D0161632

DC/AC: THE BASICS

Robert L. Boylestad

Merrill Publishing Company
A Bell & Howell Information Company
Columbus Toronto London Melbourne

To a good friend and caring educator,
Professor Joseph Aidala

Cover Photo: Jim Osburn, Osburn Photographic Illustration, Inc.

Published by Merrill Publishing Company
A Bell & Howell Information Company
Columbus, Ohio 43216

This book was set in Century Schoolbook

Administrative Editor: Steve Helba
Production Coordinator: Rex Davidson
Art Coordinator: Raydelle Clement
Cover Designer: Cathy Watterson
Text Designer: Cynthia Brunk

Copyright © 1989 by Merrill Publishing Company. All rights reserved. No part of this book may be reproduced in any form, electronic or mechanical, including photocopy, recording, or any information storage and retrieval system, without permission in writing from the publisher. "Merrill Publishing Company" and "Merrill" are registered trademarks of Merrill Publishing Company.

Library of Congress Catalog Card Number: 88-61048
International Standard Book Number: 0-675-20918-8
Printed in the United States of America
1 2 3 4 5 6 7 8 9—93 92 91 90 89

Preface

In order to become a competent contributor to any technical field, a firm, clear understanding of the basics of that field must first be established. Basics smooth the path through more complex material; their importance at the introductory level of study cannot be overstressed.

The primary goal of this presentation is to ensure that the basics of electric circuit analysis are correctly and clearly understood. Innovative techniques are used to demonstrate the impact of each basic concept, with superfluous material kept to an absolute minimum. The wealth of examples include functional use of a second color to enhance understanding. Important statements surrounding the characteristics of a basic principle are printed in boldface type with a large directional arrow for easy identification. The level of mathematics is in keeping with that typically encountered in the working environment. Whenever possible, standard element values are employed.

Each chapter begins with a set of objectives to introduce students to the content of the chapter and highlight the skills they should develop. Each chapter concludes with a formula summary including those equations of primary importance in the chapter. In most cases the formula summary lists all the equations that should be put to memory. The chapter summary includes many of the comments I offer in the classroom to guide students' efforts and emphasize the important facets of each concept. The glossary defines all those terms with which students should be familiar. Although some terms will receive limited application, all are fundamental to the field of electric circuit analysis. The problems are keyed to each section of the chapter, progressing from the simple to the challenging. Most of the problems should be within the realm of capability for the majority of students. Problems of particular difficulty are denoted by an asterisk.

A full chapter on mathematical operations with powers of ten emphasizes the importance of both becoming familiar with the notation and developing proficiency in the required operations. While mathematical complexity is kept to a minimum, there are a number of fundamental operations students must be very familiar with; these become part of a technician's set of tools.

The use of calculators is an integral feature of the text, and will help develop students' expertise. The calculator can be an invaluable tool to the practicing technician; hence, it deserves the coverage it receives. Meters and their use are introduced at the earliest opportunity to develop the practical skills that permit proper measurement of the fundamental quantities and verification of the basic laws introduced in the text. To the degree possible, the numbers reflect the actual use of instruments in the laboratory. Once introduced, the meter is referenced frequently to demonstrate how it is used in a variety of situations.

I have used BASIC for the computer programs, since it is the most "readable" language and the easiest with which to develop some familiarity in the necessarily limited coverage of an introductory text. Brackets are employed to reveal the operations performed by each sequence of steps, and a run is provided to demonstrate the capability and accuracy of the program.

Probably one of the most distinguishing features of *DC/AC: The Basics* appears in the ac section. Whereas complex algebra is introduced early in most basic texts and is used throughout the analyses, this presentation makes a bold attempt to separate the use of complex algebra from the introduction of the basic concepts of sinusoidal circuits. The analysis of ac networks is first performed without the use of complex algebra with an emphasis on certain basic relationships and the pythagorean theorem. The last sections of each ac chapter introduce complex algebra and demonstrate its impact on the calculations performed earlier. In other words, the entire ac section can be covered *without* the use of complex algebra, if desired. For those instructors who prefer an extended use of complex algebra, the necessary sections are available.

To support the text, an accompanying laboratory manual follows the text presentation in content and level. The early experiments center on the basic series and parallel configurations (for both dc and ac) to ensure the correct use of meters and verification of the concepts under investigation.

The preparation of a text emphasizing the important basics of electric circuits using a level of mathematics typically encountered in the working environment has been a challenging experience that has resulted in numerous modifications of the original manuscript. I am grateful for the support and encouragement offered by Professors Aidala and Katz, and the staff at Merrill Publishing, especially Rex Davidson and Steve Helba. I also wish to thank the following individuals who offered many helpful suggestions in the early stages of the manuscript's development: V. S. Anandu, Southwest Texas State University; Ed Bowling, Johnstone Community College; Robert Campbell, Northshore Community College; Ellsworth McGuigan, Waterbury State Technical College; Larry Oliver, Columbus State Technical College; Roy Powell, Chattanooga State Technical Institute; Burton Richardson, Longview Community College; Herb Singer, Gulf Coast Community College; Darren Whitcher, Heald College; and Ulrich Zeisler, Utah Technical Institute.

Robert L. Boylestad

Contents

APPENDIX C

APPENDIX D

APPENDIX E

APPENDIX F

APPENDIX G

1

Introduction

OBJECTIVES

- ☐ Establish a brief history of the electrical/electronics industry.
- ☐ Become aware of how developments in the industry were initiated and utilized by others to foster new ideas and directions.
- ☐ Develop a familiarity of the units of measurement used in this and related fields and the importance of using the proper unit of measurement for a quantity.
- ☐ Establish a technique for converting between units of measurement.
- ☐ Become aware of the accuracy level to be employed throughout the text and its impact on the measurements to be made.

1.1 A GROWING INDUSTRY

It is difficult, if not impossible, to identify a part of one's life-style that remains unaffected by developments in the electrical/electronics industry. This industry continues to generate interest and excitement through advances and discoveries that appeared virtually impossible a few short decades ago. Television sets are now small enough to be hand-held and have a battery capability that allows them to be more portable. Computers with significant memory capacity are now as small as portable typewriters. The size of radios is limited simply by the readability of the numbers on the faces of the dials. Hearing aids are no longer visible, and pacemakers are significantly smaller and more reliable. All this reduction in size is due primarily to a marvelous development of the last few decades—the *integrated circuit* (IC), which was first developed in the late 1950s. The IC industry has now reached a point where it can cut 1-micrometer lines. Consider that some 25,000 of these lines would fit within 1 inch. Try to visualize breaking down an inch into 100 divisions and then consider 1000 or 25,000 divisions—an incredible achievement.

The integrated circuit of Fig. 1.1 has over 68,000 transistors in addition to thousands of other elements, yet it is only about 1/4 inch on each side.

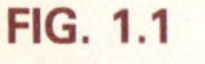

FIG. 1.1
Integrated circuit. (Courtesy of Motorola Semiconductor Products)

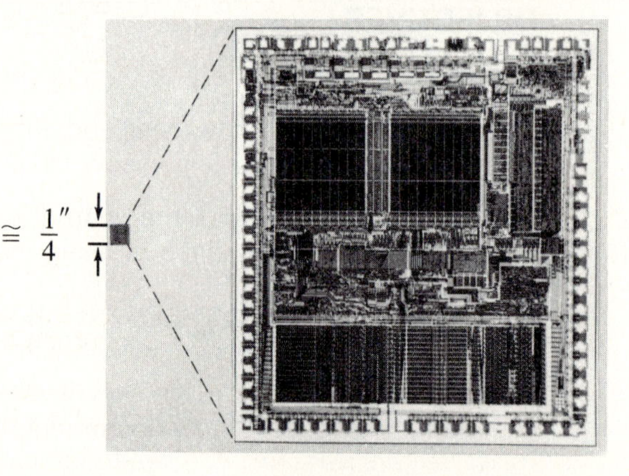

There are many publications that review the history of the field and can prove to be very interesting reading. In the late 1700s and early 1800s, discoveries and new possible areas of interest and research came fast and furious, spurred on by individuals such as Charles Coulomb, Georg Ohm, Heinrich Hertz, and Gustav Kirchhoff, to name just a few. Those listed are currently recognized through the use of their surnames in identifying important laws or units of measurement.

There is a tendency when reading about the great scientists, inventors, and innovators to believe their contributions were totally individual efforts. In many instances, this was not the case. In fact, many of the great contributors were friends or associates and provided support and encouragement in their efforts to investigate various theories. At the very least, they were aware of one another's efforts to the degree possible in the days when a letter was often the best form of communication. In particular, note the closeness of the dates during periods of rapid development in Fig. 1.2. One contributor seemed to spur on the efforts of the others or possibly provided the key needed to continue with an area of interest.

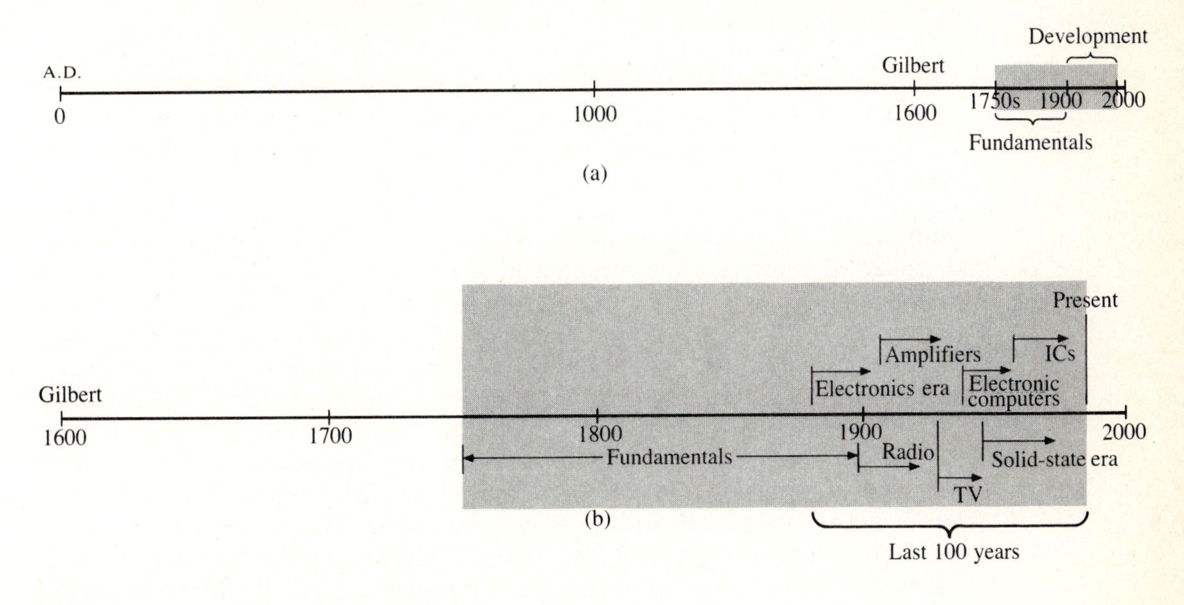

FIG. 1.2
Time charts. (a) Long-range; (b) expanded.

In the early stages, the contributors were not electrical, electronic, or computer engineers as we know them today. In most cases, they were physicists, chemists, mathematicians, or even philosophers. In addition, they were not from one or two communities of the Old World. Studying the home countries of the major contributors demonstrates that almost every established community had some impact on the development of the fundamental laws of electrical circuits. The time charts of Fig. 1.2 include a limited number of major developments, primarily to identify specific periods of rapid development and to reveal how far we have come in the last few decades. In essence, the current state of the art is a result of efforts that began in earnest some 250 years ago, with progress in the last 100 years almost exponential.

Radio The beginning of the electronics era is often traced to the efforts of Thomas Edison in the late 1800s, when he established a flow of charge between two elements in an evacuated (vacuum) glass tube. In 1904, John Ambrose Fleming expanded on the efforts of Edison to develop the first diode, commonly called *Fleming's valve,* which had a profound impact on the transmission of radio waves. By 1915 radio signals were being transmitted across the United States, and in 1918 Edwin Armstrong applied for a patent for the superheterodyne circuit employed in virtually every modern radio and TV. The major components of today's radios were in place, and sales in radios grew from a few million dollars in the early 1920s to over $1 billion by the 1930s. The 1930s were truly the golden years of radio, with a wide range of productions for the listening audience.

The 1930s were also the true beginning of the television era. In 1932, NBC installed the first commercial TV antenna on top of the Empire State Building, and RCA began regular broadcasting in 1939. The war slowed development and sales, but in the mid-1940s the number of sets grew from a few thousand to a few million. Color TV became popular in the early 1960s.

Computers The earliest computer system can be traced back to Blaise Pascal in 1642, who developed a mechanical machine for adding and subtracting numbers. In 1673 Gottfried Wilhelm von Leibniz used the *Leibniz wheel* to add multiplication and division to the range of operations, and in 1823 Charles Babbage developed the *difference engine* to add mathematical operations such as sine, cosine, and log. In the years to follow, improvements were made, but systems remained primarily mechanical until the 1930s, when electromechanical systems using components such as relays were introduced. It was not until the 1940s that totally electronic systems became the new wave. It is interesting to note that even though IBM was formed in 1924, it did not enter the computer industry until 1937. An entirely electronic system known as ENIAC was dedicated at the University of Pennsylvania in 1946. It contained 18,000 tubes and weighed 30 tons but was several times faster than most electromechanical systems. Although other vacuum-tube systems were built, it was not until the birth of the solid-state era that computer systems experienced a major change in size, speed, and capability.

The Solid-State Era In 1947, physicists William Shockley, John Bardeen, and Walter H. Brattain of Bell Telephone Laboratories demonstrated the point-contact *transistor* (Fig. 1.3), an amplifier constructed entirely of solid-state materials with no requirement for a vacuum, glass envelope, or heater voltage for the filament. Although reluctant at first due to the vast amount of material available on the design, analysis, and synthesis of tube networks, the industry eventually accepted this new technology as the wave of the future. In 1958 the first IC was developed at Texas Instruments, and in 1961 the first commercial IC was manufactured by the Fairchild Corporation.

It is impossible to review in a proper fashion the entire history of the electrical and electronics field in a few pages. Rather, through both the discussion and the time graphs of Fig. 1.2, we attempt to reveal the amazing progress of this field in the last 50 years. The growth appears to be truly exponential since the early 1900s, raising the interesting question as to where we go from here. The time chart suggests that the next few decades will probably contain a number of important innovative contributions that may cause an even faster growth curve than we are now experiencing.

FIG. 1.3
The first transistor. (Courtesy of AT&T, Bell Laboratories)

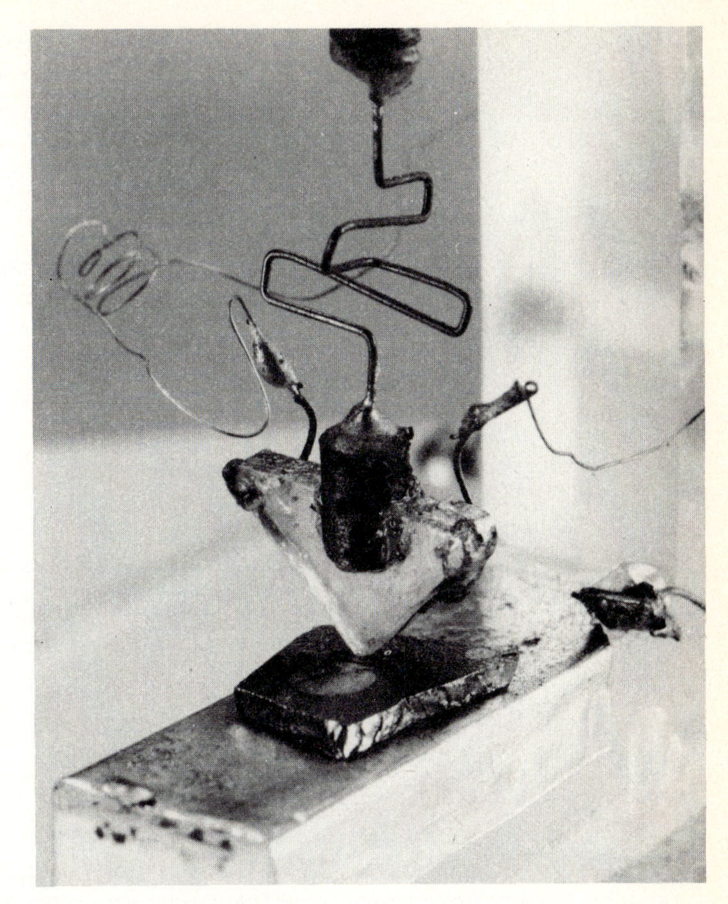

1.2 SYSTEMS OF UNITS

Throughout the world there are three *systems of units* used to give some relative meaning to measured quantities. In the United States the system used most commonly is the *English system,* with distance measured in yards (inches or feet), force, in pounds, and temperature, in degrees Fahrenheit, as shown in Table 1.1. However, in research labs and many industries in the United States and throughout the world the MKS and CGS systems are used, as defined in Table 1.1. The MKS and CGS systems draw their names from the units of measurement used with each system; the MKS system uses meters, kilograms, and seconds, whereas the CGS system uses centimeters, grams, and seconds.

Understandably, the use of more than one system of units in a world that finds itself continually shrinking in size due to advanced technical developments in communications and transportation introduces unnecessary complications to the basic understanding of any technical data. The need for a standard set of units to be adopted by all nations has become increasingly obvious. The International Bureau of Weights and Measures located at Sèvres, France, was the host for the General Conference of Weights and Measures, attended by representatives from all nations of the world. In 1960, the General Conference

TABLE 1.1
Comparison of the English and metric systems of units.

English	Metric		
	MKS	CGS	SI
Length:			
Yard (yd) (0.914 m)	Meter (m) (39.37 in.) (100 cm)	Centimeter (cm) (2.54 cm = 1 in.)	Meter (m)
Mass:			
Slug (14.6 kg)	Kilogram (kg) (1000 g)	Gram (g)	Kilogram (kg)
Force:			
Pound (lb) (4.45 N)	Newton (N) (100,000 dyn)	Dyne (dyn)	Newton (N)
Temperature:			
Fahrenheit (°F)	Celsius or Centigrade (°C)	Centigrade (°C)	Kelvin (K)
$°F = \frac{9}{5}°C + 32°$	$°C = \frac{5}{9}(°F - 32°)$		$K = 273.15 + °C$
Energy:			
Foot-pound (ft-lb) (1.356 J)	Newton-meter (N-m) or joule (J) (0.7378 ft-lb)	Dyne-centimeter or erg (1 J = 10^7 ergs)	Joule (J)
Time:			
Second (s)	Second (s)	Second (s)	Second (s)

adopted a system called Le Système International d'Unités (International System of Units), which has the international abbreviation SI. Since then, it has been adopted by the Institute of Electrical and Electronic Engineers, Inc. (IEEE) in 1965 and by the United States of America Standards Institute in 1967 as a standard for all scientific and engineering literature.

For comparison, the SI units of measurement and their abbreviations appear in Table 1.1. These abbreviations are the ones usually applied to each unit of measurement, which have been carefully chosen to be the most effective. Therefore, it is important that they be used whenever applicable to ensure universal understanding. Note the similarities of the SI system to the MKS system. Whenever possible and practical, this text will employ all the major units and abbreviations of the SI system in an effort to support the need for a universal system. For those readers requiring further information on the SI system, a complete kit has been assembled for general distribution by the American Society for Engineering Education (ASEE).*

*American Society for Engineering Education (ASEE), 1 Dupont Circle, Suite 400, Washington, D.C. 20036.

A standard exists for each unit of measurement of each system. The standards of some units are quite interesting. The *meter* was defined in 1960 as 1,650,763.73 wavelengths of the orange-red light of krypton 86. It was originally defined in 1790 to be 1/10,000,000 the distance between the equator and either pole at sea level. The current standard is preserved on a platinum-iridium bar at the International Bureau of Weights and Measures at Sèvres, France. The *kilogram* is defined as a mass equal to 1000 times the mass of 1 cm^3 of pure water at 4°C. This standard is preserved in the form of a platinum-iridium cylinder in Sèvres. The *second* was originally defined as 1/86,400 of the mean solar day. It was redefined in 1960 as 1/31,556,925.9747 of the tropical year 1900.

Do not be concerned if some of the units of measurement of Table 1.1 are not familiar or fail to provide any real feeling for the measure of the quantity. Most will appear on one or more occasions in the chapters to follow and will be described in detail as the need for the unit of measurement arises.

For the moment, carefully review the units of measurement under the heading "Length" of Fig. 1.4. Note that a meter is only slightly longer than a yard and that there are more than $2\frac{1}{2}$ cm in 1 in.

For quick recall (due to their frequent use), it is strongly suggested that the following conversion factors be memorized.

$$\begin{aligned} 1\text{ m} &= 100\text{ cm} = 39.37\text{ in.} \\ 2.54\text{ cm} &= 1\text{ in.} \end{aligned} \tag{1.1}$$

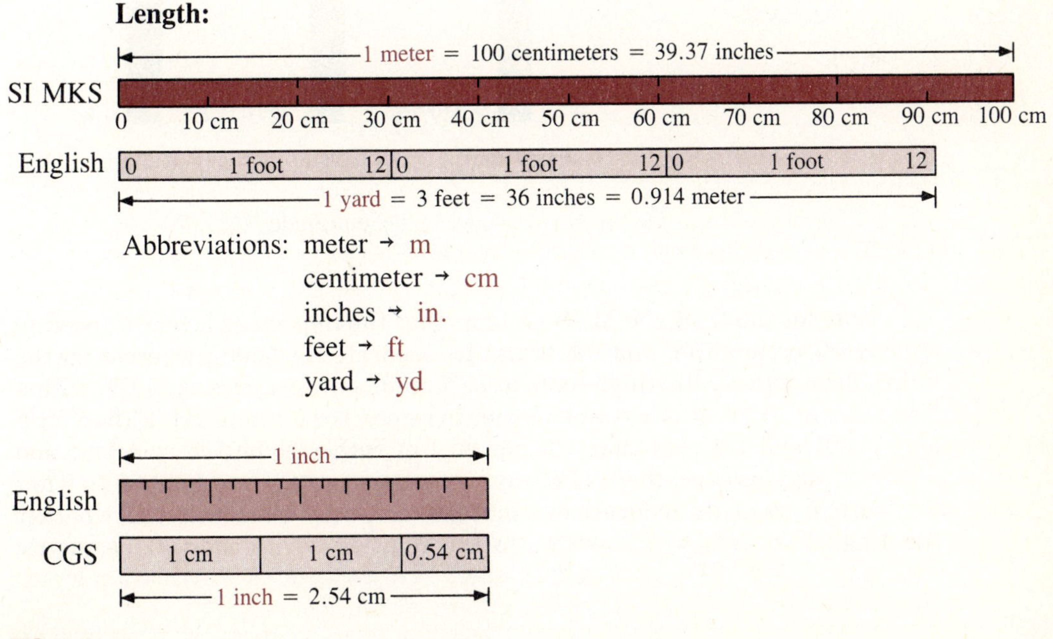

FIG. 1.4
Units of length.

The actual lengths of a centimeter and inch appear in Fig. 1.4 for comparison. The relative lengths of the meter, yard, and foot are depicted in the same figure for comparison purposes.

Figure 1.5 provides a comparison of systems for measuring temperature, with the Fahrenheit (English) scale the most familiar in North America. The Celsius scale has become more familiar in recent years. The columns were drawn to demonstrate equivalent temperatures in each system of units. For future reference, the following should also be memorized.

$$\boxed{\begin{array}{lrcl} \text{Freezing:} & 32°\text{F} & = & 0°\text{C} \\ \text{Boiling:} & 212°\text{F} & = & 100°\text{C} \end{array}} \qquad (1.2)$$

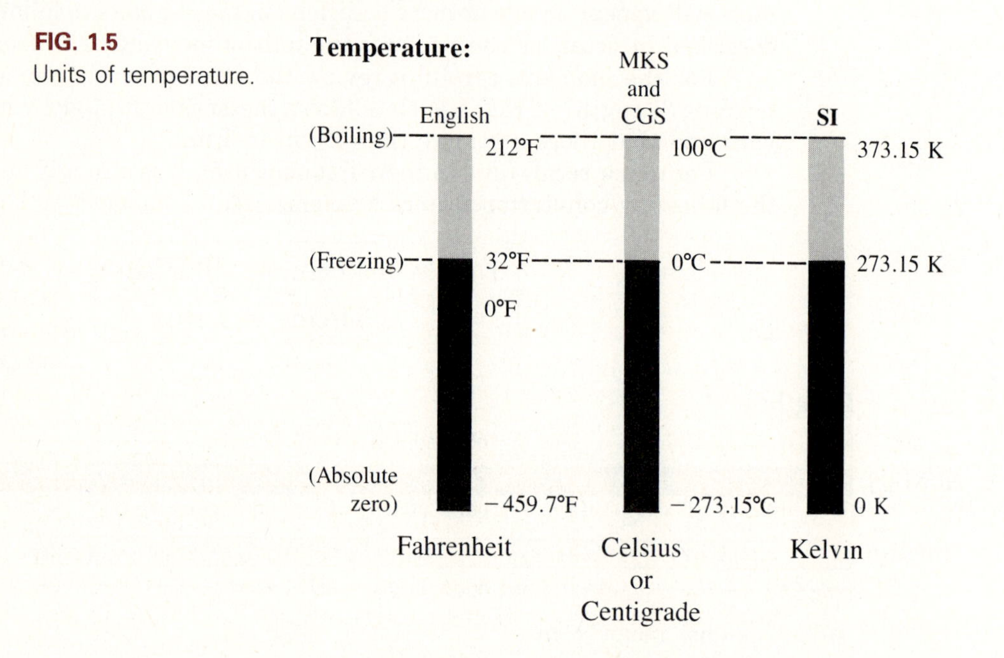

FIG. 1.5
Units of temperature.

Note for the CGS and MKS systems that the important levels of freezing and boiling occur at 0°C and 100°C and are separated by 100°C, whereas for the Fahrenheit system, freezing occurs at 32°F and boiling occurs at 212°F, with a difference of 180°F. It is certainly easier to remember 0°C and 100°C than 32°F and 212°F and the fact that 100 cm = 1 m rather than 1 ft = 12 in. and 3 ft = 1 yd. However, there is always a tendency to try and hold on to what we learn first in the educational process, and only time will tell whether the English system will succumb to the relative advantages of the metric system. The Kelvin system of Fig. 1.5 will be introduced and discussed as the need arises, but observe that absolute zero (the temperature at which all molecular activity ceases) is an extremely cold temperature (−459.7°F), which has corresponding levels of −273.15°C and 0 K. Obviously, the Kelvin scale uses absolute zero as its reference level.

The equations relating the temperature scales are as follows.

$$°F = \frac{9}{5}°C + 32° \tag{1.3}$$

$$°C = \frac{5}{9}(°F - 32°) \tag{1.4}$$

$$K = 273.15 + °C \tag{1.5}$$

The above equations have limited use in the chapters to follow but are provided here for completeness.

The units of measurement for mass, force, and energy are provided in Fig. 1.6, with graphic comparisons of the units in different systems. Although many units may be unfamiliar to those with limited scientific backgrounds, a number of them are introduced in the chapters to follow. In time, any concerns regarding units of measurement will be laid to rest by frequent exposure to the units of measurement and comparisons with more familiar units. For the moment, however, note in Fig. 1.6 that the units of measurement for the CGS system are many times smaller than those applied in the other two systems, which is logical considering its application in fields where very low levels of mass, force, and energy are encountered. At the other end of the spectrum, the English system has the largest unit of measure for each quantity. The SI system (and the MKS for these quantities) appears to lie somewhere between the two, providing some obvious evidence for its choice as the universal standard.

For the future it is paramount that the following rule be followed.

Every measured or measurable quantity must have an associated unit of measurement.

For example, to say that a length is 20 without indicating whether it is in centimeters, yards, or miles is completely meaningless. Similarly, a time measurement of 14 is meaningless unless a unit of measurement such as seconds, hours, or days is applied.

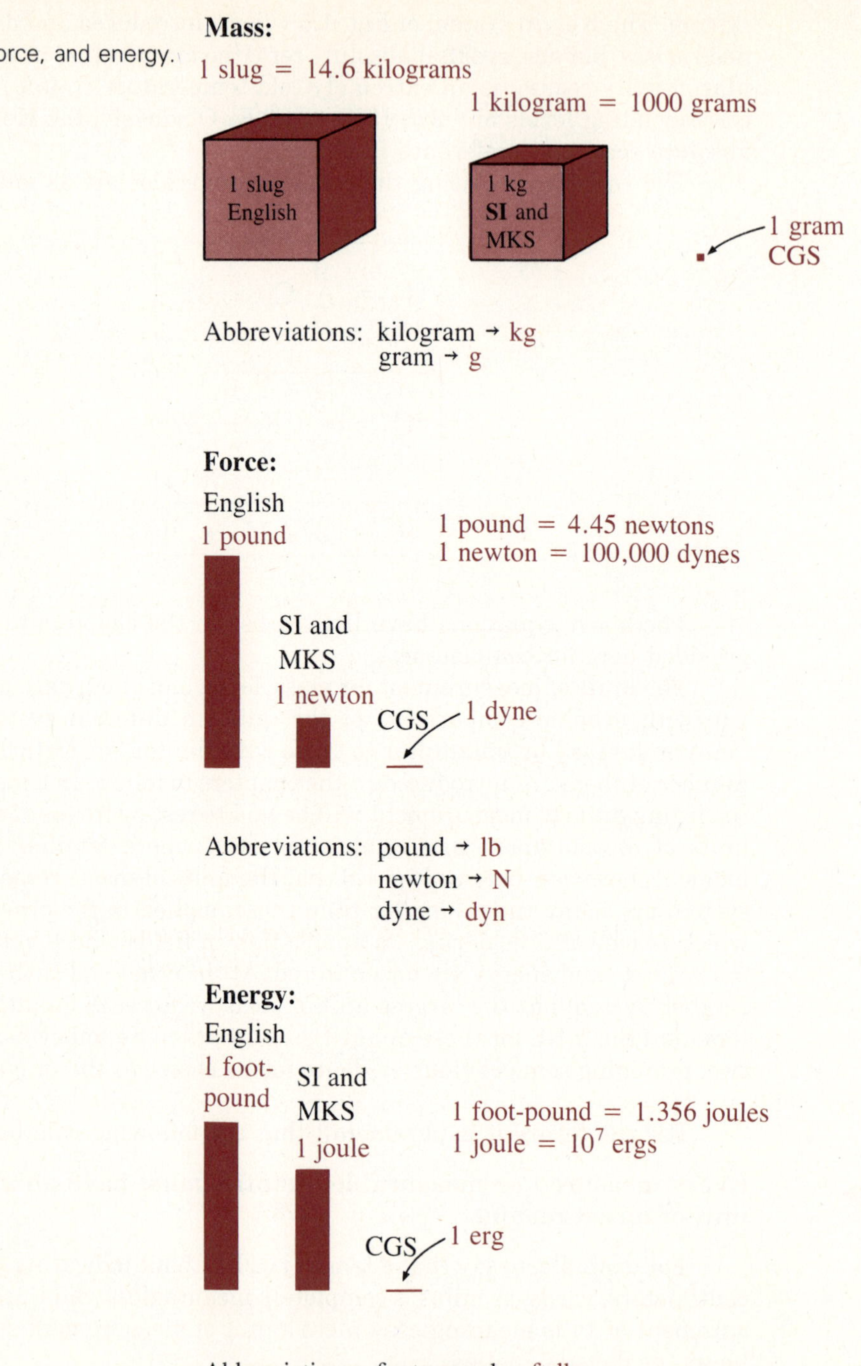

FIG. 1.6
Units of mass, force, and energy.

1.3 CONVERSION WITHIN AND BETWEEN SYSTEMS OF UNITS

The conversion within and between systems of units is a process that cannot be avoided in the study of any technical field. It is an operation, however, that is performed incorrectly so often that this section was included to provide one approach that, if applied properly, will lead to the correct result.

There is more than one way to perform the conversion process. In fact, some people prefer to determine mentally whether the conversion factor is multiplied or divided. For instance, if you were asked to determine how many meters are in 5 ft (60 in.), the first step would be to look up the conversion factor of 1 m = 39.37 in. Then you would divide 60 in. by the factor to determine the result:

$$\frac{60\ \text{in.}}{39.37\ \dfrac{\text{in.}}{\text{m}}} = 1.524\ \text{m}$$

This approach is acceptable for some elementary conversions but is risky with more complex operations.

The procedure to be described here is best introduced by examining a relatively simple problem, such as converting inches to meters. Specifically, let us convert 60 in. (5 ft) to meters.

If we multiply 60 in. by the factor 1, the result is obviously the same, 60 in. Any number multiplied by 1 remains the same. However, let us now set up the product in the following form:

$$60\ \text{in.} = (60\ \text{in.})(1)$$

Now let us look up the relationship between inches and meters. That is, how many inches are equivalent to 1 m? The result, called the *conversion factor,* appears below:

$$1\ \text{m} = 39.37\ \text{in.}$$

If we divide both sides of the conversion factor by 39.37 in., we obtain

$$\frac{1\ \text{m}}{39.37\ \text{in.}} = \frac{39.37\ \text{in.}}{39.37\ \text{in.}} = 1$$

Note the same factor of (1) appearing in the earlier equation. If we now replace the earlier (1) by the conversion factor in the form just developed, we obtain

$$60\ \text{in.} = (60\ \text{in.})(1) = (60\ \text{in.})\left(\frac{1\ \text{m}}{39.37\ \text{in.}}\right)$$

The unit of measurement inches will cancel in the numerator and denominator, leaving the desired unit of measurement meters with the proper measure:

$$\frac{60}{39.37}\ \text{m} = 1.219\ \text{m}$$

For all the examples to follow, note that the conversion factors are set up so that the unit of measurement of the denominator cancels the given unit of measurement. In general, however, follow the following sequence of steps:

1. Multiply the quantity to be converted by the factor 1.
2. Set up the conversion factor to form a numerical value of 1 with the unit of measurement to be removed in the denominator.
3. Perform the required mathematics to obtain the proper measure for the remaining unit of measurement.

EXAMPLE 1.1 Convert 6 in. to centimeters.

Solution

Step 1: 6 in. = 6 in.(1)

Step 2: $$\frac{2.54\ \text{cm}}{1\ \text{in.}} = (1)$$

Step 3: $$6\ \text{in.}\left[\frac{2.54\ \text{cm}}{1\ \text{in.}}\right] = \mathbf{15.24\ cm}$$

EXAMPLE 1.2 Determine the number of meters in a mile.

Solution

Step 1: 1 mi = 5280 ft = 5280 ft(1)

Step 2: Converting from feet to inches:

$$\frac{12\ \text{in.}}{1\ \text{ft}} = (1)$$

Step 3: $$5280\ \text{ft}\left(\frac{12\ \text{in.}}{1\ \text{ft}}\right) = 63{,}360\ \text{in.}$$

Converting from inches to meters:

Step 2: $$\frac{1\ \text{m}}{39.37\ \text{in.}} = (1)$$

Step 3: $$63{,}360\ \text{in.}\left(\frac{1\ \text{m}}{39.37\ \text{in.}}\right) = \mathbf{1609.347\ m}$$

EXAMPLE 1.3

How many seconds are there in 0.2 h?

Solution Since (1)(1) = (1), we can group the multiplying factors as shown below:

Step 1: $0.2\text{ h} = 0.2\text{ h}(1)(1)$

Step 2: $\dfrac{60\text{ min.}}{1\text{ h}} = (1), \qquad \dfrac{60\text{ s}}{1\text{ min.}} = (1)$

Step 3: $0.2\text{ h}(1)(1) = 0.2\,\cancel{\text{h}}\left(\dfrac{60\,\cancel{\text{min.}}}{1\,\cancel{\text{h}}}\right)\left(\dfrac{60\text{ s}}{1\,\cancel{\text{min.}}}\right) = \mathbf{720\text{ s}}$

1.4 ACCURACY

Inexpensive, multifunctional calculators can be used to save time and maintain a high level of accuracy. For this text, most calculations will be performed to three places (thousandths) accuracy. That is, a result such as 5.06215 will be given as 5.062 and one such as 12.00497, as 12.005. In other words, the digit in the ten-thousandths place will determine the thousandths digit. If the digit in the ten-thousandths place is greater than or equal to 5, the digit in the thousandths place is raised by one. Otherwise, it remains the same.

The following numbers are written to three-decimal-place accuracy.

$$0.00642 = 0.006$$
$$540.0579 = 540.058$$

Most good scientific calculators have the option of setting the level of accuracy using the **[2nd F]** (second function) or **[FIX]** keys. You will have to check your own calculator handbook under the heading of *rounding* or *decimal places*. For those with the [2nd F] key, the following is a common sequence for setting the result to three decimal places:

[2nd F] [TAB] [3]

Once entered, the input of a number such as 4.12345, using the keys

[4] [·] [1] [2] [3] [4] [5] [=]

results in a display of **4.123.**

An input of 1.00566, using

[1] [·] [0] [0] [5] [6] [6] [=]

results in a display of **1.006.**

For two-decimal-place and four-decimal-place accuracy, the inputs are

(2nd F) (TAB) (2)

and

(2nd F) (TAB) (4)

respectively.

For calculators with the (FIX) key, the following will set the level of accuracy to three-decimal-place accuracy.

(FIX) (3)

FORMULA SUMMARY

$$1 \text{ in.} = 2.54 \text{ cm}$$

$$1 \text{ m} = 39.37 \text{ in.} = 100 \text{ cm}$$

$$°F = \frac{9}{5}°C + 32°$$

$$°C = \frac{5}{9}(°F - 32°)$$

$$°K = 273.15 + °C$$

CHAPTER SUMMARY

It is important to remember to

1. Assign a *unit* of *measurement* to any measured or determined quantity.
2. *Check* all results to be sure they make sense based on the size (magnitude) of the quantities used to determine the result.
3. Substitute the *proper* unit of measurement into an equation to ensure the proper magnitude for the result.

The conversion process is one that will be particularly important in this course and those to follow. Make every effort to learn the technique introduced here or another procedure that will ensure the correct result.

Start to develop a close relationship with your calculator by using the keys introduced in this chapter and reviewing the manual provided with the instrument. If your calculator has different keys from those employed in this text (for the same operation), become familiar with the equivalent keys on your keyboard. The calculator is a very valuable tool of the trade and must become an integral part of your everyday calculations.

GLOSSARY

CGS system A system of units employing the centimeter, gram, and second as its fundamental units of measure.

Difference engine One of the first mechanical calculators.

Edison effect A flow of charge between two elements in an evacuated tube.

ENIAC The first totally electronic computer.

Fleming's valve The first of the electronic devices, the diode.

Integrated circuit (IC) A subminiature structure containing a vast number of electronic devices designed to perform a particular set of functions.

Joule (J) a unit of measurement for energy in the SI or MKS system equal to 0.7378 ft-lb in the English system and 10^7 ergs in the CGS system.

Kelvin (K) A unit of measurement for temperature in the SI system equal to 273.15 + °C in the MKS and CGS systems.

Kilogram (kg) A unit of measure for mass in the SI and MKS systems equal to 1000 grams in the CGS system.

Meter (m) A unit of measure for length in the SI and MKS systems equal to 1.094 yd in the English system and 100 cm in the CGS system.

MKS system A system of units employing the meter, kilogram, and second as its fundamental units of measure.

Newton (N) A unit of measurement for force in the SI and MKS systems equal to 100,000 dyn in the CGS system.

Pound (lb) A unit of measurement for force in the English system equal to 4.45 N in the SI or MKS system.

Second (s) A unit of measurement for time in the SI, MKS, English, and CGS systems.

SI system The system of units adopted by the IEEE in 1965 and the USASI in 1967 as the International System of Units (Système International d'Unités).

Slug A unit of measure for mass in the English system equal to 14.6 kg in the SI or MKS system.

Transistor The first semiconductor amplifier.

PROBLEMS

Section 1.1

1. Visit the nearest available library, find the section on electricity/electronics, and briefly describe how large an area is set aside for the subject and anything of particular interest that stands out as you review the available list of literature.
2. Choose a contributor to the field and write a brief one-page summary of his or her contributions and their impact on future developments.
3. Keeping in mind recent developments in the field, write a one-page paper on where you feel the field is moving in the near future. How do you feel we can improve on current electrical/electronics systems?

Sections 1.2–1.3

4. Which of the following lengths is closer to the height of this text, a centimeter, foot, or meter?
5. Which of the following lengths is closer to the width of a dime, a centimeter, inch, or meter?
6. Why do you believe a football field is measured in yards rather than inches? Give any reason or reasons you feel appropriate based on the action of the game.
7. What is unique about the relationship among centimeters, meters, and kilometers that is not true for inches, yards, and miles?
8. The scientific community employs the metric system almost exclusively. Why do you think it is so hard to motivate people to switch from the English to the metric system? What methods would you put in place to establish the metric system as the only system of units?
9. Why do you think the speed of a car is measured in miles per hour or kilometers per hour rather than miles per second or kilometers per second? How about miles per day or kilometers per day?
10. Why is the length of a carpenter's nail measured in inches or centimeters rather than feet, yards, or meters?
11. What unit of measurement is the same for the English and metric systems?
12. Compare the numerical values of the freezing and boiling temperatures of water in the Celsius system with the numerical values in the Fahrenheit system. Which seems more practical and easier to remember? Why do you believe we continue to use both?
13. How would you describe the difference in mass between 1 g and 1 slug?

*14. How would you compare a force of 1 N to a force of 1 lb? Can you make a good comparison between a force of 1 dyn and a force of 1 lb?

15. Condense all the *important* relationships and information of Sections 1.2 and 1.3 to a number that would fit comfortably on both sides of one 3-in. by 5-in. index card. This card can then be used for review purposes in the future and as a handy reference when trying to memorize the equations or reviewing for an examination.
16. When you solve for a quantity using an equation and you find you are not sure of the proper unit of measurement, is it better to leave the unit of measurement out or

make an educated guess as to the proper unit of measurement? What other alternatives are there?

17. When using a calculator, should you always trust the results of the calculator or should you have some idea of the magnitude of the result before plugging in the numbers? In other words, is it best to always assume the calculator result is correct? When would you be hesitant to trust the calculator result?

18. Find the number of centimeters in
a. 1 ft
b. 1 yd
c. 1/2 m

19. Find the number of inches in
a. 10 cm
b. 2.5 ft
c. 0.2 m

***20.** Find the number of feet in
a. 1/4 in.
b. 1 km
c. 10 m

21. What is the length in miles of a 10-km run?

22. How many kilometers in a marathon (26.2 mi)?

23. **a.** What is the average speed in miles per hour of a runner completing a 4-min mile?
b. Repeat part a for an average runner with an 8-min/mi pace.

24. **a.** At 55 mi/h, how many days would it take to cross the North Atlantic continent (about 3000 mi)?
b. Repeat part a for a speed of 65 mi/h.
c. Based on the results, comment on whether you feel the time saved is worth the additional risk.

***25.** Find the number of seconds in
a. 2.8 min
b. 1/4 h
c. 0.01 d

26. Convert 100 km/h to miles per hour. What is the approximate conversion factor between the two? How about from miles per hour to kilometers per hour?

27. What is the typical room temperature of 68°F in the Celsius and Kelvin temperature scales?

28. If the conversion rate between the British pound and the U.S. dollar is $1.86 = 1 British pound, how many British pounds are equivalent to U.S. $100? If 6.4 Norwegian krone equal $1, how many U.S. dollars would you need to purchase 100 Norwegian krone?

29. Convert 2500 grams to
a. kilograms
b. slugs

30. Convert 100 newtons of force to
a. pounds
b. dynes

2

Mathematics: An Important Element

OBJECTIVES

- ☐ Become aware of the importance of developing basic mathematical skills.
- ☐ Understand the usefulness and applications of powers of ten.
- ☐ Be able to write both small and large numbers in scientific notation.
- ☐ Develop calculator skills as applied to numbers in the power-of-ten format.
- ☐ Be able to perform the basic mathematical operations of addition, subtraction, multiplication, and division using numbers in the power-of-ten format.
- ☐ Find powers of powers of ten using longhand and calculator techniques.
- ☐ Learn the prefixes and abbreviations applied to specific powers of ten.
- ☐ Be able to convert from one power of ten to another.
- ☐ Understand how to use a conversion table.

2.1 INTRODUCTION

It is important for a new student of electricity/electronics to appreciate that there is a set of mathematical operations with which he or she must become proficient. This is not to imply that an advanced level of understanding of a subject area such as calculus or statistics is required but simply that a clear understanding of some very basic operations is necessary.

In general, accept mathematics as the foundation required to understand and apply the basic concepts to be introduced in the chapters to follow. Without basic math skills it is impossible to understand fully the impact of certain fundamental relationships.

Once the basic concepts and the related mathematics are understood, the application of the theory to a number of exercises should develop a level of confidence that will remove most concerns about the mathematics involved. In time, the application of the electrical/electronic concepts will be the overriding issue and the mathematics will be simply a tool needed to obtain the numerical result. In other words, mathematics will be a tool employed to demonstrate the impact and breadth of a particular concept—not the primary objective of the sections and chapters to follow.

2.2 NUMBERS—BOTH LARGE AND SMALL

The range of numbers that one encounters in this and related fields extends from the very small to the very large. It is not atypical to refer to something measured in terms of *picos* (1 pico $= 10^{-12} = 0.000000000001$) in the same sentence with another quantity measured in terms of *megas* (1 meg $= 10^{+6} = 1{,}000{,}000$). This wide range makes it important to get away from constantly dealing with all the zeros associated with such numbers. The number of zeros involved might possibly lead to errors in both calculations and the communication of results to others. To solve the problem, *powers of ten* are employed to place the numbers in a more convenient form that is relatively easy to deal with when basic mathematical maneuvers are required. The technique is employed throughout the industry. It is not a difficult technique to learn, but numerous examples are provided for clarity.

2.3 POWERS OF TEN

Large Numbers For starters let us consider the number 1000. As shown below, it is really the product of 10 taken three times:

$$10 \times 10 \times 10 = 1000$$

In the power-of-ten format the number is written as

$$10^3 = 1000 \qquad \text{(1 thousand)}$$

with the power indicating the number of times the number 10 occurs as a factor in the multiplication.

The value of 10^6 is

$$10^6 = (10)(10)(10)(10)(10)(10) = 1{,}000{,}000$$

or 1 million.

This is, therefore, a shorthand method for writing large numbers that are powers of ten. The following is a partial list extending from 1 to 1 million.

$$
\begin{aligned}
10^0 &= 1 & & \\
10^1 &= 10 & & \text{(ten)} \\
10^2 &= 100 & & \text{(1 hundred)} \\
10^3 &= 1000 & & \text{(1 thousand)} \\
10^4 &= 10{,}000 & & \text{(10 thousand)} \\
10^5 &= 100{,}000 & & \text{(1 hundred thousand)} \\
10^6 &= 1{,}000{,}000 & & \text{(1 million)}
\end{aligned}
$$

Note that $10^0 = 1$. Any number such as 2^0, 5^0, or x^0 is, by definition, equal to 1. It may seem somewhat strange at this point but it is a mathematical axiom that will prove necessary and useful in the calculations to follow.

It is a very simple process to determine the power of ten from the longhand form. For example, consider the number 100,000, rewritten below. The number of places from the *right* of the number 1 to the decimal point is the resulting power of ten. Note that

$$1\underbrace{0\,0,\,0\,0\,0.}_{\text{5 places}} = 10^5$$

The power of ten is also the number of zeros in the number, but we will find in the next example that powers of ten can also be applied to numbers that are not simply powers of ten, thereby negating this simpler technique.

For a number such as 25,000, the same procedure can be followed, but the power of ten is determined by the number of places counted. For instance, if we start to the right of the number 2, as shown below, the power of ten is 10^4:

$$2\underbrace{5,\,0\,0\,0.}_{\text{4 places}} = 2.5 \times 10^4$$

The same result can be obtained in another way:

$$25{,}000 = (2.5)(10{,}000) = 2.5 \times 10^4$$

If the count begins to the right of the number 5, the result is

$$2\,5,\underbrace{0\,0\,0.}_{\text{3 places}} = 25 \times 10^3$$

which is also correct.

In other words, the power of ten is a direct function of where the count begins, and the results (if the counting is done correctly) are equivalent. That is,

$$\begin{aligned} 25{,}000 &= 0.25 \times 10^5 = 2.5 \times 10^4 = 25 \times 10^3 = 250 \times 10^2 \\ &= 2500 \times 10^1 = 25{,}000 \times 10^0 \end{aligned}$$

The choice of which format to use is usually determined by the operation to be performed or the notation that is most convenient.

When the operation requires that we go from the power-of-ten format to the long form, we simply let the power dictate how many places the decimal point is moved to the *right* of its present position. Always keep in mind that *positive powers of ten increase the size of the multiplying factor.*

For example,

$$2.2 \times 10^3 = 2.\underbrace{200}_{\text{3 places}} = 2200$$

and

$$0.56 \times 10^5 = 0.\underbrace{56000}_{\text{5 places}} = 56{,}000$$

Small Numbers The case for small numbers is similar. For instance, if we divide 1000 into 1, we obtain

$$\frac{1}{1000} = 0.\underbrace{001}_{\text{3 places}} = 1 \times 10^{-3}$$

The power-of-ten form is written with a negative sign to indicate that the decimal point is moved three places to the *left* of the right side of the number 1 rather than to the right, as for positive powers of ten. Note for both the positive and negative powers of ten that the count begins to the right of the number 1. The following is a partial list of numbers less than 1 that have negative powers of ten:

$$\frac{1}{10} = 0.\underbrace{1}_{\text{1 place}} = 1 \times 10^{-1}$$

$$\frac{1}{100} = 0.\underbrace{01}_{\text{2 places}} = 1 \times 10^{-2}$$

$$\frac{1}{1000} = 0.\underbrace{001}_{\text{3 places}} = 1 \times 10^{-3}$$

$$\frac{1}{10{,}000} = 0.\underbrace{0001}_{4\text{ places}} = 1 \times 10^{-4}$$

$$\frac{1}{100{,}000} = 0.\underbrace{00001}_{5\text{ places}} = 1 \times 10^{-5}$$

$$\frac{1}{1{,}000{,}000} = 0.\underbrace{000001}_{6\text{ places}} = 1 \times 10^{-6}$$

For numbers other than negative powers of ten, the same procedure can be followed, with the final location of the decimal point determining the negative power of ten.

For example:

$$0.\underbrace{0045}_{4\text{ places}} = 45 \times 10^{-4}$$

or

$$0.\underbrace{00}_{2\text{ places}}45 = 0.45 \times 10^{-2}$$

or

$$0.\underbrace{00450}_{5\text{ places}} = 450 \times 10^{-5}$$

Keep in mind that all the preceding equations are equivalent, yet one form may be more convenient for a particular application.

For the reverse process, simply move the decimal point to the *left* of its current position a number of places equal to the magnitude (without regard to sign) of the power of ten.

For example:

$$60 \times 10^{-3} = \underbrace{060}_{3\text{ places}}. = 0.06$$

$$3.8 \times 10^{-4} = \underbrace{0003}_{4\text{ places}}.8 = 0.00038$$

When converting from one form to the other, do not be swayed by the magnitude of the multiplying factor. For instance, in the first example that follows, the relatively large magnitude of 4000 might lead you to make the number smaller by moving the decimal point to the left. However, positive powers of ten always make the multiplying factor larger.

$$4000 \times 10^4 = 4\,0,0\,0\underbrace{0,0\,0\,0}_{4\text{ places}},$$

and

$$0.07 \times 10^{-3} = 0.\underbrace{0\,0\,0}_{3\text{ places}}0\,7$$

Similarly, in the second example, the relatively small multiplying factor may suggest that the decimal point be moved three places to the right to make the number larger. However, negative powers of ten always reduce the size of the multiplying factor.

It is interesting to note that

$$\frac{1}{10} = \frac{1}{10^1} = 0.1 = 10^{-1}$$

$$\frac{1}{10^2} = 10^{-2}$$

$$\frac{1}{10^3} = 10^{-3}$$

and so on. In general, therefore,

$$\boxed{\frac{1}{10^n} = 10^{-n}} \tag{2.1}$$

and

$$\boxed{\frac{1}{10^{-n}} = 10^{+n}} \tag{2.2}$$

In words, both equations reveal that a power of ten can be moved from the denominator to the numerator of a fraction simply by changing the sign of the power.

Further examples include the following:

$$\frac{1}{10^{+6}} = 10^{-6} \quad \text{and} \quad \frac{1}{10^{-3}} = 10^{+3}$$

2.4 SCIENTIFIC NOTATION

When using powers of ten, if the decimal point is placed to the right of the first significant digit of the multiplying factor, the number is said to be in *scientific notation*. A few examples of numbers placed in scientific notation are

$$22{,}000 = 2.2 \times 10^{+4}$$
$$0.006 = 6 \times 10^{-3}$$
$$5{,}100{,}000 = 5.1 \times 10^{+6}$$

Calculator The power-of-ten format can be set on most calculators using the

(EXP) or (EE)

key. The format on the display will vary from calculator to calculator. However, once employed, the format will be quite clear and in time the advantage gained by being able to perform the operations on a calculator should be appreciated.

For one calculator the input of the number

$$33 \times 10^{3}$$

is

(3) (3) (EXP) (3)

resulting in

33E3

Another calculator may result in the display

33. 03

with the additional space (actually representing a positive sign) the only indication of a power-of-ten format.

For a number such as 0.1×10^{-4}, the keying sequence is

(·) (1) (EXP) (−) (4) **.1E − 4**

for one calculator. It is

(·) (1) (EXP) (+/−) (4) **.1 − 04**

for calculators with the (+/−) key.

2.5 ADDITION

The addition of two numbers in the power-of-ten format requires the following:

Addition:

The power of ten of each number must be the same.

For instance, consider the addition of 45,000 and 7000:

$$\begin{array}{rcl} 45{,}000 & = & 45 \times 10^3 \\ +\ 7{,}000 & = & +\ 7 \times 10^3 \\ \hline & & \mathbf{52 \times 10^3} \end{array} \quad \text{same!}$$

As just noted, the power of ten for the result is the same as that of each number. It is *not* the sum of the two powers, which would result in a solution of 10^6. The remaining integer values are then simply added and the power of ten is included in the solution. You will find in all the mathematical operations to be introduced in this chapter that the following is true.

Once the power-of-ten format is established, the multiplying factors can be operated on independently of the power of ten.

A further example:

$$\begin{array}{rcl} 0.006 & = & 600 \times 10^{-5} \\ +\ 0.00012 & = & +\ 12 \times 10^{-5} \\ \hline & & \mathbf{612 \times 10^{-5}} \end{array}$$

In the preceding example the power of ten chosen was based on the number 0.00012 and the desire to have a nondecimal multiplier such as 0.12 or 1.2. However, if a power of ten such as 10^{-3} were preferred, a result in terms of 10^{-3} could have been obtained that would be absolutely correct.

Calculator For most calculators, once the numbers are input in a power-of-ten format the calculator will internally set up its chosen power of ten for each and provide a result that may not include a power of ten. For instance, the calculator input for the first of the above examples is

resulting in **52,000.**

For the second example,

results in **0.00612.**

For most calculators the size of the numbers will determine whether the result is in power-of-ten or decimal form.

2.6 SUBTRACTION

The subtraction of two numbers in the power of ten format requires the following.

Subtraction:
The power of ten of each number must be the same.

For instance, let us subtract 500 from 6000 using powers of ten:

$$\begin{array}{rcl} 6000 & = & 6.0 \times 10^3 \\ -\ 500 & = & -\ 0.5 \times 10^3 \\ \hline & & \mathbf{5.5 \times 10^3} \end{array} \quad \text{same!}$$

Note the same power of ten throughout the calculation and the fact that the subtraction is performed using only the multiplying factors.

A second example is as follows:

$$\begin{array}{rcl} 0.012 & = & 120 \times 10^{-4} \\ -0.0005 & = & -\ \ 5 \times 10^{-4} \\ \hline & & \mathbf{115 \times 10^{-4}} \end{array}$$

Calculator As with addition, the calculator will perform its own internal maneuvers to generate a result in the power-of-ten or decimal format.

For the first example, the input would be

6 EXP 3 − · 5 EXP 3 =

with the result **5500.**

For the second example, key in

1 2 0 EXP +/− 4 − 5 EXP +/− 4 =

with the result **0.0115.**

2.7 MULTIPLICATION

Before we describe the procedure for multiplication of power-of-ten numbers, it is important that you realize that the following notations for multiplication are equivalent and, depending on the situation, may be used interchangeably:

$$2 \times 4 = (2)(4) = 2 \cdot 4$$

The real convenience associated with powers of ten surfaces with the multiplication operation, since we are now spared the worry about the extensive number of zeros in decimal form.

The multiplication process hinges on the following mathematical relationship, where the letters n and m can be any positive *or* negative number.

$$(10^n)(10^m) = 10^{n+m} \tag{2.3}$$

To best describe the use of the equation, the following example is provided in an expanded form:

$$(100)(1000) = \underbrace{(10^2)(10^3)}_{(10^n)(10^m)} = \underbrace{10^{2+3}}_{10^{n+m}} = 10^5 = 100{,}000$$

$$n = +2, \quad m = +3$$

Note the similarities between the middle steps and Eq. 2.3. The similarities clearly reveal that for this operation, $n = 2$, $m = 3$, and $n + m = 5$. It is therefore necessary first to place both numbers in a power of ten format (there is no need to have the same power of ten, as required for addition and subtraction) and perform the required addition of exponents.

EXAMPLES

$$(10{,}000)(100{,}000) = (10^4)(10^5) = \mathbf{10^9}$$

$$(1000)(0.0001) = (10^3)(10^{-4}) = 10^{3+(-4)} = 10^{3-4} = \mathbf{1 \times 10^{-1}}$$

In the second example note the need to carry the signs of n and m through the calculations.

For numbers that are not exact powers of ten, the multiplication process can be separated from the operation with the powers of ten.

For example,

$$\begin{aligned}(5000)(0.004) &= (5 \times 10^3)(4 \times 10^{-3}) \\ &= (5)(4) \times (10^3)(10^{-3}) \\ &= 20 \times 10^0 = 20 \times 1 = \mathbf{20}\end{aligned}$$

Note the importance of the axiom $10^0 = 1$ and the avoidance of having to deal with all the zeros of the initial format of the numbers.

A second example involves three numbers:

$$\begin{aligned}(0.0001)(50 \times 10^4)(0.06 \times 10^{-3}) &= (1 \times 10^{-4})(50 \times 10^4)(0.06 \times 10^{-3}) \\ &= (1)(50)(0.06) \times (10^{-4})(10^{+4})(10^{-3}) \\ &= 3 \times 10^{(-4+4-3)} \\ &= \mathbf{3 \times 10^{-3}}\end{aligned}$$

Calculator For the first example above, key

(5) (EXP) (3) (×) (4) (EXP) (+/−) (3) (=)

with the result **20.**

For the second example, key

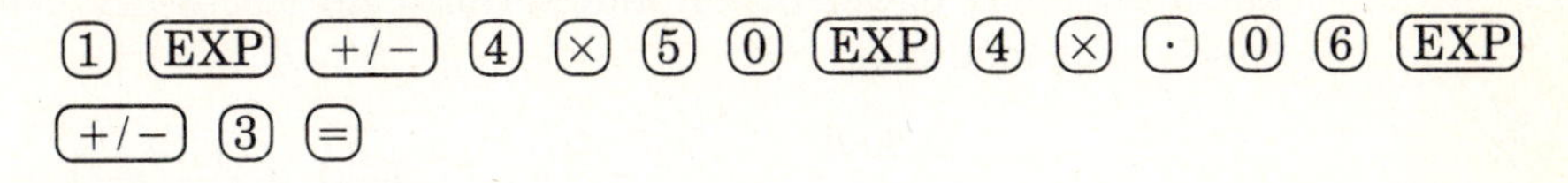

with the result **0.003.**

2.8 DIVISION

The division operation with powers of ten is defined by the following equation:

$$\frac{10^n}{10^m} = 10^{n-m} \tag{2.4}$$

When using this equation we must be especially sensitive to the signs of n and m, as demonstrated in the examples that follow. For all division operations, always keep in mind that dividing a number by a number less than 1 results in a number of larger magnitude. Similarly, when a small number is divided by a larger number, the result is less than the magnitude of the numerator. In other words, be aware of the magnitude of the result you expect from the operation to be performed so you have some check against the result. Don't be oblivious to the numbers involved or the result obtained. Too often the use of calculators has produced ridiculous results (due to incorrect entries) that are accepted as correct because the user had no idea of what to expect.

For example:

$$\frac{10{,}000}{100} = \underbrace{\frac{10^4}{10^2}}_{\frac{10^n}{10^m}} = \underbrace{10^{4-2}}_{10^{n-m}} = 10^2 = \mathbf{100}$$

and

$$\frac{0.001}{1000} = \frac{10^{-3}}{10^3} = 10^{-3-(+3)} = 10^{-3-3} = \mathbf{1 \times 10^{-6}}$$

Note in the last example that $n = -3$ and $m = +3$; the negative sign of Eq. 2.4 resulted in a change of sign for the exponent m.

As a further example,

$$\frac{100}{0.0001} = \frac{10^{+2}}{10^{-4}} = 10^{+2-(-4)} = 10^{2+4} = \mathbf{10^6}$$

In this example the negative power of the denominator became positive due to the operation $n - m$.

For numbers other than powers of ten, simply separate the multiplying factor from the power of ten and perform the necessary operations independently. For example,

$$\frac{0.0005}{0.002} = \frac{5 \times 10^{-4}}{2 \times 10^{-3}} = \left(\frac{5}{2}\right) \times \left(\frac{10^{-4}}{10^{-3}}\right)$$
$$= 2.5 \times 10^{-4-(-3)} = 2.5 \times 10^{-4+3}$$
$$= 2.5 \times 10^{-1} = \mathbf{0.25}$$

As noted earlier, the division $5 \div 2$ is performed separately from the power-of-ten operation.

Also,

$$\frac{800}{0.0004} = \frac{8 \times 10^{2}}{4 \times 10^{-4}} = \left(\frac{8}{4}\right) \times \left(\frac{10^{2}}{10^{-4}}\right)$$
$$= 2 \times 10^{2-(-4)} = 2 \times 10^{2+4}$$
$$= \mathbf{2 \times 10^{6}}$$

Calculator The division $(5 \times 10^{-4})/(2 \times 10^{-3})$ is entered as

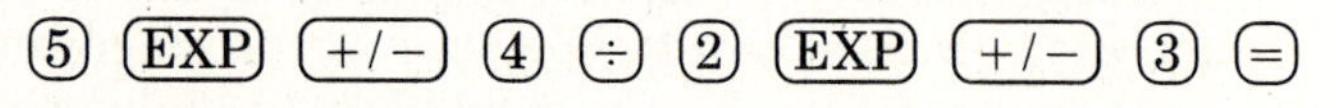

with the result **0.25.**

The division $(8 \times 10^{2})/(4 \times 10^{-4})$ is entered as

8 EXP 2 ÷ 4 EXP +/− 4 =

with the result **2,000,000 = 2 × 10⁶.**

2.9 POWERS OF POWERS OF TEN

If a power of ten is taken to a power, the result is raised to the product of the two powers. That is,

$$(10^{n})^{m} = 10^{n \times m} \tag{2.5}$$

Keep in mind that the operation with the powers of ten can be isolated from the remaining calculations. For example:

$$(1000)^{2} = (10^{3})^{2} = 10^{3 \times 2} = \mathbf{10^{6}}$$
$$(0.0001)^{4} = (10^{-4})^{4} = 10^{(-4)(+4)} = \mathbf{10^{-16}}$$
$$(40{,}000)^{3} = (4 \times 10^{4})^{3} = (4)^{3} \times (10^{4})^{3}$$
$$= (4)(4)(4) \times 10^{4 \times 3} = \mathbf{64 \times 10^{12}}$$

Calculator To perform the last two examples on a calculator requires the use of the (y^x) key in the following manner:

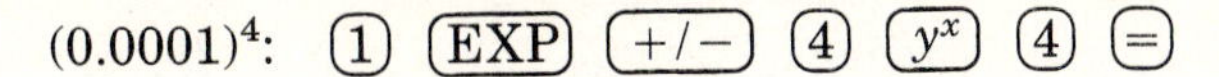

with a display **1.–16.**

$(40{,}000)^3$: (4) (EXP) (4) (y^x) (3) (=)

with a display **6.4 13.**

2.10 MIXED OPERATIONS

You will often have to use more than one of the above operations within one equation to obtain the desired result. The order in which the operations are performed is the same as with decimal numbers and is usually fairly obvious. For instance, in the first example that follows, the summation in the numerator is determined before the division is performed:

$$\frac{6.2 \times 10^4 + 9.6 \times 10^4}{0.2 \times 10^5} = \frac{(6.2 + 9.6) \times 10^4}{0.2 \times 10^5}$$

$$= \frac{15.8 \times 10^4}{0.2 \times 10^5} = \left(\frac{15.8}{0.2}\right) \times \left(\frac{10^4}{10^5}\right)$$

$$= 79 \times 10^{4-5} = 79 \times 10^{-1}$$

$$= \mathbf{7.9}$$

Calculator:

(6) (·) (2) (EXP) (4) (+) (9) (·) (6) (EXP) (4) (=) (÷) (·) (2) (EXP) (5) (=)

with a result of **7.9.**

In the next example the division operation is performed before addition, keeping in mind that the addition operation requires equal powers of ten.

$$\frac{0.0006}{0.02} + 0.008 = \frac{6 \times 10^{-4}}{2 \times 10^{-2}} + 8 \times 10^{-3}$$

$$= \frac{6}{2} \times \frac{10^{-4}}{10^{-2}} + 8 \times 10^{-3}$$

$$= 3 \times 10^{-4-(-2)} + 8 \times 10^{-3}$$

$$= 3 \times 10^{-4+2} + 8 \times 10^{-3} = 3 \times 10^{-2} + 8 \times 10^{-3}$$

$$= 30 \times 10^{-3} + 8 \times 10^{-3} = \mathbf{38 \times 10^{-3}}$$

Using the calculator, key in

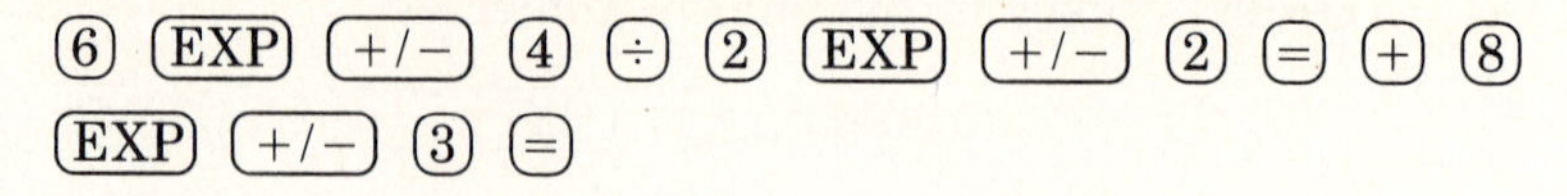

This gives a result of **0.038.** Note that the calculator internally determines a common power of ten for the addition operation.

The last example reveals the type of numbers that may be encountered in a power calculation to be performed in the following chapters.

$$\begin{aligned}(0.005)^2(4000) &= (5 \times 10^{-3})^2(4 \times 10^3) = ((5)^2 \times (10^{-3})^2)(4 \times 10^3)\\ &= (25 \times 10^{-6})(4 \times 10^3) = (25)(4) \times (10^{-6})(10^3)\\ &= 100 \times 10^{-3} = \mathbf{0.1}\end{aligned}$$

To use the calculator, key in

(5) (EXP) (+/−) (3) (x^2) (×) (4) (EXP) (3) (=)

resulting in **0.1.**

2.11 ADDITIONAL CALCULATOR FUNCTIONS

Most scientific calculators permit a choice of floating decimal point (such as 0.00065) or scientific notation (6.5×10^{-4}) by simply pressing the key (F <-> E) or one similar in notation. Pressing the key once switches from floating point to scientific notation, and pressing the key again returns the display to the original format.

For example, $0.03 = 3 \times 10^{-2}$. Using a calculator, the keys

(·) (0) (3) (=) (F <-> E)

result in a display of **3.–02**. For

$$0.0003 = 3 \times 10^{-4}$$

using a calculator, press the keys

(·) (0) (0) (0) (3) (=) (F <-> E)

This results in a display of **3.–04.**

A large number such as $450{,}000 = 4.5 \times 10^5$ can be registered using a calculator:

(4) (5) (0) (0) (0) (0) (=) (F <-> E)

This results in a display of **4.5 05.**

The quantity π is also usually available on most scientific calculators. It can be displayed by simply pressing

(2nd F) (π)

with a display of **3.141592654.**

Using the (TAB) key to limit the display to three significant places, as when pressing

(2nd F) (TAB) (3) (2nd F) (π)

provides a display of **3.142.**

The product 2π is formed by

(2) (×) (2nd F) (π) (=)

with a display of **6.283.**

The quantity π^2 can be obtained in the following manner:

(2nd F) (π) (x^2) (=)

with a display of **9.870.**

The square root of a number can be determined using the ($\sqrt{\ }$) key or the (y^x) function.

For instance, $\sqrt{16}$ is determined by

(1) (6) ($\sqrt{\ }$)

with a display of **4.** You can also use (y^x), since $\sqrt{16} = (16)^{1/2} = (16)^{0.5}$:

(1) (6) (y^x) (·) (5) (=)

with a display of **4.**

To find a cube root, (y^x) or ($x\sqrt{y}$) can be employed. That is, $\sqrt[3]{27} = (27)^{1/3}$;

(2) (7) (y^x) (( ) (1) (÷) (3) ( )) (=)

gives a display of **3.**

Note the use of the brackets in the preceding example to set x as 1/3 rather than 0.33 or 0.333, which would introduce a loss of accuracy.

Using ($x\sqrt{y}$) yields

(2) (7) (2nd F) ($x\sqrt{y}$) (3) (=)

with a display of **3.**

More will be said about the use of brackets in the chapters to follow.

2.12 PREFIXES

Since some powers of ten are used so frequently, prefixes and abbreviations have been defined for a few in order to improve levels of communication and simplicity. Those used most frequently in this and related fields appear in Table 2.1.

TABLE 2.1

Power of 10	Prefix	Abbreviation
10^{12}	Tera	T
10^{9}	Giga	G
10^{6}	Mega	M
10^{3}	kilo	k
10^{-3}	milli	m
10^{-6}	micro	μ
10^{-9}	nano	n
10^{-12}	pico	p

First note that the powers of ten are separated by a factor of 10^3 and that the abbreviations employ capital letters, lowercase letters, and the Greek letter μ (mu). Since the table reflects an international standard, it is important to use a lowercase and not a capital letter k and a capital, not a lowercase, letter M for mega; the lowercase letter m is used only for the prefix milli.

EXAMPLES

$$10{,}000 \text{ m} = 10 \times 10^3 \text{ m} = \mathbf{10\ km}$$
$$0.001 \text{ s} = 1 \times 10^{-3} \text{ s} = \mathbf{1\ ms}$$
$$5{,}000{,}000 \text{ Hz} = 5 \times 10^6 \text{ Hz} = \mathbf{5\ MHz}$$

As demonstrated in the examples, the prefix or abbreviation replaces the power of ten.

We must also be able to use the power-of-ten notation and the decimal form instead of an abbreviation, as shown in the next few examples.

$$0.05 \text{ km} = 0.05 \times 10^3 \text{ m} = \mathbf{50\ m}$$
$$400 \text{ ms} = 400 \times 10^{-3} \text{ s} = \mathbf{0.4\ s}$$
$$600\ \mu\text{s} = 600 \times 10^{-6} \text{ s} = \mathbf{0.0006\ s}$$

It is often necessary to convert from one abbreviated form to the other. The process is not difficult if you simply keep in mind that a reduction in the power of ten requires an equal shift to the right of the decimal point in the multiplying factor. Similarly, an increase in the power of ten requires an equal shift to the left of the decimal point of the multiplying factor. The process is best described with a few examples.

Convert: 0.01 ms to μs

Solution: $\underbrace{0.01}_{\substack{\text{multiplying factor} \\ \text{(increase by 3)}}} \times \overbrace{10^{-3}\ \text{s} \rightarrow \underline{\ ?\ } \times 10^{-6}}^{\text{reduce by 3}}\ \text{s}$

Since the power of ten will be *reduced* by a factor of three, the multiplying factor must be increased by moving the decimal point three places to the right. That is,

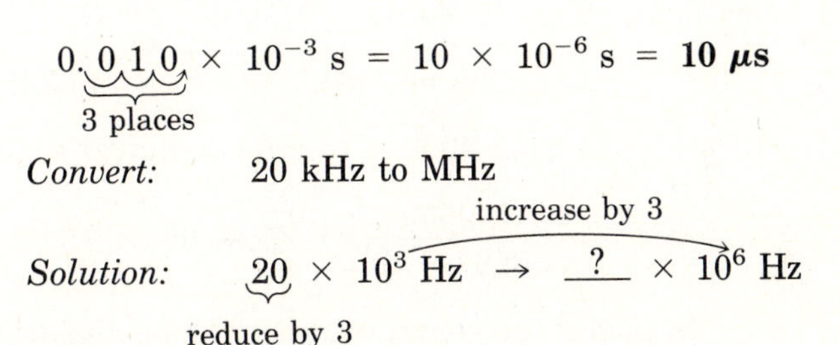

$$\underbrace{0.010}_{\text{3 places}} \times 10^{-3}\ \text{s} = 10 \times 10^{-6}\ \text{s} = \mathbf{10\ \mu s}$$

Convert: 20 kHz to MHz

Solution: $\underbrace{20}_{\text{reduce by 3}} \times \overbrace{10^{3}\ \text{Hz} \rightarrow \underline{\ ?\ } \times 10^{6}}^{\text{increase by 3}}\ \text{Hz}$

The power of ten is being increased by a factor of three, requiring a reduction in the multiplying factor as follows:

$$\underbrace{0.020}_{\text{3 places}} \times 10^{6}\ \text{Hz} = \mathbf{0.02\ MHz}$$

Convert: 4000 μm to mm

Solution: $\underbrace{4000}_{\text{reduce by 3}} \times \overbrace{10^{-6}\ \text{m} \rightarrow \underline{\ ?\ } \times 10^{-3}}^{\text{increase by 3}}\ \text{m}$

We must be careful in this conversion because the tendency is to read the change in power of ten as a reduction by a factor of three. However, keep in mind that moving from 10^{-6} to 10^{-3} is actually an increase in size, requiring that the multiplying factor be reduced as follows:

$$\underbrace{4.000}_{\text{3 places}} \times 10^{-3}\ \text{m} = 4 \times 10^{-3}\ \text{mm} = \mathbf{4\ mm}$$

2.13 CONVERSION TABLES

Conversion tables such as those appearing in Appendix A can be very useful when time does not permit the application of methods described in this chapter. However, even though such tables appear easy to use, frequent errors occur because the operations appearing at the head of the table are not properly performed. In any case, when using such tables, try to establish mentally some order of magnitude for the quantity to be determined as compared to the magnitude of the given quantity. This simple process should prevent a number of the impossible results that may occur if the conversion operation is improperly applied.

For example, consider the following from such a conversion table:

$$\frac{\text{To convert from}}{\text{Miles}} = \frac{\text{To}}{\text{Meters}} = \frac{\text{Multiply by}}{1.609 \times 10^3}$$

A conversion of 2.5 mi to meters requires that we multiply 2.5 by the conversion factor. That is,

$$2.5 \text{ mi}(1.609 \times 10^3) = 4.0225 \times 10^3 \text{ m}$$

A conversion from 4000 m to miles requires a division process:

$$\frac{4000 \text{ m}}{1.609 \times 10^3} = 2486.02 \times 10^{-3} = 2.48602 \text{ mi}$$

In each of the above, you should have little difficulty realizing that 2.5 mi would convert to a few thousand meters and 4000 m would be only a few miles. As indicated above, this kind of prior thinking will eliminate the possibility of ridiculous conversion results.

FORMULA SUMMARY

$$\frac{1}{10^n} = 10^{-n}, \qquad \frac{1}{10^{-n}} = 10^n$$

$$(10^n)(10^m) = 10^{n+m}$$

$$\frac{10^n}{10^m} = 10^{n-m}$$

$$(10^n)^m = 10^{n \times m}$$

Prefixes:

10^{12}	Tera	T
10^{9}	Giga	G
10^{6}	Mega	M
10^{3}	kilo	k
10^{-3}	milli	m
10^{-6}	micro	μ
10^{-9}	nano	n
10^{-12}	pico	p

CHAPTER SUMMARY

There are fundamental mathematical calculations in this and related fields that must be performed with speed and accuracy. In this chapter the use of powers of ten was introduced with an eye toward developing those skills necessary to proceed through the text. An entire chapter was devoted to the subject to emphasize the importance of being proficient in its use. Be assured, however, that by the time you have completed the next few chapters, you will be quite adept at using powers of ten and their prefixes.

In addition, do not become overly concerned that the field has a mathematics orientation far beyond what you expected. Yes, it does require the frequent use of basic algebraic operations, but an advanced mathematics level is not necessary to understand and apply the basic laws and theorems of electrical and electronic circuits.

When using the equations appearing in the formula summary, pay particular attention to the *sign* of n or m when substituting into the equation. The loss of a single negative sign can change a result from the micro (μ) to mega (M) range.

GLOSSARY

Scientific notation A method for describing very large and small numbers through the use of powers of ten.

PROBLEMS

Section 2.3

1. Write the following numbers as a product in which one factor is 10^3.
   a. 4000 **b.** 50,000 **c.** 104,500
   d. 1,500,000 **e.** 600 **f.** 50
2. Write the following numbers as a product in which one factor is 10^6.
   a. 5,000,000 **b.** 600,000 **c.** 40,000
3. Write the following numbers in the decimal form.
   a. 48×10^5 **b.** 0.7×10^3 **c.** 8.2×10^4
   d. 0.02×10^2 **e.** 500×10^4 **f.** 3.3×10^2
4. Write the following numbers with a power of ten of 10^{-3}.
   a. 0.005 **b.** 0.0025 **c.** 0.00009
   d. 0.02 **e.** 0.1 **f.** 20
5. Write the following numbers with a power of ten of 10^{-6}.
   a. 0.000005 **b.** 0.007 **c.** 0.000000081
6. Write the following numbers in the long-hand fashion.
   a. 0.2×10^{-3} **b.** 55×10^{-4} **c.** 2000×10^{-3}
   d. 0.009×10^{-2} **e.** 0.01×10^{-4} **f.** 4×10^{-4}

7. Write the following numbers in the $10^{\pm n}$ format (nonfractional form).

a. $\frac{1}{10^{+4}}$ b. $\frac{1}{10^{-7}}$

c. $\frac{1}{10^{+100}}$ d. $\frac{1}{10^{-0.01}}$

Section 2.4

8. Write the following numbers in scientific notation.

a. 5000 b. 0.0008 c. 1,500,000

d. 42,000 e. 150,000 f. 0.02

9. Write the numbers in Problem 1 in scientific notation.

10. Write all the numbers in Problem 4 in scientific notation.

Section 2.5

11. Perform the following additions using a power of ten of 10^3.

a. 2000 + 45,000 b. 127,000 + 8000

c. 12,000 + 600 d. 500,000 + 50

12. Perform the following additions using a power of ten of 10^{-3}.

a. 0.002 + 0.0055 b. 0.0004 + 0.006

c. 0.01 + 0.00007 d. 0.0009 + 0.000008

Section 2.6

13. Perform the following subtractions using a power of ten of 10^3.

a. 52,000 − 8.000 b. 4000 − 600

c. 108,000 − 76,500 d. 1,080,000 − 4900

14. Perform the following subtractions using a power of ten of 10^{-3}.

a. 0.009 − 0.004 b. 0.0010 − 0.0006

c. 0.050 − 0.007 d. 0.008 − 0.000072

Section 2.7

15. Perform the following operations and leave the result in scientific notation.

a. (4000) (0.004) b. (500) (0.0008)

c. (3000) (52,000) d. (0.007) (0.03)

e. $(0.01 \times 10^{-4})(50{,}000)$ f. $(5 \times 10^{+5})(0.006)$

g. $(0.009)(0.01)(4 \times 10^{+6})$ h. $(0.05 \times 10^{+5})(0.007)(4400)$

Section 2.8

16. Perform the following operations and leave the result in scientific notation.

a. $\frac{200}{0.05}$ b. $\frac{5000}{22{,}000}$

c. $\frac{0.06}{0.004}$ d. $\frac{1000}{0.05}$

e. $\frac{0.0002}{0.008}$ f. $\frac{1 \times 10^{+5}}{5000}$

g. $\frac{0.02 \times 10^{4}}{5 \times 10^{+3}}$ h. $\frac{0.002}{5 \times 10^{4}}$

Section 2.9

17. Perform the following operations and leave the result in scientific notation.

a. $(2000)^5$ **b.** $(0.006)^4$
c. $(0.05)^2$ **d.** $(4 \times 10^5)^{-3}$

Section 2.10

18. Perform the following operations and leave the result in scientific notation.

a. $(0.006)^2(4000)$ **b.** $\dfrac{(50 \times 10^{-3})^2}{2000}$

c. $\dfrac{(0.0004)^2(2200)}{5000}$ **d.** $\dfrac{200}{0.004} + 50{,}000$

e. $\dfrac{2 \times 10^3}{500 + 4 \times 10^4}$ **f.** $\dfrac{(0.003)^{-4}}{(5 \times 10^4)(0.004)}$

Section 2.11

19. Write the following quantities in the most convenient (or logical) format.

a. 22,000 m **b.** 0.0005 s
c. 95,000,000 Hz **d.** 0.000009 m
e. 0.0000000048 F **f.** 5,080,000,000 Hz
g. 0.02 s **h.** 0.0004 m

20. Convert the following as indicated.

a. 4000 μs to s **b.** 0.006 ms to μs
c. 20 kHz to MHz **d.** 0.008 MHz to kHz
e. 0.1 ms to μs **f.** 5000 μs to ms
g. 0.00004 km to mm **h.** 2200 pF to nF
i. 50,000 pF to μF **j.** 0.01 μF to nF

3

Current and Voltage

OBJECTIVES

- ☐ Become familiar with the basic atomic structure.
- ☐ Learn why copper is the most frequently employed conductor material.
- ☐ Become acquainted with Coulomb's law.
- ☐ Understand how current is established, calculated, and measured.
- ☐ Develop an understanding of voltage and how it is calculated and measured.
- ☐ Become aware of the various types of voltage sources, how they function, where they are used, and their life span.
- ☐ Understand how a current source is different from a voltage source.
- ☐ Differentiate between conductors and insulators.

3.1 INTRODUCTION

Most students of this and related fields have had some exposure to the use and meaning of the two quantities to be introduced in this chapter. Every electrical appliance has some voltage rating (usually 120 V ac) and some current rating. It may be easier to develop a sense for one than the other, but in time both will be terms that will be used and applied with equal confidence.

3.2 ATOMS AND THEIR STRUCTURE

A basic understanding of the fundamental concepts of current and voltage requires a degree of familiarity with the atom and its structure. The simplest of all atoms is the hydrogen atom, made up of two basic particles, the *proton* and the *electron,* in the relative positions shown in Fig. 3.1(a). The *nucleus* of the hydrogen atom is the proton, a positively charged particle. *The orbiting electron carries a negative charge that is equal in magnitude to the positive charge of the proton.* In all other elements, the nucleus also contains *neutrons,* which are slightly heavier than protons and have no electrical charge. The helium atom, for example, has two neutrons in addition to two electrons and two protons, as shown in Fig. 3.1(b). *In all neutral atoms the number of electrons is equal to the number of protons.* The mass of the electron is 9.11×10^{-28} g, and that of the proton and neutron is 1.672×10^{-24} g.* The

*Imagine having to write both these numbers with all the zeros required to place the decimal point properly.

FIG. 3.1
The hydrogen and helium atoms.

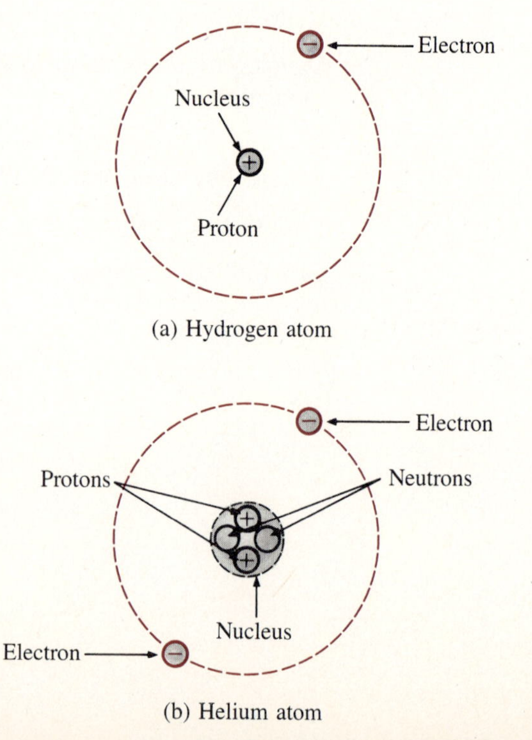

mass of the proton (or neutron) is therefore approximately 1836 times that of the electron.

For the hydrogen atom, the radius of the smallest orbit followed by the electron is about 5×10^{-11} m. The radius of this orbit is approximately 25,000 times that of the basic constituents of the atom. This is approximately equivalent to a sphere the size of a dime rotating about another sphere of similar size more than a quarter of a mile away.

Different atoms have various numbers of electrons in the concentric shells about the nucleus. The first shell, which is closest to the nucleus, can contain only 2 electrons. If an atom should have three electrons, the third must go to the next shell. The second shell can contain a maximum of 8 electrons, the third, 18, and the fourth, 32, as determined by the equation $2n^2$, where n is the shell number.

Since *copper* is the most commonly used metal in the electrical/electronics industry, let us examine its atomic structure and identify why it has such widespread application. Note in the shell structure diagram of Fig. 3.2 that the copper atom has one more electron than needed to complete the first three shells. The fact that this outermost shell is incomplete and has only one electron in the shell (the full shell is $2n^2 = 2(4)^2 = 32$ electrons) places this electron in a fairly volatile state—that is, in a position to leave the parent atom if exposed to sufficient external attractive forces.

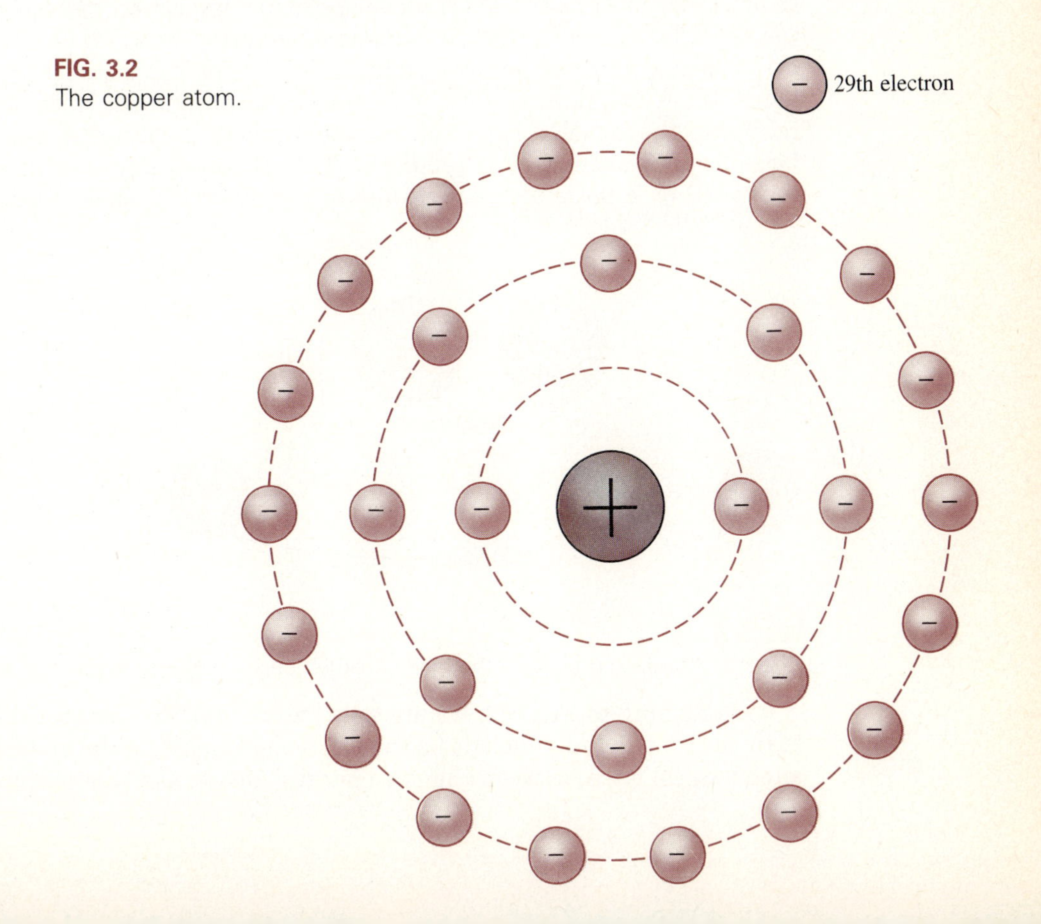

FIG. 3.2
The copper atom.

In addition, it has been determined through experimentation that

Unlike charges attract and like charges repel.

The force of attraction or repulsion between two charged bodies Q_1 and Q_2 can be determined by Coulomb's law:

$$F \text{ (attraction or repulsion)} = \frac{kQ_1Q_2}{r^2} \tag{3.1}$$

where F is the force in newtons, Q_1 and Q_2 are the charges in coulombs (to be introduced in Section 3.3), and r is the distance between the two charges in meters. The constant k is equal to 9×10^9 N-m^2/C^2 and is needed to establish a relationship among the quantities of the equation that permits the indicated units of measurement. The equation is not the simplest of nature to be introduced at this stage, but it will not be applied extensively in the analysis to follow. It was introduced primarily to reveal and support some very fundamental concepts. First, since the numerator is the product of the two charges, the larger the charges, the larger the force of attraction or repulsion. In addition, since the distance function is in the denominator, the larger the distance between the charges, the less the attractive or repulsive force. In fact, because the distance is squared, it has an even greater impact on the resulting force. Consider, for example, that doubling the distance $[(2)^2 = 4]$ reduces the force by a factor of four for the same charges. Increasing the distance by a factor of four $[(4)^2 = 16]$ reduces the force by a factor of sixteen. Note how Fig. 3.3 reinforces the fact that large charges separated by a short distance result in a large force of attraction or repulsion. The adjoining figure of small charges separated by a large distance results in a reduced measure of force.

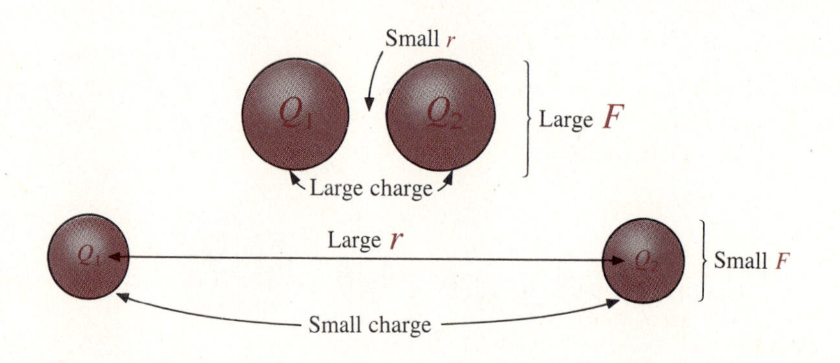

FIG. 3.3
Effect of charge and distance between charges on the force between the charges.

Returning to Fig. 3.2, we are now aware that the distance between the 29th electron and the nucleus has a significant impact on the attractive (opposite charges) force between the two charges. Due to the weaker binding force between the nucleus and the 29th electron, it is natural to expect that it would require less energy to remove this electron from the parent atom than it would take to remove any other electron of the structure.

In summary, therefore, the combination of factors just listed reveals that the 29th electron of the copper atom is *loosely* bound to the atomic structure and could leave the parent atom if sufficient energy were available from the surrounding medium or if the proper attractive forces were applied externally. The result is that this particular electron is commonly referred to as a *free electron*. These properties of the atom that permit the removal of electrons under certain conditions are essential if motion of charge is to be created. Without this motion, this discussion could venture no further—our basic quantities rely on it. In 1 in.^{3} of copper at room temperature there are approximately $1.4 \times 10^{+24}$ free electrons. Copper also has the advantage of being able to be drawn into long, thin wires (ductility) or worked into many different shapes (malleability). Other metals that exhibit the same properties as copper, but to a different degree, are silver, gold, platinum, and aluminum. Gold is used extensively in integrated circuits, where the performance level and amount of material required balance the cost factor. Aluminum has found some commercial use but suffers from being more temperature sensitive (expansion and contraction) than copper.

As an exercise in the use of powers of ten and to demonstrate the impact of changing the distance between the two charges, consider the following example.

EXAMPLE 3.1

a. Find the attractive force between two positive charges $Q_1 = 2\ \mu C$ and $Q_2 = 4\ \mu C$ separated by a distance of 10 cm.

b. Repeat part a if the distance is increased to 1 m.

Solutions

a. First note before applying Eq. 3.1 that the distance between charges is given in centimeters rather than meters, as required by the units of measurement listed with the equation. The first step, therefore, is to convert 10 cm to meters:

$$(10\ \text{cm})(1) = (10\ \cancel{\text{cm}})\left(\frac{1\ \text{m}}{100\ \cancel{\text{cm}}}\right) = 0.1\ \text{m}$$

Applying Eq. 3.1:

$$F = \frac{kQ_1Q_2}{r^2} = \frac{(9 \times 10^9\ \text{N}\cdot\text{m}^2/\text{C}^2)(2 \times 10^{-6}\ \text{F})(4 \times 10^{-6}\ \text{F})}{(0.1\ \text{m})^2}$$

$$= \frac{(9)(2)(4) \times (10^9)(10^{-6})(10^{-6})}{(10^{-1})^2}$$

$$F = \frac{72 \times 10^{9-6-6}}{10^{-2}} = 72 \times \frac{10^{-3}}{10^{-2}} = 72 \times 10^{-1} = \mathbf{7.2\ N}$$

Calculator

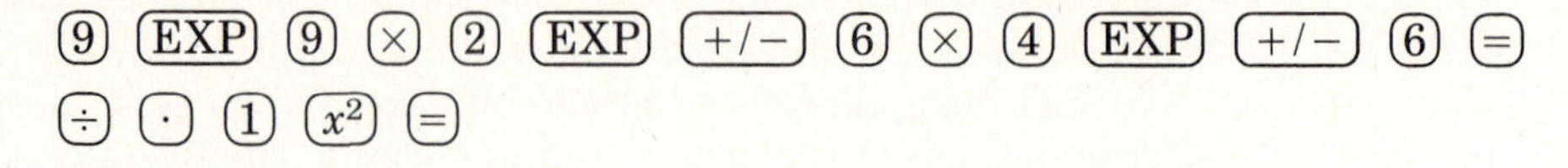

with a display of 7.2.

b. For part b the distance is given in meters:

$$F = \frac{kQ_1Q_2}{r^2} = \frac{(9 \times 10^9)(2 \times 10^{-6}\,\text{F})(4 \times 10^{-6}\,\text{F})}{(1\text{ m})^2}$$

$$= \frac{0.072}{1} = \mathbf{0.072\ N}$$

Note the significant difference in the result due to moving the charges from a 0.1-m (≅ 4-in.) to 1-m (≅ 40-in.) separation. The force was 7.2 N/0.072 N = 100 times stronger when separated by only a few inches as compared to being separated by 1 m (approximately 3 ft).

In addition, note how each step of the required calculation was performed before moving on to the next mathematical maneuver. Losing just one factor of 10 or a plus or minus sign would have resulted in a totally erroneous result.

3.3 CURRENT

Consider a short length of copper wire cut with an imaginary perpendicular plane, producing the circular cross section shown in Fig. 3.4. At room temperature with no external forces applied, there exists within the copper wire the random motion of free electrons created by the thermal energy that the electrons gain from the surrounding medium. When an atom loses its free electron, it acquires a net positive charge and is referred to as a *positive ion*. To be sure the concept of a positive ion is understood, consider the copper atom of Fig. 3.5(a). Within the imaginary boundary the *net* charge is zero, since the number of positive charges (protons) equals the number of negative charges (electrons).

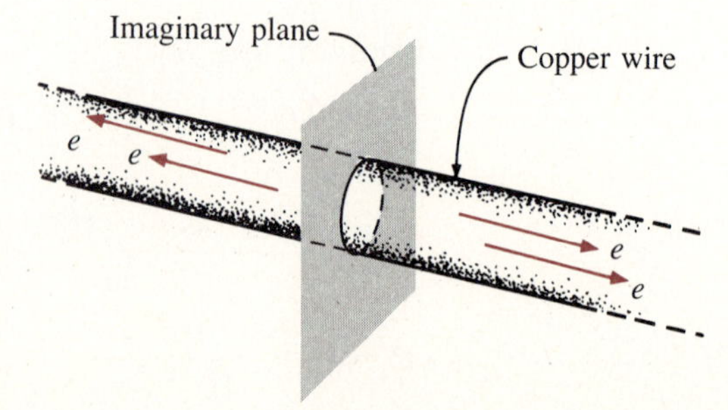

FIG. 3.4
Copper conductor with no external forces applied.

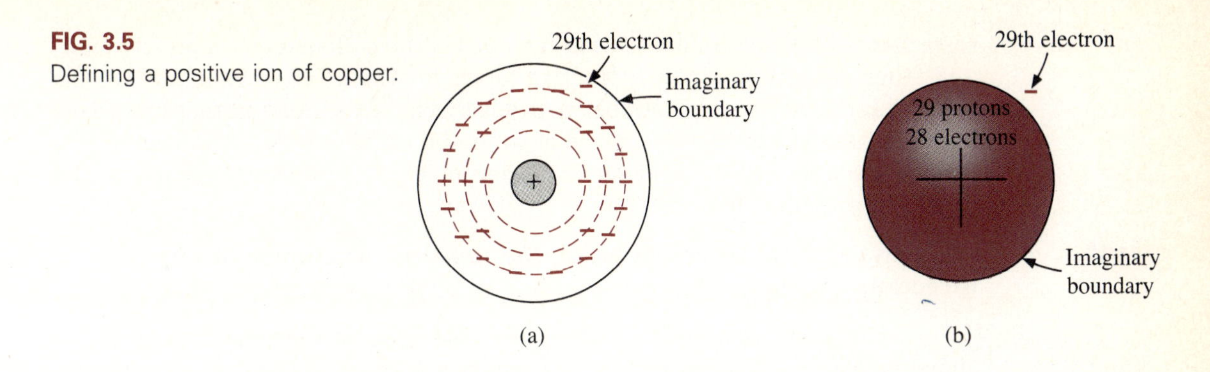

FIG. 3.5
Defining a positive ion of copper.

In other words, when we stand outside the imaginary boundary the entire structure appears to be neutral (no net charge). However, if we disassociate the 29th electron (the *free* electron) from the total structure, as shown in Fig. 3.5(b), by moving the imaginary boundary within the last shell, we find that within the boundary there are 29 protons (in the nucleus) and only 28 electrons, resulting in a net positive charge as shown in the figure. The *free* electron continues to carry its own negative charge, as shown in the same figure. In other words, by disassociating the *free* electron from the parent atom shown in Fig. 3.5(b), we create a positive ion and a *free* negative carrier. The *free* electron is able to move within these positive ions and leave the general area of the parent atom, whereas the positive ions only oscillate in a mean fixed position. For this reason,

The free electron is the charge carrier in a copper wire or in any other solid conductor of electricity.

An array of positive ions and *free* electrons is depicted in Fig. 3.6. Within this array, the *free* electrons find themselves continually gaining or losing energy by virtue of their changing direction and velocity. Some of the factors

FIG. 3.6
Random motion of free electrons in an atomic structure.

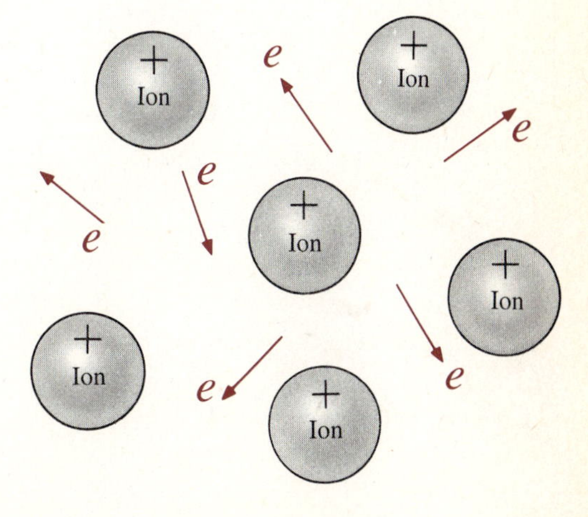

responsible for this random motion include (1) the collisions with positive ions and other electrons, (2) the attractive forces for the positive ions, and (3) the force of repulsion that exists between electrons. This random motion of free electrons is such that over a period of time, the number of electrons moving to the right across the circular cross section of Fig. 3.4 is exactly equal to the number passing over to the left.

With no external forces applied, the net flow of charge in any one direction is zero.

Let us now connect this copper wire between two battery terminals, as shown in Fig. 3.7. The battery, at the expense of chemical energy, places a net positive charge on one terminal and a net negative charge on the other. The instant the wire is connected between these two terminals, the free electrons of the copper wire will drift toward the positive terminal, whereas the positive ions will simply oscillate in a mean fixed position. The negative terminal is a supply of electrons to be drawn from when the electrons of the copper wire drift toward the positive terminal. The chemical activity of the battery absorbs the electrons at the positive terminal and maintains a steady supply of electrons at the negative terminal.

If 6.242×10^{18} *electrons* (1 C of charge) drift at uniform velocity through the imaginary circular cross section of Fig. 3.7 in 1 s, the flow of charge, or

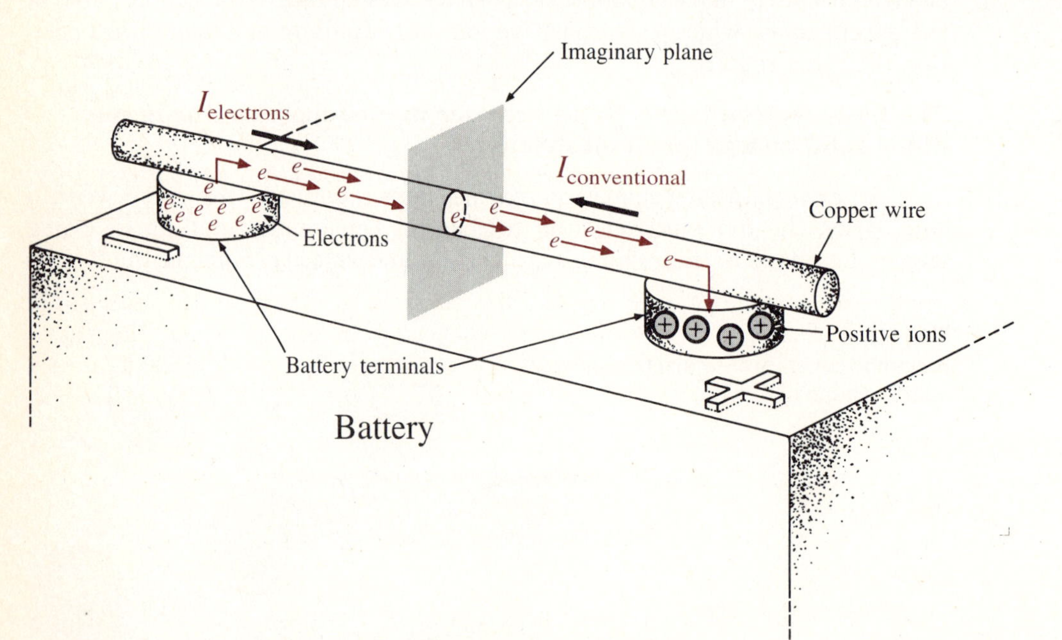

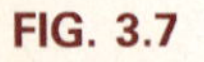
FIG. 3.7
Conduction established by a dc battery.

current, is said to be 1 *ampere* (A). The discussion of Chapter 2 revealed that this is an enormous number of electrons passing through the surface in 1 s. In fact, in floating-point form, this is 6,242,000,000,000,000,000 electrons in 1 s—further demonstrating the usefulness of scientific notation in not having to include all the zeros. The current associated with only a few electrons per second would be inconsequential and of little practical value. To establish numerical values that permit immediate comparisons between levels, 1 C of charge was defined as the total charge associated with 6.242×10^{18} electrons. The charge associated with one electron can then be determined from

$$\text{Charge/electron} = Q_e = \frac{1\text{ C}}{6.242 \times 10^{18}} = \mathbf{1.6 \times 10^{-19}\ C}$$

The current in amperes can now be calculated using the following equation:

$$I = \frac{Q}{t} \qquad \begin{aligned} I &= \text{amperes (A)} \\ Q &= \text{coulombs (C)} \\ t &= \text{seconds (s)} \end{aligned} \tag{3.2}$$

The capital letter I was chosen from the French word for current: *intensité*. The SI abbreviation for each quantity in Eq. 3.2 is provided to the right of the equation.

It is important when being exposed to an equation for the first time to take a moment to evaluate its impact rather than to simply memorize it and apply it with absolutely no feeling for the equation whatsoever. Equation 3.2 clearly indicates that since charge is in the numerator, the more charge that passes through a wire in the same time interval, the larger the current. Also, since time t is in the denominator, the longer (larger t) it takes the charge to pass through the wire, the less the resulting current. The impact of Q and t is depicted in Fig. 3.8. One of the analogies frequently used to describe current is the flow of water through a pipe. The more water that goes through a pipe during the same time period, the heavier the flow. Or, the longer the time it takes a given amount of water to flow through the pipe, the less the flow rate.

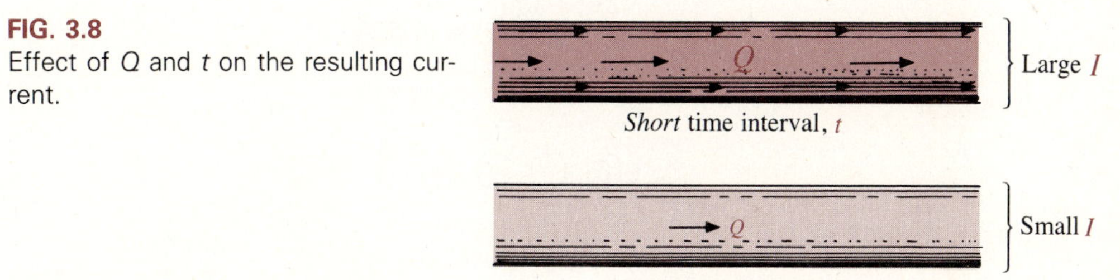

FIG. 3.8
Effect of Q and t on the resulting current.

If we multiply both sides of Eq. 3.2 by the variable t, we obtain

$$I(t) = \frac{Q}{\not{t}}(\not{t})$$

Canceling the variable t from the right side of the equation will result in the following equation for charge:

$$\boxed{Q = It} \qquad \text{(coulombs, C)} \qquad (3.3)$$

Equation 3.3 provides the amount of charge that has passed through a wire for a particular current level and time interval. Since the product of I and t determines the resulting charge, the larger the current or the longer the flow of charge is permitted to continue, the larger the amount of charge that will pass through the wire—an understandable conclusion.

And finally, if we divide both sides of Eq. 3.3 by the variable I, we obtain

$$\frac{Q}{I} = \frac{\not{I}t}{\not{I}}$$

and

$$\boxed{t = \frac{Q}{I}} \qquad \text{(seconds, s)} \qquad (3.4)$$

revealing that the time it takes a particular charge Q to pass through a wire is *inversely* related to the current level. That is, the larger the current, the less time (and vice versa) it takes a particular measure of charge to pass through the wire, since I is in the denominator of Eq. 3.4.

EXAMPLE 3.2 There are 6 C of charge passing through a wire every 4 s. Determine the current in amperes.

Solution Using Eq. 3.2:

$$I = \frac{Q}{t} = \frac{6\ \text{C}}{4\ \text{s}} = \mathbf{1.5\ A}$$

Calculator

(6) (÷) (4) (=)

with a display of 1.5.

EXAMPLE 3.3

How much charge has passed through a wire in 1/2 min if the current is 5 mA?

Solution Using Eq. 3.3:

$$Q = It = (5 \times 10^{-3}\text{ A})(30\text{ s}) = 150 \times 10^{-3}\text{ C}$$
$$= \mathbf{0.15\ C}$$

Note that we first have to convert the time interval to seconds.

Calculator

(5) (EXP) (+/−) (3) (×) (3) (0) (=)

with a display of 0.15.

EXAMPLE 3.4

Determine how long it will take 30×10^{20} electrons to pass through a wire if the current is 4 A.

Solution First we must determine the charge associated with 30×10^{20} electrons:

$$30 \times 10^{20}\ \cancel{\text{electrons}}\left(\frac{1\text{ C}}{6.242 \times 10^{18}\ \cancel{\text{electrons}}}\right) = 480.615\text{ C}$$

Calculator

(3) (0) (EXP) (2) (0) (÷) (6) (·) (2) (4) (2) (EXP) (1) (8) (=)

with a display of 480.615.

Then, using Eq. 3.4:

$$t = \frac{Q}{I} = \frac{480.615\text{ C}}{4\text{ A}} = \mathbf{120.154\ s}$$

or

$$120.154\ \cancel{\text{s}}\left(\frac{1\text{ min}}{60\ \cancel{\text{s}}}\right) = \mathbf{2.003\ min}$$

Calculator

(4) (8) (0) (·) (6) (1) (5) (÷) (4) (=)

with a display of 120.154 and

(1) (2) (0) (·) (1) (5) (4) (÷) (6) (0) (=)

with a display of 2.003.

A second glance at Fig. 3.7 reveals that two directions of charge flow have been indicated. One is called *conventional flow*, and the other is called *electron flow*. This text deals only with conventional flow for a variety of reasons, including the fact that it is the most widely used at educational institutions and in industry, is employed in the design of all electronic device symbols, and is the popular choice for all major computer software packages. The flow controversy is a result of an assumption made at the time electricity was discovered that the positive charge was the moving particle in metallic conductors. Be assured that the choice of conventional flow will not create great difficulty and confusion in the chapters to follow. Once the direction of I is established, the issue is dropped and the analysis can continue.

3.4 VOLTAGE

The current through a coil of wire or any electrical system is zero until some external source of energy is applied to establish a drift of charge in one direction. In an electrical system, the source of "pressure" is the applied voltage available from a battery, an outlet, or the like.

In order to establish this pressure, a *positioning* of charge must be established that will create a net positive charge at one end of the source and a net negative charge at the other. For the battery of Fig. 3.7 the accumulation of negative and positive charge at the respective terminals is established through internal chemical action. The energy that a body (such as an electron) has by virtue of its *position* is referred to as *potential energy*. Since *energy* is by definition the *capacity to do work*, those electrons at the negative terminal have inherited sufficient potential energy to overcome collisions with other particles in a conductor and the repulsion of other similar charges to reach the positive terminal to which they are attracted. In other words, through the expense of chemical energy, the electrons are placed in a position that provides sufficient potential energy for them to overcome opposing forces in the wire as they establish a net flow of charge or current in the conductor.

Charge can be raised to a higher potential level through the expenditure of energy from an external source, or it can lose potential energy as it travels through an electrical system. In any case, by definition:

A potential difference of 1 V exists between two points if 1 J of energy is exchanged in moving 1 C of charge between the two points.

In general, the potential difference between two points is determined by

$$V = \frac{W}{Q} \qquad \begin{aligned} V &= \text{volts (V)} \\ W &= \text{joules (J)} \\ Q &= \text{coulombs (C)} \end{aligned} \tag{3.5}$$

Through algebraic manipulations like those described for Eq. 3.2, we have

$$\boxed{W = QV} \quad \text{(joules, J)} \tag{3.6}$$

and

$$\boxed{Q = \frac{W}{V}} \quad \text{(coulombs, C)} \tag{3.7}$$

EXAMPLE 3.5 Find the potential difference between two points in an electrical system if 60 J of energy are expended by a charge of 20 C between these two points.

Solution Using Eq. 3.5:

$$V = \frac{W}{Q} = \frac{60\text{ J}}{20\text{ C}} = \mathbf{3\ V}$$

EXAMPLE 3.6 Determine the energy expended moving a charge of 50 μC through a potential difference of 6 V.

Solution Using Eq. 3.6:

$$W = QV = (50 \times 10^{-6}\text{ C})(6\text{ V}) = 300 \times 10^{-6}\text{ J} = \mathbf{300\ \mu J}$$

Calculator

(5) (0) (EXP) (+/−) (6) (×) (6) (=)

with a display of 0.0003 = 300 μJ.

By adding the (F <−> E) key to the preceding sequence, the result would be 3.−04 (scientific notation).

Notation plays a very important role in the analysis of electrical and electronic systems. To distinguish between sources of voltage (batteries and the like) and losses in potential across dissipative elements, the following notation will be used:

E for voltage sources (volts)
V for voltage drops (volts)

In summary, the applied potential difference (in volts) of a voltage source in an electric circuit is the "pressure" to set the system in motion and "cause" the flow of charge or current through the electrical system. A mechanical analogy of the applied voltage is the pressure applied to the water in a main. The resulting flow of water through the system is likened to the flow of charge through an electric circuit.

3.5 FIXED (DC) SUPPLIES

The notation *dc* employed in the heading of this section is an abbreviation for *direct current,* which encompasses the various electrical systems in which there is a *unidirectional* ("one direction") flow of charge. A great deal more will be said about this terminology in the chapters to follow. For now, we will consider only those supplies that provide a fixed voltage or current.

Dc Voltage Sources Since the dc voltage source is the more familiar of the two types of supplies, it is examined first. The symbol used for all dc voltage supplies in this text appears in Fig. 3.9. The relative lengths of the bars indicate the terminals they represent.

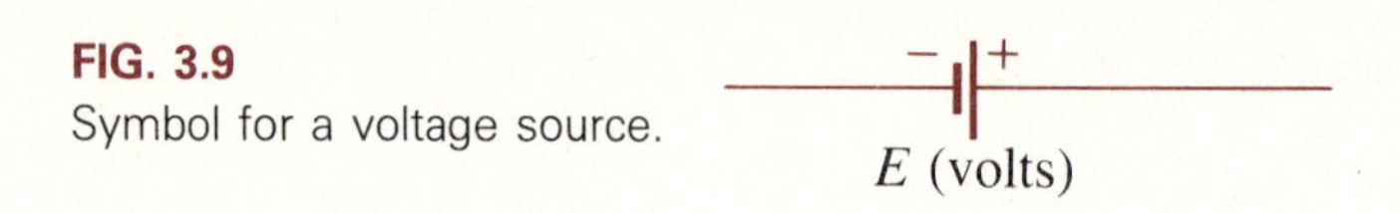

FIG. 3.9
Symbol for a voltage source.

Dc voltage sources can be divided into three broad categories: (1) batteries (chemical action), (2) generators (electromechanical), and (3) power supplies (rectification).

Batteries For the layperson, a battery is the most common of the dc sources. By definition, a battery (derived from the expression "battery of cells") consists of a combination of two or more similar *cells,* a cell being the fundamental source of electrical energy developed through the conversion of chemical or solar energy. All cells can be divided into *primary* or *secondary* types. The secondary type is rechargeable, whereas the primary is not. That is, the chemical reaction of the secondary cell can be reversed to restore its capacity. The two most common rechargeable batteries include the lead-acid unit (used primarily in automobiles) and the nickel-cadmium battery (used in calculators, tools, photoflash units, shavers, and so on). The obvious advantage of the rechargeable unit is the reduced costs associated with not having continually to replace discharged primary cells.

All the cells appearing in this chapter except the *solar cell,* which absorbs energy from incident light in the form of photons, establish a potential difference at the expense of chemical energy. In addition, each has a positive and a negative *electrode* and an *electrolyte* to complete the circuit between electrodes within the battery. The electrolyte is the contact element and the source of ions for conduction between the terminals.

The popular carbon-zinc primary battery uses a zinc can as its negative electrode, a manganese dioxide mix and carbon rod as its positive electrode, and an electrolyte that is a mix of ammonium and zinc chlorides, flour, and starch, as shown in Fig. 3.10. Figure 3.11 shows a number of other types of primary units with an area of application and a rating to be considered later in this section.

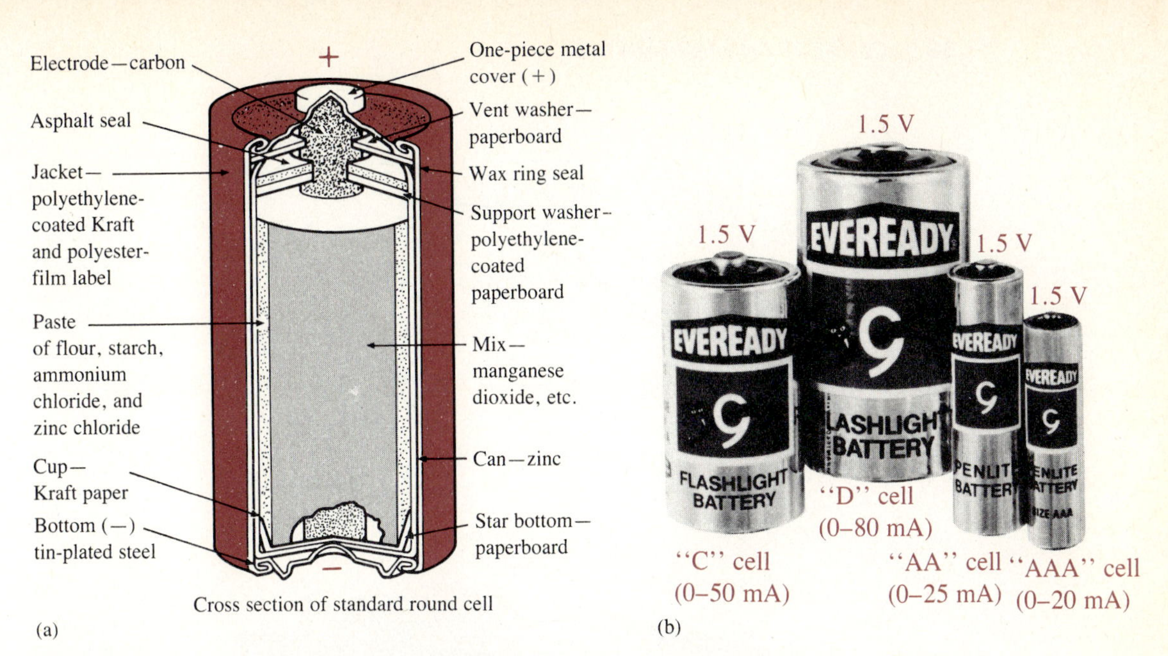

FIG. 3.10

Carbon-zinc primary battery. (a) Construction; (b) appearance and ratings. (Courtesy of Eveready Batteries)

(a) Lithiode™ lithium-iodine cell 2.8 V, 870 mAh
Long-life power sources with printed circuit board mounting capability

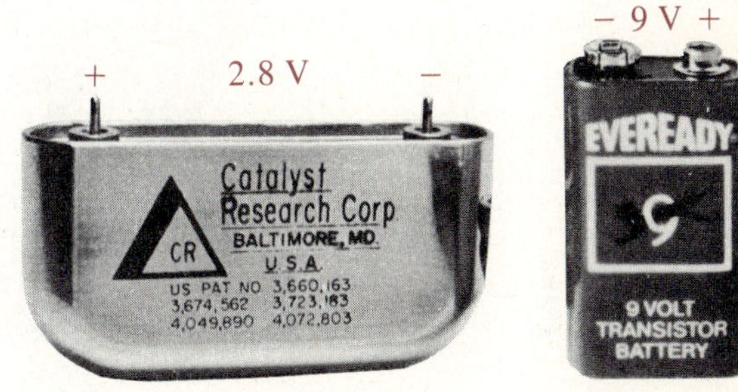

(b) Lithium-iodine pacemaker cell 2.0 Ah

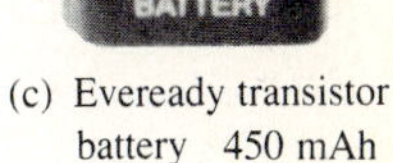

(c) Eveready transistor battery 450 mAh

FIG. 3.11

Primary cells. (Parts (a) and (b) courtesy of Catalyst Research Corp.; part (c) courtesy of Eveready Batteries)

For the secondary lead-acid unit in Fig. 3.12, the electrolyte is sulfuric acid and the electrodes are spongy lead (Pb) and lead peroxide (PbO_2). When a load is applied to the battery terminals, there is a transfer of electrons from the spongy lead electrode to the lead peroxide electrode through the load. This transfer of electrons continues until the battery is completely discharged. The discharge time is determined by how diluted the acid has become and how

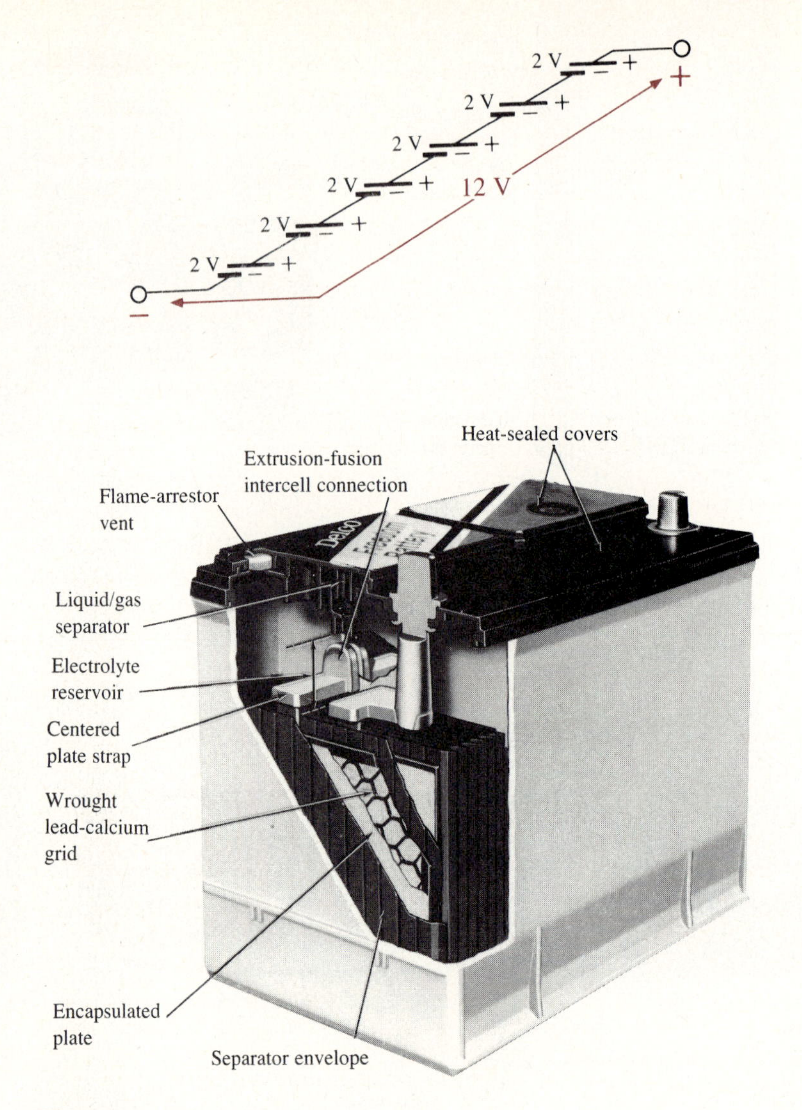

FIG. 3.12
Maintenance-free 12-V lead-acid battery. (Courtesy of Delco-Remy, a division of General Motors Corp.)

heavy the coating of lead sulfate is on each plate. The state of discharge of a lead storage cell can be determined by measuring the specific gravity of the electrolyte with a hydrometer. The specific gravity of a substance is defined to be the ratio of the weight of a given volume of the substance to the weight of an equal volume of water at 4°C. For fully charged batteries, the specific gravity should be somewhere between 1.28 and 1.30. When the specific gravity drops to about 1.1, the battery should be recharged.

Since the lead storage cell is a secondary cell, it can be recharged at any point during the discharge phase simply by applying an external dc source across the cell that passes current through the cell in a direction opposite to that in which the cell supplied current to the load. This removes the lead sulfate from the plates and restores the concentration of sulfuric acid.

The output of a lead storage cell over most of the discharge phase is about 2 V. In the commercial lead storage batteries used in the automobile, the 12 V can be produced by six cells in series, as shown in Fig. 3.12. The use of a grid made from a wrought lead-calcium alloy strip rather than the lead-antimony cast grid commonly used has resulted in maintenance-free batteries such as the one shown in the same figure. The lead-antimony structure was susceptible to corrosion, overcharge, gassing, water usage, and self-discharge. Improved design with the lead-calcium grid has either eliminated or substantially reduced most of these problems.

The nickel-cadmium battery is a rechargeable battery that has been the subject of enormous interest and development in recent years. A number of such batteries manufactured by Union Carbide Corporation and General Electric Company appear in Fig. 3.13. The internal construction of the cylindrical-

(a)

Eveready® BH 500 cell
1.2 V, 500 mAh
App: Where vertical height is severe limitation
(b)

Printed circuit board mountable battery
2.4 V, 70 mAh
(c)

FIG. 3.13
Rechargeable nickel-cadmium batteries. (Parts (a) and (b) courtesy of Eveready Batteries; part (c) courtesy of General Electric Company.)

type cell is shown in Fig. 3.14. In the fully charged condition the positive electrode is nickel hydroxide [$Ni(OH)_2$]; the negative electrode is metallic cadmium (Cd); and the electrolyte is potassium hydroxide (KOH). The oxidation (increased oxygen content) of the negative electrode occurring simultaneously

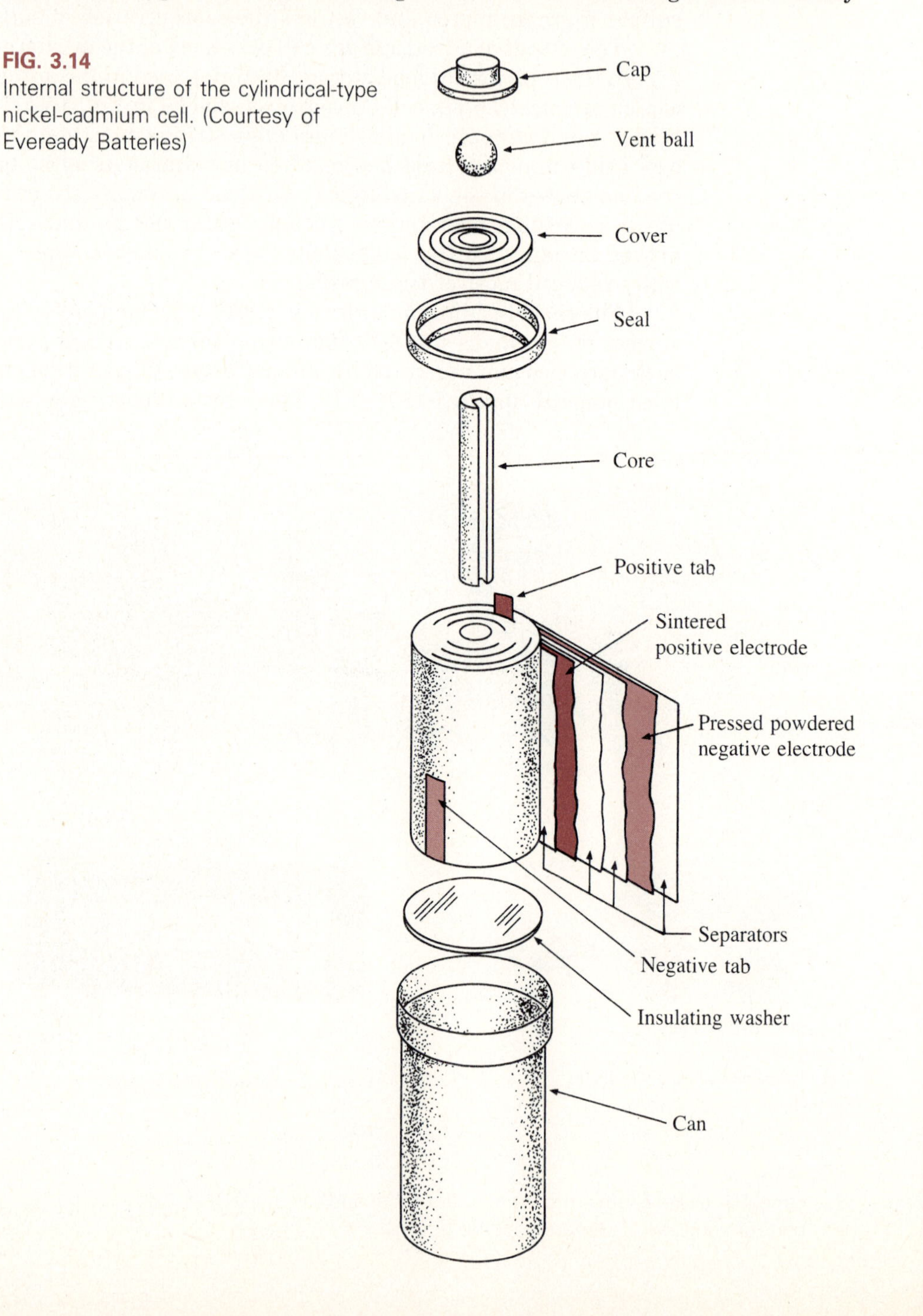

FIG. 3.14
Internal structure of the cylindrical-type nickel-cadmium cell. (Courtesy of Eveready Batteries)

with the reduction of the positive electrode provides the required electrical energy. The separator is required to isolate the two electrodes and maintain the location of the electrolyte. The advantage of such cells is that the active materials go through a change in oxidation state necessary to establish the required ion level without a change in the physical state. This establishes an excellent recovery mechanism for the recharging phase.

A high-density, 40-W solar cell appears in Fig. 3.15 with some of its associated data and areas of application. Since the maximum available wattage in an average bright sunlit day is 100 mW/cm^2 and conversion efficiencies are currently between 10% and 14%, the maximum available power per square

FIG. 3.15
Solar module. (Courtesy of Motorola Semiconductor Products)

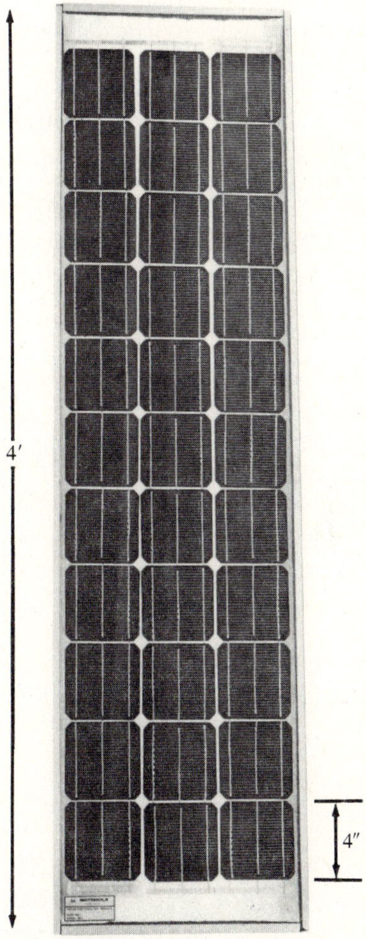

4′

4″

40-W, high-density solar module
100-mm × 100-mm (4″ × 4″) square cells are used to provide maximum power in a minimum of space. The 33 series cell module provides a strong 12-V battery charging current for a wide range of temperatures (−40°C to 60°C)

centimeter from most commercial units is between 10 mW and 14 mW. For a square meter, however, the return would be 100 W to 140 W. A more detailed description of the solar cell will be given in your electronics courses. For now, it is important to realize that a fixed illumination of a solar cell provides a fairly steady dc voltage for driving various loads, from watches to automobiles.

3.6 AMPERE-HOUR RATING

Batteries have a capacity rating given in ampere-hours (Ah) or milliampere-hours (mAh). Some of these ratings are included in earlier figures of this chapter. A battery with an ampere-hour rating of 100 Ah will theoretically provide a steady current of 1 A for 100 h, 2 A for 50 h, 10 A for 10 h, and so on, as determined by the following equation:

$$\text{Life (hours)} = \frac{\text{Ampere-hour rating (Ah)}}{\text{Amperes drawn (A)}} \tag{3.8}$$

Two factors that affect this rating, however, are the temperature and the rate of discharge. The disc-type EVEREADY® BH 500 cell shown in Fig. 3.13 has the terminal characteristics given in Fig. 3.16. Note that for the 1-V unit,

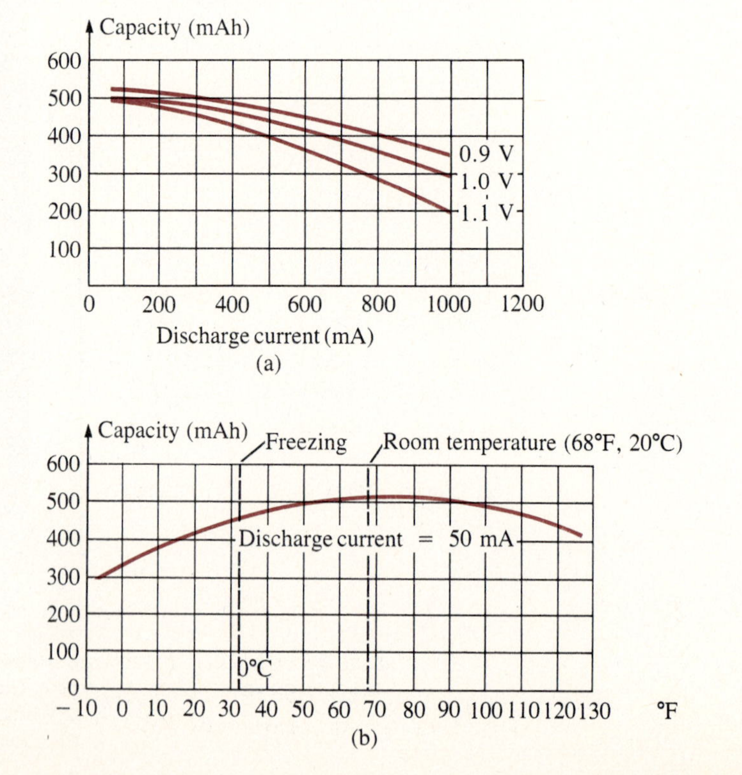

FIG. 3.16
EVEREADY® BH 500 cell characteristics. (a) Capacity vs. discharge current; (b) capacity vs. temperature. (Courtesy of Eveready Batteries)

the rating is above 500 mAh at a discharge current of 100 mA [Fig. 3.16(a)] but drops to 300 mAh at about 1 A. For a unit that is less than $1\frac{1}{2}$ in. in diameter and less than 1/2 in. in thickness, however, these are excellent terminal characteristics. Figure 3.16(b) reveals that the maximum mAh rating (at a current drain of 50 mA) occurs at about 75°F (≅ 24°C), or just above average room temperature. Note how the curve drops to the right and left of this maximum value. We are all aware of the reduced "strength" of a battery at low temperatures. Note that it has dropped to almost 300 mAh at −20°C.

Another curve of interest appears in Fig. 3.17. It provides the expected cell voltage at a particular drain over a period of hours of use. It is noteworthy that the loss in hours between 50 mA and 100 mA is much greater than between 100 mA and 150 mA, even though the increase in current is the same between levels.

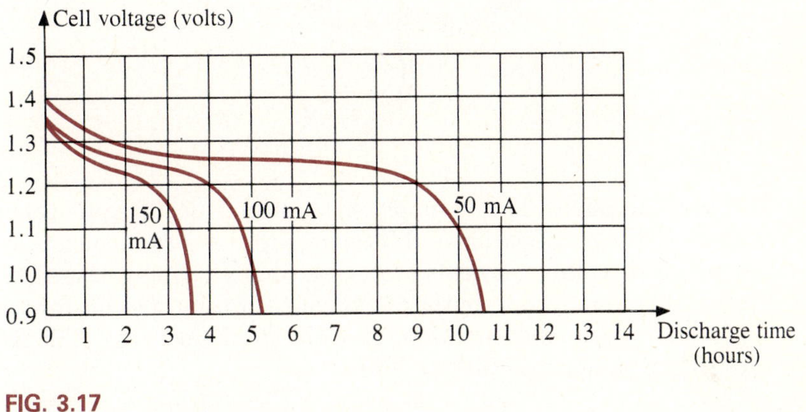

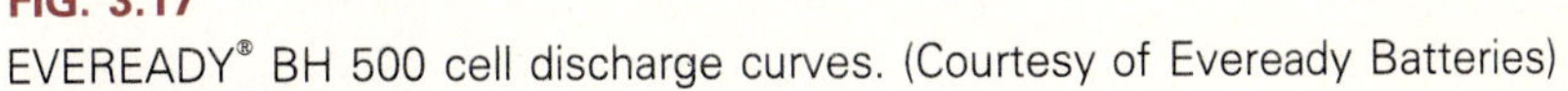

FIG. 3.17
EVEREADY® BH 500 cell discharge curves. (Courtesy of Eveready Batteries)

EXAMPLE 3.7

a. Determine the capacity in milliampere-hours for the 0.9-V BH 500 cell of Fig. 3.16(a) if the discharge current is 600 mA.

b. At what temperature will the milliampere-hour rating of the cell of Fig. 3.16(b) be 90% of its maximum value if the discharge current is 50 mA?

Solutions

a. From Fig. 3.16(a), the capacity at 600 mA is about 450 mAh. Thus, from Eq. 3.8,

$$\text{Life} = \frac{450 \text{ mAh}}{600 \text{ mA}} = 0.75 \text{ h} = \mathbf{45\ min}$$

b. From Fig. 3.16(b), the maximum is approximately 520 mAh. The 90% level is therefore 468 mAh, which occurs just above freezing, or **1°C,** and at the higher temperature of **45°C.**

Generators The dc generator is quite different, both in construction (Fig. 3.18) and in mode of operation, from the battery. When the shaft of a generator is rotating at the nameplate speed due to the applied torque of some external source of mechanical power, a voltage of rated value will appear across the external terminals. The terminal voltage and power-handling capabilities of the dc generator are typically higher than those of most batteries, and its lifetime is determined only by its construction. Commercially used dc generators are typically of the 120-V or 240-V variety. As pointed out earlier in this section, for the purposes of this text no distinction is made between the symbols for a battery and a generator.

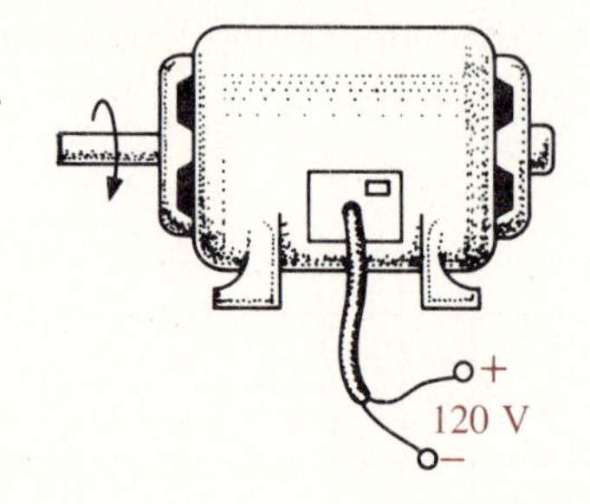

FIG. 3.18
Dc generator.

Power Supplies The dc supply encountered most frequently in the laboratory employs the rectification and filtering processes as its means toward obtaining a steady dc voltage. By this process, a time-varying voltage (such as ac voltage available from a home outlet) is converted to one of a constant magnitude. This process is covered in detail in basic electronics courses. A dc laboratory supply of this type appears in Fig. 3.19.

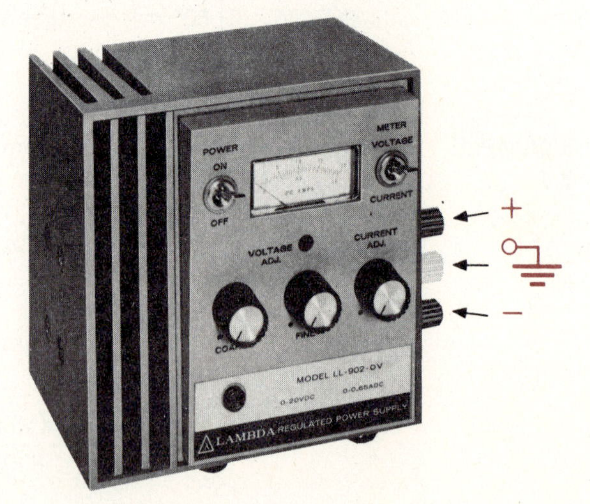

FIG. 3.19
Dc laboratory supply. (Courtesy of Lambda Electronics Corp.)

Most dc laboratory supplies have a regulated, adjustable voltage output with three available terminals, as indicated in Figs. 3.19 and 3.20(a). The symbol for ground or zero potential (the reference) is also shown in Fig. 3.20(a). If 10 V above ground potential are required, then the connections are made as shown in Fig. 3.20(b). If 15 V below ground potential are required, then the connections are made as shown in Fig. 3.20(c). If connections are as shown in Fig. 3.20(d), we say we have a "floating" voltage of 5 V, since the reference level is not included. The configuration of Fig. 3.20(d) is seldom employed, since it fails to protect the operator by providing a direct low-resistance path to ground and to establish a common ground for the system. In any case, the positive and negative terminals must be part of any circuit configuration.

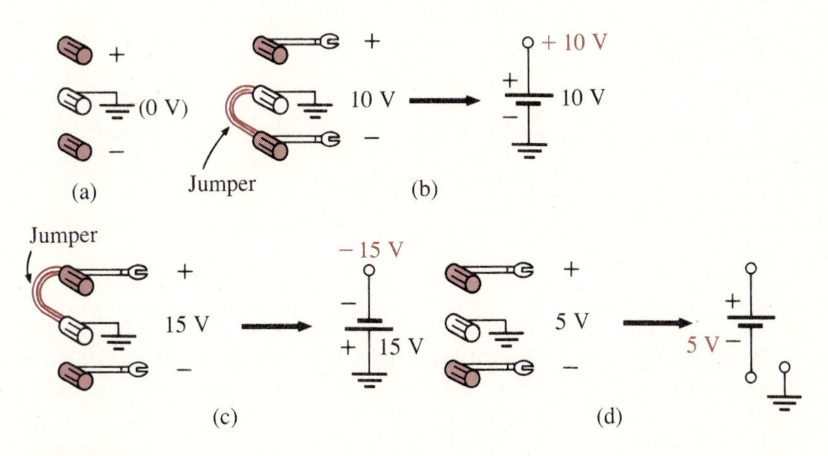

FIG. 3.20
Possible output connections of a dc laboratory supply.

Dc Current Sources The wide variety of types of and applications for the dc voltage source have resulted in its becoming a rather familiar device, the characteristics of which are understood, at least basically, by the layperson. For example, it is common knowledge that a 12-V car battery has a terminal voltage (at least approximately) of 12 V even though the current drain by the automobile may vary under different operating conditions. In other words, *a dc voltage source ideally provides a fixed terminal voltage even though the current drain may vary,* as depicted in Fig. 3.21(a). A dc current source is the dual of the voltage source. That is, *the current source, ideally, supplies a fixed current to a load even though there are variations in the terminal voltage as determined by the load,* as depicted in Fig. 3.21(b). (Do not become alarmed if the concept of a current source is strange and somewhat confusing at this point. It will be covered in great detail in later chapters.)

FIG. 3.21
Terminal characteristics. (a) Ideal voltage source; (b) ideal current source.

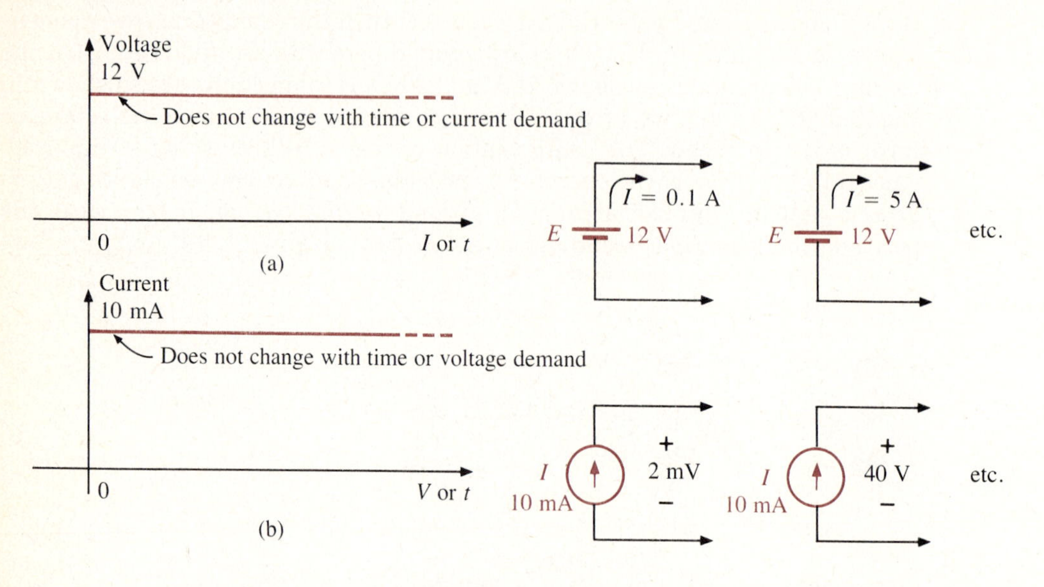

The introduction of semiconductor devices such as the transistor has accounted in large measure for the increasing interest in current sources. A representative commercially available dc current source appears in Fig. 3.22.

3.7 CONDUCTORS AND INSULATORS

Different wires placed across the same two battery terminals allow different amounts of charge to flow between the terminals. Many factors, such as the stability, density, and mobility of the material, account for these variations in charge flow. *Conductors are those materials that permit a generous flow of electrons with very little electromotive force applied.* Since copper is used most frequently, it serves as the standard of comparison for the relative conductivity in Table 3.1. Note that aluminum, which has lately seen some commercial use, has only 61% of the conductivity level of copper, but keep in mind that this must be weighed against the cost and weight factors.

Materials that have very few free electrons, high stability and density, and low mobility are called *insulators,* since a *large potential difference is required to produce any sizable current* through such materials. A common use of insulating material is for covering current-carrying wire, which, if uninsulated, could cause dangerous side effects. For example, workers on high-

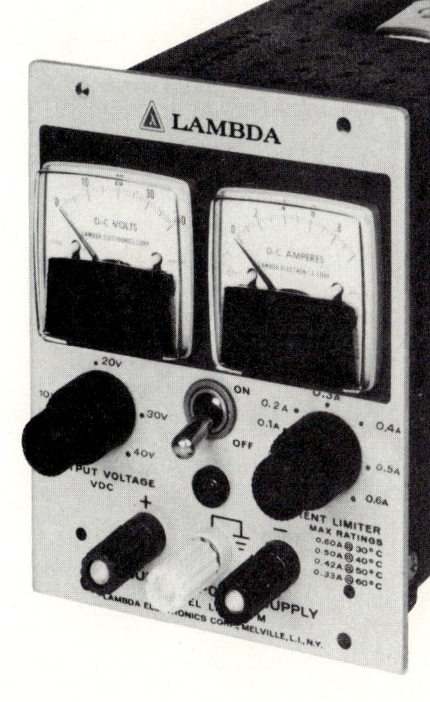

FIG. 3.22
Dc current source. (Courtesy of Lambda Electronics Corp.)

TABLE 3.1
Relative conductivity of various materials.

Metal	Relative Conductivity (%)
Silver	105
Copper	100
Gold	70.5
Aluminum	61
Tungsten	31.2
Nickel	22.1
Iron	14
Constantan	3.52
Nichrome	1.73
Calorite	1.44

voltage power lines wear rubber gloves and stand on rubber mats as additional safety measures. A number of different types of insulators and their applications appear in Fig. 3.23.

FIG. 3.23
Insulators. (a) Insulated thru-panel bushings; (b) antenna strain indicators; (c) porcelain stand-off insulators. (Courtesy of Herman H. Smith, Inc.)

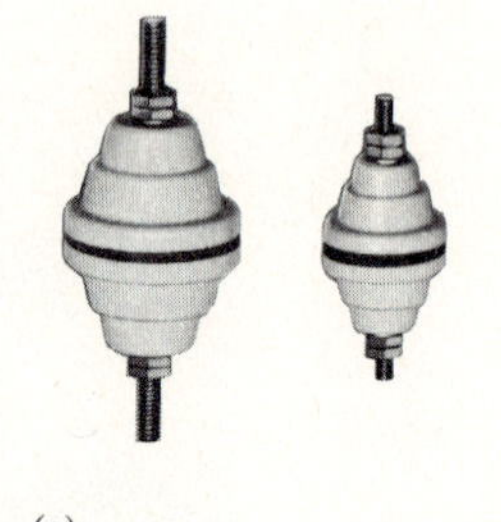

(a)

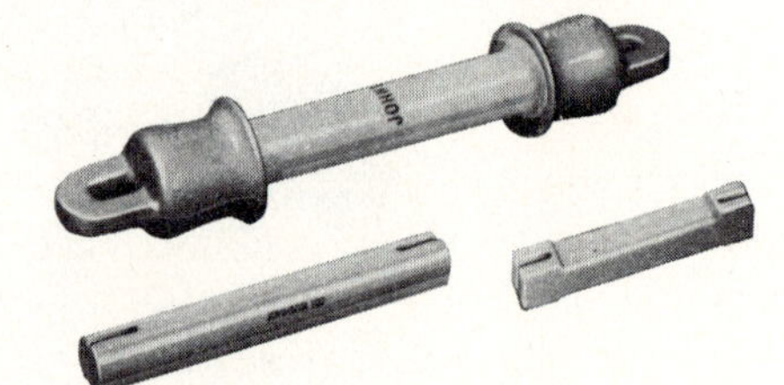

(b)

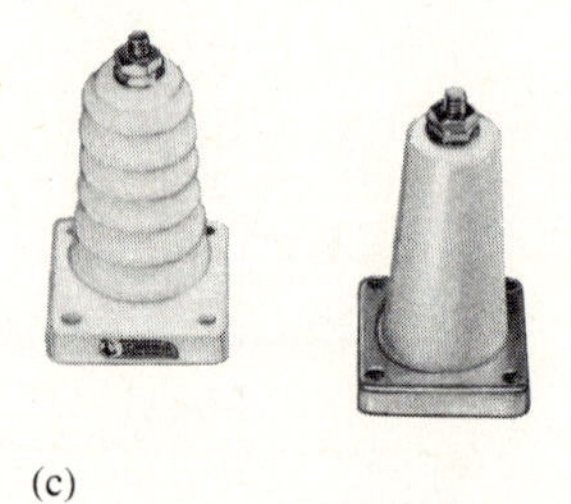

(c)

It must be pointed out, however, that even the best insulator will break down (permit charge to flow around and through it) if a sufficiently large potential is applied across it. The breakdown strengths of some common insulators are listed in Table 3.2.

According to this table, for insulators with the same geometric shape, it would require 270/30 = 9 times as much potential to pass current through rubber as through air and approximately 67 times as much voltage to pass current through mica as through air.

TABLE 3.2
Breakdown strengths of some common insulators.

Material	Average Breakdown Strength (kV/cm)
Air	30
Porcelain	70
Oils	140
Bakelite	150
Rubber	270
Paper (paraffin-coated)	500
Teflon	600
Glass	900
Mica	2000

3.8 SEMICONDUCTORS

Between the class of elements called *insulators* and those exhibiting conductor properties, there exists a group of elements of significant importance called *semiconductors*. The entire electronics industry is dependent on these materials since the diodes, transistors, and integrated circuits (ICs) that we hear so much about are constructed of semiconductor materials. Although *silicon* is the most extensively employed, *germanium* is also used in a number of devices. Both of these materials will be examined in some detail in your electronics courses. It will then be demonstrated why they are so appropriate for the applications noted above.

3.9 AMMETERS AND VOLTMETERS

It is important to be able to measure the current and voltage levels in the network in order to check its operation, isolate malfunctions, and investigate effects impossible to predict on paper. As the names imply, *ammeters* are used to measure current levels, and *voltmeters* measure the potential difference between two points. If the current levels are usually of the order of milliamperes, the instrument is referred to as a *milliammeter,* and if it is in the microampere range, as a *microammeter*. Similar statements can be made for voltage levels. Throughout the industry there are two types of scales used to display the results of a measurement—*analog* and *digital*. The volt-ohm-milliammeter (VOM) of Fig. 3.24 incorporates an analog scale, whereas the digital multimeter (DMM) of Fig. 3.25 utilizes a digital scale. Both instruments are capable of measuring current, voltage, and resistance (to be introduced in the next chapter) at a number of levels.

The digital meter has the advantage of displaying the magnitude of the measurement with the decimal point in place. For instance, a measurement of

FIG. 3.24
Volt-ohm-milliammeter (VOM). (Courtesy of Simpson Electric Co.)

FIG. 3.25
Digital multimeter (DMM). (Courtesy of John Fluke Mfg. Co. Inc.)

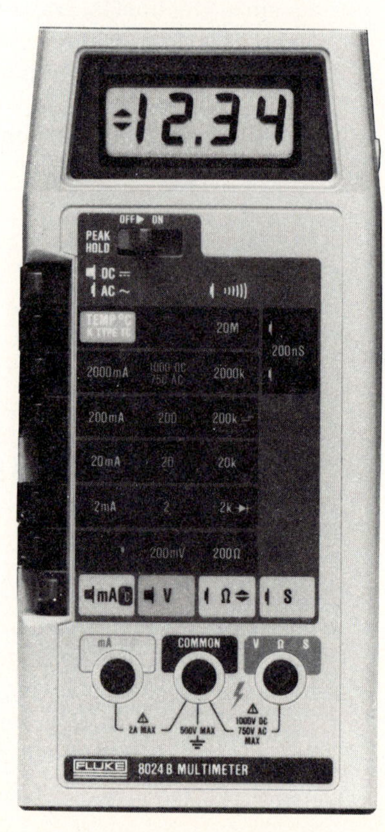

12.34 (Fig. 3.25) when the meter is set on the 20-mA scale is obviously a current of 12.34 mA. For the highest accuracy, the chosen scale should have a maximum value closest to the level being measured.

For the analog instrument, the chosen scale determines the placement of the decimal point for the numerical values of the scale. For instance, the 250 of the voltage scale must be read as 2.5 V when the dial is on the 2.5-V scale and 250 V when the 250-V scale is chosen. One must then be able to interpret the reading of the instrument from the scale divisions, as demonstrated in Fig. 3.26. The 10 divisions between 50 V and 100 V reveal that each division corresponds to a 5-V difference. The result is the levels indicated within the brackets and an actual reading of 82.5 V. If the meter were on the 2.5-V scale, the reading would be 0.825 V. When using an analog scale one must also be sure to read the meter "head-on." Reading the scale from an angle (to either side) can create a *parallax* error that can result in a reading above or below the correct level, depending on the direction from which the reading is made. In Fig. 3.27 the analog meter is on the 10-mA scale and each division carries a weight of 0.2 V. The pointer is between the short markers of 6.6 and 6.8 but leans more toward 6.8 mA. In fact, since 6.7 is exactly halfway between 6.6 mA and 6.8 mA, a reading of 6.75 mA seems in order.

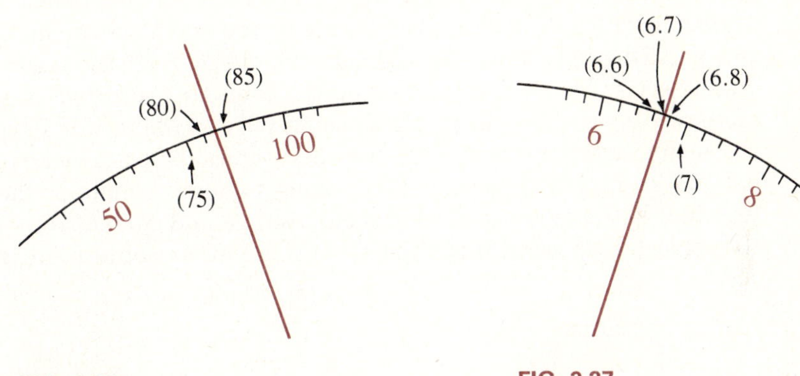

FIG. 3.26
Analog meter on the 250-V scale.

FIG. 3.27
Analog meter on the 10-mA scale.

The potential difference between two points is measured by simply connecting the leads of the meter *across the two points,* as indicated in Fig. 3.28. An up-scale reading is obtained by placing the positive lead of the meter to the point of higher potential of the network and the common or negative lead to the point of lower potential. The reverse connection results in a negative reading, or a below-zero indication.

Throughout the industry, voltage levels are measured more frequently than current levels primarily because the former does not require that the network connections be disturbed. In addition, a current level can usually be determined by simply measuring the voltage across a resistor and using Ohm's law.

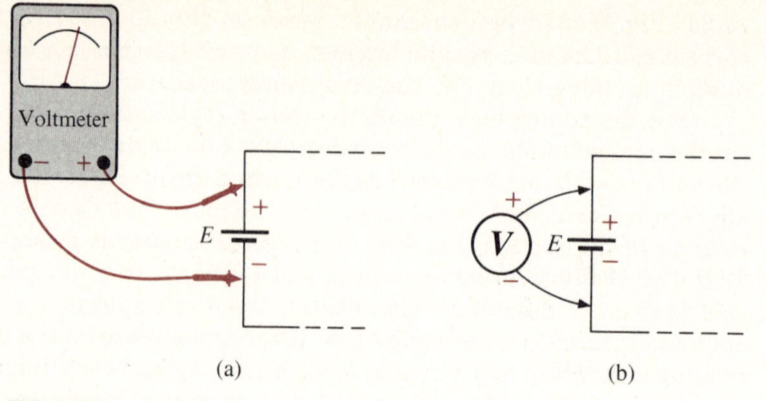

FIG. 3.28
(a) Voltmeter connection for an up-scale reading (analog or digital); schematic representation.

Ammeters are connected in the *same branch* in which the current is to be measured, as shown in Fig. 3.29. Since ammeters measure the rate of flow of charge, the meter must be placed in the network such that the charge will flow through the meter. The only way this can be accomplished is to open the branch in which the current is to be measured and place the meter between the two resulting terminals. For the network of Fig. 3.29, the source lead must be disconnected from the network and the ammeter inserted as shown. An up-scale reading will be obtained if the polarities on the terminals of the ammeter are such that the current of the network enters the positive terminal or terminal at the higher potential. For analog or digital meters, the common (−) terminal of the meter is used for both voltage and current readings (the negative side for up-scale indications), but the other connection is usually in a

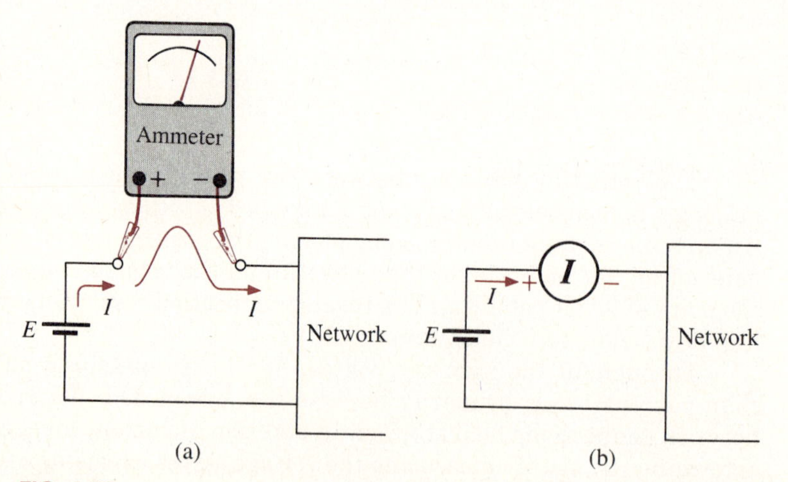

FIG. 3.29
Ammeter connection for an up-scale reading (analog or digital).

voltage terminal (+ or V) for a voltage measurement and in a current terminal (mA, A) for a current measurement. The terminals just mentioned are visible in Figs. 3.24 and 3.25.

The introduction of any meter into the network raises a concern about whether the meter will affect the behavior of the network. This question and others are examined in Chapters 6 and 7 after additional terms and concepts have been introduced.

Two final comments are in order regarding the use of instruments. Before measuring any voltage or current level, be sure *the meter is in the proper mode.* Hooking up a meter as a voltmeter with the instrument in the milliampere mode will most likely damage the meter. And second, when unsure of the voltage or current level, always *start with the highest scale* and work your way down to the scale that provides the highest degree of accuracy. Comments on the characteristics and use of a variety of meters are made throughout the text. However, the major study of meters is left for the laboratory sessions.

FORMULA SUMMARY

$$F = \frac{kQ_1Q_2}{r^2}$$

$$I = \frac{Q}{t}, \qquad Q = It, \qquad t = \frac{Q}{I}$$

$$V = \frac{W}{Q}, \qquad W = QV, \qquad Q = \frac{W}{V}$$

$$\text{Life (h)} = \frac{\text{Ampere-hour rating (Ah)}}{\text{Amperes drawn (A)}}$$

CHAPTER SUMMARY

For the basic equations introduced in this chapter, be absolutely sure the magnitude of the quantity substituted is as defined by the units of measurement of the equation. That is, the magnitude of time (t) in the equation $I = Q/t$ is in seconds and not minutes or hours. For the equation $V = W/Q$, Q is in coulombs and not electrons or another unit of measure. The improper substitution of a single magnitude can generate a solution of impossible dimensions.

In addition, for an equation such as $I = Q/t$ or $V = W/Q$, do not memorize all three forms for each at this early stage of development. Rather, develop some confidence in the algebraic manipulations required to solve for Q, t, or W and limit your memorization to the most common form (as just given).

Finally, be sure to include the proper unit of measurement with each calculated or measured quantity. It is insufficient and perhaps misleading to end a calculation with $I = 5$, when the result should appear as $I = 5$ A or perhaps $I = 5$ mA, which is a magnitude of a totally different level.

GLOSSARY

Ammeter An instrument designed to read the current through elements in series with the meter.

Ampere (A) The SI unit of measurement applied to the flow of charge through a conductor.

Ampere-hour rating The rating applied to a source of energy that will reveal how long a particular level of current can be drawn from that source.

Cell A fundamental source of electrical energy developed through the conversion of chemical or solar energy.

Conductors Materials that permit a generous flow of electrons with very little voltage applied.

Copper A material possessing physical properties that make it particularly useful as a conductor of electricity.

Coulomb (C) The fundamental SI unit of measure for charge. It is equal to the charge carried by 6.242×10^{18} electrons.

Coulomb's law An equation defining the force of attraction or repulsion between two charges.

Dc current source A source that will provide a fixed current level even though the load to which it is applied may cause its terminal voltage to change.

Dc generator A source of dc voltage available through the turning of the shaft of the device by some external means.

Direct current Current in which the magnitude does not change over a period of time.

Ductility The property of a material that allows it to be drawn into long, thin wires.

Electrolytes The contact element and the source of ions between the electrodes of the battery.

Electron The particle with negative polarity that orbits the nucleus of an atom.

Free electron An electron unassociated with any particular atom, relatively free to move through a crystal lattice structure under the influence of external forces.

Insulators Materials in which a very high voltage must be applied to produce any measurable current flow.

Malleability The property of a material that allows it to be worked into many different shapes.

Neutron A particle having no electrical charge, found in the nucleus of the atom.

Nucleus The structural center of an atom that contains both protons and neutrons.

Positive ion An atom having a net positive charge due to the loss of one of its negatively charged electrons.

Potential difference The difference in potential between two points in an electrical system.

Potential energy The energy that a mass possesses by virtue of its position.

Primary cell Sources of voltage that cannot be recharged.

Proton The particle of positive polarity found in the nucleus of the atom.

Rectification The process by which an ac signal is converted to one that has an average dc level.

Secondary cell Sources of voltage that can be recharged.

Semiconductor A material having a conductance value between that of an insulator and that of a conductor. Of significant importance in the manufacture of semiconductor electronic devices.

Solar cell Sources of voltage available through the conversion of light energy (photons) into electrical energy.

Specific gravity The ratio of the weight of a given volume of a substance to the weight of an equal volume of water at 4°C.

Volt (V) The unit of measurement applied to the difference in potential between two points. If 1 J of energy is required to move 1 C of charge between two points, the difference in potential is said to be 1 V.

Voltmeter An instrument designed to read the voltage across an element or between any two points in a network.

PROBLEMS

Section 3.2

1. Aluminum and gold are two other materials used commercially as conductors for specialized applications. Research the number of electrons in their shells, draw the shell structure, and comment on one reason they are good conductors.

2. What is the force of repulsion between two negatively charged bodies $Q_1 = 1\ \mu C$ and $Q_2 = 2\ \mu C$ if the distance between them is 0.01 m (less than 1/2 in.)?

***3.** **a.** Repeat Problem 2 if the distance is reduced to 1/1000 m (slightly less than 1/16 in.).

b. Repeat Problem 2 if the distance is increased to 1 m (more than 3 ft).

c. Compare the results obtained in parts a and b.

4. Determine the force of attraction between a positive charge of 5 μC and a negative charge of 20 μC if the distance between the charges is 1 in.

***5.** Determine the distance (in meters) between two equal negative charges of 1 μC if the force of attraction between them is 0.1 N.

Section 3.3

6. How many coulombs of charge are carried by 200×10^{16} electrons?

7. How many electrons are required to establish a charge of 8 μC?

8. If 20 μC of charge pass through a wire in 0.5 s, determine the current.

9. If 12.726×10^{18} electrons pass through a wire in 2 s, what is the current?

10. Find the current if 0.08 μC of charge passes through a conductor in 32 ms.
11. How much charge must pass through a conductor in 8 s if the current is to be 4 A?
12. If a current of 32 mA exists in a conductor for 2 min, how much charge has passed through the conductor?
13. If a current of 2 A exists for a period of 10 s, how many electrons have passed through the conductor?
14. How long will it take 15 C of charge to pass through a conductor if the current is 2.5 A?
15. How much time is required to pass 20 mC of charge through a conductor if the current is 5 μA?
16. How many minutes will it take a current of 4 A to pass 480 C of charge through a conductor?

Section 3.4

17. What is the potential difference between two points if 240 J of energy are required to move a charge of 6 C between the two points?
18. Determine the voltage between two points in a network if 16 mJ of energy are required to move 1/100 C of charge between the two points.
19. What is the potential difference between two points if 1 pJ of energy is expended moving 1000 electrons between the two points?
20. How much energy is required to move a charge of 5 C through a potential difference of 60 V?
21. Determine the energy expended moving a charge of 10 mC through a potential drop of 6 mV.
22. If 420 J of energy are required to move a charge through a potential difference of 40 V, how much charge is involved?
23. If 300 mJ of energy are required to move a charge through a potential difference of 120 V, find the charge involved.

Section 3.5

24. Describe the basic difference between the three common dc voltage sources: a battery, a generator, and a power supply.
25. Discuss the difference between primary and secondary batteries.
26. Using your local library as a resource, write a short paper on solar cells, describing how they work, their characteristics, and when they are utilized most effectively.
27. For the carbon-zinc battery of Fig. 3.10, is the voltage rating directly related to size? That is, is the available voltage larger for the larger-size units? In addition, discuss whether the current rating is directly related to size.

Section 3.6

28. Theoretically, how many hours will the transistor battery of Fig. 3.11(c) provide a current of 5 mA?
29. How long will the printed circuit board battery of Fig. 3.13(c) provide a current of 2.5 mA? At what voltage?

30. Referring to the characteristics of the EVEREADY® BH 500 cell of Fig. 3.16, determine each of the following.
    a. The capacity of the 1.1-V cell at a discharge current of 500 mA.
    b. The temperature at which the capacity will drop to 80% of its level at 10°C (50°F).
    c. Which cell (0.9 V, 1.0 V, or 1.1 V) is the most sensitive to discharge current.
31. a. Referring to Fig. 3.17, determine the cell voltage at a drain current of 100 mA and a discharge period of 1 h.
    b. Repeat part a for a period of 4 h and compare to the results of part a.
    c. Repeat part a for a period of 5 h and compare to the results of part a.
    d. What general conclusions can you make about the discharge period versus the resulting cell voltage (at a fixed drain current)?
32. Repeat Problem 31 for a 50-mA drain current. Is the conclusion of part d the same?

Sections 3.7–3.8

33. Describe in your own words the general characteristics of conductors, insulators, and semiconductors.
34. Referring to Table 3.1, which materials would appear to have conduction levels that would support their use in a wide variety of electrical applications? Even though gold has a lower conductivity level than silver, gold is used extensively in integrated circuit design, whereas silver is not. Can you think of any reasons for the use of gold rather than silver?
35. For insulators of the same geometry, how much more voltage must be applied across a mica sample compared to rubber to establish a flow of charge through the materials?
36. What voltage must be applied across a mica sheet 0.2 mm thick to establish a flow of charge if the breakdown strength is 2000 kV/cm?
37. Repeat Problem 26 for a rubber glove with a breakdown strength of 270 kV/cm if the thickness of the glove is 1/1000 m = 1 mm.

Section 3.9

38. a. Does a network have to be disturbed to measure voltage levels? How are the connections made?
    b. Does a network have to be disturbed to measure current levels with an ammeter? How are the connections made?
39. a. Find the pointer location on Fig. 3.26 for a reading of 55 V.
    b. Repeat part a for 95 V.
    c. Repeat part a for 67.5 V.
40. a. Find the pointer location on Fig. 3.27 for a reading of 7.8 mA.
    b. Repeat part a for 7.5 mA.
    c. Repeat part a for 8.15 mA.

4

Resistance

OBJECTIVES

- ☐ Become familiar with the factors that affect the resistance of a conductor.
- ☐ Understand and be able to apply the standard *wire tables.*
- ☐ Be able to apply the basic resistance equation in the English and metric systems.
- ☐ Understand the effects of temperature on resistance and be able to apply the related equations.
- ☐ Become aware of the standard types of resistors and where they are applied.
- ☐ Learn the color-coding system for resistors and become aware of the standard values.
- ☐ Understand the concept of conductance and how it relates to the resistance of an element.

4.1 INTRODUCTION

The flow of charge through any material encounters an opposing force similar in many respects to mechanical friction. This opposition, which is due to the collisions between electrons and between electrons and other atoms in the material, and *which converts electrical energy into heat,* is called the *resistance* of the material. The unit of measurement of resistance is the *ohm*, for which the symbol is Ω, the capital Greek letter omega. The circuit symbol for resistance appears in Fig. 4.1 with the graphic abbreviation for resistance (R).

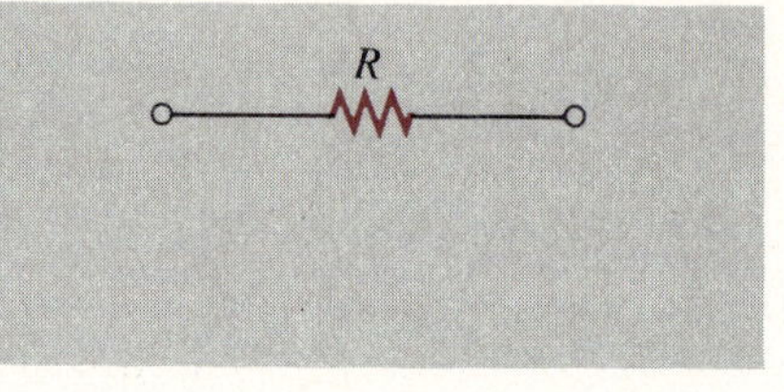

FIG. 4.1
Resistance symbol and notation.

The resistance of any material with a uniform cross-sectional area is determined by the following four factors:

1. **Material**
2. **Length**
3. **Cross-sectional area**
4. **Temperature**

The chosen material, with its unique molecular structure, will react to pressures to establish current through its core. Conductors permit a generous flow of charge with little external pressure and have low resistance levels. Insulators are designed to reduce the charge flow to its lowest levels through high resistance characteristics.

As one might expect, the longer the path the charge must pass through, the higher the resistance level, whereas the larger the area (and therefore available room), the lower the resistance. Resistance is thus directly proportional to length and inversely proportional to area.

As the temperature of most conductors and resistive elements increases, the increased motion of the particles within the molecular structure makes it increasingly difficult for the "free" carriers to pass through, and the resistance level increases.

At a fixed temperature of 20°C (room temperature), the resistance is related to the other three factors by

$$R = \rho \frac{l}{A} \qquad \text{(ohms, } \Omega\text{)} \qquad (4.1)$$

where ρ (the Greek letter rho) is a characteristic of the material called the *resistivity*, l is the length of the sample, and A is the cross-sectional area of the sample.

The units of measurement substituted into Eq. 4.1 are related to the application. For circular wires, units of measurement are usually defined as in Section 4.2. For most other applications involving important areas, such as integrated circuits, the units are as defined in Section 4.4.

4.2 RESISTANCE: CIRCULAR WIRES

For a circular wire, the quantities appearing in Eq. 4.1 are defined by Fig. 4.2. For two wires at the same temperature, as shown in Fig. 4.3, the shorter wire with the larger diameter has the lower resistance level. For similar physical dimensions, the higher the temperature, the more the resistance, as shown in Fig. 4.4.

FIG. 4.2

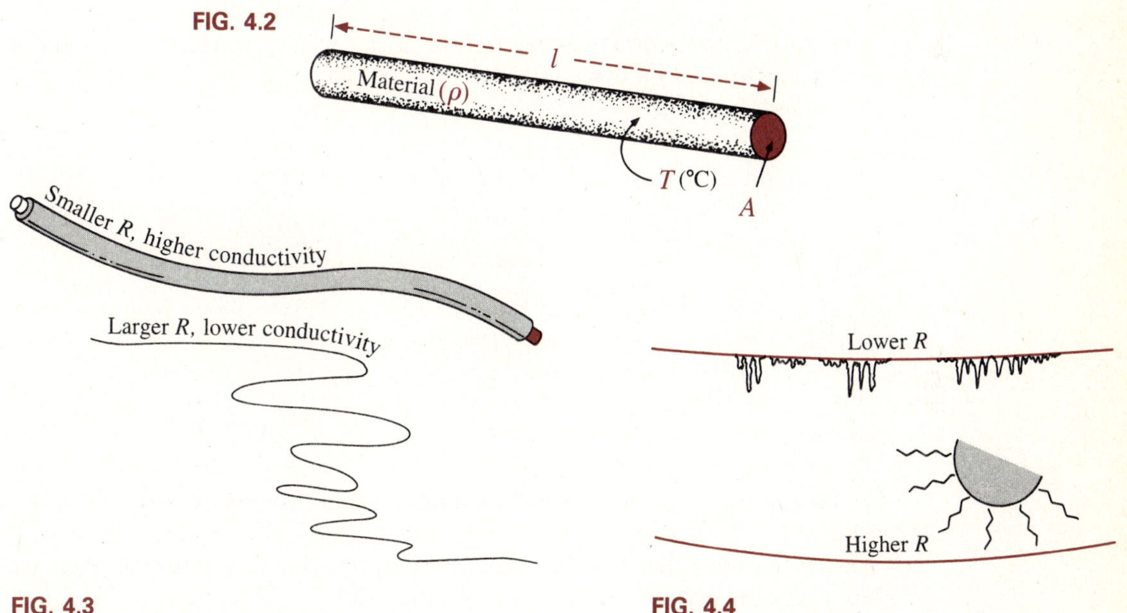

FIG. 4.3
Effect of area and length on resistance and conductivity.

FIG. 4.4
Effect of temperature on resistance.

For circular wires, the quantities of Eq. 4.1 have the following units.

ρ	CM-Ω/ft at $T = 20°C$
l	feet
A	circular mils (CM)

Note that the area of the conductor is measured in *circular mils* and *not* in square meters, inches, and so on, as determined by the familiar equation

$$\text{Area (circle)} = \pi r^2 = \frac{\pi d^2}{4} \qquad \begin{aligned} r &= \text{radius} \\ d &= \text{diameter} \end{aligned} \tag{4.2}$$

By definition:

$$1 \text{ mil} = \frac{1}{1000} \text{ in.} = 0.001 \text{ in.} = 1 \times 10^{-3} \text{ in.} \tag{4.3}$$

or

$$1000 \text{ mils} = 1 \text{ in.} \tag{4.4}$$

Figure 4.5 is provided to develop some familiarity with the relative size of a mil. The thickness of a dime is about 50 mils, whereas the thickness of this paper is about 1.5 mils.

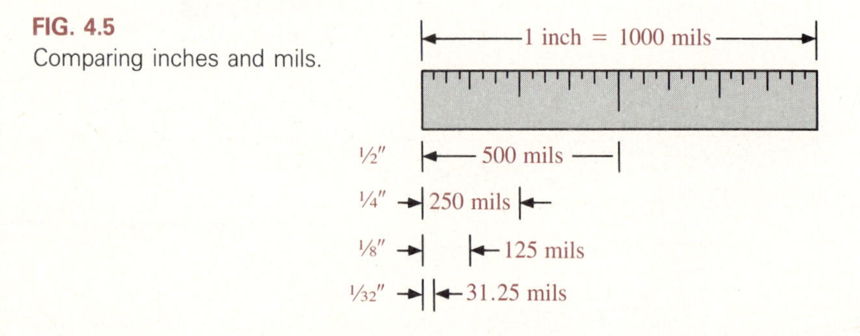

FIG. 4.5
Comparing inches and mils.

A square mil is as shown in Fig. 4.6(a). By definition, *a wire that has a diameter of 1 mil, as shown in Fig. 4.6(b), has an area of 1 CM*. One square mil was superimposed on the 1-CM area of Fig. 4.6(b) to demonstrate that a square mil has a larger surface area than a circular mil.

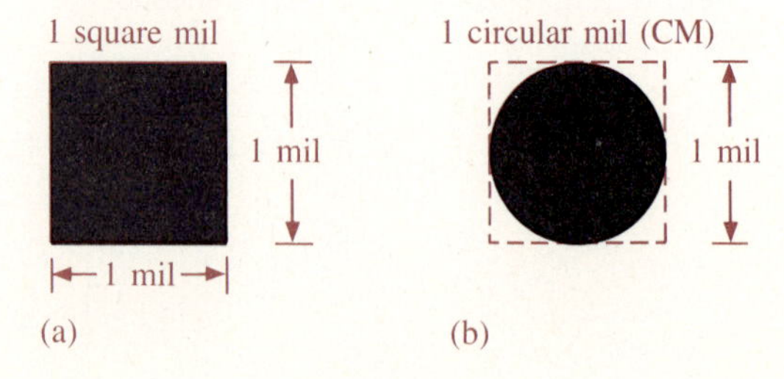

FIG. 4.6
Comparing the area of a circular mil to that of a square mil.

For conversion purposes,

$$\text{CM} = \frac{4}{\pi} \times (\text{no. of square mils})$$
$$\text{square mils} = \frac{\pi}{r} \times (\text{no. of CM}) \tag{4.5}$$

Through algebraic manipulations it can be shown that:

$$A_{\text{CM}} = (d_{\text{mils}})^2 \tag{4.6}$$

Therefore, in order to find the area in circular mils, the diameter must first be converted to mils. Since 1 mil = 0.001 in., if the diameter is given in inches, simply *move the decimal point three places to the right.* For example,

$$0.123 \text{ in.} = 123.0 \text{ mils}$$

If the number is in fractional form, first convert it to decimal form and then proceed as before. For example,

$$\frac{1}{8} \text{ in.} = 0.125 \text{ in.} = 125 \text{ mils}$$

The constant ρ (resistivity) is different for every material. Its value is the resistance of a length of wire 1 ft by 1 mil in diameter, measured at 20°C (Fig. 4.7). Some typical values of ρ are listed in Table 4.1.

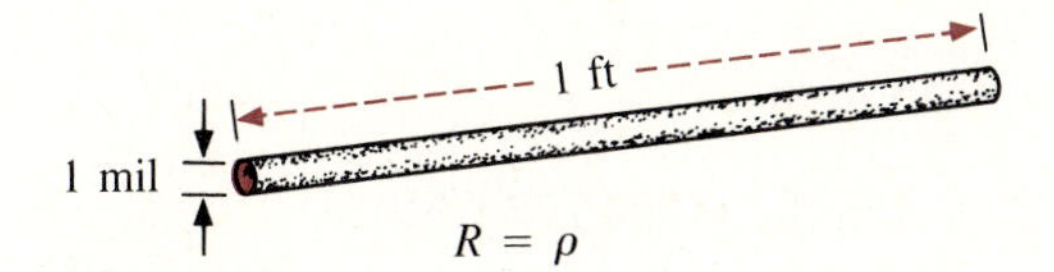

FIG. 4.7
Defining ρ.

TABLE 4.1
The resistivity of various materials.

Material	ρ at 20°C
Silver	9.9
Copper	10.37
Gold	14.7
Aluminum	17.0
Tungsten	33.0
Nickel	47.0
Iron	74.0
Constantan	295.0
Nichrome	600.0
Calorite	720.0
Carbon	21,000.0

EXAMPLE 4.1 What is the resistance of a 100-ft length of copper wire with a diameter of 0.020 in. at 20°C?

Solution

$$\rho = 10.37 \qquad 0.020 \text{ in.} = 20 \text{ mils}$$

$$A_{CM} = (d_{mils})^2 = (20 \text{ mils})^2 = 400 \text{ CM}$$

$$R = \rho\frac{l}{A} = \frac{(10.37)(100 \text{ ft})}{400 \text{ CM}}$$

$$= \mathbf{2.593\ \Omega}$$

Calculator

(1) (0) (·) (3) (7) (×) (1) (0) (0) (=) (÷) (4) (0) (0) (=)

with a display of 2.593.

EXAMPLE 4.2 An undetermined number of feet of wire have been used from the carton of Fig. 4.8. Find the length of the remaining copper wire if it has a diameter of 1/16 in. and a resistance of 0.5 Ω.

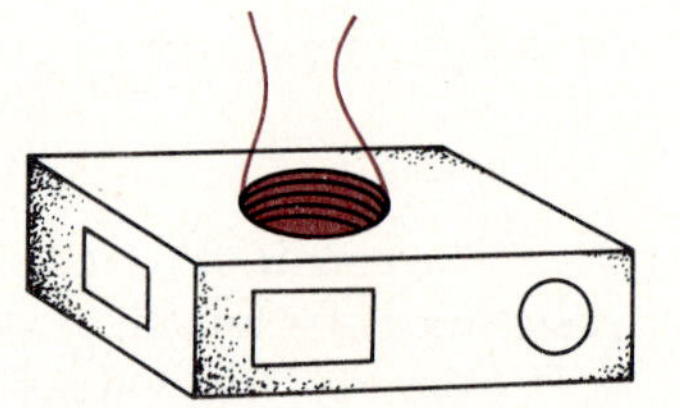

FIG. 4.8
Determining the length of conductor in the container.

Solution For copper, $\rho = 10.37$. To determine the area in circular mils:

$$\frac{1}{16} \text{ in.} = 0.0625 \text{ in.} = 62.5 \text{ mils}$$

$$A_{CM} = (d_{mils})^2 = (62.5 \text{ mils})^2 = 3906.25 \text{ CM}$$

Equation 4.1 is

$$R = \rho\frac{l}{A}$$

We must now use algebraic manipulation to obtain an equation for l.

Multiplying both sides of the equation by the area A, we have

$$R(A) = \rho\frac{l}{\cancel{A}}(\cancel{A})$$

Then dividing both sides by the resistivity ρ gives

$$\frac{RA}{\rho} = \frac{\cancel{\rho}l}{\cancel{\rho}}$$

and finally

$$\boxed{l = \frac{RA}{\rho}} \tag{4.7}$$

Substituting gives

$$l = \frac{(0.5\ \Omega)(3906.25\ \text{CM})}{10.37} = \frac{1953.125}{10.37}\ \text{ft}$$
$$= \mathbf{188.34\ ft}$$

Calculator

⊙ ⑤ ⊗ ③ ⑨ ⓪ ⑥ ⊙ ② ⑤ ⊜ ÷ ① ⓪ ⊙ ③ ⑦ ⊜

with a display of 188.344.

EXAMPLE 4.3 Find the diameter of a copper wire that has a resistance of 1 Ω and a length of 100 yd.

Solution For copper, $\rho = 10.37$. The length in feet is

$$100\ \text{yd}\left(\frac{3\ \text{ft}}{1\ \text{yd}}\right) = 300\ \text{ft}$$

Equation 4.1 is

$$R = \rho\frac{l}{A}$$

Multiplying both sides of Eq. 4.1 by the area A gives

$$R(A) = \rho\frac{l(A)}{A}$$

Dividing both sides by the resistance R yields

$$\frac{RA}{R} = \frac{\rho l}{R}$$

and

$$\boxed{A = \frac{\rho l}{R}} \tag{4.8}$$

Substituting,

$$A = \frac{(10.37)(300\ \text{ft})}{1\ \Omega} = 3111\ \text{CM}$$

To determine the diameter in mils we must first be aware of the following mathematical relationship:

$$\boxed{\text{If } y = x^2, \quad \text{then} \quad x = \sqrt{y}.} \tag{4.9}$$

For Eq. 4.6,

$$\text{If } \underbrace{A_{CM}}_{y} = \underbrace{(d_{\text{mils}})^2}_{x^2} \quad \text{then} \quad \underbrace{d_{\text{mils}}}_{x} = \underbrace{\sqrt{A_{CM}}}_{y}$$

and

$$d_{\text{mils}} = \sqrt{3111}\ \text{CM} = 55.776\ \text{mils}$$

To convert from mils to inches, the decimal point must be moved *three places to the left* as follows:

$$55.776\ \text{mils} = 0.055776\ \text{in.} = \mathbf{0.056\ in.}$$

This is slightly smaller than 1/16 in. = 0.0625 in.

Calculator

with a display of 3111.

d_{mils}: 3 1 1 1 √

with a display of 55.776.

4.3 WIRE TABLES

The *wire table* was designed primarily to standardize the size of wire produced by manufacturers throughout the United States. As a result, the manufacturer has a larger market and the consumer knows that standard wire sizes will always be available. The table was designed to assist the user in every way possible; it usually includes such data as the cross-sectional area in circular mils, diameter in mils, ohms per 1000 ft at 20°C, and weight per 1000 ft.

The American Wire Gage (AWG) sizes are given in Table 4.2 for solid round copper wire. A column indicating the maximum allowable current in amperes, as determined by the National Fire Protection Association, has also been included.

The chosen sizes have an interesting relationship: For every drop in 3 gage numbers, the area is doubled, and for every drop in 10 gage numbers, the area increases by a factor of 10.

Examining Eq. 4.1, we also note that *doubling the area cuts the resistance in half, and increasing the area by a factor of 10 decreases the resistance to 1/10 the orginal,* everything else kept constant.

TABLE 4.2
American Wire Gage (AWG) sizes.

	AWG #	Area (CM)	Ω/1000 ft at 20°C	Maximum Allowable Current for RHW Insulation (A)*
(4/0)	0000	211,600	0.0490	230
(3/0)	000	167,810	0.0618	200
(2/0)	00	133,080	0.0780	175
(1/0)	0	105,530	0.0983	150
	1	83,694	0.1240	130
	2	66,373	0.1563	115
	3	52,634	0.1970	100
	4	41,742	0.485	85
	5	33,102	0.3133	—
	6	26,250	0.3951	65
	7	20,816	0.4982	—
	8	16,509	0.6282	45
	9	13,094	0.7921	—
	10	10,381	0.9989	30
	11	8,234.0	1.260	—
	12	6,529.0	1.588	20
	13	5,178.4	2.003	—
	14	4,106.8	2.525	15
	15	3,256.7	3.184	
	16	2,582.9	4.016	
	17	2,048.2	5.064	
	18	1,624.3	6.385	
	19	1,288.1	8.051	
	20	1,021.5	10.15	
	21	810.10	12.80	
	22	642.40	16.14	
	23	509.45	20.36	
	24	404.01	25.67	
	25	320.40	32.37	
	26	254.10	40.81	
	27	201.50	51.47	
	28	159.79	64.90	
	29	126.72	81.83	
	30	100.50	103.2	
	31	79.70	130.1	
	32	63.21	164.1	
	33	50.13	206.9	
	34	39.75	260.9	
	35	31.52	329.0	
	36	25.00	414.8	
	37	19.83	523.1	
	38	15.72	659.6	
	39	12.47	831.8	
	40	9.89	1049.0	

INCREASING AWG # ↓

DECREASING DIAMETER (AND AREA) ↓

Reprinted by permission from NFPA 70–1975, National Electrical Code®, copyright © 1974, National Fire Protection Association, Quincy, MA 02269. This reprinted material is not the complete and official position of the NFPA on the referenced subject which is represented only by the standard in its entirety. *National Electrical Code* is a registered trademark of the National Fire Protection Association, Inc., Quincy, MA for a triennial electrical publication. The term *National Electrical Code,* as used herein means the triennial publication constituting the National Electrical Code and is used with permission of the National Fire Protection Association.

*Not more than three conductors in raceway, cable, or direct burial.

The actual sizes of some of the gage wires listed in Table 4.2 are shown in Fig. 4.9 with a few of their areas of application. A few examples using Table 4.2 follow.

FIG. 4.9

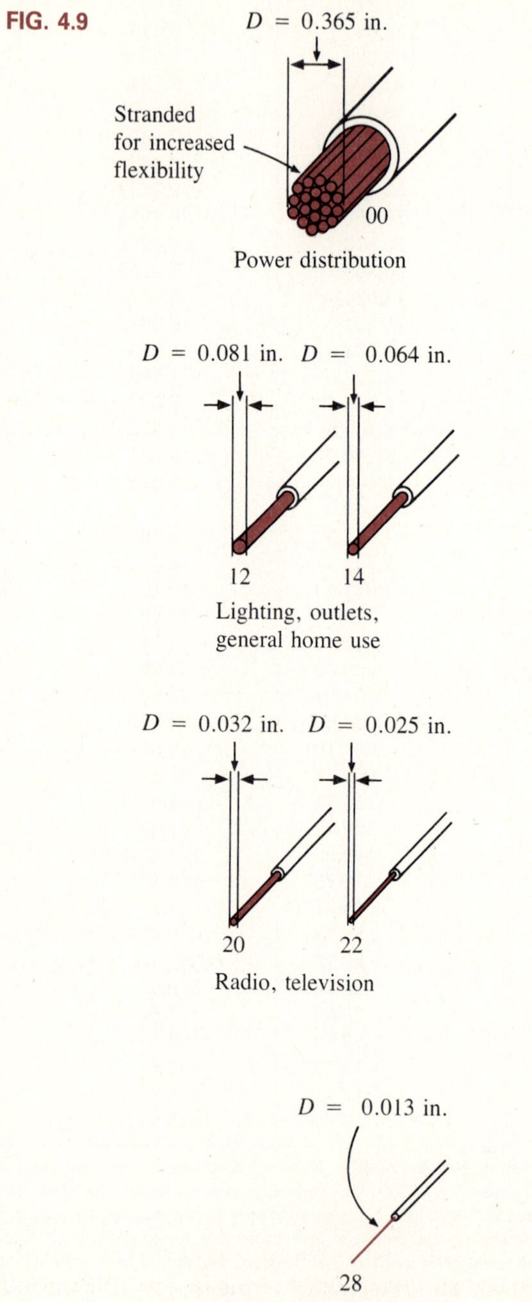

EXAMPLE 4.4 Find the resistance of 650 ft of #8 copper wire ($T = 20°C$).

Solution For #8 copper wire (solid), Ω/1000 ft at 20°C = 0.6282 Ω, and

$$650\ \text{ft}\left(\frac{0.6282\ \Omega}{1000\ \text{ft}}\right) = \mathbf{0.408\ \Omega}$$

EXAMPLE 4.5 What is the diameter, in inches, of #12 copper wire?

Solution For #12 copper wire (solid), $A = 6529.9$ CM, and

$$d_{\text{mils}} = \sqrt{A_{\text{CM}}} = \sqrt{6529.9} \cong 80.81\ \text{mils}$$
$$d = \mathbf{0.0808\ in.} \qquad \text{(or close to 1/12 in.)}$$

EXAMPLE 4.6 For the system of Fig. 4.10, the total resistance of *each* power line cannot exceed 0.025 Ω, and the maximum current to be drawn by the load is 95 A. What gage wire should be used?

FIG. 4.10

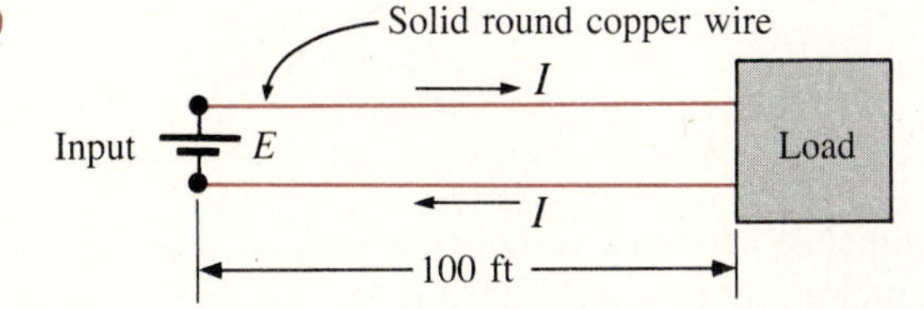

Solution

$$R = \rho\frac{l}{A} \Rightarrow A = \rho\frac{l}{R} = \frac{(10.37)(100\ \text{ft})}{0.025\ \Omega} = \mathbf{41{,}480\ CM}$$

Using the wire table, we choose the wire with the next largest area, which is #4, to satisfy the resistance requirement. We note, however, that 95 A must flow through the line. This specification requires that #3 wire be used, since the #4 wire can carry a maximum current of only 85 A.

4.4 RESISTANCE: METRIC UNITS

The design of resistive elements for a variety of areas of application including thin-film resistors and integrated circuits uses metric units for the quantities of Eq. 4.1. In SI units, the resistivity is measured in ohm-meters, the area is measured in square meters, and the length is measured in meters. However, the meter is generally too large a unit of measure for most applications, so the centimeter is usually employed. The resulting dimensions for Eq. 4.1 are, therefore,

ρ ohm-centimeters
l centimeters
A square centimeters

The resistivity of a material is actually the resistance of a sample, as shown in Fig. 4.11. Table 4.3 provides a list of values of ρ in ohm-centimeters.

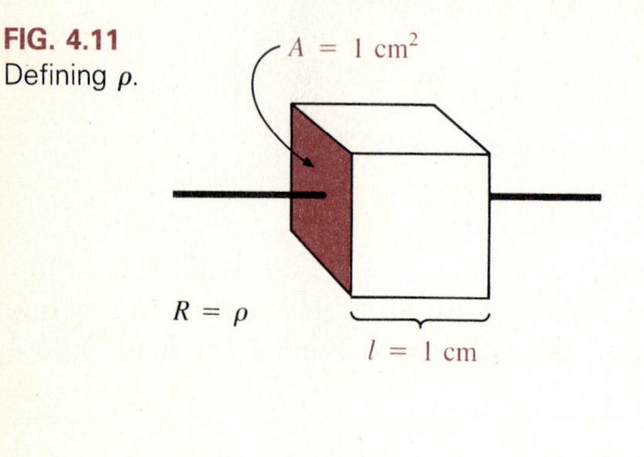

FIG. 4.11
Defining ρ.

TABLE 4.3
Resistivity (ρ) of various materials in ohm-centimeters.

Material	ρ
Silver	1.629×10^{-6}
Copper	1.724×10^{-6}
Gold	2.44×10^{-6}
Aluminum	2.688×10^{-6}
Tungsten	5.5×10^{-6}
Nickel	7.8×10^{-6}
Iron	9.8×10^{-6}
Tantalum	15.5×10^{-6}
Nichrome	100×10^{-6}
Tin oxide	250×10^{-6}
Carbon	3500×10^{-6}

Note that the area now is expressed in square centimeters, which can be determined using the basic equation $A = \pi d^2/4$, *eliminating* the need to work with circular mils, the special unit of measure associated with circular wires.

EXAMPLE 4.7 Determine the resistance of 100 ft of #28 copper telephone wire if the diameter is 0.0126 in.

Solution First, we must make some unit conversions.

l: required in centimeters

$$l = 100\ \text{ft}\left(\frac{12\ \text{in.}}{1\ \text{ft}}\right)\left(\frac{2.54\ \text{cm}}{1\ \text{in.}}\right) = 3048\ \text{cm}$$

Diameter of wire: required in centimeters

$$d = 0.0126\ \text{in.}\left(\frac{2.54\ \text{cm}}{1\ \text{in.}}\right) = 0.032\ \text{cm}$$

Therefore, the area in square centimeters is

$$A = \frac{\pi d^2}{4} = \frac{(3.1416)(0.032\ \text{cm})^2}{4} = 8.04 \times 10^{-4}\ \text{cm}^2$$

Equation 4.1 gives

$$R = \rho\frac{l}{A} = \frac{(1.724 \times 10^{-6}\ \Omega\text{-cm})(3048\ \text{cm})}{8.04 \times 10^{-4}\ \text{cm}^2} \cong \mathbf{6.5\ \Omega}$$

Using the units for circular wires and Table 4.2 for the area of a #28 wire, we find

$$R = \rho\frac{l}{A} = \frac{(10.37)(100\ \text{ft})}{159.79\ \text{CM}} = \mathbf{6.5\ \Omega}$$

Calculator

l: (1) (0) (0) (×) (1) (2) (×) (2) (·) (5) (4) (=)

with a display of 3048.

Note that an equals sign is not required between each successive multiplication:

d: (·) (0) (1) (2) (6) (×) (2) (·) (5) (4) (F <-> E) (=)

with a display of 0.032.

A: In this case let us first determine the squared term of $A = \pi d^2/4$ and then multiply by π (3.1416) and divide by 4.

(·) (0) (3) (2) (x^2) (×) (2nd F) (π) (÷) (4) (=)

with a display of 8.043–04.

$R = \rho l/A$: (1) (·) (7) (2) (4) (EXP) (+/−) (6) (×) (3) (0) (4) (8) (=) (÷) (8) (·) (0) (4) (EXP) (+/−) (4) (=)

with a display of 6.536.

EXAMPLE 4.8 Determine the resistance of the thin-film resistor of Fig. 4.12 if the sheet resistance R_s (defined by $R_s = \rho/d$) is 100 Ω.

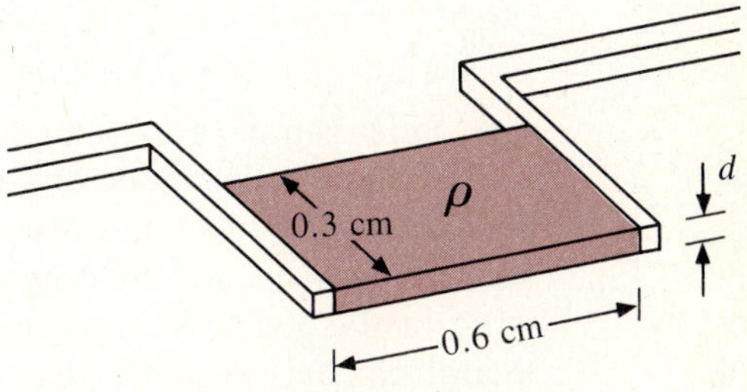

FIG. 4.12
Thin-film resistor (note Fig. 4.20).

Solution For deposited materials of the same thickness, the sheet-resistance factor is usually employed in the design of thin-film resistors.

Equation 4.1 can be written

$$R = \rho\frac{l}{A} = \rho\frac{l}{dw} = \left(\frac{\rho}{d}\right)\left(\frac{l}{w}\right) = R_s\frac{l}{w}$$

where l is the length of the sample and w is the width. Substituting into this equation yields

$$R = R_s\frac{l}{w} = \frac{(100)(0.6\ \text{cm})}{0.3\ \text{cm}} = \mathbf{200\ \Omega}$$

as one might expect, since $l = 2w$.

4.5 TEMPERATURE EFFECTS

For most conductors, the resistance increases with increase in temperature due to the increased molecular movement within the conductor, which hinders the flow of charge. Figure 4.13 reveals that for copper (and most other metallic conductors), the resistance increases almost linearly (in a straight-line relationship) with increase in temperature. For the range of semiconductor materials employed in transistors, diodes, and so on, the resistance decreases with increase in temperature.

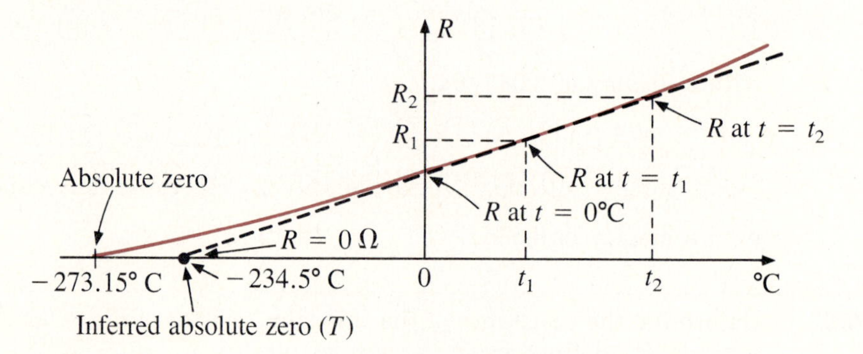

FIG. 4.13
Effect of temperature on the resistance of copper.

Since temperature can have such a pronounced effect on the resistance of a conductor, it is important that we have some method of determining the resistance at any temperature within operating limits. An equation for this purpose can be obtained by approximating the curve of Fig. 4.13 by the straight dashed line that intersects the temperature scale at −234.5°C. Although the actual curve extends to *absolute zero* (−273.15°C), the straight-line approximation is quite accurate for the normal operating temperature range. At two different temperatures, t_1 and t_2, the resistance of copper is R_1 and R_2, respectively, as indicated on the curve. Using a property of similar triangles, we may develop the following mathematical relationship between resistance and temperature.

$$\boxed{\frac{234.5 + t_1}{R_1} = \frac{234.5 + t_2}{R_2}} \quad (4.10)$$

The temperature of −234.5°C is called the *inferred absolute temperature* of copper. For different conducting materials, the intersection of the straight-line approximation occurs at different temperatures. A few typical values are listed in Table 4.4.

TABLE 4.4
Inferred absolute temperatures.

Material	Temperature (°C)
Silver	−243
Copper	−234.5
Gold	−274
Aluminum	−236
Tungsten	−204
Nickel	−147
Iron	−162
Nichrome	−2250
Constantan	−125,000

Equation 4.10 can easily be adapted to any material by inserting the proper inferred absolute temperature. It may therefore be written as follows:

$$\boxed{\frac{|T| + t_1}{R_1} = \frac{|T| + t_2}{R_2}} \quad (4.11)$$

where $|T|$ indicates that the inferred absolute temperature of the material involved is inserted as a positive value in the equation. The temperature t is inserted as a positive number if greater than 0°C and as a negative number if less than 0°C.

EXAMPLE 4.9

If the resistance of a copper wire is 50 Ω at 20°C (room temperature), what is its resistance at 100°C (boiling point of water)?

Solution Equation 4.11 yields

$$\frac{234.5 + 20}{50} = \frac{234.5 + 100}{R_2}$$

so,

$$R_2 = \frac{(50)(334.5)}{254.5} = \mathbf{65.717\ \Omega}$$

Calculator

⑤ ⓪ ⊗ ③ ③ ④ ⊙ ⑤ ⊜ ⊕ ② ⑤ ④ ⊙ ⑤ ⊜

with a display of 65.717.

EXAMPLE 4.10

If the resistance of a copper wire at freezing (0°C) is 30 Ω, what is its resistance at −40°C?

Solution Equation 4.11 gives

$$\frac{234.5 + 0}{30} = \frac{234.5 - 40}{R_2}$$

$$R_2 = \frac{(30)(194.5)}{234.5} = \mathbf{24.883\ \Omega}$$

Calculator

③ ⓪ ⊗ ① ⑨ ④ ⊙ ⑤ ⊜ ⊕ ② ③ ④ ⊙ ⑤ ⊜

with a display of 24.883.

Example 4.11 determines the temperature at which a particular resistance is obtained. The most difficult part of the exercise is simply to obtain the proper equation for the desired temperature.

EXAMPLE 4.11

If the resistance of a copper sample is 100 Ω at room temperature (68°C), at what temperature will it be 120 Ω?

Solution In Eq. 4.11, multiplying both sides by R_2 gives

$$\frac{(R_2)(|T| + t_1)}{R_1} = \frac{(\not{R}_2)(|T| + t_2)}{\not{R}_2}$$

Subtracting $|T|$ from both sides gives

$$\frac{R_2(|T| + t_1)}{R_1} - |T| = |\not{T}| + t_2 - |\not{T}|$$

and

$$t_2 = \frac{R_2}{R_1}(|T| + t_1) - |T| \tag{4.12}$$

Substituting values yields

$$t_2 = \frac{120}{100}(234.5 + 68) - (234.5)$$
$$= 1.2(302.5) - 234.5$$
$$t_2 = \mathbf{128.5°C}$$

Calculator

Using the format

$$t_2 = \frac{(234.5 + 68)(120)}{100} - 234.5$$

(2) (3) (4) (.) (5) (+) (6) (8) (=) (×) (1) (2) (0) (=)
(÷) (1) (0) (0) (=) (−) (2) (3) (4) (.) (5) (=)

with a display of 128.5.

Of course, the calculator can be used to perform the individual operations of the longhand solution.

There is a second popular equation for calculating the resistance of a conductor at different temperatures. Defining

$$\alpha_1 = \frac{1}{|T| + t_1}$$

as the *temperature coefficient of resistance* at a temperature t_1, we have

$$R_2 = R_1[1 + \alpha_1(t_2 - t_1)] \tag{4.13}$$

The values of α_1 for different materials at a temperature of 20°C have been evaluated, and a few are listed in Table 4.5. As indicated in the table, the family of *semiconductor materials have negative temperature coefficients*. In other words, the resistance of the material drops with increase in temperature and vice versa.

TABLE 4.5
Temperature coefficient of resistance for various materials at 20°C.

Material	Temperature Coefficient (α_1)
Silver	0.0038
Copper	0.00393
Gold	0.0034
Aluminum	0.00391
Tungsten	0.005
Nickel	0.006
Iron	0.0055
Constantan	0.000008
Nichrome	0.00044
Carbon	−0.0005

In addition, *the higher the value of* α_1, *the greater the rate of change of resistance with temperature*. Referring to Table 4.5, we find that copper is more sensitive to temperature variations than silver, gold, or aluminum, although the differences are relatively small.

4.6 TYPES OF RESISTORS

Resistors are made in many forms, but all belong in either of two groups: fixed or variable. The most common of the low-wattage, fixed-type resistors is the molded carbon-composition resistor. The basic construction is shown in Fig. 4.14.

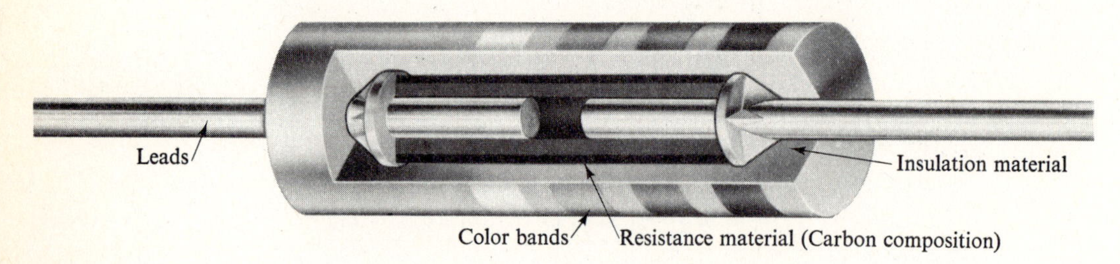

FIG. 4.14
Fixed composition resistor. (Courtesy of Ohmite Manufacturing Co.)

The relative sizes of all fixed and variable resistors change with the wattage (power) rating, increasing in size for increased wattage ratings in order to withstand the higher currents and dissipation losses. The relative sizes of the molded composition resistors for different wattage ratings are shown in Fig. 4.15. Resistors of this type are readily available in values ranging from 2.7 Ω to 22 MΩ.

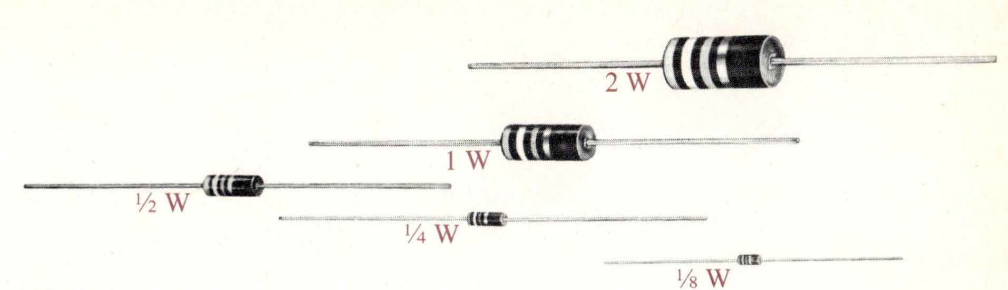

FIG. 4.15
Fixed composition resistors of different wattage ratings. (Courtesy of Ohmite Manufacturing Co.)

The temperature-versus-resistance curve for a 10,000-Ω and 0.5-MΩ composition-type resistor is shown in Fig. 4.16. Note the small percentage resistance change in the normal-temperature operating range. Several other types of fixed resistors are shown in Fig. 4.17.

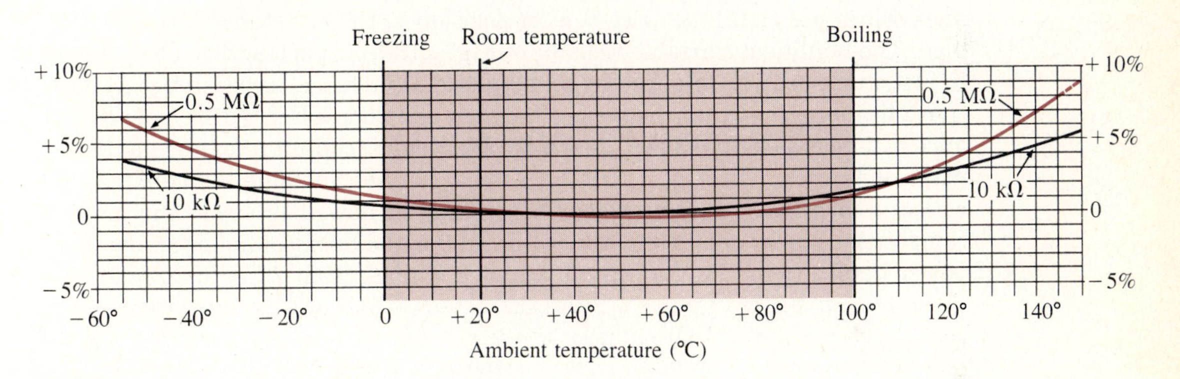

FIG. 4.16
Curves showing percent temporary resistance changes from +25°C values. (Courtesy of Allen-Bradley Co.)

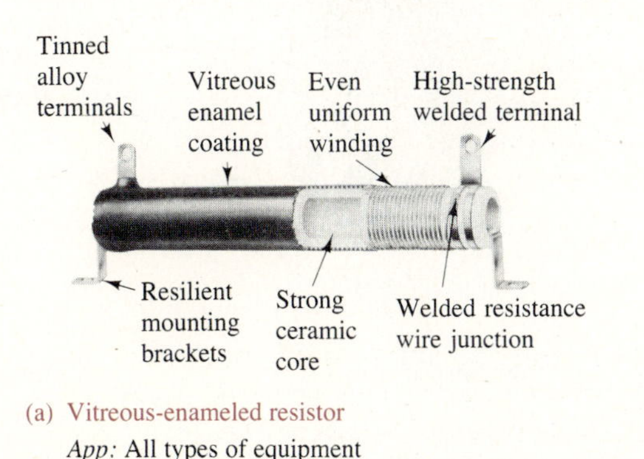

(a) Vitreous-enameled resistor
App: All types of equipment

(b) Molded vitreous-enameled wire-wound axial lead resistor
App: For low-wattage applications in electronic and similar circuits

(c) Metal-film precision resistors
App: Where high stability, low temperature coefficient, and low noise level desired

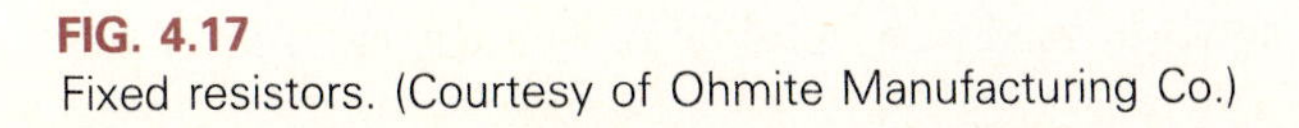
FIG. 4.17
Fixed resistors. (Courtesy of Ohmite Manufacturing Co.)

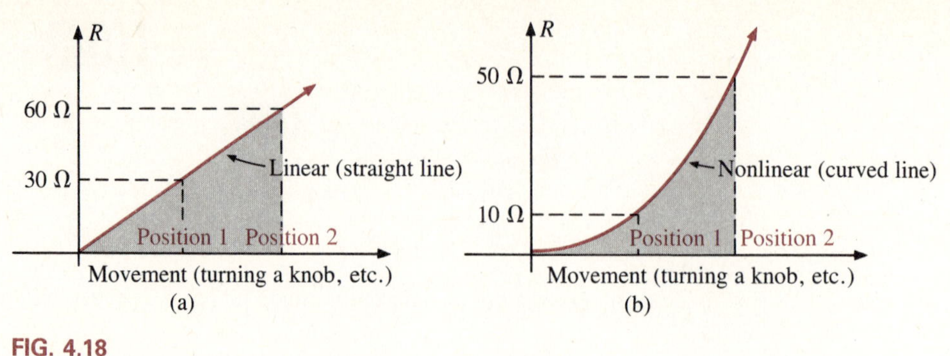

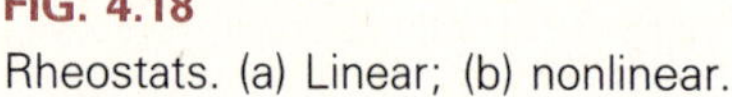

FIG. 4.18
Rheostats. (a) Linear; (b) nonlinear.

Variable resistors come in many forms, but basically they can be separated into the linear or nonlinear types. As shown in Fig. 4.18(a), a linear variable resistor has a straight-line relationship between position and resistance. Since position 2 is twice that of position 1, the resistance has also doubled. The nonlinear variable resistor of Fig. 4.18(b) experiences a slow change in resistance at first, followed by an accelerated increase in resistance as position 2 is approached. For the same change in position as Fig. 4.18(a), the resistance has increased fivefold. The symbol for a two-terminal linear or nonlinear *rheostat* is shown in Fig. 4.19. The three-terminal variable resistor may be called a *rheostat* or *potentiometer,* depending on how it is used. The molded composition linear potentiometer of Fig. 4.20 is the most common for low-level power applications. Its maximum value can range from 20 Ω to 20 MΩ.

FIG. 4.19
Rheostat.

R

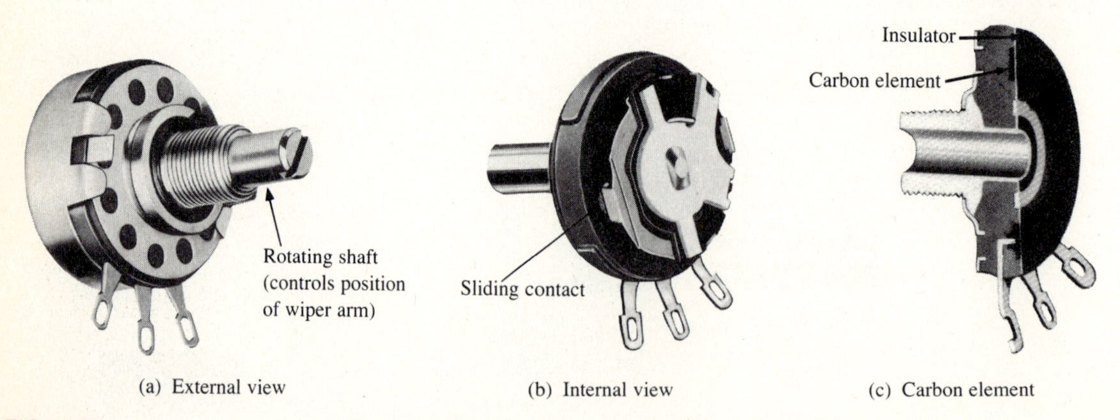

FIG. 4.20
Molded composition-type potentiometer. (Courtesy of Allen-Bradley Co.)

The symbol for a three-terminal variable resistor is shown in Fig. 4.21, along with the connections for its use as a rheostat or potentiometer. The arrow in the symbol of Fig. 4.21(a) is a contact that is movable on a continuous resistive element. As shown in Fig. 4.21(b), if the lug connected to the moving contact and a stationary lug are the only terminals used, the variable resistor is being used as a rheostat. The moving contact will determine whether R_{ab} is a minimum (0 Ω) or maximum value (R). If all three lugs are connected in the circuit, as shown in Fig. 4.21(c), it is being employed as a potentiometer. The terminology *potentiometer* refers to the fact that the moving contact (wiper arm) controls by its position the *potential* differences V_{ab} and V_{bc} of Fig. 4.21(c).

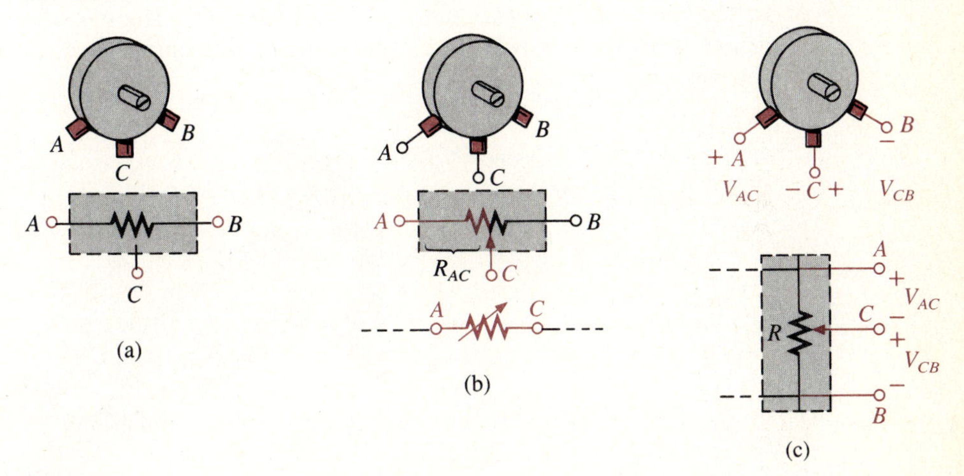

FIG. 4.21
Potentiometer. (a) Graphic symbol; (b) as a rheostat; (c) as a potentiometer.

Figure 4.22 shows both a linear and a tapered type of potentiometer. In the linear type of Fig. 4.22(a), the number of turns of the high-resistance wire per unit length of the core is uniform; therefore, the resistance varies linearly with the position of the rotating contact. One-half turn results in half the total

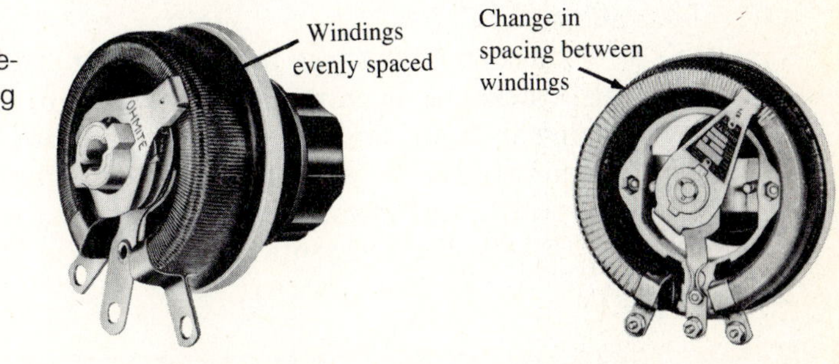

FIG. 4.22
Wirewound vitreous enamel potentiometers. (Courtesy of Ohmite Manufacturing Co.)

resistance between either stationary lug and the moving contact. Three-quarters turn establishes three-quarters of the total across two terminals and one-quarter between the other stationary lug and the moving contact. If the number of turns is not uniform, as in the tapered unit of Fig. 4.22(b), the resistance varies nonlinearly with the position of the rotating contact. That is, a quarter turn may result in more or less than one-quarter of the total resistance between a stationary lug and the moving contact. Potentiometers of both types in Fig. 4.22 are made in all sizes, with a range of maximum values from 200 Ω to 50 MΩ.

The resistance of the screw-drive linear variable resistor of Fig. 4.23 is determined by the position of the contact arm, which can be moved by using the handwheel. The stationary terminal used with the movable contact determines whether the resistance increases or decreases with movement of the contact arm.

FIG. 4.23
Screw-drive rheostat. (Courtesy of James G. Biddle Co.)

The miniaturization of parts—used quite extensively in computers—requires that resistances of different values be placed in very small packages. Two steps leading to the packaging of three resistors in a single module are shown in Fig. 4.24.

Resistor networks in a variety of configurations are available in miniature packages, such as the one shown in Fig. 4.25 with a photograph of the casing and pins, for use with printed circuit boards. The LDP is a coding for the production series, whereas the second number, 14, is the number of pins. The last two digits indicate the internal circuit configuration. The resistance range for the discrete elements in each chip is 10 Ω to 10 MΩ.

4.7 COLOR CODING AND STANDARD RESISTOR VALUES

A wide variety of resistors, fixed or variable, are large enough to have their resistance in ohms printed on the casing. There are some, however, that are too

FIG. 4.24
Placement of resistors on a module. (Courtesy of International Business Machines Corp.)

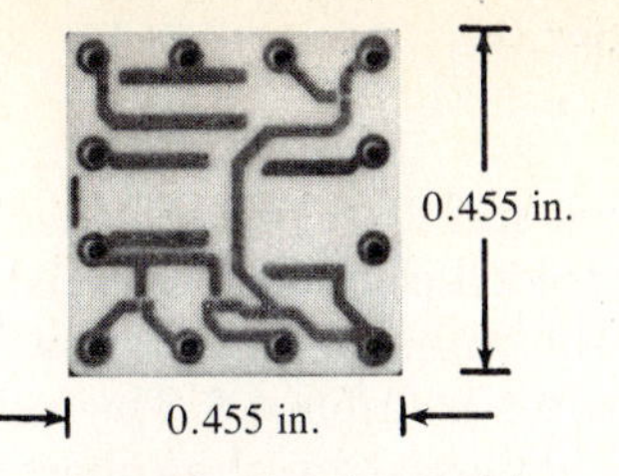

(a) Electrodes placed on module

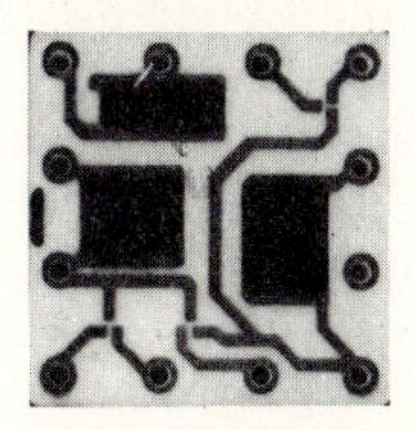

(b) Resistance applied and adjusted to desired value by air-abrasion techniques

(c) Module completely encased

FIG. 4.25
Resistor configuration microcircuit. (Courtesy of Dale Electronics, Inc.)

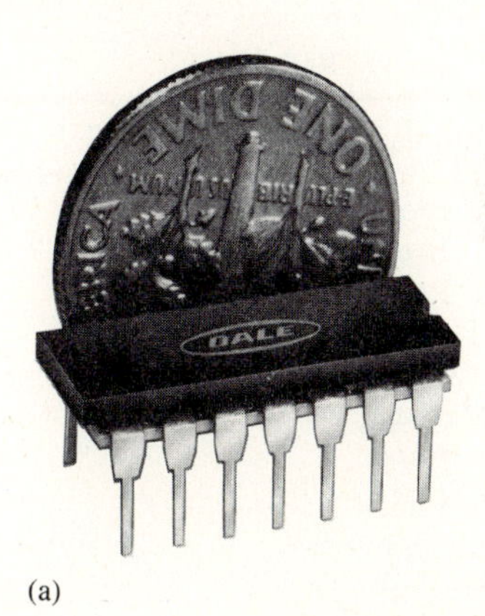

(a)

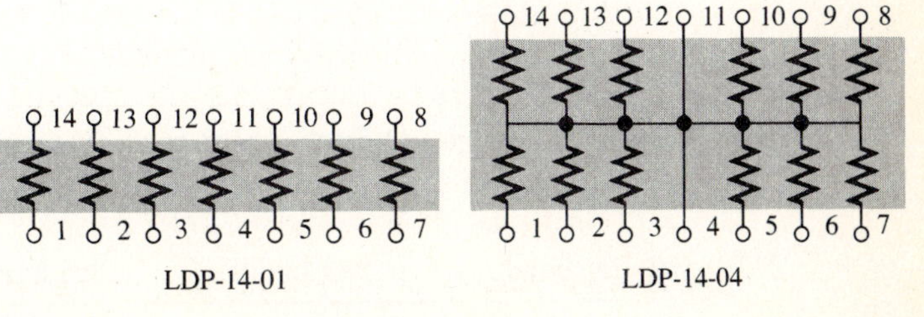

(b) (c)

small to have numbers printed on them, so a system of color coding is used. For the fixed molded composition resistor, four or five color bands are printed on one end of the outer casing, as shown in Fig. 4.26. Each color has the numerical value indicated in Table 4.6. The color bands are always read from the end that has the band closest to it, as shown in Fig. 4.26. The first and second bands represent the first and second digits, respectively. The third band determines the power-of-ten multiplier for the first two digits. The fourth band is the manufacturer's tolerance, which is an indication of the precision by which the resistor was made. If the fourth band is omitted, the tolerance is assumed to be ±20%. The fifth band is a reliability factor, which gives the percentage of failure per 1000 h of use. For instance, a 1% failure rate reveals that, on the average, one out of every 100 will fail to fall within the tolerance range after 1000 h of use.

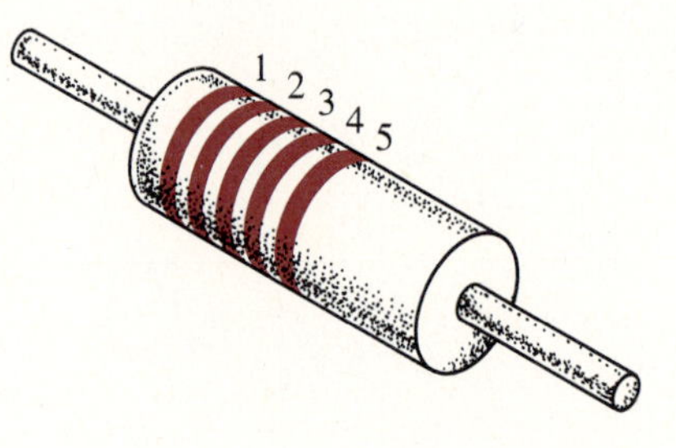

FIG. 4.26
Color coding—fixed molded composition resistor.

TABLE 4.6
Resistor color coding.

Bands 1–3	Band 3		Band 4	Band 5
0 Black	0.1 Gold	Multiplying factors	5% Gold	1% Brown
1 Brown	0.01 Silver		10% Silver	0.1% Red
2 Red			20% No band	0.01% Orange
3 Orange				0.001% Yellow
4 Yellow				
5 Green				
6 Blue				
7 Violet				
8 Gray				
9 White				

EXAMPLE 4.12

Find the range in which a resistor having the following color bands must exist to satisfy the manufacturer's tolerance:

a.

1st band	2nd band	3rd band	4th band	5th band
Gray	Red	Black	Gold	Brown
8	2	0	±5%	1%

b.

1st band	2nd band	3rd band	4th band	5th band
Orange 3	White 9	Gold 0.1	Silver ±10%	No color

Solutions

a. 82 Ω ± 5% (1% reliability). Since 5% of 82 = 4.10, the resistor should be within the range 82 Ω ± 4.10 Ω, or **between 77.90 and 86.10 Ω.**

b. 3.9 Ω ± 10% = 3.9 ± 0.39 Ω. The resistor should lie somewhere **between 3.51 and 4.29 Ω.**

Throughout the text material, resistor values in the network are typically not *standard values* to reduce the mathematical complexity of finding the solution. It was felt that the procedure or analysis technique was of primary importance and the mathematical exercise was secondary. In the problem sections, however, standard values were frequently employed to make them more familiar and demonstrate their effects on the required calculations. A list of readily available standard values appears in Table 4.7. All the resistors ap-

TABLE 4.7
Standard values of commercially available resistors.

Ohms (Ω)					Kilohms (kΩ)		Megohms (MΩ)	
0.10	1.0	10	100	1000	10	100	1.0	10.0
0.11	1.1	11	110	1100	11	110	1.1	11.0
0.12	**1.2**	**12**	**120**	**1200**	**12**	**120**	**1.2**	**12.0**
0.13	1.3	13	130	1300	13	130	1.3	13.0
0.15	1.5	15	150	1500	15	150	1.5	15.0
0.16	1.6	16	160	1600	16	160	1.6	16.0
0.18	**1.8**	**18**	**180**	**1800**	**18**	**180**	**1.8**	**18.0**
0.20	2.0	20	200	2000	20	200	2.0	20.0
0.22	2.2	22	220	2200	22	220	2.2	22.0
0.24	2.4	24	240	2400	24	240	2.4	
0.27	**2.7**	**27**	**270**	**2700**	**27**	**270**	**2.7**	
0.30	3.0	30	300	3000	30	300	3.0	
0.33	3.3	33	330	3300	33	330	3.3	
0.36	3.6	36	360	3600	36	360	3.6	
0.39	**3.9**	**39**	**390**	**3900**	**39**	**390**	**3.9**	
0.43	4.3	43	430	4300	43	430	4.3	
0.47	4.7	47	470	4700	47	470	4.7	
0.51	5.1	51	510	5100	51	510	5.1	
0.56	**5.6**	**56**	**560**	**5600**	**56**	**560**	**5.6**	
0.62	6.2	62	620	6200	62	620	6.2	
0.68	6.8	68	680	6800	68	680	6.8	
0.75	7.5	75	750	7500	75	750	7.5	
0.82	**8.2**	**82**	**820**	**8200**	**82**	**820**	**8.2**	
0.91	9.1	91	910	9100	91	910	9.1	

Key: Color: Most common: 5%, 10%, and 20% tolerances. Bold type: Fairly common: 5% and 10% tolerances. Regular type: More specialized: 5% tolerance only.

pearing in Table 4.7 are available with 5% tolerance. Those in boldface are available with 5% and 10% tolerances, whereas those in color are available with 5%, 10%, and 20% tolerances.

4.8 CONDUCTANCE

By finding the reciprocal of the resistance of a material, we have a measure of how well the material will conduct electricity. The quantity is called *conductance*, has the symbol G, and is measured in *siemens* (S).

In equation form, conductance is

$$\boxed{G = \frac{1}{R}} \quad \text{(siemens, S)} \tag{4.14}$$

A resistance of 1 MΩ is equivalent to a conductance of 10^{-6} S, and a resistance of 10 Ω is equivalent to a conductance of 10^{-1} S. The larger the conductance, therefore, the less the resistance and the greater the conductivity.

In equation form, the conductance is determined by

$$\boxed{G = \frac{A}{\rho l}} \quad \text{(S)} \tag{4.15}$$

indicating that increasing the area or decreasing either the length or the resistivity will increase the conductance.

EXAMPLE 4.13

What is the new level of conductance of a conductor if the area is cut in half and the length is increased by a factor of 4? The resistivity is fixed and the original conductance level was 0.04 S.

Solution Starting out,

$$G = \frac{A}{\rho l} = 0.04 \text{ S}$$

If the area, which appears in the *numerator* of the equation, is *cut in half,* the conductance drops to one-half the original level also, and $G' = (1/2)(0.04 \text{ S}) = 0.02 \text{ S}$.

If the length, which appears in the *denominator* of the equation, is *increased* by a factor of four, the conductance drops to one-fourth, as follows:

$$G'' = \frac{1}{4}(0.02 \text{ S}) = \mathbf{0.005\ S}$$

4.9 OHMMETERS

The *ohmmeter* is an instrument used to perform the following tasks and a number of other useful functions:

1. Measure the resistance of individual or combined elements
2. Detect open-circuit (high-resistance) and short-circuit (low-resistance) situations
3. Check continuity of network connections and identify wires of a multi-lead cable
4. Test some semiconductor (electronic) devices

For most applications, the ohmmeters used most frequently appear as part of a VOM or DMM, such as those in Figs. 3.24 and 3.25. The ohmmeter scale of the VOM is a nonlinear scale, as shown in Fig. 3.24 and Fig. 4.27. For the highest level of accuracy, a scale should be chosen to place the pointer in the middle or high end of the scale. Note in Fig. 4.27 the various levels provided in the brackets and the weight of the division in each section. The nonlinearity is quite obvious from the fact that the distance from 0 to 2 almost matches the distance from 5 to 10 even though the change in resistance is quite different.

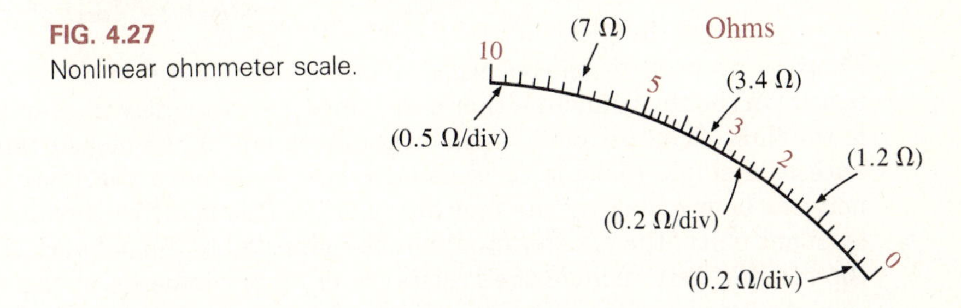

FIG. 4.27
Nonlinear ohmmeter scale.

Before using an analog VOM, the pointer must first be adjusted to reflect 0 Ω when the leads are touching, as shown in Fig. 4.28. This is accomplished using the *zero-adjust* knob on the face of the instrument. Any changes in resistance scale requires a resetting of the zero-adjust control. Only one scale is

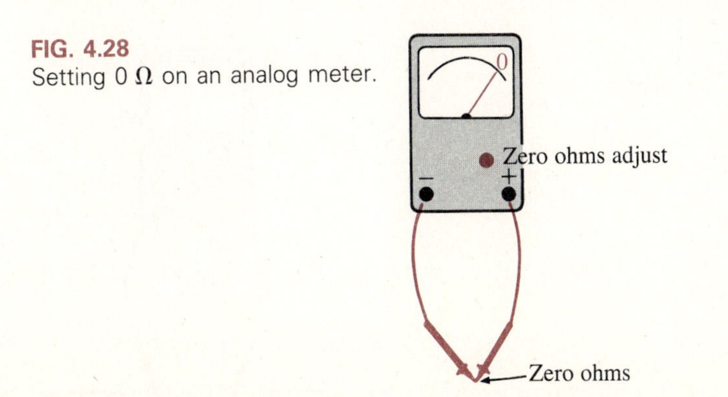

FIG. 4.28
Setting 0 Ω on an analog meter.

provided for a range of resistance levels. If the $R \times 100$ scale is chosen, any scale indication must be multiplied by 100. For instance, a reading of 4.6 actually reflects a resistance level of (4.6)(100), or 460 Ω. The same indication on the $R \times 1$ scale results in a 4.6-Ω level.

When using a DMM it is important to recognize that a display of 10 on the 20-MΩ scale reflects a 10-MΩ resistance level, whereas on the 200-kΩ scale, it is 10 kΩ, and on the 200-Ω scale, it is only 10 Ω.

In general, the resistance of a resistor can be measured by simply connecting the two leads of the meter across the resistor, as shown in Fig. 4.29.

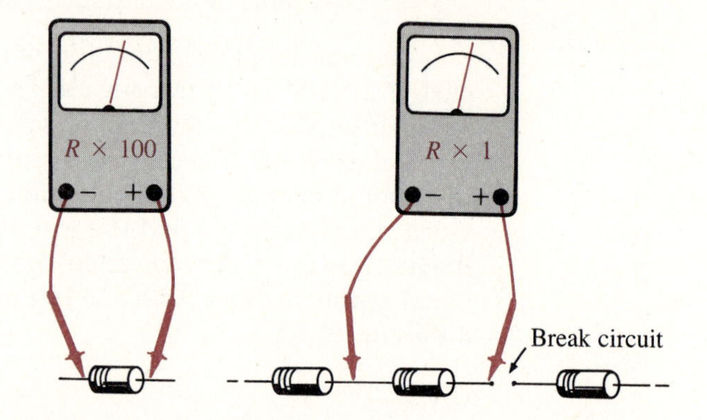

FIG. 4.29
Measuring the resistance of a single element.

There is no need to be concerned about which lead goes on which end; the result will be the same in either case, since resistors offer the same resistance to the flow of charge (current) in either direction. When measuring the resistance of a single resistor, it is usually best to remove the resistor from the network before making the measurement. If this is difficult or impossible, at least one end of the resistor must not be connected to the network (Fig. 4.29) or the reading may include the resistance of other elements in the system.

As noted earlier, if the two leads of the meter are touching in the ohmmeter mode, the resulting resistance is 0 Ω. A connection can, therefore, be checked as shown in Fig. 4.30 by simply hooking up the meter to either side of the connection. If the resistance is zero, the connection is secure. If it is other than zero, it could be a weak connection; if it is infinite, there is no connection at all.

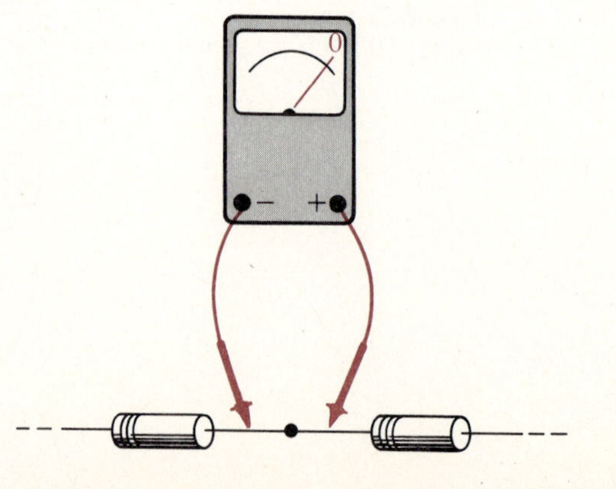

FIG. 4.30
Checking continuity of a connection.

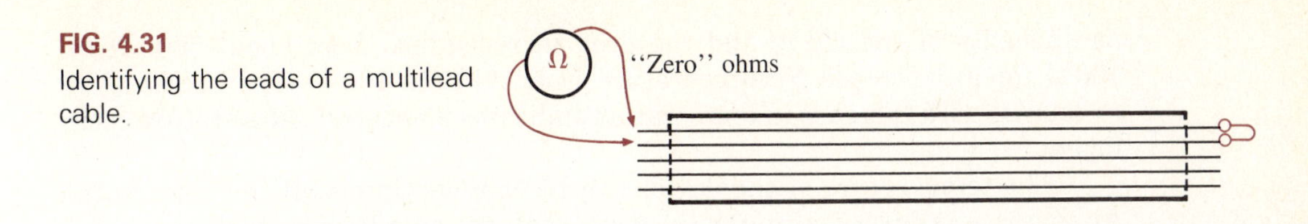

FIG. 4.31
Identifying the leads of a multilead cable.

If one wire of a harness is known, a second can be found, as shown in Fig. 4.31. Simply connect the end of the known lead to the end of any other lead. When the ohmmeter indicates 0 Ω (or very low resistance), the second lead has been identified. This procedure can also be used to determine the first known lead by simply connecting the meter to any wire at one end and then touching all the leads at the other end until a 0-Ω indication is obtained.

Preliminary measurements of the condition of some electronic devices such as the diode and transistor can be made using the ohmmeter. The meter can also be used to identify the terminals of such devices.

One important note about the use of any ohmmeter: *Never hook up an ohmmeter to a live circuit!* The reading will be meaningless and you may damage the instrument. The ohmmeter section of any meter is designed to pass a small sensing current through the resistance to be measured. Too large a current could damage the movement and would certainly throw off the calibration of the instrument. In addition, *never store an ohmmeter in the resistance mode.* The two leads of the meter could touch and the small sensing current could drain the internal battery. VOMs should be stored with the selector switch on the highest voltage range, and DMMs should be stored in the off position.

4.10 THERMISTORS

The *thermistor* is a two-terminal semiconductor device whose resistance, as the name suggests, is temperature sensitive. A representative characteristic appears in Fig. 4.32, along with the graphic symbol for the device. Note the

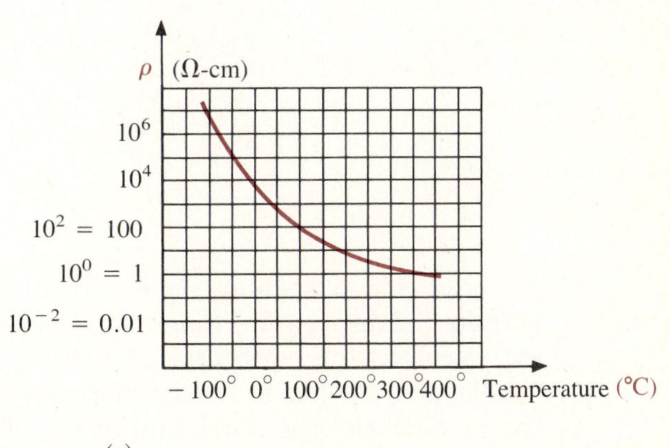

FIG. 4.32
Thermistor. (a) Characteristics; (b) symbol.

nonlinearity of the curve and the drop in resistance from about 5000 Ω to 100 Ω for an increase in temperature from 20°C to 100°C. The decrease in resistance with increase in temperature indicates a negative temperature coefficient.

The temperature of the device can be changed internally or externally. An increase in current through the device will raise its temperature, causing a drop in its terminal resistance. Any externally applied heat source will result in an increase in its body temperature and a drop in resistance. This type of action (internal or external) lends itself well to control mechanisms. A number of different types of thermistors are shown in Fig. 4.33. Materials employed in the manufacture of thermistors include oxides of cobalt, nickel, strontium, and manganese.

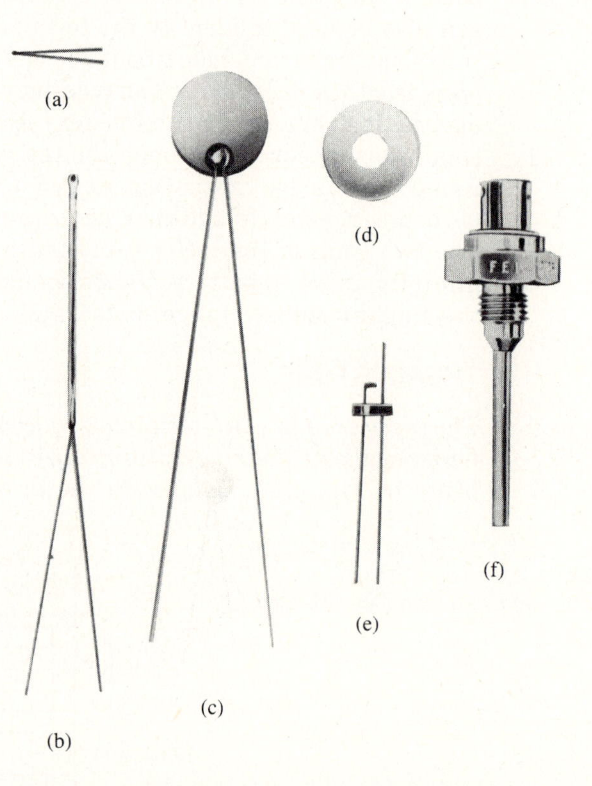

FIG. 4.33
Thermistors. (a) Beads; (b) glass probe; (c) disc; (d) washer; (e) specially mounted bead; (f) special probe assembly. (Courtesy of Fenwal Electronics, Inc.)

Note the use of a log scale in Fig. 4.32 for the vertical axis. The log scale permits the display of a wider range of specific resistance levels than a linear scale such as the horizontal axis. Note that it extends from 0.0001 Ω-cm to 100,000,000 Ω-cm over a very short interval. A log scale is also used for both the vertical and the horizontal axis of Fig. 4.34, which appears in the next section.

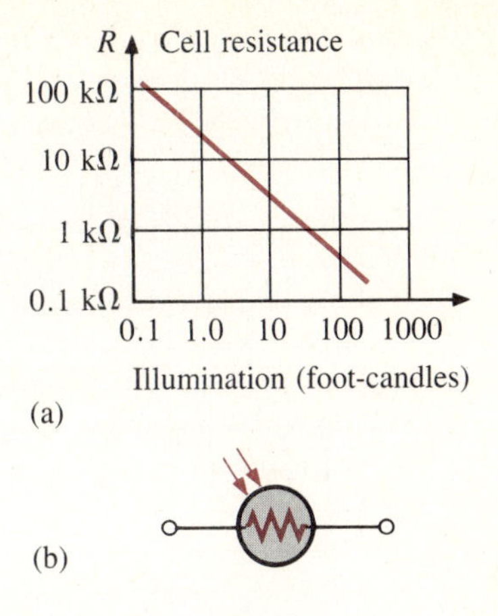

FIG. 4.34
Photoconductive cell. (a) Characteristics; (b) symbol.

4.11 PHOTOCONDUCTIVE CELL

The *photoconductive cell* is a two-terminal semiconductor device whose terminal resistance is determined by the intensity of the incident light on its exposed surface. As the applied illumination increases in intensity, the energy state of the surface electrons and atoms increases, with a resultant increase in the number of "free carriers" and a corresponding drop in resistance. A typical set of characteristics and its graphic symbol appear in Fig. 4.34. Note (as for the thermistor, which is also a semiconductor device) the negative temperature coefficient. A number of cadmium sulfide photoconductive cells are shown in Fig. 4.35.

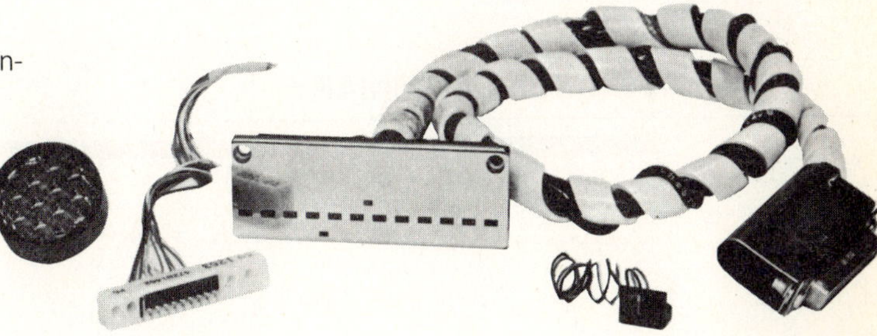

FIG. 4.35
Photoconductive cells. (Courtesy of International Rectifier)

4.12 VARISTORS

Varistors are voltage-dependent, nonlinear resistors used to suppress high-voltage transients. That is, their characteristics are such that they limit the voltage that can appear across the terminals of a sensitive device or system. A typical set of characteristics appears in Fig. 4.36(a), along with a linear resistance characteristic for comparison purposes. Note that at a particular "firing voltage," the current rises rapidly but the voltage is limited to a level just above this firing potential. In other words, the magnitude of the voltage that

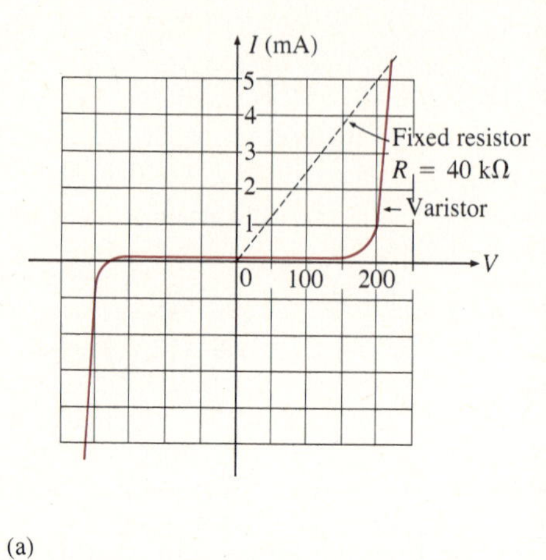

(a)

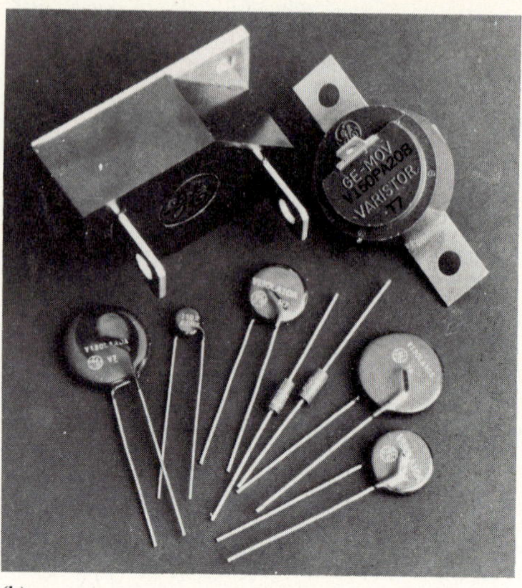

(b)

FIG. 4.36
Varistors. (a) Characteristics; (b) photograph. (Courtesy of General Electric Co.)

can appear across this device cannot exceed that level defined by its characteristics. Through proper design techniques, this device can therefore limit the voltage appearing across sensitive regions of a network. The current is simply limited by the network to which it is connected. A photograph of a number of commercial units appears in Fig. 4.36(b).

FORMULA SUMMARY

$$R = \rho\frac{l}{A}$$

$$A_{CM} = (d_{mils})^2$$

$$\frac{|T| + t_1}{R_1} = \frac{|T| + t_2}{R_2}$$

$$R_2 = R_1[1 + \alpha_1(t_2 - t_1)]$$

$$G = \frac{1}{R}$$

Color Code:

0—black, 1—brown, 2—red, 3—orange
4—yellow, 5—green, 6—blue, 7—violet
8—gray, 9—white, 5%—gold, 10%—silver

CHAPTER SUMMARY

Although a number of equations were introduced in this chapter, each area of investigation has one primary equation. For instance, the equation $R = \rho l/A$ is applied to determine R, l, or A at a *fixed temperature,* and the equation $(|T| + t_1)/R_1 = (|T| + t_2)/R_2$ is applied for temperature variations. In other words, when encountering a question related to resistance, one of the two preceding equations will usually provide the desired unknown. The equations for other quantities such as l, A, or t should be determined from algebraic manipulations and not memorization.

When using the equation $R = \rho l/A$, be sure the length is inserted in feet and the area in circular mils. In addition, convert the diameter of a wire to decimal form before converting to mils. For metric calculations, be sure each quantity is substituted with the proper unit of measurement.

When using the equation $(|T| + t_1)/R_1 = (|T| + t_2)/R_2$, be sure to insert the inferred absolute temperature T without its negative sign. As a check on the results obtained using this equation, keep in mind that for conductors the resistance will increase with increase in temperature and decrease with drop in temperature.

GLOSSARY

Absolute zero The temperature at which all molecular motion ceases; −273.15°C.

Circular mil (CM) The cross-sectional area of a wire having a diameter of one mil.

Color coding A technique employing bands of color to indicate the resistance levels and tolerance of resistors.

Conductance (G) An indication of the relative ease with which current can be established in a material. It is measured in siemens (S).

Inferred absolute temperature The temperature through which a straight-line approximation for the actual resistance-versus-temperature curve will intersect the temperature axis.

Negative temperature coefficient of resistance A value that reveals that the resistance of a material will decrease with increase in temperature.

Ohm (Ω) The unit of measurement applied to resistance.

Ohmmeter An instrument for measuring resistance levels.

Photoconductive cell A two-terminal semiconductor device whose terminal resistance is determined by the intensity of the incident light on its exposed surface.

Positive temperature coefficient of resistance A value that reveals that the resistance of a material will increase with increase in temperature.

Potentiometer A three-terminal device through which potential levels can be varied in a linear or nonlinear manner.

Resistance A measure of the opposition to the flow of charge through a material.

Resistivity (ρ) A constant of proportionality between the resistance of a material and its physical dimensions.

Rheostat An element whose terminal resistance can be varied in a linear or nonlinear manner.

Sheet resistance Defined by ρ/d for thin-film and integrated circuit design.

Thermistor A two-terminal semiconductor device whose resistance is temperature sensitive.

Varistor A voltage-dependent, nonlinear resistor used to suppress high-voltage transients.

PROBLEMS

Section 4.1

1. How would you expect the resistance of a conductor to change if the length is made three times longer with all the other factors held the same?

2. How would you expect the resistance of a conductor to change if the area is doubled and all the other factors are held the same?

3. How would you expect the resistance of a conductor to change if the resistivity is cut in half with all the other factors held the same?

4. How would you expect the resistance of a conductor to change if the area is cut in half and the length doubled if the other factors remain the same?

***5.** How would you expect the resistance of a conductor to change if the length is increased by a factor of four, the area cut in half, and the resistivity doubled?

Section 4.2

6. Convert 0.060 in. to mils.

7. Convert 1/2 in. to mils.

8. Convert 1/32 in. to mils.

9. Convert 0.1 ft to mils.

10. Calculate the area in circular mils of a wire with a diameter of 50 mils.

11. Calculate the area in circular mils of a wire with a diameter of 1/64 in.

12. **a.** Calculate the area in circular mils and square mils of a wire with a diameter of 1/4 in.
b. Compare the number of circular mils to square mils of the same-diameter wire. What is the conversion factor?

13. Calculate the area in circular mils of a wire with a diameter of 1 cm.

14. **a.** If the area of a copper wire is 20,000 CM, what is the diameter of the wire in mils and inches?
b. Repeat part a for an aluminum wire.
c. What conclusion can you draw from the results of parts a and b?

15. If the area of a copper wire is 100 CM, what is the diameter of the wire in mils and inches?

16. Calculate the resistance of a 100-ft length of copper wire with a diameter of 1/10 in.

17. Calculate the resistance of 100 yd of aluminum wire having a diameter of 0.08 in.

18. Calculate the resistance of 1/2 mi of copper telephone wire having a diameter of 0.013 in.

19. Find the length of a copper wire having a resistance of 2 Ω and a diameter of 0.05 in.

20. Find the length of a copper wire having a resistance of 1.8 Ω and a diameter of 1/32 in.

21. Find the length of a copper wire having a resistance of 5 Ω and a diameter of 0.1 cm.

22. Find the length of a gold wire in millimeters having a resistance of 2.5 mΩ and a diameter of 0.02 in.

23. Find the length of a copper wire with a resistance of 0.4 Ω and a diameter of 4 mm.

24. Find the diameter in inches of a 200-ft copper wire that has a resistance of 0.1 Ω.

25. Find the diameter in mils of a copper wire the length of a football field that has a resistance of 250 mΩ.

***26.** Find the diameter in inches of a copper wire 1/4 mi in length that has a resistance of 0.5 Ω.

Section 4.3

27. Using Table 4.2, determine the resistance of 250 ft of #14 house wire.

28. Using Table 4.2, determine the resistance of 1 mi of 2/0 wire.

29. Using Table 4.2, determine the diameter of #20 TV wire in mils and inches.

30. Using Table 4.2, determine the diameter of #40 fine instrument wire in mils, inches, and millimeters.

31. How would you compare the resistance of 200 ft of #12 house wire to #28 telephone wire of the same length?

32. If 20 A is the maximum legal limit for #12 house wire, how long a length of wire would you need to stay within the 20-A limit if the wire were connected across 120 V?

33. **a.** If 1 ft of #18 wire were used to connect a 10-Ω resistor to a network, would the resistance of the wire be of any concern compared to that of the 10-Ω resistor?
b. Answer part a for a #28 wire?
c. Answer part a for a #40 wire?

34. **a.** Repeat Example 4.6 if the length is increased to 200 ft.
b. Repeat Example 4.6 if the length is cut in half.

Section 4.4

35. Determine the resistance of 300 ft of #12 copper wire using metric units.

36. Repeat Problem 35 using Table 4.2 and compare results.

37. Determine the resistance of a circular copper conductor 100 m in length with a diameter of 1 mm using metric units.

38. **a.** Determine the resistance of a 5-cm-long copper conductor rectangular in shape with dimensions 4 mm by 1 mm.
b. Repeat part a for carbon material.

39. Using metric units, determine the length of a copper wire that has a resistance of 0.2 Ω and a diameter of 1/10 in.

40. Repeat Example 4.8 if the length is doubled and the width reduced to one-third its original dimension.

41. If the sheet resistance of a tin oxide sample is 100 Ω, what is the thickness of the oxide layer?

42. Determine the width of a carbon resistor having a sheet resistance of 150 Ω if the length is 1/2 in. and the resistance is 500 Ω.

Section 4.5

43. The resistance of a copper wire is 1 Ω at 0°C (freezing temperature of water). What is its resistance at 100°C (boiling temperature of water)?

44. The resistance of a copper wire is 0.5 Ω at −20°C. What is its resistance at +20°C?

45. The resistance of a copper wire is 0.02 Ω at room temperature (68°F = 20°C). What is its resistance if the surrounding temperature should increase to 120°F?

46. If the resistance of a silver wire is 0.04 Ω at −30°C, what is its resistance at 0°C?

47. **a.** If the resistance of a copper conductor is 5 Ω at 0°C, what is its resistance at +10°C?
b. Determine the resistance of the copper conductor of part a at +20°C.
c. Determine the resistance of the copper conductor of part a at +30°C.
d. Looking at the results of parts a, b, and c, how much does the resistance increase for each 10°C?
e. What do the results of part d suggest about the curve relating resistance and temperature? Is it linear or nonlinear?

48. If the resistance of a copper conductor is 10 Ω at 10°C, at what temperature will it be 20 Ω?

***49.** If the resistance of a 100-m length of copper conductor is 0.5 Ω at room temperature (20°C), at what temperature will it be 1 Ω?

50. If the resistance of a silver wire is 0.04 Ω at +20°C, what is its resistance at 40°C?

51. Using the defining equation for α_1, determine α_1 for copper and aluminum at +20°C and compare it to those given in Table 4.5.

52. Using Eq. 4.13, find the resistance of a copper wire at 50°C if its resistance at 20°C is 4 Ω.

Section 4.6

53. **a.** What is the approximate increase in size from a 1-W to a 2-W carbon resistor?
b. What is the approximate increase in size from a 1/2-W to a 2-W carbon resistor?

54. If the 10-kΩ resistor of Fig. 4.16 is exactly 10 kΩ at room temperature, what is its approximate resistance at 100°C?

55. Repeat Problem 54 at a temperature of −30°C.

56. If the resistance between the outside terminals of a linear potentiometer is 10 kΩ, what is the resistance between the wiper (moveable) arm and an outside terminal if the resistance between the wiper arm and the other outside terminal is 3.5 kΩ?

57. If the wiper arm of a linear potentiometer is one-quarter of the way around the contact surface, what is the resistance between the wiper arm and each outside terminal if the total resistance is 25 kΩ between outside terminals?

Section 4.7

58. Find the range in which the following resistor must face when measured if it is to satisfy the manufacturer's tolerance.

1st band:	green;	2nd band:	blue
3rd band:	orange;	4th band:	gold

59. Repeat Problem 58 for red, red, brown, silver.
60. Repeat Problem 58 for brown, black, blue, no fourth band.
61. Is there coverage between 20% resistors? That is, determine the tolerance range for a 10-Ω, 20% resistor and a 15-Ω, 20% resistor and note whether the coverage is continuous.
62. Repeat Problem 61 for 10% resistors of the same values.

Section 4.8

63. Find the conductance level of each of the following resistances.
    a. 40 Ω **b.** 10,000 Ω
    c. 0.05 Ω **d.** 5 mΩ
64. Find the conductance level of 100 ft of #18 AWG copper wire.

*65. The conductance of a wire is 100 S. If the area is cut to one-fourth its original size and the length is decreased by a factor of two, what is the new conductance level?

Section 4.9

66. How would you check the status of a fuse with an ohmmeter?
67. How would you determine the *on* and *off* states of a switch using an ohmmeter?
68. How would you use an ohmmeter to check the status of a light bulb?

Section 4.10

69. Find the resistivity of the thermistor having the characteristics of Fig. 4.32 at −50°C, 50°C, and 200°C. Note that it is a log scale. If necessary, consult a reference with an expanded log scale.

Section 4.11

70. Using the characteristics of Fig. 4.34, determine the resistance of the photoconductive cell at 10 and 100 foot-candles illumination. As in Problem 69, note that it is a log scale.

Section 4.12

71. **a.** Referring to Fig. 4.36(a), find the terminal voltage of the device at 0.5, 1, 3, and 5 mA.
    b. What is the total change in voltage for the range $I = 0.5$ mA to 5 mA?
    c. Compare the ratio of maximum to minimum current levels in parts a and b to the corresponding ratio of voltage levels.

5

Ohm's Law, Power, and Energy

OBJECTIVES

- ☐ Become familiar with the three forms of Ohm's law.
- ☐ Be able to plot Ohm's law and understand how to read the graph.
- ☐ Learn the three forms of the power equation and when to use each.
- ☐ Understand the meaning of the efficiency level of a system and be able to calculate its level using energy or power levels.
- ☐ Become aware of the difference between energy and power levels and how to calculate the energy level in joules, watthours, or kilowatthours.
- ☐ Understand the function and operation of a circuit breaker and fuse.

5.1 OHM'S LAW

Consider the following relationship:

$$\text{Effect} = \frac{\text{Cause}}{\text{Opposition}} \tag{5.1}$$

Every conversion of energy from one form to another can be related to this equation. In electric circuits, the *effect* we are trying to establish is the flow of charge, or *current*. The *potential difference* or voltage between two points is the *cause* ("pressure"), and the opposition is the *resistance* encountered.

Substituting these terms into Eq. 5.1 results in

$$\text{Current} = \frac{\text{Voltage}}{\text{Resistance}}$$

and

$$I = \frac{E}{R} \qquad \text{(amperes, A)} \tag{5.2}$$

Equation 5.2, known as *Ohm's law,* clearly reveals that the greater the voltage across a resistor, the more the current, and the more the resistance for the same voltage, the less the current. In other words, the current is proportional to the applied voltage and inversely proportional to the resistance.

By simple mathematical manipulations such as described in earlier sections, the voltage and resistance can be found in terms of the other two quantities:

$$E = IR \qquad \text{(volts, V)} \tag{5.3}$$

$$R = \frac{E}{I} \qquad \text{(ohms, } \Omega\text{)} \tag{5.4}$$

Recall from Chapter 3 that for voltage, the symbol E represents all sources of voltage such as the battery, and the symbol V represents the potential drop across a resistor or any other energy-converting device. In any case, E and V are interchangeable in Eqs. 5.2 through 5.4.

The three quantities of Eqs. 5.2 through 5.4 are defined by Fig. 5.1. The current I of Eq. 5.2 results from applying a dc supply of E volts across a network having a resistance R. Equation 5.3 determines the voltage E required to establish a current I through a network with a total resistance R, and Eq. 5.4 provides the resistance of a network that results in a current I due to an impressed voltage E.

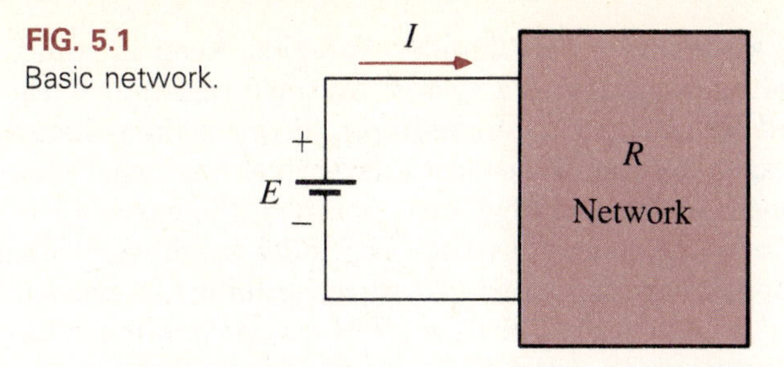

FIG. 5.1
Basic network.

Note in Fig. 5.1 that a clockwise current I was established by the battery. For single-source dc networks the direction of conventional current established by a voltage source can be determined using the mechanical analogy of Fig. 5.2. The long bar of the battery symbol can be likened to the blade of a bulldozer or snowplow. Quite obviously, if dirt (or snow) were to be moved by the blade, it would be pushed in the direction indicated.

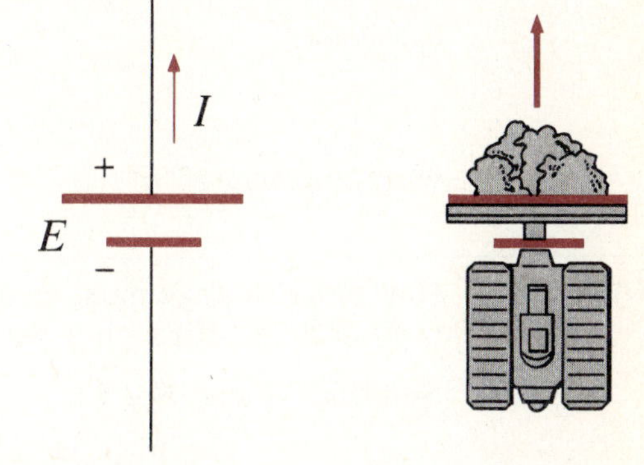

FIG. 5.2
Remembering the direction of established conventional current.

For the new student of the field, the diagram of Fig. 5.3 can be helpful in the initial memorization of the three forms of Ohm's law. If I is isolated in the diagram, the remaining letters indicate the quotient $I = E/R$. When solving for R, the remaining letters also indicate the quotient $R = E/I$, and, finally, the solution for the voltage E results in a product of $E = I \cdot R$. In any event, Ohm's law is such an important basic relationship in the electrical/electronic field that a student should be able to provide the equation for I, R, or E without a second's hesitation. A few minutes spent now to memorize all three forms would be time well spent.

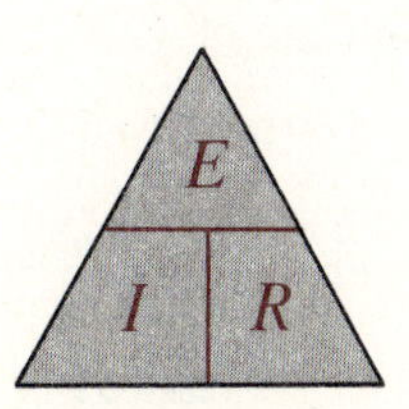

FIG. 5.3
Helpful figure for remembering Ohm's law.

Before attacking a few simple examples, keep in mind that the more pressure (voltage) applied to a system (of fixed resistance), the more the current that will result (Eq. 5.2). In addition, the more the resistance for the same applied voltage, the less the current. Everyone has experienced the fact that the less opposition there is to what is planned, the easier the "flow" (and vice versa). Further, the more flow that has to be established against a fixed or increasing opposition, the more "pressure," or effort, is needed (Eq. 5.3). And, finally, the less the current, or flow, for the same applied voltage, or pressure, the more opposition, or resistance, the system must have (Eq. 5.4).

EXAMPLE 5.1

Determine the current resulting from the application of 12 V across a resistance of 4 Ω.

Solution Using Eq. 5.2:

$$I = \frac{E}{R} = \frac{12\text{ V}}{4\ \Omega} = \mathbf{3\ A}$$

Calculator

(1) (2) (÷) (4) (=)

with a display of 3.

EXAMPLE 5.2

How much voltage must be applied to an electrical system to establish a current of 20 mA through a resistance of 4 kΩ?

Solution Using Eq. 5.3:

$$\begin{aligned} E = IR &= (20 \times 10^{-3}\text{ A})(4 \times 10^{+3}\ \Omega) \\ &= (20)(4) \times (10^{-3})(10^{+3})\text{ V} \\ &= 80 \times 10^{0}\text{ V} = 80 \times 1\text{ V} \\ &= \mathbf{80\ V} \end{aligned}$$

Calculator

(2) (0) (EXP) (+/−) (3) (×) (4) (EXP) (3) (=)

with a display of 80.

EXAMPLE 5.3

Determine the resistance of a light bulb if a current of 500 mA results from an applied voltage of 120 V.

Solution Using Eq. 5.4:

$$\begin{aligned} R = \frac{E}{I} &= \frac{120\text{ V}}{500 \times 10^{-3}\text{ A}} = \frac{120}{500} \times 10^{+3}\ \Omega = 0.240 \times 10^{+3}\ \Omega \\ &= \mathbf{240\ \Omega} \end{aligned}$$

Calculator

with a display of 240.

For the resistive element, the polarity of the voltage drop is as shown in Fig. 5.4(a) for the indicated current direction. A reversal in current reverses the polarity, as shown in Fig. 5.4(b). In general, the flow of charge is from a high (+) to a low (−) potential. In other words, if the current direction is known, then the polarities can be determined and added to the diagram. Similarly, if the polarities are known, then the current direction can be determined. The situation is likened to a runner approaching a hill from either direction. No matter which approach taken, there is a loss of energy when overcoming the obstacle. The preceding maneuvers are described in the next two examples.

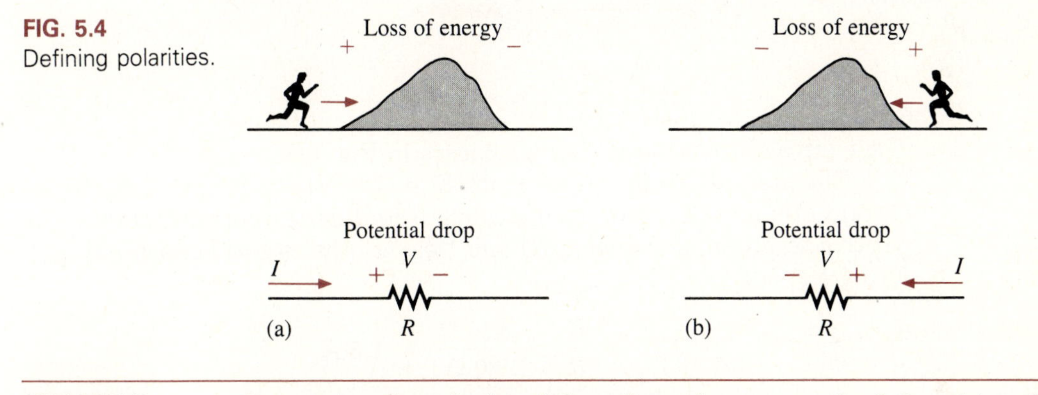

FIG. 5.4
Defining polarities.

EXAMPLE 5.4

a. Determine the polarity of the voltage drop across the 2-Ω resistor of Fig. 5.5 for the indicated current direction.

b. Calculate the voltage drop across the 2-Ω resistor.

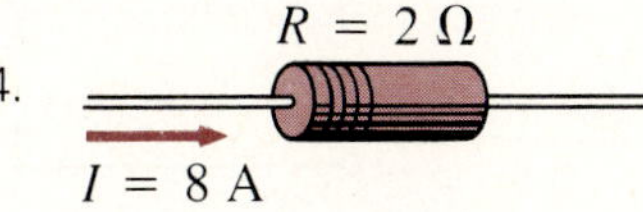

FIG. 5.5
Example 5.4.

Solutions

a. The polarities are added in Fig. 5.6, as determined by the current direction. Remember that the current direction establishes the end of the resistor it enters as the higher potential (+) side.

FIG. 5.6
Solution to Example 5.4.

$V = 16$ V, $+$ $-$, $I = 8$ A, $R = 2\ \Omega$

b. From Eq. 5.3:

$$V = IR = (8\ \text{A})(2\ \Omega) = \mathbf{16\ V}$$

as shown in Fig. 5.5.

Calculator

with a display of 16.

EXAMPLE 5.5

a. Determine the current direction for the soldering iron of Fig. 5.7 for the applied voltage.
b. Calculate the magnitude of the resulting current.

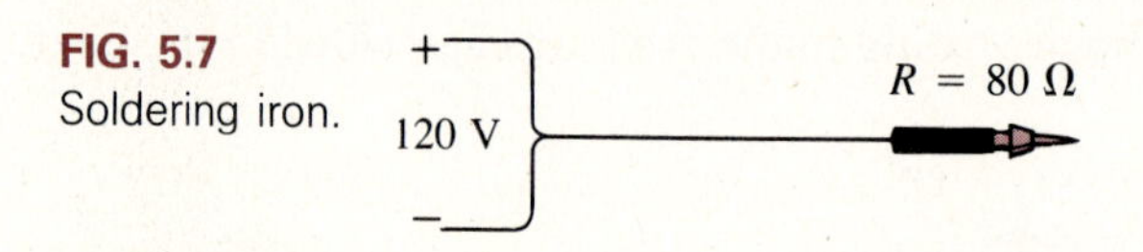

FIG. 5.7
Soldering iron.

Solutions

a. The equivalent electrical circuit appears in Fig. 5.8.
The pressure (from – to + for E = 120 V) establishes a clockwise current around the circuit, as indicated. Note that the current "enters" the positive terminal of the load (R) and "leaves" the negative terminal.

b. Using Eq. 5.2:

$$I = \frac{E}{R} = \frac{120\ \Omega}{80\ \Omega} = \mathbf{1.5\ A}$$

FIG. 5.8
Equivalent electrical circuit of Fig. 5.7.

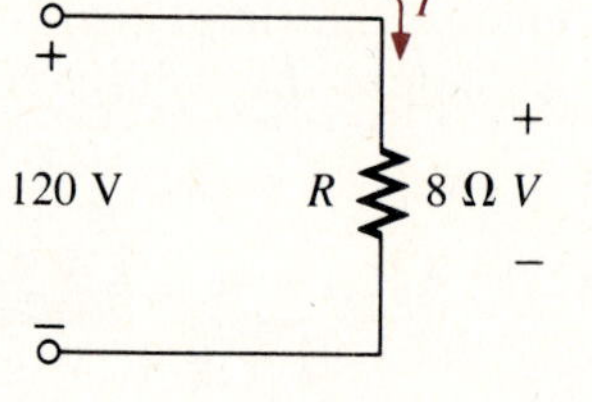

Calculator

1 2 0 ÷ 8 0 =

with a display of 1.5.

5.2 PLOTTING OHM'S LAW

Graphs, characteristics, plots, and the like play an important role in every technical field as a mode through which the broad picture of the behavior or response of a system can be conveniently displayed. It is therefore critical to develop the skills necessary both to read data and to plot them in such a manner that they can be interpreted easily.

For most sets of characteristics of electronic devices, the current is represented by the vertical axis (ordinate), or y-axis, and the voltage is represented by the horizontal axis (abscissa), or x-axis, as shown in Fig. 5.9. First note that the vertical axis is in amperes and the horizontal axis is in volts. For some plots, I may be in milliamperes, microamperes, or whatever is appropriate for the range of interest. The same is true for the levels of voltage on the horizontal axis. Note also that the chosen parameters require that the spacing between numerical values of the vertical axis be different from that of the horizontal axis.

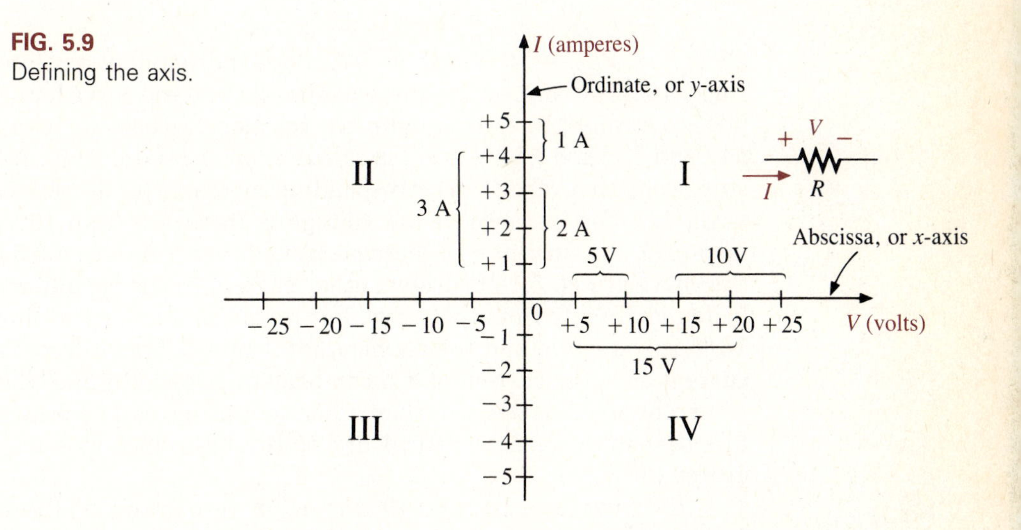

FIG. 5.9
Defining the axis.

The defining voltage polarities and current direction included with the resistor in the same figure determine which part of the graph should be examined. If the voltage across the resistor has the polarity of Fig. 5.9 and the current direction of the same figure, then the first quadrant is the region under examination. In other words, if the voltage polarity and current direction match those of Fig. 5.9, then both quantities are assigned a positive sign and region I is the region of interest. If the polarity of the applied voltage and the current direction are opposite to those appearing in Fig. 5.9, then both are assigned a negative sign and region III is the area of interest. The elements introduced in this text employ regions I and III almost exclusively. When you examine electronic devices, regions II and IV will come into play. For the remainder of this chapter and, in fact, for the majority of this text, region I will be employed almost exclusively.

Let us now plot the characteristics of a 5-Ω resistor on the characteristics of Fig. 5.9 and examine the resulting curve. At 0 V, the current as determined by Ohm's law will be zero also. That is, $I = V/R = 0\text{ V}/5\ \Omega = 0\text{ A}$. This plot point now appears on Fig. 5.10. At 10 V, $I = V/R = 10\text{ V}/5\ \Omega = 2\text{ A}$, and at

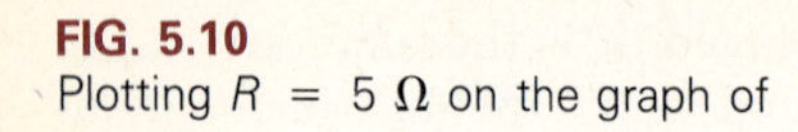
FIG. 5.10
Plotting $R = 5\ \Omega$ on the graph of

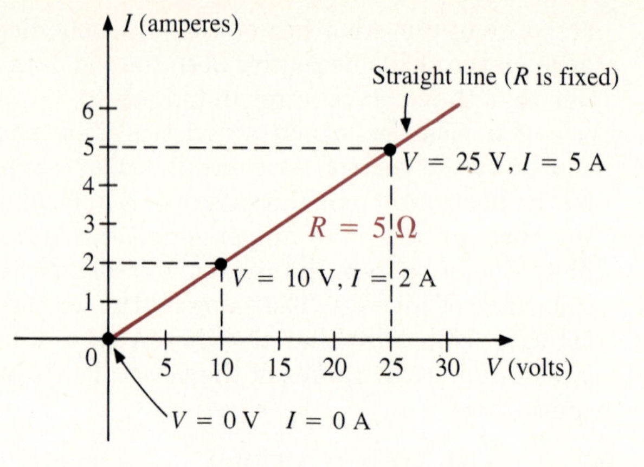

25 V, $I = V/R = 25\text{ V}/5\ \Omega = 5$ A. Both additional plot points also appear on Fig. 5.10. If we connect the three points, we find the curve for a fixed resistor of 5 Ω is a straight line. A straight-line relationship between two quantities such as I and V, I and t, or V and t is called a *linear* relationship. A linear relationship is one that results in corresponding increases (or decreases) in the related variables. For instance, if the voltage is increased from 10 V to 25 V (a 2.5 increase), the current will increase from 2 A to 5 A (also a 2.5 increase). Once the curve of Fig. 5.10 is drawn, other current levels for different voltages can be found directly from the curve. For instance, if a vertical line is drawn from 15 V to the curve and then a horizontal line is drawn from the curve to the current axis, the current of 3 A can be determined. Similarly, if the voltage at a current of 4 A is desired, then a horizontal line can be drawn at 4 A until it hits the curve; then a vertical line to the horizontal axis will reveal the required 20 V.

If curves for a 1-Ω and 10-Ω resistor were drawn on the same graph, the curves of Fig. 5.11 would result. Note that the slope is less for larger resistances and more for smaller resistances. In other words, higher resistance levels are associated with curves closer to the horizontal axis, whereas lesser resistance levels occur closer to the vertical or current axis. This fact will be very useful when we examine characteristic curves in the future.

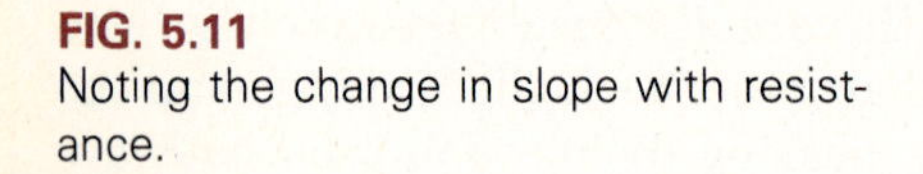
FIG. 5.11
Noting the change in slope with resistance.

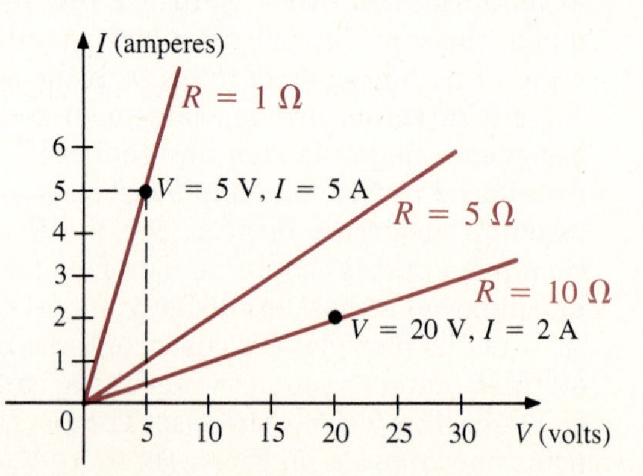

The resistance of the curves is related to the change in voltage and current by

$$R = \frac{\Delta V}{\Delta I} \quad \text{(ohms, } \Omega\text{)} \tag{5.5}$$

where Δ signifies finite change. To examine the use of Eq. 5.5, consider the curve for the 5-Ω resistor of Fig. 5.12. For a change in current from 2 A to 3 A, or $\Delta I = 1\ \text{A}\ (3\ \text{A} - 2\ \text{A} = 1\ \text{A})$, the corresponding change in voltage is from 10 V to 15 V, or $\Delta V = 5\ \text{V}\ (15\ \text{V} - 10\ \text{V} = 5\ \text{V})$.

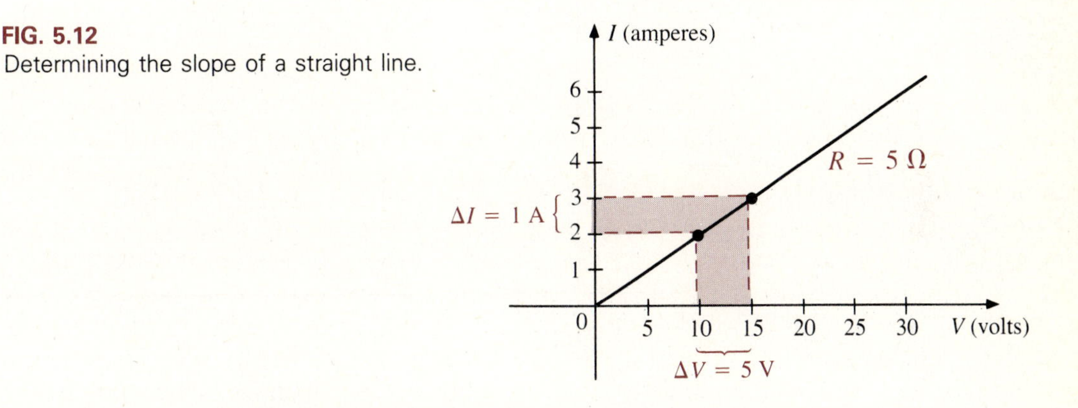

FIG. 5.12
Determining the slope of a straight line.

Substituting into Eq. 5.5:

$$R = \frac{\Delta V}{\Delta I} = \frac{5\ \text{V}}{1\ \text{A}} = \mathbf{5\ \Omega}$$

Any chosen ΔI and its corresponding ΔV will result in the same solution for R.

If a ΔV is chosen for the 10-Ω resistor of Fig. 5.11 such that $\Delta V = 20\ \text{V} - 10\ \text{V} = 10\ \text{V}$, the resulting $\Delta I = 2\ \text{A} - 1\ \text{A} = 1\ \text{A}$, and

$$R = \frac{\Delta V}{\Delta I} = \frac{10\ \text{V}}{1\ \text{A}} = \mathbf{10\ \Omega}$$

Since the slope of a curve is defined by

$$\text{Slope } m = \frac{\Delta y}{\Delta x} \tag{5.6}$$

with y the vertical axis and x the horizontal axis, substituting results in

$$m = \frac{\Delta y}{\Delta x} = \frac{\Delta I}{\Delta V} = \frac{1}{\Delta V/\Delta I} = \frac{1}{R} \tag{5.7}$$

revealing that the slope of the curves of Figs. 5.10 through 5.12 is *inversely* proportional to the resistance. That is, the *larger* the resistance, the *less* the slope (and vice versa), as concluded earlier in our introduction.

EXAMPLE 5.6 Given the axis of Fig. 5.13 plot the curves of 500-Ω (0.5-kΩ) and 2-kΩ resistors.

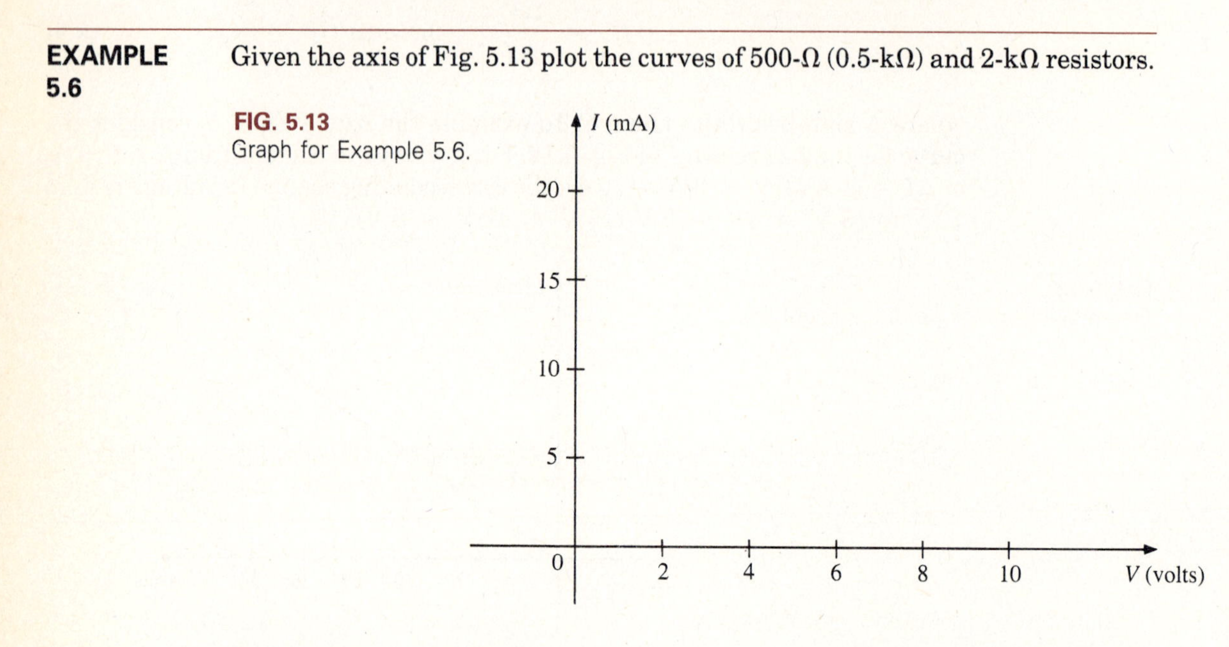

FIG. 5.13
Graph for Example 5.6.

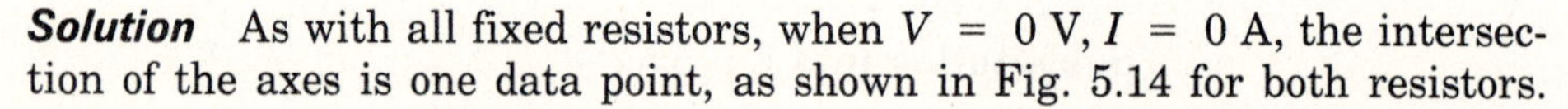

Solution As with all fixed resistors, when $V = 0$ V, $I = 0$ A, the intersection of the axes is one data point, as shown in Fig. 5.14 for both resistors.

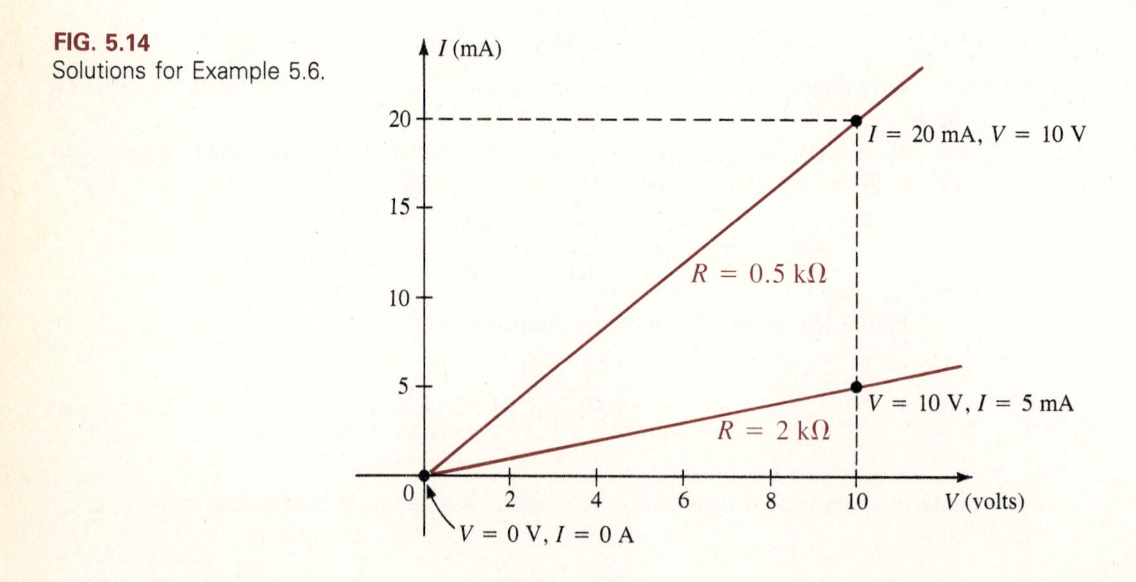

FIG. 5.14
Solutions for Example 5.6.

For the 0.5-kΩ resistor, if $I = 20$ mA, the required voltage is determined by Eq. 5.3:

$$\begin{aligned} V = IR &= (20 \times 10^{-3}\ \text{A})(0.5 \times 10^{+3}\ \Omega) \\ &= 10 \times 10^{0}\ \text{V} \\ V &= 10\ \text{V} \end{aligned}$$

A second data point is found at $I = 20$ mA and $V = 10$ V, as shown in Fig. 5.14.

Connecting a straight line between the data points provides the curve for $R = 0.5$ kΩ.

For the 2-kΩ resistor, if $V = 10$ V is chosen, the resulting current is determined by Eq. 5.2:

$$\begin{aligned} I &= \frac{V}{R} = \frac{10\ \text{V}}{2 \times 10^{3}\ \Omega} = \frac{10}{2} \times 10^{-3}\ \text{A} \\ &= 5 \times 10^{-3}\ \text{A} = 5\ \text{mA} \end{aligned}$$

which is also shown as a data point on Fig. 5.14. Connecting a straight line from the origin (intersection of the two axes) to the data point of $V = 10$ V and $I = 5$ mA provides the curve for $R = 2$ kΩ.

Note that the slope of the 2-kΩ resistor is much less ($m = 1/R = 1/2\ \text{k}\Omega = 0.5$ mS) than the slope of the 0.5-kΩ resistor ($m = 1/R = 1/0.5\ \text{k}\Omega = 0.2$ mS).

EXAMPLE 5.7

Given the plot of Fig. 5.15 determine the resistance level.

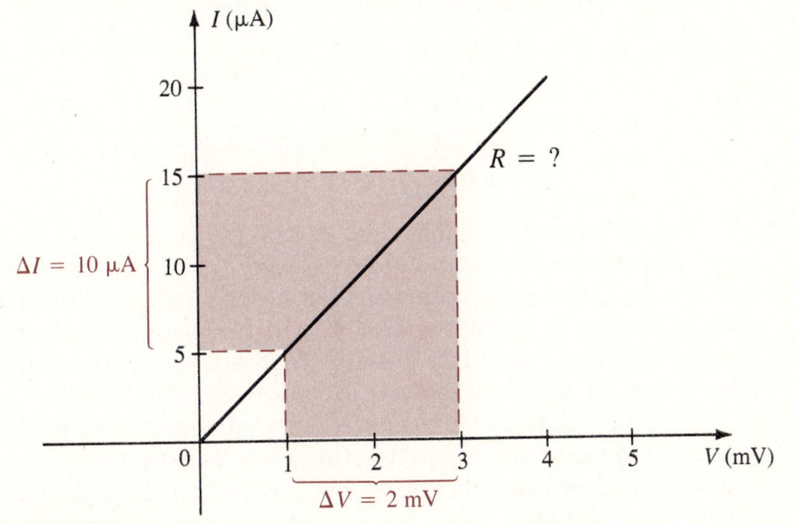

FIG. 5.15
Graph for Example 5.7.

Solution Use Eq. 5.5:

$$R = \frac{\Delta V}{\Delta I} = \frac{(3 - 1)\ \text{mV}}{(15 - 5)\ \mu\text{A}} = \frac{2 \times 10^{-3}\ \text{V}}{10 \times 10^{-6}\ \text{A}}$$
$$= \left(\frac{2}{10}\right) \times \left(\frac{10^{-3}}{10^{-6}}\right)\ \Omega = 0.2 \times 10^{-3-(-6)}\ \Omega$$
$$= 0.2 \times 10^{-3+6}\ \Omega = 0.2 \times 10^{+3}\ \Omega$$
$$R = \mathbf{0.2\ k\Omega}$$

For devices whose characteristics pass through the origin, such as the preceding resistors, the resistance can be determined using an equation simpler in application than Eq. 5.5. It is derived by simply choosing the origin as one of the data points for the application of Eq. 5.5. The result is that

$$\boxed{R = \frac{V}{I}} \quad (V, I = 0 \text{ a data point}) \qquad (5.8)$$

where V and I are voltage and current levels at any data point on the characteristic curve. For instance, for the resistor of Fig. 5.15 at $V = 3$ mV with $I = 15\ \mu$A, substituting gives

$$R = \frac{V}{I} = \frac{3\ \text{mV}}{15\ \mu\text{A}} = \frac{3 \times 10^{-3}\ \text{V}}{15 \times 10^{-6}\ \text{A}} = \left(\frac{3}{15}\right) \times \left(\frac{10^{-3}}{10^{-6}}\right)\ \Omega$$
$$= 0.2 \times 10^{+3}\ \Omega = \mathbf{0.2\ k\Omega}$$

Similarly, at $V = 1$ mV, $I = 5\ \mu$A, and

$$R = \frac{V}{I} = \frac{1\ \text{mV}}{5\ \mu\text{A}} = \frac{1 \times 10^{-3}\ \text{V}}{5 \times 10^{-6}\ \text{A}} = \left(\frac{1}{5}\right) \times \left(\frac{10^{-3}}{10^{-6}}\right)\ \Omega$$
$$= 0.2 \times 10^{+3}\ \Omega = \mathbf{0.2\ k\Omega}$$

Before leaving the subject, let us first investigate the characteristics of a very important semiconductor device called a *diode,* which is examined in detail in basic electronics courses. This device ideally acts like a low-resistance path to current in one direction and a high-resistance path to current in the reverse direction, much like a switch that will pass current in only one direction. A typical set of characteristics appears in Fig. 5.16.

First note that the vertical axis is the current axis and the horizontal axis is the voltage axis, as discussed in the previous paragraphs. The steep slope in the first quadrant reveals a very low resistance region (like a closed switch). The horizontal portion below 0.7 V suggests a very high resistance region (like an open switch). In other words, above 0.7 V the diode is like a closed switch, and below 0.7 V it is like an open switch.

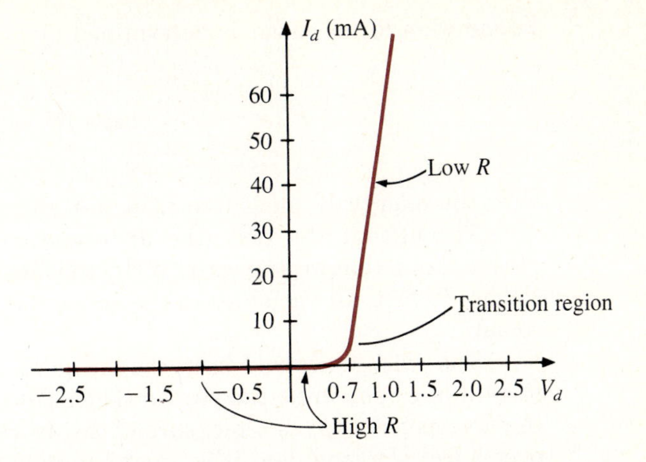

FIG. 5.16
Semiconductor diode characteristics.

Be aware that if the vertical scale is the voltage axis and the horizontal scale is the current axis, some of the conclusions described in this section must be modified. Equations 5.5 ($R = \Delta V/\Delta I$), 5.6 ($m = \Delta y/\Delta x$), and 5.7 ($R = V/I$) are the same, but Eq. 5.7 becomes

$$m = \frac{\Delta y}{\Delta x} = \frac{\Delta V}{\Delta I} = R \qquad \begin{array}{l} V\text{—vertical axis} \\ I\text{—horizontal axis} \end{array} \tag{5.9}$$

revealing that the slope is directly proportional to the resistance. In other words, the steeper the slope, the larger the resistance, and the smaller the slope, the less the resistance—just the opposite of the conclusions obtained for Fig. 5.15.

Keep in mind, however, that most device characteristics utilize current as the vertical axis and voltage as the horizontal axis, reducing any concerns about the confusion introduced by reversing the roles of each.

5.3 POWER

Power is an indication of how much work (the conversion of energy from one form to another) can be accomplished in a specified amount of time, that is, a *rate* of doing work. Obviously, therefore, the more power a system has, the more work it can accomplish in a specified amount of time.

Since converted energy is measured in *joules* (J) and time is measured in seconds (s), power is measured in joules/second (J/s). The electrical unit of measurement for power is the watt (W), defined by

$$1 \text{ watt (W)} = 1 \text{ joule/second (J/s)} \tag{5.10}$$

In equation form, power is determined by

$$P = \frac{W}{t} \quad \text{(watts, W, or joules/second, J/s)} \tag{5.11}$$

with the energy W measured in joules and the time t measured in seconds.

Throughout the text, the abbreviation for energy (W) can be distinguished from that for the watt (W) by the fact that one is italic and the other is roman. In fact, all variables in the dc section are italic, whereas the units are roman.

The watt is derived from the surname of James Watt, who was instrumental in establishing the standards for power measurements. He introduced the *horsepower* (hp) as a measure of the average power of a strong dray horse over a full working day. It is approximately 50% more than can be expected from the average horse. Horsepower and watts are related in the following manner:

$$1 \text{ hp} \cong 746 \text{ W}$$

The power delivered to, or absorbed by, an electrical device or system can be found in terms of the current and voltage by first substituting Eq. 3.6 and Eq. 5.11:

$$P = \frac{W}{t} = \frac{QV}{t} = V\frac{Q}{t}$$

But

$$I = \frac{Q}{t} \quad \text{(Eq. 3.2)}$$

so that

$$P = VI \quad \text{(watts)} \tag{5.12}$$

By direct substitution of Ohm's law, the equation for power can be obtained in two other forms:

$$P = VI = V\left(\frac{V}{R}\right)$$

and

$$P = \frac{V^2}{R} \quad \text{(watts)} \tag{5.13}$$

or

$$P = VI = (IR)I$$

and

$$\boxed{P = I^2R} \qquad \text{(watts)} \tag{5.14}$$

The result is that the power to the resistor of Fig. 5.17 can be found directly, depending on the information available. In other words, if the current and resistance are known, it pays to use Eq. 5.14 directly, and if V and I are known, Eq. 5.12 is appropriate. It saves having to apply Ohm's law before determining the power. When applying Eq. 5.13, be sure to remember that the voltage V is the voltage across the resistor and not the applied voltage. Such a misunderstanding is a source of numerous errors in the early use of Eqs. 5.12 through 5.14.

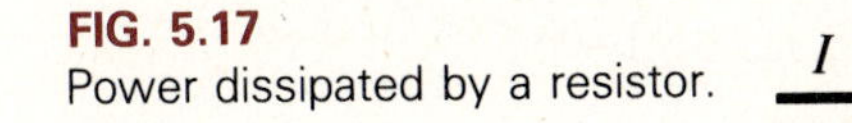

FIG. 5.17
Power dissipated by a resistor.

EXAMPLE 5.8 Find the power delivered to the dc motor of Fig. 5.18.

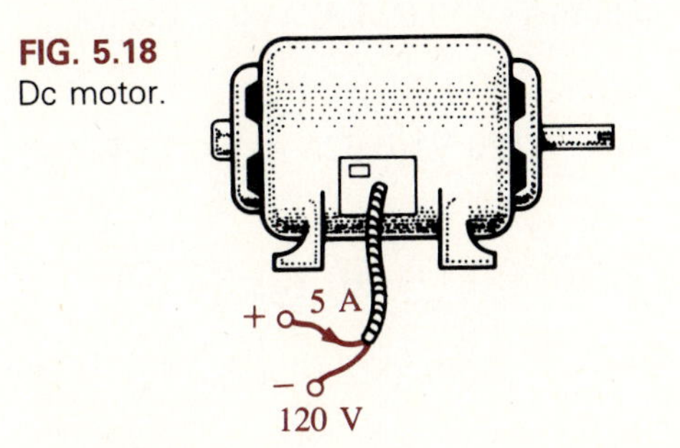

FIG. 5.18
Dc motor.

Solution $P = VI = (120\text{ V})(5\text{ A}) = 600\text{ W} = \mathbf{0.6\ kW}$

Calculator

① ② ⓪ ⊗ ⑤ ⊜

with a display of 600.

EXAMPLE 5.9

What is the power dissipated by a 5-Ω resistor if the current is 4 A?

Solution

$$P = I^2R = (4 \text{ A})^2 \cdot 5 \ \Omega = \mathbf{80 \ W}$$

Calculator

(4) (x^2) (×) (5) (=)

with a display of 80.

EXAMPLE 5.10

The *I-V* characteristics of a light bulb are provided in Fig. 5.19. Note the nonlinearity of the curve, indicating a wide range in resistance of the bulb with applied voltage. If the rated voltage is 120 V, find the wattage rating of the bulb. Also, calculate the resistance of the bulb under rated conditions.

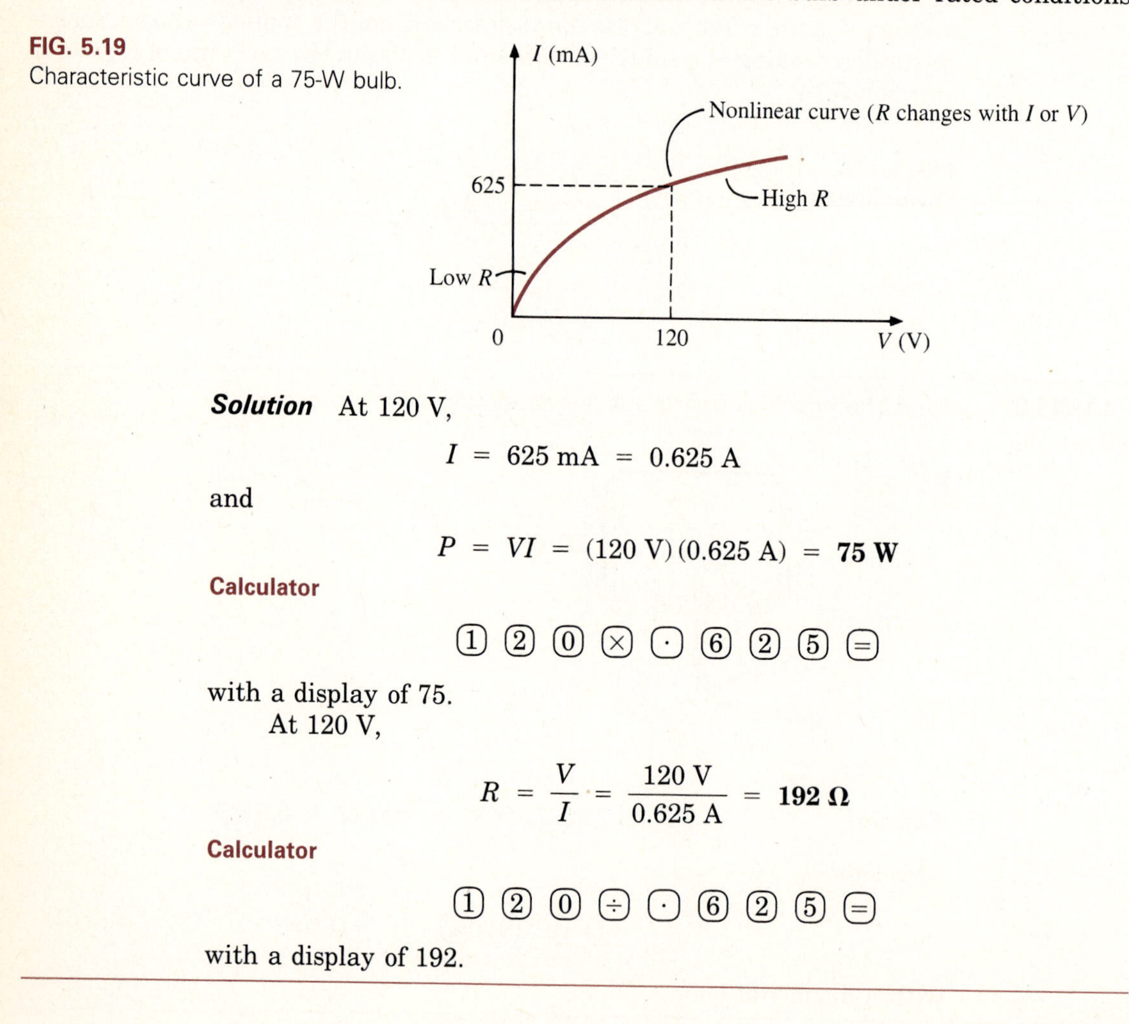

FIG. 5.19
Characteristic curve of a 75-W bulb.

Solution At 120 V,

$$I = 625 \text{ mA} = 0.625 \text{ A}$$

and

$$P = VI = (120 \text{ V})(0.625 \text{ A}) = \mathbf{75 \ W}$$

Calculator

(1) (2) (0) (×) (·) (6) (2) (5) (=)

with a display of 75.

At 120 V,

$$R = \frac{V}{I} = \frac{120 \text{ V}}{0.625 \text{ A}} = \mathbf{192 \ \Omega}$$

Calculator

(1) (2) (0) (÷) (·) (6) (2) (5) (=)

with a display of 192.

Taking a more careful look at the characteristics of Fig. 5.19, we find that for low values of current and voltage, the slope is quite steep, indicating a low-resistance region. For increasing levels of current and voltage, the curve levels off, indicating a higher resistance region. In other words, as the current level through the bulb increases, the resistance increases also—an expected result due to the heating up of the filament with the higher current levels. The power delivered by an energy source such as a battery or generator is given by

$$\boxed{P = EI} \quad \text{(W)} \tag{5.15}$$

where E is the source voltage and I is the current drain from the source, as shown in Fig. 5.20.

FIG. 5.20

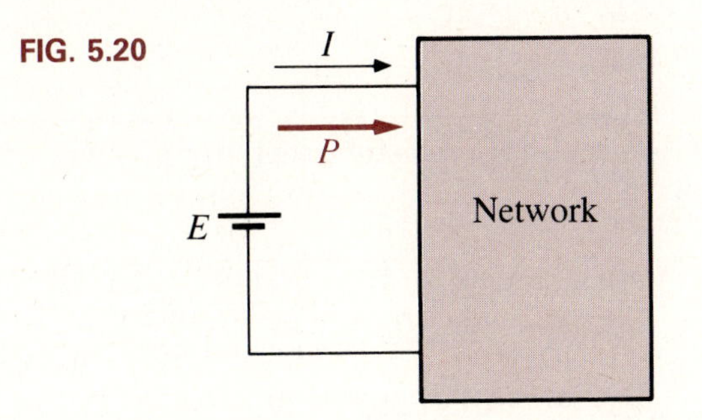

Sometimes the power is given and the current or voltage must be determined. Through algebraic manipulations, an equation for each variable is derived as follows:

$$P = I^2R \Rightarrow I^2 = \frac{P}{R}$$

and

$$\boxed{I = \sqrt{\frac{P}{R}}} \quad \text{(A)} \tag{5.16}$$

$$P = \frac{V^2}{R} \Rightarrow V^2 = PR$$

and

$$\boxed{V = \sqrt{PR}} \quad \text{(V)} \tag{5.17}$$

EXAMPLE 5.11 Determine the current through a 5-kΩ resistor when the power dissipated by the element is 20 mW.

Solution Use Eq. 5.16:

$$I = \sqrt{\frac{P}{R}} = \sqrt{\frac{20 \times 10^{-3}\ \text{W}}{5 \times 10^{3}\ \Omega}} = \sqrt{4 \times 10^{-6}}\ \text{A} = 2 \times 10^{-3}\ \text{A}$$

$$= \mathbf{2\ mA}$$

Calculator

(2) (0) (EXP) (+/−) (3) (÷) (5) (EXP) (3) (=) (√)

with a display of 0.002. Pressing the (F <−> E) key results in 2.–03 (= 2 mA).

EXAMPLE 5.12 Find the voltage across a 10-mΩ resistor receiving a power of 5 mW.

Solution Use Eq. 5.17:

$$\begin{aligned} V = \sqrt{PR} &= \sqrt{(5 \times 10^{-3}\ \text{W})(10 \times 10^{6}\ \Omega)} \\ &= \sqrt{(5)(10) \times (10^{-3})(10^{6})}\ \text{V} \\ &= \sqrt{50 \times 10^{+3}}\ \text{V} = \sqrt{5 \times 10^{4}}\ \text{V} \\ &= \sqrt{5} \times (10^{4})^{1/2}\ \text{V} \\ &= 2.236 \times 10^{2}\ \text{V} \\ V &= \mathbf{223.6\ V} \end{aligned}$$

Calculator

with a display of 236.61.

5.4 WATTMETERS

As one might expect, instruments exist that can measure the power delivered by a source and to a dissipative element. One such instrument appears in Fig. 5.21. Since power is a function of both the current and voltage levels, four terminals are required, as shown in Fig. 5.21. The potential terminals (*PC*) must be connected in parallel (across) the load, as shown in Fig. 5.22, whereas the current terminals (*CC*) are connected "in line with" (in *series* with, as will be defined in Chapter 6) the load. In addition to hooking them up as just indicated, note that the positive terminal of the *PC* connection is hooked up to the positive side of the voltage *V*, whereas the positive terminal of the *CC* connection is connected such that the current centers the terminal. The result of the preceding will be an up-scale deflection. A reversal of *either* set of connections will result in a below-zero indication.

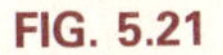

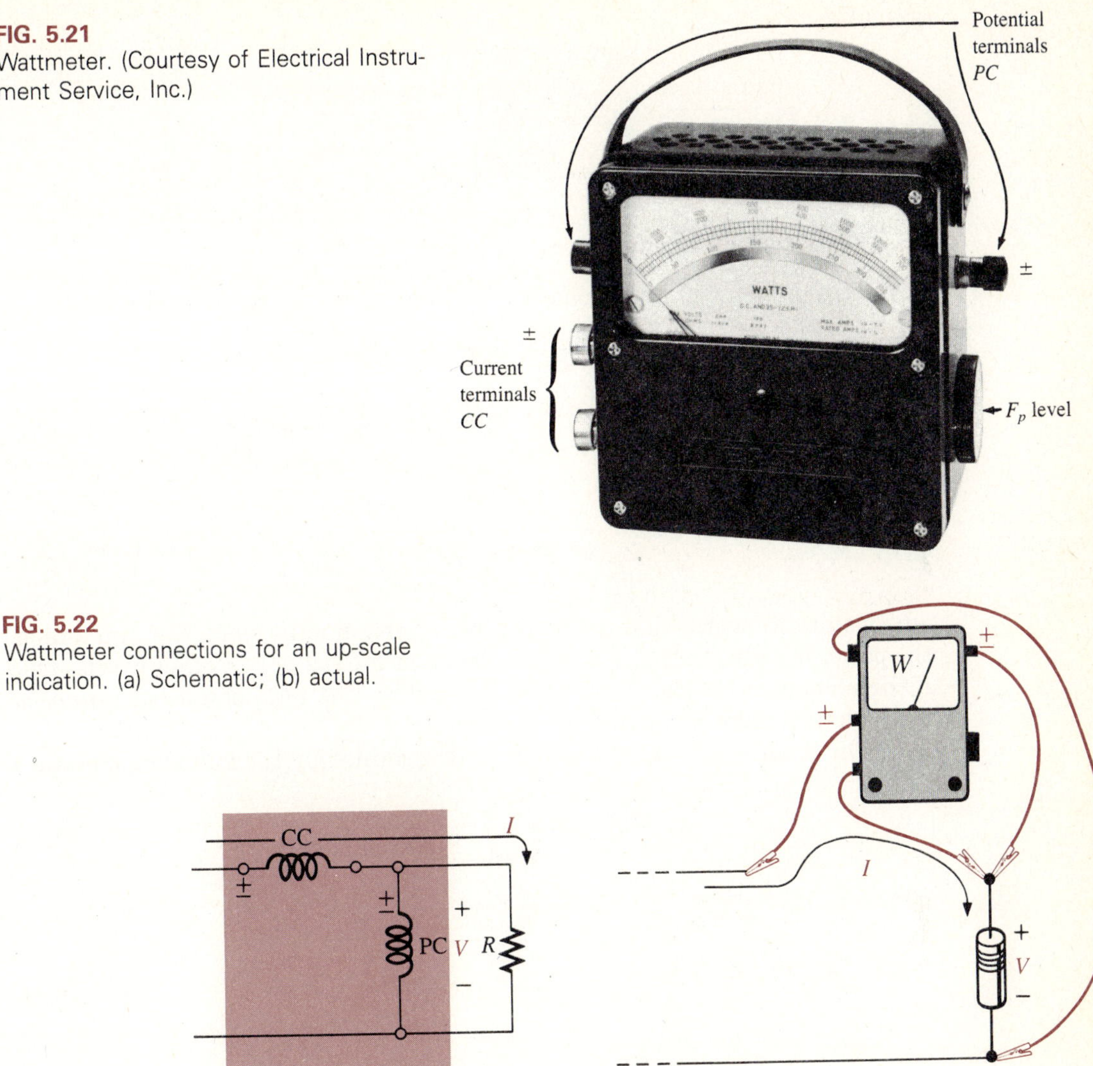

FIG. 5.21
Wattmeter. (Courtesy of Electrical Instrument Service, Inc.)

FIG. 5.22
Wattmeter connections for an up-scale indication. (a) Schematic; (b) actual.

Three voltage terminals may be available on the voltage side to permit a choice of voltage levels. On most wattmeters, the current terminals are physically larger than the voltage terminals for safety reasons and to ensure a solid connection.

5.5 EFFICIENCY

Any electrical system that converts energy from one form to another can be represented by the block diagram of Fig. 5.23 with an energy input and output terminal.

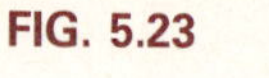

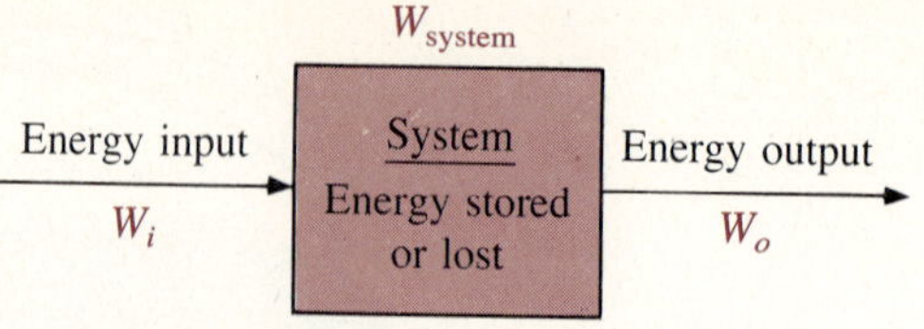

Conservation of energy requires that

$$\text{Energy input} = \text{Energy output} + \text{Energy lost or stored in the system}$$

Dividing both sides of the relationship by t gives

$$\frac{W_{\text{in}}}{t} = \frac{W_{\text{out}}}{t} + \frac{W_{\text{lost or stored by the system}}}{t}$$

Since $P = W/t$, we have the following:

$$P_i = P_o + P_{\text{lost or stored}} \tag{5.18}$$

which states in words that the power into a system must equal the sum total of the power out plus that either lost or stored in the system. In other words, even if there is no power lost or stored in the system, the total power out *cannot* be more than the power applied to the system.

The efficiency (η) of the system is determined by the following equation:

$$\text{Efficiency} = \frac{\text{Power output}}{\text{Power input}}$$

and

$$\eta = \frac{P_o}{P_i} \tag{5.19}$$

where η (lowercase Greek eta) is a decimal number. Expressed as a percentage,

$$\eta\% = \frac{P_o}{P_i} \times 100\% \tag{5.20}$$

Again, since the output can *never* be more than the input power, the efficiency can never be more than 100%. The greater the loss or storage within the system, the less will be the net efficiency.

There is no requirement that the input and output energy or power be electrical in nature. For a motor the input power is electrical, whereas the output power is mechanical. For a generator the input energy is mechanical, and the output energy is electrical. Of course, in an electronic amplifier system the input and output power levels are electrical.

EXAMPLE 5.13

A 2-hp motor operates at an efficiency of 75%.

a. What is the input power in watts?

b. If the input current is 9.05 A, what is the input voltage?

Solutions The 2-hp level of a motor is an indication of its output power level, and

$$P_o = (2\ \text{hp})\left(\frac{746\ \text{W}}{1\ \text{hp}}\right) = 1492\ \text{W}$$

a. Use Eq. 5.19:

$$\eta = \frac{P_o}{P_i}$$

Multiplying both sides by P_i gives

$$\eta(P_i) = \frac{P_o}{\cancel{P_i}}(\cancel{P_i})$$

and dividing both sides by η

$$\frac{\cancel{\eta}P_i}{\cancel{\eta}} = \frac{P_o}{\eta}$$

with

$$\boxed{P_i = \frac{P_o}{\eta}} \tag{5.21}$$

Substituting:

$$P_i = \frac{1492\ \text{W}}{0.75} = \mathbf{1989.33\ W}$$

b. $P_i = EI$. Dividing both sides by the current I gives

$$\frac{P_i}{I} = \frac{EI}{I}$$

so that

$$\boxed{E = \frac{P_i}{I}} \tag{5.22}$$

Substituting:

$$E = \frac{1989.33\ \text{W}}{9.05\ \text{A}} = 219.82\ \text{V} \cong \mathbf{220\ V}$$

EXAMPLE 5.14 What is the output in horsepower of a motor with an efficiency of 80% and an input current of 8 A at 120 V?

Solution The input power is

$$P_i = EI$$
$$= (120 \text{ V})(8 \text{ A})$$
$$P_i = 960 \text{ W}$$

From Eq. 5.19:

$$\eta = \frac{P_o}{P_i}$$

Multiplying both sides by P_i gives

$$\eta(P_i) = \frac{P_o(\cancel{P_i})}{\cancel{P_i}}$$

and

$$\boxed{P_o = \eta P_i} \qquad (5.23)$$

Substituting yields

$$P_o = (0.80)(960 \text{ W})$$
$$= 768 \text{ W}$$

In horsepower,

$$P_o = (768 \cancel{\text{W}})\left(\frac{1 \text{ hp}}{746 \cancel{\text{W}}}\right)$$
$$= \mathbf{1.029 \text{ hp}}$$

The basic components of a generating (voltage) system are depicted in Fig. 5.24. The source of mechanical power is a structure such as a paddlewheel that is turned by the water rushing over the dam. The gear train will then ensure that the rotating member of the generator is turning at rated speed. The output voltage must then be fed through a transmission system to the load. For each component of the system, an input and output power have been indicated. The efficiency of each system is given by

$$\eta_1 = \frac{P_{o_1}}{P_{i_1}} \qquad \eta_2 = \frac{P_{o_2}}{P_{i_2}} \qquad \eta_3 = \frac{P_{o_3}}{P_{i_3}}$$

If we form the product of these three efficiencies, we have

$$\eta_1 \cdot \eta_2 \cdot \eta_3 = \frac{P_{o_1}}{P_{i_1}} \cdot \frac{P_{o_2}}{P_{i_2}} \cdot \frac{P_{o_3}}{P_{i_3}}$$

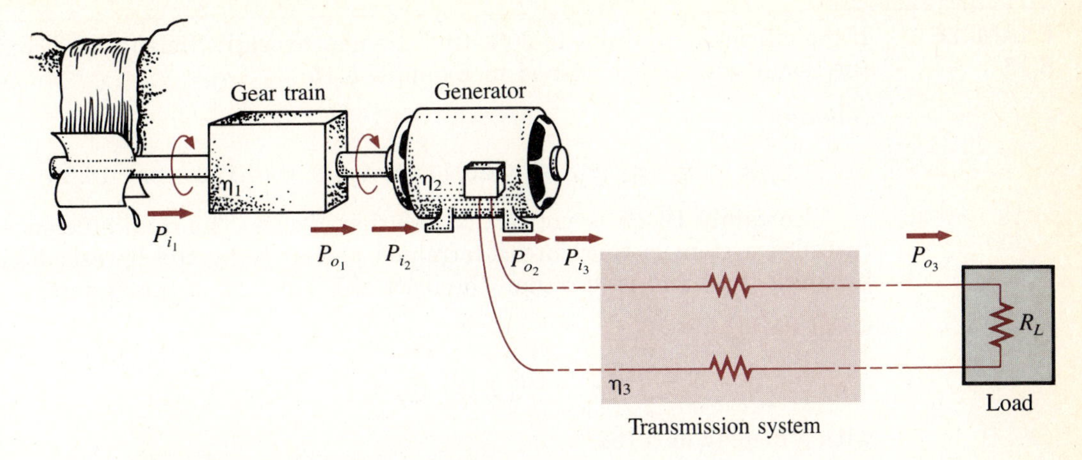

FIG. 5.24
Basic components of a generating system.

If we then substitute the fact that $P_{i_2} = P_{o_1}$ and $P_{i_3} = P_{o_2}$, we find that the quantities indicated in the last equation will cancel, resulting in P_{o_3}/P_{i_1}, which is a measure of the efficiency of the entire system. In general, for the representative cascaded system of Fig. 5.25,

$$\eta_T = \eta_1 \cdot \eta_2 \cdot \eta_3 \cdot \ldots \cdot \eta_n \qquad \textbf{(5.24)}$$

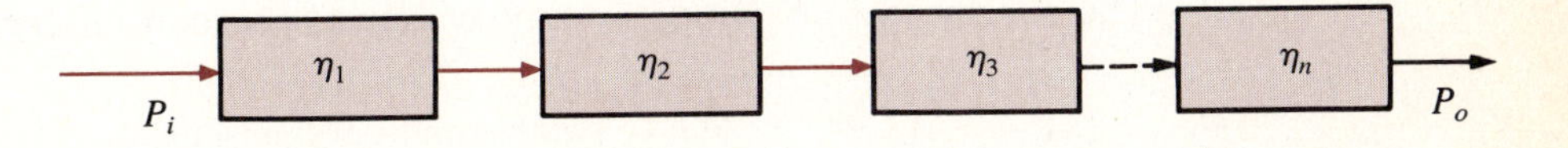

FIG. 5.25
Cascaded system.

EXAMPLE 5.15 Find the overall efficiency of the system of Fig. 5.25 if $\eta_1 = 90\%$, $\eta_2 = 85\%$, and $\eta_3 = 95\%$.

Solution

$$\eta_T = \eta_1 \cdot \eta_2 \cdot \eta_3 = (0.90)(0.85)(0.95) = 0.727, \quad \text{or} \quad \mathbf{72.7\%}$$

Calculator

[·] [9] [×] [·] [8] [5] [×] [·] [9] [5] [=]

with a display of 0.727.

EXAMPLE 5.16 If the efficiency η_1 drops to 20%, find the new overall efficiency and compare the result with that obtained in Example 5.15.

Solution

$$\eta_T = \eta_1 \cdot \eta_2 \cdot \eta_3 = (0.20)(0.85)(0.95) = 0.162, \quad \text{or} \quad \mathbf{16.2\%}$$

Certainly 16.2% is noticeably less than 72.7%. The total efficiency of a cascaded system is therefore determined primarily by the lowest efficiency (weakest link) and is less than the least efficient link of the system.

Calculator

(·) (2) (×) (·) (8) (5) (×) (·) (9) (5) (=)

with a display of 0.162.

5.6 ENERGY

In order for power, which is the rate of doing work, to produce an energy conversion of any form, it must be *used over a period of time*. For example, a motor may have the horsepower to run a heavy load, but unless the motor is *used* over a period of time, there will be no energy conversion. In addition, the longer the motor is used to drive the load, the greater will be the energy expended.

The energy lost or gained by any system is therefore determined by

$$\boxed{W = Pt} \quad \text{(wattseconds, Ws, or joules)} \qquad \mathbf{(5.25)}$$

Since power is measured in watts (or joules per second) and time is measured in seconds, the unit of energy is the *wattsecond* or *joule*, as just indicated. The wattsecond, however, is too small a quantity for most practical purposes, so the *watthour* (Wh) and *kilowatthour* (kWh) were defined. If the energy is desired in watthours, the power is substituted in watts and the time in hours as follows:

$$\boxed{W = Pt} \quad \begin{array}{l}\text{(watthours, Wh)}\\ P\text{—watts, } t\text{—hours}\end{array} \qquad \mathbf{(5.26)}$$

In order to convert the answer to kilowatthours, Eq. 5.26 is simply divided by 1000 (1 kilo = 1000).

That is,

$$\boxed{W = \frac{Pt}{1000}} \quad \begin{array}{l}\text{(kilowatthours, kWh)}\\ P\text{—watts, } t\text{—hours}\end{array} \qquad \mathbf{(5.27)}$$

To develop some sense for the kilowatthour energy level, consider that 1 kWh is the energy dissipated by a 100-W bulb in 10 h.

Further, an appreciation for the energy level associated with 1 J can be established by noting that

$$1\text{ kWh} = 1\ \cancel{\text{kWh}}\left(\frac{60\ \cancel{\text{min}}}{1\ \cancel{\text{h}}}\right)\left(\frac{60\text{ s}}{1\ \cancel{\text{min}}}\right)\left(\frac{1000\text{ W}}{1\ \cancel{\text{kW}}}\right)$$

$$= 3.6 \times 10^6\text{ Ws} = 3.6 \times 10^6\text{ J}$$

$$1\text{ kWh} = 3.6\text{ million joules}$$

Quite obviously, therefore, the joule is a very small unit of measure for energy and is inappropriate for most power applications.

The *kilowatthour meter* is an instrument for measuring the energy supplied to the residential or commercial user of electricity. It is normally connected directly to the lines at a point just prior to entering the power distribution panel of the building. A typical set of dials is shown in Fig. 5.26 with a photograph of a kilowatthour meter. As indicated, each power of 10 below a

FIG. 5.26
Kilowatthour meter. (Courtesy of Westinghouse Electric Corp.)

dial is in kilowatthours. The more rapidly the aluminum disc rotates, the greater the energy demand. The dials are connected through a set of gears to the rotation of this disc.

EXAMPLE 5.17 For the dial positions of Fig. 5.26, calculate the electricity bill if the previous reading was 4650 kWh and the average cost is 7¢ per kilowatthour.

Solution

$$5360 \text{ kWh} - 4650 \text{ kWh} = 710 \text{ kWh used}$$

$$710 \cancel{\text{kWh}}\left(\frac{7¢}{\cancel{\text{kWh}}}\right) = \mathbf{\$49.70}$$

EXAMPLE 5.18 How much energy (in kilowatthours) is required to light a 60-W bulb continuously for 1 y (365 d)?

Solution

$$1 \text{ y} = 365 \cancel{\text{d}}\left(\frac{24 \text{ h}}{1 \cancel{\text{d}}}\right) = 8760 \text{ h}$$

and

$$W = \frac{Pt}{1000} = \frac{(60 \text{ W})(8760 \text{ h})}{1000}$$
$$= \mathbf{525.6 \text{ kWh}}$$

Calculator

(6) (0) (×) (8) (7) (6) (0) (=) (÷) (1) (0) (0) (0) (=)

with a display of 525.6.

EXAMPLE 5.19 How long can a 205-W television set be on before it uses more than 4 kWh of energy?

Solution Use Eq. 5.27:

$$W = \frac{Pt}{1000}$$

Multiplying both sides by 1000 gives

$$W(1000) = \frac{Pt}{\cancel{1000}}(\cancel{1000})$$

and dividing both sides by the power P yields

$$\frac{1000\,W}{P} = \frac{\cancel{P}t}{\cancel{P}}$$

so that

$$t = \frac{1000W}{P} \quad \text{(hours, h)} \qquad (5.28)$$

Substituting gives

$$t = \frac{1000(4\ \text{k}\not{\text{W}}\text{h})}{205\ \not{\text{W}}} = \frac{4000\ \text{h}}{205}$$

$$= \mathbf{19.512\ h}$$

Note in the preceding substitution that 4 is substituted rather than 4000 to reflect the fact that W is measured in kilowatthours in Eq. 5.28.

Calculator

1 0 0 0 × 4 ÷ 2 0 5 =

with a display of 19.512.

EXAMPLE 5.20 Determine the maximum power rating of a system that will not use more than 5 kWh in 8 h.

Solution From Eq. 5.27:

$$W = \frac{Pt}{1000}$$

Multiplying both sides by 1000 and dividing both sides by t results in

$$\frac{(1000)W}{t} = \frac{P\not{t}}{\not{1000}} \frac{(\not{1000})}{\not{t}}$$

and

$$P = \frac{1000W}{t} \quad \text{(watts, W)} \qquad (5.29)$$

Substituting:

$$P = \frac{(1000)\,(5\ \text{kWh})}{8\ \text{h}} = \frac{5000\ \text{W}}{8}$$

$$= \mathbf{625\ W}$$

Calculator

1 0 0 0 × 5 ÷ 8 =

with a display of 625.

EXAMPLE 5.21

What is the total cost of using the following at 7¢ per kilowatthour?

a. A 1200-W toaster for 30 min
b. Six 50-W bulbs for 4 h
c. A 400-W washing machine for 45 min
d. A 4800-W electric clothes dryer for 20 min

Solution For problems of this type, there is no need to calculate the individual cost of each appliance and total the result. Rather, since the denominator of 1000 is the same for each equation, Eq. 5.27 can be written as

$$W = \frac{P_1t_1 + P_2t_2 + P_3t_3 + \cdots + P_nt_n}{1000} \quad \text{(kWh)} \qquad \textbf{(5.30)}$$

where each Pt product reflects a particular load.

For this example,

$$P_1 = 1200 \text{ W}, \quad t_1 = 30 \not{\text{min}} \left(\frac{1 \text{ h}}{60 \not{\text{min}}}\right) = 0.5 \text{ h}$$

$$P_2 = 6(50 \text{ W}) = 300 \text{ W}, \quad t_2 = 4 \text{ h}$$

$$P_3 = 400 \text{ W}, \quad t_3 = 45 \not{\text{min}} \left(\frac{1 \text{ h}}{60 \not{\text{min}}}\right) = 0.75 \text{ h}$$

$$P_4 = 4800 \text{ W}, \quad t = 20 \not{\text{min}} \left(\frac{1 \text{ h}}{60 \not{\text{min}}}\right) = \frac{1}{3} \text{ h}$$

Note that time must be converted to hours for insertion in Eq. 5.30. Substituting gives

$$W = \frac{(1200 \text{ W})(0.5 \text{ h}) + (300 \text{ W})(4 \text{ h}) + (400 \text{ W})(0.75 \text{ h}) + (4800 \text{ W})(1/3 \text{ h})}{1000}$$

$$= \frac{600 \text{ Wh} + 1200 \text{ Wh} + 300 \text{ Wh} + 1600 \text{ Wh}}{1000}$$

$$W = 3700 \text{ Wh} = 3.7 \text{ kWh}$$

$$\text{Cost} = (3.7 \not{\text{kWh}})\left(\frac{7¢}{\not{\text{kWh}}}\right) = \mathbf{25.9¢}$$

Calculator

W: 1 2 0 0 × . 5 + 3 0 0 × 4 + 4 0 0 × . 7 5 + 4 8 0 0 ÷ 3 = ÷ 1 0 0 0 =

with a display of 3.7.

Cost: (3) (·) (7) (×) (7) (=)

with a display of 25.9.

The chart in Fig. 5.27 shows the average cost per kilowatthour as compared to the kilowatthours used per customer. Note that the cost today is about the same as in 1926 but that the average customer uses more than 20 times as much electrical energy in a year. Keep in mind that the chart of Fig. 5.27 is the average cost across the nation. Some states have average rates of 3¢ or 4¢ per kilowatthour, and others pay over 11¢ per kilowatthour.

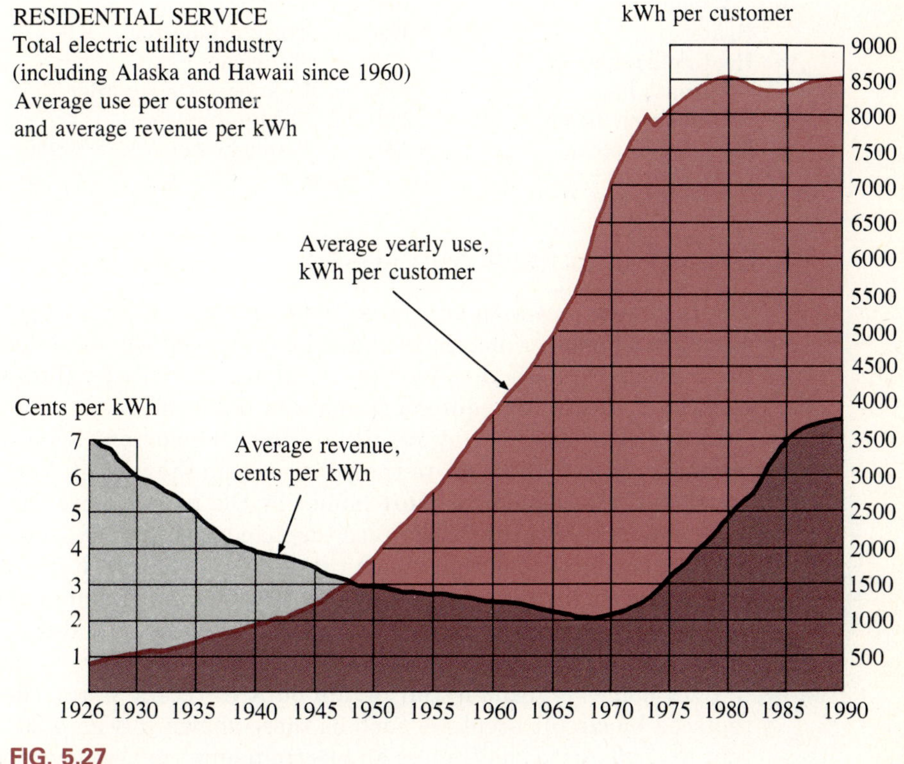

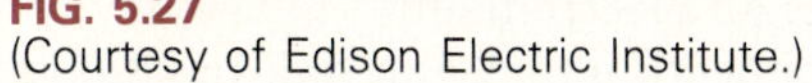

FIG. 5.27
(Courtesy of Edison Electric Institute.)

Table 5.1 lists some common household items with their typical wattage ratings. It might prove interesting for the reader to calculate the cost of operating some of these appliances over a period of time using the preceding chart to find the cost per kilowatthour.

TABLE 5.1
Typical wattage ratings of some common household appliances.

Appliance	Wattage Rating	Appliance	Wattage Rating
Air conditioner	860	Microwave oven	800
Blow dryer	1300	Phonograph	75
Cassette player/recorder	5	Projector	1200
Clock	2	Radio	70
Clothes dryer (electric)	4800	Range (self-cleaning)	12,200
Coffee maker	900	Refrigerator (automatic defrost)	1800
Dishwasher	1200	Shaver	15
Fan:		Stereo equipment	110
Portable	90	Sun lamp	280
Window	200	Toaster	1200
Heater	1322	Trash compactor	400
Heating equipment:		TV (color)	150
Furnace fan	320	Videocassette recorder	110
Oil-burner motor	230	Washing machine	400
Iron, dry or steam	1100	Water heater	2500

Courtesy of General Electric Co.

5.7 CIRCUIT BREAKERS AND FUSES

The incoming power to any large industrial plant, heavy equipment, simple circuit in a home, or meters used in a laboratory must be limited to ensure that the current through the lines is not above the rated value. Otherwise, the electric or electronic equipment may be damaged or dangerous side effects such as fire or smoke may result. To limit the current level, fuses or circuit breakers are installed where the power enters the installation, such as in the panel in the basement of most homes at the point where the outside feeder lines enter the dwelling. The fuse (depicted in Fig. 5.28) has an internal bimetallic conductor through which the current will pass; it begins to melt if the current through the system exceeds the rated value printed on the casing. Of course, if it melts through, the current path is broken and the load in its path protected.

In homes and industrial plants built in recent years, the fuse has been replaced by circuit breakers such as those shown in Fig. 5.29. When the current exceeds rated conditions, an electromagnet in the device will have sufficient strength to draw the connecting metallic link in the breaker out of the circuit and open the current path. When conditions have been corrected, the breaker can be reset and used again, unlike the fuse, which has to be replaced.

It is especially important to understand that if a fuse or circuit breaker does open up to protect the load, *do not* replace it with a fuse or circuit breaker higher in rating than the one in use. The result may be current levels that could damage the equipment or cause fire or smoke. In addition, do not increase the fuse rating if the application of another unit such as a power saw or

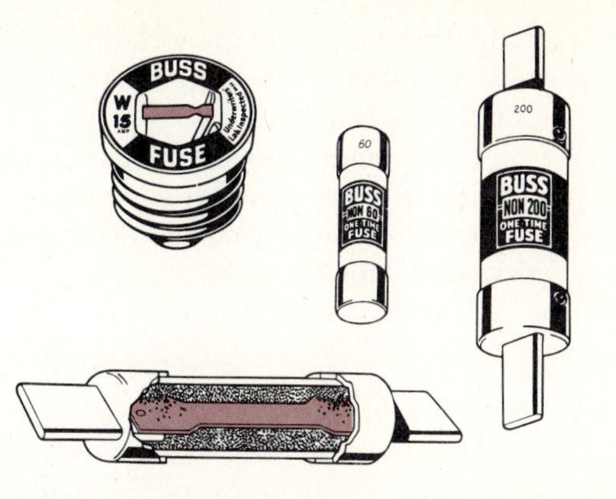

(a)

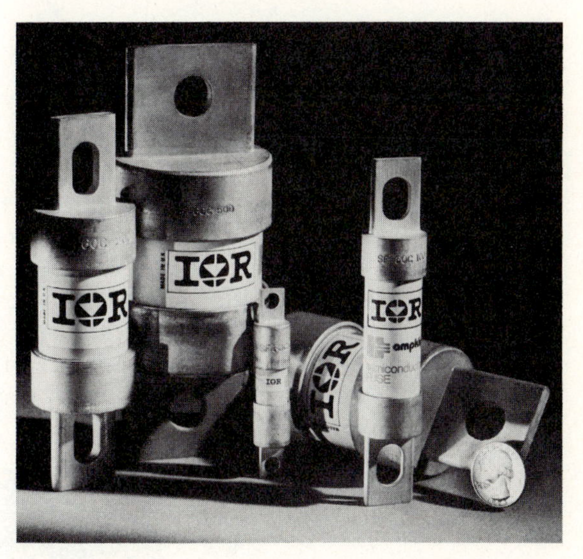

(b)

FIG. 5.28
Bimetallic fuses. (Part (a) courtesy of Bussman Manufacturing Co.; part (b) courtesy of International Rectifier Corp.)

FIG. 5.29
Circuit breakers. (Courtesy of Potter and Brumfield Division, AMF, Inc.)

hair dryer seems to blow the fuse. The wire in the interior walls may not be able to handle the higher current levels of the larger fuse, and dangerous conditions could result.

The condition of a fuse can be checked very easily by using the ohmmeter section of a multimeter, as shown in Fig. 5.30. An indication of 0 Ω usually indicates a fuse in good operating order, whereas an infinite reading indication reveals a "blown" fuse.

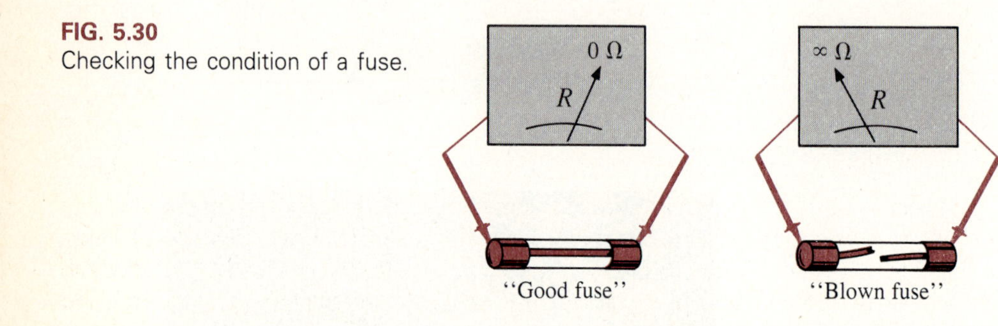

FIG. 5.30
Checking the condition of a fuse.

The same results can be applied to the circuit breaker terminals to see if it is in good operating condition. The "on" position should result in an indication of 0 Ω, and the "off" position is an open-circuit indication. The internal construction of a circuit breaker appears in Fig. 5.31. The input and output terminals normally just snap into place on a circuit breaker panel. If current levels are too high, the electromagnet will draw the movable arm to the magnet and open the circuit to protect the system.

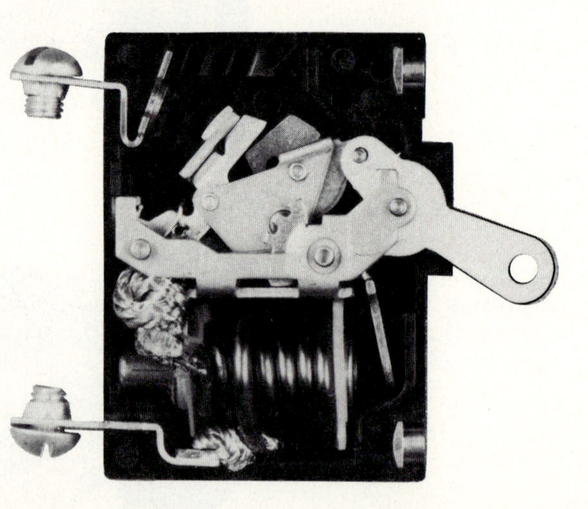

FIG. 5.31
Circuit breaker. (Courtesy of Thomas Sundheim Inc.)

Fuses and circuit breakers are hooked up in-line with the load, as shown in Fig. 5.32 for a fuse and circuit breaker. In either case a blown fuse will obviously prevent the current from reaching the load R.

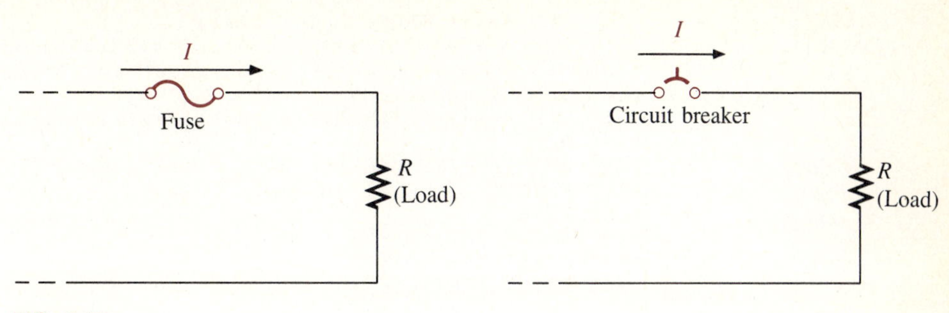

FIG. 5.32
Fuse and circuit breaker installation.

5.8 COMPUTER ANALYSIS

In recent years the development of the personal computer (PC) has placed the system within financial reach of most individuals with a need for its capabilities and scope. Texts are beginning to surface dedicated solely to the software (written programs) that perform a wide variety of tasks with a high degree of accuracy in a very short period of time. Space constraints here do not permit an in-depth review of computers and the available software. However, a few typical programs have been interspersed throughout the text to demonstrate their range of application. In case you have a personal computer or computer facilities are available for your use, exercises have been provided at the end of each chapter to test your programming skills.

The BASIC language is the communication link between the user (operator) and the computer in a wide range of personal computers manufactured today. You will note in Program 5.1 (Fig. 5.33) that the commands, statements, and so on found on each line use English words and phrases to indicate the operation to be performed. The REM statement (from the word *REMark*) simply indicates that a descriptive statement is being made about the program. The PRINT statement tells the system to print the characters between the quotes on the screen. The INPUT commands request the value of the variable appearing on the same line. Each equation on lines 440, 640, and 840 carries out the operation to be performed in each module defined by the brackets at the left of the program. Lines 400 through 460 request I and R and calculate the voltage. Line 450 prints the solution, whereas line 460 returns the program to line 210 to determine if a second calculation is to be performed. The module from line 600 to line 660 calculates the current I, and the module from line 800 to line 860 determines the resistance from the input voltage and current. Lines 100 through 200 permit a selection of the form of Ohm's law to be applied.

Three runs of the program are provided in the figure to reveal the format of the request for data and the output response.

The preceding description is by no means an attempt to make the reader an expert in the use of computer systems. The intent is simply to introduce the format of a program, reveal the readability of a frequently applied language, and demonstrate the ability of a computer to perform calculations in an accurate, rapid, and efficient manner.

FIG. 5.33
Program 5.1.

```
10 REM ***** PROGRAM 5-1 *****
20 REM ******************************************
30 REM Program demonstrates selecting various forms
40 REM of equations
50 REM ******************************************
60 REM
```

Body of Program:

Equation Selection

```
100 PRINT:PRINT "Select which form of Ohm's law equation "
110 PRINT "you wish to use."
120 PRINT
130 PRINT TAB(10);"(1) V=I*R"
140 PRINT TAB(10);"(2) I=V/R"
150 PRINT TAB(10);"(3) R=V/I"
160 PRINT TAB(20);
170 INPUT "choice=";C
180 IF C<1 OR C>3 THEN GOTO 100
190 ON C GOSUB 400,600,800
200 PRINT:PRINT
```

Continue?

```
210 INPUT "More (YES or NO)";A$
220 IF A$="YES" THEN 100
230 PRINT "Have a good day"
240 END
```

$V = IR$

```
400 REM Accept input of I,R and output V
410 PRINT:PRINT "Enter the following data:"
420 INPUT "I=";I
430 INPUT "R=";R
440 V=I*R
450 PRINT "Voltage is ";V;"volts"
460 RETURN
```

$I = \frac{V}{R}$

```
600 REM Accept input of V,R and output I
610 PRINT "Enter the following data:"
620 INPUT "V=";V
630 INPUT "R=";R
640 I=V/R
650 PRINT "Current is ";I;"amperes"
660 RETURN
```

$R = \frac{V}{I}$

```
800 REM Accept input of V,I and output R
810 PRINT "Enter the following data:"
820 INPUT "V=";V
830 INPUT "I=";I
840 R=V/I
850 PRINT "Resistance is ";R;"ohms"
860 RETURN
```

Program Runs:

```
READY

RUN


Select which form of Ohm's law equation
you wish to use.

          (1) V=I*R
          (2) I=V/R
          (3) R=V/I
                    choice=? 2
```

FIG. 5.33
(continued)

```
Enter the following data:
V=? 12

R=? 4E3

Current is  3E-03 amperes


More (YES or NO)? YES


Select which form of Ohm's law equation
you wish to use.

            (1) V=I*R
            (2) I=V/R
            (3) R=V/I
                        choice=? 1


Enter the following data:
I=? 2E-3

R=? 5.6E3

Voltage is  11.2 volts


More (YES or NO)? YES


Select which form of Ohm's law equation
you wish to use.

            (1) V=I*R
            (2) I=V/R
            (3) R=V/I
                        choice=? 3

Enter the following data:
V=? 48

I=? 0.025

Resistance is  1920 ohms


More (YES or NO)? NO

Have a good day


READY
```

FORMULA SUMMARY

$$I = \frac{E}{R}, \quad E = IR, \quad R = \frac{E}{I}$$

$$P = \frac{W}{t}, \quad VI, \quad I^2R, \quad \frac{V^2}{R}$$

$$\eta\% = \frac{P_o}{P_i} \times 100\%$$

$$\eta_T = \eta_1 \cdot \eta_2 \cdot \eta_3 \cdot \cdots \cdot \eta_n$$

$$W = Pt, \quad W\ (\text{kWh}) = \frac{Pt}{1000}$$

CHAPTER SUMMARY

Fewer equations are listed in the formula summary than are introduced in the chapter. However, the formula review includes those equations that should be memorized. The other forms should be derived from algebraic manipulations.

The equations introduced in this chapter begin to tie the three elements of current, voltage, and resistance together. The completion of this chapter is likened to reaching a first plateau in circuit analysis—a number of important basic elements have been introduced, and through the basic equations of this chapter the impact of one quantity on the others can be appreciated. It is also important to realize that the equations introduced in this chapter will not be dropped for a more advanced set in later chapters. The Ohm's law equation is applied in the most advanced levels of research and the power equations have the same appearance when we examine power sources other than dc.

In using each equation, be sure the proper magnitude is substituted into the equation as determined by the associated power of ten and be sure to include the proper unit of measurement with each result. For Ohm's law and the power equations, keep in mind that the voltage is the voltage across the load and the current is the current through the load. In the equation for the total kilowatt hours the time must be in hours and not minutes, days, or months.

Ohm's law and the power equations should be memorized in all their forms. They will be used so frequently that one should not hesitate to apply an equation such as $P = I^2R$ without deriving it from the fundamental form.

GLOSSARY

Circuit breaker A two-terminal device designed to ensure that current levels do not exceed safe levels. If tripped, it can be reset with a switch or a reset button.

Diode A semiconductor device whose behavior is much like that of a simple switch; that is, it passes current ideally in only one direction when operating within specified limits.

Efficiency (η) A ratio of output to input power that provides immediate information about the energy-converting characteristics of a system.

Energy (W) A quantity whose change in state is determined by the product of the rate of conversion (P) and the period involved (t). It is typically measured in joules (J) or wattseconds (Ws), watthours (Wh), or kilowatthours (kWh).

Fuse A two-terminal device whose sole purpose is to ensure that current levels in a circuit do not exceed safe levels.

Horsepower (hp) Equivalent to 746 W in the electrical system.

Kilowatthour meter An instrument for measuring kilowatthours of energy supplied to a residential or commercial user of electricity.

Ohm's law An equation that establishes a relationship among the current, voltage, and resistance of an electrical system.

Power An indication of how much work can be done in a specified amount of time; a *rate* of doing work. It is typically measured in joules per second (J/s), watts (W), or horsepower (hp).

Wattmeter An instrument capable of measuring the power delivered to an element by sensing both the voltage across the element and the current through the element.

PROBLEMS

Section 5.1

1. What is the potential drop across a 6-Ω resistor if the current through it is 2.5 A?
2. What is the current through a 72-Ω resistor if the voltage drop across it is 12 V?
3. How much resistance is required to limit the current to 1.5 mA if the potential drop across the resistor is 6 V?
4. Find the current through a 2.4-MΩ resistance placed across a 120-V source.
5. If the current through a 0.02-MΩ resistor is 3.6 μA, what is the voltage drop across the resistor?
6. If a voltmeter has an internal resistance of 15 kΩ, find the current through the meter when it reads 62 V.
7. If a refrigerator draws 2.2 A at 120 V, what is its resistance?
8. If a clock has an internal resistance of 7.5 kΩ, find the current through the clock if it is plugged into a 120-V outlet.
9. What voltage is required to pass 42 mA through a resistance of 0.04 MΩ?
10. If a soldering iron draws 0.76 A at 120 V, what is its resistance?
11. If an electric heater draws 9.5 A when connected to a 120-V supply, what is the internal resistance of the heater?
12. The input current to a transistor is 20 μA. If the applied (input) voltage is 24 mV, determine the input resistance of the transistor.
13. The internal resistance of a dc generator is 0.5 Ω. Determine the loss in terminal voltage across this internal resistance if the current is 15 A.

Section 5.2

14. Plot the curve of a 4-Ω resistor on a replica of the graph of Fig. 5.9.
15. a. Plot the curve of a 0.5-Ω resistor on a replica of the graph of Fig. 5.9.
    b. Plot the curve of a 100-Ω resistor on the graph of part a.
    c. Compare the results of parts a and b (location, slope, etc.).
16. Plot the curve of a 1-kΩ resistor on a replica of the graph of Fig. 5.13.
17. a. Plot the curve of a 10-kΩ resistor on a replica of the graph of Fig. 5.13.
    b. Plot the curve of a 100-Ω (0.1-kΩ) resistor on the graph of part a.
    c. Compare the results of parts a and b (location, slope, etc.).
18. Referring to Fig. 5.16, determine the resistance level at a drain current of 30 mA using Eq. 5.8.
19. Referring to Fig. 5.16, determine the resistance level at a diode voltage of 1 V using Eq. 5.8.

Section 5.3

20. If 420 J of energy are absorbed by a resistor in 7 min, what is the power to the resistor?
21. The power to a device is 40 J/s. How long will it take to deliver 640 J?
22. a. How many joules of energy does a 2-W night-light dissipate in 8 h?
    b. How many kilowatthours does it dissipate?

*23. A resistor of 10 Ω has charge flowing through it at the rate of 300 C/min. How much power is dissipated?

24. How long will a steady current of 2 A have to exist in a resistor that has 3 V across it to dissipate 12 J of energy?
25. What is the power delivered by a 6-V battery if the charge flows at the rate of 48 C/min?
26. The current through a 4-Ω resistor is 7 mA. What is the power delivered to the resistor?
27. The voltage drop across a 3-Ω resistor is 9 mV. What is the power input to the resistor?
28. If the power input to a 4-Ω resistor is 64 W, what is the current through the resistor?
29. A 1/2-W resistor has a resistance of 1000 Ω. What is the maximum current that it can safely handle?
30. A 2.2-kΩ resistor in a stereo system dissipates 42 mW of power. What is the voltage across the resistor?
31. A power supply can deliver 100 mA at 400 V. What is the power rating?
32. What are the resistance and current ratings of a 120-V, 100-W bulb?
33. What are the resistance and voltage ratings of a 450-W automatic washer that draws 3.75 A?
34. A 20-kΩ resistor has a rating of 100 W. What are the maximum current and the maximum voltage that can be applied to the resistor?

*35. A home is supplied with a 120-V, 100-A service.
  a. Find the maximum power capability.
  b. Determine if the homeowner can safely operate the following loads at the same time.
    1. A 5-hp motor
    2. A 3000-W clothes dryer
    3. A 2400-W electric range
    4. A 1000-W steam iron
  c. Find the current drawn under the loaded conditions of part b.

Section 5.5

36. What is the efficiency of a motor that has an output of 0.5 hp with an input of 450 W?

37. The motor of a power saw is rated 68.5% efficient. If 1.8 hp are required to cut a particular piece of lumber, what is the current drawn from a 120-V supply?

38. What is the efficiency of a dryer motor that delivers 1 hp when the input current and voltage are 4 A and 220 V, respectively?

39. If an electric motor having an efficiency of 87% and operating off a 220-V line delivers 3.6 hp, what input current does the motor draw?

40. A motor is rated to deliver 2 hp.
  a. If it runs on 110 V and is 90% efficient, how many watts does it draw from the power line?
  b. What is the input current?
  c. What is the input if the motor is only 70% efficient?

41. An electric motor used in an elevator system has an efficiency of 90%. If the input voltage is 220 V, what is the input current when the motor is delivering 15 hp?

42. A 2-hp motor drives a sanding belt. If the efficiency of the motor is 87% and that of the sanding belt is 75%, due to slippage, what is the overall efficiency of the system?

43. If two systems in cascade each have an efficiency of 80% and the input energy is 60 J, what is the output energy?

44. The overall efficiency of two systems in cascade is 72%. If the efficiency of one is 0.9, what is the efficiency in percent of the other?

*45. If the total input and output power of two systems in cascade are 400 W and 128 W, respectively, what is the efficiency of each system if one has twice the efficiency of the other?

46. a. What is the total efficiency of three systems in cascade with efficiencies of 0.98, 0.87, and 0.21?
  b. If the system with the least efficiency (0.21) were removed and replaced by one with an efficiency of 0.90, what would be the percent increase in total efficiency?

Section 5.6

47. A 10-Ω resistor is connected across a 15-V battery.
  a. How many joules of energy will it dissipate in 1 min?
  b. If the resistor is left connected for 2 min instead of 1 min, will the energy used increase? Will the power increase?

*48. How much energy in kilowatthours is required to keep a 230-W oil-burner motor running 12 h a week for 5 mo (1 mo = 30 d)?

49. How long can a 1500-W heater be on before using more than 10 kWh of energy?

50. How much does it cost to use a 30-W radio for 3 h at 7¢ per kilowatthour?

51. What is the total cost of using the following at 7¢ per kilowatthour?
  a. An 860-W air conditioner for 24 h
  b. A 4800-W clothes dryer for 30 min
  c. A 400-W washing machine for 1 h
  d. A 1200-W dishwasher for 45 min

52. What is the total cost of using the following at 7¢ per kilowatthour?
  a. A 110-W stereo set for 4 h
  b. A 1200-W projector for 3 h
  c. A 60-W tape recorder for 2 h
  d. A 150-W color television set for 6 h

Computer Problems

53. Given the voltage across and current through a resistor, write a program to determine the resistance of the resistor, the power dissipated by the resistor, and the energy absorbed every minute.

54. Request I, R, and t and determine V, P, and W. Print out the results with the proper units.

55. Given a resistance in kilohms, tabulate the voltage and power to the resistor for a range of current extending from 1 mA to 10 mA in increments of 1 mA.

56. Write a program to calculate the cost of using five different appliances for varying lengths of time if the cost is 7¢ per kilowatthour.

57. Tabulate the time (in hours) and the cost (at 7¢ per kilowatthour) for the use of a particular system for $T = 1$ to N hours in increments of 1 h. N is an input quantity in the range 2 to 10.

6

Series Circuits

OBJECTIVES

- ☐ Understand the conditions that must be satisfied for elements to be in series.
- ☐ Be able to calculate the total resistance, source current, voltage, and power levels of a series circuit.
- ☐ Understand and be able to apply Kirchhoff's voltage law.
- ☐ Learn the voltage divider rule and where it can be applied.
- ☐ Develop an awareness of the different types of notation employed on electrical schematics.
- ☐ Understand the impact of the internal resistance of voltage sources on the terminal characteristics.
- ☐ Be able to calculate the voltage regulation level of a supply from its terminal characteristics.
- ☐ Understand the loading effect of meters on an electrical system and how to interpret the readings.

6.1 INTRODUCTION

Two types of current are readily available to the consumer today. One is *direct current* (dc), in which, ideally, the flow of charge (current) does not change in magnitude or direction. The other is *sinusoidal alternating current* (ac), in which the flow of charge is continually changing in magnitude and direction. The next few chapters provide an introduction to circuit analysis purely from a dc approach. The methods and concepts are discussed in detail for direct current; a short discussion suffices to cover any variations we might encounter when we consider ac in the later chapters.

The supply of Fig. 6.1, by virtue of the potential difference between its terminals, has the ability to cause (or pressure) charge to flow through the simple circuit. The positive terminal attracts the electrons through the wire at the same rate at which electrons are supplied by the negative terminal. As long as the supply is connected and maintains its terminal characteristics, the current (dc) through the circuit does not change in magnitude or direction.

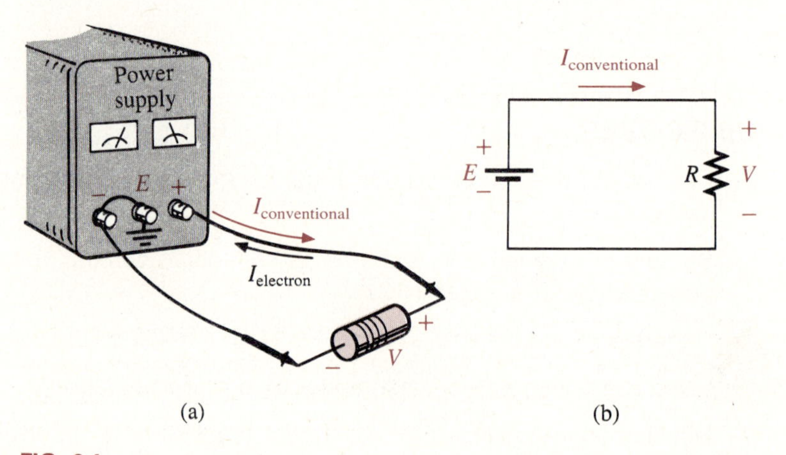

FIG. 6.1
Simplest electrical circuit. (a) Construction; (b) schematic representation.

If we consider the wire to be an ideal conductor (that is, having no opposition to flow), the potential difference V across the resistor equals the applied voltage of the battery: V (volts) $=$ E (volts).

The current is limited only by the resistor R. The higher the resistance, the less the current, and conversely, as determined by Ohm's law.

By convention, as discussed in Chapter 3, the direction of I as shown in Fig. 6.1 is opposite to that of electron flow. Also, the uniform flow of charge dictates that the direct current I be the same everywhere in the circuit. By following the direction of conventional flow, we notice that there is a rise in potential across the battery ($-$ to $+$) and a drop in potential across the resistor ($+$ to $-$). Recall from Chapter 5 that for single-voltage-source dc circuits, conventional flow always passes from a low potential to a high potential when passing through a voltage source, as shown in Fig. 6.2(a). In other words, the pressure applied by the battery is such that it will establish a current that flows from the negative to positive terminal of the battery. Reversing the battery, as shown in Fig. 6.2(b), reverses the direction of the current I, as required by this convention. Keep in mind, however, that this rule is only for single-source networks. For networks with more than one source, it is possible for the current through one or more batteries to be opposite to that shown in Fig. 6.2. Chapter 5 also pointed out that for a load such as that shown in Figs. 6.1 and 6.3(a), conventional flow is always from the high to the low potential. Reversing the current direction reverses the polarity, as shown in Fig. 6.3(b). However, conventional flow always passes from a high to a low potential when passing through a resistor for any number of voltage sources in the same circuit, as shown in Fig. 6.3.

FIG. 6.2
Current direction for one-voltage-source dc circuit.

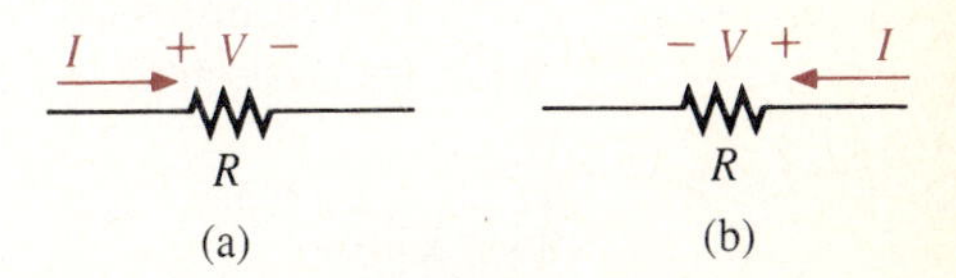

FIG. 6.3
Polarity of the voltage drop across a resistor as determined by the current direction (for any combination of voltage sources in the same network).

The circuit of Fig. 6.1 is the simplest possible configuration. This chapter and the chapters to follow add elements to the system in a very specific manner to introduce a range of concepts that will form a major part of the foundation required to analyze the most complex system. Be aware that the laws and rules introduced in Chapters 6 and 7 will be used throughout your studies of electrical, electronic, or computer systems. They will not be dropped for a more advanced set as you progress to more sophisticated material. It is therefore critical that you understand the concepts thoroughly and that you be able to apply the various procedures and methods with confidence.

6.2 SERIES CIRCUITS

There are two electrical configurations that the student of this field must fully understand and be able to analyze with confidence. One is referred to as a *series circuit* and the other is called a *parallel circuit*. Series circuits are analyzed in this chapter, with parallel circuits reserved for the next chapter. Chapter 8 then introduces methods to analyze networks that have both series and parallel configurations.

The simplest and clearest mechanical analogy for a series configuration is a connection of hoses, as shown in Fig. 6.4. Note that hose A is connected to

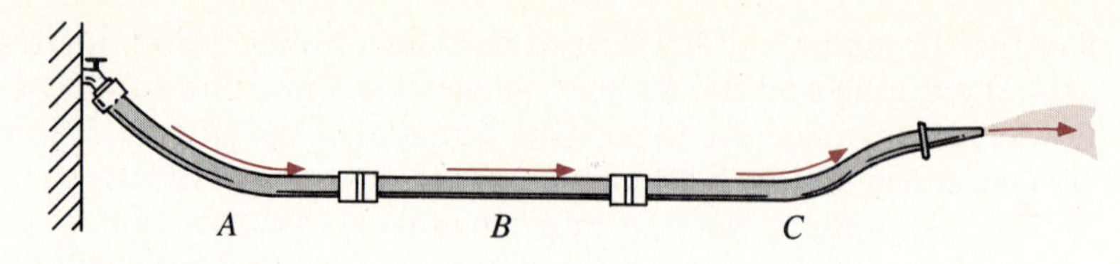

FIG. 6.4
Mechanical analogy of a series circuit.

hose B at only *one* point. In other words, both ends are not connected together. The same is true for hoses B and C. It should also be reasonable to accept that the flow of water through one hose is the same as through another hose once a steady-state flow has been established. In other words, the flow is the *same* through each section.

The simplest of series electrical configurations is shown in Fig. 6.5(a) with its electrical diagram in Fig. 6.5(b).

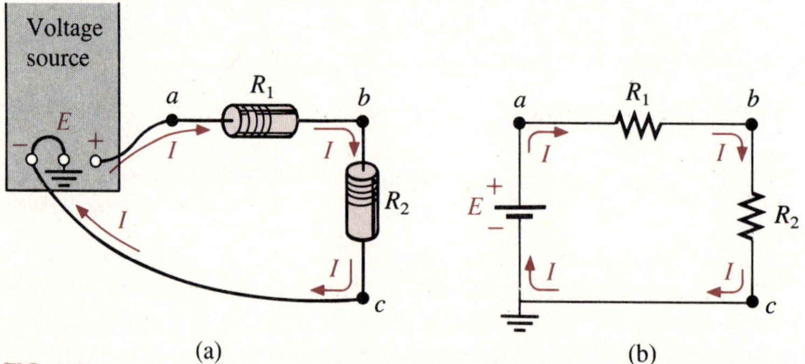

FIG. 6.5
Simple series circuit. (a) Construction; (b) electrical schematic.

First note in Fig. 6.5(b) that the single-source network has established a clockwise current direction, as determined by the polarity of the battery. Note also that the current I (like the flow of water in Fig. 6.4) is the *same* everywhere. That is, the current through R_1, R_2, and the battery E is the same.

In a series configuration, the current is the same through each series element.

By definition,

Two elements are in series if they have only one point in common that is not connected to other current-carrying elements of the network.

In Fig. 6.5(b) the resistors R_1 and R_2 are in series because they have *only* point b in common. The other ends of the resistors are connected elsewhere in the circuit. For the same reason, the battery E and resistor R_1 are in series (terminal a in common) and the resistor R_2 and the battery E are in series (terminal c in common). Since all the elements are in series, the network is called a *series circuit*.

The primary difference between the electrical system of Fig. 6.5 and the mechanical analogy of Fig. 6.4 is the fact that there is a closed loop for the flow of charge in Fig. 6.5, whereas the water leaving the hose of Fig. 6.4 is not returned to the source. In other words, the electrical system of Fig. 6.5(b) has a continuous path that permits a "looping" of the current that essentially exists as long as the battery has sufficient energy to pressure the flow of charge through the system. In order for current to exist in *any* electrical system, there must be at least one closed path that will permit current to leave in one direction and return in another. By definition,

A circuit consists of any number of elements joined at terminal points, providing at least one closed path through which charge can flow.

If a third element, R_3, is added at terminal b, as shown in Fig. 6.6, the resistors R_1 and R_2 are no longer in series because an additional element has been connected between the common terminal and some other point in the network.

FIG. 6.6

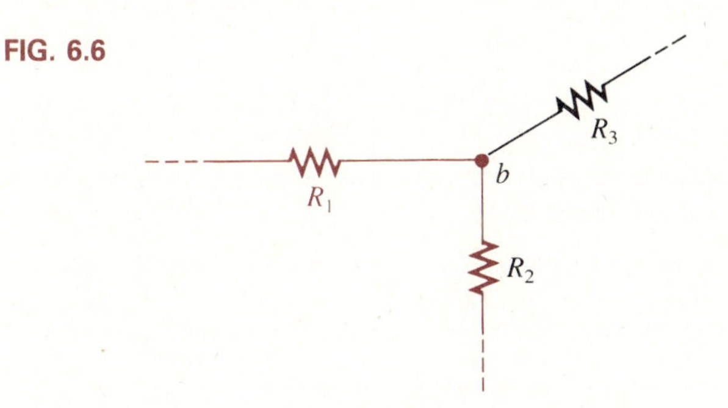

In general, to check if two elements are in series, be sure that the following hold:

1. The two elements are connected at *only* one end.
2. The common point is *not* connected to additional current-carrying elements.
3. The current through each element is the same.

A *branch* of a circuit is any portion of the circuit that has one or more elements in series. In Fig. 6.5(b), the resistor R_1 forms one branch of the circuit, the resistor R_2 forms another, and the battery E forms a third. In a larger network the series combination of E, R_1, and R_2 may form only one branch of the system.

The total resistance of the circuit is determined by simply adding the values of the various resistors. In Fig. 6.5(b), for example, the total resistance

(R_T) is equal to $R_1 + R_2$. Note that the total resistance is actually the resistance "seen" by the battery as it "looks" into the series combination of elements, as shown in Fig. 6.7.

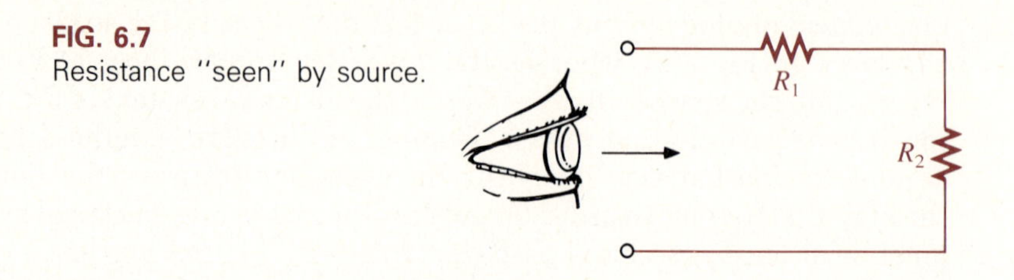

FIG. 6.7
Resistance "seen" by source.

For networks with more than two elements in series, the total resistance continues to be the sum of the resistors connected in series. In general, to find the total resistance of N resistors in series (where N can be any positive whole number such as 2, 3, or 10), the following equation is applied:

$$R_T = R_1 + R_2 + R_3 + \cdots + R_N \tag{6.1}$$

Once the total resistance is known, the current can be determined from Ohm's law:

$$I = \frac{E}{R_T} \tag{6.2}$$

and the voltage across each element can also be determined using Ohm's law in the following manner:

$$V_1 = IR_1,\ V_2 = IR_2,\ V_3 = IR_3,\ \cdots,\ V_N = IR_N \tag{6.3}$$

EXAMPLE 6.1

a. Find the total resistance of the series circuit of Fig. 6.8.
b. Calculate the current I.
c. Determine the voltages V_1 and V_2.

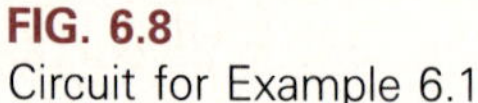

FIG. 6.8
Circuit for Example 6.1

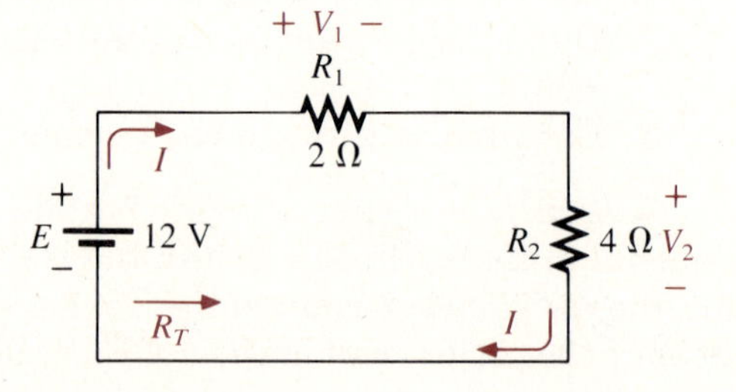

Solutions

a. $R_T = R_1 + R_2 = 2\,\Omega + 4\,\Omega = \mathbf{6\,\Omega}$

b. $I = \dfrac{E}{R_T} = \dfrac{12\text{ V}}{6\,\Omega} = \mathbf{2\text{ A}}$

c. $V_1 = IR_1 = (2\text{ A})(2\,\Omega) = \mathbf{4\text{ V}}$

$V_2 = IR_2 = (2\text{ A})(4\,\Omega) = \mathbf{8\text{ V}}$

Note in part c of Example 6.1 that the current I is the same for each voltage calculation. In addition, note the polarity of the potential drop across each resistor as determined by the current direction.

EXAMPLE 6.2

a. Find the total resistance for the series circuit of Fig. 6.9.
b. Calculate the current I.
c. Determine the voltages V_1, V_2, and V_3.

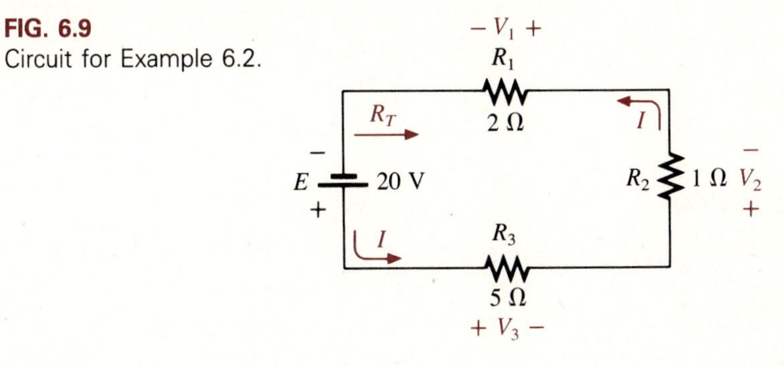

FIG. 6.9
Circuit for Example 6.2.

Solutions

a. $R_T = R_1 + R_2 + R_3 = 2\,\Omega + 1\,\Omega + 5\,\Omega = \mathbf{8\,\Omega}$

b. $I = \dfrac{E}{R_T} = \dfrac{20\text{ V}}{8\,\Omega} = \mathbf{2.5\text{ A}}$

c. $V_1 = IR_1 = (2.5\text{ A})(2\,\Omega) = \mathbf{5\text{ V}}$

$V_2 = IR_2 = (2.5\text{ A})(1\,\Omega) = \mathbf{2.5\text{ V}}$

$V_3 = IR_3 = (2.5\text{ A})(5\,\Omega) = \mathbf{12.5\text{ V}}$

In this example of three series elements, note that the direction of the current I was reversed because the voltage source E was reversed, and note the polarity of V_1 and V_2 as determined by the current direction.

If a situation should arise where the resistors in series are the same value, then Eq. 6.1 can be written as

$$\boxed{R_T = NR} \tag{6.4}$$

where N is the number of resistors R with the same value.

EXAMPLE 6.3

a. Find the total resistance of the series circuit of Fig. 6.10.
b. Determine the current I.
c. Calculate the voltage V_2.
d. Determine the power delivered to each 2-kΩ resistor.
e. Calculate the power supplied by the battery.

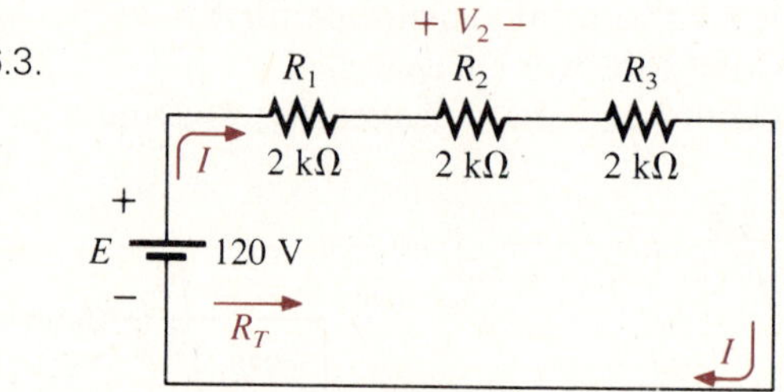

FIG. 6.10
Circuit for Example 6.3.

Solutions

a. $R_T = NR = (3)(2\text{ k}\Omega) = \mathbf{6\text{ k}\Omega}$

Calculator

(3) (×) (2) (EXP) (3) (=)

with a display of 6000.

b. $I = \dfrac{E}{R_T} = \dfrac{120\text{ V}}{6\text{ k}\Omega} = \dfrac{120\text{ V}}{6 \times 10^3\ \Omega} = \dfrac{120}{6} \times 10^{-3}\text{ A}$
$= \mathbf{20\text{ mA}}$

Calculator

(1) (2) (0) (÷) (6) (EXP) (3) (=)

with a display of 0.02.

c. $V_2 = IR_2 = (20 \times 10^{-3}\text{ A})(2 \times 10^{+3}\ \Omega)$
$= 40 \times 10^0\text{ V} = \mathbf{40\text{ V}}$

Calculator

(2) (0) (EXP) (+/−) (3) (×) (2) (EXP) (3) (=)

with a display of 40.

d. $P = I^2R = (20 \times 10^{-3}\text{ A})^2\ 2\text{ k}\Omega$

$= (20)^2 \times (10^{-3})^2 \times 2 \times 10^3\text{ W}$

$= (400 \times 10^{-6})(2 \times 10^3)\text{ W}$

$= (400)(2) \times (10^{-6})(10^3)\text{ W}$

$= 800 \times 10^{-3}\text{ W}$

$\mathbf{P = 800\ mW = 0.8\ W}$

Calculator

(2) (0) (EXP) (+/−) (3) (x^2) (×) (2) (EXP) (3) (=)

with a display of 0.8.

e. $P = EI = (120\text{ V})(20 \times 10^{-3}\text{ A})$

$= (120)(20) \times 10^{-3}\text{ W}$

$= 2400 \times 10^{-3}\text{ W}$

$\mathbf{P = 2400\ mW = 2.4\ W}$

Note that the power supplied is exactly equal to that dissipated by all three resistors. That is,

$$P_{\text{supplied}} = 2.4\text{ W} = 3(P_{2\text{k}\Omega}) = 3(0.8\text{ W}) = 2.4\text{ W}$$

There are also occasions where a series circuit can have a combination of resistors that are the same with resistors of different values. The equations for the total resistance are then combined as shown in the next example.

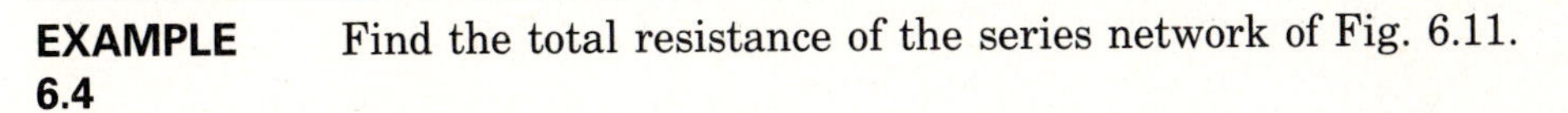

EXAMPLE 6.4 Find the total resistance of the series network of Fig. 6.11.

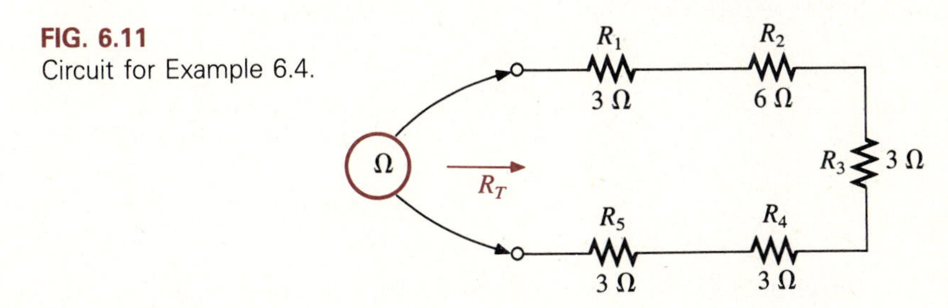

FIG. 6.11
Circuit for Example 6.4.

Solution It is important to realize that resistive elements in series can be interchanged without affecting the total resistance reading or current level. Since we have four 3-Ω resistors, let us redraw Fig. 6.11, as shown in Fig. 6.12.

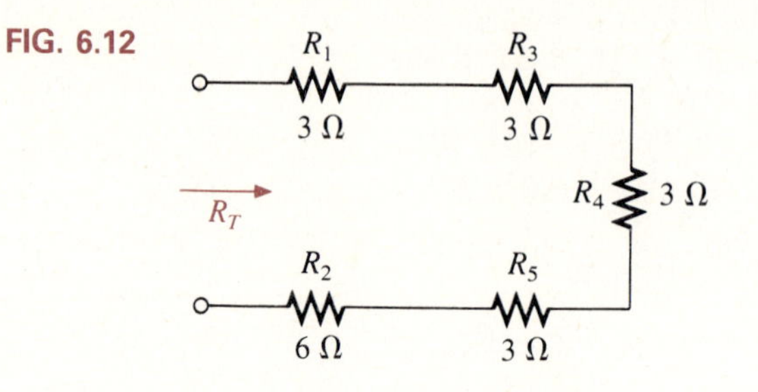

FIG. 6.12

The total resistance R_T is then

$$\begin{aligned} R_T &= NR + R_2 \\ &= (4)(3\ \Omega) + 6\ \Omega \\ &= 12\ \Omega + 6\ \Omega \\ R_T &= \mathbf{18\ \Omega} \end{aligned}$$

Examples 6.1 through 6.4 are straightforward substitution-type problems that are relatively easy to solve with some practice. Example 6.5, however, is evidence of another type of problem, which requires a firm grasp of the fundamental equations and an ability to identify which equation to use first. The best preparation for this type of exercise is simply to work through as many problems of this kind as possible.

EXAMPLE 6.5 Given the total resistance of the series circuit of Fig. 6.13, determine the unknown resistance R_3.

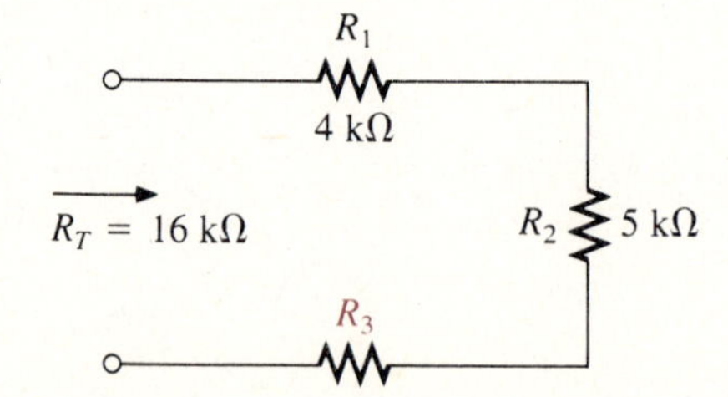

FIG. 6.13
Circuit for Example 6.5.

Solution The best path toward a solution to a problem of this type is first to identify the equation or equations that apply. Since the total resistance is provided, let us start by writing Eq. 6.3 as follows:

$$R_T = R_1 + R_2 + R_3$$

Substituting what we know,

$$16\ \text{k}\Omega = 4\ \text{k}\Omega + 5\ \text{k}\Omega + R_3$$
$$16\ \text{k}\Omega = 9\ \text{k}\Omega + R_3$$

Subtracting 9 kΩ from both sides gives

$$16\ \text{k}\Omega - 9\ \text{k}\Omega = \cancel{9\ \text{k}\Omega} + R_3 - \cancel{9\ \text{k}\Omega}$$
$$R_3 = \mathbf{7\ k\Omega}$$

In other words, find an equation that results in one unknown when all the quantities are substituted and then solve for that unknown using basic algebra.

EXAMPLE 6.6

Given V_2 and I for the network of Fig. 6.14, determine R_2, R_T, and E.

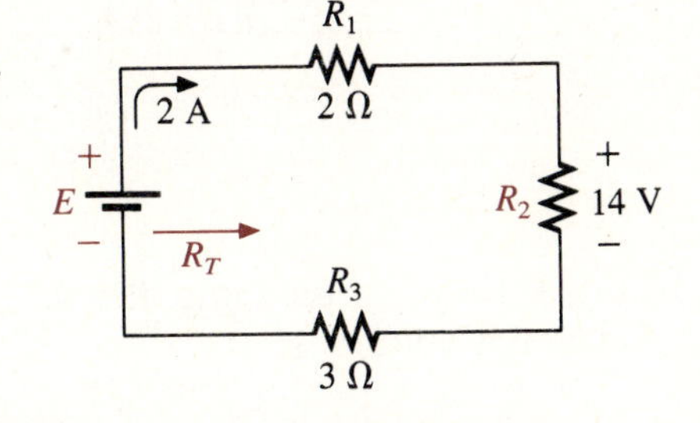

FIG. 6.14
Circuit for Example 6.6.

Solution When asked to find a resistance level, writing Ohm's law in the following form is usually a good starting point:

$$R_2 = \frac{V_2}{I}$$

It tells us that the resistance R_2 can be determined by dividing the voltage across R_2 by the current through R_2. Remember that the current is the same everywhere in a series circuit. Proceeding,

$$R_2 = \frac{V_2}{I} = \frac{14\ \text{V}}{2\ \text{A}} = \mathbf{7\ \Omega}$$

Now that R_2 is known, we can calculate the total resistance from

$$R_T = R_1 + R_2 + R_3$$
$$= 2\ \Omega + 7\ \Omega + 3\ \Omega$$
$$R_T = \mathbf{12\ \Omega}$$

Using Ohm's law,

$$E = IR_T$$
$$= (2\ \text{A})(12\ \Omega)$$
$$E = \mathbf{24\ V}$$

6.3 VOLTAGE SOURCES IN SERIES

Voltage sources can be connected in series as shown in Fig. 6.15 to increase or decrease the total voltage applied to a system. The net voltage is determined simply by summing the sources with the same polarity and subtracting the total of the sources with the opposite pressure. The net polarity is the polarity of the larger sum.

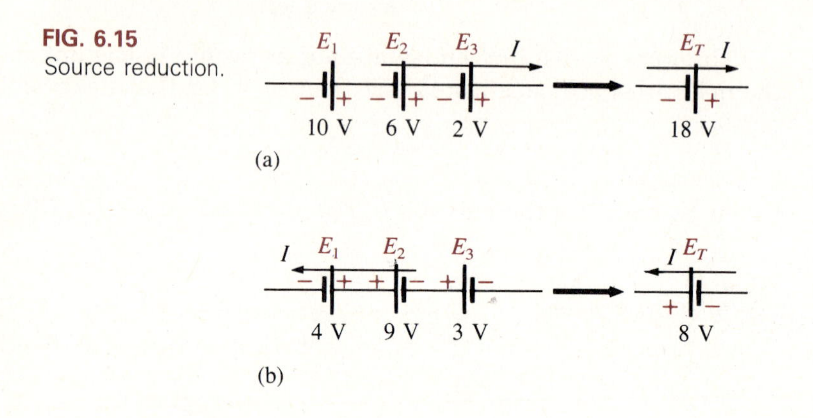

FIG. 6.15
Source reduction.

In Fig. 6.15(a), for example, the sources are all pressuring current to the right; so the net voltage is

$$E_T = E_1 + E_2 + E_3 = 10 + 6 + 2 = \mathbf{18\ V}$$

as shown in the figure. In Fig. 6.15(b), however, the greater pressure is to the left, with a net voltage of

$$E_T = E_2 + E_3 - E_1 = 9 + 3 - 4 = \mathbf{8\ V}$$

and the polarity shown in the figure.

6.4 KIRCHHOFF'S VOLTAGE LAW

One of the most important laws of electric circuits is *Kirchhoff's voltage law.* It can be used to find unknown voltage levels and required battery levels or to check voltage calculations. It states that

The algebraic sum of the potential (voltage) rises and drops around a closed loop (or path) is zero.

The term *algebraic* requires that the change in voltage levels be assigned a specific sign to determine whether they are added or subtracted when the law is applied. A positive sign is assigned to a voltage rise (− to +) and a negative sign is given to a voltage drop (+ to −) as we follow *one direction* around the entire closed loop. Since the clockwise direction is one we use frequently in the chapters to follow, let us choose this direction as our standard for the examples to follow. This is not to say, however, that the counterclockwise direction will generate an incorrect result. Both directions yield the same result, but using one direction as a standard eliminates the need to wonder about which way to proceed and will generate similar equations for anyone applying the law. Finally, the algebraic sum is zero only if the application of the law ends at the same point in the network that it was initiated. In Fig. 6.16, for example, if point a is the starting point, then point a must also be the ending point.

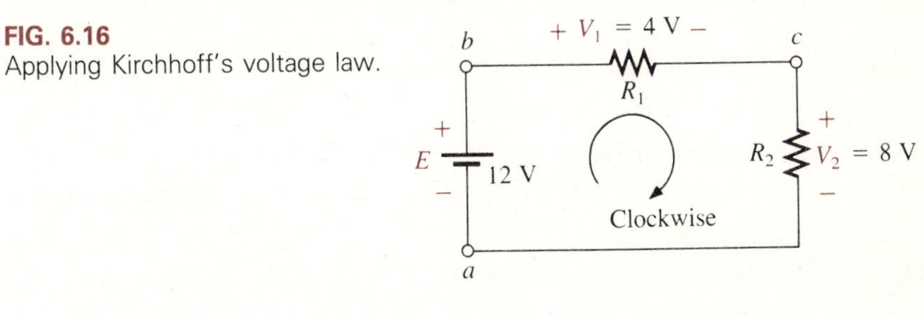

FIG. 6.16
Applying Kirchhoff's voltage law.

Starting at point a and moving in the clockwise direction to point b, we encounter a rise (− to + terminal of the battery) and therefore assign a positive sign to the 12-V change in potential level, as follows:

$$+E - V_1 - V_2 = 0$$
$$+12\text{ V} - 4\text{ V} - 8\text{ V} = 0$$

When determining the proper sign, always restrict your analysis to the change in voltage level of interest. In other words, when looking at the source voltage E, ignore the polarities surrounding both V_1 and V_2. It is also not difficult to remember which sign to apply if you simply keep in mind that going from something negative (−) to something positive (+) is a positive experience, thereby resulting in a positive sign for the 12-V change.

As we continue around the loop in the clockwise direction, we encounter a drop in potential (+ to −) across the resistor R_1 and the resistor R_2, resulting in negative signs, as just shown. We are now back at point a, and the resulting series of terms can be set equal to zero, as required by Kirchhoff's voltage law.

Looking at the resulting equation, it is quite obvious that the algebraic sum of the terms is zero. That is,

$$12\text{ V} - (4\text{ V} + 8\text{ V}) = 0$$
$$12\text{ V} - 12\text{ V} = 0$$
$$0\text{ V} = 0\text{ V}$$

If we transfer V_1 and V_2 to the right side of the equals sign by subtracting each from both sides, we obtain

$$E = V_1 + V_2$$

revealing that in single-source series networks, the applied voltage (battery voltage) equals the sum of the drops across the loads.

To demonstrate that the application of the law in the counterclockwise direction will result in the same conclusion, let us apply the law in the counterclockwise direction starting at point a. You will now find that the transition for V_1 and V_2 is a positive one (− to +), and the transition is a negative one for E (+ to −). It is therefore very important not to look at whether the change in voltage is across a load or source of voltage. It is important to note only whether the change is a positive or negative transition.

Applying Kirchhoff's voltage law gives

$$+V_1 + V_2 - E = 0$$

or

$$E = V_1 + V_2$$

as obtained earlier.

In the next example we use the law to find an unknown voltage.

EXAMPLE 6.7 Find the voltage V_3 of Fig. 6.17 using Kirchhoff's voltage law.

FIG. 6.17
Circuit for Example 6.7.

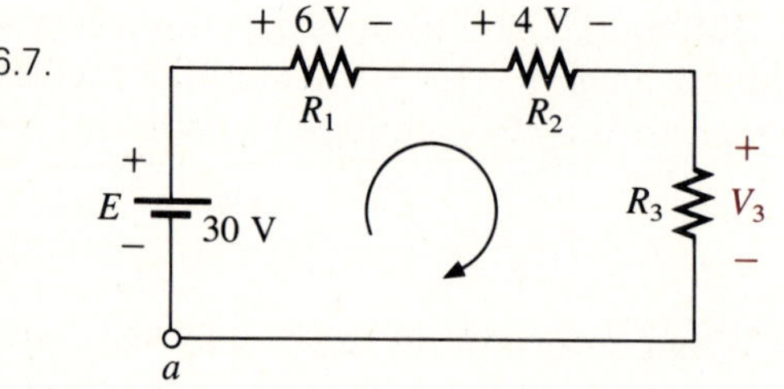

Solution Note that there is no need to know the resistor values or the current in the circuit. The law deals solely with voltage levels and their relationships around a closed loop.

Starting at a and moving in the clockwise direction, we find a rise (− to +) of 30 V and drops (+ to −) of 6 V, 4 V, and V_3. Applying Kirchhoff's voltage law,

$$+30\text{ V} - 6\text{ V} - 4\text{ V} - V_3 = 0$$

or

$$30\text{ V} - 10\text{ V} - V_3 = 0$$

and

$$20\text{ V} - V_3 = 0$$

Subtracting 20 V from both sides:

$$\cancel{20\text{ V}} - V_3 - \cancel{20\text{ V}} = -20\text{ V}$$
$$-V_3 = -20\text{ V}$$

Multiplying through by −1 on both sides gives

$$-(-V_3) = -(-20\text{ V})$$
$$V_3 = \mathbf{20\text{ V}}$$

EXAMPLE 6.8

Given the voltage levels indicated in Fig. 6.18, determine the voltage level V.

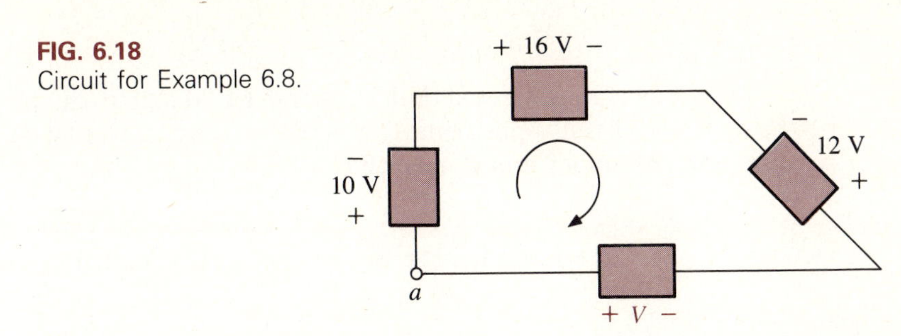

FIG. 6.18
Circuit for Example 6.8.

Solution This example demonstrates that it is unnecessary to know whether there is a source of electrical energy or a load in each container. The law is specifically interested in voltage levels and their polarity and is not concerned about the source of the potential difference.

Applying Kirchhoff's voltage law in the clockwise direction starting at point a,

$$-10\text{ V} - 16\text{ V} + 12\text{ V} + V = 0$$
$$-26\text{ V} + 12\text{ V} + V = 0$$
$$-14\text{ V} + V = 0$$

Adding +14 V to both sides gives

$$-\cancel{14\text{ V}} + \cancel{14\text{ V}} + V = +14\text{ V}$$
$$V = \mathbf{14\text{ V}}$$

EXAMPLE 6.9

Determine V_1 and V_2 for Fig. 6.19.

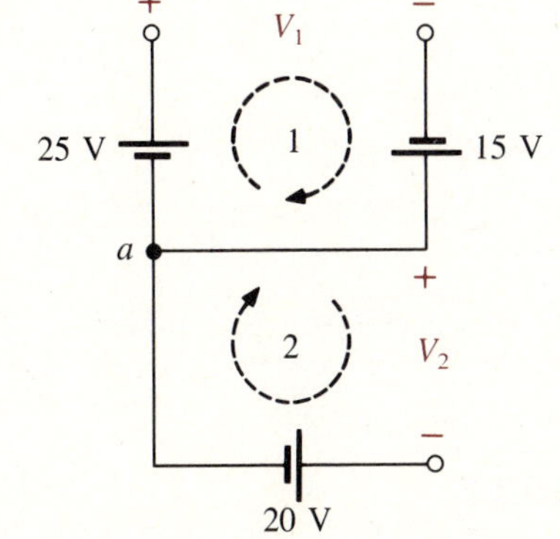

FIG. 6.19

Solution For clockwise path 1 starting at a:

$$+25\text{ V} - V_1 + 15\text{ V} = 0$$
$$40\text{ V} - V_1 = 0$$
$$V_1 = \mathbf{40\text{ V}}$$

For clockwise path 2 starting at *a:*

$$-V_2 - 20\text{ V} = 0$$
$$V_2 = \mathbf{-20\text{ V}}$$

The minus sign indicates that V_2 is indeed 20 V, but the polarity should be the opposite of that appearing in Fig. 6.19. That is, the lower end of V_2 is positive and the upper end is negative.

EXAMPLE 6.10 Determine the voltage V_5 for the network of Fig. 6.20 using loops 1 and 2.

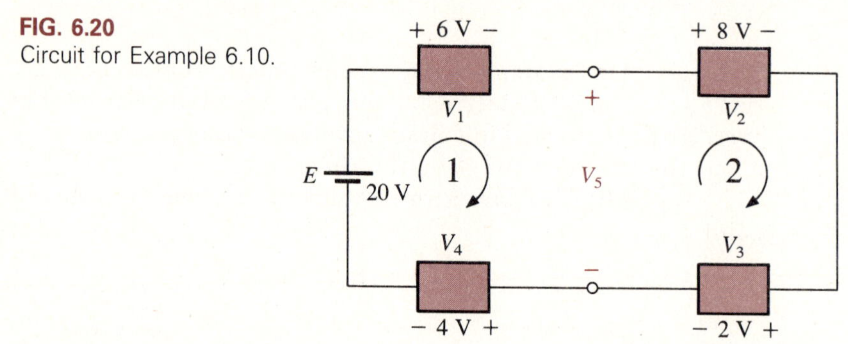

FIG. 6.20
Circuit for Example 6.10.

Solution First note that V_5 is not across a particular element. It demonstrates that potential differences can be determined between any two points in a network. However, the analysis to follow reveals that only *one* potential difference can exist between any two points of a network.

Loop 1: Starting at *a*,

$$+20\text{ V} - 6\text{ V} - V_5 - 4\text{ V} = 0$$
$$10\text{ V} - V_5 = 0$$
$$V_5 = \mathbf{10\text{ V}}$$

Loop 2: Starting at *a*,

$$+V_5 - 8\text{ V} - 2\text{ V} = 0$$
$$V_5 - 10\text{ V} = 0$$
$$V_5 = \mathbf{10\text{ V}}$$

verifying the preceding conclusion.

6.5 VOLTAGE DIVIDER RULE

The voltage divider rule provides a method of determining the voltage across a load without having to first find the network current. In other words, a voltage can be determined in one step rather than by applying methods described earlier.

In words, the rule states that

The voltage across a resistor in a series circuit is equal to the value of that resistor times the impressed voltage divided by the total resistance of the series elements.

For the circuit of Fig. 6.21, the rule results in the following equation for V_1:

$$V_1 = \frac{R_1 E}{R_T} \tag{6.5}$$

and for V_2,

$$V_2 = \frac{R_2 E}{R_T} \tag{6.6}$$

Note that the numerator always includes the resistor value of the resistor across which the voltage is to be found. The remaining part of the equation is the same for each.

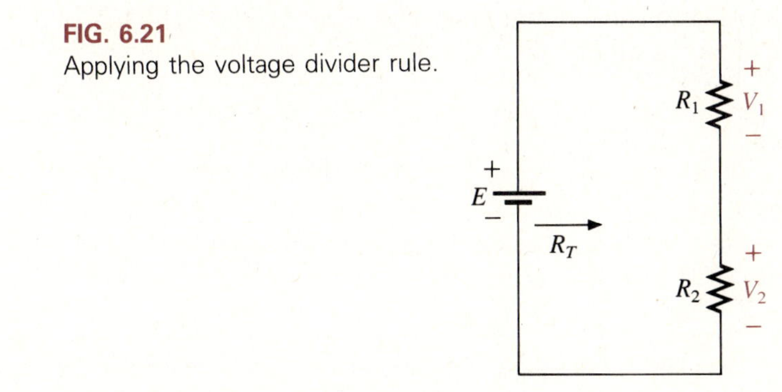

FIG. 6.21
Applying the voltage divider rule.

EXAMPLE 6.11 Determine V_1 and V_2 for the circuit of Fig. 6.22.

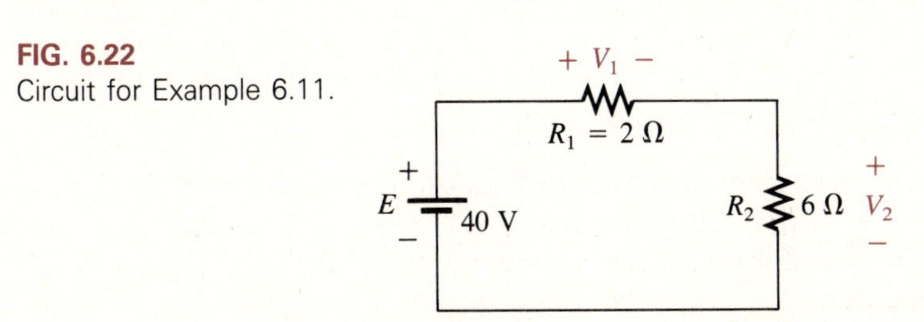

FIG. 6.22
Circuit for Example 6.11.

Solution By Eq. 6.5:

$$V_1 = \frac{R_1 E}{R_T}$$

$$R_T = R_1 + R_2 = 2\,\Omega + 6\,\Omega = 8\,\Omega$$

and

$$V_1 = \frac{(2\,\Omega)(40\text{ V})}{8\,\Omega} = \frac{80}{8}\text{ V} = \mathbf{10\text{ V}}$$

By Eq. 6.7:

$$V_2 = \frac{R_2 E}{R_T} = \frac{(6\,\Omega)(40\text{ V})}{8\,\Omega}$$

$$V_2 = \frac{240}{8}\text{ V} = \mathbf{30\text{ V}}$$

Due to the similarities between Eqs. 6.5 and 6.6, the voltage divider rule can be written in the following more general format, which is applicable to any number of series elements.

$$\boxed{V_x = \frac{R_x E}{R_T}} \tag{6.7}$$

where V_x is the voltage to be found across resistor R_x and R_T is the total resistance of the series circuit.

EXAMPLE 6.12

a. Determine the voltage V_1 for the series circuit of Fig. 6.23.
b. Repeat part a for V_2 and V_3.

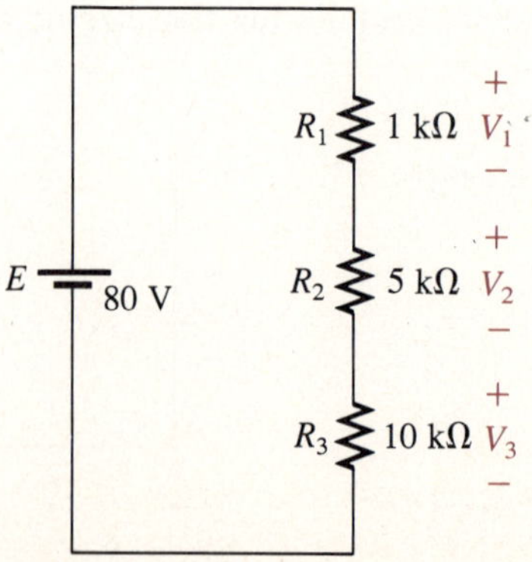

FIG. 6.23
Circuit for Example 6.12.

Solutions Using Eq. 6.7:

$$V_x = \frac{R_x E}{R_T}$$

$$V_1 = \frac{R_1 E}{R_T} \qquad (x = 1)$$

$$= \frac{(1\ \text{k}\Omega)(80\ \text{V})}{1\ \text{k}\Omega + 5\ \text{k}\Omega + 10\ \text{k}\Omega} = \frac{(1\ \text{k}\Omega)(80\ \text{V})}{16\ \text{k}\Omega}$$

$$= \frac{80}{16}\ \text{V}$$

$$V_1 = \mathbf{5\ V}$$

$$V_2 = \frac{R_2 E}{R_T} \qquad (x = 2)$$

$$= \frac{(5\ \text{k}\Omega)(80\ \text{V})}{16\ \text{k}\Omega}$$

$$= \frac{400}{16}\ \text{V}$$

$$V_2 = \mathbf{25\ V}$$

Note that the value of R_T was simply substituted from the equation for V_1, since it does not change for each voltage level.

$$V_3 = \frac{R_3 E}{R_T} \qquad (x = 3)$$

$$= \frac{(10\ \text{k}\Omega)(80\ \text{V})}{16\ \text{k}\Omega}$$

$$= \frac{800}{16}\ \text{V}$$

$$V_3 = \mathbf{50\ V}$$

Before leaving this example, note that the voltage levels across R_1, R_2, and R_3 are direct functions of their relative magnitudes. That is, R_2 is five times as large as R_1, and therefore V_2 is five times as large as V_1. In addition, R_3 is 10 times as large as R_1, and therefore V_3 is 10 times as large as V_1. This is a result of the fact that Eq. 6.7 can be written as

$$V_x = R_x\left(\frac{E}{R_T}\right)$$

clearly showing that E/R_T is fixed for all levels of V_x, so the larger R_x, the larger the resulting level of V_x.

In general, it is helpful to remember that

In a series circuit, the ratio of the voltage levels across series elements is directly related to the ratio of their resistive levels.

For example, without analyzing the network of Fig. 6.24, we know that the vast majority of the applied voltage E will appear across R_3. In fact, $V_3 = 100V_1$ and $V_3 = (100/2)V_2 = 50V_2$. The voltages V_1 and V_2 are related by $V_2 = 2V_1$ but are of a much lower level than V_3.

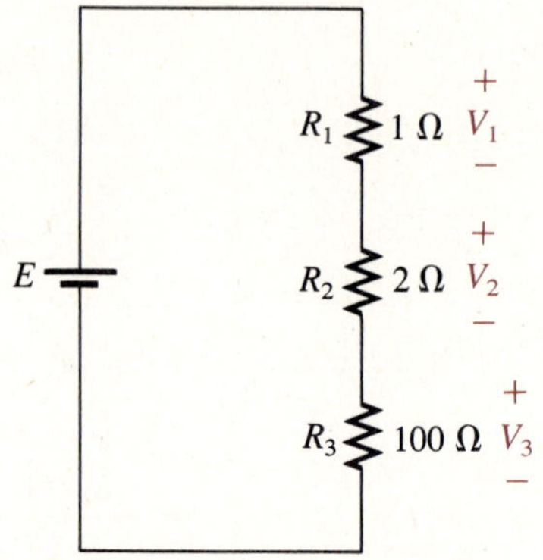

FIG. 6.24
Examining the effects of the voltage divider rule.

Example 6.13 is of design variety in which the elements of the network are determined to fit a given set of specifications—often referred to as the *synthesis* process.

EXAMPLE 6.13 Determine the resistance R_2 of Fig. 6.25 required to establish a 12-V drop across R_2.

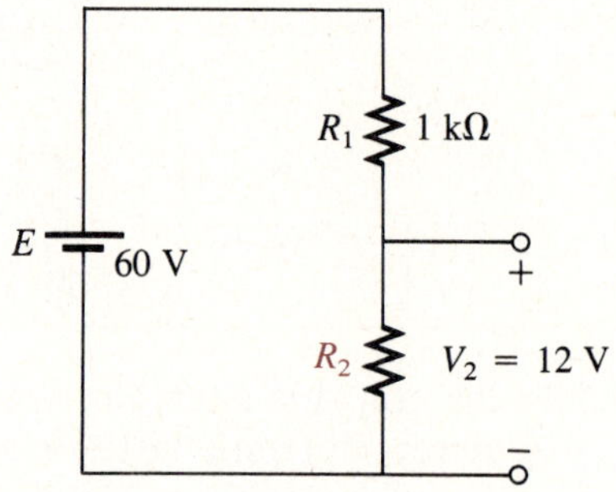

FIG. 6.25
Circuit for Example 6.13.

Solution Applying Eq. 6.8,

$$V_x = \frac{R_x E}{R_T}$$

and substituting the known quantities,

$$12\text{ V} = \frac{(R_2)(60\text{ V})}{1\text{ k}\Omega + R_2}$$

Multiplying both sides by $(1\ \text{k}\Omega + R_2)$ gives

$$(1\ \text{k}\Omega + R_2)12\ \text{V} = \frac{(60\ \text{V})(R_2)}{\cancel{(1\ \text{k}\Omega + R_2)}}\cancel{(1\ \text{k}\Omega + R_2)}$$

Multiplying through, we have

$$(1\ \text{k}\Omega)(12\ \text{V}) + (12\ \text{V})(R_2) = (60\ \text{V})(R_2)$$

Subtracting $(12\ \text{V})(R_2)$ from both sides gives

$$(1\ \text{k}\Omega)(12\ \text{V}) + \cancel{(12\ \text{V})(R_2)} - \cancel{(12\ \text{V})(R_2)} = (60\ \text{V})(R_2) - (12\ \text{V})(R_2)$$
$$(1\ \text{k}\Omega)(12\ \text{V}) = (48\ \text{V})(R_2)$$

Dividing both sides by 48 V yields

$$\frac{(1\ \text{k}\Omega)(12\ \text{V})}{48\ \text{V}} = \frac{\cancel{(48\ \text{V})}(R_2)}{\cancel{48\ \text{V}}}$$

$$R_2 = \frac{12}{48}(1\ \text{k}\Omega) = 0.25\ \text{k}\Omega$$
$$= \mathbf{250\ \Omega}$$

A less mathematical method would be to determine the voltage across R_1 using Kirchhoff's voltage law:

$$+E - V_1 - V_2 = 0$$

or

$$V_1 = E - V_2 = 60\ \text{V} - 12\ \text{V} = 48\ \text{V}$$

and

$$I = \frac{V_1}{R_1} = \frac{48\ \text{V}}{1\ \text{k}\Omega} = 48\ \text{mA}$$

so that

$$R_2 = \frac{V_2}{I} = \frac{12\ \text{V}}{48 \times 10^{-3}\ \text{A}} = \frac{12}{48} \times 10^3\ \Omega$$
$$= \frac{1}{4} \times 10^3\ \Omega = 0.25 \times 10^3\ \Omega$$
$$R_2 = \mathbf{250\ \Omega}$$

The latter solution is preferred because it requires a broader and more confident understanding of basic electrical concepts. However, the first solution includes algebraic manipulations that are very basic in nature and the application of only one equation.

Before starting the analysis of the circuit of Fig. 6.25, we should expect that R_2 would be considerably less in magnitude than R_1 because R_2 is getting only 12 V of the total of 60 V. In fact, since the ratio of V_1 to V_2 is 48 V/12 V = 4:1, we should expect that R_2 will be 1/4 of R_1, or 250 Ω.

6.6 NOTATION

Notation plays an increasingly important role in the analysis to follow. It is important, therefore, that we begin to consider some of the notation used throughout the industry.

Except for a few special cases, electrical and electronic systems are grounded for reference and safety purposes. The symbol for the ground connection appears in Fig. 6.26 with its defined potential level—0 V. None of the circuits discussed thus far have contained the ground connection. If Fig. 6.25 were redrawn with a grounded supply, it might appear as shown in Fig. 6.27(a) or (b). In either case, it is understood that the negative terminal of the battery and the bottom of the resistor R_2 are at ground potential. Although Fig. 6.27(b) shows no connection between the two grounds, it is recognized that such a connection exists for the continuous flow of charge. If E = 12 V, then point a is 12 V positive with respect to ground potential, and 12 V exist across the series combination of resistors R_1 and R_2. If a voltmeter placed from point b to ground reads 4 V, then the voltage across R_2 is 4 V, with the higher potential at point b.

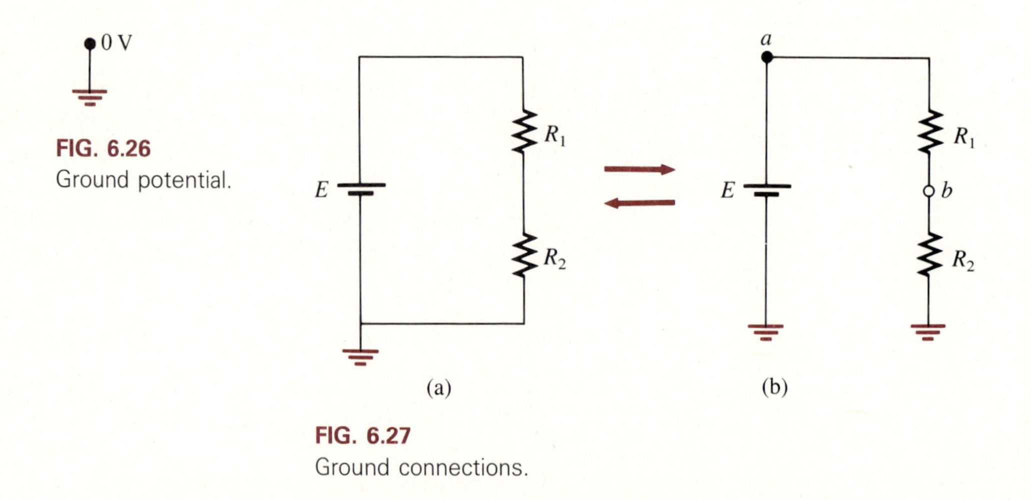

FIG. 6.26
Ground potential.

FIG. 6.27
Ground connections.

On large schematics where space is at a premium and clarity is important, voltage sources may be indicated as shown in Figs. 6.28(a) and 6.29(a) rather than as illustrated in Figs. 6.28(b) and 6.29(b). In addition, potential levels may be indicated as in Fig. 6.30 to permit a rapid check of the potential levels at various points in a network with respect to ground to ensure the system is operating properly.

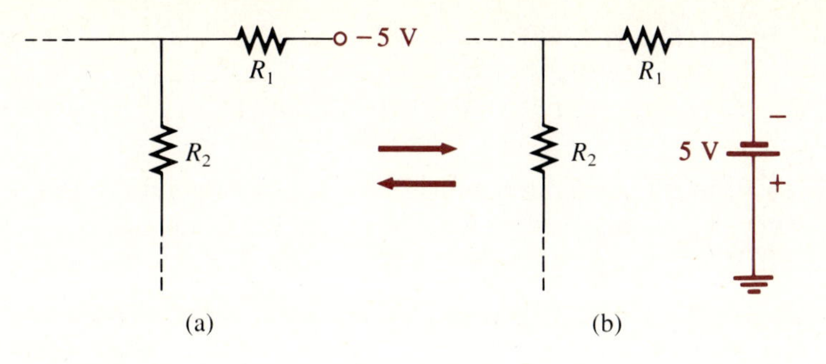

FIG. 6.28

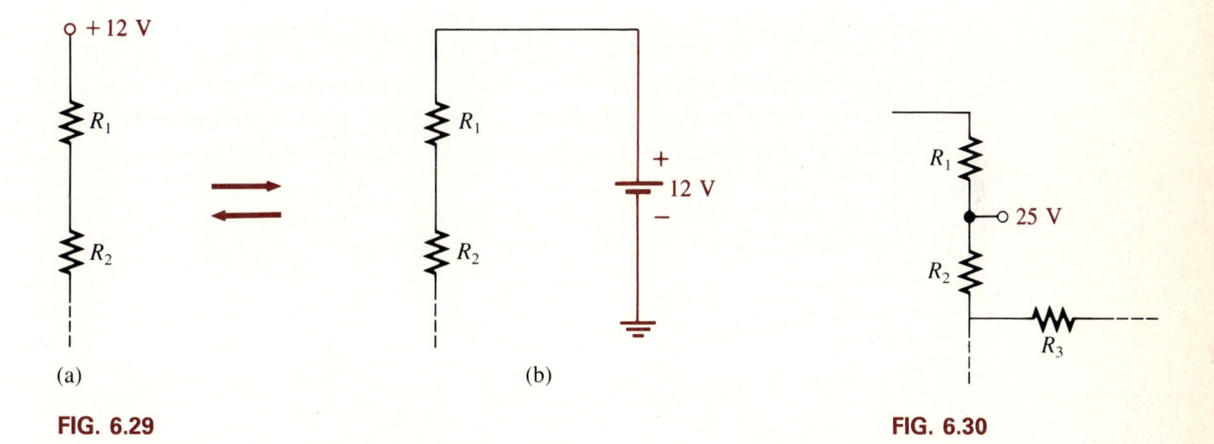

FIG. 6.29

FIG. 6.30

The fact that voltage is an *across* variable and exists between two points has resulted in a double-subscript notation that defines the first subscript as the higher potential. The notation V_{ab} for the potential difference appearing in Fig. 6.31 specifies that the potential at a is higher than the potential at b by 6 V. In fact, $V_{ab} = V_a - V_b = 6$ V. For a situation such as that shown in Fig. 6.32, where a is a lesser potential than b, a negative sign is employed to reveal the polarity difference.

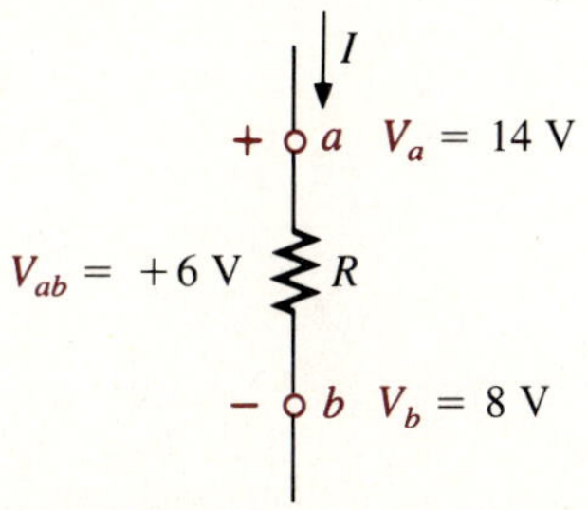

FIG. 6.31
Double-subscript notation (+ value, $V_a > V_b$).

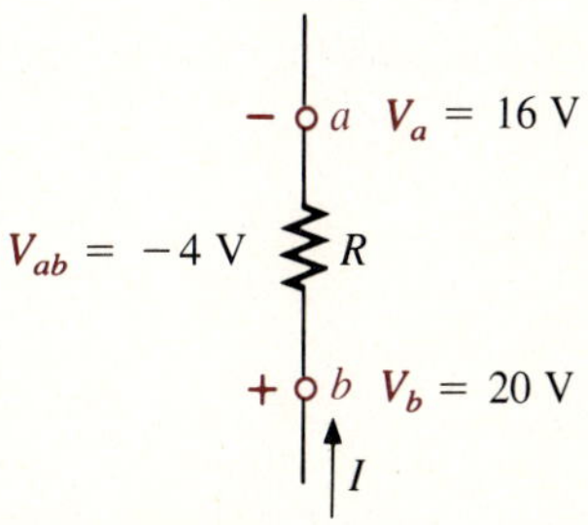

FIG. 6.32
Double-subscript notation (− value, $V_b > V_a$).

In summary,

$$V_{xy} = V_x - V_y \tag{6.8}$$

where V_x and V_y are the potential levels at two points in the network. If $V_x > V_y$, then V_{xy} is positive. If $V_x < V_y$, then V_{xy} is negative.

If V_y is ground potential, or 0 V, then Eq. 6.8 becomes

$$V_{xy} = V_x - V_y = V_x - 0 = V_x$$

revealing that a single-subscript notation such as V_x or V_y provides the level at x or y with respect to ground.

V_x is the voltage at point x with respect to ground (0 V).

Three examples of single-subscript notation appear in Fig. 6.33. Note the negative sign for Fig. 6.33(b), which reveals that point a is 7 V below the ground, or 0-V, reference level.

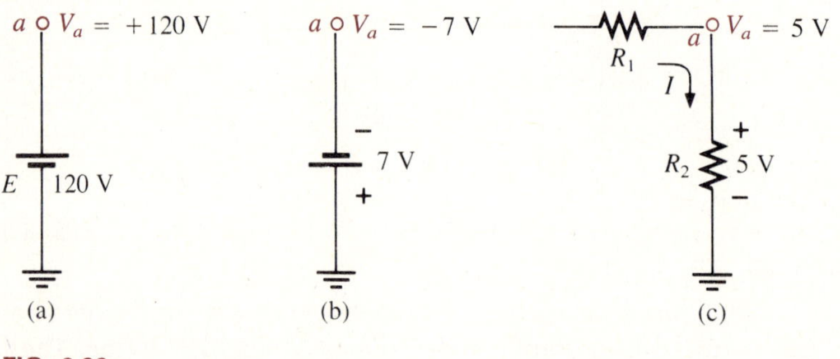

FIG. 6.33
Single-subscript notation.

EXAMPLE 6.14 Using the voltage divider rule, determine the voltages V_1 and V_2 of Fig. 6.34.

FIG. 6.34
Circuit for Example 6.14.

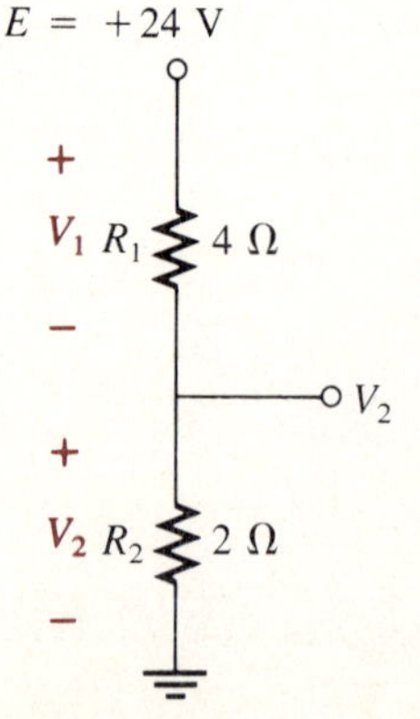

$E = +24$ V

+ V_1 R_1 4 Ω −

V_2

+ V_2 R_2 2 Ω −

Solution Redrawing the network with the standard battery symbol results in the network of Fig. 6.35. Applying the voltage divider rule,

$$V_1 = \frac{R_1 E}{R_1 + R_2} = \frac{(4\ \Omega)(24\ \text{V})}{4\ \Omega + 2\ \Omega} = \mathbf{16\ V}$$

$$V_2 = \frac{R_2 E}{R_1 + R_2} = \frac{(2\ \Omega)(24\ \text{V})}{4\ \Omega + 2\ \Omega} = \mathbf{8\ V}$$

FIG. 6.35

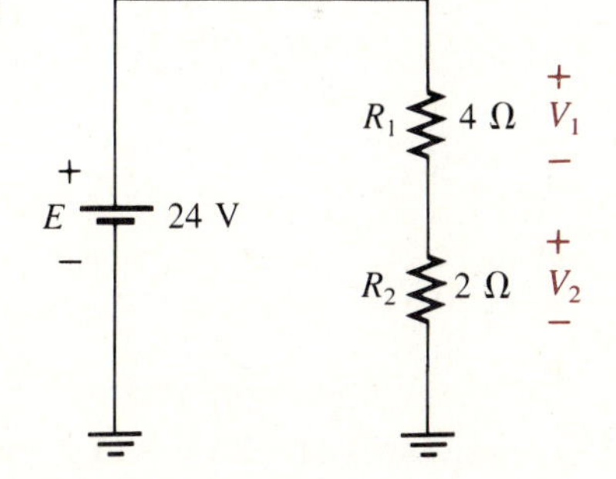

EXAMPLE 6.15 For the network of Fig. 6.36,

a. Calculate V_{ab}.
b. Determine V_b.
c. Calculate V_c.

FIG. 6.36

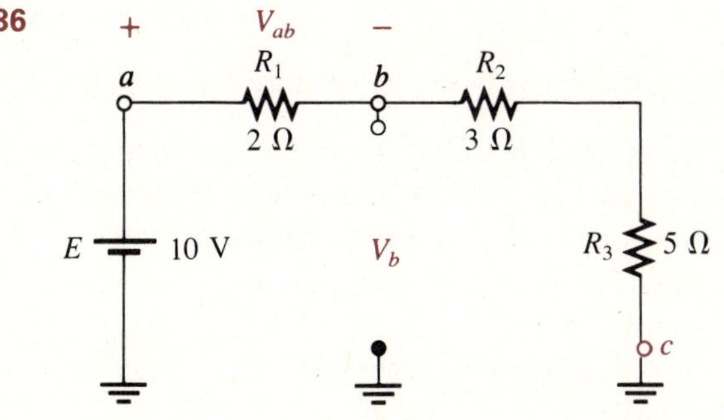

Solutions

a. Using the voltage divider rule,

$$V_{ab} = \frac{R_1 E}{R_T} = \frac{(2\ \Omega)(10\ \text{V})}{2\ \Omega + 3\ \Omega + 5\ \Omega} = \mathbf{2\ V}$$

b. Using the voltage divider rule,

$$V_b = V_{R_2} + V_{R_3} = \frac{(R_2 + R_3)E}{R_T} = \frac{(3\ \Omega + 5\ \Omega)(10\ \text{V})}{10\ \Omega} = \mathbf{8\ V}$$

or

$$V_b = E - V_{ab} = 10 \text{ V} - 2 \text{ V} = \mathbf{8 \text{ V}}$$

c. V_c = ground potential = **0 V.**

EXAMPLE 6.16 Determine V_a, I_1, and V_2 for the network of Fig. 6.37.

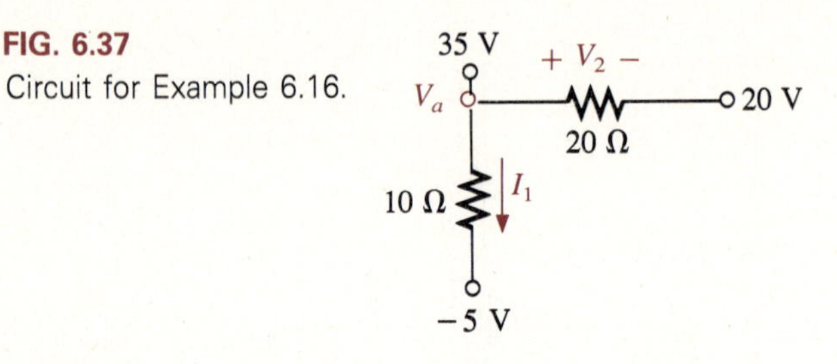

FIG. 6.37
Circuit for Example 6.16.

Solution

$$V_a = \mathbf{35 \text{ V}}$$

One end of the 10-Ω resistor is at +35 V and the other is at −5 V, creating a difference of 35 + 5 = 40 V across the 10-Ω resistor. The current I_1 is, therefore,

$$I_1 = \frac{40 \text{ V}}{10 \ \Omega} = \mathbf{4 \text{ A}}$$

The voltage V_2 is 35 V − 20 V = **15 V.**

6.7 INTERNAL RESISTANCE OF VOLTAGE SOURCES

Every source of voltage, whether it be a generator, battery, or laboratory supply, as shown in Fig. 6.38(a), has some internal resistance. The equivalent circuit of any source of voltage therefore appears as shown in Fig. 6.38(b). In

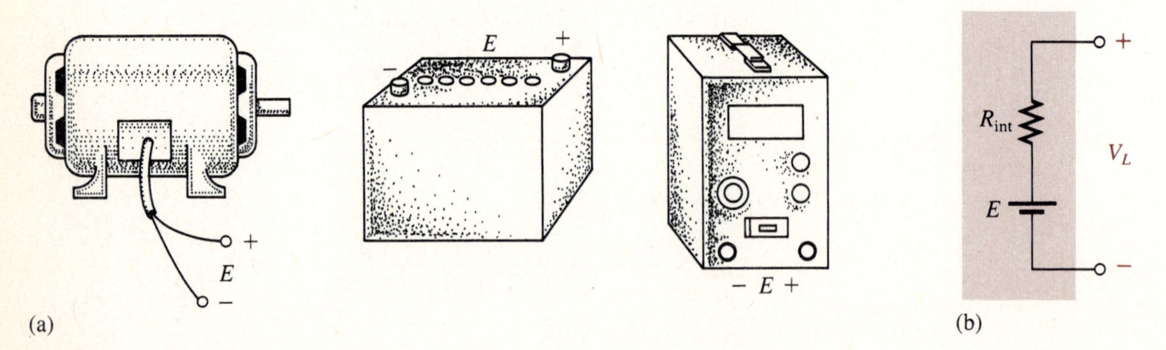

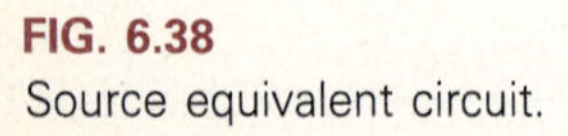

FIG. 6.38
Source equivalent circuit.

this section, we examine the effect of the internal resistance on the output voltage so that any unexpected changes in terminal characteristics can be explained.

In all the analyses to this point, we have assumed the voltage source was an ideal device with no internal resistance, as shown in Fig. 6.39. The ideal source of Fig. 6.39 has an output of E volts with R_L connected or not (no-load).

FIG. 6.39
Ideal voltage source.

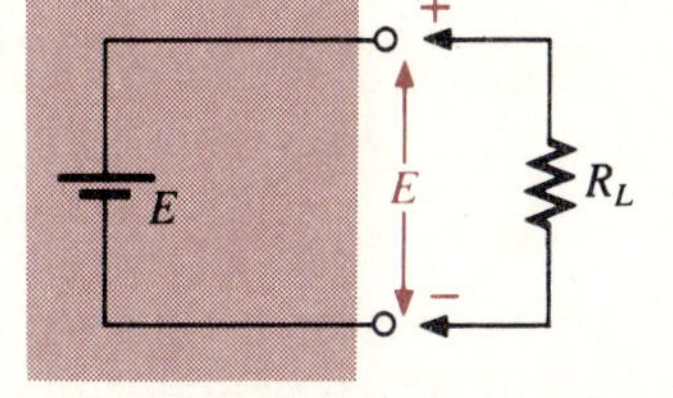

In Fig. 6.40(a) we find that the internal resistance does not affect the output voltage under no-load conditions because the source current is zero. A complete path does not exist for charge to flow. In other words, there is no drop in voltage ($V_{int} = I_L R_{int} = 0 \cdot R_{int} = 0$ V) across R_{int}, and the output voltage, or no-load voltage, is $V_{NL} = E$ volts. However, if we connect a load R_L as shown in Fig. 6.40(b), current will exist in the series circuit and a drop in voltage will occur across the R_{int} resistance.

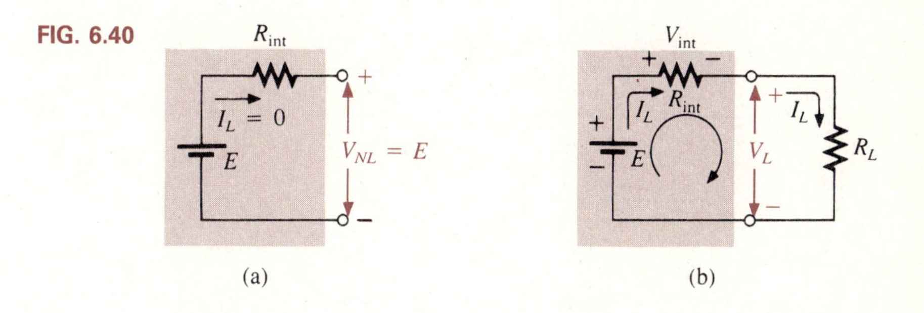

FIG. 6.40

If we apply Kirchhoff's voltage law around the closed path of Fig. 6.40(b), we obtain

$$E - V_{int} - V_L = 0$$

or

$$\boxed{V_L = E - V_{int}} \tag{6.9}$$

Substituting $V_{int} = I_L R_{int}$, we obtain

$$\boxed{V_L = E - I_L R_{int}} \tag{6.10}$$

Equation 6.9 clearly shows that the terminal voltage (V_2) of a supply will be less than the ideal source voltage (E) by the drop across the internal resistance. If we assume a fixed internal resistance, Eq. 6.10 reveals that the higher the load current, the less the terminal voltage (V_L) and the less ideal the supply.

Through algebraic manipulations, an equation for the internal resistance can be developed as follows:

$$R_{\text{int}} = \frac{V_{NL}}{I_L} - R_L \qquad (6.11)$$

In words, Eq. 6.11 requires that we measure the terminal voltage of a battery with no load attached to determine V_{NL} and then apply a known load of R_L and measure the current to determine I_L.

A plot of the output voltage versus current appears in Fig. 6.41(b) for the circuit in Fig. 6.41(a). Note that an increase in load demand (I_L) increases the current through, and thereby the voltage drop across, the internal resistance of the source, resulting in a decrease in terminal voltage. Eventually, as the load resistance approaches 0 Ω, all the generated voltage will appear across the internal resistance and none will appear at the output terminals. The steeper the slope of the curve of Fig. 6.41(b), the greater the internal resistance.

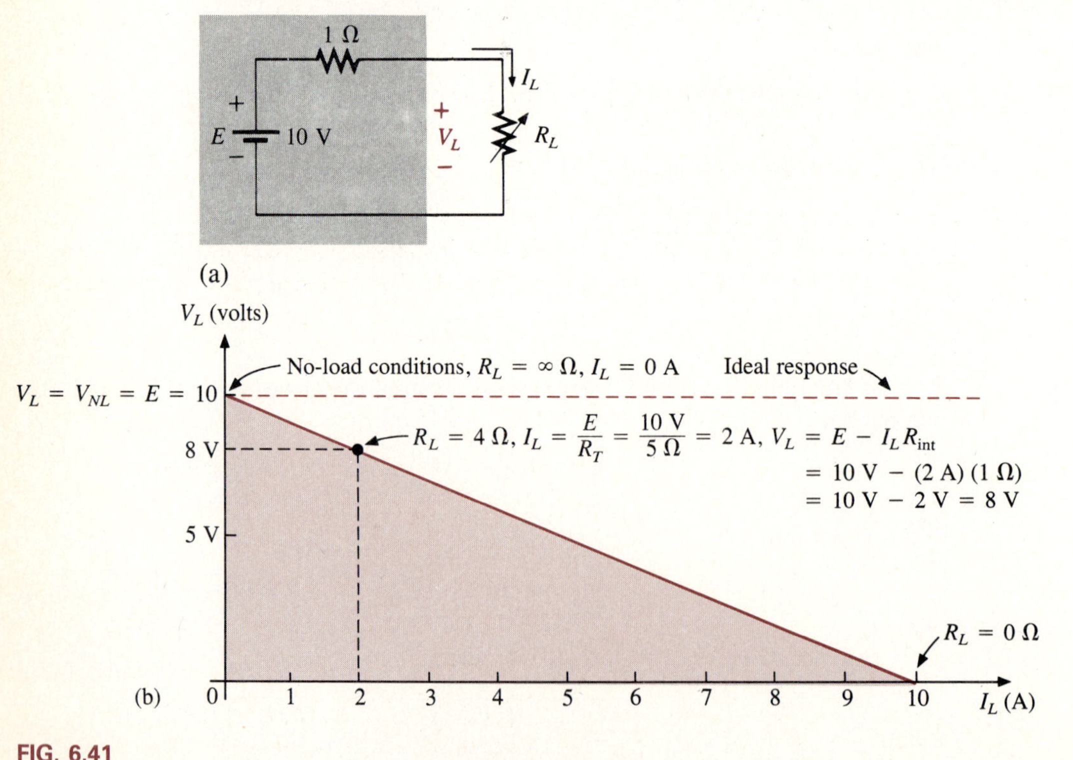

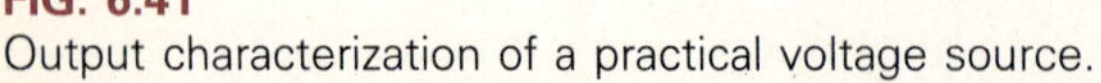

FIG. 6.41
Output characterization of a practical voltage source.

EXAMPLE 6.17

Before a load is applied, the terminal voltage of the power supply of Fig. 6.42(a) is set to 40 V. When a load of 500 Ω is attached, as shown in Fig. 6.42(b), the terminal voltage drops to 36 V. What happened to the remainder of the no-load voltage, and what is the internal resistance of the source?

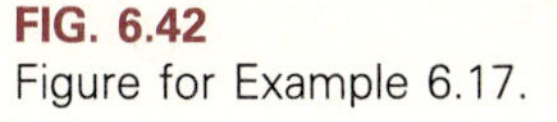

FIG. 6.42
Figure for Example 6.17.

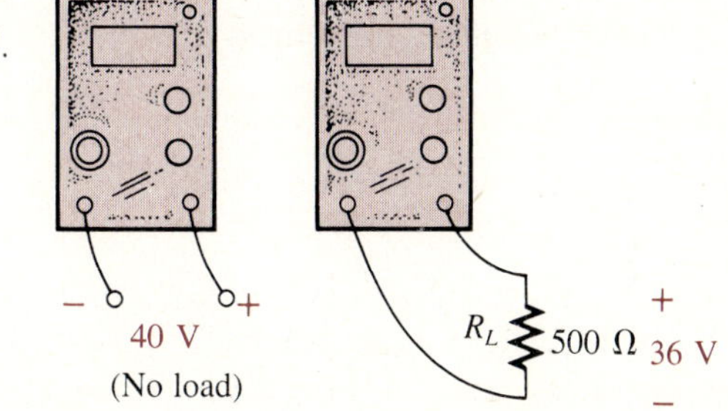

Solution The difference of 40 V − 36 V = 4 V now appears across the internal resistance of the source. The load current is V_L/R_L = 36 V/0.5 kΩ = 72 mA. Applying Eq. 6.12,

$$R_{\text{int}} = \frac{V_{NL}}{I_L} - R_L = \frac{40\text{ V}}{72\text{ mA}} - 0.5\text{ k}\Omega$$

$$= 555.55\ \Omega - 500\ \Omega = \mathbf{55.55\ \Omega}$$

Calculator

(4) (0) (÷) (7) (2) (EXP) (+/−) (3) (−) (·) (5) (EXP) (3) (=)

with a display of 55.555.

EXAMPLE 6.18

The battery of Fig. 6.43 has an internal resistance of 2 Ω. Find the voltage V_L and the power lost to the internal resistance if the applied load is a 13-Ω resistor.

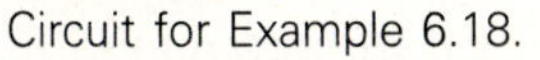

FIG. 6.43
Circuit for Example 6.18.

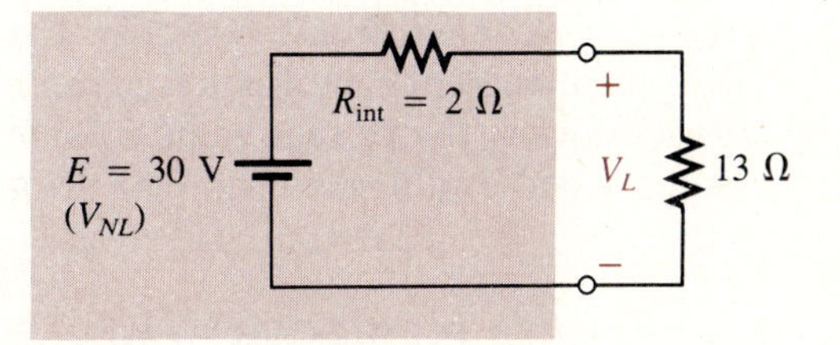

Solution

$$I_L = \frac{30\text{ V}}{2\ \Omega + 13\ \Omega} = \frac{30\text{ V}}{15\ \Omega} = \mathbf{2\ A}$$

$$V_L = V_{NL} - I_L R_{\text{int}} = 30\text{ V} - (2\text{ A})(2\ \Omega) = \mathbf{26\ V}$$

$$P_{\text{lost}} = I_L^2 R_{\text{int}} = (2\text{ A})^2(2\ \Omega) = (4)(2) = \mathbf{8\ W}$$

EXAMPLE 6.19

The terminal characteristics of a dc generator appear in Fig. 6.44. Rated (full-load) conditions are indicated at 120 V, 8 A. Calculate the average internal resistance of the supply.

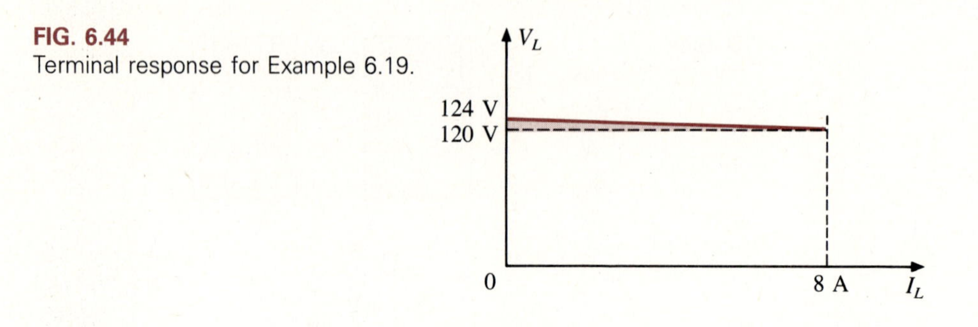

FIG. 6.44
Terminal response for Example 6.19.

Solution

$$R_L = \frac{V_L}{I_L} = \frac{120\text{ V}}{8\text{ A}} = 15\ \Omega$$

Using Eq. 6.12,

$$R_{\text{int}} = \frac{V_{NL}}{I_L} - R_L$$

$$= \frac{124\text{ V}}{8\text{ A}} - 15\ \Omega$$

$$= 15.5\ \Omega - 15\ \Omega$$

$$R_{\text{int}} = \mathbf{0.5\ \Omega}$$

6.8 VOLTAGE REGULATION

For any supply, ideal conditions dictate that for the range of load demand (I_L), the terminal voltage remain fixed in magnitude. In other words, if a supply is set for 12 V, it is desirable that it maintain this terminal voltage even though the current demand on the supply may vary. A measure of how close a supply will come to ideal conditions is given by the voltage-regulation characteristic. By definition, the voltage regulation of a supply between the limits of full-load and no-load conditions (Fig. 6.45) is given by the following:

$$\text{Voltage regulation (VR)\%} = \frac{V_{NL} - V_{FL}}{V_{FL}} \times 100\% \qquad \textbf{(6.12)}$$

For ideal conditions, $V_{FL} = V_{NL}$ and VR% $= 0$. Therefore, *the smaller the voltage regulation, the less the variation in terminal voltage with change in load.*

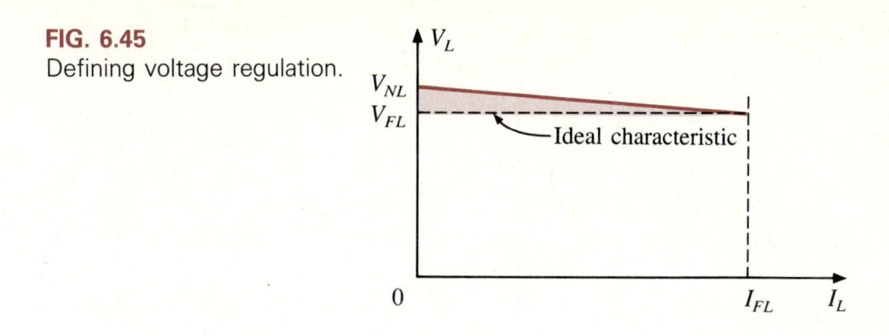

FIG. 6.45
Defining voltage regulation.

It can be shown by a short derivation that the voltage regulation is also given by

$$\text{VR\%} = \frac{R_{\text{int}}}{R_L} \times 100\% \tag{6.13}$$

In other words, the smaller the internal resistance for the same load, the smaller the regulation and more ideal the output.

EXAMPLE 6.20 Calculate the voltage regulation of a supply having the characteristics of Fig. 6.44.

Solution

$$\text{VR\%} = \frac{V_{NL} - V_{FL}}{V_{FL}} \times 100\% = \frac{124\text{ V} - 120\text{ V}}{120\text{ V}} \times 100\%$$

$$= \frac{4}{120} \times 100\% \cong \mathbf{3.33\%}$$

EXAMPLE 6.21 Determine the voltage regulation of the supply of Fig. 6.43.

Solution

$$\text{VR\%} = \frac{R_{\text{int}}}{R_L} \times 100\% = \frac{2\ \Omega}{13\ \Omega} \times 100\% \cong \mathbf{15.38\%}$$

6.9 INSTRUMENTATION

In Chapter 3, it was noted that ammeters are inserted in the branch in which the current is to be measured. We now realize that such a condition specifies that *ammeters be placed in series with the branch in which the current is to be measured,* as shown in Fig. 6.46(b) for the circuit of Fig. 6.46(a).

If the ammeter is to have minimal impact on the behavior of the network, its resistance should be very small (ideally 0 Ω) compared to the other series elements (R_1, R_2, and R_3) of the circuit of Fig. 6.46(b). It is also noteworthy

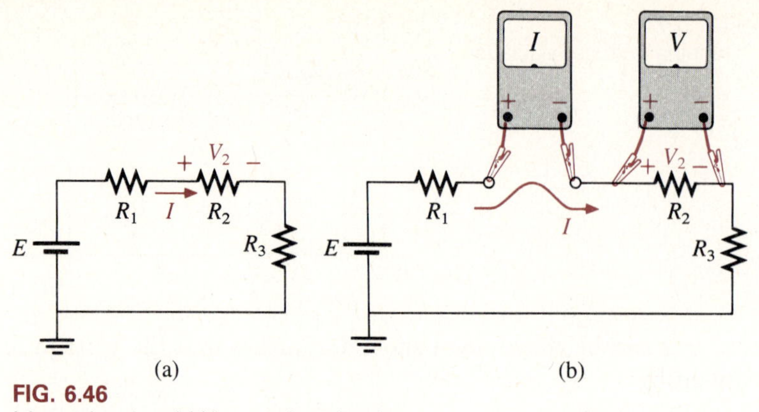

FIG. 6.46
Measuring I and V in a series circuit.

that the resistances of the separate current scales of the same meter are usually not the same. In fact, the meter resistance normally increases with decreasing current levels. However, for the majority of situations, we can simply assume that the internal ammeter resistance is small enough compared to the other circuit elements that it can be ignored.

Note in Fig. 6.46(b) that the voltmeter is placed "across" the resistor R_2 to measure V_2 with the positive (+) lead of the meter connected to the high side of the voltage V_2 and the negative (or common) lead to the other end for a positive, up-scale indication. We will find in the next chapter that the voltmeter and R_2 are in parallel and that the internal resistance of the meter should be much larger than R_2 to have minimum impact on the circuit.

In Fig. 6.47 a wattmeter has been applied to measure the power to the resistor R_2. Note the need to sense the current and voltage levels to R_2. In addition, an up-scale (positive) reading is obtained if the current I enters the (±) terminal of the current terminals and the high (+) side of V_2 is connected to the (±) terminal of the voltage terminals.

FIG. 6.47
Measuring the power to R_2.

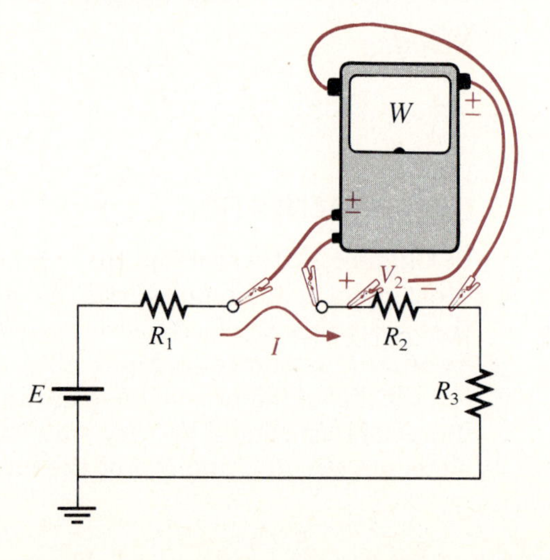

Figure 6.48 demonstrates how the voltage levels V_a and V_b are measured in a series circuit with a voltmeter. In each case the negative (common) terminal of the voltmeter is connected to ground potential (0 V) for an up-scale reading.

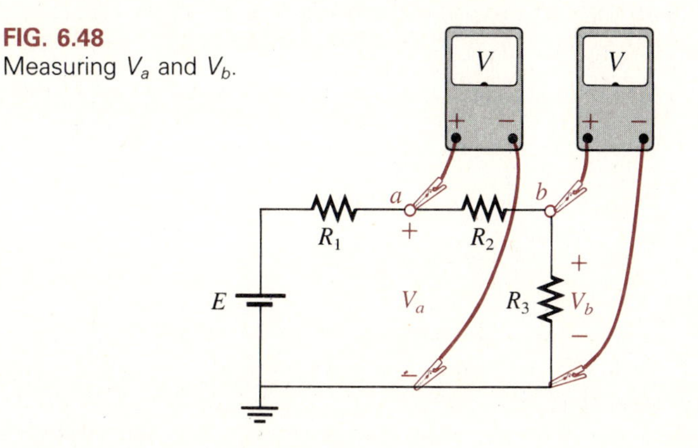

FIG. 6.48
Measuring V_a and V_b.

6.10 COMPUTER SOLUTIONS

The true value of computer methods can be clearly demonstrated by Program 6.1 of Fig. 6.50, which calculates every quantity of significance for the network of Fig. 6.49 once the battery voltage and resistance values have been entered.

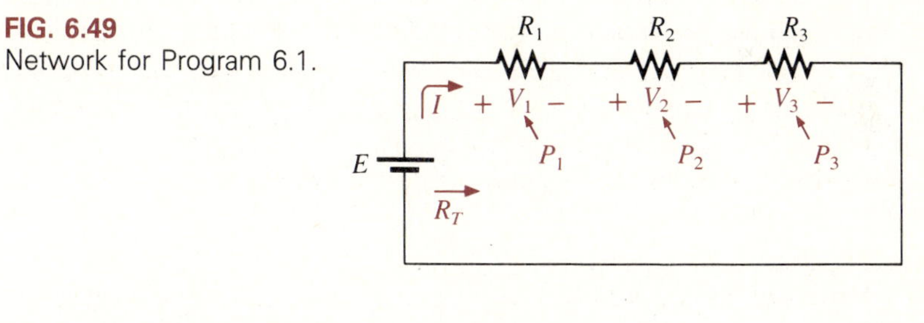

FIG. 6.49
Network for Program 6.1.

```
10 REM ***** PROGRAM 6-1 *****
20 REM ***********************************************
30 REM Analysis of a series resistor network
40 REM ***********************************************
50 REM
```

Body of Program:

```
100 PRINT:PRINT "Enter resistor values for up to 3 resistors"
110 PRINT "in series (enter 0 if no resistor):"
120 INPUT "R1=";R1
130 INPUT "R2=";R2
140 INPUT "R3=";R3
150 RT=R1+R2+R3
160 PRINT:PRINT "The total resistance is RT=";RT;"ohms"
170 PRINT:INPUT "Enter value of supply voltage, E=";E
180 I=E/RT
190 PRINT
200 PRINT "Supply current is, I=";I;"amperes"
```

```
     210 PRINT
     220 PRINT "The voltage drop across each resistor is:"
Vx   230 V1=I*R1:V2=I*R2 :V3=I*R3
     240 PRINT "V1=";V1;"volts   V2=";V2;"volts   V3=";V3;"volts"
Px   250 P1=I^2*R1 :P2=I^2*R2 :P3=I^2*R3
     260 PRINT
     270 PRINT "The power dissipated by each resistor is:"
     280 PRINT "P1=";P1;"watts","P2=";P2;"watts","P3=";P3;"watts"
     290 PRINT
     300 PRINT "Total voltage around loop is, V1+V2+V3=";V1+V2+V3;"volts"
     310 PRINT "and total power dissipated, P1+P2+P3=";P1+P2+P3;"watts"
     320 END
```

Program Runs:

```
READY

RUN


Enter resistor values for up to 3 resistors
in series (enter 0 if no resistor):

R1=? 6

R2=? 7

R3=? 5


The total resistance is RT= 18 ohms

Enter value of supply voltage, E=? 54


Supply current is, I= 3 amperes

The voltage drop across each resistor is:
V1= 18 volts   V2= 21 volts   V3= 15 volts

The power dissipated by each resistor is:
P1= 54 watts    P2= 63 watts    P3= 45 watts

Total voltage around loop is, V1+V2+V3= 54 volts
and total power dissipated, P1+P2+P3= 162 watts

READY

RUN


Enter resistor values for up to 3 resistors
in series (enter 0 if no resistor):


R1=? 1E3
```

```
R2=? 4E3

R3=? 0


The total resistance is RT= 5000 ohms

Enter value of supply voltage, E=? 50


Supply current is, I= .01 amperes

The voltage drop across each resistor is:
V1= 10 volts    V2= 40 volts    V3= 0 volts

The power dissipated by each resistor is:
P1= .1 watts    P2= .4 watts    P3= 0 watts

Total voltage around loop is, V1+V2+V3= 50 volts
and total power dissipated, P1+P2+P3= .5 watts


READY
```

FIG. 6.50
Program 6.1.

Note that R_T is calculated on line 150, I, on line 180, the voltage across each resistor, on line 230, and the power to each resistor, on line 250. In addition, Kirchhoff's voltage law is applied on line 300 to show that the applied voltage equals the sum of the voltage drops. The total power supplied or dissipated is then calculated on line 310.

In the second run of the program, only two resistors were entered, with R_3 given a resistance of 0 Ω (short circuit). Note that V_3 and P_3 are 0 V and 0 W, respectively.

FORMULA SUMMARY

$$R_T = R_1 + R_2 + R_3 + \cdots + R_N$$

$$R_T = NR$$

$$I_T = \frac{E}{R_T}$$

$$V_x = \frac{R_x E}{R_T}$$

$$V_L = E - I_L R_{\text{int}}$$

$$\text{VR\%} = \frac{V_{NL} - V_{FL}}{V_{FL}} \times 100\%$$

CHAPTER SUMMARY

This is the first chapter of a series that deal with the interconnection of elements in an electrical system. A firm knowledge of the concepts introduced in this and the next few chapters will establish a foundation that will carry through to the most difficult and complex system. Be absolutely sure the definitions introduced are read very carefully with the impact of each term correctly understood. For instance, keep in mind that one and only one end of two elements can be connected to satisfy the definition of two series elements. In addition, no other elements (whether they be resistors or batteries) can be connected to the junction point if the two elements are to be in series. Since the current is the same through series elements, we can check to see if the current is the same through two adjoining elements to determine if they are in series. This latter approach is true only if a third element is not connected to the junction point.

For series configurations, the current is *always* the same throughout the system, but the voltage will divide as the size of the resistors. The larger the resistor, the more of the applied voltage it will capture. If one resistor is twice another, it will have twice the voltage, and so on. As far as power is concerned, since the current is the same through series elements and $P = I^2R$, the power delivered to one resistor as compared to another is a direct function of the resistor's value.

Since ammeters are always connected in series with the elements through which the current is to be measured, the circuit must be *opened* at some point and the meter inserted. Keep the water analogy for current in mind to ensure that the meter is measuring the correct flow. Voltages can be measured *without* disturbing the circuit simply by placing the meter across the elements. An up-scale (or positive) indication is obtained only if the positive (+) lead of the meter is connected to the high (+) side of the potential drop across the load.

GLOSSARY

Branch The portion of a circuit consisting of one or more elements in series.

Circuit A combination of a number of elements joined at terminal points providing at least one closed path through which charge can flow.

Closed loop Any continuous connection of branches that allows tracing of a path that leaves a point in one direction and returns to that same point from another direction without leaving the circuit.

Conventional current flow A defined direction for the flow of charge in an electrical system that is opposite to that of the motion of electrons.

Electron flow The flow of charge in an electrical system having the same direction as the motion of electrons.

Internal resistance The inherent resistance found internal to any source of energy.

Kirchhoff's voltage law Law that states that the algebraic sum of the potential rises and drops around a closed loop (or path) is zero.

Series circuit A circuit configuration in which the elements have only one point in common and each terminal is not connected to a third, current-carrying element.

Voltage divider rule A method by which a voltage in a series circuit can be determined without first calculating the current in the circuit.

Voltage regulation (VR) A value, given as a percent, that provides an indication of the change in terminal voltage of a supply with change in load demand.

PROBLEMS

Section 6.2

1. Find the total resistance of the configurations of Fig. 6.51.

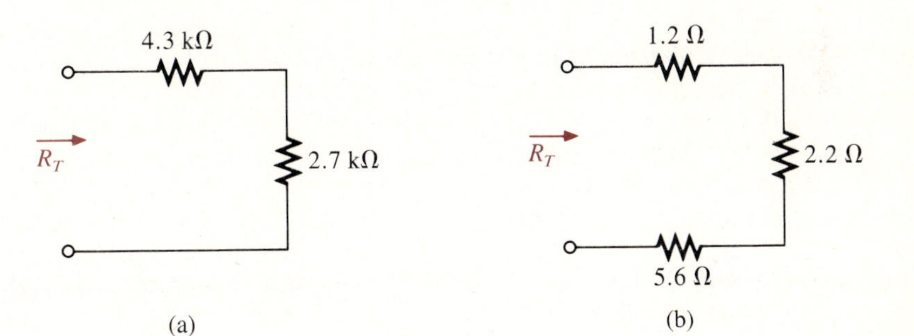

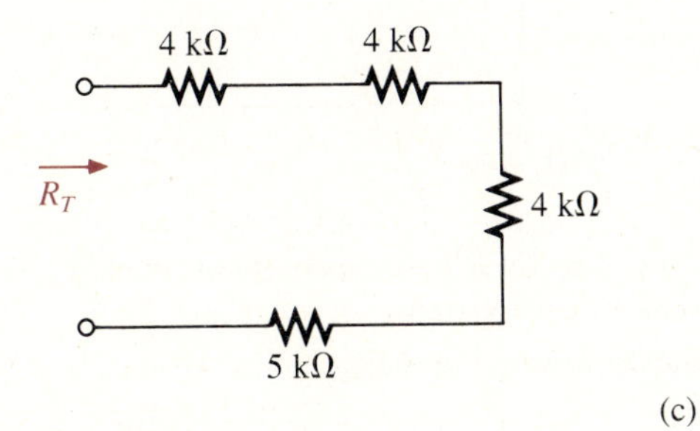

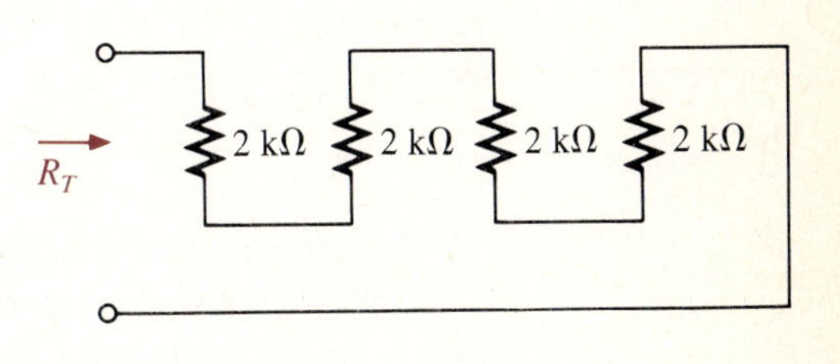

(c)

FIG. 6.51

2. For the circuit of Fig. 6.52, determine
   a. R_T
   b. I
   c. V_1 and V_2
   d. P_1 and P_2 (power to each resistor)
   e. P_{del} (power delivered by the source)

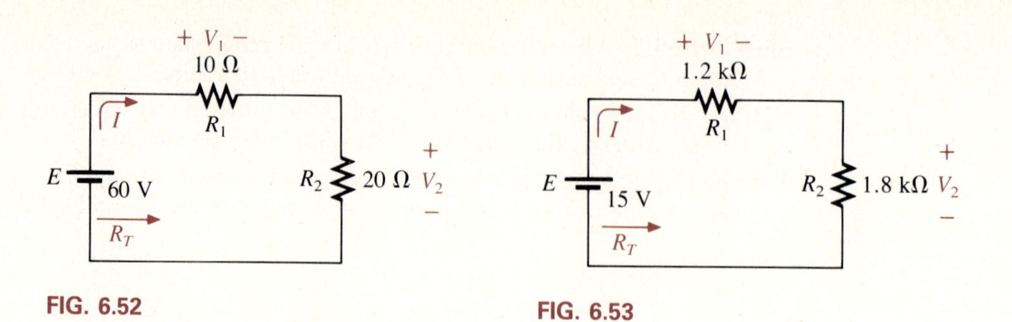

FIG. 6.52

FIG. 6.53

3. Repeat Problem 2 for the network of Fig. 6.53.
4. For the series circuit of Fig. 6.54, determine
   a. R_T
   b. I
   c. V_1
   d. P_2
5. For the series circuit of Fig. 6.55, determine
   a. R_T
   b. I
   c. V_3

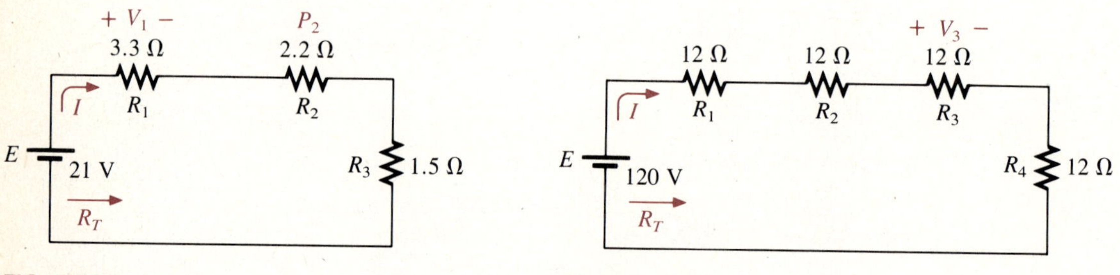

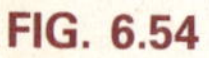
FIG. 6.54

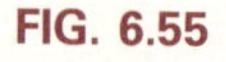
FIG. 6.55

6. For the circuit of Fig. 6.56, the total resistance is specified. Find the unknown resistance and the current in each circuit.
7. Repeat Problem 6 for the circuit of Fig. 6.57.

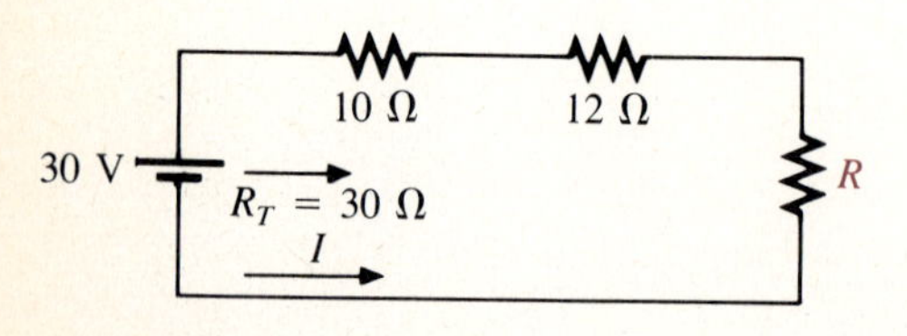

FIG. 6.56

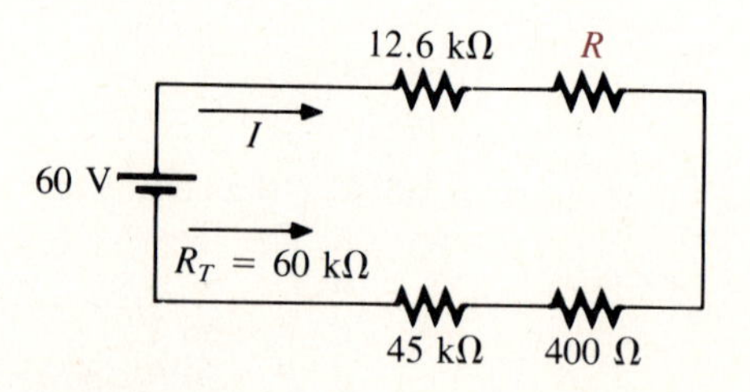

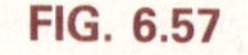
FIG. 6.57

8. Repeat Problem 6 for the circuit of Fig. 6.58.

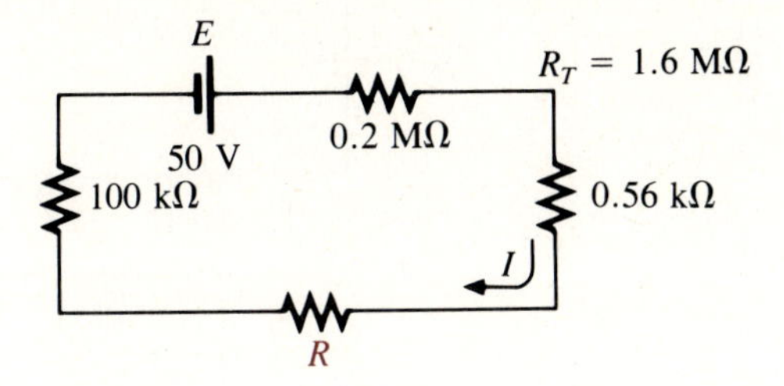

FIG. 6.58

9. Find the applied voltage E necessary to develop the current specified in each circuit of Fig. 6.59.

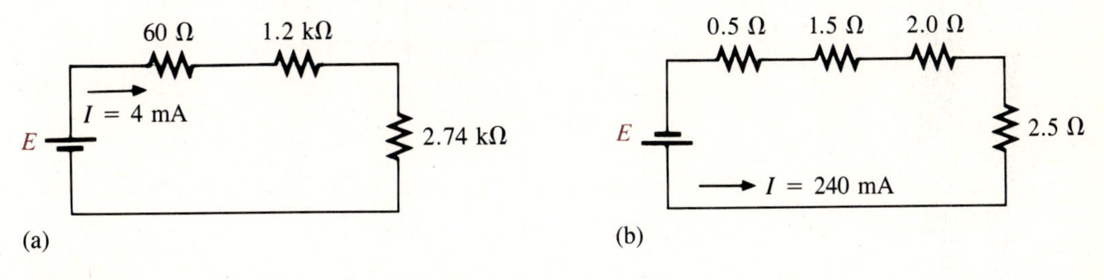

FIG. 6.59

*10. For each circuit of Fig. 6.60, determine the current I, the unknown resistance, and the voltage across each element.

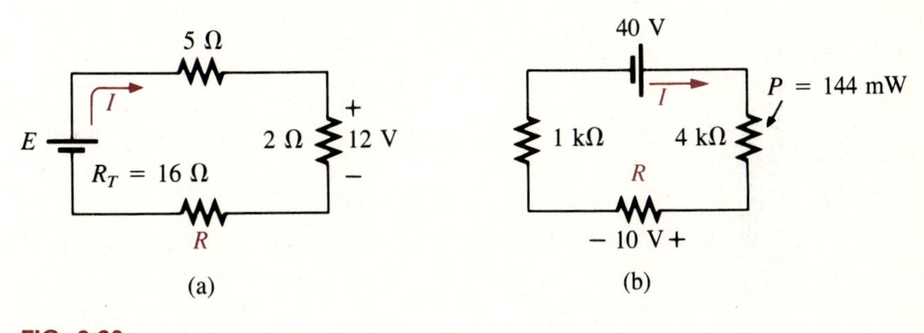

FIG. 6.60

Section 6.3

11. Determine the current I for each circuit of Fig. 6.61. Before solving for I, redraw each network with a single voltage source.

12. Find the voltage V for the circuits of Fig. 6.62.

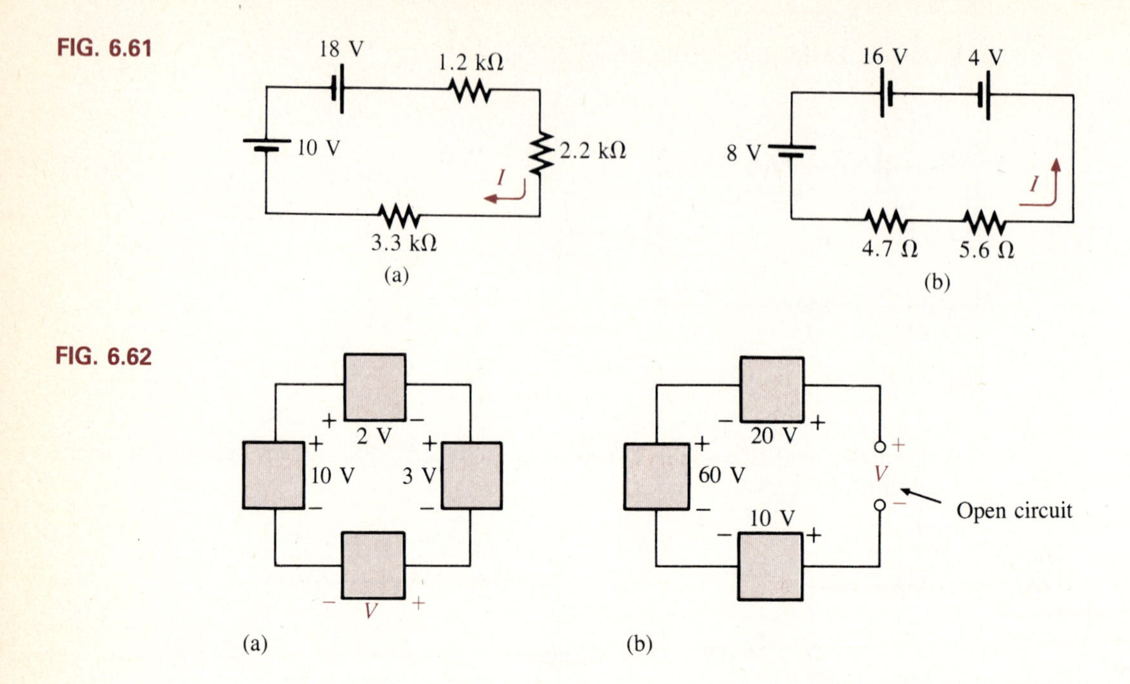

FIG. 6.61

FIG. 6.62

13. Find the voltage V_1 for the circuit of Fig. 6.63 using the given voltage levels.
14. Find the battery voltage E for the network of Fig. 6.64 using the given information.

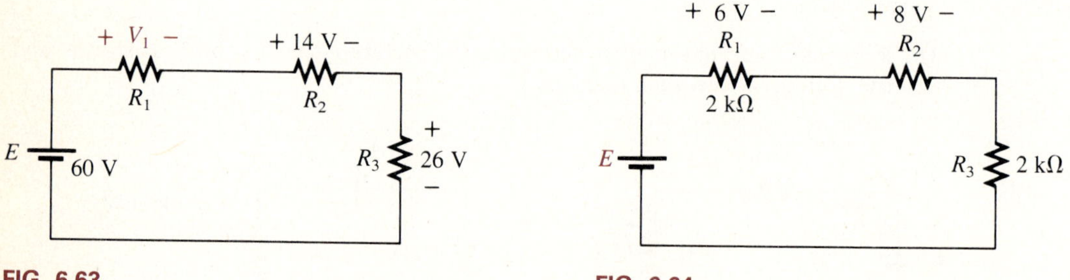

FIG. 6.63

FIG. 6.64

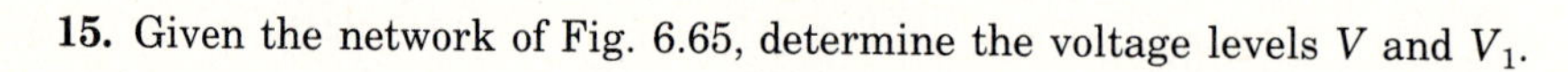
15. Given the network of Fig. 6.65, determine the voltage levels V and V_1.

FIG. 6.65

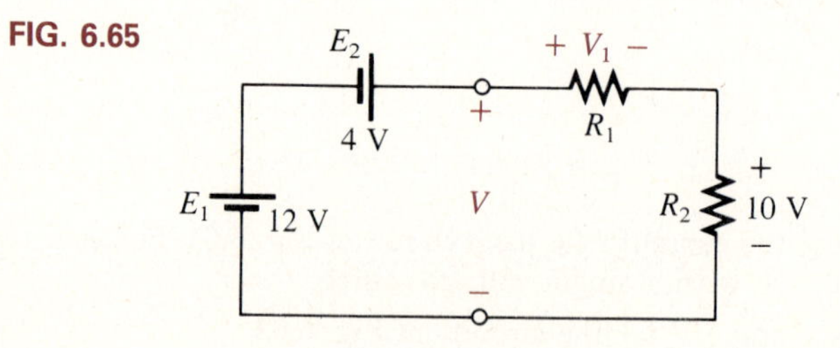

*16. Find the unknown quantities for the circuits of Fig. 6.66.

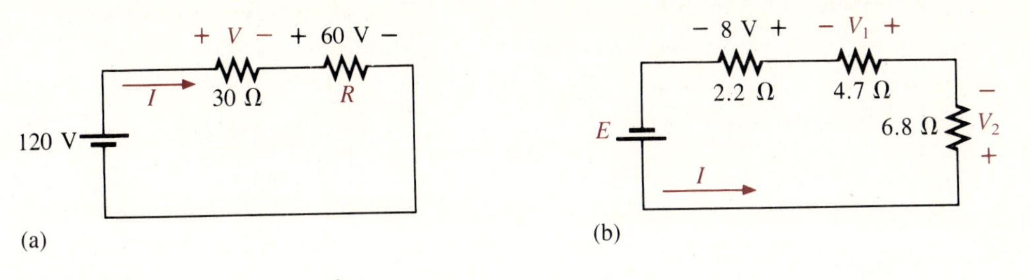

FIG. 6.66

*17. Find the unknown quantities for the circuits of Fig. 6.67.

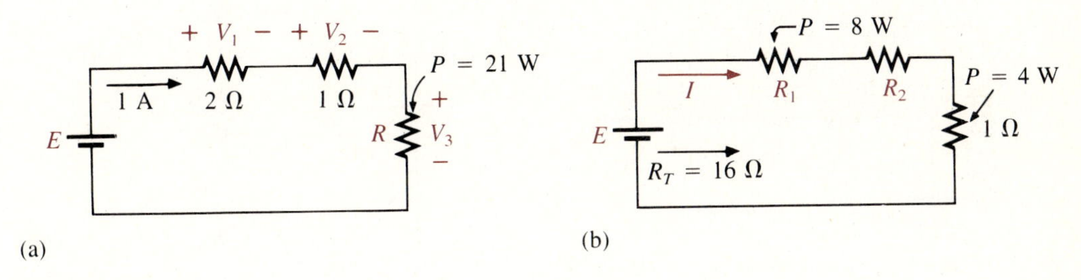

FIG. 6.67

Section 6.4

18. Using the voltage divider rule, find the voltage V_2 for the networks of Fig. 6.68.

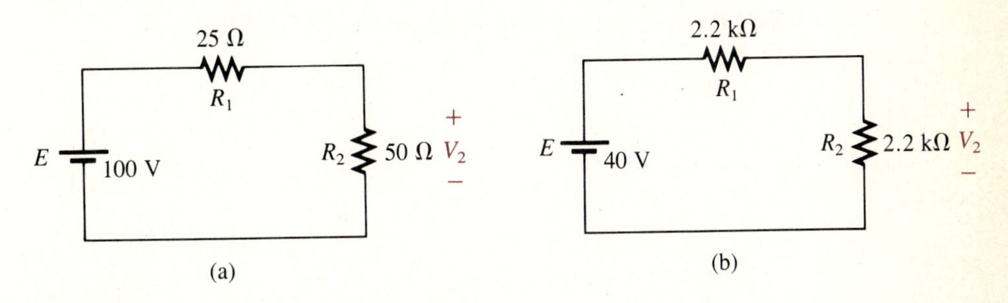

FIG. 6.68

19. Using the voltage divider rule, find the voltage V_2 for the networks of Fig. 6.69.

*20. Using the voltage divider rule, determine the resistance R_1 of Fig. 6.70 such that the voltage across R_1 is 40 V.

*21. Given the information appearing in Fig. 6.71, determine the voltages V_1, V_2, and E by inspection.

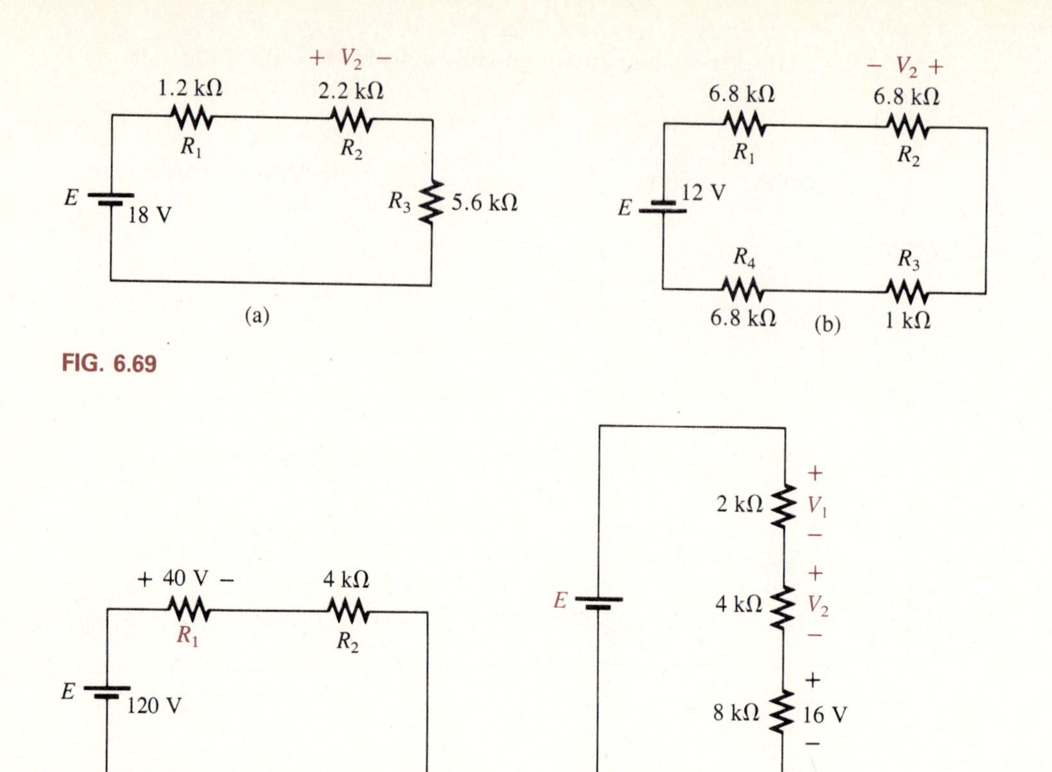

FIG. 6.69

FIG. 6.70

FIG. 6.71

Section 6.5

22. For the configuration of Fig. 6.72, find the source current I and the voltage V_2.

***23.** For the configuration of Fig. 6.73, find the voltages

a. V_a

b. V_b

c. V_{ab}

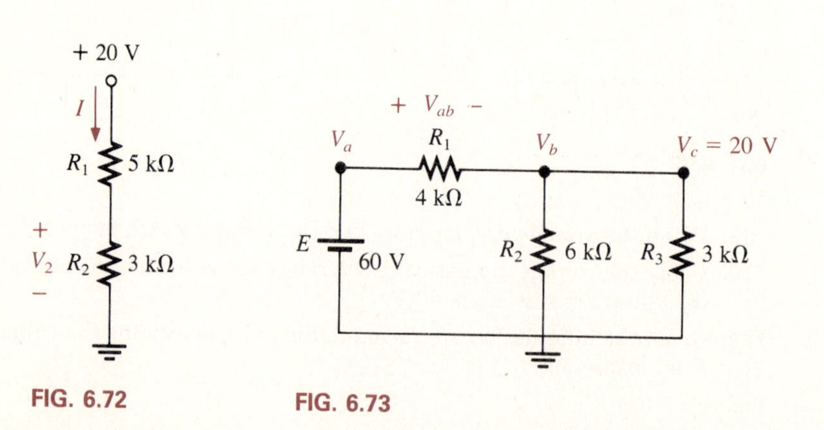

FIG. 6.72

FIG. 6.73

*24. For the two-source network of Fig. 6.74, find
  a. The voltages V_a and V
  b. The voltages V_b and V_{ab}
  c. The currents I_1 and I_2

*25. For the system of Fig. 6.75, determine
  a. The voltages V_a, V_b, and V_c
  b. The voltages V_{ac}, V_{ab}, and V_{bc}
  c. The currents I_1, I_2, and I_3

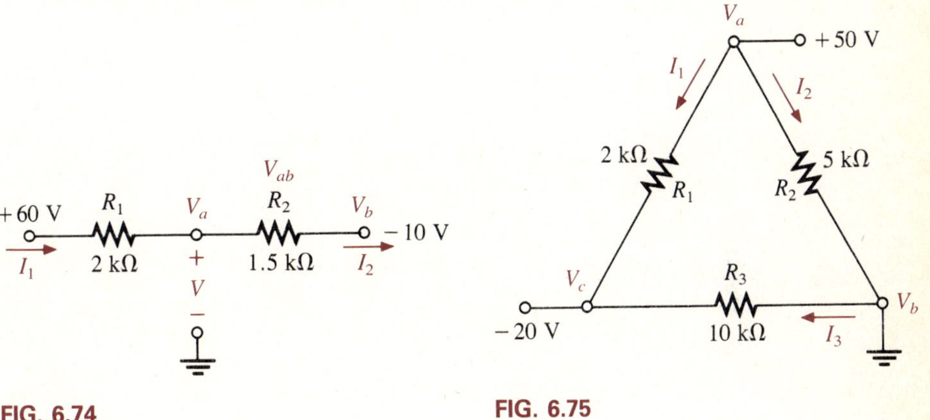

FIG. 6.74

FIG. 6.75

Section 6.6

26. For the battery of Fig. 6.76,
  a. Determine the no-load terminal voltage V_L.
  b. Determine the loaded (R_L = 33 kΩ) terminal voltage V_L.
  c. Determine the voltage lost across R_{int}.
  d. Determine the load current with R_L = 33 kΩ.
  e. Determine the power dissipated by R_{int}.
  f. Plot V_L versus I_L for R_L extending from ∞ Ω (no load) to 0 Ω (short-circuit across output terminals).

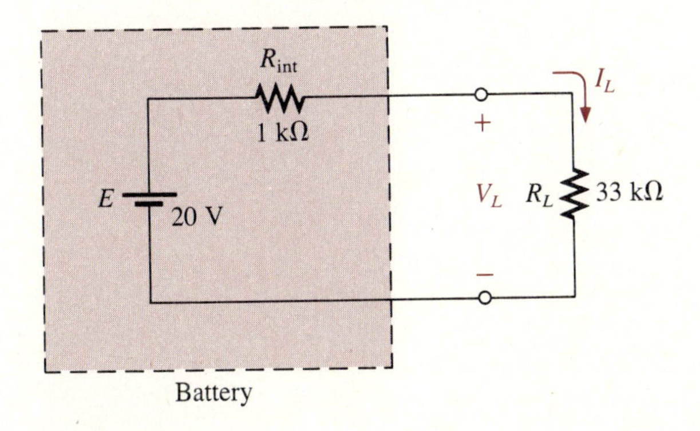

FIG. 6.76

27. For a dc generator having the characteristics of Fig. 6.45, where $V_{FL} = 110$ V, $V_{NL} = 112$ V, and $I_{FL} = 4$ A, determine
a. R_L (for full-load conditions)
b. R_{int}

28. Given the terminal characteristics appearing in Fig. 6.77, determine E, R_L, and R_{int}.

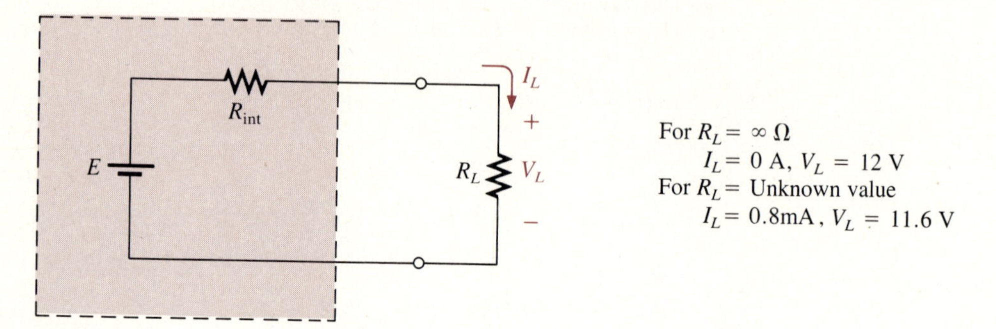

FIG. 6.77

Section 6.8

29. Determine the voltage regulation of a dc generator having the characteristics of Problem 27.

30. Repeat Problem 29 for the supply of Problem 28.

Computer Problems

31. Write a program to determine the total resistance of any number of resistors in series.

32. Write a program that will apply the voltage divider rule to either resistor of a series circuit with a single source and two series resistors.

33. Write a program to tabulate the current and power to the resistor R_L of the network of Fig. 6.41 for a range of values for R_L from 0 Ω to 20 Ω in increments of 1 Ω.

7

Parallel Circuits

OBJECTIVES

- ☐ Understand the conditions that must be satisfied for elements to be in parallel.
- ☐ Be able to calculate the total resistance of two or more parallel branches.
- ☐ Perform a complete analysis of a parallel configuration including the total resistance, source, and branch currents and power levels in the most efficient manner.
- ☐ Be able to apply Kirchhoff's current law.
- ☐ Learn the current divider rule and how to apply it to a parallel configuration with two or three parallel branches.
- ☐ Understand the practical use of placing voltage sources in parallel.
- ☐ Develop an awareness of how open and short circuits can affect the analysis of series or parallel networks.
- ☐ Understand the loading effect of voltmeters and ammeters on a parallel network.

7.1 INTRODUCTION

There are two network configurations that form the framework for some of the most complex network structures. A clear understanding of each will pay enormous dividends as more complex methods and networks are examined. The series connection was discussed in detail in the last chapter. We now examine the *parallel* connection and all the methods and laws associated with this important configuration.

7.2 PARALLEL ELEMENTS

Two elements, branches, or networks are in parallel if they have two points (and only two points) in common.

In Fig. 7.1(a), (b), and (c), for example, the two resistors have terminals a and b in common; the resistors are, therefore, in parallel. The electrical schematic of Fig. 7.1(b) is a closer representation of the actual connections, but Fig. 7.1(c) is the format normally employed.

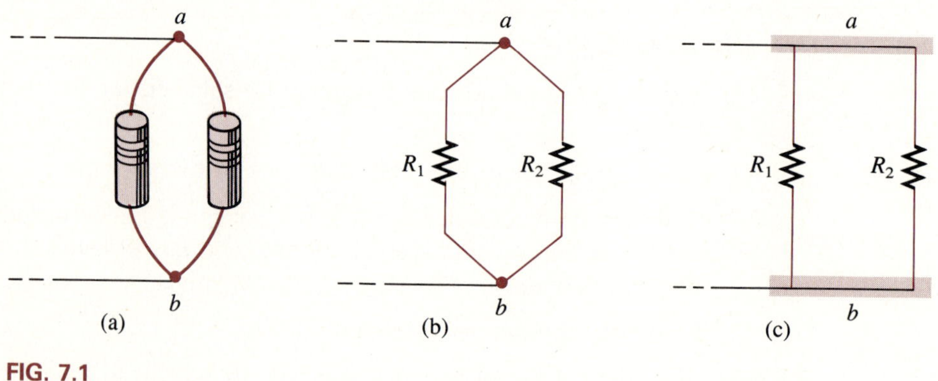

FIG. 7.1
Parallel elements.

In Fig. 7.2, all the elements are in parallel because they satisfy the preceding criterion. Two electrical schematics are provided to demonstrate how the parallel networks can be drawn. Do not let the squaring of the connections at the top and bottom of Fig. 7.2(b) and (c) cloud the fact that all the elements are connected to one terminal point at top and bottom, as shown in Fig. 7.2(a).

In Fig. 7.3, resistors R_1 and R_2 are in parallel because they have terminals a and b in common. The parallel combination of R_1 and R_2 is then in series with the resistor R_3 due to the common terminal point b.

In Fig. 7.4, resistors R_1 and R_2 are in series due to the common point c, but the series combination of R_1 and R_2 is in parallel with R_3, as defined by the common terminal connections at a and b.

Common examples of parallel elements include the rungs of a ladder, the tying of more than one rope between two points to increase the strength of the

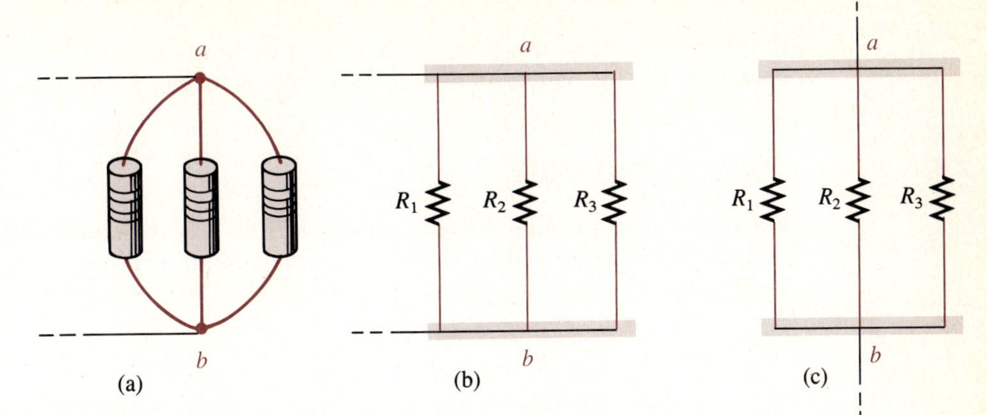

FIG. 7.2
Three parallel elements.

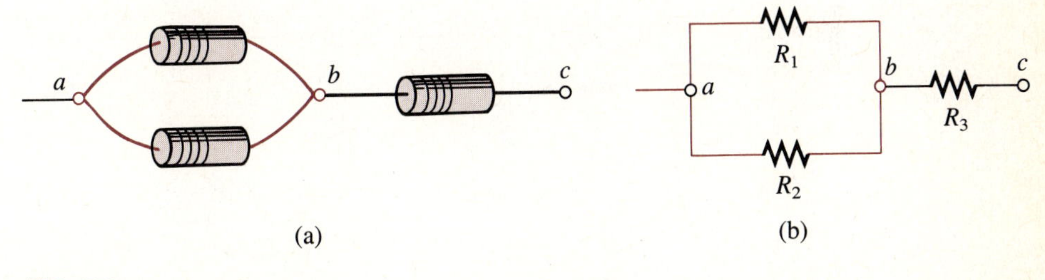

FIG. 7.3
Identifying parallel elements.

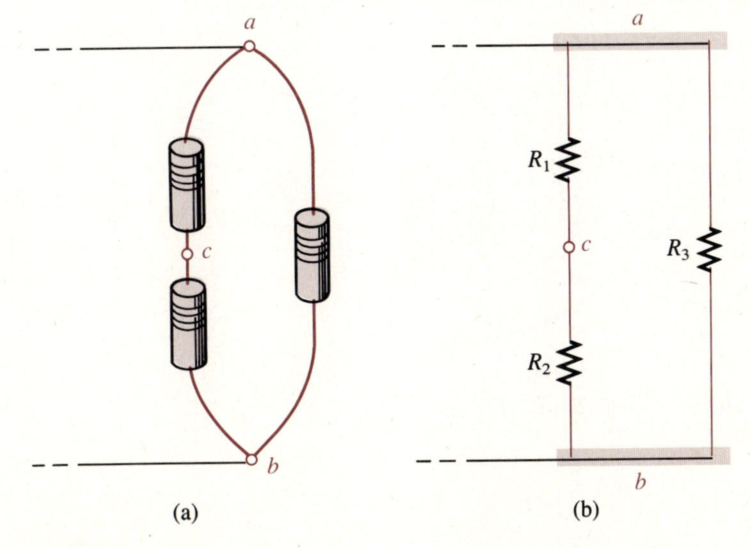

FIG. 7.4
Parallel branches.

connection, and the use of pipes between two points to split the water flowing between the two points at a ratio determined by the areas of the pipes.

7.3 TOTAL RESISTANCE

For resistors in parallel, such as shown in Fig. 7.5, the equation for the total resistance R_T measured between the two left-hand terminals of the system is given by

$$\frac{1}{R_T} = \frac{1}{R_1} + \frac{1}{R_2} + \cdots + \frac{1}{R_N} \tag{7.1}$$

where N is the number of parallel elements. Note that the equation is for 1 divided by the total resistance rather than for R_T directly.

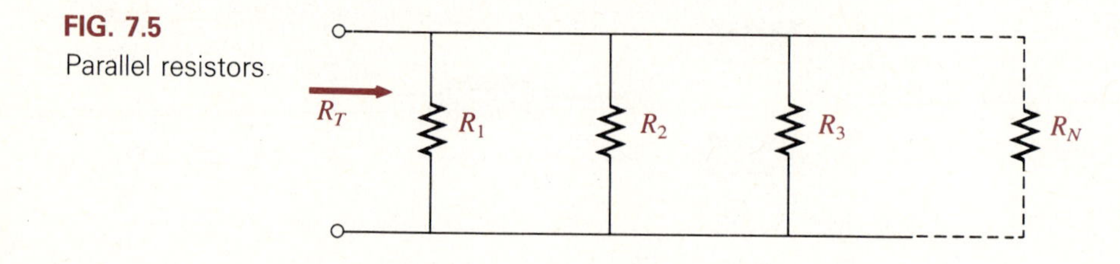

FIG. 7.5
Parallel resistors.

EXAMPLE 7.1 Determine the total resistance of the parallel network of Fig. 7.6(a) having the electrical schematic equivalent of Fig. 7.6(b).

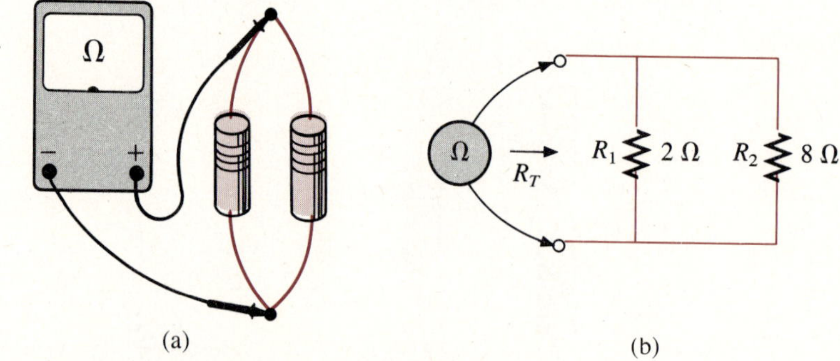

FIG. 7.6
Network for Example 7.1.

Solution First find $1/R_T$ using Eq. 7.1:

$$\frac{1}{R_T} = \frac{1}{R_1} + \frac{1}{R_2}$$

$$= \frac{1}{2\ \Omega} + \frac{1}{8\ \Omega} = 0.5\ \text{S} + 0.125\ \text{S} = \mathbf{0.625\ S}$$

Calculator

(2) (1/x) (+) (8) (1/x) (=)

with a display of 0.625.

Since

$$\frac{1}{R_T} = 0.625 \text{ S}$$

then

$$R_T = \frac{1}{0.625 \text{ S}} = \mathbf{1.6\ \Omega}$$

Calculator

(·) (6) (2) (5) (1/x) (=)

with a display of 1.6.

Note the need to add the fractional numbers in decimal form before finding R_T.

EXAMPLE 7.2 Determine the total resistance of the parallel network of Fig. 7.7.

FIG. 7.7
Network for Example 7.2.

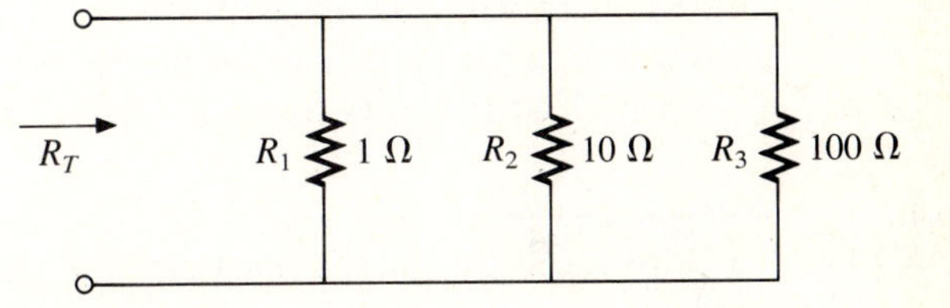

Solution Use Eq. 7.1 to find $1/R_T$:

$$\frac{1}{R_T} = \frac{1}{R_1} + \frac{1}{R_2} + \frac{1}{R_3}$$

$$= \frac{1}{1\ \Omega} + \frac{1}{10\ \Omega} + \frac{1}{100\ \Omega} = 1 \text{ S} + 0.1 \text{ S} + 0.01 \text{ S} = \mathbf{1.11\ S}$$

Calculator

(1) (1/x) (+) (1) (0) (1/x) (+) (1) (0) (0) (1/x) (=)

with a display of 1.11.

Then find R_T:

$$\frac{1}{R_T} = 1.11$$

$$R_T = \frac{1}{1.11} \cong \mathbf{0.901\ \Omega}$$

Calculator

[1] [.] [1] [1] [1/x] or [1] [÷] [1] [.] [1] [1] [=]

with a display of 0.901.

The last two examples clearly reveal a very interesting and useful (for checking purposes) characteristic of parallel resistors.

The total resistance of parallel resistors is *always* less than the value of the smallest resistor.

In Example 7.1 the total resistance is 1.6 Ω, compared with the value of 2 Ω for the smallest resistor, whereas the total resistance in Example 7.2 is 0.9 Ω, compared with the value of 1 Ω for the smallest resistor. This fact gives an excellent way to make an initial check on your calculations. The conclusion also permits a quick judgment (on an approximate basis) of the total resistance of a parallel network. Without any calculations, we know the total resistance of the network of Fig. 7.7 will be close to 1 Ω and that of Fig. 7.6 will be close to 2 Ω.

In addition, the wider the spread in numerical value between parallel resistors, the closer the total resistance will be to that of the smallest resistor. For instance, 1 Ω in parallel with 2 Ω has a total resistance of 2/3 Ω = 0.667 Ω, whereas 1 Ω in parallel with 10 Ω has 10/11 Ω ≅ 0.909 Ω and 1 Ω in parallel with 100 Ω has 100/101 Ω ≅ 0.99 Ω.

EXAMPLE 7.3

Determine the total resistance for the network of Fig. 7.8(a).

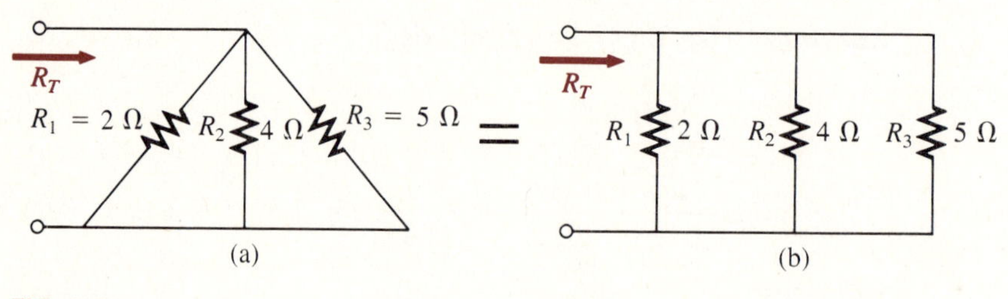

FIG. 7.8

Solution In this case note that redrawing the network as shown in Fig. 7.8(b) clarifies the analysis to be performed. Again, use Eq. 7.1:

$$\frac{1}{R_T} = \frac{1}{R_1} + \frac{1}{R_2} + \frac{1}{R_3}$$

$$= \frac{1}{2\ \Omega} + \frac{1}{4\ \Omega} + \frac{1}{5\ \Omega} = 0.5\ \text{S} + 0.25\ \text{S} + 0.2\ \text{S}$$

$$\frac{1}{R_T} = \mathbf{0.95\ S}$$

Calculator

(2) (1/x) (+) (4) (1/x) (+) (5) (1/x) (=)

with a display of 0.95.

Therefore,

$$R_T = \frac{1}{0.95\text{ S}} \cong \mathbf{1.053\ \Omega}$$

Calculator

(·) (9) (5) (1/x) or (1) (÷) (·) (9) (5) (=)

with a display of 1.053.

For *equal* resistors in parallel, the total resistance can be determined by

$$\boxed{R_T = \frac{R}{N}} \tag{7.2}$$

where N is the number of resistors with value R in parallel. In other words, the total resistance of N equal parallel resistors is the resistance of *one* resistor divided by the number (N) of parallel elements.

EXAMPLE 7.4 Determine the total resistance of the network of Fig. 7.9.

FIG. 7.9
Network for Example 7.4.

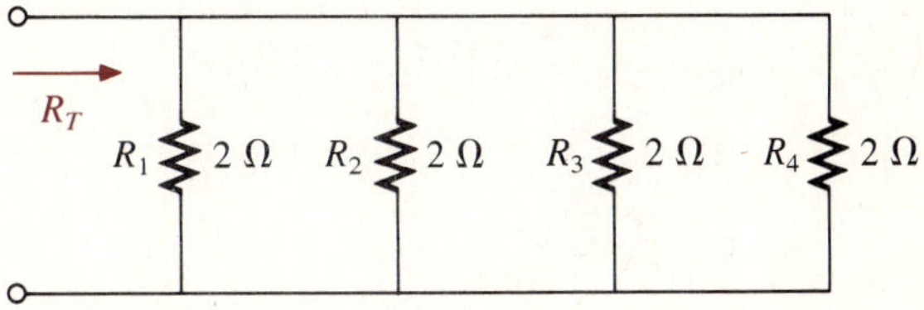

Solution $N = 4$ (four 2-Ω resistors in parallel). So, using Eq. 7.2:

$$R_T = \frac{R}{N} = \frac{2\ \Omega}{4} = \mathbf{0.5\ \Omega}$$

EXAMPLE 7.5 Determine the total resistance of the network of Fig. 7.10.

FIG. 7.10
Network for Example 7.5.

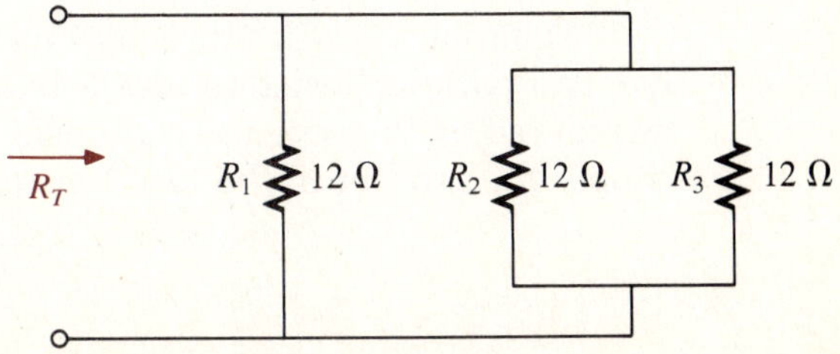

Solution For clarity, the network is redrawn as shown in Fig. 7.11.

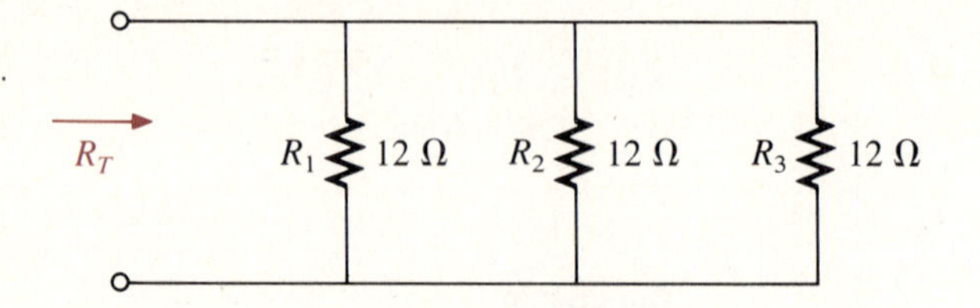

FIG. 7.11
Redrawn network of Fig. 7.10.

Here, $N = 3$ (three 12-Ω resistors in parallel). Using Eq. 7.2:

$$R_T = \frac{R}{N} = \frac{12\ \Omega}{3} = \mathbf{4\ \Omega}$$

In the vast majority of situations, only two or three parallel resistive elements need be combined. With this in mind, the following equations were developed to simplify the computation when determining R_T.

For two parallel resistors, we write

$$\frac{1}{R_T} = \frac{1}{R_1} + \frac{1}{R_2}$$

Multiplying the top and bottom of each term of the right side of the equation by the other resistor results in

$$\frac{1}{R_T} = \left(\frac{R_2}{R_2}\right)\frac{1}{R_1} + \left(\frac{R_1}{R_1}\right)\frac{1}{R_2} = \frac{R_2}{R_1R_2} + \frac{R_1}{R_1R_2}$$

$$= \frac{R_2 + R_1}{R_1R_2}$$

Hence,

$$\boxed{R_T = \frac{R_1R_2}{R_1 + R_2}} \tag{7.3}$$

In words,

The total resistance of two parallel resistors is the product of the two resistances divided by their sum.

Equation 7.3 will prove very useful in the analysis of parallel elements. For two parallel resistors, there is no longer a need to worry about errors introduced by the inverse relationships of Eq. 7.1. The total resistance is now simply the product of the two resistors divided by the sum.

EXAMPLE 7.6

Determine the total resistance of the network of Fig. 7.12.

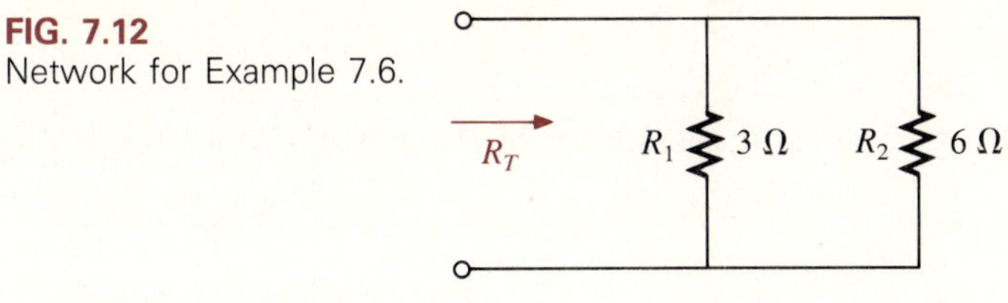

FIG. 7.12
Network for Example 7.6.

Solution Using Eq. 7.3:

$$R_T = \frac{R_1 R_2}{R_1 + R_2} = \frac{(3\ \Omega)(6\ \Omega)}{3\ \Omega + 6\ \Omega}$$

$$= \frac{18}{9}\ \Omega = \mathbf{2\ \Omega}$$

Calculator

$R_1 + R_2$: [3] [+] [6] [=] with a display of 9

R_T: [3] [×] [6] [÷] [9] [=] with a display of 2

You can also use parentheses to group the addition in the denominator:

$$R_T = \frac{(3\ \Omega)(6\ \Omega)}{(3\ \Omega + 6\ \Omega)}$$

R_T: [3] [×] [6] [÷] [(] [3] [+] [6] [)] [=]

with a display of 2.

EXAMPLE 7.7

Repeat Example 7.1 using Eq. 7.3.

Solution

$$R_T = \frac{R_1 R_2}{R_1 + R_2} = \frac{(2\ \Omega)(8\ \Omega)}{2\ \Omega + 8\ \Omega} = \frac{16}{10}\ \Omega$$

$$= \mathbf{1.6\ \Omega}$$

as obtained earlier.

For three parallel resistors, the equation becomes

$$R_T = \frac{R_1 R_2 R_3}{R_1 R_2 + R_1 R_3 + R_2 R_3} \tag{7.4}$$

with a numerator equal to the product of the three resistors and a denominator that includes all the possible product combinations taken two resistors at a time.

EXAMPLE 7.8

Determine the total resistance of the parallel network of Fig. 7.13.

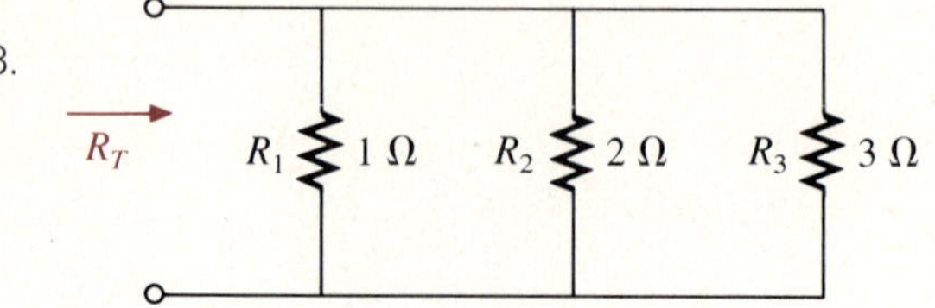

FIG. 7.13
Network for Example 7.8.

Solution Using Eq. 7.4:

$$R_T = \frac{R_1R_2R_3}{R_1R_2 + R_1R_3 + R_2R_3}$$
$$= \frac{(1\ \Omega)(2\ \Omega)(3\ \Omega)}{(1\ \Omega)(2\ \Omega) + (1\ \Omega)(3\ \Omega) + (2\ \Omega)(3\ \Omega)}$$
$$R_T = \frac{6}{2 + 3 + 6}\ \Omega = \frac{6}{11}\ \Omega \cong \mathbf{0.545\ \Omega}$$

Calculator

Using parentheses, R_T can be expressed as

$$R_T = \frac{1 \times 2 \times 3}{(1 \times 2 + 1 \times 3 + 2 \times 3)}$$

1 × 2 × 3 ÷ (1 × 2 + 1 × 3 + 2 × 3) =

with a display of 0.545.

EXAMPLE 7.9

Repeat Example 7.2 and compare the results.

Solution Using Eq. 7.4:

$$R_T = \frac{R_1R_2R_3}{R_1R_2 + R_1R_3 + R_2R_3}$$
$$= \frac{(1\ \Omega)(10\ \Omega)(100\ \Omega)}{(1\ \Omega)(10\ \Omega) + (1\ \Omega)(100\ \Omega) + (10\ \Omega)(100\ \Omega)}$$
$$= \frac{1000\ \Omega}{10 + 100 + 1000} = \frac{1000}{1110}\ \Omega$$
$$R_T \cong \mathbf{0.901\ \Omega}$$

as obtained earlier.

Recall that series elements can be interchanged without affecting the magnitude of the total resistance or current.

In parallel networks, parallel elements can be interchanged without changing the total resistance or input current.

In other words, redrawing the network with the parallel elements in a different order can often make the analysis simpler and more direct.

EXAMPLE 7.10 Determine the total resistance of the network of Fig. 7.14.

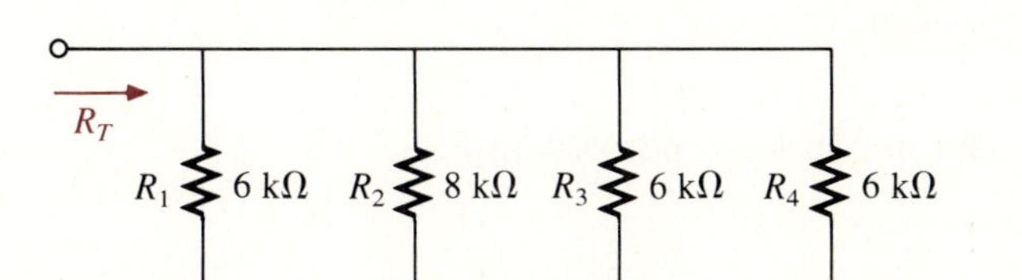

FIG. 7.14
Parallel network for Example 7.10.

Solution Redrawing the network as shown in Fig. 7.15 reveals three equal resistors of 6 kΩ, permitting the use of Eq. 7.2:

$$R'_T = \frac{R}{N} = \frac{6\ \text{k}\Omega}{3} = 2\ \text{k}\Omega$$

This now results in the configuration of Fig. 7.16. Using Eq. 7.3 with this network gives

$$R_T = \frac{R_2 R'_T}{R_2 + R'_T} = \frac{(8\ \text{k}\Omega)(2\ \text{k}\Omega)}{8\ \text{k}\Omega + 2\ \text{k}\Omega}$$

$$= \frac{16}{10}\ \text{k}\Omega = \mathbf{1.6\ k\Omega}$$

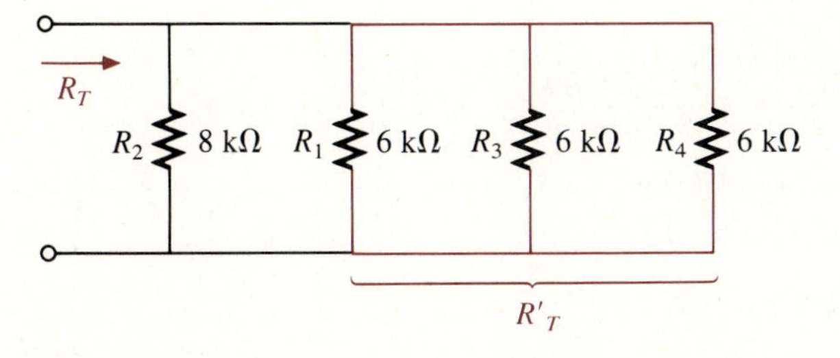

FIG. 7.15
Redrawn network of Fig. 7.14.

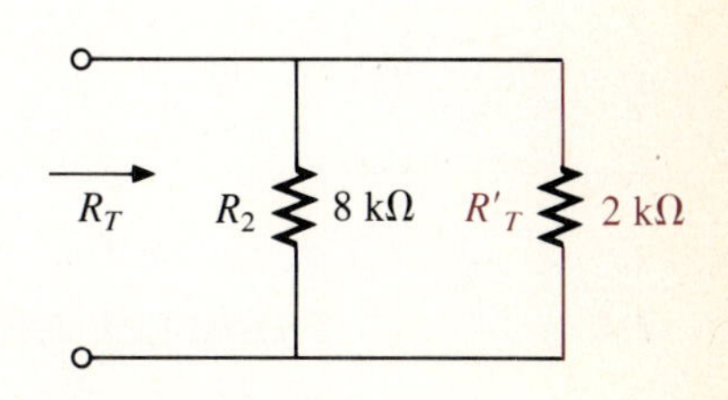

FIG. 7.16
Redrawn network of Fig. 7.15.

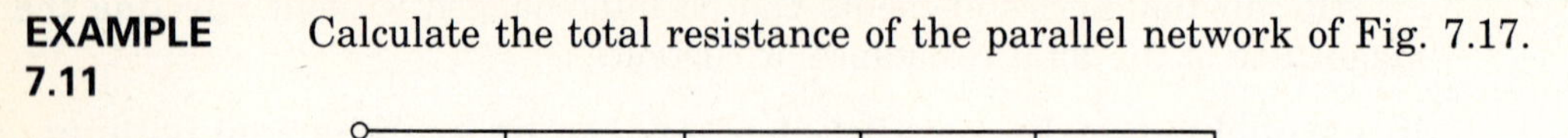

EXAMPLE 7.11 Calculate the total resistance of the parallel network of Fig. 7.17.

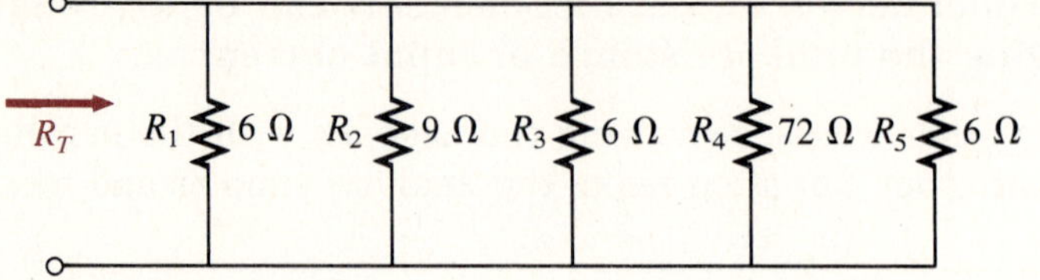

FIG. 7.17
Network for Example 7.11.

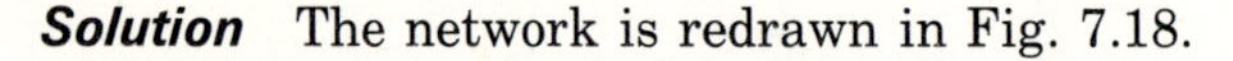

Solution The network is redrawn in Fig. 7.18.

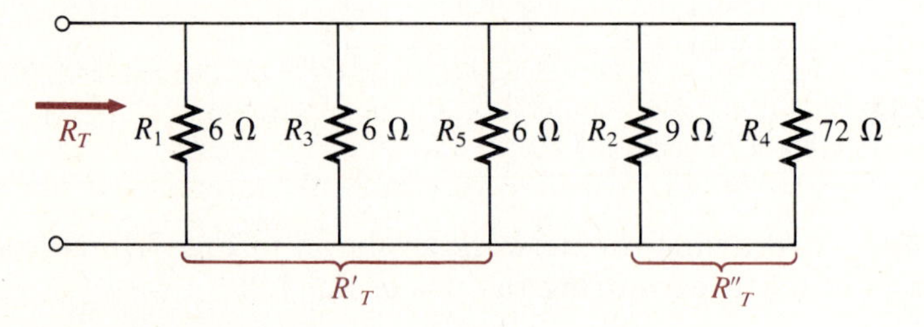

FIG. 7.18
Redrawn network of Fig. 7.17.

$$R'_T = \frac{R}{N} = \frac{6\ \Omega}{3} = 2\ \Omega$$

$$R''_T = \frac{R_1R_2}{R_1 + R_2} = \frac{(9\ \Omega)(72\ \Omega)}{9\ \Omega + 72\ \Omega} = \frac{648\ \Omega}{81} = 8\ \Omega$$

Finally,

$$R_T = R'_T \parallel R''_T \quad (\text{in parallel with})$$

$$= \frac{R'_TR''_T}{R'_T + R''_T} = \frac{(2\ \Omega)(8\ \Omega)}{2\ \Omega + 8\ \Omega} = \frac{16\ \Omega}{10} = \mathbf{1.6\ \Omega}$$

7.4 PARALLEL NETWORKS

The network of Fig. 7.19 is the simplest of parallel networks. All the elements have terminals a and b in common. The total resistance is determined by $R_T = R_1R_2/(R_1 + R_2)$, and the source or total current is determined by

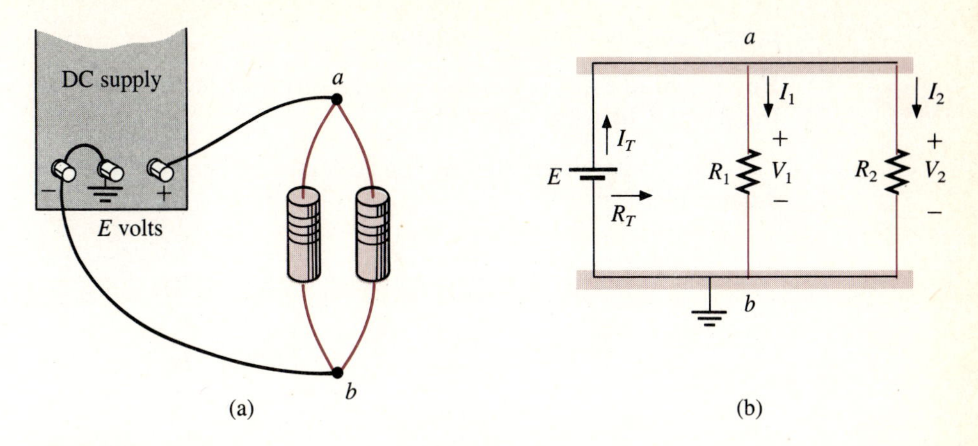

FIG. 7.19
Parallel network. (a) Construction; (b) electrical schematic.

$I_T = E/R_T$. Since the terminals of the battery are connected directly across the resistors R_1 and R_2, the following should be fairly obvious:

The voltage across parallel elements is the same.

Using this fact results in

$$V_1 = V_2 = E$$

and

$$I_1 = \frac{V_1}{R_1} = \frac{E}{R_1}$$

with

$$I_2 = \frac{V_2}{R_2} = \frac{E}{R_2}$$

If we take the equation for the total resistance and multiply both sides by the applied voltage, we obtain

$$E\left(\frac{1}{R_T}\right) = E\left(\frac{1}{R_1} + \frac{1}{R_2}\right)$$

and

$$\frac{E}{R_T} = \frac{E}{R_1} + \frac{E}{R_2}$$

Substituting the preceding Ohm's law relationships, we find that

$$\boxed{I_T = I_1 + I_2} \qquad \textbf{(7.5)}$$

permitting the following conclusion:

For parallel networks, the source current is equal to the sum of the individual branch currents.

The power dissipated by the resistors and delivered by the source can be determined from

$$P_1 = V_1I_1 = I_1^2R_1 = \frac{V_1^2}{R_1}$$

$$P_2 = V_2I_2 = I_2^2R_2 = \frac{V_2^2}{R_2}$$

$$P_s = EI_T = I_T^2R_T = \frac{E^2}{R_T}$$

EXAMPLE 7.12

For the parallel network of Fig. 7.20:

a. Calculate R_T.
b. Determine I_T.
c. Calculate I_1 and I_2 and demonstrate that $I_T = I_1 + I_2$.
d. Determine the power to each resistive load.
e. Determine the power delivered by the source and compare to the total power dissipated by the resistive elements.

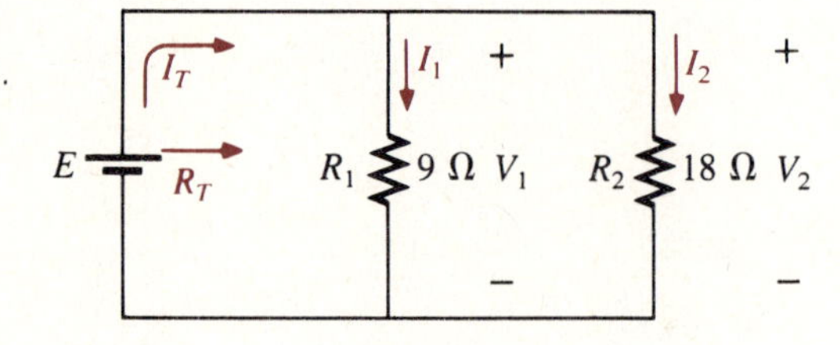

FIG. 7.20
Network for Example 7.12.

Solutions

a. $R_T = \dfrac{R_1R_2}{R_1 + R_2} = \dfrac{(9\ \Omega)(18\ \Omega)}{9\ \Omega + 18\ \Omega} = \dfrac{162\ \Omega}{27} = \mathbf{6\ \Omega}$

b. $I_T = \dfrac{E}{R_T} = \dfrac{27\text{ V}}{6\ \Omega} = \mathbf{4.5\text{ A}}$

c. $I_1 = \dfrac{V_1}{R_1} = \dfrac{E}{R_1} = \dfrac{27\text{ V}}{9\ \Omega} = \mathbf{3\text{ A}}$

$I_2 = \dfrac{V_2}{R_2} = \dfrac{E}{R_2} = \dfrac{27\text{ V}}{18\ \Omega} = \mathbf{1.5\text{ A}}$

$I_T = I_1 + I_2$
$4.5\text{ A} = 3\text{ A} + 1.5\text{ A}$
$4.5\text{ A} = 4.5\text{ A}$ (checks)

d. $P_1 = V_1I_1 = EI_1 = (27\text{ V})(3\text{ A}) = \mathbf{81\ W}$
$P_2 = V_2I_2 = EI_2 = (27\text{ V})(1.5\text{ A}) = \mathbf{40.5\ W}$
e. $P_s = EI_T = (27\text{ V})(4.5\text{ A}) = \mathbf{121.5\ W}$
$P_s = P_1 + P_2 = 81\text{ W} + 40.5\text{ W} = \mathbf{121.5\ W}$

EXAMPLE 7.13

Given the information provided in Fig. 7.21:

a. Determine R_3.
b. Calculate E.
c. Find I_T.
d. Find I_2.
e. Determine P_2.

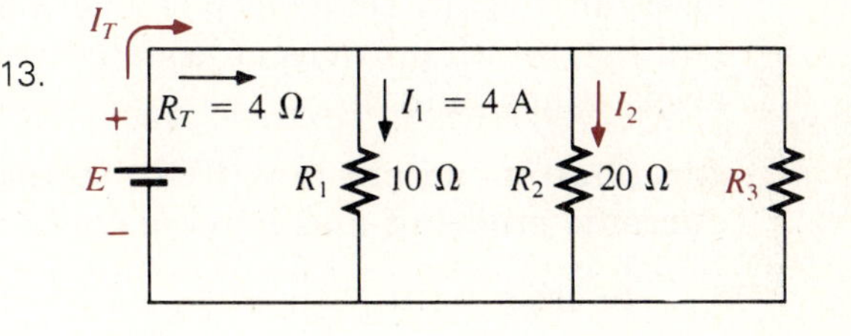

FIG. 7.21
Network for Example 7.13.

Solutions

a.
$$\frac{1}{R_T} = \frac{1}{R_1} + \frac{1}{R_2} + \frac{1}{R_3}$$
$$\frac{1}{4\ \Omega} = \frac{1}{10\ \Omega} + \frac{1}{20\ \Omega} + \frac{1}{R_3}$$
$$0.25\text{ S} = 0.1\text{ S} + 0.05\text{ S} + \frac{1}{R_3}$$
$$0.25\text{ S} = 0.15\text{ S} + \frac{1}{R_3}$$

Subtracting 0.15 S from both sides results in

$$\frac{1}{R_3} = 0.1\text{ S}$$

so

$$R_3 = \frac{1}{0.1\text{ S}} = \mathbf{10\ \Omega}$$

b. $E = V_1 = I_1R_1 = (4\text{ A})(10\ \Omega) = \mathbf{40\ V}$

c. $I_T = \dfrac{E}{R_T} = \dfrac{40\text{ V}}{4\ \Omega} = \mathbf{10\ A}$

d. $I_2 = \dfrac{V_2}{R_2} = \dfrac{E}{R_2} = \dfrac{40\text{ V}}{20\ \Omega} = \mathbf{2\ A}$

e. $P_2 = I_2^2 R_2 = (2\text{ A})^2(20\ \Omega) = \mathbf{80\ W}$

Calculator

with a display of 80.

7.5 KIRCHHOFF'S CURRENT LAW

Kirchhoff's voltage law provides an important relationship between voltage levels around any closed loop of a network. We now consider Kirchhoff's current law, which provides an equally important relationship between current levels at any junction.

Kirchhoff's current law (KCL) states that the algebraic sum of the currents entering and leaving a junction is zero.

In other words,

The sum of the currents entering a junction must equal the sum of the currents leaving the junction.

In equation form:

$$\boxed{\Sigma I_{\text{entering}} = \Sigma I_{\text{leaving}}} \quad \textbf{(7.6)}$$

In Fig. 7.22,

$$\begin{aligned} \Sigma I_{\text{entering}} &= \Sigma I_{\text{leaving}} \\ 6\text{ A} &= 2\text{ A} + 4\text{ A} \\ 6\text{ A} &= 6\text{ A} \quad \text{(checks)} \end{aligned}$$

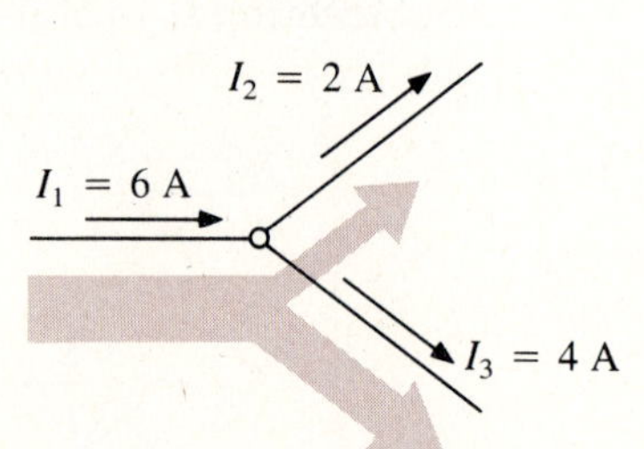

FIG. 7.22
Demonstrating Kirchhoff's current law.

In the next two examples unknown currents can be determined by applying Kirchhoff's current law. Simply remember to place all current levels entering a junction to the left of the equals sign and the sum of all currents leaving a junction to the right of the equals sign. The water-in-the-pipe analogy is an excellent one for supporting and clarifying the preceding law. Quite obviously, the sum total of the water entering a junction must leave through the remaining pipes.

EXAMPLE 7.14

Determine the current I_4 of Fig. 7.23 using Kirchhoff's current law.

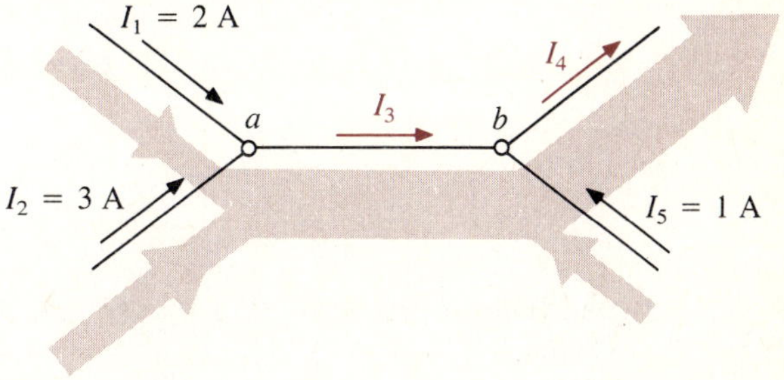

FIG. 7.23
Figure for Example 7.14.

Solution We must first work with junction a, since the only unknown is I_3. At junction b there are two unknowns and both cannot be determined from one application of the law.

At a:

$$\begin{aligned} \Sigma I_{\text{entering}} &= \Sigma I_{\text{leaving}} \\ I_1 + I_2 &= I_3 \\ 2\text{ A} + 3\text{ A} &= I_3 \\ I_3 &= \mathbf{5\text{ A}} \end{aligned}$$

At b:

$$\begin{aligned} \Sigma I_{\text{entering}} &= \Sigma I_{\text{leaving}} \\ I_3 + I_5 &= I_4 \\ 5\text{ A} + 1\text{ A} &= I_4 \\ I_4 &= \mathbf{6\text{ A}} \end{aligned}$$

EXAMPLE 7.15

Determine I_1, I_3, I_4, and I_5 for the network of Fig. 7.24.

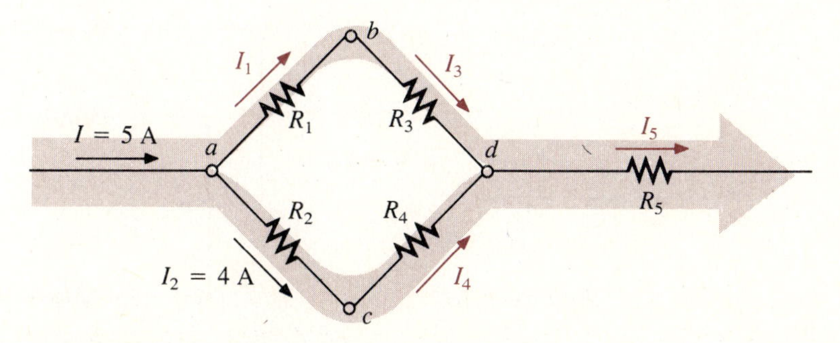

FIG. 7.24
Network for Example 7.15.

Solution At a:

$$\Sigma I_{\text{entering}} = \Sigma I_{\text{leaving}}$$
$$I = I_1 + I_2$$
$$5\text{ A} = I_1 + 4\text{ A}$$

Subtracting 4 A from both sides gives

$$5\text{ A} - 4\text{ A} = I_1 + \cancel{4\text{ A}} - \cancel{4\text{ A}}$$
$$I_1 = 5\text{ A} - 4\text{ A} = \mathbf{1\text{ A}}$$

At b:

$$\Sigma I_{\text{entering}} = \Sigma I_{\text{leaving}}$$
$$I_1 = I_3 = \mathbf{1\text{ A}}$$

as it should, since R_1 and R_3 are in series and the current is the same in series elements.

At c:

$$I_2 = I_4 = \mathbf{4\text{ A}}$$

for the same reasons given for junction b.

At d:

$$\Sigma I_{\text{entering}} = \Sigma I_{\text{leaving}}$$
$$I_3 + I_4 = I_5$$
$$1\text{ A} + 4\text{ A} = I_5$$
$$I_5 = \mathbf{5\text{ A}}$$

If we enclose the entire network, we find that the current entering is $I = 5$ A; the net current leaving from the far right is $I_5 = 5$ A. The two must be equal, since the net current entering any system must equal that leaving.

EXAMPLE 7.16 Determine the currents I_3 and I_5 of Fig. 7.25 through applications of Kirchhoff's current law.

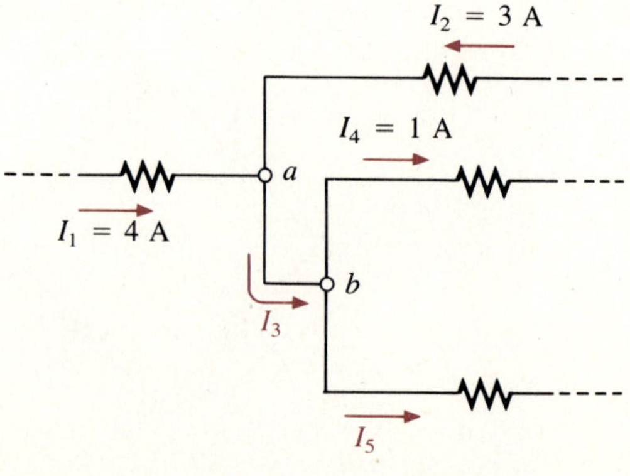

FIG. 7.25
Network for Example 7.16.

Solution Note that since point b has two unknown quantities and a only one, we must first apply Kirchhoff's current law to a. The result can then be applied to junction b. For a,

$$I_1 + I_2 = I_3$$
$$4\text{ A} + 3\text{ A} = I_3$$

and

$$I_3 = \mathbf{7\text{ A}}$$

For b,

$$I_3 = I_4 + I_5$$
$$7\text{ A} = 1\text{ A} + I_5$$

and

$$I_5 = 7\text{ A} - 1\text{ A} = \mathbf{6\text{ A}}$$

EXAMPLE 7.17

Find the magnitude and direction of the currents I_3, I_4, I_6, and I_7 for the network of Fig. 7.26. Even though the elements are not simply in series or parallel, Kirchhoff's current law can be applied to determine all the unknown currents.

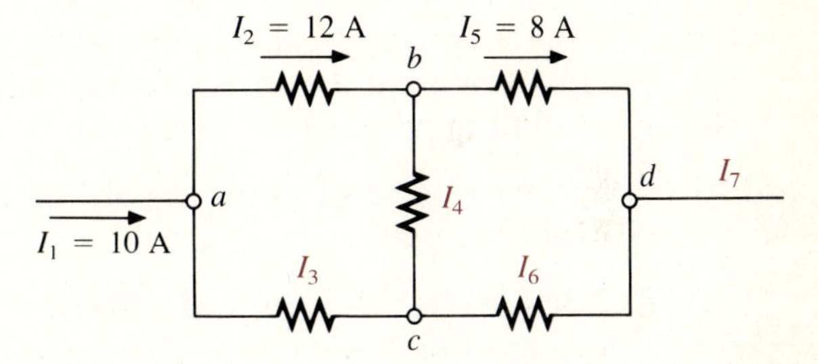

FIG. 7.26
Network for Example 7.17.

Solution Considering the overall system, we know that the current entering must equal that leaving. Therefore,

$$I_7 = I_1 = \mathbf{10\text{ A}}$$

Since 10 A are entering point a and 12 A are leaving, I_3 must be supplying current to the node. Applying Kirchhoff's current law at a,

$$I_1 + I_3 = I_2$$
$$10\text{ A} + I_3 = 12\text{ A}$$

and

$$I_3 = 12\text{ A} - 10\text{ A} = \mathbf{2\text{ A}}$$

At b, since 12 A are entering and 8 A are leaving, I_4 must be leaving. Therefore,

$$I_2 = I_4 + I_5$$
$$12\text{ A} = I_4 + 8\text{ A}$$

and

$$I_4 = 12\text{ A} - 8\text{ A} = \mathbf{4\text{ A}}$$

At c, I_3 is leaving at 2 A and I_4 is entering at 4 A, requiring that I_6 be leaving. Applying Kirchhoff's current law at c,

$$I_4 = I_3 + I_6$$
$$4\text{ A} = 2\text{ A} + I_6$$

and

$$I_6 = 4\text{ A} - 2\text{ A} = \mathbf{2\text{ A}}$$

As a check at d,

$$I_5 + I_6 = I_7$$
$$8\text{ A} + 2\text{ A} = 10\text{ A}$$
$$10\text{ A} = 10\text{ A} \quad \text{(checks)}$$

7.6 CURRENT DIVIDER RULE

As the name suggests, the *current divider rule* (CDR) determines how the current entering a set of parallel branches is split between the elements. Before examining the law in detail, there are a few conclusions that one should keep in mind:

For two parallel elements of equal value, the entering current will divide equally.

For parallel elements of different values, the *smaller* the resistance, the greater the current through that branch. Current always seeks the path of least resistance.

In Fig. 7.27 the current I will split into I_1 and I_2 and then return to its full magnitude at the other end of the parallel branches.

The *current divider rule* states that

note the difference in subscripts

$$I_1 = \frac{R_2 I}{R_1 + R_2} \tag{7.7}$$

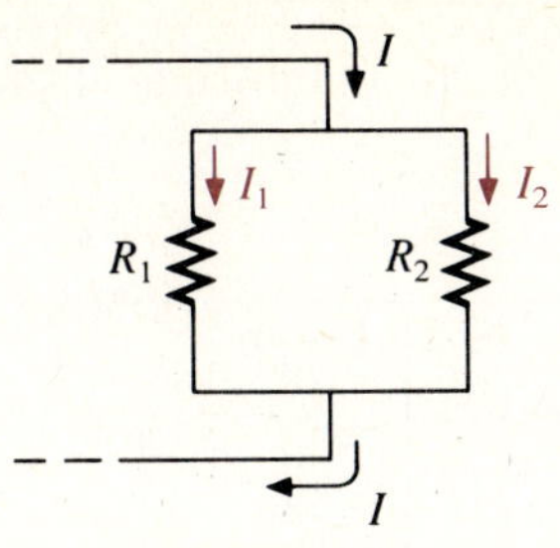

FIG. 7.27
Applying the current divider rule.

and

$$I_2 = \frac{R_1 I}{R_1 + R_2} \tag{7.8}$$

Note in each case that the resistor in the numerator has the opposite subscript as the current to be determined. The remainder of each equation is exactly the same. In words,

For two parallel branches, the current through either branch is equal to the product of the *other* parallel resistor and the input current divided by the *sum* (not the total parallel resistance) of the two parallel resistances.

EXAMPLE 7.18 Determine the current I_2 for the network of Fig. 7.28 using the current divider rule.

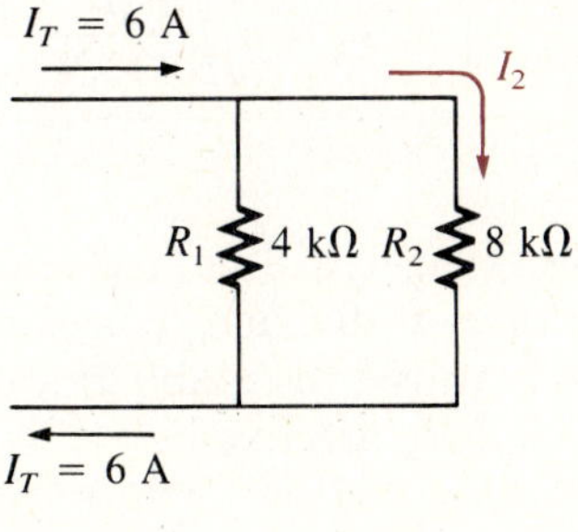

FIG. 7.28
Network for Example 7.18.

$I_T = 6$ A, I_2, R_1 4 kΩ, R_2 8 kΩ, $I_T = 6$ A

Solution

$$I_2 = \frac{R_1 I_T}{R_1 + R_2} = \frac{(4\text{ k}\Omega)(6\text{ A})}{4\text{ k}\Omega + 8\text{ k}\Omega} = \frac{4}{12}(6\text{ A}) = \frac{1}{3}(6\text{ A})$$
$$= \mathbf{2\ A}$$

EXAMPLE 7.19 Determine the magnitude of the currents I_1 and I_2 for the network of Fig. 7.29.

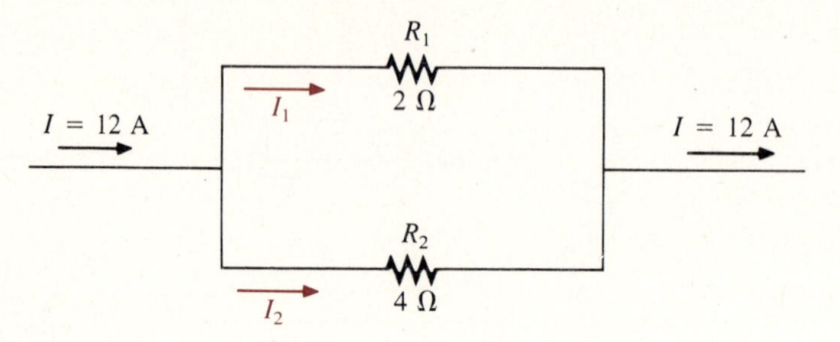

FIG. 7.29
Network for Example 7.19.

Solution By Eq. 7.7, the current divider rule,

$$I_1 = \frac{R_2 I}{R_1 + R_2} = \frac{(4\,\Omega)(12\text{ A})}{2\,\Omega + 4\,\Omega} = \mathbf{8\text{ A}}$$

Applying Kirchhoff's current law,

$$I = I_1 + I_2$$

so

$$I_2 = I - I_1 = 12\text{ A} - 8\text{ A} = \mathbf{4\text{ A}}$$

or, using the current divider rule, Eq. 7.8 again:

$$I_2 = \frac{R_1 I}{R_1 + R_2} = \frac{(2\,\Omega)(12\text{ A})}{2\,\Omega + 4\,\Omega} = \mathbf{4\text{ A}}$$

On occasion it will be necessary to find the resistance levels to effect a particular current division. For example, in Fig. 7.30 assume that resistance level R_1 is required to establish a current I_1 through the resistance R_1. The levels of I and R_2 are fixed.

FIG. 7.30
Determining R_1 and R_2.

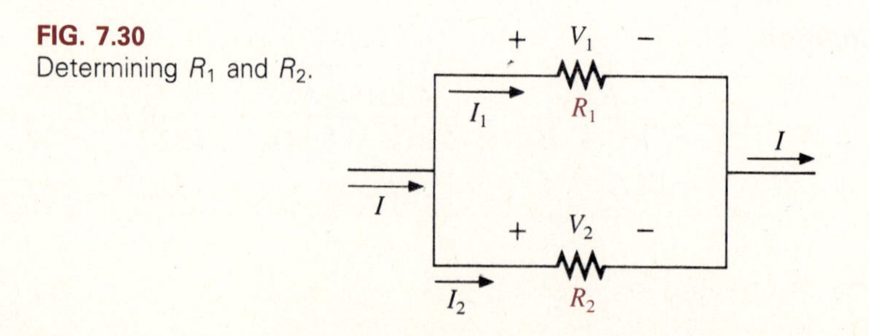

$$\begin{aligned} &\textit{Step 1:} & I_2 &= I - I_1 \\ &\textit{Step 2:} & V_2 &= I_2R_2 \\ &\textit{Step 3:} & V_1 &= V_2 \\ &\textit{Step 4:} & R_1 &= \frac{V_1}{I_1} \quad \text{(solution)} \end{aligned}$$

The resistance R_1 can also be determined through algebraic manipulations of Eq. 7.7:

$$\boxed{R_1 = \frac{R_2}{I_1}(I - I_1)} \tag{7.9}$$

Similarly for R_2,

$$\boxed{R_2 = \frac{R_1}{I_2}(I - I_2)} \tag{7.10}$$

EXAMPLE 7.20 Determine the resistance R_1 to effect the division of current in Fig. 7.31.

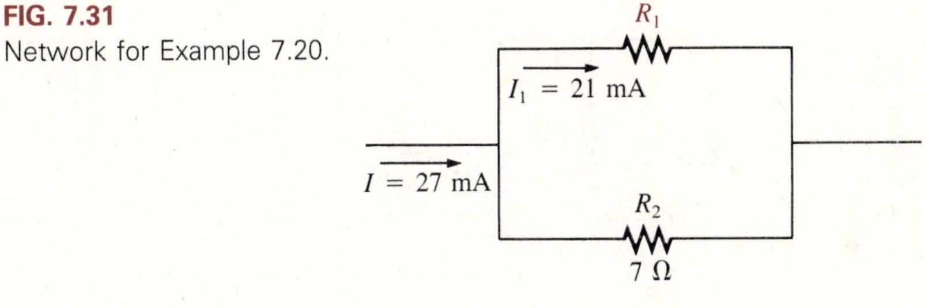

FIG. 7.31
Network for Example 7.20.

Solution

$$\begin{aligned} &\textit{Step 1:} & I_2 &= I - I_1 = 27\text{ mA} - 21\text{ mA} = 6\text{ mA} \\ &\textit{Step 2:} & V_2 &= I_2R_2 = (6\text{ mA})(7\ \Omega) = 42\text{ mV} \\ &\textit{Step 3:} & V_1 &= V_2 = 42\text{ mV} \\ &\textit{Step 4:} & R_1 &= \frac{V_1}{I_1} = \frac{42\text{ mV}}{21\text{ mA}} = \mathbf{2\ \Omega} \end{aligned}$$

Using Eq. 7.9:

$$\begin{aligned} R_1 &= \frac{R_2}{I_1}(I - I_1) \\ &= \frac{7\ \Omega}{21\text{ mA}}(27\text{ mA} - 21\text{ mA}) = \frac{7\ \Omega}{21\text{ mA}}(6\text{ mA}) \\ R_1 &= \frac{42}{21}\ \Omega = \mathbf{2\ \Omega} \end{aligned}$$

Calculator

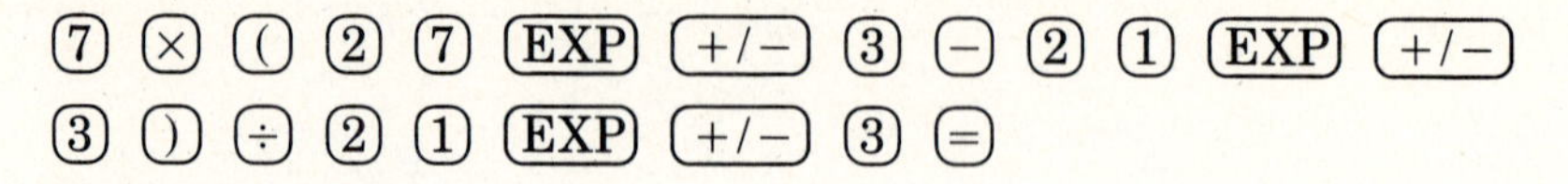

with a display of 2.0.

Obviously, the second solution seems simpler and more direct, but Eq. 7.9 is one that will not be used that often and may be difficult to remember. In addition, the sequence of steps of the first solution utilizes some very basic relationships of electronic circuits that cannot be forgotten or misunderstood if any expertise in this field is to be developed. From the examples just described, note the following:

1. More current passes through the smaller of two parallel resistors.
2. The current entering any number of parallel resistors divides as the inverse ratio of their ohmic values. This relationship is depicted in Fig. 7.32.

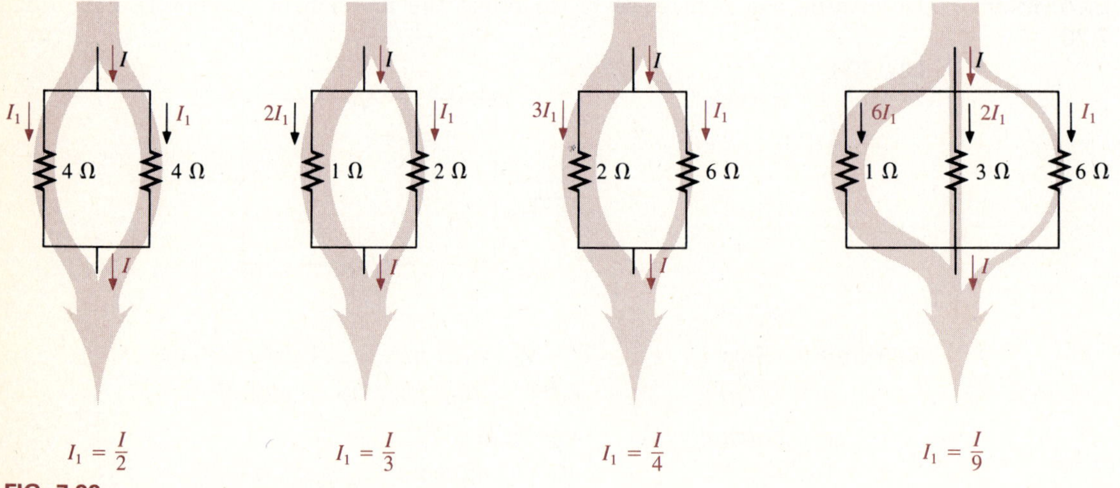

FIG. 7.32
Current division through parallel branches.

For parallel systems with more than two parallel resistors, as shown in Fig. 7.33, the current through any branch can be determined using the following equation:

$$\boxed{I_x = \frac{R_T}{R_x} I} \tag{7.11}$$

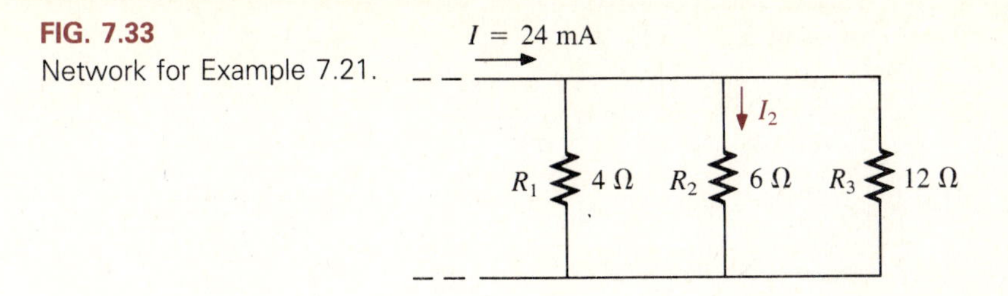

FIG. 7.33
Network for Example 7.21.

where x is the branch in which the current is to be determined and R_T is the total resistance of the parallel network. This equation can also be applied to two parallel resistors but Eqs. 7.7 and 7.8 are more direct. The following example demonstrates the use of Eq. 7.11.

EXAMPLE 7.21

Determine the current I_2 of Fig. 7.33 using the current divider rule.

Solution The parallel resistance of 6 Ω in parallel with 12 Ω is

$$R'_T = 6\ \Omega \| 12\ \Omega = \frac{(6\ \Omega)(12\ \Omega)}{6\ \Omega + 12\ \Omega}$$

$$= \frac{72\ \Omega}{18} = 4\ \Omega$$

and

$$R_T = R_1 \| R'_T = 4\ \Omega \| 4\ \Omega = \frac{4\ \Omega}{2} = 2\ \Omega$$

or

$$\frac{1}{R_T} = \frac{1}{R_1} + \frac{1}{R_2} + \frac{1}{R_3} = \frac{1}{4\ \Omega} + \frac{1}{6\ \Omega} + \frac{1}{12\ \Omega}$$

$$= 0.25\ \mathrm{S} + 0.167\ \mathrm{S} + 0.083\ \mathrm{S}$$

$$\frac{1}{R_T} = 0.5\ \mathrm{S}$$

and

$$R_T = \frac{1}{0.5\ \mathrm{S}} = 2\ \Omega$$

Calculator

(4) (1/x) (+) (6) (1/x) (+) (1) (2) (1/x) (=) (1/x)

with a display of 2.0.

Using Eq. 7.11:

$$I_x = \frac{R_T}{R_x} I$$

$$I_2 = \frac{R_T}{R_2} I = \frac{(2\ \Omega)(24\ \text{mA})}{(6\ \Omega)}$$

$$= \frac{48\ \text{mA}}{6} = \mathbf{8\ mA}$$

Calculator

(2) (×) (2) (4) (EXP) (+/−) (3) (=) (÷) (6) (=) (F <−> E)

with a display of 8.−03.

Since $R_3 = 12\ \Omega$ is twice $R_2 = 6\ \Omega$, we can deduce that $I_3 = (1/2)(8\ \text{mA}) = 4\ \text{mA}$. Since $R_1 = 4\ \Omega$ is 1/3 of $R_3 = 12\ \Omega$, we can surmise that $I_1 = 3(4\ \text{mA}) = 12\ \text{mA}$, always keeping in mind that the heaviest current exists in the smallest resistance levels. Checking our results with Kirchhoff's current law, we find

$$I = I_1 + I_2 + I_3$$
$$24\ \text{mA} = 12\ \text{mA} + 8\ \text{mA} + 4\ \text{mA}$$
$$24\ \text{mA} = 24\ \text{mA} \quad \text{(checks)}$$

EXAMPLE 7.22

Repeat Example 7.18 using Eq. 7.11 and compare results.

Solution

$$R_T = 4\ \text{k}\Omega \| 8\ \text{k}\Omega = \frac{(4\ \text{k}\Omega)(8\ \text{k}\Omega)}{4\ \text{k}\Omega + 8\ \text{k}\Omega}$$

$$= \frac{32\ \text{k}\Omega}{12} = 2.667\ \text{k}\Omega$$

Using Eq. 7.11:

$$I_x = \frac{R_T}{R_x} I$$

$$I_2 = \frac{R_T I}{R_2}$$

$$= \frac{(2.667\ \text{k}\Omega)(6\ \text{A})}{8\ \text{k}\Omega}$$

$$= \frac{16\ \text{A}}{8}$$

$$I_2 = \mathbf{2\ A}$$

as obtained earlier.

Calculator

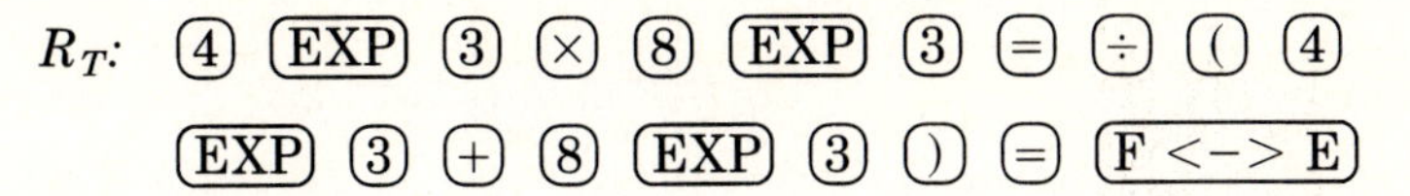

with a display of 2.667 03.

I_2: 2 . 6 6 7 EXP 3 × 6 = ÷ 8 EXP 3 =

with a display of 2.0.

A mechanical analogy often used to describe this division of current is the flow of water through pipes. The water represents the flow of charge, and the tubes or pipes represent conductors. In this analogy, the greater the resistance of the corresponding electrical element, the smaller the area of the tubing.

The total current I in Fig. 7.34(a) divides equally between the two equal resistors. The analogy just described is shown to the right. Obviously, for two pipes of equal diameter, the water will divide equally.

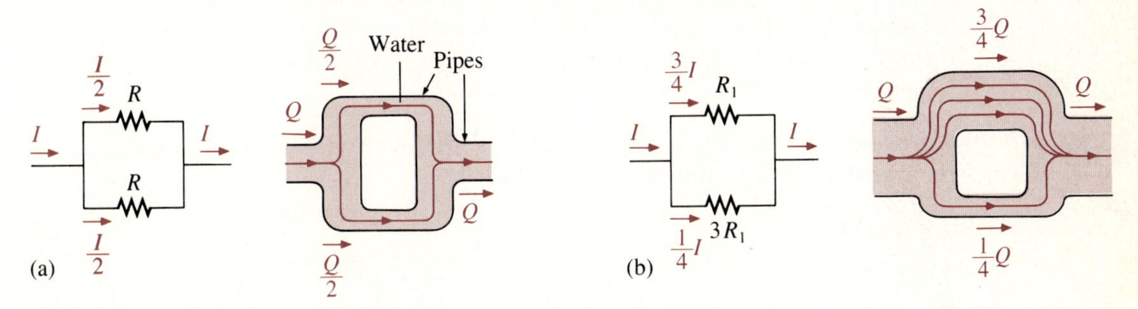

FIG. 7.34
Mechanical analogy of current division in a parallel network.

In Fig. 7.34(b), one resistor is three times the other, resulting in the current dividing as shown. Its mechanical analogy is shown in the adjoining figure. There is three times as much water (current) passing through one pipe as through the other. In both Fig. 7.34(a) and (b), the total water (current) entering the parallel systems from the left equals that leaving to the right.

7.7 VOLTAGE SOURCES IN PARALLEL

Voltage sources are placed in parallel, as shown in Fig. 7.35, only if they have the same voltage rating. Otherwise, Kirchhoff's voltage law would be violated around the internal loop of the two batteries and one battery would discharge through the other. The primary reason for placing two or more batteries in parallel with the same terminal voltage is to increase the current rating of the

FIG. 7.35
Parallel voltage sources.

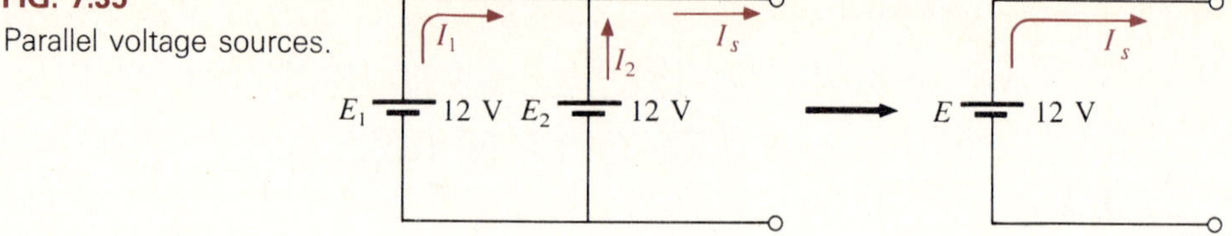

source. As shown in Fig. 7.35, the current rating of the combination is determined by $I_s = I_1 + I_2$ at the same terminal voltage. The resulting power rating is twice that available with one supply.

7.8 OPEN AND SHORT CIRCUITS

Open circuits and short circuits can often cause more confusion and difficulty in the analysis of a system than standard series or parallel configurations. This will become more obvious in the chapters to follow, when we apply some of the methods and theorems.

An *open circuit* is simply two isolated terminals not connected by an element of any kind. Consider the battery of Fig. 7.36. An open circuit exists between terminals *a* and *b*. There is a voltage of *E* volts between the two terminals, but the current between the two is zero due to the absence of a closed path for the flow of charge. In general:

An open circuit can have a potential difference (voltage) across its terminals, but the current is always 0 A.

FIG. 7.36
Open-circuit characteristics.

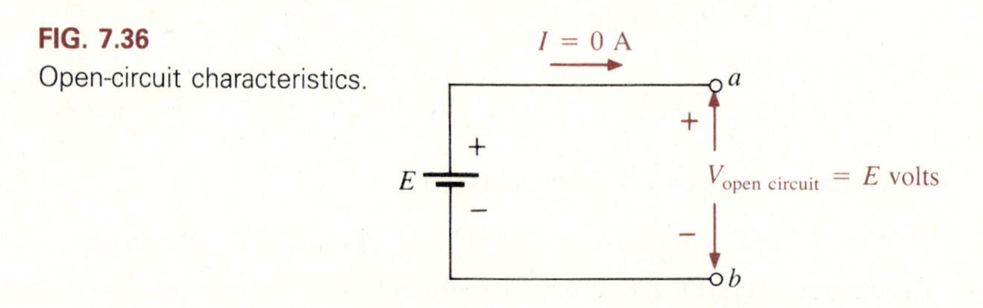

A *short circuit* is a direct connection of 0 Ω across an element or combination of elements. In Fig. 7.37(a), the current through the 2-Ω resistor is 5 A. If a short circuit is established across the 2-Ω resistor, as shown in Fig. 7.37(b), due to a faulty wire, connection, or other unexpected circumstance, the total resistance of the circuit R_T is now $R_T = (2)(0)/(2 + 0) = 0\ \Omega$, and the current will rise to very high levels. The 2-Ω resistor has effectively been "shorted out" by the low-resistance connection. The maximum current is limited only by the circuit breaker or fuse in series with the source. The resulting high current is often the cause of fire or smoke if the protective device fails to respond quickly enough. Since the resistance of a short circuit is 0 Ω, there is no volt-

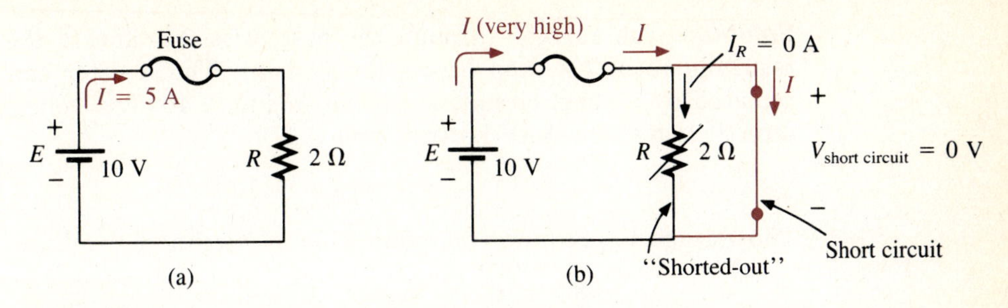

FIG. 7.37
Effect of a short circuit on the source current.

age drop across a short circuit, as determined by Ohm's law ($V = IR$). In general:

A short circuit can carry a current of any level, but the potential difference (voltage) across its terminals is always 0 V.

EXAMPLE 7.23 Determine the voltage V_{ab} for the network of Fig. 7.38.

FIG. 7.38
Current for Example 7.23.

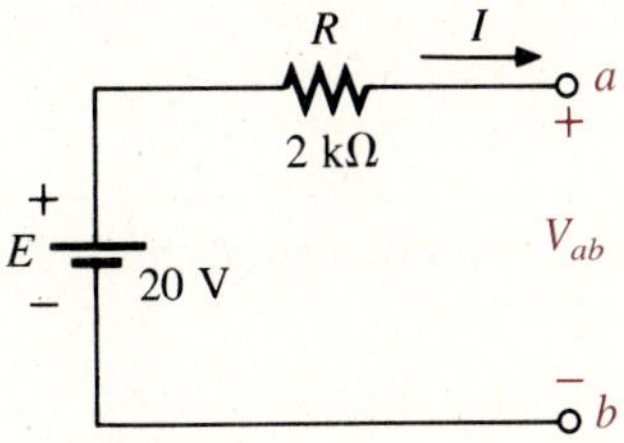

Solution The open circuit requires that I be 0 A. The voltage drop across the resistor is therefore 0 V, since $V = IR = (0)R = 0$ V. Applying Kirchhoff's voltage law around the closed loop,

$$V_{ab} = E = \mathbf{20\ V}$$

EXAMPLE 7.24 Determine the voltages V_{ab} and V_{cd} for the network of Fig. 7.39.

FIG. 7.39
Circuit for Example 7.24.

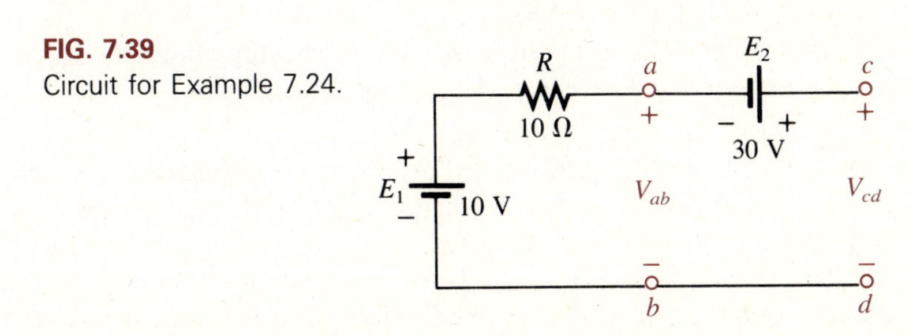

Solution The current through the system is 0 A due to the open circuit resulting in a 0-V drop across the resistor. The resistor can therefore be replaced by a short circuit, as shown in Fig. 7.40. The voltage V_{ab} is then directly across the 10-V battery, and

$$V_{ab} = E_1 = \mathbf{10\ V}$$

The voltage V_{cd} requires an application of Kirchhoff's voltage law:

$$+E_1 + E_2 - V_{cd} = 0$$

or

$$V_{cd} = E_1 + E_2 = 10\text{ V} + 30\text{ V} = \mathbf{40\ V}$$

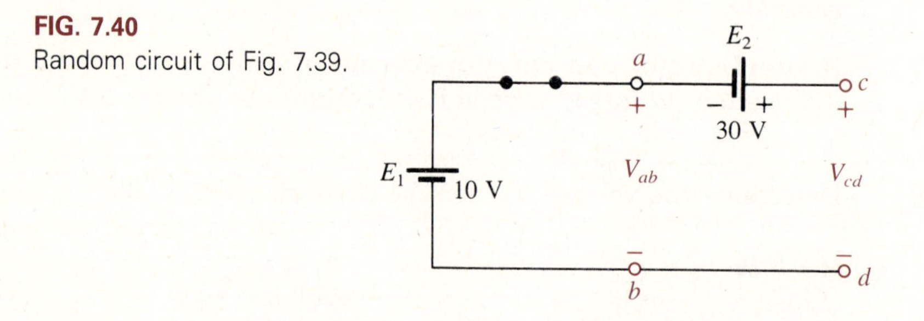

FIG. 7.40
Random circuit of Fig. 7.39.

EXAMPLE 7.25 Calculate the current I and the voltage V for the network of Fig. 7.41.

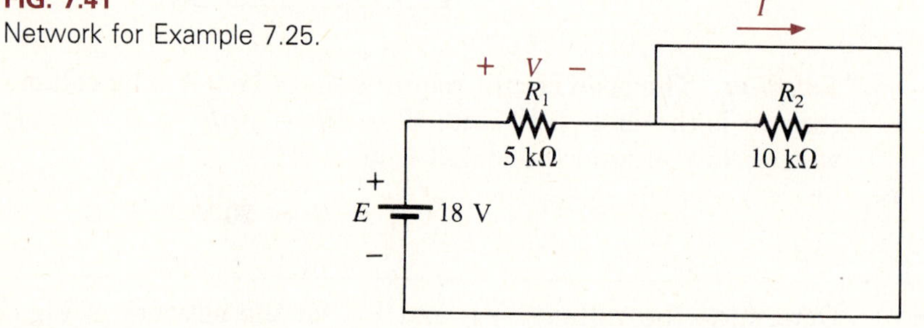

FIG. 7.41
Network for Example 7.25.

Solution The 10-Ω resistor has been effectively shorted out, resulting in the equivalent network of Fig. 7.42. Using Ohm's law,

$$I = \frac{E}{R_1} = \frac{18\text{ V}}{5\text{ k}\Omega} = \mathbf{3.6\ mA}$$

and

$$V = E = \mathbf{18\ V}$$

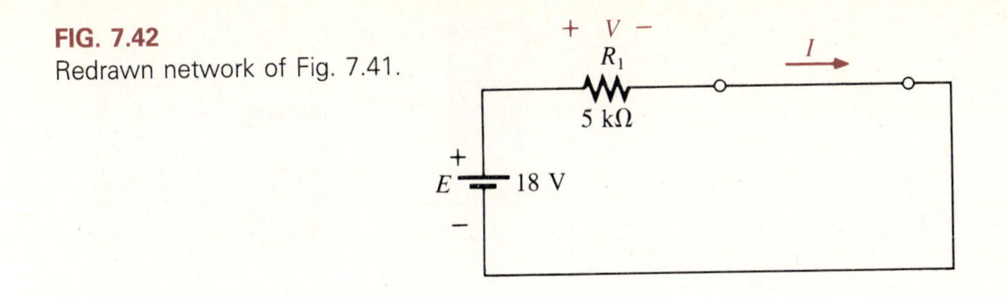

FIG. 7.42
Redrawn network of Fig. 7.41.

EXAMPLE 7.26 Determine V and I for the network of Fig. 7.43 if the resistor R_2 is shorted out.

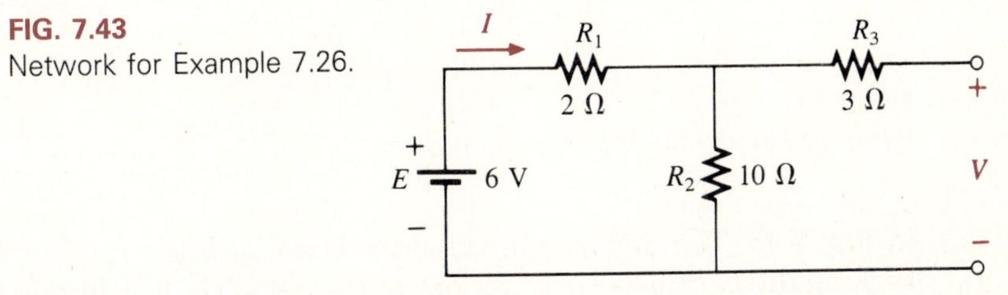

FIG. 7.43
Network for Example 7.26.

Solution The redrawn network appears in Fig. 7.44. The current through the 3-Ω resistor is zero due to the open circuit, causing all the current I to pass through the short circuit. Since $V_{3\Omega} = IR = (0)R = 0$ V, the voltage V is directly across the short.

$$V = \mathbf{0\ V}$$

and

$$I = \frac{E}{R_1} = \frac{6\text{ V}}{2\ \Omega} = \mathbf{3\ A}$$

FIG. 7.44
Redrawn network of Fig. 7.43.

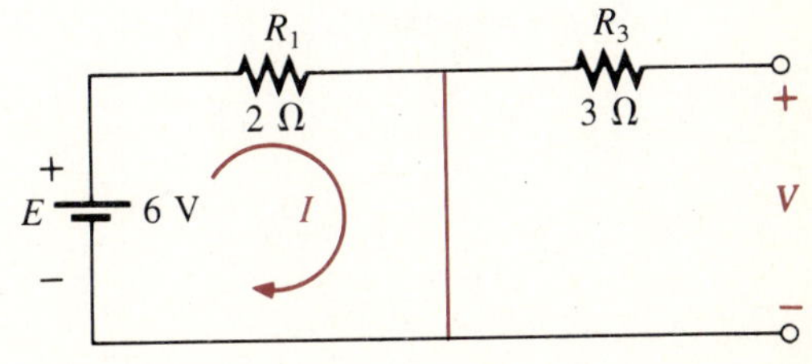

7.9 INSTRUMENTATION

The use of meters to measure current and voltage levels in a parallel network requires more attention to detail than encountered for the basic series configuration. In particular, note the connections required to measure the current I_1

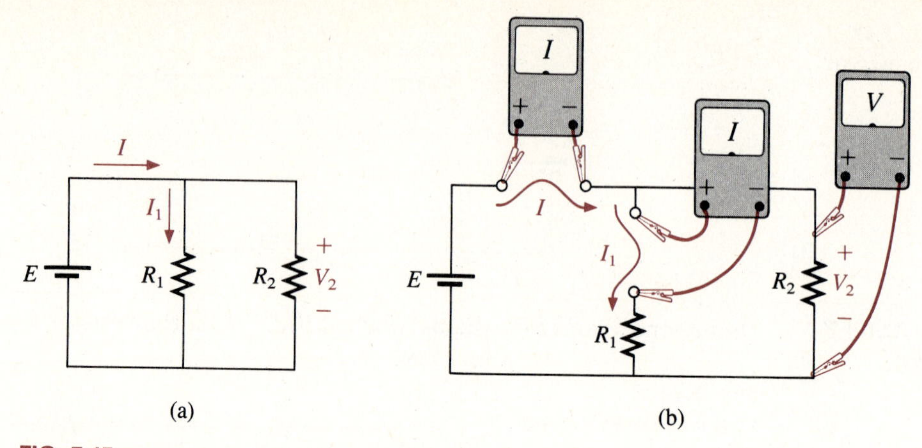

FIG. 7.45
Basic measurements for a parallel network.

in Fig. 7.45. The meter is inserted in series with the resistor R_1 without disturbing the parameter connections of the network. In addition, note the direction of I_1 and the polarities of the meter to ensure an up-scale reading.

The remaining meters of Fig. 7.45 are measuring the source current I and the voltage across the resistor R_2. Note, in general, how current measurements require that the network be disturbed (by insertion of the meter), whereas voltage measurements are simply made between the points of interest.

Figure 7.46 depicts the connections required to measure the total resistance of a parallel network. Note the absence of the supply, since resistance measurements are not made on a "live" network.

FIG. 7.46
Measuring the total resistance of two parallel resistors.

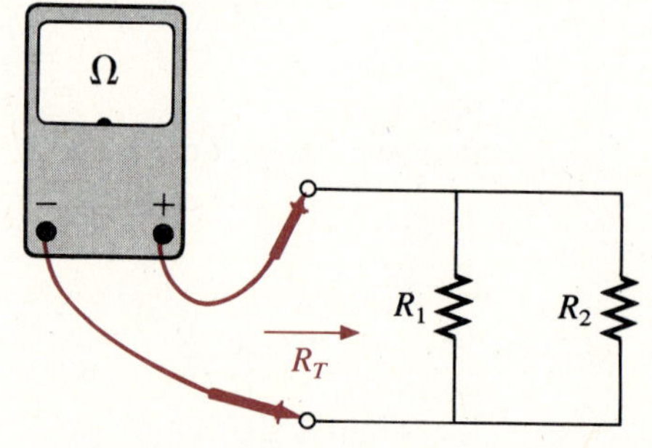

Recall from the previous chapter that the resistance of an ammeter is placed in series with the circuit, as shown in Fig. 7.47. For most instances, the resistance R_m is sufficiently small compared to R to be ignored. However, for voltmeters where the *loading effect* is a meter connected in parallel with the network resistance, care must be taken to be sure that R_m is sufficiently large compared to R, as depicted in Fig. 7.48. In fact, the ideal situation would be

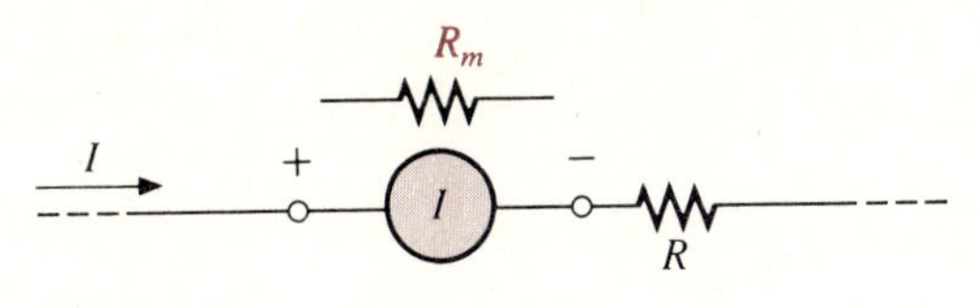

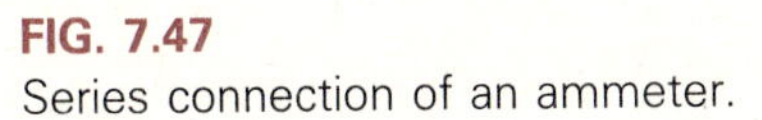

FIG. 7.47
Series connection of an ammeter.

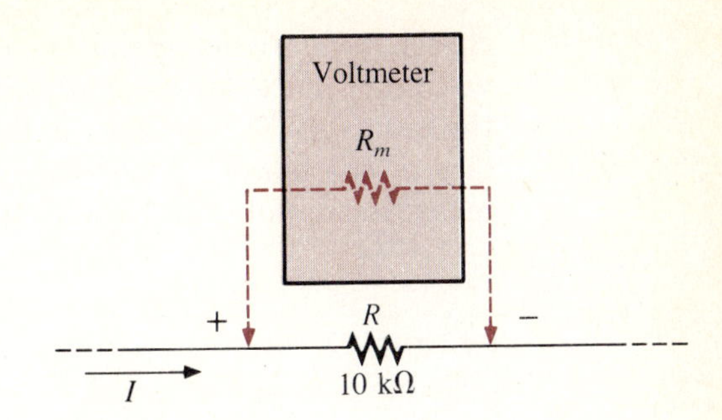

FIG. 7.48
Voltmeter loading.

established if R_m were infinite in magnitude, corresponding to an open circuit. The total resistance of the combination is

$$R_T = 10\text{ k}\Omega \| 11\text{ M}\Omega = \frac{(10^4\ \Omega)(11 \times 10^6\ \Omega)}{10^4\ \Omega + (11 \times 10^6\ \Omega)} = 9.991\text{ k}\Omega$$

Calculator

(1) (EXP) (4) (×) (1) (1) (EXP) (6) (=) (÷) (( ) (1) (EXP) (4)
(+) (1) (1) (EXP) (6) ( )) (=) (F <−> E)

with a display of 9.991 03. We find that the network is essentially undisturbed, since 9.991 kΩ ≅ 10 kΩ. However, if we use a VOM with an internal resistance of 50 kΩ on the 2.5-V scale, the parallel resistance is

$$R_T = 10\text{ k}\Omega \| 50\text{ k}\Omega = \frac{(10^4\ \Omega)(50 \times 10^3\ \Omega)}{10^4\ \Omega + (50 \times 10^3\ \Omega)} = 8.333\text{ k}\Omega$$

Calculator

(1) (EXP) (4) (×) (5) (0) (EXP) (3) (=) (÷) (( ) (1) (EXP) (4)
(+) (5) (0) (EXP) (3) ( )) (=) (F <−> E)

with a display of 8.333 03; the behavior of the network will be altered somewhat, since the 10-kΩ resistor will now appear to be 8.33 kΩ to the rest of the network.

The loading of a network by the insertion of meters is not to be taken lightly, especially in research efforts where accuracy is a primary consideration. It is good practice always to check the meter-resistance level against the resistive elements of the network before making measurements. A factor of at least 10 between resistance levels will usually provide fairly accurate meter readings for a wide range of applications.

Most DMMs have internal resistance levels in excess of 10 MΩ on all voltage scales, whereas the internal resistance of VOMs is sensitive to the chosen scale. To determine the resistance of each scale setting of a VOM in the voltmeter mode, simply multiply the maximum voltage of the scale setting by the ohm/volt (Ω/V) rating of the meter, normally found at the bottom of the face of the meter, as shown in Fig. 7.49. Note that it is 20,000 Ω/V for dc measurements but only 5000 Ω/V for ac measurements.

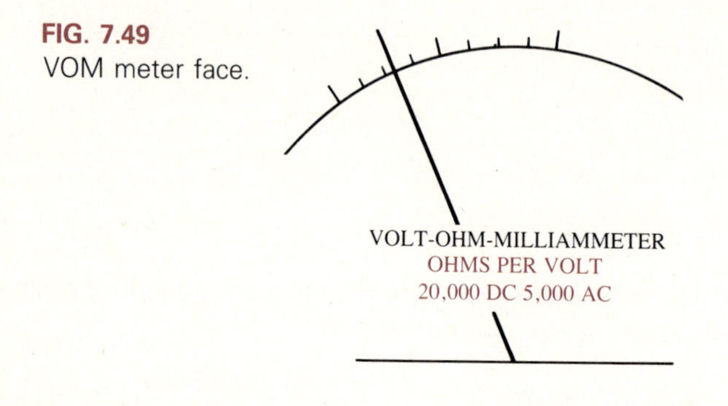

FIG. 7.49
VOM meter face.

For a typical ohm/volt rating of 20,000, the 2.5-V scale would have an internal resistance of

$$(2.5\ \text{V})(20{,}000\ \Omega/\text{V}) = 50\ \text{k}\Omega$$

Calculator

(2) (·) (5) (×) (2) (EXP) (4) (=) (F <-> E)

with a display of 5. 04. For the 100-V scale, it would be

$$(100\ \text{V})(20{,}000\ \Omega/\text{V}) = 2\ \text{M}\Omega$$

Calculator

(1) (0) (0) (×) (2) (EXP) (4) (=) (F <-> E)

with a display of 2. 06. For the 250-V scale,

$$(250\ \text{V})(20{,}000\ \Omega/\text{V}) = 5\ \text{M}\Omega$$

Calculator

(2) (5) (0) (×) (2) (EXP) (4) (=) (F <-> E)

with a display of 5. 06.

FORMULA SUMMARY

$$\frac{1}{R_T} = \frac{1}{R_1} + \frac{1}{R_2} + \frac{1}{R_3} + \cdots + \frac{1}{R_N}$$

$$R_T = \frac{R}{N}$$

$$R_T = \frac{R_1 R_2}{R_1 + R_2}$$

$$\Sigma I_{\text{entering}} = \Sigma I_{\text{leaving}}$$

$$I_1 = \frac{R_2 I}{R_1 + R_2}, \qquad I_2 = \frac{R_1 I}{R_1 + R_2}$$

$$I_x = \frac{R_T}{R_x} I$$

CHAPTER SUMMARY

This chapter introduced a second very important method of connecting two or more elements. At first glance the definitions of series and parallel elements do not appear that difficult to comprehend. However, due to the various ways in which networks can be drawn, a fair amount of exposure is usually required before one can consider himself or herself an "expert" on the subject. For parallel elements the two ends of one element must be connected directly across the other element. Leaving either end of one element will result in a connection point connected to one end of the other element. Oftentimes simply redrawing the network will help reveal whether two elements are indeed in series or parallel.

The equation for two parallel resistors, $R_T = R_1R_2/(R_1 + R_2)$, is a very useful one that avoids getting involved with the reciprocal terms that often result in incorrect numerical results. As pointed out in this chapter, this formula can also be used to find the total resistance of more than two parallel resistors by simply using the equation for each set of two parallel resistors and then again for the intermediate results.

For series circuits the current is the same through each element, whereas for parallel elements the voltage is the same. Furthermore, for series circuits the magnitude of the voltage across a resistor is *directly* related to the magnitude of the resistor. For parallel networks, the current through a parallel branch is *inversely* related to the resistance; that is, the smaller the parallel resistor the larger the current through the resistor for the same terminal voltage.

Kirchhoff's current law is not that difficult to apply if you simply keep in mind that the net sum of the currents entering a junction (or region) must equal the net sum of the currents leaving the junction (or region). In addition, always start at the location that results in one unknown quantity, since an application of the law can provide only one unknown quantity. When applying the current divider rule, the denominator of the equation for either current is simply the *sum* of the two parallel resistances. The numerator of each equation is formed by taking the product of the total current entering the junction times the magnitude of the other resistor.

GLOSSARY

Current divider rule A method by which the current through parallel elements can be determined without first finding the voltage across those parallel elements.

Kirchhoff's current law The law that states that the algebraic sum of the currents entering and leaving a node is zero.

Node A junction of two or more branches.

Ohm/volt rating A rating used to determine both the current sensitivity of the movement and the internal resistance of the meter.

Open circuit The absence of a direct connection between two points in a network.

Parallel circuit A circuit configuration in which the elements have two points (and only two points) in common.

Short circuit A direct connection of low resistive value that can significantly alter the behavior of an element or system.

PROBLEMS

Section 7.2

1. For each configuration of Fig. 7.50, determine which elements are in series or parallel.

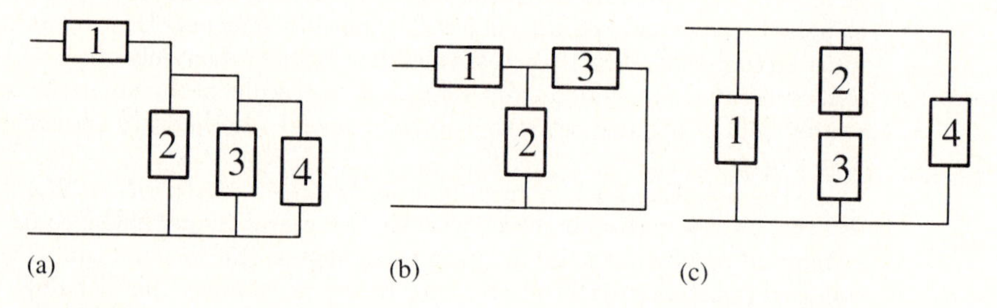

FIG. 7.50

2. Which elements of the configurations of Fig. 7.51 are in parallel?

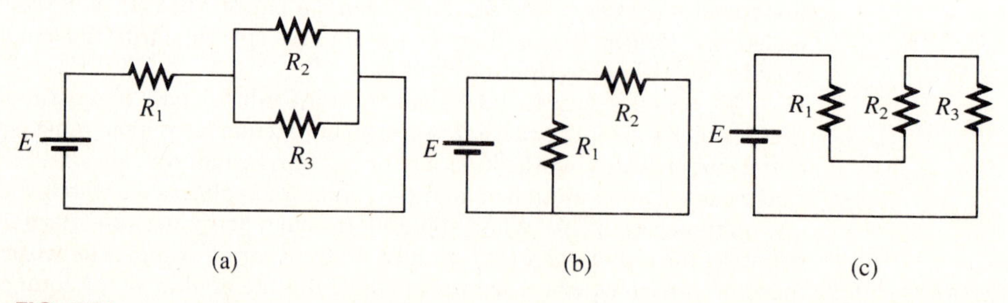

FIG. 7.51

Section 7.3

3. Determine the total resistance of the networks of Fig. 7.52.

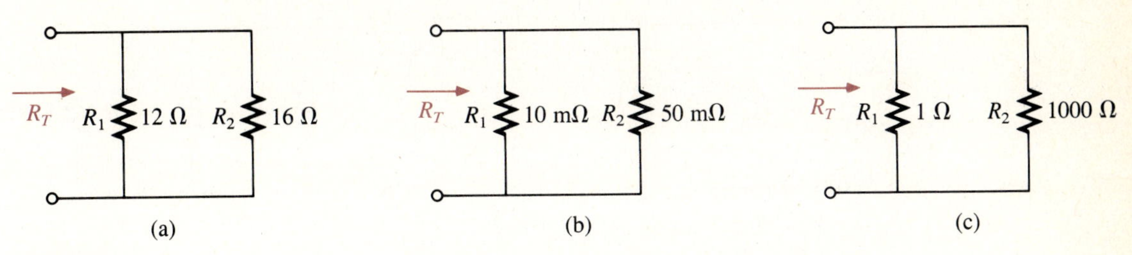

FIG. 7.52

4. Determine the total resistance of the networks of Fig. 7.53.

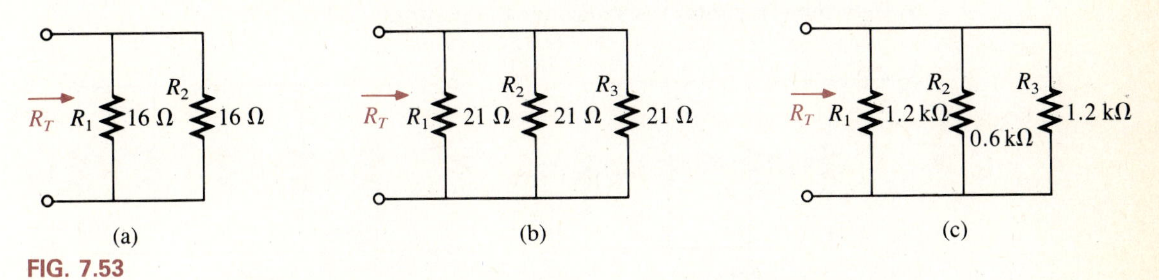

FIG. 7.53

5. Determine the total resistance of the networks of Fig. 7.54.

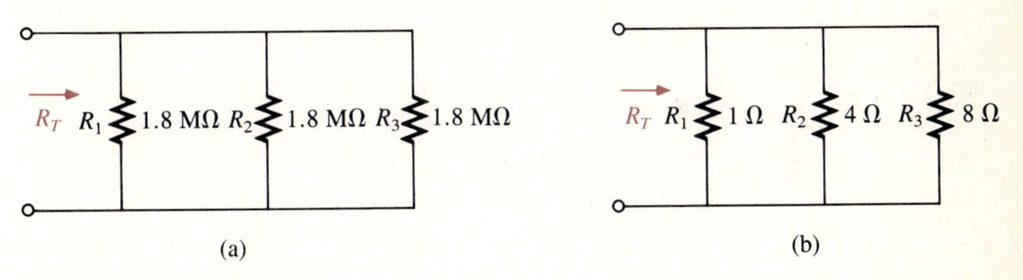

FIG. 7.54

***6. a.** By inspection, determine the approximate total resistance of the networks of Fig. 7.55.

b. Verify the conclusion of part a by making the necessary calculations.

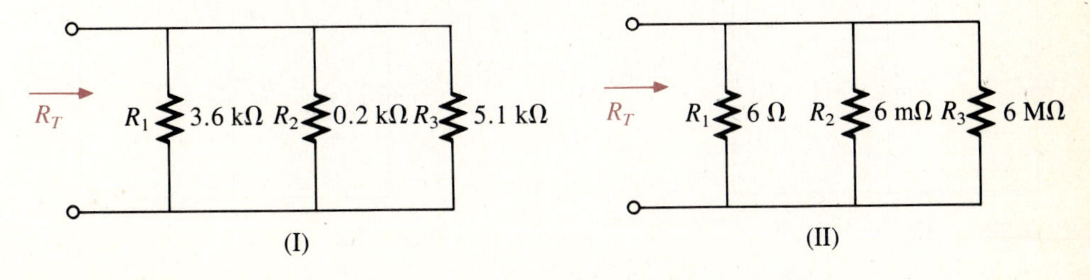

FIG. 7.55

*7. Given the total resistance R_T, determine the unknown resistors for the configurations of Fig. 7.56.

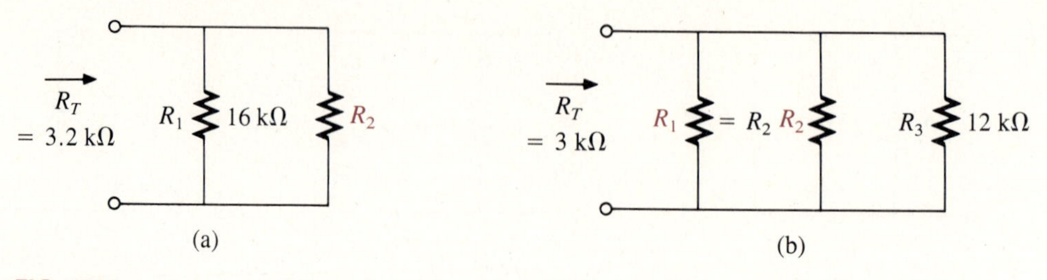

FIG. 7.56

8. Determine the total resistance for the networks of Fig. 7.57.

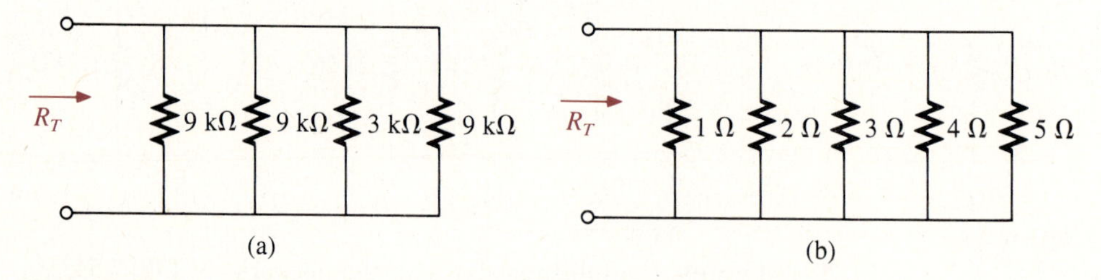

FIG. 7.57

Section 7.4

9. For the network of Fig. 7.58, determine each value.
   a. R_T
   b. I_T
   c. V_1 and V_2
   d. I_1 and I_2
   e. P_1, P_2, and P_{del} (power delivered by the supply)
   f. Does $I_T = I_1 + I_2$?
   g. Does $P_{del} = P_1 + P_2$?

10. Repeat Problem 9 for the network of Fig. 7.59.

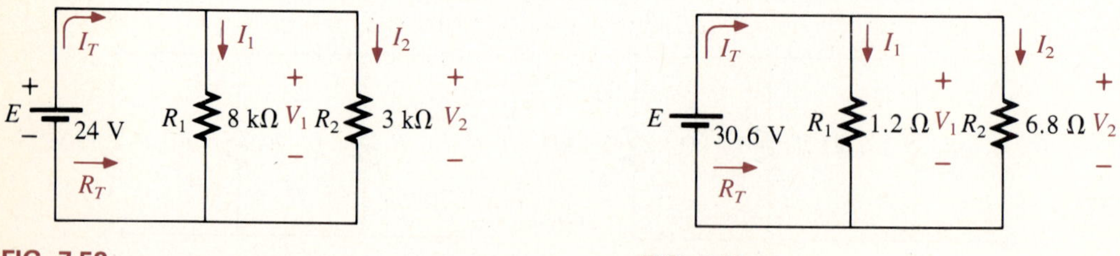

FIG. 7.58

FIG. 7.59

*11. For the network of Fig. 7.60, determine each value.
a. R_T
b. I_T
c. V_1, V_2, and V_3
d. I_1, I_2, and I_3
e. Does $I_T = I_1 + I_2 + I_3$?
f. P_3

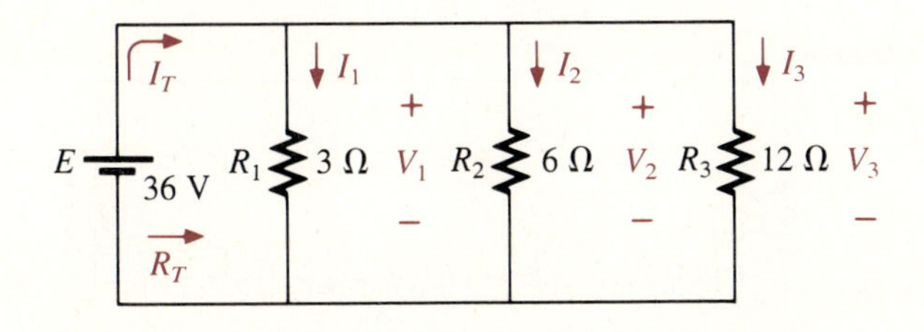

FIG. 7.60

*12. Repeat Problem 11 for the network of Fig. 7.61.

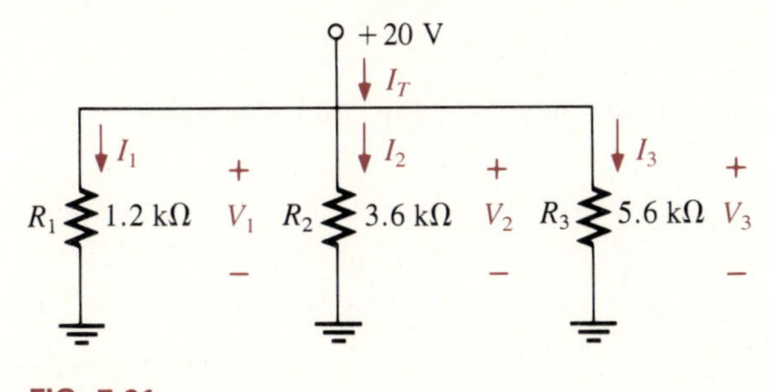

FIG. 7.61

13. There are eight lights connected in parallel, as shown in Fig. 7.62.
a. If the set is connected to a 120-V source, what is the current through each bulb if each bulb has an internal resistance of 1.8 kΩ?
b. Determine the total resistance of the network.
c. Find the power delivered to each bulb.
d. If one bulb burns out (that is, the filament opens), what is the effect on the remaining bulbs?
e. Compare the parallel arrangement of Fig. 7.62 to a series arrangement. What are the relative advantages and disadvantages of the parallel system as compared to the series arrangement?

FIG. 7.62

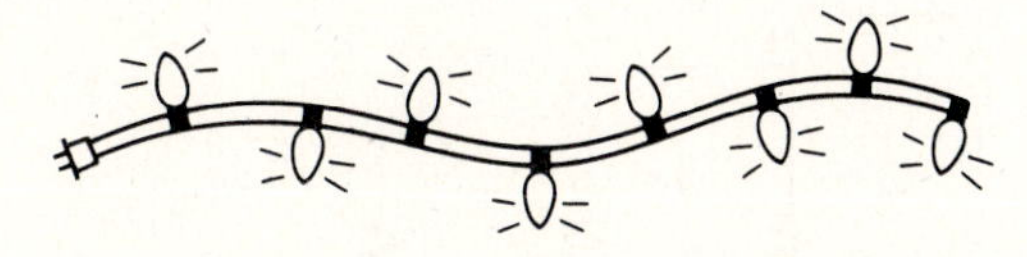

14. A portion of a residential service to a home is depicted in Fig. 7.63.
   a. Determine the current through each parallel branch of the network.
   b. Calculate the current drawn from the 120-V source. Will the 20-A circuit breaker trip?
   c. What is the total resistance of the network?
   d. Determine the power supplied by the 120-V source. How does it compare to the total power of the load?

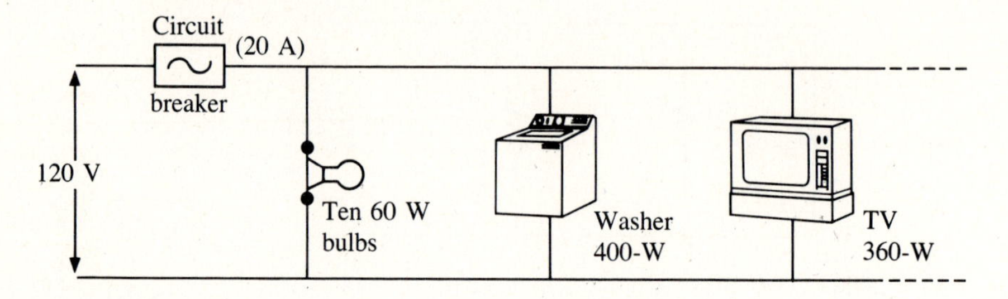

FIG. 7.63

15. Given the information provided in Fig. 7.64, determine
   a. E
   b. R_1

***16.** Given the information provided in Fig. 7.65, determine
   a. R_2
   b. E

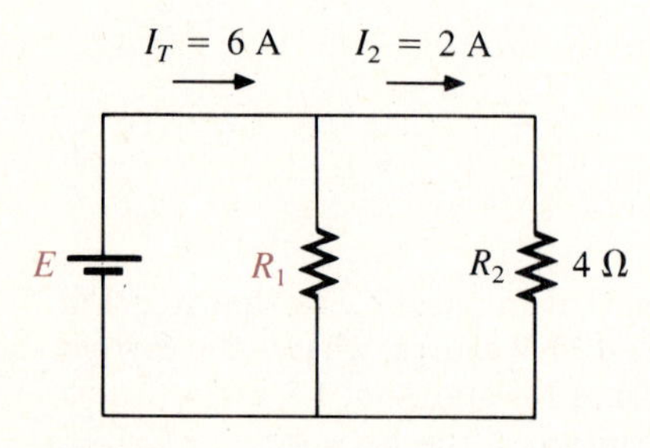

FIG. 7.64

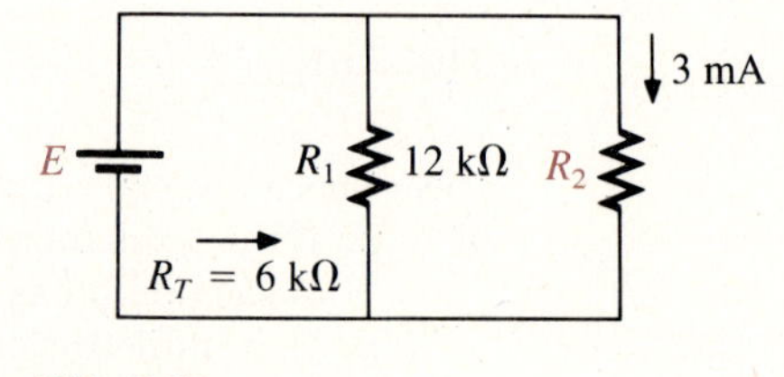

FIG. 7.65

***17.** Given the information provided in Fig. 7.66, determine
   a. E
   b. R_2 and R_3

FIG. 7.66

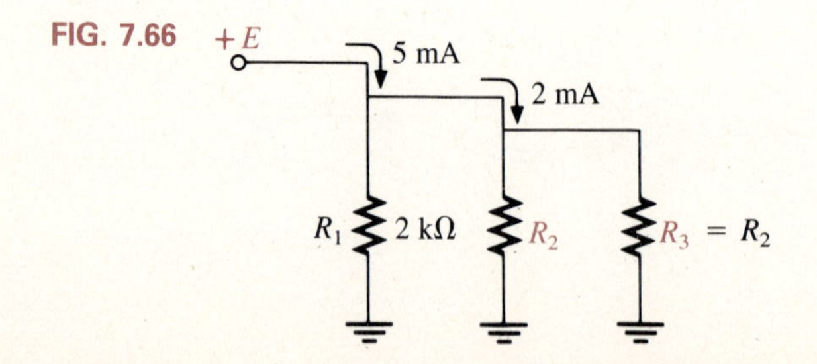

Section 7.5

18. For the configurations of Fig. 7.67, determine the unknown currents.

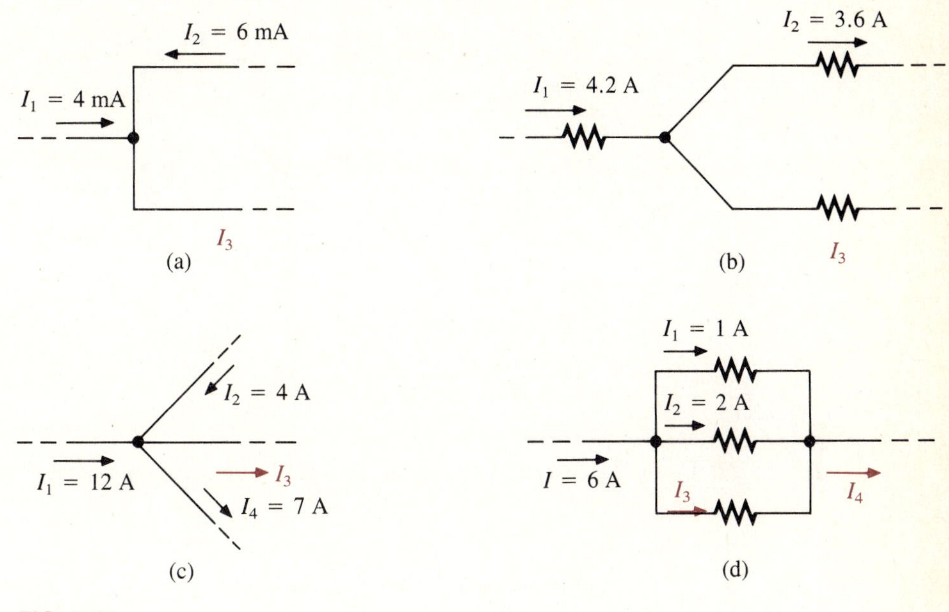

FIG. 7.67

***19.** Determine the unknown currents for the networks of Fig. 7.68.

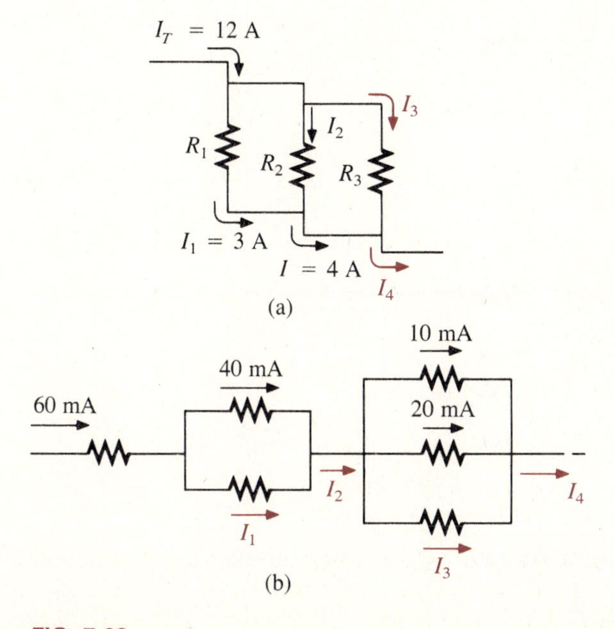

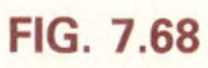
FIG. 7.68

Section 7.6

20. Using the current divider rule, find the unknown currents for the networks of Fig. 7.69.

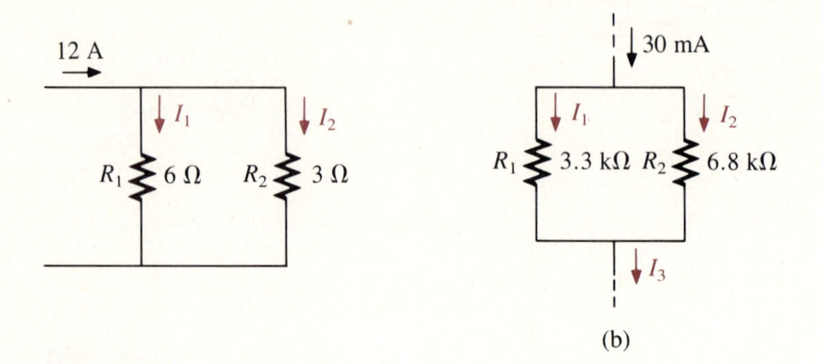

FIG. 7.69

21. Using the current divider rule, find the unknown currents for the networks of Fig. 7.70.

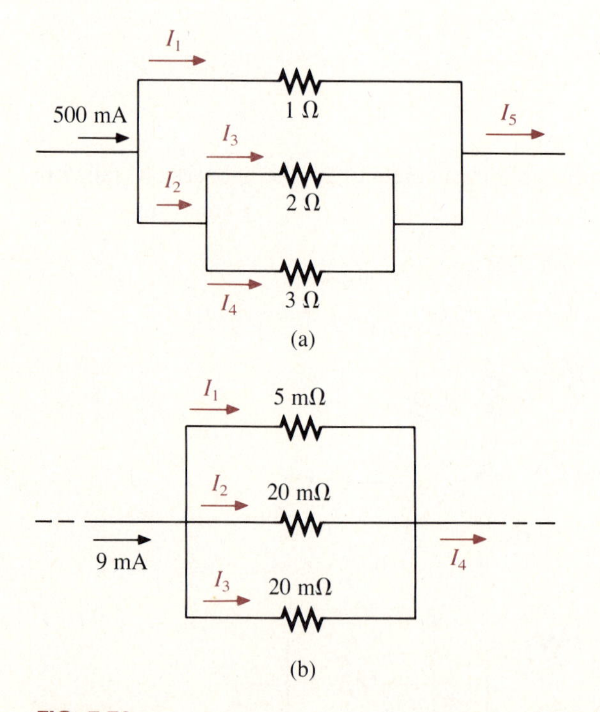

FIG. 7.70

22. Determine the resistance R_1 necessary to effect the current division shown in Fig. 7.71.

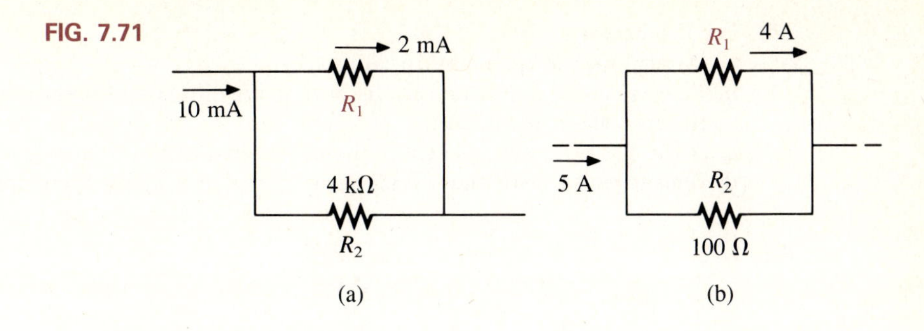

FIG. 7.71

***23.** Find the unknown quantities using the information provided for the networks of Fig. 7.72.

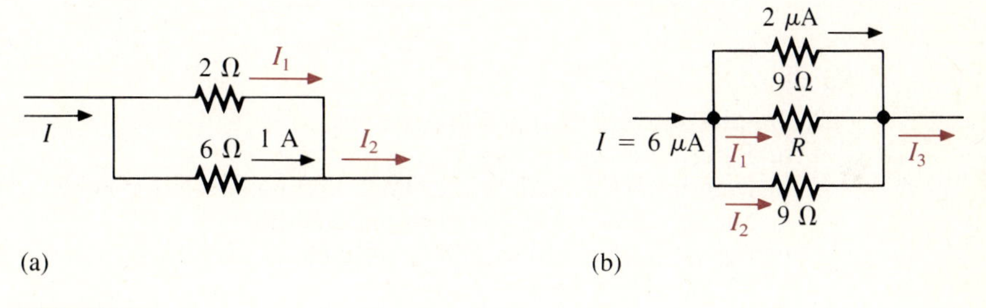

FIG. 7.72

Section 7.7

24. Determine the currents I_1 and I_2 for the network of Fig. 7.73.

FIG. 7.73

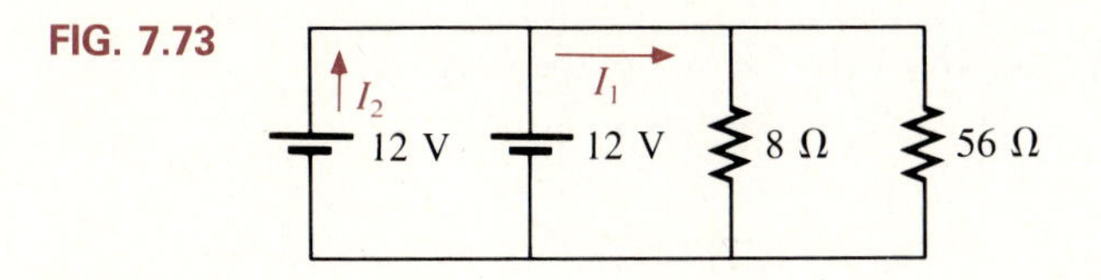

Section 7.8

25. For the network of Fig. 7.74, determine
a. I_T and V_L
b. I_T if R_L is shorted out
c. V_L if R_L is replaced by an open circuit

FIG. 7.74

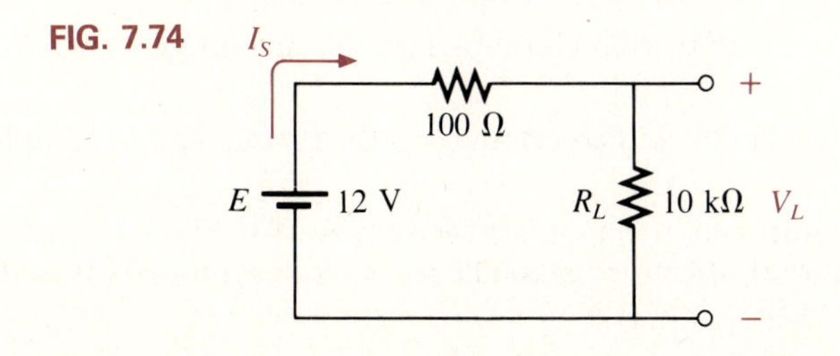

***26.** For the network of Fig. 7.75,

a. Determine the open-circuit voltage V_L.

b. Place a short circuit across the output terminals and determine the current through the short circuit.

c. If the 2.2-kΩ resistor is short-circuited, what is the new value of V_L?

d. Repeat part b with the 4.7-kΩ resistor replaced by an open circuit.

FIG. 7.75

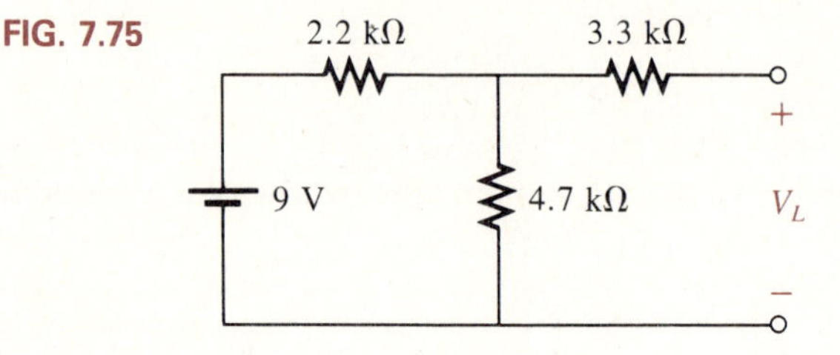

Section 7.9

27. For the network of Fig. 7.76,

a. Determine the voltage V_2.

b. Determine the reading of a DMM (voltmeter section) having an internal resistance of 11 MΩ when used to measure V_2.

c. Repeat part b with a VOM (voltmeter section) having an ohm/volt rating of 20,000 using the 10-V scale.

d. Repeat part c with $R_1 = 100$ kΩ and $R_2 = 200$ kΩ.

FIG. 7.76

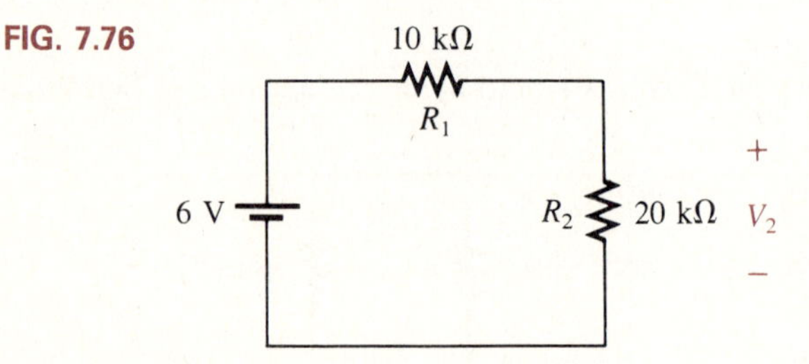

Computer Problems

28. Write a program to determine the total resistance of any number of parallel resistors.

29. Write a program to provide a complete solution of parallel network with a single source and two parallel resistors. That is, print out the total resistance, source current, branch currents, and power to each element.

30. Write a program that will determine how the current splits between two parallel elements.

31. Write a program that will determine how the current splits through three parallel elements.

32. Write a program that will tabulate the voltage V_2 of Fig. 7.76 measured by a VOM with an internal resistance of 200 kΩ as R_2 varies from 10 kΩ to 200 kΩ in increments of 10 kΩ.

8

Series-Parallel Networks

OBJECTIVES

- ☐ Identify series and parallel configurations in a series-parallel network.
- ☐ Develop an approach to series-parallel networks that will provide a path toward the desired solution.
- ☐ Recognize a ladder network and be able to solve for any unknown quantity.

8.1 INTRODUCTION

Series-parallel networks are networks that have both series and parallel configurations in the same structure. The first step in analyzing networks of this type usually is to determine which elements are in series or parallel. Once this is determined, elements can be combined and the network can usually be reduced to a form that will permit determining quantities such as the total resistance and source current. The analysis can then work backward until the desired quantities of the network have been determined. In some cases, none of the elements may be in series or parallel, and other techniques—such as those described in the chapters to follow—must be applied.

Since a determination of which elements are in series or parallel is the first critical step in the analysis, keep the following in mind.

Elements or branches are in series *only if* they are connected at one end and that end is *not* connected to any other current-carrying elements. In addition, if two adjoining elements or branches have the same current and a third element is not connected to the junction point, then they are in series.

Elements or branches are in parallel *only if* they are connected at both ends. In addition, if two elements or branches have the same voltage across them, chances are that they are in parallel.

In general, a firm understanding of the basic principles associated with series and parallel circuits is a sufficient background to approach most complicated series-parallel networks with *one* source of voltage. Multisource networks are considered in Chapters 9 and 10.

Aside from making an initial determination of which elements are in series or parallel, the following is a natural sequence:

1. Study the problem and make a brief mental "sketch" of the overall approach you plan to use. The result may be time- and energy-saving shortcuts.
2. After you have determined the overall approach, examine each branch independently before tying them together in series-parallel combinations. This will eliminate many of the errors that might develop due to the lack of a systematic approach.
3. When you have a solution, check that it is reasonable by considering the magnitudes of the energy source and the elements in the network. If it does not seem reasonable, either solve the circuit using another approach or check over your work very carefully.

The block-diagram approach is employed to emphasize the fact that combinations of elements, not simply single resistive elements, can be in series or parallel. The approach also reveals the number of seemingly different networks that have the same basic structure and therefore can involve similar analysis techniques.

8.2 ILLUSTRATIVE EXAMPLES

In Fig. 8.1, blocks B and C are in parallel (points b and c in common), and the voltage source E is in series with block A (point a in common). The parallel combination of B and C is also in series with A and the voltage source E due to the common points b and c, respectively.

FIG. 8.1

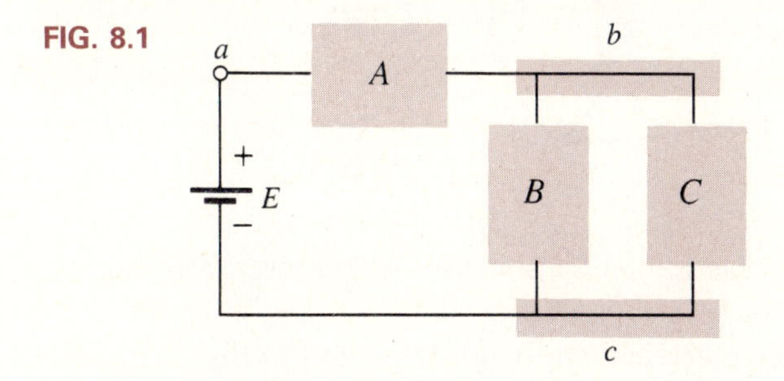

If blocks A, B, and C were resistors, the series elements of the battery E and block A could not be combined, since they are different types of elements. Blocks B and C could be combined since they are both resistors. When combined, they would be replaced by a single block D, such as shown in Fig. 8.2, and then block A could be combined with the resulting resistance, since they are now in series. Note in Fig. 8.2 that terminals a, b, and c still exist and that the voltage from b to c is the same in Figs. 8.1 and 8.2. Once the series combination of A and D is determined, the current through E, A, and D can be determined, and the voltage from b to c can be calculated using Ohm's law.

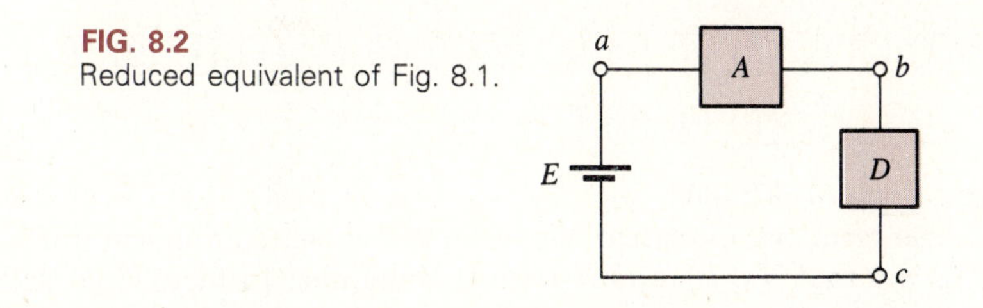

FIG. 8.2
Reduced equivalent of Fig. 8.1.

EXAMPLE 8.1 For the network of Fig. 8.3(a), determine

a. R_T
b. I_T
c. V_1 and V_3
d. Power to R_3
e. I_3

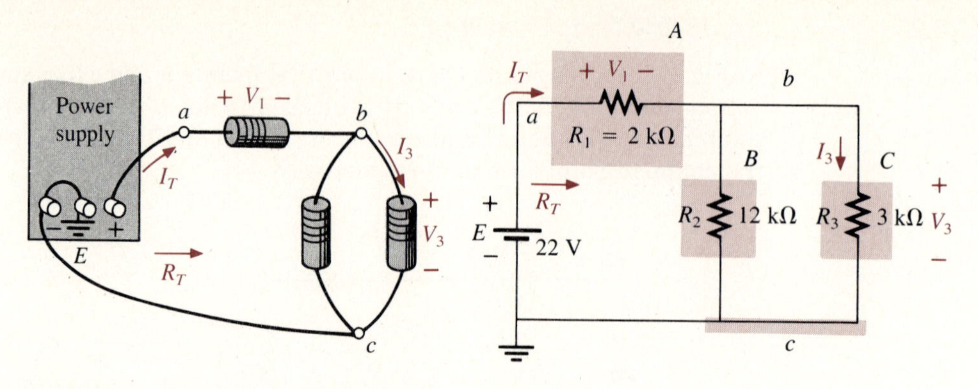

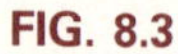
FIG. 8.3
Network for Example 8.1. (a) Construction; (b) electrical schematic.

Solutions The electrical equivalent appears in Fig. 8.3(b). All resistor values are commercially available values.

a. R_2 and R_3 are in parallel, since both ends of one (b and c) are connected to both ends of the other. Combining the two:

$$R' = R_2 \| R_3 = \frac{R_2 R_3}{R_2 + R_3}$$

$$= \frac{(12\text{ k}\Omega)(3\text{ k}\Omega)}{12\text{ k}\Omega + 3\text{ k}\Omega} = \frac{36\text{ k}\Omega}{15} = 2.4\text{ k}\Omega$$

Calculator

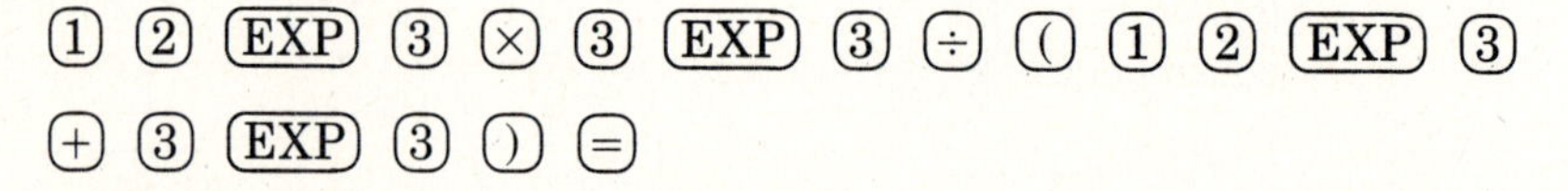

with a display of 2400.

The resulting electrical network then appears as shown in Fig. 8.4. Note in Fig. 8.4 that V_1, I_T, and R_T are still defined. In fact, V_3 is still the same because the voltage across parallel elements is the same even if the parallel combination is reduced to its single equivalent. We can therefore use the network of Fig. 8.4 to determine V_3, even though it was initially defined by Fig. 8.3.

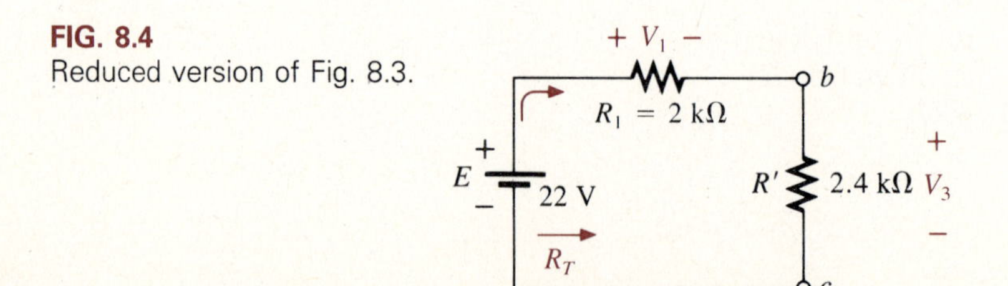

FIG. 8.4
Reduced version of Fig. 8.3.

Since R_1 and R' are now in series (only point b in common),

$$R_T = R_1 + R' = 2\text{ k}\Omega + 2.4\text{ k}\Omega = \mathbf{4.4\text{ k}\Omega}$$

Calculator

(2) (EXP) (3) (+) (2) (·) (4) (EXP) (3) (=)

with a display of 4400.

b. The source current I_T can then be determined using Ohm's law:

$$I_T = \frac{E}{R_T} = \frac{22\text{ V}}{4.4\text{ k}\Omega} = \frac{2.2\text{ A}}{4.4 \times 10^3} = \frac{22}{4.4} \times 10^{-3}\text{ A}$$
$$= 5 \times 10^{-3}\text{ A} = \mathbf{5\text{ mA}}$$

Calculator

(2) (2) (÷) (4) (·) (4) (EXP) (3) (=) (F <-> E)

with a display of 5.−03.

c. In Fig. 8.4 the source current I_T is the same through R_1 and R', and the voltages V_1 and V_3 can be determined using Ohm's law.

$$\begin{aligned} V_1 = I_1R_1 &= I_TR_1 = (5\text{ mA})(2\text{ k}\Omega) \\ &= (5)(2) \times (10^{-3})(10^{+3})\text{ V} \\ &= 10 \times 10^0\text{ V} \\ V_1 &= \mathbf{10\text{ V}} \end{aligned}$$

In general, for calculations of this type that will appear quite frequently, keep in mind that the product of milliamps and kilohms will result in a multiplying factor of 10^0, or 1. In other words, the units balance each other and do not affect the magnitude of the product of multipliers.

Calculator

(5) (EXP) (+/−) (3) (×) (2) (EXP) (3) (=)

with a display of 10.

$$\begin{aligned} V_3 = I_TR' &= (5\text{ mA})(2.4\text{ k}\Omega) \\ &= (5)(2.4)\text{ V} \\ V_3 &= \mathbf{12\text{ V}} \end{aligned}$$

Calculator

(5) (EXP) (+/−) (3) (×) (2) (·) (4) (EXP) (3) (=)

with a display of 12.

d. The power to R_3 can be found using the equations

$$P_3 = I_3^2R_3 \quad \text{or} \quad P_3 = \frac{V_3^2}{R_3}$$

Since V_3 is known and not I_3, the latter equation will be used.

$$P_3 = \frac{V_3^2}{R_3} = \frac{(12\text{ V})^2}{3\text{ k}\Omega}$$

$$= \frac{(12)^2}{3 \times 10^3}\text{ W}$$

$$P_3 = \frac{144}{3} \times 10^{-3}\text{ W} = \mathbf{48\text{ mW}}$$

Calculator

(1) (2) (x^2) (÷) (3) (EXP) (3) (=)

with a display of 0.048.

e. The current I_3 can be determined by Ohm's law, $I_3 = V_3/R_3$, or the current divider rule, $I_3 = R_2 I_T/(R_2 + R_3)$. Since Ohm's law is more direct,

$$I_3 = \frac{V_3}{R_3} = \frac{12\text{ V}}{3\text{ k}\Omega} = \frac{12}{3 \times 10^3}\text{ A}$$

$$= \frac{12}{3} \times 10^{-3}\text{ A} = 4 \times 10^{-3}\text{ A}$$

$$I_3 = \mathbf{4\text{ mA}}$$

Calculator

(1) (2) (÷) (3) (EXP) (3) (=) (F <-> E)

with a display of 4.−03.

Checking:

$$I_3 = \frac{R_2 I_T}{R_2 + R_3}$$

$$= \frac{(12\text{ k}\Omega)(5\text{ mA})}{12\text{ k}\Omega + 3\text{ k}\Omega} = \frac{(12)(5)}{15} \times \frac{(10^{+3})(10^{-3})}{10^{+3}}\text{ A}$$

$$= \frac{60}{15} \times \frac{10^0}{10^{+3}}\text{ A} = 4 \times 10^{-3}\text{ A}$$

$$I_3 = \mathbf{4.0\text{ mA}}$$

as above.

Calculator

(1) (2) (EXP) (3) (×) (5) (EXP) (+/−) (3) (÷) (( ) (1) (2) (EXP) (3) (+) (3) (EXP) (3) ( )) (=) (F <-> E)

with a display of 4.−03.

In Fig. 8.1 the first combination to be made was the parallel configuration of blocks B and C followed by the series combination of block A with the resulting parallel equivalent. In Fig. 8.5, blocks B and C are in series (point c in common), with block A in parallel with source E and the branch containing blocks B and C (points a and b in common).

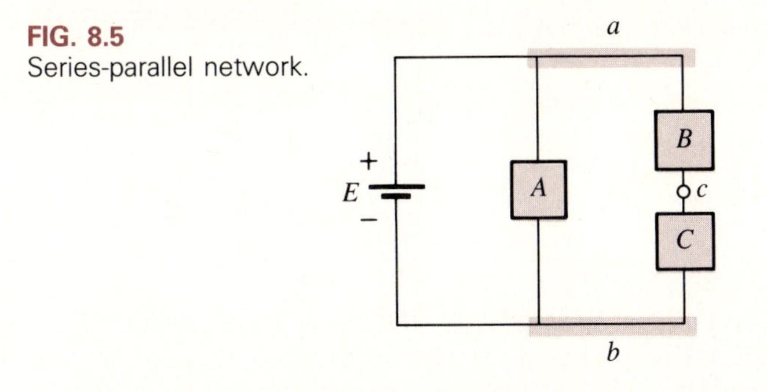

FIG. 8.5
Series-parallel network.

For situations where A, B, and C are resistive elements, the parallel elements of the battery E and block A cannot be combined, since they are different elements. However, blocks B and C can be combined as series elements, and the resulting resistance can be combined with the parallel block A. The next example includes a network of the configuration of Fig. 8.5.

EXAMPLE 8.2

For the network of Fig. 8.6(a), determine

a. R_T
b. I_T
c. I_2
d. V_3

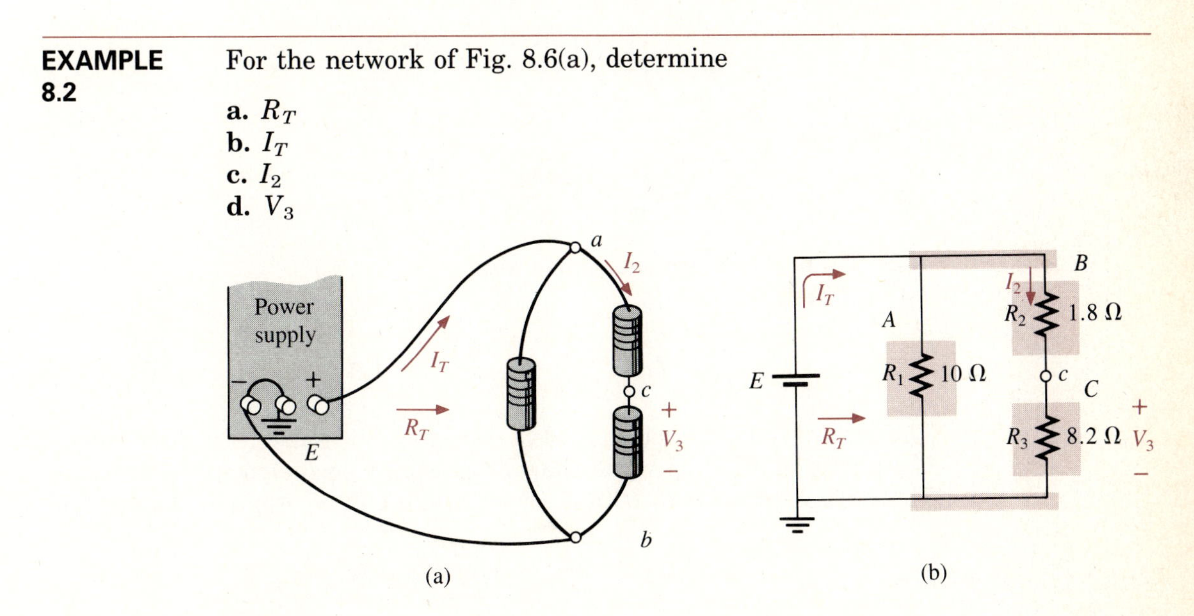

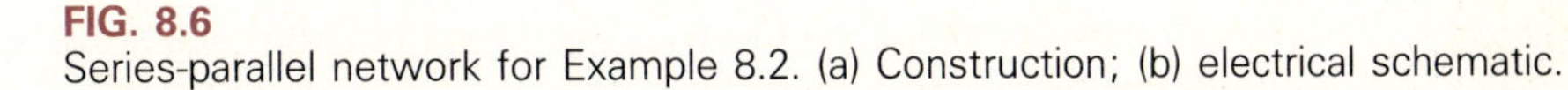

FIG. 8.6
Series-parallel network for Example 8.2. (a) Construction; (b) electrical schematic.

Solutions The electrical equivalent of Fig. 8.6(a) appears in Fig. 8.6(b). All resistor values are commercially available values, as listed in Table 4.7.

a. Since resistors R_2 and R_3 are in series, they can be combined as

$$R' = R_2 + R_3 = 1.8\ \Omega + 8.2\ \Omega = 10\ \Omega$$

The network then appears as shown in Fig. 8.7.

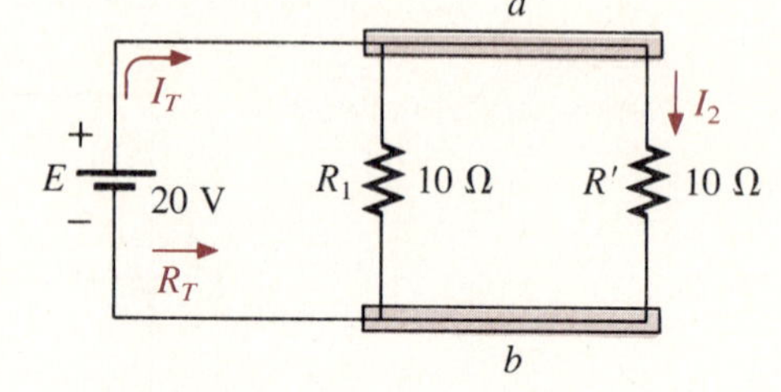

FIG. 8.7
Reduced equivalent of Fig. 8.6.

Note in Fig. 8.7 that I_2 is still the current through R', since the current through series elements is the same. The voltage V_3 was lost, however, since it is the voltage across the resistor R_3 and not the series combination of R_2 and R_3.

The total resistance R_T is now the parallel combination of R_1 and R', and

$$R_T = R_1 \| R' = \frac{R_1 R'}{R_1 + R'}$$

$$= \frac{(10\ \Omega)(10\ \Omega)}{10\ \Omega + 10\ \Omega} = \frac{100\ \Omega}{20} = \mathbf{5\ \Omega}$$

or

$$R_T = \frac{R}{N} = \frac{10\ \Omega}{2} = \mathbf{5\ \Omega}$$

since the resistors are the same value.

b. Using Ohm's law,

$$I_T = \frac{E}{R_T} = \frac{20\ \text{V}}{5\ \Omega} = \mathbf{4\ A}$$

c. The current I_2 can be determined in two ways. Since the network of Fig. 8.7 is a totally parallel network, the voltage across each element is the same. The voltage across both R_1 and R' is therefore 20 V. Applying Ohm's law:

$$I_2 = \frac{E}{R'} = \frac{20\ \text{V}}{10\ \Omega} = \mathbf{2\ A}$$

However, the current entering *equal* parallel resistors splits equally, and therefore

$$I_2 = \frac{I_T}{2} = \frac{4\ \text{A}}{2} = \mathbf{2\ A}$$

d. For V_3 we must return to Fig. 8.6(b), where Ohm's law provides

$$V_3 = I_3R_3 = I_2R_3$$
$$= (2\text{ A})(8.2\ \Omega)$$
$$V_3 = \mathbf{16.4\ V}$$

By the voltage divider rule:

$$V_3 = \frac{R_3E}{R_3 + R_2} = \frac{(8.2\ \Omega)(20\text{ V})}{8.2\ \Omega + 1.8\ \Omega}$$
$$= \frac{164\text{ V}}{10} = \mathbf{16.4\ V}$$

The preceding networks are only two of an infinite variety of possibilities for series-parallel combinations. In each block of Fig. 8.1 or 8.5 there could have been a series, parallel, or series-parallel combination. In the next example, the network of Fig. 8.6 is modified to have two series branches in parallel. The analysis is similar to that just introduced, but there are a few additional considerations to be aware of.

EXAMPLE 8.3

For the network of Fig. 8.8, determine

a. R_T
b. I_1
c. V_2
d. V_4
e. V_{ab}

FIG. 8.8
Network for Example 8.3.

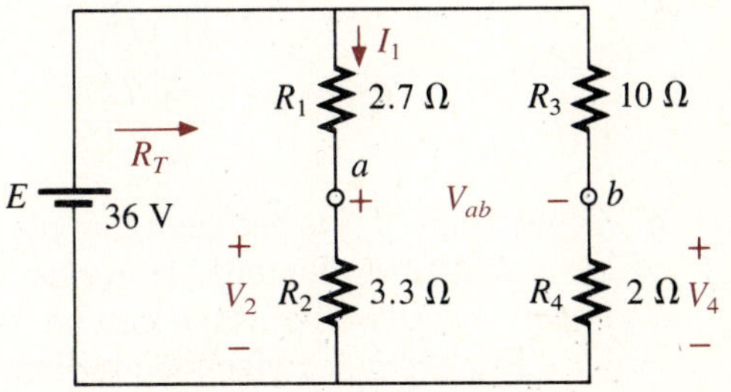

Solutions

a. By inspection, we note that R_1 and R_2 are in series and R_3 and R_4 are in series. The total resistance for each branch is then determined by

$$R' = R_1 + R_2 = 2.7\ \Omega + 3.3\ \Omega = 6\ \Omega$$

and

$$R'' = R_3 + R_4 = 10\ \Omega + 2\ \Omega = 12\ \Omega$$

resulting in the configuration of Fig. 8.9.

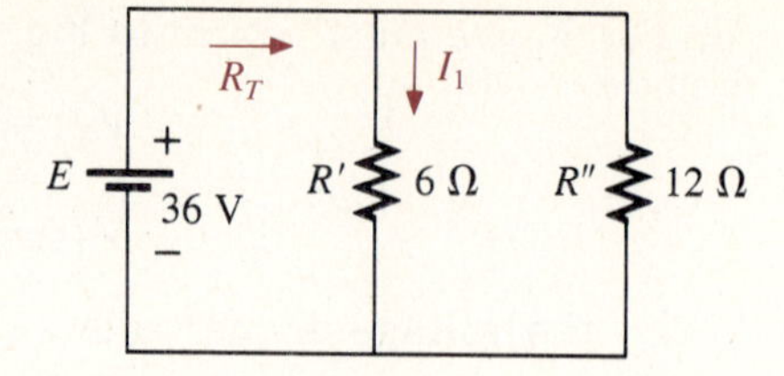

FIG. 8.9
Reduced equivalent of Fig. 8.8.

Note that the redrawing of the network has resulted in the loss of V_2, V_4, and V_{ab}, but R_T and I_1 are still defined.

Since R' and R'' are in parallel,

$$R_T = R' \| R'' = \frac{R'R''}{R' + R''}$$

$$= \frac{(6\ \Omega)(12\ \Omega)}{6\ \Omega + 12\ \Omega} = \frac{72\ \Omega}{18} = \mathbf{4\ \Omega}$$

b. The voltage is the same across parallel elements, resulting in

$$I_1 = \frac{E}{R'} = \frac{36\ \text{V}}{6\ \Omega} = \mathbf{6\ A}$$

c. For the remaining unknowns, we must return to the configuration of Fig. 8.8, where

$$V_2 = I_2R_2 = I_1R_2 = (6\ \text{A})(3.3\ \Omega)$$
$$= \mathbf{19.8\ V}$$

d. Applying the voltage divider rule:

$$V_4 = \frac{R_4E}{R_4 + R_3} = \frac{(2\ \Omega)(36\ \text{V})}{2\ \Omega + 10\ \Omega}$$

$$= \frac{72\ \text{V}}{12} = \mathbf{6\ V}$$

e. The voltage V_{ab} is between two points in the network that are not connected by a resistive element. However, a difference in potential does exist between these two points; it can be determined using Kirchhoff's voltage law.

For the clockwise loop indicated in Fig. 8.10, an application of the law results in

$$+V_2 - V_{ab} - V_4 = 0$$
$$V_{ab} = V_2 - V_4$$

Substituting gives

$$V_{ab} = 19.8\ \text{V} - 6\ \text{V}$$
$$= \mathbf{13.8\ V}$$

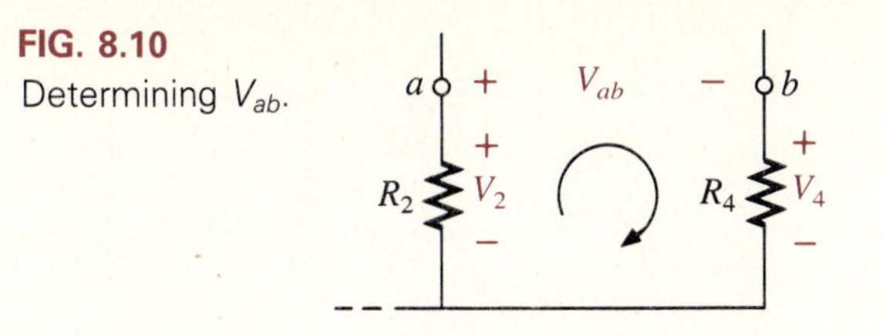

FIG. 8.10
Determining V_{ab}.

8.3 ILLUSTRATIVE EXAMPLES USING NOTATIONAL ABBREVIATIONS

The examples that follow use the type of notation introduced in Chapter 6, which is often employed on schematics of electrical systems. It permits a cleaner-looking diagram with a reduced amount of clutter.

EXAMPLE 8.4 For the network of Fig. 8.11, determine

a. I_1
b. I_3
c. V_4

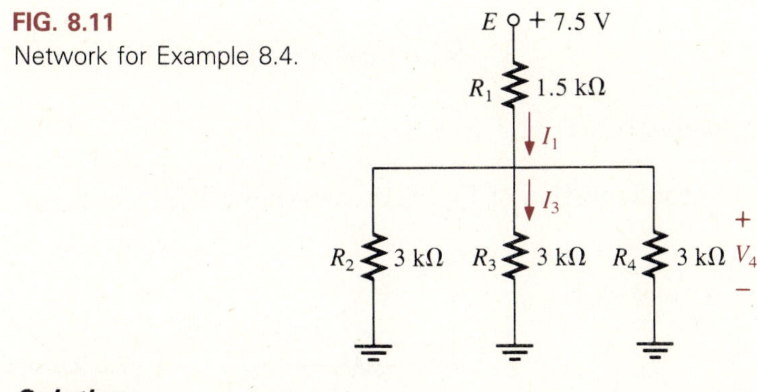

FIG. 8.11
Network for Example 8.4.

Solutions

a. For clarity, the network is redrawn in Fig. 8.12. The three resistors R_2, R_3, and R_4 are in parallel, with a total resistance

$$R' = R_2 \| R_3 \| R_4 = \frac{R}{N} = \frac{3\text{ k}\Omega}{3} = 1\text{ k}\Omega$$

and

$$\begin{aligned} R_T &= R_1 + R' = 1.5\text{ k}\Omega + 1\text{ k}\Omega \\ &= 2.5\text{ k}\Omega \end{aligned}$$

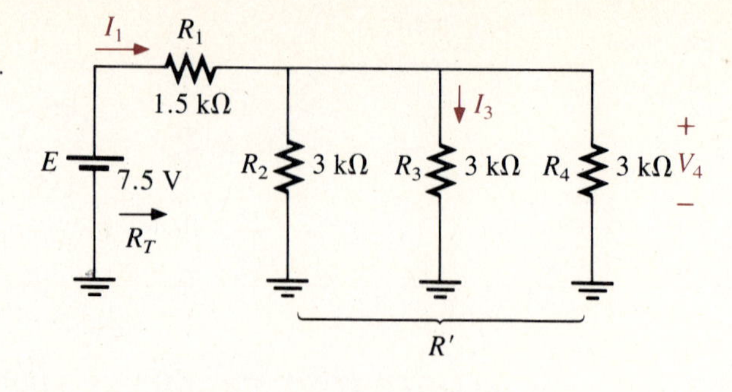

FIG. 8.12
Network of Fig. 8.11 redrawn.

Thus

$$I_1 = \frac{E}{R_T} = \frac{7.5 \text{ V}}{2.5 \text{ k}\Omega} = \mathbf{3 \text{ mA}}$$

b. Since $R_2 = R_3 = R_4$, the current I_1 entering the three parallel branches of Fig. 8.10 splits equally, and

$$I_3 = \frac{I_1}{3} = \frac{3 \text{ mA}}{3} = \mathbf{1 \text{ mA}}$$

c.

$$I_4 = I_3 = I_2 = 1 \text{ mA}$$

and

$$V_4 = I_4 R_4 = (1 \text{ mA})(3 \text{ k}\Omega) = \mathbf{3 \text{ V}}$$

EXAMPLE 8.5 For the configuration of Fig. 8.13, determine V_1, I_1, and V_2.

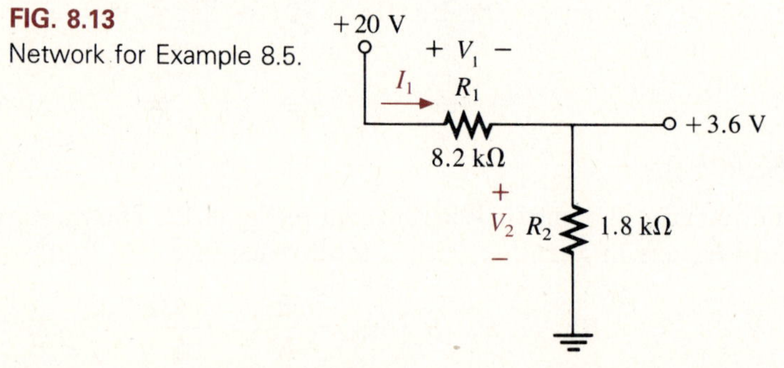

FIG. 8.13
Network for Example 8.5.

Solution In this case there are 20 V at one end of R_1 and +3.6 V tied to the other end. The voltage across R_1 is the difference between these two, with the positive end at the 20-V level.

That is,

$$V_1 = 20 \text{ V} - 3.6 \text{ V} = \mathbf{16.4 \text{ V}}$$

The current I_1 is

$$I_1 = \frac{V_1}{R_1} = \frac{16.4\ \text{V}}{8.2\ \text{k}\Omega} = \mathbf{2\ mA}$$

The voltage V_2 is the voltage across the 1.8-kΩ resistor. A second look also reveals that the resistor R_2 is connected to a 3.6-V supply at one end and ground potential (0 V) at the other end. The result is that the voltage across R_2 is the same as the 3.6-V supply level and

$$V_2 = \mathbf{3.6\ V}$$

8.4 LADDER NETWORKS

Ladder networks are so named due to their similarity in structure to a ladder, as shown in Fig. 8.14. Once the analysis of two or three sections is understood, the analysis of a longer section is quite similar. A detailed analysis of a two-section network is provided in the next example.

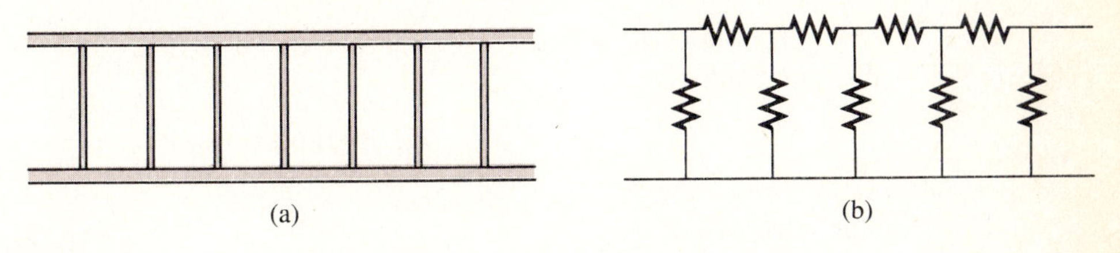

FIG. 8.14
Ladder network. (a) Mechanical analogy; (b) electrical schematic.

EXAMPLE 8.6 Determine the current I_4 for the network of Fig. 8.15.

FIG. 8.15
Network for Example 8.6.

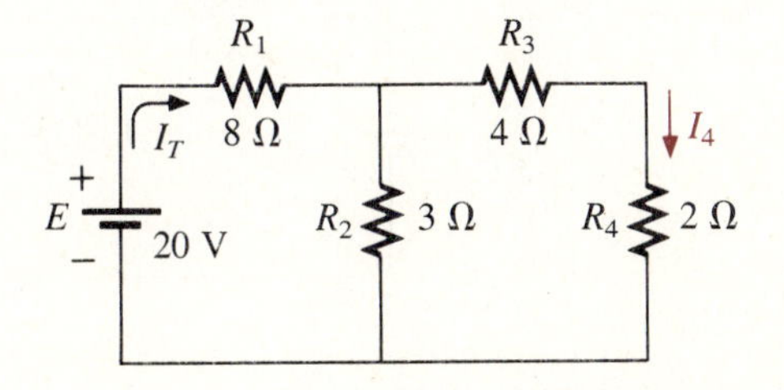

Solution The analysis of ladder networks is typically one where we first work back to the source to determine R_T and I_T and then work back to the desired quantity. A review of the network reveals that the only series resistors are R_3 and R_4. There are parallel branches but no parallel elements. The result is that only R_3 and R_4 can be combined and replaced by a single element.

$$R' = R_3 + R_4 = 4\ \Omega + 2\ \Omega = 6\ \Omega$$

The network then appears as shown in Fig. 8.16. Note in the figure that we have lost the desired unknown I_4 but have reduced the network to a simpler form.

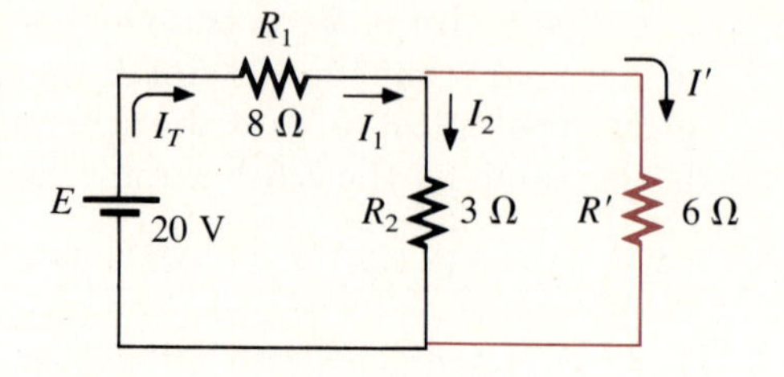

FIG. 8.16
Reduced version of Fig. 8.15.

The resistors R_2 and R' are now in parallel, and

$$R'' = R_2 \| R' = \frac{R_2 R'}{R_2 + R'}$$
$$= \frac{(3\ \Omega)(6\ \Omega)}{3\ \Omega + 6\ \Omega} = \frac{18\ \Omega}{9} = 2\ \Omega$$

The network will then appear as shown in Fig. 8.17. It is now fairly obvious that

$$R_T = R_1 + R''$$
$$= 8\ \Omega + 2\ \Omega = 10\ \Omega$$

The current is

$$I_T = \frac{E}{R_T} = \frac{20\ \text{V}}{10\ \Omega} = \mathbf{2\ A}$$

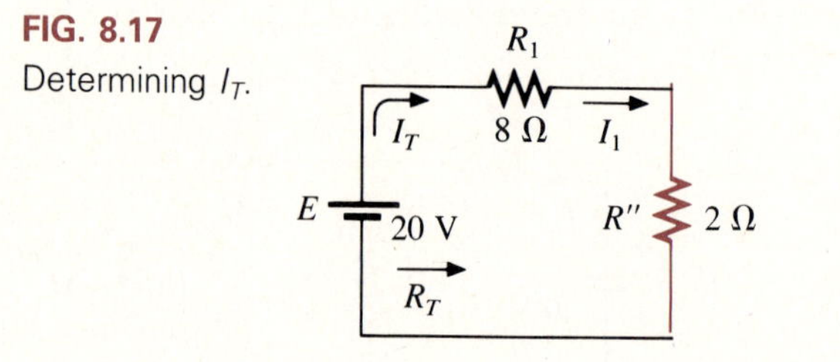

FIG. 8.17
Determining I_T.

We must now work back to I_4 by first noting that $I_1 = I_T = 2$ A; as shown in Fig. 8.16, the current I_1 splits up into I_2 and I'.

Applying the current divider rule to Fig. 8.16 gives

$$I' = \frac{R_2 I_1}{R_2 + R'} = \frac{(3\ \Omega)(2\ \text{A})}{3\ \Omega + 6\ \Omega} = \frac{6\ \text{A}}{9}$$
$$= \frac{2}{3}\ \text{A}$$

Looking back at Fig. 8.15, we find

$$I_4 = I_3 = I' = \frac{2}{3}\ \mathbf{A}$$

8.5 AMMETER, VOLTMETER, AND OHMMETER DESIGN

Now that the fundamentals of series, parallel, and series-parallel networks have been introduced, we are prepared to investigate the fundamental design of an ammeter, a voltmeter, and an ohmmeter. Our design of each employs the d'Arsonval movement of Fig. 8.18. The movement consists basically of an iron-core coil mounted on bearings between a permanent magnet. The helical springs limit the turning motion of the coil and provide a path for the current to reach the coil. When a current is passed through the movable coil, the fluxes of the coil and permanent magnet interact to develop a torque on the coil that will cause it to rotate on its bearings. The movement is adjusted to indicate zero deflection on a meter scale when the current through the coil is zero. The direction of current through the coil will then determine whether the pointer will display an up-scale or below-zero indication. For this reason, ammeters and voltmeters have an assigned polarity on their terminals to ensure an up-scale reading.

FIG. 8.18
D'Arsonval movement. (Courtesy of Weston Instruments, Inc.)

D'Arsonval movements are usually rated by current and resistance. The specifications of a typical movement might be 1 mA, 50 Ω. The 1 mA is the *current sensitivity (CS)* of the movement, which is the current required for a full-scale deflection. It will be denoted by the symbol I_{CS}. The 50 Ω represents the internal resistance (R_m) of the movement. A common notation for the movement and its specifications is provided in Fig. 8.19.

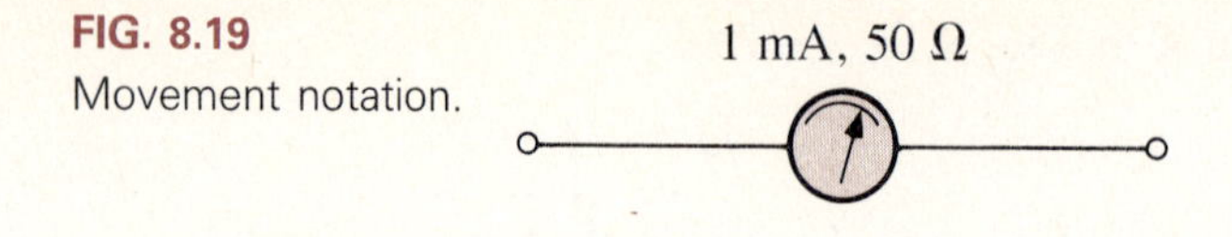

FIG. 8.19
Movement notation.

The Ammeter The maximum current that the d'Arsonval movement can read independently is equal to the current sensitivity of the movement. However, higher currents can be measured if additional circuitry is introduced. This additional circuitry, as shown in Fig. 8.20, results in the basic construction of an ammeter.

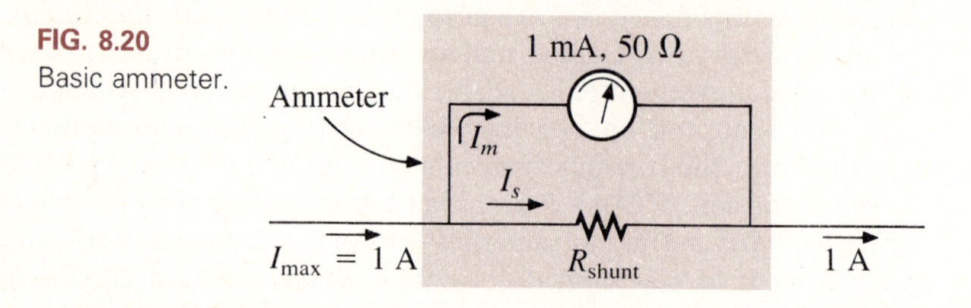

FIG. 8.20
Basic ammeter.

The resistance R_{shunt} is chosen for the ammeter of Fig. 8.20 to allow 1 mA to flow through the movement when a maximum current of 1 A enters the ammeter. If less than 1 A should flow through the ammeter, the movement will have less than 1 mA flowing through it and will indicate less than full-scale deflection.

Since the voltage across parallel elements must be the same, the potential drop across the movement in Fig. 8.20 must equal that across the resistor; that is,

$$\begin{aligned} V_{movement} &= V_{R_{shunt}} \\ (1 \text{ mA})(50\ \Omega) &= I_s R_{shunt} \end{aligned}$$

Also, I_s must equal 1 A − 1 mA = 999 mA if the current is to be limited to 1 mA through the movement (Kirchhoff's current law). Therefore,

$$\begin{aligned} (1 \times 10^{-3}\text{ A})(50\ \Omega) &= (999 \times 10^{-3}\text{ A})(R_{shunt}) \\ R_{shunt} &= \frac{(1 \times 10^{-3}\text{ A})(50\ \Omega)}{999 \times 10^{-3}\text{ A}} \\ &\cong \mathbf{0.05\ \Omega} \end{aligned}$$

Calculator

(1) (EXP) (+/−) (3) (×) (5) (0) (÷) (9) (9) (9) (EXP) (+/−) (3) (=)

with a display of 0.050.

In general,

$$R_{shunt} = \frac{R_m I_{CS}}{I_{max} - I_{CS}} \tag{8.1}$$

One method of constructing a multirange ammeter is shown in Fig. 8.21, where the rotary switch determines the R_{shunt} to be used for the maximum current indicated on the face of the meter. Most meters employ the same scale for various values of maximum current. If you read 375 on the 0–5 mA scale with the switch on the 5 setting, the current is 3.75 mA; on the 50 setting, the current is 37.5 mA; and so on.

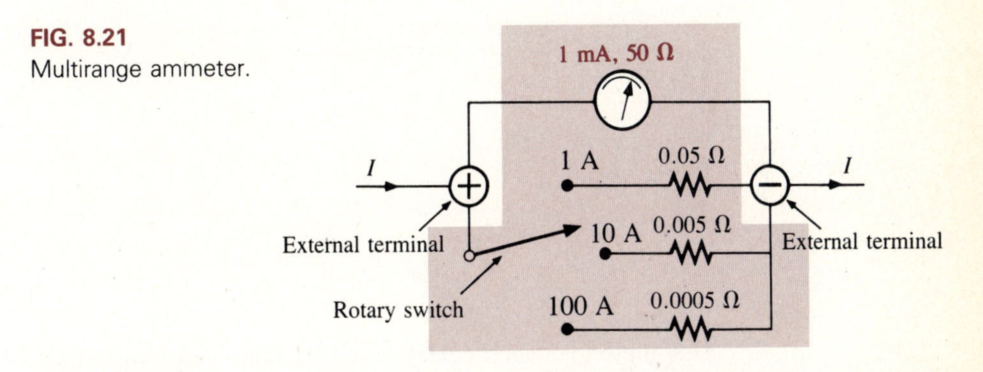

FIG. 8.21
Multirange ammeter.

The Voltmeter A variation in the additional circuitry permits the use of the d'Arsonval movement in the design of a voltmeter. The 1-mA, 50-Ω movement can also be rated as a 50-mV (1 mA × 50 Ω), 50-Ω movement, indicating that the maximum voltage that the movement can measure independently is 50 mV. The millivolt rating is sometimes referred to as the *voltage sensitivity (VS)*. The basic construction of the voltmeter is shown in Fig. 8.22.

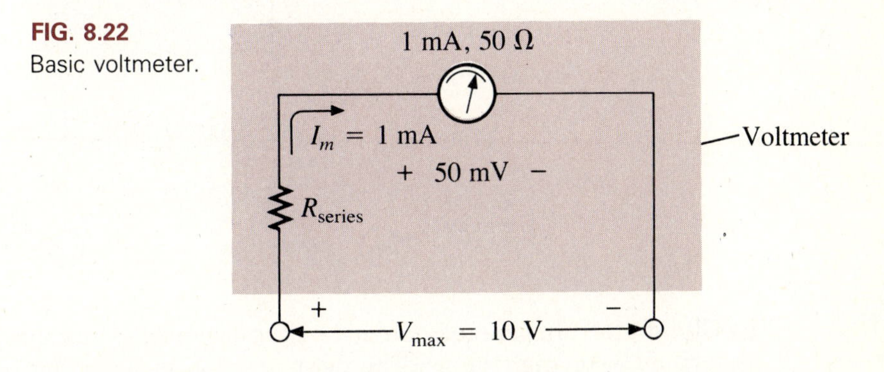

FIG. 8.22
Basic voltmeter.

The R_{series} is adjusted to limit the current through the movement to 1 mA when the maximum voltage is applied across the voltmeter. A lesser voltage would simply reduce the current in the circuit and, thereby, the deflection of the movement.

Applying Kirchhoff's voltage law around the closed loop of Fig. 8.21, we obtain

$$V_{max} - I_m R_{series} - VS = 0$$
$$[10\text{ V} - (1 \times 10^{-3}\text{ A})(R_{series})] - 50 \times 10^{-3}\text{ V} = 0$$

or

$$R_{series} = \frac{10\text{ V} - (50 \times 10^{-3}\text{ V})}{1 \times 10^{-3}\text{ A}} = \mathbf{9950\ \Omega}$$

Calculator

(1) (0) (−) (5) (0) (EXP) (+/−) (3) (=) (÷) (1) (EXP) (+/−) (3) (=)

with a display of 9950.

In general,

$$R_{series} = \frac{V_{max} - V_{VS}}{I_{CS}} \tag{8.2}$$

One method of constructing a multirange voltmeter is shown in Fig. 8.23. If the rotary switch is at 10 V, R_{series} = 9.950 kΩ; at 50 V, R_{series} = 40 kΩ + 9.950 kΩ = 49.950 kΩ; and at 100 V, R_{series} = 50 kΩ + 40 kΩ + 9.950 kΩ = 99.950 kΩ.

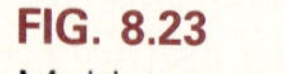

FIG. 8.23
Multirange voltmeter.

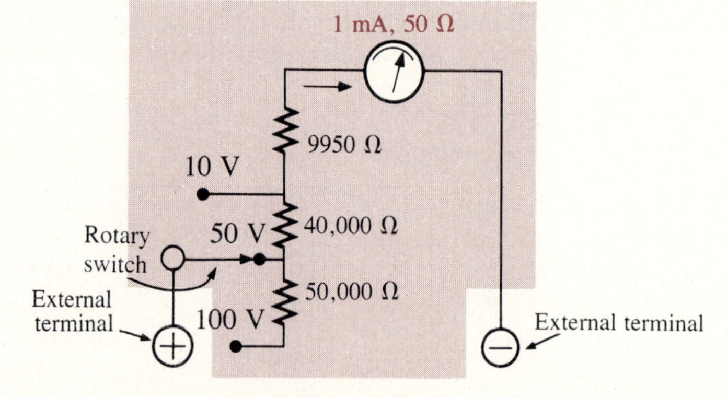

The Ohmmeter In general, ohmmeters are designed to measure resistance in the low, mid-, or high range. The most common is the *series ohmmeter,* designed to read resistance levels in the midrange. It employs the series configuration of Fig. 8.24. The design is quite different from that of the ammeter or voltmeter in that it will show a full-scale deflection for 0 Ω and no deflection for infinite resistance.

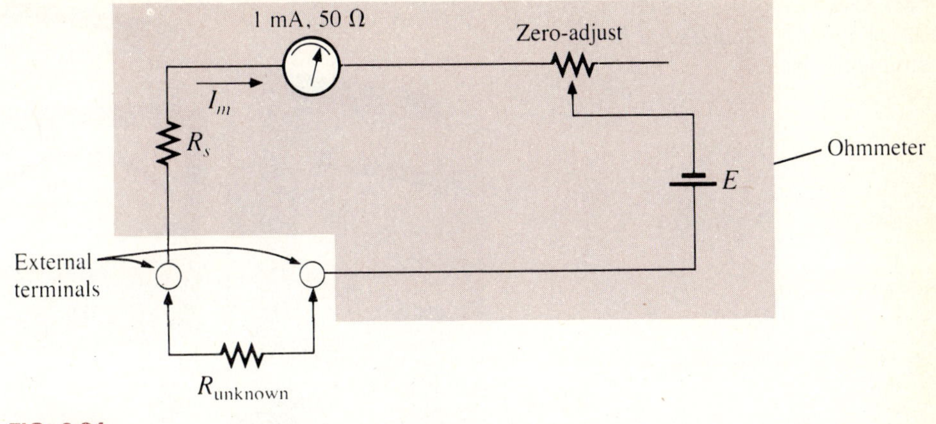

FIG. 8.24
Series ohmmeter.

To determine the series resistance R_s, the external terminals are shorted (a direct connection of 0 Ω between the two) to simulate 0 Ω, and the zero-adjust is set to half its maximum value. The resistance R_s is then adjusted to allow a current equal to the current sensitivity of the movement (1 mA) to flow in the circuit. The zero-adjust is set to half its value so that any variation in the components of the meter that may produce a current more or less than the current sensitivity can be compensated for. The current I_m is, by Ohm's law,

$$I_m \text{ (full scale)} = I_{CS} = \frac{E}{R_s + R_m + \dfrac{\text{zero-adjust}}{2}} \tag{8.3}$$

Using some basic algebra,

$$R_s = \frac{E}{I_{CS}} - R_m - \frac{\text{zero-adjust}}{2} \tag{8.4}$$

If an unknown resistance is then placed between the external terminals, the current will be reduced, causing a deflection less than full scale. If the terminals are left open, simulating infinite resistance, the pointer will not deflect, since the current through the circuit is zero.

An instrument designed to read very low values of resistance appears in Fig. 8.25. Because of its low-range capability, the network design must be a great deal more sophisticated than described earlier. It employs electronic components that eliminate the inaccuracies introduced by lead and contact resistances. It is similar to the preceding system in the sense that it is completely portable and does require a dc battery to establish measurement conditions. Note that special leads designed to limit any introduced resistance levels. The maximum scale setting can be set as low as 0.00352 (3.52 mΩ).

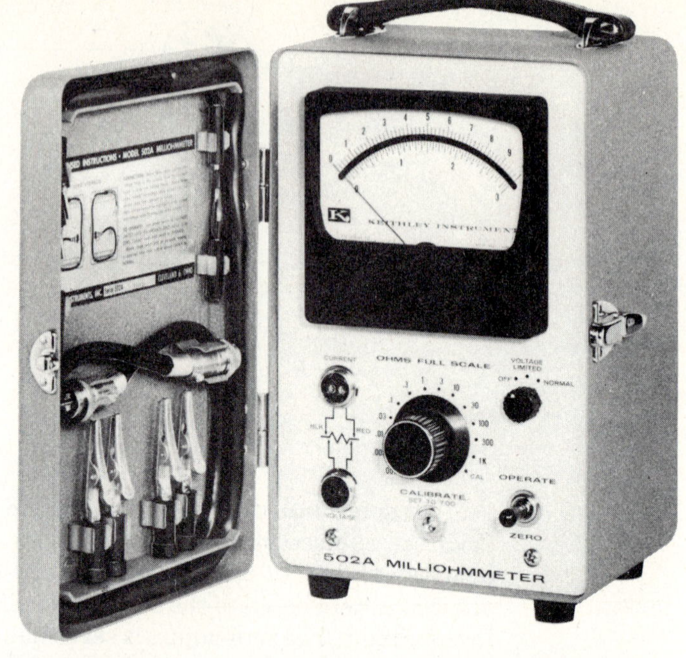

FIG. 8.25
Milliohmmeter. (Courtesy of Keithley Instruments, Inc.)

The Megger® tester is an instrument for measuring very high resistance values. The term *Megger* is derived from the fact that the device measures resistance values in the megohm range. Its primary function is to test the insulation found in power-transmission systems, electrical machinery, transformers, and so on. To measure the high-resistance values, a high dc voltage is

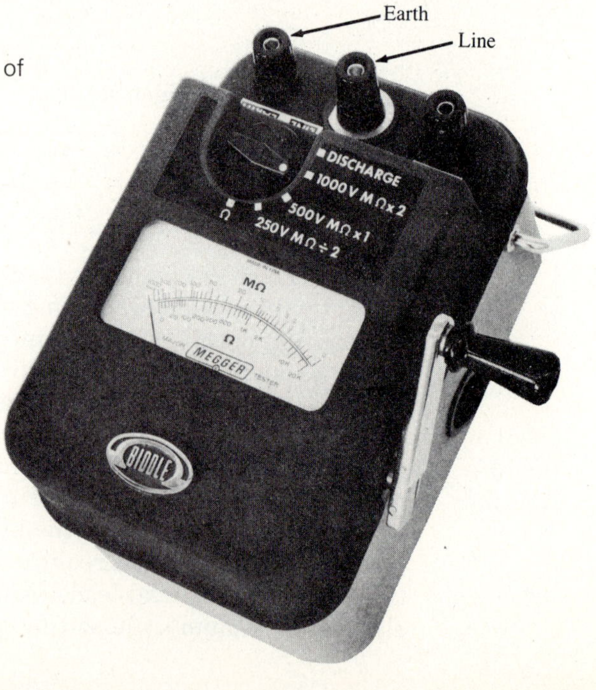

FIG. 8.26
The Megger® tester. (Courtesy of James G. Biddle Co.)

established by a hand-driven generator. If the shaft is rotated above some set value, the output of the generator will be fixed at one selectable voltage, typically 250, 500, or 1000 V. A photograph of the commercially available Megger® tester is shown in Fig. 8.26. The unknown resistance is connected between the terminals marked *line* and *earth*. For this instrument, the range is zero to 2000 MΩ.

CHAPTER SUMMARY

The key words in the analysis of series-parallel networks are *preparation, reduction,* and *practice*.

Preparation

It is critical that a plan of attack be carefully developed before diving into a network to avoid approaches that lead to dead ends and to identify channels that can save time and energy. It is usually possible to establish mentally the "best" approach to a complete solution of the exercise without making a single calculation. Leave the numerical calculations for the *next* phase of the exercise—first choose the approach.

Reduction

Once the approach is established, take the time to redraw the network after each combination of elements and include in the reduced network all those unknowns unaffected by the reduction. Often the reduced networks will be utilized in the "return trip" to provide unknown quantities using equations simpler than those required with the original network.

Practice

It is virtually impossible to become adept at solving series-parallel networks without attacking a number and variety of problems. Read the provided examples carefully, or, in fact, try the examples before looking at the solutions and then compare notes. Do as many problems at the end of the chapter as time permits. Solutions to all the odd-numbered problems are provided at the end of the text.

Finally, the key to successful analysis of series-parallel networks is an understanding of series and parallel networks, a conclusion strongly suggesting a review of Chapters 6 and 7 as you proceed through this chapter.

GLOSSARY

d'Arsonval movement An iron-core coil mounted on bearings between a permanent magnet. A pointer connected to the movable core indicates the strength of the current passing through the coil.

Ladder network A network that consists of a cascaded set of series-parallel combinations and has the appearance of a ladder.

Megger® tester An instrument for measuring very high resistance levels, such as in the megohm range.

Series ohmmeter A resistance-measuring instrument in which the movement is placed in series with the unknown resistance.

Series-parallel network A network consisting of a combination of both series and parallel branches.

Shunt ohmmeter A resistance-measuring instrument in which the movement is placed in parallel with the unknown resistance.

PROBLEMS

Section 8.2

1. For the network of Fig. 8.27, determine
   - a. R_T
   - b. I_T
   - c. I_1, I_2, and I_3
   - d. V_1, V_2, and V_3
   - e. Power to R_3 (P_3)
2. For the network of Fig. 8.28, determine
   - a. R_T
   - b. I_T
   - c. I_1, I_3, and I_4
   - d. Power delivered by battery (P_{del})

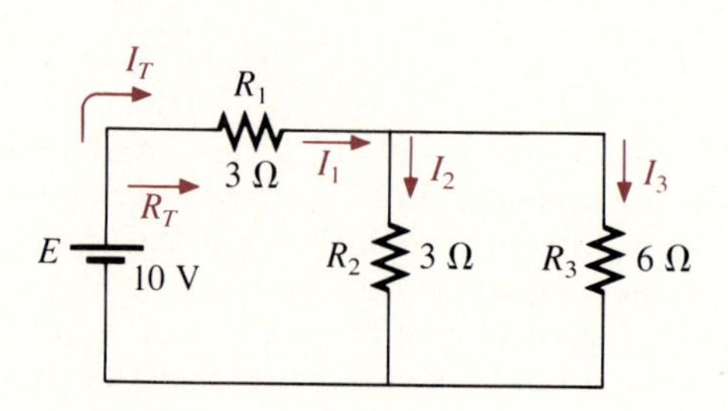

FIG. 8.27

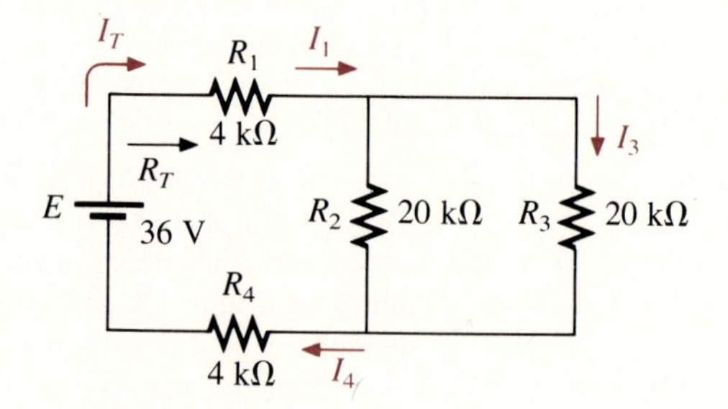

FIG. 8.28

3. For the network of Fig. 8.29, determine
   - a. V_1
   - b. I_1, I_3, and I_T

*4. For the network of Fig. 8.30, determine
   - a. I_T
   - b. V_2 and V_4

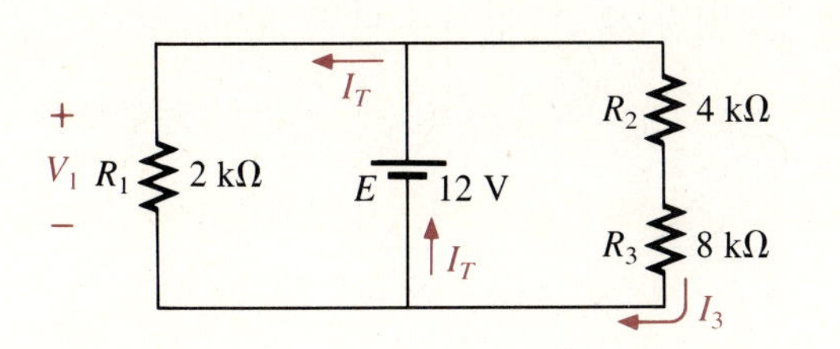

FIG. 8.29

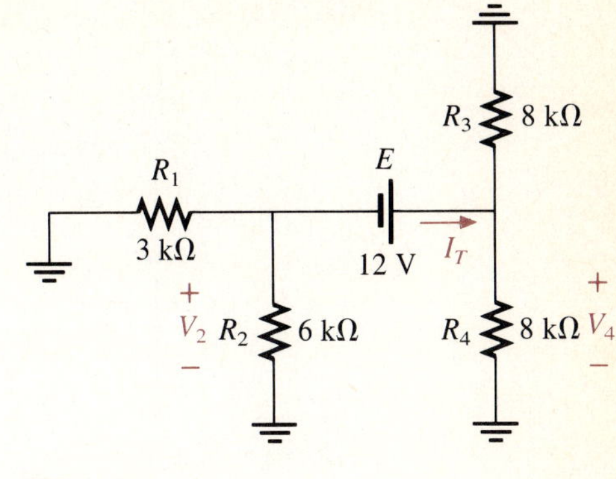

FIG. 8.30

*5. For the network of Fig. 8.31, determine
 a. I_1 and I_3
 b. V_1 and V_3
 c. V_{ab}

6. For the network of Fig. 8.32, determine
 a. R_T
 b. I_T
 c. I_2 and I_3

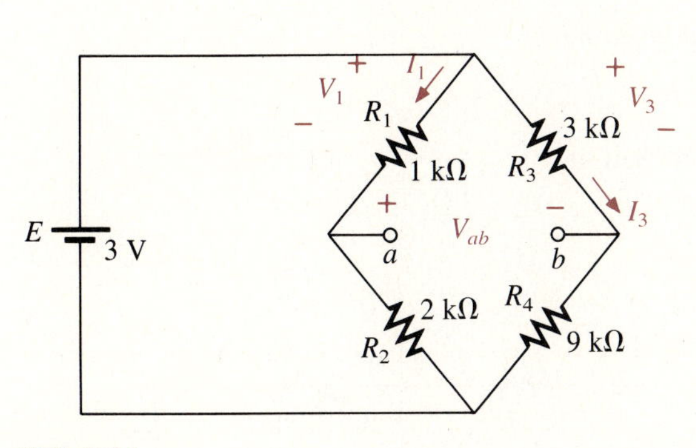

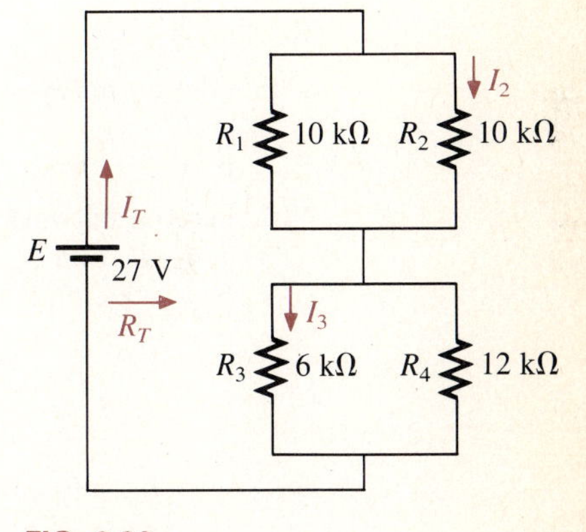

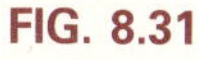
FIG. 8.31

FIG. 8.32

7. For the network of Fig. 8.33, determine
 a. R_T
 b. I_T
 c. I_1, I_2, and I_4
 d. V_a, V_b, and V_c

*8. For the network of Fig. 8.34, find the resistance R_3 if the current through it is 2 A.

FIG. 8.33

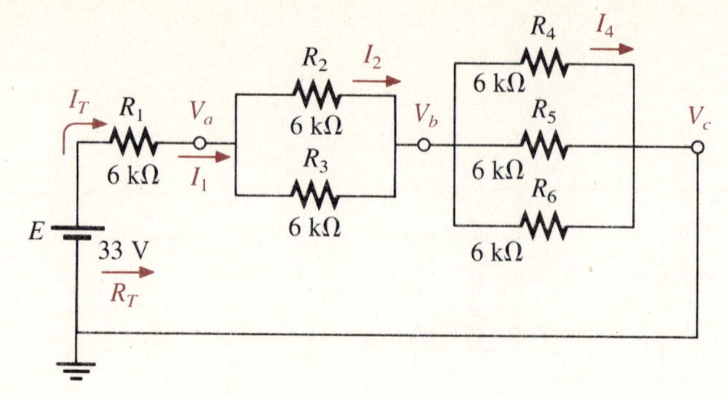

FIG. 8.34

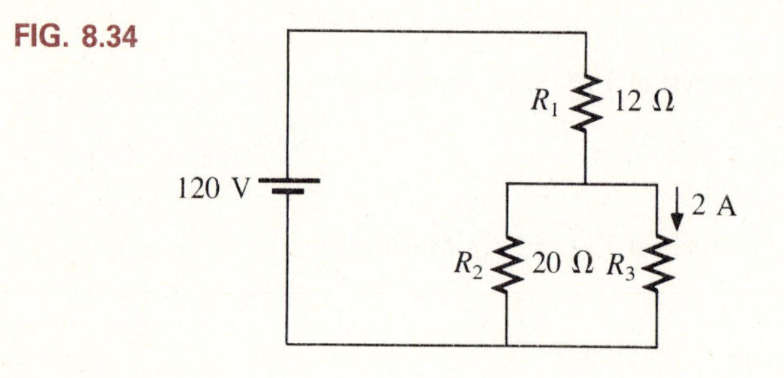

Section 8.3

***9.** For the network of Fig. 8.35, determine
   a. I_2
   b. V_a

***10.** For the network of Fig. 8.36, determine
   a. I_1, I_2, and I_3
   b. V_1, V_2, and V_3

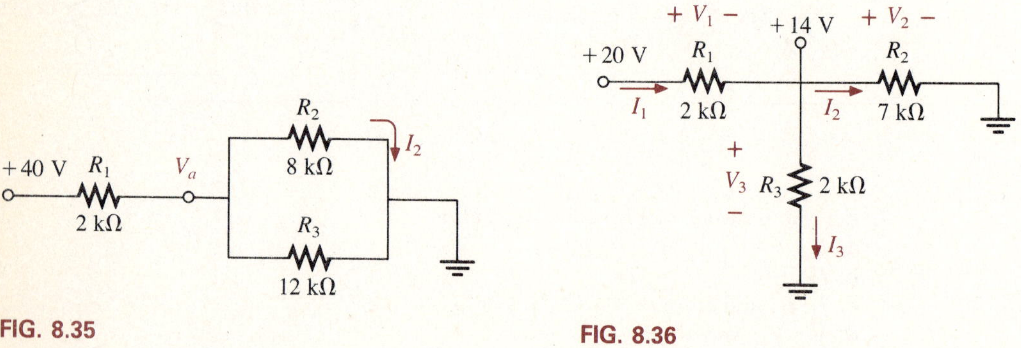

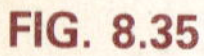
FIG. 8.35

FIG. 8.36

***11.** For the network of Fig. 8.37, determine
   a. I_1 and I_3
   b. V and V_4

FIG. 8.37

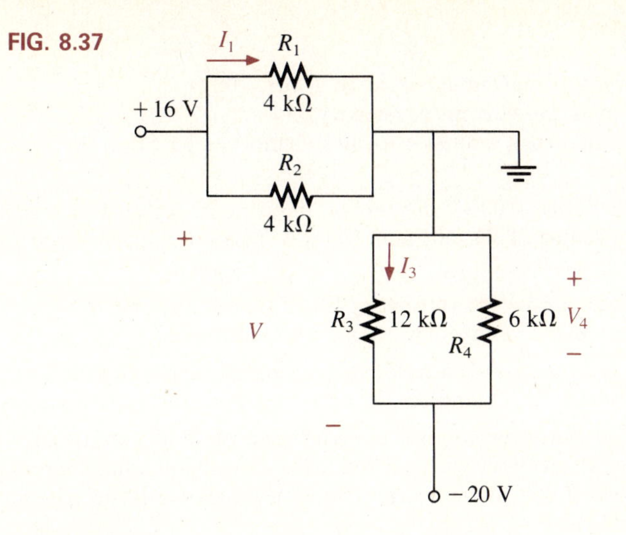

Section 8.4

12. For the network of Fig. 8.38, determine

a. R_T

b. I_T

c. I_2 and I_4

d. P_4 (power to R_4)

e. Power delivered by source (P_{del}); compare it to P_4.

13. For the network of Fig. 8.39, determine V_2 and V_5.

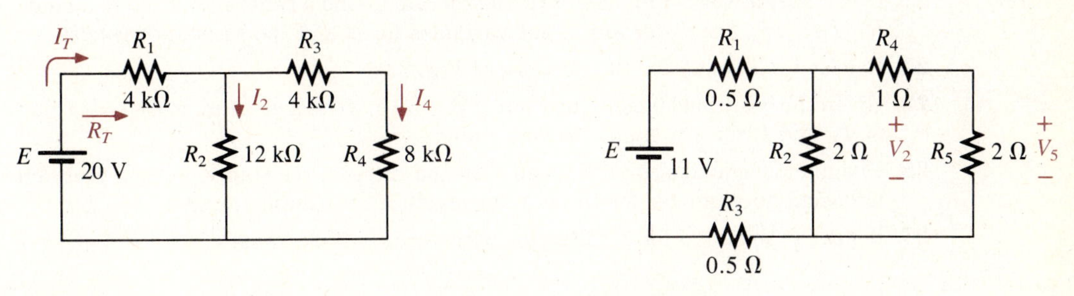

FIG. 8.38

FIG. 8.39

***14.** For the ladder network of Fig. 8.40, determine I_6.

FIG. 8.40

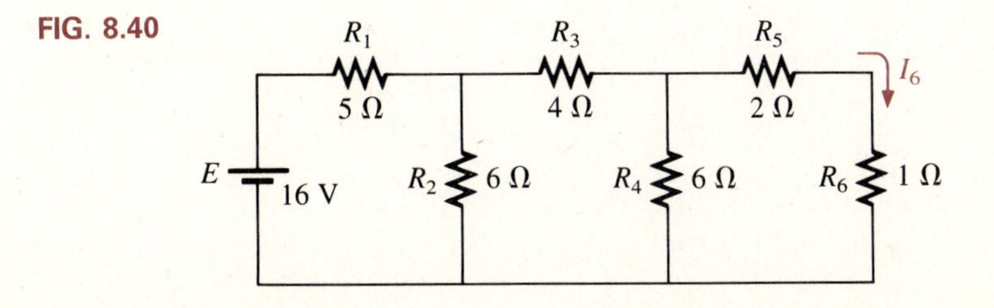

Section 8.5

15. A d'Arsonval movement is rated 1 mA, 100 Ω.
    a. What is the current sensitivity?
    b. Design a 20-A ammeter using this movement. Show the circuit and component values.
16. Using a 50-μA, 1000-Ω d'Arsonval movement, design a multirange milliammeter having scales of 25, 50, and 100 mA. Show the circuit and component values.
17. A d'Arsonval movement is rated 50 μA, 1000 Ω.
    a. Design a 15-V dc voltmeter. Show the circuit and component values.
    b. What is the ohm/volt rating of the voltmeter?
18. Using a 1-mA, 100-Ω d'Arsonval movement, design a multirange voltmeter having scales of 5, 50, and 500 V. Show the circuit and component values.
19. A digital meter has an internal resistance of 10 MΩ on its 0.5-V range. If you had to build a voltmeter with a d'Arsonval movement, what current sensitivity would you need if the meter were to have the same internal resistance on the same voltage scale?
20. a. Design a series ohmmeter using a 100-μA, 1000-Ω movement, a zero-adjust with a maximum value of 2 kΩ, a battery of 3 V, and a series resistor whose value is to be determined.
    b. Find the resistance required for full scale, 3/4 scale, 1/2 scale, and 1/4 scale deflection.
    c. Using the results of part b, draw the scale to be used with the ohmmeter.
21. Describe the basic construction and operation of the Megger®.

Computer Problems

22. Given the network of Fig. 8.3, write a program to find a general solution to include R_T, I_T, I_1, I_2, and I_3 for any input variables for E and the resistance levels.
23. Repeat Problem 22 for the network of Fig. 8.6.
24. Given the basic ladder configuration of Fig. 8.15, write a program to find R_T, I_T, I_1, I_2, I_3, and I_4 for any input network values.
25. Given a movement's current sensitivity and internal resistance, write a program to design an ammeter for three input levels of maximum current.
26. Repeat Problem 25 for a voltmeter with three voltage ranges.

9

Methods of Analysis and Network Theorems

OBJECTIVES

- ☐ Become familiar with the characteristics of a current source.
- ☐ Apply superposition to a network with more than one voltage or current source.
- ☐ Understand the impact of Thevenin's theorem and learn how to apply the theorem.
- ☐ Understand the conditions and equations associated with the maximum power transfer theorem.
- ☐ Be able to apply branch-current analysis to a multiloop network.
- ☐ Understand how mesh analysis is applied to a multiloop network.
- ☐ Become aware of the nodal-analysis technique and the significance of the results obtained.
- ☐ Be able to perform a Δ-Y or Y-Δ conversion.

9.1 INTRODUCTION

The circuits described in the previous chapters had only one source or two or more sources in series or parallel present. The step-by-step procedure outlined in those chapters cannot be applied if two or more sources in the same network are not in series or parallel. There will be an interaction of sources that will not permit the reduction technique used in Chapter 8 to find such quantities as the total resistance and source current.

Fortunately, there are several network theorems that not only permit an analysis of multiloop networks but also provide further insight into the characteristics of a configuration with one or more sources. The theorems to be included are the *superposition*, *Thevenin*, and *maximum power transfer theorems*.

Methods of analysis have also been developed that allow us to approach, in a systematic manner, a network with any number of sources in any arrangement. Fortunately, these methods can also be applied to networks with only one source. The methods to be discussed in detail in this chapter include *branch-current analysis, mesh analysis,* and *nodal analysis.* Each can be applied to the same network. The "best" method cannot be defined by a set of rules but can be determined only by acquiring a firm understanding of the relative advantages of each.

The chapter concludes with a look at bridge networks and Δ-Y (and Y-Δ) conversions.

9.2 CURRENT SOURCES

The concept of the current source was introduced in Section 3.4 with the photograph of a commercially available unit. We must now investigate its characteristics in greater detail so that we can properly determine its effect on the networks to be examined in this chapter.

The current source is often referred to as the *dual* of the voltage source. That is, whereas a battery supplies a *fixed* voltage and the source current can vary, the current source supplies a *fixed* current to the branch in which it is located, and its terminal voltage may vary as determined by the network to which it is applied. Note from this discussion that *duality* simply implies an interchange of current and voltage to distinguish the characteristics of one source from the other.

The increasing interest in the current source is due fundamentally to semiconductor devices such as the transistor. In basic electronics courses, you will find that the transistor is a current-controlled device. In the physical model (equivalent circuit) of a transistor used in the analysis of transistor networks, there appears a current source, as indicated in Fig. 9.1. The symbol for a current source appears in Fig. 9.1. The direction of the arrow within the circle indicates the direction in which current is being supplied.

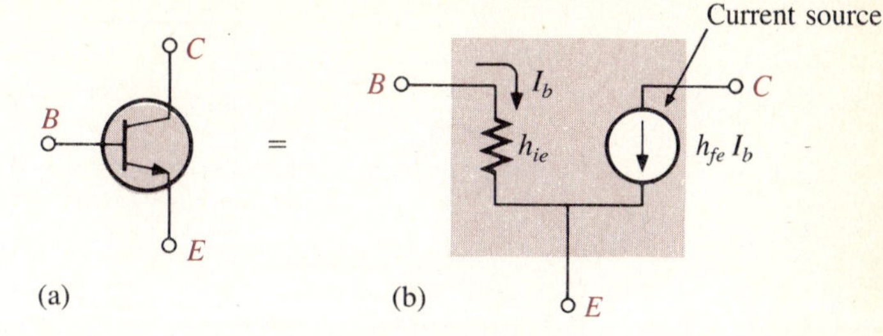

FIG. 9.1
Transistor. (a) Symbol; (b) ac equivalent circuit.

For further comparison, the terminal characteristics of a dc voltage and current source are presented in Fig. 9.2. Note that for the voltage source, the terminal voltage is fixed at E volts for the range of current values. The characteristics of the current source, shown in Fig. 9.2(b), indicate that the current source will supply a fixed current, even though the voltage across the source may vary in magnitude. This is indicated in the associated figure of Fig. 9.2(b). For the voltage source, the current direction will be determined by the remaining elements of the network. For all one-voltage-source networks, it will have the direction indicated to the right of the battery in Fig. 9.2(a). For the current source, the network to which it is connected will also determine the magnitude and polarity of the voltage across the source. For all single-current-source networks, it will have the polarity indicated to the right of the current source in Fig. 9.2(b).

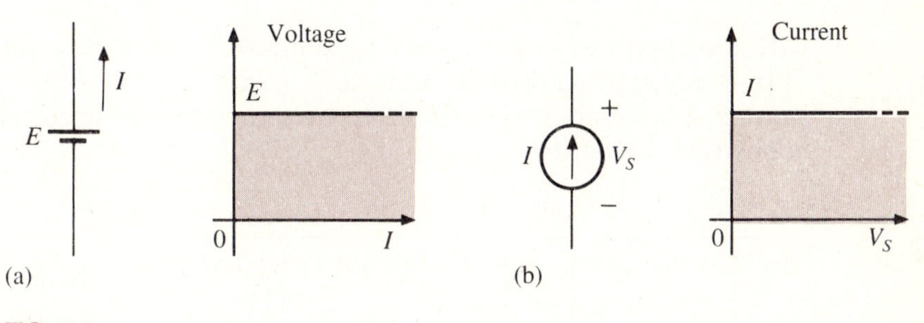

FIG. 9.2
Voltage vs. current source.

In Fig. 9.3, a current source of 5 mA has been applied to three *different* networks. In Fig. 9.3(a) the applied current source establishes a 12-V drop across the network. Since parallel elements have the same voltage, the voltage V_s across the current source has the magnitude and polarity shown in Fig. 9.3(a). In Fig. 9.3(b), the same current source has been applied across a different network. The entering current is still 5 mA, but the network is such that the voltage across it is now only 1.6 V. The result is that V_s has dropped to this new level but with the same polarity. In Fig. 9.3(c) the network contains *additional* current or voltage sources, resulting in a drop of 6 V across the network

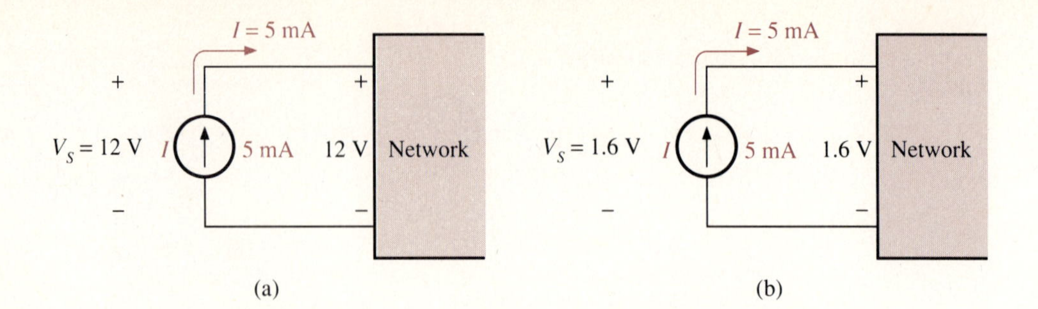

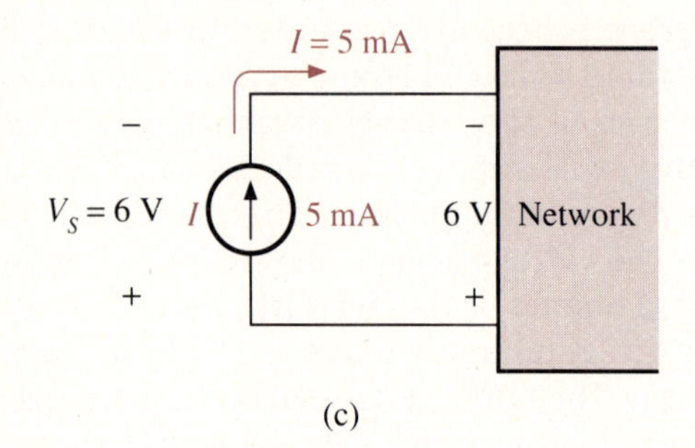

FIG. 9.3
Possible levels of V_s for a current source.

with the opposite polarity. Thus V_s is also 6 V but with a polarity to match. This is a situation very similar to a series circuit with opposing voltage sources, where the current through one source may be opposite the "pressure" applied by the battery.

EXAMPLE 9.1 Find the voltage V_s and the current I_1 for the circuit of Fig. 9.4.

FIG. 9.4
Circuit for Example 9.1.

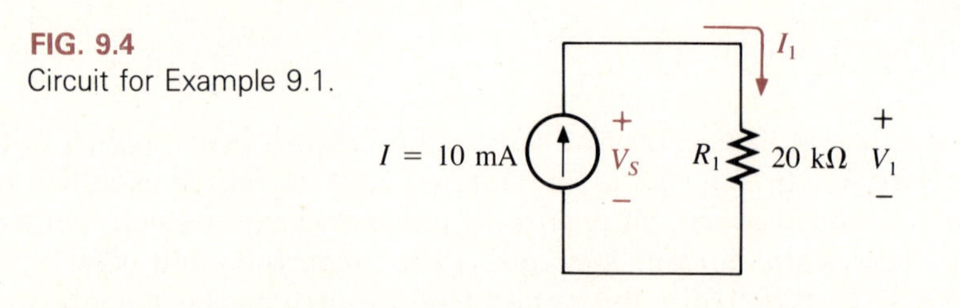

Solution

$$I_1 = I = \mathbf{10\ mA}$$

$$V_s = V_1 = I_1R_1 = (10\text{ mA})(20\text{ k}\Omega) = \mathbf{200\ V}$$

EXAMPLE 9.2

Repeat Example 9.1 with $R_1 = 5\ \text{k}\Omega$.

Solution

$$I_1 = I = \mathbf{10\ mA} \qquad (I_1 \text{ unaffected by change in } R_1)$$

$$V_s = V_1 = I_1R_1 = (10\ \text{mA})(5\ \text{k}\Omega) = \mathbf{50\ V} \qquad (V_s \text{ affected by the magnitude of load applied})$$

EXAMPLE 9.3

Calculate the voltages V_1, V_2, and V_s for the circuit of Fig. 9.5.

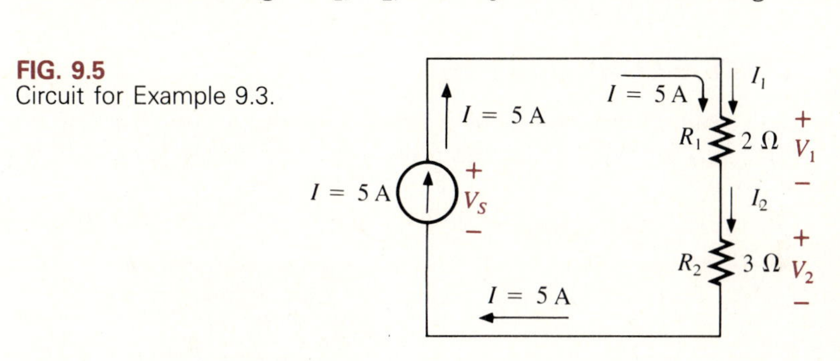

FIG. 9.5
Circuit for Example 9.3.

Solution

$$V_1 = I_1R_1 = IR_1 = (5\ \text{A})(2\ \Omega) = \mathbf{10\ V}$$

$$V_2 = I_2R_2 = IR_2 = (5\ \text{A})(3\ \Omega) = \mathbf{15\ V}$$

Applying Kirchhoff's voltage law in the clockwise direction, we obtain

$$+V_s - V_1 - V_2 = 0$$

or

$$V_s = V_1 + V_2 = 10\ \text{V} + 15\ \text{V} = \mathbf{25\ V}$$

Note the polarity of V_s for the single-source circuit.

EXAMPLE 9.4

For the parallel configuration of Fig. 9.6, determine I_1, I_2, and V_s.

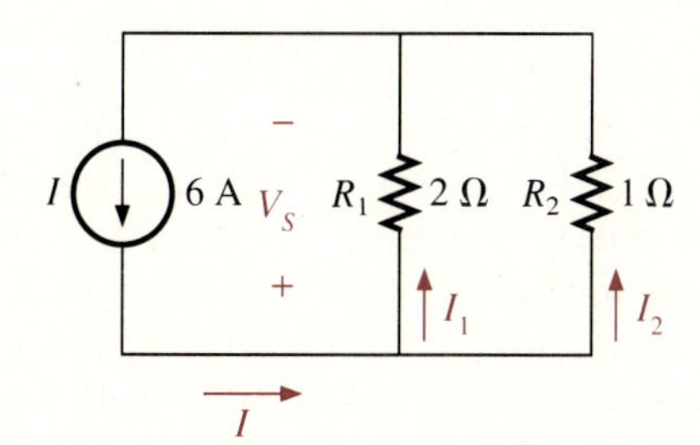

FIG. 9.6
Network for Example 9.4.

Solution Using the current divider rule:

$$I_1 = \frac{R_2 I}{R_2 + R_1} = \frac{(1\ \Omega)(6\ \text{A})}{1\ \Omega + 2\ \Omega} = \frac{1}{3}(6\ \text{A}) = \mathbf{2\ A}$$

Using Kirchhoff's current law:

$$I_2 = I - I_1 = 6\ \text{A} - 2\ \text{A} = \mathbf{4\ A}$$

$$V_s = V_1 = V_2 = I_2 R_2 = (4\ \text{A})(1\ \Omega) = \mathbf{4\ V}$$

with the polarity shown in Fig. 9.6.

9.3 SOURCE CONVERSIONS

The basic arrangement of a practical voltage and current source appears in Figs. 9.7 and 9.8 with their internal resistance (present in every practical source) R_s.

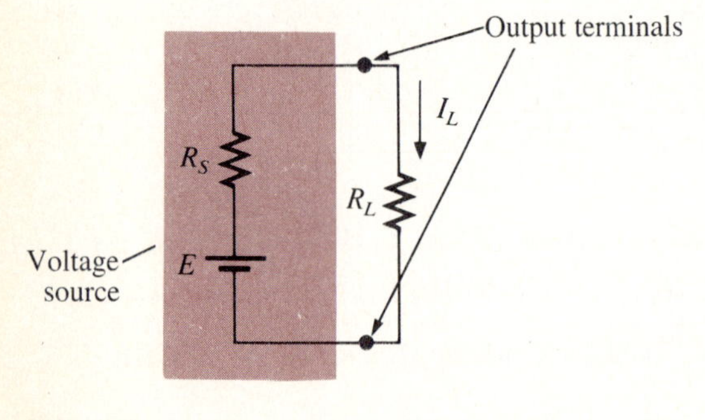

FIG. 9.7
Practical voltage source.

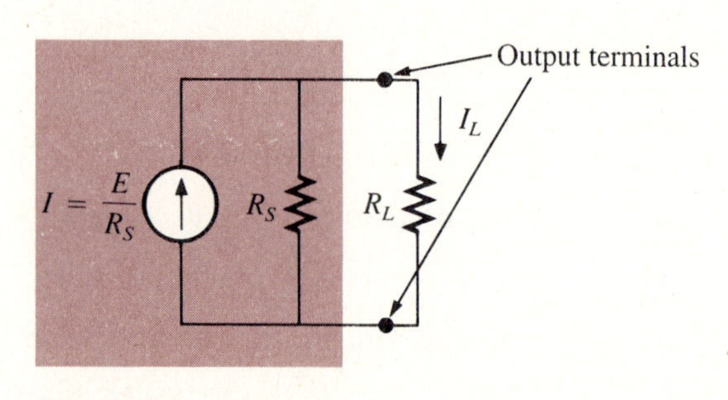

FIG. 9.8
Practical current source.

Note that for the voltage source, the internal resistance is placed in series with the voltage source, whereas the internal resistance is placed in parallel for the current source. Since it is often necessary to have a voltage source rather than a current source or vice versa, a method must be developed to convert from one type of source to the other. In order for the sources of Figs. 9.7 and 9.8 to be equivalent, they must have an equivalence at the source terminals. That is, the applied load R_L must not be able to detect whether it is attached to the configuration of Fig. 9.7 or Fig. 9.8. In Fig. 9.9, for example, a load R_L has been attached to a supply that may contain a voltage or current source. If the internal supplies are equivalent, the load R_L will receive the same current I_L (and, therefore, power, since $P_L = I_L^2 R_L$) and will not be aware of whether it is connected to a voltage or current source.

Fortunately, the configurations of Figs. 9.7 and 9.8 are related by some rather simple equations. In fact, the conversion from one form to the other is

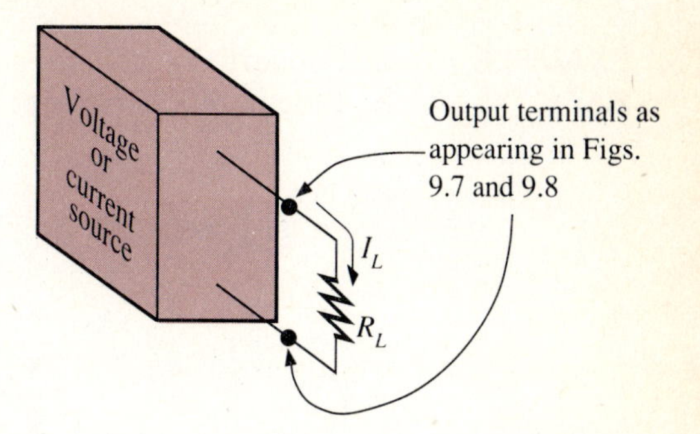

FIG. 9.9
Demonstrating "equivalence" between a voltage and current source.

completely described by Fig. 9.10. Note that the internal resistance of one is the same for the other—only their placement is changed. If converting to a current source, the source current is determined by Ohm's law, $I = E/R_s$, where the voltage E is that of the voltage source and the resistance is the internal resistance of the voltage source. For a current-to-voltage-source conversion, the source voltage is again determined by Ohm's law, $E = IR_s$, where I is the source current of the current source and R_s is the parallel internal resistance of the current source. Any load connected between a and b of either configuration will have the same voltage across it and the same current through it.

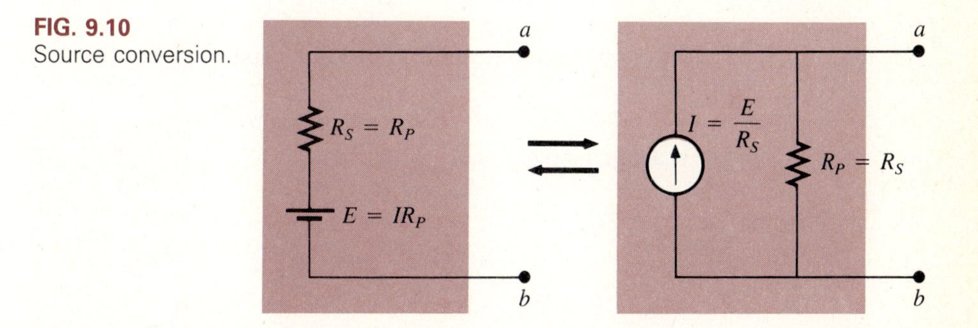

FIG. 9.10
Source conversion.

It was pointed out in some detail in Chapter 7 that every source of voltage has some internal series resistance. *For the current source, some internal parallel resistance will always exist in the practical world.* However, in many cases, it is an excellent approximation to drop the internal resistance of a source due to the magnitude of the elements of the network to which it is applied. For this reason, in the analysis to follow, voltage sources may appear without a series resistor, and current sources may appear without a parallel resistance. Realize, however, that in order to perform a conversion from one type of source to another, a voltage source must have a resistor in series with it, and a current source must have a resistor in parallel.

EXAMPLE 9.5

a. Convert the voltage source of Fig. 9.11 to a current source.
b. Calculate the current through the load for each source.

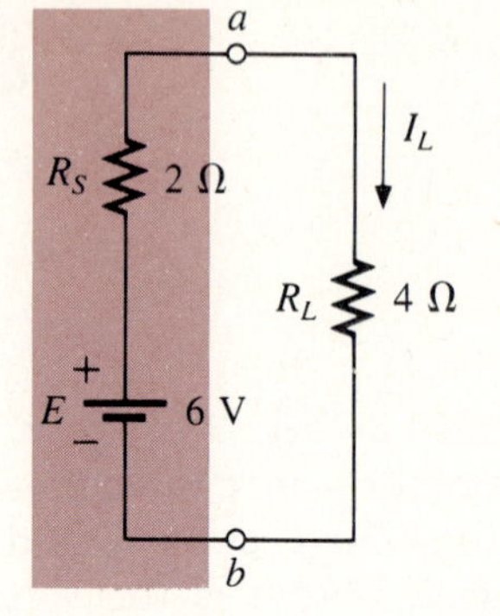

FIG. 9.11
Circuit for Example 9.5.

Solutions

a. R_s (current source) = R_s (voltage source) = **2 Ω**

and

$$I = \frac{E}{R_s} = \frac{6\text{ V}}{2\ \Omega} = \mathbf{3\ A}$$

The current source and applied load appear in Fig. 9.12.

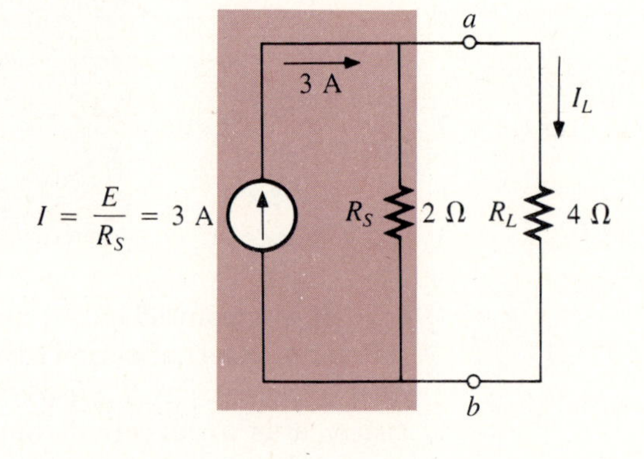

FIG. 9.12
Current source equivalent of Fig. 9.11.

b. For Fig. 9.11,

$$I_L = \frac{E}{R_s + R_L} = \frac{6\text{ V}}{2\ \Omega + 4\ \Omega} = \frac{6\text{ V}}{6\ \Omega} = \mathbf{1\ A}$$

For Fig. 9.12, using the current divider rule:

$$I_L = \frac{R_s I}{R_s + R_L} = \frac{(2\ \Omega)(3\text{ A})}{2\ \Omega + 4\ \Omega} = \frac{1}{3}(3\text{ A}) = \mathbf{1\ A}$$

EXAMPLE 9.6 Convert the current source of Fig. 9.13 to a voltage source and find the current through the load for each source.

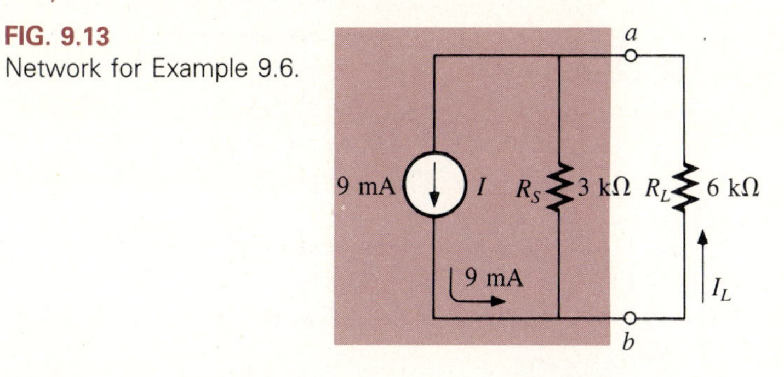

FIG. 9.13
Network for Example 9.6.

Solution Note Fig. 9.14, where $R_s = 3\text{ k}\Omega$ and $E = IR_s = (9\text{ mA})(3\text{ k}\Omega) = 27\text{ V}$.

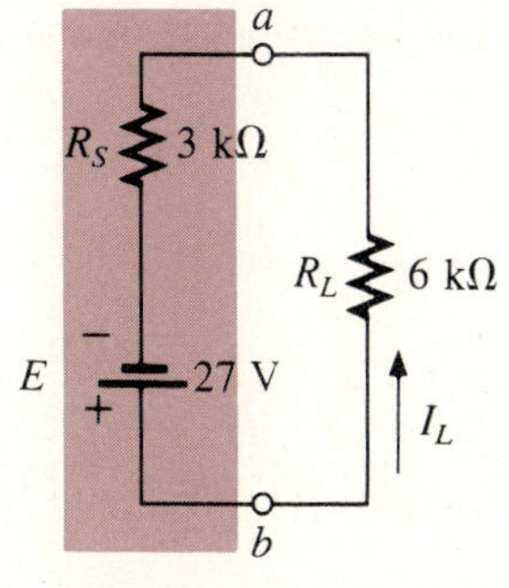

FIG. 9.14
Voltage source equivalent of Fig. 9.13.

For Fig. 9.13, using the current divider rule:

$$I_L = \frac{R_s I}{R_s + R_L} = \frac{(3\text{ k}\Omega)(9\text{ mA})}{3\text{ k}\Omega + 6\text{ k}\Omega} = \frac{1}{3}(9\text{ mA}) = \mathbf{3\text{ mA}}$$

For Fig. 9.14,

$$I_L = \frac{E}{R_s + R_L} = \frac{27\text{ V}}{3\text{ k}\Omega + 6\text{ k}\Omega} = \frac{27\text{ V}}{9\text{ k}\Omega} = \mathbf{3\text{ mA}}$$

9.4 CURRENT SOURCES IN PARALLEL

If two or more current sources are in parallel, they may all be replaced by one current source having the magnitude and direction of the resultant, which can be found by summing the currents in one direction and subtracting the sum of the currents in the opposite direction. The new parallel resistance is determined by methods described in the discussion of parallel resistors in Chapter 6. Consider the following examples.

EXAMPLE 9.7

Reduce the network of Fig. 9.15 to its simplest form and find the voltage across the resistor R_1.

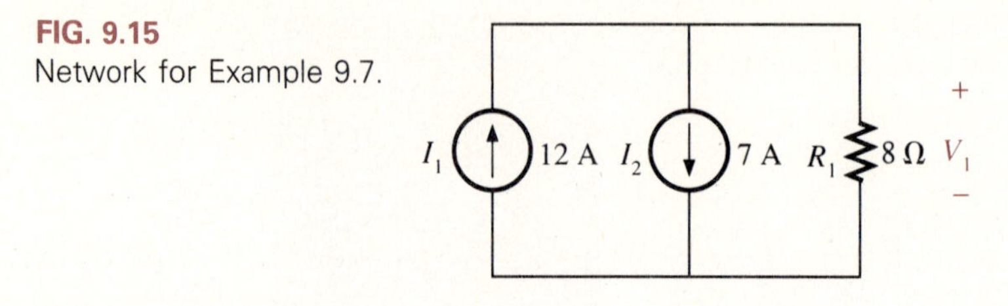

FIG. 9.15
Network for Example 9.7.

Solution The net result of 12 A "up" and 7 A "down" is 5 A "up," as shown in Fig. 9.16, with

$$V_1 = IR_1 = (5\text{ A})(8\ \Omega) = \mathbf{40\ V}$$

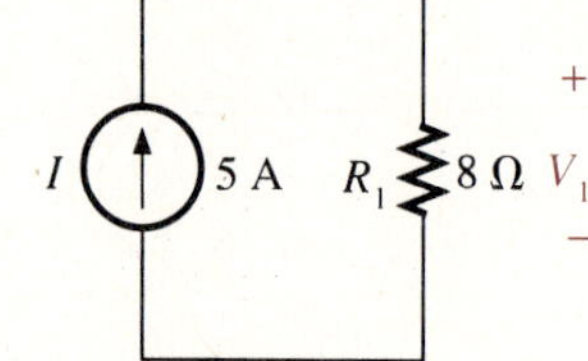

FIG. 9.16
Reduced equivalent network of Fig. 9.15.

EXAMPLE 9.8

Reduce the left side of Fig. 9.17 to a minimum number of elements.

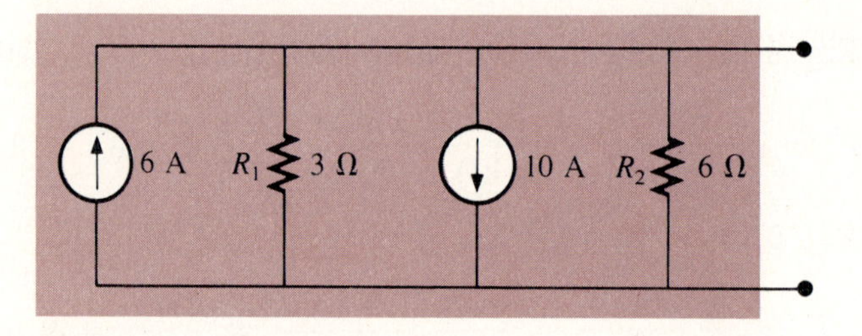

FIG. 9.17
Network for Example 9.8.

Solution

$$I_s = 10\text{ A} - 6\text{ A} = \mathbf{4\ A}$$
$$R_s = R_1 \| R_2 = 3\ \Omega \| 6\ \Omega = \mathbf{2\ \Omega}$$

as shown in Fig. 9.18.

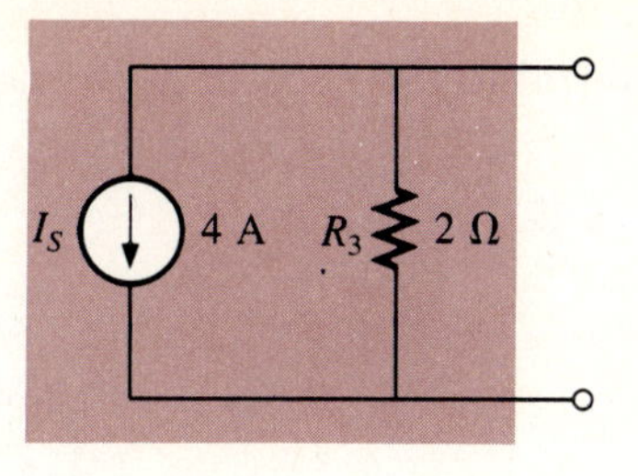

FIG. 9.18
Reduced equivalent of Fig. 9.17.

EXAMPLE 9.9 Reduce the network of Fig. 9.19 to a single current source and calculate the current through R_L.

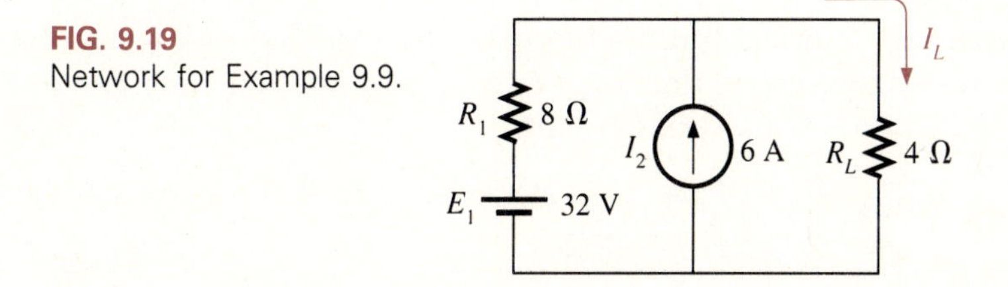

FIG. 9.19
Network for Example 9.9.

Solution In this example, the voltage source will first be converted to a current source, as shown in Fig. 9.20. Combining current sources,

$$I_s = I_1 + I_2 = 4\text{ A} + 6\text{ A} = \mathbf{10\text{ A}}$$

and

$$R_s = R_1 = \mathbf{8\ \Omega}$$

Applying the current divider rule to the network of Fig. 9.21,

$$I_L = \frac{R_s I_s}{R_s + R_L} = \frac{(8\ \Omega)(10\text{ A})}{8\ \Omega + 4\ \Omega} = \frac{8\ \Omega}{12\ \Omega}(10\text{ A}) = \mathbf{6.667\text{ A}}$$

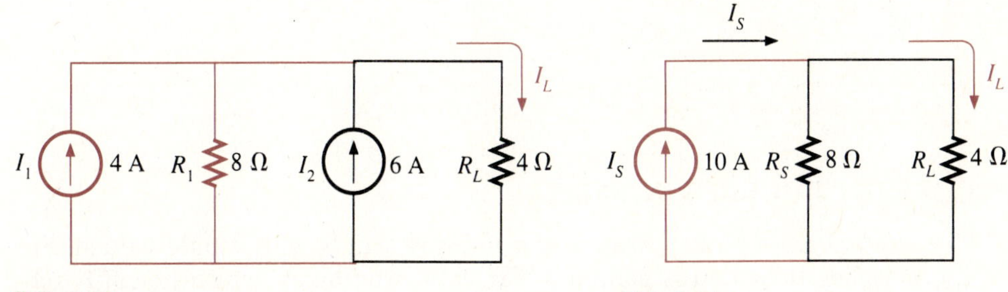

FIG. 9.20
Network of Fig. 9.19 redrawn with current source equivalent.

FIG. 9.21
Reduced form of Fig. 9.20.

EXAMPLE 9.10 Determine the current I_2 in the network of Fig. 9.22.

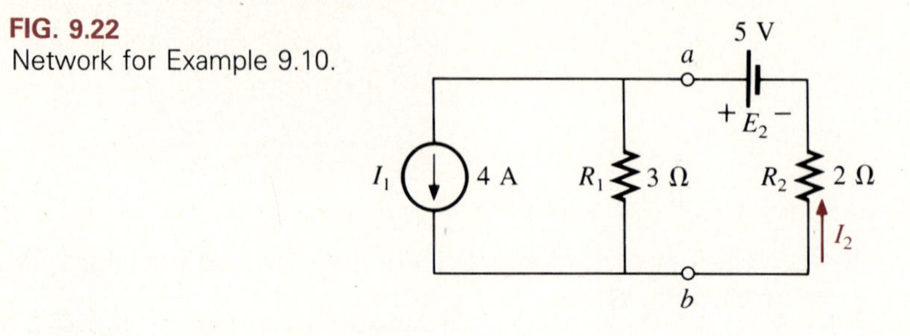

FIG. 9.22
Network for Example 9.10.

Solution Although it might appear that the network cannot be solved using methods introduced thus far, one source conversion, as shown in Fig. 9.23, will result in a simple series circuit:

$$E_s = I_1R_1 = (4\ \text{A})(3\ \Omega) = \mathbf{12\ V}$$
$$R_s = R_1 = \mathbf{3\ \Omega}$$

and

$$I_2 = \frac{E_s + E_2}{R_s + R_2} = \frac{12\ \text{V} + 5\ \text{V}}{3\ \Omega + 2\ \Omega} = \frac{17\ \text{V}}{5\ \Omega} = \mathbf{3.4\ A}$$

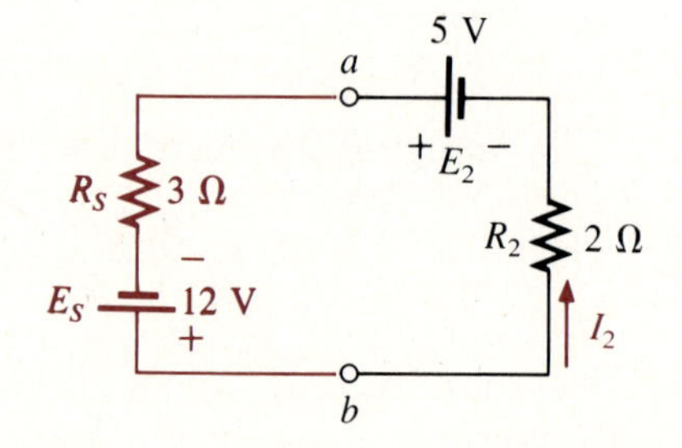

FIG. 9.23
Network of Fig. 9.22 with a voltage source equivalent of the current source.

9.5 CURRENT SOURCES IN SERIES

The current through any branch of a network can be only single-valued. For the situation indicated at point a in Fig. 9.24, we find by application of Kirchhoff's current law that the current leaving that point is greater than that entering—an impossible situation. Therefore, *current sources of different current ratings are not connected in series,* just as voltage sources of different voltage ratings are not connected in parallel.

FIG. 9.24
Current sources of different magnitude are never hooked up in series.

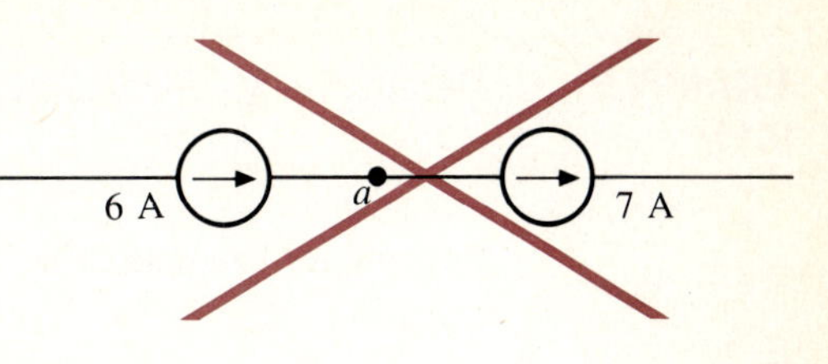

9.6 SUPERPOSITION THEOREM

The superposition theorem, like the methods just described, can be used to find the solution to networks with two or more sources that are not in series or parallel. The most obvious advantage of this method is that it does not require the use of a mathematical technique such as substitution or determinants to find the required voltages or currents. Instead, each source is treated independently, and the algebraic (paying attention to sign) sum is found to determine a particular unknown quantity of the network. In other words, for a network with two sources, two independent series-parallel networks would have to be considered before a solution could be obtained. For three sources, three independent networks would be examined, and so on.

The superposition theorem states the following:

The current through, or voltage across, an element in a series-parallel network is equal to the algebraic sum of the currents or voltages produced independently by each source.

To consider the effects of each source independently requires that sources be removed and replaced without affecting the final result. To remove a voltage source when applying this theorem, the difference in potential between the terminals of the voltage source must be set to zero (short circuit); removing a current source requires that its terminals be opened (open circuit). Any internal resistance or conductance associated with the displaced sources is not eliminated but must still be considered.

The total current through any portion of the network is equal to the algebraic sum of the currents produced independently by each source. That is, for a two-source network, if the current produced by one source is in one direction and that produced by the other is in the opposite direction through the same resistor, *the resulting current is the difference of the two and has the direction of the larger*. If the individual currents are in the same direction, *the resulting current is the sum of the two in the direction of either current.* This rule holds true for the voltage across a portion of a network as determined by polarities, and it can be extended to networks with any number of sources.

The superposition principle is not applicable to power effects, since the power loss in a resistor varies as the square (nonlinear) of the current or voltage. For this reason, the power to an element cannot be calculated until the total current through (or voltage across) the element has been determined by superposition. This is demonstrated in Example 9.13.

EXAMPLE 9.11 Determine V_1 for the network of Fig. 9.25 using superposition.

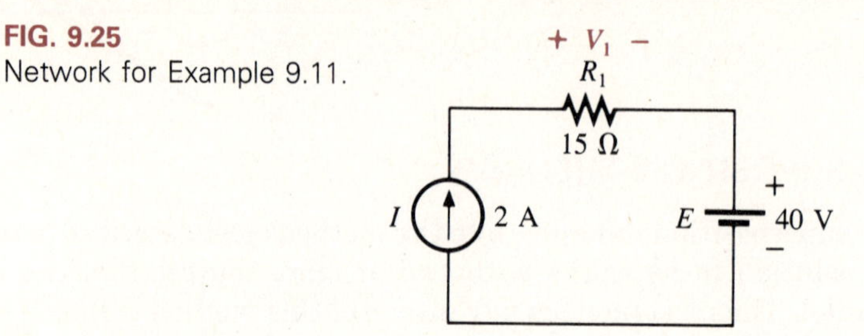

FIG. 9.25
Network for Example 9.11.

Solution For the voltage source set to 0 V (short circuit), the circuit of Fig. 9.26 will result, and

$$V'_1 = I_1R_1 = IR_1 = (2\text{ A})(15\ \Omega) = 30\text{ V}$$

The network of Fig. 9.27 will result for the current source set to zero (open circuit), and

$$V''_1 = I_1R_1 = (0)R_1 = 0\text{ V}$$

Note that V'_1 and V''_1 have opposite polarities. The voltage V_1, however, has the same polarity as V'_1, and, therefore,

$$V_1 = V'_1 - V''_1 = 30\text{ V} - 0\text{ V} = \mathbf{30\text{ V}}$$

Note that the 40-V supply has no effect on V_1, since the current source determined the current through the 15-Ω resistor.

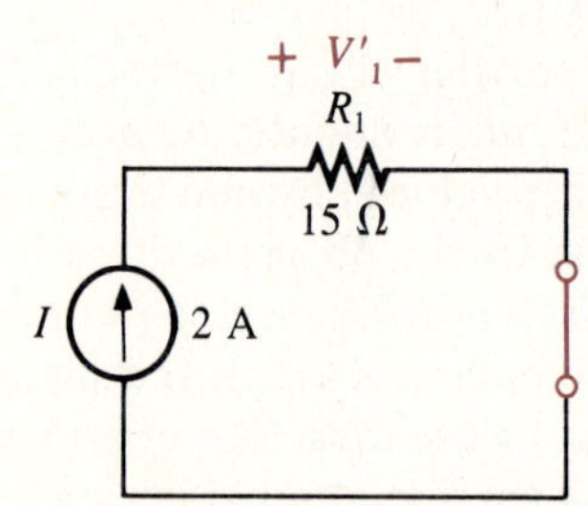

FIG. 9.26
Considering the effects of the current source of Fig. 9.25.

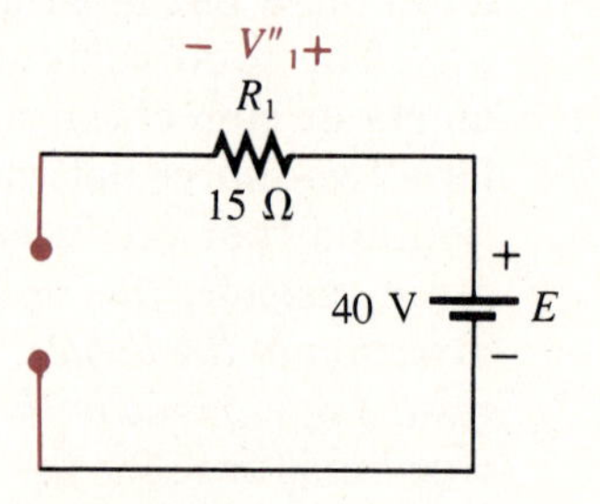

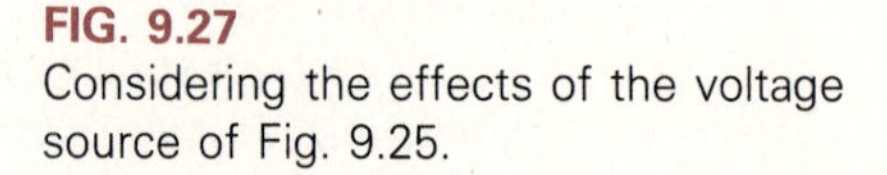

FIG. 9.27
Considering the effects of the voltage source of Fig. 9.25.

EXAMPLE 9.12

Determine I_1 for the network of Fig. 9.28.

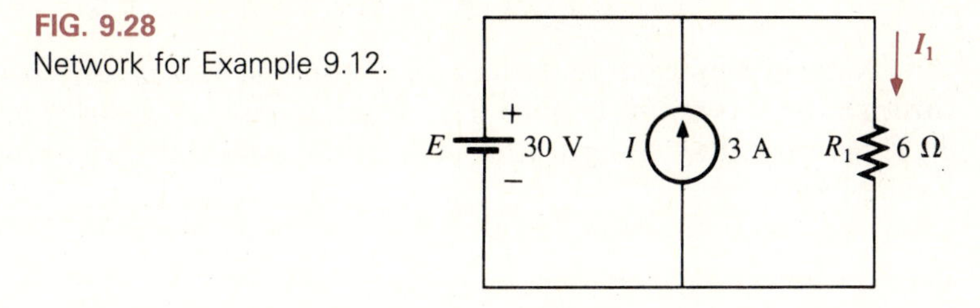

FIG. 9.28
Network for Example 9.12.

Solution Setting $E = 0$ V for the network of Fig. 9.28 results in the network of Fig. 9.29, where a short-circuit equivalent has replaced the 30-V source.

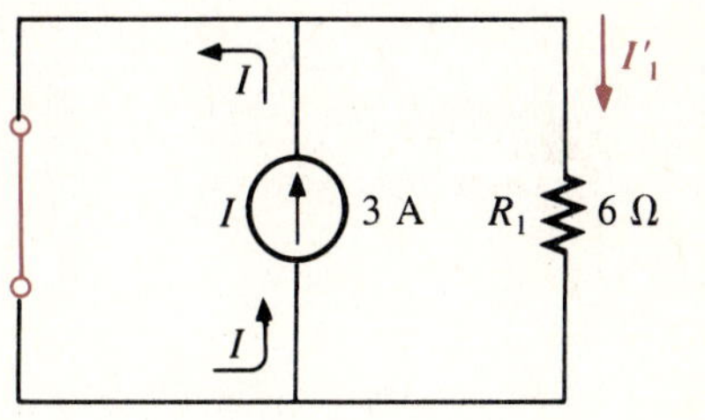

FIG. 9.29
Considering the effects of the current source of Fig. 9.28.

As shown in Fig. 9.29, the source current chooses the short-circuit path, and $I'_1 = 0$ A. If we apply the current divider rule,

$$I'_1 = \frac{R_{sc}I}{R_{sc} + R_1} = \frac{(0\ \Omega)I}{0\ \Omega + 6\ \Omega} = 0\ \text{A}$$

Setting I to 0 A will result in the network of Fig. 9.30 with the current source replaced by an open circuit. Applying Ohm's law,

$$I''_1 = \frac{E}{R_1} = \frac{30\ \text{V}}{6\ \Omega} = 5\ \text{A}$$

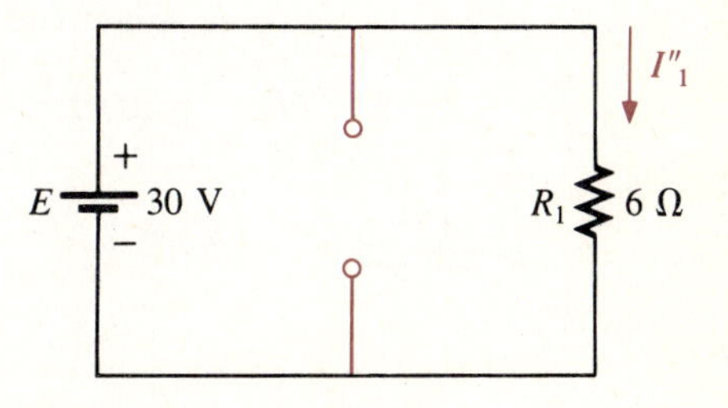

FIG. 9.30
Considering the effects of the voltage source of Fig. 9.28.

Since I'_1 and I''_1 have the same defined direction in Figs. 9.29 and 9.30, the current I_1 is the sum of the two, and

$$I_1 = I'_1 + I''_1 = 0 \text{ A} + 5 \text{ A} = \mathbf{5\ A}$$

Note in this case that the current source has no effect on the current through the 6-Ω resistor, since the voltage across the resistor must be fixed at 30 V because they are parallel elements—essentially the dual situation of Example 9.11.

EXAMPLE 9.13 Using superposition, find the current through the 6-Ω resistor of the network of Fig. 9.31.

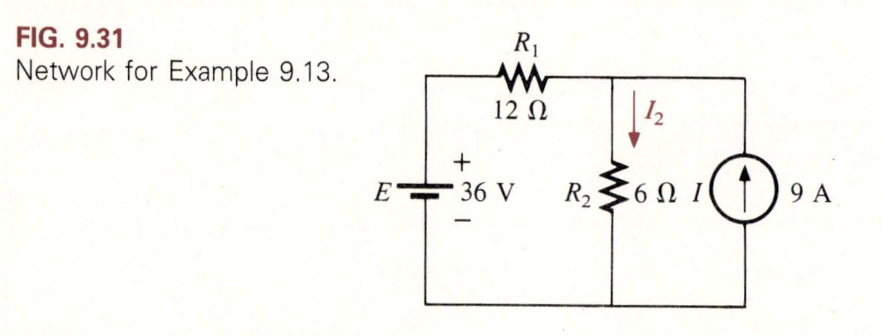

FIG. 9.31
Network for Example 9.13.

Solution Considering the effect of the 36-V source (Fig. 9.32),

$$I'_2 = \frac{E}{R_T} = \frac{E}{R_1 + R_2} = \frac{36 \text{ V}}{12\ \Omega + 6\ \Omega} = 2 \text{ A}$$

Considering the effect of the 9-A source (Fig. 9.33), by applying the current divider rule,

$$I''_2 = \frac{R_1 I}{R_1 + R_2} = \frac{(12\ \Omega)(9 \text{ A})}{12\ \Omega + 6\ \Omega} = \frac{108 \text{ A}}{18} = 6 \text{ A}$$

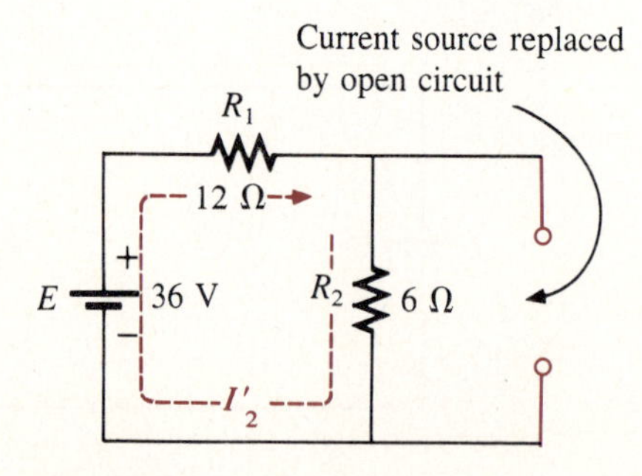

FIG. 9.32
Considering the effects of the voltage source of Fig. 9.31.

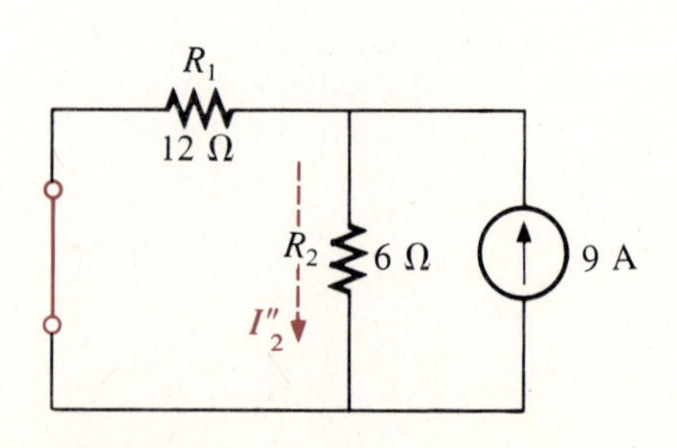

FIG. 9.33
Considering the effects of the current source of Fig. 9.31.

The total current through the 6-Ω resistor (Fig. 9.34) is

$$I_2 = I'_2 + I''_2 = 2\text{ A} + 6\text{ A} = \mathbf{8\text{ A}}$$

The power to the 6-Ω resistor is

$$\text{Power} = I^2R = (8\text{ A})^2(6\ \Omega) = \mathbf{384\text{ W}}$$

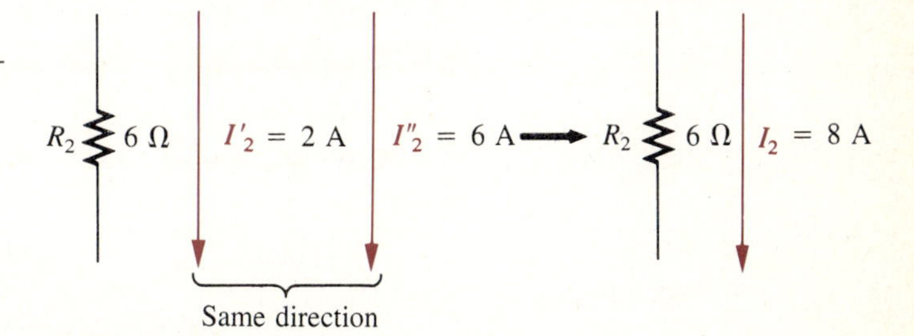

FIG. 9.34
Finding the algebraic sum of the currents established by each source.

The calculated power to the 6-Ω resistor due to each source, *misusing* the principle of superposition, is

$$P_1 = (I'_2)^2R = (2\text{ A})^2(6\ \Omega) = 24\text{ W}$$
$$P_2 = (I''_2)^2R = (6\text{ A})^2(6\ \Omega) = 216\text{ W}$$
$$P_T = P_1 + P_2 = 240\text{ W} \neq 384\text{ W}$$

This results because 2 + 6 = 8, but

$$(2)^2 + (6)^2 \neq (8)^2$$

As mentioned previously, the superposition principle is not applicable to power effects, since power is proportional to the square of the current or voltage (I^2R or V^2/R).

Figure 9.35 is a plot of the power delivered to the 6-Ω resistor versus current.

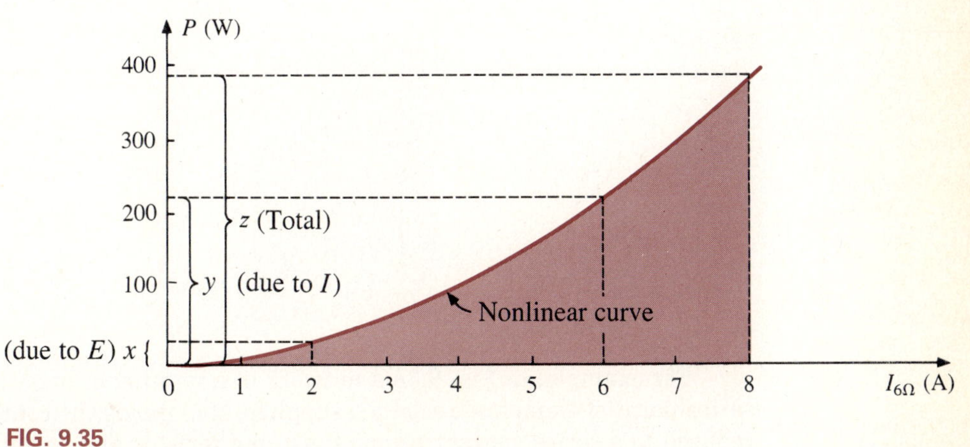

FIG. 9.35
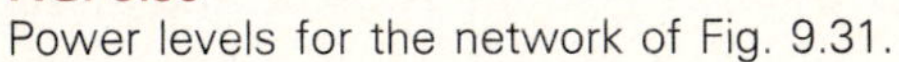
Power levels for the network of Fig. 9.31.

Obviously, $x + y \neq z$, or $24 + 216 \neq 384$, and superposition does not hold. However, for a linear relationship, such as that between the voltage and current of the fixed-type 6-Ω resistor, superposition can be applied, as demonstrated by the graph of Fig. 9.36, where $a + b = c$, or $2 + 6 = 8$.

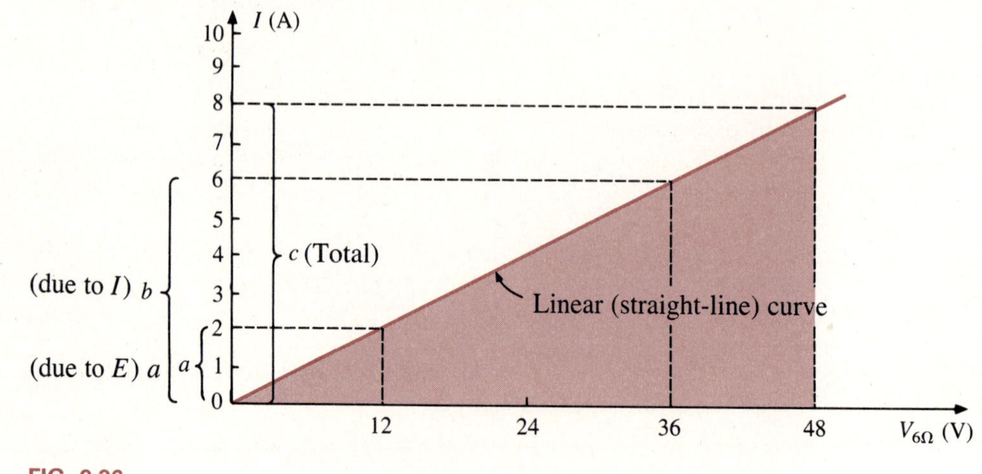

FIG. 9.36
Current levels (vs. voltage) for the network of Fig. 9.31.

9.7 THEVENIN'S THEOREM

Thevenin's theorem states the following:

Any two-terminal dc network can be replaced by an equivalent circuit consisting of a voltage source and a series resistor, as shown in Fig. 9.37.

FIG. 9.37
Thevenin equivalent circuit.

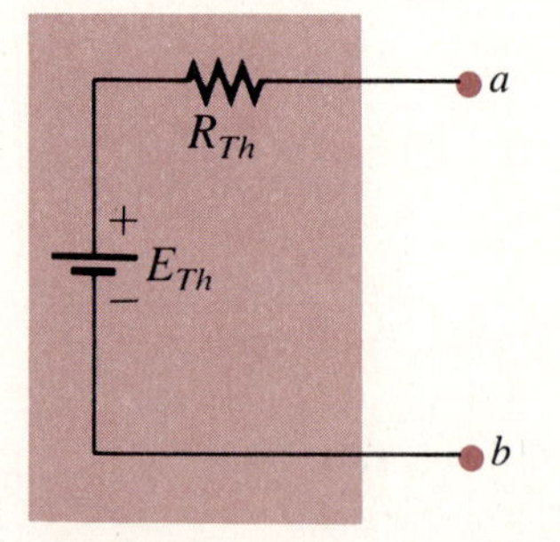

For example, in Fig. 9.38 a network of any number of series, parallel, or series-parallel elements (resistors, supplies, etc.) inside the container has been reduced to a series configuration of just one resistor and dc supply. The theo-

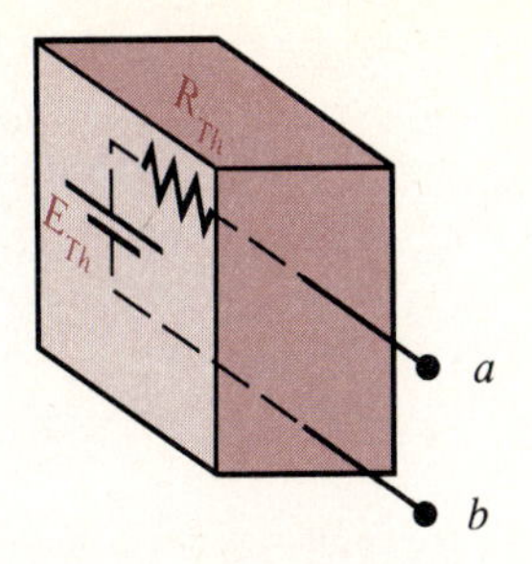

FIG. 9.38
Defining the Thevenin equivalent circuit.

rem states that by choosing the *proper* value of E_{Th} and R_{Th}, the voltage measured across terminals a and b will be the same with the original network or the series configuration of Fig. 9.38.

In Fig. 9.39, the network between terminals a and b can be replaced by one resistor of 10 Ω and a single battery of 8 V, as shown in the adjoining figure. Note the reduction in the network and the remaining series supply-resistor configuration while maintaining the same terminal voltage (open circuit = 8 V) between terminals a and b. The procedure described next allows us to extend the procedure just applied to more complex configurations and still end up with the relatively simple network of Fig. 9.38.

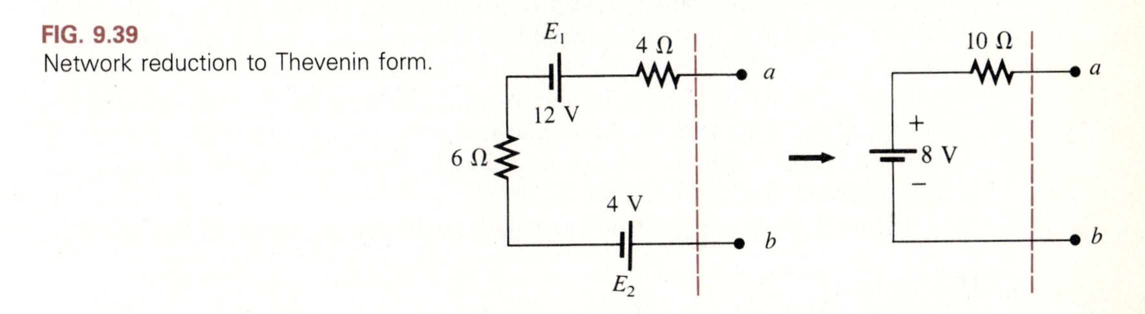

FIG. 9.39
Network reduction to Thevenin form.

In most cases, there will be other elements connected to the right of terminals a and b in Fig. 9.39. To apply the theorem, however, the network to be reduced to the Thevenin equivalent form must be isolated, as shown in Fig. 9.39, and the two "holding" terminals must be identified. Once the proper Thevenin equivalent circuit has been determined, the voltage, current, or resistance readings between the two holding terminals will be the same whether the original or Thevenin equivalent circuit is connected to the left of terminals a and b in Fig. 9.39. Any load connected to the right of terminals a and b of Fig. 9.39 will receive the same voltage or current with either network.

This theorem achieves two important objectives. First, as was true for all the methods previously described, it allows us to find any particular voltage or current in a linear network with one, two, or any other number of sources. Second, we can concentrate on a specific portion of a network by replacing the remaining network by an equivalent circuit. In Fig. 9.40(a), for example, by finding the Thevenin equivalent circuit for the network in the enclosed area,

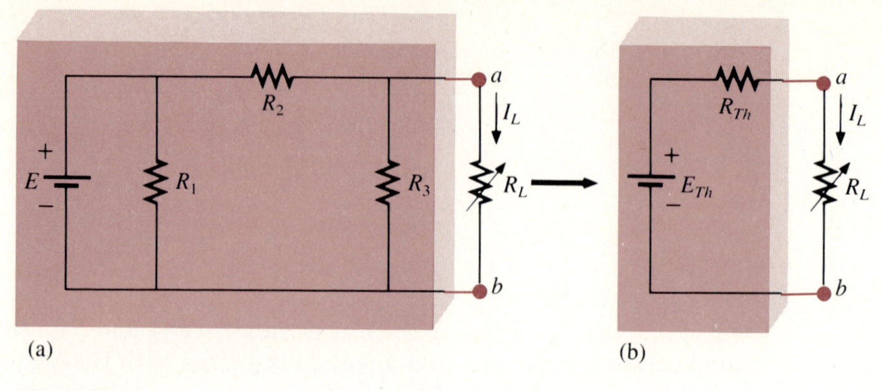

FIG. 9.40

Demonstrating the effect of the application of Thevenin's theorem.

we can quickly calculate the change in current through or voltage across the variable resistor R_L for the various values that it may assume. This is demonstrated in Example 9.14.

Before we can examine the steps involved in applying this theorem, it is important that an additional word be included here to ensure that the implications of the Thevenin equivalent circuit are clear. In Fig. 9.40(a), the entire network, except R_L, is to be replaced by a single series resistor and battery, as shown in Fig. 9.40(b). The values of these two elements of the Thevenin equivalent circuit must be chosen to ensure that the resistor R_L will react to the network of Fig. 9.40(a) in the same manner as to the network of Fig. 9.40(b). In other words, the current through or voltage across R_L must be the same for either network for any value of R_L.

The following sequence of steps leads to the proper values of R_{Th} and E_{Th}.

Preliminary

1. Remove that portion of the network across which the Thevenin equivalent circuit is to be found. In Fig. 9.40(a), this requires that the load resistor R_L be temporarily removed from the network.
2. Mark the terminals of the remaining two-terminal network. (The importance of this step will become obvious as we progress through some examples.)

R_{Th}

3. Calculate R_{Th} by first setting all sources to zero (voltage sources are replaced by short circuits and current sources by open circuits) and then finding the resultant resistance between the two marked terminals. (If the internal resistance of the voltage and/or current sources is included in the original network, it must remain when the sources are set to zero.)

E_{Th}

4. Calculate E_{Th} by first returning all sources to their original positions and finding the *open-circuit* voltage between the marked terminals. (This step is invariably the one that leads to the most confusion and errors. In *all* cases, keep in mind that it is the *open-circuit* potential between the two terminals marked in step 2.

Conclusion

5. Draw the Thevenin equivalent circuit with the portion of the circuit previously removed replaced between the terminals of the equivalent circuit. This step is indicated by the placement of the resistor R_L between the terminals of the Thevenin equivalent circuit as shown in Fig. 9.40(b).

EXAMPLE 9.14

Find the Thevenin equivalent circuit for the network in the shaded area of the network of Fig. 9.41. Then find the current through R_L for values of 2, 10, and 100 Ω.

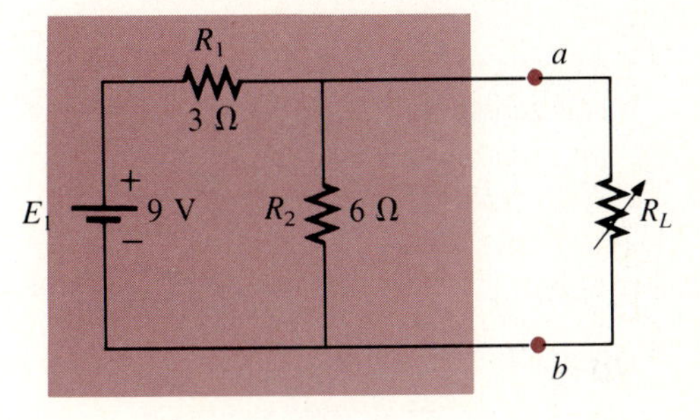

FIG. 9.41
Network for Example 9.14.

Solution

Steps 1 and 2 produce the network of Fig. 9.42. Note that the load resistor R_L has been removed and the two "holding" terminals have been defined as *a* and *b*.

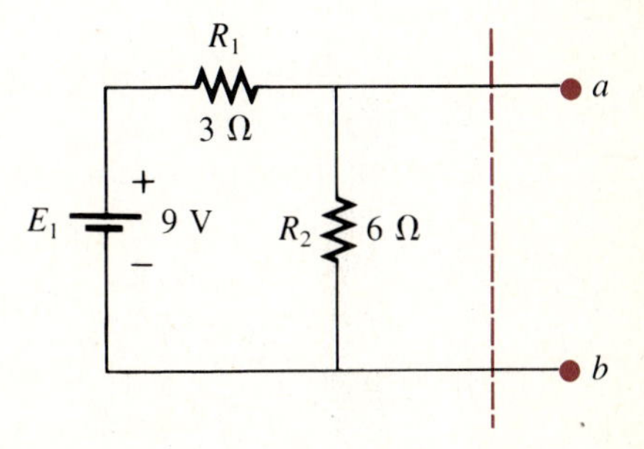

FIG. 9.42
Effect of removing the load resistor from Fig. 9.41.

Step 3: Replacing the voltage source E_1 by a short-circuit equivalent yields the network of Fig. 9.43, where

$$R_{Th} = R_1 \| R_2 = \frac{(3\ \Omega)(6\ \Omega)}{3\ \Omega + 6\ \Omega} = \mathbf{2\ \Omega}$$

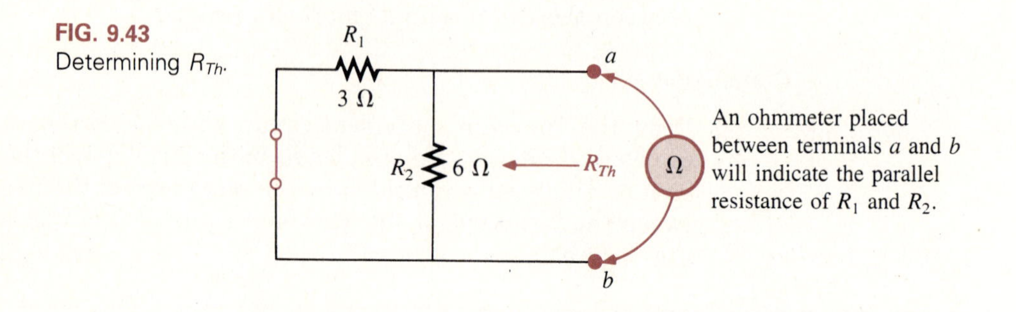

FIG. 9.43
Determining R_{Th}.

Experimentally, R_{Th} is determined using an ohmmeter, as shown in Fig. 9.44. Note that the source has been *replaced* by a short-circuit equivalent—a short circuit was *not* placed across the output terminals of the supply.

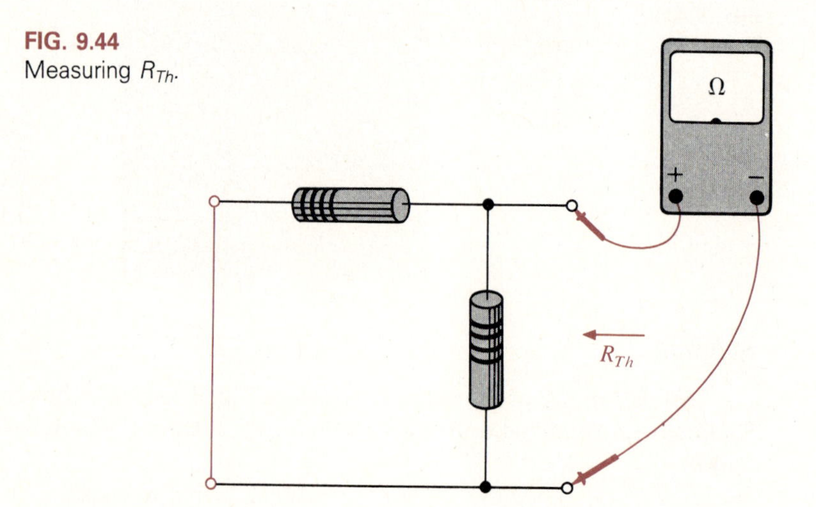

FIG. 9.44
Measuring R_{Th}.

Step 4: Replace the voltage source (Fig. 9.45). For this case, the open-circuit voltage E_{Th} is the same as the voltage drop across the 6-Ω resistor. Applying the voltage divider rule,

$$E_{Th} = \frac{R_2 E_1}{R_2 + R_1} = \frac{(6\ \Omega)(9\ \Omega)}{6\ \Omega + 3\ \Omega} = \frac{54}{9}\text{ V} = \mathbf{6\ V}$$

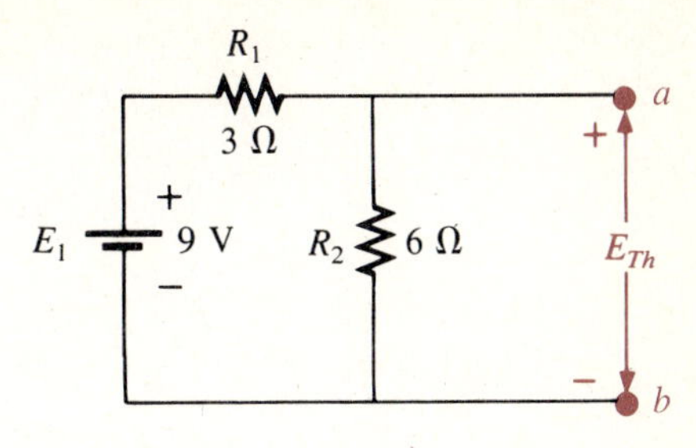

FIG. 9.45
Determining E_{Th} for the network of Fig. 9.41.

E_{Th} is determined experimentally using a voltmeter, as shown in Fig. 9.46. The supply has been returned to the network and E_{Th} is simply the voltmeter reading across the 6-Ω resistor.

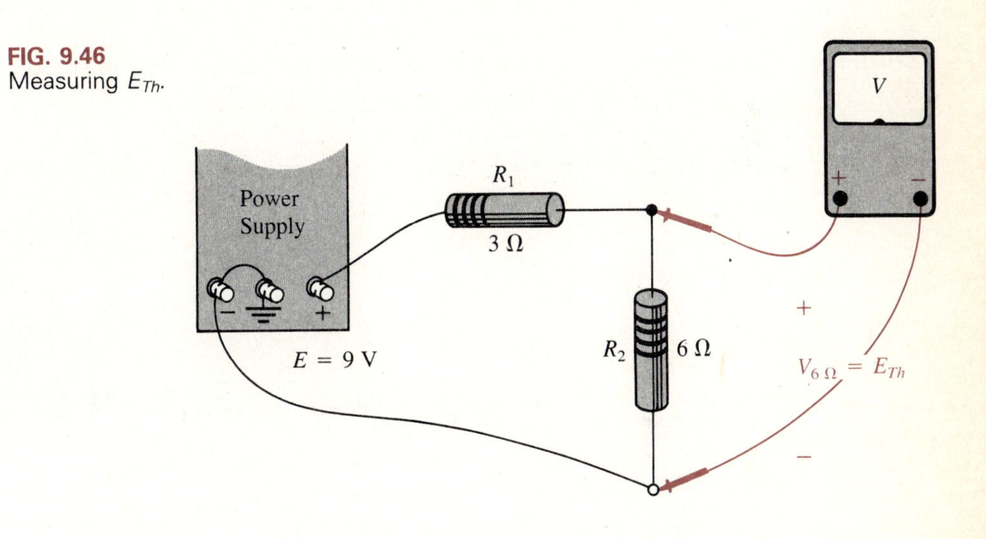

FIG. 9.46
Measuring E_{Th}.

Step 5: See Fig. 9.47.

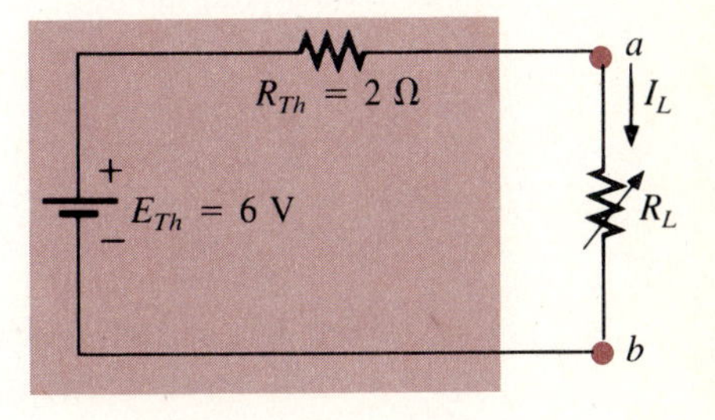

FIG. 9.47
Constructing the Thevenin equivalent circuit.

$$I_L = \frac{E_{Th}}{R_{Th} + R_L}$$

$$R_L = 2\ \Omega: \quad I_L = \frac{6\ \text{V}}{2\ \Omega + 2\ \Omega} = \mathbf{1.5\ A}$$

$$R_L = 10\ \Omega: \quad I_L = \frac{6\ \text{V}}{2\ \Omega + 10\ \Omega} = \mathbf{0.5\ A}$$

$$R_L = 100\ \Omega: \quad I_L = \frac{6\ \text{V}}{2\ \Omega + 100\ \Omega} = \mathbf{0.059\ A}$$

If Thevenin's theorem were unavailable, each change in R_L would require that the entire network of Fig. 9.41 be reexamined to find the new value of R_L.

EXAMPLE 9.15 Find the Thevenin equivalent circuit for the network in the shaded area of the network of Fig. 9.48.

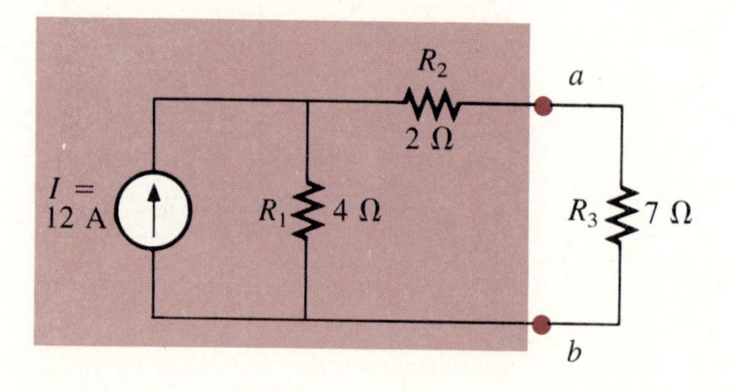

FIG. 9.48
Network for Example 9.15.

Solution

Steps 1 and 2 are shown in Fig. 9.49.

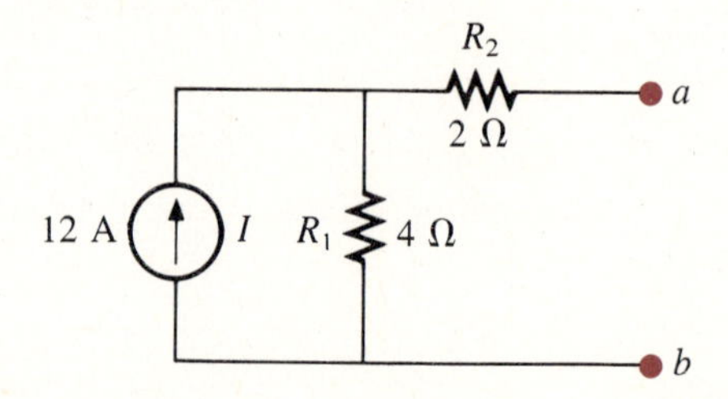

FIG. 9.49
Removing the element not included in the Thevenin equivalent circuit.

Step 3 is shown in Fig. 9.50. The current source has been replaced by an open-circuit equivalent and the resistance determined between terminals *a*

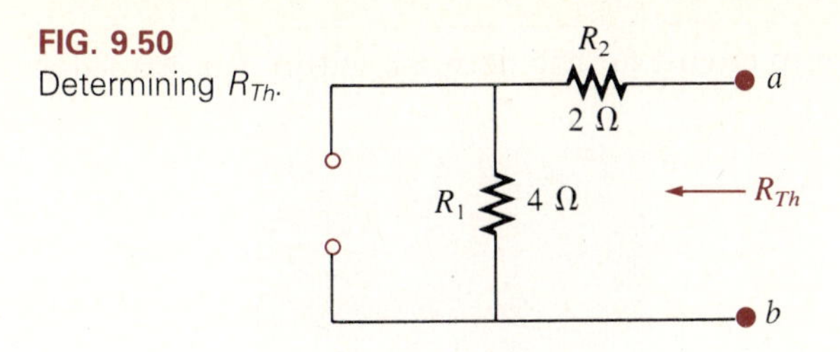

FIG. 9.50
Determining R_{Th}.

and b. In this case, an ohmmeter placed across terminals a-b would read the series resistance of R_1 and R_2.

$$R_{Th} = R_1 + R_2 = 4\ \Omega + 2\ \Omega = \mathbf{6\ \Omega}$$

Step 4 (Fig. 9.51): In this case, since an open circuit exists between the two marked terminals, the current is zero between these terminals and through the 2-Ω resistor. The voltage drop across R_2 is, therefore,

$$V_2 = I_2R_2 = (0)R_2 = 0\ \text{V}$$

and

$$E_{Th} = V_1 = I_1R_1 = IR_1 = (12\ \text{A})(4\ \Omega) = \mathbf{48\ V}$$

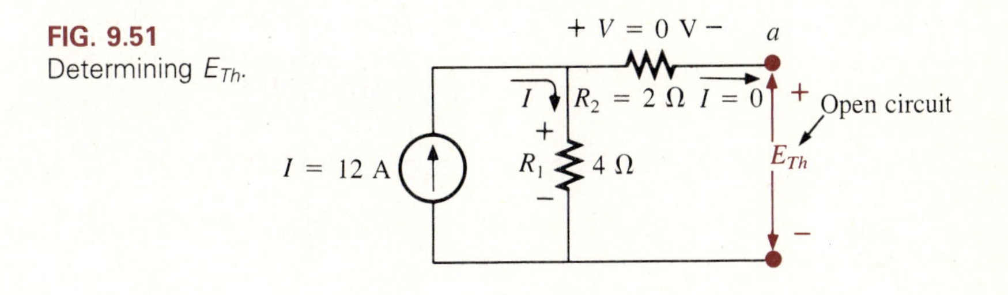

FIG. 9.51
Determining E_{Th}.

Step 5 is shown in Fig. 9.52.

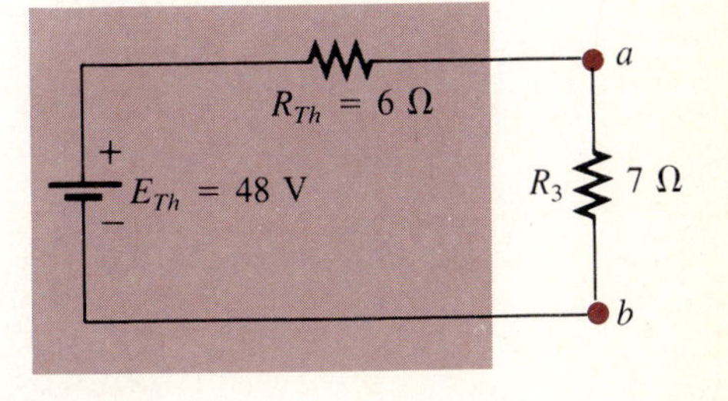

FIG. 9.52
Constructing the Thevenin circuit.

EXAMPLE 9.16 Find the Thevenin circuit for the network within the shaded area of Fig. 9.53.

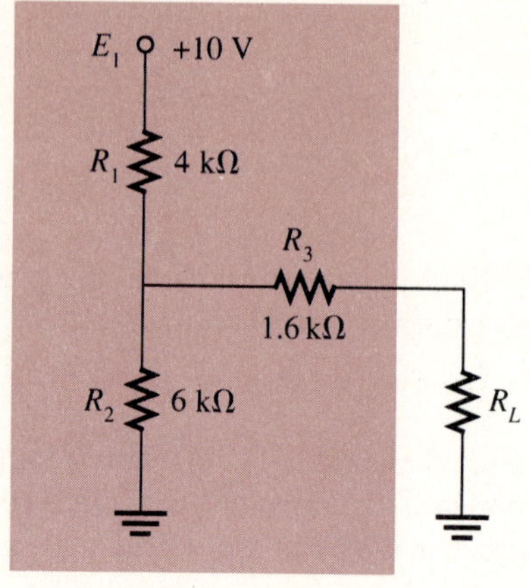

FIG. 9.53
Network for Example 9.16.

Solution The network is redrawn and *Steps 1 and 2* are applied as shown in Fig. 9.54.

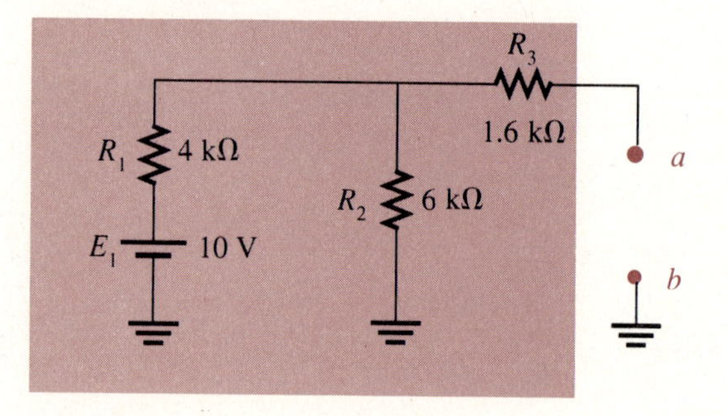

FIG. 9.54
Network of Fig. 9.53 following the removal of R_2.

Step 3: See Fig. 9.55.

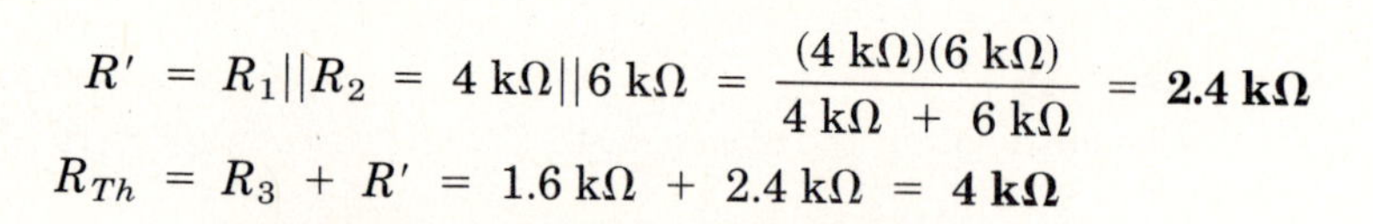

$$R' = R_1 \| R_2 = 4\text{ k}\Omega \| 6\text{ k}\Omega = \frac{(4\text{ k}\Omega)(6\text{ k}\Omega)}{4\text{ k}\Omega + 6\text{ k}\Omega} = \mathbf{2.4\text{ k}\Omega}$$

$$R_{Th} = R_3 + R' = 1.6\text{ k}\Omega + 2.4\text{ k}\Omega = \mathbf{4\text{ k}\Omega}$$

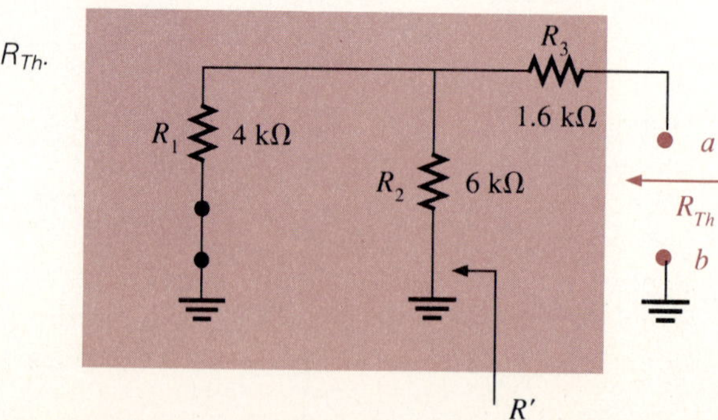

FIG. 9.55
Determining R_{Th}.

Step 4: Refer to Fig. 9.56. Use the voltage divider rule:

$$V_2 = \frac{R_2 E_1}{R_2 + R_1} = \frac{(6\ \text{k}\Omega)(10\ \text{V})}{6\ \text{k}\Omega + 4\ \text{k}\Omega}$$

$$= \frac{6}{10}(10\ \text{V}) = \mathbf{6\ V}$$

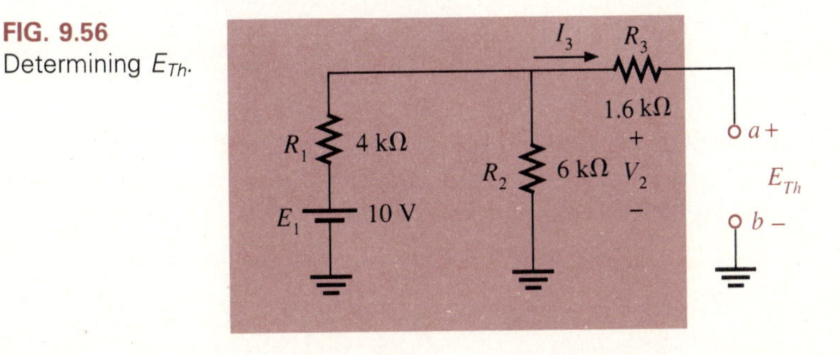

FIG. 9.56
Determining E_{Th}.

The current I_3 is 0 A due to the series open circuit from a to b. The result is

$$V_3 = I_3 R_3 = (0)R_3 = 0\ \text{V}$$

and

$$E_{Th} = V_2 - V_3 = 6\ \text{V} - 0\ \text{V} = \mathbf{6\ V}$$

Step 5: See Fig. 9.57.

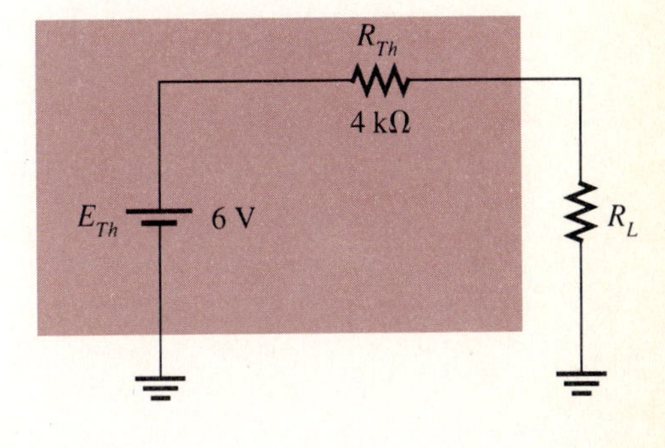

FIG. 9.57
Constructing the Thevenin equivalent circuit.

As noted before, Thevenin's theorem resulted in an equivalent series circuit consisting of a dc voltage source and resistor. Another approach, referred to as Norton's theorem, results in a parallel current source equivalent such as the one shown in Fig. 9.58.

The resistance R_N is found in exactly the same manner as R_{Th}, but I_N is the short-circuit current between the two "holding" terminals labeled a and b in the preceding analysis. However, rather than go into depth on how I_N is

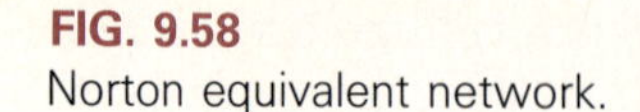
FIG. 9.58
Norton equivalent network.

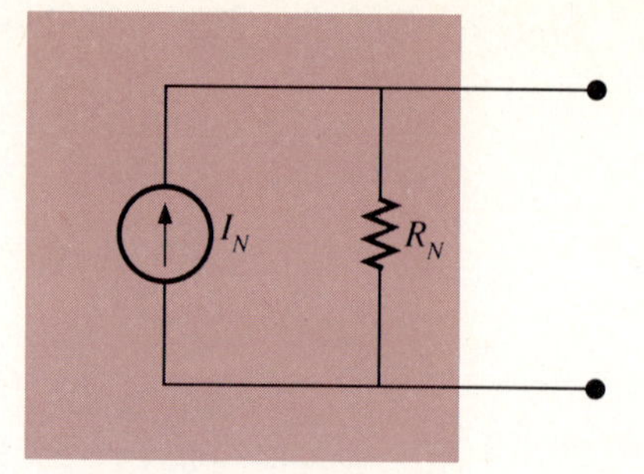

determined, we will simply remember that the current source model of Fig. 9.58 can be obtained directly from the Thevenin model just described using the standard source conversion equations:

$$I_N = \frac{E_{Th}}{R_{Th}} \quad \text{and} \quad R_N = R_{Th} \tag{9.1}$$

In other words, whenever the Norton equivalent is desired, we first determine the Thevenin equivalent and perform a source conversion.

For the network of Example 9.16,

$$I_N = \frac{E_{Th}}{R_{Th}} = \frac{6\ \text{V}}{4\ \text{k}\Omega} = \mathbf{1.5\ mA}$$

and

$$R_N = R_{Th} = \mathbf{4\ k\Omega}$$

as shown in Fig. 9.59.

FIG. 9.59
Norton equivalent network for the network of Example 9.16.

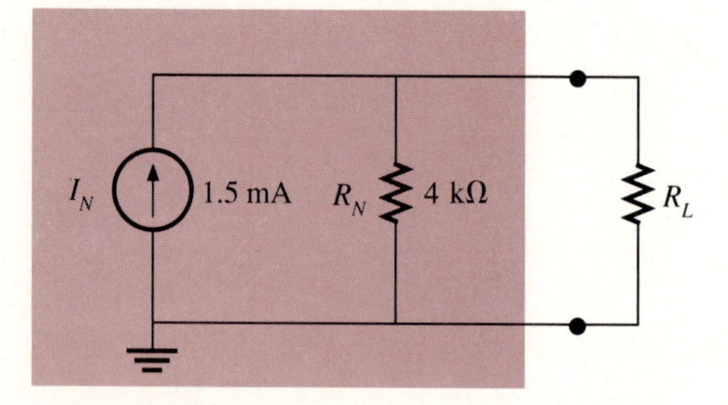

9.8 MAXIMUM POWER TRANSFER THEOREM

The maximum power transfer theorem states the following:

A load will receive maximum power from a dc network when its total resistive value is exactly equal to the Thevenin resistance of the network as "seen" by the load.

For the network of Fig. 9.60, maximum power is, therefore, delivered to the load when

$$\boxed{R_L = R_{Th}} \tag{9.2}$$

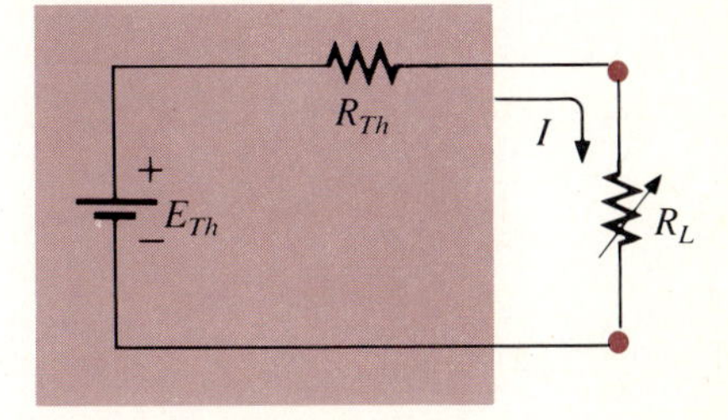

FIG. 9.60
Applying a variable load to the Thevenin equivalent circuit.

In other words, if we want to apply a load R_L to the network of Fig. 9.60 such that R_L will receive maximum power, then the value of R_L should equal R_{Th}.

For transistor configurations, the Norton equivalent circuit of Fig. 9.61 will appear more frequently. In this case maximum power transfer to the load R_L will result when the following condition is satisfied:

$$\boxed{R_L = R_N} \tag{9.3}$$

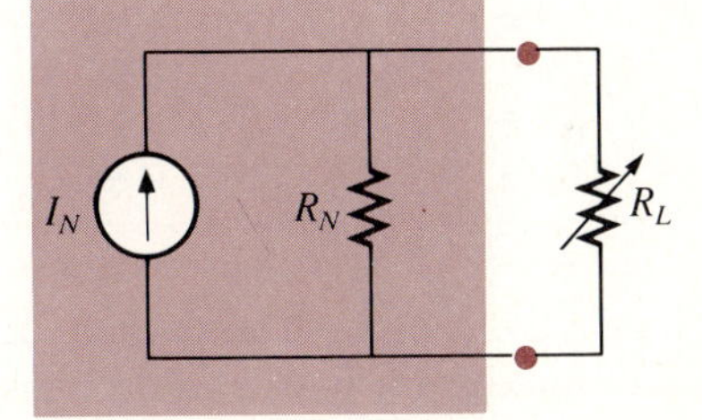

FIG. 9.61
Applying a variable load to the Norton equivalent circuit.

From the past discussions, we realize that a Thevenin equivalent circuit can be found across any element or group of elements in a dc network. Therefore, if we consider the case of the Thevenin equivalent circuit with respect to the maximum power transfer theorem, we are, in essence, considering the *total* effects of any network across a resistor R_L, such as in Fig. 9.60.

The initial tendency when examining maximum power transfer is to believe that a larger value of R_L will result in more power to R_L, since $P = V_L^2/R_L$ and V_L will be larger. However, keep in mind that R_L in the denominator will be larger and will thereby reduce the net value of P. If it is believed that a smaller R_L will result in a larger current and therefore power, since $P = I_L^2 R_L$, we must note that the multiplier of R_L will be less and reduce the level of P.

If we solve for the current through the load of Fig. 9.60, we find

$$I_L = \frac{E_{Th}}{R_{Th} + R_L}$$

Plugging this value into the power equation gives

$$P_L = I_L^2 R_L = \left(\frac{E_{Th}}{R_{Th} + R_L}\right)^2 R_L$$

and

$$P_L = \frac{E_{Th}^2 R_L}{(R_{Th} + R_L)^2} \tag{9.4}$$

For discussion purposes let us assume $E_{Th} = 6$ V and $R_{Th} = 9\ \Omega$ and calculate the power levels for different values of R_L using Eq. 9.4. The results are tabulated in Table 9.1.

TABLE 9.1

R_L (ohms)	$P_L = \dfrac{36R_L}{(9 + R_L)^2}$ (watts)	
1	0.36	Increase ↓
3	0.75	
6	0.96	
9	1	← Maximum
12	0.98	Decrease ↓
15	0.94	
18	0.89	

A plot of the data of Table 9.1 as provided in Fig. 9.62 clearly reveals that maximum power is delivered to R_L when it is exactly equal to R_{Th}.

For any physical network, the value of E_{Th} can be determined experimentally by measuring the open-circuit voltage across the load terminals, as shown in Fig. 9.63; $E_{Th} = V_{ab}$. The value of R_{Th} can then be determined by completing the network with an R_L such as the potentiometer of Fig. 9.64(b), page 300. R_L can then be varied until the voltage appearing across the load is one-half the open-circuit value, or $V_L = E_{Th}/2$. For the series circuit of Fig. 9.64(a), when the load voltage is reduced to one-half the open-circuit level, the voltage across R_{Th} and R_L must be the same. If we read the value of R_L [as shown in Fig. 9.64(c)] that resulted in the preceding calculations, we will also have the value of R_{Th}, since $R_L = R_{Th}$ if V_L equals the voltage across R_{Th}.

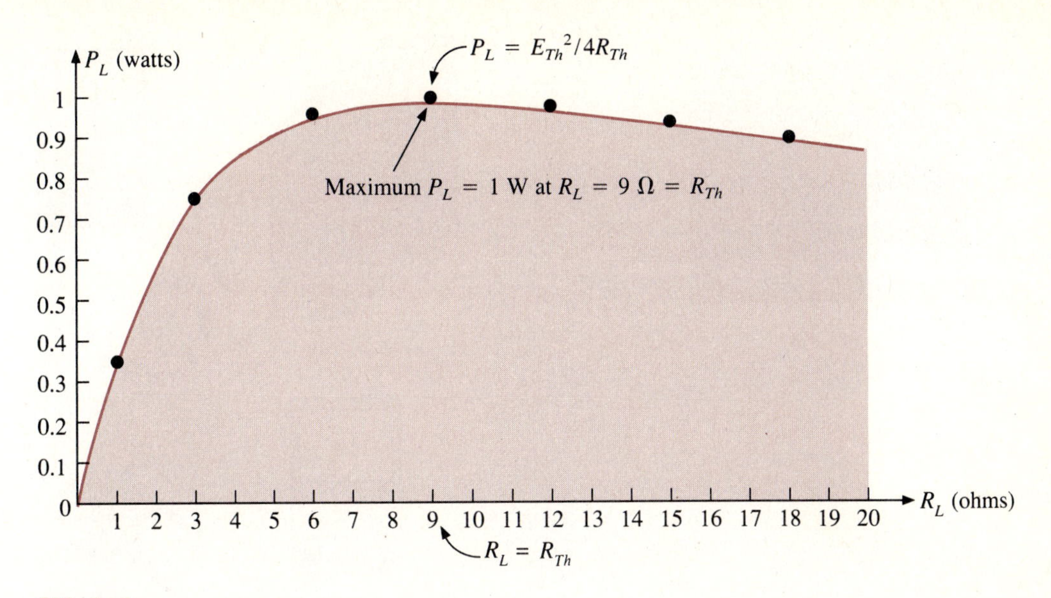

FIG. 9.62
P_L + R_L for the Thevenin equivalent circuit with E_{Th} = 6 V and R_{Th} = 9 Ω.

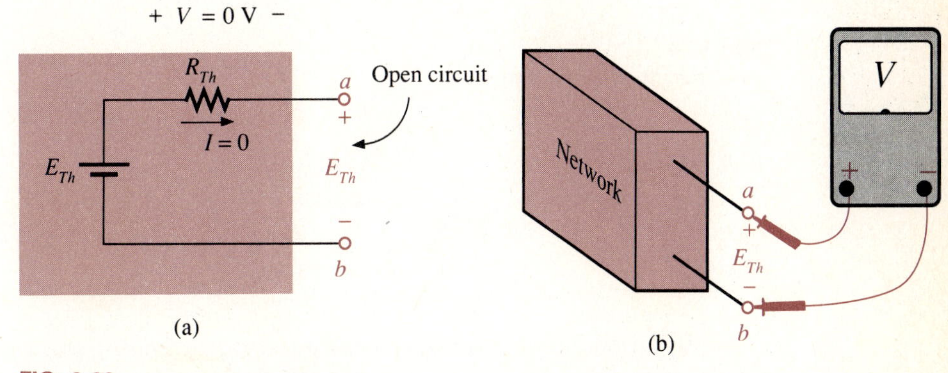

FIG. 9.63
Determining E_{Th} experimentally.

If we apply a load $R_L = R_{Th}$ to the circuit of Fig. 9.64(a), we obtain

$$I_L = \frac{E_{Th}}{R_{Th} + R_L} = \frac{E_{Th}}{R_{Th} + R_{Th}} = \frac{E_{Th}}{2R_{Th}}$$

and the maximum power to the load is

$$P_{L_{\max}} = I_L^2 R_L = \left(\frac{E_{Th}}{2R_{Th}}\right)^2 R_{Th} = \frac{E_{Th}^2}{4R_{Th}^2} \cdot R_{Th}$$

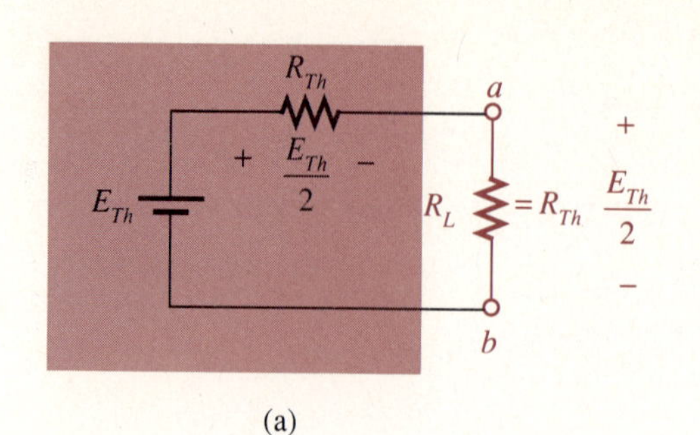

(a)

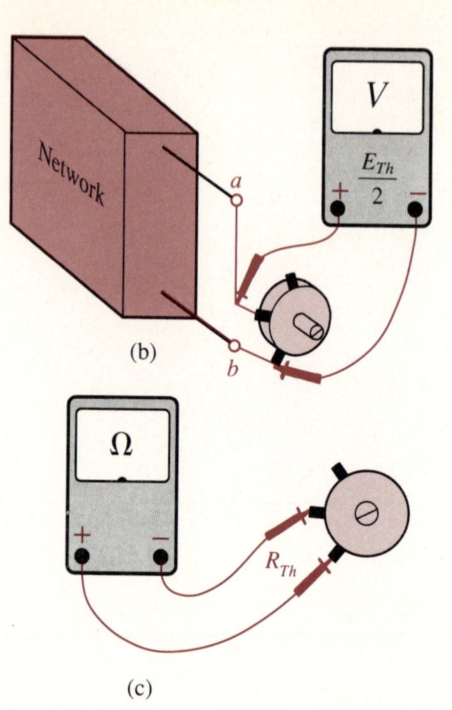

(b)

(c)

FIG. 9.64
Determining R_{Th} experimentally.

and

$$P_{L_{max}} = \frac{E_{Th}^2}{4R_{Th}} \tag{9.5}$$

Equation 9.5 provides a direct route toward finding the maximum power that can be delivered to a load from a particular network. Simply find E_{Th} and R_{Th} at the point where the load is applied and substitute into Eq. 9.5.

EXAMPLE 9.17

For each configuration of Fig. 9.65 with $E = 20$ V, determine

a. The value of R_L for maximum power to R_L.
b. The maximum power to each load.

Solutions

a. Fig. 9.65(a): $R_L = R_{int} = \mathbf{2.5\ \Omega}$
Fig. 9.65(b): $R_L = R_{int} = \mathbf{0.5\ \Omega}$
Fig. 9.65(c): $R_L = R_{int} = \mathbf{40\ \Omega}$

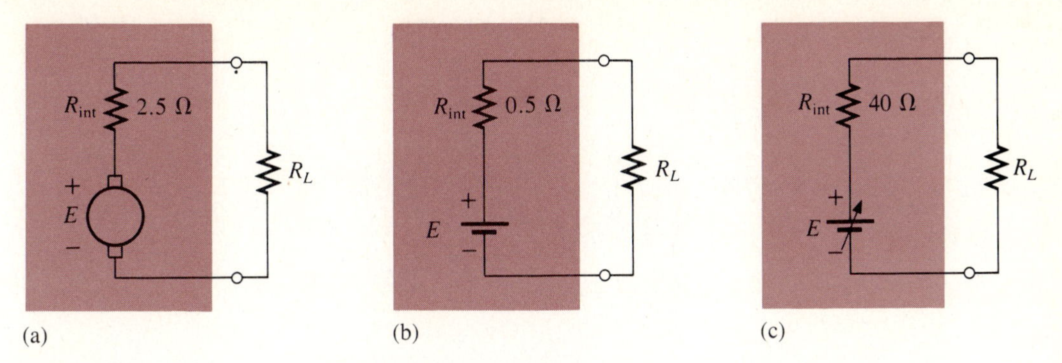

FIG. 9.65
Dc sources. (a) Generator; (b) battery; (c) laboratory supply.

b. Fig. 9.65(a):

$$E_{Th} = E = 20 \text{ V}, \qquad R_{Th} = 2.5 \ \Omega$$
$$P_{max} = \frac{E_{Th}^2}{4R_{Th}} = \frac{(20 \text{ V})^2}{4(2.5 \ \Omega)} = \mathbf{40 \ W}$$

Fig. 9.65(b):

$$E_{Th} = E = 20 \text{ V}, \qquad R_{Th} = 0.5 \ \Omega$$
$$P_{max} = \frac{E_{Th}^2}{4R_{Th}} = \frac{(20 \text{ V})^2}{4(0.5 \ \Omega)} = \mathbf{200 \ W}$$

Fig. 9.65(c):

$$E_{Th} = E = 20 \text{ V}, \qquad R_{Th} = 40 \ \Omega$$
$$P_{max} = \frac{E_{Th}^2}{4R_{Th}} = \frac{(20 \text{ V})^2}{4(40 \ \Omega)} = \mathbf{2.5 \ W}$$

For the same no-load voltage, note the impact of the R_{int} on the maximum power that can be delivered to a load.

EXAMPLE 9.18

Analysis of a transistor network resulted in the reduced configuration of Fig. 9.66. Determine the R_L necessary for maximum power transfer to R_L, and calculate the power to R_L under these conditions.

FIG. 9.66
Network for Example 9.18.

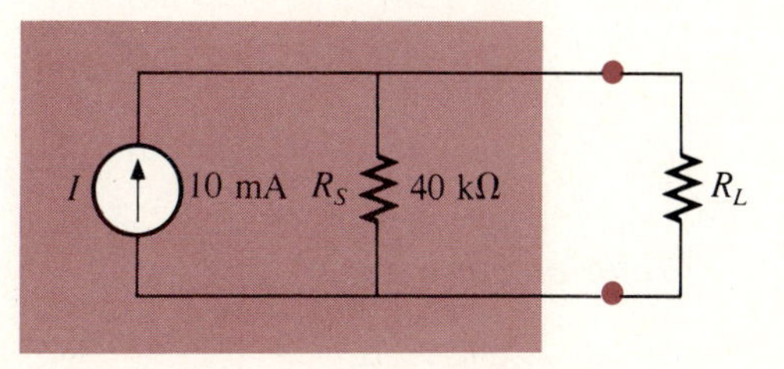

Solution

R_{Th}: See Fig. 9.67.

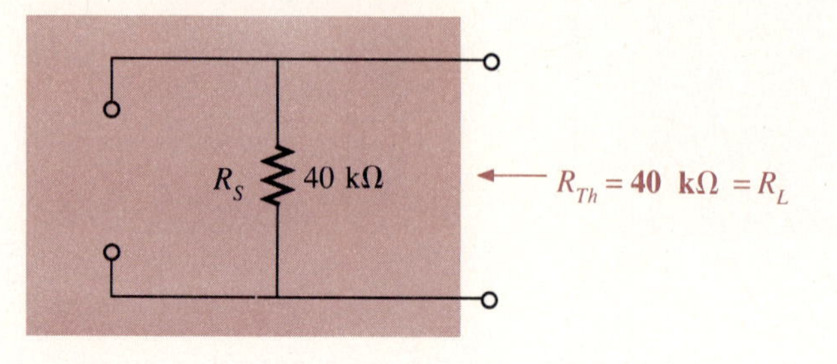

FIG. 9.67
Determining R_{Th} for the network of Fig. 9.66.

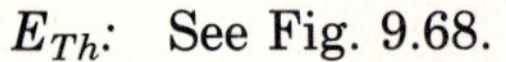

E_{Th}: See Fig. 9.68.

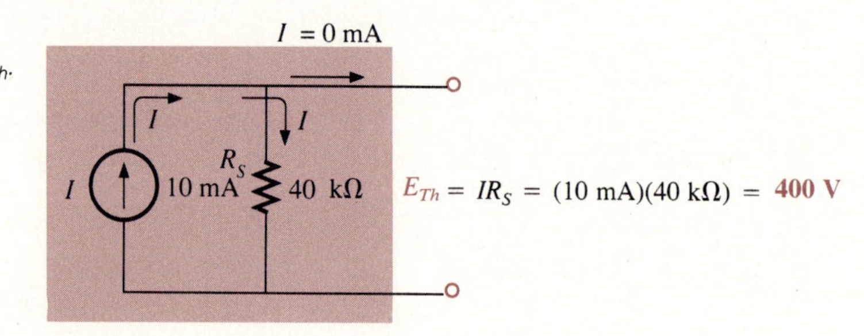

FIG. 9.68
Determining E_{Th}.

Using Eq. 9.5:

$$P_{L_{max}} = \frac{E_{Th}^2}{4R_{Th}} = \frac{(400\text{ V})^2}{4(40\text{ k}\Omega)} = \mathbf{1\text{ W}}$$

EXAMPLE 9.19 For the network of Fig. 9.69, determine the value of R for maximum power to R, and calculate the power delivered under these conditions.

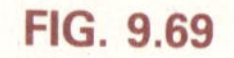

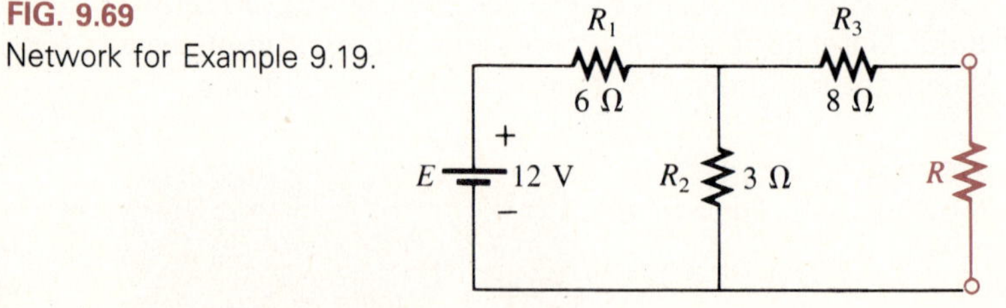

FIG. 9.69
Network for Example 9.19.

Solution See Fig. 9.70;

$$R_{Th} = R_3 + R_1 \| R_2 = 8\,\Omega + \frac{(6\,\Omega)(3\,\Omega)}{6\,\Omega + 3\,\Omega} = 8\,\Omega + 2\,\Omega$$

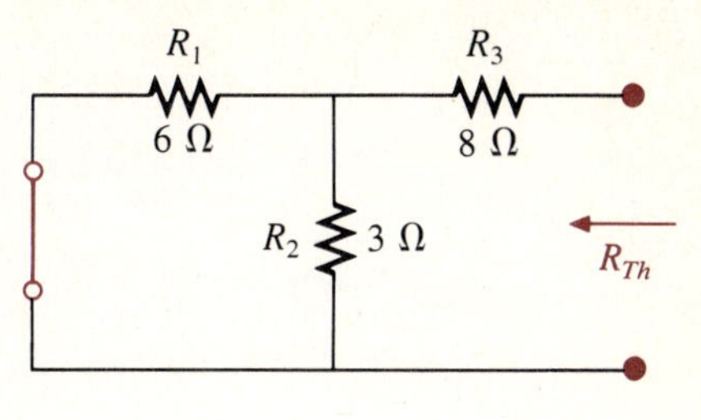

FIG. 9.70
Determining R_{Th} for the network of Fig. 9.69.

and

$$R = R_{Th} = \mathbf{10\ \Omega}$$

See Fig. 9.71;

$$E_{Th} = \frac{R_2 E}{R_2 + R_1} = \frac{(3\ \Omega)(12\ \text{V})}{3\ \Omega + 6\ \Omega} = \frac{36\ \text{V}}{9} = \mathbf{4\ V}$$

and, by Eq. 9.5,

$$P_{L_{\max}} = \frac{E_{Th}^2}{4R_{Th}} = \frac{(4\ \text{V})^2}{4(10\ \Omega)} = \mathbf{0.4\ W}$$

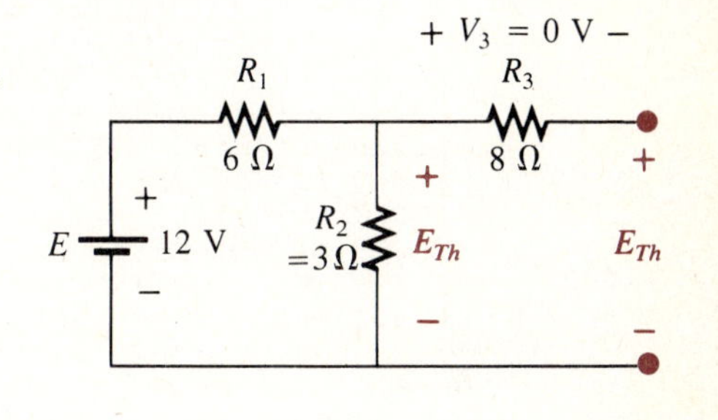

FIG. 9.71
Determining E_{Th} for the network of Fig. 9.69.

9.9 COMPUTER ANALYSIS

Program 9.1 (Fig. 9.73) provides a complete Thevenin analysis of the network of Fig. 9.72, including a tabulation of power and efficiency versus load resistance.

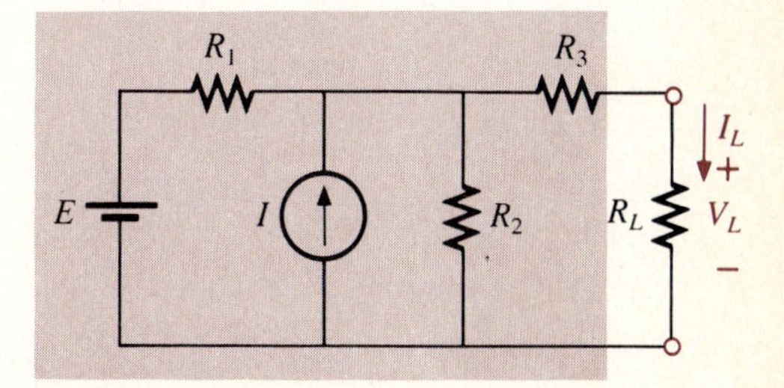

FIG. 9.72
Network to be analyzed by Program 9.1.

Body of Program:

```
 10 REM ***** PROGRAM 9-1 *****
 20 REM ******************************************
 30 REM Program to tabulate changes in load levels for
 40 REM a range of load values using Thevenin's theorem
 50 REM ******************************************
 60 REM
100 PRINT "For the network of Fig. 9.84"
110 PRINT "enter the following data:"
120 PRINT
130 INPUT "R1=";R1 :REM Enter 0 if resistor non-existant      [Input Data]
140 INPUT "R2=";R2 :REM Enter 1E30 if resistor non-existant
150 INPUT "R3=";R3 :REM Enter 0 if resistor non-existant
160 INPUT "RL=";RL
170 INPUT "Supply voltage, E=";E
180 INPUT "and supply current, I=";I
190 PRINT
200 REM Determine Rth                                          [R_Th]
210 RT=R3+R1*R2/(R1+R2)
220 REM Use superposition to determine Eth                     [E_Th]
230 E1=R2*E/(R1+R2)
240 I2=R2*I/(R1+R2)
250 E2=R1*R2*I/(R1+R2)
260 ET=E1+E2
270 PRINT "Using Thevenin's theorem:"                          [Output R_Th & E_Th]
280 PRINT "Rth=";RT;"ohms"
290 PRINT "and Eth=";ET;"volts"
300 PRINT
310 PRINT TAB(7);"RL";TAB(15);"IL";TAB(25);"VL";              [Table Heading]
320 PRINT TAB(35);"PL";TAB(45);"PD";TAB(55);"n%"
330 FOR RL=RT/4 TO 4*RT STEP RT/4                              [Calc.]
340 IL=ET/(RT+RL)
350 VL=IL*RL
360 PL=IL^2*RL
370 PD=ET*IL
380 N=100*PL/PD
390 IF RL=RT THEN PRINT "Rth=";                                [Output Control]
400 PRINT TAB(5);RL;TAB(13);IL;TAB(23);VL;
410 PRINT TAB(33);PL;TAB(43);PD;TAB(53);N
420 NEXT RL
430 END
```

Program Runs:

```
READY

RUN


For the network of Fig. 9.84
enter the following data:

R1=? 20

R2=? 1E30

R3=? 5

RL=? 25

Supply voltage, E=? -10

and supply current, I=? 4
```

```
Using Thevenin's theorem:
Rth= 25 ohms
and Eth= 70 volts

       RL        IL        VL        PL        PD        n%
       6.25      2.24      14        31.36     156.8     20
       12.5      1.8667    23.3333   43.5556   130.6667  33.3333
       18.75     1.6       30        48        112       42.8571
Rth=   25        1.4       35        49        98        50
       31.25     1.2444    38.8889   48.3951   87.1111   55.5556
       37.5      1.12      42        47.04     78.4      60
       43.75     1.0182    44.5455   45.3554   71.2727   63.6364
       50        .9333     46.6667   43.5556   65.3333   66.6667
       56.25     .8615     48.4615   41.7515   60.3077   69.2308
       62.5      .8        50        40        56        71.4286
       68.75     .7467     51.3333   38.3289   52.2667   73.3333
       75        .7        52.5      36.75     49        75
       81.25     .6588     53.5294   35.2664   46.1176   76.4706
       87.5      .6222     54.4444   33.8765   43.5556   77.7778
       93.75     .5895     55.2632   32.5762   41.2632   78.9474
       100       .56       56        31.36     39.2      80

READY
```

FIG. 9.73
Program 9.1.

The Thevenin resistance is determined by lines 200 and 210, and E_{Th} is determined using superposition on lines 220 through 260. E_1 on line 230 is the contribution to E_{Th} due to the voltage source E, and E_2 on line 250 is the contribution due to the current source I. E_{Th} is then determined on line 260. The results are printed out by lines 270 through 290. The details of the tabulating routine are left for your first computer course.

9.10 BRANCH-CURRENT ANALYSIS

We now consider the first in a series of methods for solving networks with two or more sources that are not in series or parallel. Keep in mind that networks with two isolated sources cannot be solved using the approach of Chapter 8. There is an interaction of sources that does not permit finding such quantities as the total resistance R_T and a singular source current. For further evidence of this fact, try solving for the unknown quantities of Example 9.11 using the methods of Chapter 8.

The branch-current analysis approach requires that we apply both Kirchhoff's voltage and current laws a number of times determined by the size of the network. Before looking at the steps required to apply this method, be sure it is understood that the quantities to be found are the *branch* currents of the network. They are the currents of each branch of the network (remember that two elements in series have the same current and constitute a branch of the network) having one or more elements in series. Once all the branch currents of a network are known, all other quantities—such as voltage or power—can be determined.

The most direct introduction to a method of this type is to list the series of steps required for its application and then carefully examine each step individually.

Step 1: Assign a distinct current of arbitrary direction to each branch of the network.

This step requires that you first determine how many different series branches the network has. Actually, simply determine how many different currents the network actually has (keeping in mind that series elements have the same current) and you have the number of branch currents. When you assign a current to each branch, give it a name such as I_1, I_2, or I_3, as often determined by the resistors in the branch. The term *arbitrary* reveals that *any* direction can be chosen for each branch. A wrong choice simply results in a negative sign in the answer (as we will see in some of the examples). In Fig. 9.74 three possible choices for branch-current direction are provided for a two-loop network.

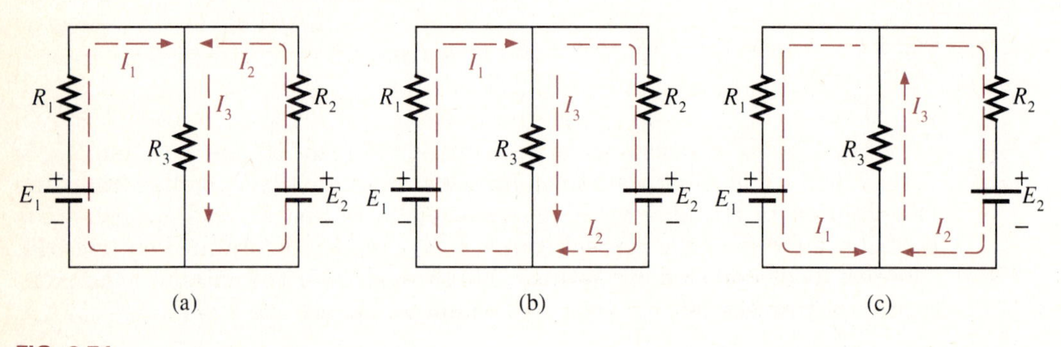

FIG. 9.74
Defining the branch currents.

Step 2: Indicate the polarities for each resistor, as determined by the assumed branch-current direction.

When applying this step, be as mechanical as possible. That is, keep in mind that current entering a resistor establishes a positive (+) potential on that side and a negative (−) potential on the side that it leaves. Do not be influenced by batteries in the network or the other assumed branch currents of the network. The polarity of all voltage sources is unaffected by the direction of the assumed branch current in the branch in which it is located. In Fig. 9.75 the polarities as defined by the current directions of Fig. 9.74 are introduced. The polarity for each resistor is defined solely by the current direction with no concern for the polarity of the voltage sources.

Step 3: Apply Kirchhoff's voltage law around each *closed loop and Kirchhoff's current law at the minimum number of junction points that will be sure to include all the branch currents of the network.*

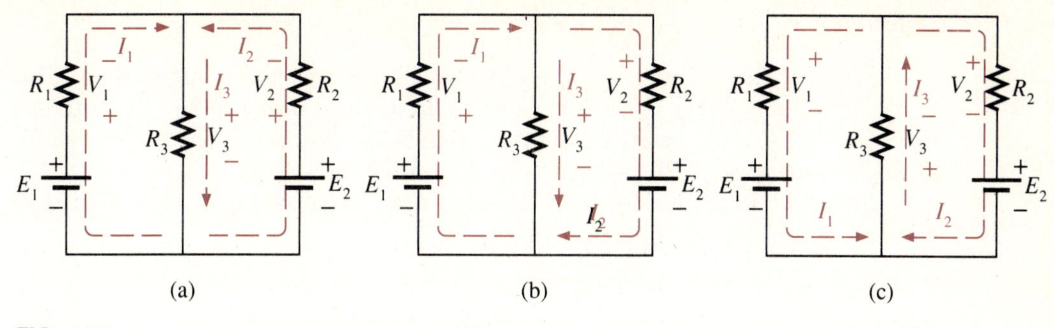

FIG. 9.75
Inserting polarities across each resistive element.

The easiest way to determine how many independent closed loops a network has is to count the number of "windows" in the configuration. A window is simply any area of the network completely enclosed on all sides by branches of the network. Usually, the number of junction points at which Kirchhoff's current law has to be applied is one less than the number of windows, or independent closed paths. For the network of Fig. 9.75 there are two windows, requiring two applications of Kirchhoff's voltage law and one application of Kirchhoff's current law.

Step 4: Solve the resulting equations for the branch currents of the network.

There are a number of mathematical techniques, such as determinants (Appendix B) and matrices, that can be used to solve for the unknown quantities of a set of equations, but our attention will be limited to networks that can be solved in a fairly direct manner by the substitution process. If time permits, however, it would be time well spent to investigate the more advanced mathematical techniques.

EXAMPLE 9.20 Apply the branch-current method to the network of Fig. 9.76.

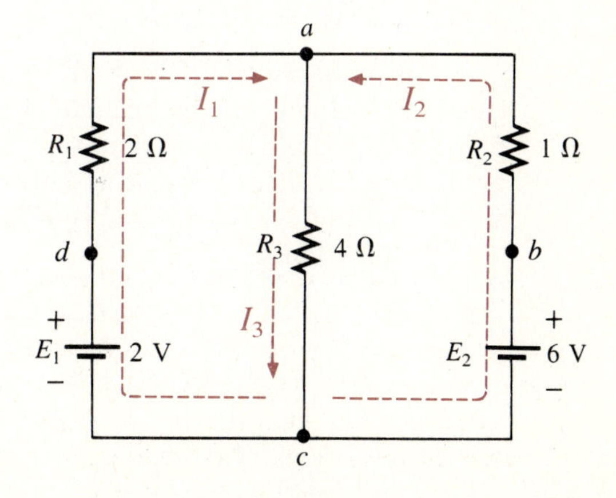

FIG. 9.76
Network for Example 9.20.

Solution

Step 1: Since there are three distinct branches, three currents of arbitrary directions (I_1, I_2, I_3) are chosen, as indicated in Fig. 9.76. The current directions for I_1 and I_2 were chosen to match the pressure applied by sources E_1 and E_2, respectively. Since both I_1 and I_2 enter node a, I_3 is leaving.

Step 2: Polarities for each resistor are drawn to agree with assumed current directions, as indicated in Fig. 9.77.

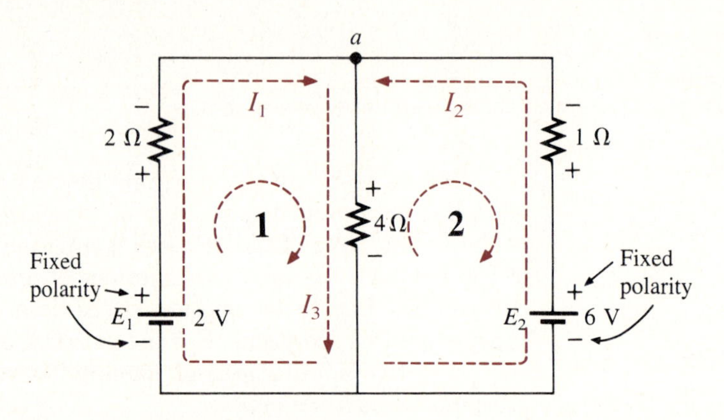

FIG. 9.77
Defining polarities and applying Kirchhoff's voltage law.

Step 3: It should be fairly obvious from Fig. 9.76 that there are two windows (enclosed areas) in the configuration, requiring two applications of Kirchhoff's voltage law. They are labeled as 1 and 2 in Fig. 9.77.

Kirchhoff's voltage law is applied around each closed loop (1 and 2) in the clockwise direction starting at the bottom of the network:

	rise in potential		drop in potential			
Loop 1:	$+2$	$-$	$2I_1$	$-$	$4I_3$	$= 0$
	battery potential		voltage drop across 2-Ω resistor		voltage drop across 4-Ω resistor	
Loop 2:	$4I_3$	$+$	$1I_2$	$-$	6	$= 0$

An application of Kirchhoff's current law at junction a results in an equation that includes all the branch currents of the network. There is, therefore, no need to write a second equation using this law. Note also that since there are two independent loops, if we subtract one from this number (as described in step 3), we find we need to apply Kirchhoff's current law only to one junction.

At junction *a:*

$$\Sigma I_{\text{entering}} = \Sigma I_{\text{leaving}}$$
$$I_1 + I_2 = I_3$$

Step 4: There are then three equations and three unknowns, permitting a solution for the three unknown branch currents.

$$\begin{aligned} 2 - 2I_1 - 4I_3 &= 0 \\ 4I_3 + 1I_2 - 6 &= 0 \\ I_1 + I_2 &= I_3 \end{aligned}$$

Lining up all the unknown branch currents:

$$\begin{aligned} 2I_1 + 0 + 4I_3 &= +2 \\ 0 + I_2 + 4I_3 &= +6 \\ I_1 + I_2 - I_3 &= 0 \end{aligned}$$

If we knew determinants (Appendix B), matrices, or computer techniques, the solutions for I_1, I_2, and I_3 could be found quite directly. However, solving for I_1 in the third equation and substituting into the first equation results in two equations and two unknowns (I_2 and I_3). That is, $I_1 = I_3 - I_2$ and Eq. 1 becomes

$$2(I_3 - I_2) + 4I_3 = +2$$

or

$$-2I_2 + 6I_3 = +2$$

With Eq. 2:

$$\begin{aligned} -2I_2 + 6I_3 &= +2 \\ I_2 + 4I_3 &= +6 \end{aligned}$$

Solving for I_2 in the second equation and substituting into the first equation results in

$$I_2 = 6 - 4I_3$$

and

$$\begin{aligned} -2(6 - 4I_3) + 6I_3 &= +2 \\ -12 + 14I_3 &= +2 \\ 14I_3 &= 14 \\ I_3 &= \mathbf{1\ A} \end{aligned}$$

Now that we have I_3, we can find I_1 and I_2 by substituting back into the original equations.

For I_1:

$$\begin{aligned} 2I_1 + 4I_3 &= +2 \\ 2I_1 + 4(1) &= +2 \\ 2I_1 &= -2 \\ I_1 &= \mathbf{-1\ A} \end{aligned}$$

The minus simply reveals that the actual direction of the branch current I_1 is opposite to that assumed. However, the magnitude of 1 A is correct.

For I_2:

$$\begin{aligned} I_2 + 4I_3 &= +6 \\ I_2 + 4(1) &= +6 \\ I_2 &= \mathbf{2\ A} \end{aligned}$$

The network is redrawn once more, with the results in Fig. 9.78. The current through E_1 and R_1 is 1 A in the direction shown (opposite to that assumed). The current through R_3 is also 1 A, with the current through R_2 and E_2 at a level of 2 A. If desired, the voltage across R_1 is

$$V_1 = I_1R_1 = (1\ \text{A})(2\ \Omega) = \mathbf{2\ V}$$

and the voltage across R_2 is

$$V_2 = I_2R_2 = (2\ \text{A})(1\ \Omega) = \mathbf{2\ V}$$

with

$$V_3 = I_3R_3 = (1\ \text{A})(4\ \Omega) = \mathbf{4\ V}$$

The power to R_3 is

$$\begin{aligned} P_3 &= I_3^2R_3 \\ &= (1\ \text{A})^2(4\ \Omega) \\ &= (1)(4) = \mathbf{4\ W} \end{aligned}$$

clearly indicating that once the branch currents of a network are known, all the remaining quantities can be found quite directly.

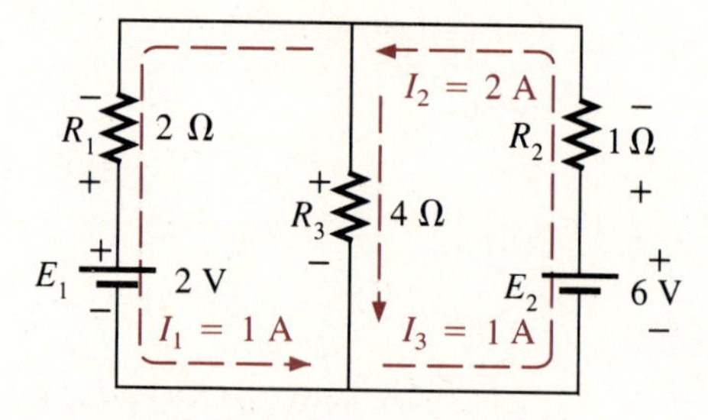

FIG. 9.78
Understanding the results obtained.

EXAMPLE 9.21 Apply branch-current analysis to the network of Fig. 9.79.

Solution Again, the current directions are chosen to match the pressure of each battery. The polarities are then added and Kirchhoff's voltage law is applied around each closed loop in the clockwise direction. The result is as follows:

$$\begin{aligned} &\text{Loop 1:} \quad +15 - 4I_1 + 10I_3 - 20 = 0 \\ &\text{Loop 2:} \quad +20 - 10I_3 - 5I_2 + 40 = 0 \end{aligned}$$

FIG. 9.79
Network for Example 9.21.

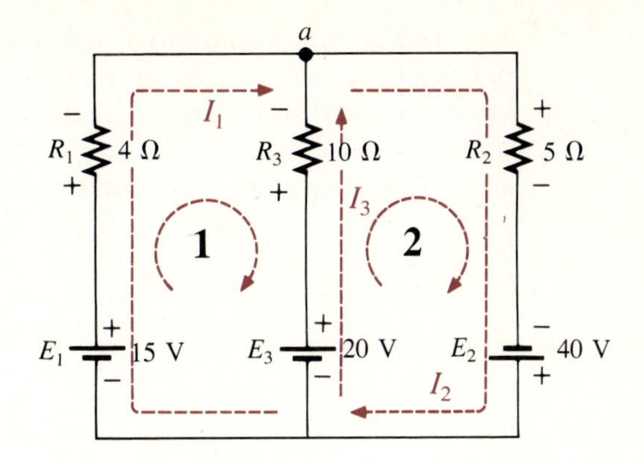

Applying Kirchhoff's current law at node a,

$$I_1 + I_3 = I_2$$

Substituting the third equation into the other two yields

$$15 - 4I_1 + 10I_3 - 20 = 0$$
$$20 - 10I_3 - 5(I_1 + I_3) + 40 = 0$$

substituting for I_2 (since it occurs only once in the two equations)

or

$$-4I_1 + 10I_3 = 5$$
$$5I_1 + 15I_3 = 60$$

Using the substitution technique, we find that

$$I_1 = \mathbf{4.773\ A}, \quad I_2 = \mathbf{7.182\ A}, \quad \text{and} \quad I_3 = \mathbf{2.409\ A}$$

revealing that we chose the correct direction for each. It is also clear that the results will not always be whole integer values.

9.11 MESH ANALYSIS

The second method of analysis to be described is called *mesh analysis*. The term *mesh* is derived from the similarities in appearance between the closed loops of a network and a wire mesh fence. Although this approach is on a more sophisticated plane than the branch-current method, it incorporates many of the ideas just developed. Of the two methods, mesh analysis is the one more frequently applied today. Branch-current analysis is introduced as a stepping stone to mesh analysis because branch currents are initially more "real" to the student than the loop currents employed in mesh analysis. Essentially, the mesh-analysis approach simply eliminates the need to substitute the results of Kirchhoff's current law into the equations derived from Kirchhoff's voltage law. The substitution is now accomplished in the initial writing of the equations.

In Fig. 9.80 two mesh or loop currents I_1 and I_2 have been drawn. Note in particular that I_1 and I_2 are not the branch currents of the configuration but are drawn within the complete enclosure, or window, of the network. In addition, note that they are continuous loops starting in one direction and ending at the starting point while maintaining one direction.

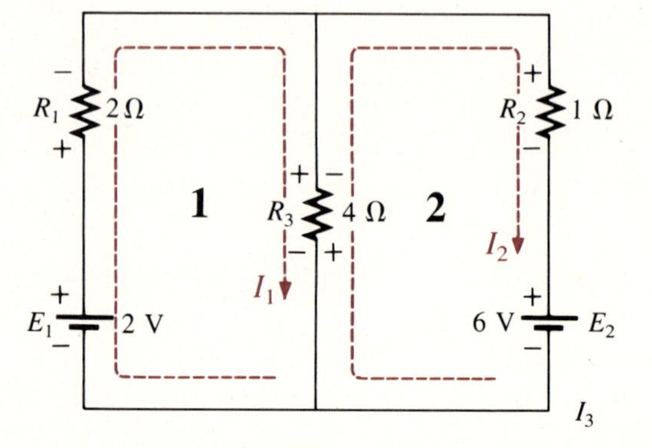

FIG. 9.80
Defining the loop (mesh) currents.

Taking a closer look at the loop current I_1, we find it is the only current through R_1 and E_1. Therefore, when a solution for I_1 is determined, we have the branch current for R_1 and E_1. For similar reasons, the current I_2 is the branch current for the branch including R_2 and E_2. The resistor R_3 includes both I_1 and I_2 in opposite directions. The magnitude of the current through R_3 is, therefore, the difference of the two mesh currents $I_1 - I_2$.

An application of Kirchhoff's voltage law to each window results in two equations with two unknowns, I_1 and I_2. Essentially, the application of Kirchhoff's current law has been incorporated by using loop or mesh currents. Of course, one must realize that the current through R_3 is not one or the other but their difference.

The true advantage of mesh analysis is not simply that one step of branch-current analysis has been eliminated but rather that a method can be developed for writing the mesh equations that is direct, error saving, simple, and relatively easy to apply using computer techniques. The technique is often referred to as the *format* approach to mesh analysis and can be applied by following these steps:

Step 1: *Assign a loop or mesh current to each independent closed loop (the windows) in the clockwise direction.*

Normally, the loop current is simply drawn as a closed loop within each enclosure, as shown in Fig. 9.80.

Step 2: *For each loop, multiply the loop current by the sum of the resistances that the loop current "touches."*

Simply write the loop current of interest and multiply it by the sum of the resistances in the loop.

Step 3: *For each loop, subtract the mutual terms composed of the product of any other current that passes through a resistor associated with the loop of interest and the resistance value.*

Take careful note of each resistor having the loop current of interest and any loop current. The product of the mutual resistance and the other loop current is subtracted from the term developed by step 2. There may be more than one mutual term, but they are all subtracted from the term of step 2.

Step 4: The column to the right of the equals sign is the algebraic sum of the voltage sources through which the loop current of interest passes.

Positive signs are assigned to those sources of voltage having a polarity such that the loop current passes from the negative to the positive terminal. A negative sign is assigned to those potentials for which the reverse is true.

Step 5: Solve the resulting simultaneous equations for the desired loop currents.

EXAMPLE 9.22

Solve for the mesh currents of the network of Fig. 9.80.

Solution *Step 2:* $I_1(R_1 + R_3)$ (I_1 does not "touch" R_2)
$I_2(R_2 + R_3)$ (I_2 does not "touch" R_1)

Step 3: $I_1(R_1 + R_3) - I_2R_3$

The resistor R_2 is not mutual with I_1. That is, I_1 does not pass through R_2.

$$I_2(R_2 + R_3) - I_1R_3$$

The resistor R_1 is not mutual with I_2.

Step 4:

$$\begin{aligned} I_1(R_1 + R_3) - I_2R_3 &= +E_1 \\ I_2(R_2 + R_3) - I_1R_3 &= -E_2 \end{aligned}$$

Note that E_1 is positive, since we progress from a negative to a positive sign as we move clockwise, whereas a negative sign is applied to E_2, since we move from a positive sign to a negative sign as we progress in the clockwise direction.

Substituting numbers:

$$\begin{aligned} I_1(2 + 4) - I_2(4) &= 2 \\ I_2(1 + 4) - I_1(4) &= -6 \end{aligned}$$

or

$$\begin{aligned} 6I_1 - 4I_2 &= 2 \\ -4I_1 + 5I_2 &= -6 \end{aligned}$$

Solving for I_1 and I_2 using techniques described in the last section gives

$$I_1 = \mathbf{-1\ A} \quad \text{and} \quad I_2 = \mathbf{-2\ A}$$

The negative sign for I_1 reveals that the branch current through R_1 and E_1 is 1 A but in the other direction. The negative sign for I_2 reveals that the branch current through R_2 and E_2 is 2 A but in the other direction. For the resistor R_3, we can refer to Fig. 9.80 and note that the current through R_3 is

$I_1 - I_2$ with the direction of I_1, since we gave it a positive sign in the equation. That is,

$$I_3 = I_1 - I_2 \qquad \text{(direction of } I_1\text{)}$$

or

$$I_3 = I_2 - I_1 \qquad \text{(direction of } I_2\text{)}$$

Using $I_3 = I_1 - I_2$:

$$\begin{aligned} I_3 &= (-1) - (-2) \\ &= -1 + 2 = \mathbf{1\ A} \end{aligned}$$

and I_3 is indeed 1 A in the direction of I_1. Note the careful substitution of the values for I_1 and I_2 with their respective minus signs.

EXAMPLE 9.23 Solve for the mesh currents of the network of Fig. 9.81.

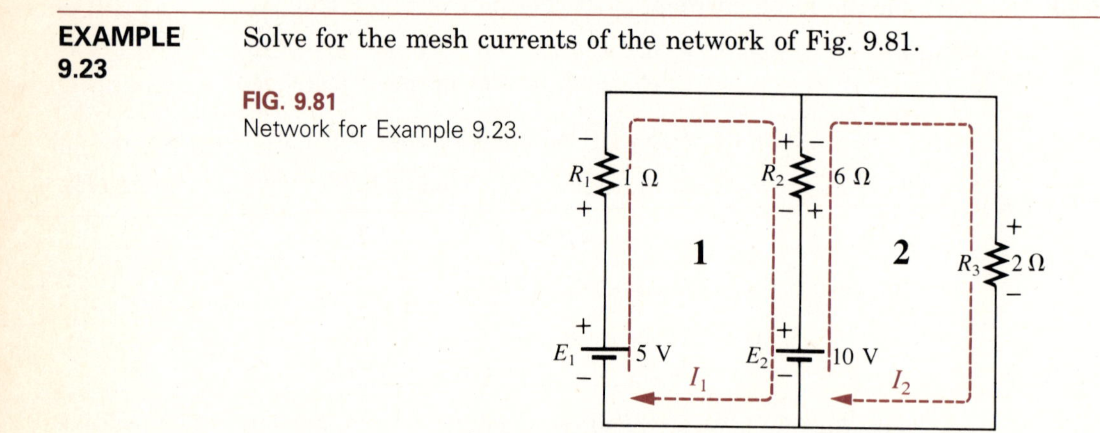

FIG. 9.81
Network for Example 9.23.

Solution

Step 1: This step is shown in Fig. 9.81.
Step 2: $I_1(R_1 + R_2)$ and $I_2(R_2 + R_3)$
Step 3: $I_1(R_1 + R_2) - I_2R_2$
$I_2(R_2 + R_3) - I_1R_2$

So far, the results are similar to those of Example 9.22, but take note of the results of the next step.

Step 4:
$$\begin{aligned} I_1(R_1 + R_2) - I_2R_2 &= E_1 - E_2 \\ I_2(R_2 + R_3) - I_1R_2 &= E_2 \end{aligned}$$

Substituting numbers gives

$$\begin{aligned} I_1(1 + 6) - I_26 &= 5 - 10 \\ I_2(6 + 2) - I_16 &= 10 \end{aligned}$$

$$\begin{aligned} 7I_1 - 6I_2 &= -5 \\ 8I_2 - 6I_1 &= 10 \end{aligned}$$

$$\begin{aligned} 7I_1 - 6I_2 &= -5 \\ -6I_1 + 8I_2 &= 10 \end{aligned}$$

resulting in

$$I_1 = \mathbf{1\ A} \quad \text{and} \quad I_2 = \mathbf{2\ A}$$

with the current through R_2 equal to

$$\begin{aligned} I\ (\text{of } R_2) &= I_1 - I_2 \qquad (\text{direction of } I_1) \\ &= 1\ \text{A} - 2\ \text{A} \\ &= \mathbf{-1\ A} \end{aligned}$$

The result reveals that the current through R_2 is 1 A but in the opposite direction.

It may happen that current sources are present in the network to which we wish to apply mesh analysis. The first step is then to convert all current sources to voltage sources and then proceed as before.

As noted earlier, the natural sequence of steps makes the mesh-analysis approach one that can easily be programmed on a computer, as demonstrated by Program 9.2 in Fig. 9.83. The network analyzed appears in Fig. 9.82, and the mesh equations obtained are the following:

$$\begin{aligned} I_1(R_1 + R_3) - I_2R_3 &= E_1 \\ -I_1R_3 + I_2(R_2 + R_3) &= -E_2 \end{aligned}$$

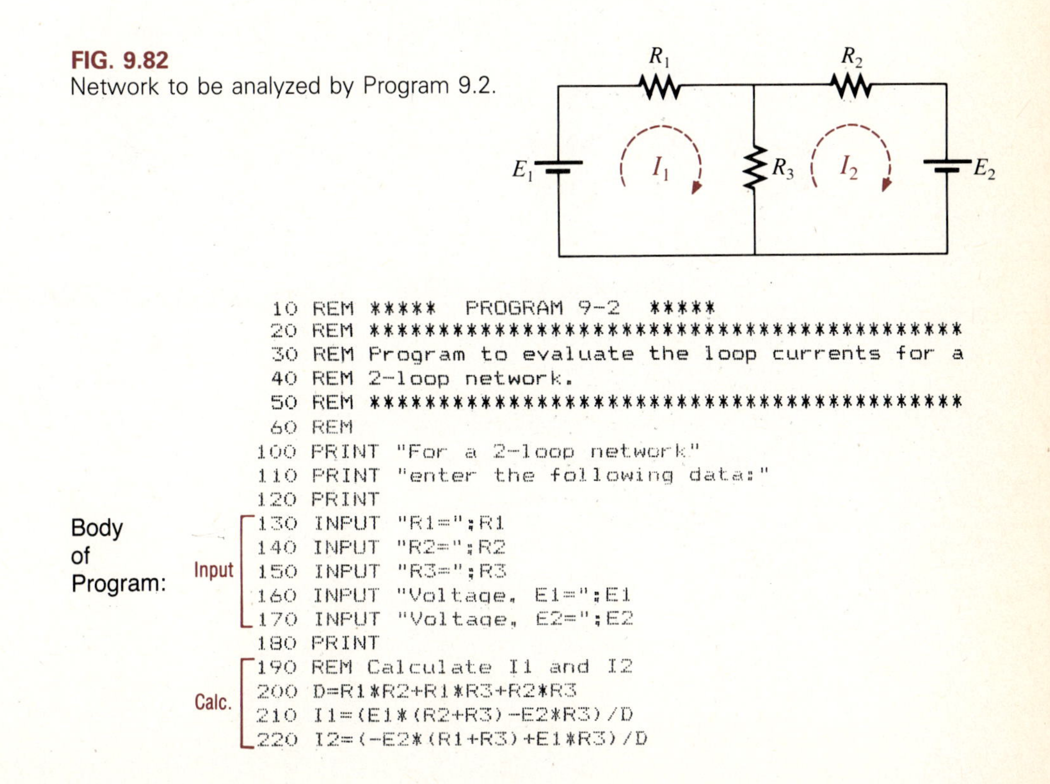

FIG. 9.82
Network to be analyzed by Program 9.2.

```
10 REM *****   PROGRAM 9-2   *****
20 REM ******************************************
30 REM Program to evaluate the loop currents for a
40 REM 2-loop network.
50 REM ******************************************
60 REM
100 PRINT "For a 2-loop network"
110 PRINT "enter the following data:"
120 PRINT
130 INPUT "R1=";R1
140 INPUT "R2=";R2
150 INPUT "R3=";R3
160 INPUT "Voltage, E1=";E1
170 INPUT "Voltage, E2=";E2
180 PRINT
190 REM Calculate I1 and I2
200 D=R1*R2+R1*R3+R2*R3
210 I1=(E1*(R2+R3)-E2*R3)/D
220 I2=(-E2*(R1+R3)+E1*R3)/D
```

Output

```
230 PRINT "The loop currents are:"
240 PRINT "I1=";I1;"amps"
250 PRINT "I2=";I2;"amps"
260 END
```

```
READY


For a 2-loop network
enter the following data:

R1=? 2

R2=? 1

R3=? 4

Voltage, E1=? 2

Voltage, E2=? 6


The loop currents are:
I1=-1 amps
I2=-2 amps


READY
```

Program Runs:

```
RUN

For a 2-loop network
enter the following data:

R1=? 1E3

R2=? 2.2E3

R3=? 3.3E3

Voltage, E1=? -5.4

Voltage, E2=? 8.6


The loop currents are:
I1=-4.5517E-03 amps
I2=-4.2947E-03 amps


READY
```

FIG. 9.83
Program 9.2.

The equations of lines 200, 210, and 220 can be obtained using the substitution, determinant, or matrix mathematical techniques. The second run includes standard resistor values in the kilohm range, resulting in currents in the milliampere range. Note that both I_1 and I_2 are negative for the values chosen, revealing that the current through R_1 and R_2 has the opposite direction from that shown in Fig. 9.82.

9.12 NODAL ANALYSIS

Recall from the development of loop analysis that the general network equations were obtained by applying Kirchhoff's voltage law around each closed loop. We will now employ Kirchhoff's current law to develop a method referred to as *nodal analysis*.

A *node* is defined as a junction of two or more branches (a connection point for the elements of the network). If we define one node of any network as a reference of 0 V, the remaining nodes of the network will all have a fixed potential (voltage) relative to this reference. For a network of three nodes, therefore, there will exist two nodes with a fixed potential relative to the assigned reference node. A set of equations will result that can be solved in a manner similar to that described for mesh analysis; these equations will provide all the voltage levels from the nodes to the reference. All the currents and power levels for the network can be found using Ohm's law and the standard power equations.

To facilitate the writing of the network equations, all voltage sources within the network must first be converted to current sources before Kirchhoff's current law is applied.

In general, the nodal-analysis method is applied as follows:

1. Convert all voltage sources to current sources.
2. Determine the number of nodes within the network.
3. Pick a reference node and label each remaining node with a subscripted value of voltage: V_1, V_2, and so on.
4. Write Kirchhoff's current law at each node except the reference.
5. Solve the resulting equations for nodal voltages.

EXAMPLE 9.24 Apply nodal analysis to the network of Fig. 9.84.

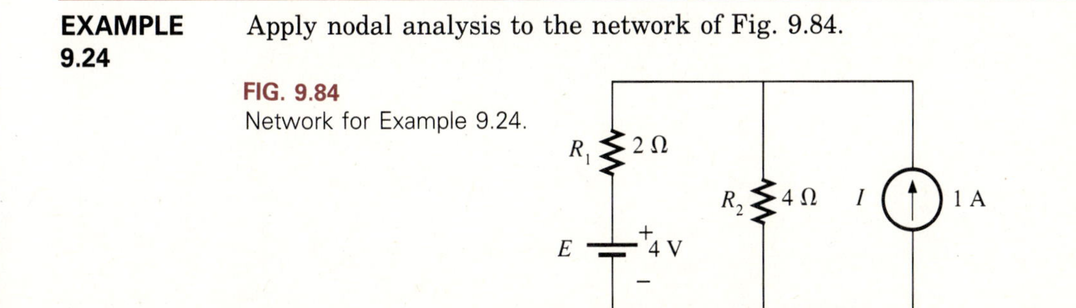

FIG. 9.84
Network for Example 9.24.

Solution

Step 1: Convert the voltage source to a current source (Fig. 9.85).

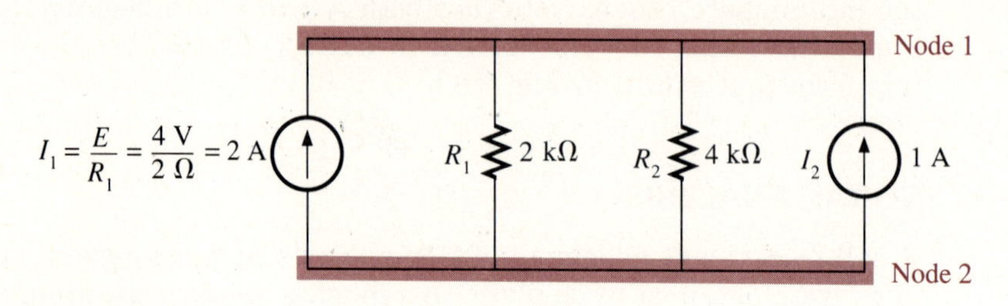

FIG. 9.85
Network of Fig. 9.84 following the source conversion.

Steps 2 and 3: See Fig. 9.86.

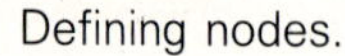

FIG. 9.86
Defining nodes.

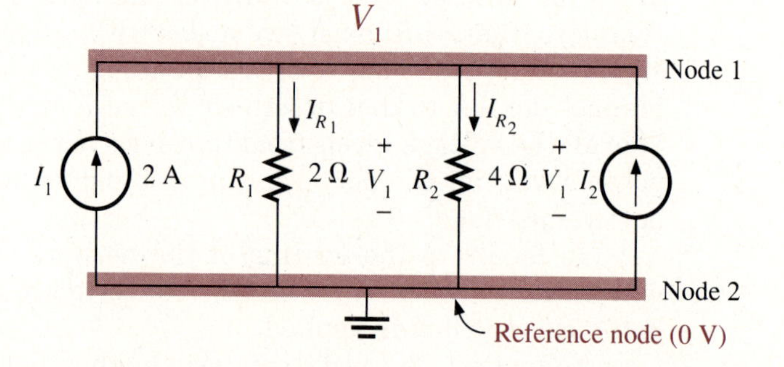

Note in Fig. 9.85 that there are two junction points (nodes), one on top and one on the bottom of the network. In Fig. 9.86 the bottom node was chosen as the reference, and the node at the top was labeled V_1. When the analysis is complete and V_1 is determined, we will know the voltage from the top node to the reference node (assigned a level of 0 V). In other words, we will have the voltage across each of the elements, since they are all in parallel.

Step 4: Applying Kirchhoff's current law at the top node will result in

$$\Sigma I_{\text{entering}} = \Sigma I_{\text{leaving}}$$
$$I_1 + I_2 = I_{R_1} + I_{R_2}$$

or

$$I_1 + I_2 = \frac{V_1}{R_1} + \frac{V_1}{R_2}$$
$$= V_1\left(\frac{1}{R_1} + \frac{1}{R_2}\right)$$

Step 5: Substituting numbers and rearranging gives

$$V_1\left(\frac{1}{2} + \frac{1}{4}\right) = 2 + 1$$

$$V_1\left(\frac{3}{4}\right) = 3$$

$$V_1 = \frac{12}{3}\,\text{V} = \mathbf{4\ V}$$

The currents I_{R_1} and I_{R_2} can then be found using Ohm's law:

$$I_{R_1} = \frac{V_1}{R_1} = \frac{4\ \text{V}}{2\ \Omega} = \mathbf{2\ A}$$

$$I_{R_2} = \frac{V_1}{R_2} = \frac{4\ \text{V}}{1\ \Omega} = \mathbf{4\ A}$$

EXAMPLE 9.25

Determine the nodal voltages for the network of Fig. 9.87.

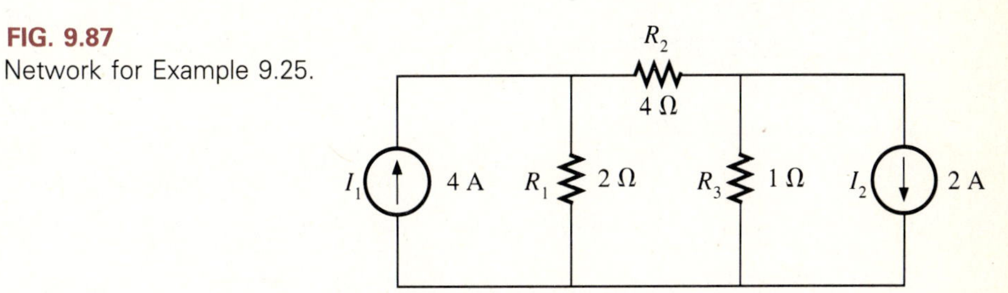

FIG. 9.87
Network for Example 9.25.

Solution

Step 1: All sources are current sources. No source conversions required.

Steps 2 and 3: As shown in Fig. 9.88, there are three nodes, with the bottom node (or junction) chosen as the reference. The other two are labeled V_1 and V_2. The voltage across the 4-A source and R_1 is therefore V_1, and the voltage across the 2-A source and R_2 is V_2. The voltage across R_3 is the difference between V_1 and V_2 $(V_1 - V_2)$. The meters are measuring the voltages $V_1 = V_{R_1}$ and $(V_1 - V_2) = V_{R_3}$.

Step 4: Applying Kirchhoff's current law at node V_1:

$$\Sigma I_{\text{entering}} = \Sigma I_{\text{leaving}}$$

$$4\ \text{A} = I_1 + I_3$$

$$4\ \text{A} = \frac{V_1}{R_1} + \frac{V_1 - V_2}{R_3}$$

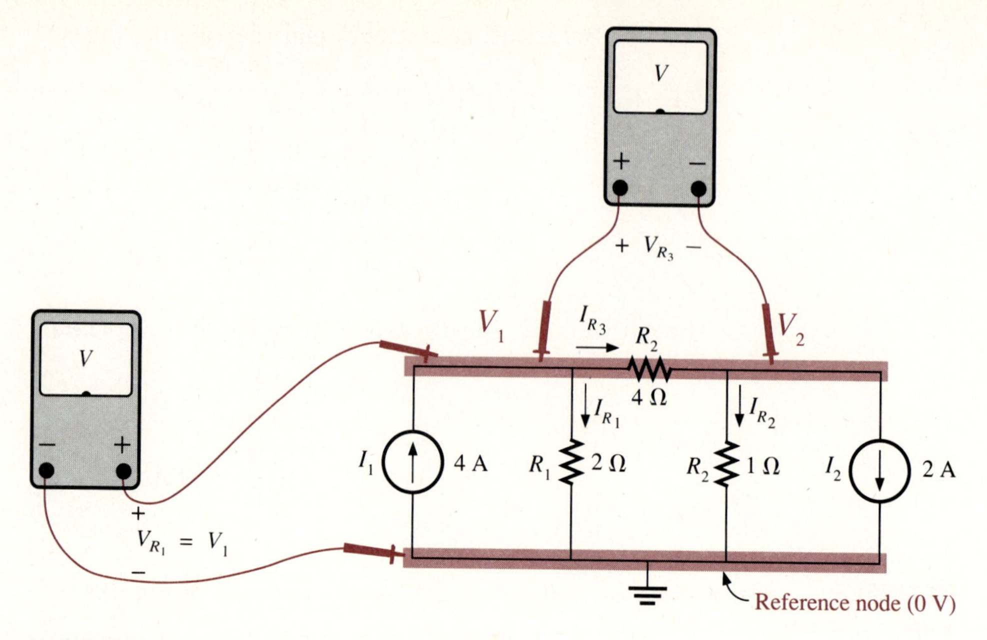

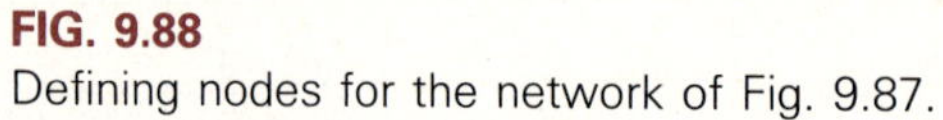
FIG. 9.88
Defining nodes for the network of Fig. 9.87.

At node V_2:

$$I_3 = I_2 + 2\text{ A}$$

$$\frac{V_1 - V_2}{R_3} = \frac{V_2}{R_2} + 2\text{ A}$$

Rewriting the two equations with the current source values to the right of the equals sign results in

$$V_1\left(\frac{1}{R_1} + \frac{1}{R_3}\right) - V_2\left(\frac{1}{R_3}\right) = 4\text{ A}$$

$$V_1\left(\frac{1}{R_3}\right) - V_2\left(\frac{1}{R_2} + \frac{1}{R_3}\right) = 2\text{ A}$$

Substituting numbers gives

$$V_1\left(\frac{1}{2\ \Omega} + \frac{1}{4\ \Omega}\right) - V_2\left(\frac{1}{4\ \Omega}\right) = 4\text{ A}$$

$$V_1\left(\frac{1}{4\ \Omega}\right) - V_2\left(\frac{1}{1\ \Omega} + \frac{1}{4\ \Omega}\right) = 2\text{ A}$$

$$\frac{3}{4}V_1 - \frac{1}{4}V_2 = 4$$
$$\frac{1}{4}V_1 - \frac{5}{4}V_2 = 2$$

resulting in two equations and two unknowns.

Using the substitution technique results in V_1 = **5.143 V** and V_2 = **0.571 V.**

The voltage across R_3 is

$$V_{R_3} = V_1 - V_2 = 5.143\text{ V} - (-0.571\text{ V}) = \mathbf{5.714\ V}$$

The current is

$$I_{R_1} = \frac{V_1}{R_1} = \frac{5.143\text{ V}}{2\ \Omega} = \mathbf{2.572\ A}$$

and

$$I_{R_2} = \frac{V_2}{R_2} = \frac{-0.571\text{ V}}{1\ \Omega} = \mathbf{-0.571\ A}$$

The minus sign simply indicates that I_{R_2} has the opposite direction of that indicated in Fig. 9.88.

The current I_{R_3} is

$$I_{R_3} = \frac{V_{R_3}}{R_3} = \frac{5.714\text{ V}}{4\ \Omega} = \mathbf{1.429\ A}$$

with the positive sign revealing that the assumed direction of I_{R_3} in Fig. 9.88 is correct.

It should be obvious from the preceding example that the mathematics associated with nodal analysis is more demanding and time consuming than that encountered for mesh analysis. Fortunately, however, there is a format approach to nodal analysis that will at least make the writing of the nodal equations more direct than just experienced.

Probably the most direct path toward writing the nodal equations using the format approach is to follow the next sequence of steps.

Step 1: Choose a reference node and assign a subscripted voltage label to the remaining nodes of the network.

Step 2: The number of equations required for a complete solution is equal to the number of subscripted nodes. To write the equations, first take each nodal voltage and multiply by the sum of the conductances ($G = 1/R$) connected to that node. Any conductances not connected to the node of interest are not included in the equation.

Step 3: From each equation subtract the mutual terms composed of the product of any conductance tied directly to another node and the other nodal voltage. In other words, if a conductance is connected directly from the node of interest to another node, a term composed of the product of the conductance and the other subscripted nodal voltage should be subtracted from the term generated in step 2.

Step 4: The column to the right of the equality sign is the algebraic sum of the current sources tied to the node of interest. A current source is assigned a positive sign if it supplies current to a node and a negative sign if it draws current from the node.

Step 5: Solve the resulting simultaneous equations for the desired voltages.

Let us now consider a few examples.

EXAMPLE 9.26 Write the nodal equations for the network of Fig. 9.89.

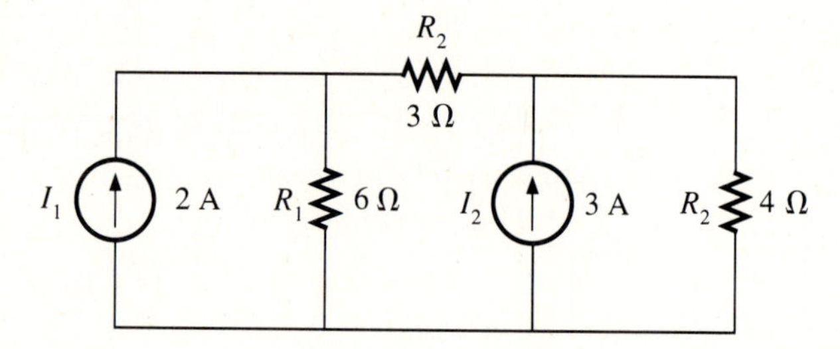

FIG. 9.89
Network for Example 9.26.

Solution

Step 1: The figure is redrawn with assigned subscripted voltages in Fig. 9.90.

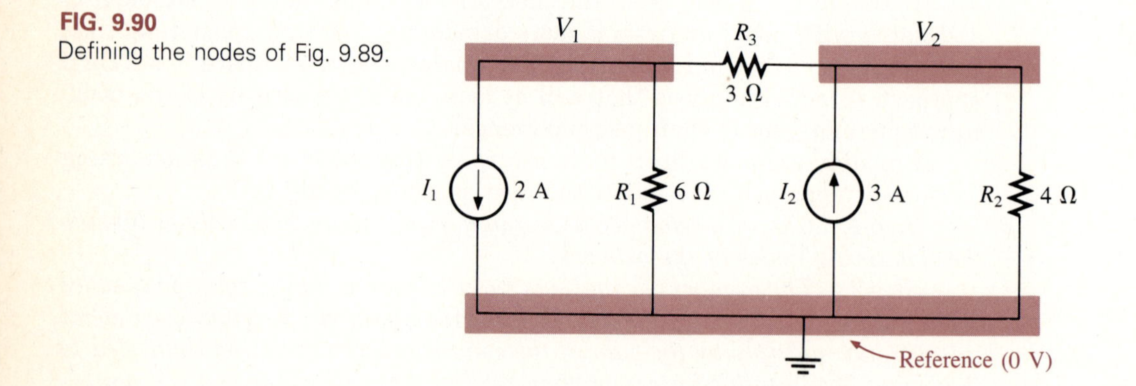

FIG. 9.90
Defining the nodes of Fig. 9.89.

Steps 2 to 4:

$$V_1: \quad \underbrace{\left(\frac{1}{6} + \frac{1}{3}\right)}_{\substack{\text{sum of}\\ \text{conductances}\\ \text{connected}\\ \text{to node 1}}} V_1 - \underbrace{\left(\frac{1}{3}\right)}_{\substack{\text{mutual}\\ \text{conductance}}} V_2 = \overset{\substack{\text{drawing current}\\ \text{from node 1}}}{-2}$$

$$V_2: \quad \underbrace{\left(\frac{1}{4} + \frac{1}{3}\right)}_{\substack{\text{sum of}\\ \text{conductances}\\ \text{connected}\\ \text{to node 2}}} V_2 - \underbrace{\left(\frac{1}{3}\right)}_{\substack{\text{mutual}\\ \text{conductance}}} V_1 = \overset{\substack{\text{supplying current}\\ \text{to node 2}}}{+3}$$

and

$$\begin{aligned} \frac{1}{2}V_1 - \frac{1}{3}V_2 &= -2 \\ \frac{1}{3}V_1 + \frac{7}{12}V_2 &= 3 \end{aligned}$$

EXAMPLE 9.27 Write the nodal equations for the network of Fig. 9.91.

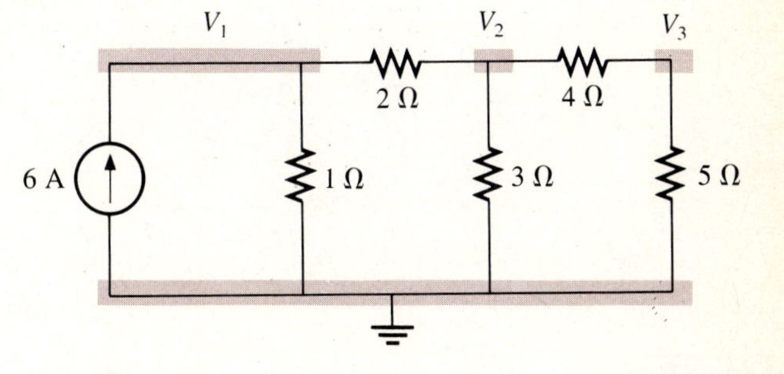

FIG. 9.91
Network for Example 9.27.

Solution Node V_1:

$$V_1\left[\frac{1}{1} + \frac{1}{2}\right] - \frac{1}{2}V_2 = +6$$

Node V_2:

$$V_2\left[\frac{1}{2} + \frac{1}{3} + \frac{1}{4}\right] - \frac{1}{2}V_1 - \frac{1}{4}V_3 = 0$$

Node V_3:

$$V_3\left[\frac{1}{4} + \frac{1}{5}\right] - \frac{1}{4}V_2 = 0$$

and

$$\begin{aligned} 1.5V_1 - 0.5V_2 + 0 &= 6 \\ -0.5V_1 + 1.083V_2 - 0.25V_3 &= 0 \\ 0 - 0.25V_2 + 0.75V_3 &= 0 \end{aligned}$$

Note the zero at the right of the equals sign for the last two equations due to the absence of a supply connected to the nodes V_2 and V_3.

9.13 Y-Δ (T-π) AND Δ-Y (π-T) CONVERSIONS

Circuit configurations are often encountered in which the resistors do not appear to be in series or parallel. Under these conditions, it may be necessary to convert the circuit from one form to another in order to solve for any unknown quantities if mesh or nodal analysis is not applied. Two circuit configurations that often account for these difficulties are the wye (Y) and delta (Δ), depicted in Fig. 9.92(a). They are also referred to as the tee (T) and pi (π) networks, respectively, as indicated in Fig. 9.92(b). Note that the pi is actually an inverted delta.

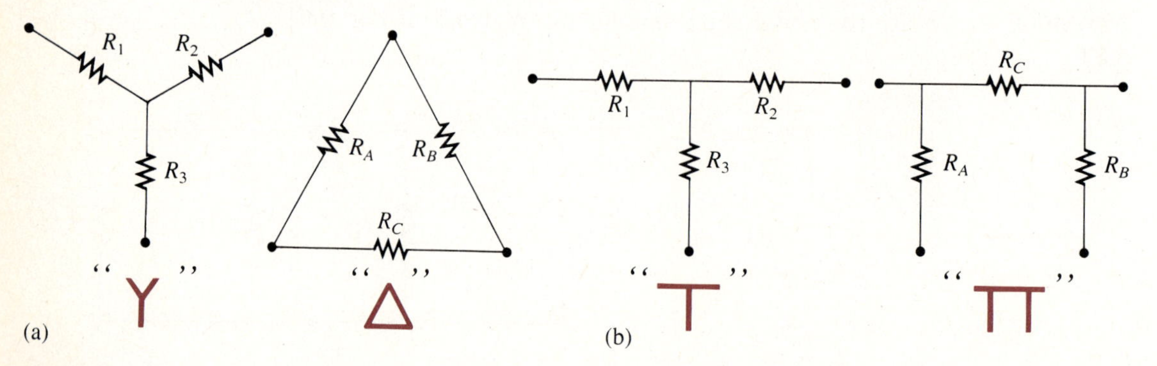

FIG. 9.92
Introducing a number of important network configurations.

The purpose of this section is to develop the equations for converting from Δ to Y or vice versa. This type of conversion normally leads to a network that can be solved using techniques such as described in Chapter 8. In other words, in Fig. 9.93, with terminals a, b, and c held fast, if the wye (Y) configuration (R_1, R_2, and R_3) were desired *instead of* the inverted delta (Δ) configuration, all that would be necessary is a direct application of the equations to be derived. The phrase *instead of* is emphasized to ensure that it is understood that only one of these configurations is to appear at one time between the indicated terminals.

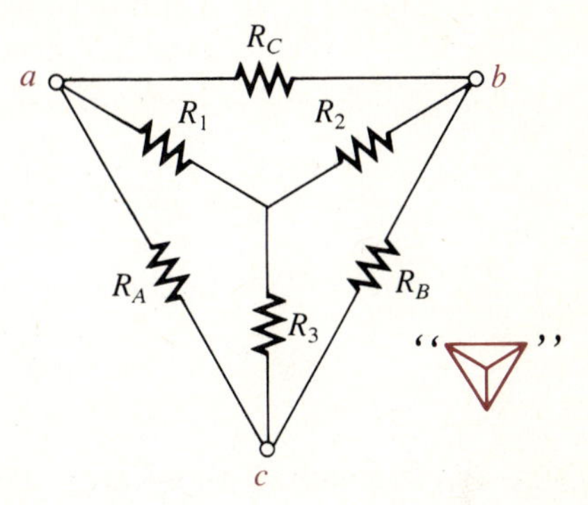

FIG. 9.93
Inserting the Y in an inverted Δ configuration.

If one configuration is to be replaced by the other, a series of equations must be found for each resistor in terms of the resistors of the other configuration.

Converting the Δ of Fig. 9.93 to a Y:

$$R_1 = \frac{R_A R_C}{R_A + R_B + R_C} \quad \textbf{(9.6a)}$$

$$R_2 = \frac{R_B R_C}{R_A + R_B + R_C} \quad \textbf{(9.6b)}$$

$$R_3 = \frac{R_A R_B}{R_A + R_B + R_C} \quad \textbf{(9.6c)}$$

Note that each resistor of the Y is equal to the product of the resistors in the two closest branches of the Δ divided by the sum of the resistors in the Δ.

Converting the Y of Fig. 9.92 to a Δ:

$$R_A = \frac{R_1 R_2 + R_1 R_3 + R_2 R_3}{R_2} \quad \textbf{(9.7a)}$$

$$R_B = \frac{R_1 R_2 + R_1 R_3 + R_2 R_3}{R_1} \quad \textbf{(9.7b)}$$

$$R_C = \frac{R_1R_2 + R_1R_3 + R_2R_3}{R_3} \tag{9.7c}$$

Note that the value of each resistor of the Δ is equal to the sum of the possible product combinations of the resistances of the Δ divided by the resistance of the Y farthest from the resistor to be determined.

Let us consider what would occur if all the values of a Δ or Y were the same. If $R_A = R_B = R_C$, Eq. 9.6a would become (using R_A only)

$$R_3 = \frac{R_AR_B}{R_A + R_B + R_C} = \frac{R_AR_A}{R_A + R_A + R_A} = \frac{R_A^2}{3R_A} = \frac{R_A}{3}$$

and, following the same procedure,

$$R_1 = \frac{R_A}{3} \qquad R_2 = \frac{R_A}{3}$$

In general, therefore,

$$R_Y = \frac{R_\Delta}{3} \tag{9.8a}$$

or

$$R_\Delta = 3R_Y \tag{9.8b}$$

which indicates that *for a Y of three equal resistors, the value of each resistor of the Δ is equal to three times the value of any resistor of the Y*. If only two elements of a Y or a Δ are the same, the corresponding Δ or Y of each will also have two equal elements.

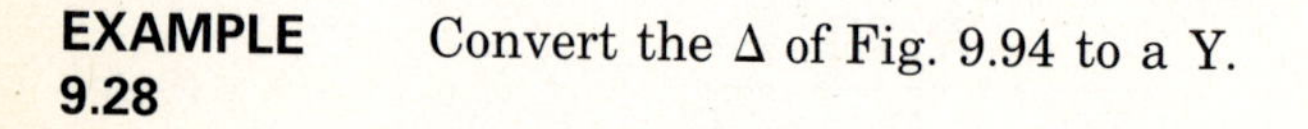

EXAMPLE 9.28 Convert the Δ of Fig. 9.94 to a Y.

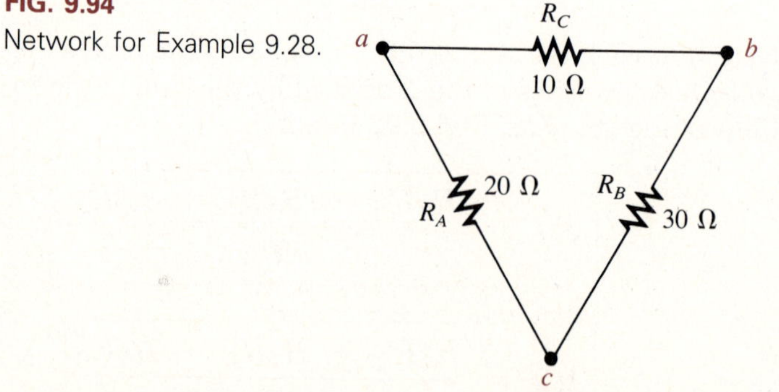

FIG. 9.94
Network for Example 9.28.

Solution

$$R_1 = \frac{R_A R_C}{R_A + R_B + R_C} = \frac{(20\ \Omega)(10\ \Omega)}{20\ \Omega + 30\ \Omega + 10\ \Omega} = \frac{200}{60}\ \Omega = \mathbf{3\tfrac{1}{3}\ \Omega}$$

$$R_2 = \frac{R_B R_C}{R_A + R_B + R_C} = \frac{(30\ \Omega)(10\ \Omega)}{60\ \Omega} = \frac{300}{60}\ \Omega = \mathbf{5\ \Omega}$$

$$R_3 = \frac{R_A R_B}{R_A + R_B + R_C} = \frac{(20\ \Omega)(30\ \Omega)}{60\ \Omega} = \frac{600}{60}\ \Omega = \mathbf{10\ \Omega}$$

The equivalent network is shown in Fig. 9.95.

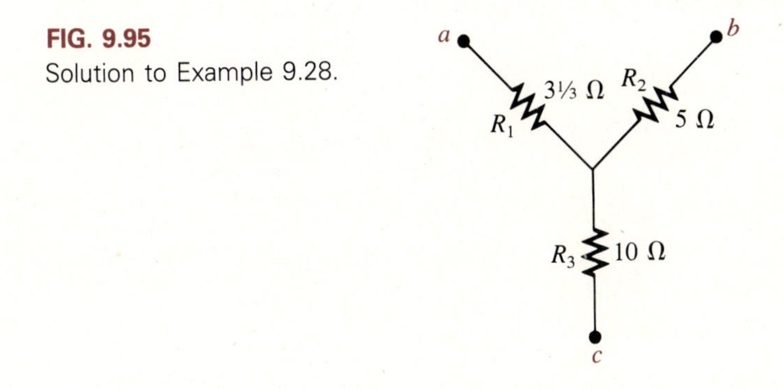

FIG. 9.95
Solution to Example 9.28.

EXAMPLE 9.29 Convert the Y of Fig. 9.96 to a Δ.

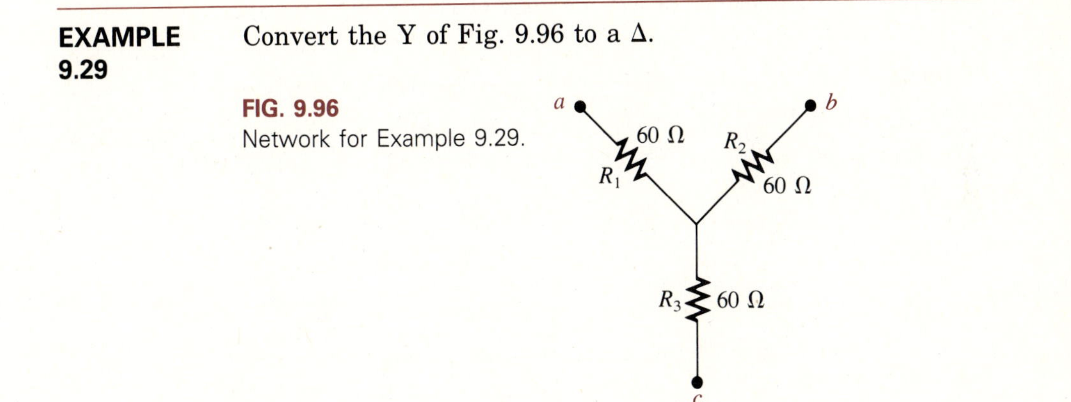

FIG. 9.96
Network for Example 9.29.

Solution

$$R_A = \frac{R_1R_2 + R_1R_3 + R_2R_3}{R_2}$$

$$= \frac{(60\ \Omega)(60\ \Omega) + (60\ \Omega)(60\ \Omega) + (60\ \Omega)(60\ \Omega)}{60\ \Omega}$$

$$= \frac{3600\ \Omega + 3600\ \Omega + 3600\ \Omega}{60\ \Omega} = \frac{10{,}800}{60}\ \Omega$$

$$R_A = \mathbf{180\ \Omega}$$

However, the three resistors for the Y are equal, permitting the use of Eq. 9.8b, which yields

$$R_{\Delta} = 3R_{Y} = 3(60\ \Omega) = 180\ \Omega$$

and

$$R_B = R_C = \mathbf{180\ \Omega}$$

The equivalent network is shown in Fig. 9.97.

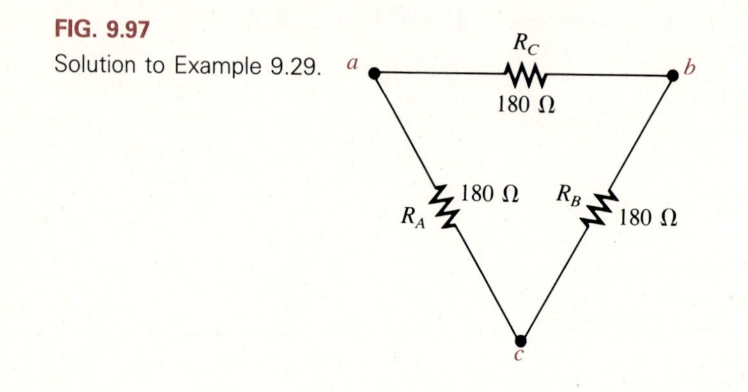

FIG. 9.97
Solution to Example 9.29.

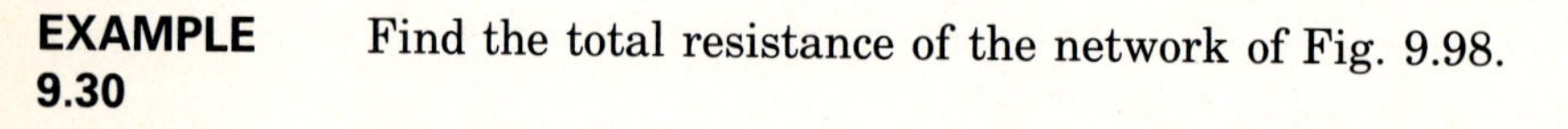

EXAMPLE 9.30

Find the total resistance of the network of Fig. 9.98.

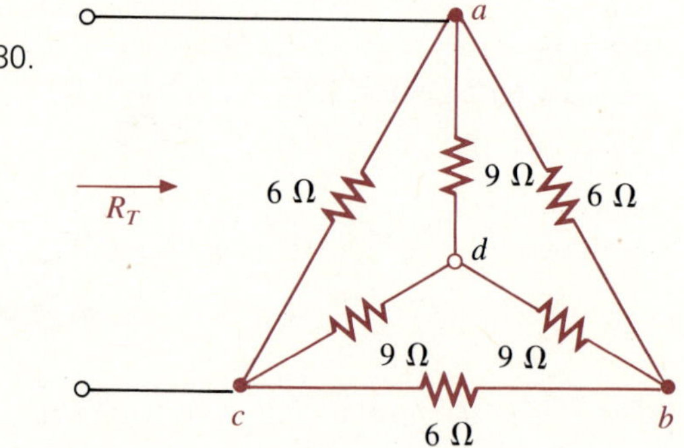

FIG. 9.98
Network for Example 9.30.

Solutions Since all the resistors of the Δ or Y are the same, Eqs. 9.8a and 9.8b can be used to convert either form to the other.

a. *Converting the Δ to a Y. Note:* When this is done, the resulting d' of the new Y will be the same as the point d shown in the original figure because both systems are balanced. That is, the resistance in each branch of each system has the same value:

$$R_Y = \frac{R_{\Delta}}{3} = \frac{6}{3}\ \Omega = 2\ \Omega \qquad \text{(Fig. 9.99)}$$

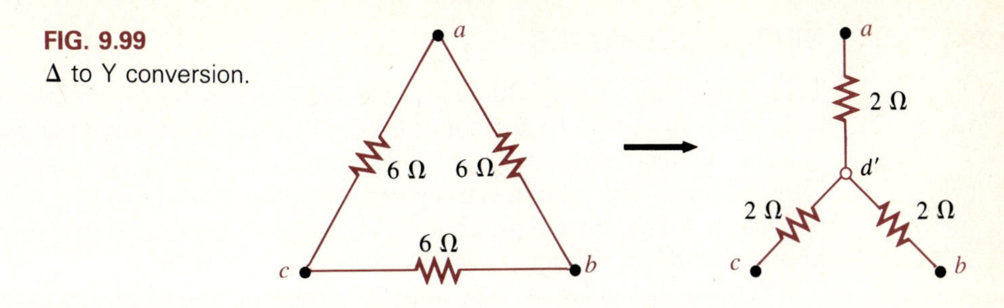

FIG. 9.99
Δ to Y conversion.

The original circuit then appears as shown in Fig. 9.100 and

$$R_T = 2\left[\frac{(2\ \Omega)(9\ \Omega)}{2\ \Omega + 9\ \Omega}\right] = \mathbf{3.273\ \Omega}$$

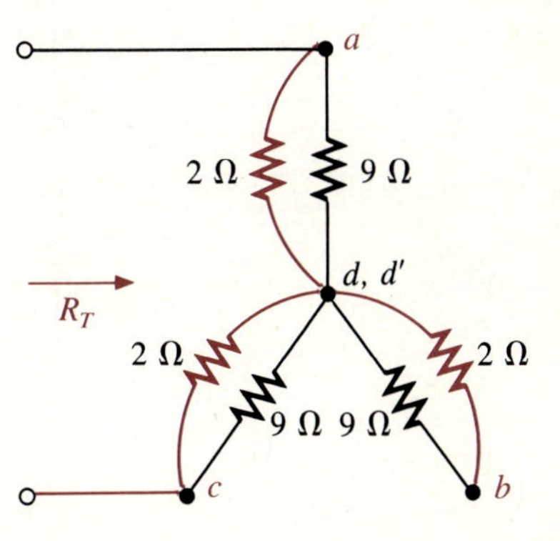

FIG. 9.100
Inserting the new Y configuration into the network of Fig. 9.98.

b. *Converting the Y to a Δ.*

$$R_\Delta = 3R_Y = (3\ \Omega)(9\ \Omega) = 27\ \Omega \qquad \text{(Fig. 9.101)}$$

$$R'_T = \frac{(6\ \Omega)(27\ \Omega)}{6\ \Omega + 27\ \Omega} = \frac{162}{33}\ \Omega = 4.9091\ \Omega$$

$$R_T = \frac{R'_T(R'_T + R'_T)}{R'_T + (R'_T + R'_T)} = \frac{R'_T 2R'_T}{3R'_T} = \frac{2R'_T}{3}$$

$$= \frac{2(4.9091\ \Omega)}{3} = \mathbf{3.273\ \Omega}$$

which checks with the previous solution.

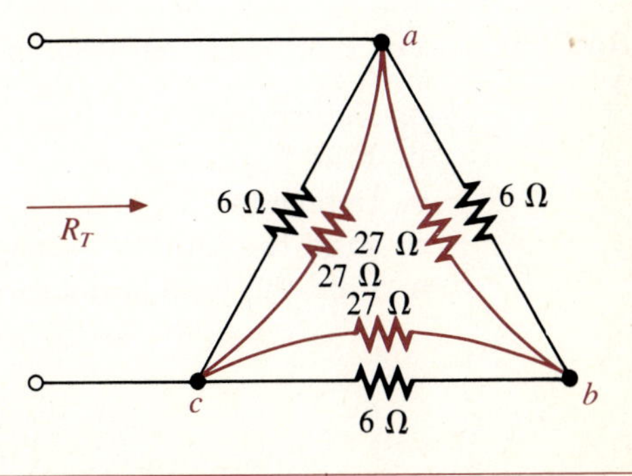

FIG. 9.101
Inserting a Δ configuration into the network of Fig. 9.98.

9.14 BRIDGE NETWORKS

This section introduces the bridge network, a configuration that has a multitude of applications. In the chapters to follow, it is employed in both dc and ac meters. In electronics courses it is encountered early in the discussion of rectifying circuits employed in converting a varying signal to one of a steady nature (such as dc). There are a number of other areas of application that require some knowledge of ac networks, which are discussed later.

The bridge network may appear in one of three forms, as indicated in Fig. 9.102. The network of Fig. 9.102(c) is also called a *symmetrical lattice network* if $R_2 = R_3$ and $R_1 = R_4$. Note how resistor R_5 forms a "bridge" between opposite branches of the configuration.

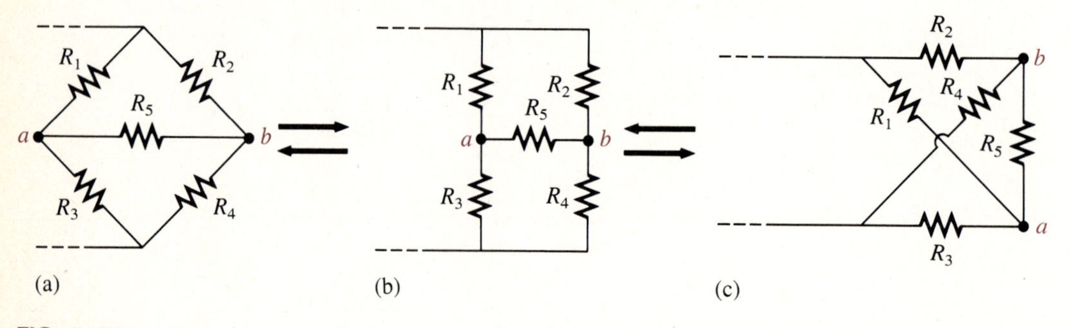

FIG. 9.102
Bridge network.

If the resistors of the configuration are such that

$$\frac{R_1}{R_3} = \frac{R_2}{R_4} \tag{9.9}$$

the voltage across R_5 will be 0 V, resulting in a current of 0 A through the same resistor. In addition, $V_1 = V_2$ and $V_3 = V_4$. If Eq. 9.9 is not satisfied, a method of analysis to be described in a later chapter must be employed.

EXAMPLE 9.31 For the bridge network of Fig. 9.103,

a. Determine the voltage V_5 and current I_5.
b. Find V_1 and V_3.
c. Determine V_2 and V_4.
d. Find the total resistance using the balance condition.
e. Find the total resistance using a Δ-Y conversion and compare with part d.

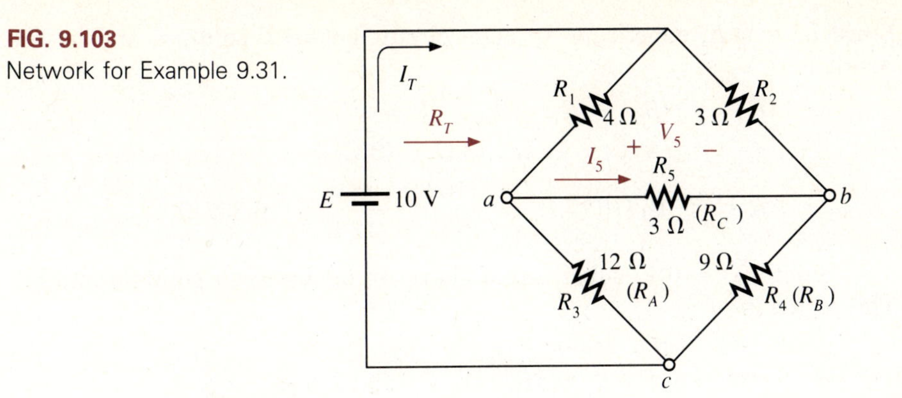

FIG. 9.103
Network for Example 9.31.

Solutions

a. Eq. 9.9:

$$\frac{R_1}{R_3} = \frac{R_2}{R_4}$$

$$\frac{4\ \Omega}{12\ \Omega} = \frac{3\ \Omega}{9\ \Omega}$$

$$\frac{1}{3} = \frac{1}{3} \quad \text{(balanced)}$$

Therefore, $V_5 = 0$ V and $I_5 = 0$ A.

b. If $I_5 = 0$ A, $I_1 = I_3$ and

$$I_1 = I_3 = \frac{E}{R_1 + R_3} = \frac{10\ \text{V}}{4\ \Omega + 12\ \Omega} = \frac{10\ \text{V}}{16\ \text{V}} = 0.625\ \text{A}$$

with

$$V_1 = I_1R_1 = (0.625\ \text{A})(4\ \Omega) = \mathbf{2.5\ V}$$

and

$$V_3 = I_3R_3 = (0.625\ \text{A})(12\ \Omega) = \mathbf{7.5\ V}$$

c. When balanced, $V_1 = V_2$ and $V_3 = V_4$. Therefore,

$$V_2 = \mathbf{2.5\ V}$$

and

$$V_4 = \mathbf{7.5\ V}$$

d. Since $I_5 = 0$ A, substitute an open circuit between points a and b. The result is

$$R_T = (R_1 + R_3)||(R_2 + R_4)$$
$$= (4\ \Omega + 12\ \Omega)||(3\ \Omega + 9\ \Omega)$$
$$R_T = 16\ \Omega||12\ \Omega = \frac{(16\ \Omega)(12\ \Omega)}{16\ \Omega + 12\ \Omega} = \mathbf{6.857\ \Omega}$$

Since $V_5 = 0$ V, substitute a short circuit between points a and b. The result is

$$R_T = (R_1||R_2) + (R_3||R_4)$$
$$= (4\ \Omega||3\ \Omega) + (12\ \Omega||9\ \Omega)$$
$$= 1.714\ \Omega + 5.143\ \Omega$$
$$R_T = \mathbf{6.857\ \Omega}$$

as above.

e. R_A, R_B, and R_C are labeled in Fig. 9.103. The equivalent Y configuration appears in Fig. 9.104, with

$$R_1 = \frac{R_A R_C}{R_A + R_B + R_C} = \frac{(12\ \Omega)(3\ \Omega)}{12\ \Omega + 9\ \Omega + 3\ \Omega} = 1.5\ \Omega$$
$$R_2 = \frac{R_B R_C}{R_A + R_B + R_C} = \frac{(9\ \Omega)(3\ \Omega)}{12\ \Omega + 9\ \Omega + 3\ \Omega} = 1.125\ \Omega$$
$$R_3 = \frac{R_A R_B}{R_A + R_B + R_C} = \frac{(12\ \Omega)(9\ \Omega)}{12\ \Omega + 9\ \Omega + 3\ \Omega} = 4.5\ \Omega$$

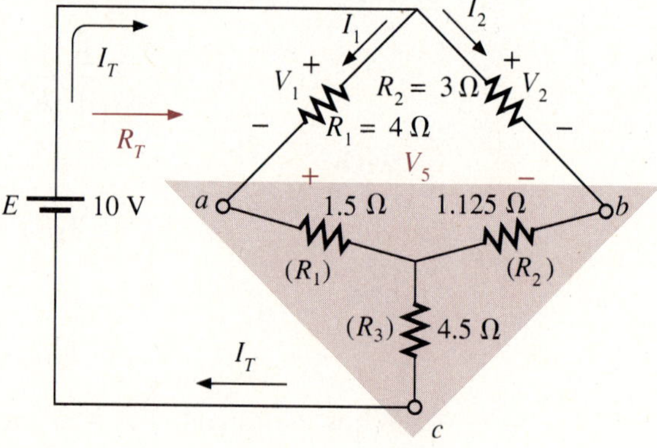

FIG. 9.104
Inserting the Y configuration in Fig. 9.103.

The resulting network appears in Fig. 9.105, where the total resistance is

$$R_T = (4\ \Omega + 1.5\ \Omega)||(3\ \Omega + 1.125\ \Omega) + 4.5\ \Omega$$
$$= 5.5\ \Omega||4.125\ \Omega + 4.5\ \Omega$$
$$= 2.357\ \Omega + 4.5\ \Omega$$
$$R_T = \mathbf{6.857\ \Omega}$$

as obtained in part d.

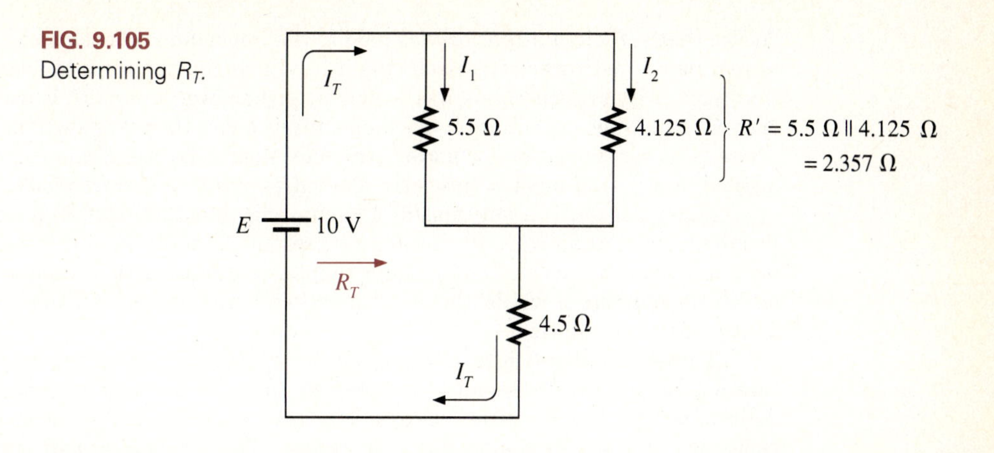

FIG. 9.105
Determining R_T.

FORMULA SUMMARY

Source Conversions

$$I_s = \frac{E_s}{R_s}, \qquad E_s = I_s R_s, \qquad R_s \text{ the same}$$

Maximum Power Transfer

$$R_L = R_{Th} = R_N$$

$$P_{L_{\max}} = \frac{E_{Th}^2}{4R_{Th}}$$

Δ-Y Conversions

$$R_Y = \frac{R_\Delta}{3}, \qquad R_\Delta = 3R_Y$$

Bridge Networks (for Balance)

$$\frac{R_1}{R_3} = \frac{R_2}{R_4}$$

CHAPTER SUMMARY

As evident by the formula list, there are very few equations to memorize in this chapter. The content is primarily an introduction to theorems and methods of analysis that will provide a solution to networks with interactive sources and more than one series or parallel path. The sequence of steps for each subject area was carefully outlined to ensure that sufficient attention was paid to the importance of each step. Take the time to redraw the networks after each step if it will clarify the steps to follow and ensure that a quantity is properly defined—no matter how well the math is performed, if the original equations are incorrect (even one minus sign), the results will be meaningless. There is no compensation for performing four steps correctly if one step is improperly applied. *Every* step must be properly applied to obtain a correct result.

Superposition is a very useful technique in that the network for each source is usually simpler than with all the sources present. In addition to using the theorem to isolate dc sources, we will find in a later chapter that it can also be used to isolate dc and ac solutions. In other words, the dc solution to a network can be found totally independently of the ac solution.

Thevenin's theorem is particularly interesting in its ability to reduce a two-terminal network to its simplest form. In most cases the Thevenin resistance is easier to determine than the Thevenin voltage. The key to success in this latter situation is to remember that E_{Th} is the open-circuit voltage. The current through any elements in series with E_{Th} must be 0 A, and E_{Th} may simply be the voltage across a single element of the network. The maximum power transfer theorem is a natural extension of Thevenin's theorem, since $R_L = R_{Th}$ and $P_{L_{max}} = E_{Th}^2/4R_{Th}$.

The branch-current method is closely linked to fundamental laws of electric circuits, whereas the mesh- and nodal-analysis techniques provide a more direct approach that normally raises the level of confidence with multiloop networks.

Bridge configurations are commonplace in instrument and power supply design due to conditions that we established for the arm of the bridge. For the "balance" condition, simply keep in mind that $I = 0$ A and $V = 0$ V for the bridge element. The result is that an open circuit or short circuit can be substituted for the bridge element when investigating the remaining current and voltage levels of the network.

GLOSSARY

Branch-current method A technique for determining the branch currents of a multiloop network.

Current sources Sources that supply a fixed current to a network and have a terminal voltage dependent on the network to which they are applied.

Delta (Δ), pi (π) configuration A network constructed of three branches that has the appearance of the Greek letter delta (Δ) or pi (π).

Maximum power transfer theorem A theorem used to determine the load resistance necessary to ensure maximum power transfer to the load.

Mesh analysis A technique for determining the mesh (loop) currents of a network that results in a reduced set of equations compared to the branch-current method.

Mesh (loop) current A labeled current assigned to each distinct closed loop of a network that can individually or in combination with other mesh currents define all the branch currents of a network.

Nodal analysis A technique for determining the nodal voltages of a network.

Node A junction of two or more branches in a network.

Norton's theorem A theorem that permits the reduction of any two-terminal linear dc network to one having a single current source and parallel resistor.

Superposition theorem A network theorem that permits considering the effects of each source independently. The resulting current and/or voltage is the algebraic sum of the currents and/or voltages developed by each source independently.

Thevenin's theorem A theorem that permits the reduction of any two-terminal linear dc network to one having a single voltage source and series resistor.

Wye (Y), tee (T) configuration A network constructed of three branches that has the appearance of the capital letter Y or T.

PROBLEMS

Section 9.2

1. Find the voltage V_{ab} (with polarity) for the circuit of Fig. 9.106.
2. For the network of Fig. 9.107:
   a. Find the voltages V_s and V_3.
   b. Find the current I_2.

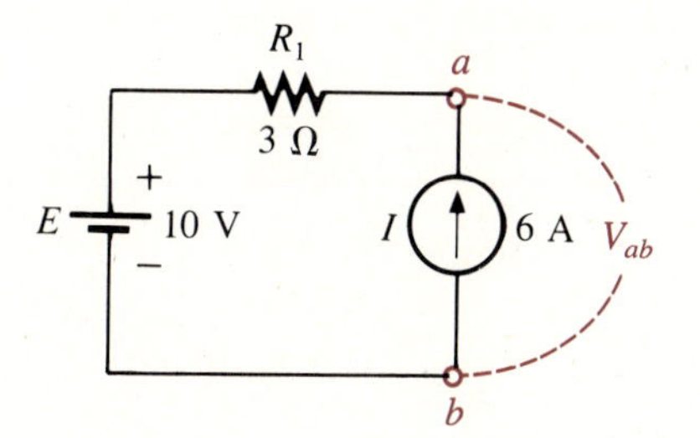

FIG. 9.106

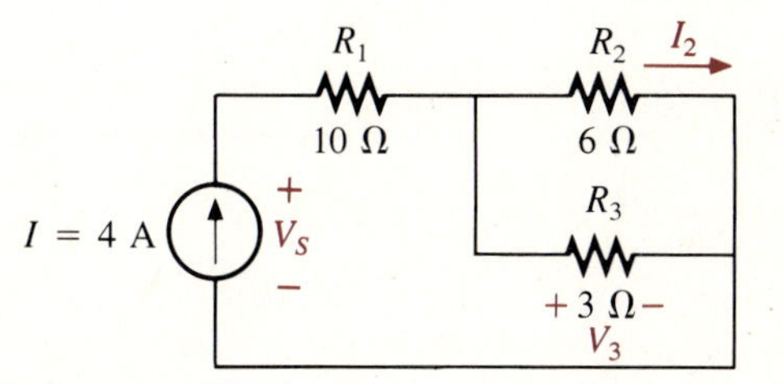

FIG. 9.107

3. For the network of Fig. 9.108:
   a. Find the currents I_T and I_2.
   b. Find the voltage V_s.

FIG. 9.108

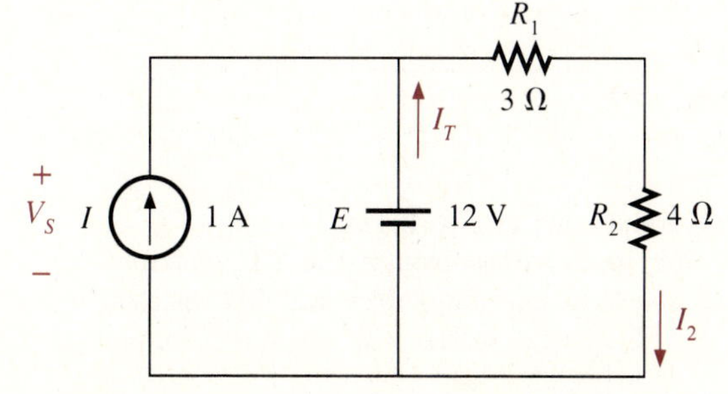

4. For the network of Fig. 9.109:
   a. Find the current I_2.
   b. Find the voltage V_3.

FIG. 9.109

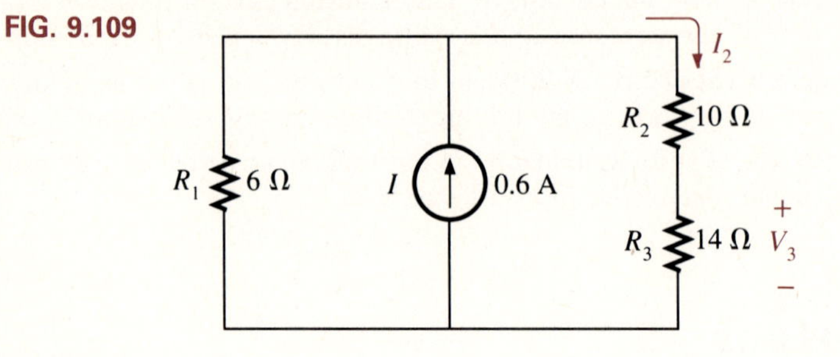

Section 9.3

5. Convert the voltage sources of Fig. 9.110 to current sources.

FIG. 9.110

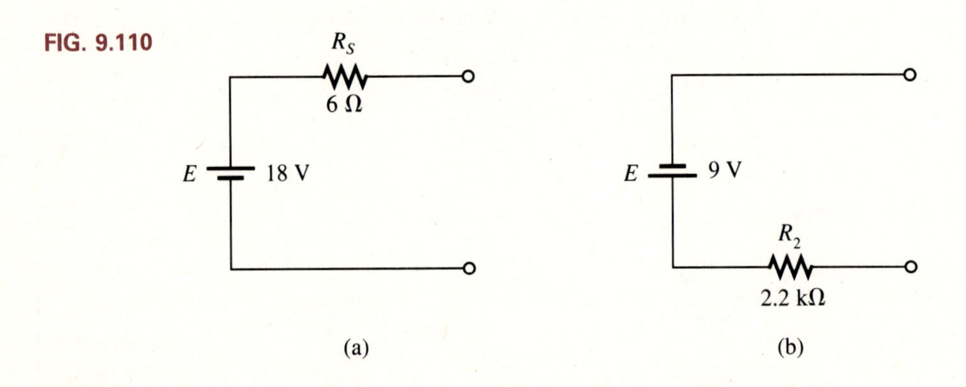

6. Convert the current sources of Fig. 9.111 to voltage sources.

FIG. 9.111

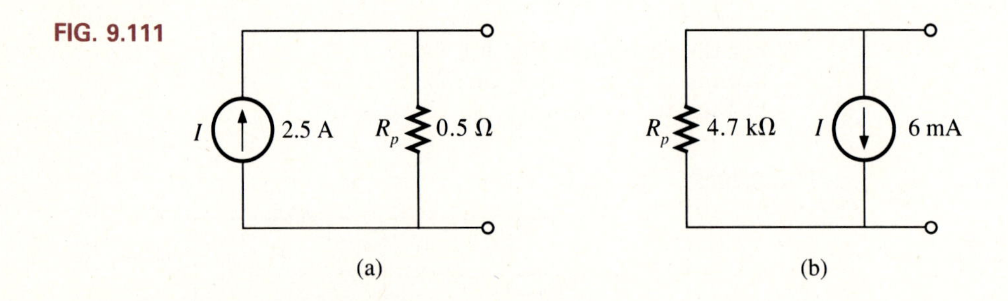

7. For the network of Fig. 9.112:
   a. Find the current through the 2-Ω resistor.
   b. Convert the current source and 4-Ω resistor to a voltage source, and again solve for the current in the 2-Ω resistor. Compare the results.

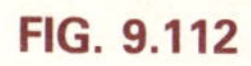
FIG. 9.112

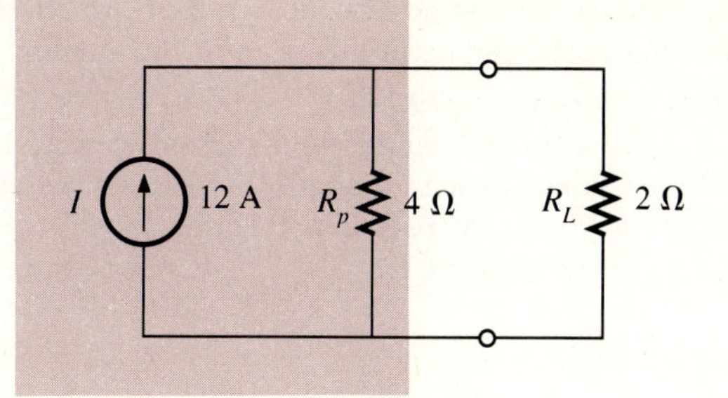

8. Convert the voltage sources of Fig. 9.113 to current sources.

FIG. 9.113

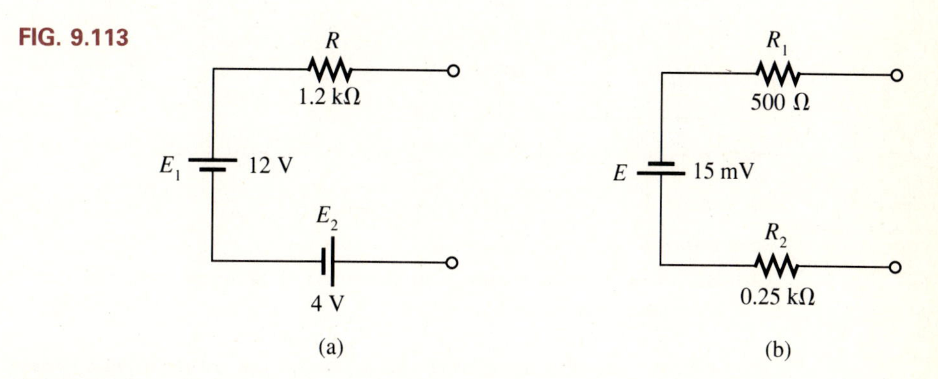

Section 9.4

9. Reduce the sources of Fig. 9.114 to a single current source.

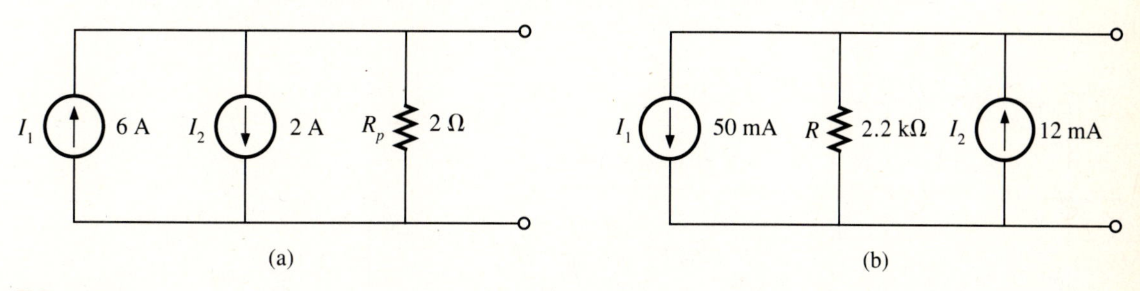

FIG. 9.114

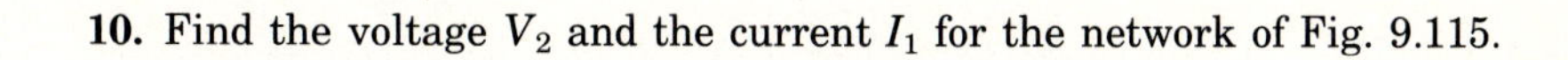
10. Find the voltage V_2 and the current I_1 for the network of Fig. 9.115.

FIG. 9.115

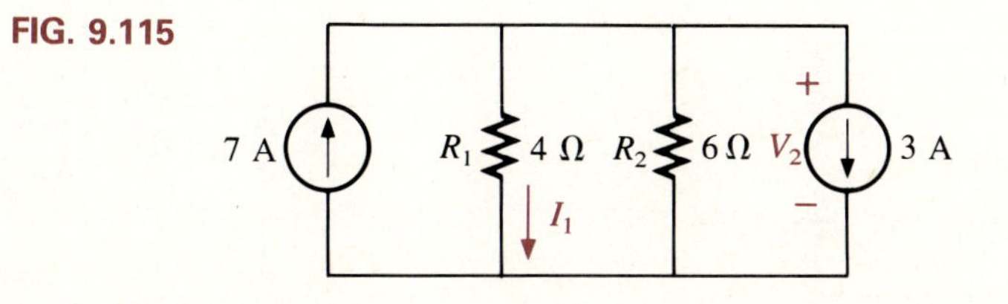

11. a. Convert the voltage sources of Fig. 9.116 to current sources.
b. Find the magnitude and direction of the current I.

12. For the network of Fig. 9.117:
a. Convert the voltage source to a current source.
b. Reduce the network to a single current source and determine the voltage V_1.
c. Using the results of part b, determine V_2.
d. Calculate the current I_2.

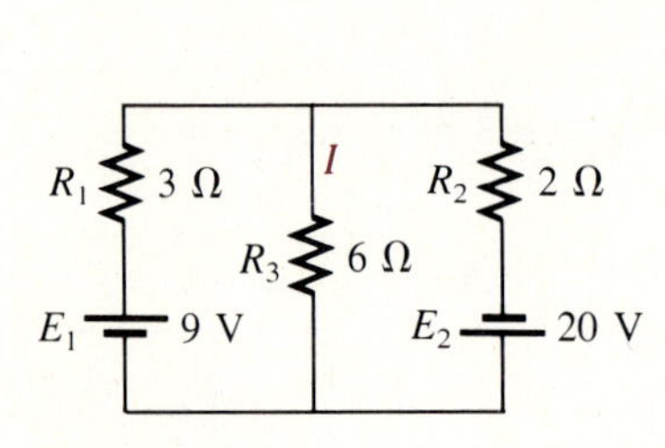

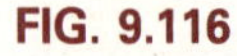
FIG. 9.116

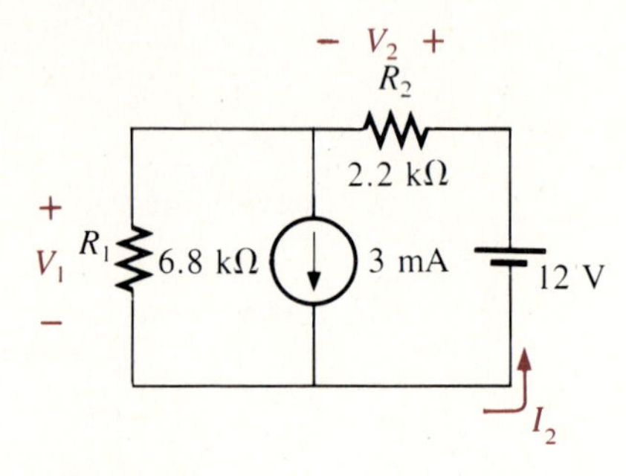

FIG. 9.117

Section 9.6

13. a. Using superposition, find the current through each resistor of the network of Fig. 9.118.
b. Find the power delivered to R_1 for each source.
c. Find the power delivered to R_1 using the total current through R_1.
d. Does superposition apply to power effects? Explain.

14. Using superposition, find the current through R_1 for the network of Fig. 9.119.

15. Using superposition, find the voltage V_2 for the network of Fig. 9.120.

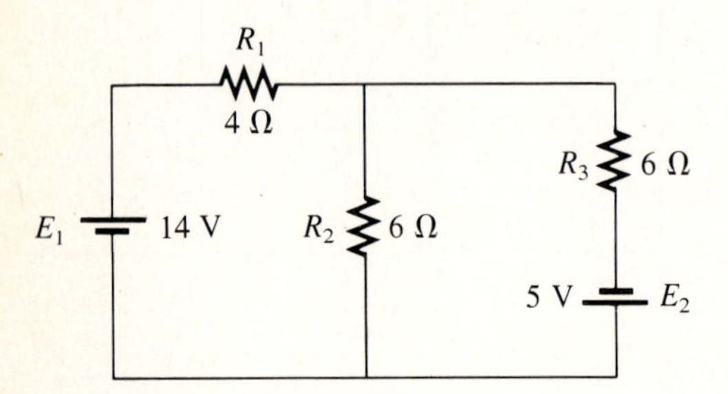

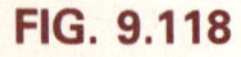
FIG. 9.118

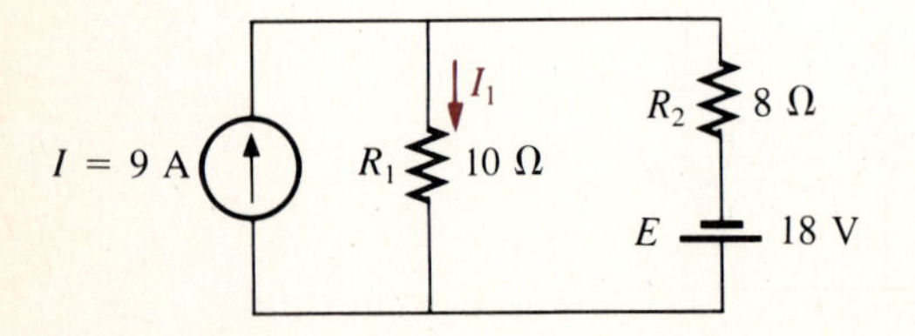

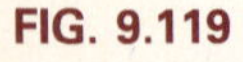
FIG. 9.119

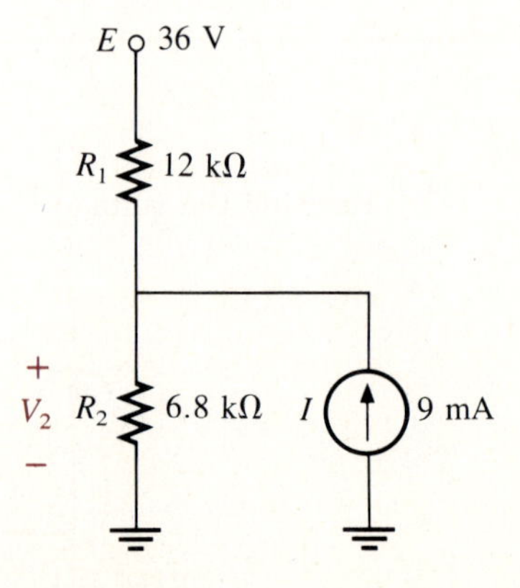

FIG. 9.120

Section 9.7

16. **a.** Find the Thevenin equivalent circuit for the network external to the resistor R of Fig. 9.121.
    b. Find the current through R when R is 2, 30, and 100 Ω.

17. **a.** Find the Thevenin equivalent circuit for the network external to the resistor R for the network of Fig. 9.122.
    b. Find the power delivered to R when R is 2 Ω and 100 Ω.

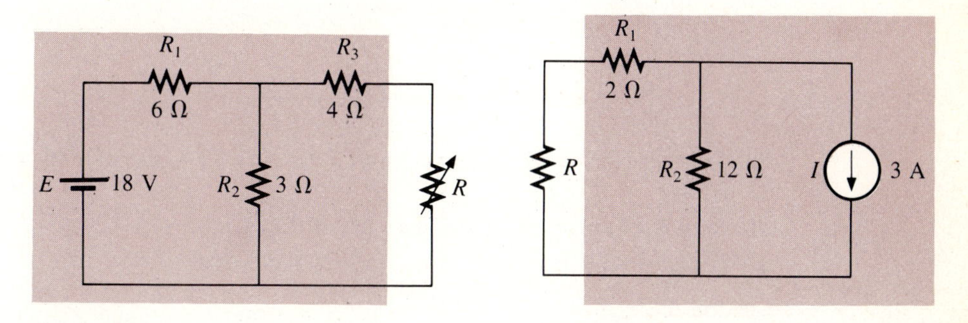

FIG. 9.121

FIG. 9.122

18. Repeat Problem 17 for the network of Fig. 9.123.

19. Find the Thevenin equivalent circuit for the network external to the resistor R of Fig. 9.124.

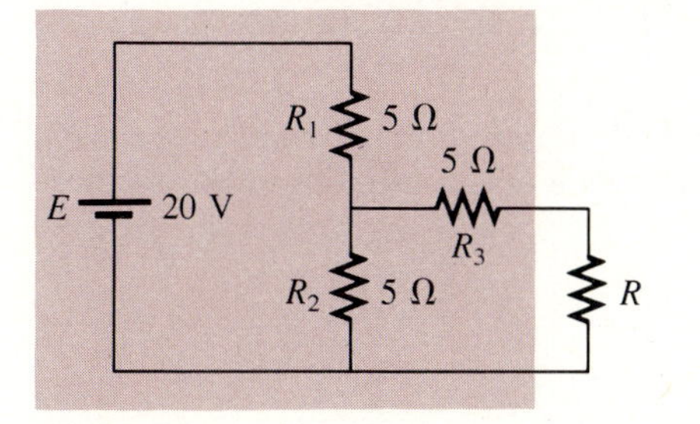

FIG. 9.123

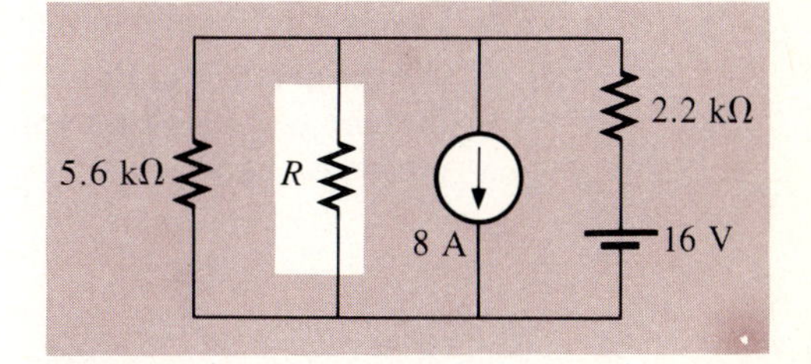

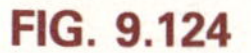

FIG. 9.124

20. Find the Thevenin equivalent circuit for the network external to the resistor R of Fig. 9.125.

FIG. 9.125

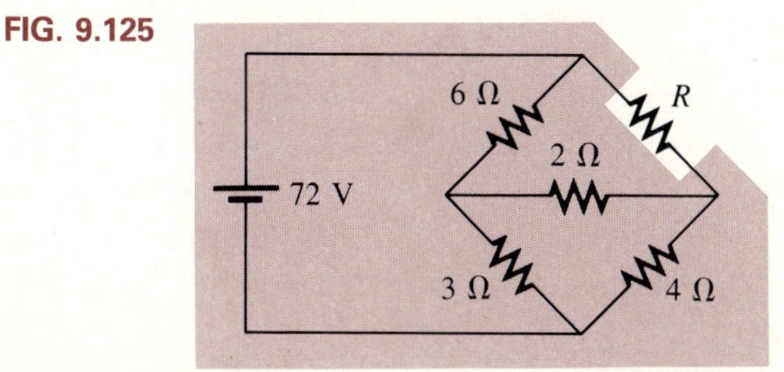

21. Determine the Thevenin equivalent circuit for the network external to the resistor R in the network of Fig. 9.126.

FIG. 9.126

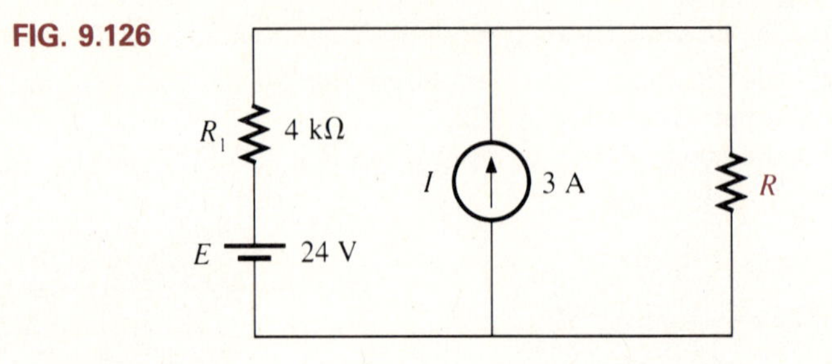

22. Find the Norton equivalent circuit for the network external to the resistor R in Fig. 9.121.

23. Repeat Problem 22 for the network of Fig. 9.123.

24. Repeat Problem 22 for the network of Fig. 9.125.

Section 9.8

25. a. For the network of Fig. 9.121, determine the value of R for maximum power to R.
    b. Determine the maximum power to R.

26. Repeat Problem 25 for the network of Fig. 9.123.

27. Repeat Problem 25 for the network of Fig. 9.125.

28. For the network of Fig. 9.127:
    a. Determine the value of R for maximum power to R.
    b. Determine the maximum power to R.

29. a. For the network of Fig. 9.128, determine the value R_2 for maximum power to R_4.
    b. Is there a general statement that can be made about situations such as those presented here?

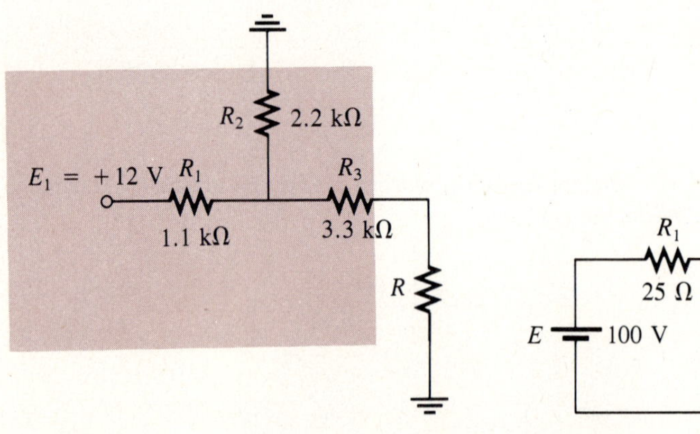

FIG. 9.127

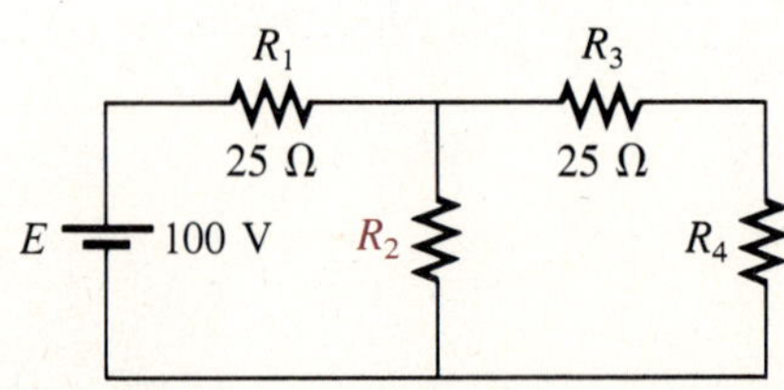

FIG. 9.128

Section 9.10

30. Determine the current through each resistor of Fig. 9.129 using branch-current analysis.

31. Determine the current through each resistor of Fig. 9.130 using branch-current analysis.

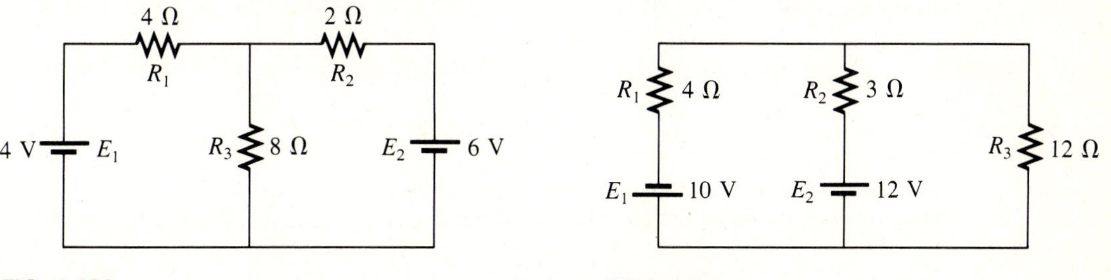

FIG. 9.129

FIG. 9.130

32. Determine the current through each resistor of Fig. 9.131 using branch-current analysis.

33. Using branch-current analysis, determine the current through each resistor of Fig. 9.132.

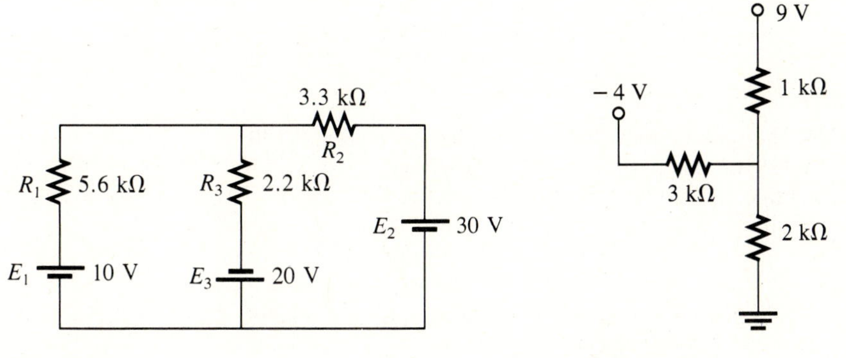

FIG. 9.131

FIG. 9.132

Section 9.11

34. Repeat Problem 30 using mesh analysis.

35. Repeat Problem 31 using mesh analysis.

36. Repeat Problem 32 using mesh analysis.

37. Repeat Problem 33 using mesh analysis.

38. Determine the current I_3 for the network of Fig. 9.133 using mesh analysis.

39. **a.** Write the mesh equations for the network of Fig. 9.134.
b. Solve for the current I_4.

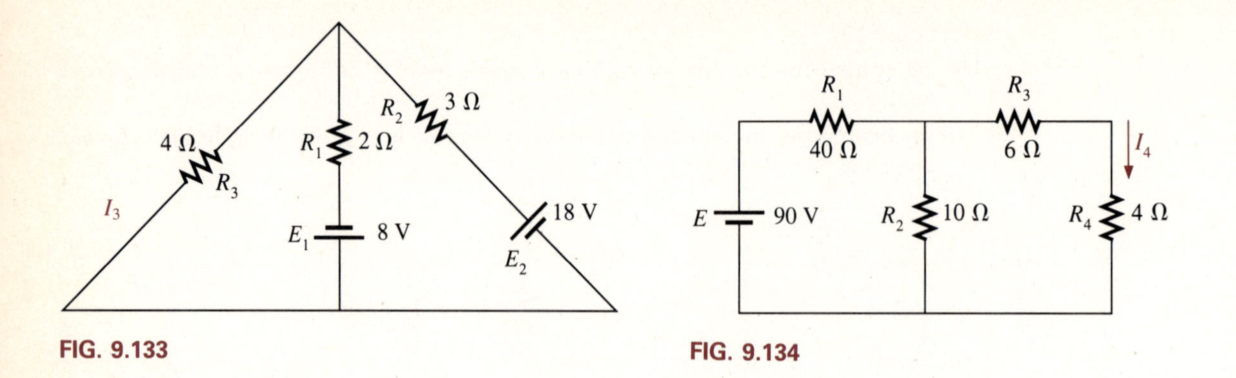

FIG. 9.133

FIG. 9.134

40. Write the mesh equations for the network of Fig. 9.135.

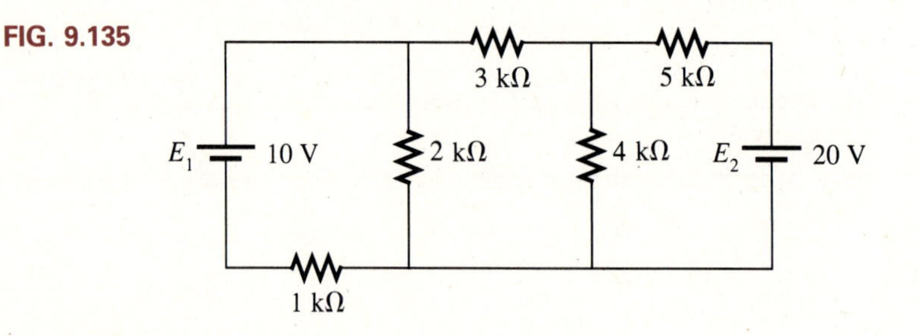

FIG. 9.135

Section 9.12

41. a. Write the nodal equations for the network of Fig. 9.136.
b. Solve for the nodal voltages of the network.
c. Determine the current through R_3.

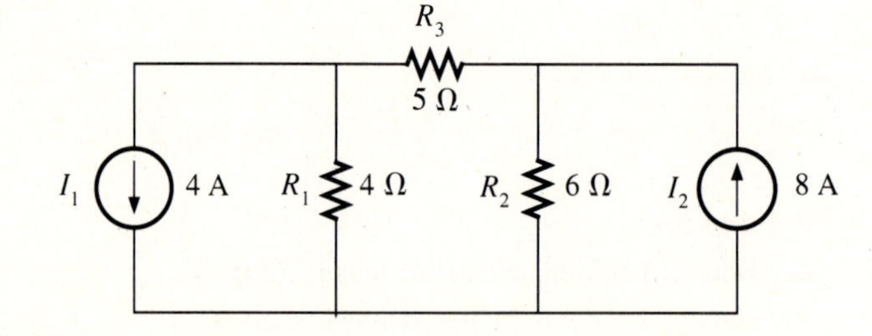

FIG. 9.136

42. a. Write the nodal equations for the network of Fig. 9.137.
b. Solve for the nodal voltages of the network.
c. Determine the voltage across R_1.

43. a. Convert the voltage source of Fig. 9.138 to a current source.
b. Write the nodal equation for the network.
c. Solve for the nodal voltage.
d. Determine the current I_1.

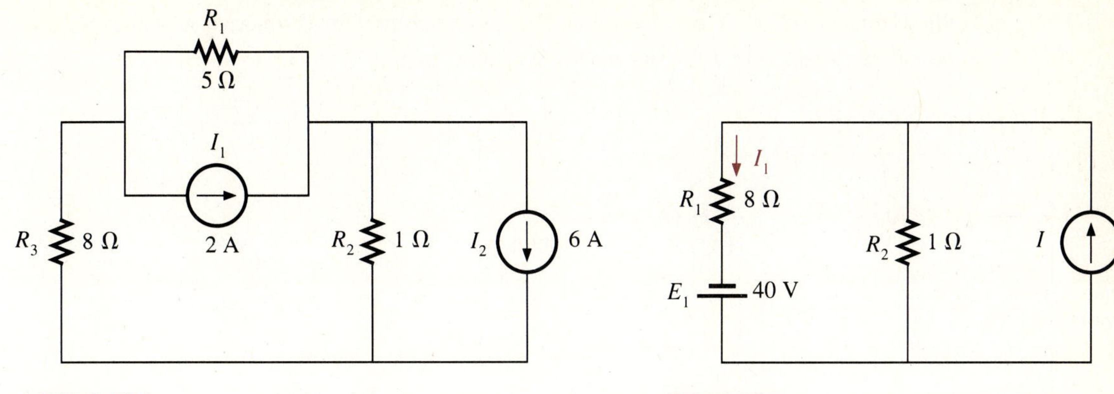

FIG. 9.137

FIG. 9.138

44. **a.** Write the nodal equations for the network of Fig. 9.134.
    b. Solve for the voltage across the resistor R_4.
45. **a.** Convert the voltage sources of Fig. 9.135 to current sources.
    b. Write the nodal equations.
46. Write the nodal equations for the network of Fig. 9.139.
47. **a.** Write the nodal equations for the bridge network of Fig. 9.140. Do not solve.
    b. Write the mesh equations for the same network. Do not solve.

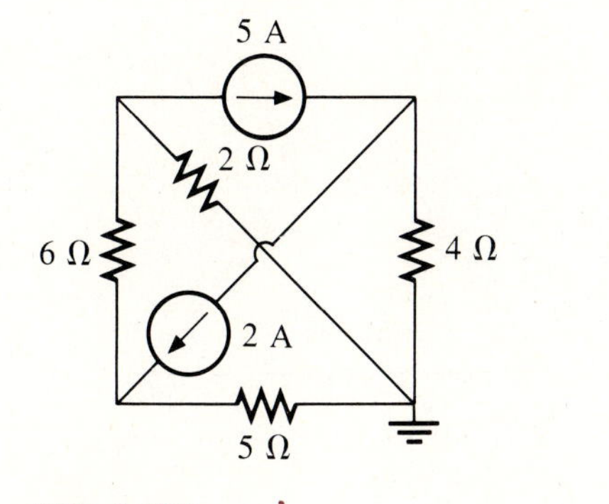

FIG. 9.139

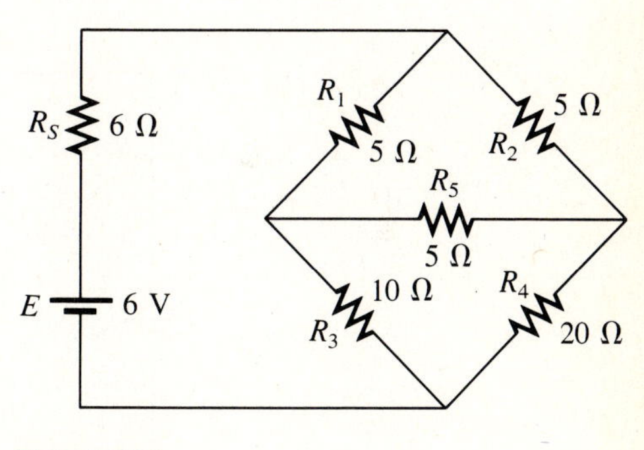

FIG. 9.140

Section 9.13

48. Using a Δ-Y or Y-Δ conversion, find the current I in the network of Fig. 9.141.

FIG. 9.141

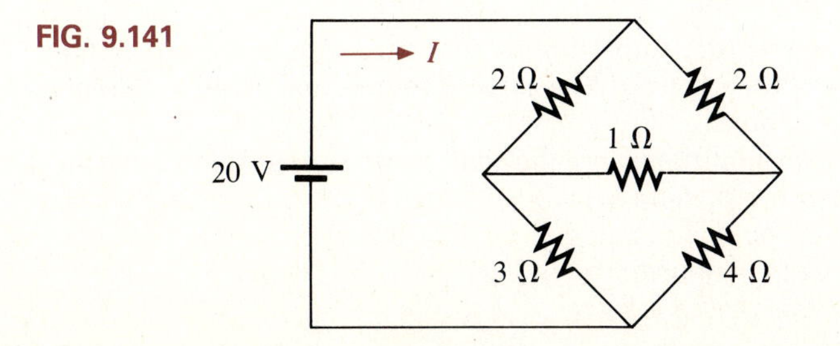

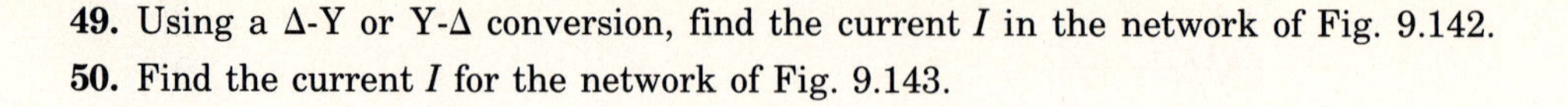

49. Using a Δ-Y or Y-Δ conversion, find the current I in the network of Fig. 9.142.

50. Find the current I for the network of Fig. 9.143.

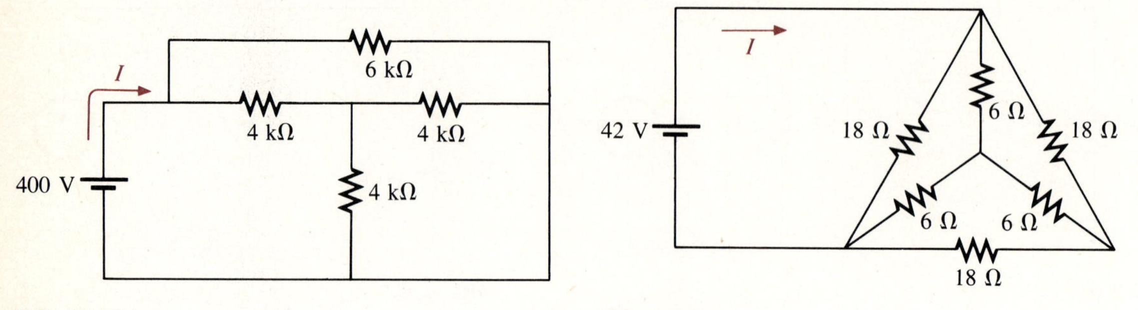

FIG. 9.142

FIG. 9.143

Section 9.14

51. For the bridge configuration of Fig. 9.141:
  a. Determine whether the system is balanced.
  b. Find the total resistance R_T.

52. For the bridge configuration of Fig. 9.144:
  a. Is the system balanced?
  b. Find the total resistance R_T.
  c. Calculate I_5, I_1, and I_2.

FIG. 9.144

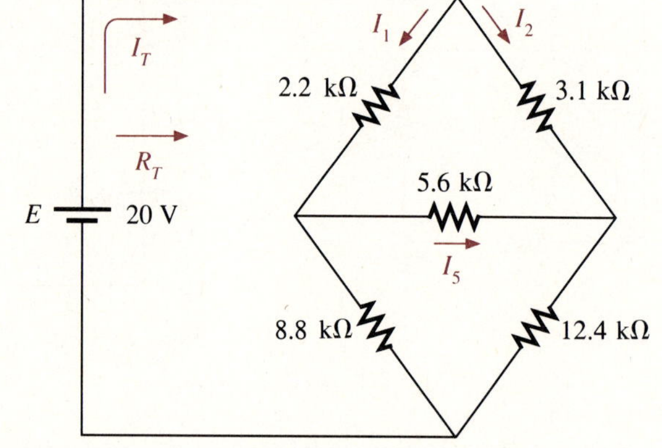

Computer Problems

53. Write a program to perform a source conversion of either type. That is, given a voltage source, convert to a current source, and given a current source, convert to a voltage source.

54. Given two simultaneous equations, write a program to solve for the unknown variables.

55. Write a program to perform mesh analysis on the network of Fig. 9.129 for any element values. That is, the program should generate the mesh currents of the network.

56. Write a program that will determine E_{Th} for the network of Fig. 9.121 for any element values.

57. Expand on Problem 56 to include a tabulation of power to R_L for values of R_L extending from $R_{Th}/10$ to $2R_{Th}$ in increments of $0.1R_{Th}$.

10

Capacitors

OBJECTIVES

- ☐ Become familiar with the characteristics of an electric field.
- ☐ Become aware of the construction of a capacitor and the parameters that determine the level of capacitance.
- ☐ Be able to calculate the capacitance of a capacitor from its dimensions and the type of dielectric.
- ☐ Become familiar with the various types of capacitors, their construction, and their areas of application.
- ☐ Understand the transient behavior of a capacitive network and how the network parameters affect the shape of the response curves.
- ☐ Be able to combine series or parallel capacitors and calculate the voltage across each capacitor, the charge stored on the plates, and the energy stored by each capacitor.

10.1 INTRODUCTION

Our analysis thus far has been limited solely to networks having only resistive elements. We now consider two additional elements, called the *capacitor* and *inductor,* which are quite different from the resistor in purpose, operation, and construction. All three are similar in that they have two terminals (two ends) for connection purposes.

Unlike the standard fixed-value resistor, the capacitor and inductor display their total characteristics only when a *change* in voltage or current is made in the circuit in which they are connected. In other words, they show a response to a change in voltage or current totally different from that obtained for a resistor. In addition, if we consider the *ideal* situation, they do not dissipate energy like the resistor but have the capability to store energy in a form that can be returned to the system whenever required by the circuit design.

Proper treatment of each requires that we devote this entire chapter to capacitors and the major portion of the next chapter to inductors.

10.2 THE ELECTRIC FIELD

Recall from Chapter 3 that a force of attraction or repulsion exists between two charged bodies. We now examine this phenomenon in greater detail by considering the electric field that exists in the region around any charged body. This electric field is represented by electric flux lines, which are drawn to indicate the strength of the electric field at any point around the charged body; that is, the denser (closer together) the lines of flux, the stronger the electric field. In Fig. 10.1, the electric field strength is stronger at position a than at position b because the flux lines are denser at a than at b. That is, for the same area at a

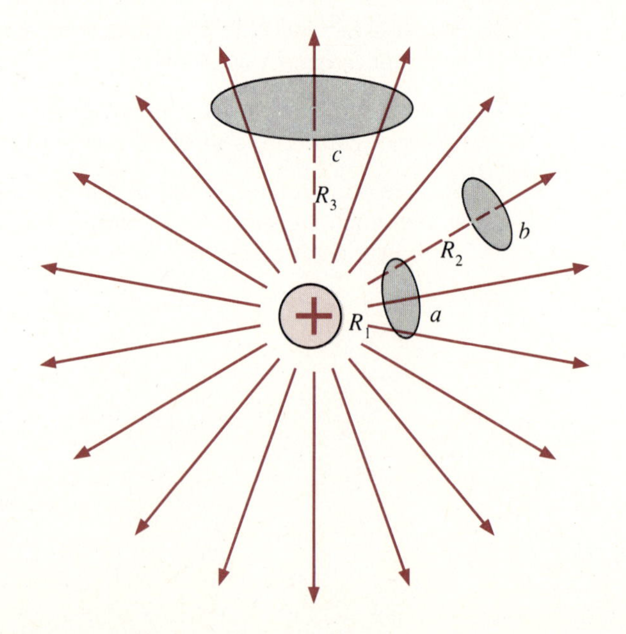

FIG. 10.1
Radiating electric flux lines from a positive charge.

and b, there are more flux lines through a than through b. Region c has the same number of flux lines as region a, but note the increased area required at distance R_3, resulting in a reduced density at c compared with a.

By definition, the electric field strength ($\mathscr{E}$) at a point in an electric field is the force acting on a charge Q at that point; that is,

$$\mathscr{E} = \frac{F}{Q} \qquad \text{(newtons/coulomb, N/C)} \qquad (10.1)$$

For an isolated charge Q the electric field strength at any point distant from the charge is given by

$$\mathscr{E} = \frac{kQ}{r^2} \qquad \text{(N/C)} \qquad (10.2)$$

where $k = 9 \times 10^9$ (constant) and r is the distance from the charge in meters. The squared distance term in the denominator clearly reveals that the strength of the electric field drops off very quickly with distance from the charge. Doubling the distance reduces the field strength by $1/(2)^2$, or 1/4. Tripling the distance reduces the strength by $1/(3)^2$, or 1/9.

Flux lines always extend from a positively charged to a negatively charged body, always extend or terminate perpendicular to the charged surfaces, and never intersect. For two charges of similar and opposite polarities, the flux distribution appears as shown in Fig. 10.2.

The attraction and repulsion between charges can now be explained in terms of the electric field and its flux lines. In Fig. 10.2(a), the flux lines are not interlocked but tend to act as a buffer, preventing attraction and causing

FIG. 10.2
Electric flux lines. (a) Similar charges; (b) opposite charges.

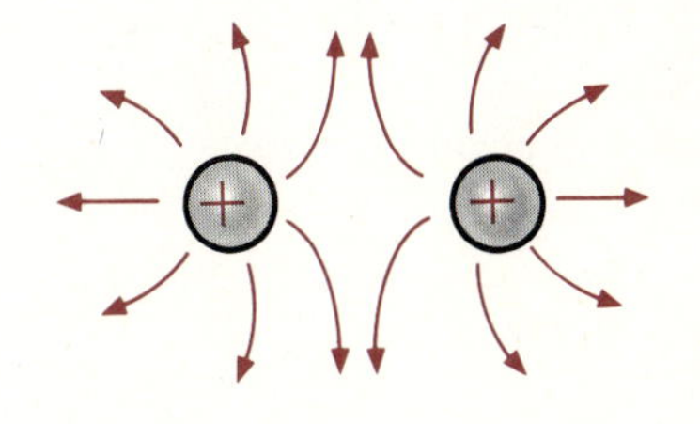

(a)

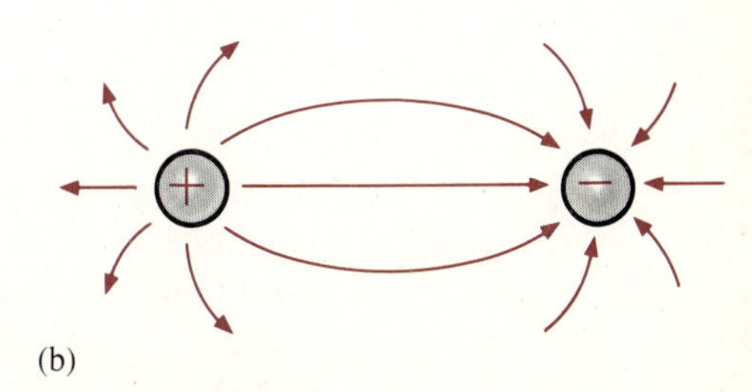

(b)

repulsion. The electric field strength is stronger (flux lines denser) for each charge the closer we are to the charge, so the more we try to bring the two charges together, the stronger will be the force of repulsion between them. In Fig. 10.2(b), the flux lines extending from the positive charge are terminated at the negative charge. A basic law of physics states that electric flux lines always tend to be as short as possible. The two charges will therefore be drawn to each other. Again, the closer the two charges, the stronger the attraction between the two charges due to the increased field strengths.

10.3 CAPACITANCE

Up to this point we have considered only isolated positive and negative spherical charges, but the analysis can be extended to charged surfaces of any shape and size. In Fig. 10.3, for example, two parallel plates of a conducting material separated by an air gap have been connected through a switch and a resistor to a battery. If the parallel plates are initially uncharged and the switch is left open, no net positive or negative charge will exist on either plate. The instant the switch is closed, however, electrons are drawn from the upper plate through the resistor due to their attraction for the positive terminal of the battery. There will be a surge of current at first, limited in magnitude by the resistance present. The level of flow will then decline in a preset manner to be demonstrated in the sections to follow. This action creates a net positive charge on the top plate. Electrons are being repelled by the negative terminal through the lower conductor to the bottom plate at the same rate they are being drawn to the positive terminal. This transfer of electrons continues until the potential difference across the parallel plates is exactly equal to the battery voltage. The final result is a net positive charge on the top plate and a negative charge on the bottom plate, very similar in many respects to the two isolated charges of Fig. 10.2(b).

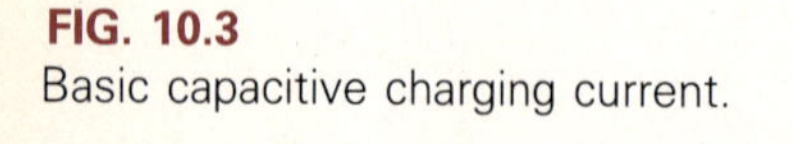

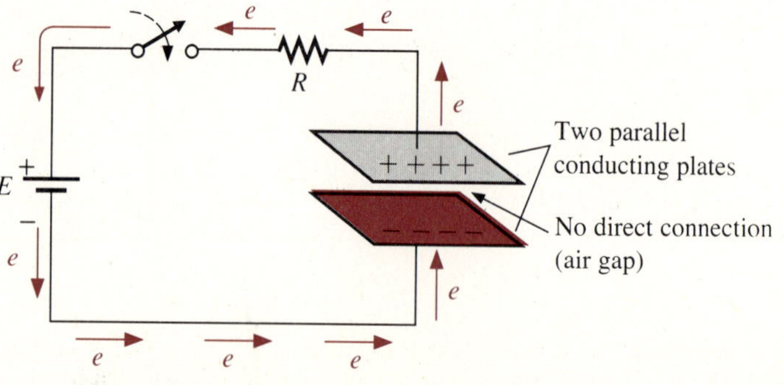

FIG. 10.3
Basic capacitive charging current.

This element, constructed simply of two parallel conducting plates separated by an insulating material (in this case, air), is called a *capacitor*. *Capacitance* is a measure of a capacitor's ability to store charge on its plates—in other words, its storage capacity. A capacitor has a capacitance of 1 farad (F) if 1 C of charge is deposited on the plates by a potential difference of 1 V across the plates. The farad is named after Michael Faraday, a nineteenth-century English chemist and physicist. The farad, however, is generally too large a measure of capacitance for most practical applications, so the microfarad (μF, 10^{-6}) or picofarad (pF, 10^{-12}) is more commonly used. Expressed as an equation, the capacitance is determined by

$$C = \frac{Q}{V} \qquad \begin{aligned} C &= \text{farads (F)} \\ Q &= \text{coulombs (C)} \\ V &= \text{volts (V)} \end{aligned} \tag{10.3}$$

Different capacitors having the same voltage across their plates will acquire greater or lesser amounts of charge on their plates. Hence the capacitors have a greater or lesser capacitance, respectively.

A cross-sectional view of the parallel plate is shown with the distribution of electric flux lines in Fig. 10.4(a). The number of flux lines per unit area between the two plates is quite uniform. At the edges, the flux lines extend outside the common surface area of the plates, producing an effect known as *fringing*. This effect, which reduces the capacitance somewhat, can be neglected for most practical applications. For the analysis to follow, we assume that all the flux lines leaving the positive plate pass directly to the negative plate within the common surface area of the plates [Fig. 10.4(b)].

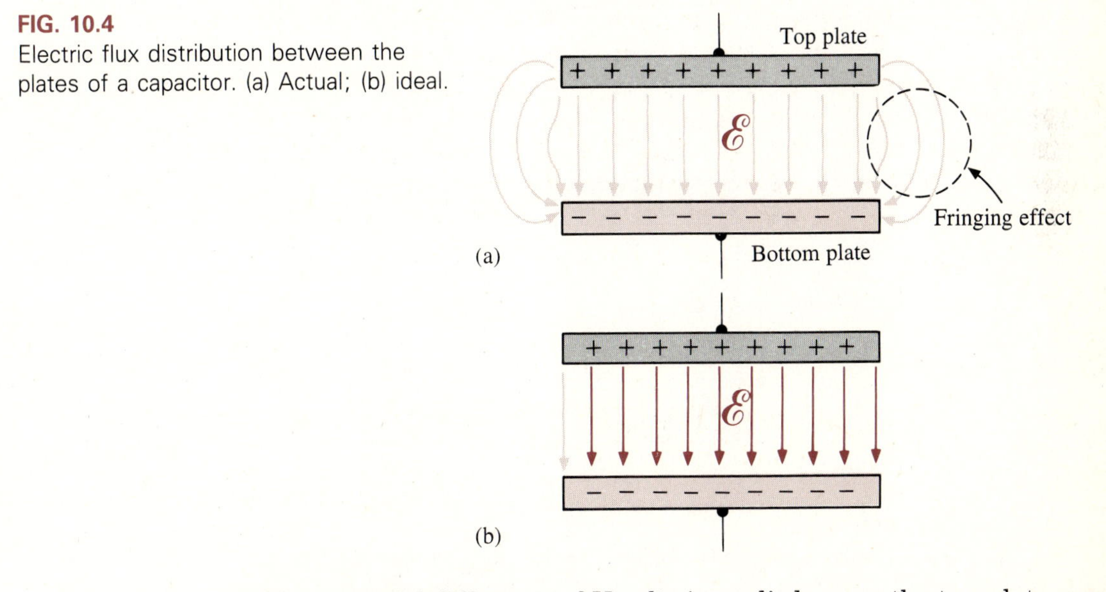

FIG. 10.4
Electric flux distribution between the plates of a capacitor. (a) Actual; (b) ideal.

If a potential difference of V volts is applied across the two plates separated by a distance of d (in meters), the electric field strength between the plates is determined by

$$\mathscr{E} = \frac{V}{d} \qquad \text{(V/m)} \tag{10.4}$$

The uniformity of the flux distribution in Fig. 10.4(b) also indicates that the electric field strength is the same at any point between the two plates.

Many values of capacitance can be obtained for the same set of parallel plates by the addition of certain insulating materials between the plates. In Fig. 10.5(a), an insulating material has been placed between a set of parallel plates having a potential difference of V volts across them.

Since the material is an insulator, the electrons within the insulator are unable to leave the parent atom and travel to the positive plate. The positive components (protons) and negative components (electrons) of each atom do shift, however [as shown in Fig. 10.5(a)], to form *dipoles*.

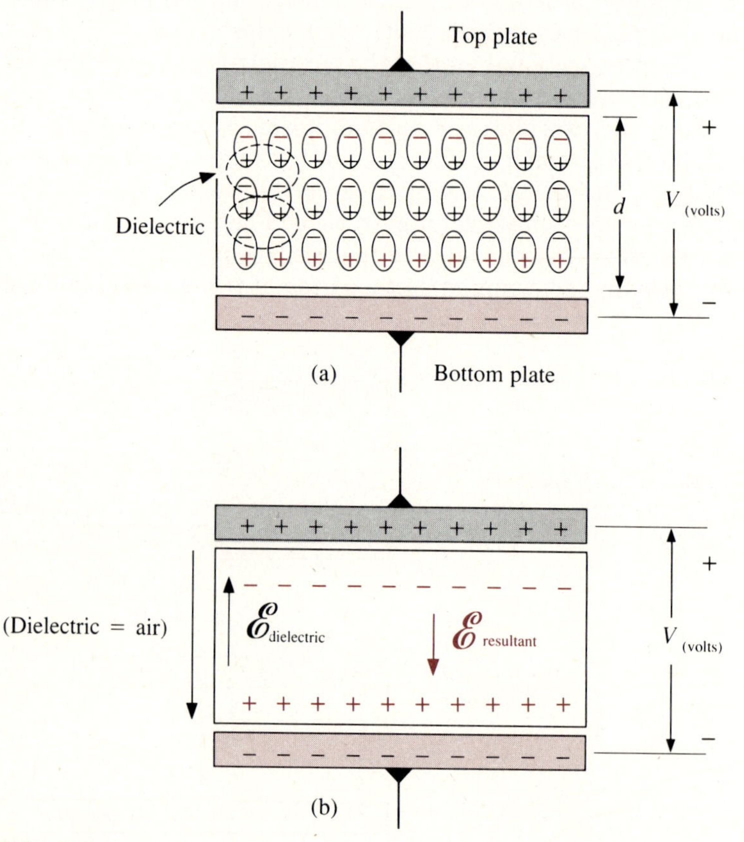

FIG. 10.5
Electric field strength. (a) Dielectric polarization; (b) net electric field.

When the dipoles align themselves as shown in Fig. 10.5(a), the material is *polarized*. A close examination within this polarized material indicates that the positive and negative components of adjoining dipoles are neutralizing the effects of each other [note the dashed areas in Fig. 10.5(a)]. The layer of positive charge on one surface and the negative charge on the other are not neutralized, however, resulting in the establishment of an electric field within the insulator [$\mathscr{E}_{\text{dielectric}}$, Fig. 10.5(b)]. The net electric field between the plates [$\mathscr{E}_{\text{resultant}} = \mathscr{E}_{\text{air}} - \mathscr{E}_{\text{dielectric}}$] is therefore reduced due to the insertion of the dielectric.

The purpose of the dielectric, therefore, is to establish an electric field that will oppose the electric field set up by free charges on the parallel plates. For this reason, the insulating material is referred to as a *dielectric, di* for *opposing* and *electric* for *electric field.*

In either case—with or without the dielectric—if the potential across the plates is kept constant and the distance between the plates is fixed, the net electric field within the plates must remain the same, as determined by the equation $\mathscr{E} = V/d$. We just ascertained, however, that the net electric field between the plates decreases with insertion of the dielectric for a fixed amount of free charge on the plates. To compensate and keep the net electric field equal to the value determined by V and d, more charge must be deposited on the plates. This additional charge for the same potential across the plates increases the capacitance, as determined by Eq. 10.5:

$$C\uparrow = \frac{Q\uparrow}{V\ (\text{fixed})}$$

For different dielectric materials between the same two parallel plates, different amounts of charge will be deposited on the plates. The *permittivity* of a dielectric is a measure of how easily the dielectric will "permit" the establishment of flux lines within the dielectric. The greater its value, the greater the amount of charge deposited on the plates and, consequently, the greater the flux density for a fixed area.

For a vacuum, the permittivity, denoted by ϵ_o is 8.85×10^{-12} F/m. The ratio of the permittivity of any dielectric to that of a vacuum is called the *relative permittivity,* ϵ_r. In equation form,

$$\boxed{\epsilon_r = \frac{\epsilon}{\epsilon_o}} \tag{10.5}$$

The value of ϵ for any material, therefore, is

$$\epsilon = \epsilon_r \epsilon_o$$

Note that ϵ_r is a dimensionless quantity. The relative permittivity, or *dielectric constant,* as it is often called, is given in Table 10.1 for various dielectric materials.

In terms of the area of the plates, distance between the plates, and permittivity, the capacitance of a capacitor is given by:

$$\boxed{C = \epsilon\frac{A}{d}} \tag{10.6}$$

TABLE 10.1
Relative permittivity (dielectric constant) of various dielectrics.

Dielectric	ϵ_r (Average Values)
Vacuum	1.0
Air	1.0006
Teflon®	2.0
Paper, paraffined	2.5
Rubber	3.0
Transformer oil	4.0
Mica	5.0
Porcelain	6.0
Bakelite	7.0
Glass	7.5
Distilled water	80.0
Barium-strontium titanite (ceramic)	7500.0

or

$$C = \epsilon_o \epsilon_r \frac{A}{d}$$

and

$$C = 8.85 \times 10^{-12} \epsilon_r \frac{A}{d} \tag{10.7}$$

where A is in square meters, d is in meters, and ϵ_r is as provided in Table 10.1. The capacitance, therefore, is greater if the area of the plates is increased, the distance between the plates is decreased, or the dielectric is changed so that ϵ_r is increased.

Defining C_o as the capacitance of a capacitor with a dielectric of air, the capacitance of the same capacitor with a dielectric having a relative permittivity of ϵ_r is determined by

$$C = \epsilon_r C_o \tag{10.8}$$

In other words, for the same set of parallel plates, the capacitance using a dielectric of mica ($\epsilon_r = 5$) is five times that obtained for a vacuum (or air, approximately) between the plates. This relationship between ϵ_r and the capacitances provides an excellent experimental method for finding the value of ϵ_r for various dielectrics.

EXAMPLE 10.1 Determine the capacitance of each capacitor on the right side of Fig. 10.6 using the information provided on the left side.

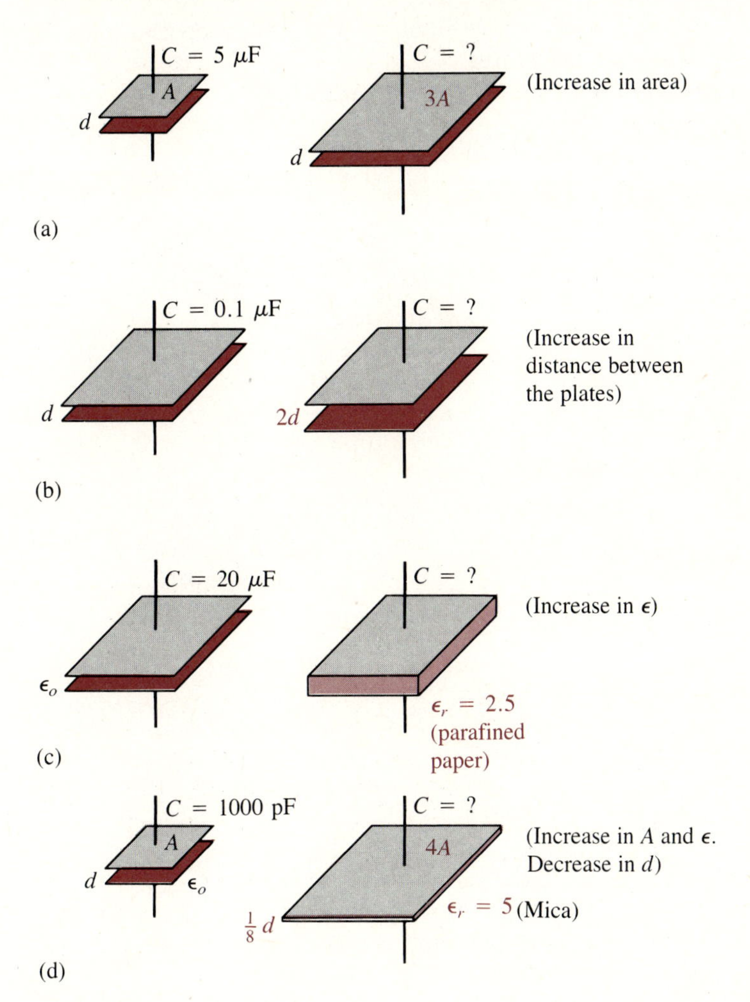

FIG. 10.6
Capacitors for Example 10.1.

Solutions

a. $C = 3(5\ \mu\text{F}) = \mathbf{15\ \mu F}$

b. $C = \frac{1}{2}(0.1\ \mu\text{F}) = \mathbf{0.05\ \mu F}$

c. $C = 2.5(20\ \mu\text{F}) = \mathbf{50\ \mu F}$

d. $C = (5)\dfrac{4}{1/8}\ 1000\ \text{pF} = (160)(1000\ \text{pF}) = \mathbf{0.16\ \mu F}$

EXAMPLE 10.2

For the capacitor of Fig. 10.7:

a. Determine the capacitance.
b. Determine the electric field strength between the plates if 450 V are applied across the plates.
c. Find the resulting charge on each plate.

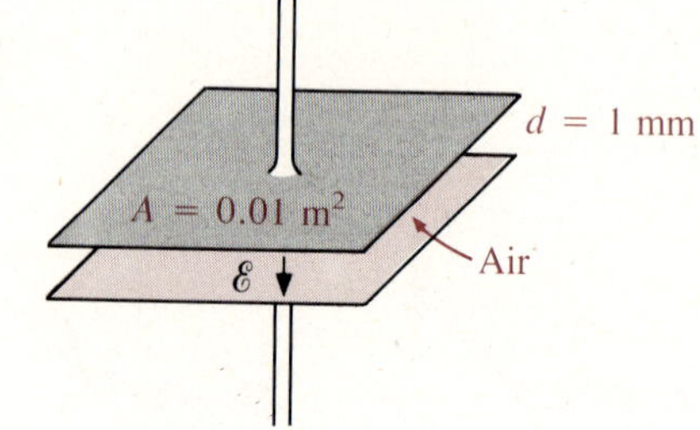

FIG. 10.7
Capacitor for Example 10.2.

Solutions

a. $$C_o = \epsilon_o \frac{A}{d} = 8.85 \times 10^{-12}\ \text{F/m}\,\frac{0.01\ \text{m}^2}{1 \times 10^{-3}\ \text{m}}$$
$$= \mathbf{88.5\ pF}$$

Calculator

(8) (·) (8) (5) (EXP) (+/−) (1) (2) (×) (·) (0) (1) (=) (÷) (1) (EXP) (+/−) (3) (=)

with a display of 8.85−11.

b. $$\mathscr{E} = \frac{V}{d} = \frac{450\ \text{V}}{1 \times 10^{-3}\ \text{m}}$$
$$\cong \mathbf{450 \times 10^3\ V/m}$$

Calculator

(4) (5) (0) (÷) (1) (EXP) (+/−) (3) (=) (F <−> E)

with a display of 4.5 05.

c. $C = Q/V$, so

$$Q = CV = (88.5 \times 10^{-12}\ \text{F})(450\ \text{V})$$
$$= 39.825 \times 10^{-9}\ \text{C}$$
$$= \mathbf{39.825\ nC}$$

Calculator

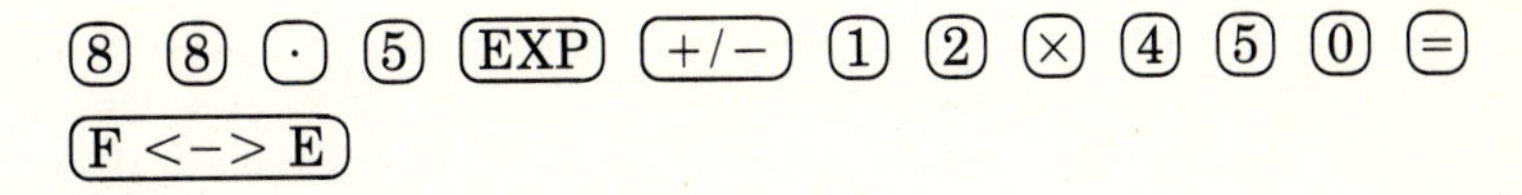

with a display of 3.983 – 08.

EXAMPLE 10.3 A sheet of mica 1 mm thick having the same area as the plates is inserted between the plates of Example 10.2.

a. Find the electric field strength between the plates.
b. Find the capacitance.
c. Find the charge on each plate.

Solutions

a. $\mathscr{E}$ is fixed by

$$\mathscr{E} = \frac{V}{d} = \frac{450\text{ V}}{1 \times 10^{-3}\text{ m}}$$
$$\cong \mathbf{450 \times 10^{3}\ V/m}$$

Calculator

(4) (5) (0) (÷) (1) (EXP) (+/−) (3) (=) (F <-> E)

with a display of 4.5 05.

b. $C = \epsilon_r C_o$
$= (5)(88.5 \times 10^{-12}\text{ F}) = \mathbf{442.5\ pF}$

Calculator

(5) (×) (8) (8) (·) (5) (EXP) (+/−) (1) (2) (=)

with a display of 4.425 – 10.

c. $Q = CV = (442.5 \times 10^{-12}\text{ F})(450\text{ V})$
$= \mathbf{199.125\ nC}$

Calculator

(4) (4) (2) (·) (5) (EXP) (+/−) (1) (2) (×) (4) (5) (0) (=)

with a display of 1.991–07.

10.4 DIELECTRIC STRENGTH

For every dielectric there is a potential that if applied across the dielectric will break the bonds within the dielectric and cause charge to flow. The voltage required per unit length (electric field intensity) to establish conduction in a dielectric is an indication of its *dielectric strength* and is called the *breakdown voltage*. When breakdown occurs, the capacitor has characteristics very similar to those of a conductor. A typical example of breakdown is lightning, which occurs when the potential between the clouds and the earth is so high that charge can pass from one to the other through the atmosphere, which acts as the dielectric.

The average dielectric strengths for various dielectrics are tabulated in volts/mil in Table 10.2 (1 mil = 0.001 in.). The relative permittivity appears in parentheses to emphasize the importance of considering both factors in the design of capacitors. Take particular note of barium-strontium titanite and mica.

TABLE 10.2
Dielectric strength of some dielectric materials.

Dielectric	Dielectric Strength (Average Value), in Volts/mil	(ϵ_r)
Air	75	(1.006)
Barium-strontium titanite (ceramic)	75	(7500.0)
Porcelain	200	(6.0)
Transformer oil	400	(4.0)
Bakelite	400	(7.0)
Rubber	700	(3.0)
Paper, paraffined	1300	(2.5)
Teflon®	1500	(2.0)
Glass	3000	(7.5)
Mica	5000	(5.0)

EXAMPLE 10.4

If 24,000 V establishes conduction through a glass dielectric, determine the thickness of the glass in inches.

Solution From Table 10.2, dielectric strength = 3000 V/mil.

$$\text{Thickness} = \frac{24{,}000\text{ V}}{3000\text{ V/mil}} = 8\text{ mils}$$

$$8\ \cancel{\text{mils}}\left(\frac{0.001\text{ in.}}{1\ \cancel{\text{mil}}}\right) = \mathbf{0.008\text{ in.}}$$

or eight thousandths of an inch.

EXAMPLE 10.5 What voltage would have to be applied across the thickness of a rubber glove to establish conduction if the gloves are 8 mils thick?

Solution From Table 10.2, dielectric strength = 700 V/mil.

$$8 \text{ mils}\left(\frac{700 \text{ V}}{1 \text{ mil}}\right) = \mathbf{5600 \text{ V}}$$

10.5 LEAKAGE CURRENT

Up to this point, we have assumed that the flow of electrons will occur in a dielectric only when the breakdown voltage is reached. This is the ideal case. In actuality, there are free electrons in every dielectric. These free electrons are due in part to impurities in the dielectric and to forces within the material itself.

When a voltage is applied across the plates of a capacitor, a leakage current due to the free electrons flows from one plate to the other. The current is usually so small, however, that it can be neglected for most practical applications. This effect is represented by a resistor in parallel with the capacitor, as shown in Fig. 10.8(a). Its value is usually in the order of 1000 megohms (MΩ). There are some capacitors, however, such as the electrolytic type, that have high leakage currents. When charged and then disconnected from the charging circuit, these capacitors lose their charge in a matter of seconds because of the flow of charge (leakage current) from one plate to the other [Fig. 10.8(b)].

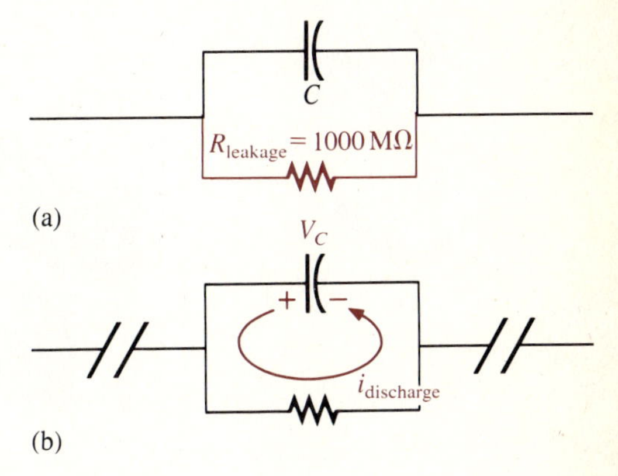

FIG. 10.8
Leakage resistance. (a) Equivalent network; (b) capacitive discharge.

10.6 TYPES OF CAPACITORS

Like resistors, all capacitors can be included under either of two general headings: fixed or variable. The symbol for a fixed capacitor is ; for a variable capacitor the symbol is . The curved line represents the plate that is usually connected to the point of lower potential.

Many types of fixed capacitors are available today. Some of the most common are the mica, ceramic, electrolytic, tantalum, and polyester-film capacitors. The typical *mica capacitor* consists basically of mica sheets separated by sheets of metal foil. The plates are connected to two electrodes, as shown in Fig. 10.9. The total area is the area of one sheet times the number of dielectric sheets. The entire system is encased in a plastic insulating material, as shown in Fig. 10.10(a). The mica capacitor exhibits excellent characteristics under stress of temperature variations and high-voltage applications (its dielectric strength is 5000 V/mil). Its leakage current is also very small (R_{leakage} is about 1000 MΩ).

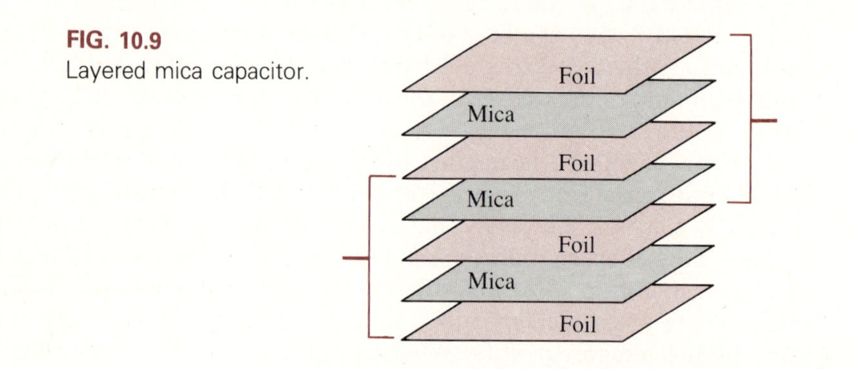

FIG. 10.9
Layered mica capacitor.

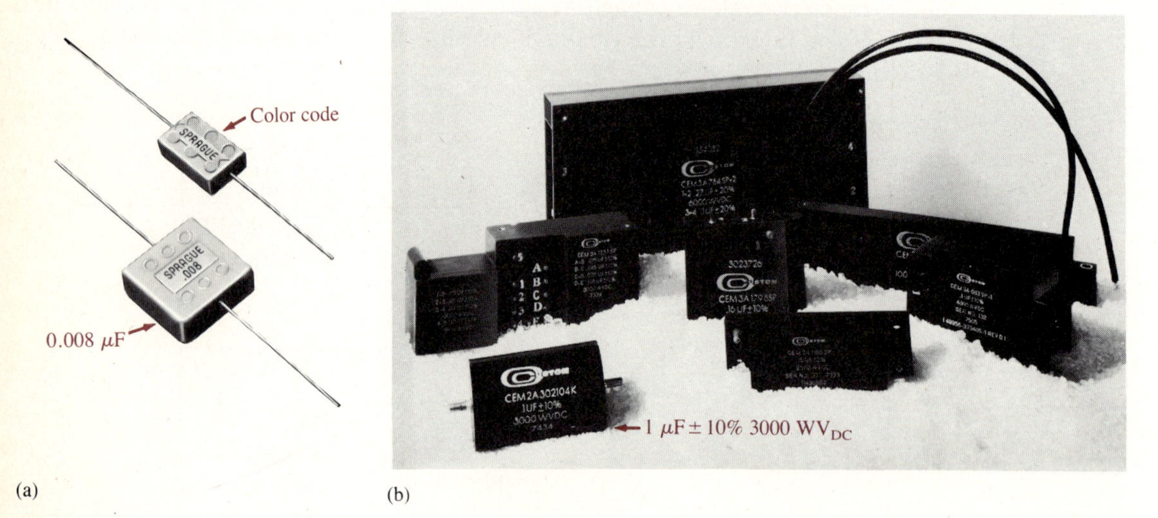

FIG. 10.10
Mica capacitors. (Courtesy of Custom Electronics Inc.)

Mica capacitors are typically between a few picofarads and 0.2 μF, with voltages of 100 V or more. The color code for the mica capacitors of Fig. 10.10(a) can be found in Appendix C.

A second type of mica capacitor appears in Fig. 10.10(b). Note in particular the cylindrical unit in the bottom left-hand corner of the figure. The ability to "roll" the mica to form the cylindrical shape is due to a process whereby the soluble contaminants in natural mica are removed, leaving a paperlike structure due to the cohesive forces in natural mica. It is commonly referred to as *reconstituted mica,* although the terminology does not mean "recycled" or "second-hand" mica. For some of the units in the photograph, different levels of capacitance are available between different sets of terminals.

Ceramic capacitors are made in many shapes and sizes, some of which are shown in Fig. 10.11. The basic construction, however, is about the same for each, as shown in Fig. 10.12. A ceramic base is coated on two sides with a metal, such as copper or silver, to act as the two plates. The leads are then attached through electrodes to the plates. An insulating coating of ceramic or plastic is then applied over the plates and dielectric. Ceramic capacitors also have a very low leakage current (R_{leakage} is about 1000 MΩ) and can be used in both dc and ac networks. They can be found in values ranging from a few picofarads to perhaps 2 μF, with very high working voltages such as 5000 V or more.

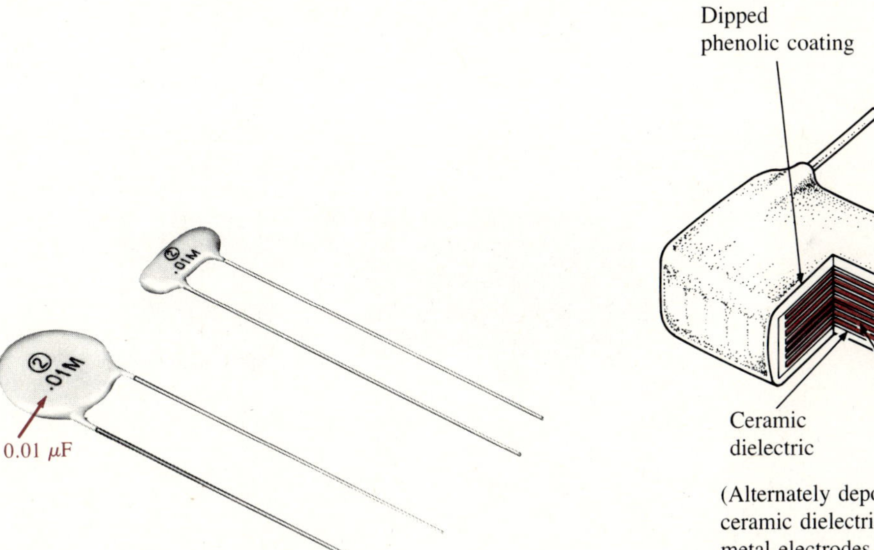

FIG. 10.11
Ceramic disc capacitors. (Courtesy of Sprague Electric Co.)

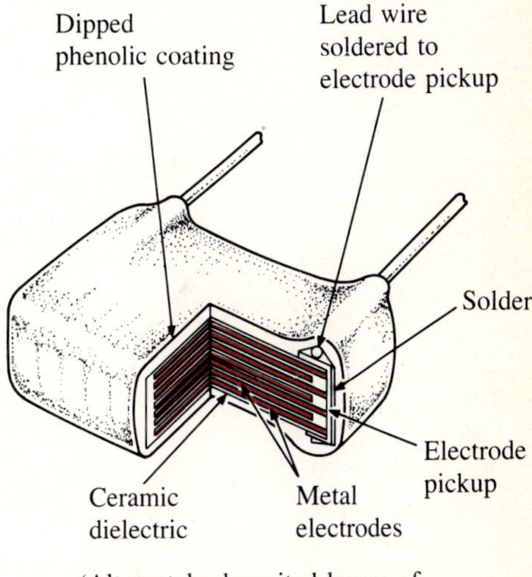

FIG. 10.12
Multilayer, radial-lead ceramic capacitor.

In recent years there has been increasing interest in monolithic (single-structure) chip capacitors, such as those shown in Fig. 10.13(a), due to their application on hybrid circuitry (networks using both discrete and IC compo-

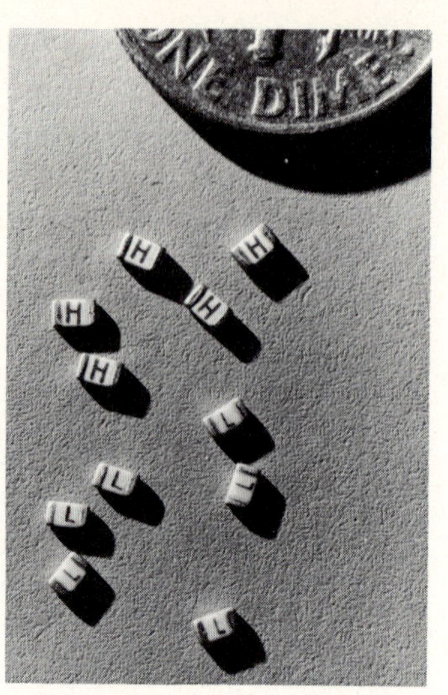

(a)

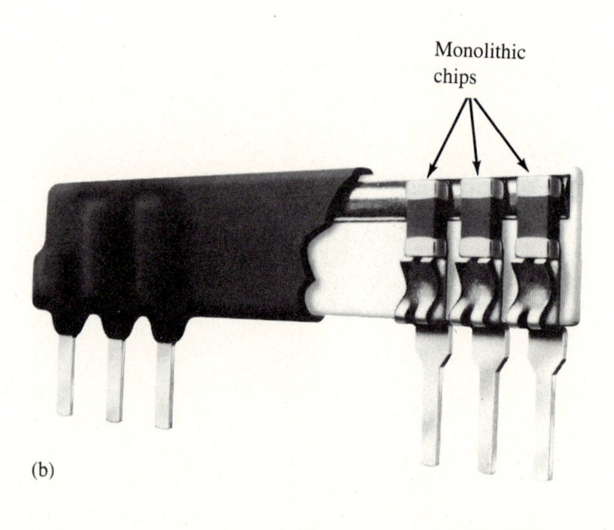

(b)

FIG. 10.13
Monolithic chip capacitors. (Courtesy of Vitramon, Inc.)

nents). There has also been increasing use of microstrip (strip-line) circuitry such as that shown in Fig. 10.13(b). Note the small chips in this cutaway section. The L and H of Fig. 10.13(a) indicate the level of capacitance. For example, the letter H in black letters represents 16 units of capacitance (in picofarads), or 16 pF. If blue ink is used, a multiplier of 100 is applied, resulting in 1600 pF. Although the size is similar, the type of ceramic material controls the capacitance level.

The *electrolytic capacitor* is used most commonly in situations where capacitances of the order of one to several thousand microfarads are required. They are designed primarily for use in networks where only dc voltages will be applied across the capacitor. There are electrolytic capacitors available that can be used in ac circuits (for starting motors) and in cases where the polarity of the dc voltage will reverse across the capacitor for short periods of time.

The basic construction of the electrolytic capacitor consists of a roll of aluminum foil coated on one side with an aluminum oxide, the aluminum being the positive plate and the oxide the dielectric. A layer of paper or gauze saturated with an electrolyte is placed over the aluminum oxide on the positive plate. Another layer of aluminum without the oxide coating is then placed over this layer, to assume the role of the negative plate. In most cases the negative plate is connected directly to the aluminum container, which then serves as the negative terminal for external connections. Because of the size of the roll of aluminum foil, the overall area of this capacitor is large; and due to the use of an oxide as the dielectric, the distance between the plates is extremely small. The negative terminal of the electrolytic capacitor is usually the one with no visible identification on the casing. The positive terminal is usually indicated by such designs as +, △, □, and so on. Due to the polarity requirement, the symbol for an electrolytic normally appears as $\overset{+}{\not\,\!\!=}$.

Associated with each electrolytic capacitor are the dc working voltage and the surge voltage. The *working voltage* is the voltage that can be applied across the capacitor for long periods of time without breakdown. The *surge voltage* is the maximum dc voltage that can be applied for a short period of time. Electrolytic capacitors are characterized as having low breakdown voltages and high leakage currents (R_{leakage} of about 1 MΩ). Various types of electrolytic capacitors are shown in Fig. 10.14. They can be found in values extending from a few microfarads to several thousand microfarads and working voltages as high as 500 V. However, increased levels of voltage are normally associated with lower values of available capacitance.

There are fundamentally two types of *tantalum capacitors:* the *solid* and the *wet-slug*. In each case, tantalum powder of high purity is pressed into a rectangular or cylindrical shape, as shown in Fig. 10.15. The anode (+) connection is then simply pressed into the resulting structures, as shown in the figure. The resulting unit is then sintered (baked) in a vacuum at very high temperatures to establish a very porous material. The result is a structure with a very large surface area in a limited volume. Through immersion in an acid solution, a very thin manganese dioxide (MnO_2) coating is established on the large, porous surface area. An electrolyte is then added to establish contact between the surface area and the cathode, producing a solid tantalum capacitor. If an appropriate "wet" acid is introduced, the result is called a *wet-slug* tantalum capacitor.

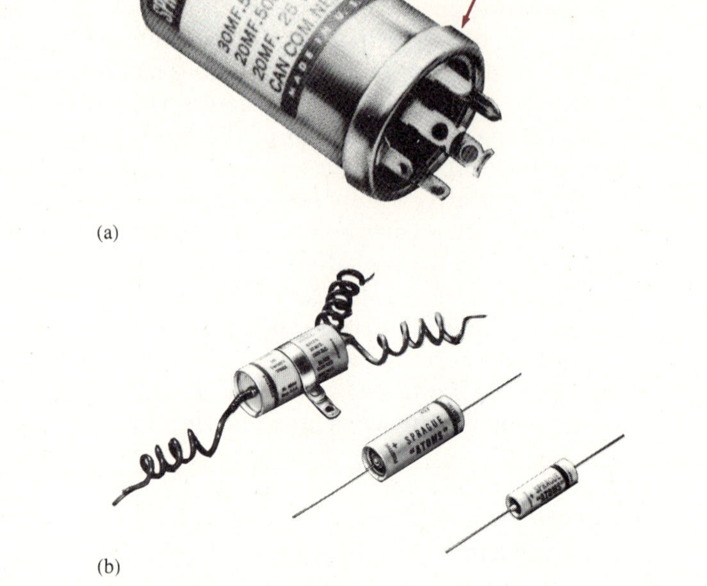

FIG. 10.14
Electrolytic capacitors. (Courtesy of Sprague Electric Co.)

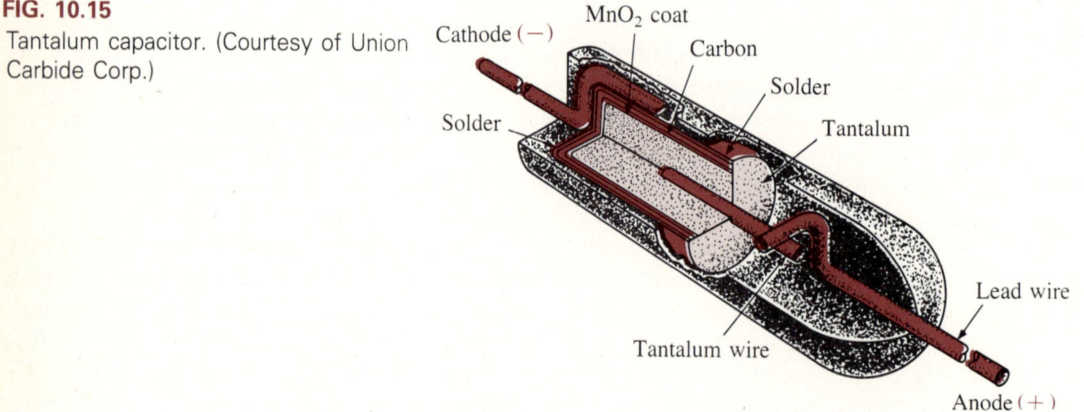

FIG. 10.15
Tantalum capacitor. (Courtesy of Union Carbide Corp.)

The last type of fixed capacitor to be introduced is the *polyester-film capacitor,* the basic construction of which is shown in Fig. 10.16. It consists simply of two metal foils separated by a strip of polyester material such as Mylar®.

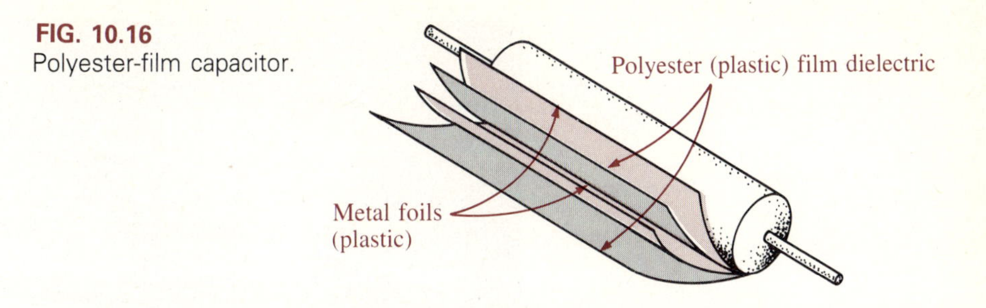

FIG. 10.16
Polyester-film capacitor.

The outside layer of polyester is applied to act as an insulating jacket. Each metal foil is connected to a lead, which extends either axially or radially from the capacitor. The rolled construction results in a large surface area, and the use of the plastic dielectric results in a very thin layer between the conducting surfaces.

Data such as capacitance and working voltage are printed on the outer wrapping if the polyester capacitor is large enough. Color coding is used on smaller devices (see Appendix C). A band (usually black) is sometimes printed near the lead that is connected to the other metal foil. The lead nearest this band should always be connected to the point of lower potential. This capacitor can be used for both dc and ac networks. Its leakage resistance is of the order of 100 MΩ. A typical polyester capacitor appears in Fig. 10.17. Polyester capacitors range in value from a few hundred picofarads to 10 to 20 μF, with working voltages as high as a few thousand volts.

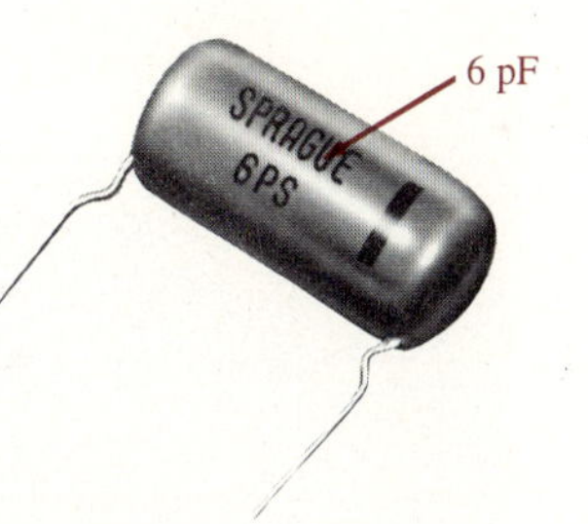

FIG. 10.17
Orange Drop® tabular capacitor. (Courtesy of Sprague Electric Co.)

The most common of the variable-type capacitors is shown in Fig. 10.18. The dielectric for each capacitor is air. The capacitance in Fig. 10.18(a) is changed by turning the shaft at one end to vary the common area of the movable and fixed plates. The greater the common area, the larger the capacitance, as determined by Eq. 10.7. The capacitance of the trimmer capacitor in Fig. 10.18(b) is changed by turning the screw, which varies the distance between the plates and, thereby, the capacitance. A digital reading capacitance meter appears in Fig. 10.19.

FIG. 10.18
Variable air capacitors. (Part (a) courtesy of James Millen Manufacturing Co.; part (b) courtesy of Johnson Manufacturing Co.)

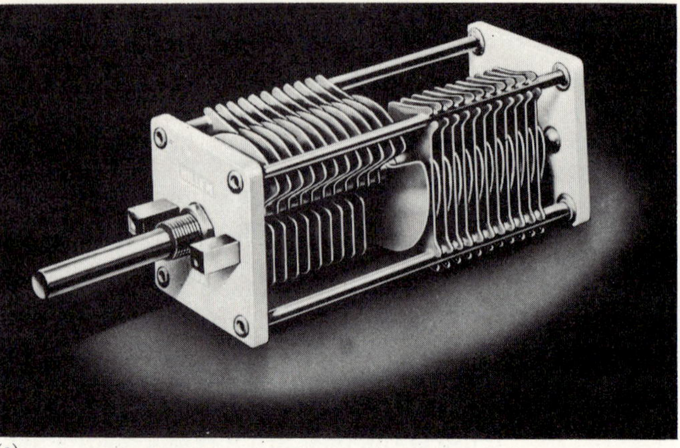

(a)

(b)

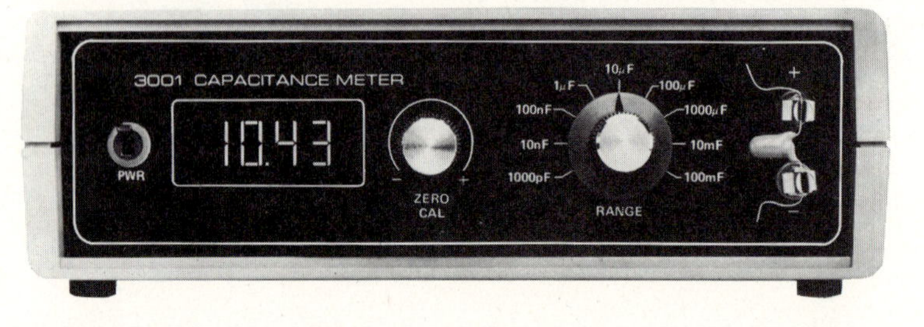

FIG. 10.19
Digital reading capacitance meter. (Courtesy of Global Specialties Corp.)

10.7 TRANSIENTS IN CAPACITIVE NETWORKS: CHARGING PHASE

Section 10.3 described how a capacitor acquires its charge. Let us now extend this discussion to include the potentials and current developed within the network of Fig. 10.20 following the closing of the switch (to position 1).

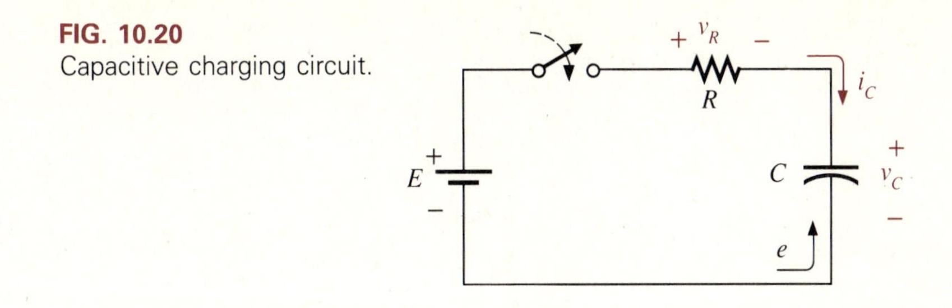

FIG. 10.20
Capacitive charging circuit.

You will recall that the instant the switch is closed, electrons are drawn from the top plate and deposited on the bottom plate by the battery, resulting in a net positive charge on the top plate, and a negative charge on the bottom plate. The transfer of electrons is very rapid at first, slowing down as the potential across the capacitor approaches the applied voltage of the battery. When the voltage across the capacitor equals the battery voltage, the transfer of electrons will cease and the plates will have a net charge determined by $Q = CV_C = CE$.

Plots of the changing current and voltage are shown in Figs. 10.21 and 10.22, respectively. When the switch is closed at $t = 0$ s, the current jumps to a value limited only by the resistance of the network and then decays to zero as the plates are charged. Note the rapid decay in current level, revealing that the amount of charge deposited on the plates per unit time is rapidly decaying also. Since the voltage across the plates is directly related to the charge on the plates by $v_C = q/C$, the rapid rate with which charge is initially deposited on the plates results in a rapid increase in v_C. Obviously, as the rate of flow of charge (i_C) decreases, the rate of change in voltage will follow suit. Eventually, the flow of charge will stop, the current i_C will be zero, and the voltage will cease to change in magnitude—the *charging phase* has passed. At this point the capacitor takes on the characteristics of an open circuit: a voltage drop across the plates without a flow of charge "between" the plates. As demonstrated in Fig. 10.23, the voltage across the capacitor is the source voltage, since $i = i_C = i_R = 0$ A and $v_R = i_R R = (0)R = 0$ V. For all future analysis:

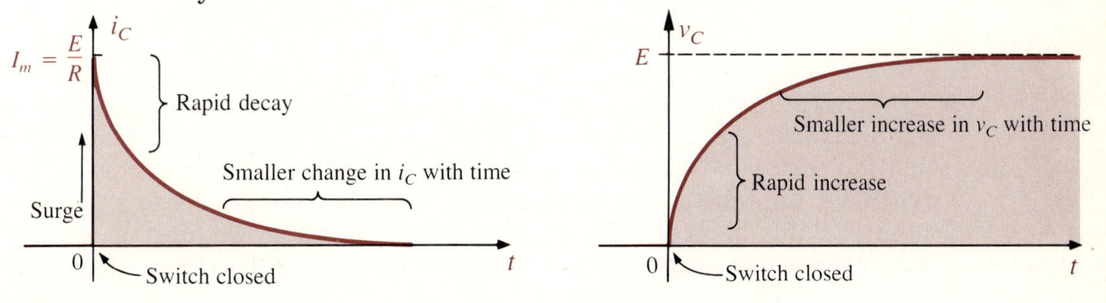

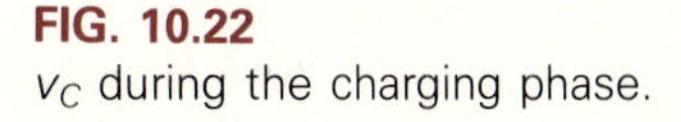

FIG. 10.21
i_C during the charging phase.

FIG. 10.22
v_C during the charging phase.

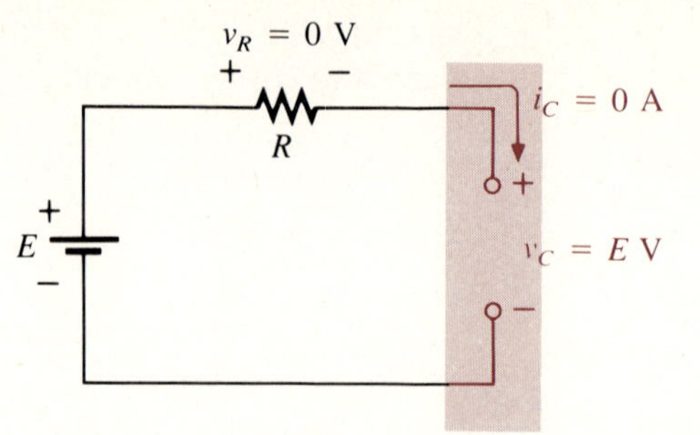

FIG. 10.23
Open-circuit equivalent for a capacitor (t = long interval).

since $i = i_C = i_R = 0$ A and $v_R = i_R R = (0)R = 0$ V. For all future analysis:

A capacitor can be replaced by an open-circuit equivalent once the charging phase in a dc network has passed.

This characteristic of the capacitor is extremely useful in the design of electronic systems where it is desirable to isolate dc levels of the network. Due to the preceding characteristic, capacitors can be treated as open circuits for dc levels in a network containing both fixed and varying voltage levels.

Looking back at the instant the switch is closed, we can also surmise that a capacitor behaves like a short circuit the moment the switch is closed in a dc charging network, as shown in Fig. 10.24. The current $i = i_C = i_R = E/R$, and the voltage $v_C = E - v_R = E - I_R R = E - (E/R)R = E - E = 0$ V at $t = 0$ s.

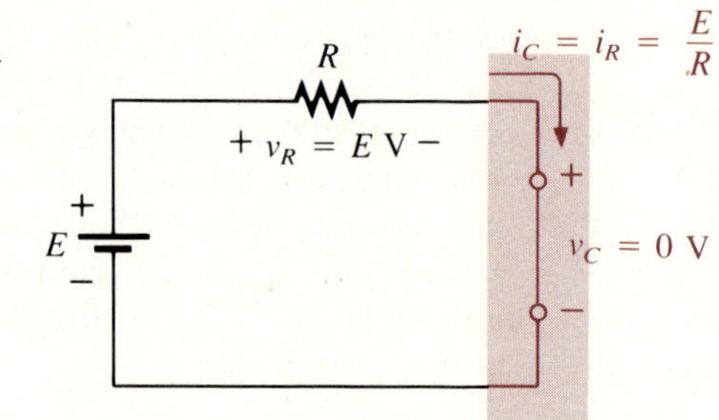

FIG. 10.24
Short-circuit equivalent for a capacitor ($t = 0$ s).

The curved shape of i_C and v_C in Figs. 10.21 and 10.22 can be described by mathematical equations that employ exponential functions. Although the analysis of this and succeeding chapters does not stress the use of exponential functions, the function is defined in the next few paragraphs to ensure its recognizability and use when required. It is a mathematical function appearing frequently in the mathematical expression of a number of important electrical variables.

The exponential function e^x appears on most scientific calculators. The numerical value of e is 2.718281828 if we limit ourselves to nine-place accuracy, although 2.718 is frequently employed. For $x = 1$ we have

$$e^x = e^1 \cong (2.718)^1 = 2.718$$

On a calculator, the procedure is

(1) (2nd F) (e^x)

with a display of 2.718.

For $x = 2$, $e^x = e^2$.

Calculator

(2) (2nd F) (e^x)

with a display of 7.389. Note that doubling x results in a much larger increase for e^x.

Similarly, for $x = 10$, $e^x = e^{10} \cong 22{,}026.466$.

Calculator

(1) (0) (2nd F) (e^x)

with a display of 22,026.466, revealing how fast the function increases.

An equation for the current i_C of Fig. 10.21 from $t = 0$ s to its final value of 0 A is given by

$$\boxed{i_C = I_m e^{-x}} \qquad \textbf{(10.9)}$$

Note in this case that the exponential function is e^{-x} rather than e^x. The result is a function that will decrease with increasing values of x at the same rate e^x increased with increasing values of x.

Our current interest in the function e^{-x} is limited to values of x greater than zero, as noted in the curve of Fig. 10.25. To obtain e^{-x} on the calculator, the sign of x must be changed using the sign key (not the subtraction key) before the exponential function is keyed in. For example, e^{-2} is obtained as follows:

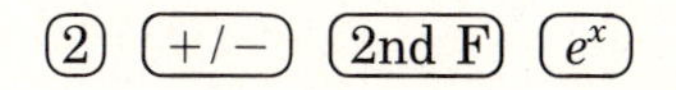

with a display of 0.135.

The values of e^{-x} are listed in Table 10.3 for a range of values of x. Note the rapidly decreasing magnitude of e^{-x} with increasing value of x.

Referring to Fig. 10.25, we find that the greatest change in e^{-x} occurs between $x = 0$ and $x = 1$ with only a slight change from $x = 4$ to $x = 5$.

TABLE 10.3
Selected values of e^{-x}

$x = 0$	$e^{-x} = e^{-0} = \frac{1}{e^0} = \frac{1}{1} = 1$
$x = 1$	$e^{-1} = \frac{1}{e} = \frac{1}{2.71828...} = 0.3679$
$x = 2$	$e^{-2} = \frac{1}{e^2} = 0.1353$
$x = 5$	$e^{-5} = \frac{1}{e^5} = 0.00674$
$x = 10$	$e^{-10} = \frac{1}{e^{10}} = 0.0000454$
$x = 100$	$e^{-100} = \frac{1}{e^{100}} = 3.72 \times 10^{-44}$

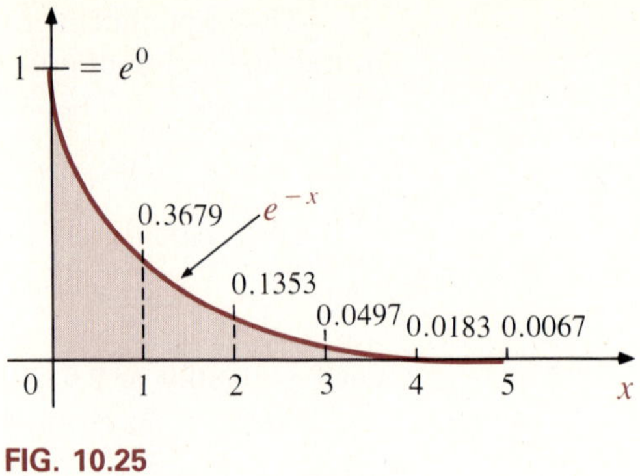

FIG. 10.25
The e^{-x} function ($x \geq 0$).

Note the similarities with the curve of Fig. 10.21. Substituting the maximum value of Fig. 10.21 and replacing x by the network factors that affect the duration of the curve results in

$$i_C = \frac{E}{R} e^{-t/RC} \tag{10.10}$$

The factor RC in Eq. 10.10 is called the *time constant* of the system and is measured in seconds. Its symbol is the Greek letter τ (tau), as shown in Eq. 10.11.

$$\tau = RC \quad \text{(seconds, s)} \tag{10.11}$$

If we substitute τ into Eq. 10.10, we obtain

$$i_C = \frac{E}{R} e^{-t/\tau} \tag{10.12}$$

When $t = \tau$, $E^{-t/\tau}$ becomes $e^{-\tau/\tau} = e^{-1} = 0.3679$, as obtained in Table 10.3. When $t = 2\tau$, $e^{-t/\tau} = e^{-2\tau/\tau} = e^{-2} = 0.1353$, and so on.

The magnitude of $e^{-t/\tau}$ and the percent change between time constants have been tabulated in Tables 10.4 and 10.5, respectively. Note that the current has dropped 63.2% (100% − 36.8%) in the first time constant but only 0.4% between the fifth and sixth time constants. The rate of change of i_C is therefore quite sensitive to the time constant determined by the network parameters R and C. For this reason, the universal time constant chart of Fig. 10.26 is provided to permit a more accurate estimate of the value of the

TABLE 10.4
i_C vs. τ (charging phase).

t	Magnitude	
0	100%	
1τ	36.8%	
2τ	13.5%	
3τ	5.0%	
4τ	1.8%	
5τ	0.67% ←	Less than 1% of maximum
6τ	0.24%	

TABLE 10.5
Change in i_C between time constants.

$(0 \rightarrow 1)\tau$	63.2%	
$(1 \rightarrow 2)\tau$	23.3%	
$(2 \rightarrow 3)\tau$	8.6%	
$(3 \rightarrow 4)\tau$	3.0%	
$(4 \rightarrow 5)\tau$	1.2%	
$(5 \rightarrow 6)\tau$	0.4% ←	Less than 1%

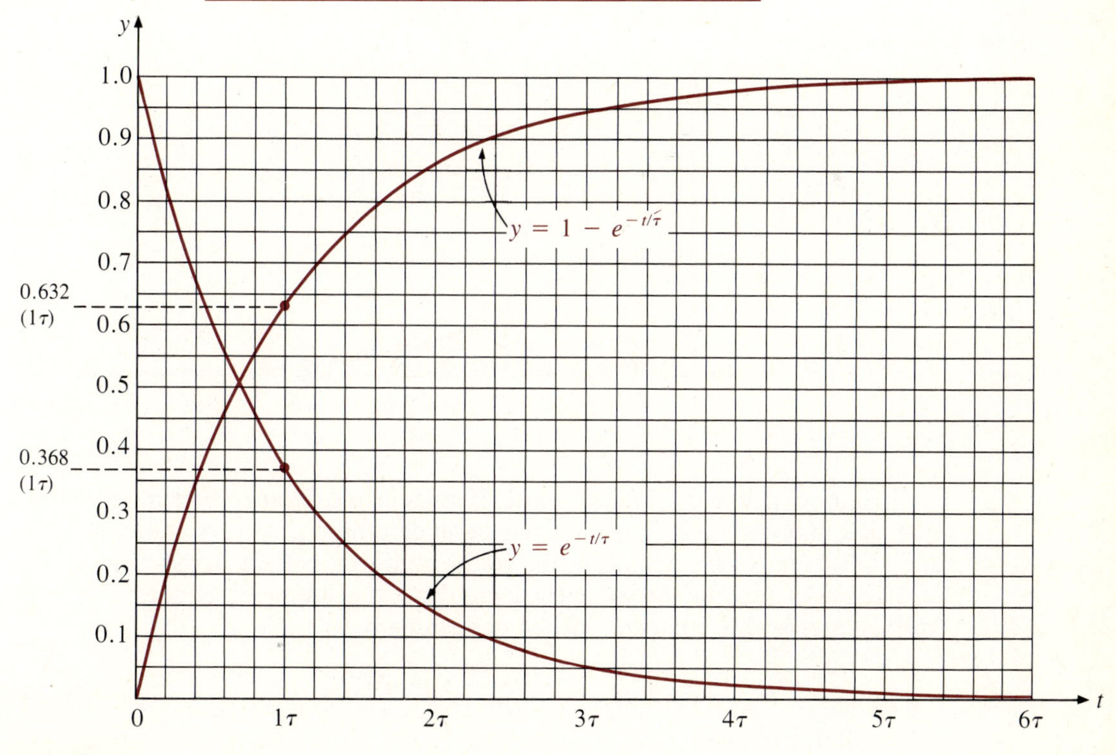

FIG. 10.26
Universal time constant chart.

function e^{-x} for specific time intervals related to the time constant. The term *universal* is used because the axes are not scaled to specific values. The function $y = 1 - e^{-t/\tau}$ is employed to describe the voltage v_C mathematically. Note how $e^{-t/\tau}$ and $1 - e^{-t/\tau}$ have a similar shape, but one is increasing at the same rate the other is decreasing.

Returning to Eq. 10.11, we find that the multiplying factor E/R is the maximum value the current i_C can attain, as shown in Fig. 10.21. Substituting $t = 0$ s into Eq. 10.10 yields

$$i_C = \frac{E}{R}e^{-t/RC} = \frac{E}{R}e^{-0} = \frac{E}{R}$$

verifying our earlier conclusion.

For increasing values of t, the magnitude of $e^{-t/\tau}$, and therefore the value of i_C, decreases, as shown in Fig. 10.27. Since the magnitude of i_C is less than 1% of its maximum after five time constants, we assume for future analysis that:

The current i_C of a capacitive network is effectively zero amperes after five time constants of the charging phase have passed in a dc network.

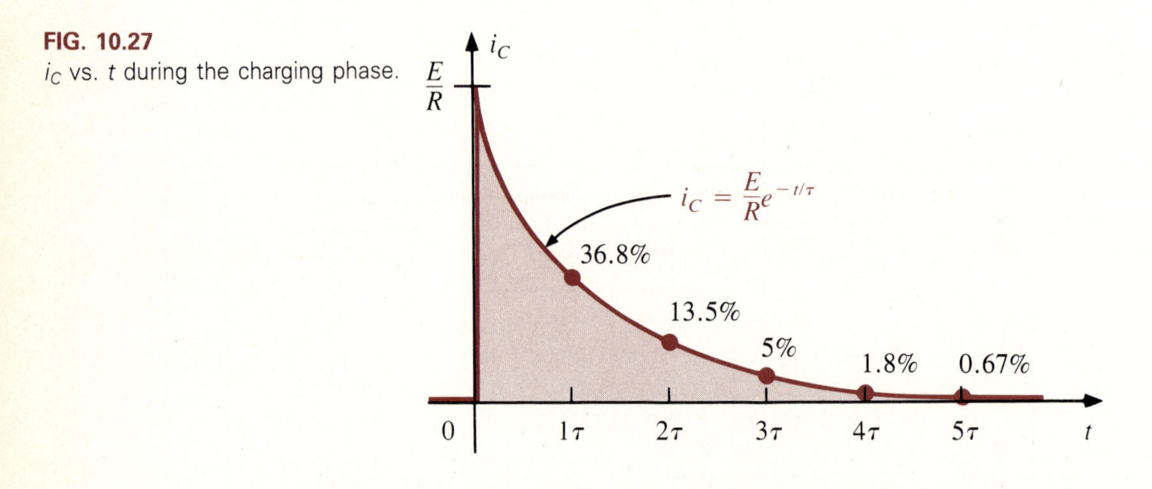

FIG. 10.27
i_C vs. t during the charging phase.

Since C is usually found in microfarads or picofarads, the time constant $\tau = RC$ is never greater than a few seconds unless R is very large.

Let us now turn our attention to the charging voltage across the capacitor. Through further mathematical analysis, the following equation for the voltage across the capacitor can be determined:

$$\boxed{v_C = E(1 - e^{-t/RC})} \qquad \textbf{(10.13)}$$

Note the presence of the same factor $e^{-t/RC}$ and the function $1 - e^{-t/RC}$ appearing in Fig. 10.26. Since $e^{-t/\tau}$ is a decaying function, the factor $1 - e^{-t/\tau}$

grows toward a maximum value of 1 with time, as shown in Fig. 10.26. In addition, since E is the multiplying factor, we can conclude that for all practical purposes the voltage v_C is E volts after five time constants of the charging phase. A plot of v_C versus t is provided in Fig. 10.28.

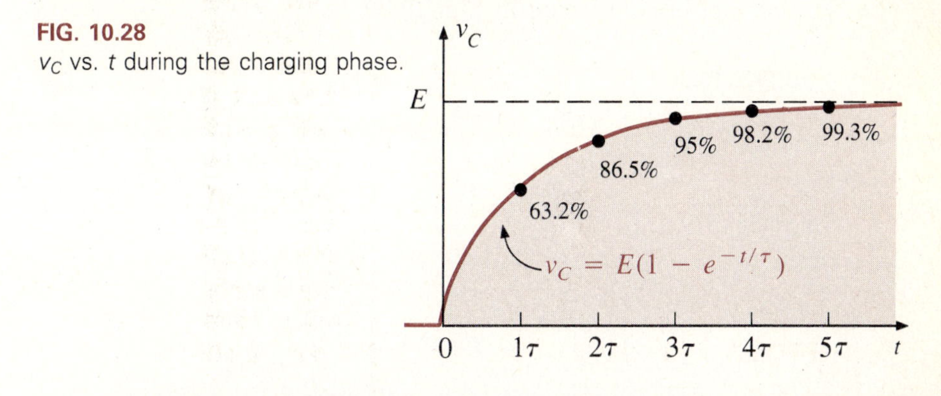

FIG. 10.28
v_C vs. t during the charging phase.

If we keep R constant and reduce C, the product RC will decrease, and the rise time of five time constants will decrease. The change in transient behavior of the voltage v_C is plotted in Fig. 10.29 for various values of C. The product RC will always have some numerical value, even though it may be very small in some cases. For this reason,

The voltage across a capacitor cannot change instantaneously.

In fact, the capacitance of a network is also a measure of how much it will oppose a change in voltage across the network. The larger the capacitance, the larger the time constant and the longer it takes to charge up to its final value (curve of C_3 in Fig. 10.29). A lesser capacitance would permit the voltage to build up more quickly, since the time constant is less (curve of C_1 in Fig. 10.29).

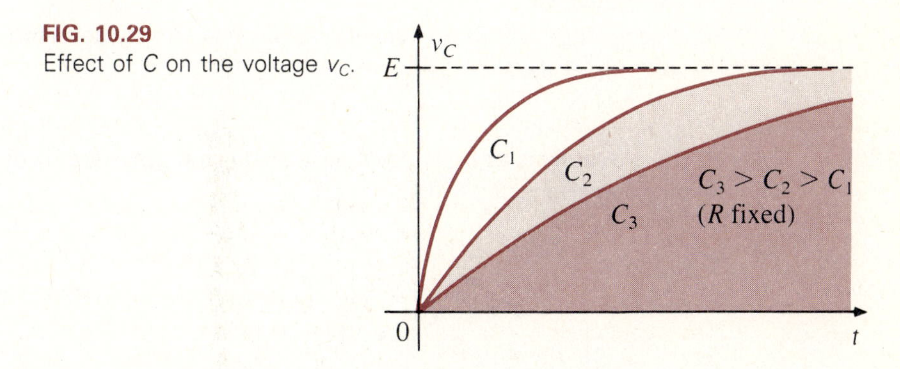

FIG. 10.29
Effect of C on the voltage v_C.

The voltage across the resistor is determined by Ohm's law:

$$v_R = i_R R = i_C R = R \frac{E}{R} e^{-t/\tau}$$

or

$$v_R = Ee^{-t/\tau} \tag{10.14}$$

A plot of v_R appears in Fig. 10.30.

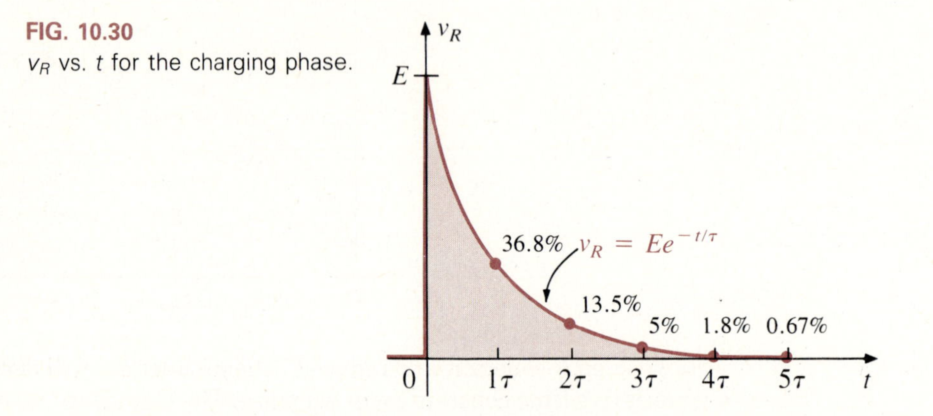

FIG. 10.30
v_R vs. t for the charging phase.

For the analysis to follow, a great deal of the mathematical complexity can be avoided by simply keeping the following in mind:

1. The voltages and current of the charging network will all essentially reach their final state (steady-state condition) after five time constants.
2. To find the maximum current level, simply replace the capacitor by a short circuit (since $v_C = 0$ V) the instant the switch is closed, and determine the resulting current level.
3. To determine the final steady-state voltage, simply replace the capacitor by an open circuit and determine the open-circuit voltage across the capacitor.
4. The voltage across any series resistors is simply determined by Ohm's law: $v_R = i_R R = i_C R$.

EXAMPLE 10.6 Sketch the waveforms of i_C, v_C, and v_R for the charging circuit of Fig. 10.31 if the switch is closed at $t = 0$ s.

FIG. 10.31

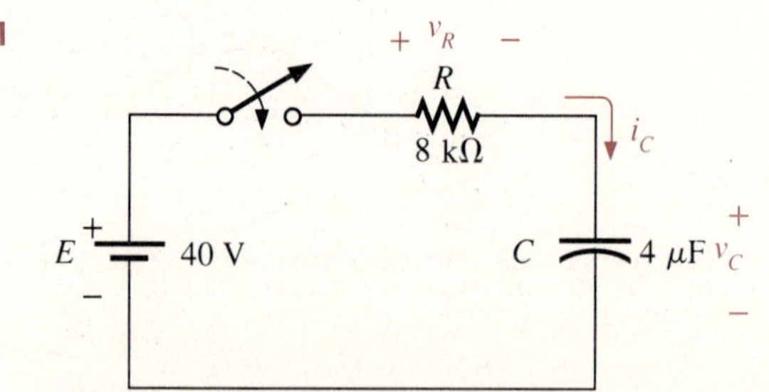

Solution

$$i_C\text{:}\quad \tau = RC = (8 \times 10^3\ \Omega)(4 \times 10^{-6}\ \text{F})$$
$$= (8)(4) \times (10^3)(10^{-6})\ \text{s}$$
$$\tau = 32 \times 10^{-3}\ \text{s} = 32\ \text{ms}$$

Calculator

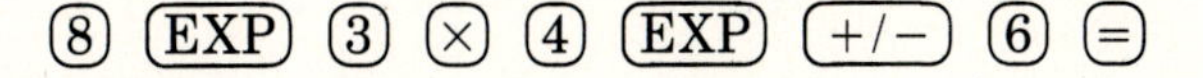

with a display of 0.032.

Therefore, $5\tau = 5(32\ \text{ms}) = 160\ \text{ms}$.

Replacing the capacitor by a short circuit results in the configuration of Fig. 10.32. Then

$$I_m = \frac{E}{R} = \frac{40\ \text{V}}{8\ \text{k}\Omega} = 5 \times 10^{-3}\ \text{A} = 5\ \text{mA}$$

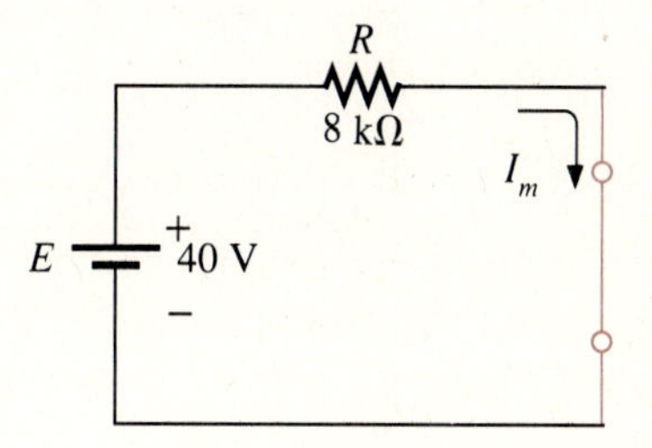

FIG. 10.32
Circuit of Fig. 10.31 at the instant the switch is closed.

Calculator

with a display of 5. −03.

Plotting the curve yields Fig. 10.33.

FIG. 10.33
i_C for the circuit of Fig. 10.31 following the closing of the switch.

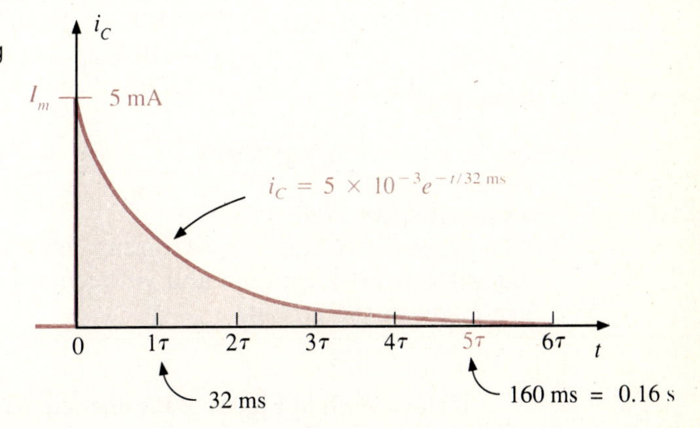

v_C: For steady-state conditions, the capacitor is replaced by its open-circuit equivalent, as shown in Fig. 10.34, and the open-circuit voltage is determined.

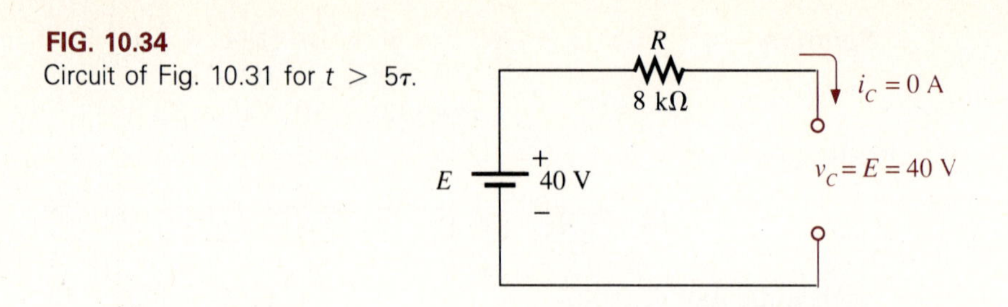

FIG. 10.34
Circuit of Fig. 10.31 for $t > 5\tau$.

Using the same τ as just determined, the curve for v_C can be drawn as shown in Fig. 10.35.

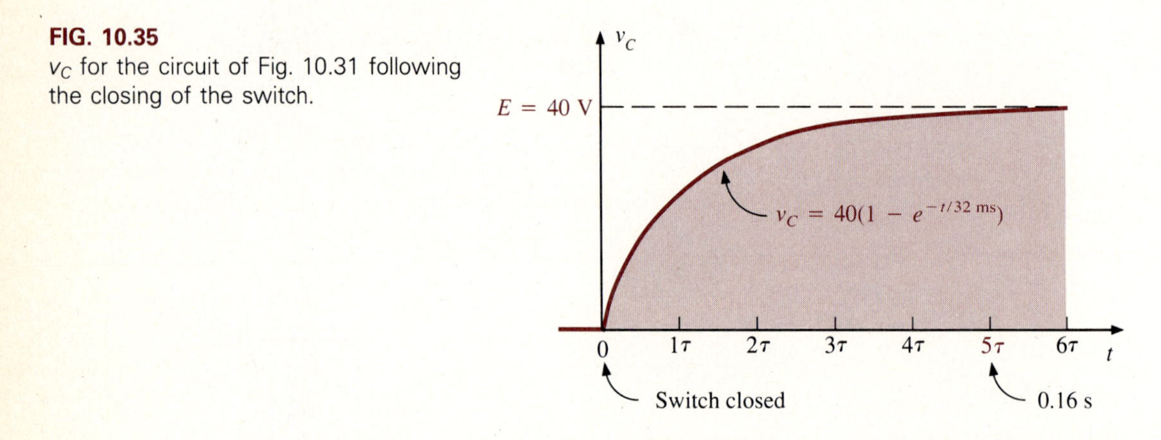

FIG. 10.35
v_C for the circuit of Fig. 10.31 following the closing of the switch.

v_R: Since v_L and i_C (= i_R) are related by a constant (R), they have the same shape. The maximum value of v_R is determined using Ohm's law:

$$
\begin{aligned}
V_{R_{\max}} &= I_m R \\
&= (5 \times 10^{-3}\ \text{A})(8 \times 10^{3}\ \Omega) \\
&= (5)(8) \times (10^{-3})(10^{3}) \\
V_{R_{\max}} &= \mathbf{40\ V}
\end{aligned}
$$

Calculator

(5) (EXP) (+/−) (3) (×) (8) (EXP) (3) (=)

with a display of 40.

Once the voltage across the capacitor has reached the input voltage E, the capacitor is fully charged and will remain in this state if no further changes are made in the circuit.

If the switch of Fig. 10.20 is opened, as shown in Fig. 10.36(a), the capacitor will retain its charge for a period of time determined by its leakage current. For capacitors such as the mica and ceramic, the leakage current ($i_{\text{leakage}} = v_C/R_{\text{leakage}}$) is very small, enabling the capacitor to retain its charge—and hence the potential difference across its plates—for a longer pe-

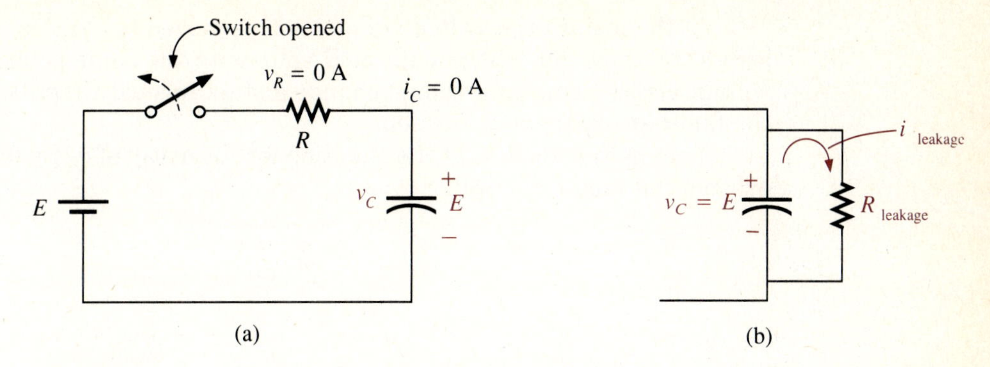

FIG. 10.36
(a) Opening the switch on the circuit of Fig. 10.20 when the capacitor is fully charged; (b) discharge of the capacitor through the leakage path.

riod of time. For electrolytic capacitors, which have very high leakage currents, the capacitor discharges more rapidly, as shown in Fig. 10.36(b). In any event, to ensure that they are completely discharged, capacitors should be shorted by a lead or a screwdriver before they are handled.

10.8 DISCHARGE PHASE

The network of Fig. 10.20 can be modified to charge and discharge the capacitor by adding a second terminal, as shown in Fig. 10.37, and moving the switch to position 2 when the capacitor is fully charged.

FIG. 10.37
Capacitive discharge circuit.

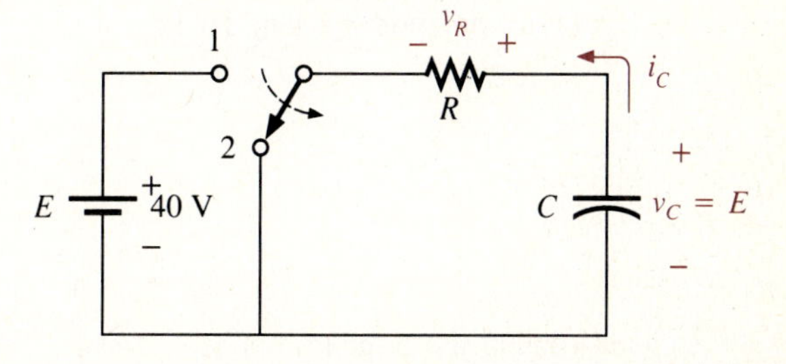

By moving the switch to position 2, the battery E is removed from the capacitive circuit and the voltage across the capacitor (E volts) becomes the source of pressure to establish a current i_C in the direction shown in Fig. 10.37 (opposite to that of the charging current). The major differences between the source voltage E and the voltage across the capacitor is that as the discharge current begins to flow, the voltage across the capacitor will decrease at the same rate. Since R and C of the circuit are the same as the charging network, the time constant of the discharging elements is the same. As with the charging phase, it takes (on a practical basis) five time constants to complete the discharge phase.

At the instant the switch of Fig. 10.37 is moved to position 2, the voltage across the capacitor will remain at E volts with the same polarity, since the voltage across a capacitor cannot change instantaneously. It will then decay in the familiar exponential fashion.

Note in Fig. 10.38 that the equation for the decay of v_C includes the same exponential factor e^{-x} and

$$v_C = Ee^{-t/\tau} \tag{10.15}$$

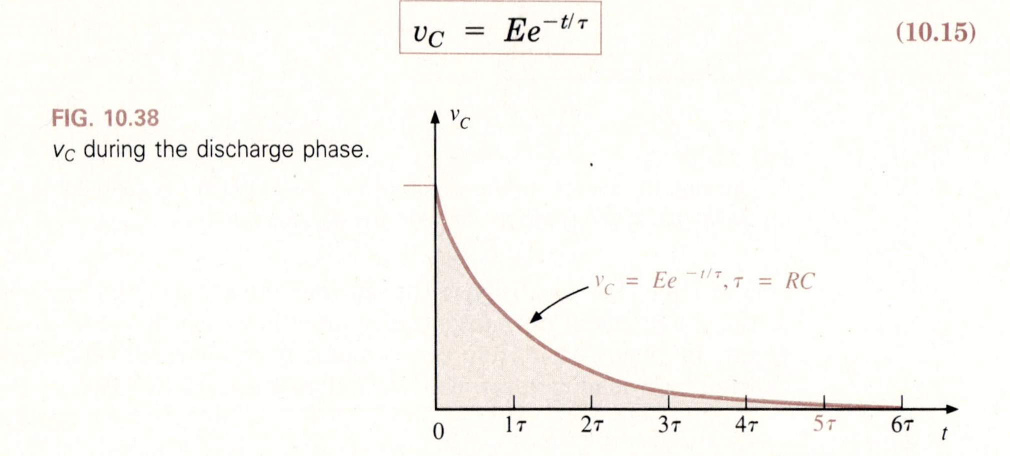

FIG. 10.38
v_C during the discharge phase.

The discharge current i_C flows heavily at first in a direction (opposite to that of the charging current) established by the polarity of v_C. The rate of flow then decreases in an exponential manner, as depicted in Fig. 10.39. Since the voltage across the resistor R at the instant the switch is closed is the voltage across the capacitor, the current at this instant can be determined by Ohm's law: $I_m = E/R$. The current i_C then decays toward zero in an exponential manner as defined by Eq. 10.16.

$$i_C = \frac{E}{R}e^{-t/\tau} \tag{10.16}$$

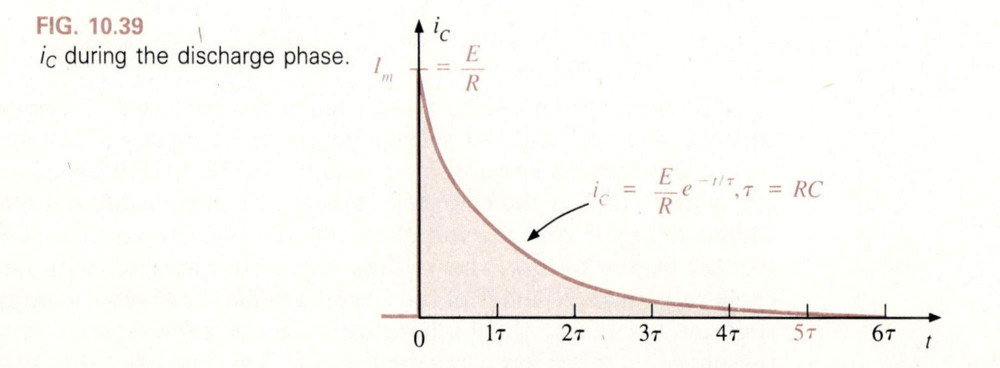

FIG. 10.39
i_C during the discharge phase.

In Fig. 10.37 it is quite clear that $v_C = v_R$, and therefore

$$v_R = v_C = Ee^{-t/\tau} \tag{10.17}$$

during the discharge phase.

EXAMPLE 10.7 The switch of Fig. 10.40 has been in position 1 for a long period of time (much greater than 5τ). Sketch v_C and i_C if the switch is thrown into position 2.

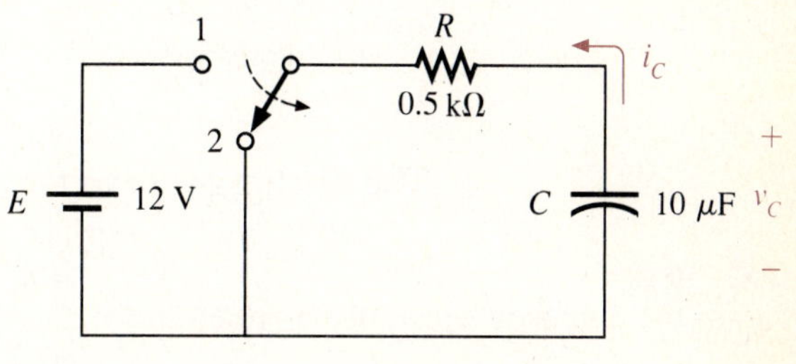

FIG. 10.40
Discharge circuit for Example 10.7.

Solution

v_C: In position 1 the voltage across the capacitor has reached a steady-state level of $E = 12$ V. The instant the switch is moved to position 2, the voltage v_C will remain at 12 V with the same polarity.

The time constant of the discharge phase is

$$\begin{aligned}\tau = RC &= (0.5 \times 10^3\ \Omega)(10 \times 10^{-6}\ \text{F})\\ &= (0.5)(10) \times (10^3)(10^{-6})\ \text{s} = 5 \times 10^{-3}\ \text{s}\\ \tau &= \mathbf{5\ ms}\end{aligned}$$

Calculator

(.) (5) (EXP) (3) (×) (1) (0) (EXP) (+/−) (6) (=) (F <−> E)

with a display of 5.−03.

For a full discharge,

$$5\tau = 5(5 \times 10^{-3}\ \text{s}) = 25 \times 10^{-3}\ \text{s} = \mathbf{25\ ms}$$

The discharge curve for v_C appears in Fig. 10.41.

i_C: The peak current is

$$\begin{aligned}I_m = \frac{E}{R} &= \frac{12\ \text{V}}{0.5\ \text{k}\Omega} = \frac{12\ \text{V}}{0.5 \times 10^3\ \Omega}\\ &= \mathbf{24\ mA}\end{aligned}$$

Calculator

(1) (2) (÷) (.) (5) (EXP) (3) (=)

with a display of 0.024.

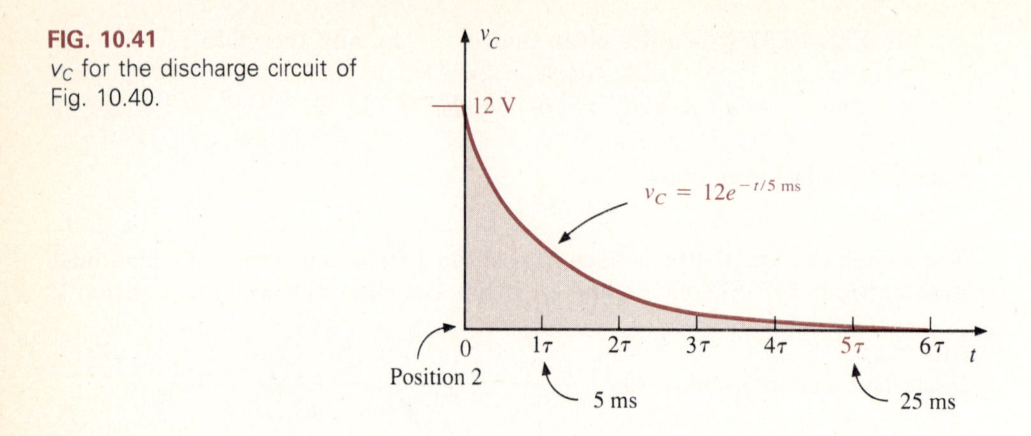

FIG. 10.41
v_C for the discharge circuit of Fig. 10.40.

The discharge curve for i_C appears in Fig. 10.42.

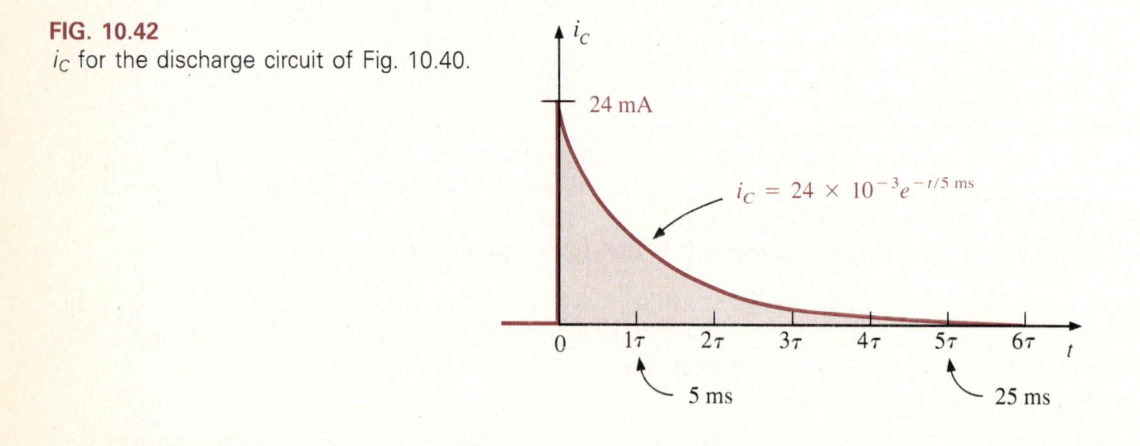

FIG. 10.42
i_C for the discharge circuit of Fig. 10.40.

10.9 INSTANTANEOUS VALUES

On occasion it will be necessary to determine the voltage or current at a particular instant of time. For example, if

$$v_C = 20(1 - e^{-t/2 \times 10^{-3}})$$

the voltage v_C may be required at $t = 5$ ms, which does not correspond to a particular value of τ. Figure 10.26 reveals that $1 - e^{-t/\tau}$ is approximately 0.93 at $t = 5\text{ ms} = 2.5\tau$, resulting in $v_C = 20(0.93) = 18.6$ V. Additional accuracy can be obtained simply by substituting $t = 5$ ms into the equation and solving for v_C, using a calculator or table to determine $e^{-2.5}$. Thus,

$$\begin{aligned} v_C &= 20(1 - e^{-5\text{ ms}/2\text{ ms}}) \\ &= 20(1 - e^{-2.5}) \\ &= 20(1 - 0.082) \\ &= 20(0.918) \\ v_C &= \mathbf{18.36\ V} \end{aligned}$$

The results are close, but accuracy beyond the tenths' place is suspect using Fig. 10.26. This procedure can also be applied to any other equation introduced in this chapter for currents or other voltages.

Calculator

$e^{-2.5}$: 2 · 5 +/− 2nd F e^x

with a display of 0.0821.

$20(1 - e^{-2.5})$: 2 0 × (1 − 2 · 5 +/− 2nd F e^x) =

with a display of 18.358.

There are also occasions when the time to reach a particular voltage or current is required. The procedure is complicated somewhat by the use of natural logs ($\log_e$, or ln), but today's calculators are equipped to handle the operation with ease.

For the charging phase of v_C, the time to reach a level v_C with a source voltage of E volts is determined by

$$t = \tau \log_e\left(1 - \frac{v_C}{E}\right) \qquad \text{(seconds, s)} \qquad (10.18)$$

EXAMPLE 10.8 Determine how long it will take v_C to reach 20 V if $E = 40$ V, $R = 8$ kΩ, and $C = 4$ μF, as in Example 10.6.

Solution

$$\tau = RC = (8 \times 10^3\ \Omega)(4 \times 10^{-6}\ \text{F}) = \mathbf{32\ ms}$$

$$t = -\tau \log_e\left(1 - \frac{v_C}{E}\right) = -(32\ \text{ms})\log_e\left(1 - \frac{20\ \text{V}}{40\ \text{V}}\right)$$

$$= \mathbf{-(32\ ms)\log_e 0.5}$$

Calculator

· 5 ln

with a display of −0.691. Hence

$$t = -(32\ \text{ms})(-0.691)$$
$$= \mathbf{22.112\ ms}$$

For the charging current, $i_C = (E/R)e^{-t/\tau}$, and

$$t = -\tau \log_e\left(\frac{i_C}{I_m}\right) \qquad (10.19)$$

and for the function $v_C = Ee^{-t/\tau}$, we have

$$t = -\tau \log_e \frac{v_C}{E} \tag{10.20}$$

10.10 THEVENIN'S THEOREM $\tau = R_{Th}C$

Occasions will arise in which the network does not have the simple series form of Fig. 10.20. It will then be necessary first to find the Thevenin equivalent circuit for the network external to the capacitive element. E_{Th} will then be the source voltage E of Eqs. 10.13 through 10.16 and R_{Th} will be the resistance R. The time constant is then $\tau = R_{Th}C$.

EXAMPLE 10.9 For the network of Fig. 10.43, find the mathematical expression for the transient behavior of the voltage v_C following the closing of the switch.

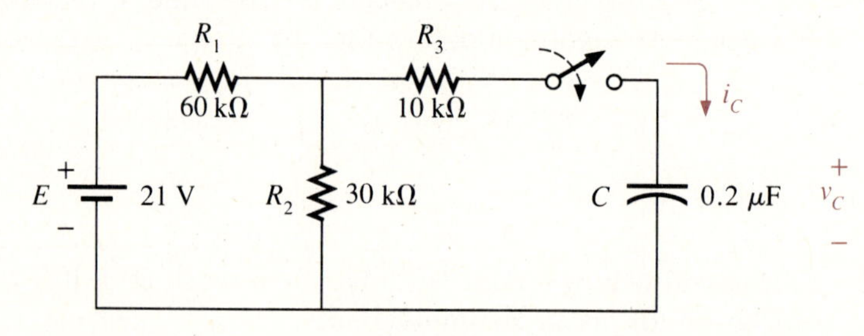

FIG. 10.43
Network for Example 10.9.

Solution Applying Thevenin's theorem to the 0.2-μF capacitor, we obtain Fig. 10.44 for R_{Th}:

$$R_{Th} = (R_1 || R_2) + R_3 = \frac{(60\text{ k}\Omega)(30\text{ k}\Omega)}{90\text{ k}\Omega} + 10\text{ k}\Omega$$

$$= 20\text{ k}\Omega + 10\text{ k}\Omega$$

$$R_{Th} = \mathbf{30\text{ k}\Omega}$$

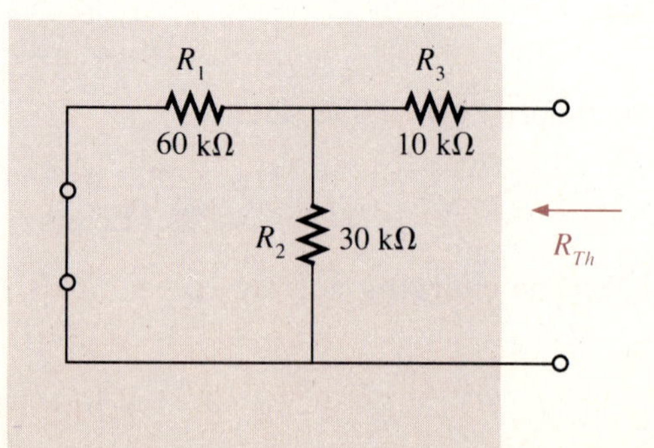

FIG. 10.44
Determining R_{Th} for the network of Fig. 10.43.

For E_{Th} (Fig. 10.45):

$$E_{Th} = \frac{R_2E}{R_2 + R_1} = \frac{(30\ \text{k}\Omega)(21\ \text{V})}{30\ \text{k}\Omega + 60\ \text{k}\Omega} = \frac{1}{3}(21) = \mathbf{7\ V}$$

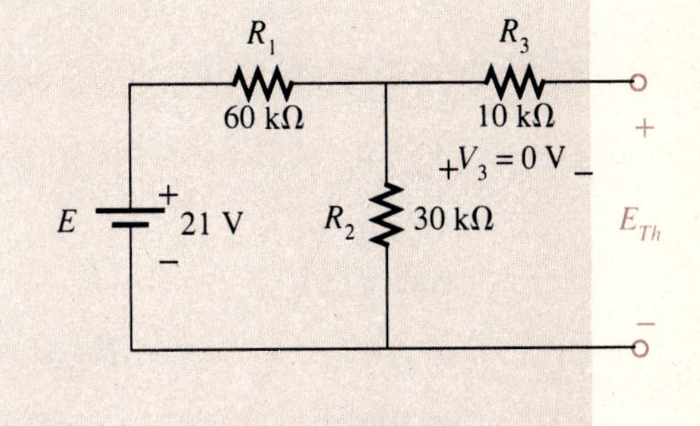

FIG. 10.45
Determining E_{Th} for the network of Fig. 10.43.

The resultant Thevenin equivalent circuit with the capacitor replaced is shown in Fig. 10.46.

$$\begin{aligned}\tau = RC &= (30 \times 10^3\ \Omega)(0.2 \times 10^{-6}\ \text{F}) \\ &= (30)(0.2) \times (10^3)(10^{-6})\ \text{s} = 6 \times 10^{-3}\ \text{s} \\ \tau &= 6\ \text{ms}\end{aligned}$$

Calculator

(3) (0) (EXP) (3) (×) (·) (2) (EXP) (+/−) (6) (=) (F <-> E)

with a display of 6.−03 and

$$v_C = \mathbf{7(1 - e^{-t/6ms})}$$

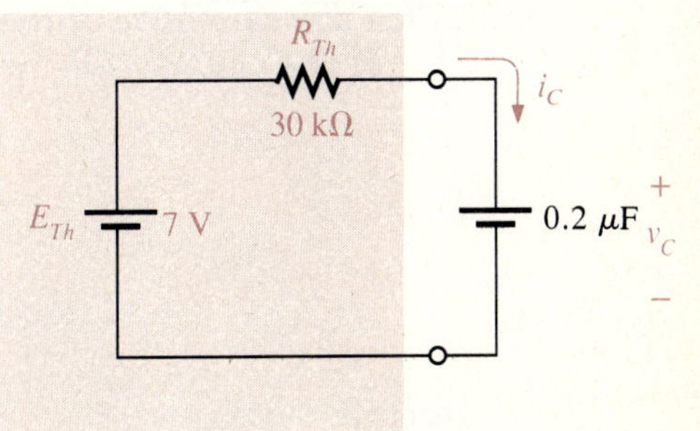

FIG. 10.46
Substituting the Thevenin equivalent circuit.

EXAMPLE 10.10 For the network of Fig. 10.47, find the mathematical expression for the voltage v_C after the closing of the switch (at $t = 0$).

Solution

$$\begin{aligned}R_{Th} &= R_1 + R_2 = 6\ \Omega + 10\ \Omega = 16\ \Omega \\ E_{Th} &= V_1 + V_2 = IR_1 + 0 \\ &= (20 \times 10^{-3}\ \text{A})(6\ \Omega) = 120 \times 10^{-3}\ \text{V} = 0.12\ \text{V}\end{aligned}$$

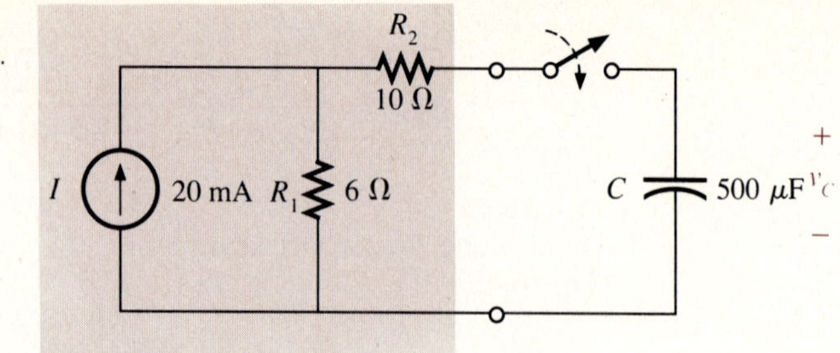

FIG. 10.47
Circuit for Example 10.10.

and

$$\tau = R_{Th}C = (16\ \Omega)(500 \times 10^{-6}\ \text{F}) = 8\ \text{ms}$$

so that

$$v_C = \mathbf{0.12(1 - e^{-t/(8 \times 10^{-3})})}$$

10.11 THE CURRENT i_C

The current i_C associated with a capacitance C is related to the voltage across the capacitor by

$$i_C = C\frac{dv_C}{dt} \tag{10.21}$$

where dv_C/dt is a measure of the rate of change of v_C. The function dv_C/dt is called the *derivative* of the voltage v_C with respect to time t.

If the voltage fails to change at a particular instant, then

$$dv_C = 0$$

and

$$i_C = C\frac{dv_C}{dt} = 0\ \text{A}$$

In other words,

If the voltage across a capacitor fails to change with time, the current i_C associated with the capacitor is zero.

To take this a step further, the equation also states that *the more the change in voltage across the capacitor, the greater the resulting current.*

In an effort to develop a clearer understanding of Eq. 10.21, let us calculate the average current associated with a capacitor for various voltages impressed across the capacitor. The average current is defined by the equation

$$i_{C_{av}} = C\frac{\Delta v_C}{\Delta t} \tag{10.22}$$

where Δ indicates a finite (measurable) change in charge, voltage, or time.

In the following example, the change in voltage Δv_C is considered for each slope of the voltage waveform. If the voltage increases with time, the average current is the change in voltage divided by the change in time, with a positive sign. If the voltage decreases with time, the average current is again the change in voltage divided by the change in time, but with a negative sign.

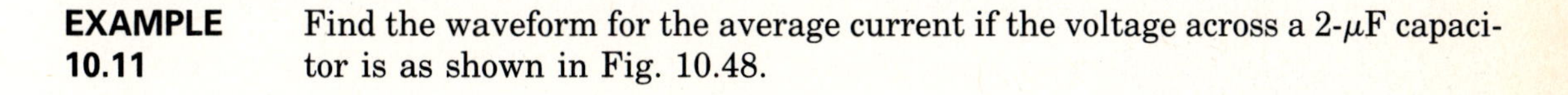

EXAMPLE 10.11 Find the waveform for the average current if the voltage across a 2-μF capacitor is as shown in Fig. 10.48.

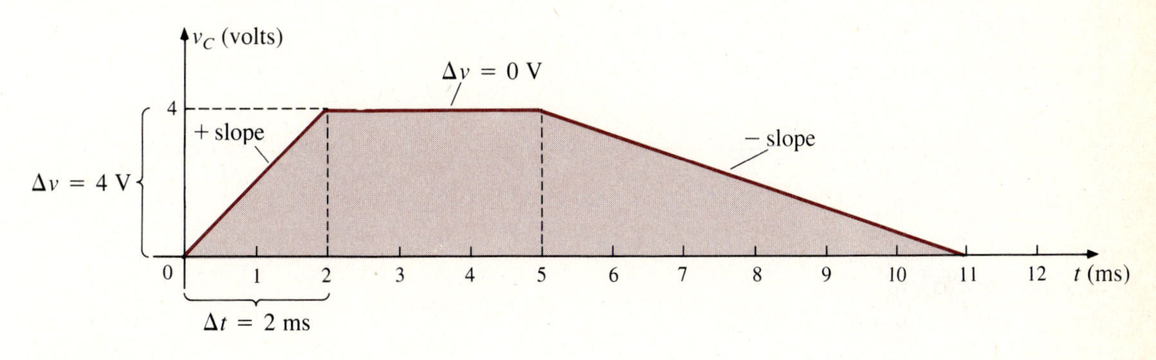

FIG. 10.48
Applied voltage v_C for Example 10.11.

Solutions

a. From 0 to 2 ms, the voltage increases linearly from 0 to 4 V, so the change in voltage $\Delta v = 4\text{ V} - 0\text{ V} = 4\text{ V}$ (with a positive sign, since the voltage increases with time). The change in time is $\Delta t = 2\text{ ms} - 0\text{ ms} = 2\text{ ms}$, and

$$i_{C_{av}} = C\frac{\Delta v_C}{\Delta t} = (2 \times 10^{-6}\text{ F})\left(\frac{4\text{ V}}{2 \times 10^{-3}\text{ s}}\right)$$
$$= 4 \times 10^{-3}\text{ A} = \mathbf{4\ mA}$$

b. From 2 to 5 ms, the voltage remains constant at 4 V; the change in voltage is $\Delta v = 0$ V. The change in time is $\Delta t = 3$ ms, and

$$i_{C_{av}} = C\frac{\Delta v_C}{\Delta t} = C\frac{0}{\Delta t} = 0 \text{ A}$$

c. From 5 to 11 ms, the voltage decreases from 4 V to 0 V. The change in voltage Δv is therefore 4 V − 0 V = 4 V (with a negative sign, since the voltage is decreasing with time). The change in time is $\Delta t = 11$ ms − 5 ms = 6 ms, and

$$i_{C_{av}} = C\frac{\Delta v_C}{\Delta t} = -(2 \times 10^{-6} \text{ F})\left(\frac{4 \text{ V}}{6 \times 10^{-3} \text{ s}}\right)$$
$$= -1.33 \times 10^{-3} \text{ A} = \mathbf{-1.33 \text{ mA}}$$

d. From 11 ms on, the voltage remains constant at 0 and $\Delta v = 0$ V, so $i_{C_{av}} = 0$ A. The waveform for the average current for the impressed voltage is as shown in Fig. 10.49.

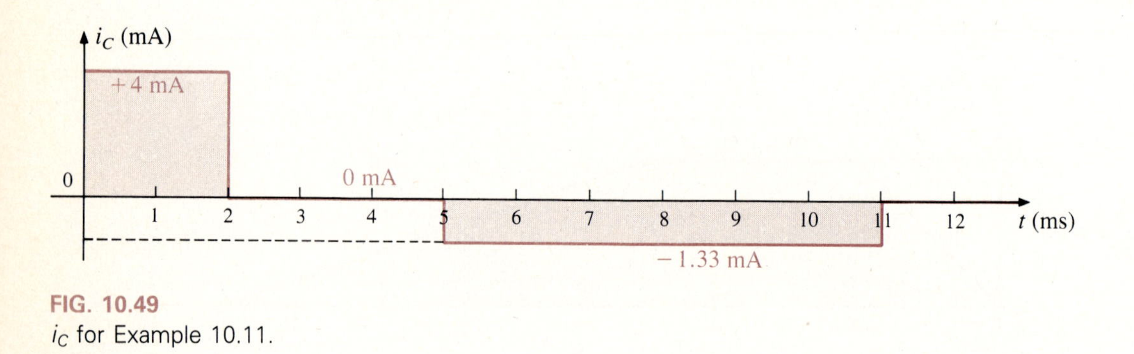

FIG. 10.49
i_C for Example 10.11.

In general, note in Example 10.11 that the steeper the slope, the greater the current, and when the voltage fails to change, the current is zero. In addition, the average value is the same as the instantaneous value at any point along the slope over which the average value was found. If we consider the limit as $\Delta t \Rightarrow 0$, the slope will still remain the same, and therefore $i_{C_{av}} = i_{C_{inst}}$ at any instant of time between 0 and t_1. The same can be said about any portion of the voltage waveform that has a constant slope.

An important point to be noted from this discussion is that it is not the magnitude of the voltage across a capacitor that determines the current but rather how quickly the voltage *changes* across the capacitor. An applied steady dc voltage of 10,000 V (ideally) does not create any flow of charge (current), but a change in voltage of 1 V in a very brief period of time could create a significant current.

The method just described is only for waveforms with straight-line (linear) segments. For nonlinear (curved) waveforms, a method of calculus (differentiation) must be employed.

10.12 CAPACITORS IN SERIES AND PARALLEL

Capacitors, like resistors, can be placed in series and parallel. Increasing levels of capacitance can be obtained by placing capacitors in parallel, and decreasing levels can be obtained by placing capacitors in series.

For capacitors in series, the charge is the same on each capacitor (Fig. 10.50):

$$Q_T = Q_1 = Q_2 = Q_3 \tag{10.23}$$

and the total capacitance is determined by

$$\frac{1}{C_T} = \frac{1}{C_1} + \frac{1}{C_2} + \frac{1}{C_3} \tag{10.24}$$

which is similar to the manner in which we found the total resistance of a parallel resistive circuit. The total capacitance of two capacitors in series is

$$C_T = \frac{C_1 C_2}{C_1 + C_2} \tag{10.25}$$

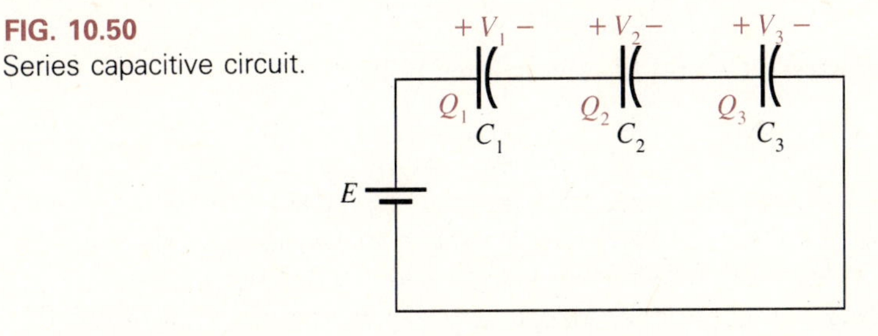

FIG. 10.50 Series capacitive circuit.

The voltages V_1, V_2, and V_3 are related to the charge on a capacitor and the capacitance by $V = Q/C$, or $V_1 = Q_1/C_1$, $V_2 = Q_2/C_2$, and $V_3 = Q_3/C_3$. Of course, $E = V_1 + V_2 + V_3$.

For capacitors in parallel, as shown in Fig. 10.51, the voltage is the same across each capacitor, and the total charge is the sum of that on each capacitor:

$$Q_T = Q_1 + Q_2 + Q_3 \tag{10.26}$$

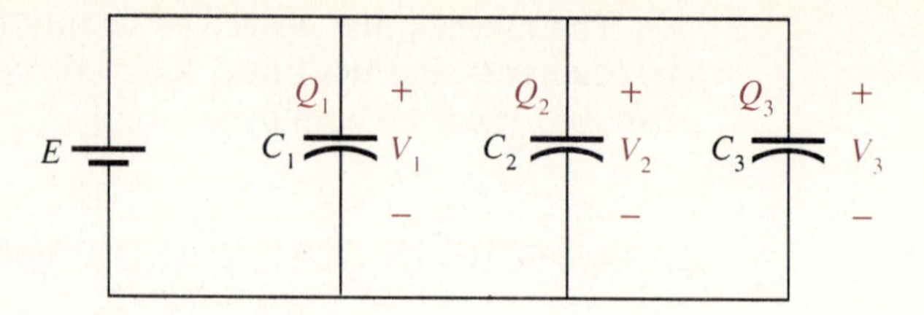

FIG. 10.51
Parallel capacitive circuit.

The total capacitance is determined by

$$C_T = C_1 + C_2 + C_3 \tag{10.27}$$

which is similar to the manner in which the total resistance of a series circuit is found.

EXAMPLE 10.12 Determine the total capacitance of the network of Fig. 10.52.

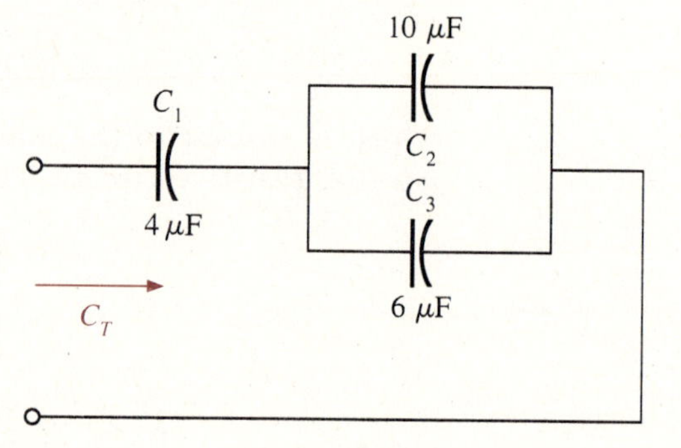

FIG. 10.52
Capacitive network for Example 10.12.

Solution C_2 and C_3 are in parallel:

$$\begin{aligned} C'_T &= C_2 + C_3 \\ &= 10\ \mu\text{F} + 6\ \mu\text{F} \\ C'_T &= 16\ \mu\text{F} \end{aligned}$$

C_1 is in series with C'_T:

$$\begin{aligned} C_T &= \frac{C_1 C'_T}{C_1 + C'_T} \\ &= \frac{(4\ \mu\text{F})(16\ \mu\text{F})}{4\ \mu\text{F} + 16\ \mu\text{F}} = \frac{64\ \mu\text{F}}{20} = \mathbf{3.2\ \mu F} \end{aligned}$$

EXAMPLE 10.13 For the circuit of Fig. 10.53:

a. Find the total capacitance.
b. Determine the charge on each plate.
c. Find the voltage across each capacitor.

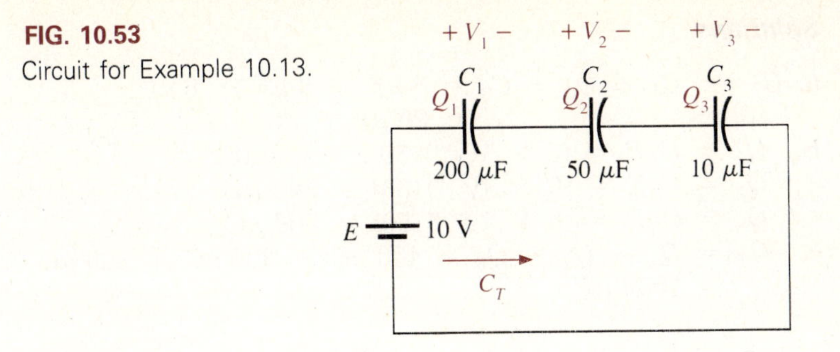

FIG. 10.53
Circuit for Example 10.13.

Solutions

a. $$\frac{1}{C_T} = \frac{1}{C_1} + \frac{1}{C_2} + \frac{1}{C_3}$$

$$= \frac{1}{200 \times 10^{-6}\text{ F}} + \frac{1}{50 \times 10^{-6}\text{ F}} + \frac{1}{10 \times 10^{-6}\text{ F}}$$

$$= 0.005 \times 10^6 + 0.02 \times 10^6 + 0.1 \times 10^6$$

$$\frac{1}{C_T} = 0.125 \times 10^6$$

$$C_T = \frac{1}{0.125 \times 10^6} = \mathbf{8\ \mu F}$$

b. $$Q_T = Q_1 = Q_2 = Q_3$$
$$= C_T E = (8 \times 10^{-6}\text{ F})(10\text{ V}) = \mathbf{80\ \mu C}$$

c. $$V_1 = \frac{Q_1}{C_1} = \frac{80 \times 10^{-6}\text{ C}}{200 \times 10^{-6}\text{ F}} = \mathbf{0.4\ V}$$

$$V_2 = \frac{Q_2}{C_2} = \frac{80 \times 10^{-6}\text{ C}}{50 \times 10^{-6}\text{ F}} = \mathbf{1.6\ V}$$

$$V_3 = \frac{Q_3}{C_3} = \frac{80 \times 10^{-6}\text{ C}}{10 \times 10^{-6}\text{ F}} = \mathbf{8.0\ V}$$

EXAMPLE 10.14

For the network of Fig. 10.54:

a. Find the total capacitance.
b. Determine the charge on each plate.
c. Find the total charge.

FIG. 10.54
Network for Example 10.14.

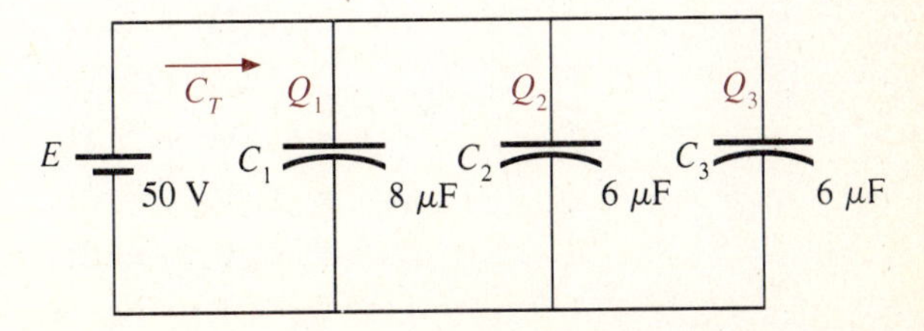

Solutions

a. $C_T = C_1 + C_2 + C_3 = 8\ \mu\text{F} + 6\ \mu\text{F} + 6\ \mu\text{F}$
$= \mathbf{20\ \mu F}$

b. $Q_1 = C_1E = (8 \times 10^{-6})(50) = \mathbf{400\ \mu C}$
$Q_2 = C_2E = (6 \times 10^{-6})(50) = \mathbf{300\ \mu C}$
$Q_3 = C_3E = (6 \times 10^{-6})(50) = \mathbf{300\ \mu C}$

c. $Q_T = Q_1 + Q_2 + Q_3 = 400\ \mu\text{C} + 300\ \mu\text{C} + 300\ \mu\text{C}$
$= \mathbf{1000\ \mu C}$

EXAMPLE 10.15 Find the voltage across and charge on capacitor C_1 of Fig. 10.55 after it has charged up to its final value.

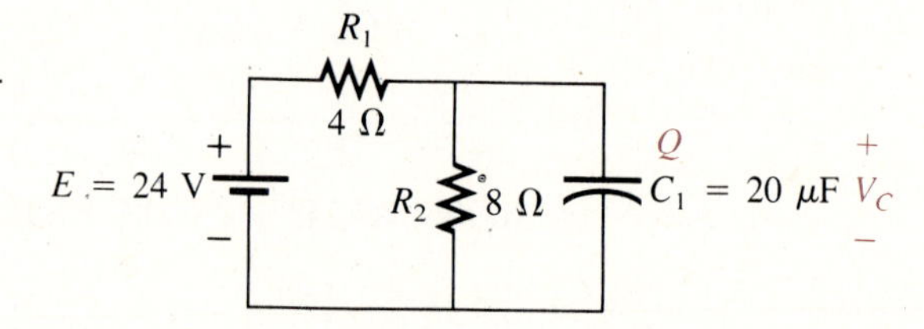

FIG. 10.55
Network for Example 10.15.

Solution As previously discussed, the capacitor is effectively an open circuit for dc after charging up to its final value (Fig. 10.56). Therefore,

$$V_C = \frac{(8\ \Omega)(24\ \text{V})}{4\ \Omega + 8\ \Omega} = \mathbf{16\ V}$$

$$Q_1 = C_1V_C = (20 \times 10^{-6}\ \text{F})(16\ \text{V})$$
$$= \mathbf{320\ \mu C}$$

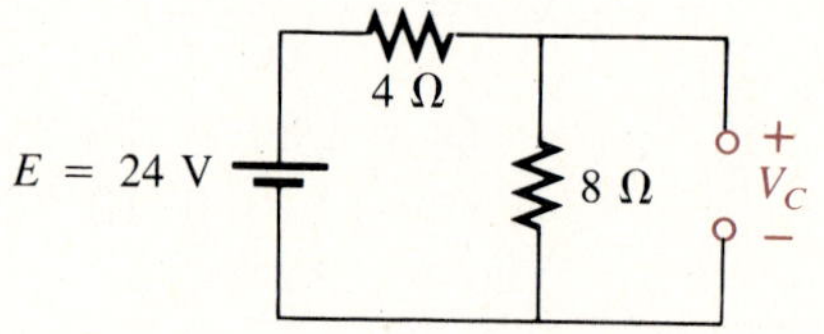

FIG. 10.56
Determining the steady-state value of V_C.

EXAMPLE 10.16 Find the voltage across and charge on each capacitor of the network of Fig. 10.57 after each has charged up to its final value.

Solution

$$V_{C_2} = \frac{(7\ \Omega)(72\ \text{V})}{7\ \Omega + 2\ \Omega} = \mathbf{56\ V}$$

$$V_{C_1} = \frac{(2\ \Omega)(72\ \text{V})}{2\ \Omega + 7\ \Omega} = \mathbf{16\ V}$$

$$Q_1 = C_1V_{C_1} = (2 \times 10^{-6}\ \text{F})(16\ \text{V}) = \mathbf{32\ \mu C}$$
$$Q_2 = C_2V_{C_2} = (3 \times 10^{-6}\ \text{F})(56\ \text{V}) = \mathbf{168\ \mu C}$$

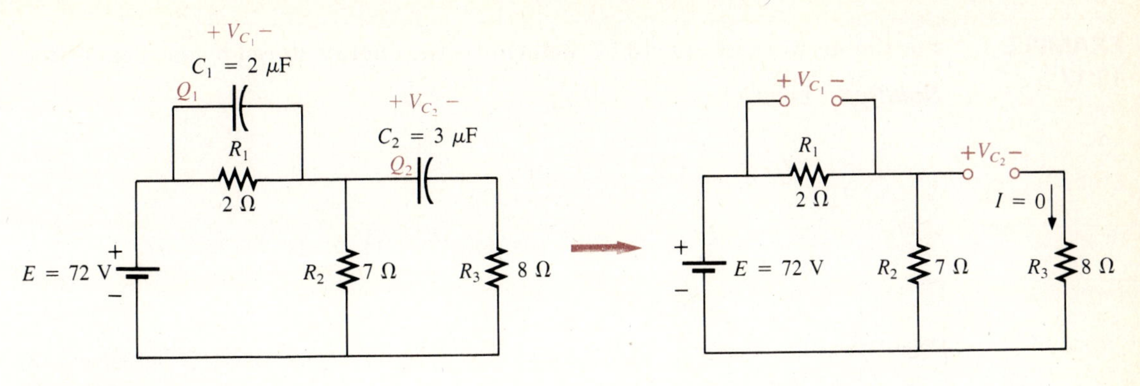

FIG. 10.57
Network for Example 10.16.

10.13 ENERGY STORED BY A CAPACITOR

The ideal capacitor does not dissipate any of the energy supplied to it. It stores the energy in the form of an electric field between the conducting surfaces. A plot of the voltage, current, and power to a capacitor during the charging phase is shown in Fig. 10.58. The power curve can be obtained by finding the product of the voltage and current at selected intervals of time and connecting the points obtained. The energy stored is represented by the shaded area under the power curve. Using calculus, we can determine the area under the curve:

$$W_C = \frac{1}{2}CE^2$$

In general,

$$W_C = \frac{1}{2}CV^2 \tag{10.28}$$

where V is the steady-state voltage across the capacitor.

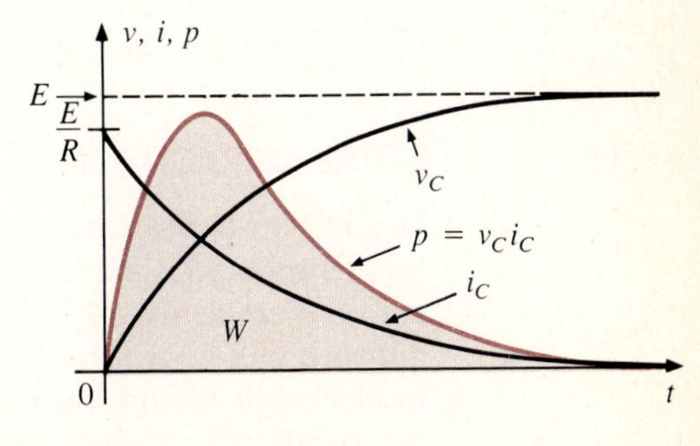

FIG. 10.58
Determining the energy stored by a capacitor.

EXAMPLE 10.17 For the network of Fig. 10.57, determine the energy stored by each capacitor.

Solution For C_1,

$$W_C = \frac{1}{2}CV^2$$

$$= \frac{1}{2}(2 \times 10^{-6}\text{ F})(16\text{ V})^2 = (1 \times 10^{-6})(256)\text{ J}$$

$$W_C = \mathbf{256\ \mu J}$$

Calculator

(1) (÷) (2) (×) (2) (EXP) (+/−) (6) (×) (1) (6) (x^2) (=) (F <−> E)

with a display of 2.56−04.

For C_2,

$$W_C = \frac{1}{2}CV^2$$

$$= \frac{1}{2}(3 \times 10^{-6}\text{ F})(56\text{ V})^2 = (1.5 \times 10^{-6})(3136)\text{ J}$$

$$W_C = \mathbf{4704\ \mu J}$$

Calculator

(1) (÷) (2) (×) (3) (EXP) (+/−) (6) (×) (5) (6) (x^2) (=) (F <−> E)

with a display of 4.704−03.

Due to the squared term, note the difference in energy stored due to a higher voltage.

Since capacitors can store the energy accumulated for a period of time determined by the leakage resistance of the capacitor, it is often important to release this energy before picking up or using the element. This is easily accomplished using a spare lead or screwdriver, as shown in Fig. 10.59.

10.14 STRAY CAPACITANCES

In addition to the capacitors discussed so far in this chapter, there are stray capacitances that exist not through design but simply because two conducting surfaces are relatively close to each other. Two conducting wires in the same network will have a capacitive effect between them, as shown in Fig. 10.60(a). In electronic circuits, capacitance levels exist between conducting surfaces of the transistor as shown in Fig. 10.60(b). In Chapter 12 we discuss another

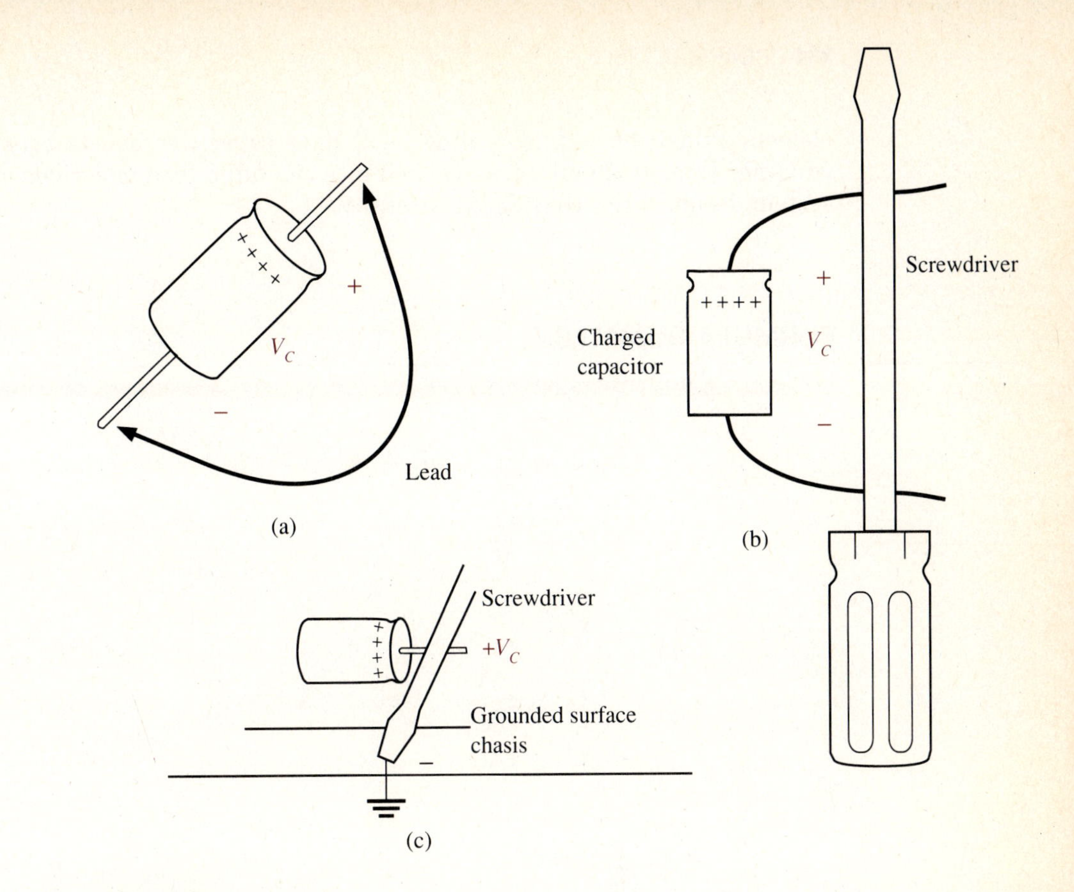

FIG. 10.59
Discharging a capacitor.

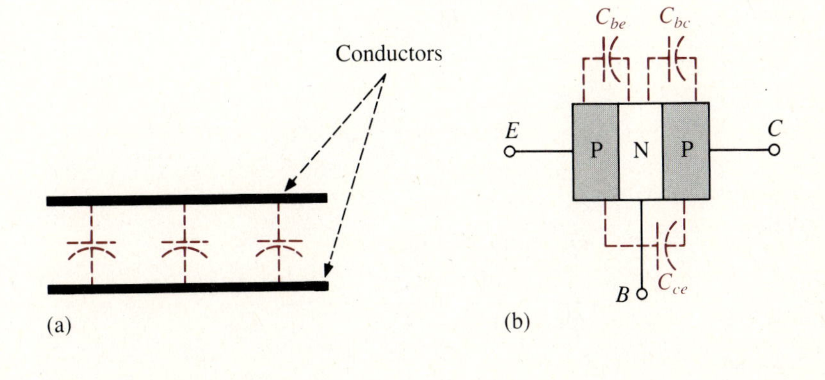

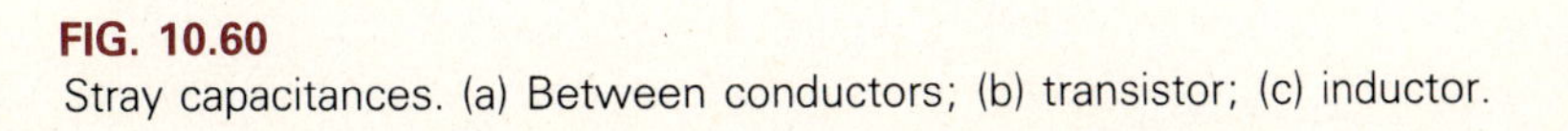
FIG. 10.60
Stray capacitances. (a) Between conductors; (b) transistor; (c) inductor.

element called the *inductor,* which will have capacitive effects between the windings [Fig. 10.60(c)]. Stray capacitance can often lead to serious errors in system design if not considered carefully.

FORMULA SUMMARY

$$\mathscr{E} = \frac{F}{Q}, \qquad \mathscr{E} = k\frac{Q}{r^2}, \qquad \mathscr{E} = \frac{V}{d}$$

$$C = \frac{Q}{V}$$

$$\epsilon_r = \frac{\epsilon}{\epsilon_o}$$

$$C = \epsilon\frac{A}{d} = 8.85 \times 10^{-12}\epsilon_r\frac{A}{d}$$

$$C = \epsilon_r C_o$$

Charging Phase

$$i_C = \frac{E}{R}e^{-t/RC}, \qquad v_C = E(1 - e^{-t/RC}), \qquad v_R = i_C R$$

$$\tau = RC$$

Discharge Phase

$$i_C = \frac{E}{R}e^{-t/\tau}, \qquad v_C = v_R = Ee^{-t/\tau}$$

Capacitive Current

$$i_C = C\frac{dv_C}{dt}, \qquad i_{C_{av}} = C\frac{\Delta v_C}{\Delta t}$$

Series Capacitors

$$Q_T = Q_1 = Q_2 = Q_3$$

$$\frac{1}{C_T} = \frac{1}{C_1} + \frac{1}{C_2} + \frac{1}{C_3} + \cdots + \frac{1}{C_N}, \qquad C_T = \frac{C_1 C_2}{C_1 + C_2}$$

Parallel Capacitors

$$Q_T = Q_1 + Q_2 + Q_3$$

$$C_T = C_1 + C_2 + C_3 + \cdots + C_N$$

$$W_C = \frac{1}{2}CV^2$$

CHAPTER SUMMARY

The ideal capacitor is quite different from the resistor in that the capacitor stores the energy delivered to it rather than dissipating it in the form of heat. In addition, the voltage across a capacitor cannot change instantaneously, although it can change instantaneously across a resistor. Further, capacitive current is directly related to the *rate* of change of voltage across the capacitor, whereas the current of a resistor is simply a direct function of the magnitude of the applied voltage. It is also important to keep in mind that a capacitor has the characteristics of an open circuit when steady-state conditions are established—an important characteristic for isolating dc stages of an electrical system.

The transient behavior of a capacitor can be plotted with an acceptable level of accuracy without getting deeply involved with the exponential equations. Simply recall the open-circuit equivalent at steady state to determine the final values and utilize the fact that steady-state conditions are established after five time constants. Using the fact that the transition is 63.2% complete in one time constant and effectively zero or the maximum value after 5τ usually permits the drawing of a fairly accurate response curve. Instantaneous values may be determined using the universal curve or equations provided in the text. No need to memorize the equations for the instantaneous time t, since they can be easily generated using basic algebraic manipulations or simply referencing the text when required.

As noted in the examples, the capacitive current is a maximum when the slope of the curve for v_C is a maximum (increasing or decreasing with time). If v_C fails to change, the current is zero, and if v_C drops with time, a negative sign is applied to the resulting current. Too frequently students get so involved with the $\Delta v_C/\Delta t$ term that they forget to multiply the result by the capacitance of the capacitor.

Finally, keep in mind that the equations for capacitors in series match those for resistors in parallel, whereas for capacitors in parallel the equations for series resistors are similar. However, note that Q is the same for series capacitors (like current for series resistors) with the total Q_T of parallel capacitors equal to the sum of the charge levels of each capacitor (like $I_T = I_1 + I_2 + I_3$ for parallel resistors).

GLOSSARY

Breakdown voltage Another term for *dielectric strength.*

Capacitance A measure of a capacitor's ability to store charge; measured in farads (F).

Capacitive time constant The product of resistance and capacitance that establishes the required time for the charging and discharging phases of a capacitive transient.

Capacitive transient The waveforms for the voltage and current of a capacitor that result during the charging and discharging phases.

Capacitor A fundamental electrical element having two conducting surfaces separated by an insulating material and having the capacity to store charge on its plates.

Dielectric The insulating material between the plates of a capacitor that can have a pronounced effect on the charge stored on the plates of a capacitor.

Dielectric constant Another term for *relative permittivity.*

Dielectric strength An indication of the voltage required for unit length to establish conduction in a dielectric.

Electric field strength The force acting on a unit positive charge in the region of interest.

Electric flux lines Lines drawn to indicate the strength and direction of an electric field in a particular region.

Fringing An effect established by flux lines that do not pass directly from one conducting surface to another.

Leakage current The current that will result in the total discharge of a capacitor if the capacitor is disconnected from the charging network for a sufficient length of time.

Permittivity A measure of how well a dielectric will *permit* the establishment of flux lines within the dielectric.

Relative permittivity The permittivity of a material compared to that of air.

Stray capacitance Capacitances that exist not through design but simply because two conducting surfaces are relatively close to each other.

Surge voltage The maximum voltage that can be applied across the capacitor for very short periods of time.

Working voltage The voltage that can be applied across a capacitor for long periods of time without concern for dielectric breakdown.

PROBLEMS

Section 10.2

1. Find the electric field strength at a point 2 m from a charge of 4 μC.
2. The electric field strength is 36 N/C at a point r meters from a charge of 0.064 μC. Find the distance r.

Section 10.3

3. Find the capacitance of a parallel plate capacitor if 1400 μC of charge are deposited on its plates when 20 V are applied across the plates.
4. How much charge is deposited on the plates of a 0.05-μF capacitor if 45 V are applied across the capacitor?
5. Find the electric field strength between the plates of a parallel plate capacitor if 100 mV are applied across the plates and the plates are 2 mm apart.
6. Repeat Problem 5 if the plates are separated by 4 mils.
7. A 4-μF parallel plate capacitor has 160 μC of charge on its plates. If the plates are 5 mm apart, find the electric field strength between the plates.
8. Find the capacitance of a parallel plate capacitor if the area of each plate is 0.08 m^2 and the distance between the plates is 2 mm. The dielectric is air.

9. Repeat Problem 8 if the dielectric is paraffin-coated paper.

10. Find the distance in mils between the plates of a 2-μF capacitor if the area of each plate is 0.09 m^2 and the dielectric is transformer oil.

11. The capacitance of a capacitor with a dielectric of air is 1200 pF. When a dielectric is inserted between the plates, the capacitance increases to 0.006 μF. Of what material is the dielectric made?

12. The plates of a parallel plate air capacitor are 0.2 mm apart and have an area of 0.08 m^2, and 200 V are applied across the plates.

a. Determine the capacitance.
b. Find the electric field intensity between the plates.
c. Find the charge on each plate if the dielectric is air.

13. A sheet of Bakelite 0.2 mm thick having an area of 0.08 m^2 is inserted between the plates of Problem 12.

a. Find the electric field strength between the plates.
b. Determine the capacitance.
c. Determine the charge on each plate.

Section 10.4

14. Find the maximum voltage ratings of the capacitors of Problems 12 and 13 assuming a linear relationship between the breakdown voltage and the thickness of the dielectric.

***15.** Find the maximum voltage that can be applied across a parallel plate capacitor of 0.006 μF. The area of one plate is 0.02 m^2 and the dielectric is mica. Assume a linear relationship between the dielectric strength and the thickness of the dielectric.

16. Find the distance in millimeters between the plates of a parallel plate capacitor if the maximum voltage that can be applied across the capacitor is 1250 V. The dielectric is mica. Assume a linear relationship between the breakdown strength and the thickness of the dielectric.

Section 10.7

17. For the circuit of Fig. 10.61:

a. Determine the time constant of the circuit.
b. Write the mathematical equation for the voltage v_C following the closing of the switch.
c. Determine the voltage v_C after one, three, and five time constants.
d. Sketch the waveforms for v_C and i_C.
e. Write the equations for the current i_C and the voltage v_R.

FIG. 10.61

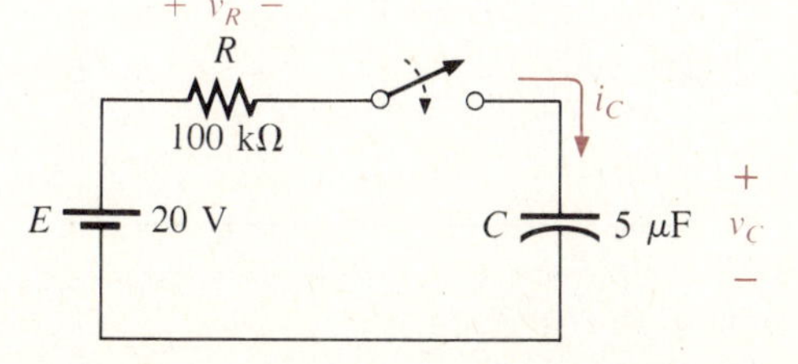

18. Repeat Problem 17 for $R = 1\ \text{M}\Omega$ and compare results.

19. Repeat Problem 17 for the network of Fig. 10.62.

*20. For the circuit of Fig. 10.63:

a. Determine the time constant of the circuit.

b. Write the mathematical expression for the voltage v_C following the closing of the switch.

c. Write the mathematical expression for the current i_C following the closing of the switch.

d. Sketch the waveforms of v_C and i_C.

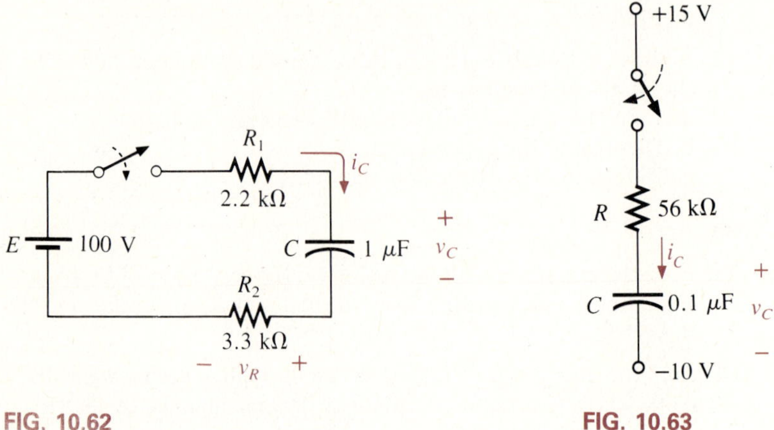

FIG. 10.62

FIG. 10.63

Section 10.8

21. For the circuit of Fig. 10.64:

a. Determine the time constant of the circuit when the switch is thrown into position 1.

b. Sketch the waveform of v_C and i_C during the charging phase.

c. Write the mathematical expressions for v_C and i_C.

d. Determine the time constant when the switch is thrown into position 2.

e. Sketch the discharge waveforms for v_C and i_C.

f. Write the mathematical expressions for v_C and i_C during the discharge phase.

FIG. 10.64

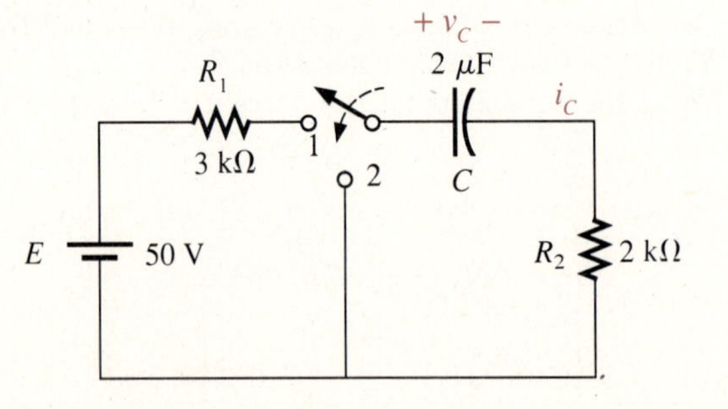

*22. The capacitor of Fig. 10.65 is initially charged to 40 V before the switch is closed. Sketch the voltages v_C and v_R and the current i_C for the decay phase.

23. The 1000-μF capacitor of Fig. 10.66 is charged to 6 V. To discharge the capacitor before further use, a wire with a resistance of 0.002 Ω is placed across the capacitor.
 a. How long will it take to discharge the capacitor?
 b. What is the peak value of the current?
 c. Based on the result of part b, is a spark expected when contact is made with both ends of the capacitor?

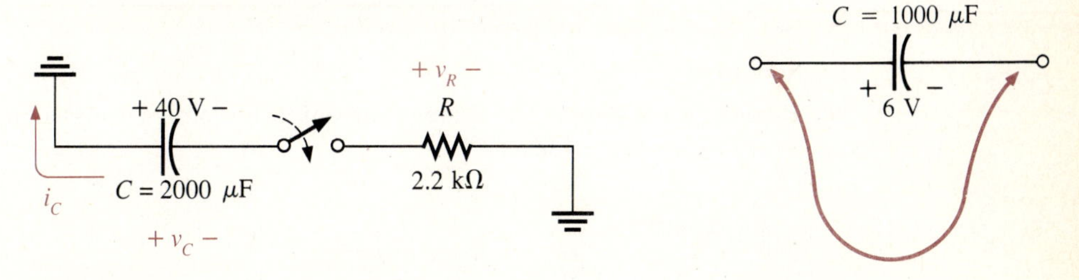

FIG. 10.65

FIG. 10.66

Section 10.9

24. Given the expression $v_C = 8(1 - e^{-t/(20 \times 10^{-6})})$:
 a. Determine v_C after 5 time constants.
 b. Determine v_C after 10 time constants.
 c. Determine v_C at $t = 5\ \mu$s.

25. For the situation of Problem 23, determine when the discharge current is one-half its maximum value if contact is made at $t = 0$ s.

26. Given $i_C = 10 \times 10^{-3} e^{-t/20\text{ ms}}$, determine the time required for i_C to drop to one-half its peak value.

27. Given $v_C = 20e^{-t/10\ \mu\text{s}}$, determine when v_C will drop to 2 V.

Section 10.10

28. For the circuit of Fig. 10.67:
 a. Sketch the waveforms of v_C and i_C following the closing of the switch at $t = 0$ s.
 b. How long will it take v_C to reach 6 V?

FIG. 10.67

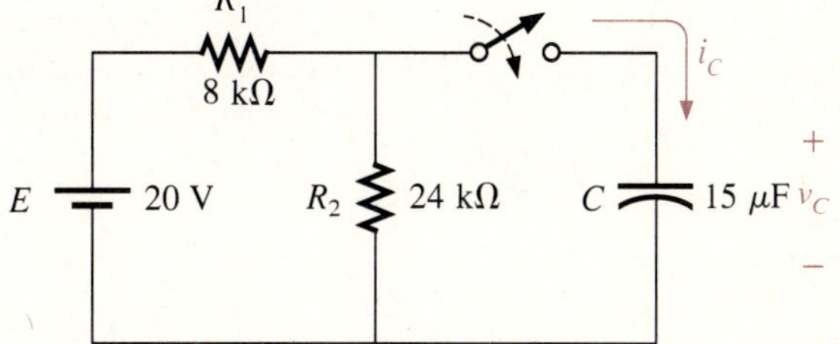

*29. Repeat Problem 28 for the circuit of Fig. 10.68.

FIG. 10.68

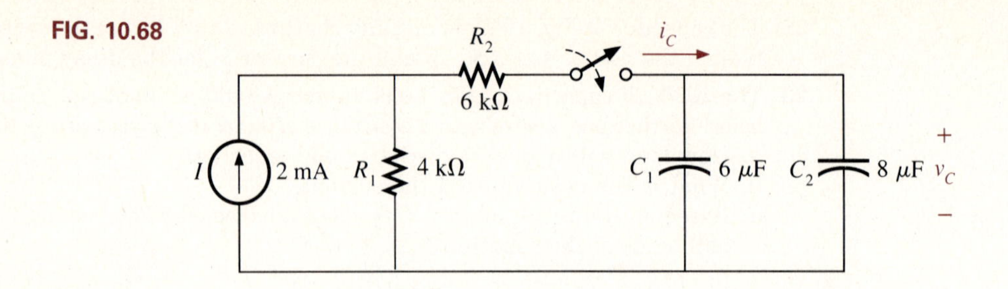

Section 10.11

30. Find the waveform for the average current if the voltage across a 6-μF capacitor is as shown in Fig. 10.69.

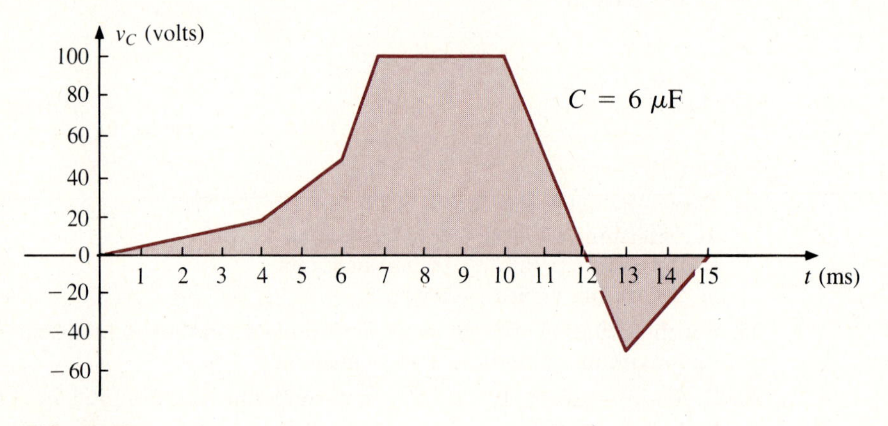

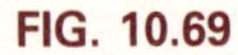
FIG. 10.69

31. Find the waveform for the average current if the voltage across a 20-μF capacitor is as shown in Fig. 10.70.

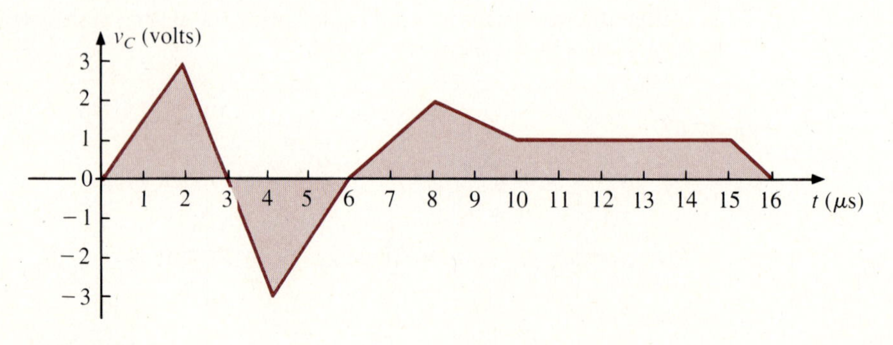

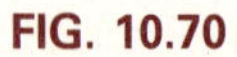
FIG. 10.70

Section 10.12

32. Find the total capacitance C_T between points a and b of the circuits of Fig. 10.71.

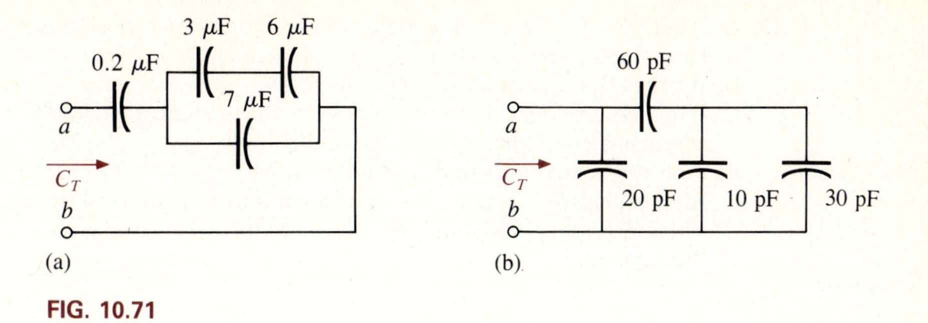

FIG. 10.71

33. Find the voltage across and charge on each capacitor for the circuits of Fig. 10.72.

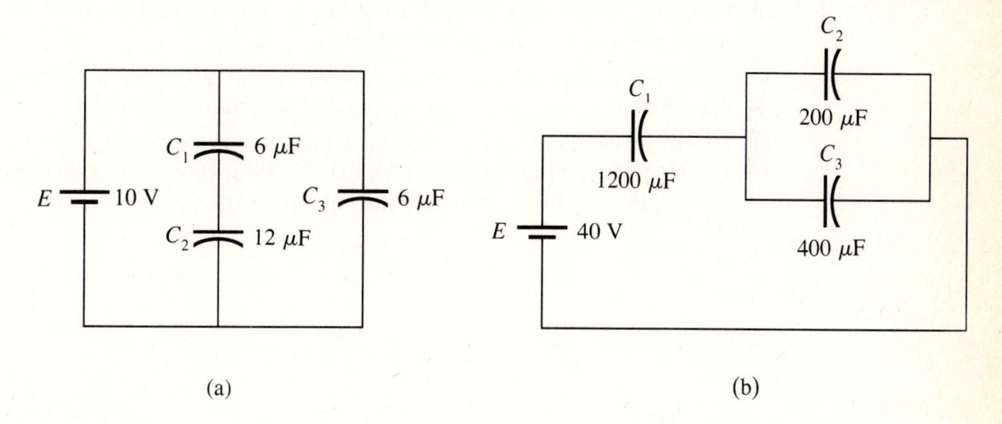

FIG. 10.72

34. For the circuits of Fig. 10.73, find the voltage across and charge on each capacitor after each capacitor has charged to its final value.

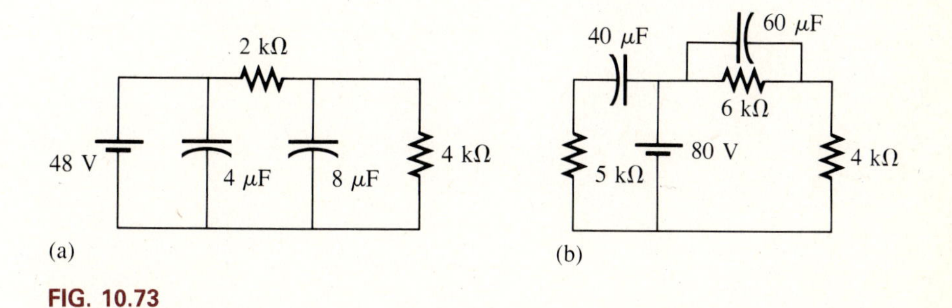

FIG. 10.73

Section 10.13

35. Find the energy stored by a 120-pF capacitor with 12 V across its plates.

36. If the energy stored by a 6-μF capacitor is 1200 J, find the charge Q on each plate of the capacitor.

*37. An electronic flashgun has a 1000-μF capacitor, which is charged to 100 V.
   a. How much energy is stored by the capacitor?
   b. What is the charge on the capacitor?
   c. When the photographer takes a picture, the flash fires for 1/2000 s. What is the average current through the flashtube?
   d. Find the power delivered to the flashtube.
   e. After a picture is taken, the capacitor has to be recharged by a power supply, which delivers a maximum current of 10 mA. How long will it take to charge the capacitor?

Computer Problems

38. Given the area of a capacitor in square meters, the distance between the plates in *meters*, and the permittivity of the dielectrics, write a program to determine the capacitance of the capacitor in microfarads.

39. Write a program to tabulate the voltage v_C if $v_C = 10(1 - e^{-t/1\text{ ms}})$ for each time constant from 1 to 20 time constants.

40. Given a standard configuration such as Fig. 10.20, write a program to generate the mathematical expression for v_C and i_C given E, R, and C.

41. Given two capacitors in any series or parallel arrangement with the applied voltage E, write a program to determine the total capacitance, voltage across each capacitor, and charge on the plates of each capacitor.

11

Magnetic Circuits and Inductors

OBJECTIVES

- ☐ Become familiar with the use of magnetic flux lines to reveal the strength of a magnetic field and how magnetic materials interact.
- ☐ Understand how an electromagnet is established and how its properties match those of a permanent magnet.
- ☐ Become aware of the general characteristics of magnetic materials.
- ☐ Determine the required *NI* or the resultant flux Φ for series magnetic circuits.
- ☐ Understand the impact of Faraday's and Lenz's laws.
- ☐ Become aware of the construction, characteristics, and types of inductors.
- ☐ Be able to sketch the important transient response curves of inductive networks.
- ☐ Understand how both the capacitor and inductor behave in dc networks and how the energy stored by each is determined.

11.1 INTRODUCTION

Magnetism plays an integral part in almost every electrical device used today in industry, research, or the home. Generators, motors, transformers, circuit breakers, televisions, computers, tape recorders, and telephones all employ magnetic effects to perform a variety of important tasks. This chapter reviews the basic concept of magnetism and how it is employed in the construction and design of an important two-terminal device called the *inductor* (or *coil*). The response characteristics of the inductor are similar in some respects to those of the capacitor, although its construction and areas of application are quite different from those of the capacitor. The inductor will join the resistor and capacitor as elements of primary concern in the chapters to follow.

There is a great deal of similarity between the analyses of electric circuits and magnetic circuits. This is demonstrated late in this chapter when we compare the basic equations and methods used to solve magnetic circuits with those used for electric circuits.

11.2 MAGNETIC FIELDS

In the region surrounding a permanent magnet, there exists a magnetic field, which can be represented by magnetic flux lines similar to electric flux lines. Magnetic flux lines, however, do not have origins or terminating points like electric flux lines but exist in continuous loops, as shown in Fig. 11.1. The symbol for magnetic flux is Φ (capital Greek phi).

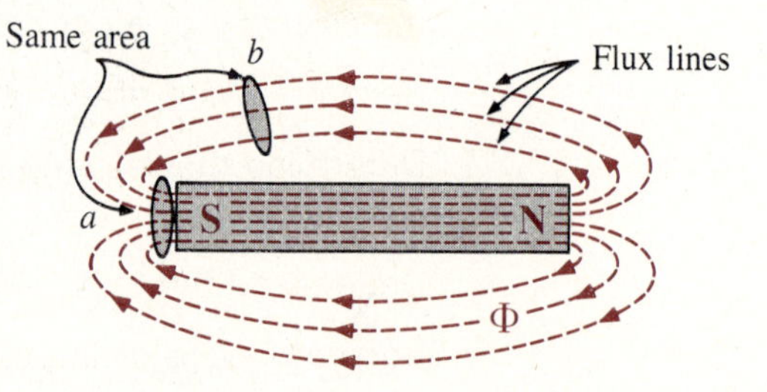

FIG. 11.1
Distribution of a permanent magnet.

The magnetic flux lines radiate from the north pole to the south pole, returning to the north pole through the metallic bar. Note the equal spacing between the flux lines within the core and the symmetric distribution outside the magnetic material. It is also important to realize that the continuous magnetic flux line will strive to occupy as small an area as possible. This will result in magnetic flux lines of minimum length between the like poles, as shown in Fig. 11.2. The strength of a magnetic field in a particular region is directly related to the density of flux lines in that region. In Fig. 11.1, for example, the magnetic field strength at a is twice that at b, since there are twice as many magnetic flux lines associated with the perpendicular area at a than at b. Recall from childhood experiments how the strength of permanent magnets was always stronger near the poles.

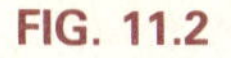
FIG. 11.2
Flux distribution of two permanent magnets with adjoining opposite poles.

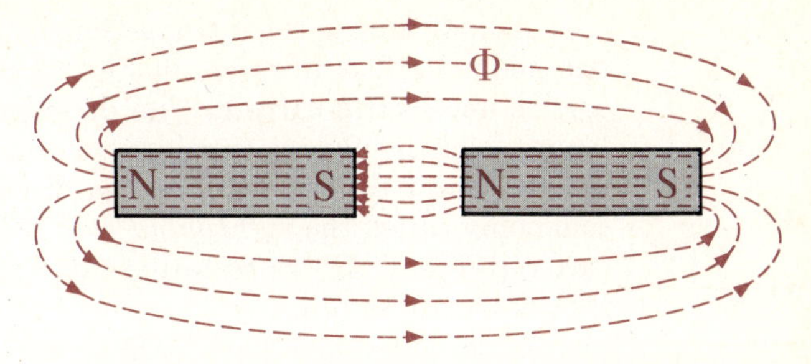

If unlike poles of two permanent magnets are brought together, the magnets will attract, and the flux distribution will be as shown in Fig. 11.2. If like poles are brought together, the magnets will repel, and the flux distribution will be as shown in Fig. 11.3.

FIG. 11.3
Flux distribution of two permanent magnets with adjoining similar poles.

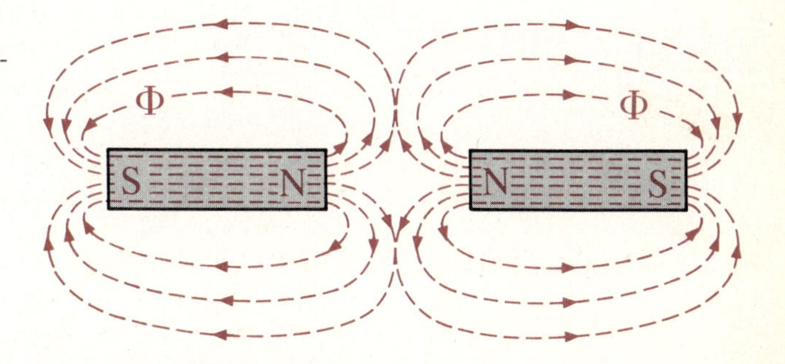

If a nonmagnetic material, such as glass or copper, is placed in the flux paths surrounding a permanent magnet, there will be an almost unnoticeable change in the flux distribution (Fig. 11.4). However, if a magnetic material, such as soft iron, is placed in the flux path, the flux lines will pass through the soft iron rather than through the surrounding air because flux lines pass with greater ease through magnetic materials than through air. This principle is put to use in the shielding of sensitive electrical elements and instruments that can be affected by stray magnetic fields (Fig. 11.5).

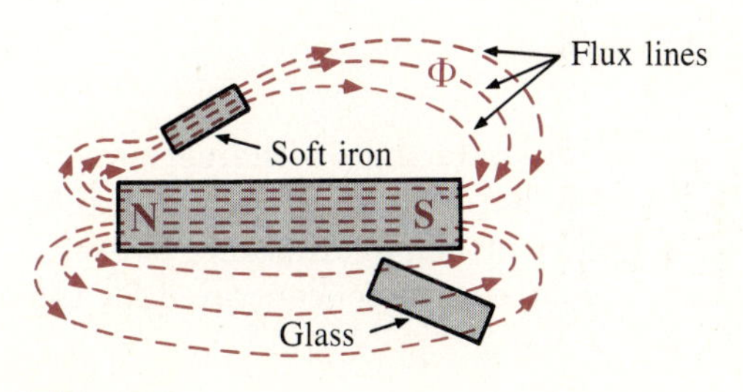

FIG. 11.4

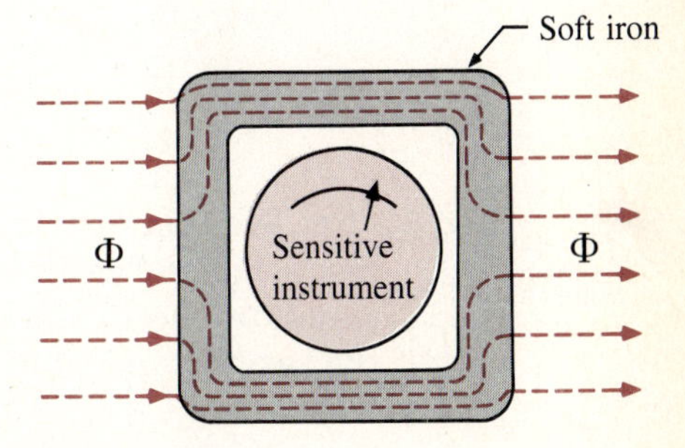

FIG. 11.5
Magnetic shield.

As indicated in the introduction, a magnetic field (represented by concentric magnetic flux lines, as in Fig. 11.6) is present around every wire that carries an electric current. The direction of the magnetic flux lines can be found simply by placing the thumb of the *right* hand in the direction of *conventional* current flow and noting the direction of the fingers. (This method is commonly called the *right-hand rule*.) If the conductor is wound in a single-turn coil (Fig. 11.7), the resulting flux will flow in a common direction through the center of the coil.

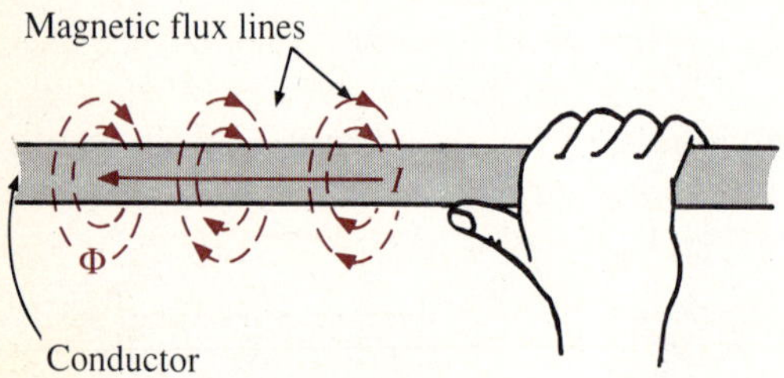

Magnetic flux lines

I

Φ

Conductor

FIG. 11.6
Applying the right-hand rule.

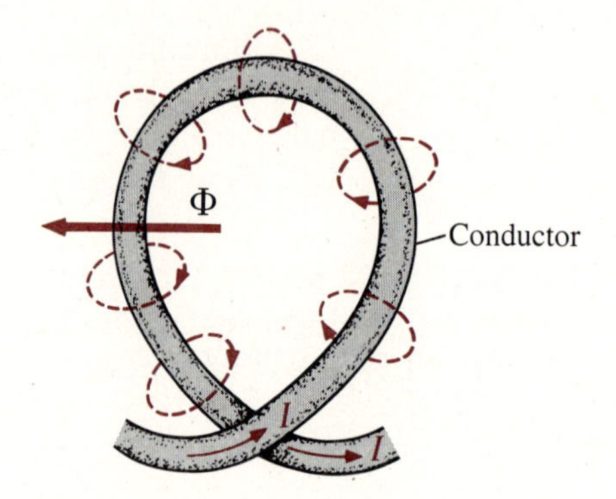

FIG. 11.7
Net flux direction within a winding.

A coil of more than one turn produces a magnetic field that exists in a continuous path through and around the coil (Fig. 11.8).

FIG. 11.8
Flux distribution through a coil.

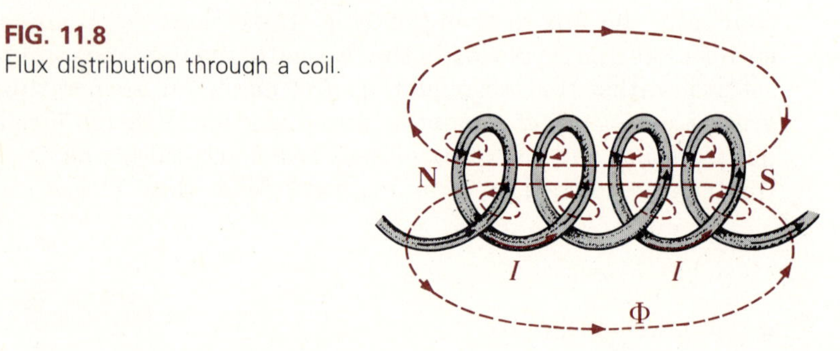

The flux distribution of a coil is quite similar to that of the permanent magnet. The flux lines leaving the coil from the left and entering to the right simulate a north and south pole, respectively. The principal difference between the two flux distributions is that the flux lines are more concentrated for the permanent magnet than for the coil. Also, since the strength of a magnetic field is determined by the density of the flux lines, the coil has a weaker field strength. The field strength of the coil can be effectively increased by placing certain materials, such as iron, steel, or cobalt, within the coil to increase the flux density within the coil. By increasing the field strength with the addition of the core, we have devised an *electromagnet* (Fig. 11.9), which, in addition to having all the properties of a permanent magnet, also has a field strength that can be varied by changing one of the component values (current, turns, and so on). Of course, current must pass through the coil of the electromagnet in order for magnetic flux to be developed, whereas there is no need for the coil or current in the permanent magnet. The direction of flux lines can be determined for the electromagnet (or in any core with a wrapping of turns) by placing the fingers of the right hand in the direction of current flow around the core. The thumb will then point in the direction of the north pole of the induced magnetic flux. This is demonstrated in Fig. 11.10. A cross section of the same electromagnet is included in the figure to introduce the convention for directions perpendicular to the page. The cross and dot refer to the tail and head of the arrow, respectively.

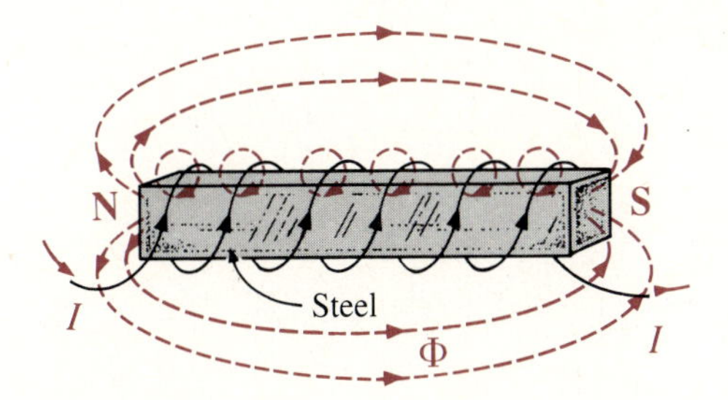

FIG. 11.9
Flux distribution of an electromagnet.

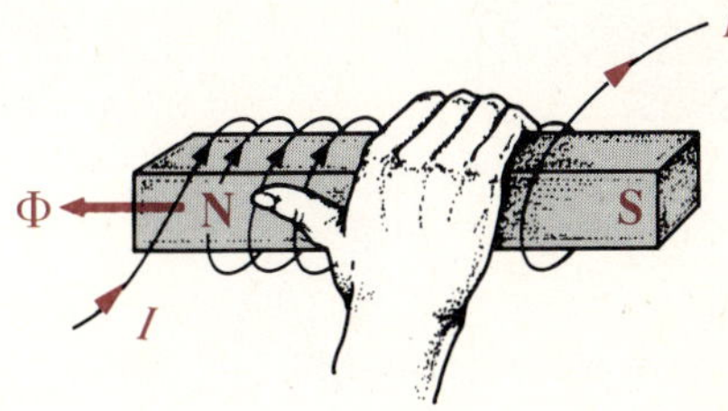

(a)

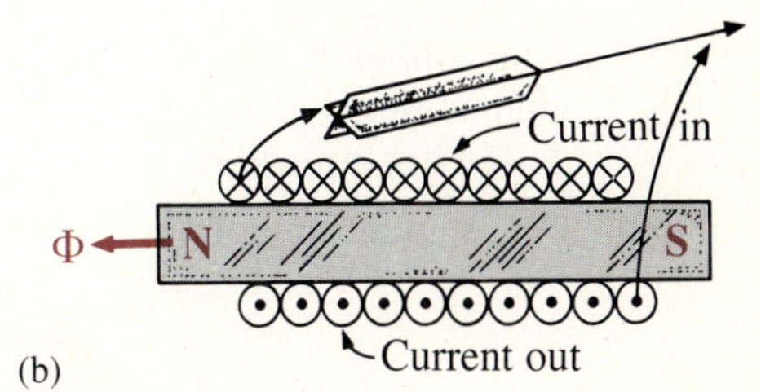

(b)

FIG. 11.10
Determining the flux direction of an electromagnet.

Other areas of application for electromagnetic effects are shown in Fig. 11.11. The flux path for each is indicated in each figure.

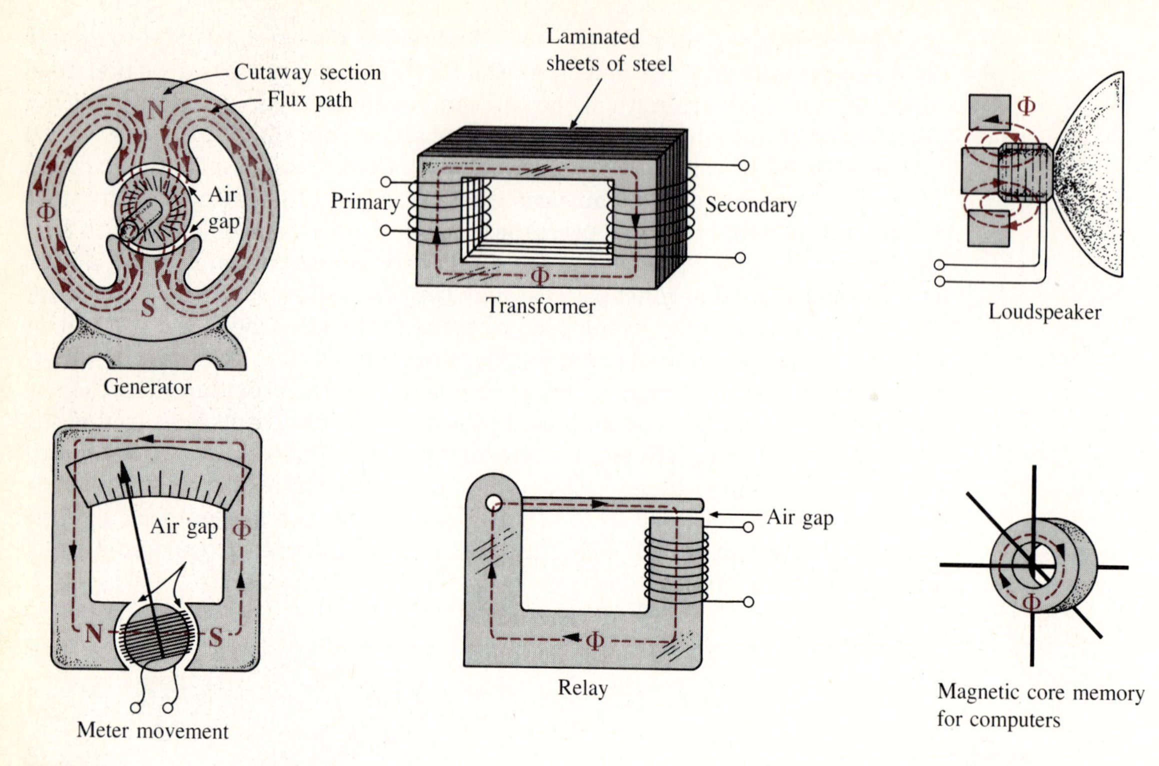

FIG. 11.11
Applications of magnetic effects.

11.3 FLUX DENSITY

In the SI system of units, magnetic flux is measured in webers (Wb) and has the symbol Φ. The number of flux lines per unit area is called the *flux density* and is denoted by the capital letter B. Its magnitude is determined by the following equation:

$$\boxed{B = \frac{\Phi}{A}} \qquad \begin{aligned} B &= \text{teslas (T)} \\ \Phi &= \text{webers (Wb)} \\ A &= \text{square meters (m}^2\text{)} \end{aligned} \tag{11.1}$$

where Φ is the number of flux lines passing through the area A (Fig. 11.12). The flux density at position a in Fig. 11.1 is twice that at b because twice as many flux lines are passing through the same area.

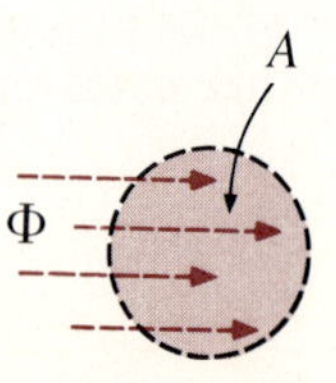

FIG. 11.12
The parameters that determine flux density.

As noted in Eq. 11.1, magnetic flux density in the SI system of units is measured in *teslas,* for which the symbol is T. By definition,

$$1\ \text{T} = 1\ \text{Wb/m}^2$$

EXAMPLE 11.1 For the core of Fig. 11.13, determine the flux density B in teslas.

FIG. 11.13
Flux path for Example 11.1.

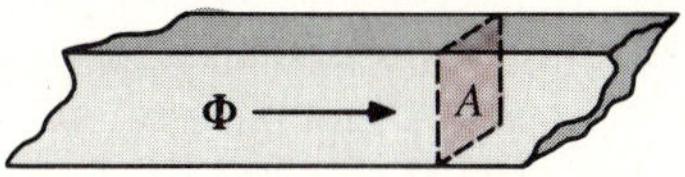

$\Phi \longrightarrow$ A

$$\Phi = 6 \times 10^{-5}\ \text{Wb}$$
$$A = 1.2 \times 10^{-3}\ \text{m}^2$$

Solution

$$B = \frac{\Phi}{A} = \frac{6 \times 10^{-5}\ \text{Wb}}{1.2 \times 10^{-3}\ \text{m}^2} = \mathbf{5 \times 10^{-2}\ T}$$

An instrument designed to measure flux density in gauss (CGS system) is shown in Fig. 11.14. Appendix F indicates that 1 Wb/m^2 = 10^4 gauss. The magnitude of the reading appearing on the face of the meter in Fig. 11.14 is, therefore,

$$1.964\ \cancel{\text{gauss}}\left(\frac{1\ \text{Wb/m}^2}{10^4\ \cancel{\text{gauss}}}\right) = 1.964 \times 10^{-4}\ \text{Wb/m}^2$$
$$= \mathbf{1.964 \times 10^{-4}\ T}$$

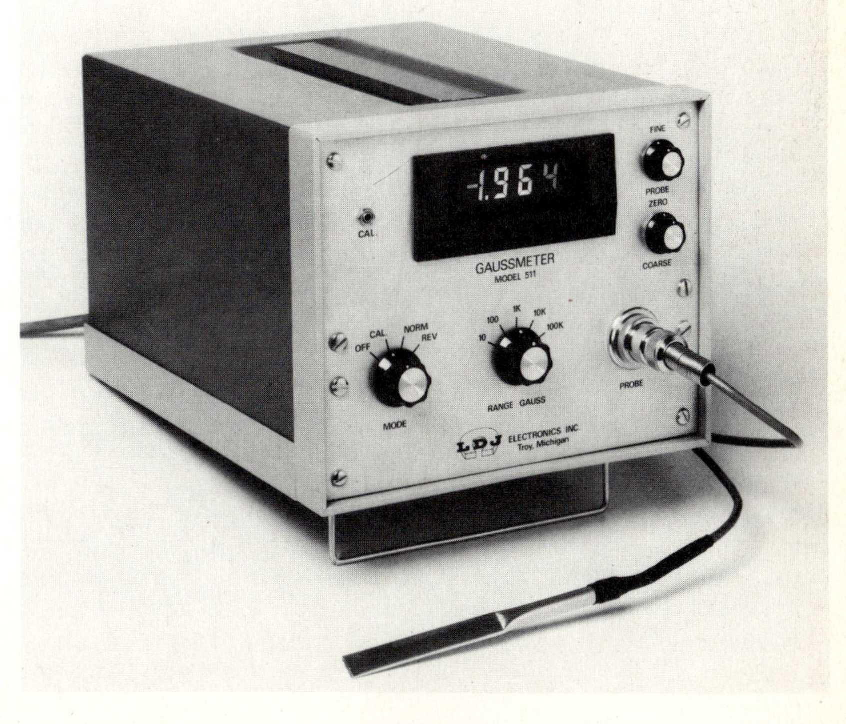

FIG. 11.14
Digital display gaussmeter. (Courtesy of LDJ Electronics, Inc.)

11.4 PERMEABILITY

If cores of different materials with the same physical dimensions are used in the electromagnet described in Section 11.2, the strength of the magnet will vary in accordance with the core used. This variation in strength is due to the greater or lesser number of flux lines passing through the core. Materials in which flux lines can readily be set up are said to be *magnetic* and to have *high permeability*. The permeability (μ) of a material, therefore, is a measure of the ease with which magnetic flux lines can be established in the material. It is similar in many respects to conductivity in electric circuits. The permeability of free space, μ_o (vacuum), is

$$\mu_o = 4\pi \times 10^{-7} \text{ Wb/A·m}$$

As indicated, μ has the units of webers per ampere-meter. Practically speaking, the permeability of all nonmagnetic materials, such as copper, aluminum, wood, glass, and air, is the same as for free space. Materials that have permeabilities slightly less than that of free space are said to be *diamagnetic,* and those with permeabilities slightly greater than that of free space are said to be *paramagnetic*. Magnetic materials, such as iron, nickel, steel, cobalt, and alloys of these metals, have permeabilities hundreds and even thousands of times that of free space. Materials with these very high permeabilities are referred to as *ferromagnetic*.

The ratio of the permeability of a material to that of free space is called its *relative permeability;* that is,

$$\boxed{\mu_r = \frac{\mu}{\mu_o}} \tag{11.2}$$

In general, for ferromagnetic materials, $\mu_r \geq 100$, and for nonmagnetic materials, $\mu_r = 1$.

Since μ_r is a variable dependent on other quantities of the magnetic circuit, values of μ_r are not tabulated. Methods of calculating μ_r from the data supplied by manufacturers are considered in a later section.

11.5 RELUCTANCE

The resistance of a material to the flow of charge (current) is determined for electric circuits by the equation

$$R = \rho\frac{l}{A} \qquad \text{(ohms, } \Omega\text{)}$$

The *reluctance* of a material to the setting up of magnetic flux lines in the material is determined by the following equation:

$$\mathcal{R} = \frac{l}{\mu A} \qquad \text{(rels or A·t/Wb)} \tag{11.3}$$

where $\mathcal{R}$ is the reluctance, l is the length of the magnetic path, and A is the material's cross-sectional area. The "t" in the units ampere-turns per weber (A·t/Wb) stands for the number of turns of the applied winding. More will be said about ampere-turns in the next section. Note that the resistance and reluctance are inversely proportional to the area, indicating that an increase in area will result in a reduction in each and an *increase* in the desired result: current and flux. For an increase in length the opposite is true, and the desired effect is reduced. The reluctance, however, is inversely proportional to the permeability, whereas the resistance is directly proportional to the resistivity. The larger the μ or the smaller the ρ, the smaller the reluctance and resistance, respectively. Obviously, therefore, materials with high permeability, such as the ferromagnetics, have very small reluctances and will result in an increased measure of flux through the core. There is no widely accepted unit for reluctance, although the rel and the ampere-turn per weber are usually applied.

11.6 OHM'S LAW AND MAGNETIC CIRCUITS

Recall the equation

$$\text{Effect} = \frac{\text{Cause}}{\text{Opposition}}$$

from Chapter 4, which introduced Ohm's law for electric circuits. For magnetic circuits, the effect desired is the flux Φ. The cause is the *magnetomotive force* (mmf) $\mathcal{F}$, which is the external force (or "pressure") required to set up the magnetic flux lines within the magnetic material. The opposition to the setting up of the flux Φ is the reluctance $\mathcal{R}$.

Substituting, we have

$$\Phi = \frac{\mathcal{F}}{\mathcal{R}} \tag{11.4}$$

The magnetomotive force is proportional to the product of the number of turns around the core (in which the flux is to be established) and the current through the turns of wire (Fig. 11.15). In equation form,

$$\boxed{\mathscr{F} = NI} \quad \text{(ampere-turns, A·t)} \qquad (11.5)$$

The equation clearly indicates that an increase in the number of turns, or the current through the wire, will result in an increased "pressure" on the system to establish flux lines through the core.

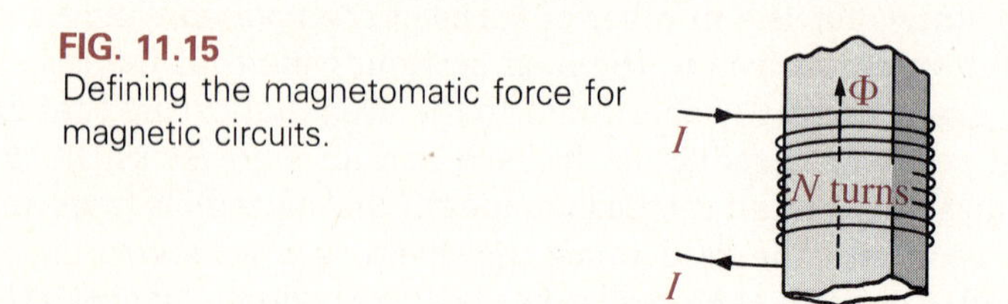

FIG. 11.15
Defining the magnetomatic force for magnetic circuits.

Although there is a great deal of similarity between electric and magnetic circuits, one must continue to realize that the flux Φ is not a "flow" variable, such as current in an electric circuit. Magnetic flux is established in the core through the alteration of the atomic structure of the core due to external pressure and is not a measure of the flow of some charged particles through the core.

In Fig. 11.16 two magnetic circuits are provided to emphasize the effect of certain parameters in the magnitude of the flux in the core. In Fig. 11.16(a), a large flux Φ is established due to a large magnetomotive force (both N and I large) and a large area with a small l. In Fig. 11.16(b), a small flux is induced by a small magnetomotive force (both N and I small) and a small area with a large l.

11.7 MAGNETIZING FORCE AND HYSTERESIS

The magnetomotive force per unit length is called the *magnetizing force* (H). In equation form,

$$\boxed{H = \frac{\mathscr{F}}{l}} \quad \text{(A·t/m)} \qquad (11.6)$$

Substituting for the magnetomotive force results in

$$\boxed{H = \frac{NI}{l}} \quad \text{(A·t/m)} \qquad (11.7)$$

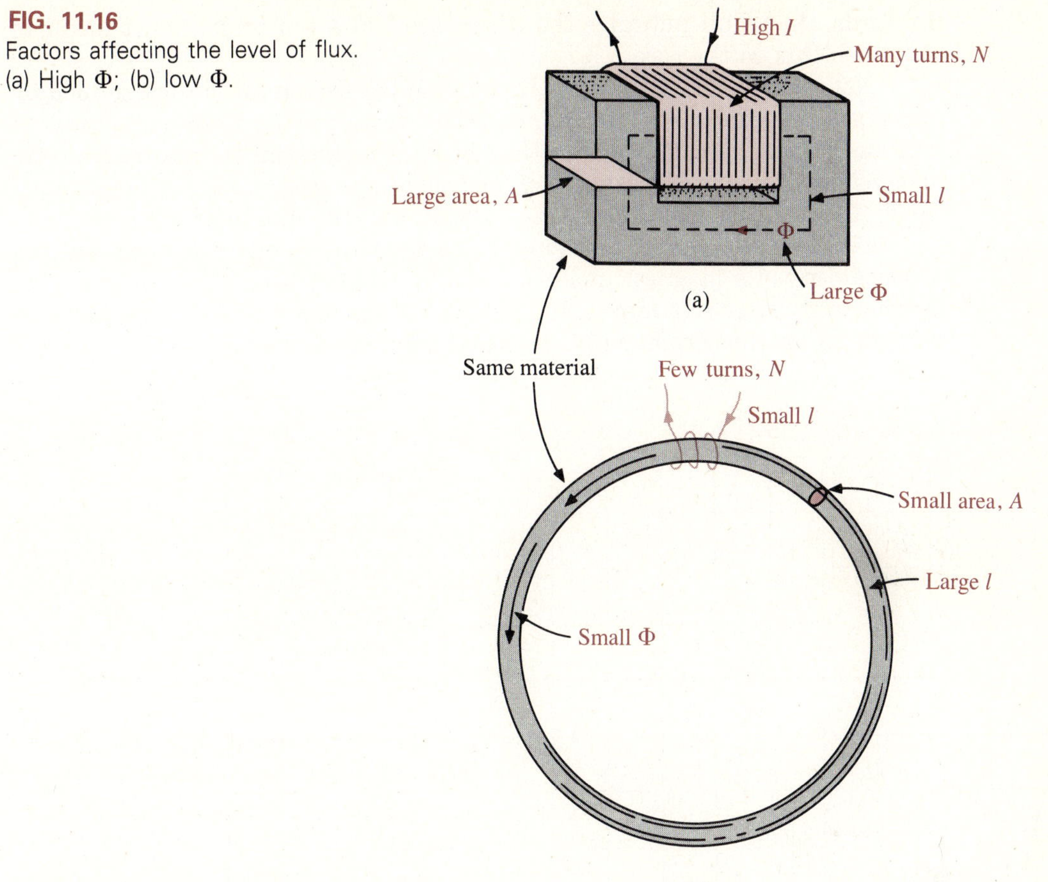

FIG. 11.16
Factors affecting the level of flux. (a) High Φ; (b) low Φ.

For the magnetic circuit of Fig. 11.17, if $I = 1$ A, $N = 40$ turns (t), and $l = 0.2$ m, then

$$H = \frac{NI}{l} = \frac{(40 \text{ t})(1 \text{ A})}{0.2 \text{ m}} = \frac{40 \text{ A·t}}{0.2 \text{ m}} = 200 \text{ A·t/m}$$

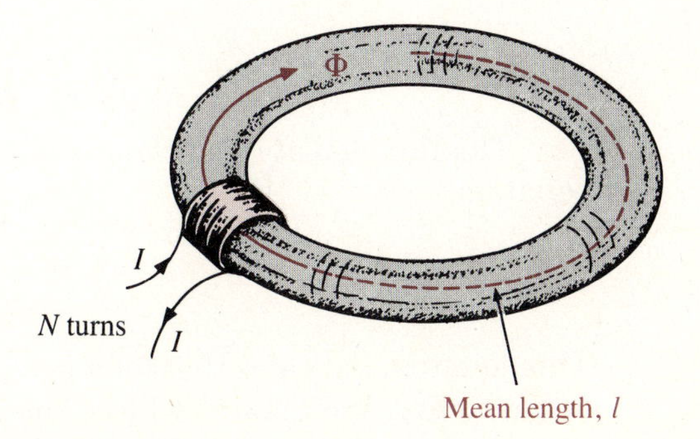

FIG. 11.17
Defining magnetizing force.

In words, the result indicates that there are 200 A·t of pressure per meter to establish flux in the core.

Note in Fig. 11.17 that the direction of the flux Φ can be determined by placing the fingers of the right hand in the direction of current around the core and noting the direction of the thumb. It is interesting to realize that *the magnetizing force is independent of the type of core material*—it is determined solely by the number of turns, the current, and the length of the core.

The applied magnetizing force has a pronounced effect on the resulting permeability of a magnetic material. As the magnetizing force increases, the permeability rises to a maximum and then drops to a minimum, as shown in Fig. 11.18 for three commonly employed magnetic materials.

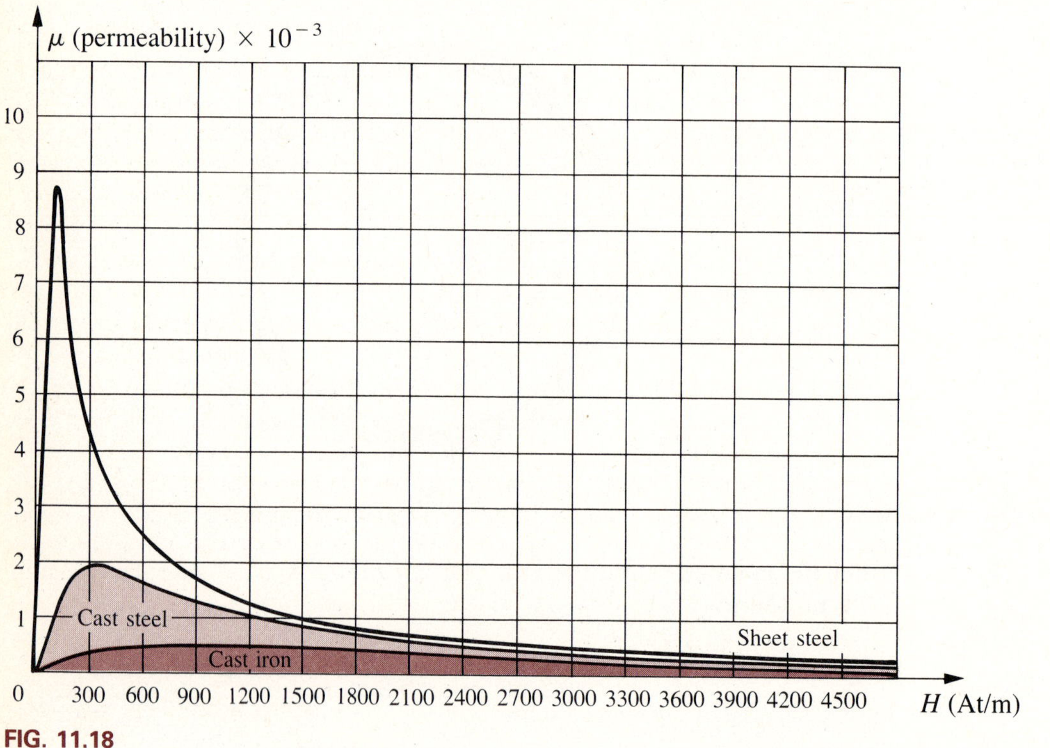

FIG. 11.18
Variation of μ with H.

The flux density and the magnetizing force are related by the following equation:

$$B = \mu H \tag{11.8}$$

This equation indicates that for a particular magnetizing force, the greater the permeability, the greater will be the induced flux density.

A curve of the flux density B versus the magnetizing force H of a material is of particular importance to the engineer. Curves of this type can usually be found in manuals and descriptive pamphlets and brochures published by manufacturers of magnetic materials. A typical B-H curve for a ferromagnetic material such as steel appears in Fig. 11.19. The curve provides a closed curve that shows the variation in the flux density $B = \Phi/A$ with the applied magnetizing force $H = NI/l$. For an unmagnetized sample, increasing the level of H will result in an increase in B, as shown by the dark colored curve of Fig. 11.19. As H is increased, however, the level of B will eventually level off at the saturation level, revealing that no matter how much more we increase the magnetizing force (or current through the turns of wire), the flux density (flux in the core) will increase only slightly. If we then reduce the level of the magnetizing force to zero (by letting the current I drop to 0 A), there will still be a level of B remaining, denoted by B_R in Fig. 11.19. The subscript R is derived from the word *retentivity* or the word *residual* from residual flux density. It is a measure of the flux density (and therefore flux in the core) remaining when the magnetizing force has been removed. Recall the permanent magnet that has magnetic characteristics even though there is no current-carrying coil wrapped around the core. Obviously, therefore, without this retentivity, permanent magnets would not be a possibility. If the current of the magnetizing force is reversed, a flux is established that opposes the flux due to the retentivity; this eventually results in a net flux of zero (with $B = \Phi/A = 0$ Wb also). Eventually, the flux density saturates with the flux in the opposite direction, as shown by the bottom left-hand corner of Fig. 11.19. As the applied current is then reversed, the flux density again decreases to the retentivity level and works toward positive saturation again. A repeat of the changing current (between maximums) results in a repeat of the same outside curve of Fig. 11.19. The entire curve of Fig. 11.19 is called the *hysteresis curve* for the ferromagnetic material, from the Greek *hysterein,* meaning "to lag behind." The flux density B *lagged* behind the magnetizing force H during the entire plotting of

FIG. 11.19
B-H (hysteresis) curve.

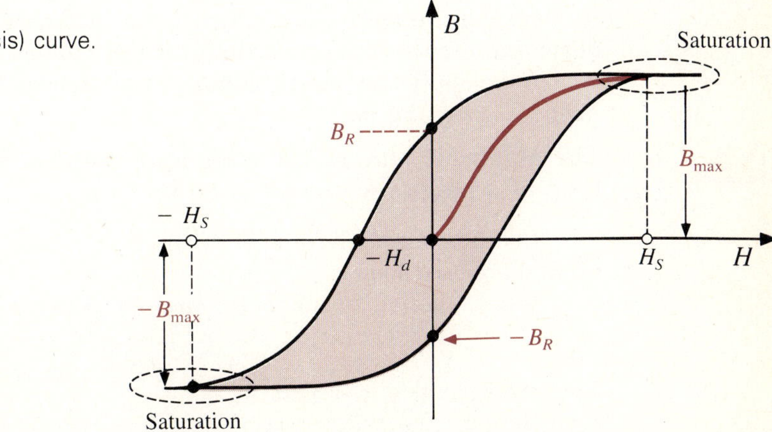

the curve. When H was zero at B_R, B was not zero but had only begun to decline. Long after H had passed through zero and had become equal to $-H_d$ did the flux density B finally become equal to zero. If the process is repeated for maximum levels of H less than the saturation levels, the curves of Fig. 11.20 will result.

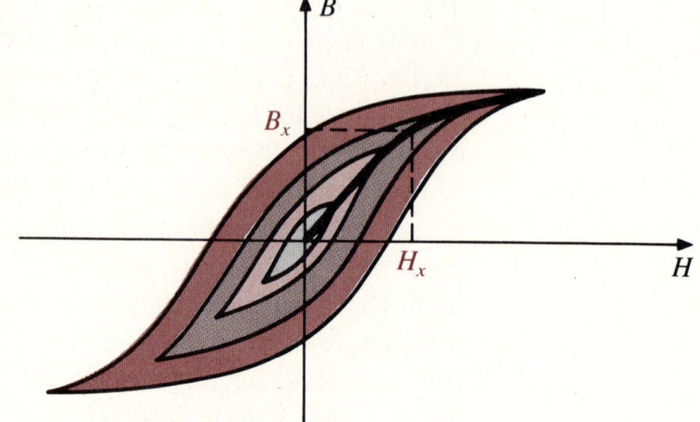

FIG. 11.20
Defining a "normal magnetization curve."

Note from the various curves that for a particular value of H, say, H_x, the value of B can vary widely, as determined by the history of the core. In an effort to assign a particular value for B to each value of H, we compromise by connecting the tips of the hysteresis loops. The resulting curve, shown by the heavy, solid line in Fig. 11.20 and for various materials in Fig. 11.21, is called the *normal magnetization curve.*

An instrument that will provide a plot of the B-H curve for a magnetic sample appears in Fig. 11.22, page 418.

11.8 AMPÈRE'S CIRCUITAL LAW

We mentioned in the introduction to this chapter that there is a broad similarity between the analyses of electric and magnetic circuits. We have already demonstrated this to some extent for the quantities in Table 11.1, page 418.

If we apply the "cause" analogy to Kirchhoff's voltage law, we establish *Ampère's circuital law:*

The algebraic sum of the rises and drops of the mmf around a closed loop of a magnetic circuit is equal to zero.

In other words, the sum of the mmf rises equals the sum of the mmf drops around a closed loop.

When applied to magnetic circuits, sources of mmf are determined by

$$\boxed{\mathscr{F} = NI} \quad \text{(A·t)} \qquad \textbf{(11.9)}$$

revealing that the pressure on a magnetic system to establish the magnetic flux is determined by the product of the number of turns and the applied

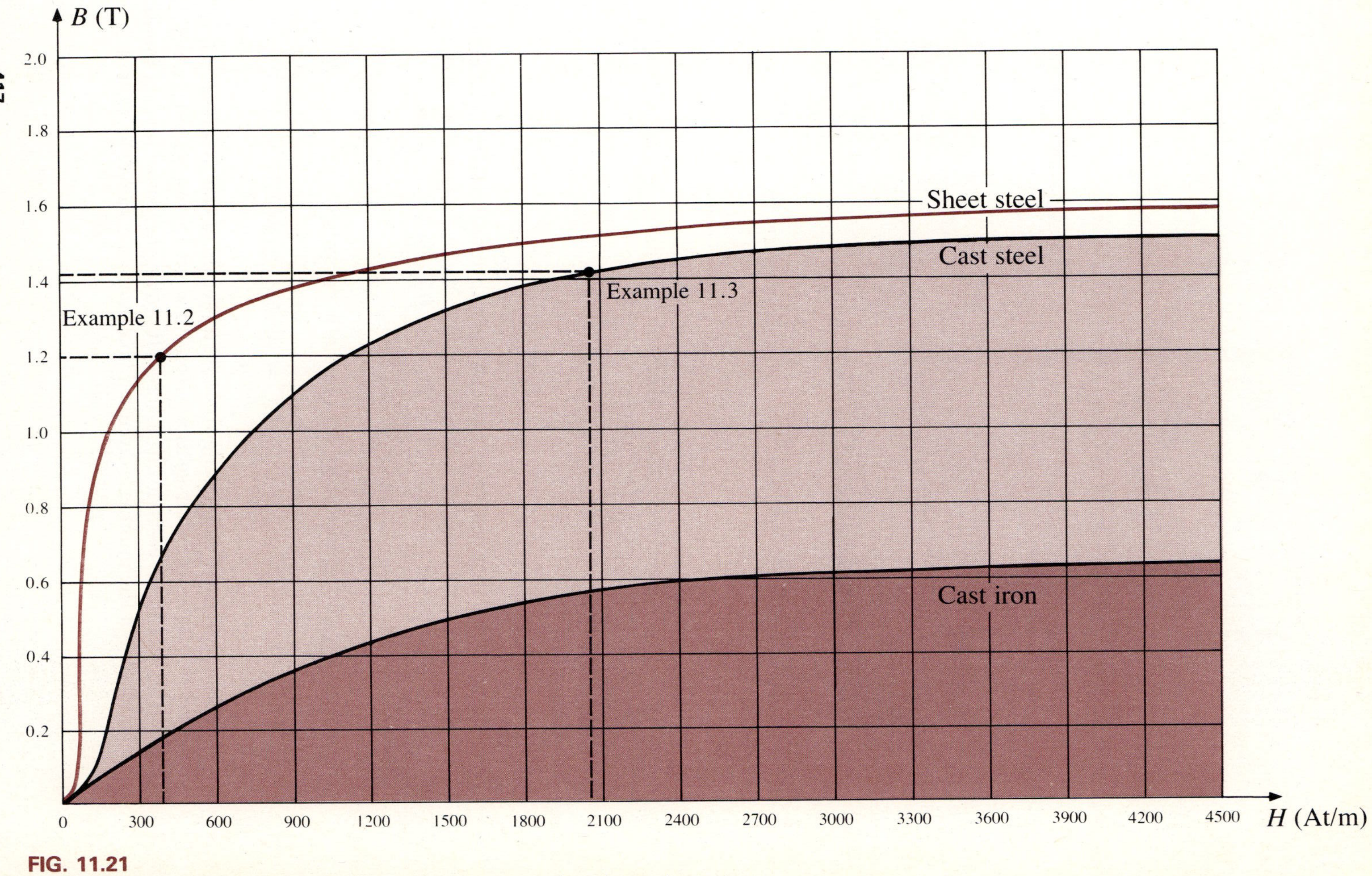

FIG. 11.21
Normal magnetization curves for sheet steel, cast steel, and cast iron.

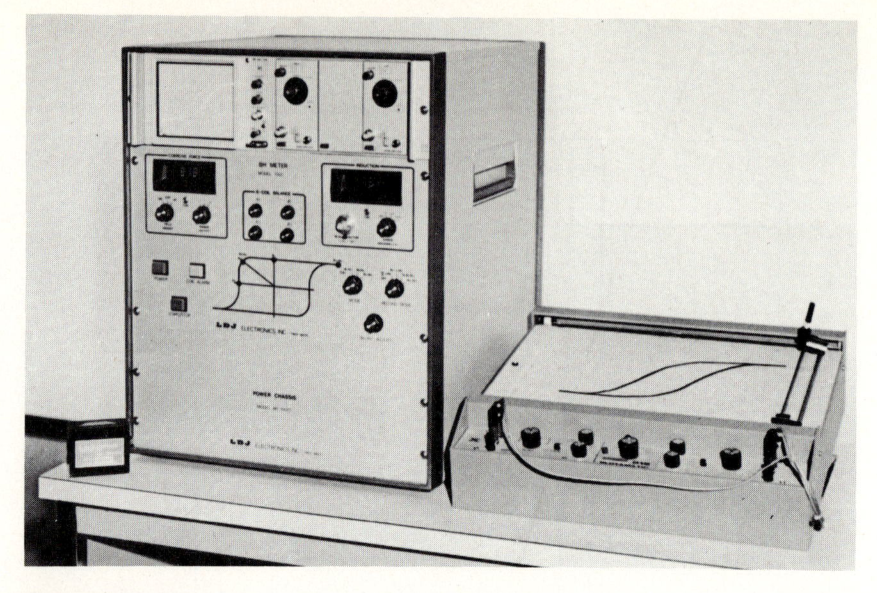

FIG. 11.22
Hysteresis curve plotter. (Courtesy of LDJ Electronics, Inc.)

TABLE 11.1

	Electric Circuits	Magnetic Circuits
Cause	E	$\mathscr{F}$
Effect	I	Φ
Opposition	R	$\mathscr{R}$

current. An increase in either one or both increases the resulting flux in the core. The mmf drops of a magnetic circuit can be found by applying the relationships listed in Table 11.1. That is, for electric circuits,

$$V = IR$$

resulting in the following for magnetic circuits:

$$\boxed{\mathscr{F} = \Phi\mathscr{R}} \quad \text{(A·t)} \qquad \textbf{(11.10)}$$

where Φ is the flux passing through a section of the magnetic circuit and $\mathscr{R}$ is the reluctance of that section. Note how an increased reluctance level (like resistance in an electric circuit) results in an increased level of mmf loss across a section of the magnetic circuit. The reluctance, however, is seldom calculated in the analysis of magnetic circuits. A more practical equation for the mmf drop is

$$\boxed{\mathscr{F} = Hl} \quad \text{(A·t)} \tag{11.11}$$

as derived from Eq. 11.6, where H is the magnetizing force on a section of a magnetic circuit and l is the mean length of the section.

For the magnetic circuit of Fig. 11.23, the pressure on the system is the magnetomotive force $\mathscr{F} = NI$ and the result is the flux Φ in the core. Since the algebraic sum of the rises and drops in mmf ($\mathscr{F}$) around a closed loop must equal zero, the applied mmf must equal the mmf drop across the core ($\mathscr{F} = Hl$) and

$$\mathscr{F}\text{ (pressure)} = \mathscr{F}\text{ (drop across load)}$$

and

$$\boxed{NI = Hl} \tag{11.12}$$

The length l is the mean length of the core, as shown in Fig. 11.23.

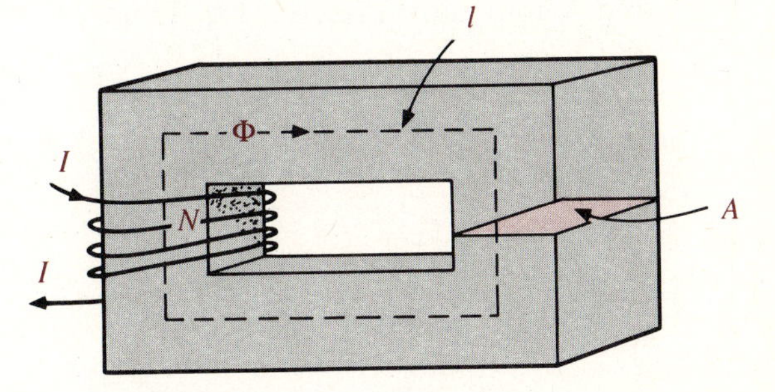

FIG. 11.23
Basic magnetic circuit.

Given N, I, l, and A, the magnetizing force H can be determined from $H = NI/l$. The flux density B can then be determined from Fig. 11.21 and the flux in the core can be found from $\Phi = BA$. Given Φ, Nl, and A, the flux density can be determined from $B = \Phi/A$ followed by H from Fig. 11.21. The required I can then be determined from $I = Hl/N$. In other words, given NI, the resulting flux can be determined, or, given the desired level of Φ, the required NI can be determined.

The preceding conclusions can be applied to a variety of electromagnetic elements, devices, and systems such as transformers, relays, generators, motors, and the like. In general, keep in mind that the more the turns or current, the greater the pressure on the system to increase the flux level. Further, the shorter the core length, the larger the core area, and the higher the relative permeability, the greater the resulting flux, as indicated in Fig. 11.16.

EXAMPLE 11.2 Determine the current I required to establish the flux Φ indicated in Fig. 11.24.

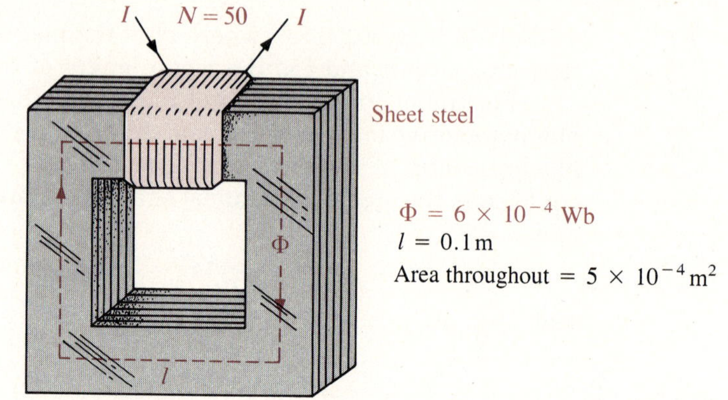

FIG. 11.24
Magnetic circuit for Example 11.2.

Solution Since all the units are in the SI system (Φ in webers, l in meters, and A in square meters), Eq. 11.12 can be applied directly:

$$NI = Hl$$

and

$$I = \frac{Hl}{N}$$

The flux density is

$$B = \frac{\Phi}{A} = \frac{6 \times 10^{-4}\ \text{Wb}}{5 \times 10^{-4}\ \text{m}^2}$$

$$B = 1.2\ \text{Wb/m}^2 = 1.2\ \text{T}$$

From Fig. 11.21,

$$H \cong 400\ \text{A·t/m}$$

Substituting into the preceding equation gives

$$I = \frac{Hl}{N} = \frac{(400\ \text{A·t/m})(0.1\ \text{m})}{50\ \text{t}} = \mathbf{0.8\ A}$$

EXAMPLE 11.3 Determine the flux Φ developed in the core of Fig. 11.25 due to the applied mmf.

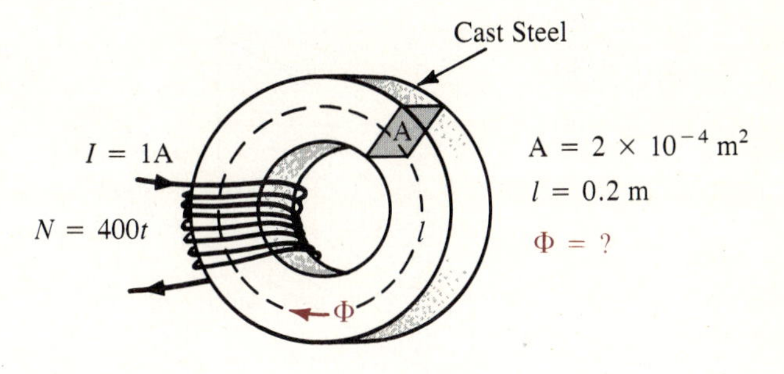

FIG. 11.25
Magnetic circuit for Example 11.3.

Solution From Eq. 11.12:

$$H = \frac{NI}{l} = \frac{(400\text{ t})(1\text{ A})}{0.2\text{ m}}$$

$$H = 2000\text{ A}\cdot\text{t/m}$$

From Fig. 11.21,

$$B \cong 1.42\text{ T} \quad \text{and} \quad B = \frac{\Phi}{A}$$

results in

$$\Phi = BA = (1.42\text{ T})(2 \times 10^{-4}\text{ m}^2) = \mathbf{2.84 \times 10^{-4}\ Wb}$$

11.9 FARADAY'S LAW OF ELECTROMAGNETIC INDUCTION

If a conductor is moved through a magnetic field so that it cuts magnetic lines of flux, a voltage will be induced across the conductor, as shown in Fig. 11.26. The greater the number of flux lines cut per unit time (by increasing the speed with which the conductor passes through the field) or the stronger the magnetic field strength (for the same traversing speed), the greater will be the induced voltage across the conductor. If the conductor is held fixed and the magnetic field is moved so that its flux lines cut the conductor, the same effect will be produced.

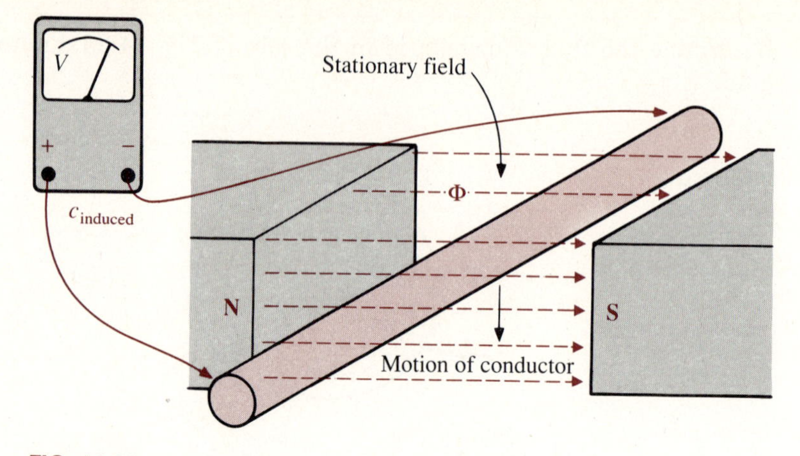

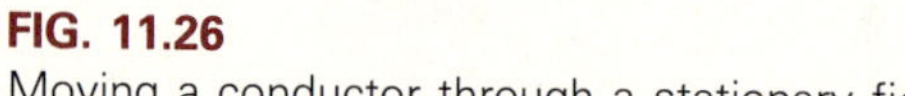

FIG. 11.26
Moving a conductor through a stationary field.

If a coil of N turns is placed in the region of a changing flux, as in Fig. 11.27, a voltage will be induced across the coil as determined by *Faraday's law:*

$$\boxed{e = N\frac{d\Phi}{dt}} \qquad \text{(volts, V)} \qquad \textbf{(11.13)}$$

where N represents the number of turns of the coil and $d\Phi/dt$ is the instantaneous change in flux (in webers) linking the coil. The term *linking* refers to the flux within the turns of wire. The term *changing* simply indicates that either the strength of the field linking the coil changes in magnitude or the coil is moved through the field in such a way that the number of flux lines through the coil changes with time.

FIG. 11.27
A stationary coil in a changing field.

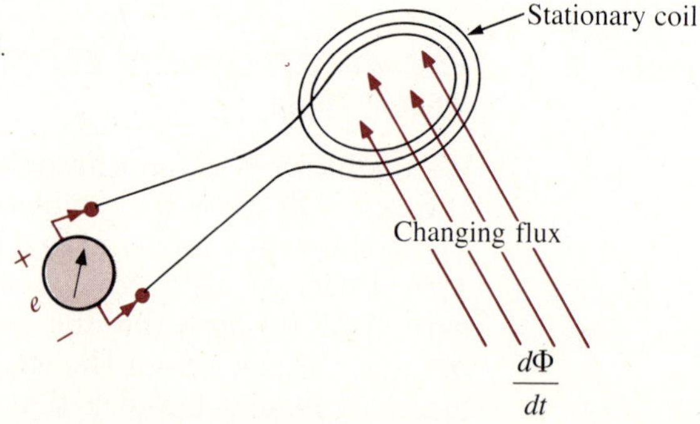

If the flux linking the coil ceases to change, such as when the coil simply sits still in a magnetic field of fixed strength, $d\Phi/dt = 0$, and the induced voltage is $e = N(d\Phi/dt) = N(0) = 0$ V.

11.10 LENZ'S LAW

In Section 11.2 it was shown that the magnetic flux linking a coil of N turns with a current I has the distribution of Fig. 11.28.

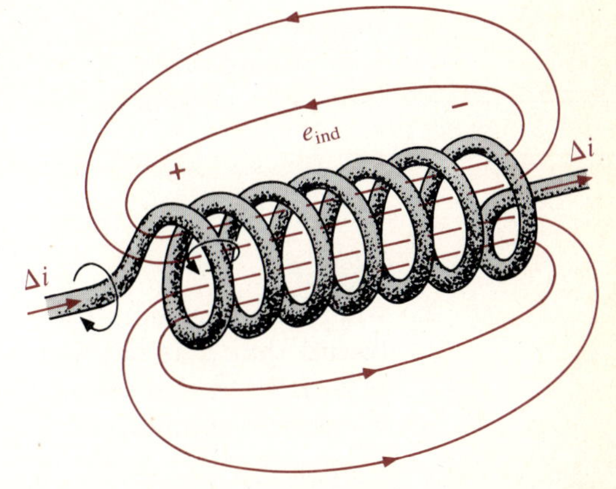

FIG. 11.28
Induced voltage across a coil due to a changing current.

If the current increases in magnitude, the flux linking the coil also increases. It was shown in Section 11.9, however, that a changing flux linking a coil induces a voltage across the coil. For this coil, therefore, an induced voltage is developed *across* the coil due to the change in current *through* the coil. The polarity of this induced voltage tends to establish a current in the coil that produces a flux that will oppose any change in the original flux. In other words, the induced effect (e_{ind}) is a result of the increasing current through the coil. However, the resulting induced voltage will tend to establish a current that will oppose the increasing change in current through the coil. Keep in mind that this is all occurring simultaneously. The instant the current begins to increase in magnitude, there will be an opposing effect trying to limit the change. This effect is "choking" the change in current through the coil. Hence, the term *choke* is often applied to the inductor or coil. In fact, we will find shortly that the current through a coil cannot change instantaneously. A period of time determined by the coil and the resistance of the circuit is required before the inductor discontinues its opposition to a momentary change in current. Recall a similar situation for the voltage across a capacitor in Chapter 10. This reaction is true for increasing or decreasing levels of current through the coil. This effect is an example of a general principle known as *Lenz's law*, which states that

An induced effect is always such as to oppose the cause that produced it.

11.11 SELF-INDUCTANCE

The ability of a coil to oppose any change in current is a measure of the *self-inductance L* of the coil. For brevity, the prefix *self* is usually dropped. Inductance is measured in henries (H), after the American physicist Joseph Henry.

Inductors are coils of various dimensions designed to introduce specified amounts of inductance into a circuit. The inductance of a coil varies directly with the magnetic properties of the coil. Ferromagnetic materials, therefore, are frequently employed to increase the inductance by increasing the flux linking the coil.

A close approximation, in terms of physical dimensions, for the inductance of the coils of Fig. 11.29 can be found using the following equation:

$$L = \frac{N^2 \mu A}{l} \quad \text{(henries, H)} \tag{11.14}$$

where N represents the number of turns, μ is the permeability of the core (recall that μ is not a constant but depends on the level of B and H, since $\mu = B/H$), A is the area of the core in square meters, and l is the mean length of the core in meters.

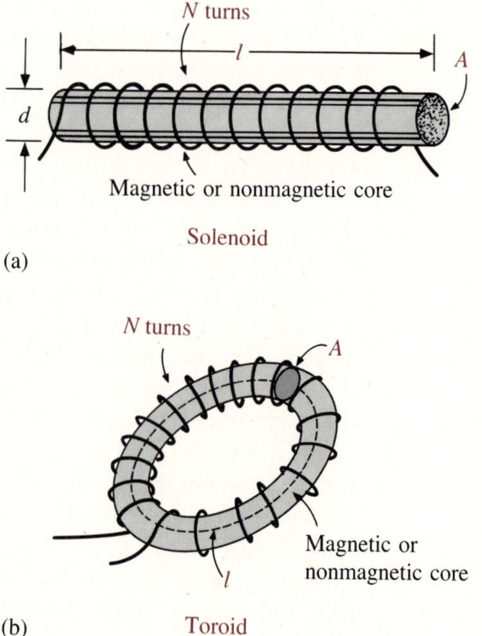

FIG. 11.29
Inductors. (a) Solenoid; (b) toroid.

In another form:

$$L = \mu_r L_o \tag{11.15}$$

where L_o is the inductance of the coil with an air core. In other words, the inductance of a coil with a ferromagnetic core is the relative permeability of the core times the inductance achieved with an air core.

Equations for the inductance of coils different from those just shown can be found in reference handbooks. Most of the equations are more complex than those just described.

EXAMPLE 11.4 Find the inductance of the air-core coil of Fig. 11.30.

FIG. 11.30
Inductor for Example 11.4.

$l = 50$ mm
Air core (μ_0)
$A = 100 \times 10^{-6}$ m^2
200 turns

Solution

$$\mu = \mu_r\mu_o = (1)(\mu_o) = \mu_o = 4\pi \times 10^{-7} \text{ Wb/A·m}$$

$$L = \frac{N^2\mu_o A}{l} = \frac{(200 \text{ t})^2(4\pi \times 10^{-7} \text{ Wb/A·m})(100 \times 10^{-6} \text{ m}^2)}{50 \times 10^{-3} \text{ m}}$$

$$= \frac{(4 \times 10^4)(12.566 \times 10^{-7})(100 \times 10^{-6})}{50 \times 10^{-3}}$$

$$= \frac{(4)(12.566)(100)}{50} \times \frac{(10^4)(10^{-7})(10^{-6})}{10^{-3}}$$

$$L = 100.528 \times 10^{-6} \cong 0.1 \times 10^{-3} = \mathbf{0.1 \text{ mH}}$$

Calculator

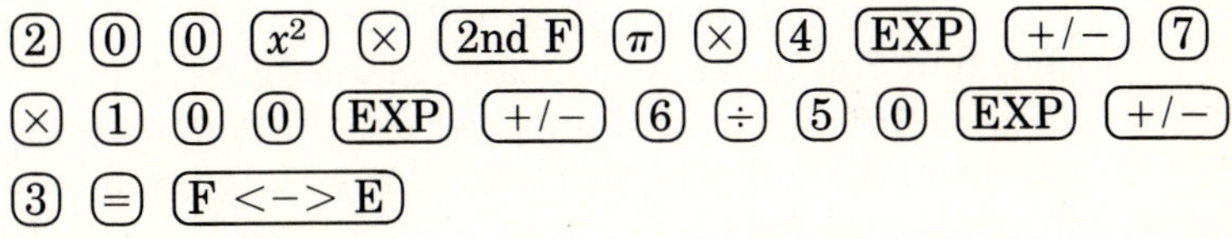

with a display of 1.005−04.

EXAMPLE 11.5 Repeat Example 11.4 with an iron core and conditions such that $\mu_r = 2000$.

Solution By Eq. 11.15,

$$L = \mu_r L_o = (2000)(0.1 \times 10^{-3} \text{ H}) = \mathbf{200 \text{ mH}}$$

11.12 TYPES OF INDUCTORS

Associated with every inductor are a resistance equal to the resistance of the turns and a stray capacitance due to the capacitance between the turns of the coil. To include these effects, the equivalent circuit for the inductor is as shown in Fig. 11.31. However, for most applications considered in this text, the stray capacitance appearing in Fig. 11.31 can be ignored, resulting in the equivalent model of Fig. 11.32.

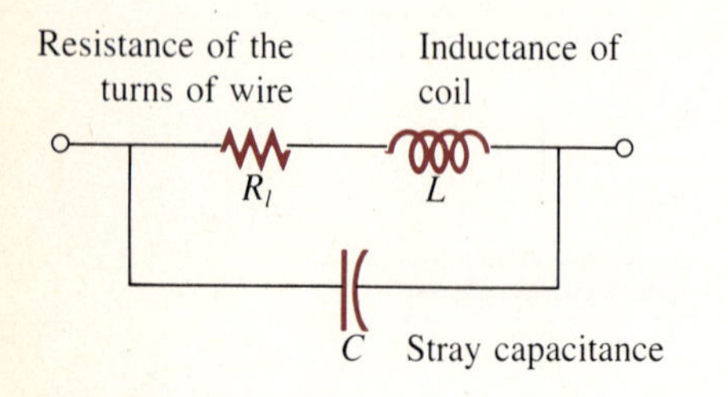

FIG. 11.31
Complete equivalent model for an inductor.

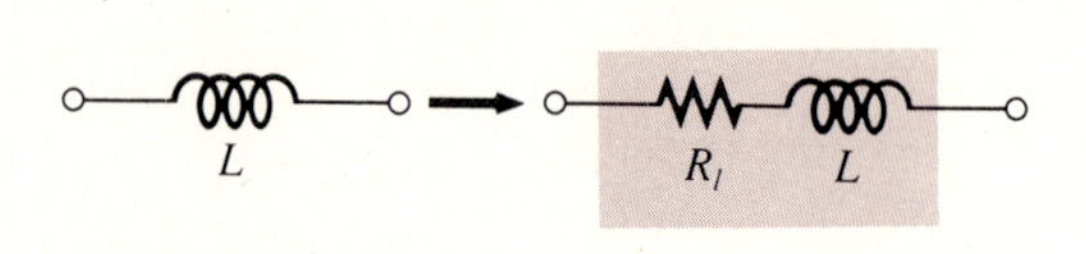

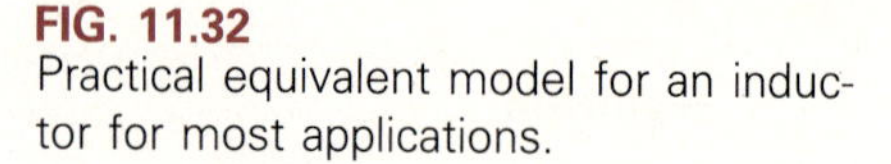

FIG. 11.32
Practical equivalent model for an inductor for most applications.

The resistance R_l can play an important role in the analysis of networks with inductive elements. For most applications, we have been able to treat the capacitor as an ideal element and maintain a high degree of accuracy. For the inductor, however, R_l must often be included in the analysis and can have a pronounced effect on the response of a system (see Chapter 18). The level of R_l can extend from a few ohms to a few hundred ohms. Keep in mind that the longer or thinner the wire used in the construction of the inductor, the greater will be the dc resistance as determined by $R = \rho l/A$. Our initial analysis will treat the inductor as an ideal element. Once a general feeling for the response of the element is established, the effects of R_l will be included.

The dc resistance of an inductor can be measured by an ohmmeter, as shown in Fig. 11.33. As indicated in the figure, typical values of R_l extend from 10 Ω to 100 Ω. A reading of very high resistance typically reveals an open circuit in the coil, whereas a very low reading could indicate that the windings have shorted together.

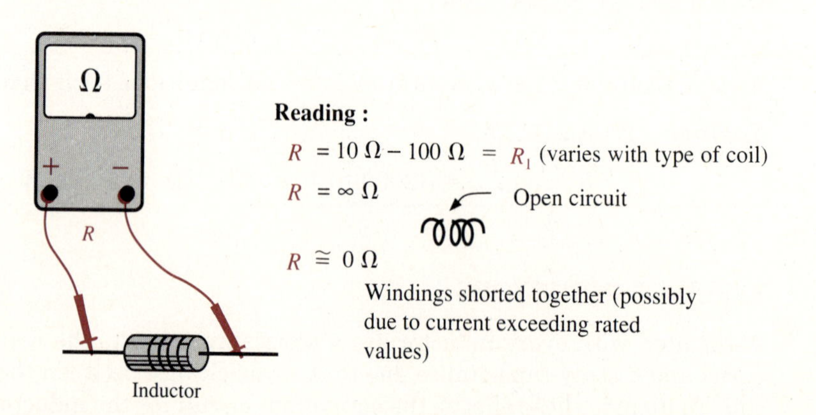

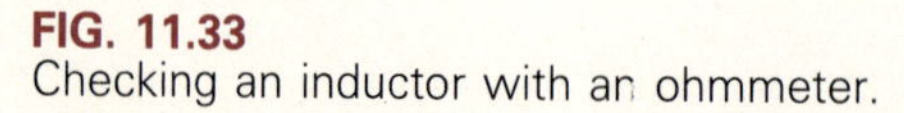

FIG. 11.33
Checking an inductor with an ohmmeter.

The primary function of the inductor, however, is to introduce inductance—not resistance or capacitance—into the network. For this reason, the symbols employed for inductance are as shown in Fig. 11.34.

FIG. 11.34
Graphics symbols for inductors.

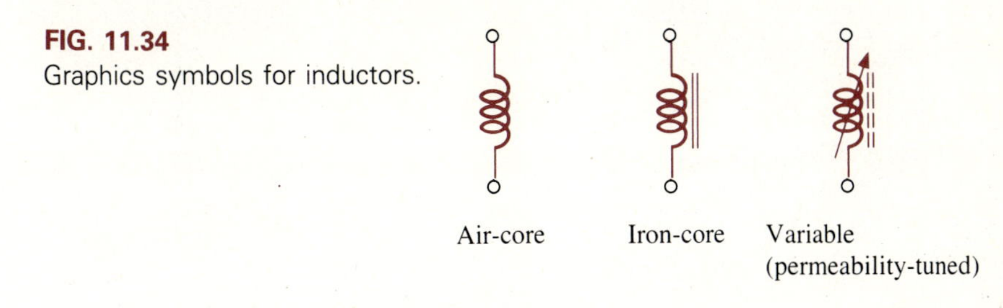

All inductors, like capacitors, can be listed under two general headings: *fixed* and *variable*. The fixed air-core and iron-core inductors were described in the last section. The permeability-tuned variable coil has a ferromagnetic shaft that can be moved within the coil to vary the flux linkages of the coil and thereby its inductance. Several fixed and variable inductors appear in Fig. 11.35.

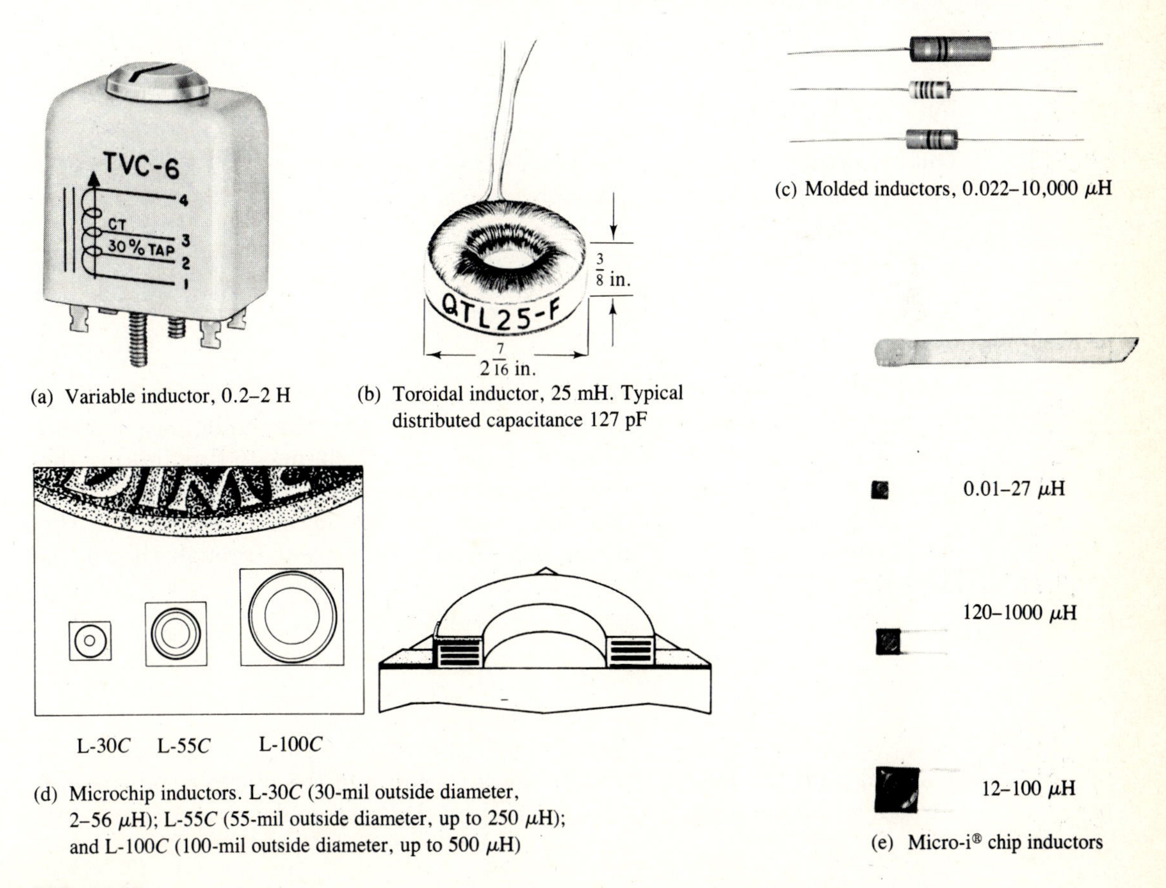

(a) Variable inductor, 0.2–2 H

(b) Toroidal inductor, 25 mH. Typical distributed capacitance 127 pF

(c) Molded inductors, 0.022–10,000 μH

(d) Microchip inductors. L-30*C* (30-mil outside diameter, 2–56 μH); L-55*C* (55-mil outside diameter, up to 250 μH); and L-100*C* (100-mil outside diameter, up to 500 μH)

(e) Micro-i® chip inductors

FIG. 11.35

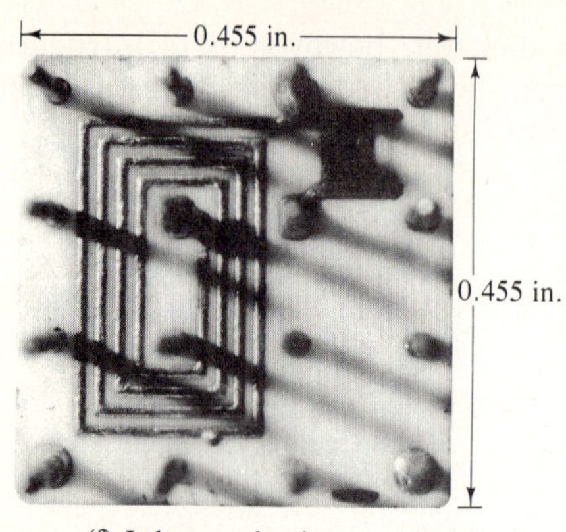

(f) Inductor and resistor on a module

(g) 1.0 H at 8 A, 8 kV working voltage

FIG. 11.35 (continued)
Various types of inductors. (Part (a) courtesy of United Transformer Corp.; part (b) courtesy of Microtan Company, Inc.; part (c) courtesy of Delevan, Division of American Precision Industries, Inc.; part (d) courtesy of Thinco Division, Hull Corp.; part (e) courtesy of Delevan, Division of American Precision Industries, Inc.; part (f) courtesy of International Business Machines Corp.; part (g) courtesy of Basler Electric Co.)

11.13 INDUCED VOLTAGE

The inductance of a coil is also a measure of the change in flux linking a coil due to a change in current through the coil; that is,

$$L = N\frac{d\Phi}{di} \quad \text{(H)} \tag{11.16}$$

where N is the number of turns, Φ is the flux in webers, and i is the current through the coil. The equation reveals that the larger the inductance of a coil (with N fixed), the larger will be the instantaneous change in flux linking the coil due to an instantaneous change in current through the coil.

Using the preceding definition for inductance, it can be shown that the voltage induced across a coil due to a change in current through the coil is given by

$$v_L = L\frac{di_L}{dt} \tag{11.17}$$

revealing that the larger the induction of a coil and the quicker the current through a coil changes, the greater will be the voltage developed across the coil.

If a current through the coil fails to change at a particular instant, the induced voltage across the coil will be zero. For dc applications, after the transient effect has passed, $di_L/dt = 0$, and the induced voltage is

$$v_L = L\frac{di_L}{dt} = L(0) = 0\text{ V}$$

Recall that the equation for the current of a capacitor is

$$i_C = C\frac{dv_C}{dt}$$

Note the similarity between this equation and Eq. 11.17. In fact, if we apply the dualities $v \rightleftarrows i$ (that is, interchange the two) and $L \rightleftarrows C$ for capacitance and inductance, each equation can be derived from the other.

The average voltage across the coil is defined by the equation

$$\boxed{v_{L_{av}} = L\frac{\Delta i}{\Delta t}} \quad \text{(V)} \qquad \textbf{(11.18)}$$

where Δ signifies finite change (a measurable change). Compare this to $i_C = C(\Delta v/\Delta t)$; the meaning of Δ and application of Eq. 11.18 should be clarified from Chapter 10. An example follows.

EXAMPLE 11.6 Find the waveform for the average voltage across the coil if the current through a 4-mH coil is shown in Fig. 11.36.

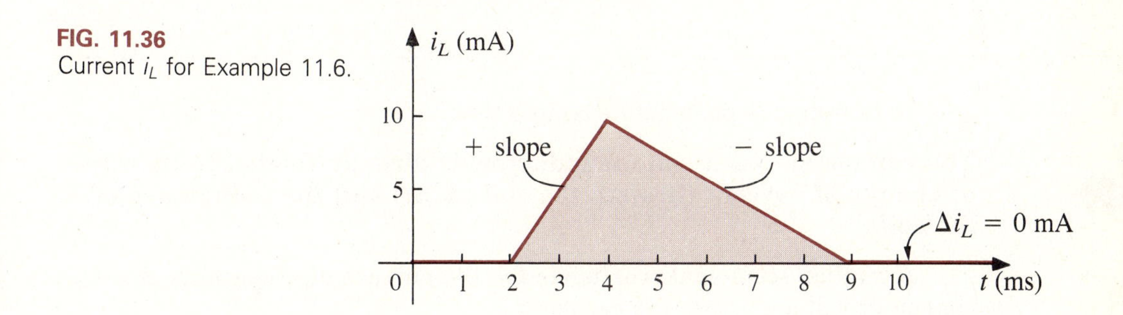

FIG. 11.36
Current i_L for Example 11.6.

Solution

a. 0 to 2 ms: Since there is no change in current through the coil, there is no voltage induced across the coil; that is,

$$v_L = L\frac{\Delta i}{\Delta t} = L\frac{0}{\Delta t} = \mathbf{0\ V}$$

b. 2 ms to 4 ms:

$$v_L = L\frac{\Delta i}{\Delta t} = (4 \times 10^{-3}\text{ H})\left(\frac{10 \times 10^{-3}\text{ A}}{2 \times 10^{-3}\text{ s}}\right) = 20 \times 10^{-3}\text{ V}$$
$$= \mathbf{20\ mV}$$

c. 4 ms to 9 ms:

$$v_L = L\frac{\Delta i}{\Delta t} = -(4 \times 10^{-3}\text{ H})\left(\frac{10 \times 10^{-3}\text{ A}}{5 \times 10^{-3}\text{ s}}\right) = -8 \times 10^{-3}\text{ V}$$
$$= \mathbf{-8\ mV}$$

d. 9 ms to ∞:

$$v_L = L\frac{\Delta i}{\Delta t} = L\frac{0}{\Delta t} = \mathbf{0\ V}$$

The waveform for the average voltage across the coil is shown in Fig. 11.37.

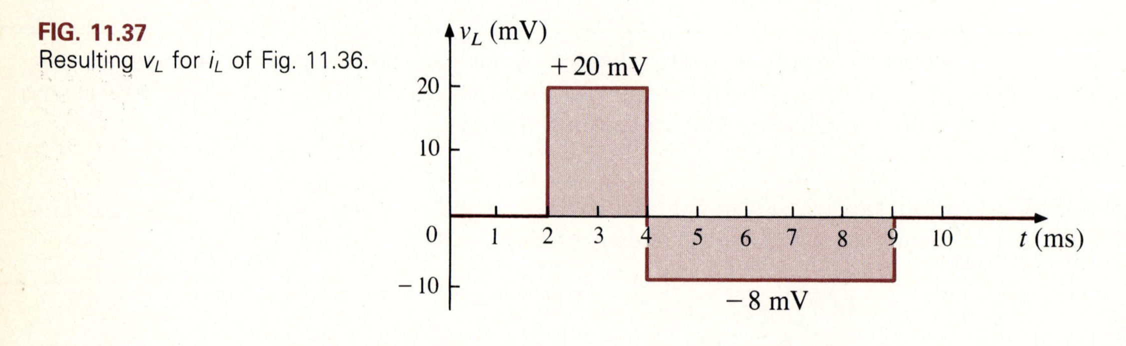

FIG. 11.37
Resulting v_L for i_L of Fig. 11.36.

Note from the preceding example that

The voltage across a coil (or inductor) is directly related to the rate of change of current through the coil ($\Delta i/\Delta t$) and the inductance of the coil.

A similar statement was made for the current of a capacitor due to a change in voltage across the capacitor.

A careful examination of Fig. 11.37 also reveals that the area under the positive pulse from 2 ms to 4 ms equals the area under the negative pulse from 4 ms to 9 ms. In Section 11.20, we will find that the area under the curves represents the energy stored or released by the inductor. From 2 ms to 4 ms, the inductor is storing energy, whereas from 4 ms to 9 ms, the inductor is releasing the energy stored. For the full period 0 to 10 ms, energy has simply been stored and released; there has been no dissipation as experienced for the resistive elements. Over a full cycle, both the ideal capacitor and inductor do not consume energy but simply store and release it in their respective forms.

11.14 *R-L* TRANSIENTS: STORAGE CYCLE

The changing voltages and current that result during the storing of energy in the form of a magnetic field by an inductor in a dc circuit can best be described by the network of Fig. 11.38. At the instant the switch is closed, the inductance of the coil will prevent an instantaneous change in current through the coil, and the current will remain at 0 A, as shown in Fig. 11.39. The potential drop across the coil, v_L, will equal the impressed voltage E as determined by Kirchhoff's voltage law, since $v_R = iR = (0)R = 0$ V and the coil will take on the characteristics of an open circuit at the instant the switch is closed. The current i_L will then build up from zero, establishing a voltage drop across the resistor and a corresponding drop in v_L. The current will continue to increase until the voltage across the inductor drops to 0 V and the full impressed voltage appears across the resistor. Initially, the current i_L increases quite rapidly, followed by a continually decreasing rate until it reaches its maximum value of E/R.

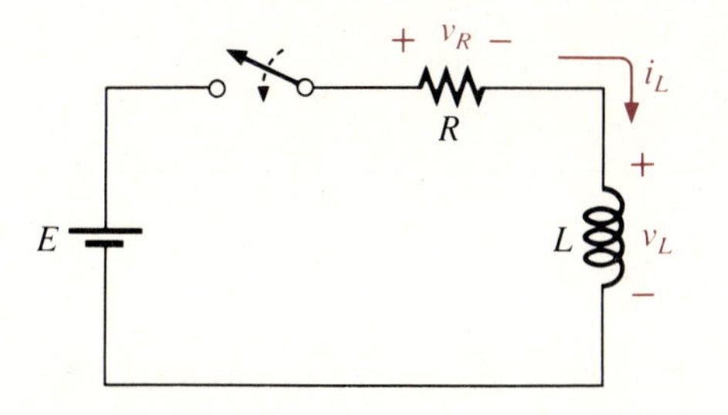

FIG. 11.38
Closing the switch to establish a storage cycle for the inductor.

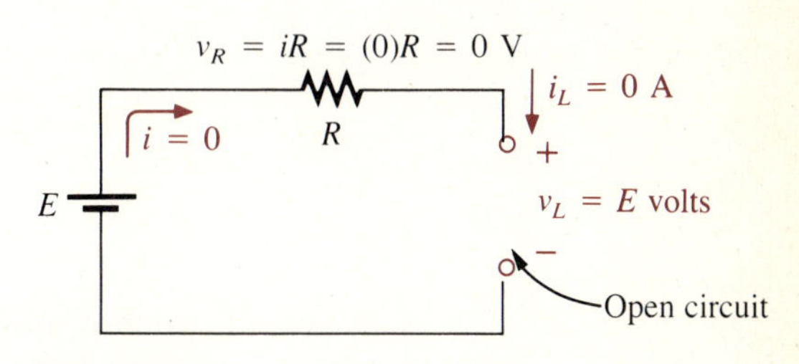

FIG. 11.39
Circuit of Fig. 11.38 the instant the switch is closed.

When steady-state conditions have been established and the storage phase is complete, the "equivalent" network will appear, as shown in Fig. 11.40. The network clearly reveals that

An ideal inductor assumes a short-circuit equivalent in a dc network once steady-state conditions have been established.

FIG. 11.40
Circuit of Fig. 11.38 under steady-state conditions ($t > 5\tau$).

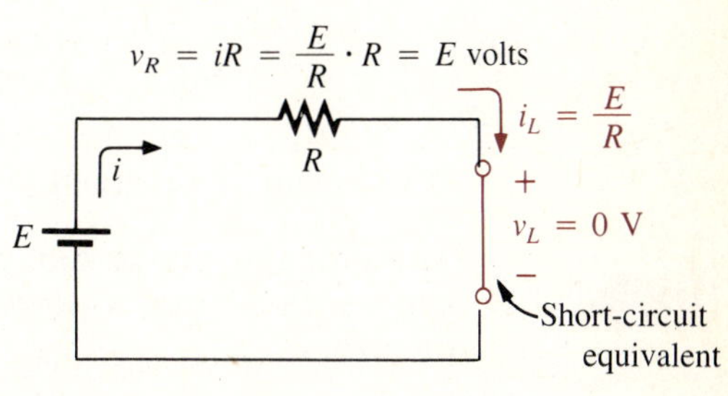

Fortunately, the resulting curves for the voltages and current for the storage phase are similar in many respects to those encountered for the *R-C*

network. The experience gained with these equations in Chapter 10 will undoubtedly make the analysis of R-L networks somewhat easier to understand. A plot of the charging current is provided in Fig. 11.41, clearly revealing that the maximum steady-state value of i_L is E/R and that the rate of change in current decreases as time passes. The abscissa is scaled in time constants, with τ for inductive circuits defined by the following:

$$\boxed{\tau = \frac{L}{R}} \quad \text{(seconds, s)} \qquad \textbf{(11.19)}$$

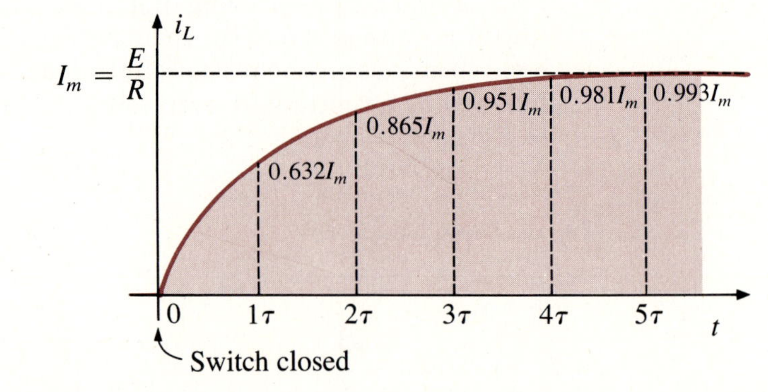

FIG. 11.41
i_L versus time following the closing of the switch in Fig. 11.38.

If we keep R constant and increase L, the ratio L/R increases and the rise time increases. The change in transient behavior for the current i_L is plotted in Fig. 11.42 for various values of L. Note again the duality between these curves and those obtained for the R-C network in Fig. 10.29.

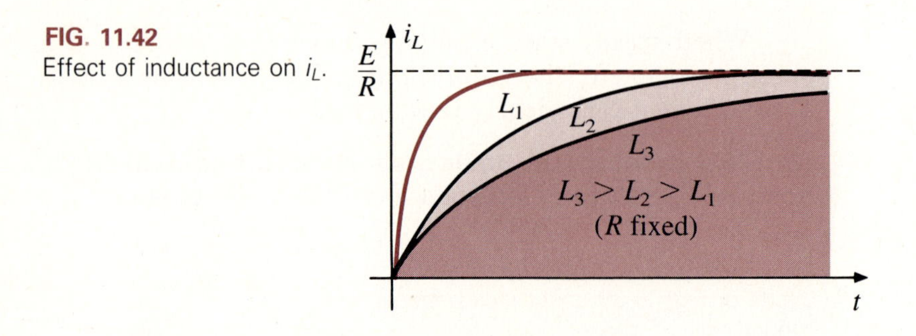

FIG. 11.42
Effect of inductance on i_L.

For most practical applications, we will assume that

The storage phase has passed and steady-state conditions have been established once a period of time equal to five time constants has occurred.

In addition, since L/R will always have some numerical value even though it may be very small, the period 5τ will always be greater than zero, confirming the fact that

The current cannot change instantaneously in an inductive network.

In fact, the larger the inductance, the more the circuit will oppose a rapid buildup in current level.

The equation for the current i_L during the storage phase is the following:

$$i_L = I_m(1 - e^{-t/\tau})$$

$$i_L = \frac{E}{R}(1 - e^{-t/(L/R)}) \qquad \textbf{(11.20)}$$

Note the factor $(1 - e^{-t/\tau})$, which also appeared for the voltage v_C of a capacitor during the charging phase.

Our experience with the factor $(1 - e^{-t/\tau})$ verifies the level of 63.2% after one time constant, 86.5% after two time constants, and so on. For convenience, Fig. 10.21 is repeated as Fig. 11.43 to evaluate the functions $(1 - e^{-t/\tau})$ and $e^{-t/\tau}$ at various values of τ.

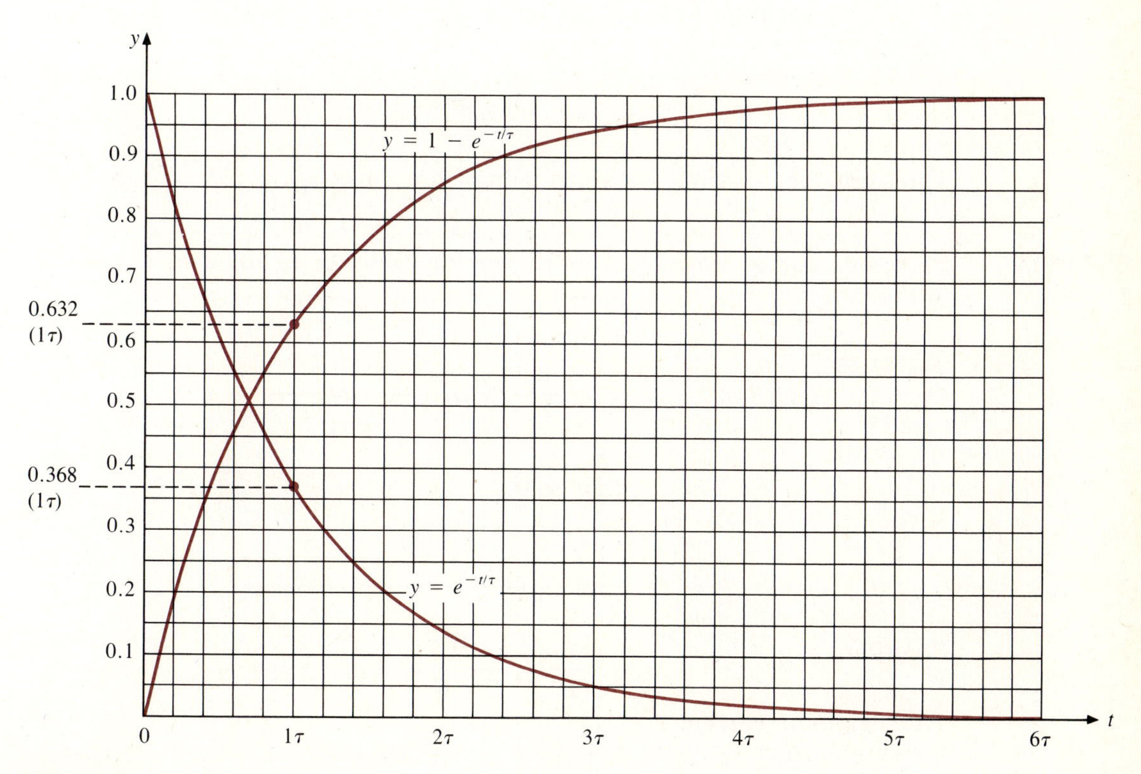

FIG. 11.43

The voltage v_L of Fig. 11.38 jumps to E volts at the closing of the switch and then decays in an exponential manner to a final value of 0 V, as shown in Fig. 11.44. Again note that the voltage has effectively dropped to 0 V in about five time constants (5τ).

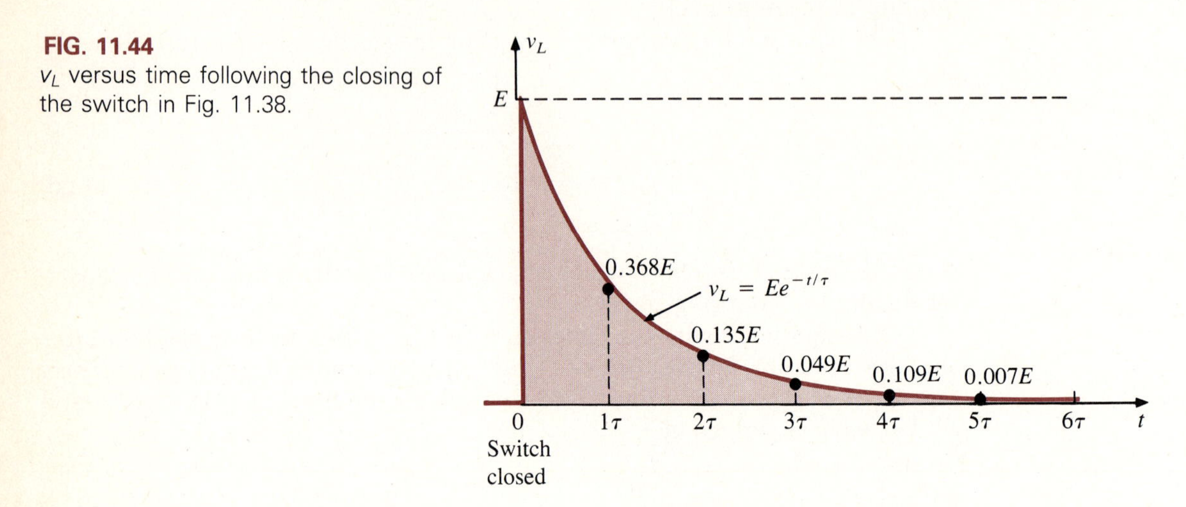

FIG. 11.44
v_L versus time following the closing of the switch in Fig. 11.38.

In equation form, the decaying voltage v_L can be mathematically represented by

$$v_L = Ee^{-t/\tau} \tag{11.21}$$

The curve for v_R can be obtained using Ohm's law: $v_R = i_R R = i_L R$.

EXAMPLE 11.7 Sketch the curves of i_L and v_L for the circuit of Fig. 11.45 following the closing of the switch.

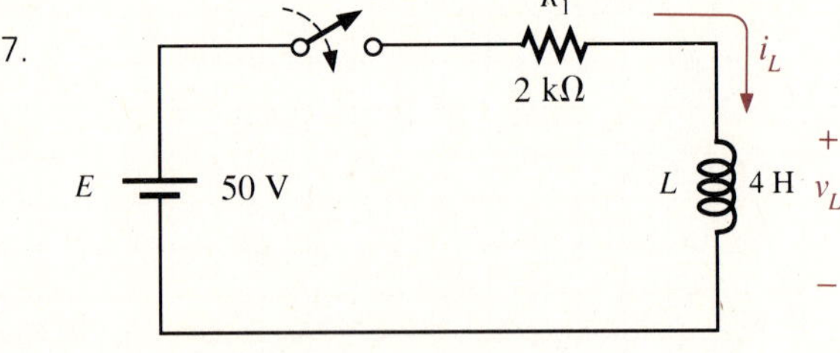

FIG. 11.45
Circuit for Example 11.7.

Solution

$$\tau = \frac{L}{R} = \frac{4\text{ H}}{2\text{ k}\Omega} = 2\text{ ms}$$

The maximum current is

$$I_m = \frac{E}{R} = \frac{50\text{ V}}{2\text{ k}\Omega} = 25 \times 10^{-3}\text{ A} = 25\text{ mA}$$

A sketch of i_L appears in Fig. 11.46(a).

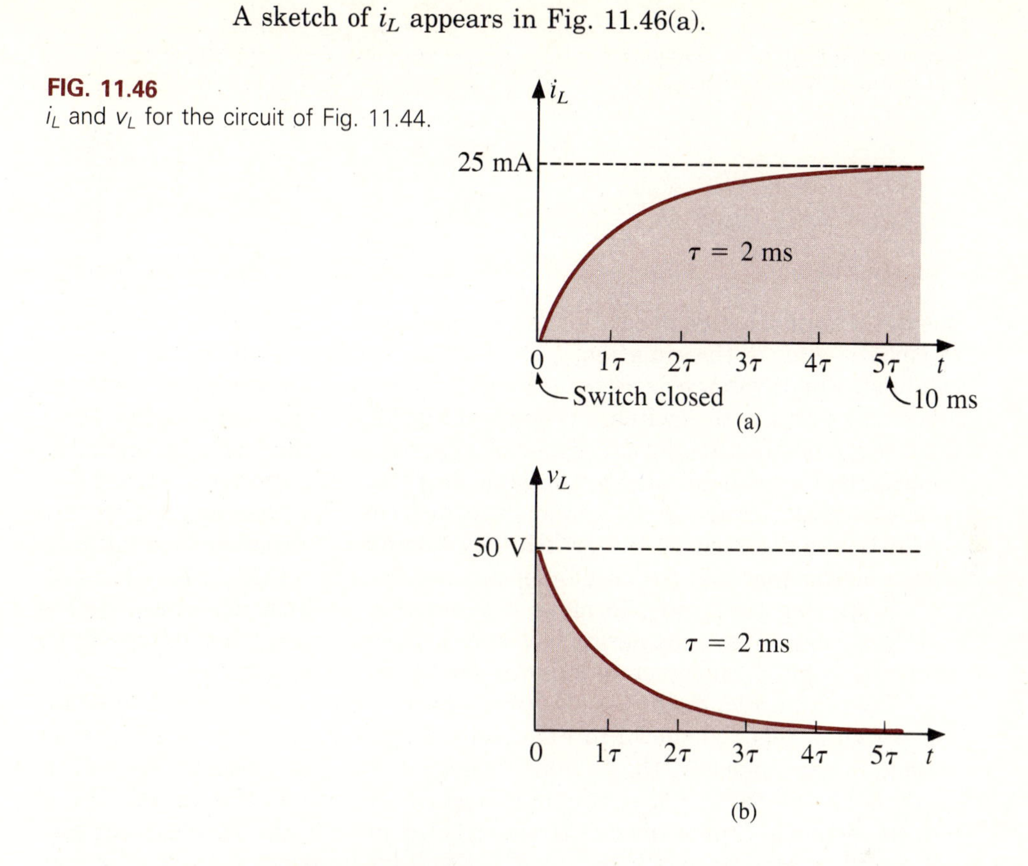

FIG. 11.46
i_L and v_L for the circuit of Fig. 11.44.

Mathematically,

$$i_L = (25 \times 10^{-3})(1 - e^{-t/(2 \times 10^{-3})})$$

The maximum voltage is $E = 50$ V with an exponential decay, as shown in Fig. 11.46(b).

Mathematically,

$$v_L = 50e^{-t/(2 \times 10^{-3})}$$

11.15 R-L TRANSIENTS: DECAY PHASE

In the analysis of R-C circuits, we found that the capacitor could hold its charge and store energy in the form of an electric field for a period of time determined by the leakage factors. In R-L circuits, the energy is stored in the form of a magnetic field established by the current through the coil. Unlike the isolated capacitor, however, an isolated inductor cannot continue to store energy, since the absence of a closed path causes the current to drop to zero, releasing the energy stored in the form of a magnetic field. If the switch of Fig. 11.47 were opened quickly, a spark would probably occur across the contacts

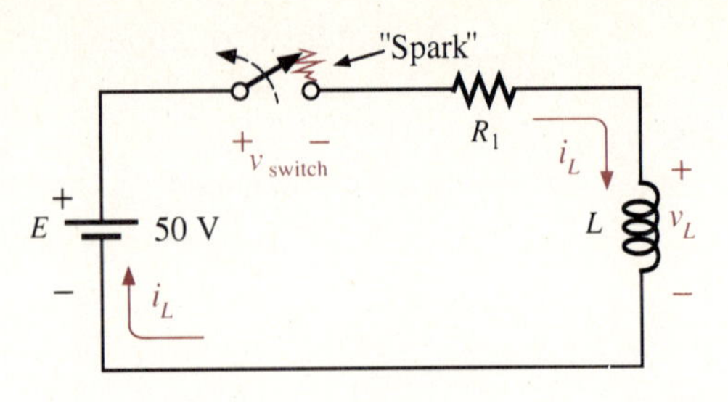

FIG. 11.47
Opening the switch on a series *R-L* circuit in a steady-state mode.

due to the rapid change in current from a maximum of E/R to 0 A. The change in current di/dt of the equation $v_L = L(di/dt)$ would be sufficiently large to establish a high voltage v_L across the coil that would contribute to the voltage across the points of the switch, as shown in Fig. 11.47. This is the same mechanism as applied in the ignition system of a car to ignite the fuel in the cylinder. Some 25,000 V are generated by the rapid decrease in ignition coil current that occurs when the switch in the system is opened. (In older systems, the points in the distributor served as the switch.) This inductive reaction is significant when you consider that the only independent source in a car is a 12-V battery.

If opening the switch to move it to another position will cause such a rapid discharge in stored energy, how can the decay phase of an *R-L* circuit be analyzed in much the same manner as for the *R-C* circuit?

The most direct method is to apply a resistor R_2 across the coil the instant the switch is open, as shown in Fig. 11.48. Since the current of a coil cannot change instantaneously, the current through the coil and resistor R_2 will be $I_m = E/R_2$ the instant the switch is opened. Note that the direction if i_L (which obviously cannot reverse direction if the magnitude of the current cannot change instantaneously) establishes a voltage drop across R_2 with the polarity shown in Fig. 11.48. The magnetic field surrounding the inductor will now begin to collapse at a rapid rate at first, followed by a continuously slower rate until the current through the coil is zero. The curve for the decaying current appears in Fig. 11.49. Note again the use of τ to define the decay period (5τ). However, since the field is collapsing through a load R_2, the time constant is

$$\boxed{\tau = \frac{L}{R_2}} \tag{11.22}$$

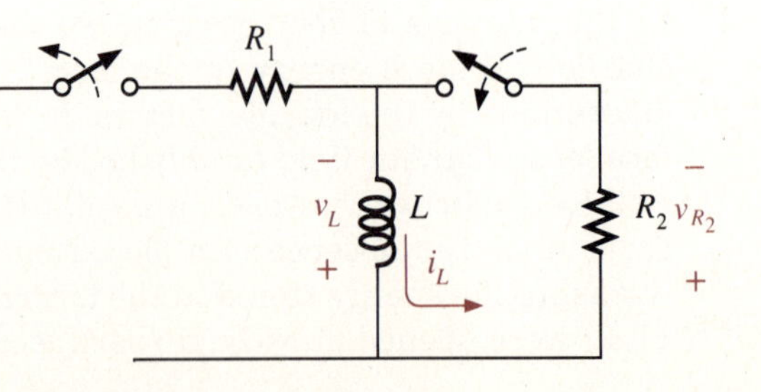

FIG. 11.48
Establishing a controlled discharge of the energy stored by the inductor.

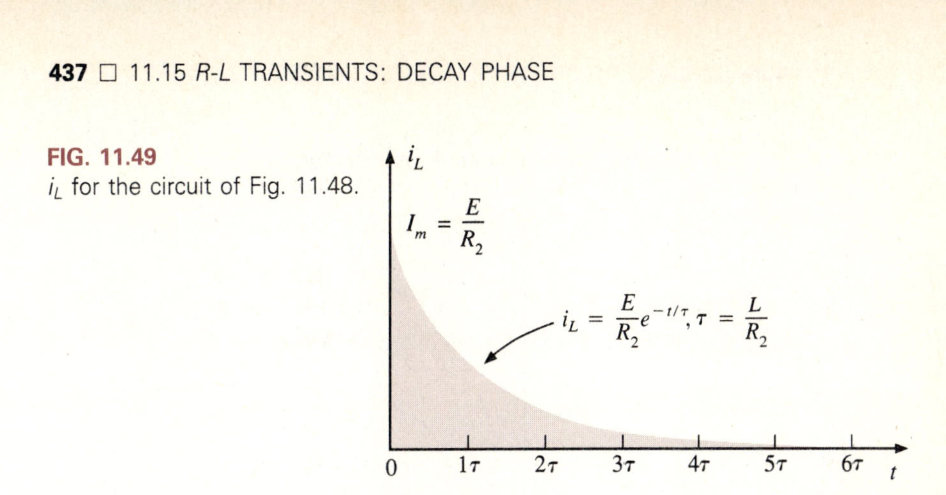

FIG. 11.49
i_L for the circuit of Fig. 11.48.

The maximum current for the curve of Fig. 11.49 is determined by the resistor R_2 of the storage phase: $I_m = E/R_2$.

The resistor R_2 will also determine the level of the voltage v_R at $t = 0$ s: $V_{R_{max}} = I_{max}R_2$ (Fig. 11.50).

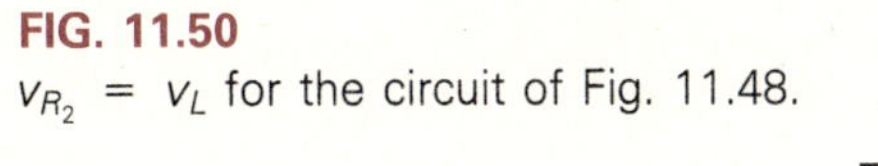

FIG. 11.50
$v_{R_2} = v_L$ for the circuit of Fig. 11.48.

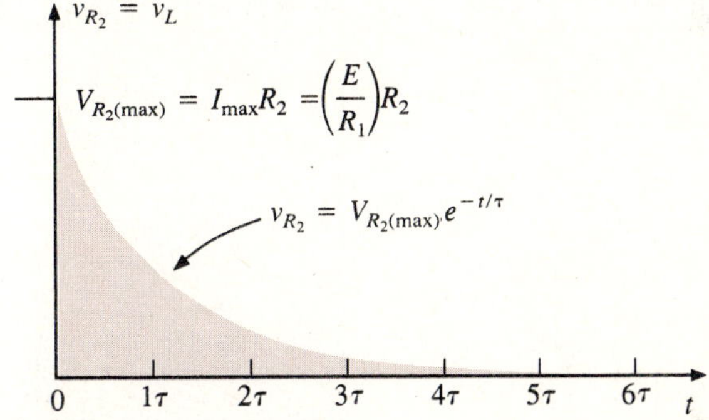

In Fig. 11.48, since $v_{R_2} = v_L$ (in parallel), the shape and characteristics of v_L match those of v_{R_2} in Fig. 11.50. In other words, when the storage phase has passed, v_L drops to 0 V, but during the decay phase just described, the voltage v_L jumps to $V_{R_{max}} = V_{L_{max}}$ from 0 V and then decays at the same rate as the current i_L.

EXAMPLE 11.8 If a load $R_2 = 10\text{ k}\Omega$ is applied across the 4-H inductor of the circuit of Fig. 11.45 the instant the switch is opened (after 5τ of the storage cycle), sketch the resulting waveforms of i_L and v_L.

Solution

$$I_m = \frac{E}{R} = \frac{50\text{ V}}{2\text{ k}\Omega} = \mathbf{25\text{ mA}} \qquad \text{(as obtained earlier)}$$

$$\tau = \frac{L}{R_d} = \frac{4\text{ H}}{10\text{ k}\Omega} = 0.4 \times 10^{-3}\text{ s} = \mathbf{0.4\text{ ms}}$$

Note that even though R_2 is five times the size of R, the time constant is significantly less, since $\tau = L/R$, and the discharge time is much quicker than the storage phase.

The curve for i_L appears in Fig. 11.51.

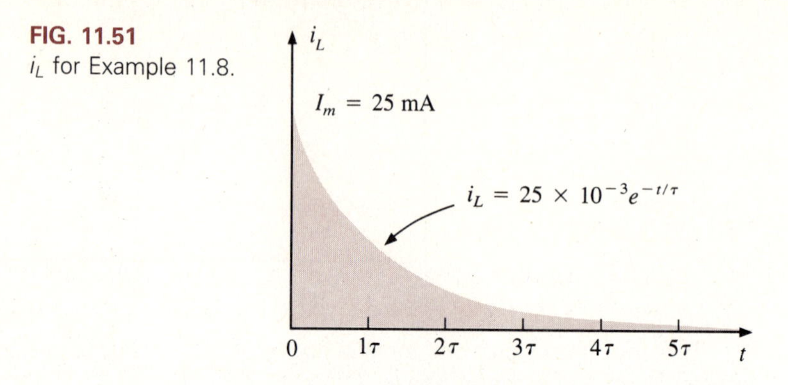

FIG. 11.51
i_L for Example 11.8.

$V_{L_{max}} = V_{R_{max}} = I_m R_2 = (25\ \text{mA})(10\ \text{k}\Omega) = 250$ V. (Note the magnitude of $V_{L_{max}}$ compared to the source voltage $E = 50$ V.)

The curve for v_L appears in Fig. 11.52.

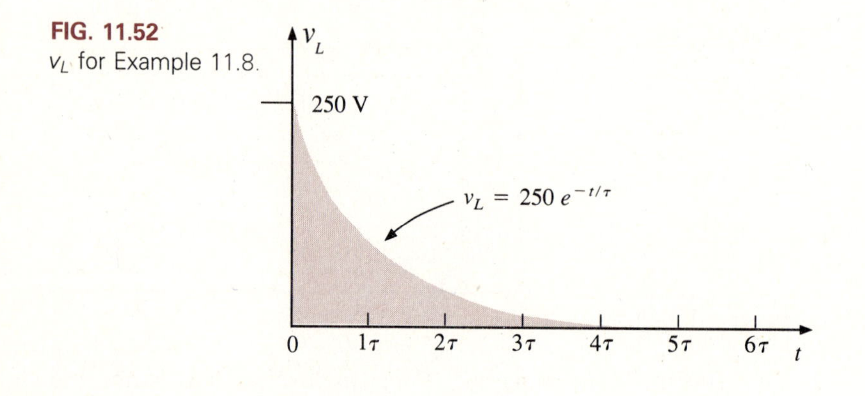

FIG. 11.52
v_L for Example 11.8.

11.16 INSTANTANEOUS VALUES

The development presented in Section 10.9 for capacitive networks can also be applied to *R-L* networks to determine instantaneous voltages, currents, and time. The instantaneous values of any voltage or current can be determined by simply inserting *t* into the equation and using a calculator or table to determine the magnitude of the exponential term.

The similarity between the equations $v_C = E(1 - e^{-t/\tau})$ and $i_L = I_m(1 - e^{-t/\tau})$ results in a derivation of the following for *t* that is identical to the one used to obtain Eq. 10.19:

$$t = -\tau \log_e\left(1 - \frac{i_L}{I_m}\right) \qquad \textbf{(11.23)}$$

For the other form, the equation $v_C = Ee^{-t/\tau}$ is a close match with $v_L = Ee^{-t/\tau}$, permitting a derivation similar to that employed for Eq. 10.20:

$$t = -\tau \log_e \frac{v_L}{E} \tag{11.24}$$

11.17 THEVENIN'S THEOREM $\tau = L/R_{Th}$

In Chapter 10, we found that there are occasions when the circuit does not have the basic form of Fig. 11.38. The same is true for inductive networks. Again, it is necessary to find the Thevenin equivalent circuit before proceeding in the manner described in this chapter. Consider the following example.

EXAMPLE 11.9 For the network of Fig. 11.53, sketch the waveforms of i_L and v_L if the switch is closed at $t = 0$ s.

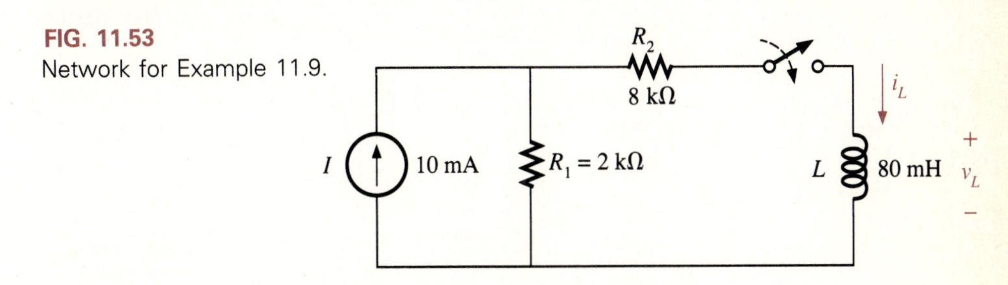

FIG. 11.53
Network for Example 11.9.

Solution To find R_{Th} (Fig. 11.54):

$$R_{Th} = R_1 + R_2 = 2\ \text{k}\Omega + 8\ \text{k}\Omega = 10\ \text{k}\Omega$$

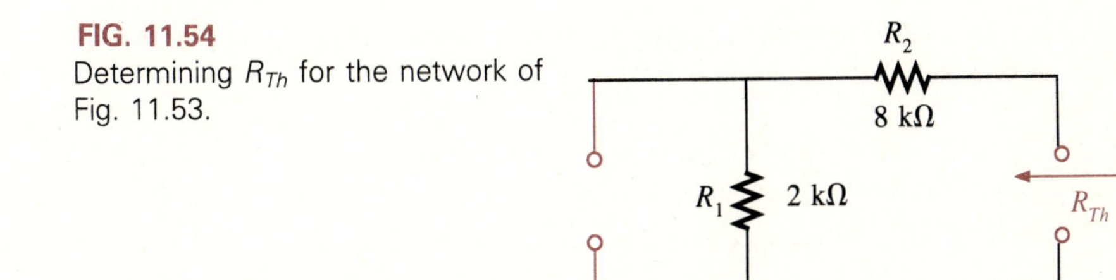

FIG. 11.54
Determining R_{Th} for the network of Fig. 11.53.

To find E_{Th} (Fig. 11.55):

$$I_2 = 0\ \text{A} \quad \text{and} \quad E_{Th} = V_1 - V_2$$

We have

$$V_2 = I_2R = (0)R_2 = 0\ \text{V}$$

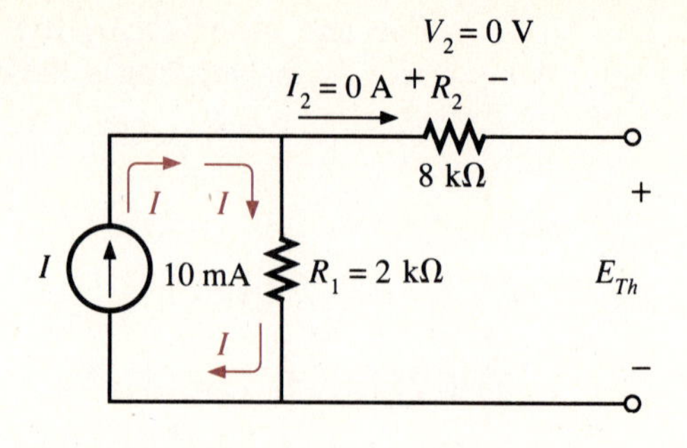

FIG. 11.55
Determining E_{Th} for the network of Fig. 11.53.

and

$$V_1 = I_1R_1 = IR_1 = (10 \text{ mA})(2 \text{ k}\Omega) = 20 \text{ V}$$

Therefore, $E_{Th} = V_1 - V_2 = 20 \text{ V} - 0 \text{ V} = 20 \text{ V}$.

Replacing the 80-mH inductor (Fig. 11.56):

$$I_m = \frac{E}{R} = \frac{E_{Th}}{R_{Th}} = \frac{20 \text{ V}}{10 \text{ k}\Omega} = \mathbf{2\ mA}$$

$$\tau = \frac{L}{R} = \frac{L}{R_{Th}} = \frac{80 \text{ mH}}{10 \text{ k}\Omega} = \mathbf{8\ \mu s} \qquad \text{(fast storage cycle)}$$

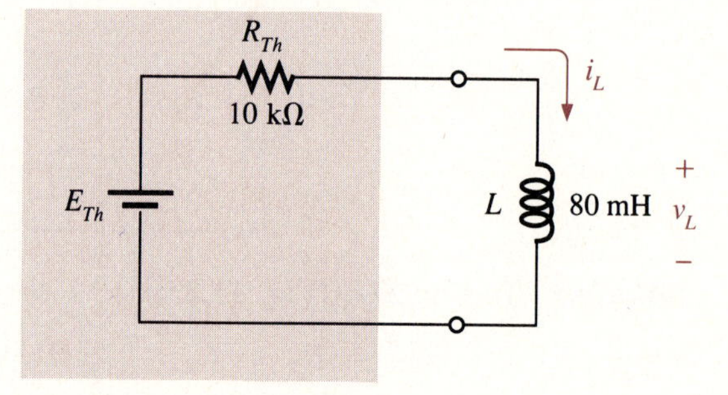

FIG. 11.56
Substituting the Thevenin equivalent circuit.

The curve for i_L appears in Fig. 11.57, and the curve for v_L appears in Fig. 11.58.

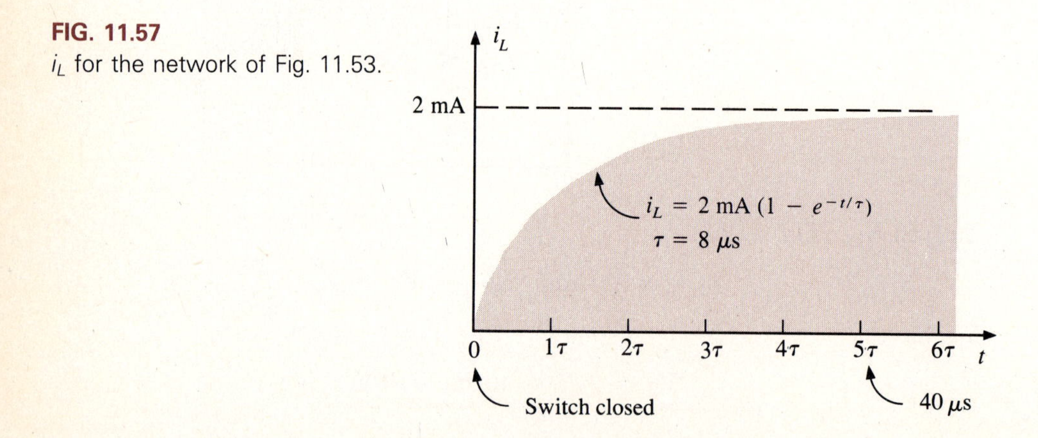

FIG. 11.57
i_L for the network of Fig. 11.53.

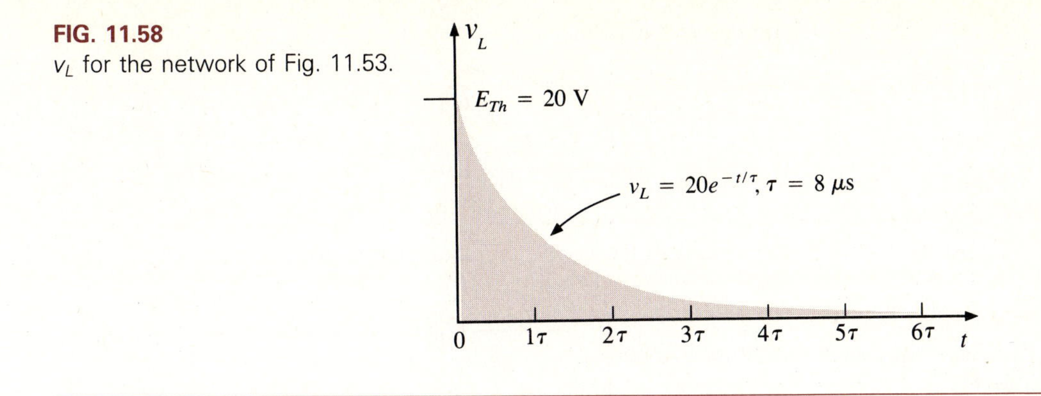

FIG. 11.58
v_L for the network of Fig. 11.53.

11.18 INDUCTORS IN SERIES AND PARALLEL

Inductors, like resistors and capacitors, can be placed in series or parallel. Increasing levels of inductance can be obtained by placing inductors in series, whereas decreasing levels can be obtained by placing inductors in parallel.

For inductors in series, the total inductance is found in the same manner as the total resistance of resistors in series (Fig. 11.59):

$$L_T = L_1 + L_2 + L_3 + \cdots + L_N \quad (11.25)$$

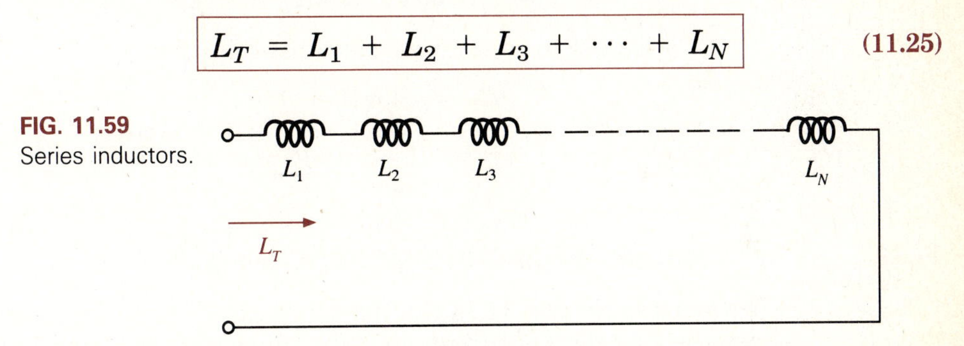

FIG. 11.59
Series inductors.

For inductors in parallel, the total inductance is found in the same manner as the total resistance of resistors in parallel (Fig. 11.60):

$$\frac{1}{L_T} = \frac{1}{L_1} + \frac{1}{L_2} + \frac{1}{L_3} + \cdots + \frac{1}{L_N} \quad (11.26)$$

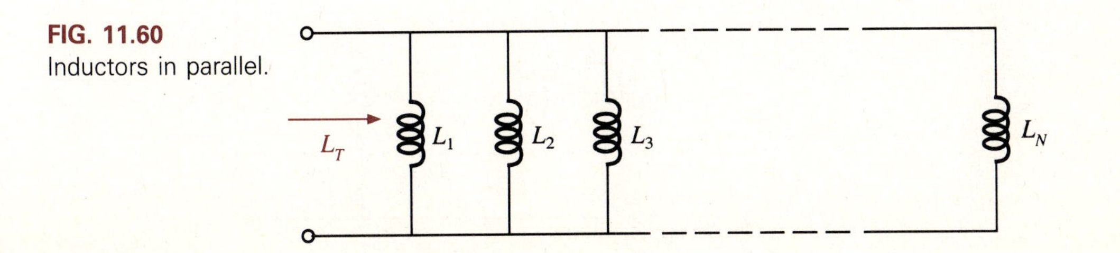

FIG. 11.60
Inductors in parallel.

For two inductors in parallel,

$$L_T = \frac{L_1 L_2}{L_1 + L_2} \tag{11.27}$$

EXAMPLE 11.10 Determine the total inductance L_T of the configuration of Fig. 11.61.

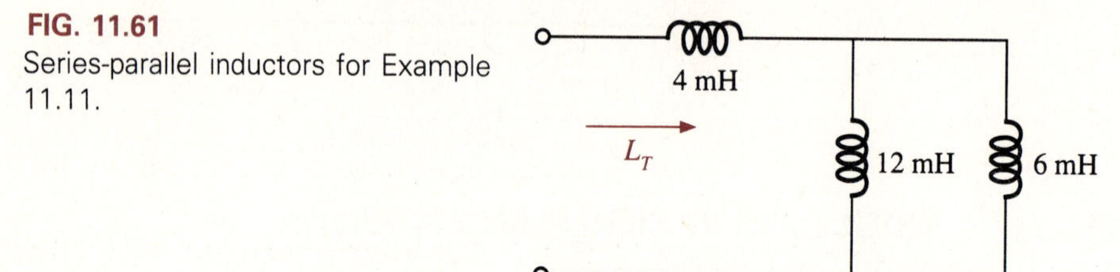

FIG. 11.61
Series-parallel inductors for Example 11.11.

Solution The 12-mH and 6-mH coils are in parallel, resulting in

$$L'_T = \frac{(12\text{ mH})(6\text{ mH})}{12\text{ mH} + 6\text{ mH}} = \frac{72\text{ mH}}{18} = 4\text{ mH}$$

L'_T is then in series with the 4-mH coil, and

$$\begin{aligned} L_T &= 4\text{ mH} + L'_T \\ &= 4\text{ mH} + 4\text{ mH} = \mathbf{8\text{ mH}} \end{aligned}$$

11.19 *R-L* AND *R-L-C* CIRCUITS WITH DC INPUTS

We found in Section 11.14 that for all practical purposes, an inductor can be replaced by a short circuit in a dc circuit after a period of time greater than five time constants has passed. If in the following circuits we assume that all the currents and voltages have reached their final values, the current through each inductor can be found by replacing each inductor by a short circuit. For the circuit of Fig. 11.62, for example, the short-circuit equivalent has been

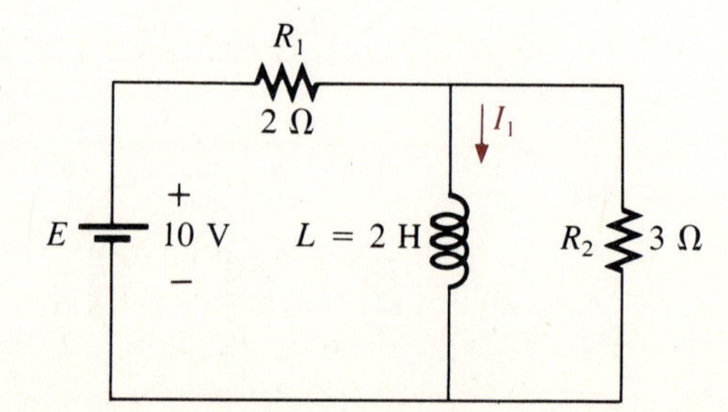

FIG. 11.62
Steady-state inductive network.

inserted for the inductor in Fig. 11.63, and

$$I_1 = \frac{E}{R_1} = \frac{10\text{ V}}{2\ \Omega} = \mathbf{5\ A}$$

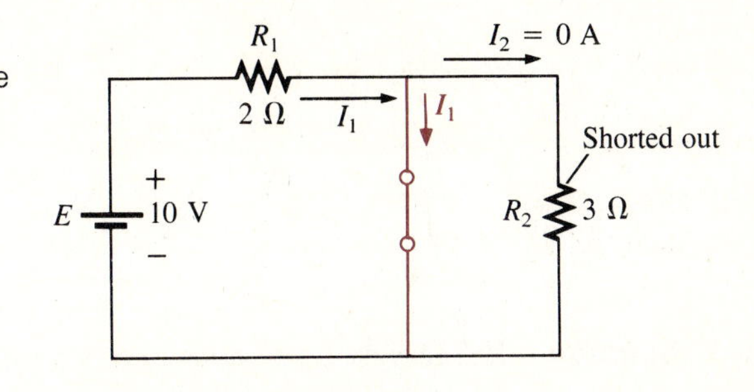

FIG. 11.63
Determining the steady-state I_1 for the network of Fig. 11.62.

For the circuit of Fig. 11.64, the short-circuit equivalent has been inserted for each inductor in Fig. 11.65, and

$$I = \frac{E}{R_2 \| R_3} = \frac{21\text{ V}}{2\ \Omega} = \mathbf{10.5\ A}$$

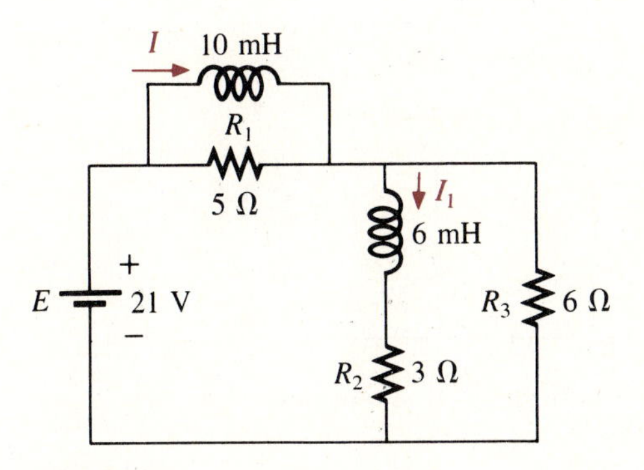

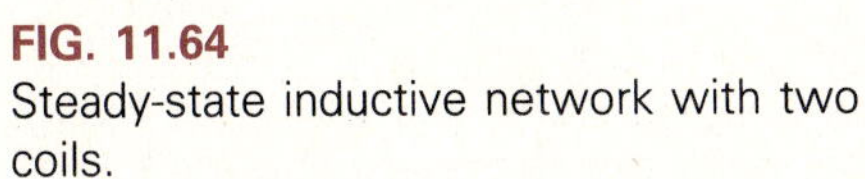
FIG. 11.64
Steady-state inductive network with two coils.

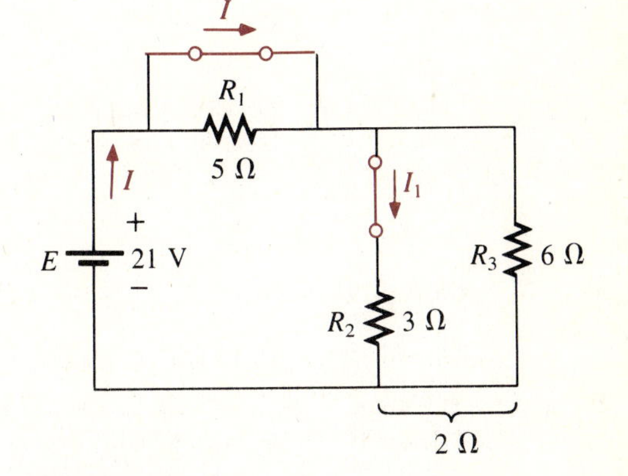

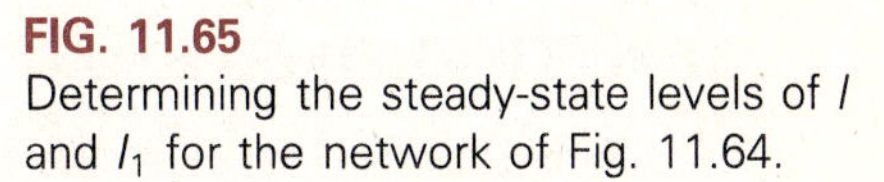
FIG. 11.65
Determining the steady-state levels of I and I_1 for the network of Fig. 11.64.

Applying the current divider rule,

$$I_1 = \frac{R_3 I}{R_3 + R_2} = \frac{(6\ \Omega)(10.5\text{ A})}{6\ \Omega + 3\ \Omega} = \frac{63\text{ A}}{9} = \mathbf{7\ A}$$

In the following examples we will assume that the voltage across the capacitors and the current through the inductors have reached their final values. Under these conditions, the inductors can be replaced by short circuits, and the capacitors can be replaced by open circuits.

EXAMPLE 11.11

Find the steady-state current I_L and voltage V_C for the network of Fig. 11.66.

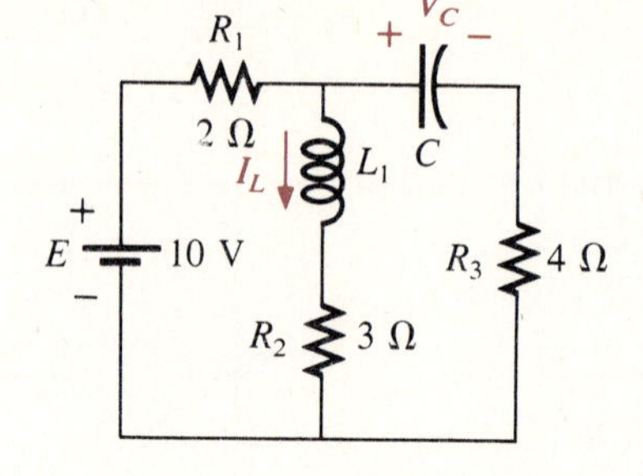

FIG. 11.66
Network for Example 11.11.

Solution The steady-state equivalent appears in Fig. 11.67.

$$I_L = \frac{E}{R_1 + R_2} = \frac{10\text{ V}}{5\ \Omega} = \mathbf{2\ A}$$

and

$$V_C = \frac{R_2E}{R_2 + R_1} = \frac{(3\ \Omega)(10\text{ V})}{3\ \Omega + 2\ \Omega} = \mathbf{6\ V}$$

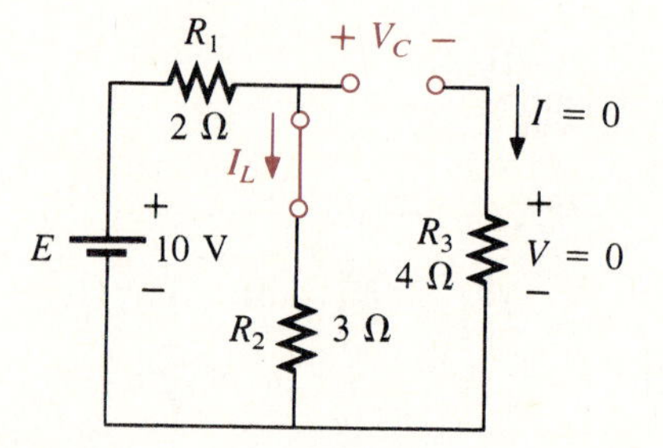

FIG. 11.67
Determining the steady-state levels of V_C and I_L for the network of Fig. 11.66.

11.20 ENERGY STORED BY AN INDUCTOR

The ideal inductor, like the ideal capacitor, does not dissipate the electrical energy supplied to it. It stores the energy in the form of a magnetic field. A plot of the voltage, current, and power to an inductor is shown in Fig. 11.68 during the buildup of the magnetic field surrounding the inductor. The energy stored is represented by the shaded area under the power curve. Using calculus, we can show that the evaluation of the area under the curve yields

$$W_{\text{stored}} = \frac{1}{2}LI_m^2 \qquad \text{(joules, J)} \qquad (11.28)$$

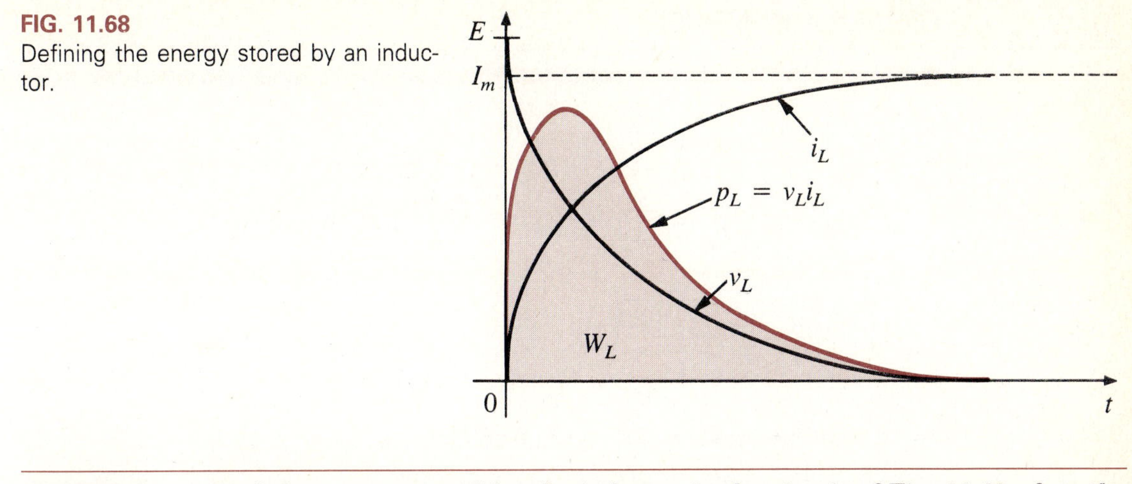

FIG. 11.68
Defining the energy stored by an inductor.

EXAMPLE 11.12 Find the energy stored by the inductor in the circuit of Fig. 11.69 when the current through it has reached its final value.

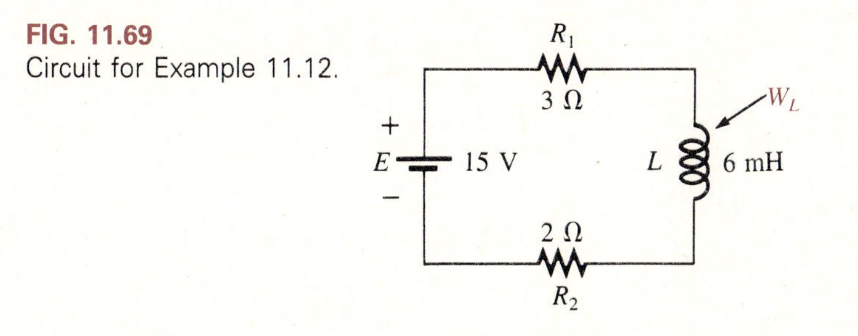

FIG. 11.69
Circuit for Example 11.12.

Solution The steady-state equivalent appears in Fig. 11.70.

$$I_m = \frac{E}{R_1 + R_2} = \frac{15\text{ V}}{3\ \Omega + 2\ \Omega} = \frac{15\text{ V}}{5\ \Omega} = 3\text{ A}$$

$$W_{\text{stored}} = \frac{1}{2}LI_m^2 = \frac{1}{2}(6 \times 10^{-3}\text{ H})(3\text{ A})^2 = \frac{54}{2} \times 10^{-3}\text{ J}$$

$$= \mathbf{27 \times 10^{-3}\ J}$$

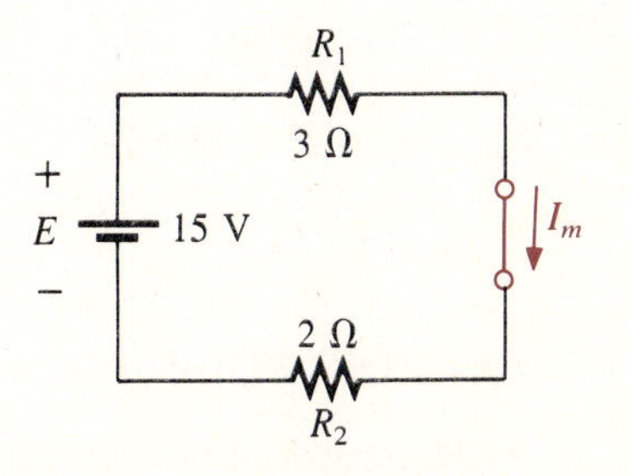

FIG. 11.70
Determining the steady-state level of i_L for the circuit of Fig. 11.69.

FORMULA SUMMARY

$$B = \frac{\Phi}{A}$$

$$\mu_r = \frac{\mu}{\mu_o}, \qquad \mu_o = 4\pi \times 10^{-7} \text{ Wb/A·m}$$

$$\mathcal{R} = \frac{l}{\mu A}$$

$$\Phi = \frac{\mathcal{F}}{\mathcal{R}}$$

$$\mathcal{F} = NI, \qquad \mathcal{F} = Hl$$

$$B = \mu H$$

$$e = N\frac{d\Phi}{dt}$$

$$L = \frac{N^2\mu A}{l}, \qquad L = \mu_r L_o$$

$$v_L = L\frac{di_L}{dt}, \qquad v_{L_{av}} = L\frac{\Delta i}{\Delta t}$$

Storage

$$i_L = \frac{E}{R}(1 - e^{-t/\tau}), \qquad \tau = \frac{L}{R}$$

$$v_L = Ee^{-t/\tau}$$

Series Inductors

$$L_T = L_1 + L_2 + L_3 + \cdots + L_N$$

Parallel Inductors

$$\frac{1}{L_T} = \frac{1}{L_1} + \frac{1}{L_2} + \frac{1}{L_3} + \cdots + \frac{1}{L_N}$$

$$L_T = \frac{L_1 L_2}{L_1 + L_2}$$

$$W_L = \frac{1}{2}LI_m^2$$

CHAPTER SUMMARY

The similarities that exist between magnetic and electric circuits should be employed to their fullest when analyzing magnetic circuits for the first time. For series magnetic circuits, the flux (flow variable) is the same throughout, whereas for parallel magnetic circuits, $\Phi_T = \Phi_1 + \Phi_2$ to match $I_T = I_1 + I_2$ for electric circuits. The applied "pressure" on the magnetic circuit is NI compared to the supply voltage E of an electric circuit. Ampère's circuital law has a direct similarity with Kirchhoff's voltage law in that the applied pressure equals the sum of the magnetomotive drops around any closed flux path of the magnetic circuit. For drops in magnetomotive force around a closed loop, the magnetizing force H can be determined from the B-H curve once B is calculated from $B = \Phi/A$. Due to the variation in μ with magnetizing force H, the reluctance of a magnetic reaction is seldom determined. Generally, the background established thus far in electric circuits will make the analysis of magnetic circuits fairly straight-forward if the analogies just described are used to their full advantage.

There are also a number of helpful similarities between the capacitive and inductive elements that should be helpful when analyzing inductive networks for the first time. The transient curves have the same shape with a decay and storage cycle of 5τ. In addition, a number of conclusions associated with the voltage of a capacitive circuit can be applied to the current of an inductive circuit. For instance, the voltage across a capacitor or the current of an inductor cannot change instantaneously. When the switch is first thrown on a storage cycle, the current of a capacitive network jumps to its peak value, and the voltage of a coil does the same. For steady-state conditions, the capacitor can be approximated by an open circuit and the open-circuit voltage determined, whereas an inductor can be replaced by a short-circuit equivalent and the short-circuit current determined. The impact of the relationship $v_{L_{av}} = L(\Delta i_L/\Delta t)$ is also very similar to that experienced with $i_{C_{av}} = C(\Delta v_C/\Delta t)$.

A common error when attacking the first few problems is to use, incorrectly, $\tau = RL$ rather than the correct form of $\tau = L/R$ due to the similar equation $\tau = RC$ for capacitive networks. The results obtained with the incorrect form will clearly reveal an error in substitution was made.

Use the similarities between the equations for series or parallel resistors and series or parallel inductors, respectively, to remember the equations employed for combining inductors. In addition note the correspondence between the equation for the stored energy of an inductor ($W_L = \frac{1}{2}LI_m^2$) and that for a capacitor ($W_C = \frac{1}{2}CV_C^2$).

In total, this chapter introduced a host of new equations and concepts. However, the similarities to the content of earlier concepts should be a tremendous aid in approaching and solving the variety of problems and technical questions associated with magnetic and inductive circuits.

GLOSSARY

Ampère's circuital law A law establishing the fact that the algebraic sum of the rises and drops of the mmf around a closed loop of a magnetic circuit is equal to zero.

Choke A term often applied to an inductor, due to the ability of a coil to resist a change in current through it.

Diamagnetic materials Materials that have permeabilities slightly less than that of free space.

Domain A group of magnetically aligned atoms.

Electromagnetism Magnetic effects introduced by the flow of charge or current.

Faraday's law A law relating the voltage induced across a coil to the number of turns in the coil and the rate at which the flux linking the coil is changing.

Ferromagnetic materials Materials having permeabilities hundreds and thousands of times greater than that of free space.

Flux density (B) A measure of the flux per unit area perpendicular to a magnetic flux path. It is measured in teslas (T) or webers per square meter (Wb/m^2).

Hysteresis The lagging effect between flux density of a material and the magnetizing force applied.

Inductor A fundamental element of electrical systems constructed of numerous turns of wire around a ferromagnetic or air core.

Lenz's law A law stating that an induced effect is always such as to oppose the cause that produced it.

Magnetic flux lines Lines of a continuous nature that reveal the strength and direction of the magnetic field.

Magnetizing force (H) A measure of the magnetomotive force per unit length of a magnetic circuit.

Magnetomotive force ($\mathcal{F}$) The pressure required to establish magnetic flux in a ferromagnetic material. It is measured in ampere-turns (A·t).

Paramagnetic materials Materials that have permeabilities slightly greater than that of free space.

Permanent magnet A material such as steel or iron that will remain magnetized for long periods of time without the aid of external means.

Permeability (μ) A measure of the ease with which magnetic flux can be established in a material. It is measured in webers per ampere-meter.

Relative permeability (μ_r) The ratio of the permeability of a material to that of free space.

Reluctance ($\mathcal{R}$) A quantity determined by the physical characteristics of a material that will provide an indication of the "reluctance" of that material to the setting up of magnetic flux lines in the material. It is measured in rels or ampere-turns per weber.

Self-inductance A measure of the ability of a coil to oppose any change in current through the coil and to store energy in the form of a magnetic field in the region surrounding the coil.

PROBLEMS

Section 11.3

1. Using Appendix F, fill in the blanks in the following table. Indicate the units for each quantity.

	Φ	B
SI	5×10^{-4} Wb	8×10^{-4} T
CGS	___	___
English	___	___

2. Repeat Problem 1 for the following table if area = 2 in.^{2}:

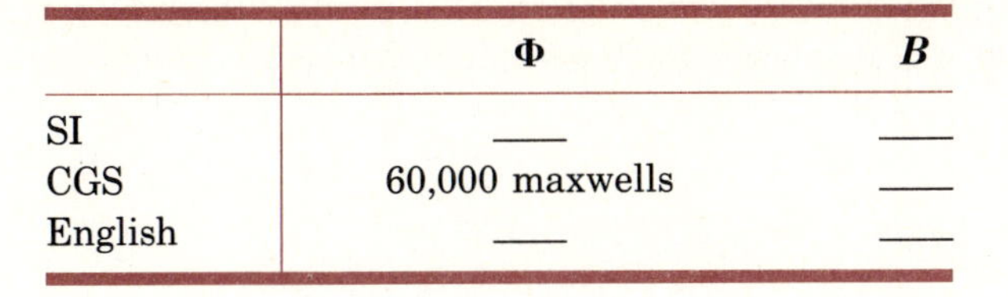

	Φ	B
SI	——	——
CGS	60,000 maxwells	——
English	——	——

3. For the electromagnet of Fig. 11.71:
  a. Find the flux density in the core.
  b. Sketch the magnetic flux lines and indicate their direction.
  c. Indicate the north and south poles of the magnet.

FIG. 11.71

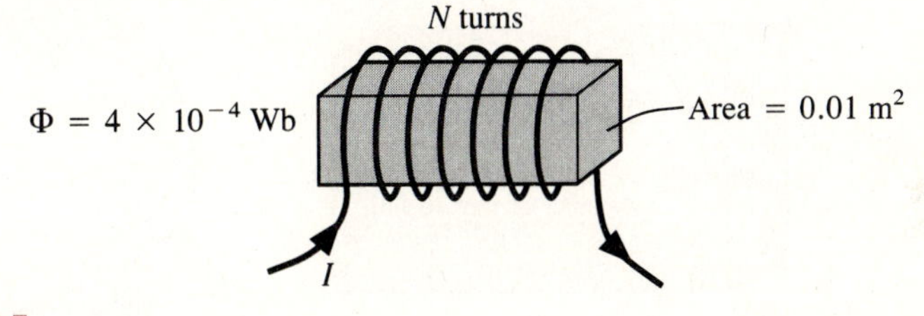

Section 11.5

4. a. Which section of Fig. 11.72 [(a), (b), or (c)] has the largest reluctance to the setting up of flux lines through its longest dimension?
  b. Which has the least reluctance (see part a).

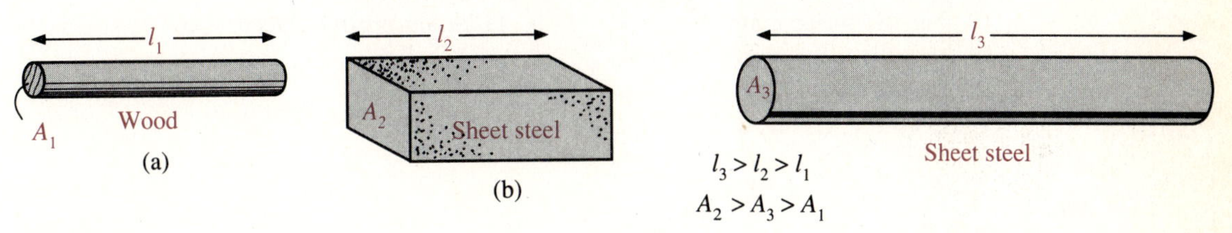

FIG. 11.72

Section 11.6

5. Find the reluctance of a magnetic circuit if a magnetic flux $\Phi = 4.2 \times 10^{-4}$ Wb is established by an impressed mmf of 400 A·t.

6. What is the magnetomotive force on a core if the coil wrapping has 250 turns and the current is 1.2 A?

Section 11.7

7. Find the magnetizing force applied to the core of Fig. 11.71 if $N = 150$ t, $I = 2$ A, and $l = 0.1$ m.

8. For the system of Fig. 11.17, if $H = 10$ A·t/m, what magnetomotive force would be required to establish the flux in the core if the mean length of the core is 6 in.?

9. a. In reference to the curves of Fig. 11.18, what is the relative permeability (μ_r) of the sheet steel sample at $H = 3000$ A·t/m?
  b. Repeat part a for $H = 300$ A·t/m.
  c. Comment on the results obtained for parts a and b.

10. If a magnetizing force H of 600 A·t/m is applied to a magnetic circuit, a flux density B of 1200×10^{-4} Wb/m^2 is established. Find the permeability μ of a material that will produce twice the original flux density for the same magnetizing force.

Section 11.8

***11.** For the magnetic circuits of Fig. 11.73:

a. Which appears to have the highest reluctance?

b. What is the reluctance of each circuit?

c. Find the flux in each core.

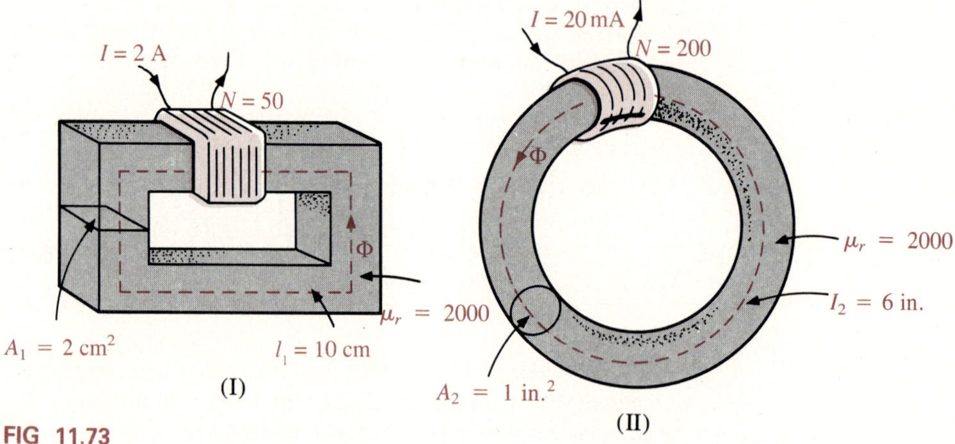

FIG 11.73

12. For the series magnetic circuit of Fig. 11.74, determine the current I necessary to establish the indicated flux.

13. Find the current necessary to establish a flux of $\Phi = 3 \times 10^{-4}$ Wb in the series magnetic circuit of Fig. 11.75.

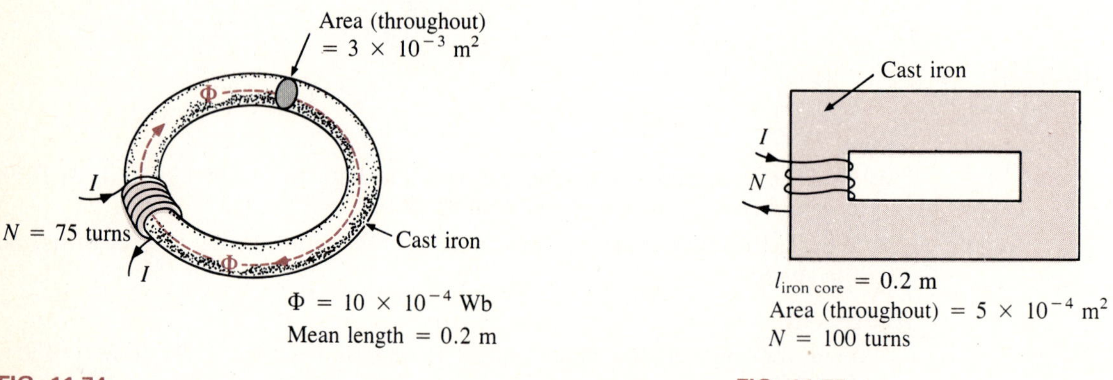

FIG. 11.74

FIG. 11.75

14. a. Find the number of turns N required to establish a flux $\Phi = 12 \times 10^{-4}$ Wb in the magnetic circuit of Fig. 11.76.

b. Find the permeability μ of the material.

15. Find the magnetic flux Φ established in the series magnetic circuit of Fig. 11.77.

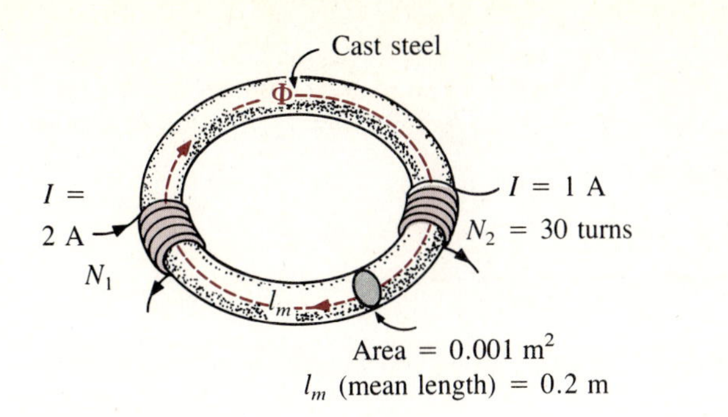

FIG. 11.76

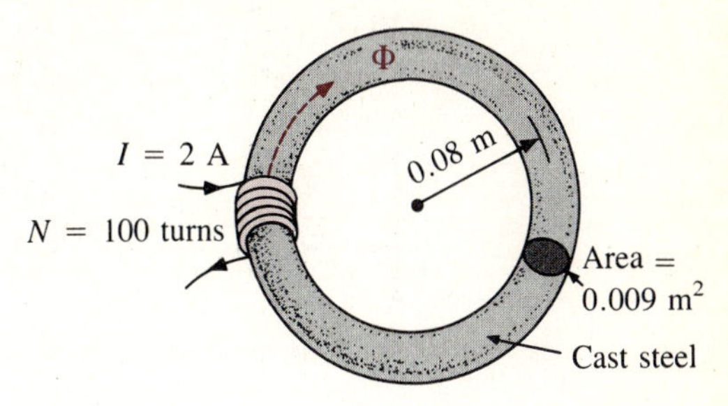

FIG. 11.77

Section 11.9

16. If the flux linking a coil of 50 turns changes at a rate of 0.085 Wb/s, what is the induced voltage across the coil?

17. Determine the rate of change of flux linking a coil if 20 V are induced across a coil of 40 turns.

18. How many turns does a coil have if 42 mV are induced across the coil by a change of flux of 0.003 Wb/s?

Section 11.11

19. Find the inductance L in henries of the inductor of Fig. 11.78.

***20.** Repeat Problem 19 with $l = 4$ in. and $d = 1/4$ in.

21. a. Find the inductance L in henries of the inductor of Fig. 11.79.
b. Repeat part a if a ferromagnetic core is added having a μ_r of 2000.

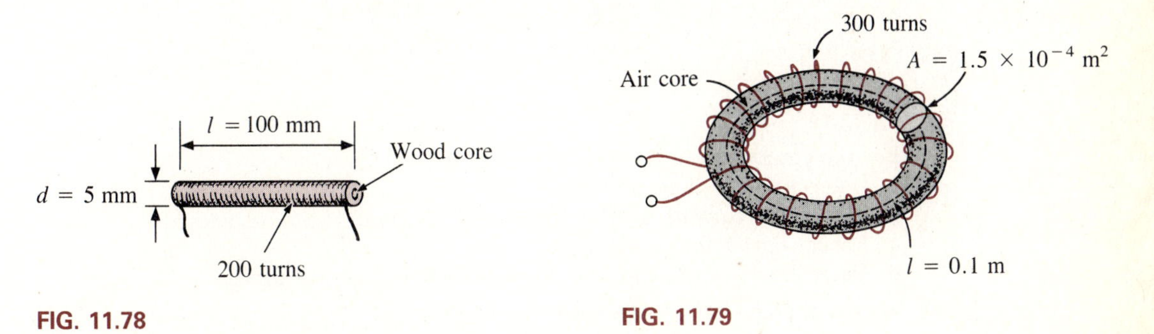

FIG. 11.78

FIG. 11.79

Section 11.13

22. Find the voltage induced across a coil of 5 H if the rate of change of current through the coil is
a. 0.5 A/s
b. 60 mA/s
c. 0.04 A/ms

23. Find the induced voltage across a 50-mH inductor if the current through the coil changes at a rate of 0.1 mA/μs.

24. Find the waveform for the voltage induced across a 200-mH coil if the current through the coil is as shown in Fig. 11.80.

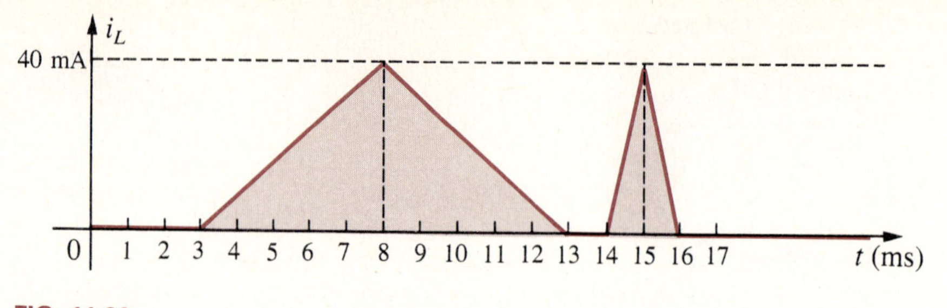

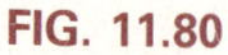
FIG. 11.80

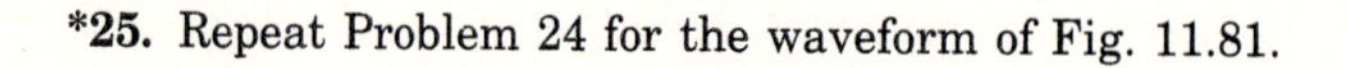
***25.** Repeat Problem 24 for the waveform of Fig. 11.81.

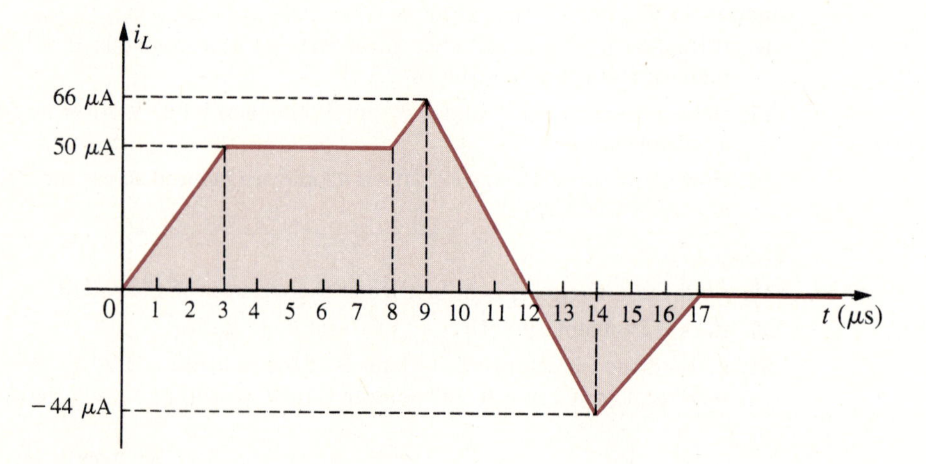

FIG. 11.81

Section 11.14

26. For the circuit of Fig. 11.82:
  a. Determine the time constant.
  b. Sketch the current i_L after the switch is closed.
  c. Repeat part b for v_L and v_R.
  d. Determine i_L and v_L at one, three, and five time constants.

27. Repeat Problem 26 for the network of Fig. 11.83.

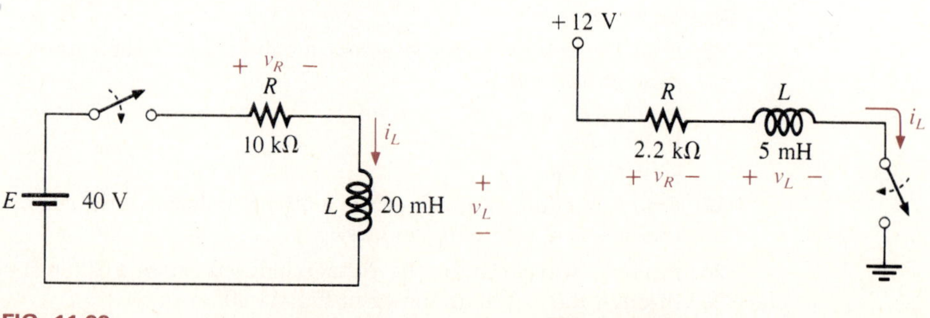

FIG. 11.82

FIG. 11.83

Section 11.15

28. If a load resistor $R_2 = 40\ \text{k}\Omega$ were applied in parallel with the 20-mH coil of Fig. 11.82, the instant the switch is opened (only after the full storage period has passed $t > 5\tau$), sketch the resulting waveforms for i_L and v_L.

***29.** **a.** If a load resistor $R_2 = 100\ \Omega$ were placed in parallel with the 5-mH coil of Fig. 11.83, the instant the switch is opened (only after the full storage period has passed $t > 5\tau$), sketch the resulting waveforms for i_L and v_L.

b. Sketch the waveforms for i_L for both the storage and decay phases of Problems 27 and 29(a) as a single continuous graph showing the effects of the differing time constants. Assume the switch is opened at $t = 5\tau$ of the storage phase.

Sections 11.16 and 11.17

30. For the network of Fig. 11.84:

a. Sketch the waveforms for the current i_L and the voltage v_L when the switch is closed.

b. Determine i_L and v_L when $t = 2\ \mu\text{s}$.

31. **a.** Sketch i_L and v_L following the closing of the switch in Fig. 11.85.

b. Determine i_L and v_L at $t = 100$ ns.

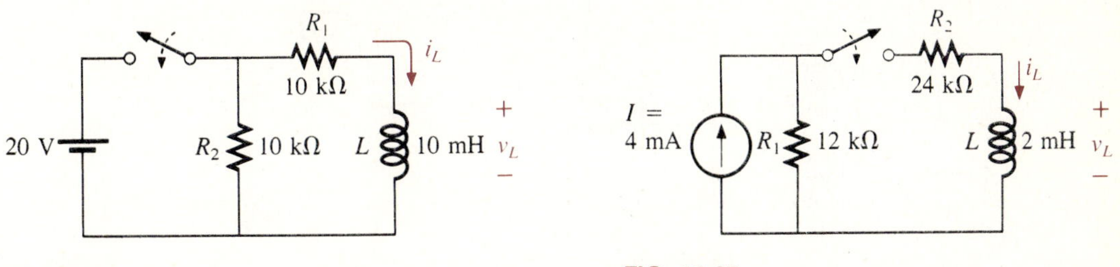

FIG. 11.84

FIG. 11.85

***32.** **a.** Sketch i_L and v_L following the closing of the switch in Fig. 11.86.

b. Calculate i_L and v_L at $t = 10\ \mu\text{s}$.

c. Sketch the current i_L and the voltage v_L if the switch is opened at $t = 10\ \mu\text{s}$.

FIG. 11.86

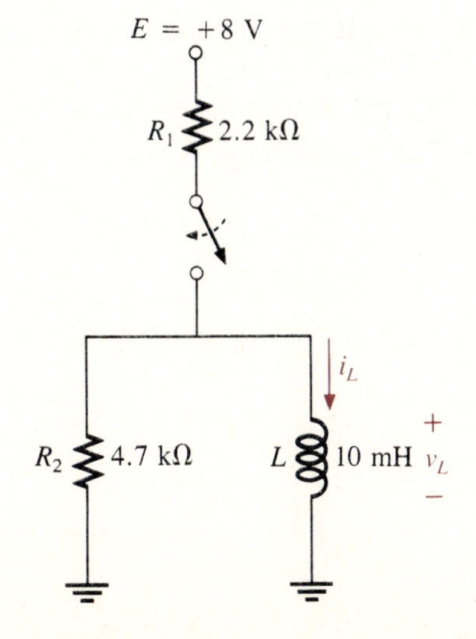

Section 11.18

33. Find the total inductance of the circuits of Fig. 11.87.

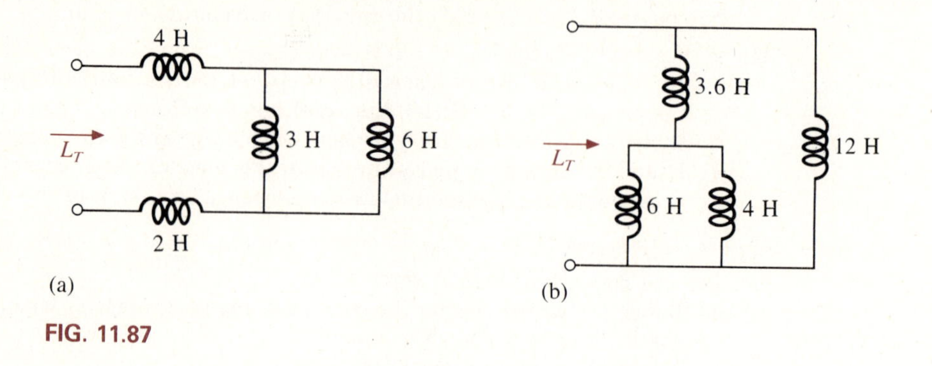

FIG. 11.87

***34.** Reduce the networks of Fig. 11.88 to the fewest elements.

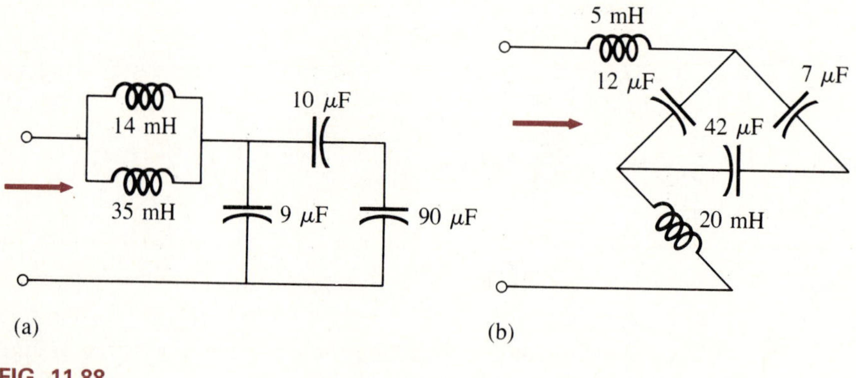

FIG. 11.88

Section 11.19

For Problems 35 and 36 assume that the voltage across each capacitor and the current through each inductor have reached their final values.

35. Find the voltages V_1 and V_2 and the current I_1 for the circuit of Fig. 11.89.

36. Find the current I_1 and the voltage V_1 for the circuit of Fig. 11.90.

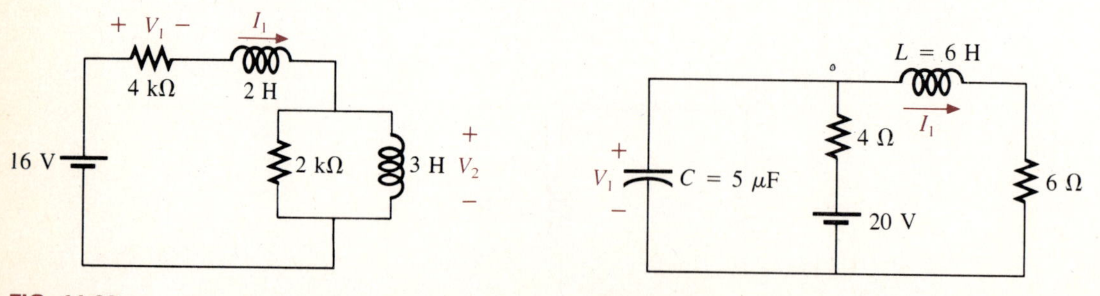

FIG. 11.89

FIG. 11.90

Section 11.20

37. Find the energy stored by each inductor of Problem 35.
38. Find the energy stored by the capacitor and inductor of Problem 36.

Computer Problems

39. Write a program that will provide the flux density in teslas (T), gauss, and lines per square inch given the flux in webers and area in square inches.
40. Research available material and write a program that will provide the value of B in webers per square meter for the sheet steel sample of Fig. 11.21 given the value of H in ampere-turns per meter.
41. Write a program to determine the inductance of a coil given the number of turns, area, length, and relative permeability of the core.
42. Given $i_L = 5\text{ mA}(1 - e^{-t/2\,\mu s})$, write a program that will tabulate i_L versus t for t from 0 to 20 μs in intervals of 1 μs. In addition, have the program label the levels of i_L at 1τ through 5τ.
43. Given E, R, and L, write a program to determine the time t when specific levels of i_L and v_L are reached (Eqs. 11.23 and 11.24).

12

Sinusoidal Alternating Waveforms

OBJECTIVES

- ☐ Understand how sinusoidal ac voltages can be generated.
- ☐ Become aware of the characteristics of alternating waveforms and how to calculate the various magnitudes, frequencies, and periods.
- ☐ Be able to plot a sinusoidal waveform against a horizonal axis of degrees, radians, or time.
- ☐ Understand how to determine the phase relationship between sinusoidal waveforms.
- ☐ Be able to calculate the average value of a variety of signals and the effective value of sinusoidal waveforms.
- ☐ Become aware of a variety of ac meters, how they are used, and what they reveal about the measured quantity.

12.1 INTRODUCTION

The analysis thus far has been limited to dc networks, networks in which the currents or voltages are fixed in magnitude except for transient effects. We now turn our attention to the analysis of networks in which the magnitude of the source varies in a set manner. Of particular interest is the time-varying voltage that is commercially available in large quantities and is commonly called the *alternating current (ac) voltage*. The term *alternating* reveals that the waveform alternates between two prescribed levels in a set time sequence (Fig. 12.1). To be absolutely correct, the term *sinusoidal, square wave,* or *triangular* must also be applied. The pattern of particular interest is the *sinusoidal* ac voltage of Fig. 12.1, generated by utilities throughout the world. Other reasons for being interested in this form include its application throughout electrical, electronic, communication, and industrial systems. In addition, the chapters to follow will reveal that the waveform itself has a number of characteristics that result in a unique response when applied to the basic electrical elements. A wide range of theorems and methods introduced for dc networks are also applied to sinusoidal ac systems. Although the application of sinusoidal signals will raise the required mathematical level, once the notation is understood, most of the concepts introduced in the dc chapters can be applied to ac networks with a minimum of added difficulty.

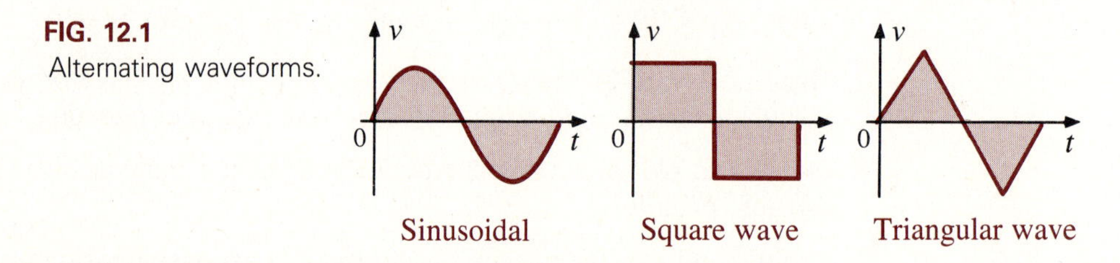

FIG. 12.1
Alternating waveforms.

12.2 SINUSOIDAL AC VOLTAGE GENERATION

The characteristics of the sinusoidal voltage and current and their effect on the basic R, L, and C elements are described in some detail in this chapter and those to follow. Of immediate interest is the generation of sinusoidal voltages.

The terminology *ac generator,* or *alternator,* should not be new to most technically oriented students. An ac generator is an electromechanical device capable of converting mechanical power to electrical power. As shown in the very basic ac generator of Fig. 12.2, it is constructed of two main components: the *rotor* (or armature, in this case) and the *stator*. As implied by the terminology, the rotor rotates within the framework of the stator, which is stationary. When the rotor is caused to rotate by some mechanical power such as is available from the forces of rushing water (dams) or steam-turbine engines, the conductors on the rotor cut magnetic lines of force established by the poles of the stator, as shown in Fig. 12.2. The poles may be those of a permanent magnet or may result from the turns of wire around the ferromagnetic core of the pole through which a dc current is passed to establish the necessary magnetomotive force for the required flux density.

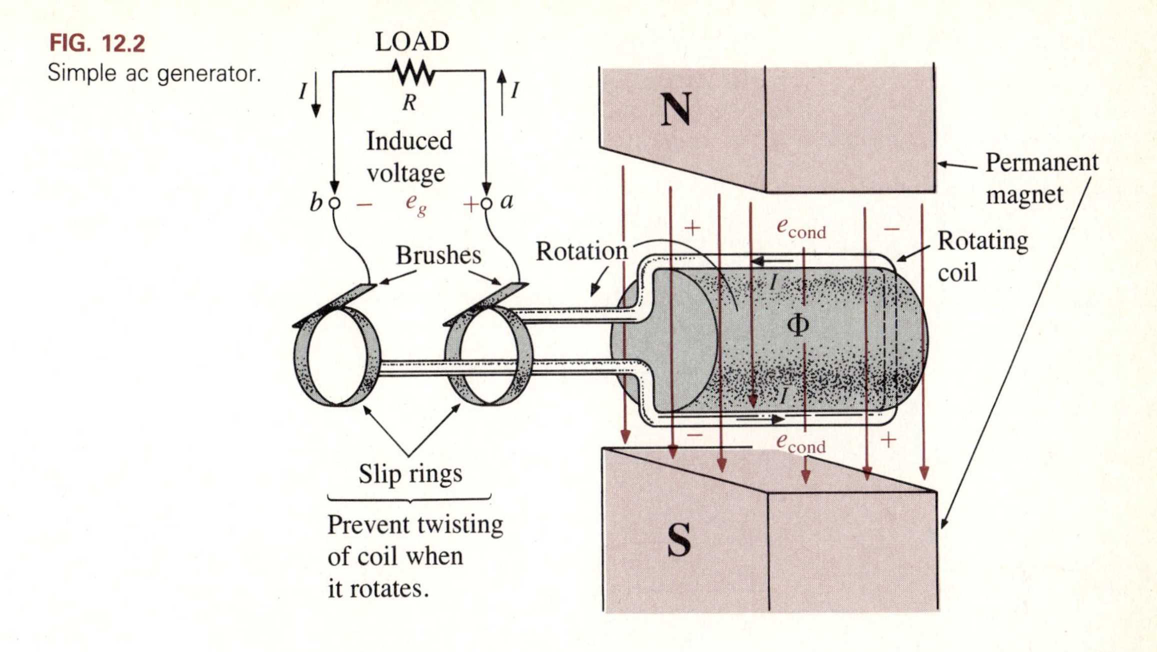

FIG. 12.2
Simple ac generator.

Dictated by Eq. 11.20, each conductor on the rotor will have a voltage induced across it, as shown in Fig. 12.2. Note that the induced voltage across each conductor is additive, so the generated terminal voltage is the sum of the two induced voltages. Since the armature of Fig. 12.2 is rotating and the output terminals a and b are connected to some fixed external load, there is the necessity for *slip rings*. Slip rings are circular conducting surfaces that provide a path of conduction from the generated voltage to the load without twisting of the coil at a and b when the coil rotates. The induced voltage has a polarity at terminals a and b that develops a current I having the direction indicated in Fig. 12.2.

Let us now consider a few representative positions of the rotating coil and determine the relative magnitude and polarity of the generated voltage at these positions. At the instant the coil passes through position 1 in Fig. 12.3(a), there are no flux lines being cut, and the induced voltage is zero. As the coil moves from position 1 to position 2, indicated in Fig. 12.3(b), the number of flux lines cut per unit time increases, resulting in an increased induced voltage across the coil. At position 3, the number of flux lines being cut per unit time is a maximum, resulting in a maximum induced voltage. The polarity of the induced voltage is the same as that of position 2.

As the coil continues to rotate toward position 4, indicated in Fig. 12.4(a), the polarity of the induced voltage remains the same, as shown in the figure, although the induced voltage drops due to the reduced number of flux lines cut per unit time. At position 4, the induced voltage is again zero, since the number of flux lines cut per unit time has dropped to zero. As the coil now turns toward position 5, the magnitude of the induced voltage again increases, but the polarity of the induced voltage is opposite to that of positions 2 and 3. The similarities between the coil positions of positions 2 and 5 and of 3 and 6 reveal

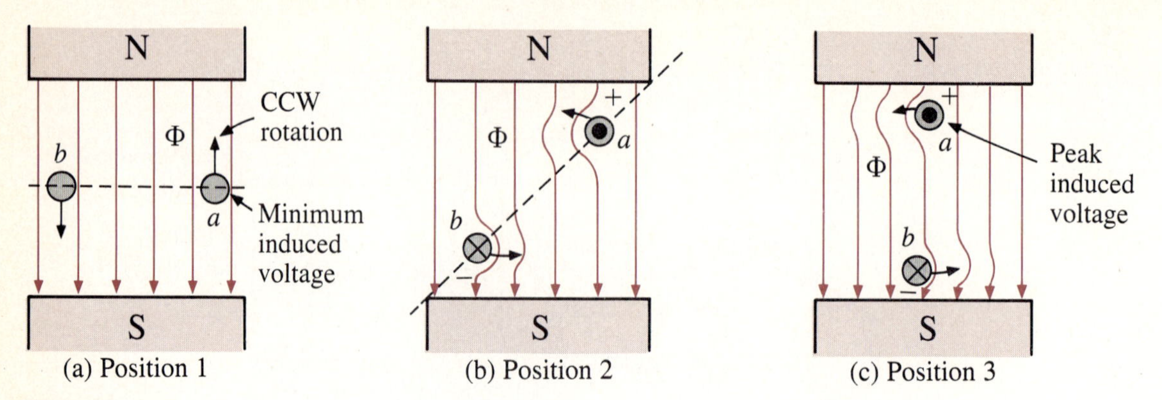

FIG. 12.3
Examining the voltage induced across the coil of Fig. 12.2 as it starts its counterclockwise rotations.

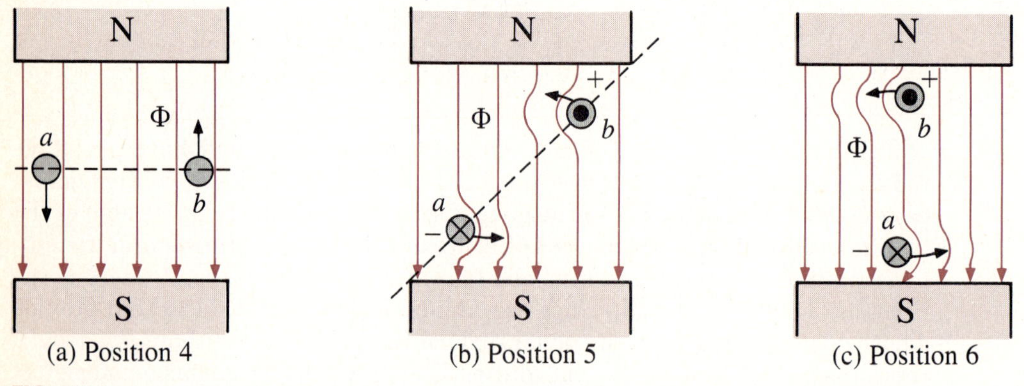

FIG. 12.4
Examining the voltage induced across the coil of Fig. 12.2 as it completes a revolution.

that the magnitude of the induced voltage is the same, although the polarity has reversed.

A continuous plot of the induced voltage e_g appears in Fig. 12.5. The polarities of the induced voltage are shown for terminals a and b to the left of the vertical axis.

Take a moment to relate the various positions to the resulting waveform of Fig. 12.5. This waveform will become very familiar in the discussions to follow. Note some of its obvious characteristics. As shown in the figure, if the coil is allowed to continue rotating, the generated voltage will repeat itself in equal intervals of time. Note also that the pattern is exactly the same below the axis as it is above and that it changes continually with time (the horizontal axis). At the risk of being repetitious, let us again state that the waveform of Fig. 12.5 is the appearance of a *sinusoidal ac voltage*.

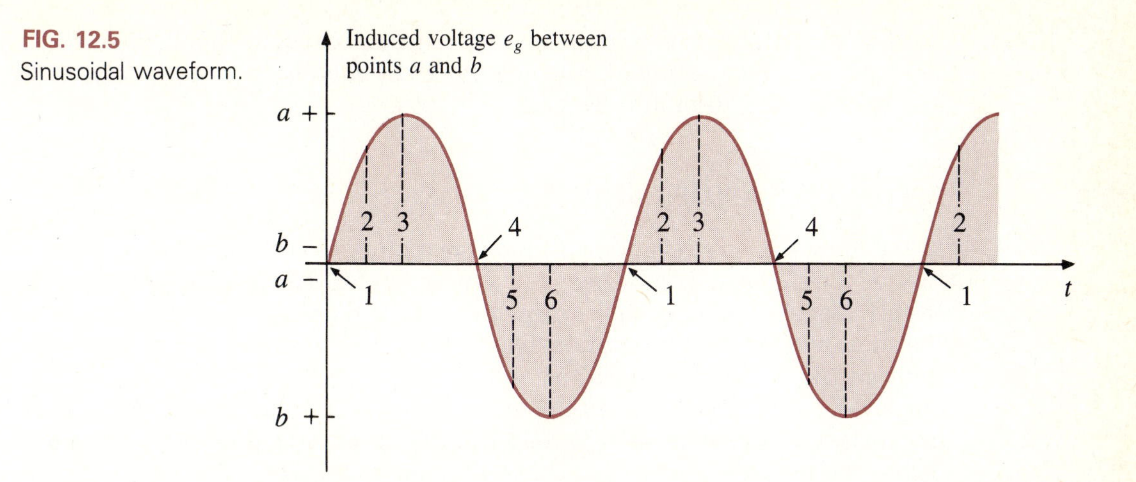

FIG. 12.5
Sinusoidal waveform.

The *function generator* of Fig. 12.6, which employs semiconductor electronic components, will provide sinusoidal, square-wave, and triangular signals over a wide frequency range determined by the dial setting and the chosen frequency range.

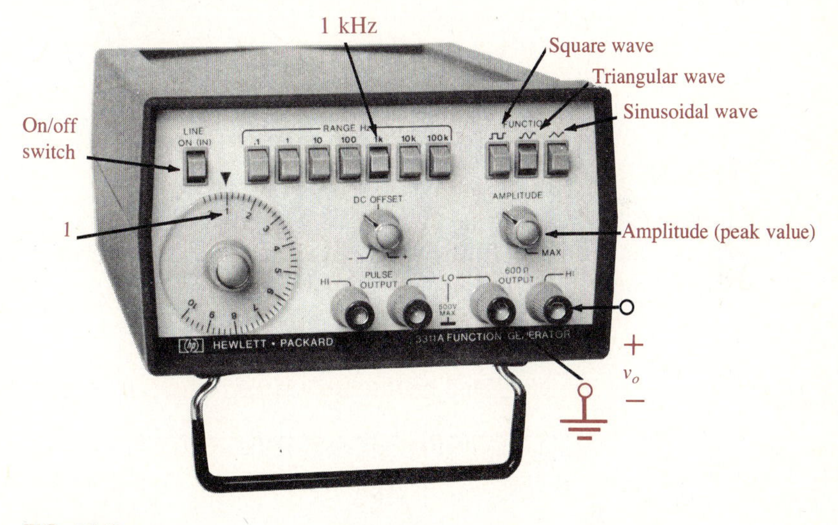

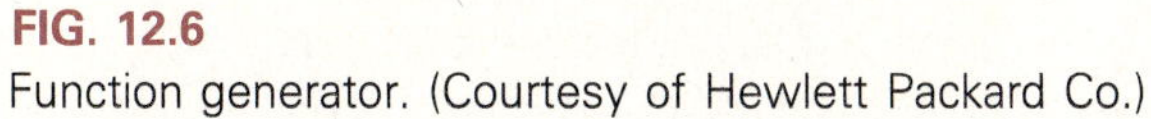
FIG. 12.6
Function generator. (Courtesy of Hewlett Packard Co.)

If a sinusoidal signal with a frequency of 1000 Hz and a peak value of 10 V were required, the sinusoidal function switch would first be depressed, as shown in the figure. Next, the dial would be set to 1 and the 1K-range switch depressed, as shown. The output frequency is the product of the dial position and the chosen range setting. The amplitude control would adjust the output until an ac voltmeter or oscilloscope indicated an output with a peak of 10 V. The same frequency could also have been set by choosing a dial position of 10 and pressing the 100-range switch.

12.3 DEFINED POLARITIES AND DIRECTION

In the following analysis, we will find it necessary to establish a set of polarities for the sinusoidal ac voltage and a direction for the sinusoidal ac current. In each case, the polarity and current direction will be in the positive portion of the sinusoidal waveform for an instant of time. This is shown in Fig. 12.7 with the symbols for the sinusoidal ac voltage and current. A lowercase letter is employed for each to indicate that the quantity is time dependent; that is, its magnitude will change with time. Note the absence of the term *sinusoidal* before the phrase *ac networks*. This will occur to an increasing degree as we progress; *sinusoidal* is to be understood unless otherwise indicated.

FIG. 12.7
(a) Sinusoidal ac voltage sources; (b) sinusoidal ac current sources.

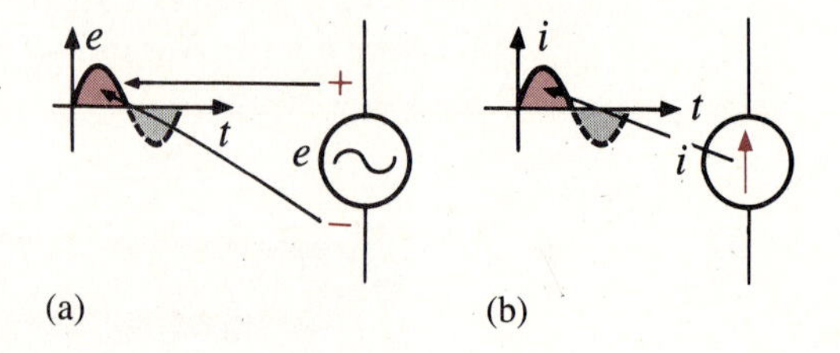

12.4 DEFINITIONS

The sinusoidal waveform of Fig. 12.8 with its additional notation will now be used as a model in defining a few basic terms. These terms can, however, be applied to any alternating waveform.

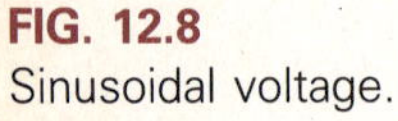

FIG. 12.8
Sinusoidal voltage.

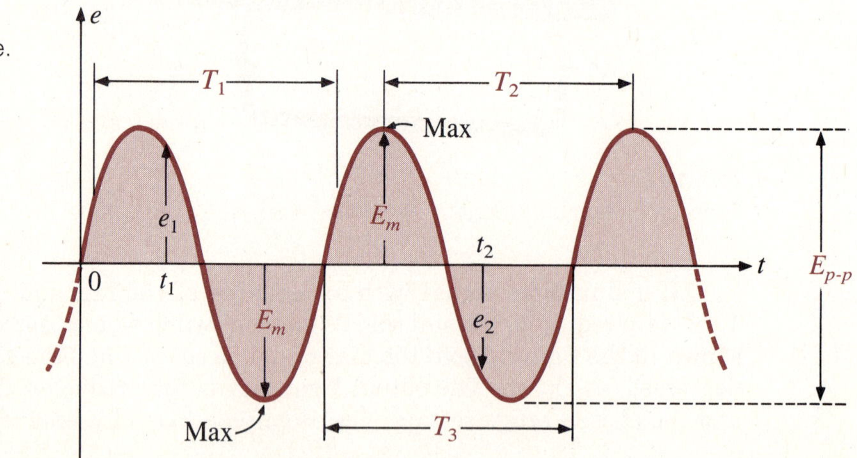

Waveform: The path traced by a quantity such as the voltage in Fig. 12.8, plotted as a function of some variable such as time (as before), position, degrees, radians, temperature, and so on.

Instantaneous value: The magnitude of a waveform at any *instant* of time, denoted by lowercase letters (e_1, e_2).

Amplitude, or *peak value:* The maximum value of a waveform, denoted by uppercase letters (E_m for sources of voltage and V_m for the voltage across common loads).

Peak-to-peak value: Denoted by $E_{p\text{-}p}$ or $V_{p\text{-}p}$, the full voltage between positive and negative peaks of the waveform—that is, the sum of the magnitude of the positive and negative peaks.

Periodic waveform: A waveform that continually repeats itself after the same time interval. The waveform of Fig. 12.8 is a periodic waveform.

Period (T)*:* The time interval between successive repetitions of a periodic waveform; the period $T_1 = T_2 = T_3$ in Fig. 12.8, as long as successive *similar points* of the periodic waveform are used in determining T.

Cycle: The portion of a waveform contained in *one period* of time. The cycles within T_1, T_2, and T_3 of Fig. 12.8 may appear differently in Fig. 12.9, but they are all bounded by one period of time and therefore satisfy the definition of a cycle.

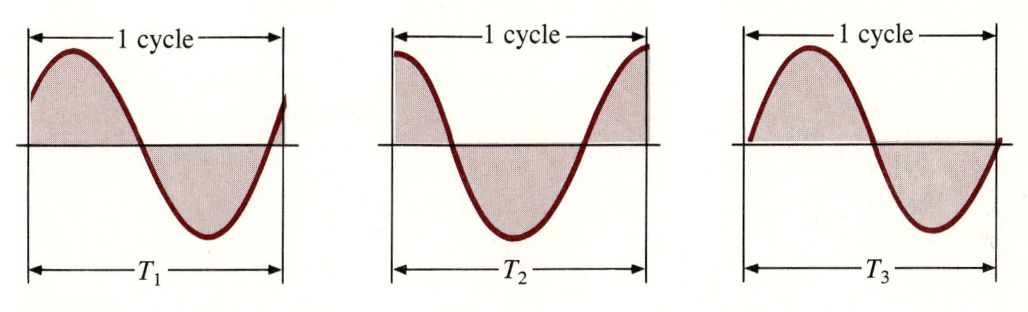

FIG. 12.9
Examining the definition of period.

Frequency (f)*:* The number of cycles that occur in 1 s. The frequency of the waveform of Fig. 12.10(a) is 1 cycle per second and for Fig. 12.10(b), $2\frac{1}{2}$

FIG. 12.10

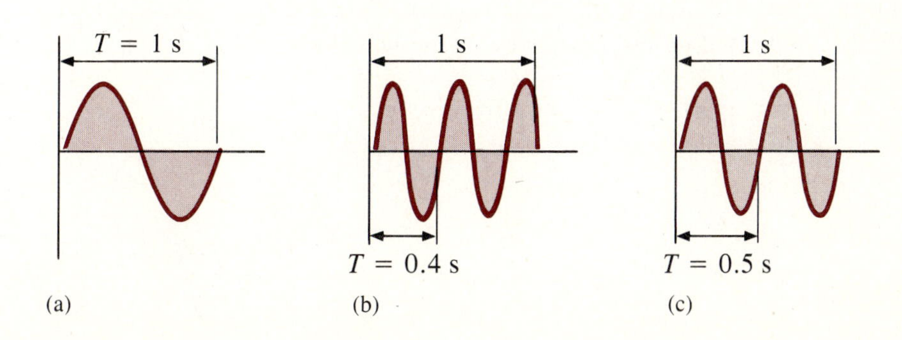

cycles per second. If a waveform of similar shape had a period of 0.5 s [Fig. 12.10(c)], the frequency would be 2 cycles per second.

By definition:

$$\boxed{1 \text{ hertz (Hz)} = 1 \text{ cycle per second (cps)}} \tag{12.1}$$

The unit hertz is derived from the surname of Heinrich Rudolph Hertz, who did original research in the area of alternating currents and voltages and their effect on the basic R, L, and C elements. The frequency standard for North America is 60 Hz, whereas for Europe it is predominantly 50 Hz.

Since the frequency is inversely related to the period—that is, as one increases the other decreases by an equal amount—the two can be related by the following equations:

$$\boxed{f = \frac{1}{T}} \qquad \begin{array}{l} f = \text{hertz, Hz} \\ T = \text{seconds, s} \end{array} \tag{12.2}$$

and

$$\boxed{T = \frac{1}{f}} \tag{12.3}$$

EXAMPLE 12.1 Find the period of a periodic waveform with a frequency of

a. 60 Hz
b. 1000 Hz

Solutions

a. $T = \frac{1}{f} = \frac{1}{60 \text{ Hz}} = 0.01667 \text{ s, or } \mathbf{16.67 \text{ ms}}$

(a recurring value, since 60 Hz is so prevalent)

b. $T = \frac{1}{f} = \frac{1}{1000 \text{ Hz}} = 10^{-3} \text{ s} = \mathbf{1 \text{ ms}}$

EXAMPLE 12.2 Determine the frequency of the waveform of Fig. 12.11.

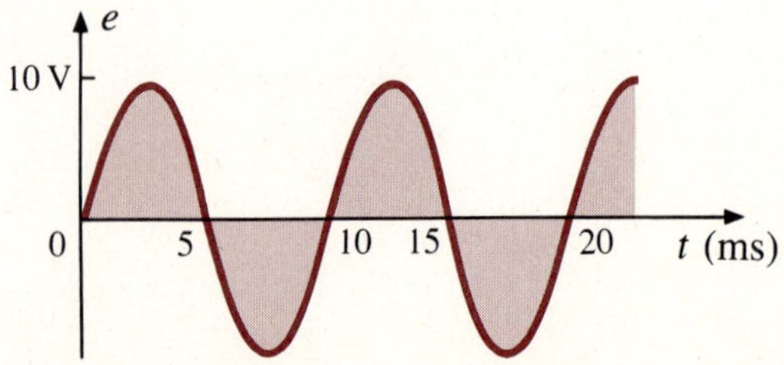

FIG. 12.11
Sinusoidal voltage for Example 12.2.

Solution From the figure, $T = 10$ ms, and

$$f = \frac{1}{T} = \frac{1}{10 \times 10^{-3}\,\text{s}} = \mathbf{100\ Hz}$$

Calculator

(1) (÷) (1) (0) (EXP) (+/−) (3)

with a display of 100.

EXAMPLE 12.3 The oscilloscope is an instrument that will display alternating waveforms such as those just described. A sinusoidal pattern appears on the oscilloscope of Fig. 12.12 with the indicated scale settings. Determine the period, frequency, and peak value of the waveform.

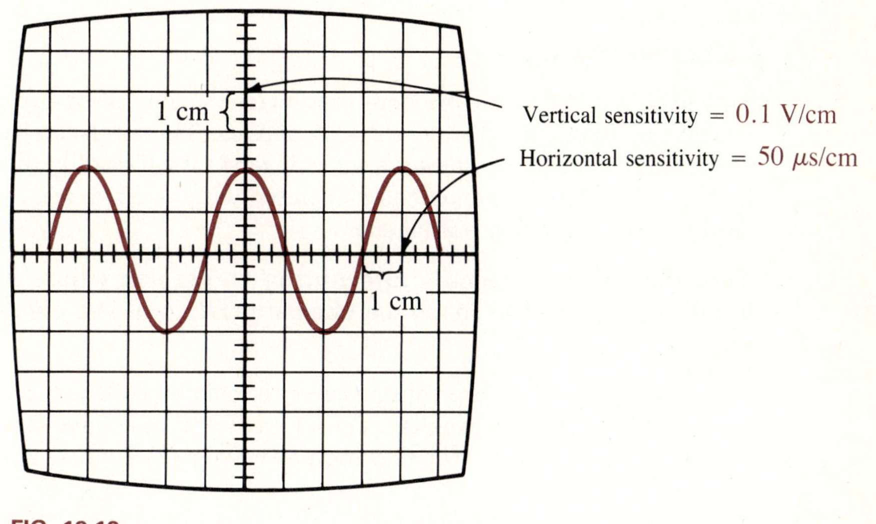

FIG. 12.12
Oscilloscope display for Example 12.3.

Solution The vertical and horizontal grid lines on a scope are separated by 1 cm. Careful examination of Fig. 12.12 reveals that it takes 4 cm (horizontal) to display a full cycle. Compare the distance between positive or negative peaks or when it crosses the horizontal axis with the same slope.

Therefore, $T = 4$ cm (in length), but the horizontal sensitivity is set on 50 μs/cm, revealing that each centimeter spacing encompasses a time interval of 50 μs. The result is

$$T = 4\,\cancel{\text{cm}}\left(\frac{50\ \mu\text{s}}{\cancel{\text{cm}}}\right) = \mathbf{200\ \mu s}$$

The frequency is

$$f = \frac{1}{T} = \frac{1}{200 \times 10^{-6}\text{ s}} = \frac{1}{200} \times \frac{1}{10^{-6}}\text{ Hz}$$
$$= 0.005 \times 10^{+6}\text{ Hz}$$
$$f = \mathbf{5000\ Hz} = \mathbf{5\ kHz}$$

Calculator

(1) (÷) (2) (0) (0) (EXP) (+/−) (6) (=)

with a display of 5000.

The vertical height above the horizontal axis is 2 cm. Therefore,

$$V_m = 2\ \text{cm}\left(\frac{0.1\text{ V}}{\text{cm}}\right) = \mathbf{0.2\ V} = \mathbf{200\ mV}$$

12.5 THE SINE WAVE

The terms defined in the previous section can be applied to any type of periodic waveform, whether smooth or discontinuous. The sinusoidal waveform is of particular importance, however, since it lends itself readily to the mathematics and the physical phenomena associated with electric circuits. Consider the power of the following statement:

The sine wave is the only alternating waveform whose appearance is unaffected by the response characteristics of the elements *R*, *L*, and *C*.

In other words, if the voltage across a resistor, coil, or capacitor is sinusoidal in nature, the resulting current for each will also have sinusoidal characteristics. If a square wave or a triangular wave were applied, such would not be the case.

The unit of measurement for the horizontal axis of Fig. 12.13 is the *degree*. Note that the waveform has a positive peak at $\alpha = 90°$ and a negative peak at $\alpha = 270°$. It crosses the horizontal axis at $\alpha = 0°, 180°$, and $360°$. A

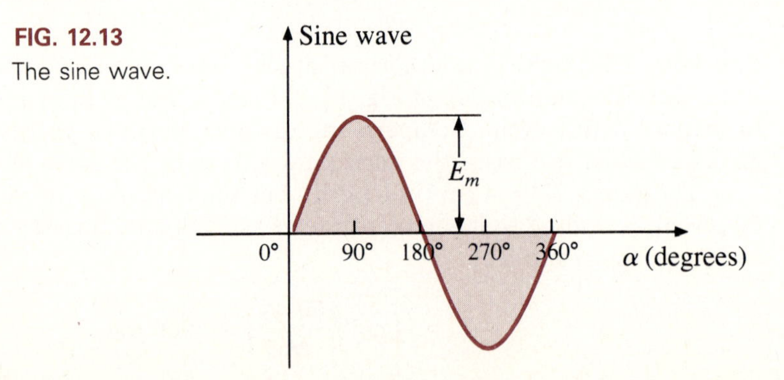

FIG. 12.13
The sine wave.

second unit of measurement frequently used is the *radian* (rad). It is defined by a quadrant of a circle, as shown in Fig. 12.14, where the distance subtended on the circumference equals the radius of the circle.

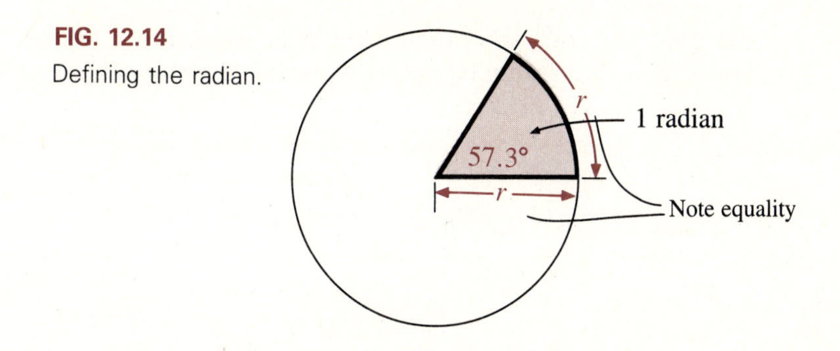

FIG. 12.14
Defining the radian.

There are 2π radians around a 360° circle, as shown in Fig. 12.15, and

$$\boxed{2\pi \text{ radians} = 360^\circ} \qquad \textbf{(12.4)}$$

with

$$\boxed{1 \text{ radian} \cong 57.3^\circ} \qquad \textbf{(12.5)}$$

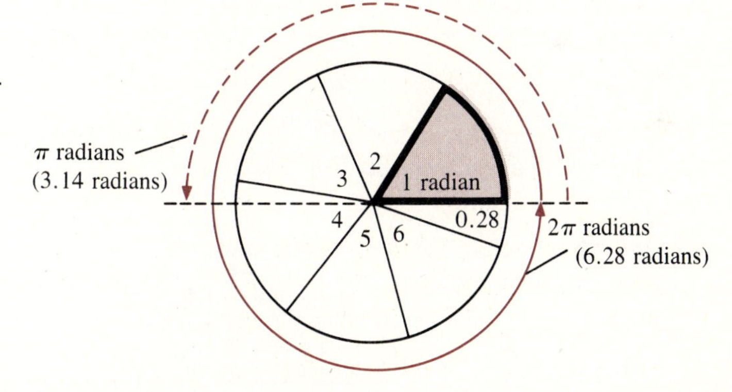

FIG. 12.15
Demonstrating that 2π rad $\cong$ 6.28 rad = 360°.

The value of π has been determined to an extended number of places for accuracy purposes and to see if a repetitive sequence of numbers appears (it does not). A few of the places the effort has yielded are:

$$\pi = 3.14159\ 26535\ 89793\ 23846\ 26433\ \ldots$$

For our purposes, the following approximation will frequently be used:

$$\boxed{\pi = 3.14} \qquad \textbf{(12.6)}$$

Using a calculator set to third-place accuracy,

[2nd F] [π]

with a display of 3.142.

For 180° and 360°, the two units of measurement are related as shown in Fig. 12.15. The conversion equations between the two are

$$\text{Radians} = \left(\frac{\pi}{180^\circ}\right) \times \text{Degrees} \tag{12.7}$$

and

$$\text{Degrees} = \left(\frac{180^\circ}{\pi}\right) \times \text{Radians} \tag{12.8}$$

Applying these equations, we find the following:

$$\mathbf{90^\circ} = \frac{\pi}{180^\circ}(90^\circ) = \frac{\pi}{2}\ \textbf{rad}$$

$$\mathbf{30^\circ} = \frac{\pi}{180^\circ}(30^\circ) = \frac{\pi}{6}\ \textbf{rad}$$

$$\frac{\pi}{3} = \frac{180^\circ}{\pi}\left(\frac{\pi}{3}\right) = \mathbf{60^\circ}$$

$$\frac{3\pi}{2} = \frac{180^\circ}{\pi}\left(\frac{3\pi}{2}\right) = \mathbf{270^\circ}$$

Using the radian as the unit of measurement for the abscissa, we would obtain the sine wave shown in Fig. 12.16.

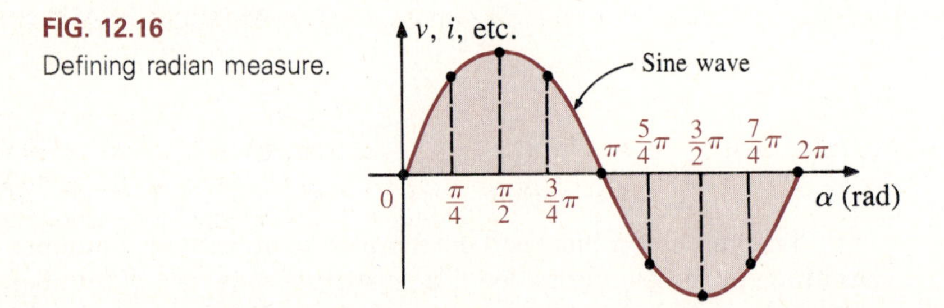

FIG. 12.16
Defining radian measure.

It is of particular interest that the sinusoidal waveform can be derived from the length of the *vertical projection* of a radius vector rotating in a uniform circular motion about a fixed point. In other words, we will plot the curve generated by the distance between the tip of the vector and the horizontal axis. The vector will have the same length as it rotates around the fixed pivot point.

The angle between the rotating vector and the horizontal axis will provide the horizontal movement from $\alpha = 0°$. For instance, when $\alpha = 45°$, a 45° interval will be marked off on the horizontal axis and the height of the curve will be determined by the distance between the tip of the vector and the horizontal axis.

As shown by Fig. 12.17 on page 470, if we let the vector complete a full rotation, a complete sinusoidal waveform is generated. A repeat of the rotation simply generates a similar sinusoidal waveform.

The velocity with which the radius vector rotates about the center, called the *angular velocity,* can be determined from the following equation:

$$\text{Angular velocity} = \frac{\text{Distance (degrees or radians)}}{\text{Time (seconds)}} \tag{12.9}$$

Substituting into Eq. 12.9 and assigning ω (the Greek letter omega) to the angular velocity, we have

$$\omega = \frac{\alpha}{t} \tag{12.10}$$

and

$$\alpha = \omega t \tag{12.11}$$

Since ω is typically given in radians per second, the angle α obtained using Eq. 12.11 is usually in radians. If α is required in degrees, Eq. 12.8 must then be applied. The importance of remembering the preceding relationship will become obvious in the examples to follow.

In Fig. 12.17, the time required to complete one revolution is equal to the period (T) of the sinusoidal waveform of Fig. 12.17(i). Since 2π radians are subtended in this time interval, we find

$$\omega = \frac{2\pi}{T} \quad \text{(rad/s)} \tag{12.12}$$

In words, this equation states that the smaller the period of the sinusoidal waveform of Fig. 12.17(i), or the smaller the time interval before one complete cycle is generated, the greater must be the angular velocity of the rotating radius vector. Certainly this statement agrees with what we have learned thus far. We can now go one step further and apply the fact that the frequency of the generated waveform is inversely related to the period of the waveform; that is, $f = 1/T$. Thus,

$$\omega = 2\pi f \quad \text{(rad/s)} \tag{12.13}$$

FIG. 12.17
Sinusoidal generation by the vertical projection of a rotating vector.

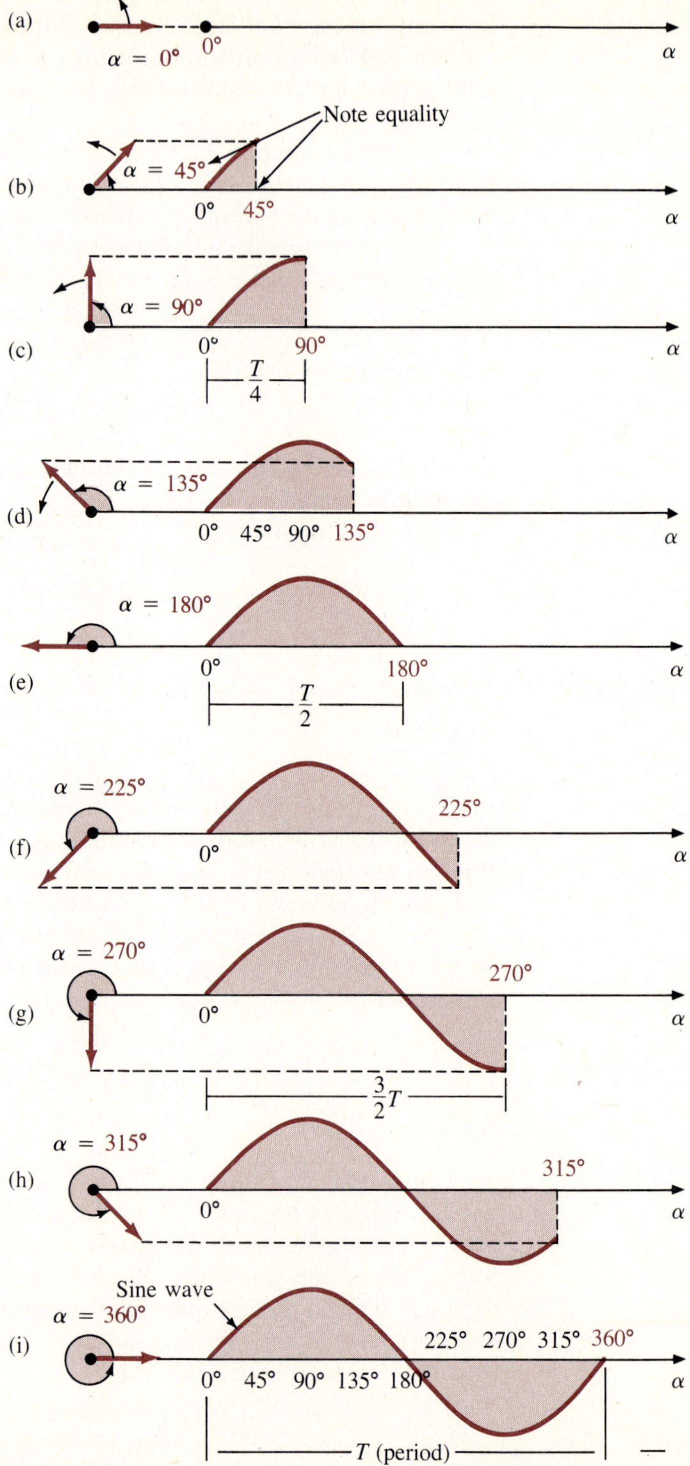

This equation states that the higher the frequency of the generated sinusoidal waveform, the higher must be the angular velocity. Equations 12.12 and 12.13 are verified somewhat by Fig. 12.18, where for the same-radius vector, $\omega = 100$ rad/s and 500 rad/s.

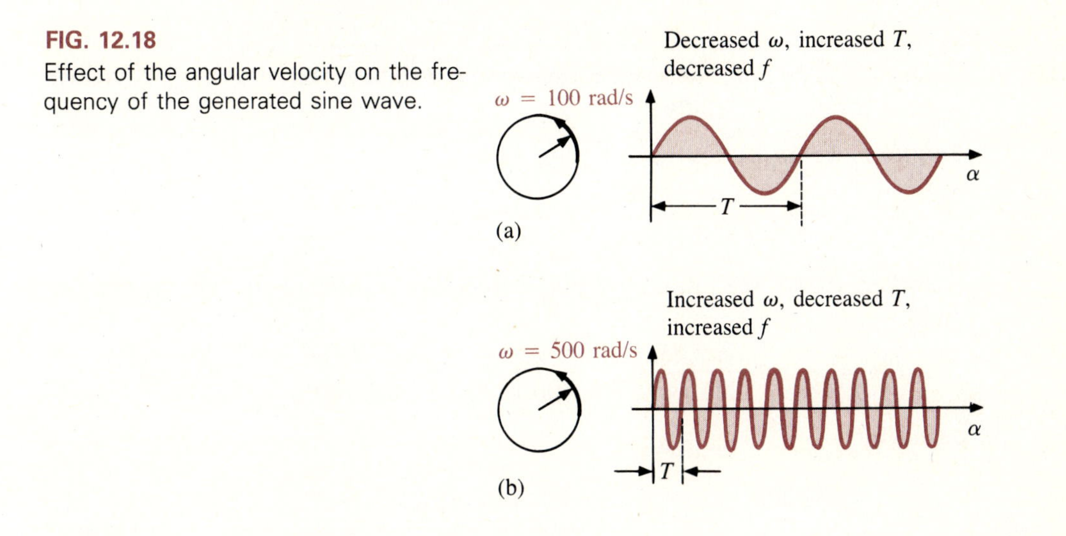

FIG. 12.18
Effect of the angular velocity on the frequency of the generated sine wave.

EXAMPLE 12.4 Determine the angular velocity of a sine wave having a frequency of 60 Hz.

Solution

$$\omega = 2\pi f = (6.28)(60 \text{ Hz}) = 376.991 \text{ rad/s} \cong \mathbf{377 \text{ rad/s}}$$

(a recurring value due to 60-Hz predominance).

Calculator

(2) (×) (2nd F) (π) (×) (6) (0) (=)

with a display of 376.991.

EXAMPLE 12.5 Determine the frequency and period of the sine wave of Fig. 12.18(b).

Solution Since $\omega = 2\pi/T$,

$$T = \frac{2\pi}{\omega} = \frac{2\pi \text{ rad}}{500 \text{ rad/s}} = \frac{6.28 \text{ rad}}{500 \text{ rad/s}} = \mathbf{12.56 \text{ ms}}$$

Calculator

(2) (×) (2nd F) (π) (÷) (5) (0) (0) (=) (F <-> E)

with a display of 1.257−02. Also,

$$f = \frac{1}{T} = \frac{1}{12.56 \times 10^{-3} \text{ s}} = \mathbf{79.617 \text{ Hz}}$$

Calculator

(1) (÷) (1) (2) (·) (5) (6) (EXP) (+/−) (3) (=)

with a display of 79.617.

EXAMPLE 12.6 Given $\omega = 200$ rad/s, determine how long it will take the sinusoidal waveform to pass through an angle of 90°.

Solution From Eq. 12.11: $\alpha = \omega t$, and

$$t = \frac{\alpha}{\omega}$$

However, $\pi/2$ (= 90°) must be substituted for α, since ω is in radians per second:

$$t = \frac{\alpha}{\omega} = \frac{\pi/2 \text{ rad}}{200 \text{ rad/s}} = \frac{\pi \text{ rad}}{400 \text{ rad/s}} = \frac{3.14 \text{ rad}}{400 \text{ rad/s}} = \mathbf{7.854 \text{ ms}}$$

Calculator

(2nd F) (π) (÷) (2) (÷) (2) (0) (0) (=) (F <−> E)

with a display of 7.854−03.

EXAMPLE 12.7 Find the angle through which a sinusoidal waveform of 60 Hz will pass in a period of 5 ms.

Solution From Eq. 12.11, $\alpha = \omega t$, or

$$\alpha = 2\pi f t = (6.28)(60 \text{ Hz})(5 \times 10^{-3} \text{ s}) \cong \mathbf{1.884 \text{ rad}}$$

If not careful, you might be tempted to interpret the answer as 1.884°. However, using Eq. 12.8,

$$\alpha\,(°) = \frac{180°}{\pi}(1.884) \cong \mathbf{108°}$$

Calculator

(1) (8) (0) (×) (1) (·) (8) (8) (4) (÷) (2nd F) (π) (=)

with a display of 107.945.

12.6 GENERAL FORMAT FOR THE SINUSOIDAL VOLTAGE OR CURRENT

The basic mathematical format for the sinusoidal waveform is

$$\boxed{A_m \sin \alpha} \qquad (12.14)$$

where A_m is the peak value of the waveform and α is the unit of measure for the horizontal axis, as shown in Fig. 12.19.

$$\text{At } \alpha = 90°, A_m \sin \alpha = A_m \sin 90° = A_m(1) = A_m$$
$$\text{At } \alpha = 180°, A_m \sin \alpha = A_m \sin 180° = A_m(0) = 0$$
$$\text{At } \alpha = 270°, A_m \sin \alpha = A_m \sin 270° = A_m(-1) = -A_m$$

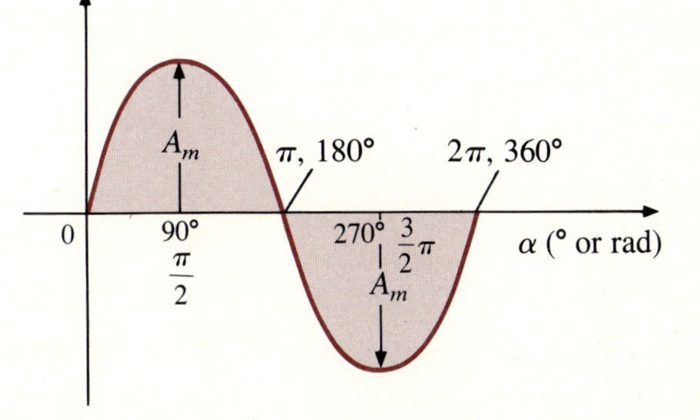

FIG. 12.19
Defining the general mathematical format for a sinusoidal waveform.

EXAMPLE 12.8

If a sinusoidal voltage $e = 20 \sin \alpha$, determine e at $\alpha = 45°$ and $\alpha = (2/3)\pi$.

Solution For $\alpha = 45°$,

$$e = 20 \sin \alpha = 20 \sin 45° = 20(0.7071) = \mathbf{14.142\ V}$$

Calculator

(2) (0) (×) (4) (5) (sin) (=)

with a display of 14.142.

For $\alpha = (2/3)\pi$,

$$\alpha° = \left(\frac{180°}{\pi}\right)\left(\frac{2}{3}\pi\right) = \frac{360°}{3} = 120°$$

$$e = 20 \sin \alpha = 20 \sin 120° = 20(0.8660) = \mathbf{17.32\ V}$$

Calculator

(2) (0) (×) (1) (2) (0) (sin) (=)

with a display of 17.321.

Due to Eq. 12.11, $\alpha = \omega t$, the general format of a sine wave can also be written

$$\boxed{A_m \sin \omega t} \qquad \textbf{(12.15)}$$

with ωt as the horizontal unit of measure.

For electrical quantities such as current and voltage, the general format is

$$i = I_m \sin \omega t = I_m \sin \alpha$$
$$e = E_m \sin \omega t = E_m \sin \alpha$$

where the capital letters with the subscript m represent the amplitude and the lowercase letters i and e represent the instantaneous value of current or voltage, respectively, at any time t.

The angle at which a particular voltage level is attained can be determined from the following equation:

$$\alpha = \sin^{-1}\left(\frac{e}{E_m}\right) \quad (12.16)$$

Similarly, for a particular current level,

$$\alpha = \sin^{-1}\left(\frac{i}{I_m}\right) \quad (12.17)$$

The function $\sin^{-1}$ is available on all scientific calculators. Both e and i are the levels at which the value of α is to be determined, with E_m and I_m representing the peak value of each sinusoidal waveform.

EXAMPLE 12.9

a. Determine the angle at which the magnitude of the sinusoidal function $e = 10 \sin 377t$ is 10 V (Fig. 12.20).
b. Repeat part a for a 4-V level.
c. Determine the instant when $e = 10$ V.
d. Determine the instant when $e = 4$ V.

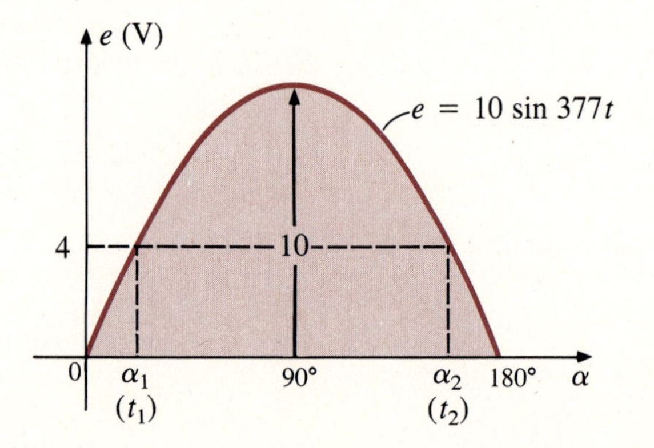

FIG. 12.20
Determining the angle and time at which a sinusoidal function has a particular magnitude.

Solutions

a. Using Eq. 12.16, at $e = 10$ V (the peak value),

$$\alpha = \sin^{-1}\left(\frac{e}{E_m}\right) = \sin^{-1}\left(\frac{10}{10}\right) = \sin^{-1}(1)$$
$$= \mathbf{90^\circ}$$

Calculator

(1) (2nd F) ($\sin^{-1}$)

with a display of 90.

b. Referring to Fig. 12.20, we note that $e = 4$ V at α_1 and α_2. For α_1,

$$\alpha_1 = \sin^{-1}\left(\frac{e}{E_m}\right) = \sin^{-1}\left(\frac{4}{10}\right) = \sin^{-1}(0.4)$$
$$= \mathbf{23.578^\circ}$$

Calculator

(4) (÷) (1) (0) (=) (2nd F) ($\sin^{-1}$)

with a display of 23.578.

The second intersection of α_2 is determined by

$$\alpha_2 = 180^\circ - \alpha_1 = 180^\circ - (23.578^\circ) = \mathbf{156.422^\circ}$$

since the distance (in time units) from 0 to α_1 is the same as from α_2 to 180°.

c. To use the equation $t = \alpha/\omega$, we must convert $\alpha = 90^\circ$ to radians:

$$\alpha = 90^\circ = \frac{\pi}{2} \text{ rad}$$

and

$$t = \frac{\alpha}{\omega} = \frac{\pi/2 \text{ rad}}{377 \text{ rad/s}} = \frac{3.14/2}{377} \text{ s} = \mathbf{4.167 \text{ ms}}$$

Calculator

(2nd F) (π) (÷) (2) (÷) (3) (7) (7) (=) (F <–> E)

with a display of 4.167−03.

d. To find t_1:

$$\alpha_1 = \frac{\pi}{180^\circ}(23.578^\circ) = 0.411 \text{ rad}$$

$$t_1 = \frac{\alpha_1}{\omega} = \frac{0.411 \text{ rad}}{377 \text{ rad/s}} = \mathbf{1.092 \text{ ms}}$$

Calculator

2nd F | π | ÷ | 1 | 8 | 0 | × | 2 | 3 | . | 5 | 7 | 8 | ÷ | 3 | 7 | 7 | = | F <-> E

with a display of 1.092−03.

To find t_2, for the second intersection,

$$T = \frac{2\pi}{\omega} = \frac{6.28 \text{ rad}}{377 \text{ rad/s}} = 16.66 \text{ ms}$$

and

$$\frac{T}{2} = \frac{16.66 \text{ ms}}{2} = 8.33 \text{ ms}$$

with

$$t_2 = \frac{T}{2} - t_1 = (8.33 - 1.09) \text{ ms} = \mathbf{7.24 \text{ ms}}$$

The sine wave can also be plotted against *time* on the horizontal axis. The time period for each interval can be determined from $t = \alpha/\omega$, but the most direct route is simply to find the period T from $T = 1/f$ and break it up into the required intervals. This latter technique is demonstrated in Example 12.10.

Before reviewing the example, take special note of the relative simplicity of the mathematical equation that can represent a sinusoidal waveform. Any alternating waveform whose characteristics differ from those of the sine wave cannot be represented by a single term but may require two, four, six, or perhaps an infinite number of terms to be represented accurately.

EXAMPLE 12.10 Sketch $e = 10 \sin 314t$ with the abscissa

a. Angle (α) in degrees
b. Angle (α) in radians
c. Time (t) in seconds

Solutions

a. See Fig. 12.21. (Note that no calculations are required.)

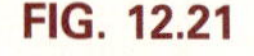

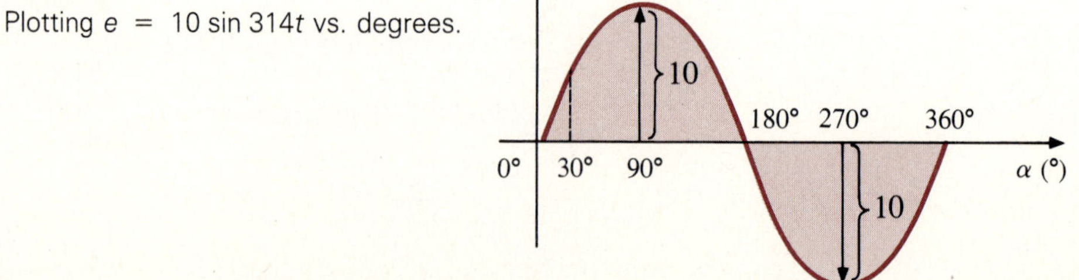

FIG. 12.21
Plotting $e = 10 \sin 314t$ vs. degrees.

b. See Fig. 12.22. (Once the relationship between degrees and radians is understood, there is again no need for calculations.)

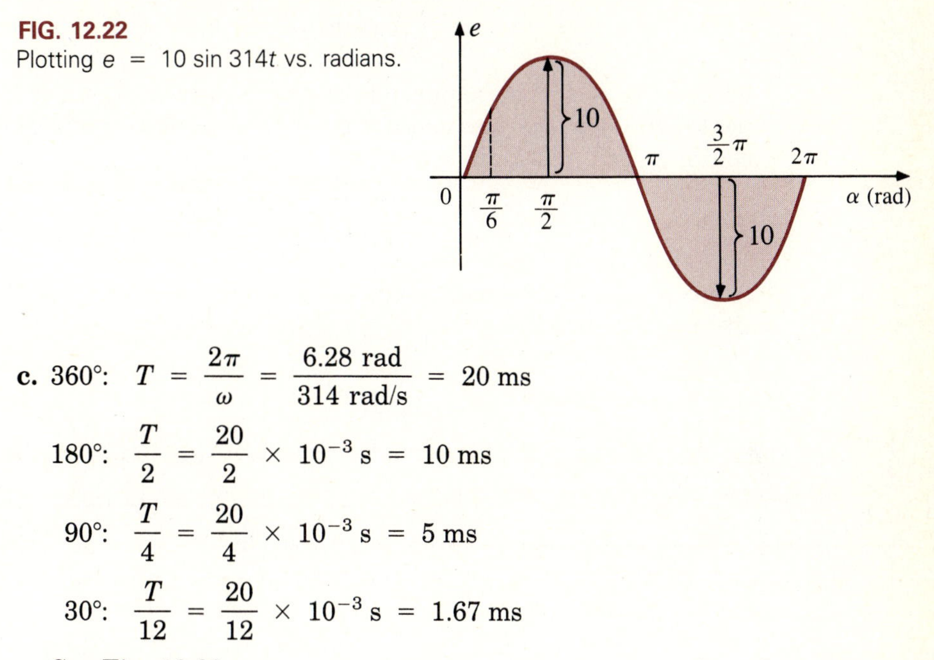

FIG. 12.22
Plotting $e = 10 \sin 314t$ vs. radians.

c. 360°: $T = \dfrac{2\pi}{\omega} = \dfrac{6.28 \text{ rad}}{314 \text{ rad/s}} = 20 \text{ ms}$

180°: $\dfrac{T}{2} = \dfrac{20}{2} \times 10^{-3} \text{ s} = 10 \text{ ms}$

90°: $\dfrac{T}{4} = \dfrac{20}{4} \times 10^{-3} \text{ s} = 5 \text{ ms}$

30°: $\dfrac{T}{12} = \dfrac{20}{12} \times 10^{-3} \text{ s} = 1.67 \text{ ms}$

See Fig. 12.23.

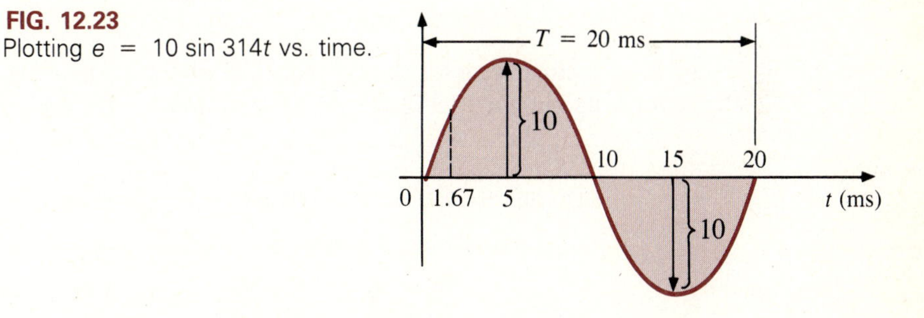

FIG. 12.23
Plotting $e = 10 \sin 314t$ vs. time.

12.7 PHASE RELATIONS

Thus far, we have considered only sine waves that have maxima at $\pi/2$ and $3\pi/2$ with a zero value at 0, π, and 2π, as shown in Fig. 12.22. If the waveform is shifted to the right or left of 0°, the expression becomes

$$A_m \sin(\omega t \pm \theta) \tag{12.18}$$

where θ is the angle in degrees or radians that the waveform has been shifted.

If the waveform passes through the horizontal axis with a *positive-going* (increasing with time) slope *before* 0°, as shown in Fig. 12.24, the expression is

$$A_m \sin(\omega t + \theta) \tag{12.19}$$

At $\omega t = \alpha = 0°$, the magnitude is determined by $A_m \sin \theta$. If the waveform passes through the horizontal axis with a positive-going slope *after* 0°, as shown in Fig. 12.25, the expression is

$$A_m \sin(\omega t - \theta) \tag{12.20}$$

At $\omega t = \alpha = 0°$, the magnitude is $A_m \sin(-\theta)$, which by a trigonometric identity is $-A_m \sin \theta$.

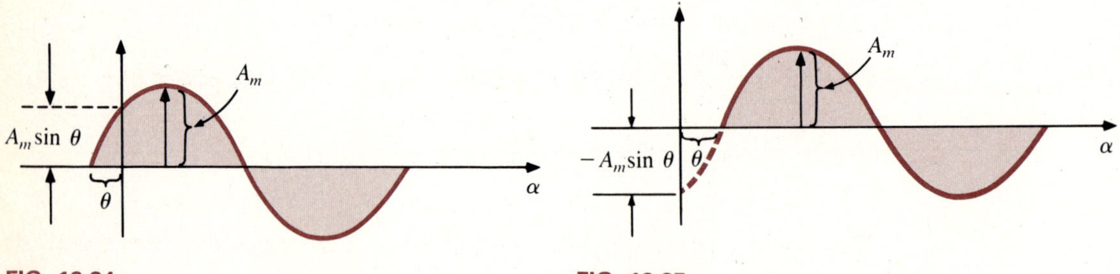

FIG. 12.24
Sinusoidal waveform with a positive phase shift.

FIG. 12.25
Sinusoidal function with a negative phase shift.

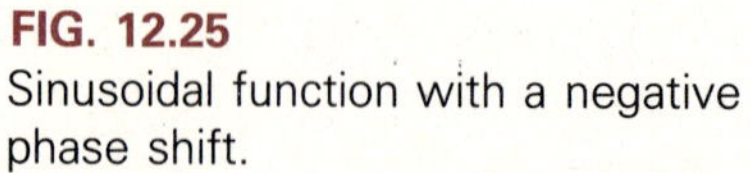

If the waveform crosses the horizontal axis with a positive-going slope 90° ($\pi/2$) sooner, as shown in Fig. 12.26, it is called a *cosine wave*. That is,

$$\sin(\omega t + 90°) = \sin\left(\omega t + \frac{\pi}{2}\right) = \cos \omega t \tag{12.21}$$

or

$$\sin \omega t = \cos(\omega t - 90°) = \cos\left(\omega t - \frac{\pi}{2}\right) \tag{12.22}$$

FIG. 12.26
Comparing a sine and cosine wave.

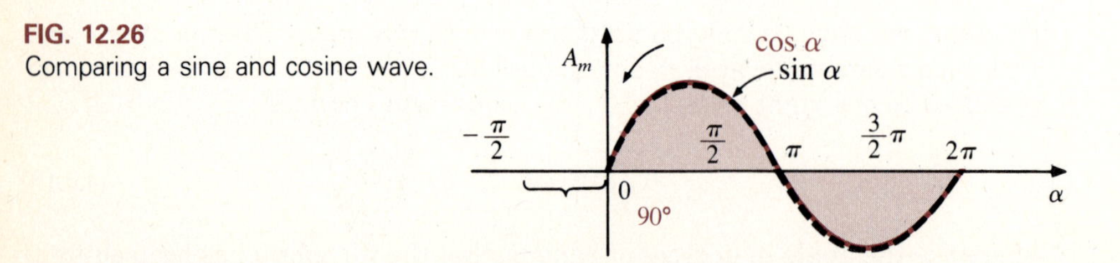

The terms *lead* and *lag* are used to indicate the relationship between two sinusoidal waveforms of the *same frequency* plotted on the same set of axes. In Fig. 12.26, the cosine curve is said to *lead* the sine curve by 90°, and the sine curve is said to *lag* the cosine curve by 90°. The 90° is referred to as the *phase angle* between the two waveforms. In language commonly applied, the waveforms are *out of phase* by 90°. Note that the phase angle between the two waveforms is measured between those two points on the horizontal axis through which each passes with the *same slope*. If both waveforms cross the axis at the same point with the same slope, they are *in phase*.

If a sinusoidal expression should appear as

$$e = -E_m \sin \omega t$$

the negative sign is associated with the sine portion of the expression, and the expression can also be written

$$e = E_m \sin(\omega t \pm 180^\circ)$$

revealing that a negative sign can be replaced by a 180° change in phase angle (+ or −). That is,

$$e = -E_m \sin \omega t = E_m \sin(\omega t + 180^\circ)$$
$$e = E_m \sin(\omega t - 180^\circ)$$

The *phase relationship* between two waveforms indicates which one leads or lags and by how many degrees or radians.

EXAMPLE 12.11 What is the phase relationship between

$$v = 10 \sin \omega t \quad \text{and} \quad i = 5 \sin(\omega t + 60^\circ)$$

Solution See Fig. 12.27: ***i* leads *v* by 60°.**

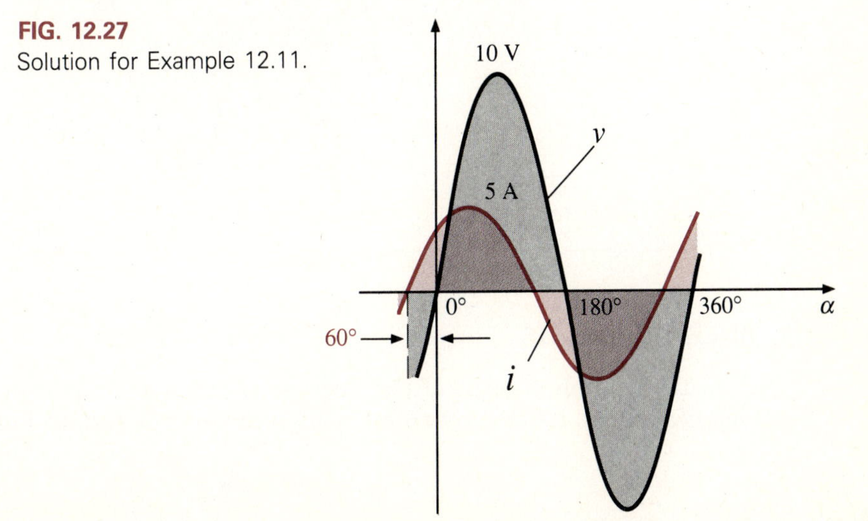

FIG. 12.27
Solution for Example 12.11.

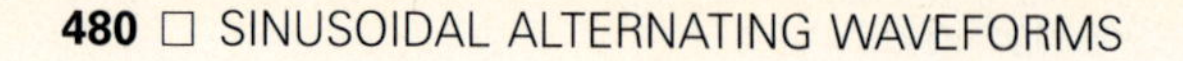

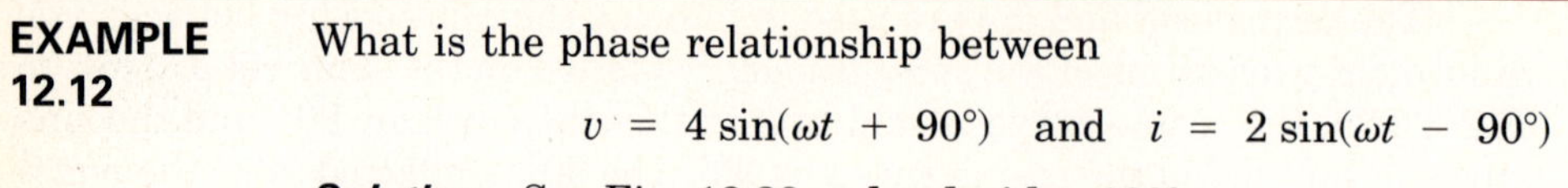

EXAMPLE 12.12

What is the phase relationship between

$$v = 4\sin(\omega t + 90°) \quad \text{and} \quad i = 2\sin(\omega t - 90°)$$

Solution See Fig. 12.28: ***v* leads *i* by 180°.**

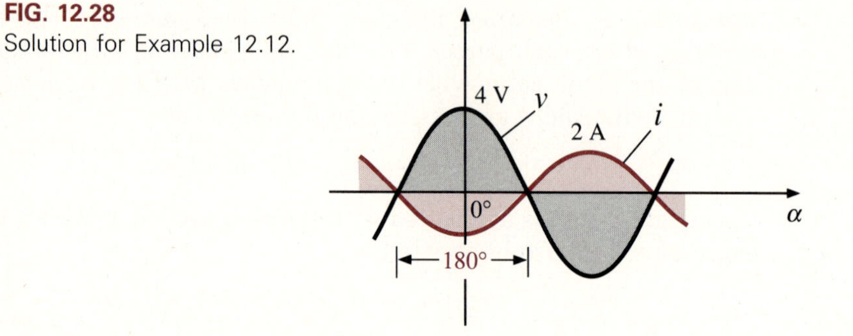

FIG. 12.28
Solution for Example 12.12.

EXAMPLE 12.13

What is the phase relationship between

$$v = 2\sin(\omega t + 30°) \quad \text{and} \quad i = -\sin\omega t$$

Solution See Fig. 12.29: ***v* leads *i* by (180° + 30°) = 210°.**

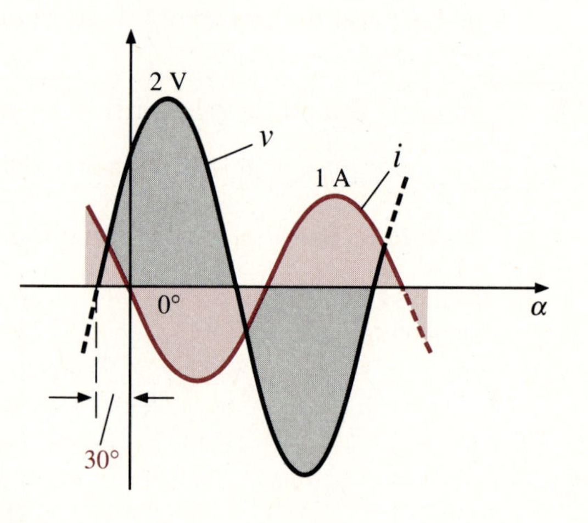

FIG. 12.29
Solution for Example 12.13.

12.8 AVERAGE VALUE

Even though the concept of the *average value* is an important one in most technical fields, its true meaning is often misinterpreted. In Fig. 12.30(a), for

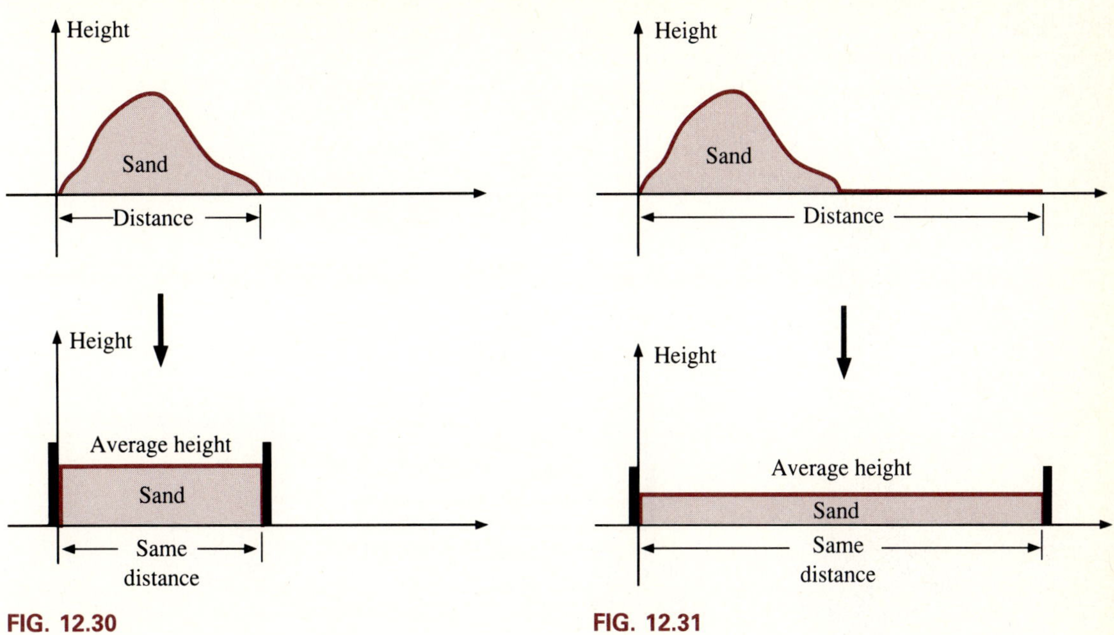

FIG. 12.30
Defining average value.

FIG. 12.31
Effect of distance on average value.

example, the average height of the sand may be required to determine the volume of sand available. The average height of the sand is that height obtained if the distance from one end to the other is maintained while the sand is leveled off, as shown in Fig. 12.30(b). The area under the mound of Fig. 12.30(a) will then equal the area under the rectangular shape of Fig. 12.30(b) as determined by $A = b \times h$. Of course, the depth (into the page) of the sand must be the same for Fig. 12.30(a) and 12.30(b) for the preceding conclusions to have any meaning.

In Fig. 12.30 the distance was measured from one end to the other. In Fig. 12.31(a) the distance extends beyond the end of the pile for the same original pile of Fig. 12.30. The situation could be one where a landscaper would like to know the average height of the sand if spread out over a distance such as defined in Fig. 12.31(a). The result of an increased distance is as shown in Fig. 12.31(b). The average height has decreased compared to Fig. 12.30. Quite obviously, therefore, the longer the distance, the lower is the average value.

If the distance parameter includes a depression, as shown in Fig. 12.32(a), some of the sand will be used to fill the depression, resulting in an even lower average value for the landscaper, as shown in Fig. 12.32(b). For a sinusoidal waveform, the depression would have the same shape as the mound of sand (over one full cycle) resulting in an average value at ground level (or 0 V for a sinusoidal voltage over one full period).

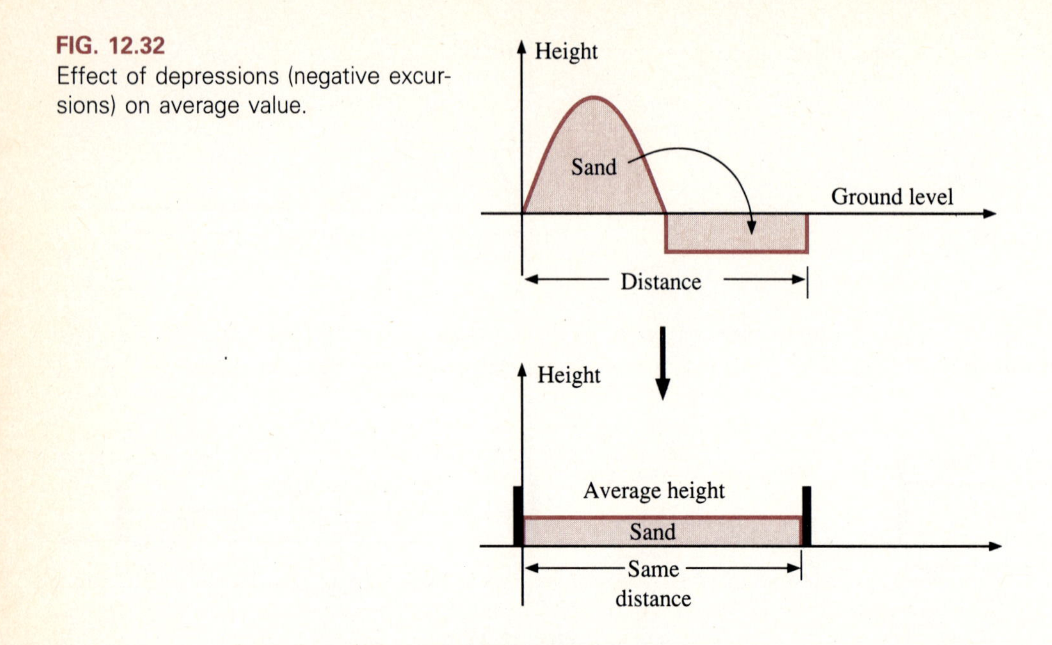

FIG. 12.32
Effect of depressions (negative excursions) on average value.

After traveling a considerable distance by car, some drivers like to calculate their average speed for the entire trip. This is usually done by dividing the miles traveled by the hours required to drive that distance. For example, if a person traveled 180 mi in 5 h, his or her average speed was 180/5, or 36, mi/h. This same distance may have been traveled at various speeds for various intervals of time, as shown in Fig. 12.33.

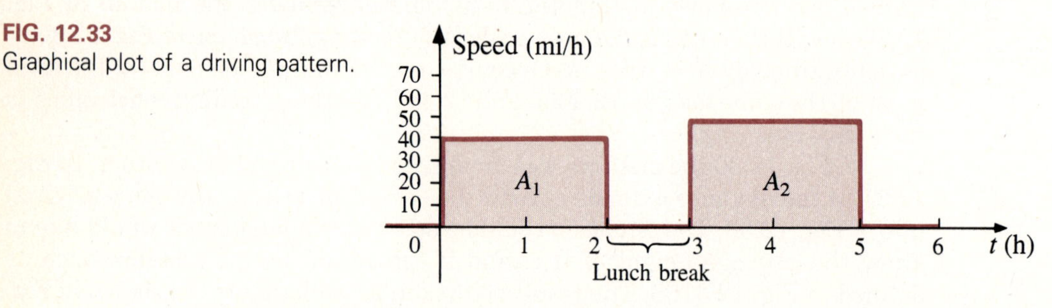

FIG. 12.33
Graphical plot of a driving pattern.

By finding the total area under the curve for 5 h and then dividing the area by 5 h (the total time for the trip), we obtain the same result of 36 mi/h; that is,

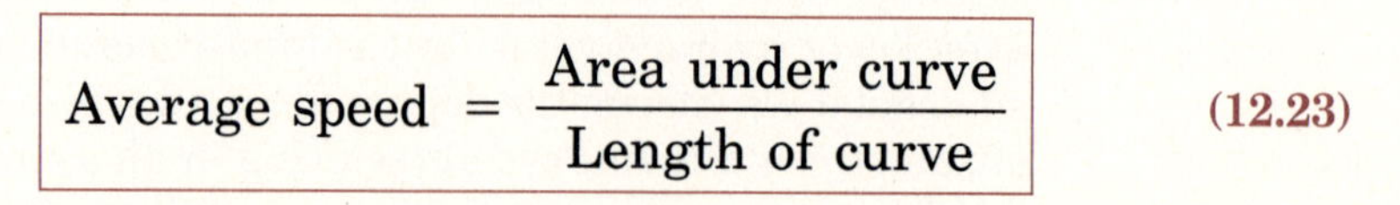

$$\text{Average speed} = \frac{\text{Area under curve}}{\text{Length of curve}} \qquad (12.23)$$

$$\text{Average speed} = \frac{A_1 + A_2}{5}$$
$$= \frac{(40 \text{ mi/h})(2 \text{ h}) + (50 \text{ mi/h})(2 \text{ h})}{5 \text{ h}}$$
$$= \frac{80 \text{ mi/h}}{5}$$
$$\text{Average speed} = \mathbf{36\ mi/h}$$

Equation 12.23 can be extended to include any variable quantity, such as current or voltage, if we let G denote the average value, as follows:

$$G \text{ (average value)} = \frac{\text{Algebraic sum of areas}}{\text{Length of curve}} \tag{12.24}$$

The algebraic (affect of sign) sum of the areas must be determined, since the contributions of some areas will be from below the horizontal axis. Areas above the axis will be assigned a positive sign, and those below, a negative sign. A positive average value will then be above the axis and a negative value, below.

The average value of *any* current or voltage is the value indicated on a dc meter. In other words, over a complete cycle the average value is the equivalent dc value. In the analysis of electronic circuits to be considered in a later course, both dc and ac sources of voltage will be applied to the same network. It will then be necessary to know or determine the dc (or average value) and ac components of the voltage or current in various parts of the system.

EXAMPLE 12.14 Find the average value of the waveform of Fig. 12.34.

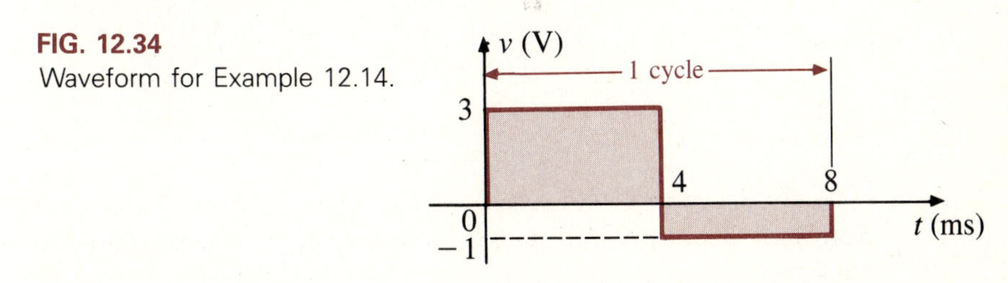

FIG. 12.34
Waveform for Example 12.14.

Solution By inspection we can see that the area above the axis is much greater than that below the axis, so a positive average value for v can be expected.

Using Eq. 12.24,

$$G = \frac{+(3\text{ V})(4\text{ ms}) - (1\text{ V})(4\text{ ms})}{8\text{ ms}}$$

$$= \frac{12\text{ V} - 4\text{ V}}{8} = \frac{8\text{ V}}{8} = \mathbf{1\ V}$$

See Fig. 12.35.

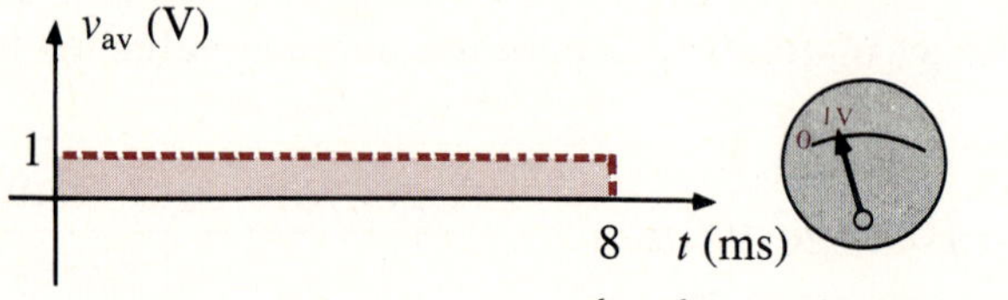

FIG. 12.35
Solution for Example 12.14.

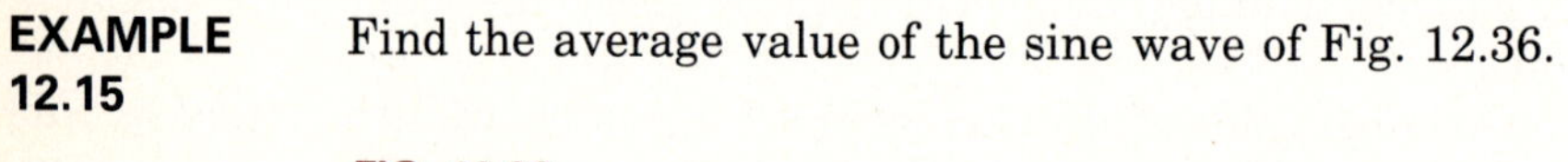

EXAMPLE 12.15 Find the average value of the sine wave of Fig. 12.36.

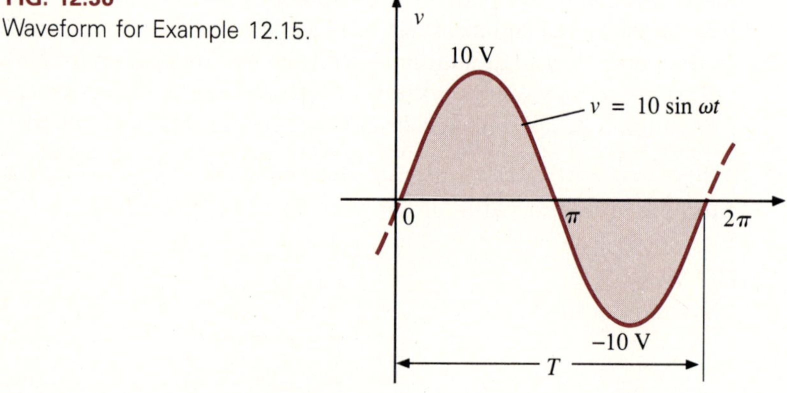

FIG. 12.36
Waveform for Example 12.15.

Solution Over one full period of Fig. 12.36, the area above the axis is exactly equal to the area below the axis. The result is that

The average value of a sinusoidal waveform over one full cycle is zero.

By a mathematical procedure known as *integration,* the exact area of a positive (or negative) pulse of a sine wave is simply $\mathbf{2A_m}$, or twice the peak value.

Using the preceding information, we can perform the analysis of Fig. 12.36 using Eq. 12.24.

$$G = \frac{2A_m - 2A_m}{2\pi}$$

$$= \frac{2(10\text{ V}) - 2(10\text{ V})}{2\pi} = \frac{20\text{ V} - 20\text{ V}}{2\pi} = \frac{0}{2\pi} = \mathbf{0\ V}$$

as determined by inspection.

EXAMPLE 12.16 Find the average value of the current waveform of Fig. 12.37.

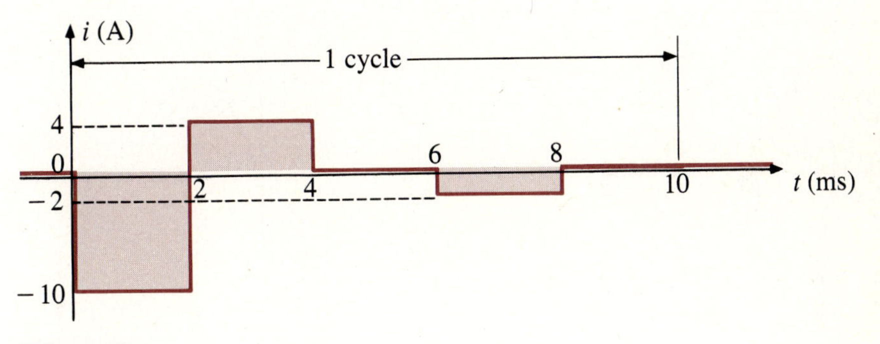

FIG. 12.37
Waveform for Example 12.16.

Solution Using Eq. 12.24,

$$G = \frac{(-10\text{ A})(2\text{ ms}) + (4\text{ A})(2\text{ ms}) - (2\text{ A})(2\text{ ms})}{10\text{ ms}}$$

$$= \frac{-20\text{ A} + 8\text{ A} - 4\text{ A}}{10}$$

$$= \frac{-24\text{ A} + 8\text{ A}}{10} = \frac{-16\text{ A}}{10}$$

$$G = \mathbf{-1.6\ A}$$

Figure 12.38 includes a plot of the average value with the response of an average reading dc ammeter.

FIG. 12.38
Solution for Example 12.16.

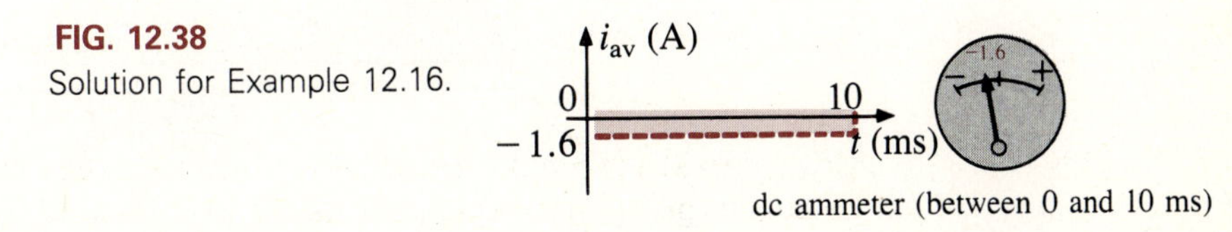

For an isolated positive pulse of a sinusoidal function, as shown in Fig. 12.39, the average value from 0 to π is

$$G = \frac{2A_m}{\pi} = \frac{2}{\pi}A_m = \frac{2}{3.1414}A_m \cong 0.637A_m$$

and for the situation of Fig. 12.40, with a period extended to 2π,

$$G = \frac{2A_m}{2\pi} = \frac{A_m}{\pi} = \frac{1}{\pi}A_m = \frac{1}{3.1414}A_m \cong 0.318A_m$$

which will prove useful in our conversion of ac to dc to be introduced in a later chapter.

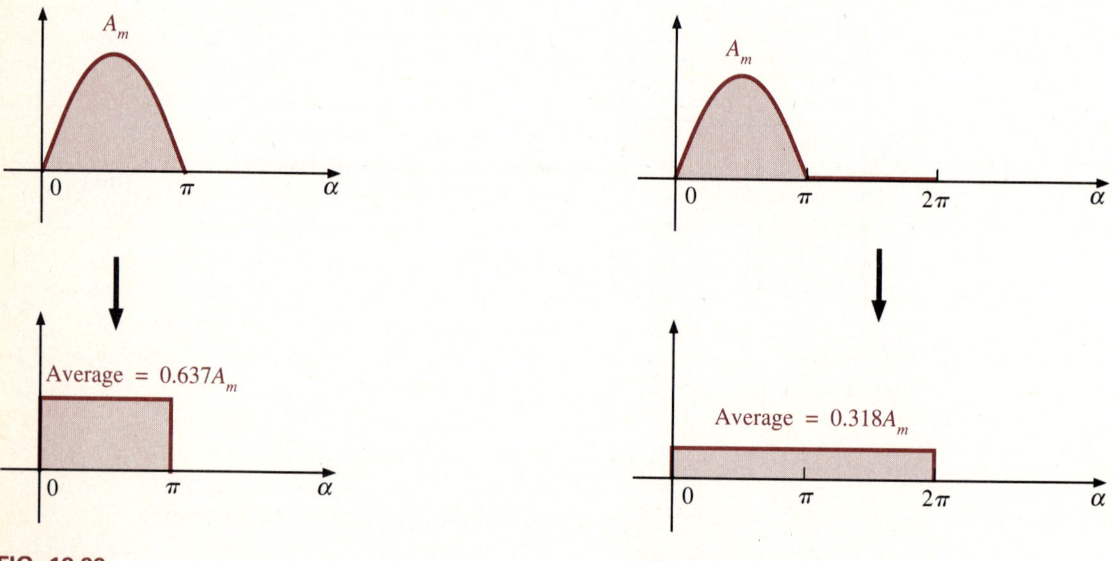

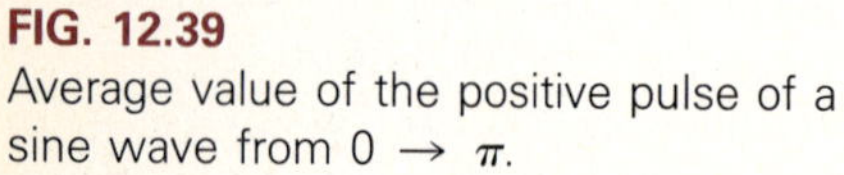

FIG. 12.39
Average value of the positive pulse of a sine wave from $0 \rightarrow \pi$.

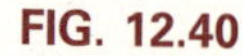

FIG. 12.40
Average value of the positive pulse of a sine wave for a period $0 \rightarrow 2\pi$.

EXAMPLE 12.17 Find an approximate value for the average level of the waveform of Fig. 12.41.

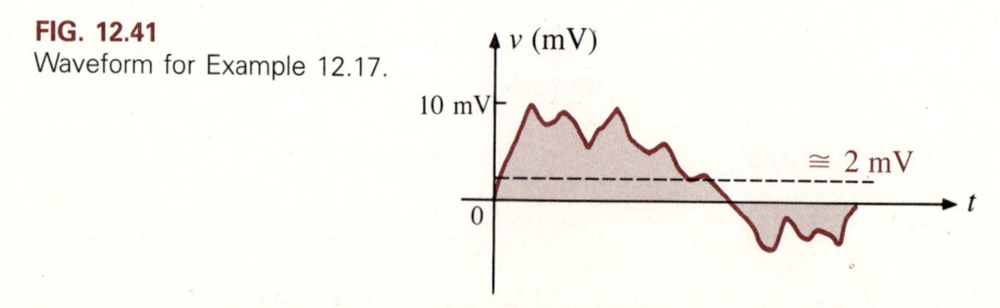

FIG. 12.41
Waveform for Example 12.17.

Solution For situations such as this, it would be almost impossible to obtain an exact value for the area above and below the axis. However, we can see that the area above the axis is much greater than that below. It would appear that a dotted line drawn as shown in Fig. 12.41 would divide the waveform such that the area above the dotted line equals the area below. The result (by visualization) is that the average value is about **2 mV.** Occasionally, judgments of this type have to be made.

12.9 EFFECTIVE VALUE

This section will begin to relate dc and ac quantities with respect to the power delivered to a load. It will help us determine the amplitude of a sinusoidal ac current required to deliver the same power as a particular dc current. The question frequently arises, How is it possible for a sinusoidal ac quantity to deliver a net power if, over a full cycle, the net current in any one direction is zero (average value 0)? It would almost appear that the power delivered during the positive portion of the sinusoidal waveform is withdrawn during the negative portion, and since the two are equal in magnitude, the net power delivered is zero. However, understand that *irrespective of direction,* current of any magnitude through a resistor will deliver power *to that resistor.* In other words, during the positive or negative portions of a sinusoidal ac current, power is being delivered at *each instant of time* to the resistor. The power delivered at each instant will, of course, vary with the magnitude of the sinusoidal ac current, but there will be a net flow during either the positive or negative pulses with net flow over the full cycle. The net power flow will equal twice that delivered by either the positive or negative regions of the sinusoidal quantity.

A fixed relationship between ac and dc voltages and currents can be derived from the experimental setup shown in Fig. 12.42. A resistor in a water bath is connected by switches to a dc and an ac supply. If switch 1 is closed, a dc current I, determined by the resistance R and battery voltage E, will be established through the resistor R. The temperature reached by the water is determined by the dc power dissipated in the form of heat by the resistor.

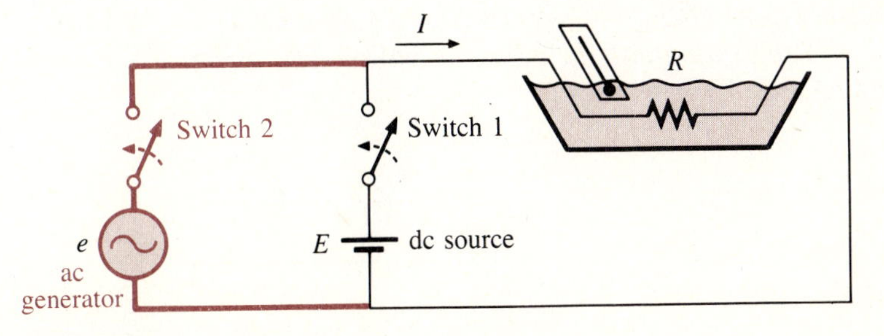

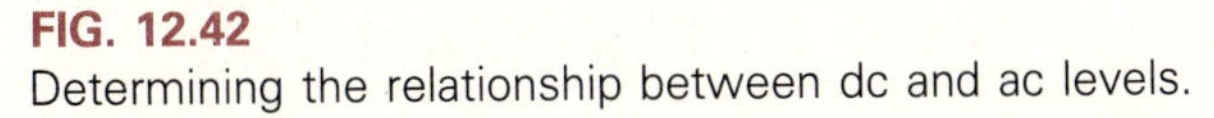
FIG. 12.42
Determining the relationship between dc and ac levels.

If switch 2 is closed and switch 1 is left open, the ac current through the resistor will have a peak value of I_m. The temperature reached by the water is now determined by the ac power dissipated in the form of heat by the resistor. The ac input is varied until the temperature is the same as that reached with the dc input. When this is accomplished, the average electrical power delivered to the resistor R by the ac source is the same as that delivered by the dc source.

For the dc source, the power delivered to the resistive load R is determined by

$$P_{dc} = I_{dc}^2 R$$

as introduced in Chapter 5.

For the ac source, let us write the equation for the power to the resistor R as

$$P_{ac} = I_{ac}^2 R$$

Since R is the same for each equation, we simply have to determine an equivalence between I_{ac} and I_{dc} to ensure $P_{dc} = P_{ac}$.

Through the preceding experimental setup (and also through calculus techniques), it can be shown that

$$\boxed{I_{ac} = \frac{I_m}{\sqrt{2}} = I_{dc}} \qquad (12.25)$$

In other words, to satisfy the condition that $P_{dc} = P_{ac}$, the peak value of the ac sinusoidal current ($i = I_m \sin \omega t$) divided by $\sqrt{2}$ must equal the magnitude of the dc current.

Writing Eq. 12.25 in the form

$$\boxed{I_m = \sqrt{2} I_{dc}} \qquad (12.26)$$

we find that the peak value of the ac current must equal $\sqrt{2}$ times the dc current.

For example, if $I_{dc} = 10$ A in Fig. 12.42, the sinusoidal current that must be applied to deliver the same power to the resistor must have a peak value $I_m = \sqrt{2} I_{dc} = \sqrt{2}(10 \text{ A}) = 1.414(10 \text{ A}) = 14.14$ A.

The *equivalent dc* value is called the *effective value* of the sinusoidal waveform. In summary,

$$\boxed{I_{eff} = \frac{I_m}{\sqrt{2}} = 0.707 I_m} \qquad (12.27)$$

or

$$\boxed{I_m = \sqrt{2} I_{eff} = 1.414 I_{eff}} \qquad (12.28)$$

Similarly, for voltage levels,

$$V_{\text{eff}} = \frac{V_m}{\sqrt{2}} = 0.707V_m \tag{12.29}$$

or

$$V_m = \sqrt{2}V_{\text{eff}} = 1.414V_{\text{eff}} \tag{12.30}$$

The effective value of a sinusoidal waveform is also the *root-mean-square,* or simply *rms,* value derived from a mathematical procedure to determine the effective value of any waveform.

Note in Eqs. 12.27 through 12.30 that the frequency of the current or voltage does not appear. Consequently,

The effective value of a sinusoidal voltage or current is independent of the frequency of the waveform.

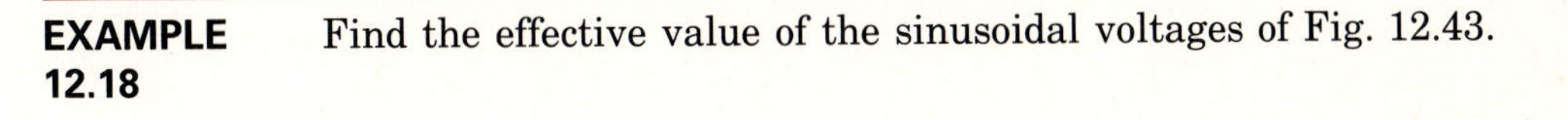

EXAMPLE 12.18 Find the effective value of the sinusoidal voltages of Fig. 12.43.

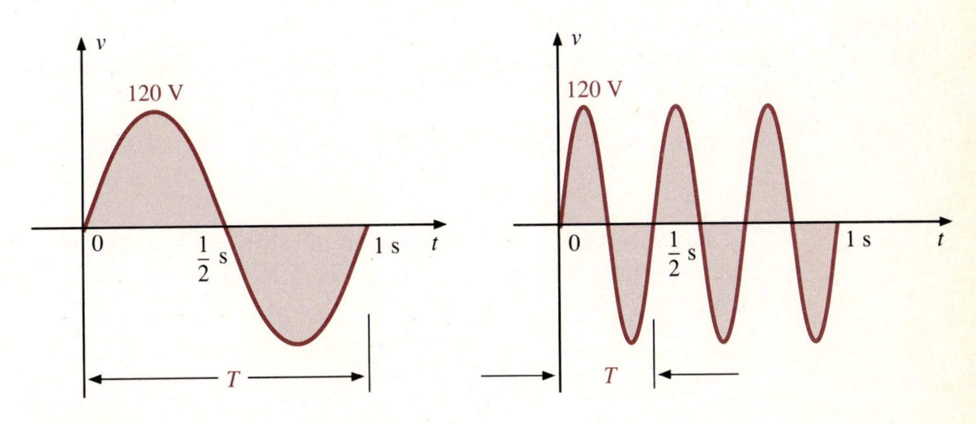

FIG. 12.43
Waveforms for Example 12.18.

Solution Since the effective value of a sinusoidal voltage is independent of the frequency, the effective values of both voltages of Fig. 12.43 are the same.

$$V_{\text{eff}} = \frac{V_m}{\sqrt{2}} = 0.707V_m = 0.707(120\text{ V})$$
$$= \mathbf{84.84\ V}$$

EXAMPLE 12.19 Sketch a sinusoidal ac current having an effective or rms value of 12 mA.

Solution The peak value of the waveform is determined by

$$I_m = \sqrt{2}I_{\text{eff}} = 1.414(12 \text{ mA}) = \mathbf{16.97 \text{ mA}}$$

The sinusoidal waveform appears in Fig. 12.44.

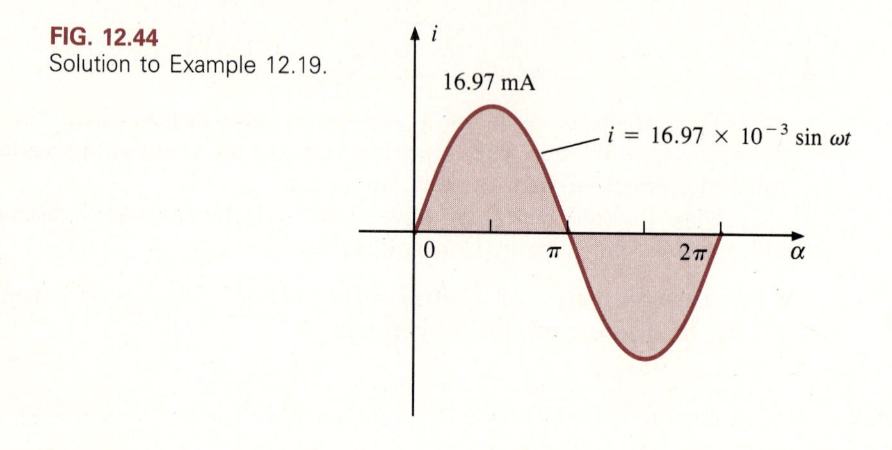

FIG. 12.44
Solution to Example 12.19.

EXAMPLE 12.20 A sinusoidal voltage $e = 60 \sin \omega t$ is delivering power to a resistive load. What dc voltage would have to be applied to deliver the same power to the load?

Solution $$E_{\text{dc}} = 0.707E_m = 0.707(6 \text{ V}) = \mathbf{42.42 \text{ V}}$$

For interest's sake, it has been determined that the effective value of a square wave such as that shown in Fig. 12.45 is equal to the peak value of the waveform. That is,

$$\boxed{V_{\text{eff}} = V_m} \quad \text{(square wave)} \tag{12.31}$$

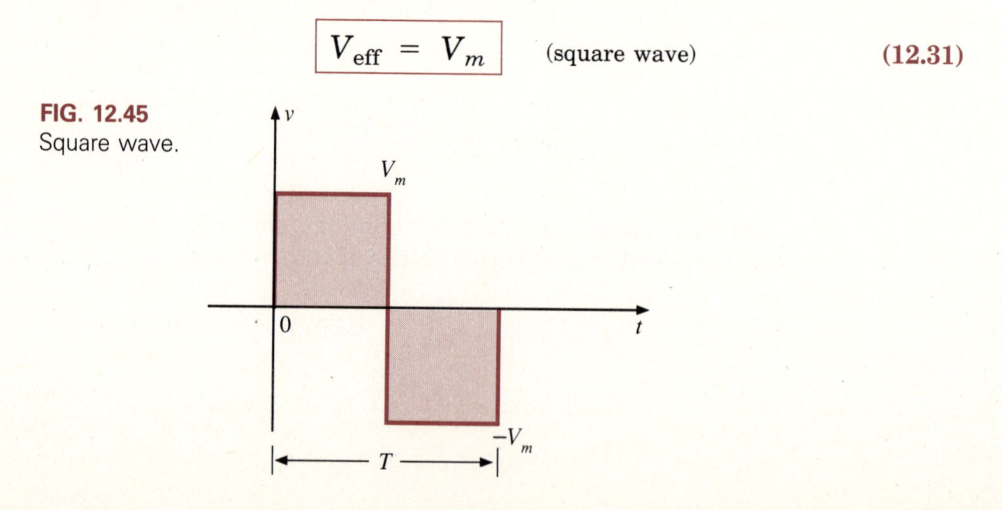

FIG. 12.45
Square wave.

The effective values of sinusoidal quantities such as voltage or current are represented by E and I. These symbols are the same as those used for dc voltages and currents. To avoid confusion, the peak value of a waveform always has a subscript m associated with it: $I_m \sin \omega t$. *Caution:* When finding the effective value of the positive pulse of a sine wave, note that the squared area is *not* simply $(2A_m)^2 = 4A_m^2$; it must be found by a completely new integration. This is always the case for any waveform that is not rectangular.

12.10 AC METERS AND INSTRUMENTS

Virtually all the VOMs and DMMs manufactured today can measure both dc and ac quantities. For current and voltage readings in the ac domain, the *effective value* is normally provided. Very few current models provide the peak-to-peak value of the sinusoidal function. For the typical VOM of Fig. 12.46, the control switch must first be placed in the A.C. position and the range selector must be set above the level to be measured. The analog scale is then carefully read to provide the effective value (37.5 V) of the sinusoidal voltage.

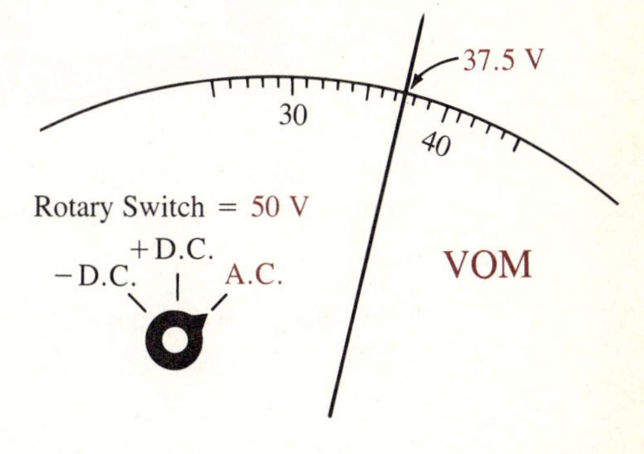

FIG. 12.46
Determining the effective value of a sinusoidal voltage using a VOM.

For the DMM of Fig. 12.47 the dial must first be set on an ac (rms) scale. For the chosen scale (20 V), the display is providing a reading of 12.34 V (effective value).

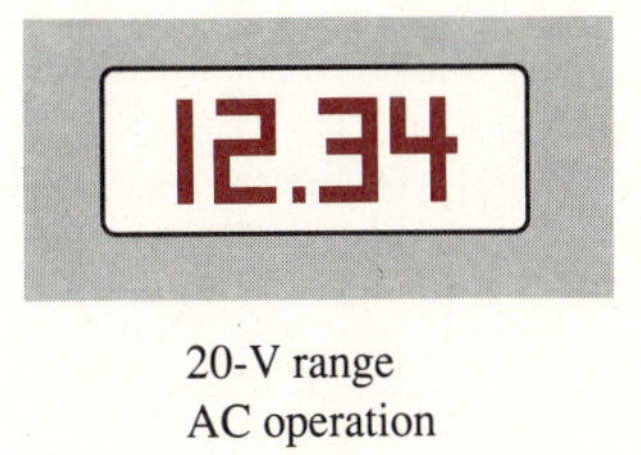

FIG. 12.47
Determining the effective value of a sinusoidal voltage using a DMM.

One of the most versatile and important instruments in the electronics industry is the oscilloscope. It provides a display of the waveform on a cathode-ray tube to permit the detection of irregularities and the determination quantities such as magnitude, frequency, period, dc component, and so on. The unit of Fig. 12.48 is particularly interesting for two reasons: It is portable (working off internal batteries), and it is very small and lightweight. It weighs only 3.5 lb and is approximately 3 in. by 5 in. by 10 in. in size. It has an input impedance of 1 MΩ and a time base that can be set for 5 μs to 500 ms per horizontal division. The vertical scale can be set to sensitivities extending from 1 mV to 50 V per division. This oscilloscope can also display two signals (dual trace) at the same time for magnitude and phase comparisons.

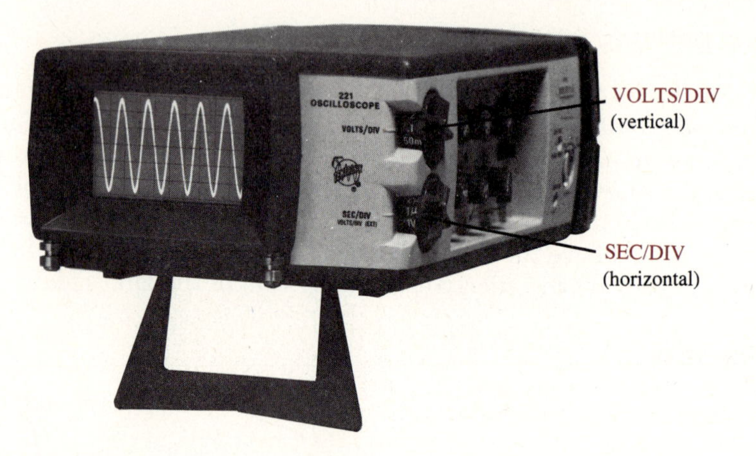

FIG. 12.48
Miniscope. (Courtesy of Tektronics, Inc.)

For the sinusoidal voltage of Fig. 12.49 appearing on the scope of Fig. 12.48, the peak value can be determined from

$$V_m = (1.6 \text{ divisions})(2 \text{ V/division}) = \mathbf{3.2 \text{ V}}$$

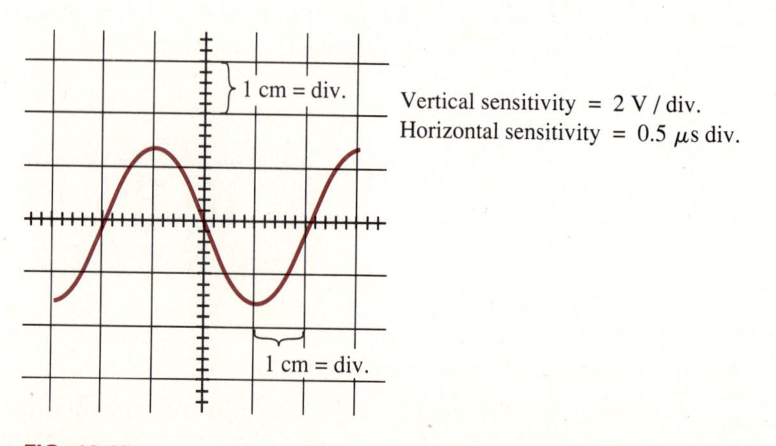

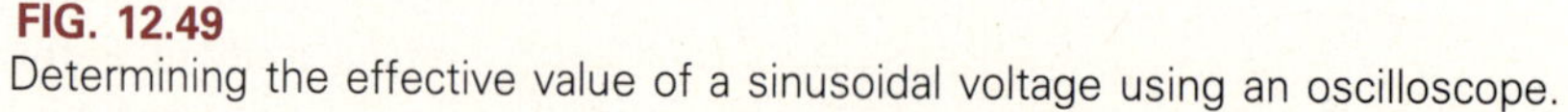
FIG. 12.49
Determining the effective value of a sinusoidal voltage using an oscilloscope.

with the peak-to-peak value determined by

$$V_{p\text{-}p} = 2V_m = 2(3.2 \text{ V}) = \mathbf{6.4 \text{ V}}$$

and an effective value of

$$V_{\text{eff}} = 0.707V_m = 0.707(3.2 \text{ V}) = \mathbf{2.262 \text{ V}}$$

Note in the preceding calculations that the horizontal time scale did not affect the determination of V_m, $V_{p\text{-}p}$, or V_{eff}.

For frequency measurements, the frequency counter of Fig. 12.50 provides a digital readout of the frequency or period of waveforms having a frequency range from 5 Hz to 80 MHz. In a period-average mode, it can average the cycle time over 10, 100, or 1000 cycles. It has an input impedance of 1 MΩ and an internal rechargeable battery for portability. Note the high degree of accuracy available from the six-digit display.

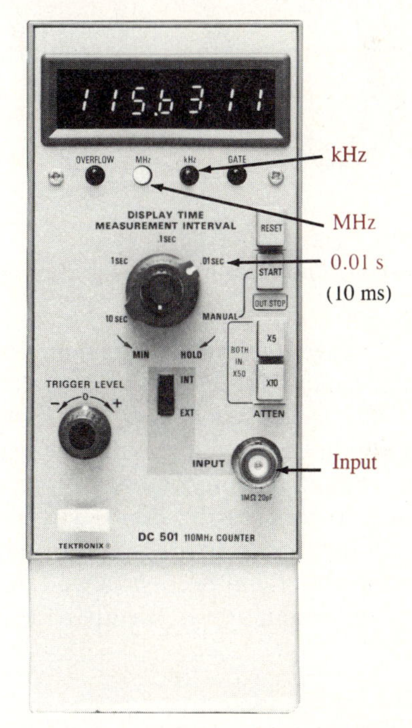

FIG. 12.50
Frequency counter. (Courtesy of Tektronix, Inc.)

The Amp-Clamp® of Fig. 12.51 is an instrument that can measure alternating current in the ampere range without having to open the circuit. The loop is opened by squeezing the "trigger"; then it is placed around the current-carrying conductor. Through transformer action, the level of current in rms units will appear on the appropriate scale. The accuracy of this instrument is ±3% of full scale at 60 Hz, and its scales have maximum values ranging from 6 A to 300 A. The addition of two leads as indicated in the figure permits its use as both a voltmeter and an ohmmeter.

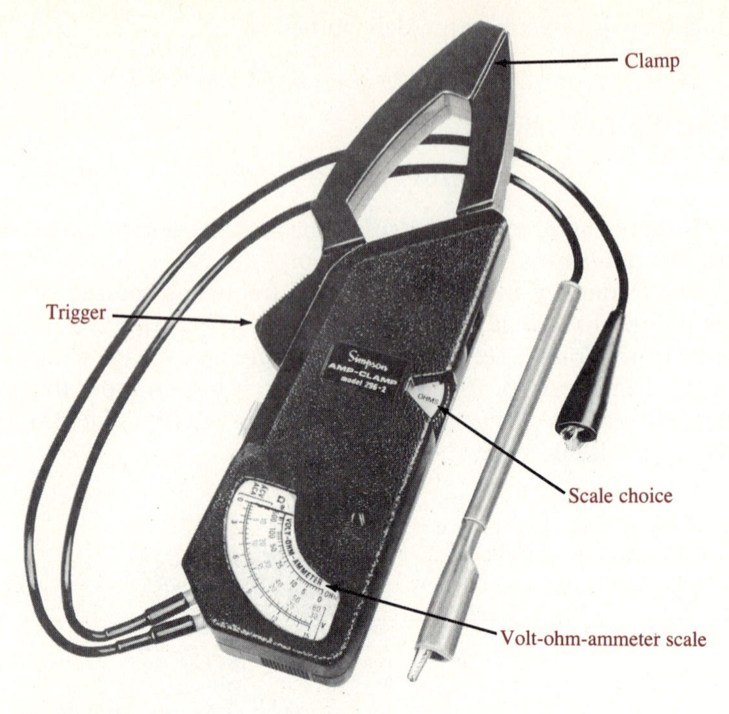

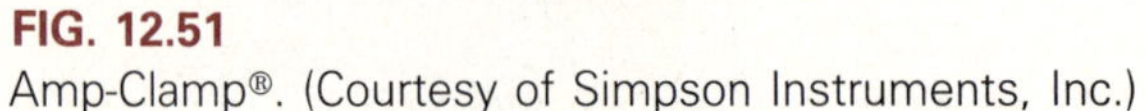

FIG. 12.51
Amp-Clamp®. (Courtesy of Simpson Instruments, Inc.)

Average reading movements such as those employed to measure dc quantities can also be used to measure ac quantities if the sinusoidal signal is first converted to one having an average value and then a *form factor* thrown in to provide a correct ac reading. For the d'Arsonval movement, the *bridge rectifier* of Fig. 12.52 constructed of four *diodes* (electronic switches that conduct current in only one direction) will convert the input signal of zero average value to one having an average value sensitive to the peak value of the input signal. The conversion process is well described in most electronics texts. Fundamentally, conduction is permitted through the diodes in such a manner as to con-

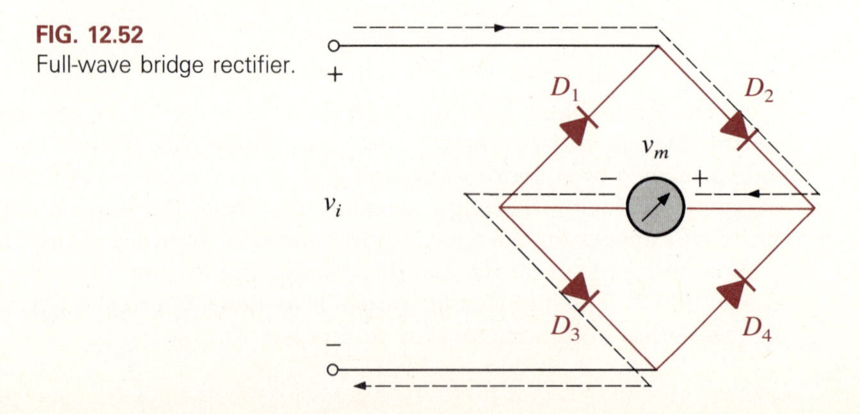

FIG. 12.52
Full-wave bridge rectifier.

vert the sinusoidal input of Fig. 12.53(a) to one having the appearance of Fig. 12.53(b). The negative portion of the input has been effectively "flipped over" by the bridge configuration. In Fig. 12.52 conduction is shown through the diodes for the positive portion of the applied sinusoidal signal. The resulting waveform of Fig. 12.53(b) is called a *full-wave rectified waveform.*

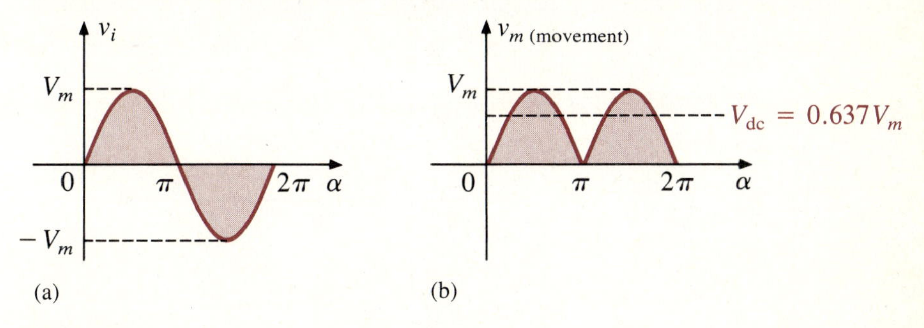

FIG. 12.53
(a) Sinusoidal input; (b) full-wave rectified signal.

The zero average value of Fig. 12.53(a) has been replaced by a pattern having an average value determined by

$$G = \frac{2V_m + 2V_m}{2\pi} = \frac{4V_m}{2\pi} = \frac{2V_m}{\pi} = 0.637V_m$$

The movement of the pointer (sensitive only to average values) is therefore directly related to the peak value of the signal by the factor 0.637.

Forming the ratio between the rms and dc levels will result in

$$\frac{V_{rms}}{V_{dc}} = \frac{0.707V_m}{0.637V_m} \cong 1.11$$

revealing that the scale indication must be 1.11 times the dc level measured by the movement. That is,

$$\boxed{\text{Meter indication} = 1.11(\text{dc, or average, value})} \quad \textit{full-wave} \qquad \textbf{(12.32)}$$

EXAMPLE 12.21 If a sinusoidal voltage $e = 120 \sin t$ is applied to the bridge rectifier of Fig. 12.52:

a. Find V_{dc} of the waveform across the movement (v_m).
b. Determine the indication on the meter.

Solutions

a. $V_{dc} = 0.636V_m = 0.636(120\text{ V}) = \mathbf{76.44\text{ V}}$
b. $V_{eff} = 1.11V_{dc} = 1.11(76.44\text{ V}) = \mathbf{84.84\text{ V}}$ (as obtained from $V_{eff} = 0.707V_m$)

Some ac meters use a half-wave rectifier arrangement that results in the waveform of Fig. 12.54, which has half the average value of Fig. 12.54(b) over one full cycle. The result is

$$\text{Meter indication} = 2.22(\text{dc, or average, value}) \quad \textit{half-wave} \tag{12.33}$$

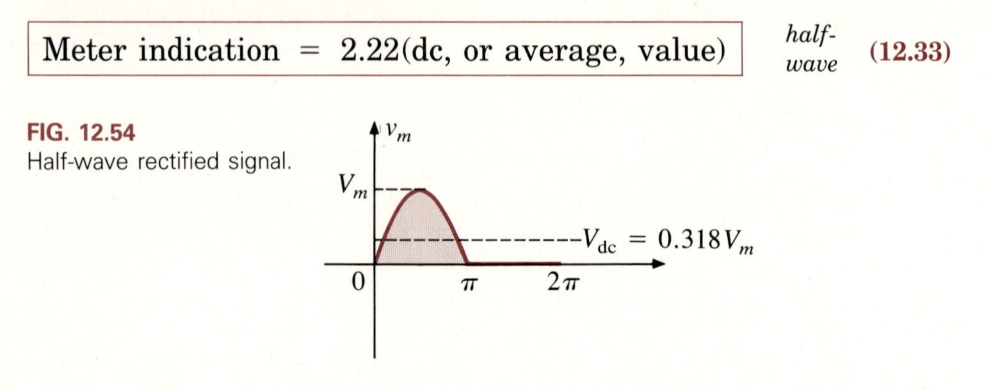

FIG. 12.54
Half-wave rectified signal.

A second movement, called the electrodynamometer movement (Fig. 12.55), can measure both ac and dc quantities without a change in internal circuitry. The movement can, in fact, read the effective value of any periodic or nonperiodic waveform because a reversal in current direction reverses the fields of both the stationary and the movable coils, so the deflection of the pointer is always up-scale. Its use is primarily limited to power measurements, however, due to cost, weight, and sensitivity considerations.

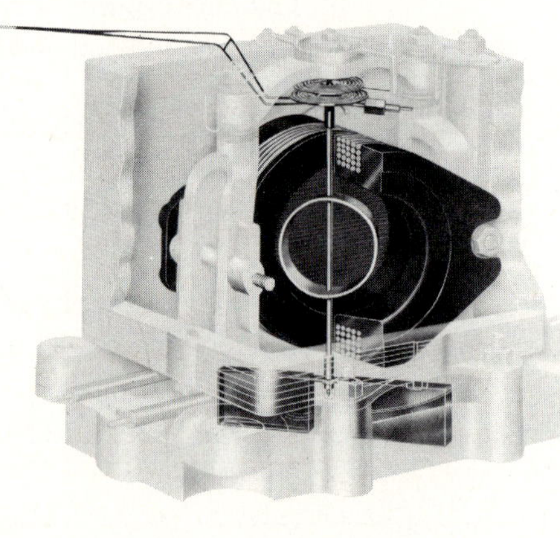

FIG. 12.55
Electrodynamometer movement. (Courtesy of Weston Instruments, Inc.)

EXAMPLE 12.22 Determine the reading of each meter for each situation of Fig. 12.56.

Solution

For part (a), situation (1), by Eq. 12.32,

$$\text{Meter indication} = 1.11(20\text{ V}) = \mathbf{22.2\text{ V}}$$

For part (a), situation (2),

$$V_{\text{rms}} = 0.707V_m = 0.707(20\text{ V}) = \mathbf{14.14\text{ V}}$$

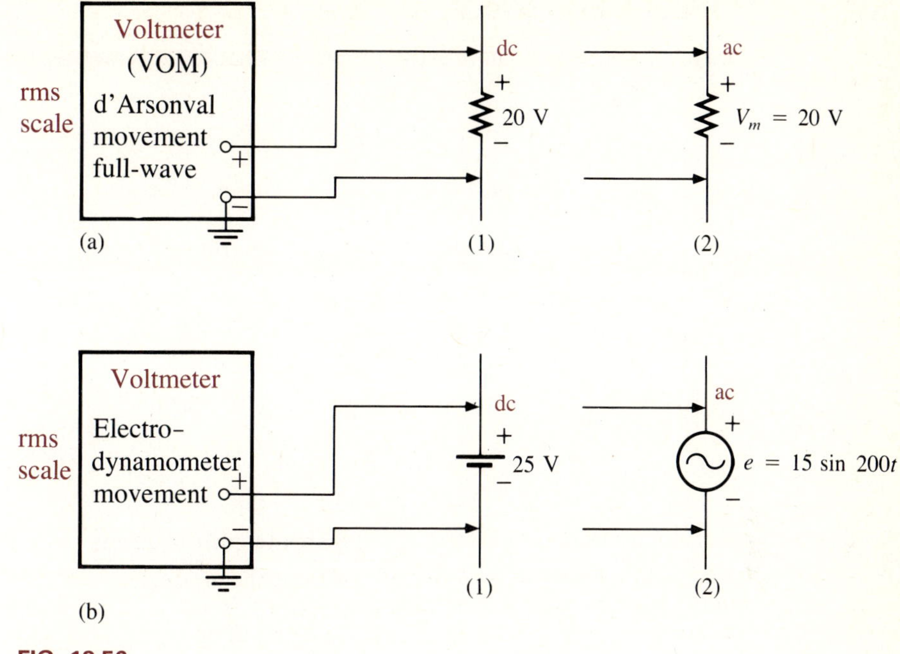

FIG. 12.56
Connections for Example 12.22.

For part (b), situation (1),

$$V_{\text{rms}} = V_{\text{dc}} = \mathbf{25\ V}$$

For part (b), situation (2),

$$V_{\text{rms}} = 0.707V_m = 0.707(15\text{ V}) \cong \mathbf{10.6\ V}$$

Most DMMs employ a full-wave rectification system to convert the input ac signal to one with an average value. In fact, for the DMM of Fig. 3.25, the same scale factor of Eq. 12.32 is employed. That is, the average value is called up by a factor of 1.11 to obtain the rms value. In the digital meters, however, there are no moving parts such as in the d'Arsonval or electrodynamometer movements to display the signal level. Rather, the average value is sensed by a multiprocessor IC, which in turn determines which digits should appear on the digital display.

Digital meters can also be used to measure nonsinusoidal signals, but the scale factor of each input waveform must first be known (normally provided by the manufacturer in the operator's manual). For instance, the scale factor for an average responding DMM on the ac rms scale will produce an indication for a square-wave input that is 1.11 times the peak value. For a triangular input, the response is 0.555 times the peak value. Obviously, for a sine wave input the response is 0.707 times the peak value.

For any instrument, it is always good practice to read (if only briefly) the operator's manual if it appears you will use the instrument on a regular basis.

FORMULA SUMMARY

$$f = \frac{1}{T}, \qquad T = \frac{1}{f}$$

$$2\pi \text{ radians} = 360°, \qquad \pi \cong 3.14$$

$$\text{Radians} = \left(\frac{\pi}{180°}\right)(\text{degrees})$$

$$\text{Degrees} = \left(\frac{180°}{\pi}\right)(\text{radians})$$

$$\omega = \frac{2\pi}{T}, \qquad \omega = 2\pi f$$

$$e = E_m \sin \omega t, \qquad i = I_m \sin 2\pi f t$$

$$\cos \alpha = \sin(\alpha + 90°)$$

$$G = \frac{\text{Algebraic sum of areas}}{\text{Length of curve}}$$

$$I_{\text{eff}} = \frac{1}{\sqrt{2}} I_m = 0.707 I_m$$

$$V_m = \sqrt{2} V_{\text{eff}} = 1.414 V_{\text{eff}}$$

CHAPTER SUMMARY

This is the first of a number of chapters introducing definitions and concepts associated with sinusoidal waveforms. Since the sinusoidal voltage is the prevalent waveform in the electrical/electronics industry, it is imperative that the basic definitions introduced in this and following chapters be correctly and clearly understood. Be comforted by the fact that once the basic characteristics of a sinusoidal voltage or current are understood, the application of most of the laws introduced for dc systems—such as Ohm's law, Kirchhoff's voltage law, and Kirchhoff's current law—can be applied with a minimum of added confusion. There will naturally be differences due to a time-varying function, but through exposure, review, and application, the equations introduced in this chapter will simply be the tools necessary to complete the analysis.

Since f and T are *inversely* related, any change in one will have the *opposite* effect on the other. That is, an increase in one will decrease the other, and vice versa. Keep in mind that radian measure is simply an extension of degree measure, with 1 rad $\cong$ 57.3° and 2π rad = 360°. In time the correspondence between 180° and π, 90° and $\pi/2$, 45° and $\pi/4$, and the like will be so familiar that the use of radian measure will occur with the same confidence as use of degrees. For units of time, it is usually best to determine the period T and then break the period down into the desired increments.

When reviewing the concept of angular velocity, keep the rotating vector in mind, since it relates well to the resulting equations. The faster the vector rotates, the shorter the time required to complete a revolution and the smaller the period $T(T = 2\pi/\omega)$. Obviously, the greater the angular velocity, the more frequently the rotating vector will complete a revolution in one second and the greater will be the frequency $(f = \omega/2\pi)$.

The phase shift between sinusoidal functions is determined from the intersections of corresponding slopes of the two waveforms. That is, both functions cross the axis increasing (or decreasing) with time. In addition, don't be afraid to state that one function leads the other by more than 90° or 180°. To state that one sine wave *leads* another by 290° is the same as saying that the other function *lags* the first by 70°(360° − 290°). Both statements are correct, with the choice dependent on which lends itself best to the general conclusions to be derived from the analysis.

The average value of a time-varying function is determined by the algebraic sum of the positive and negative excursions and the time interval under investigation. A change in time interval can have a significant impact on the average value. Be sure to include the proper sign when determining the algebraic sum of the areas and include the proper power of 10 with the variable of interest.

A mathematical equation was not provided for the effective value of any time-varying signal because the calculation is seldom required for practical applications. Simply remember the appropriate equations for the sinusoidal function $(I_{\text{eff}} = (1/\sqrt{2})I_m)$ and the square wave, where $V_{\text{eff}} = V_m$.

GLOSSARY

Alternating waveform A waveform that oscillates above and below a defined reference level.

Amp-Clamp® A clamp-type instrument that permits noninvasive current measurements.

Angular velocity The velocity with which a radius vector projecting a sinusoidal function rotates about its center.

Average value The level of a waveform defined by the condition that the area enclosed by the curve above this level is exactly equal to the area enclosed by the curve below this level for a specified interval.

Cycle A portion of a waveform contained in one period of time.

Effective value The equivalent dc value of any alternating voltage or current.

Electrodynamometer meters Instruments that can measure both ac and dc quantities without a change in internal circuitry.

Frequency (f) The number of cycles of a periodic waveform that occur in 1 s.

Frequency counter An instrument that provides a digital display of the frequency or period of a periodic time-varying signal.

Instantaneous value The magnitude of a waveform at any instant of time, denoted by lowercase letters.

Oscilloscope An instrument that will display, through the use of a cathode-ray tube, the characteristics of a time-varying signal.

Peak-to-peak value The magnitude of the total swing of a signal from positive to negative peaks. The sum of the absolute values of the positive and negative peak values.

Peak value The maximum value of a waveform, denoted by uppercase letters.

Period (T) The time interval between successive repetitions of a periodic waveform.

Periodic waveform A waveform that continually repeats itself after a defined time interval.

Phase relationship An indication of which of two waveforms leads or lags the other and by how many degrees or radians.

Radian A unit of measure used to define a particular segment of a circle. One radian is approximately equal to 57.3°; 2π radians are equal to 360°.

Rectifier-type ac meter An instrument calibrated to indicate the effective value of a current or voltage through the use of a rectifier network and d'Arsonval-type movement.

Rms value The root-mean-square or effective value of a waveform.

Sinusoidal ac waveform An alternating waveform of unique characteristics that oscillates with equal amplitude above and below a given axis.

VOM A multimeter with the capability to measure resistance and both ac and dc levels of current and voltage.

Waveform The path traced by a quantity, plotted as a function of some variable such as position, time, degrees, temperature, and so on.

PROBLEMS

Section 12.4

1. For the sinusoidal waveform of Fig. 12.57, determine each of the following.
   a. Period T
   b. Time t_1
   c. Peak value
   d. Peak-to-peak value
   e. Frequency
   f. Number of cycles shown

FIG. 12.57

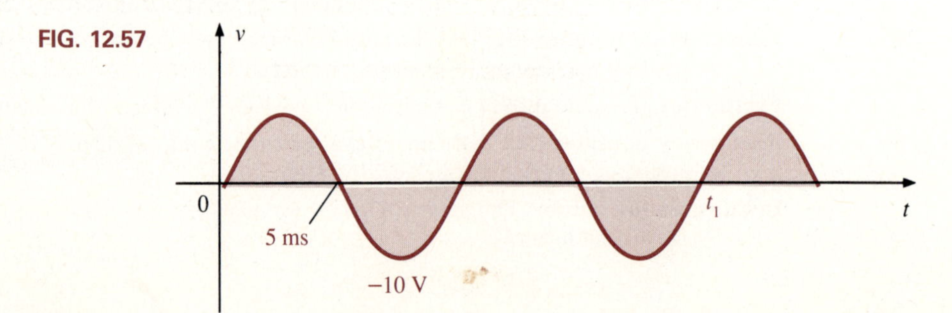

2. For the periodic waveform of Fig. 12.58, determine each of the following.
   a. Period T
   b. Frequency
   c. Number of cycles shown
   d. Peak and peak-to-peak values

FIG. 12.58

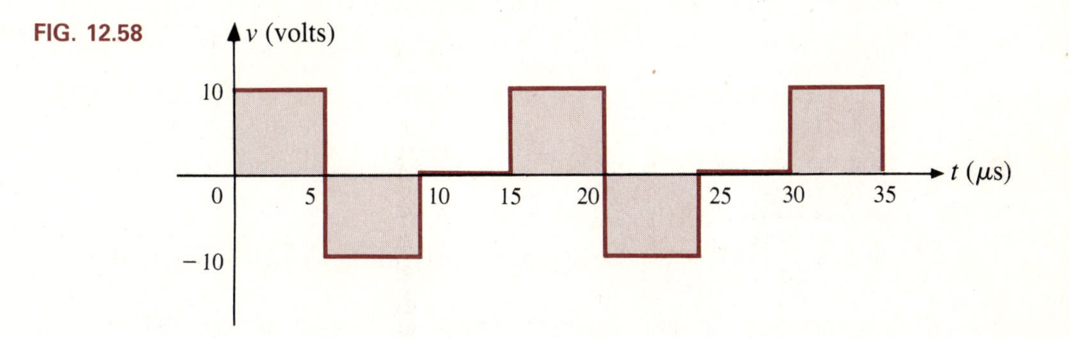

3. Sketch a sinusoidal waveform having a frequency of 2 kHz and a peak-to-peak value of 8 V. Use a time scale broken down into divisions of 1 ms.

*4. Determine the period and frequency of the sawtooth waveform of Fig. 12.59.

FIG. 12.59

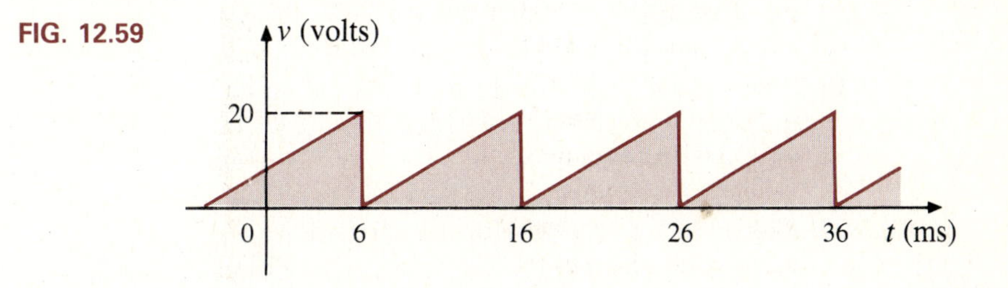

5. Find the period of a periodic waveform whose frequency is
   a. 25 Hz b. 35 MHz
   c. 55 kHz d. 1 Hz
6. Find the frequency of a repeating waveform whose period is
   a. 1/60 s b. 0.01 s
   c. 34 ms d. 25 μs
7. Find the period of a sinusoidal waveform that completes 80 cycles in 24 ms.
8. If a periodic waveform has a frequency of 20 Hz, how long (in seconds) will it take to complete 5 cycles?
9. What is the frequency of a periodic waveform that completes 42 cycles in 6 s?
10. Sketch a periodic square wave like that appearing in Fig. 12.50 with a frequency of 20,000 Hz and a peak value of 10 mV.

Section 12.5

11. Convert the following degrees to radians:
   a. 45° b. 60°
   c. 120° d. 270°
   e. 178° f. 221°

12. Convert the following radians to degrees:
a. $\pi/4$ **b.** $\pi/6$
c. $\frac{1}{10}\pi$ **d.** $\frac{7}{6}\pi$
e. 3π **f.** 0.55π

13. Find the angular velocity of a waveform with a period of
a. 2 s **b.** 0.3 ms
c. 4 μs **d.** 1/25 s

14. Find the angular velocity of a waveform with a frequency of
a. 50 Hz **b.** 600 Hz
c. 2 kHz **d.** 0.004 MHz

15. Find the frequency and period of sine waves having an angular velocity of
a. 754 rad/s **b.** 8.4 rad/s
c. 6000 rad/s **d.** 1/16 rad/s

16. Given $f = 60$ Hz, determine how long it will take the sinusoidal waveform to pass through an angle of 45°.

***17.** If a sinusoidal waveform passes through an angle of 30° in 5 ms, determine the angular velocity of the waveform.

Section 12.6

18. Find the amplitude and frequency of the following waves:
a. $20 \sin 377t$ **b.** $5 \sin 754t$
c. $10^6 \sin 10{,}000t$ **d.** $0.001 \sin 942t$
e. $-7.6 \sin 43.6t$ **f.** $(1/42)\sin 6.28t$

19. Sketch $5 \sin 754t$ with the abscissa
a. Angle in degrees
b. Angle in radians
c. Time in seconds

20. Sketch $10^6 \sin 10{,}000t$ with the abscissa
a. Angle in degrees
b. Angle in radians
c. Time in seconds

***21.** Sketch $-7.6 \sin 43.6t$ with the abscissa
a. Angle in degrees
b. Angle in radians
c. Time in seconds

22. If $e = 300 \sin 157t$, how long (in seconds) does it take this waveform to complete 1/2 cycle?

23. Given $i = 0.5 \sin \alpha$, determine i at $\alpha = 72°$.

24. Given $v = 20 \sin \alpha$, determine v at $\alpha = 1.2\pi$.

***25.** Given $v = 30 \times 10^{-3} \sin \alpha$, determine the angles at which v will be 6 mV.

***26.** If $v = 40$ V at $\alpha = 30°$ and $t = 1$ ms, determine the mathematical expression for the sinusoidal voltage.

Section 12.7

27. Sketch $\sin(377t + 60°)$ with the abscissa
a. Angle in degrees
b. Angle in radians
c. Time in seconds

28. Sketch the following waveforms:

a. $50 \sin \omega t$ **b.** $20 \sin(\omega t - 20°)$
c. $5 \sin(\omega t + 60°)$ **d.** $4 \cos \omega t$
e. $2 \cos(\omega t + 10°)$ **f.** $-5 \sin(\omega t + 30°)$

29. Find the phase relationship between the waveforms of each set:

a. $v = 4 \sin(\omega t + 50°)$
$i = 6 \sin \omega t$
b. $v = 25 \sin(\omega t - 80°)$
$i = 5 \times 10^{-3} \sin(\omega t - 80°)$
c. $v = 0.2 \sin(\omega t - 60°)$
$i = 0.1 \sin(\omega t + 20°)$
d. $v = 200 \sin(\omega t + 90°)$
$i = 25 \cos \omega t$

30. Write the analytical expression for the waveforms of Fig. 12.60 with the phase angle in degrees.

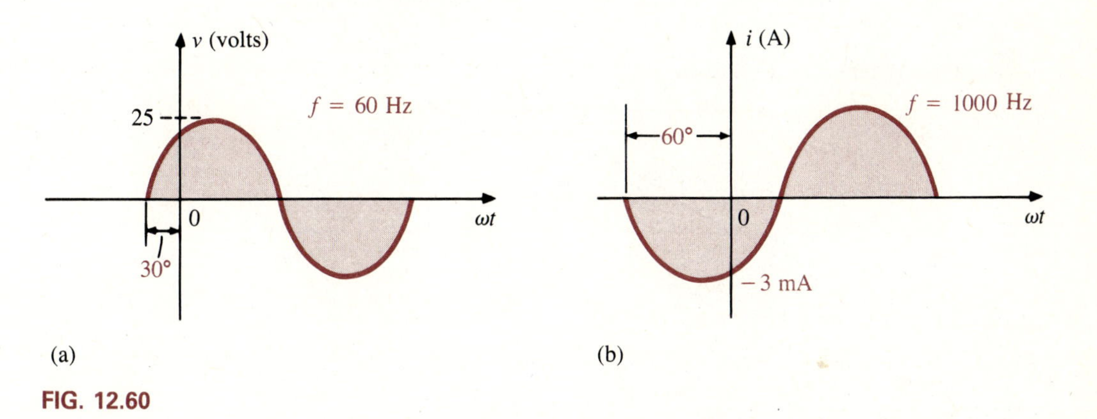

FIG. 12.60

31. Repeat Problem 30 for the waveforms of Fig. 12.61.

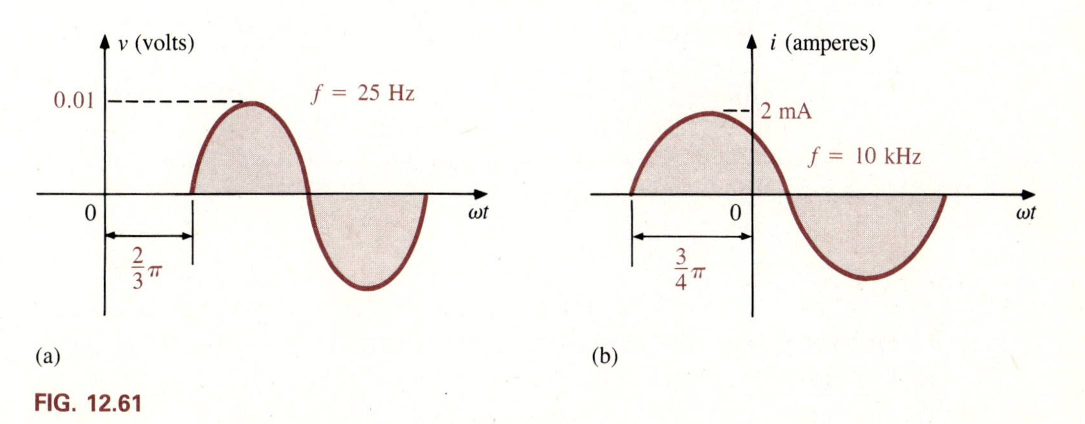

FIG. 12.61

Section 12.8

32. Find the average value of the periodic waveform of Fig. 12.62 over one full cycle.

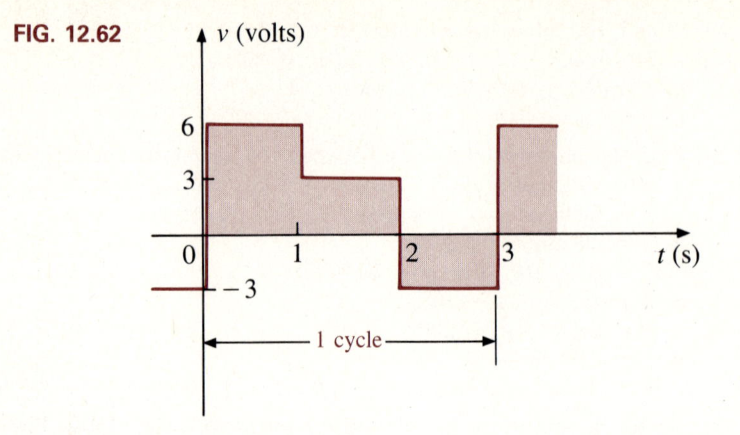

FIG. 12.62

***33.** Find the average value of the periodic waveform of Fig. 12.63.

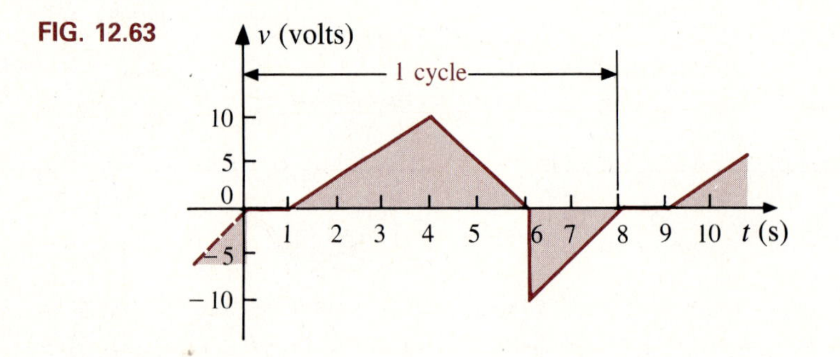

FIG. 12.63

34. Find the average value of the periodic waveform of Fig. 12.64.

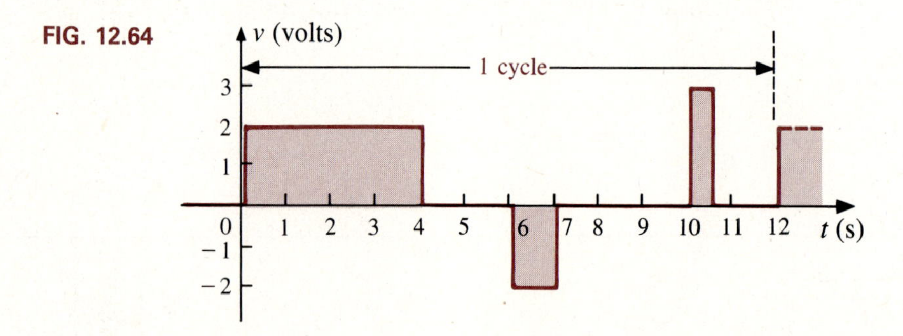

FIG. 12.64

35. Find the approximate average value for the periodic waveform of Fig. 12.65.

36. Find the average value of the periodic waveform of Fig. 12.66.

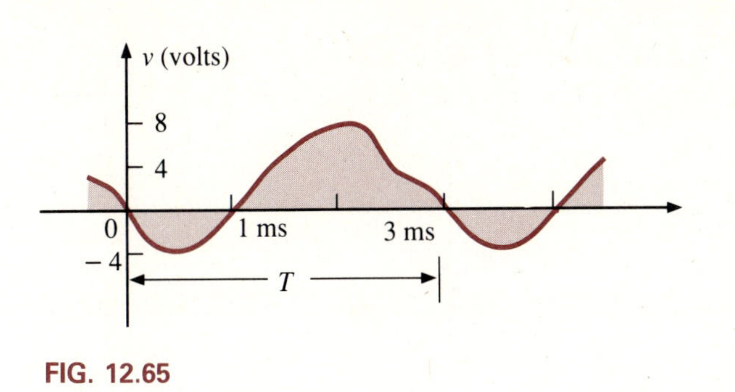

FIG. 12.65

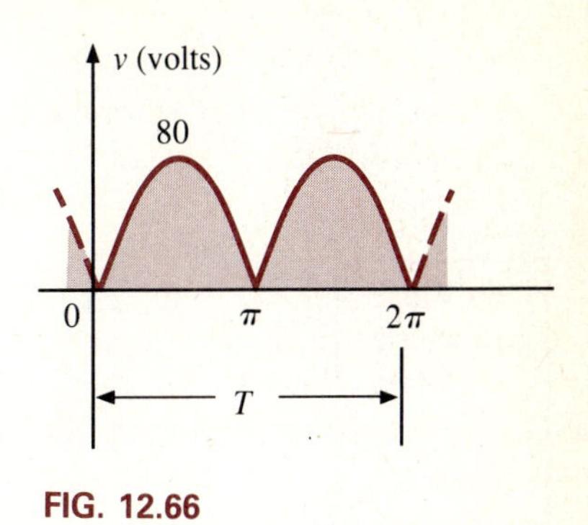

FIG. 12.66

Section 12.9

37. Find the effective values of the following sinusoidal waveforms:

a. $v = 20 \sin 754t$

b. $v = 7.07 \sin 377t$

c. $i = 0.006 \sin(400t + 20°)$

d. $i = 16 \times 10^{-3} \sin(377t - 10°)$

38. Write the sinusoidal expressions for voltages and currents having the following effective values at a frequency of 60 Hz with zero phase shift:

a. 1.414 V **b.** 70.7 V

c. 0.06 A **d.** 24 μA

39. Find the effective value of the waveform of Fig. 12.12.

40. If the current through a 5-kΩ resistor is $i = 20 \times 10^{-3} \sin(\omega t + 30°)$, determine the dc current necessary to deliver the same power to the load resistor.

41. The dc power delivered to a load resistor by a 12-V battery is 72 mW.

a. Determine the resistance R.

b. Calculate the peak value of the sinusoidal ac voltage that must be applied to deliver the same power to the resistor R.

42. What are the average and effective values of the square wave of Fig. 12.67?

FIG. 12.67

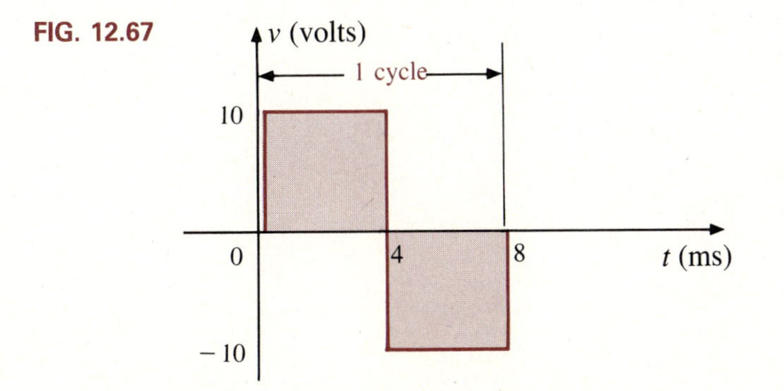

Section 12.10

***43.** Determine the reading of the meter for each situation of Fig. 12.68.

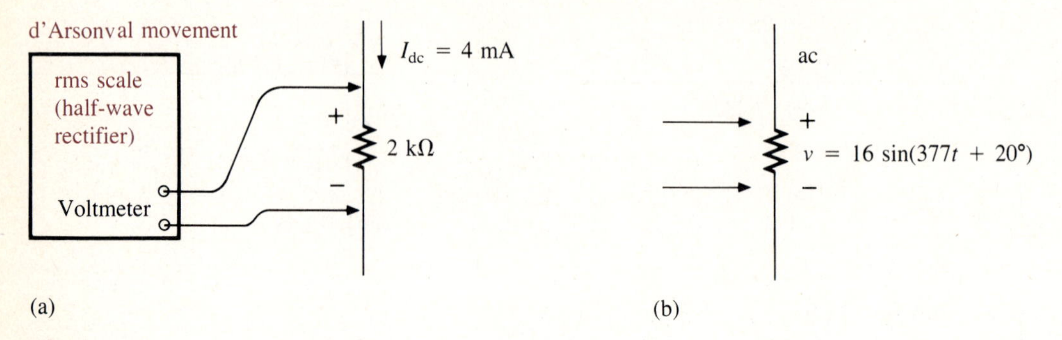

FIG. 12.68

44. Repeat Problem 43 if an electrodynamometer movement is employed in the voltmeter.

45. Repeat Problem 43 if a d'Arsonval movement is employed with a full-wave rectifier.

Computer Problems

46. Given a sinusoidal function, write a program to determine the effective value, frequency, and period.

47. Given two sinusoidal functions, write a program to determine the phase shift between the two waveforms, and indicate which is leading or lagging.

48. Given an alternating pulse waveform, write a program to determine the average value of the waveform over one complete cycle.

13

The Basic Elements and Phasors

OBJECTIVES

- ☐ Become aware of how resistors, inductors, and capacitors react to sinusoidal functions of different frequencies and phase shifts.
- ☐ Understand the difference between resistance and reactance and how inductive and capacitive reactance vary with frequency.
- ☐ Be able to determine and sketch the sinusoidal current or voltage of a resistor, capacitor, or inductor given the other quantity and the resistance or reactance of the element.
- ☐ Become familiar with the basic power equations for ac systems and how to determine the power factor of a load.
- ☐ Become aware of phasors and how they relate to sinusoidal functions and permit the addition or subtraction of sinusoidal voltages or currents.

13.1 INTRODUCTION

The response of the basic R, L, and C elements to a sinusoidal voltage and current are examined in this chapter, with special note of how frequency affects the "opposing" characteristic of each element. Phasor notation is then introduced to ensure its use in industrial labeling and application is understood.

13.2 RESISTANCE (ac DOMAIN)

For power-line frequencies (such as 60 Hz) and frequencies up to a few hundred kilohertz, resistance is, for all practical purposes, unaffected by the frequency of the applied sinusoidal voltage or current.

If we plot resistance versus frequency, we obtain the curve of Fig. 13.1, clearly revealing that R is a fixed value and is unaffected by the changing frequency. At 5 kHz, 15 kHz, or 20 kHz, the resistance is fixed at 1 kΩ.

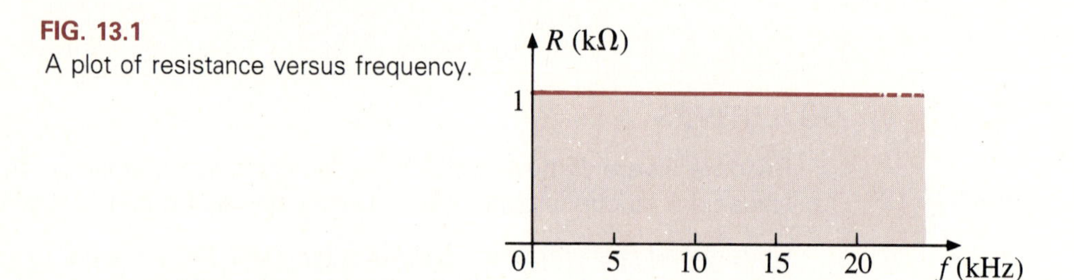

FIG. 13.1
A plot of resistance versus frequency.

Let us now apply a sinusoidal voltage across the resistor of Fig. 13.2 and use Ohm's law to find the current.

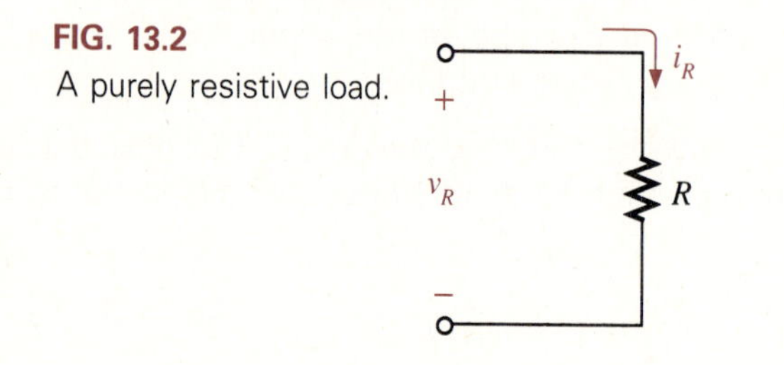

FIG. 13.2
A purely resistive load.

For $v_R = V_m \sin \omega t$,

$$i_R = \frac{v_R}{R} = \frac{V_m \sin \omega t}{R}$$

Since V_m and R are constants, we can isolate the ratio;

$$i_R = \frac{V_m}{R} \sin \omega t = I_m \sin \omega t$$

which reveals that the maximum (peak) value of the current i_R of Fig. 13.2 is determined by

$$I_m = \frac{V_m}{R} \tag{13.1}$$

In addition, note that the sinusoidal expressions for v_R and i_R do not have a phase shift associated with either function. The result is that

The voltage across a resistor is in phase with the current through the resistor.

If the voltage across a resistor is desired, given the sinusoidal current, the peak value of the voltage is determined from

$$V_m = I_m R \tag{13.2}$$

maintaining the in-phase relationship between v_R and i_R.

A plot of v_R and i_R is provided in Fig. 13.3, clearly revealing the in-phase relationship and the peak values V_m and I_m related by Ohm's law.

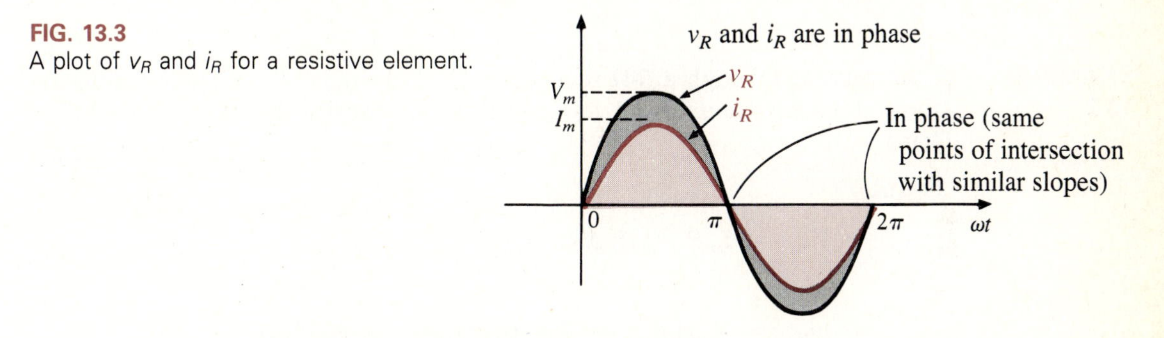

FIG. 13.3
A plot of v_R and i_R for a resistive element.

If v_R or i_R have a phase shift included in the sinusoidal expression, then the resulting i_R or v_R (respectively) will have the same phase shift as shown in Fig. 13.4 for $v_R = 10 \sin(\omega t + 60°)$.

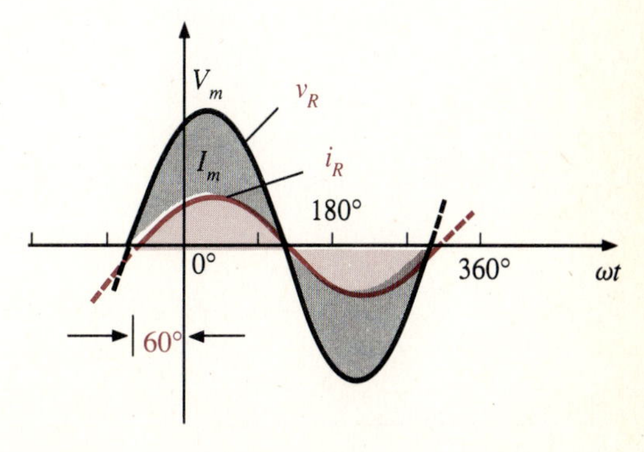

FIG. 13.4
Effect of a phase shift on a phase relationship between v_R and i_R.

EXAMPLE 13.1

Given a voltage $v_R = 20 \sin \omega t$ across a resistor of 5 kΩ, determine the sinusoidal expression for the current and sketch both waveforms.

Solution

$$I_m = \frac{V_m}{R} = \frac{20\text{ V}}{5\text{ k}\Omega} = 4\text{ mA}$$

and

$$i_R = I_m \sin \omega t \text{ (in phase)} = \mathbf{4 \times 10^{-3} \sin \omega t}$$

See Fig. 13.5.

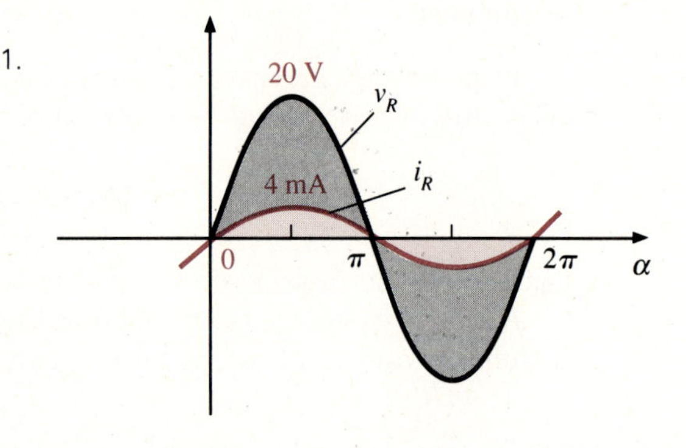

FIG. 13.5
Solution to Example 13.1.

EXAMPLE 13.2

If the current through a 5-Ω resistor is $i_R = 8 \sin(\omega t - 45°)$, determine the sinusoidal expression for the voltage across the resistor and sketch both waveforms.

Solution

$$V_m = I_m R = (8\text{ A})(5\ \Omega) = 40\text{ V}$$

and

$$v_R = V_m \sin(\omega t - 45°) \text{ (in phase)} = \mathbf{40 \sin(\omega t - 45°)}$$

See Fig. 13.6.

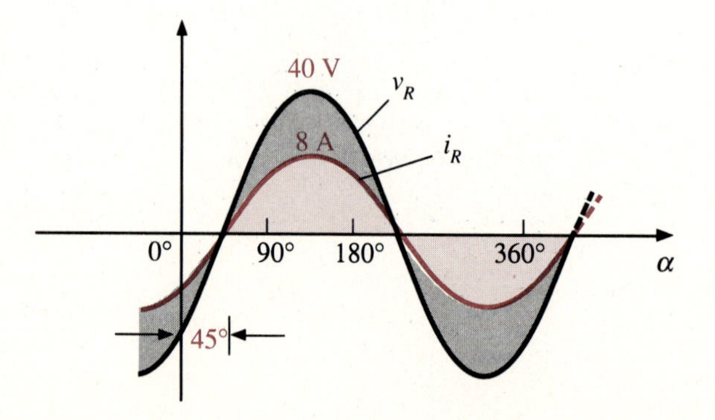

FIG. 13.6
Solution to Example 13.2.

13.3 INDUCTANCE (ac DOMAIN)

The inductor is quite different from a resistor in that it is sensitive to the frequency of the applied voltage or current. The difference stems from the relationship between the voltage across a coil v_L and the current through the coil i_L, as introduced in Chapter 11 and repeated as Eq. 13.3.

$$v_L = L\frac{di_L}{dt} \tag{13.3}$$

Both v_L and i_L are defined by Fig. 13.7. Note the polarity of v_L as defined by the direction of the current i_L.

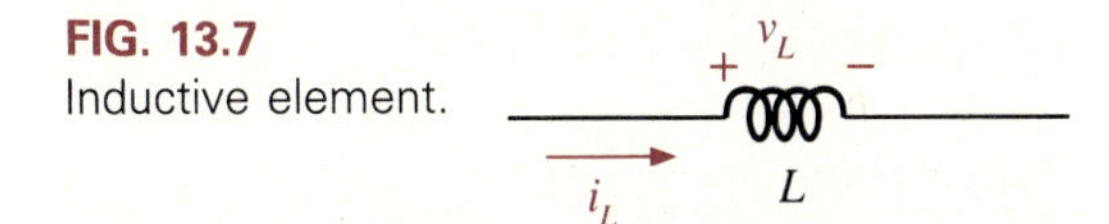

FIG. 13.7
Inductive element.

When we considered the resistive element, the voltage and current were related by a constant R for the frequency range of interest. Equation 13.3, however, states that the voltage is directly related to the magnitude of the inductance (L) and how rapidly the current through the coil is changing with time (di_L/dt). In other words, the larger the inductance or the more rapid the change in current, the greater the voltage developed across the coil. Since L is assumed constant for our frequency range of interest, we can concentrate on the effect of the derivative di_L/dt introduced in Chapter 11.

If we consider two currents of different frequencies, such as those in Fig. 13.8, we note that the *higher the frequency,* the greater the change in current per unit time. This is clearly revealed by the slope of both curves when they cross the axis. We can, therefore, conclude that the higher the frequency of the

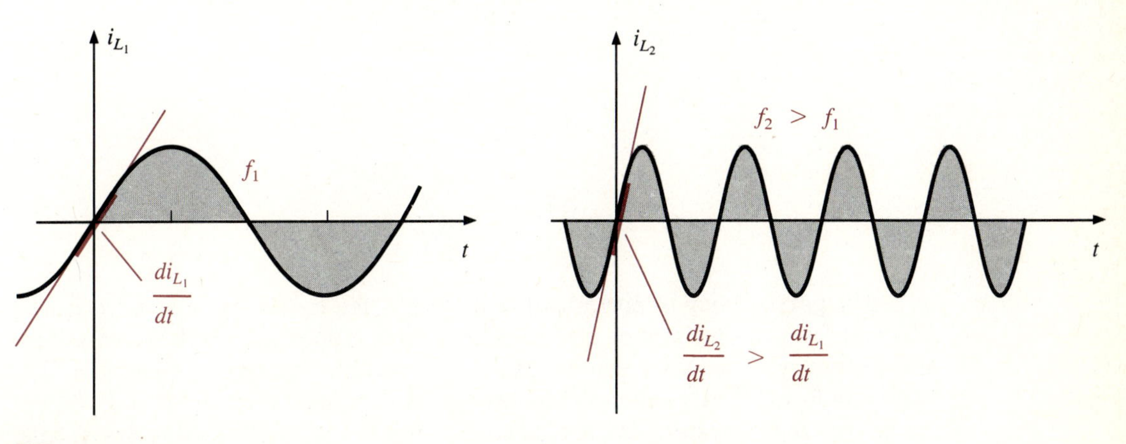

FIG. 13.8
Effect of frequency on di_L/dt.

current, the greater the factor di_L/dt of Eq. 13.3 and the greater the voltage developed across the coil. The magnitude of the voltage is, therefore, directly related to the inductance of the coil and the frequency of the current i_L. An increase in either (or both) results in an increase in the voltage across the coil.

Through experimental or mathematical techniques, it can be shown that the peak value of the voltage across the coil is related to the peak value of the current through the coil by

$$\boxed{V_m = (2\pi f L)I_m} \tag{13.4}$$

Note the multiplying factor of f and L as described before. The factor 2π is an added proportionality factor required to permit a direct calculation of one from the other.

For resistance we found in Chapter 5 that

$$\text{Opposition} = \frac{\text{Cause}}{\text{Effect}} = \frac{V_m}{I_m} = R$$

For inductance, the result is

$$\text{Opposition} = \frac{\text{Cause}}{\text{Effect}} = \frac{V_m}{I_m} = 2\pi f L$$

as derived from Eq. 13.4.

In general, the opposition of a coil is referred to as the *reactance* of the inductor and is given the notation X_L. In addition, since $\omega = 2\pi f$,

$$\boxed{X_L = \omega L} \quad \text{(ohms, } \Omega\text{)} \tag{13.5}$$

As noted by Eq. 13.5, X_L is also measured in ohms, and the peak values of v_L and i_L are related by

$$\boxed{V_m = I_m X_L} \tag{13.6}$$

with

$$\boxed{I_m = \frac{V_m}{X_L}} \tag{13.7}$$

It is particularly important to realize that the reactance of a coil is quite different from the resistance of a resistor in that it does not result in a dissipation of energy (like the resistor) but simply converts it from an electrical to a magnetic form. When called for by electrical design, the inductor can return the energy to the system. We are, of course, assuming "ideal" inductors in this discussion. Every coil has some resistance associated with the turns of wire, and some energy is always lost in changing it from one form to another. However, from an ideal viewpoint and one that is appropriate for many applications, the conversion process can be considered ideal.

The inductor has one other important impact on the relationship between v_L and i_L. It introduces a phase shift between the two of 90° that results in the following fact:

The voltage across coil v_L leads the current through the coil i_L by 90°.

The phase shift is shown in Fig. 13.9.

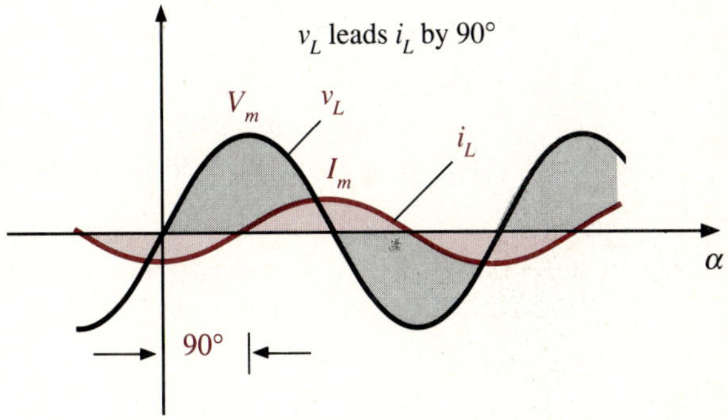

FIG. 13.9
Phase relationship between v_L and i_L for a pure inductor.

EXAMPLE 13.3 Determine the reactance of a 10-mH coil at a frequency of

a. 100 Hz
b. 5 kHz
c. 2 MHz

Solutions

a. $X_L = \omega L = 2\pi f L$
$= 2(3.14)(100\text{ Hz})(10 \times 10^{-3}\text{ H})$
$X_L =$ **6.283 Ω**

Calculator

(2) (×) (2nd F) (π) (×) (1) (0) (0) (×) (1) (0) (EXP) (+/−) (3) (=)

with a display of 6.283.

b. $X_L = \omega L = 2\pi f L$
$= 2(3.14)(5 \times 10^{3}\text{ Hz})(10 \times 10^{-3}\text{ H})$
$X_L =$ **314 Ω**

Calculator

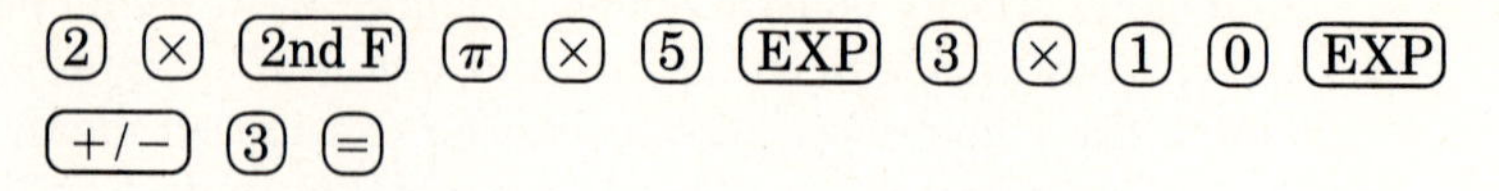

with a display of 314.159.

c. $X_L = \omega L = 2\pi fL$

$= 2(3.14)(2 \times 10^6\ \text{Hz})(10 \times 10^{-3}\ \text{H})$

$X_L = \mathbf{125.6\ k\Omega}$

Calculator

(2) (×) (2nd F) (π) (×) (2) (EXP) (6) (×) (1) (0) (EXP)
(+/−) (3) (=) (F <−> E)

with a display of 1.257 05.

Note how dramatically the change in frequency can affect the reactance of a coil from a level of 6.28 Ω at 100 Hz to over 125 kΩ at 2 MHz.

EXAMPLE 13.4

a. Determine the voltage across a 0.01-H coil if the current $i_L = 2 \sin 400t$.
b. Sketch the waveforms of v_L and i_L.

Solutions

a. Since ω is given as 400 rad/s, there is no need to determine the frequency and calculate $2\pi f$.

Therefore,

$$X_L = \omega L = (400\ \text{rad/s})(0.01\ \text{H}) = 4\ \Omega$$
$$V_m = I_m X_L = (2\ \text{A})(4\ \Omega) = 8\ \text{V}$$

Since v_L leads i_L by 90°, $v_L = V_m \sin(\omega t + 90°)$ and $v_L =$ **8 sin(400*t* + 90°).**

b. See Fig. 13.10.

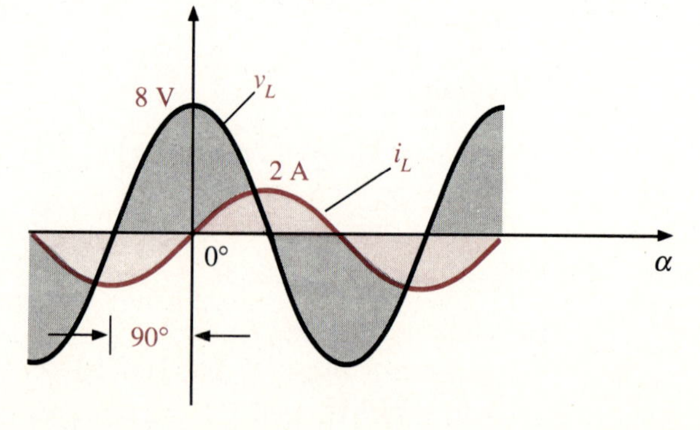

FIG. 13.10
Solution to Example 13.4.

EXAMPLE 13.5

a. Determine the current i_L of a 200-mH coil if the voltage across the coil is $v_L = 5\sin(\omega t + 45°)$ at a frequency of 60 Hz.
b. Sketch the waveforms of v_L and i_L.

Solutions

a. $X_L = 2\pi fL = 2(3.14)(60\text{ Hz})(200 \times 10^{-3}\text{ H})$
$= \mathbf{75.36\ \Omega}$

Calculator

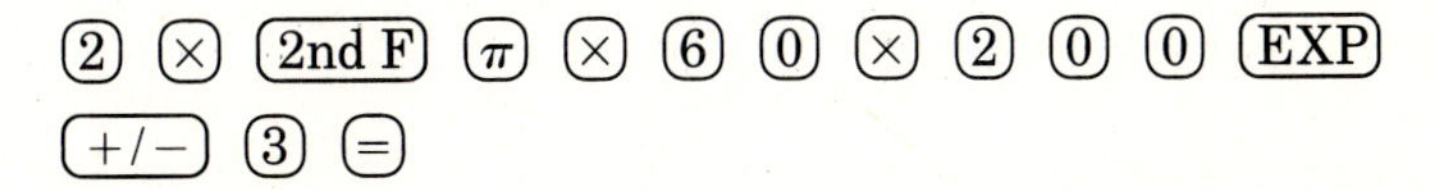

with a display of 75.398 (the difference in answers is due to the higher accuracy for π used here).

$$I_m = \frac{V_m}{X_L} = \frac{5\text{ V}}{75.36\ \Omega} = 66.35\text{ mA}$$

Since v_L leads i_L by 90°, $i_L = I_m\sin(\omega t + 45° - 90°)$ and $i_L = \mathbf{66.35 \times 10^{-3}\sin(\omega t - 45°)}$.

b. See Fig. 13.11.

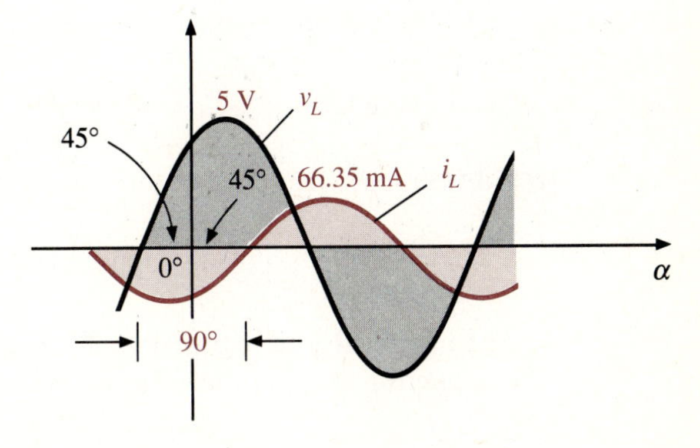

FIG. 13.11
Solution to Example 13.5.

13.4 CAPACITANCE (ac DOMAIN)

Like the inductor, the capacitor's capacitance does not dissipate (ideally) electrical energy but simply stores it temporarily in the form of an electric field. In addition, it is also sensitive to the applied frequency, as we shall see, when a relationship between the voltage across a capacitor v_C and the current i_C is established.

Recall from Chapter 10 that

$$i_C = C\frac{dv_C}{dt} \tag{13.8}$$

as defined by Fig. 13.12. Note the similarities between Eq. 13.8 and Eq. 13.3 for the inductor. In fact, Eq. 13.8 clearly reveals that the magnitude of the current i_C is directly related to the capacitance of the capacitor and the applied frequency. An increase in either one (or both) will increase the magnitude of the resulting current.

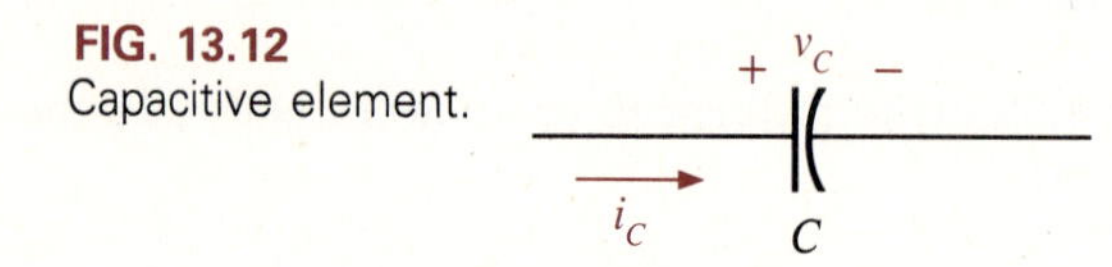

FIG. 13.12
Capacitive element.

As with the inductor, it can be shown experimentally or mathematically that I_m and V_m are related by

$$I_m = (2\pi f C)V_m \tag{13.9}$$

Note the multiplying factor of f and C as described before.
Using the equation

$$\text{Opposition} = \frac{\text{Cause}}{\text{Effect}}$$

we obtain

$$\text{Opposition} = \frac{V_m}{I_m} = \frac{1}{2\pi f C}$$

In general, the opposition of a capacitor is referred to as the *reactance* of the capacitor and is given the notation X_C. In addition, since $\omega = 2\pi f$,

$$X_C = \frac{1}{\omega C} \quad \text{(ohms, } \Omega) \tag{13.10}$$

As noted by Eq. 13.10, X_C, like X_L and R, is measured in ohms, and the peak values of i_C and v_C are related by

$$I_m = \frac{V_m}{X_C} \tag{13.11}$$

with

$$\boxed{V_m = I_m X_C} \tag{13.12}$$

It is important to note that even though Eqs. 13.11 and 13.12 have the Ohm's law format, the reactance X_C is inversely related to the frequency and capacitance. In other words, an increase in capacitance or frequency reduces the reactance, although it results in an increase in I_m for the same V_m.

The inductor created a phase shift, where v_L leads i_L by 90°. For the capacitor:

The current i_C leads the voltage v_C of a capacitor by 90°.

The phase shift is shown in Fig. 13.13.

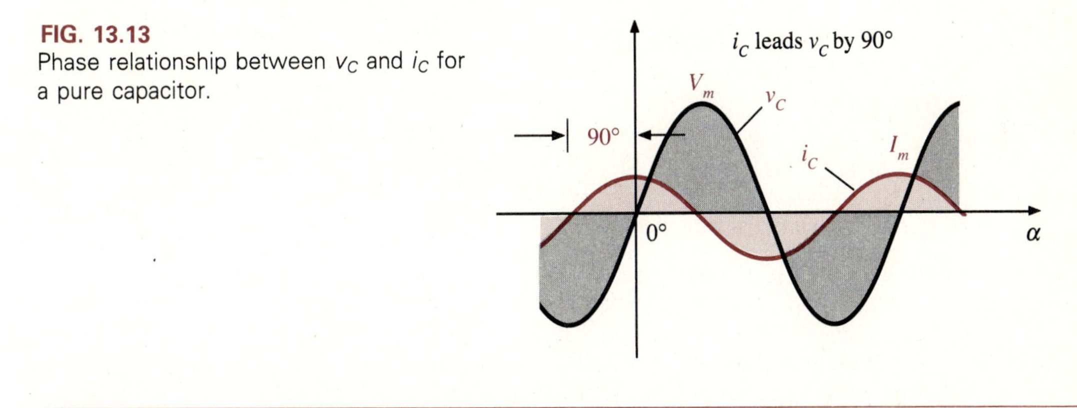

FIG. 13.13
Phase relationship between v_C and i_C for a pure capacitor.

EXAMPLE 13.6 Determine the reactance of a 0.1-μF capacitor at a frequency of

a. 100 Hz
b. 5 kHz
c. 2 MHz

Solutions

a. $$X_C = \frac{1}{2\pi f C} = \frac{1}{2(3.14)(100\text{ Hz})(0.1 \times 10^{-6}\text{ F})}$$

$$= \frac{1}{62.8 \times 10^{-6}} = 0.0159 \times 10^{6}\ \Omega = \mathbf{15.9\ k\Omega}$$

Calculator

(2) (×) (2nd F) (π) (×) (1) (0) (0) (×) (·) (1) (EXP)
(+/−) (6) (=) (1/x) (F <−> E)

with a display of 1.592 04.

b. $X_C = \dfrac{1}{2\pi fC} = \dfrac{1}{2(3.14)(5 \times 10^3\ \text{Hz})(0.1 \times 10^{-6}\ \text{F})}$

$= \dfrac{1}{3.14 \times 10^{-3}} = 0.318 \times 10^3\ \Omega = \mathbf{318\ \Omega}$

Calculator

(2) (×) (2nd F) (π) (×) (5) (EXP) (3) (×) (·) (1) (EXP)
(+/−) (6) (=) (1/x)

with a display of 318.31.

c. $X_C = \dfrac{1}{2\pi fC} = \dfrac{1}{2(3.14)(2 \times 10^6\ \text{Hz})(0.1 \times 10^{-6}\ \text{F})}$

$= \dfrac{1}{1.256} = \mathbf{0.796\ \Omega}$

Calculator

(2) (×) (2nd F) (π) (×) (2) (EXP) (6) (×) (·) (1) (EXP)
(+/−) (6) (=) (1/x)

with a display of 0.796.

Note again how dramatically the change in frequency can affect the reactance of a capacitor, from a level of 15.9 kΩ at 100 Hz to only 0.796 Ω at 2 MHz. Note also, however, that whereas the reactance of a coil increased with frequency, the reactance of a capacitor decreased with increase in frequency. Both X_L and X_C are plotted versus frequency in the next section.

EXAMPLE 13.7

a. Determine the current i_C if the voltage across a 2-μF capacitor is $v_C = 4 \sin \omega t$ at a frequency of 1 kHz.
b. Sketch the waveforms of i_C and v_C.

Solutions

a. $\omega = 2\pi f = 2(3.14)(1 \times 10^3\ \text{Hz}) = 6280\ \text{rad/s}$

$$X_C = \frac{1}{\omega C} = \frac{1}{(6280\ \text{rad/s})(2 \times 10^{-6}\ \text{F})} = 79.618\ \Omega$$

$$I_m = \frac{V_m}{X_C} = \frac{4\ \text{V}}{79.618\ \Omega} = 50.24\ \text{mA}$$

$i_C = I_m \sin(\omega t + 90°)$ (due to i_C leading v_C by 90°)

$= \mathbf{50.24 \times 10^{-3} \sin(6280t + 90°)}$

b. See Fig. 13.14.

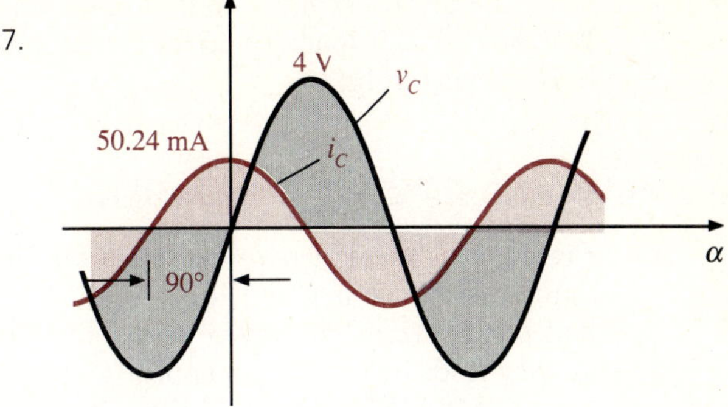

FIG. 13.14
Solution to Example 13.7.

EXAMPLE 13.8

a. Determine the voltage across a 10-pF capacitor if the current i_C is $5 \times 10^{-6}\sin(10^6 t + 30°)$.
b. Sketch v_C and i_C.

Solutions

a. $$X_C = \frac{1}{\omega C} = \frac{1}{(10^6 \text{ rad/s})(10 \times 10^{-12}\text{ F})} = \frac{1}{10} \times 10^{-6} \times 10^{+12}\ \Omega$$

$$= 100\text{ k}\Omega$$

$$V_m = I_m X_C = (5 \times 10^{-6}\text{ A})(100 \times 10^3\ \Omega) = 500 \times 10^{-3}\text{ V} = 0.5\text{ V}$$

$$v_C = V_m\sin(\omega t + \underbrace{30° - 90°)}_{i_C \text{ leads } v_C \text{ by } 90°} = V_m\sin(\omega t - 60°)$$

$$= \mathbf{0.5\sin(10^6 t - 60°)}$$

b. See Fig. 13.15.

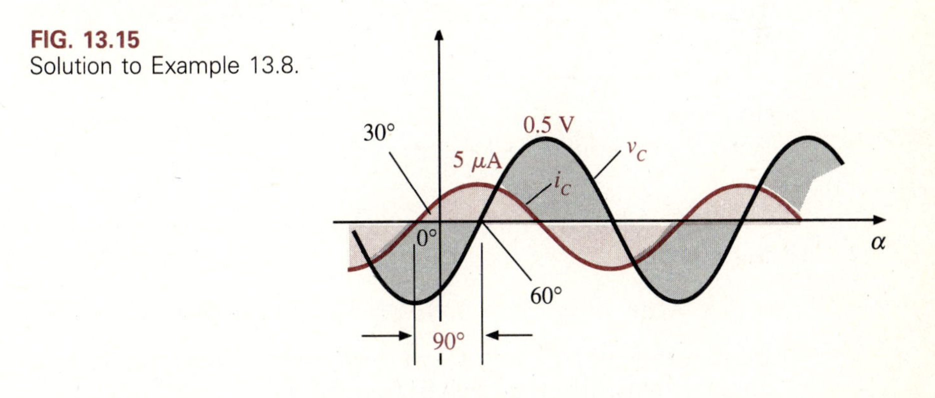

FIG. 13.15
Solution to Example 13.8.

Now that the phase relationships between the voltage and current of the inductor and capacitor have been introduced, a mnemonic phrase that can be used to remember the phase relationships is the following:

ELI the *ICE* man

The phrase seems so out of place that it is usually remembered with ease. For the coil L, E leads I and for the capacitor C, I leads E in the nouns *ELI* and *ICE*, respectively.

13.5 X_L AND X_C VERSUS FREQUENCY

Figure 13.1 clearly revealed that for the low to midrange of the frequency scale, resistance is essentially unaffected by frequency. We have found in the last two sections, however, that both X_L and X_C are quite sensitive to frequency and, in fact, have opposite reactions to changes in frequency.

For the 100-mH coil of Fig. 13.16, the reactance at $f = 0$ Hz (essentially dc conditions) is 0 Ω, and the short-circuit equivalent introduced in Chapter 11 is appropriate, as shown in Fig. 13.16(b). Note also that $v_L = 0$ V due to the short-circuit equivalent, resulting in $V_m = I_m X_C = I_m(0\ \Omega) = 0$ V. At very high frequencies, such as 100 MHz, the reactance becomes quite large: $X_L = 2\pi fL = (6.28)(100 \times 10^6\ \text{Hz})(100 \times 10^{-3}) = 62.8\ \text{M}\Omega$, and an open-circuit equivalent can often be applied, as shown in Fig. 13.16(c). In this case, $i_L = 0$ A and v_L depends on the surrounding network. In general, therefore,

For an inductor, as the frequency approaches the low end of the scale, the reactance will decrease until the short-circuit equivalent can be applied, whereas as the applied frequency increases, the inductive reactance will increase until the open-circuit equivalent is appropriate.

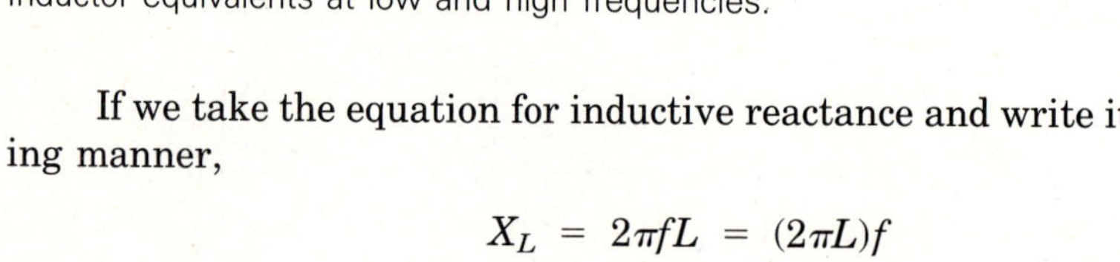

FIG. 13.16
Inductor equivalents at low and high frequencies.

If we take the equation for inductive reactance and write it in the following manner,

$$X_L = 2\pi fL = (2\pi L)f$$

we find that since π and L are constants (ideally), the inductive reactance is directly related to the applied frequency:

$$X_L = kf$$

constant

In fact, if we plot X_L (the y-axis) against f (the x-axis) as shown in Fig. 13.17 for the 100-mH coil, we find that the resulting curve is a *straight* line (implying a linear relationship) that intersects the axis at $f = 0$ since $X_L = 0\ \Omega$ and $X_L = 6.28\ \text{k}\Omega$ at $f = 10$ kHz. The reactance at any other frequency, such as $f = 6$ kHz, can then be found by simply moving vertically from the frequency axis until you hit the curve and then moving horizontally until X_L is determined. The curve clearly reveals that the inductive reactance will continue to rise so long as the frequency increases. There is no apparent limit to its magnitude.

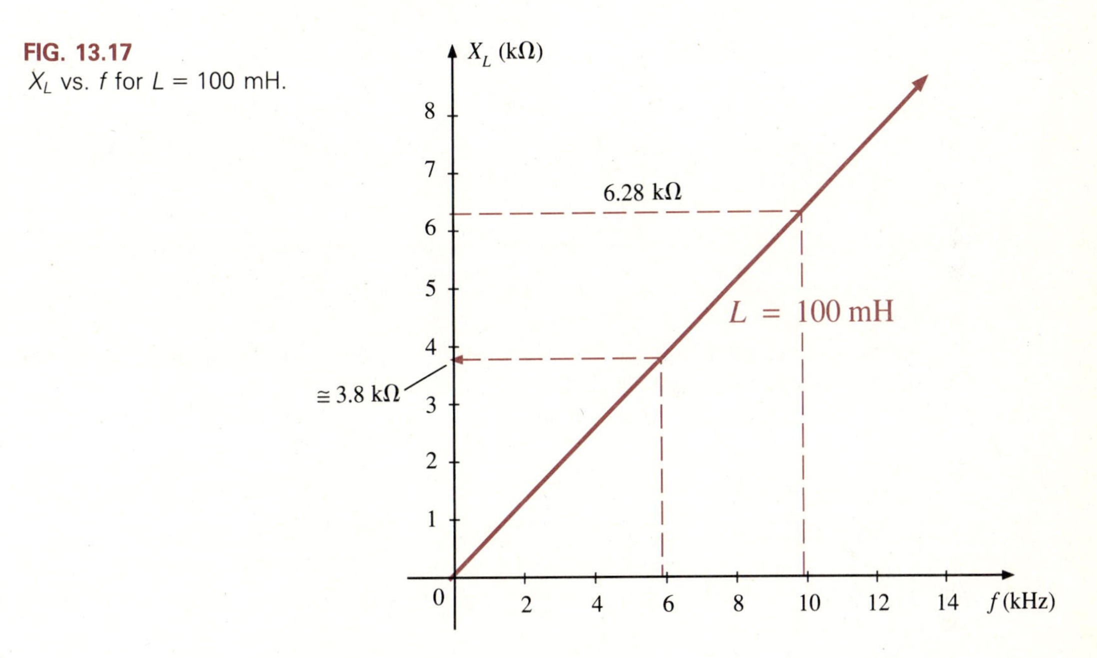

FIG. 13.17
X_L vs. f for $L = 100$ mH.

The slope of the curve defined by $m = \Delta y/\Delta x = \Delta X_L/\Delta f$ is also quite sensitive to the inductance. In fact,

$$\boxed{\text{Slope } m = 2\pi L} \tag{13.13}$$

Any increase in inductance will increase the slope, and the inductive reactance will increase that much quicker, as shown for three inductance levels in Fig. 13.18 with $L_3 > L_2 > L_1$. For example, if the inductance of Fig. 13.18 were increased to 1 H at $f = 10$ kHz, the reactance would be 10 times as great (1 H/0.1 H = 10), or 62.8 kΩ rather than 6.28 kΩ—a measurable difference.

The discussion of X_C versus frequency is somewhat complicated by the fact that X_C and f have a nonlinear relationship. The result is a curve relating X_C and f rather than the straight line of Fig. 13.17. However, let us first

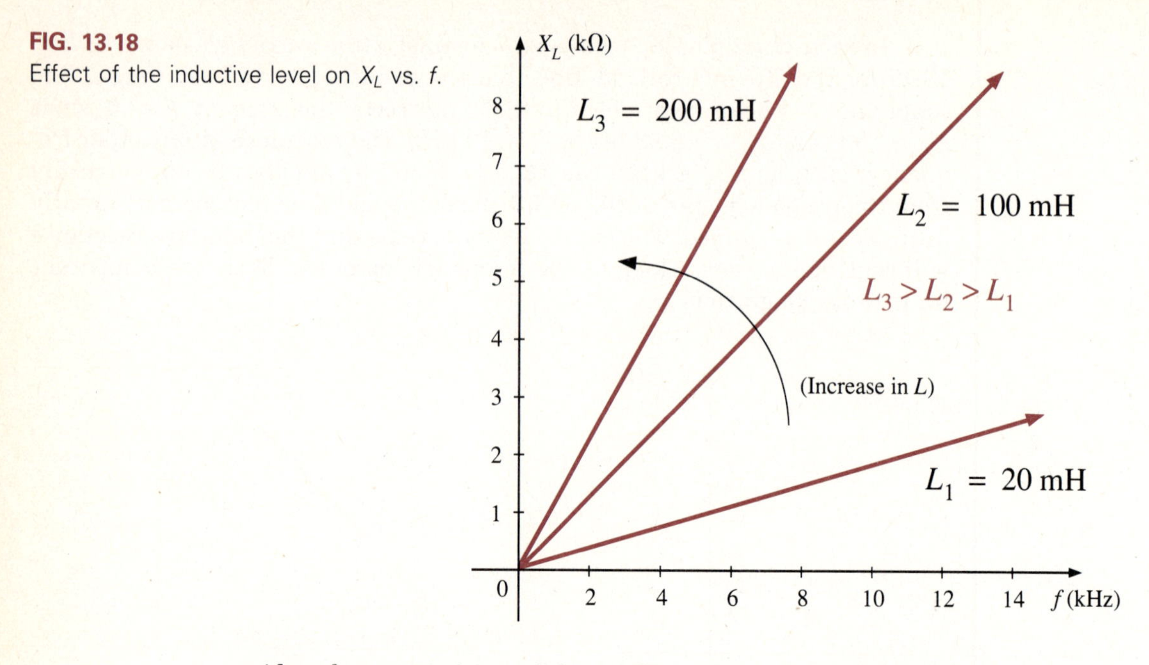

FIG. 13.18
Effect of the inductive level on X_L vs. f.

consider the extremities of $f = 0$ Hz and f = very high frequency on the 0.01-μF capacitor of Fig. 13.19(a).

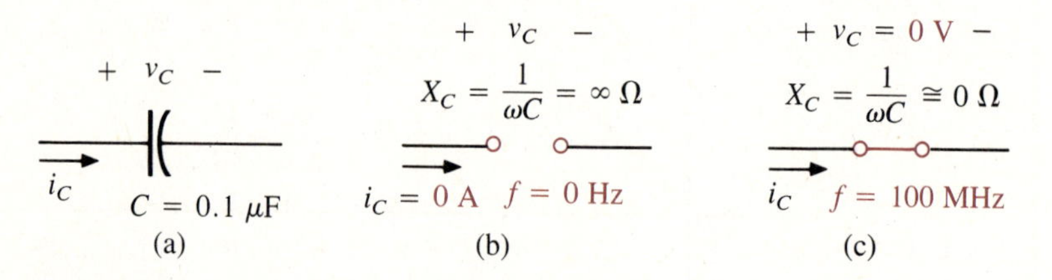

FIG. 13.19
Capacitive equivalents at low and high frequencies.

For $f = 0$ Hz, $X_C = 1/\omega C = 1/(0)C \rightarrow \infty\ \Omega$, and the open-circuit equivalent of Fig. 13.19(b) can be applied as introduced in Chapter 10 for dc conditions. The current i_C is 0 A, but the voltage can be any value, as determined by the surrounding network. At very high frequencies such as 100 MHz, $X_C = 1/\omega C = 0.16\ \Omega$, and the short-circuit equivalent becomes appropriate, as shown in Fig. 13.19(c). The voltage across the capacitor is then 0 V and i_C is determined by the surrounding network.

In general, therefore,

For a capacitor, as the frequency approaches the low end of the scale, the reactance will increase until the open-circuit equivalent can be applied, whereas as the applied frequency increases, the

capacitive reactance will decrease until the short-circuit equivalent is appropriate.

If we take the equation for capacitive reactance and write it in the following manner,

$$X_C = \frac{1}{2\pi fC} = \frac{1}{(2\pi C)f}$$

we find that since π and C are constants, the capacitive reactance is inversely related to the applied frequency as follows:

$$X_C = \frac{1}{kf}$$

The result of such a mathematical format is a parabolic curve between X_C and f, as shown in Fig. 13.20 for a 0.01-μF capacitor. At $f = 1$ kHz, $X_C \cong 15.9$ kΩ, whereas at $f = 100$ kHz, X_C will drop to about 159 Ω. Note how quickly the curve drops with increasing frequency and how it quickly levels off with increasing frequency. From a distance it would appear to change quickly from a high reactance level to a much lower level. For the inductor, the change was more gradual and occurred at a steady rate.

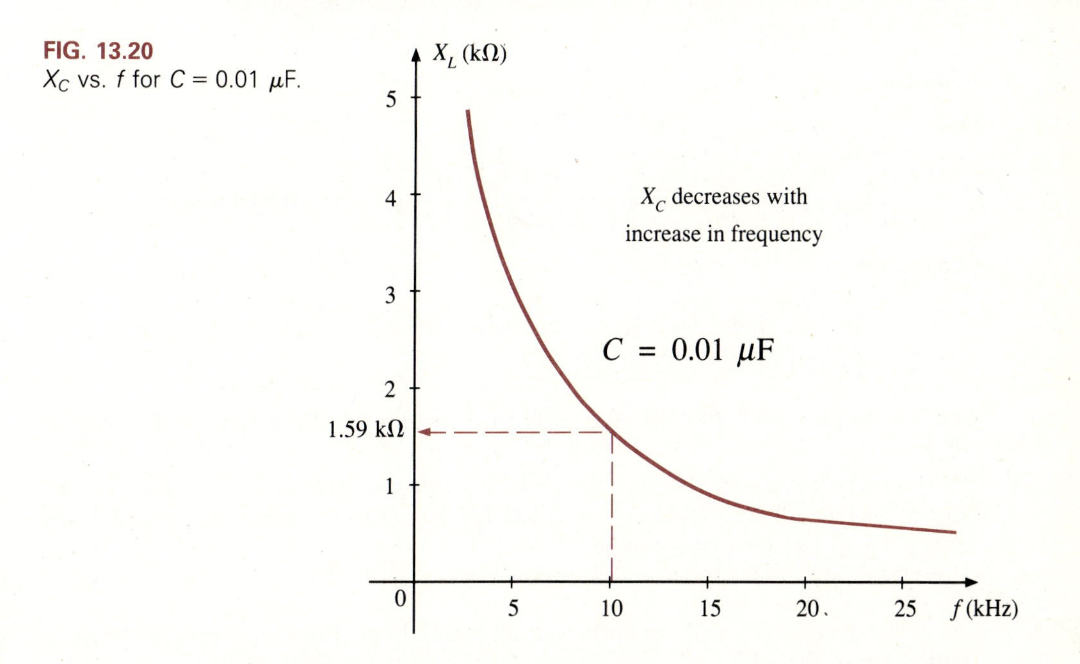

FIG. 13.20
X_C vs. f for $C = 0.01$ μF.

If the capacitance is changed to one of a greater or lesser value, the curves of Fig. 13.21 will result.

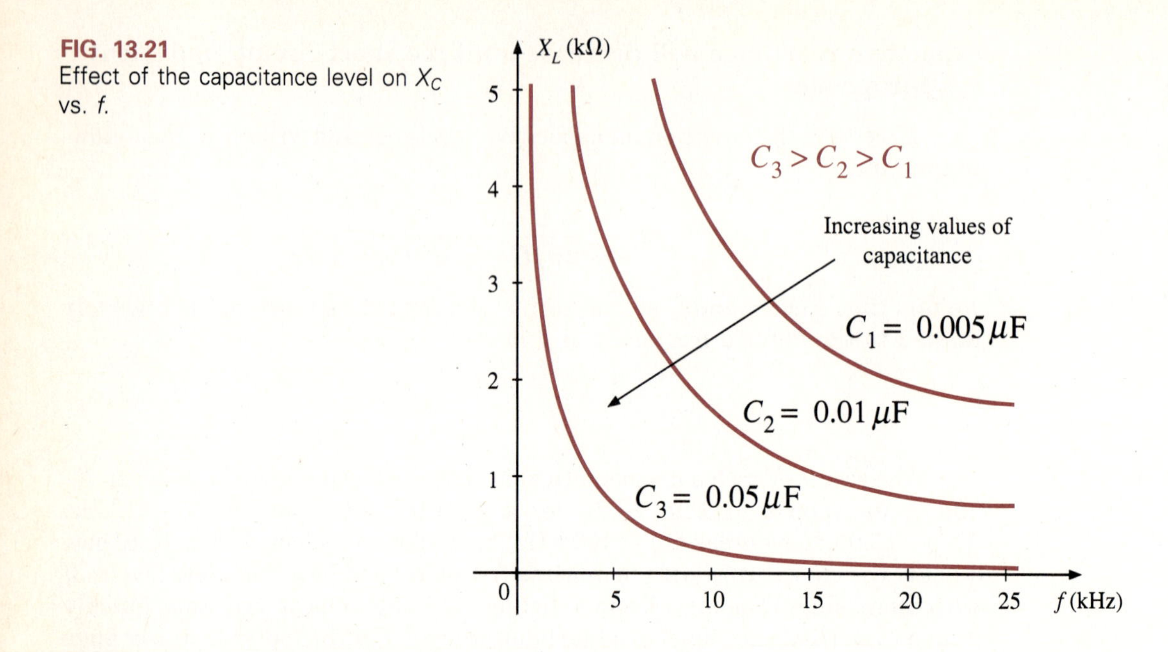

FIG. 13.21
Effect of the capacitance level on X_C vs. f.

EXAMPLE 13.9 Determine the frequency at which the 100-mH coil of Fig. 13.17 will have a reactance of 2 kΩ.

Solution $X_L = 2\pi f L$

or

$$f = \frac{X_L}{2\pi L} = \frac{2 \times 10^3\,\Omega}{2(3.14)(100 \times 10^{-3}\,\text{H})} = \mathbf{3184.713\ Hz}$$

Calculator

(2) (×) (2nd F) (π) (×) (1) (0) (0) (EXP) (+/−) (3) (=)
(1/x) (×) (2) (EXP) (3) (=)

with a display of 3183.099 (the difference is due to the differing accuracies of π).

The result could also be obtained (to a lesser degree of accuracy) by drawing a horizontal line from $X_L = 2$ kΩ (of Fig. 13.17) to the curve and then dropping to the frequency axis.

EXAMPLE 13.10 At what frequency will an inductor of 5 mH have the same reactance as a capacitor of 0.1 μF?

Solution

$$X_L = X_C$$

$$\omega L = \frac{1}{\omega C}$$

or

$$\omega^2 = \frac{1}{LC}$$

and

$$\omega = \frac{1}{\sqrt{LC}}$$

However,

$$\omega = 2\pi f$$

and

$$\boxed{f = \frac{1}{2\pi\sqrt{LC}}} \qquad \textbf{(13.14)}$$

Substituting,

$$\begin{aligned} f &= \frac{1}{2\pi\sqrt{(5 \times 10^{-3}\text{ H})(0.1 \times 10^{-6}\text{ F})}} \\ &= \frac{1}{2\pi\sqrt{(5)(0.1) \times (10^{-3})(10^{-6})}} = \frac{1}{2\pi\sqrt{0.5 \times 10^{-9}}} \\ &= \frac{1}{2\pi\sqrt{5 \times 10^{-10}}} = \frac{1}{2\pi\sqrt{5} \times (10^{-10})^{1/2}} \\ &= \frac{1}{2\pi(2.236) \times 10^{-5}} = \frac{10^{+5}}{14.049} \\ f &\cong \mathbf{7.118\ kHz} \end{aligned}$$

Calculator

(5) (EXP) (+/−) (3) (×) (·) (1) (EXP) (+/−) (6) (=) (√) (×)

(2) (×) (2nd F) (π) (=) (1/x)

with a display of 7.118 03.

Note in this example how the proper application of the calculator can significantly reduce the number of steps required to obtain a solution.

13.6 AVERAGE POWER AND POWER FACTOR

We found in our analysis of dc networks that the power delivered to a load is determined by

$$P_{\text{dc}} = V_L I_L \qquad \text{(watts)}$$

For an ac system the *instantaneous* power (the power delivered at each instant) is determined by

$$\boxed{P = vi} \qquad \text{(watts)} \qquad (13.15)$$

as defined by Fig. 13.22.

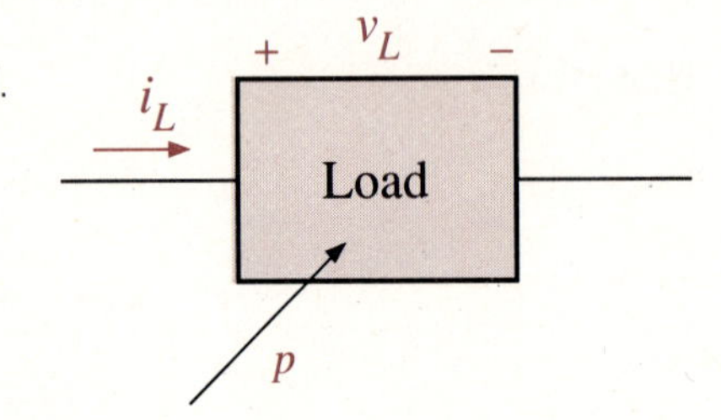

FIG. 13.22
Defining the power to the load.

The lowercase letter was employed for the power in Eq. 13.15 because the power level will continually change with changes in magnitudes of v and i. Over a complete cycle of v or i, however, the product vi will have an *average* power determined by

$$\boxed{P = \frac{V_m I_m}{2} \cos \theta} \qquad \text{(watts)} \qquad (13.16)$$

as shown in Fig. 13.23.

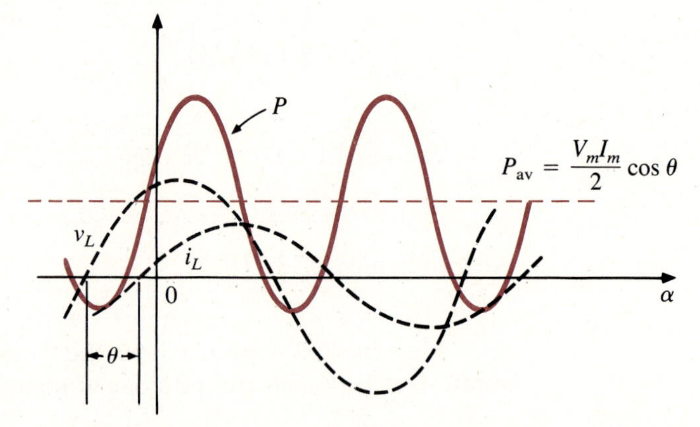

FIG. 13.23
A power curve derived from the product $p = v_L i_L$.

In other words, the average power—or *real* power, as it is sometimes called—is the net power delivered to a load during the sinusoidal variation of v and i. Quite obviously V_m and I_m are the peak values of v and i, respectively. The angle θ is the phase angle between v and i *without* a concern for whether v leads i or i leads v.

Since

$$\frac{V_m I_m}{2} = \left(\frac{V_m}{\sqrt{2}}\right)\left(\frac{I_m}{\sqrt{2}}\right) = V_{\text{eff}} I_{\text{eff}}$$

the equation for the average power can also be written as

$$\boxed{P = V_{\text{eff}} I_{\text{eff}} \cos \theta} \qquad \text{(watts)} \tag{13.17}$$

EXAMPLE 13.11 Determine the average power to the load of Fig. 13.22 if

$$v = 10 \sin \omega t$$

and

$$i = 5 \sin(\omega t + 60°)$$

Solution Using Eq. 13.16:

$$P = \frac{V_m I_m}{2} \cos \theta = \frac{(10 \text{ V})(5 \text{ A})}{2} \cos 60°$$

$$= \frac{50}{2}(0.5) = \mathbf{12.5 \text{ W}}$$

Calculator

1 0 × 5 × 6 0 cos = ÷ 2 =

with a display of 12.5.

EXAMPLE 13.12 **a.** Determine the effective values of the following voltage and current:

$$v = 12 \sin(\omega t + 60°)$$
$$i = 5 \times 10^{-3} \sin(\omega t - 20°)$$

b. Calculate the average power delivered to the load of Fig. 13.22 if the voltage and current are as defined in part a.

Solutions

a. $V_{\text{eff}} = 0.707 V_m = (0.707)(12 \text{ V}) = \mathbf{8.484 \text{ V}}$
$I_{\text{eff}} = 0.707 I_m = (0.707)(5 \times 10^{-3} \text{ A}) = \mathbf{3.535 \text{ mA}}$

b. In Eq. 13.17, $\theta = 60° + 20° = 80°$, since v intersects the horizontal axis 60° before 0° and i intersects the axis 20° after 0°.

$$P = V_{\text{eff}} I_{\text{eff}} \cos \theta$$
$$= (8.484 \text{ V})(3.535 \times 10^{-3} \text{ A}) \cos 80°$$
$$= (30 \times 10^{-3})(0.1736)$$
$$P = 5.208 \times 10^{-3} \text{ W} \cong \mathbf{5.21 \text{ mW}}$$

From Eq. 13.16,

$$P = \frac{V_m I_m}{2} \cos \theta$$

$$= \frac{(12 \text{ V})(5 \times 10^{-3} \text{ A})}{2} \cos 80°$$

$$= (30 \times 10^{-3})(0.1736)$$

$$P = \mathbf{5.21\ mW}$$

Resistor For a purely resistive network, v_R and i_R are in phase and θ of Eqs. 13.16 and 13.17 is 0°. The result is

$$\cos \theta = \cos 0° = 1$$

and

$$P = \frac{V_m I_m}{2} = V_{\text{eff}} I_{\text{eff}} \quad \text{(watts)} \tag{13.18}$$

or, since

$$I_{\text{eff}} = \frac{V_{\text{eff}}}{R}$$

then

$$P = I_{\text{eff}}^2 R = \frac{V_{\text{eff}}^2}{R} \quad \text{(watts)} \tag{13.19}$$

The result is that the equations for the power to a resistor have the same format as the equations applied for dc networks. The only difference is that the current and voltage are the effective values of the sinusoidal functions.

EXAMPLE 13.13 Find the power delivered to a 3.3-kΩ resistor by a current $i = 4 \times 10^{-3} \sin(10^6 t + 30°)$.

Solution There is no need to consider the frequency ($\omega = 2\pi f = 10^6$ rad/s) or phase angle (30°), since they do not appear in the preceding equations for the power to a resistor.

Therefore,

$$I_{\text{eff}} = (0.707)(4 \times 10^{-3} \text{ A}) = 2.828 \text{ mA}$$

and

$$\begin{aligned} P &= I_{\text{eff}}^2 R \\ &= (2.828 \times 10^{-3}\ \text{A})^2\,(3.3 \times 10^3\ \Omega) \\ &= (2.828)^2(10^{-3})^2(3.3 \times 10^3) \\ &= (8 \times 10^{-6})(3.3 \times 10^3) = (8)(3.3) \times (10^{-6})(10^{+3}) \\ P &= 26.4 \times 10^{-3}\ \text{W} = \mathbf{26.4\ mW} \end{aligned}$$

Calculator

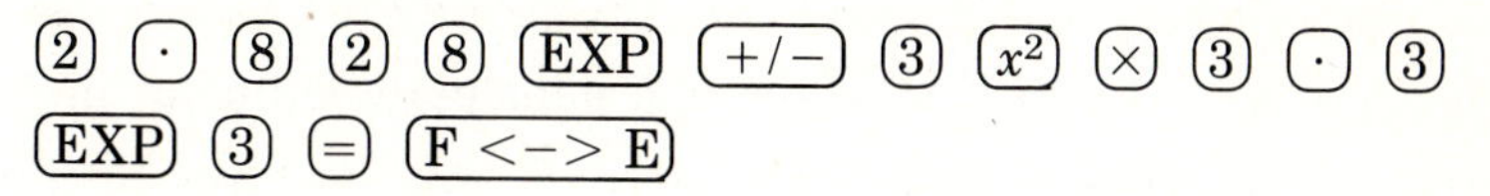

with a display of 2.639–02.

Note again the advantage of becoming astute in the use of the calculator to save time and effort and preserve the highest degree of accuracy.

Inductor and Capacitor For both the inductor and capacitor, the voltage across the element and its current are 90° out of phase, and $\theta = 90°$. The result is $\cos 90° = 0$ and

$$P = V_{\text{eff}} I_{\text{eff}} \cos\theta = V_{\text{eff}} I_{\text{eff}}(0) = \mathbf{0\ W}$$

Therefore, for the ideal inductor and capacitor, the average power to each is 0 W. This supports earlier comments that the ideal inductor and capacitor are nondissipative elements and simply convert energy from one form to another for temporary storage.

EXAMPLE 13.14

Find the average power dissipated in a circuit whose input current and voltage are the following:

$$i = 5 \sin(\omega t + 40°)$$
$$v = 10 \sin(\omega t + 40°)$$

Solution Since v and i are in phase, the circuit appears at the input terminals to be purely resistive. Therefore,

$$P = \frac{V_m I_m}{2} = \frac{(10\ \text{V})(5\ \text{A})}{2} = \mathbf{25\ W}$$

and

$$R = \frac{V_m}{I_m} = \frac{10\ \text{V}}{5\ \text{A}} = 2\ \Omega$$

$$P = \frac{V_{\text{eff}}^2}{R} = \frac{[(0.707)(10\ \text{V})]^2}{2\ \Omega} = \mathbf{25\ W}$$

or

$$P = I_{\text{eff}}^2 R = [(0.707)(5 \text{ A})]^2(2 \ \Omega) = \mathbf{25 \ W}$$

EXAMPLE 13.15

a. Find the average power dissipated by an element whose current and voltage are

$$i = 2 \sin(10^6 t + 20°)$$
$$v = 16 \sin(10^6 t - 70°)$$

b. Determine the type and nameplate value of the element.

Solutions

a. A careful examination of the phase relationship between i and v reveals that i leads v by 90°, resulting in a load with capacitive characteristics. The result is that $P = \mathbf{0 \ W}$.

b. $X_C = \dfrac{V_m}{I_m} = \dfrac{16 \text{ V}}{2 \text{ A}} = 8 \ \Omega$

Since $X_C = 1/\omega C$,

$$C = \frac{1}{\omega X_C} = \frac{1}{(10^6 \text{ rad/s})(8 \ \Omega)}$$
$$= 0.125 \times 10^{-6} \text{ F}$$
$$C = \mathbf{0.125 \ \mu F}$$

Power Factor The factor $\cos \theta$ appearing in Eqs. 13.16 and 13.17 is called the *power factor* of a load and is symbolized by F_p. That is,

$$\boxed{F_p = \cos \theta} \qquad (13.20)$$

Obviously, for a purely resistive load, $\theta = 0°$ and $F_p = \cos 0° = 1$, whereas for a purely reactive load, $\theta = 90°$ and $F_p = \cos 90° = 0$. For loads that are combinations of both reactive and resistive loads, the power factor lies somewhere between the limits of 0 and 1. The more resistive the load, the closer the power factor is to 1; the more reactive the load, the closer the power factor is to 0. In addition, the more resistive the load, the greater the average or dissipated power for the same peak values of voltage and current. For situations where the load voltage leads the load current, the system is said to be inductive and have a *lagging power factor*. For the reverse situation, established by a capacitive load (i leads v), the system is said to have a *leading power factor*. In other words, capacitive networks have *leading power factors, and inductive networks have lagging power factors*. For networks that are highly inductive, the power factor is close to 0 and is lagging in nature. For networks that are highly dissipative, but capacitive, the power factor is close to 1 and is leading in nature.

In terms of the average power and terminal voltage and current,

$$F_p = \cos\theta = \frac{P}{V_{\text{eff}} I_{\text{eff}}} \tag{13.21}$$

EXAMPLE 13.16 Determine the power factors of the following loads, and indicate whether they are leading or lagging:

a. The load of Fig. 13.24.

FIG. 13.24

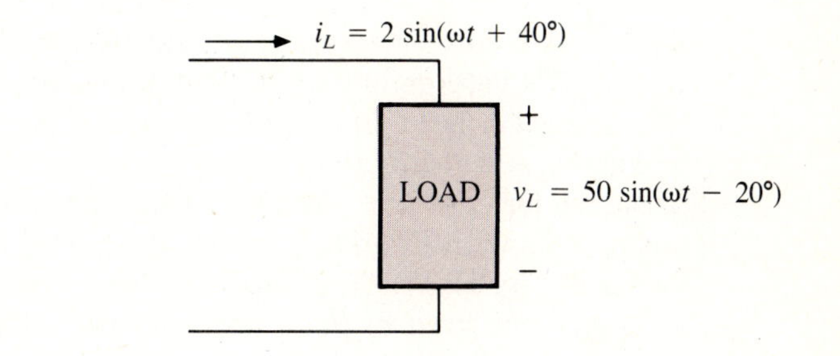

b. The load of Fig. 13.25.

FIG. 13.25

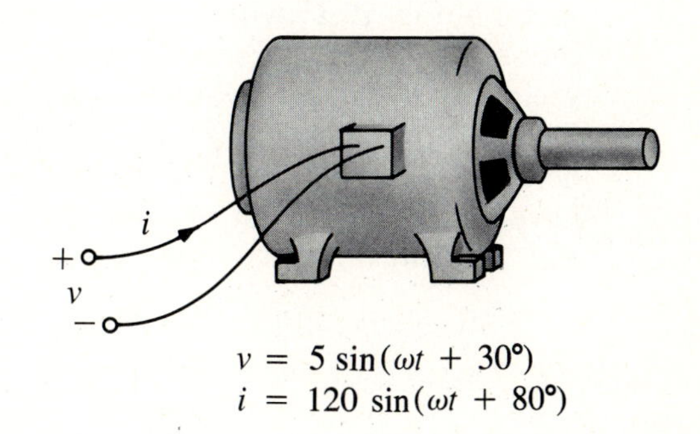

$v = 5 \sin(\omega t + 30°)$
$i = 120 \sin(\omega t + 80°)$

c. The load of Fig. 13.26.

FIG. 13.26

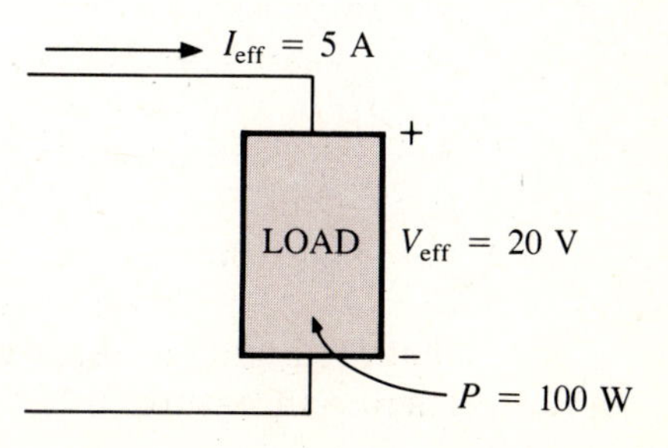

Solutions

a. $F_p = \cos\theta = \cos 60° =$ **0.5 leading** $\quad$ (i leads v)
b. $F_p = \cos\theta = \cos 50° =$ **0.6428 lagging** $\quad$ (v leads i)
c. $F_p = \cos\theta = \dfrac{P}{V_{\text{eff}}I_{\text{eff}}} = \dfrac{100}{(20)(5)} = \dfrac{100}{100} = \mathbf{1}$ $\quad$ (in phase)

The load is resistive, and F_p is neither leading nor lagging.

13.7 PHASORS

Even though the voltages and currents of an ac network are sinusoidal in nature, it is often necessary to add two voltages or currents, such as when Kirchhoff's voltage law or current law is applied. One method of performing this operation is to place both sinusoidal waveforms on the same set of axes and add algebraically the magnitudes of each at every point along the abscissa, as shown for $c = a + b$ in Fig. 13.27. This, however, can be a long and tedious process with limited accuracy. A shorter method uses the rotating radius vector of Fig. 12.17. This *radius vector,* having a *constant magnitude* (length) with *one end fixed at the origin,* is called a *phasor* when applied to electric circuits. During its rotational development of the sine wave, the phasor will, at the instant $t = 0$, have the positions shown in Fig. 13.28(a) for each waveform in Fig. 13.28(b). In other words, for the waveform $v_1 = 1 \sin \omega t$, the rotating vector of Fig. 12.17 will be horizontal at $t = 0$, since the vertical projection is zero at $t = 0$ (the waveform passes through $v = 0$ at $t = 0$.) The length of the vector is still the peak value of the sinusoidal function, as shown in Fig. 12.17. For $v_2 = 2 \sin(\omega t + 90°)$, the phasor is vertical, since the height above the horizontal axis at $t = 0$ is the full peak value of 2 V. In total, therefore, we have superimposed the rotating vectors of both v_1 and v_2 on the same axis of Fig. 13.28(a).

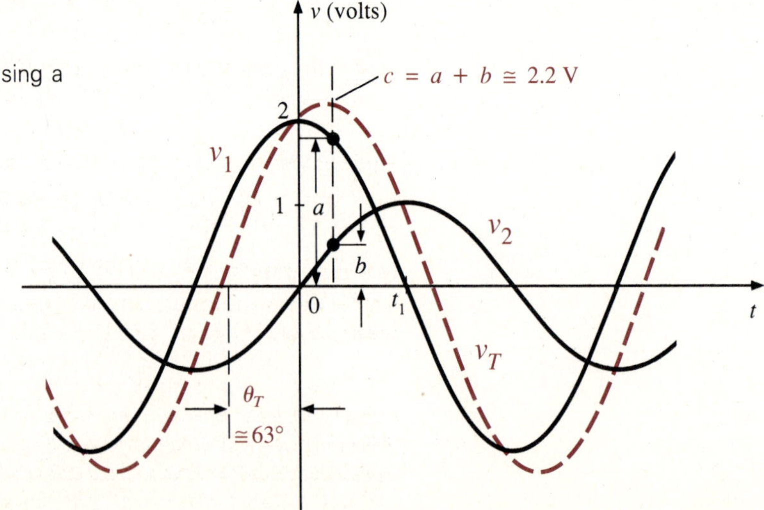

FIG. 13.27
Adding two sinusoidal voltages using a graphical approach.

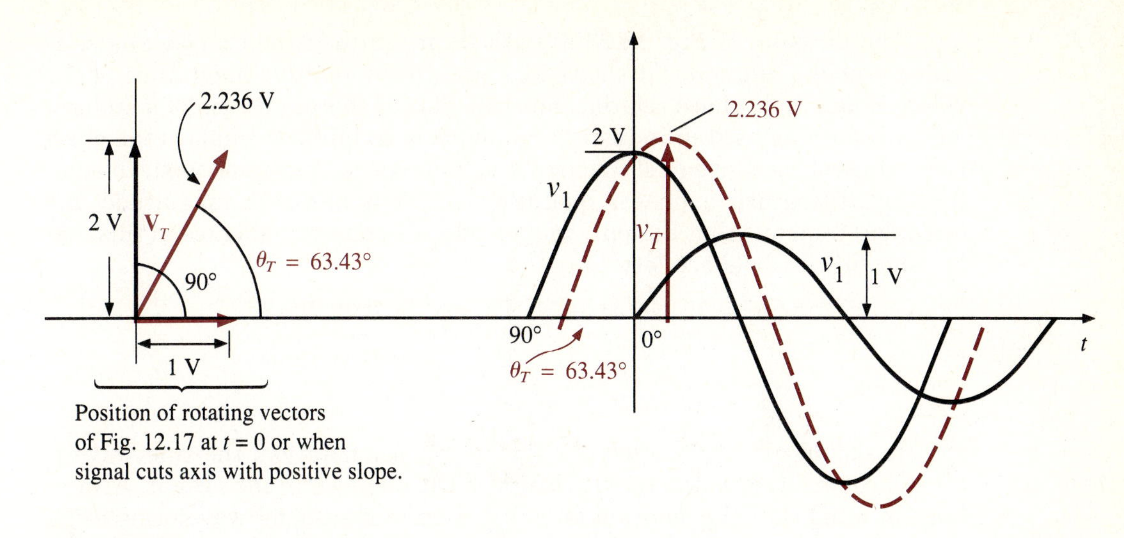

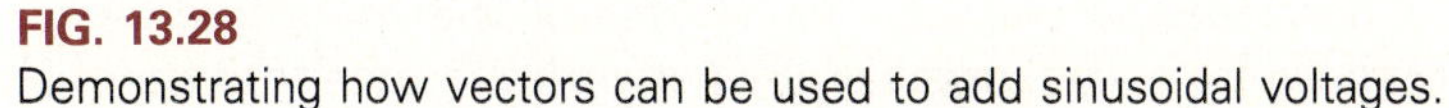

FIG. 13.28
Demonstrating how vectors can be used to add sinusoidal voltages.

The interesting result is that the sum of the two sinusoidal voltages can be determined by finding the resultant vector, as shown in Fig. 13.28(a). That is, the peak value of the sum of v_1 and v_2 is related to the peak values of v_1 and v_2 by the Pythagorean theorem:

$$V_T^2 = V_1^2 + V_2^2 \qquad \text{(all peak values)}$$

or

$$V_T = \sqrt{V_1^2 + V_2^2}$$

Substituting peak values gives

$$V_T = \sqrt{(1\text{ V})^2 + (2\text{ V})^2} = \sqrt{1 + 4}\text{ V} = \sqrt{5}\text{ V} = \mathbf{2.236\text{ V}}$$

The phase angle between the resultant sinusoid is the angle between the horizontal axis (representing $t = 0$ s or $\alpha = 0°$) and the resultant vector, denoted by θ_T in Fig. 13.28(a) and 13.28(b).

From right triangles,

$$\tan\theta_T = \frac{\text{Opposite}}{\text{Adjacent}} = \frac{V_2}{V_1} = \frac{2\text{ V}}{1\text{ V}} = 2$$

and

$$\begin{aligned}\theta_T &= \tan^{-1}2\\ &= 63.43°\end{aligned}$$

The resultant sinusoidal voltage is, therefore,

$$\begin{aligned}V_T &= V_T\sin(\omega t + \theta)\\ &= \mathbf{2.236\sin(\omega t + 63.43°)}\end{aligned}$$

with an accuracy level that would be hard to accomplish graphically.

The diagram of Fig. 13.28(a) with its magnitudes and phase angles is called a *phasor diagram.* It shows at a glance the relative magnitude of the voltages and their phase relationship. The phasor representation of a sinusoidal waveform as used in industry and analysis techniques employs the effective values of the sinusoidal function and its angle, as indicated next. In addition, to distinguish a phasor quantity (one that has both magnitude and direction) from one that has only magnitude, all phasor quantities in this text are shown with boldface lettering:

$$v = V_m \sin(\omega t + \theta) \rightarrow \mathbf{V} = \underbrace{0.707V_m \angle \theta}_{\text{effective value}}$$

$$i = I_m \sin(\omega t + \theta) \rightarrow \mathbf{I} = \overbrace{0.707I_m \angle \theta}$$

It should also be pointed out that in phasor notation, the sine wave is always the reference, and the frequency is not included in the format. Always keep in mind that the addition or subtraction of sinusoidal waveforms using the techniques just described is *only* for waveforms of the *same* frequency.

An analysis of the voltages or currents of a network requires that they all be in the phasor *or* sinusoidal form—a mix limits the extent of the investigation. All quantities in the phasor form are said to be in the *phasor domain,* whereas all quantities in the sinusoidal form are in the *time domain.*

EXAMPLE 13.17

Convert the time domains in Table 13.1 to the phasor domain.

TABLE 13.1

Time Domain	Phasor Domain
a. $\sqrt{2}(50)\sin \omega t$	$\mathbf{50 \angle 0°}$
b. $69.6 \sin(\omega t + 72°)$	$(0.707)(69.6) \angle 72° = \mathbf{49.21 \angle 72°}$
c. $45 \cos \omega t$	$(0.707)(45) \angle 90° = \mathbf{31.82 \angle 90°}$

EXAMPLE 13.18

Write the sinusoidal expression for each phasor in Table 13.2 if the frequency is 60 Hz.

TABLE 13.2

Phasor Domain	Time Domain
a. $\mathbf{I} = 10 \angle 30°$	$i = \sqrt{2}(10)\sin(2\pi 60t + 30°)$
	and $i = \mathbf{14.14 \sin(377t + 30°)}$
b. $\mathbf{V} = 115 \angle -70°$	$v = \sqrt{2}(115)\sin(377t - 70°)$
	and $v = \mathbf{162.6 \sin(377t - 70°)}$

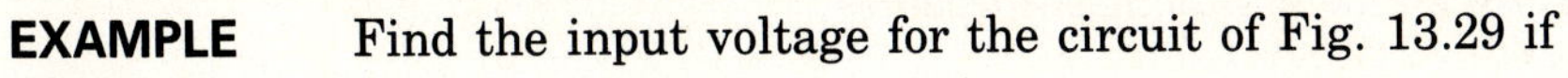

EXAMPLE 13.19

Find the input voltage for the circuit of Fig. 13.29 if

$$v_1 = 8 \sin \omega t$$

and

$$v_2 = 2 \sin \omega t$$

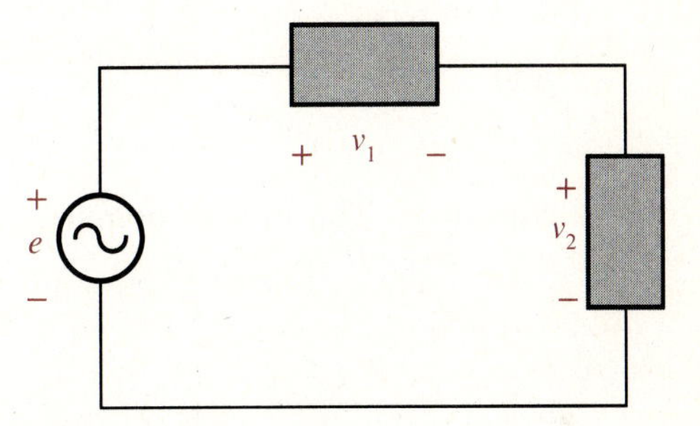

FIG. 13.29
Series circuit for Example 13.19.

Solution The phasor domain is

$$\mathbf{V}_1 = (0.707)8 \text{ V} \angle 0^\circ = 5.656 \text{ V} \angle 0^\circ$$
$$\mathbf{V}_2 = (0.707)2 \text{ V} \angle 0^\circ = 1.414 \text{ V} \angle 0^\circ$$

The phasor diagram (Fig. 13.30) clearly reveals that the two phasors have the same direction and, therefore, the total input voltage as determined by vector addition is

$$\mathbf{E} = \mathbf{V}_1 + \mathbf{V}_2 \qquad \text{(Kirchhoff's voltage law)}$$

resulting in

$$\mathbf{E} = 5.656 \text{ V} \angle 0^\circ + 1.414 \text{ V} \angle 0^\circ$$
$$= 7.07 \text{ V} \angle 0^\circ$$

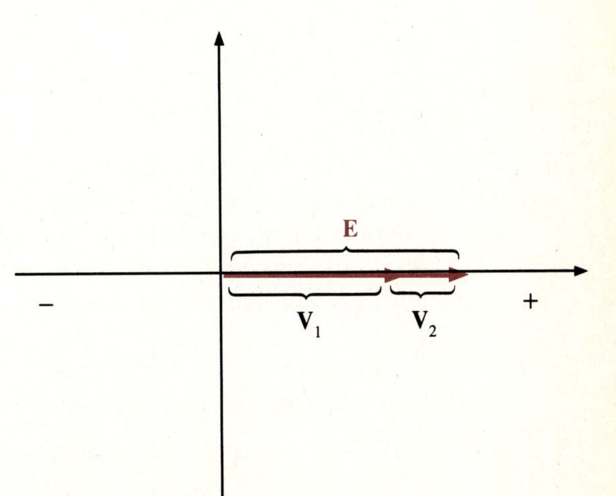

FIG. 13.30
Determining **E** for the circuit of Fig. 13.29.

which in the sinusoidal domain becomes

$$\begin{aligned} e &= (1.414)(7.07)\sin \omega t \\ &= \mathbf{10 \sin \omega t} \end{aligned}$$

EXAMPLE 13.20 Repeat Example 13.19 if

$$v_1 = (\sqrt{2})6 \sin(\omega t + 60°)$$

and

$$v_2 = (\sqrt{2})4 \sin \omega t$$

Solution The phasor domain is

$$\mathbf{V}_1 = \frac{1}{\sqrt{2}}(\sqrt{2})6\text{ V}\,\underline{/60°} = 6\text{ V}\,\underline{/60°}$$

$$\mathbf{V}_2 = \frac{1}{\sqrt{2}}(\sqrt{2})4\text{ V}\,\underline{/0°} = 4\text{ V}\,\underline{/0°}$$

The phasor diagram of Fig. 13.31 clearly reveals that $\mathbf{V}_1$ and $\mathbf{V}_2$ do not form the legs of a right triangle, thus not permitting the use of the Pythagorean theorem to find the sum of the two voltages.

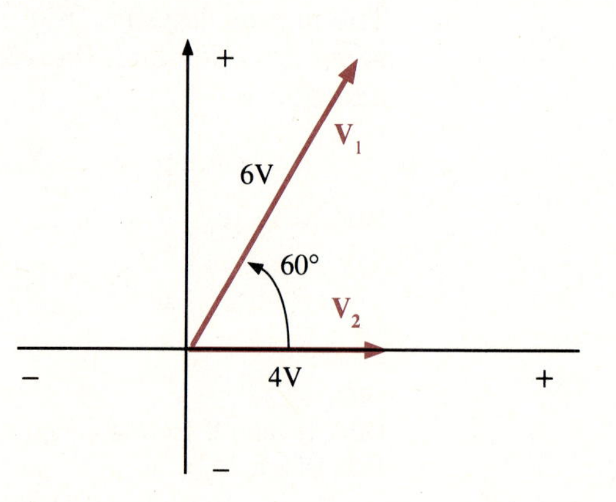

FIG. 13.31
The phasor diagram for the voltages of Example 13.20.

The solution is obtained by breaking down $\mathbf{V}_1$ into horizontal and vertical components and then adding the horizontal component to $\mathbf{V}_2$. The Pythagorean theorem can then be applied to determine **E**.

In Fig. 13.32, the horizontal and vertical projections of $\mathbf{V}_1$ are determined. The horizontal component is

$$V_1\cos\theta = 6\text{ V}\cos 60° = (6\text{ V})(0.5) = 3\text{ V}$$

The vertical component is

$$V_1\sin\theta = 6\text{ V}\sin 60° = (6\text{ V})(0.866) = 5.196\text{ V}$$

FIG. 13.32
Breaking down $\mathbf{V}_1$ into its components.

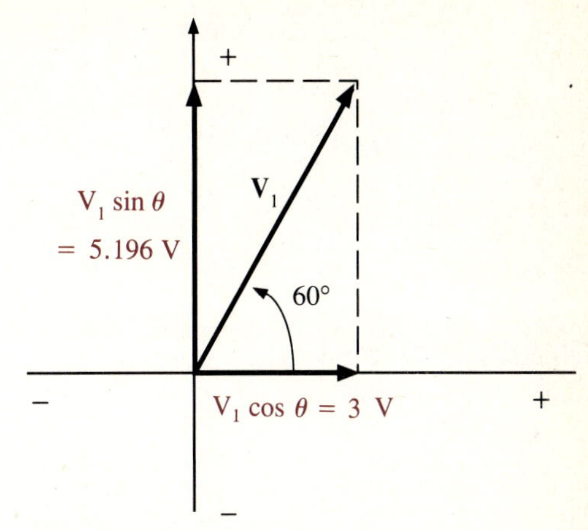

The result is the vector diagram of Fig. 13.33.

FIG. 13.33
Determining **E.**

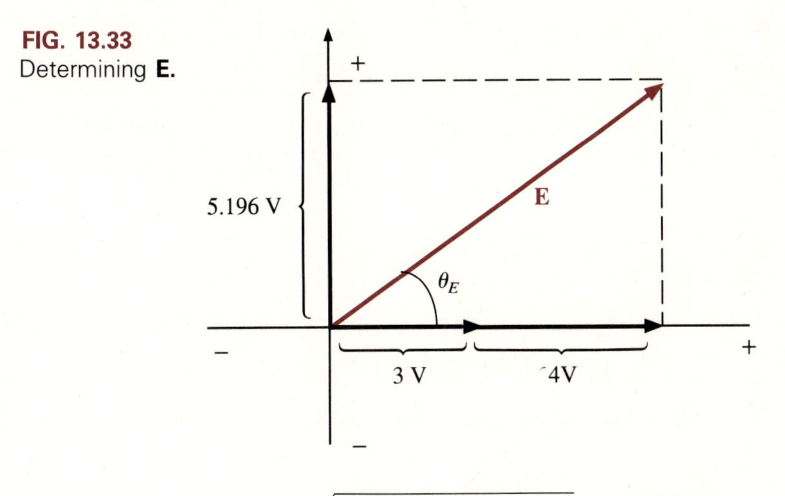

$$E = \sqrt{(7\text{ V})^2 + (5.196\text{ V})^2} = 8.718\text{ V}$$

Calculator

7 x^2 + 5 . 1 9 6 x^2 = √

with a display of 8.718.

$$\theta_E = \tan^{-1}\frac{5.196\text{ V}}{7\text{ V}} = \tan^{-1}0.742 = 36.59°$$

Calculator

with a display of 36.586.

Therefore,

$$\mathbf{E} = E \angle \theta_E = 8.718 \text{ V} \angle 36.59^\circ$$

and

$$e = (1.414)(8.718)\sin(\omega t + 36.59^\circ)$$
$$= \mathbf{12.33 \sin(\omega t + 36.59^\circ)}$$

EXAMPLE 13.21 Find the current i_1 for the network of Fig. 13.34 if $i_T = 5 \times 10^{-3}\sin(\omega t + 30^\circ)$ and $i_2 = 2 \times 10^{-3}\sin(\omega t + 210^\circ)$.

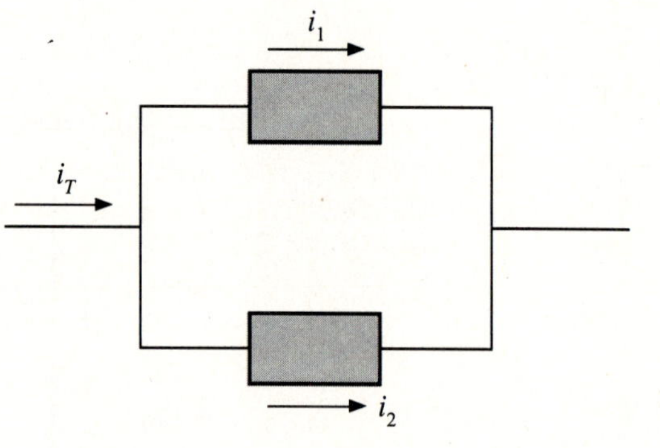

FIG. 13.34
Parallel network for Example 13.21.

Solution The phasor domain is

$$\mathbf{I}_T = (0.707)(5 \times 10^{-3} \text{ A}) \angle 30^\circ = 3.535 \text{ mA} \angle 30^\circ$$
$$\mathbf{I}_2 = (0.707)(2 \times 10^{-3} \text{ A}) \angle 210^\circ = 1.414 \text{ mA} \angle 210^\circ$$

The phasor diagram of Fig. 13.35 clearly reveals that the vectors are 180° apart.

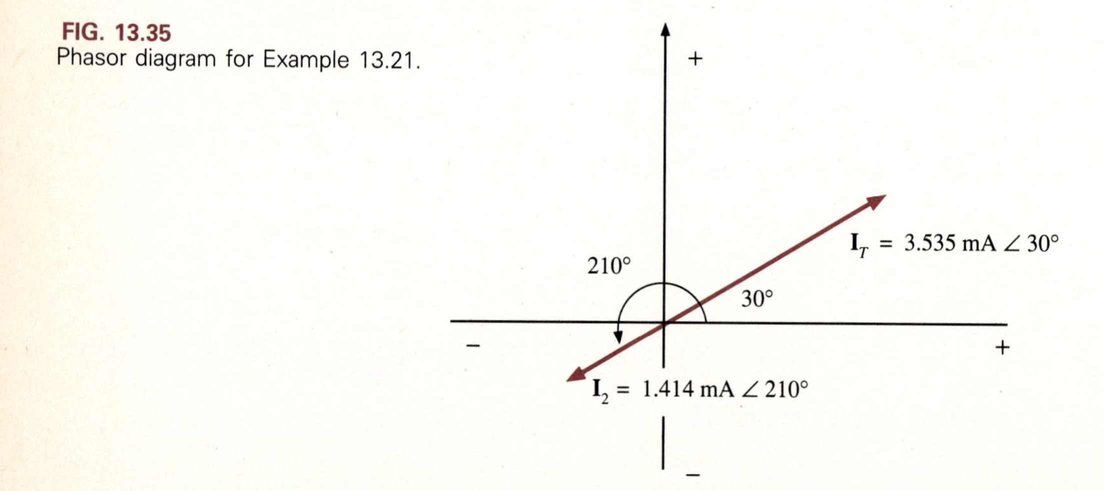

FIG. 13.35
Phasor diagram for Example 13.21.

Applying Kirchhoff's current law to the network of Fig. 13.34 gives

$$\mathbf{I}_T = \mathbf{I}_1 + \mathbf{I}_2$$
$$\mathbf{I}_1 = \mathbf{I}_T - \mathbf{I}_2$$

The minus sign in front of $\mathbf{I}_2$ reveals that the vector for $\mathbf{I}_2$ in Fig. 13.35 must be reversed, which will give it the same direction as $\mathbf{I}_T$, and

$$\mathbf{I}_1 = 3.535 \times 10^{-3}\text{ A }\underline{/30^\circ} + 1.414 \times 10^{-3}\text{ A }\underline{/30^\circ}$$
$$= 4.949 \times 10^{-3}\text{ A }\underline{/30^\circ}$$

or

$$i_1 = (1.414)(4.949) \times 10^{-3}\sin(\omega t + 30^\circ)$$
$$= \mathbf{7 \times 10^{-3}\sin(\boldsymbol{\omega} t + 30^\circ)}$$

The preceding examples reveal that sinusoidal voltages and currents can be added and subtracted in their phasor (vector) form by simply using the Pythagorean theorem once the vector components can be added or subtracted. Once the sinusoidal expression for the desired quantity has been determined, the function can easily be plotted on the same axis as the given sinusoidal functions.

13.8 THE *j*-OPERATOR

It should be clear from Example 13.20 that occasions will arise in ac systems where phasor addition or subtraction will not be accomplished with a simple application of the Pythagorean theorem. A method employing a *j*-operator, to be introduced in this section, reduces the complexity of such mathematical maneuvers. It is a method employed throughout industry and education that should be learned if time permits. It is a direct procedure that saves both time and energy.

To distinguish between the horizontal and vertical components of a vector, the horizontal axis is referred to as the *real axis* and the vertical axis is the *imaginary axis,* as shown in Fig. 13.36. The grid structure is referred to as a *complex plane.* Prior to the development of this mathematical representation of vectors, it was believed that any number not on the real axis could not exist—hence the term *imaginary* for the vertical axis. The symbol *j* is used to denote the vertical component of the vector.

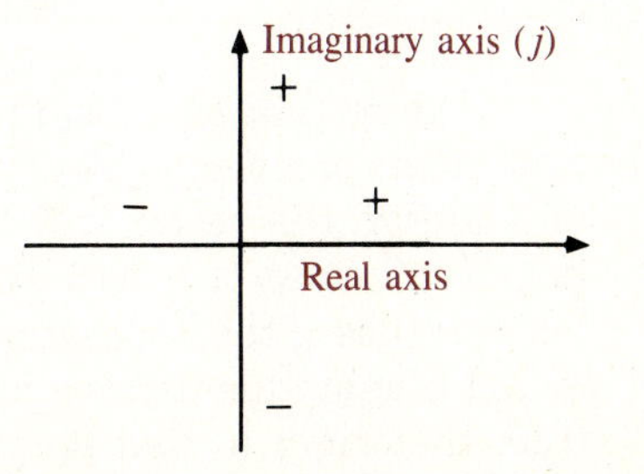

FIG. 13.36
Identifying the horizontal and vertical axes.

A vector on the grid of Fig. 13.36 can be represented by a *polar* or a *rectangular form*. The polar form is defined by the length of the vector and the angle measured from the horizontal axis, as shown in Fig. 13.37. Angles measured in the counterclockwise direction are given a positive sign, and angles measured in the clockwise direction have a minus sign. Note the similarity between the notation $\mathbf{C} = C \angle \theta$ and the phasor notation $\mathbf{E} = E \angle \theta$. In fact, all phasor quantities are in polar form. By knowing the length C and the angle θ, a point (at the head of the vector) is defined, which permits drawing a vector from the origin to that point.

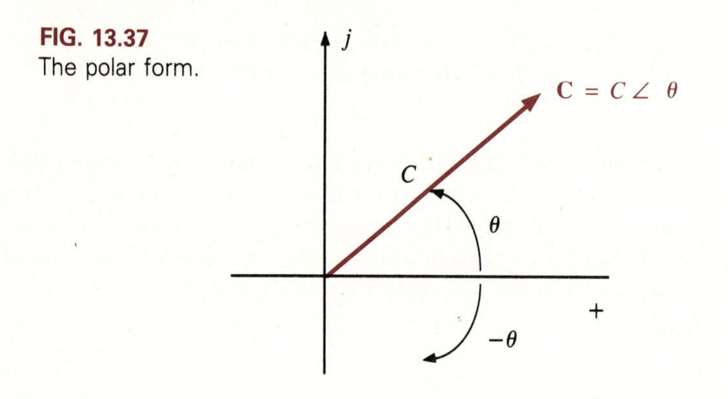

FIG. 13.37
The polar form.

EXAMPLE 13.22 Draw the vector $5 \angle 60°$ and $2 \angle -20°$ on the grid structure of Fig. 13.36.

Solution Both vectors appear in Fig. 13.38.

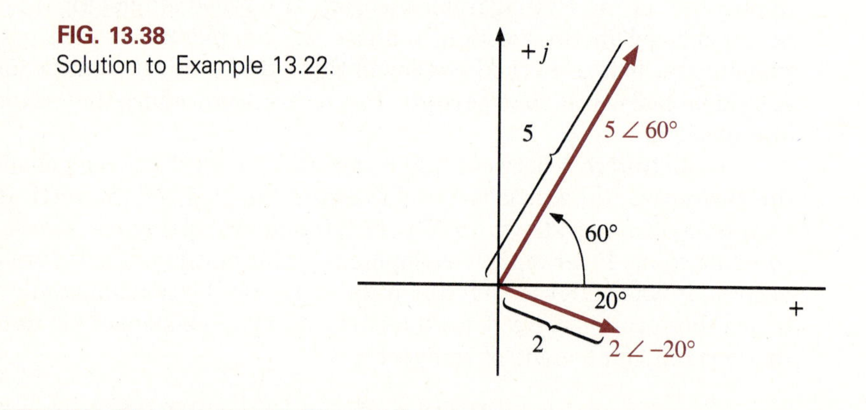

FIG. 13.38
Solution to Example 13.22.

The rectangular form requires that we know the horizontal and vertical projections of a vector, as denoted by A and B in Fig. 13.39. For positive values of A and B, the vector will be in the first quadrant. Positive values of A and negative values of B will place the vector in the fourth quadrant, and so on. Note the use of the j-operator to identify the vertical component. Knowing both A and B again identifies a point in the plane, which permits drawing a vector from the origin to that point.

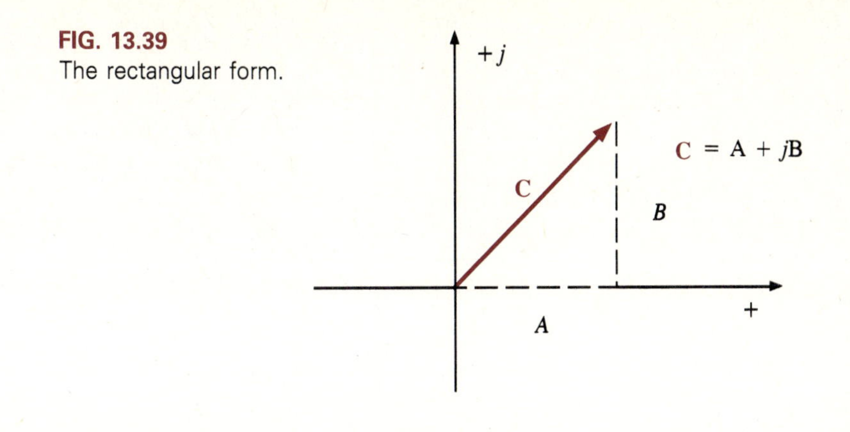

FIG. 13.39
The rectangular form.

EXAMPLE 13.23 Draw the vectors $2 + j6$ and $5 - j5$ on the same grid structure, as in Example 13.22.

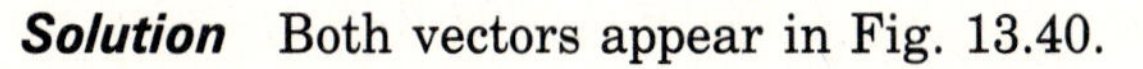

Solution Both vectors appear in Fig. 13.40.

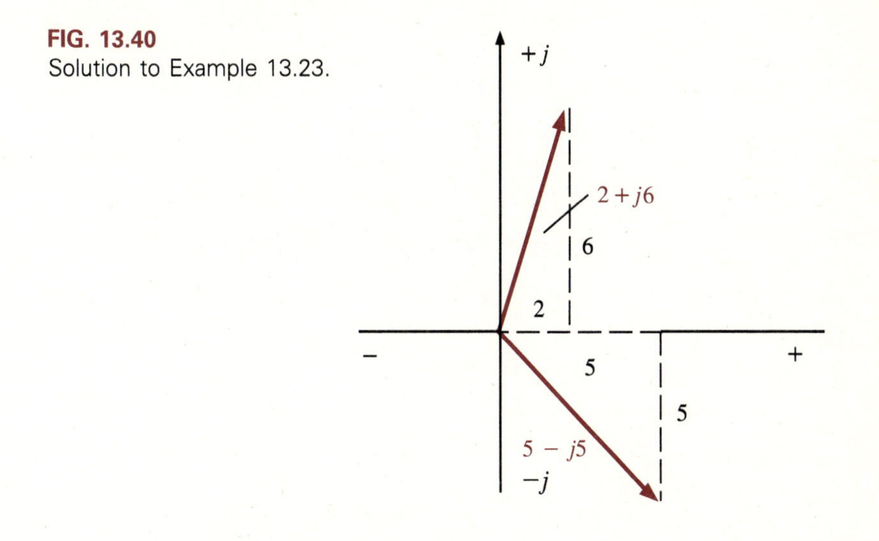

FIG. 13.40
Solution to Example 13.23.

Fortunately, one form can be derived from the other using Fig. 13.41 and the following equations employing basic geometric relationships.

Rectangular to polar:

$$C = \sqrt{A^2 + B^2}$$
$$\theta = \tan^{-1}\frac{B}{A} \quad (13.22)$$

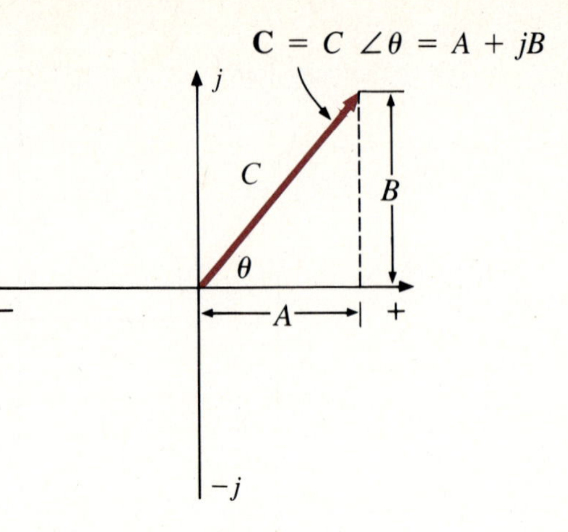

FIG. 13.41
Converting between forms.

Polar to rectangular:

$$\boxed{\begin{aligned} A &= C\cos\theta \\ B &= C\sin\theta \end{aligned}} \tag{13.23}$$

EXAMPLE 13.24 Convert the vector $\mathbf{C} = 3 + j4$ to polar form.

Solution The vector $\mathbf{C} = 3 + j4$ is shown in Fig. 13.42.

$$C = \sqrt{3^2 + 4^2} = \sqrt{25} = 5$$

$$\theta = \tan^{-1}\frac{3}{4} = \tan^{-1}0.75 = 36.87°$$

and

$$\mathbf{C} = \underline{\mathbf{5\ /36.87°}}$$

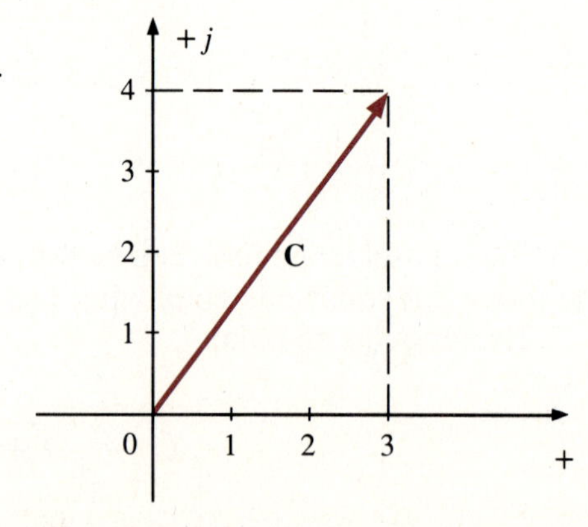

FIG. 13.42
Vector for Example 13.24.

EXAMPLE 13.25 Convert the vector **C** $= 8 \angle -30^\circ$ of Fig. 13.43 to rectangular form.

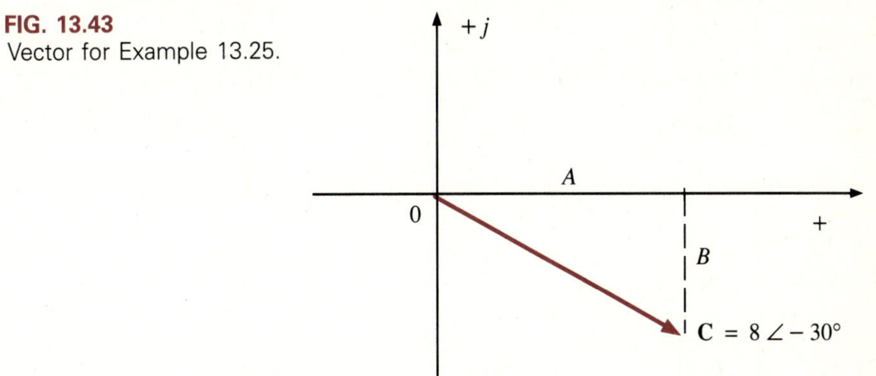

FIG. 13.43
Vector for Example 13.25.

Solution To find only the magnitude:

$$|A| = C\cos\theta = 8\cos 30^\circ = 8(0.866) = 6.93$$
$$|B| = C\sin\theta = 8\sin 30^\circ = 8(0.5) = 4$$

with

$$\mathbf{C} = \mathbf{6.93 - j4}$$

Unless the angles of the polar form are the same, *vector addition and subtraction will be performed in the rectangular form.* In the rectangular form the sum or difference is found by finding the algebraic (noting the sign) sum of the real and imaginary parts.

EXAMPLE 13.26 Find the sum of $(5 + j1) + (2 + j8)$.

Solution

$$\begin{array}{r} (5 + j1) \\ +\ (2 + j8) \\ \hline \mathbf{7 + j9} \end{array}$$

Note the use of the j-operator to identify the vertical component.

EXAMPLE 13.27 Find the difference of $10 \angle 40^\circ - (6 + j2)$.

Solution Converting the polar form to rectangular form gives

$$10 \angle 40^\circ = 7.66 + j6.43$$

Then

$$\begin{array}{r} (7.66 + j6.43) \\ -\ (6 \quad + j2) \\ \hline \mathbf{1.66 + j4.43} \end{array}$$

Since the polar form relates directly to the phasor notation, the methods just described can be applied to the addition and subtraction of sinusoidal voltages and currents.

EXAMPLE 13.28

Repeat Example 13.20 using the j-operator.

Solution

$$\mathbf{V}_1 = 6\text{ V}\,\underline{/60^\circ} = 3 + j5.196$$

$$\mathbf{V}_2 = 4\text{ V}\,\underline{/0^\circ} = 4 + j0$$

$$\mathbf{V}_1 + \mathbf{V}_2 = (3 + j5.196) + (4 + j0) = 7 + j5.196$$

The polar form is $8.718\,\underline{/36.59^\circ}$, which relates directly to $\mathbf{E} =$ **$8.718\text{ V}\underline{/36.59^\circ}$.**

EXAMPLE 13.29

Find the difference $e = v_1 - v_2$ if $v_1 = \sqrt{2}(20)\sin(\omega t + 45^\circ)$ and $v_2 = \sqrt{2}(6)\sin(\omega t + 90^\circ)$.

Solution

$$\mathbf{V}_1 = 20\text{ V}\,\underline{/45^\circ} = 14.142 + j14.142$$

$$\mathbf{V}_2 = 6\text{ V}\,\underline{/90^\circ} = +j6$$

Hence, $\mathbf{V}_1 - \mathbf{V}_2$ is

$$\begin{array}{r} (14.142 + j14.142) \\ -\quad (0 + j6) \\ \hline 14.142 + j8.142 \end{array} = 16.318\,\underline{/29.93^\circ}$$

with $\mathbf{E} = 16.318\text{ V}\,\underline{/29.93^\circ}$.

Thus

$$e = \sqrt{2}(16.138)\sin(\omega t + 29.93^\circ)$$

$$= \mathbf{23.077\,\sin(\omega t + 29.93^\circ)}$$

Since addition and subtraction of sinusoidal quantities are normally performed in rectangular form, it is necessary to convert the phasor representation of a sinusoidal function to rectangular form before the required operation can be performed. Fortunately, scientific calculators like the one illustrated in Fig. 13.44 are now available that can perform a polar-to-rectangular or rectangular-to-polar conversion with a high degree of accuracy in a matter of seconds.

A computer program designed to perform the conversion in either direction is provided in Fig. 13.45. Line 220 accepts the rectangular form and jumps to the subroutine on line 2000 to perform the conversion. Line 240 then prints out the solution. Note the use of the Pythagorean theorem on line 2020 to determine the magnitude of the polar form; lines 2030 through 2060 determine the angle theta (TH). The factor (180/3.14159) converts the radian results obtained from $\theta = \tan^{-1}y/x$ to degree measure.

FIG. 13.44
Scientific calculator. (Courtesy of Hewlett Packard Co.)

```
  10 REM *****  PROGRAM 13-1 *****
  20 REM ****************************************
  30 REM Program to perform selected conversions
  40 REM ****************************************
  50 REM
 100 PRINT
 110 PRINT "Enter  (1) for rectangular to polar conversion"
 120 PRINT "       (2) for polar to rectangular conversion"
 130 PRINT TAB(20);
 140 INPUT "Choice=";C :REM C is choice 1 or 2
 150 IF C<0 OR C>2 THEN GOTO 110
 160 ON C GOSUB 200,300
 170 PRINT:INPUT "More(YES or NO)";A$
 180 IF A$="YES" THEN GOTO 100
 190 END
 200 REM Use rectangular to polar conversion module
 210 PRINT:PRINT:PRINT "Enter rectangular data:"
```

Body of Program:

Input-(Rect.)

```
 220 INPUT "X=";X :INPUT "Y=";Y
 230 GOSUB 2000
```

Output-(Polar)

```
 240 PRINT:PRINT "Polar form is";Z;"at an angle of";TH;"degrees"
 250 RETURN
 300 REM Use polar to rectangular conversion
```

Input (Polar)

```
 310 PRINT:PRINT "Enter polar data:":PRINT:INPUT "Z=";Z
 320 INPUT "Angle(degrees), TH=";TH
 330 GOSUB 2100
```

Output (Rect.)

```
 340 PRINT:PRINT "Rectangular form is";X;
 350 IF Y>=0 THEN PRINT "+j";Y
 360 IF Y<0 THEN PRINT "-j";ABS(Y)
 370 RETURN
```

Rect. → Polar

```
2000 REM Module to convert from rectangular to polar form.
2010 REM Enter with X, Y - Return with Z, TH(eta)
2020 Z=SQR(X^2+Y^2)
2030 IF X<0 THEN TH=(180/3.14159)*ATN(Y/X)
2040 IF X<0 THEN TH=180*SGN(Y)+(180/3.14159)*ATN(Y/X)
2050 IF X=0 THEN TH=90*SGN(Y)
2060 IF Y=0 THEN IF X<0 THEN TH=180
2070 RETURN
```

Polar → Rect.

```
2100 REM Module to convert from polar to rectangular form.
2110 REM Enter with Z, TH(eta) - return with X, Y
2120 X=Z*COS(TH*3.14159/180)
2130 Y=Z*SIN(TH*3.14159/180)
2140 RETURN
```

Program Runs:

```
READY

Enter (1) for rectangular to polar conversion
      (2) for polar to rectangular conversion
                    Choice=? 2


Enter polar data:

Z=? 5

Angle(degrees), TH=? -53.13


Rectangular form is 3 -j 4

More(YES or NO)? YES


Enter (1) for rectangular to polar conversion
      (2) for polar to rectangular conversion
                    Choice=? 1


Enter rectangular data:
X=? -10

Y=? 20


Polar form is 22.3607 at an angle of 116.565 degrees

More(YES or NO)? YES


Enter (1) for rectangular to polar conversion
      (2) for polar to rectangular conversion
                    Choice=? 2


Enter polar data:

Z=? 12

Angle(degrees), TH=? 35


Rectangular form is 9.8298 +j 6.8829

More(YES or NO)? NO


READY
```

FIG. 13.45
Program 13.1.

Lines 310 and 320 accept the polar form and output the rectangular form on lines 340 through 360. Line 2120 provides the real component and line 2130 shows the imaginary, or vertical, component. Review the provided examples with an eye toward the format of the results as dictated by the program.

FORMULA SUMMARY

$$R:\quad I_m = \frac{V_m}{R}, \qquad v_R \text{ and } i_R \text{ in phase}$$

$$L:\quad X_L = \omega L, \qquad I_m = \frac{V_m}{X_L}, \qquad v_L \text{ leads } i_L \text{ by } 90^\circ$$

$$C:\quad X_C = \frac{1}{\omega C}, \qquad I_m = \frac{V_m}{X_C}, \qquad i_C \text{ leads } v_C \text{ by } 90^\circ$$

$$P = \frac{V_m I_m}{2}\cos\theta = V_{\text{eff}} I_{\text{eff}} \cos\theta$$

$$P_R = V_{\text{eff}} I_{\text{eff}} = I_{\text{eff}}^2 R = \frac{V_{\text{eff}}^2}{R}$$

$$F_p = \cos\theta = \frac{P}{V_{\text{eff}} I_{\text{eff}}}$$

$$\text{Rectangular} \rightarrow \text{polar form:}\quad C = \sqrt{A^2 + B^2}, \qquad \theta = \tan^{-1}\frac{B}{A}$$

$$\text{Polar} \rightarrow \text{rectangular form:}\quad A = C\cos\theta, \qquad B = C\sin\theta$$

CHAPTER SUMMARY

For the individual R, L, and C elements, Ohm's law can be applied in the same manner as for dc networks to determine the peak or effective values of the sinusoidal functions. The format is exactly the same for each element once R, X_L, and X_C are determined. Keep in mind that for the frequency range of interest, R is not a function of frequency, but X_L and X_C are directly and inversely related to frequency, respectively. It will be necessary to memorize the phase relationship for each element, but remembering that v_R and i_R are in phase is not particularly difficult, and once the fact that v_L leads i_L by 90° is learned, it is necessary only to reverse v and i in the preceding statement for the purely capacitive element. (Or, simply remember "*ELI* the *ICE* man.")

From a practical standpoint, it is useful to remember that the opposition of inductive elements *increases linearly* with frequency, extending from a short- to open-circuit equivalent. For the capacitor, the reactance decreases exponentially (nonlinearly) with frequency from a very high level at low frequencies to low levels at higher frequencies (progressing from an open- to short-circuit equivalent).

For resistive elements (the only dissipating elements), the power equations have the *same* format as for dc networks with the current and voltage levels replaced by the effective values of the sinusoidal functions. The equation $P = VI \cos \theta$ is usually applied to networks that are not purely resistive, inductive, or capacitive but a mixture of resistive and reactive components. For each power equation, the voltage V is the voltage across the load and the current I is the current through the load. Do not make the mistake of simply substituting the source voltage into the equation unless it happens to be directly across the load. The power factor $\cos \theta$ is very close to 1 for all networks that are primarily resistive and close to 0 for primarily reactive networks.

The j-operator approach is an option that establishes a close relationship between the analysis of dc and ac systems. It is not particularly difficult to understand and apply but does require a measure of practice and experience to be totally confident in the results obtained. The remainder of the text can be covered without becoming familiar with the j-operator approach, but it is a method familiar to all practicing electrical engineers and utilized in a number of computer software packages, which suggests the importance of at least reading the material when time permits.

GLOSSARY

Average or real power The power delivered to and dissipated by the load over a full cycle.

Derivative The instantaneous rate of change of a function with respect to time or another variable.

Leading and lagging power factors An indication of whether a network is primarily capacitive or inductive in nature. Leading power factors are associated with capacitive networks, and lagging power factors, with inductive networks.

Phasor A radius vector that has a constant magnitude at a fixed angle from the positive real axis and that represents a sinusoidal voltage or current in the vector domain.

Phasor diagram A "snapshot" of the phasors that represent a number of sinusoidal waveforms at $t = 0$.

Polar form A method of defining a point in a complex plane that includes a single magnitude to represent the distance from the origin and an angle to reflect the counterclockwise distance from the positive real axis.

Power factor (F_p) An indication of how reactive or resistive an electrical system is. The higher the power factor, the greater is the resistive component.

Reactance The opposition of an inductor or capacitor to the flow of charge that results in the continual exchange of energy between the circuit and magnetic field of an inductor or the electric field of a capacitor.

Rectangular form A method of defining a point in a complex plane that includes the magnitude of the real component and the magnitude of the imaginary component, the latter component being defined by an associated letter j.

PROBLEMS

Section 13.2

1. The voltage across a 5-Ω resistor is as indicated. Find the sinusoidal expression for the current. In addition, sketch the v and i curves with the abscissa in radians.
   a. $150 \sin 377t$ **b.** $30 \sin(377t + 60°)$
2. The current through a 7-kΩ resistor is as indicated. Find the sinusoidal expression for the voltage. In addition, sketch the v and i curves with the abscissa in radians.
   a. $3 \times 10^{-6} \sin 754t$
   b. $2 \times 10^{-3} \sin(400t + 120°)$
3. For the following pairs of voltage and current, determine the resistance level.
   a. $v = 20 \sin(\omega t + 70°)$
   $i = 4 \sin(\omega t + 70°)$
   b. $v = 6 \times 10^{-6} \sin(10^6 t - 20°)$
   $i = 4 \times 10^{-8} \sin(10^6 t - 20°)$

Section 13.3

4. Determine the inductive reactance (in ohms) of a 2-H coil for
   a. dc
   and for the following frequencies:
   b. 25 Hz **c.** 60 Hz
   d. 2000 Hz **e.** 100,000 Hz
5. Determine the inductance of a coil that has a reactance of
   a. 20 Ω at $f = 2$ Hz
   b. 1000 Ω at $f = 60$ Hz
   c. 5280 Ω at $f = 1000$ Hz
6. Determine the frequency at which a 10-H inductance has the following inductive reactances:
   a. 50 Ω **b.** 3770 Ω
   c. 15.7 kΩ **d.** 243 Ω
7. The current through a 20-Ω inductive reactance is given. What is the sinusoidal expression for the voltage? Sketch the v and i curves with the abscissa in radians.
   a. $i = 5 \sin \omega t$ **b.** $i = 0.4 \sin(\omega t + 60°)$
8. The current through a 0.1-H coil is given. What is the sinusoidal expression for the voltage?
   a. $30 \sin 30t$
   b. $0.006 \sin 377t$
   c. $5 \times 10^{-6} \sin(400t + 80°)$
9. The voltage across a 50-Ω inductive reactance is given. What is the sinusoidal expression for the current? Sketch the v and i curves with the abscissa in radians.
   a. $50 \sin \omega t$
   b. $30 \sin(\omega t + 20°)$
   c. $40 \cos(\omega t + 10°)$

10. The voltage across a 0.2-H coil is given. What is the sinusoidal expression for the current?
a. $1.5 \sin 60t$
b. $0.016 \sin(\omega t + 4°)$

Section 13.4

11. Determine the capacitive reactance (in ohms) of a 5-μF capacitor for
a. dc
and for the following frequencies:
b. 60 Hz **c.** 120 Hz
d. 1800 Hz **e.** 24,000 Hz

12. Determine the capacitance in microfarads if a capacitor has a reactance of
a. 250 Ω at $f = 60$ Hz
b. 55 Ω at $f = 312$ Hz
c. 10 Ω at $f = 25$ Hz

13. Determine the frequency at which a 50-μF capacitor has the following capacitive reactances:
a. 342 Ω **b.** 0.684 Ω
c. 171 Ω **d.** 2000 Ω

14. The voltage across a 2.5-Ω capacitive reactance is given. What is the sinusoidal expression for the current? Sketch the v and i curves with the abscissa in radians.
a. $100 \sin \omega t$ **b.** $0.4 \sin(\omega t + 20°)$

15. The voltage across a 1-μF capacitor is given. What is the sinusoidal expression for the current?
a. $30 \sin 200t$ **b.** $120 \sin(374t + 30°)$

16. The current through a 10-Ω capacitive reactance is given. Write the sinusoidal expression for the voltage. Sketch the v and i curves with the abscissa in radians.
a. $i = 50 \sin \omega t$ **b.** $i = 40 \sin(\omega t + 60°)$

17. The current through a 0.5-μF capacitor is given. What is the sinusoidal expression for the voltage?
a. $0.007 \sin 377t$ **b.** $0.08 \sin(1600t - 80°)$

18. For the following pairs of voltages and currents, indicate whether the element involved is a capacitor, inductor, or resistor, and the value of C, L, or R if sufficient data are given:
a. $v = 550 \sin(377t + 40°)$
$i = 11 \sin(377t - 50°)$
b. $v = 36 \sin(754t + 80°)$
$i = 4 \sin(754t + 170°)$
c. $v = 10.5 \sin(\omega t + 13°)$
$i = 1.5 \sin(\omega t + 13°)$

***19.** Repeat Problem 18 for the following pairs of voltages and currents:
a. $v = 2000 \sin 200t$
$i = 5 \cos 200t$
b. $v = 80 \sin(157t + 150°)$
$i = 2 \sin(157t + 60°)$
c. $v = 35 \sin(\omega t - 20°)$
$i = 7 \cos(\omega t - 110°)$

Section 13.5

20. Plot X_L versus frequency for a 5-mH coil using a frequency range of zero to 100 kHz.

21. Plot X_C versus frequency for a 1-μF capacitor using a frequency range of 0 to 10 kHz on a linear scale.

22. At what frequency will the reactance of a 1-μF capacitor equal the resistance of a 2-kΩ resistor?

23. The reactance of a coil equals the resistance of a 10-kΩ resistor at a frequency of 5 kHz. Determine the inductance of the coil.

24. Determine the frequency at which a 1-μF capacitor and a 10-mH inductor will have the same reactance.

***25.** Determine the capacitance required to establish a capacitive reactance that will match that of a 2-mH coil at a frequency of 50 kHz.

Section 13.6

26. Find the average power loss in watts for each set of Problem 18.

27. Find the average power loss in watts for each set of Problem 19.

28. Find the average power loss and power factor for each of the circuits whose input current and voltage are as follows:

a. $v = 60 \sin(\omega t + 30°)$
$i = 15 \sin(\omega t + 60°)$

b. $v = -50 \sin(\omega t - 20°)$
$i = -2 \sin(\omega t + 40°)$

c. $v = 50 \sin(\omega t + 80°)$
$i = 3 \cos(\omega t + 20°)$

d. $v = 75 \sin(\omega t - 5°)$
$i = 0.08 \sin(\omega t - 35°)$

29. If the current through and voltage across an element are $i = 8 \sin(\omega t + 40°)$ and $v = 48 \sin(\omega t + 40°)$, respectively, compute the power by I^2R, $(V_mI_m/2)\cos\theta$, and $VI \cos \Omega$, and compare answers.

30. A circuit dissipates 100 W (average power) at 150 V (effective input voltage) and 2 A (effective input current). What is the power factor? Repeat if the power is 0 W; 300 W.

***31.** The power factor of a circuit is 0.5 lagging. The power delivered in watts is 500. If the input voltage is $50 \sin(\omega t + 10°)$, find the sinusoidal expression for the input current.

32. In Fig. 13.46, $e = 30 \sin(377t + 20°)$.

a. What is the sinusoidal expression for the current?
b. Find the power loss in the circuit.
c. How long (in seconds) does it take the current to complete 6 cycles?

33. In Fig. 13.47, $e = 100 \sin(157t + 30°)$.

a. Find the sinusoidal expression for i.
b. Find the value of the inductance L.
c. Find the average power loss by the inductor.

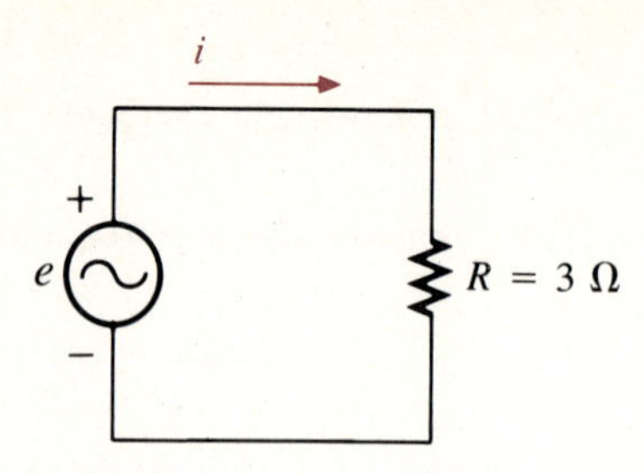

FIG. 13.46

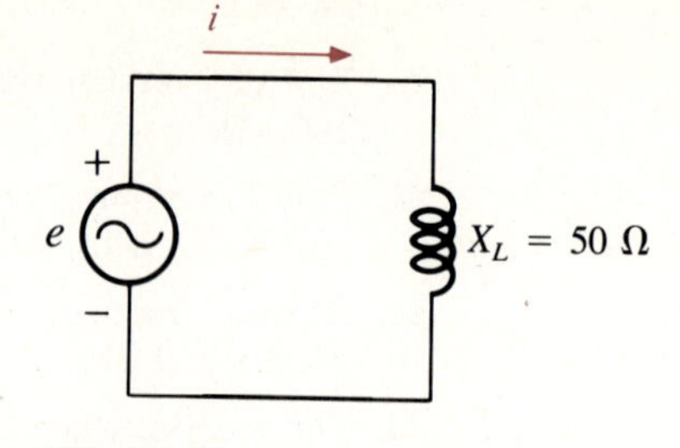

FIG. 13.47

34. In Fig. 13.48, $i = 3 \sin(377t - 20°)$.
a. Find the sinusoidal expression for e.
b. Find the value of the capacitance C in microfarads.
c. Find the average power loss in the capacitor.

FIG. 13.48

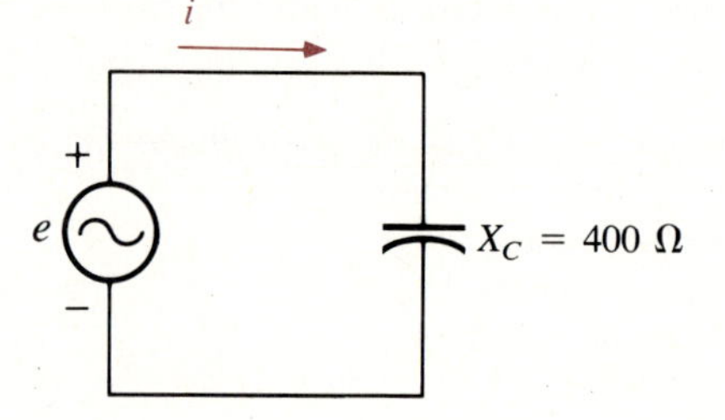

Section 13.7

35. Express the following in phasor form:
a. $\sqrt{2}(100)\sin(\omega t + 30°)$
b. $\sqrt{2}(0.25)\sin(157t - 40°)$
c. $100 \sin(\omega t - 90°)$
d. $42 \sin(377t + 0°)$
e. $6 \times 10^{-6} \cos \omega t$

36. Express the following phasor currents and voltages as sine waves if the frequency is 60 Hz:
a. $\mathbf{I} = 40 \angle 20°$ **b.** $\mathbf{V} = 120 \angle 0°$
c. $\mathbf{I} = 8 \times 10^{-3} \angle 120°$ **d.** $\mathbf{V} = 5 \angle 90°$
e. $\mathbf{I} = 1200 \angle -120°$ **f.** $\mathbf{V} = \dfrac{6000}{\sqrt{2}} \angle -180°$

37. For the system of Fig. 13.49, find the sinusoidal expression for the unknown voltage v_a if

$$e_{in} = 60 \sin(377t + 90°)$$
$$v_b = 20 \sin 377t$$
$$v_c = 0 \text{ V}$$

FIG. 13.49

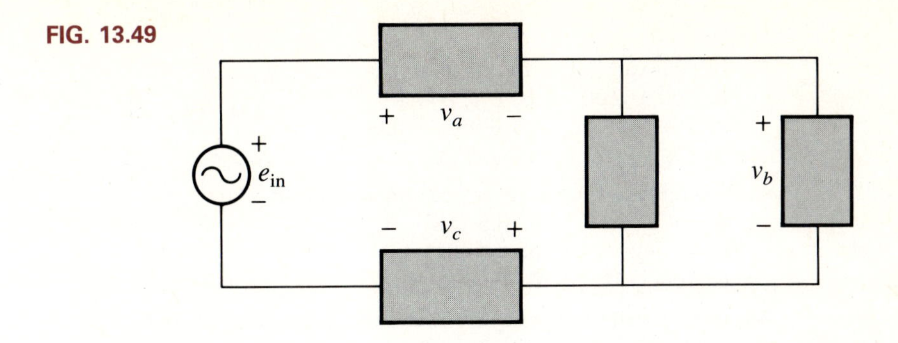

38. For the system of Fig. 13.50, find the sinusoidal expression for the unknown current i_1 if

$$i_T = 20 \times 10^{-6}\sin \omega t$$
$$i_2 = 6 \times 10^{-6}\sin(\omega t + 60°)$$

FIG. 13.50

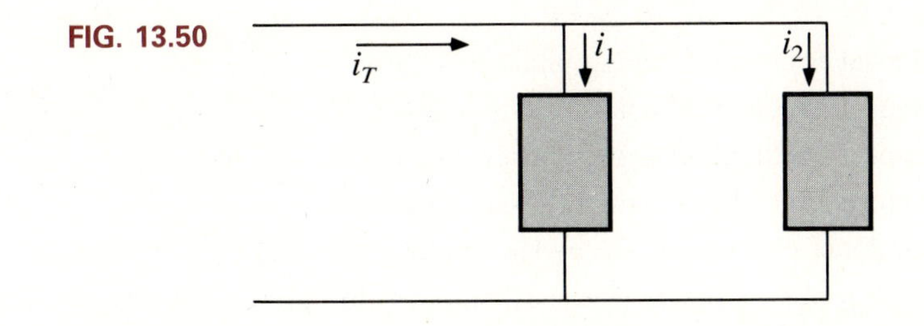

***39.** Find the sinusoidal expression for the voltage v_c for the system of Fig. 13.49 if

$$e_{in} = 120 \sin(\omega t + 45°)$$
$$v_a = 60 \sin \omega t$$
$$v_b = 30 \sin \omega t$$

***40.** Find the sinusoidal expression for the current i_T for the system of Fig. 13.51 if

$$i_1 = 6 \times 10^{-3}\sin(377t + 180°)$$
$$i_2 = 8 \times 10^{-3}\sin 377t$$
$$i_3 = 2i_2$$

FIG. 13.51

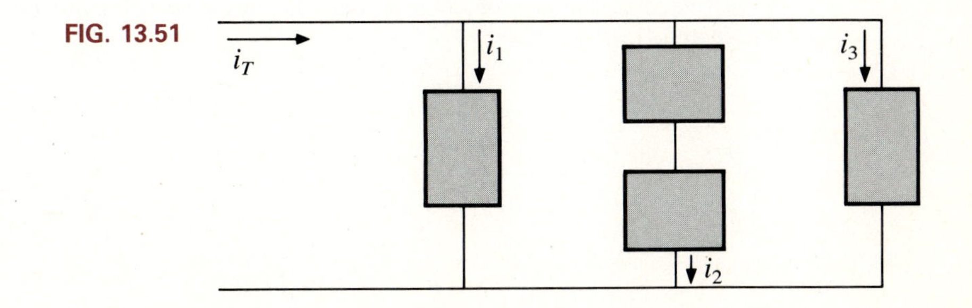

Section 13.8

41. Sketch the following vectors on a complex plane.
a. $100 + j400$ **b.** $0.1 + j0.4$
c. $10 - j2$ **d.** $0.001 + j0.006$
e. $4.6 + j1.2$ **f.** $7 + j42$

42. Sketch the following vectors on a complex plane.
a. $5 \angle 70^\circ$ **b.** $0.8 \angle 15^\circ$
c. $42 \angle 42^\circ$ **d.** $6 \angle -30^\circ$
e. $100 \angle -10^\circ$ **f.** $20{,}000 \angle 40^\circ$

43. Convert the following from the rectangular to polar form.
a. $6 + j3$ **b.** $3 + j6$
c. $1000 + j250$ **d.** $0.06 + j0.08$
e. $8 - j8$ **f.** $50 + j200$

44. Convert the following from polar to rectangular form.
a. $6 \angle 75^\circ$ **b.** $0.4 \angle -10^\circ$
c. $2000 \angle 36^\circ$ **d.** $30 \angle 30^\circ$
e. $0.01 \angle 2^\circ$ **f.** $84 \angle -15^\circ$

45. Repeat Problem 37 using phasor algebra.

46. Repeat Problem 38 using phasor algebra.

47. Repeat Problem 39 using phasor algebra.

48. Repeat Problem 40 using phasor algebra.

49. Find e of $e = v_1 + v_2$ if $v_1 = 100 \sin(\omega t + 30^\circ)$ and $v_2 = 100 \sin(\omega t - 30^\circ)$.

50. Find i_1 of $i_T = i_1 + i_2$ if $i_T = 12 \sin(\omega t + 30^\circ)$ and $i_1 = 8 \sin(\omega t + 60^\circ)$.

Computer Problems

51. Given the sinusoidal expression for the current, determine the expression for the voltage across a resistor, capacitor, or inductor, depending on the element involved. In other words, the program will ask which element is to be investigated and will then request the pertinent data to obtain the mathematical expression for the sinusoidal voltage.

52. Write a program to tabulate the reactance versus frequency for an inductor or capacitor for a specified frequency range.

53. Given the sinusoidal expression for the voltage and current of a load, write a program to determine the average power and power factor.

54. Given two sinusoidal functions (one at 90° and the other at 0°), write a program to convert each to the phasor domain, add the two, and print out the sum in the phasor and time domains.

14

Series ac Circuits

OBJECTIVES

- ☐ Be able to analyze a series sinusoidal ac circuit with resistive, inductive, and capacitive elements.
- ☐ Understand the impedance terminology and how to determine the total impedance (magnitude and angle) of a series sinusoidal ac circuit using the impedance diagram.
- ☐ Be able to draw a phasor diagram for any series ac circuit including all the voltages and current.
- ☐ Find the total power delivered to any series ac circuit with resistive and reactive elements.
- ☐ Be able to find the power factor of a series circuit using one of three methods.
- ☐ Become aware of how the voltage divider rule is applied to sinusoidal ac circuits.
- ☐ Understand how frequency can affect the behavior of series *R-L*, *R-C*, and *R-L-C* circuits.

14.1 INTRODUCTION

In this chapter the impact of placing resistive, inductive, and capacitive elements in series is investigated. The analysis initially involves a fixed frequency, and then the effect of varying the applied frequency is examined. Although the analysis may be more mathematically complex due to the presence of both resistive and reactive components, a number of basic laws, such as Ohm's law and Kirchhoff's voltage law will be applied here also.

14.2 IMPEDANCE

The resistance, reactance, or combination thereof of an ac network is referred to as the *impedance* of the network. It has the symbol Z, is measured in ohms, and is derived from the fact that it provides a measure of how much the network will "impede" (reduce or lower) the flow of charge (I) through the system. Recall for dc series circuits that the total resistance is the sum of the individual resistances. For series ac circuits composed of both resistive and reactive components, the total impedance is not simply the sum of the magnitudes of the individual impedances. The phase shift between v and i for both inductors and capacitors requires that the total impedance be found via vector addition. The magnitude of the vector format is simply the resistance or reactance in ohms. The angle associated with each is defined by the phase relationship between the voltage across the element and the current. It is measured from the positive region of the horizontal axis of Fig. 14.1. The horizontal axis is commonly referred to as the real axis, since any "real" number can be found at some point on the horizontal axis.

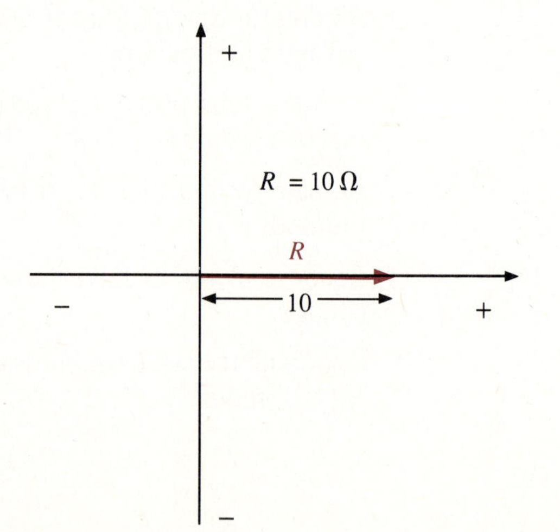

FIG. 14.1
Impedance diagram for a resistive element.

Resistor Since v_R and i_R are *in phase* (a phase shift of 0°), the angle associated with an impedance that is purely resistive is 0°. That is,

$$\mathbf{Z}_R = R \angle 0^\circ \tag{14.1}$$

The *impedance diagram* of a purely resistive network appears in Fig. 14.1 for a total resistance of 10 Ω. The length is laid off on the positive real axis ($\theta = 0^\circ$), with a length of 10 units as defined by the chosen scale.

Inductor For an inductive reactance, the voltage v_L *leads* the current i_L by 90°. The result is that the angle associated with X_L is +90°. A positive sign is associated with the angle of any impedance in which the voltage leads the current.

Therefore,

$$\mathbf{Z}_L = X_L \angle +90^\circ \tag{14.2}$$

The impedance diagram of a purely inductive network having a total reactance of 20 Ω is shown in Fig. 14.2. Note that the positive angle is measured in the counterclockwise direction from the positive real axis. Again, the length of the vector is defined by the magnitude of the impedance and the chosen scale.

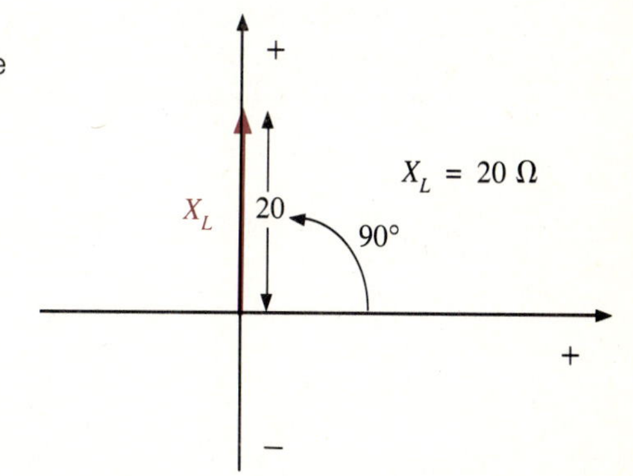

FIG. 14.2
Impedance diagram for an inductive element.

Capacitor For a capacitive reactance, the voltage v_C lags the current i_C by 90°. The result is that the angle associated with X_C is −90°, with the minus sign revealing that v_C lags i_C. Therefore,

$$\mathbf{Z}_C = X_C \angle -90^\circ \tag{14.3}$$

The impedance diagram of a purely capacitive network having a total reactance of 12 Ω appears in Fig. 14.3. Note that a negative angle is measured in the clockwise direction from the positive real axis.

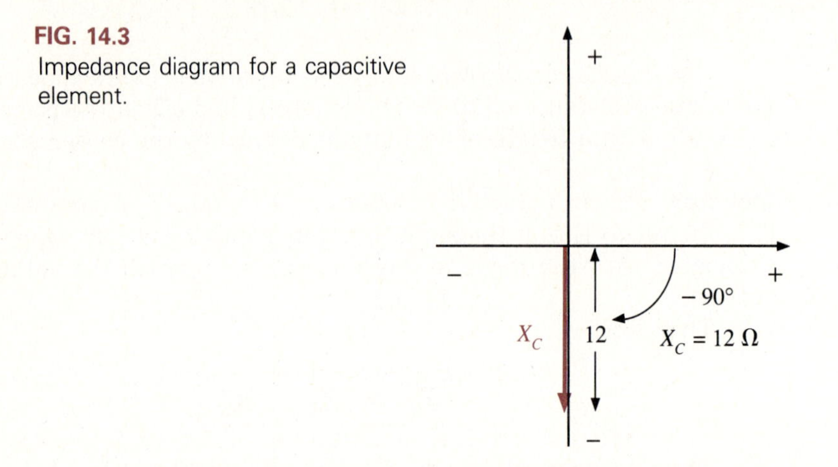

FIG. 14.3
Impedance diagram for a capacitive element.

The impact of placing the various impedances at particular angles becomes apparent in the analysis to follow. The magnitude of the total impedance is not simply the sum of the magnitudes of the individual impedances, and the resulting angle has an important impact on the phase shift between important quantities of the network.

For impedances in series, as shown in Fig. 14.4, the total impedance is the *vector sum* of the individual impedances, taking care to include the effects of the angles associated with each.

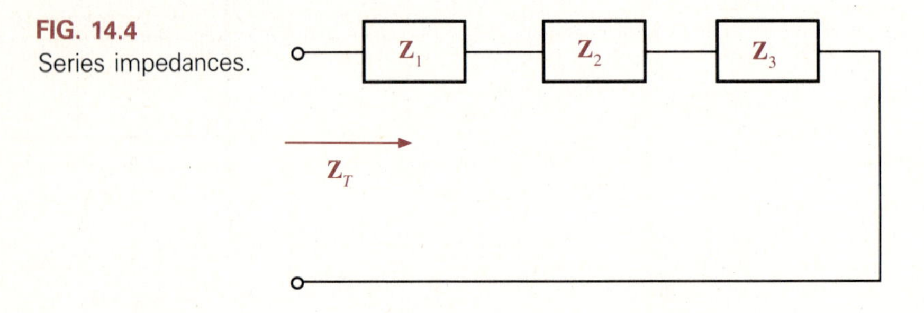

FIG. 14.4
Series impedances.

That is,

$$\mathbf{Z}_T = \mathbf{Z}_1 + \mathbf{Z}_2 + \mathbf{Z}_3 \tag{14.4}$$

A few examples demonstrate its use.

EXAMPLE 14.1 Determine the total impedance of the network of Fig. 14.5 and draw the impedance diagram.

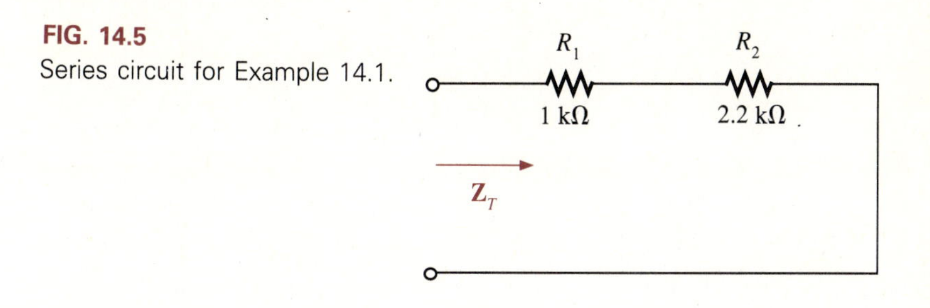

FIG. 14.5
Series circuit for Example 14.1.

Solution Using Eq. 14.4:

$$\begin{aligned} \mathbf{Z}_T &= \mathbf{Z}_1 + \mathbf{Z}_2 \\ &= 1\ \text{k}\Omega\ \angle 0^\circ + 2.2\ \text{k}\Omega\ \angle 0^\circ \\ \mathbf{Z}_T &= \mathbf{3.2\ k\Omega\ \angle 0^\circ} \end{aligned}$$

as shown in Fig. 14.6. Vectors with the same angle can be added directly.

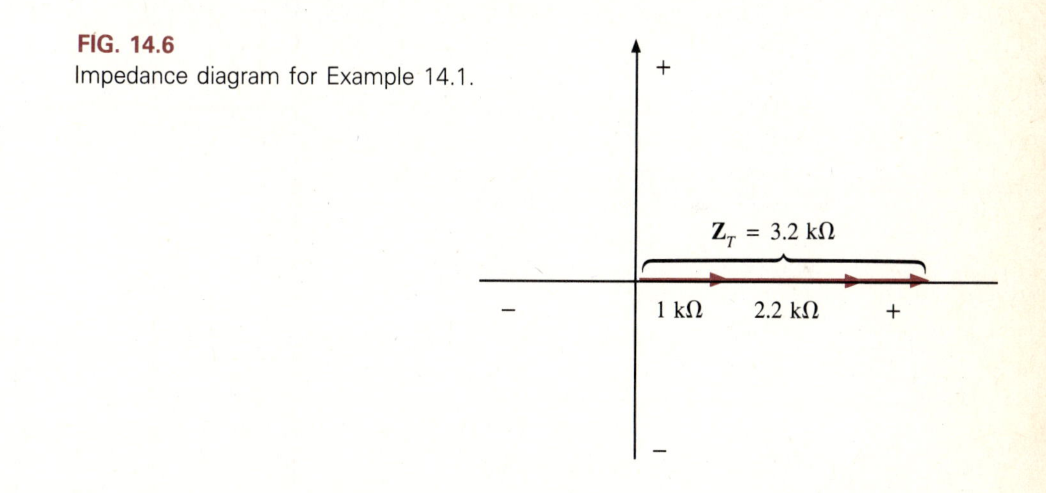

FIG. 14.6
Impedance diagram for Example 14.1.

EXAMPLE 14.2 Determine the total impedance of the network of Fig. 14.7 and draw the impedance diagram.

Solution Using Eq. 14.4:

$$\begin{aligned} \mathbf{Z}_T &= \mathbf{Z}_1 + \mathbf{Z}_2 \\ &= 10\ \Omega\ \angle 0^\circ + 20\ \Omega\ \angle 90^\circ \end{aligned}$$

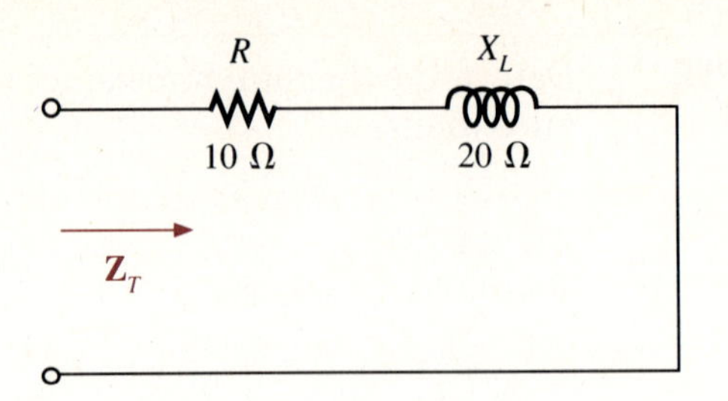

FIG. 14.7
Series ac circuit for Example 14.2.

which appears in Fig. 14.8. The result of the addition is the vector drawn as $\mathbf{Z}_T$ in the same figure.

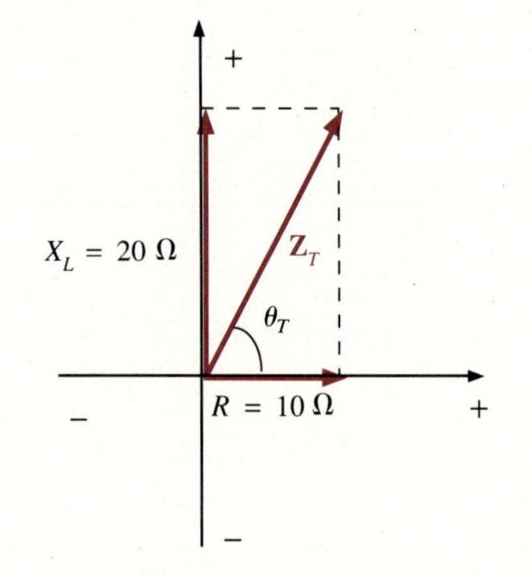

FIG. 14.8
Impedance diagram for Example 14.2.

The magnitude of $\mathbf{Z}_T$ can be determined using the Pythagorean theorem:

$$\begin{aligned} Z_T &= \sqrt{R^2 + X_L^2} \\ &= \sqrt{(10\ \Omega)^2 + (20\ \Omega)^2} = \sqrt{100 + 400} = \sqrt{500} \\ Z_T &= 22.361\ \Omega \end{aligned}$$

Calculator

(1) (0) (x^2) (+) (2) (0) (x^2) (=) ($\sqrt{\ }$)

with a display of 22.361.

The angle associated with $\mathbf{Z}_T$ is

$$\begin{aligned} \theta_T &= \tan^{-1}\frac{\text{Opposite}}{\text{Adjacent}} = \tan^{-1}\frac{X_L}{R} \\ &= \tan^{-1}\frac{20\ \Omega}{10\ \Omega} = \tan^{-1}2 = \mathbf{63.435^\circ} \end{aligned}$$

Calculator

(2) (0) (÷) (1) (0) (=) (2nd F) (tan⁻¹)

with a display of 63.435. This results in $\mathbf{Z}_T = Z_T \angle \theta_T$ = **22.36 Ω ∠63.43°.**

A careful drawing of Fig. 14.8 on an expanded scale would permit the determination of Z_T and θ_T purely from a graphical approach (ruler and protractor). The more inductive the network, the closer Z_T would approach X_L and the more θ_T would approach 90° (for a purely inductive network). Of course, the opposite is also there—the larger R is compared to X_L, the more Z_T will approach R and the more θ_T will approach 0° (for a purely resistive network). Note also that if R and X_L were simply added, Z_T would be 30 Ω, or significantly more than obtained by the required vector addition.

EXAMPLE 14.3 Determine the total impedance of the network of Fig. 14.9 and draw the impedance diagram.

FIG. 14.9
Series ac circuit for Example 14.3.

Solution From Eq. 14.4:

$$\mathbf{Z}_T = \mathbf{Z}_1 + \mathbf{Z}_2$$
$$= 12\ \Omega \angle 90^\circ + 4\ \Omega \angle -90^\circ$$

which is shown in Fig. 14.10. Since the impedances are 180° out of phase, the resultant vector is the difference of the two with the direction of the larger.

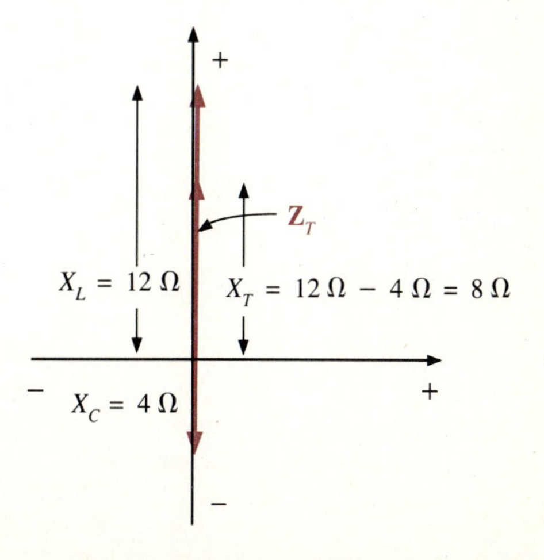

FIG. 14.10
Impedance diagram for Example 14.3.

Therefore,

$$\mathbf{Z}_T = Z_T \underline{/\theta_T} = \mathbf{8\ \Omega\ \underline{/90°}}$$

In this case, the inductive reactance is greater than the capacitive reactance, resulting in a net inductive load.

EXAMPLE 14.4 Determine the total impedance of the series R-L-C network of Fig. 14.11 and draw the impedance diagram.

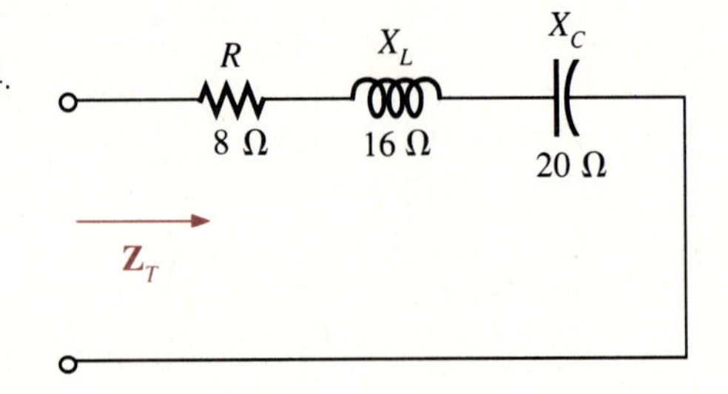

FIG. 14.11
Series ac circuit for Example 14.4.

Solution From Eq. 14.4:

$$\begin{aligned}\mathbf{Z}_T &= \mathbf{Z}_1 + \mathbf{Z}_2 + \mathbf{Z}_3 \\ &= 8\ \Omega\ \underline{/0°} + 16\ \Omega\ \underline{/90°} + 20\ \Omega\ \underline{/-90°}\end{aligned}$$

which is shown in Fig. 14.12.

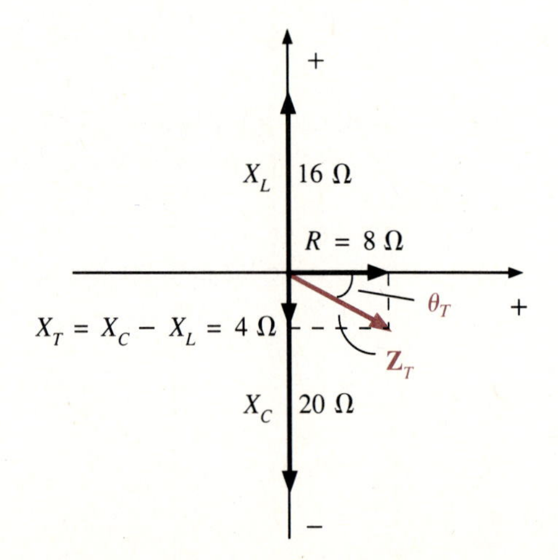

FIG. 14.12
Impedance diagram for Example 14.4.

To determine $\mathbf{Z}_T$, the first step is to find the resultant reactance, which in this case is capacitive.

$$X_C - X_L = 20\ \Omega - 16\ \Omega = 4\ \Omega \qquad \text{(capacitive)}$$

Next, the resulting capacitive reactance is combined with the resistance to determine $\mathbf{Z}_T$. For the magnitude Z_T:

$$\begin{aligned} Z_T &= \sqrt{R^2 + (X_C - X_L)^2} = \sqrt{(8\ \Omega)^2 + (4\ \Omega)^2} \\ &= \sqrt{64 + 16} = \sqrt{80} \\ Z_T &= 8.944\ \Omega \end{aligned}$$

For θ_T:

$$\begin{aligned} \theta_T &= \tan^{-1}\frac{(X_C - X_L)}{R} \\ &= \tan^{-1}\frac{4\ \Omega}{8\ \Omega} = \tan^{-1}0.5 \\ \theta_T &= 26.565° \end{aligned}$$

with $\mathbf{Z}_T = Z_T \angle \theta_T = \mathbf{8.944\ \Omega \angle -26.565°}$.

The resulting angle reveals that the circuit is capacitive in nature. A positive angle for θ_T would reveal that $X_L > X_C$. For $0° < \theta_T < 90°$, the circuit is predominantly inductive, and for $-90° < \theta_T < 0°$, the circuit is predominantly capacitive.

14.3 *R-L* SERIES CIRCUIT

Let us now perform a detailed analysis of a series *R-L* circuit with a sinusoidal voltage applied at a fixed frequency. A fixed frequency establishes a fixed value for X_L—specifically, 4 Ω for the network of Fig. 14.13.

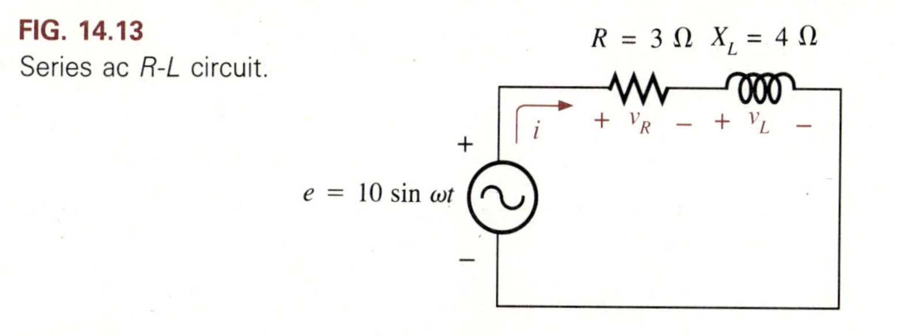

FIG. 14.13
Series ac *R-L* circuit.

$\mathbf{Z}_T$: The total impedance $\mathbf{Z}_T$ is determined by

$$\begin{aligned} \mathbf{Z}_T &= \mathbf{Z}_1 + \mathbf{Z}_2 \\ &= 3\ \Omega \angle 0° + 4\ \Omega \angle 90° \end{aligned}$$

As shown in Fig. 14.14, the magnitude of $\mathbf{Z}_T$ is determined by

$$Z_T = \sqrt{R^2 + X_L^2} = \sqrt{(3\ \Omega)^2 + (4\ \Omega)^2} = 5\ \Omega$$

and the angle θ_T is

$$\theta_T = \tan^{-1}\frac{X_L}{R} = \tan^{-1}\frac{4\ \Omega}{3\ \Omega} = \tan^{-1}1.333 = 53.13°$$

with $\mathbf{Z}_T = Z_T \angle \theta_T = \mathbf{5\ \Omega \angle 53.13°}$.

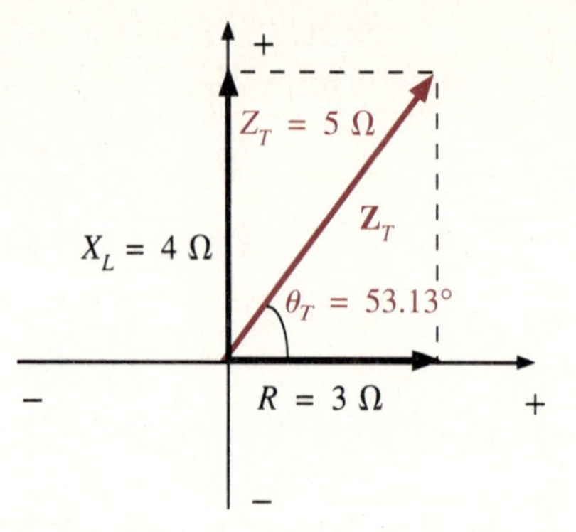

FIG. 14.14
Impedance diagram for the circuit of Fig. 14.13.

I: The current through the series circuit is related to the applied voltage and total impedance by Ohm's law:

$$I = \frac{E}{Z_T} \tag{14.5}$$

If the peak value of the voltage or current is substituted, then the peak value of the other quantity will be determined. If the effective values are employed, then effective values will result.

Employing peak values,

$$I_m = \frac{E_m}{Z_T} = \frac{10\ \text{V}}{5\ \Omega} = \mathbf{2\ A}$$

with an effective value of

$$I = (0.707)(2\ \text{A}) = \mathbf{1.414\ A}$$

Employing effective values,

$$I = \frac{E}{Z_T} = \frac{(0.707)(10\ \text{V})}{5\ \Omega} = \frac{7.07\ \text{V}}{5\ \Omega} = \mathbf{1.414\ A}$$

The angle associated with the current I can be determined using the following powerful statement:

The angle θ_T associated with the total impedance of a network is the angle by which the applied voltage leads the source current. If θ_T is negative, the current leads the voltage by the same angle.

For the preceding example, $\theta_T = +53.13°$, revealing that e leads i by 53.13°, and since $e = 10 \sin(\omega t + 0°)$,

$$i = I_m \sin(\omega t + \theta) = \mathbf{2\ sin(\boldsymbol{\omega} t - 53.13°)}$$

In phasor form, $\mathbf{I} = \mathbf{1.414\ A\ \angle -53.13°}$.

$\mathbf{V}_R$: Ohm's law can also be written in the following form:

$$\boxed{V = IZ} \tag{14.6}$$

where V and I can be peak or effective values and Z can be resistive or reactive.

For the resistor of Fig. 14.13, if we substitute peak values,

$$V_m = IZ = I_m R = (2\ \text{A})(3\ \Omega) = 6\ \text{V}$$

with an effective value of

$$V_R = (0.707)(6\ \text{V}) = \mathbf{4.242\ V}$$

Employing effective values,

$$\begin{aligned} V_R = IR &= [(0.707)(2\ \text{A})](3\ \Omega) \\ &= \mathbf{4.242\ V} \end{aligned}$$

Since v_R and i_R are *in phase,* the angle associated with v_R is the same as that associated with i. That is,

$$v_R = V_m \sin(\omega t + \theta) = \mathbf{6\ sin(\boldsymbol{\omega} t - 53.13°)}$$

In phasor form, $\mathbf{V}_R = \mathbf{4.242\ V\ \angle -53.13°}$.

$\mathbf{V}_L$: Using peak values:

$$V_m = I_m X_L = (2\ \text{A})(4\ \Omega) = 8\ \text{V}$$

with an effective value of

$$V_L = (0.707)(8\ \text{V}) = 5.656\ \text{V}$$

Since v_L leads i_L ($= i$) by 90°, the angle associated with v_L is 90° counterclockwise from i, so $\theta_L = -53.13° + 90° = +36.87°$. That is,

$$v_L = V_m \sin(\omega t + \theta) = \mathbf{8\ sin(\boldsymbol{\omega} t + 36.87°)}$$

In phasor form, $\mathbf{V}_L = \mathbf{5.656\ V\ \angle +36.87°}$.

Phasor Diagram Now that all the quantities of the network have been determined in phasor (vector) form, they can all be placed on the same figure (Fig. 14.15) to show their phase relationship.

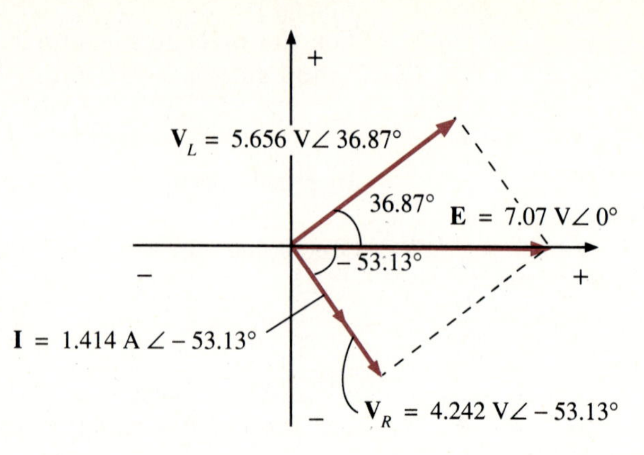

FIG. 14.15
Phasor diagram for the circuit of Fig. 14.13.

Figure 14.15 clearly reveals that $\mathbf{V}_R$ and $\mathbf{I}$ are in phase (same vector direction) and that $\mathbf{V}_L$ leads $\mathbf{I}$ by 90° ($\mathbf{V}_L$ is 90° ahead of $\mathbf{I}$ in the counterclockwise direction). In addition, note that the vector $\mathbf{E}$ equals the sum of $\mathbf{V}_R$ and $\mathbf{V}_L$, as required by Kirchhoff's voltage law. That is,

$$\mathbf{E} = \mathbf{V}_R + \mathbf{V}_L$$

Power The total power in watts delivered to the circuit is

$$\boxed{P_T = EI\cos\theta_T} \qquad (14.7)$$

where θ_T is the angle between E and I and the angle associated with $\mathbf{Z}_T$. Both E and I are effective values.

Substituting gives

$$\begin{aligned} P_T &= (7.07 \text{ V})(1.414 \text{ A})\cos 53.13^\circ \\ &= 10(0.6) = \mathbf{6\ W} \end{aligned}$$

In addition,

$$\boxed{P_T = I^2R} \qquad (14.8)$$

with I the effective value of the current.

Substituting gives

$$\begin{aligned} P_T &= (1.414 \text{ A})^2(3\ \Omega) \\ &= (2)(3) = \mathbf{6\ W} \end{aligned}$$

The following is also applicable:

$$\boxed{P_T = \frac{V_R^2}{R}} \qquad (14.9)$$

So that

$$P_T = \frac{(4.242\ \text{V})^2}{3\ \Omega} = \mathbf{6\ W}$$

or, finally,

$$\begin{aligned} P_T &= P_R + P_L \\ &= V_R I_R \cos\,\theta_R + V_L I_L \cos\,\theta_L \\ &= V_R I\,\cos\,\theta_R + V_L I\,\cos\,\theta_L \\ &= (4.242\ \text{V})(1.414\ \text{A})\cos\,0° + (5.656\ \text{V})(1.414\ \text{A})\cos\,90° \\ &= 6(1) + 8(0) \\ P_T &= \mathbf{6\ W} \end{aligned}$$

Power Factor The power factor of the network is defined by

$$F_p = \cos\,\theta_T = \frac{R}{Z_T} = \frac{P}{EI} \tag{14.10}$$

If θ_T is a positive angle, the network is said to be inductive, or to have a *lagging power factor*. If θ_T is negative, the network is capacitive and has a *leading power factor*.

For this example,

$$F_p = \cos\,53.13° = \mathbf{0.6}$$

and the network has a *lagging* power factor of 0.6.

As a check,

$$F_p = \frac{R}{Z_T} = \frac{3\ \Omega}{5\ \Omega} = \mathbf{0.6}$$

and

$$F_p = \frac{P}{EI} = \frac{6\ \text{W}}{(7.07\ \text{V})(1.414\ \text{A})} = \frac{6}{10} = \mathbf{0.6}$$

Note in the preceding calculations that effective values are used a great deal more than peak values. This is generally true for most methods of analysis applied to sinusoidal ac networks. Simply keep track of which are being used to ensure a mixed set of parameters is not employed.

Sinusoidal Domain If required by the analysis, the sinusoidal functions can also be sketched on the same set of axes. For this example,

$$\begin{aligned} e &= 10\,\sin\,\omega t \\ i &= 2\,\sin(\omega t - 53.13°) \\ v_R &= 6\,\sin(\omega t - 53.13°) \\ v_L &= 8\,\sin(\omega t + 36.87°) \end{aligned}$$

All are plotted in Fig. 14.16, which clearly reveals that v_R and i are in phase and v_L leads i by 90°. In addition, a graphical addition for v_L and v_R results in the sinusoidal waveform for e.

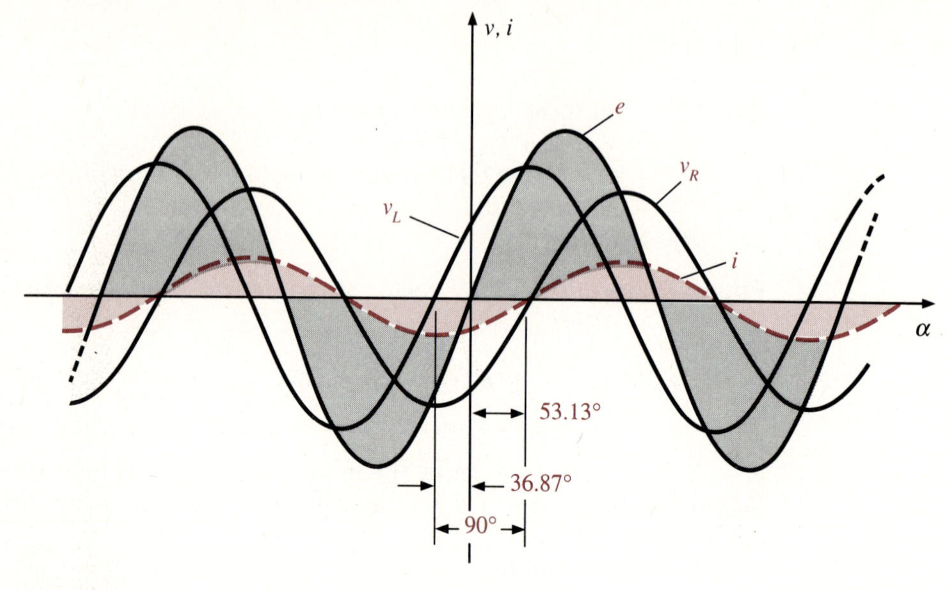

FIG. 14.16
Sinusoidal waveforms for the circuit of Fig. 14.13.

The circuit of Fig. 14.13 is redrawn in Fig. 14.17 with effective values to emphasize the fact that ac meters provide the effective rather than the peak values. In particular, note that the magnitude of **E** is not equal to the sum of the magnitudes of $\mathbf{V}_R$ and $\mathbf{V}_L$. A vector addition must be made to show that $\mathbf{E} = \mathbf{V}_R + \mathbf{V}_L$.

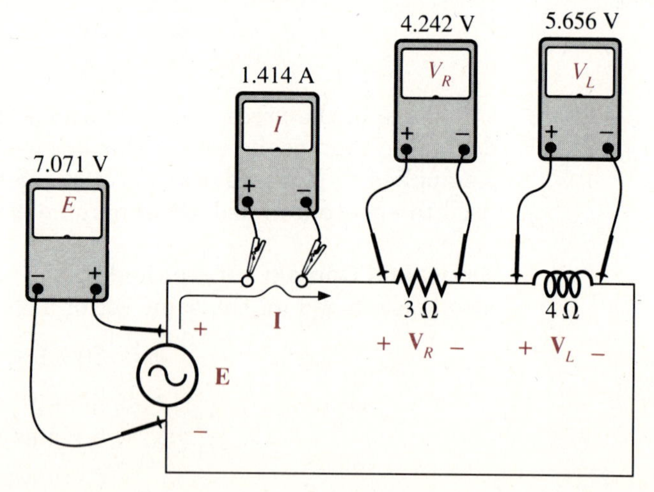

FIG. 14.17
Meter measurements for a series ac *R-L* circuit.

Also keep in mind that the current **I** is the same throughout the circuit and can be measured between any two elements.

14.4 *R-C* SERIES CIRCUIT

There is a great deal of similarity between the analysis of the R-C series circuits and the R-L series circuits. The major difference is in the fact that the source current i now leads the applied voltage e by an angle determined by the network parameters.

At the frequency of application, the reactance of the capacitor of Fig. 14.18 is 8 Ω.

FIG. 14.18
Series ac *R-L* circuit.

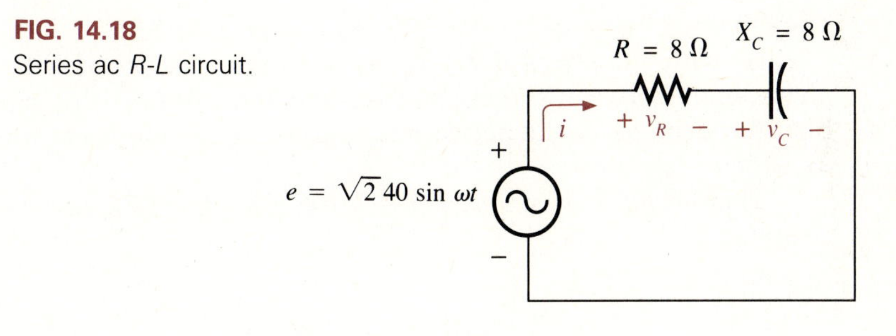

$\mathbf{Z}_T$: The total impedance $\mathbf{Z}_T$ is determined by

$$\begin{aligned}\mathbf{Z}_T &= \mathbf{Z}_1 + \mathbf{Z}_2\\ &= 8\ \Omega\ \angle 0^\circ + 8\ \Omega\ \angle -90^\circ\end{aligned}$$

as shown in Fig. 14.19.

FIG. 14.19
Impedance diagram for the circuit of Fig. 14.18.

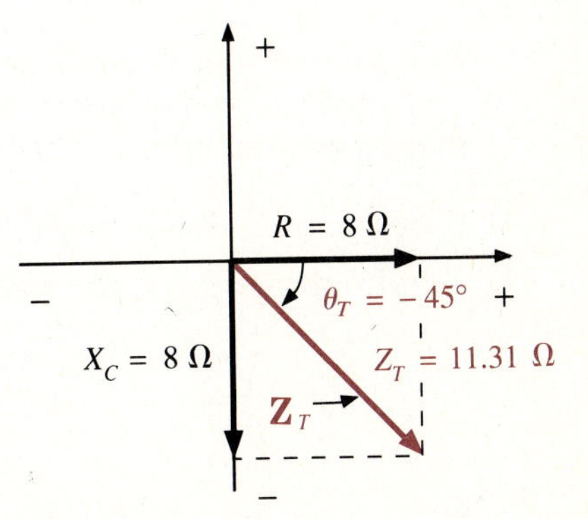

The magnitude of $\mathbf{Z}_T$ is determined by

$$Z_T = \sqrt{R^2 + X_C^2} = \sqrt{(8\ \Omega)^2 + (8\ \Omega)^2} \cong 11.314\ \Omega$$

and the angle θ_T is found by

$$\theta_T = \tan^{-1}\frac{X_C}{R} = \tan^{-1}\frac{8\ \Omega}{8\ \Omega} = \tan^{-1}1 = 45°$$

with $\mathbf{Z}_T = Z_T \angle\theta_T = \mathbf{11.314\ \Omega \angle -45°}$.

Note the negative sign associated with angles measured below the positive real axis.

I: The peak value of the current can be determined by Ohm's law:

$$I_m = \frac{E_m}{Z_T} = \frac{20\ \text{V}}{11.314\ \Omega} = 1.768\ \text{A}$$

The effective value is $I = (0.707)(1.768\ \text{A}) = \mathbf{1.25\ A}$.

As with the series R-L circuit, the angle associated with $\mathbf{Z}_T$ is also the phase angle between the applied voltage and the resulting current i. In this case, however, since the network is capacitive, the angle obtained is the angle by which i leads e.

Therefore, $i = I_m\sin(\omega t + \theta) = \mathbf{1.768\ sin(\omega t + 45°)}$, since $e = 20\ \sin(\omega t + 0°)$.

In phasor form,

$$\begin{aligned}\mathbf{I} &= I_{\text{eff}} \angle\theta \\ &= \mathbf{1.25\ A \angle +45°}\end{aligned}$$

V_R The peak value of v_R is determined using Ohm's law:

$$\begin{aligned}V_m &= I_mR \\ &= (1.768\ \text{A})(8\ \Omega) \\ V_m &= 14.14\ \text{V}\end{aligned}$$

with an effective value $V_{\text{eff}} = (0.707)(14.14\ \text{V}) = 10\ \text{V}$.

Since v_R and i_R are in phase, the angle associated with v_R is the same as that associated with i. That is,

$$v_R = V_m\sin(\omega t + \theta) = \mathbf{14.14\ sin(\omega t + 45°)}$$

In phasor form, the voltage v_R is

$$\begin{aligned}\mathbf{V}_R &= V_R \angle\theta_R \\ &= \mathbf{10\ V \angle +45°}\end{aligned}$$

V_C

$$\begin{aligned}V_m &= I_mX_C \\ &= (1.768\ \text{A})(8\ \Omega) \\ V_m &= 14.14\ \text{V}\end{aligned}$$

with an effective value $V_{\text{eff}} = (0.707)(14.14\ \text{V}) = 10\ \text{V}$.

Since v_C lags i_C by 90°, the angle associated with v_C is (+45° − 90°) = −45°. That is,

$$v_C = V_m \sin(\omega t + \theta) = \mathbf{14.14 \sin(\omega t - 45°)}$$

In phasor form,

$$\begin{aligned} \mathbf{V}_C &= V_C \angle \theta_C \\ &= \mathbf{10\ V \angle -45°} \end{aligned}$$

Phasor Diagram The phasor diagram for the voltages and current appears in Fig. 14.20. In particular, note that the magnitudes of $\mathbf{V}_R$ and $\mathbf{V}_C$ are the same, since $R = X_C = 8\Omega$—a result obtained for dc series circuits for two resistors of equal value. In addition, **I** leads $\mathbf{V}_C$ by 90° and $\mathbf{V}_R$ and **I** are in phase. Since the network is capacitive, the current **I** *leads* the applied voltage **E** by 45°.

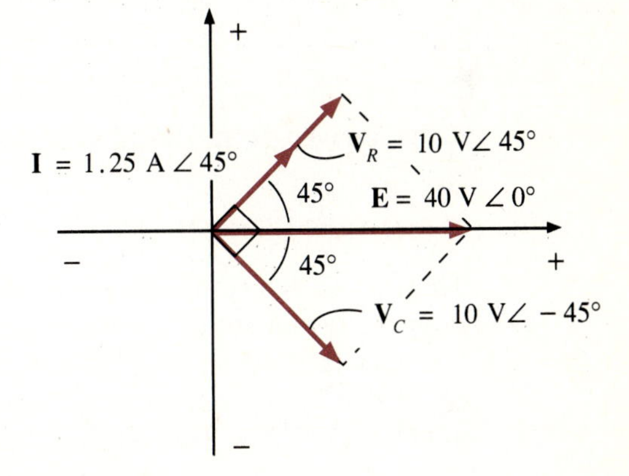

FIG. 14.20
Phasor diagram for the circuit of Fig. 14.18.

Note again that the vector addition of $\mathbf{V}_R$ and $\mathbf{V}_C$ equals the applied voltage **E,** as required by Kirchhoff's voltage law:

$$\mathbf{E} = \mathbf{V}_R + \mathbf{V}_C$$

Power The total power delivered to the circuit is

$$\begin{aligned} P_T &= EI \cos \theta_T \\ &= (14.14\ \text{V})(1.25\ \text{A})\cos 45° \\ &= (17.675)(0.707) \\ P_T &= \mathbf{12.5\ W} \end{aligned}$$

Also,

$$\begin{aligned} P_T &= I^2R = (1.25\ \text{A})^2(8\ \Omega) \\ &= \mathbf{12.5\ W} \end{aligned}$$

or

$$\begin{aligned} P_T &= \frac{V_R^2}{R} = \frac{(10\ \text{V})^2}{8\ \Omega} \\ &= \mathbf{12.5\ W} \end{aligned}$$

Power Factor

$$F_p = \cos\theta_T = \cos 45^\circ = \mathbf{0.707 \quad leading}$$

or

$$F_p = \frac{R}{Z_T} = \frac{8\ \Omega}{11.31\ \Omega} = \mathbf{0.707 \quad leading}$$

Sinusoidal Domain

$$\begin{aligned} e &= 20 \sin \omega t \\ i &= 1.768 \sin(\omega t + 45^\circ) \\ v_R &= 14.14 \sin(\omega t + 45^\circ) \\ v_C &= 14.14 \sin(\omega t - 45^\circ) \end{aligned}$$

A plot of each on the same set of axes would support the earlier conclusions derived from the phasor diagram.

14.5 *R-L-C* SERIES CIRCUIT

The last series combination to be analyzed includes all three elements, with reactive levels determined at a particular frequency. At the chosen frequency, the network of Fig. 14.21 is inductive, since $X_L > X_C$. A change in frequency could result in a net capacitive reactance ($X_C > X_L$) or total resistive appearance if $X_L = X_C$. The effects of frequency are analyzed in the next section.

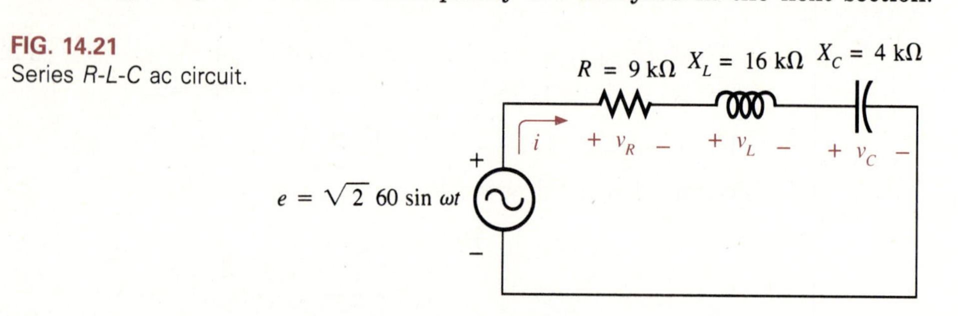

FIG. 14.21
Series *R-L-C* ac circuit.

$\mathbf{Z}_T$: The total impedance $\mathbf{Z}_T$ is determined by

$$\begin{aligned} \mathbf{Z}_T &= \mathbf{Z}_1 + \mathbf{Z}_2 + \mathbf{Z}_3 \\ &= 9\ \text{k}\Omega\ \angle 0^\circ + 16\ \text{k}\Omega\ \angle 90^\circ + 4\ \text{k}\Omega\ \angle -90^\circ \end{aligned}$$

as shown on the impedance diagram of Fig. 14.22.

The magnitude of $\mathbf{Z}_T$ is determined by

$$\begin{aligned} Z_T &= \sqrt{R^2 + (X_L - X_C)^2} = \sqrt{(9\ \text{k}\Omega)^2 + (16\ \text{k}\Omega - 4\ \text{k}\Omega)^2} \\ &= \sqrt{(9\ \text{k}\Omega)^2 + (12\ \text{k}\Omega)^2} = 15\ \text{k}\Omega \end{aligned}$$

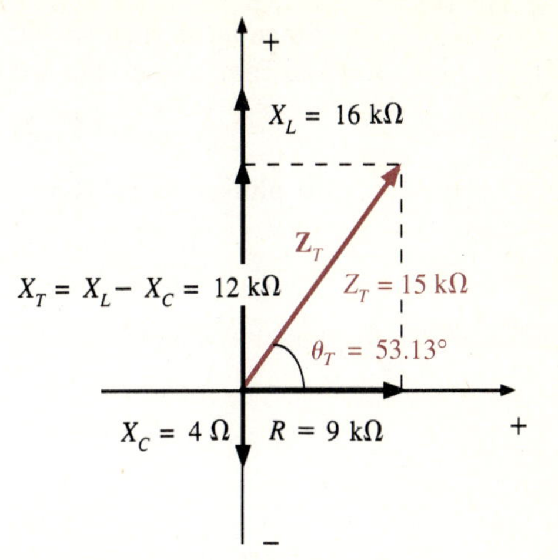

FIG. 14.22
Impedance diagram for the circuit of Fig. 14.21.

and the angle θ_T is

$$\theta_T = \tan^{-1}\frac{X_L - X_C}{R}$$
$$= \tan^{-1}\frac{12 \text{ k}\Omega}{9 \text{ k}\Omega} = \tan^{-1}1.333$$
$$\theta_T = 53.13°$$

with $\mathbf{Z}_T = Z_T \angle\theta_T = \mathbf{15\ k\Omega \angle 53.13°}$.

I:

$$I_m = \frac{E_m}{Z_T}$$
$$= \frac{60 \text{ V}}{15 \text{ k}\Omega} = 4 \text{ mA}$$

The effective value is $I = (0.707)(4 \text{ mA}) = \mathbf{2.828\ mA}$.

Since the angle θ_T is the angle by which e leads i (inductive network, $X_L > X_C$), if

$$e = 60 \sin(\omega t + 0°) \quad \text{then} \quad \mathbf{i = 4 \times 10^{-3}\sin(\omega t - 53.13°)}$$

In phasor form,

$$\mathbf{I} = I_{\text{eff}} \angle\theta$$
$$= \mathbf{2.828\ mA \angle -53.13°}$$

V_R:

$$V_m = I_m R$$
$$= (4 \times 10^{-3} \text{ A})(9 \times 10^3 \ \Omega)$$
$$V_m = 36 \text{ V}$$

with an effective value $V_{\text{eff}} = (0.707)(36 \text{ V}) = 25.45 \text{ V}$.

Since v_R and i_R are in phase, the phase angle associated with v_R is the same as that associated with i. That is,

$$v_R = V_m \sin(\omega t + \theta) = \mathbf{36 \sin(\omega t - 53.13°)}$$

In phasor form, the voltage v_R is

$$\begin{aligned} \mathbf{V}_R &= V_R \angle \theta_R \\ &= \underline{\mathbf{25.45\ V \angle -53.13°}} \end{aligned}$$

$\mathbf{V}_L$:

$$\begin{aligned} V_m &= I_m X_L \\ &= (4 \times 10^{-3}\ \text{A})(16 \times 10^3\ \Omega) \\ V_m &= 64\ \text{V} \end{aligned}$$

with effective value $V_{eff} = (0.707)(64\ \text{V}) = 45.25\ \text{V}$.

Since v_L leads $i_L = i$ by 90°, the angle associated with v_L is $(-53.13° + 90°) = +36.87°$. That is,

$$v_L = V_m \sin(\omega t + \theta) = \mathbf{64 \sin(\omega t + 36.87°)}$$

In phasor form,

$$\begin{aligned} \mathbf{V}_L &= V_L \angle \theta_L \\ &= \underline{\mathbf{45.25\ V \angle +36.87°}} \end{aligned}$$

$\mathbf{V}_C$:

$$\begin{aligned} V_m &= I_m X_C \\ &= (4 \times 10^{-3}\ \text{A})(4 \times 10^3\ \Omega) \\ V_m &= 16\ \text{V} \end{aligned}$$

with an effective value $V_{eff} = (0.707)(16\ \text{V}) = 11.31\ \text{V}$.

Since v_C lags $i_C = i$ by 90°, the angle associated with v_C is $(-53.13° - 90°) = -143.13°$. That is,

$$v_C = V_m \sin(\omega t + \theta) = \mathbf{16 \sin(\omega t - 143.13°)}$$

In phasor form, the voltage v_C is

$$\begin{aligned} \mathbf{V}_C &= V_C \angle \theta_C \\ &= \underline{\mathbf{11.31\ V \angle -143.13°}} \end{aligned}$$

The Phasor Diagram The phasor diagram for the voltages and current is shown in Fig. 14.23. In particular, note that $\mathbf{I}$ and $\mathbf{V}_R$ are in phase, that $\mathbf{V}_L$ leads $\mathbf{I}$ by 90°, that $\mathbf{V}_C$ lags $\mathbf{I}$ by 90°, and that the vector sum is $\mathbf{E} = \mathbf{V}_R + \mathbf{V}_L + \mathbf{V}_C$ (as determined by Kirchhoff's voltage law).

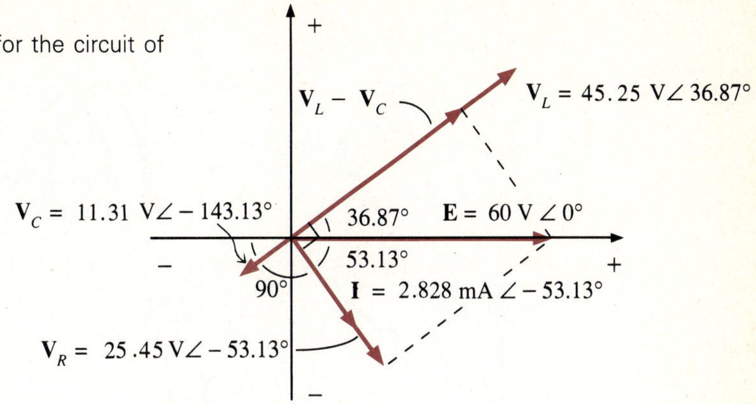

FIG. 14.23
Phasor diagram for the circuit of Fig. 14.21.

Power

$$
\begin{aligned}
P_T &= EI_T\cos\theta_T \\
&= (42.42\text{ V})(2.828\text{ mA})\cos(53.13^\circ) \\
&= (119.96 \times 10^{-3})(0.6) \\
P_T &= \mathbf{71.98\text{ mW}}
\end{aligned}
$$

or

$$
\begin{aligned}
P_T &= I^2R \\
&= (2.828\text{ mA})^2(9\text{ k}\Omega) \\
P_T &= \mathbf{71.98\text{ mW}}
\end{aligned}
$$

Power Factor

$$F_p = \cos\theta_T = \cos 53.13^\circ = \mathbf{0.6\ \ lagging}$$

or

$$F_p = \frac{R}{Z_T} = \frac{9\text{ k}\Omega}{15\text{ k}\Omega} = \mathbf{0.6\ \ lagging}$$

Sinusoidal Domain

$$
\begin{aligned}
e &= 60\sin\omega t \\
i &= 4 \times 10^{-3}\sin(\omega t - 53.13^\circ) \\
v_R &= 36\sin(\omega t - 53.13^\circ) \\
v_L &= 64\sin(\omega t + 36.87^\circ) \\
v_C &= 16\sin(\omega t - 143.13^\circ)
\end{aligned}
$$

A plot of each on the same set of axes will result in the waveforms of Fig. 14.24. Note at $\alpha = 0^\circ$ how $e = v_L - v_C - v_R = 0$ V and at $\alpha = +143.13^\circ$ that $v_L = v_C = 0$ V and $e = v_R$. In addition, note that v_R and i are in phase, with v_L leading i by 90° and v_C lagging i by 90°.

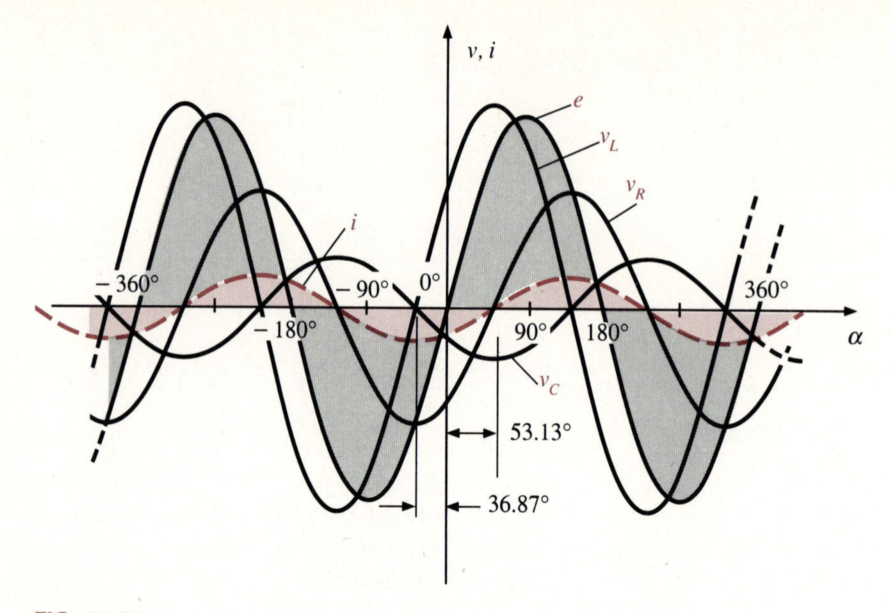

FIG. 14.24
Sinusoidal waveforms for the circuit of Fig. 14.21.

14.6 VOLTAGE DIVIDER RULE

The voltage divider rule introduced for dc circuits can also be applied to ac networks if we simply keep in mind the vector relationship between impedances and the phase relationships established by each element.

Resistive Networks For a totally resistive network such as that appearing in Fig. 14.25, all the voltage levels are the effective values of each quantity.

FIG. 14.25
Applying the voltage divider rule to a series resistive circuit.

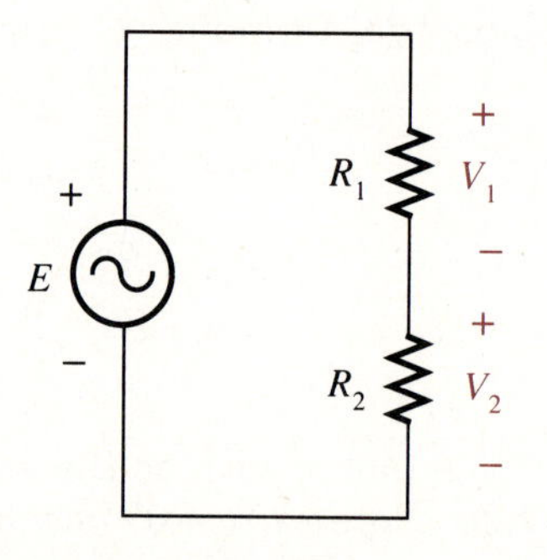

Applying the voltage divider rule results in

$$V_1 = \frac{R_1 E}{R_T} = \frac{R_1 E}{R_1 + R_2} \tag{14.11}$$

and

$$V_2 = \frac{R_2E}{R_T} = \frac{R_2E}{R_1 + R_2} \tag{14.12}$$

Note the similarities to the results obtained for dc circuits. In fact, the only change was to employ effective values of the voltage rather than the dc level.

The same format can be applied for the peak values by simply substituting the peak value for the applied voltage e. The fact that the network is purely resistive eliminates the possibility of any phase shifts, and the angle associated with v_1 or v_2 is the same as that of e.

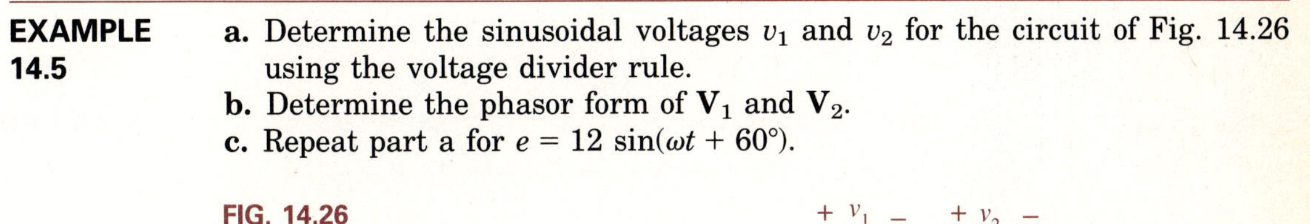

EXAMPLE 14.5

a. Determine the sinusoidal voltages v_1 and v_2 for the circuit of Fig. 14.26 using the voltage divider rule.
b. Determine the phasor form of $\mathbf{V}_1$ and $\mathbf{V}_2$.
c. Repeat part a for $e = 12 \sin(\omega t + 60°)$.

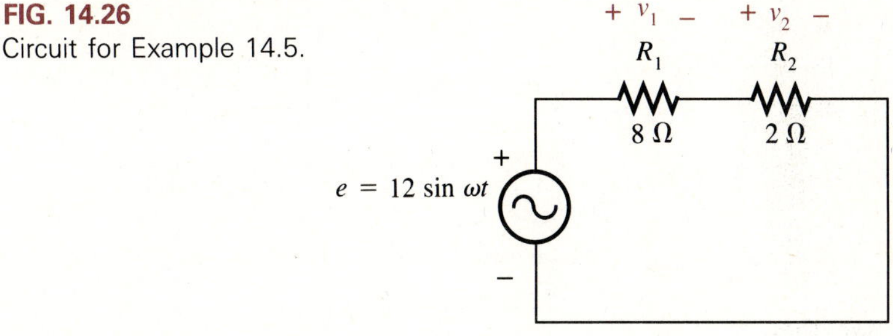

FIG. 14.26
Circuit for Example 14.5.

Solution

a. $$V_{m_1} = \frac{R_1E_m}{R_1 + R_2} = \frac{(8\ \Omega)(12\ \text{V})}{8\ \Omega + 2\ \Omega} = 0.8(12\ \text{V}) = 9.6\ \text{V}$$

with $v_1 = \mathbf{9.6 \sin \omega t}$.

$$V_{m_2} = \frac{R_2E_m}{R_1 + R_2} = \frac{(2\ \Omega)(12\ \text{V})}{8\ \Omega + 2\ \Omega} = 0.2(12\ \text{V}) = 2.4\ \text{V}$$

with $v_2 = \mathbf{2.4 \sin \omega t}$.

b. $$\mathbf{V}_1 = (0.707)(9.6\ \text{V}) \angle 0° = \mathbf{6.787\ V \angle 0°}$$

and

$$\mathbf{V}_2 = (0.707)(2.4\ \text{V}) \angle 0° = \mathbf{1.697\ V \angle 0°}$$

c. For $e = 12 \sin(\omega t + 60°)$, the only change is the angle associated with v_1 and v_2. That is,

$$v_1 = \mathbf{9.6 \sin(\omega t + 60°)}$$

and

$$v_2 = \mathbf{2.4 \sin(\omega t + 60°)}$$

***R-L* or *R-C* Networks** For a series R-L or R-C circuit such as the one in Fig. 14.27, where the reactance X can be either inductive or capacitive, the following equations provide the magnitude of V_R and V_X.

$$V_R = \frac{RE}{Z_T} = \frac{RE}{\sqrt{R^2 + X^2}} \tag{14.13}$$

$$V_X = \frac{XE}{Z_T} = \frac{XE}{\sqrt{R^2 + X^2}} \tag{14.14}$$

FIG. 14.27
Series *R-L* circuit.

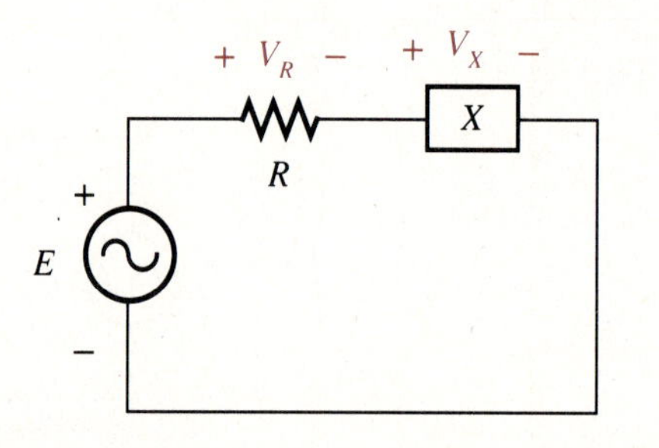

If E is the effective value, the effective value of V_R and V_X results. If E is the peak value, then V_R and V_X are the peak values.

Equations 14.13 and 14.14 do not provide the angle associated with $\mathbf{V}_R$ and $\mathbf{V}_X$, but they do provide the magnitude of the quantities normally measured by a voltmeter. If the phase angle associated with $\mathbf{V}_R$ or $\mathbf{V}_X$ is important, then the method of the previous section should be applied.

EXAMPLE 14.6 For the circuit of Fig. 14.28, determine the effective value of the voltage V_L using the voltage divider rule.

Solution Using Eq. 14.14,

$$V_L = \frac{X_L E}{\sqrt{R^2 + X_L^2}}$$

$$= \frac{(4\ \text{k}\Omega)(120\ \text{V})}{\sqrt{(2.2\ \text{k}\Omega)^2 + (4\ \text{k}\Omega)^2}} = \frac{(4\ \text{k}\Omega)(120\ \text{V})}{4.565\ \text{k}\Omega}$$

$$V_L = (0.876)(120\ \text{V}) = \mathbf{105.15\ V}$$

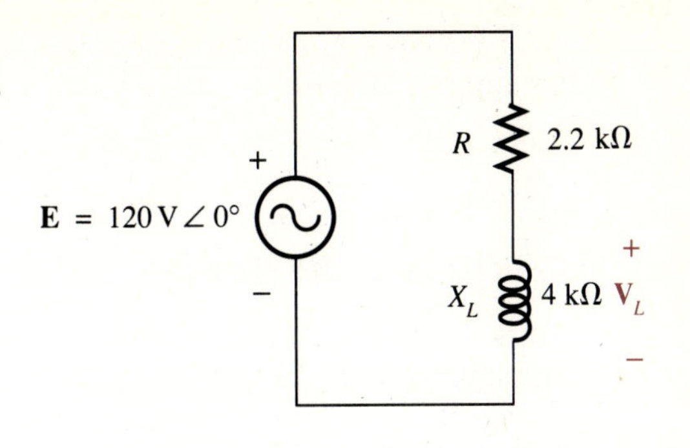

FIG. 14.28
R-L circuit for Example 14.6.

***R-L-C* or *L-C* Networks** Equations 14.13 and 14.14 can be extended to include *R-L-C* series ac circuits if X of Z_T is the net reactance of the reactive elements.

EXAMPLE 14.7 Determine the reading of the digital meter of Fig. 14.29.

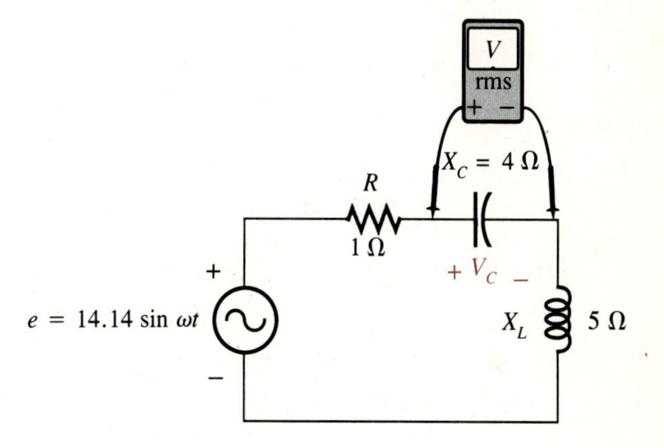

FIG. 14.29
Circuit for Example 14.7.

Solution

$$X_T = X_L - X_C = 1\ \Omega$$

and

$$V_C = \frac{X_C E}{Z_T} = \frac{4\ \Omega[(0.707)(14.14\ \text{V})]}{\sqrt{(1\ \Omega)^2 + (1\ \Omega)^2}}$$

$$= \frac{4\ \Omega(10\ \text{V})}{1.414\ \Omega} = (2.829)(10\ \text{V})$$

$$V_C = \mathbf{28.29\ V} \quad \text{(reading of the digital meter)}$$

Note in the preceding example that since the net reactance is less than either reactance, the voltage across X_C is significantly more than the effective value of the applied voltage, a situation often encountered in ac systems.

For an L-C network lacking resistive elements, Eqs. 14.13 and 14.14 can be applied, by simply remembering that

$$Z_T = X_T = X_L - X_C \text{ (or } X_C - X_L)$$

14.7 FREQUENCY RESPONSE: *R-L* CONFIGURATION

Thus far, the analysis of series circuits has been limited to a particular frequency resulting in fixed values for X_L and X_C. We now examine the effects of changing the applied frequency on the voltages and current of the series R-L circuit of Fig. 14.30. For the analysis to follow, all voltage and current levels will be effective values, since these are the quantities measured by most ac meters.

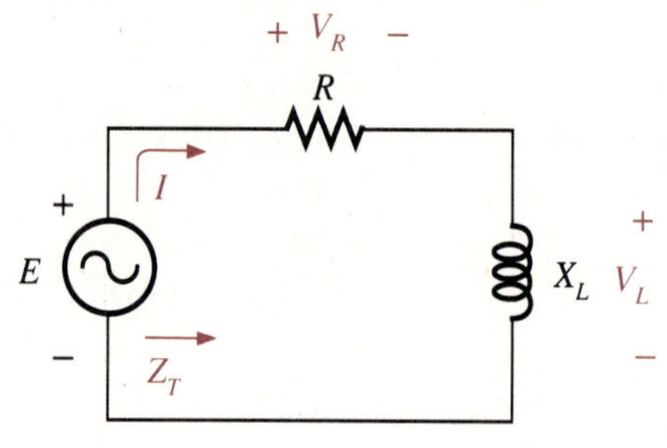

FIG. 14.30
Series R-L circuit to be analyzed for frequency response.

In the last section, we found that

$$V_L = \frac{X_L E}{Z_T} = \frac{X_L E}{\sqrt{R^2 + X_L^2}} \qquad \text{with } X_L = 2\pi f L \qquad (14.15)$$

revealing that V_L is quite sensitive to the applied frequency.

At $f = 0$ Hz, the ideal inductor will have a reactance $X_L = 2\pi f L = 2\pi(0)L = 0\ \Omega$, and the coil can assume a short-circuit equivalent on an approximate basis.

The result is that the total impedance of the network of Fig. 14.30 is

$$Z_T = R \qquad (f = 0 \text{ Hz})$$

and both **E** and **I** are in phase.

The voltage $V_L = 0$ V, since $V_L = IX_L = I(0\ \Omega) = 0$ V, or, using Eq. 14.15,

$$V_L = \frac{(0\ \Omega)E}{\sqrt{R^2 + (0\ \Omega)^2}} = 0 \text{ V}$$

In other words, at $f = 0$ Hz or low frequencies, the network appears to be highly resistive, and the phase angle between e and i is close to or equals 0°. In addition, since $V_L = 0$ V, $E = V_R$.

The frequency response for both elements of Fig. 14.30 is provided in Fig. 14.31. Note for the full frequency range that R is constant, whereas X_L in-

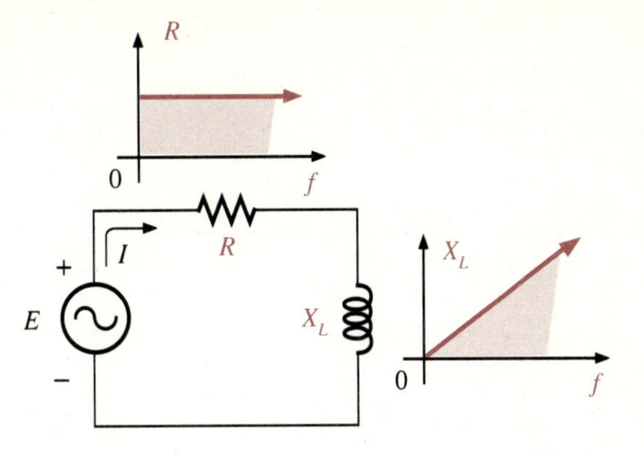

FIG. 14.31
Frequency response of each element of a series *R-L* circuit.

creases from 0 Ω with increase in frequency. Quite naturally, therefore, at low frequencies, $R > X_L$, but as the frequency increases, X_L increases; the result is that X_L will eventually be greater than R. At low frequencies the total impedance is controlled by the value of R, but at very high frequencies, X_L takes over. In the midfrequency range each contributes to the total impedance in a manner controlled by the applied frequency.

At very high frequencies X_L is much greater than R, and the coil can be replaced by an open-circuit equivalent on an approximate basis. Then

$$Z_T \cong X_L \Rightarrow \infty\ \Omega \qquad (f = \text{very high frequencies})$$

and

$$I = \frac{E}{Z_T} = 0\text{ A}$$

The result is $E = V_L$, since $V_R = IR = (0)R = 0$ V. In addition, the network is highly reactive (inductive), and e leads i by an angle equal to or close to 90°.

At low frequencies, therefore, the total impedance appears resistive, with a phase angle between E and I close to 0°, whereas at very high frequencies the impedance appears highly inductive reactive with a phase angle between e and i that approaches 90° (e leading i). Both conclusions are displayed in Fig. 14.32. The impedance diagrams appear in Fig. 14.33.

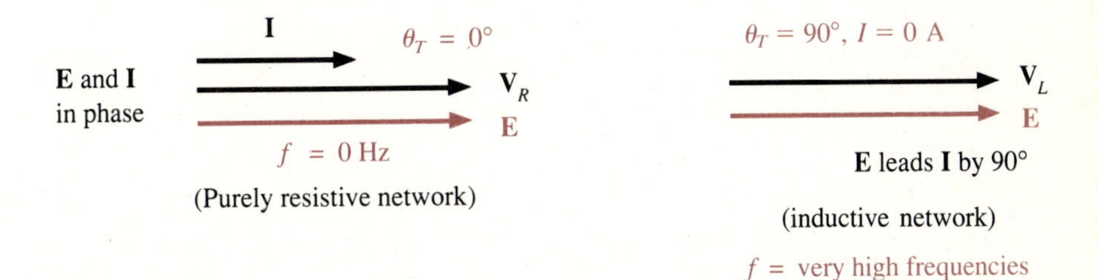

FIG. 14.32
Phasor diagrams for the circuit of Fig. 14.30 at $f = 0$ Hz and f = very high frequencies.

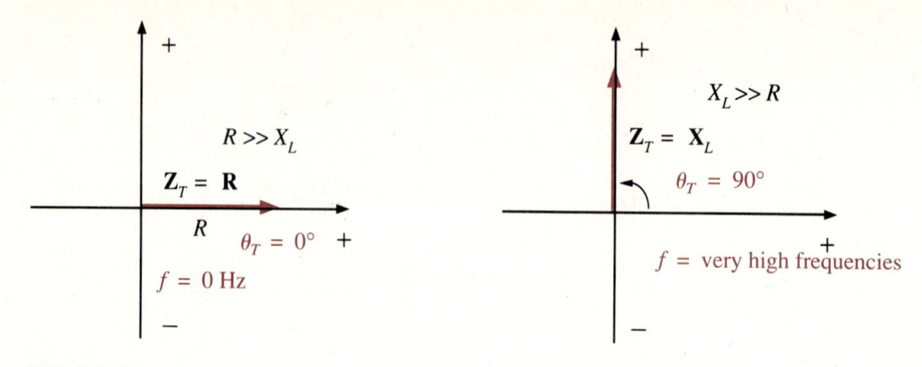

FIG. 14.33
Impedance diagrams for the circuit of Fig. 14.30 at f = 0 Hz and f = very high frequencies.

In order to investigate the effect of frequency between 0 Hz and high frequencies, the magnitude of X_L would have to be determined at each frequency in Eq. 14.15 and the value of V_L determined. The result of such an exercise for a frequency range of $f = 0$ Hz to 20 kHz for $R = 1\ \text{k}\Omega$, $L = 20$ mH, and $E = 10$ V appears in Fig. 14.34.

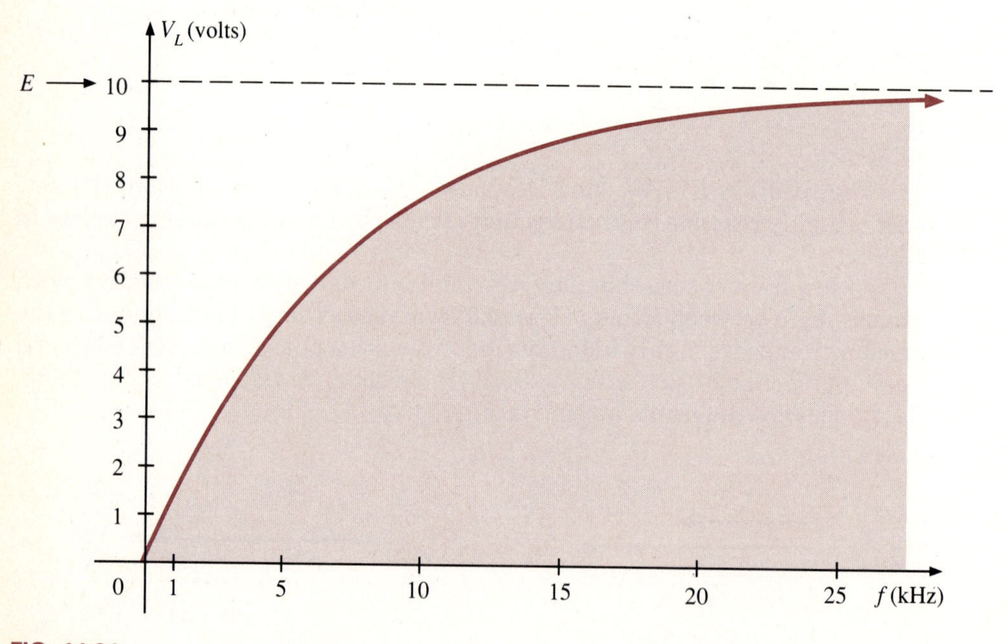

FIG. 14.34
V_L vs. frequency for the circuit of Fig. 14.30.

Note the exponential shape of the curve, revealing that the network becomes inductive very quickly as the frequency increases. Between $f = 0$ Hz and

10 kHz, V_L jumps from 0 V to 7.82 V, or 78.2% the magnitude of E. At $f =$ 10 kHz,

$$X_L = 2\pi f L = (6.28)(10 \times 10^3 \text{ Hz})(20 \times 10^{-3} \text{ H}) = 1256 \ \Omega$$

and

$$V_L = \frac{(1256 \ \Omega)(10 \text{ V})}{\sqrt{(1 \text{ k}\Omega)^2 + (1.256 \text{ k}\Omega)^2}} = 7.82 \text{ V}$$

with

$$\theta_T = \tan^{-1}\frac{X_L}{R} = \tan^{-1}\frac{1256 \ \Omega}{1000 \ \Omega} = \tan^{-1}1.256$$
$$= \mathbf{51.47^\circ} \quad (e \text{ leads } i)$$

revealing that the network is highly inductive at $f = 10$ kHz. The shape of V_R versus frequency is quite different from that of V_L, since R is not a function of frequency. For the same frequency range of Fig. 14.33, the curve of V_R in Fig. 14.35 reveals that voltage V_R drops fairly quickly with frequency. The network is obviously changing from one with resistive characteristics to one with inductive characteristics in the frequency range 0 to 25 kHz.

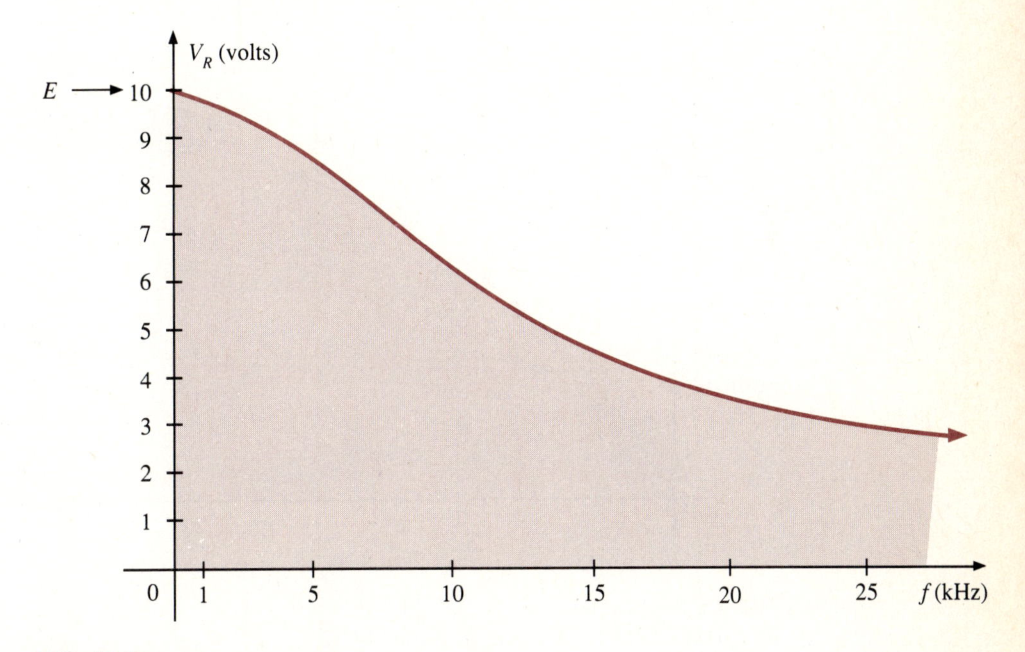

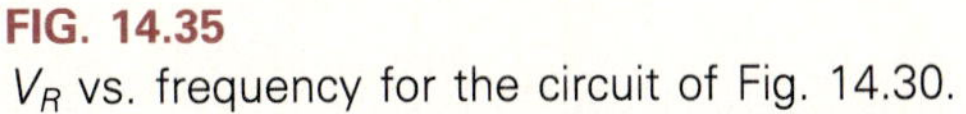
FIG. 14.35
V_R vs. frequency for the circuit of Fig. 14.30.

It is also important to note that the sum of V_L and V_R does not equal $E = 10$ V due to the vector relationship between the voltages. At $f = 5$ kHz,

$E = 10\text{ V} \neq V_R + V_L = 8.47\text{ V} + 5.32\text{ V} = 13.79\text{ V}$ and at $f = 10\text{ kHz}$, $E = 10\text{ V} \neq V_R + V_L = 1.57\text{ V} + 9.88\text{ V} = 11.45\text{ V}$.

The phasor diagram at $f = 1$ kHz and $f = 20$ kHz in Fig. 14.36 also reveals how the network is making a transition from one that is resistive to one that is primarily inductive.

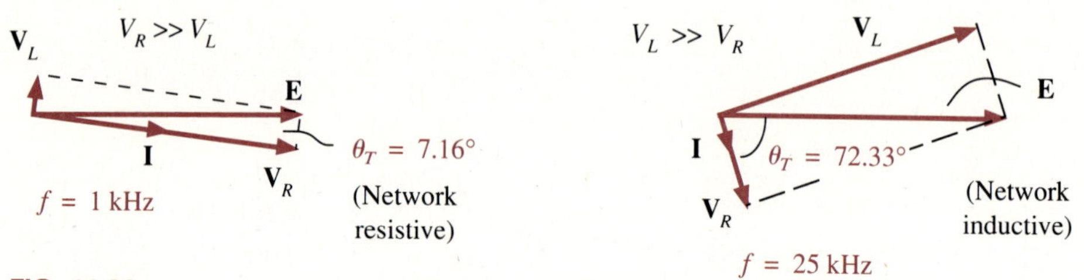

FIG. 14.36
Effect of frequency on the phasor diagram.

The angle associated with the total $\mathbf{Z}_T$ is of interest, since it reveals whether the network is predominantly resistive or inductive. A plot for the network with the parameters just introduced appears in Fig. 14.37.

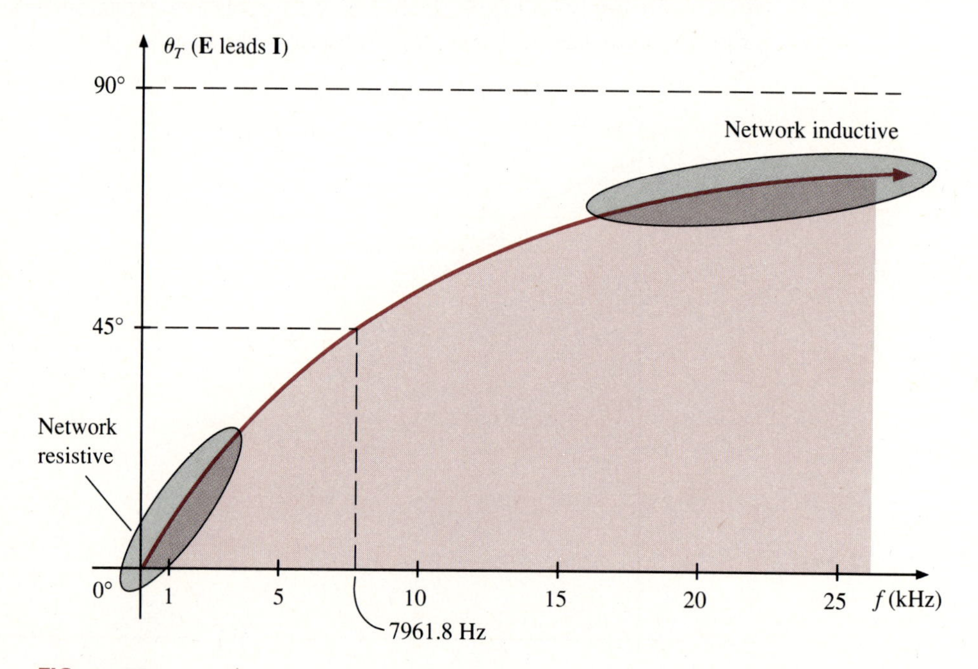

FIG. 14.37
θ_T vs. frequency for the circuit of Fig. 14.30.

For the special condition $X_L = R$,

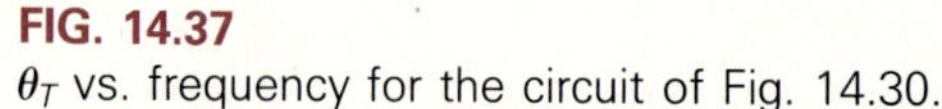

$$\theta_T = \tan^{-1}\frac{X_L}{R} = \tan^{-1}1 = 45°$$

The frequency at which this occurs can be determined from

$$X_L = R$$
$$2\pi fL = R$$

and

$$f = \frac{R}{2\pi L} = \frac{1000\ \Omega}{(6.28)(20 \times 10^{-3}\ \text{H})} = \frac{1000}{0.1256}\ \text{Hz} = \mathbf{7961.8\ Hz}$$

as shown in Fig. 14.36.

Note in Fig. 14.37 the shaded area for the frequency range in which the network is primarily resistive. The angle associated with the total impedance is in the vicinity of 0° (as required for a purely resistive network). As the angle approaches 90°, the network becomes more and more inductive, as shown in Fig. 14.37. In the midfrequency range the network is in a transition mode, with characteristics determined by the frequency of interest.

14.8 FREQUENCY RESPONSE: *R-C* CONFIGURATION

At the low- and high-frequency ends, the circuit equivalent of the capacitor of Fig. 14.38 will be opposite to that encountered for the inductor.

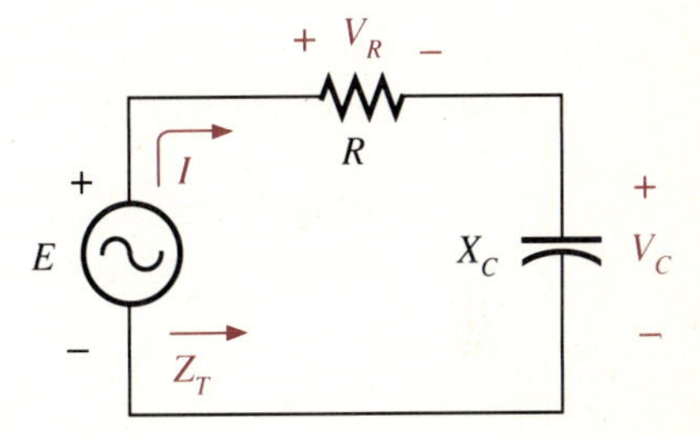

FIG. 14.38
Series *R-C* circuit to be analyzed for frequency response.

In general,

$$V_C = \frac{X_C E}{Z_T} = \frac{X_C E}{\sqrt{R^2 + X_C^2}} \qquad \text{with } X_C = \frac{1}{2\pi fC} \qquad \textbf{(14.16)}$$

clearly revealing that V_C is sensitive to the applied frequency. In fact, since X_C is inversely proportional to frequency, the numerator of Eq. 14.16 will decrease rapidly with frequency, resulting in an accelerated drop in V_C with increase in frequency.

The frequency response for both elements of Fig. 14.38 is provided in Fig. 14.39. Note for the full frequency range that R is constant, whereas X_C decreases rapidly with increase in frequency. Quite naturally, therefore, at low frequencies $X_C > R$, but as the frequency increases, X_C will decrease, with the result that R will eventually be greater than X_C. At low frequencies the total impedance is controlled by the magnitude of X_C, but at very high frequencies R

FIG. 14.39
Frequency response of each element of a series *R-C* circuit.

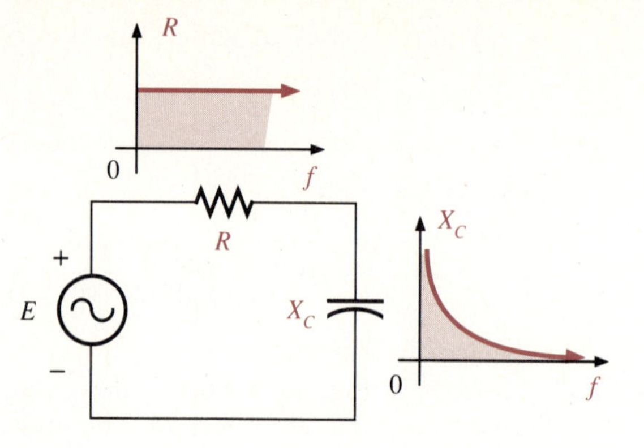

takes over. In the midfrequency range, each will contribute to the total impedance in a manner controlled by the applied frequency.

At $f = 0$ Hz, the ideal capacitor has a very large reactance $X_C = 1/2\pi fC = 1/2\pi(0)C \Rightarrow \infty\ \Omega$, and the capacitor can assume an open-circuit equivalent on an approximate basis.

The result is that all the applied voltage appears across the open-circuit equivalent, and $V_C = E$ volts (at $f = 0$ Hz) and $I = 0$ A with $V_R = IR = (0)R = 0$ V.

At low frequencies, therefore, the network is highly capacitive, since

$$Z_T = \sqrt{R^2 + X_C^2} \cong \sqrt{X_C^2} = X_C \qquad (\text{for } X_C >> R)$$

and the angle associated with $\mathbf{Z}_T$ is 90°. The result is that $i = i_R = i_C$ leads the applied voltage by 90°.

At very high frequencies $X_C = 1/2\pi fC$ approaches 0 Ω, and the short-circuit equivalent can be applied. The magnitude of the total impedance is then

$$Z_T = \sqrt{R^2 + X_C^2} = \sqrt{R^2 + (0)^2} = R$$

with $i = i_R = i_C$ in phase with the applied voltage e.

The short-circuit equivalent results in $V_C = 0$ V and $V_R = E$ volts. The phasor diagrams for the low- and high-frequency regions are provided in Fig. 14.40. The impedance diagrams appear in Fig. 14.41.

FIG. 14.40
Phasor diagrams for the circuit of Fig. 14.38 at $f = 0$ Hz and f = very high frequencies.

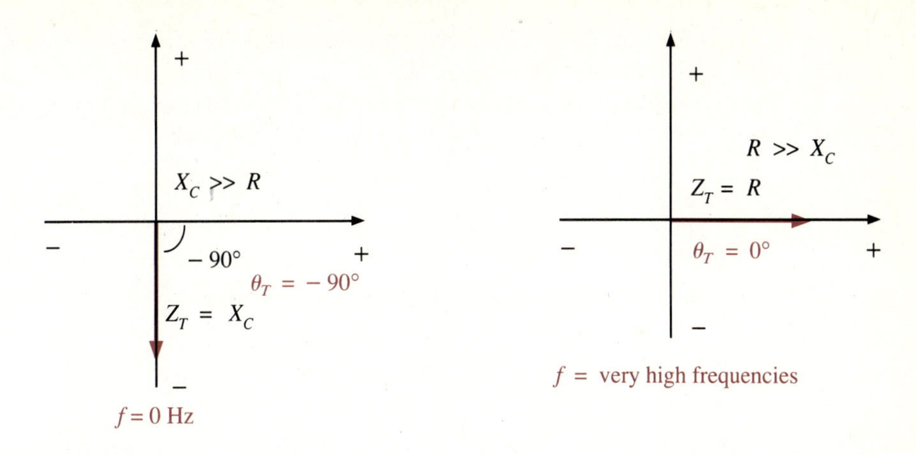

FIG. 14.41
Impedance diagrams for the circuit of Fig. 14.38 at $f = 0$ Hz and f = very high frequencies.

In general, therefore, at low frequencies the circuit of Fig. 14.38 is highly capacitive, with i leading e by an angle approaching 90°. At very high frequencies the circuit approaches resistive characteristics, with i and e leaning toward an in-phase relationship. For the region between the two frequency extremes, the voltage V_C will have to be determined using Eq. 14.16. A plot of V_C versus frequency is provided in Fig. 14.42 for $R = 5$ kΩ and $C = 0.01$ μF for a frequency range of $f = 0$ Hz to 25 kHz with $\mathbf{E} = 10\text{ V}\angle 0°$.

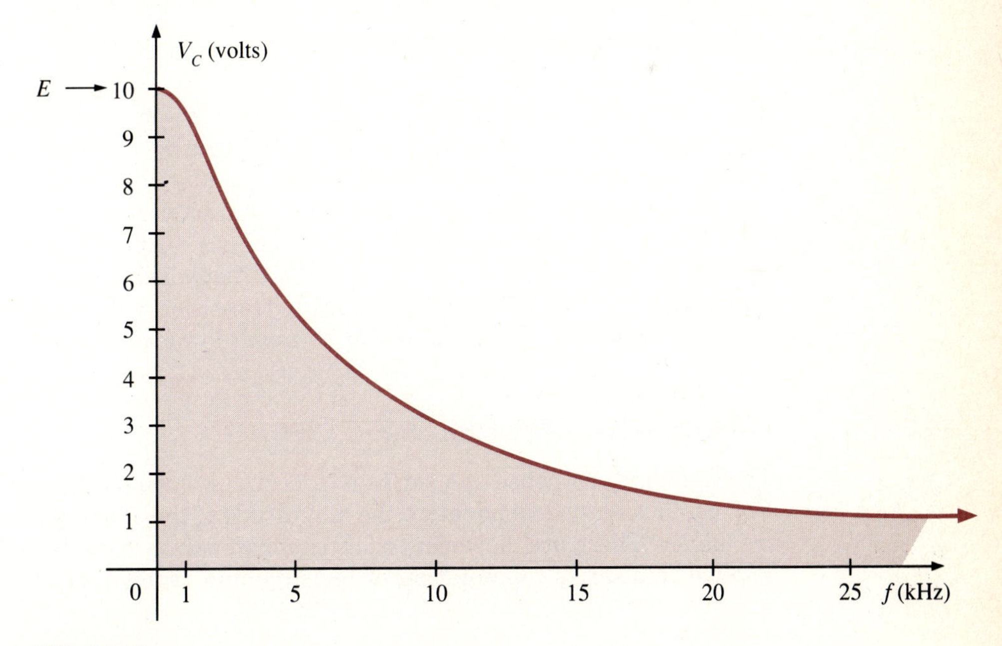

FIG. 14.42
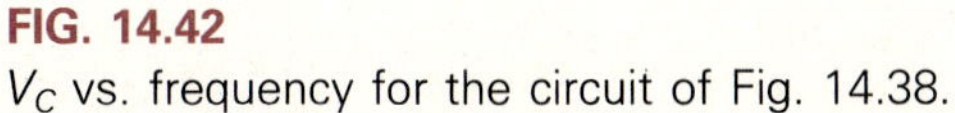
V_C vs. frequency for the circuit of Fig. 14.38.

Note that by $f = 20$ kHz, the magnitude of V_C has dropped to less than 20% of the applied voltage. A plot of V_R is provided in Fig. 14.43. In this figure note how quickly V_R rises to over 80% of E at 5 kHz, clearly revealing the rapid transition from a capacitive to a resistive network.

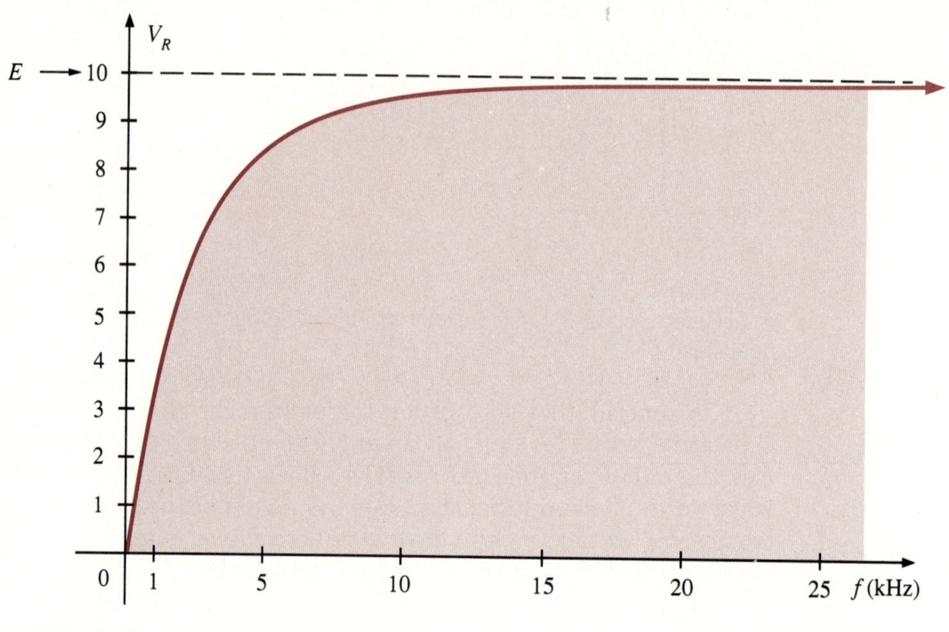

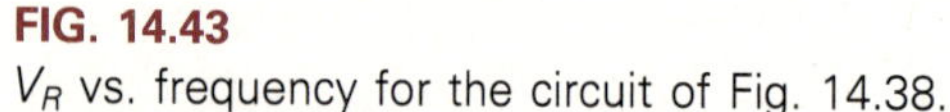
FIG. 14.43
V_R vs. frequency for the circuit of Fig. 14.38.

A plot of the angle θ_T associated with $\mathbf{Z}_T = Z_T \angle \theta_T$, representing the angle by which **I** leads **E,** is provided in Fig. 14.44. Since $\theta_T = 90°$ for a pure capacitor and $\theta_T = 0°$ for a purely resistive network, the plot clearly shows how quickly the network has changed from one input characteristic to the other.

Of course, it cannot be concluded that at $f = 20$ kHz, all R-C networks appear resistive and for $f = 1$ kHz, all R-C networks appear capacitive. The network parameters determine when the transition will occur. For some systems the transition frequencies can be much lower *or* higher.

14.9 FREQUENCY RESPONSE: *R-L-C* CIRCUIT

The frequency response of a series R-L-C circuit such as the one appearing in Fig. 14.45 is truly dependent on the magnitude of the elements and the applied frequency. There are, however, a few common parameters that will always be applied to a series combination of resistive and reactive elements.

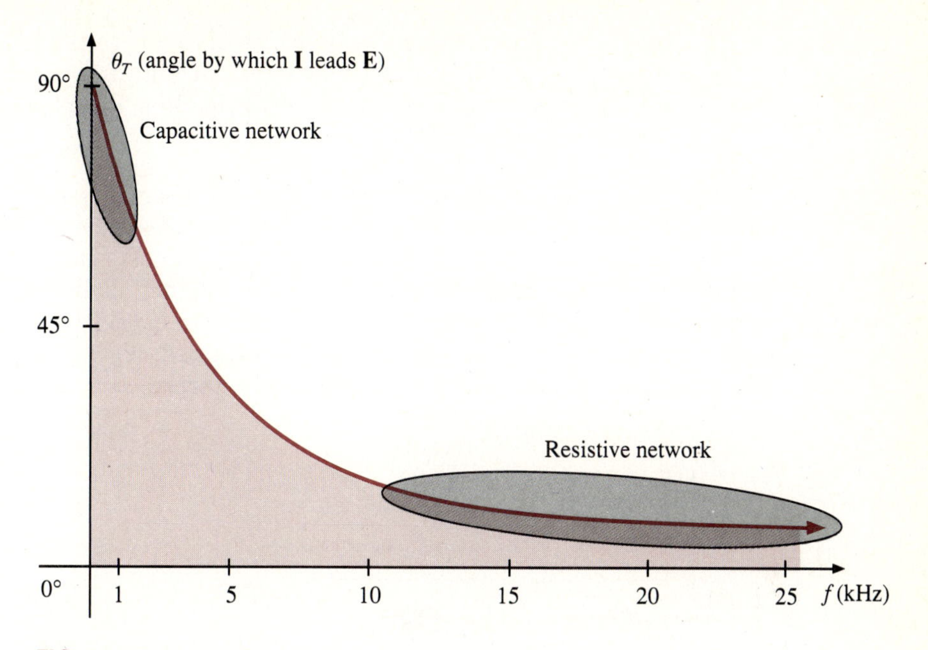

FIG. 14.44
θ_T vs. frequency for the circuit of Fig. 14.38.

FIG. 14.45
Series *R-L-C* circuit.

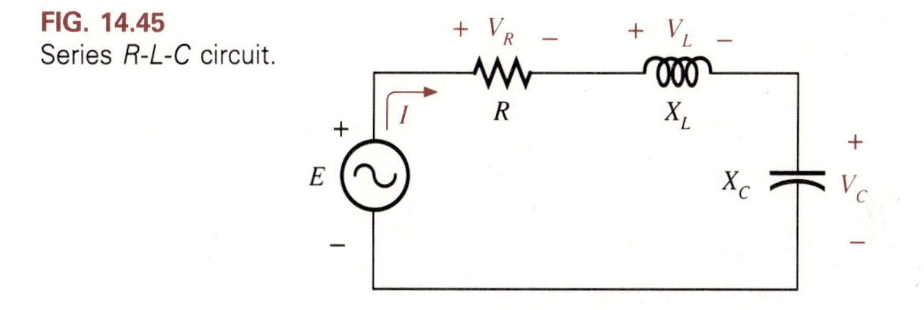

1. Assume R a constant for the frequency range of interest.
2. X_L will increase with frequency at a rate determined by the inductance L. The higher the inductance level, the more rapid the increase in X_L with frequency.
3. The reactance of a capacitor is its highest at the low end of the frequency spectrum. It also drops to low levels in an exponential manner (very rapid) rather than in a linear manner (as encountered for increasing levels of X_L).
4. At very low frequencies, any R-L-C circuit will be predominantly capacitive, whereas at very high frequencies the circuit is predominantly inductive. At some point, X_L will equal X_C, and the circuit will be purely resistive, since X_L and X_C are opposing reactance levels.

5. At very low frequencies the angle associated with the total impedance will be near 90° (i leading e), whereas at high frequencies, the angle will again approach 90° (but with e leading i). When $X_L = X_C$ the phase angle will be 0°.

The frequency response of each element of a series R-L-C circuit is provided in Fig. 14.46. At low frequencies note how X_C is larger than R or X_L, whereas at high frequencies, X_L is larger than R or X_C. In the midfrequency region the magnitude of X_L or X_C at the frequency of interest determines whether the network is predominantly inductive or capacitive. Always keep in mind that when the total impedance is determined, X_L and X_C are opposing elements and are not additive, as might be interpreted from Fig. 14.46.

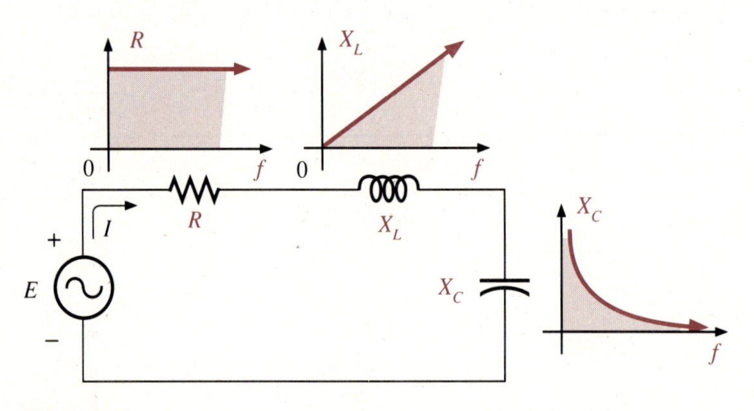

FIG. 14.46
Frequency response of each element of a series R-L-C circuit.

If both L and C are relatively large, the transition from capacitive to inductive characteristics takes place at low frequencies. If L and C are relatively small, the transition level occurs at high frequencies. The transition between capacitive and inductive characteristics occurs when $X_L = X_C$ or when

$$2\pi f L = \frac{1}{2\pi f C}$$

Solving for f yields

$$f_s = \frac{1}{2\pi\sqrt{LC}} \qquad (14.17)$$

At frequencies less than f_s the network appears capacitive. At frequencies greater than f_s the network appears inductive. At $f = f_s$ the network is resistive, since

$$Z_T = \sqrt{R^2 + (X_L - X_C)^2} = \sqrt{R^2 + (0)^2} = R$$

with

$$\theta_T \text{ of } \mathbf{Z}_T \text{ equal to } 0°$$

EXAMPLE 14.8 For the series *R-L-C* circuit of Fig. 14.44, if $R = 1\ \text{k}\Omega$, $L = 10$ mH, and $C = 0.1\ \mu\text{F}$, determine

a. The transition frequency f_s.
b. R, X_L, and X_C at $f = f_s/10$ and whether the network is capacitive, inductive, or resistive.
c. R, X_L, and X_C at $f = 10f_s$ and whether the network is capacitive, inductive, or resistive.

Solutions

a. Using Eq. 14.17:

$$f_s = \frac{1}{2\pi\sqrt{LC}} = \frac{1}{6.28\sqrt{(10 \times 10^{-3}\ \text{H})(0.1 \times 10^{-6}\ \text{F})}}$$

$$f_s = \mathbf{5035.5\ Hz}$$

Calculator

(1) (0) (EXP) (+/−) (3) (×) (·) (1) (EXP) (+/−) (6) (=) (√‾) (×) (6) (·) (2) (8) (=) (1/x)

with a display of 5035.474.

b. At $f = f_s/10 = 503.55$ Hz,

$$R = \mathbf{1\ k\Omega}$$

$$X_L = 2\pi fL = (6.28)(503.55\ \text{Hz})(10 \times 10^{-3}\ \text{H}) = \mathbf{31.62\ \Omega}$$

$$X_C = \frac{1}{2\pi fC} = \frac{1}{(6.28)(503.55\ \text{Hz})(0.1 \times 10^{-6}\ \text{F})} = \mathbf{3162.3\ \Omega}$$

Obviously, $X_C > X_L$ and $X_C > R$, resulting in a *capacitive* network.

c. At $f = 10f_s = 50{,}355$ Hz,

$$R = \mathbf{1\ k\Omega}$$

$$X_L = 2\pi fL = (6.28)(50{,}355\ \text{Hz})(10 \times 10^{-3}\ \text{H}) = \mathbf{3162.3\ \Omega}$$

$$X_C = \frac{1}{2\pi fC} = \frac{1}{(6.28)(50{,}355\ \text{Hz})(0.1 \times 10^{-6}\ \text{F})} = \mathbf{31.62\ \Omega}$$

reversing the magnitude of X_L and X_C. Since $X_L > X_C$ and $X_L > R$, the network is **inductive**.

If a frequency response curve for V_C or V_L is derived, the magnitude of each is determined by the following voltage divider rule equations:

$$V_C = \frac{X_C E}{Z_T} = \frac{X_C E}{\sqrt{R^2 + (X_L - X_C)^2}} \quad (14.18)$$

$$V_L = \frac{X_L E}{Z_T} = \frac{X_L E}{\sqrt{R^2 + (X_L - X_C)^2}} \quad (14.19)$$

A great deal more about series R-L-C configurations is introduced in Chapter 18, where series resonant circuits are examined. For the moment, the intent is simply to develop a sense for the general characteristics for a wide range of frequencies.

14.10 MEASUREMENTS—SERIES ac CIRCUITS

The majority of the measurements made on any sinusoidal ac network are made with a digital (or analog) meter or an oscilloscope. In addition, since very few meters are designed to read the peak or peak-to-peak value of a quantity, it is safe to say that

Digital and analog ac meters indicate the *effective* value of the sinusoidal voltage or current.

The meter connections are the same as those employed in the analysis of dc circuits. In addition,

Unless dealing with very low or high frequencies, it is usually unnecessary to be concerned about the effect of the applied frequency on the meter reading.

The meter connections for the voltages of a series R-L-C circuit are provided in Fig. 14.47. In addition, the connections for a *frequency counter* are also included to monitor the applied frequency. Note that the counter is simply connected directly across the supply. The connections for an ac ammeter are the same as employed for dc circuits, but the ammeter was not included because in the majority of situations for ac networks or networks that should not be disturbed, the current is determined insofar as possible by an application of Ohm's law to the resistive element. That is,

$$I_{\text{rms}} = \frac{V_{R_{\text{rms}}}}{R}$$

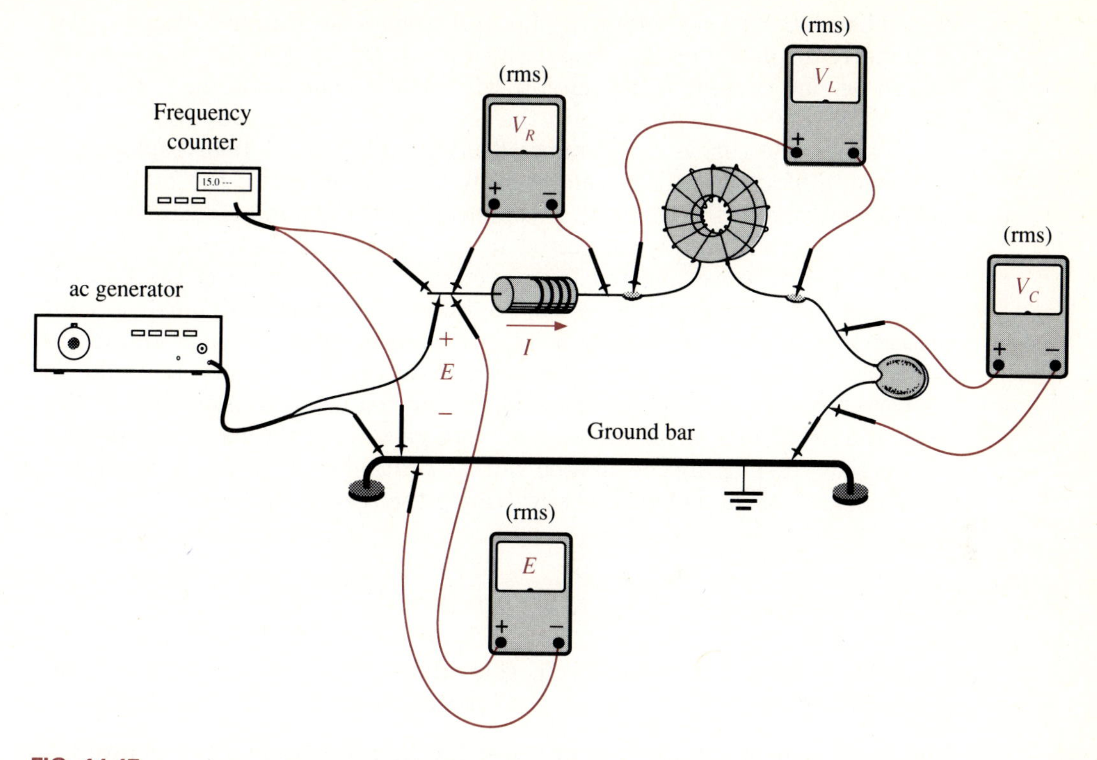

FIG. 14.47
Basic meter connections for a series *R-L-C* circuit.

The use of an oscilloscope to measure or view a voltage of a series R-L-C circuit has to be treated with more care than necessary for meters because one side of the vertical input of the scope may be grounded through the power cord. The effect of this internal grounding is best described by the series R-L circuit of Fig. 14.48. The supply and one end of the 10-kΩ inductive reactance are grounded. If the oscilloscope is hooked up as shown, there will be a common

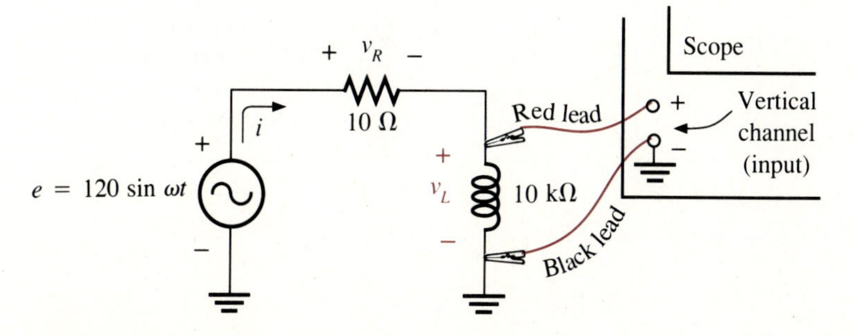

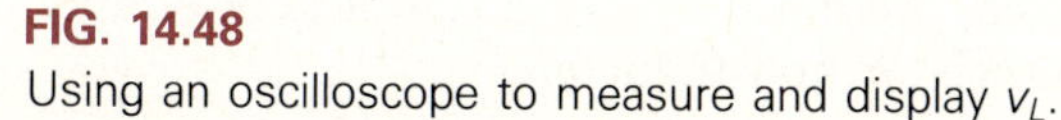
FIG. 14.48
Using an oscilloscope to measure and display v_L.

ground between the oscillator (supply), coil, and scope, and the voltage v_L will be properly displayed on the scope. Since $X_L >> R_L$ will be very close to $e =$ 120 sin ωt, the current in the circuit will have a peak value $I_m \cong 120$ V/10 kΩ = 12 mA, which is well within the limits of the circuit design.

If the oscilloscope is now connected as shown in Fig. 14.49 to display v_R on the scope, difficulties will develop due to the placement of the oscilloscope ground. One end of the inductor is connected to the ground of the generator, whereas the other end is connected to the ground of the scope. Since the grounds of each are connected to the common earth ground of the 60-cycle supply, all the grounds of the system are internally connected. Since a ground connection has a 0-V reference level, a 0-V level exists at each end of the inductor, resulting in a 0-V drop across the inductor. In actuality, the inductor has been "shorted out" by the internal ground connections. As shown in Fig. 14.49, the result is that the full applied voltage will appear across the resistor and the scope displaying the applied voltage e rather than v_R. More important, the impedance of the network is now simply the resistance R and the current

$$I_m = \frac{E_m}{R} = \frac{120 \text{ V}}{10 \ \Omega} = 12 \text{ A}$$

a level that can result in dangerous side effects if the protective fuses of the generator and scope fail to protect the instrumentation. In general, therefore,

When using an oscilloscope, pay particular attention to the ground connections of the network.

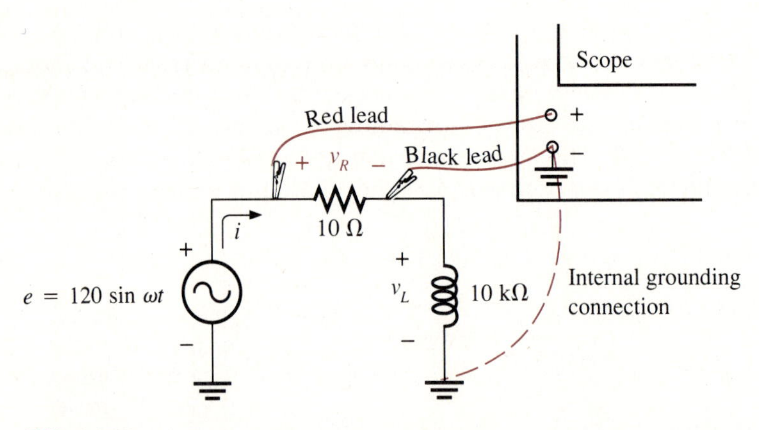

FIG. 14.49
Improper use of an oscilloscope to measure v_R.

If v_R were desired, the most direct route would be simply to interchange the positions of the resistor and inductor.

14.11 THE j-OPERATOR: MULTIPLICATION AND DIVISION

The last section of the previous chapter introduced a method for adding and subtracting vectors that included the use of the *j*-operator. The result was a technique for adding and subtracting sinusoidal voltages and currents using a rectangular format that simplified the process and reduced the possibility of errors when the desired operation is performed.

We now examine a procedure using polar form that can provide the product or division of two vectors that may be representing a voltage, current, or impedance of a network. It permits the application of Ohm's law and the voltage divider rule, which provides both the magnitude and angle of a quantity without resorting to some of the general rules appearing in the last few sections.

Division The division of two vectors is best performed in *polar form.* In fact, in this text all divisions are in the polar form.

The operation is performed in the following manner:

$$\frac{\mathbf{C}_1}{\mathbf{C}_2} = \frac{C_1 \angle \theta_1}{C_2 \angle \theta_2} = \frac{C_1}{C_2} \angle \theta_1 - \theta_2 \qquad (14.20)$$

Note that the magnitude of the resultant is simply the division of the two magnitudes and the angle of the resultant is the angle of the numerator less the angle of the denominator.

The preceding equation lends itself directly to Ohm's law, as demonstrated in the following examples.

EXAMPLE 14.9 Determine the current through a 10-Ω resistor if the voltage across the resistor is $\mathbf{V}_R = 40 \angle 30^\circ$.

Solution

$$\mathbf{I} = \frac{\mathbf{V}_R}{\mathbf{R}} = \frac{40 \text{ V} \angle 30^\circ}{10 \ \Omega \angle 0^\circ} = 4 \text{ A} \angle 30^\circ - 0^\circ$$

$$= \mathbf{4 \ A \angle 30^\circ}$$

The operation included the fact that v_R and i_R are in phase by simply associating 0° with **R** and using Eq. 14.20. In Example 14.9 both **I** and $\mathbf{V}_R$ have the same phase angle of 30°; it is not necessary to recall that v_R and i_R are in phase.

EXAMPLE 14.10 Determine the current through a 2-Ω inductive reactance if the voltage across the coil is $v_L = \sqrt{2}(20)\sin(\omega t + 60^\circ)$.

Solution

$$\mathbf{I}_L = \frac{\mathbf{V}_L}{\mathbf{X}_L} = \frac{20 \text{ V} \angle 60^\circ}{2 \ \Omega \angle 90^\circ}$$

Applying Eq. 14.18,

$$\mathbf{I}_L = 10 \text{ A } \angle 60^\circ - 90^\circ = \mathbf{10 \text{ A } \angle -30^\circ}$$

and we find that v_L at 60° does in fact lead i_L at −30° by 90°. Note again that the phase angle associated with i_L was determined by Eq. 14.18, as was the fact that 90° was associated with X_L. There was no need to remember that v_L leads i_L by 90° for a purely inductive element to find the angle of −30°.

EXAMPLE 14.11 Determine the current through a circuit with an impedance of $6\ \Omega \angle -53.13^\circ$ if the applied voltage is $120 \text{ V } \angle 0^\circ$.

Solution Using Ohm's law:

$$\mathbf{I} = \frac{\mathbf{E}}{\mathbf{Z}_T} = \frac{120 \text{ V } \angle 0^\circ}{6\ \Omega \angle -53.13^\circ} = \frac{120 \text{ V}}{6\ \Omega} \angle 0^\circ - (-53.13^\circ)$$

and

$$\mathbf{I} = \mathbf{20 \text{ A } \angle 53.13^\circ}$$

In the past we had to recall that the angle associated with $\mathbf{Z}_T$ was the angle by which the applied voltage led the resulting current. In this case the network is capacitive and **I** leads **E** by 53.13°.

Multiplication The product of two vectors is also best performed in polar form.

The operation is performed in the following manner:

$$\mathbf{C}_1\mathbf{C}_2 = (C_1 \angle \theta_1)(C_2 \angle \theta_2) = C_1C_2 \angle \theta_1 + \theta_2 \qquad (14.21)$$

Note in this case that the magnitude of the resultant is the product of the magnitudes of the two vectors, with the angle being the sum of the angles associated with each vector.

The best use of Eq. 14.21 at this juncture is in the application of Ohm's law to determine the voltage across an element, as demonstrated in the next few examples.

EXAMPLE 14.12 Determine the voltage across a capacitor of 1.5 kΩ if the current is $0.6 \text{ mA } \angle 0^\circ$.

Solution Applying Ohm's law:

$$\mathbf{V}_C = \mathbf{I}\mathbf{X}_C = (0.6 \text{ mA } \angle 0^\circ)(1.6 \text{ k}\Omega \angle -90^\circ)$$

From Eq. 14.21,

$$\begin{aligned}\mathbf{V}_C &= (0.6 \text{ mA})(1.6 \text{ k}\Omega) \angle 0^\circ - 90^\circ \\ &= \mathbf{0.96 \text{ V } \angle -90^\circ}\end{aligned}$$

The result reveals that $\mathbf{V}_C$ lags $\mathbf{I}_C$ by 90°, as it must for a capacitive element.

EXAMPLE 14.13 Determine the applied voltage necessary to establish a current $\mathbf{I} = 3\text{ A }\angle 60^\circ$ through a network with an impedance $\mathbf{Z}_T = 12\ \Omega\ \angle -45^\circ$.

Solution

$$\begin{aligned}\mathbf{E} &= \mathbf{IZ}_T \\ &= (3\text{ A }\angle 60^\circ)(12\ \Omega\ \angle -45^\circ) \\ &= 36\text{ V }\angle 60^\circ - 45^\circ \\ \mathbf{E} &= \mathbf{36\text{ V }\angle 15^\circ}\end{aligned}$$

There was again no need to remember that the angle associated with $\mathbf{Z}_T$ is the angle by which **E** leads **I.** Due to the minus sign, **I** leads **E** by 45° for this network, resulting in **E** having an angle of 15° with **I** at an angle of 60°. The whole process of finding the angle was included in the application of Eq. 14.21.

Combined Operations There are occasions such as the voltage divider rule, where both a product and division of vectors must be found to determine the unknown quantity. It is then necessary only to proceed in the same manner as if the vectors were simply real (no angle) numbers. For instance, determine the magnitude and angle of the voltage $\mathbf{V}_L$ for Example 14.6 of the previous section using the *j*-operator.

Referring to Fig. 14.28,

$$\mathbf{V}_L = \frac{\mathbf{X}_L\mathbf{E}}{\mathbf{R} + \mathbf{X}_L}$$

Note in this case that vectors—rather than just their magnitudes—are employed for all quantities.

$$\mathbf{V}_L = \frac{(4\text{ k}\Omega\ \angle 90^\circ)(120\text{ V }\angle 0^\circ)}{2.2\text{ k}\Omega\ \angle 0^\circ + 4\text{ k}\Omega\ \angle 90^\circ}$$

The numerator is a product operation:

$$\begin{aligned}(4 \times 10^3\ \angle 90^\circ)(120\ \angle 0^\circ) &= (4 \times 10^3)(120)\ \angle 90^\circ + 0^\circ \\ &= 480 \times 10^3\ \angle 90^\circ\end{aligned}$$

The denominator is a sum operation:

$$\begin{aligned}2.2\text{ k}\Omega\ \angle 0^\circ + 4\text{ k}\Omega\ \angle 90^\circ &= 2.2\text{ k}\Omega + j4\text{ k}\Omega \\ &= 4.565 \times 10^3\ \angle 61.19^\circ\end{aligned}$$

This results in

$$\mathbf{V}_L = \frac{480 \times 10^3\text{ V }\angle 90^\circ}{4.565 \times 10^3\ \angle 61.19^\circ}$$

Performing a division operation:

$$\mathbf{V}_L = \frac{480 \times 10^3\ \text{V}}{4.565 \times 10^3} \underline{/90^\circ - 61.19^\circ}$$
$$= 105.15\ \text{V}\ \underline{/28.81^\circ}$$

The magnitude of $\mathbf{V}_L$ matches the solution of Example 14.6, but now we have the angle associated with $\mathbf{V}_L$ also.

The question now is whether all the additional mathematics is worth knowing the angle associated with $\mathbf{V}_L$. How often will we have to calculate the angle associated with $\mathbf{V}_L$ in the real world? The mathematics of Example 14.6 is certainly simpler than the preceding work if only the meter reading of $\mathbf{V}_L$ is required. However, if the angle is required, a lengthy analysis—including first finding **I**—is required, using the procedure of earlier sections. In total, the question of which method to use is determined primarily by the situation. If only the magnitude is required, either method can be used, but the straightforward substitution of Example 14.6 is quicker. If the phase angle is required, then the *j*-operator approach is the more direct avenue.

All in all, there is nothing wrong with knowing both methods. Such knowledge provides options and options can often be the key to success.

As a final example, consider analyzing the series *R-L* network of Fig. 14.13 using the *j*-operator.

$$\mathbf{Z}_T\text{:}\quad \mathbf{Z}_T = \mathbf{R} + \mathbf{X}_L$$
$$= 3\ \Omega + j4\ \Omega$$
$$\mathbf{Z}_T = \mathbf{5\ \Omega\ \underline{/53.13^\circ}}$$
$$\mathbf{I}\text{:}\quad \mathbf{I} = \frac{\mathbf{E}}{\mathbf{Z}_T} = \frac{7.07\ \text{V}\ \underline{/0^\circ}}{5\ \Omega\ \underline{/53.13^\circ}} = \mathbf{1.414\ A\ \underline{/-53.13^\circ}}$$
$$\mathbf{V}_R\text{:}\quad \mathbf{V}_R = \mathbf{IR} = (1.414\ \text{A}\ \underline{/-53.13^\circ})(3\ \Omega\ \underline{/0^\circ})$$
$$= \mathbf{4.242\ V\ \underline{/-53.13^\circ}}$$
$$\mathbf{V}_L\text{:}\quad \mathbf{V}_L = \mathbf{IX}_L = (1.414\ \text{A}\ \underline{/-53.13^\circ})(4\ \Omega \underline{/90^\circ})$$
$$= \mathbf{5.656\ V\ \underline{/36.87^\circ}}$$

There is no question that the preceding analysis appears somewhat "cleaner" than the approach applied in Section 14.3. However, mathematical approaches can often eliminate the need to remember certain fundamental facts about the network behavior that can often be useful. For instance, to remember that θ_T of $\mathbf{Z}_T$ is the angle by which e leads i can quickly reveal whether the network is capacitive or inductive and to what degree.

FORMULA SUMMARY

$$\mathbf{R} = R\ \underline{/0^\circ}, \qquad \mathbf{X}_L = X_L\ \underline{/90^\circ}, \qquad \mathbf{X}_C = X_C\ \underline{/-90^\circ}$$
$$\mathbf{Z}_T = \mathbf{Z}_1 + \mathbf{Z}_2 + \mathbf{Z}_3$$

$$\theta_T = \tan^{-1}\frac{X}{R}$$

$$V_R = \frac{RE}{\sqrt{R^2 + X^2}}, \qquad V_X = \frac{XE}{\sqrt{R^2 + X^2}}$$

$$\frac{\mathbf{C}_1}{\mathbf{C}_2} = \frac{C_1}{C_2}\,\underline{/\theta_1 - \theta_2}$$

$$\mathbf{C}_1\mathbf{C}_2 = C_1C_2\,\underline{/\theta_1 + \theta_2}$$

CHAPTER SUMMARY

The similarities between the analysis of dc and ac circuits should now be fairly obvious from the content of this chapter. The major difference is the frequent use of the Pythagorean theorem to find the proper magnitude of a quantity. Ohm's law has the same format to determine the current and voltage across each element, and the power dissipated is determined by a similar equation that employs effective values. The current is the same for series ac elements, and the total impedance is the sum of the individual impedances.

Impedance and phasor diagrams should be utilized not only to display the obtained quantities but to verify that the analysis is correct. A rough sketch of the impedance of an network will quickly reveal an approximate magnitude of the total impedance and whether the network is resistive, inductive, or capacitive. The phasor diagram will help in much the same way, clearly revealing whether the applied voltage is the vector sum of the individual voltages and whether the proper phase relationship exists between the voltages and current.

For all ac systems the voltages and currents are measured using the same connections employed for dc circuits. Voltmeters are in parallel with the load and do not require disturbing the network, whereas ammeters must be placed in series with the load, requiring that the circuit be opened at some point. Unless otherwise specified, all meters provide the effective values of the measured quantities.

For the majority of situations, it is the magnitude of a voltage that is of primary importance, with the associated angle secondary. The voltage divider rule in its magnitude form is therefore a useful equation to determine quickly a voltage level to match the measured level. If the angle must be determined, the current can first be determined; this is followed by an application of Ohm's law. If the *j*-operator approach is available, the magnitude and angle of the desired quantity can be determined at the same time.

At one frequency the magnitude of X_L and X_C is fixed, and the current and voltage levels of a series circuit have a fixed magnitude and angle. As the frequency changes, X_L and X_C will change dramatically from the low- to high-frequency region. In general, keep in mind that R remains fairly constant, X_L increases linearly with frequency, and X_C decreases dramatically with increase in frequency. For series *R-L-C* configurations, X_C is the predominant factor at low frequencies and X_L is predominant at high frequencies. When $X_C = X_L$ the network impedance is simply the resistance R, and the applied voltage and current are in phase. Since $X_C \gg X_L$ and $X_C \gg R$ at low frequencies, the network is primarily capacitive, and the source current leads the applied voltage by about 90°. But since $X_L \gg X_C$ and $X_L \gg R$ at high frequencies, the applied voltage will lead the source current by about 90°. In other words, a change in phase relationship of about 180° occurs between the high and low end of the frequency spectrum with the crossover from a leading to lagging power factor occurring at $X_L = X_C$.

GLOSSARY

Impedance diagram A vector display that clearly depicts the magnitude of the impedance of the resistive, reactive, and capacitive components of a network and the magnitude and angle of the total impedance of the system.

Phasor diagram A vector display that provides at a glance the magnitude and phase relationships among the various voltages and currents of a network.

Series ac configuration A connection of elements in an ac network in which no two impedances have more than one terminal in common and the current is the same through each element.

Voltage divider rule A method by which the voltage across one element of a series of elements in an ac network can be determined without first having to find the current through the elements.

PROBLEMS

Section 14.2

1. Determine the impedance in vector form and draw the impedance diagram for each of the impedances of Fig. 14.50.

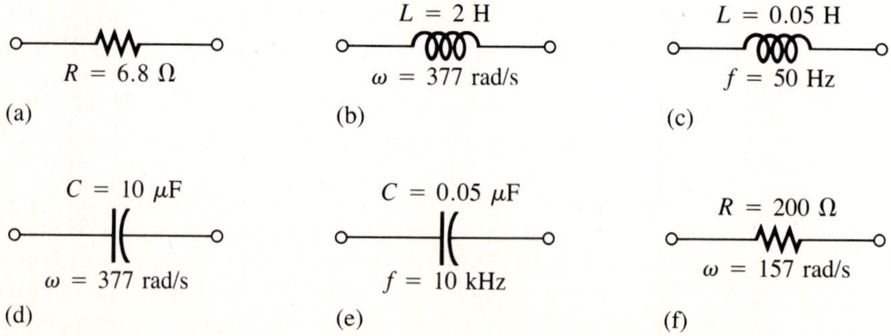

(a) $R = 6.8\ \Omega$

(b) $L = 2$ H, $\omega = 377$ rad/s

(c) $L = 0.05$ H, $f = 50$ Hz

(d) $C = 10\ \mu$F, $\omega = 377$ rad/s

(e) $C = 0.05\ \mu$F, $f = 10$ kHz

(f) $R = 200\ \Omega$, $\omega = 157$ rad/s

FIG. 14.50

2. Calculate the total impedance of the circuits of Fig. 14.51 and draw the impedance diagram.

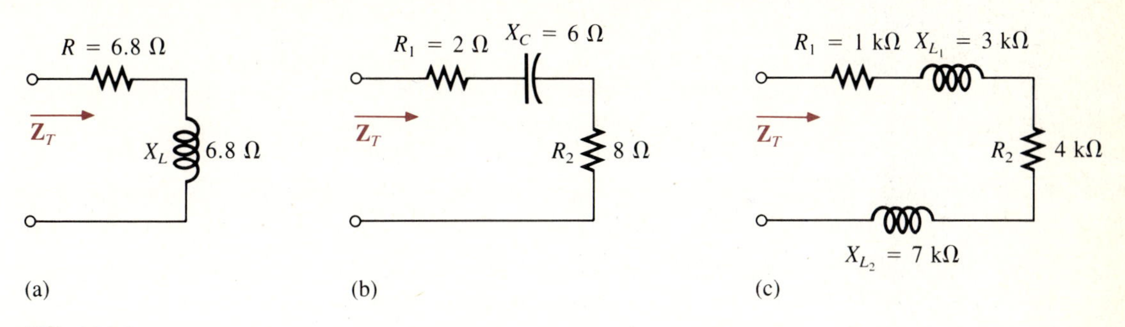

FIG. 14.51

***3.** Repeat Problem 2 for the circuits of Fig. 14.52.

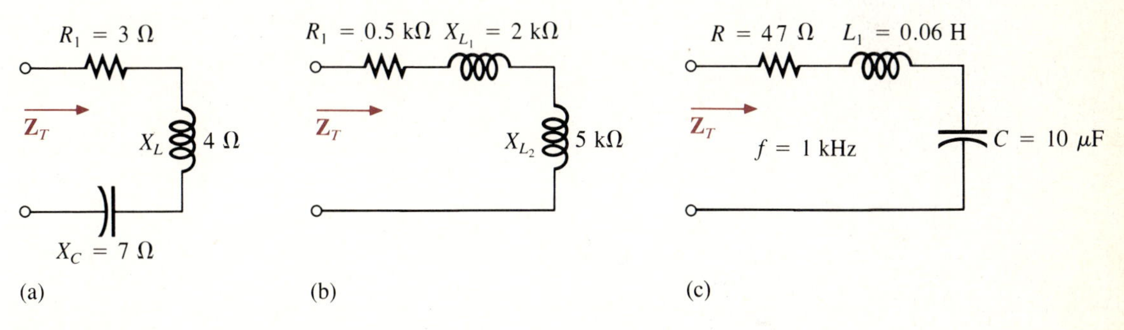

FIG. 14.52

***4.** Find the type and impedance in ohms of the series circuit elements that must be in the closed container of Fig. 14.53 in order for the indicated voltages and currents to exist at the input terminals. (Find the simplest series circuit that will satisfy the indicated conditions.)

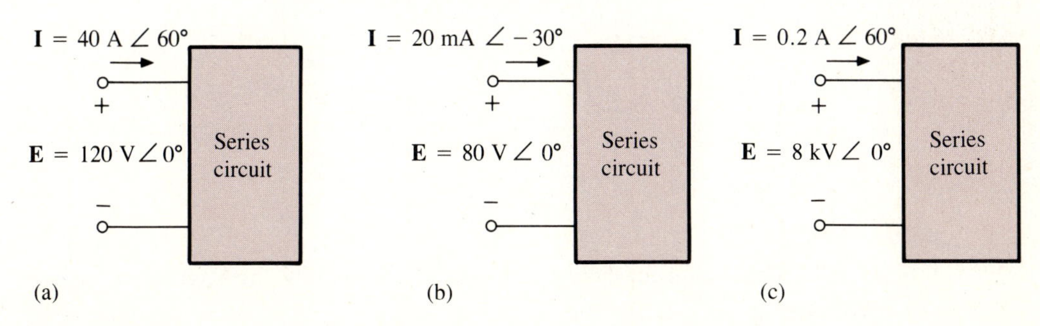

FIG. 14.53

Section 14.3

5. For the circuit of Fig. 14.54:
 a. Find the total impedance $\mathbf{Z}_T$.
 b. Draw the impedance diagram.
 c. Find the current **I** and the voltages $\mathbf{V}_R$ and $\mathbf{V}_L$ in phasor form.
 d. Draw the phasor diagram of the voltages **E**, $\mathbf{V}_R$, and $\mathbf{V}_L$ and the current **I**.
 e. Find the average power delivered to the circuit.
 f. Find the power factor of the circuit and indicate whether it is leading or lagging.
 g. Find the sinusoidal expressions for the voltage and current if the frequency is 60 Hz.
 h. Plot the waveforms for the voltages and current on the same set of axes.

6. Repeat Problem 5 for the network of Fig. 14.55.

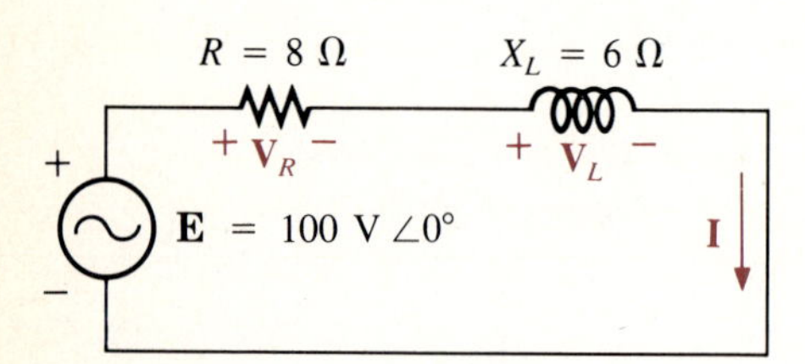

FIG. 14.54

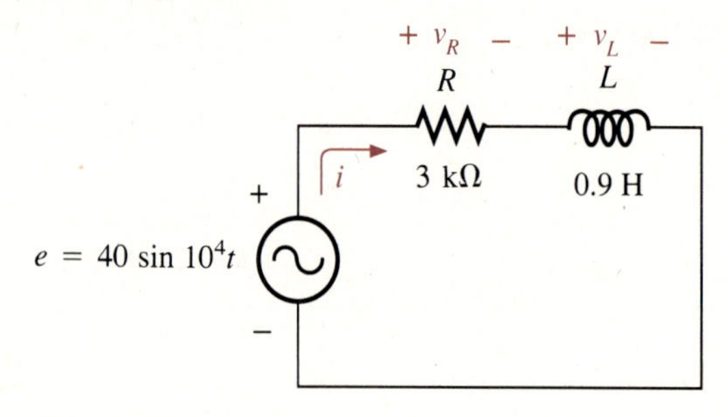

FIG. 14.55

***7.** For the circuit of Fig. 14.56:
 a. Find the total impedance $\mathbf{Z}_T$.
 b. Draw the impedance diagram.
 c. Find the current **I**.
 d. Determine the voltage $\mathbf{V}_{R_2}$.
 e. Find the average power delivered to the network.
 f. Find the sinusoidal form of i and v_{R_2}.

FIG. 14.56

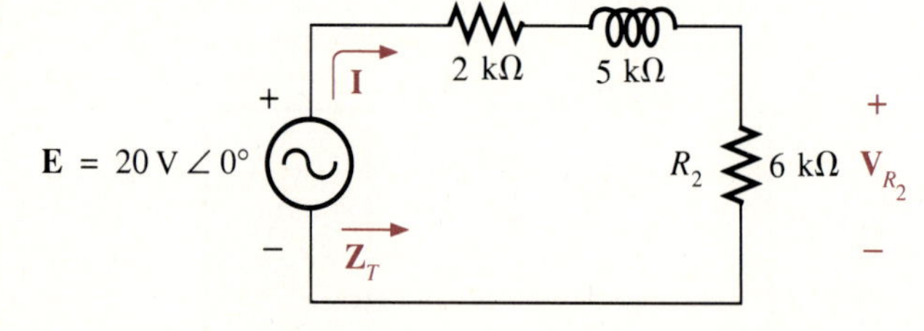

Section 14.4

8. Repeat Problem 5 for the circuit of Fig. 14.57, replacing $\mathbf{V}_L$ by $\mathbf{V}_C$ in parts c and d.

9. Given the network of Fig. 14.58:
 a. Determine $\mathbf{Z}_T$.
 b. Find **I**.
 c. Calculate $\mathbf{V}_R$ and $\mathbf{V}_L$.
 d. Find P and F_p.

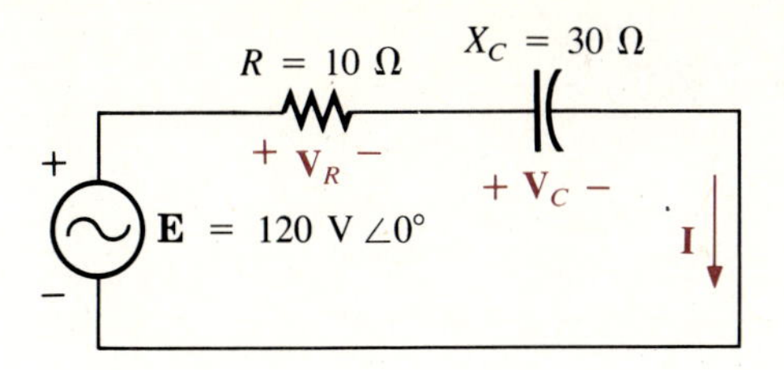

FIG. 14.57

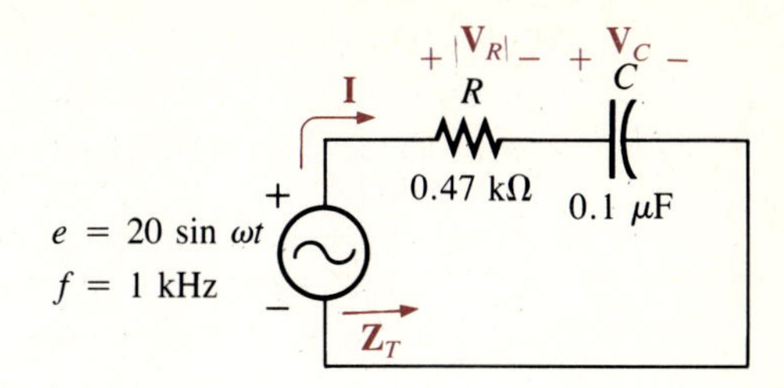

FIG. 14.58

***10.** For the circuit of Fig. 14.59:

a. Determine $\mathbf{Z}_T$.

FIG. 14.59

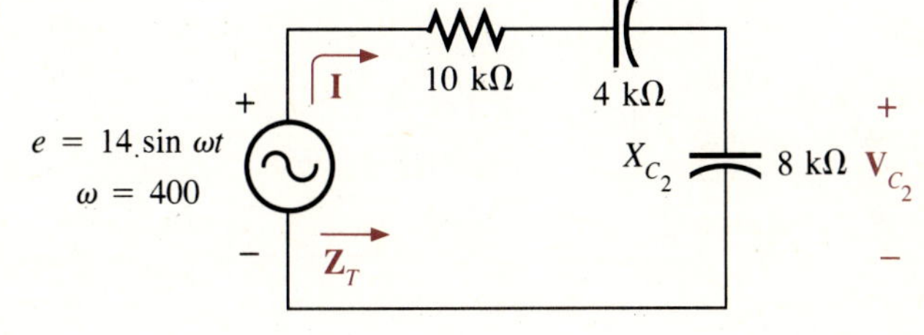

b. Find **I**.

c. Sketch the impedance diagram.

d. Calculate $\mathbf{V}_{C_2}$.

e. Find P and F_p.

Section 14.5

11. For the circuit of Fig. 14.60:

a. Find the total impedance $\mathbf{Z}_T$ in polar form.

b. Draw the impedance diagram.

c. Find the value of C in microfarads and **L** in henries.

d. Find the current i and the voltages v_R, v_L, and v_C in phasor form.

e. Draw the phasor diagram of the voltages **E**, $\mathbf{V}_R$, $\mathbf{V}_L$, and $\mathbf{V}_C$ and the current **I**.

f. Find the average power delivered to the circuit.

g. Find the power factor of the circuit and indicate whether it is leading or lagging.

h. Find the sinusoidal expressions for the voltages and current.

i. Plot the waveforms for the voltages and current on the same set of axes.

FIG. 14.60

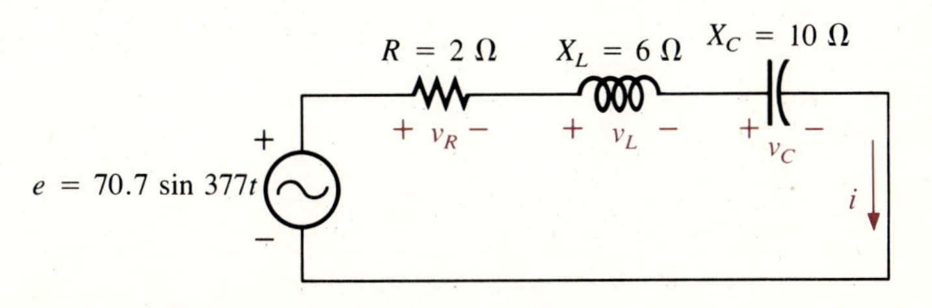

12. Repeat Problem 11 for the circuit of Fig. 14.61.

FIG. 14.61

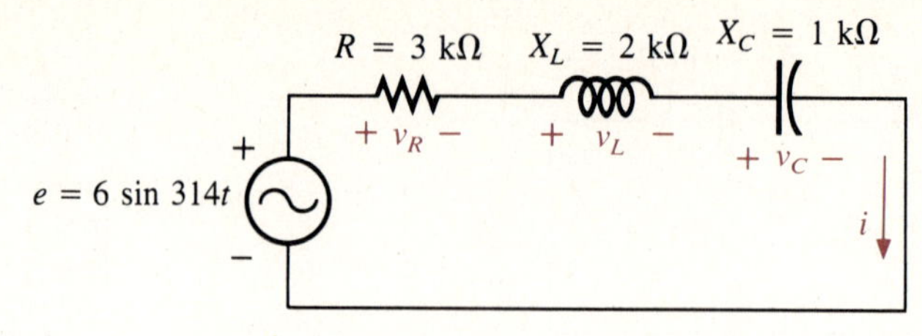

***13.** Repeat Problem 11 for the circuit of Fig. 14.62. Omit part C.

FIG. 14.62

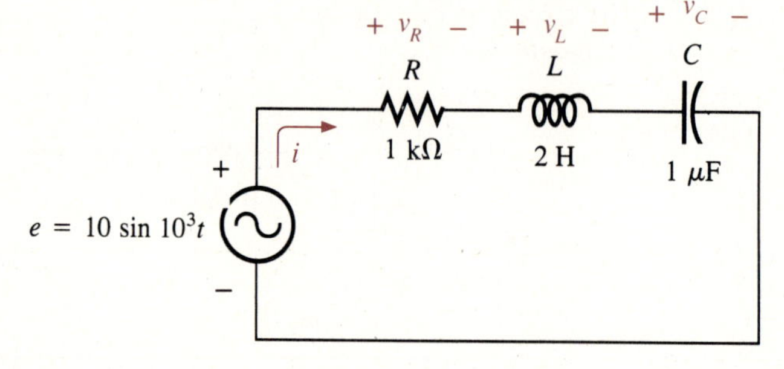

Section 14.6

14. Calculate the magnitude of the voltages $\mathbf{V}_1$ and $\mathbf{V}_2$ for the circuit of Fig. 14.63 using the voltage divider rule.

FIG. 14.63

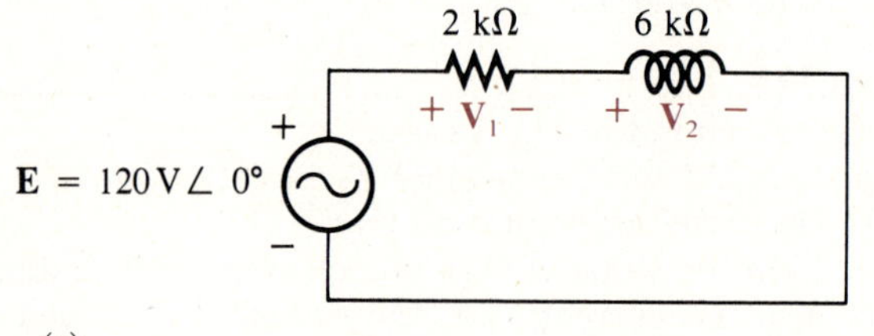

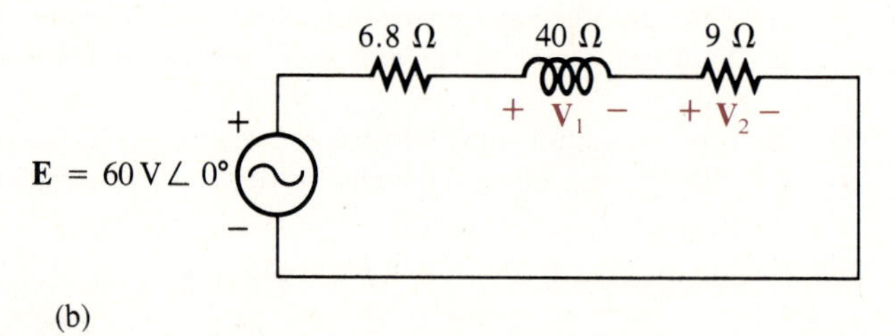

15. Repeat Problem 14 for the circuits of Fig. 14.64.

FIG. 14.64

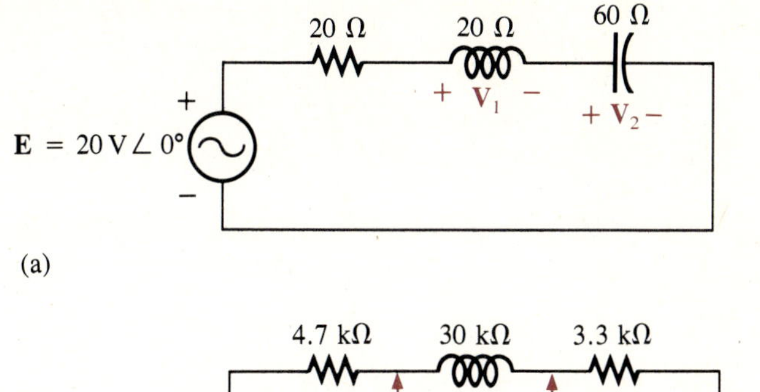

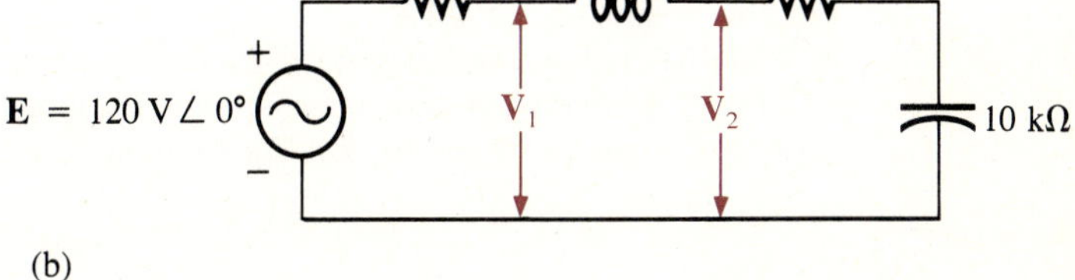

16. An electrical load has a power factor of 0.8 lagging. It dissipates 8 kW at a voltage of 200 V. Determine the resistance and reactance (type also) of the load.

***17.** Find the series elements or elements that must be in the enclosed container of Fig. 14.65 to satisfy the following conditions:
a. Average power to circuit = 300 W.
b. Circuit has a lagging power factor.

FIG. 14.65

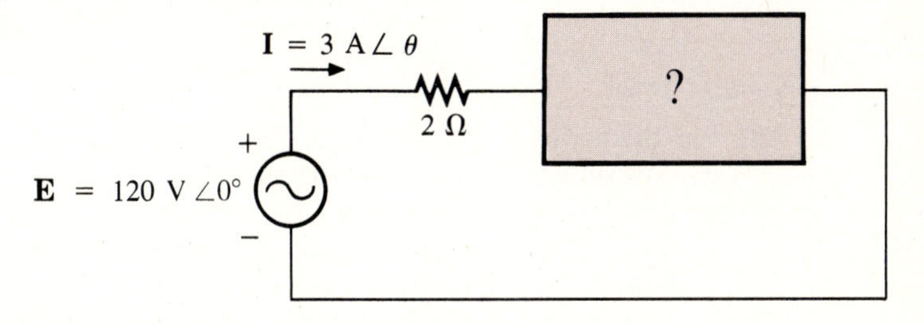

Section 14.7

18. For the circuit of Fig. 14.66:
a. Plot V_L versus frequency for a frequency range of 0 to 20 kHz.
b. Plot θ_T versus frequency for the same frequency range as in part a.
c. Plot V_R versus frequency for the frequency range of part a.

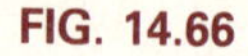
FIG. 14.66

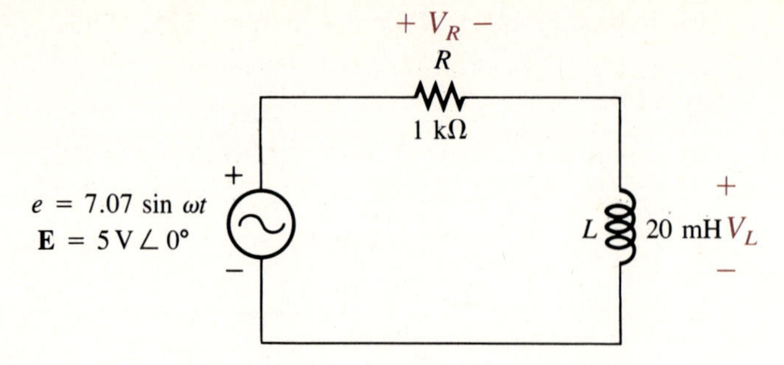

Section 14.8

19. For the circuit of Fig. 14.67:
  a. Plot V_C versus frequency for a frequency range of 0 to 10 kHz.
  b. Plot θ_T versus frequency for the same frequency range as in part a.
  c. Plot V_R versus frequency for the frequency range of part a.

FIG. 14.67

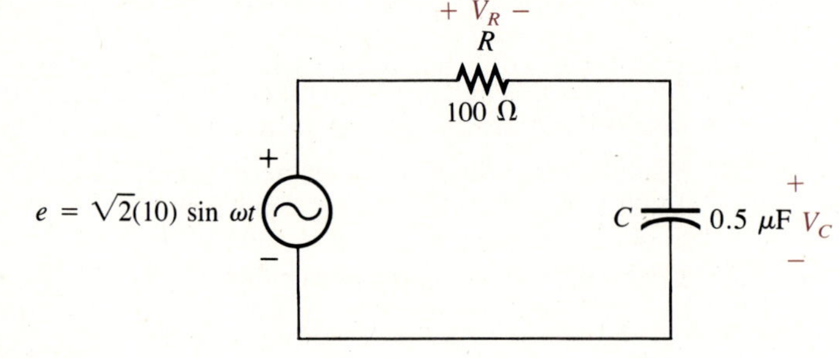

Section 14.9

***20.** For the series R-L-C circuit of Fig. 14.68:
  a. Determine the transition frequency f_s.
  b. Plot V_C versus frequency for the frequency range 0 to 20 kHz.
  c. Plot V_L versus frequency for the frequency range 0 to 20 kHz.
  d. Plot Z_T versus frequency for the frequency range 0 to 20 kHz.
  e. Looking at the results of parts b, c, and d, is there anything immediately obvious about the characteristics near or at $f = f_s$?

FIG. 14.68

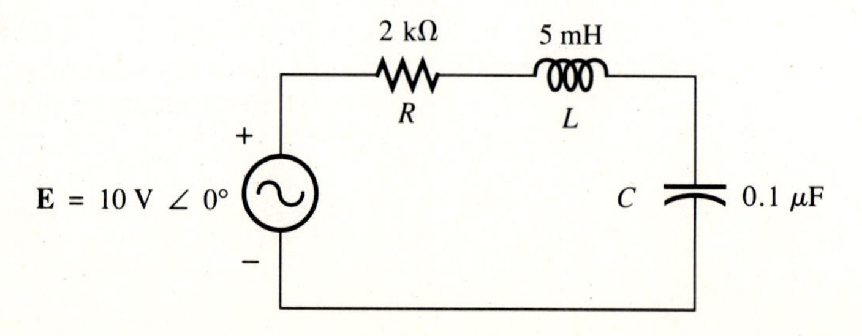

Section 14.10

21. Repeat Problem 5 using the j-operator.
22. Repeat Problem 7 using the j-operator.
23. Determine $\mathbf{I}$, $\mathbf{V}_C$, and $\mathbf{V}_R$ for the network of Fig. 14.57 using the j-operator.
24. Determine $\mathbf{I}$, $\mathbf{V}_R$, and $\mathbf{V}_C$ using the j-operator for the circuit of Fig. 14.59.
25. Determine $\mathbf{I}$, $\mathbf{V}_R$, $\mathbf{V}_L$, and $\mathbf{V}_C$ for the circuit of Fig. 14.61 using the j-operator.
26. Determine the phasor form of $\mathbf{V}_1$ and $\mathbf{V}_2$ for the circuits of Fig. 14.63 using the j-operator and the voltage divider rule.
27. Determine the phasor form of $\mathbf{V}_1$ and $\mathbf{V}_2$ for the circuits of Fig. 14.64 using the j-operator and the voltage divider rule.

Computer Problems

28. Given an R-L or R-C series network, write a program to determine the magnitude and angle of the total impedance. Assume R and L (or C) are given and the reactance has to be determined using the applied frequency.
29. Repeat Problem 28 for an R-L-C series circuit.
30. Given an R-L network such as in Fig. 14.13, write a program to determine $\mathbf{Z}_T$, $\mathbf{I}$, $\mathbf{V}_R$, $\mathbf{V}_L$, P, and F_p.
31. Repeat Problem 24 for the R-L-C circuit of Fig. 14.21, including the additional voltages $\mathbf{V}_C$.
32. Write a program to perform the frequency analysis of the network of Fig. 14.30 for any given value of R and L and a frequency of $f = 0$ Hz to 100 kHz in increments of 5 kHz. That is, tabulate V_L, V_R, and θ_T for the indicated frequencies.
33. Write a program to generate the sinusoidal expression for the current of a resistor, inductor, or capacitor given the value of R, L, or C and the applied voltage in sinusoidal form.

15

Parallel ac Circuits

OBJECTIVES

- ☐ Be able to analyze a parallel sinusoidal ac network with resistive, inductive, and capacitive elements.
- ☐ Understand the admittance terminology and how to determine the total admittance (magnitude and angle) of a parallel sinusoidal ac network using the admittance diagram.
- ☐ To draw a phasor diagram of the currents and voltage of a parallel sinusoidal ac network.
- ☐ Find the total power delivered to any parallel ac network with resistive and reactive elements.
- ☐ Be able to find the power factor of a parallel network using one of a number of methods.
- ☐ Become aware of how the current divider rule is applied and how to determine the magnitude and angle associated with each of the parallel currents.
- ☐ Understand how frequency can affect the characteristics of a parallel *R-L*, *R-C*, or *R-L-C* circuit.
- ☐ Be able to determine the equivalent parallel network of a series configuration at a fixed frequency and the equivalent series network of a parallel configuration.

15.1 INTRODUCTION

The discussion of Chapter 14 will now be extended to include parallel ac networks. Again, the introductory discussion involves a fixed frequency with the effect of frequency variation reserved for the later sections. Through the use of the impedance parameter, the similarity between equations presented here and those introduced in the dc chapters will be quite obvious. A number of new parameters such as admittance and susceptance are introduced to ensure their recognizability in the industrial arena and their use in the analysis of parallel networks.

15.2 TOTAL IMPEDANCE AND ADMITTANCE

For the individual element in a series *or* parallel configuration, the impedance is still defined by

$$\mathbf{Z}_R = R\,\underline{/0^\circ}, \qquad \mathbf{Z}_L = X_L\,\underline{/90^\circ}, \qquad \text{and} \quad \mathbf{Z}_C = X_C\,\underline{/-90^\circ}$$

However, for resistance you will recall from the dc chapters that a quantity called conductance is defined by $G = 1/R$, measured in siemens (S). For ac systems the conductance is defined by the following vector format, with 0° a result of the fact that v_R and i_R are in phase.

$$\mathbf{G} = G\,\underline{/0^\circ} = \frac{1}{R}\,\underline{/0^\circ} \qquad \text{(siemens, S)} \tag{15.1}$$

For the inductive reactance, a quantity called *inductive susceptance* is defined by

$$B_L = \frac{1}{X_L} \qquad \text{(S)} \tag{15.2}$$

The inverse relationship between susceptance and inductive reactance requires that the angle associated with the susceptance be −90°. Therefore,

$$\mathbf{B}_L = B_L\,\underline{/-90^\circ} = \frac{1}{X_L}\,\underline{/-90^\circ} \qquad \text{(S)} \tag{15.3}$$

For the capacitive reactance, a quantity called *capacitive susceptance* is defined by

$$B_C = \frac{1}{X_C} \tag{15.4}$$

The inverse relationship between susceptance and capacitive reactance requires that the angle associated with the susceptance is +90°. Therefore,

$$\mathbf{B}_C = B_C \angle 90^\circ = \frac{1}{X_C} \angle 90^\circ \quad \text{(S)} \qquad (15.5)$$

For parallel networks, each quantity just defined is, individually or in combination, a measure of the *admittance* of the network as defined by

$$\mathbf{Y} = \frac{1}{\mathbf{Z}} \quad \text{(S)} \qquad (15.6)$$

Equation 15.6 clearly reveals that the higher the impedance of a network, the lower its admittance and vice versa. A less obvious but important conclusion is that, due to the inverse relationship between the two,

The angle associated with Y is the angle by which the input current I leads the applied voltage E.

That is, for

$$\mathbf{Y} = Y \angle \theta_T \quad \text{and} \quad \mathbf{E} = E \angle 0^\circ, \qquad \mathbf{I} = I \angle \theta_T$$

A minus sign for θ_T reveals that the applied voltage **E** leads **I** by the indicated angle.

For the system of Fig. 15.1, the admittance level of each impedance can be determined as shown in Fig. 15.2, and the *total admittance* can be determined by the following equation:

$$\mathbf{Y}_T = \mathbf{Y}_1 + \mathbf{Y}_2 + \mathbf{Y}_3 \qquad (15.7)$$

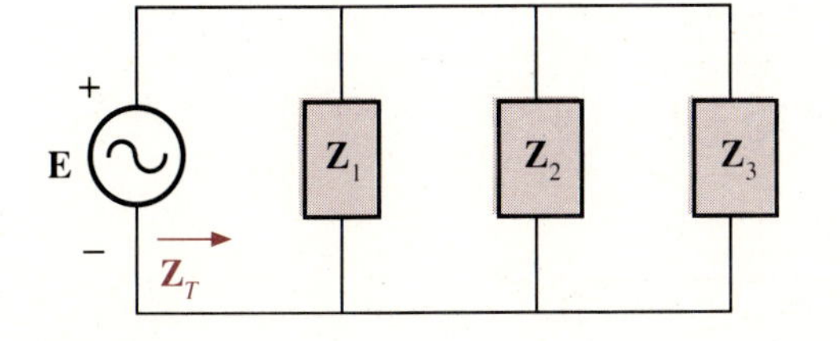

FIG. 15.1
Parallel ac network.

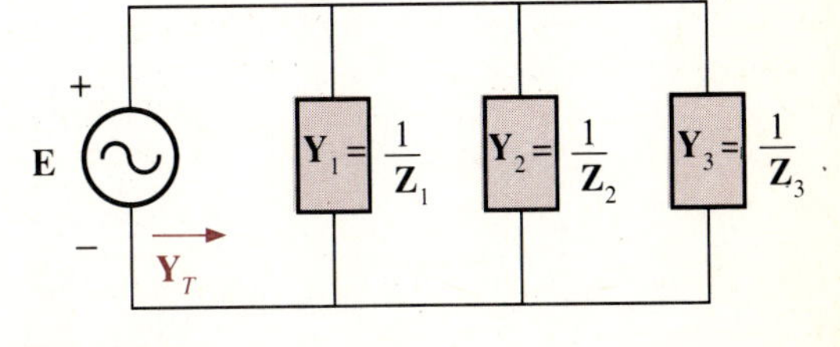

FIG. 15.2
Finding the total admittance of a parallel network.

An *admittance diagram* can be used to provide a display of the network admittances and a hint as to its general characteristics. Using the magnitudes and angles appearing in Eqs. 15.1, 15.3, and 15.4 results in the admittance diagram of Fig. 15.3.

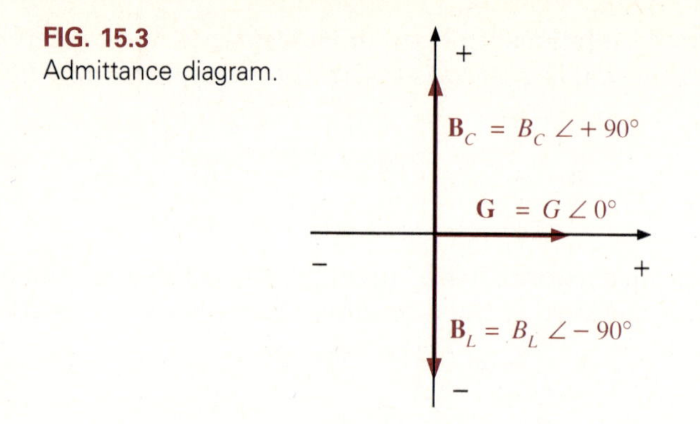

FIG. 15.3
Admittance diagram.

As with impedances in the series ac circuit, the important point to be made by Fig. 15.3 is that the total admittance of a parallel network must be determined by a vector addition and not simply by an algebraic manipulation of magnitudes.

Once the total admittance is determined, the total impedance can be determined in the following manner: For $\mathbf{Y}_T = Y_T \angle \theta_T$,

$$\mathbf{Z}_T = \frac{1}{\mathbf{Y}_T} = \frac{1}{Y_T} \angle -\theta_T \tag{15.8}$$

In particular, note the change in sign of the angle. Naturally, if θ_T of $\mathbf{Y}_T$ is negative, the angle associated with $\mathbf{Z}_T$ is positive. A new examples should clarify the use of the preceding equations.

EXAMPLE 15.1 Determine the total admittance and impedance of the parallel network of Fig. 15.4.

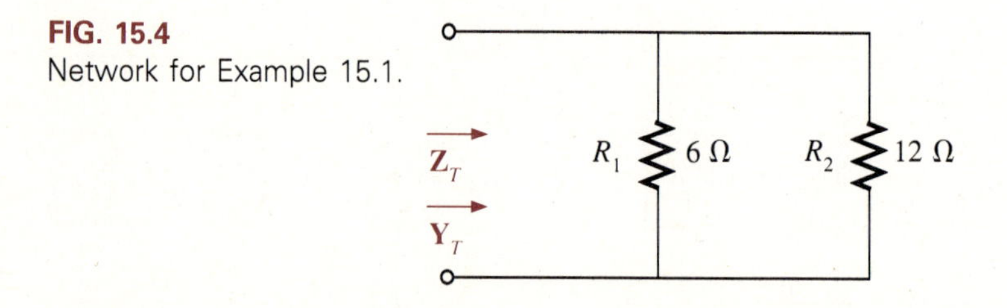

FIG. 15.4
Network for Example 15.1.

Solution For purely resistive elements, the total resistance or conductance can be determined using methods introduced for dc networks. Using the equation

$$R_T = \frac{R_1 R_2}{R_1 + R_2} = \frac{(6\ \Omega)(12\ \Omega)}{6\ \Omega + 12\ \Omega} = 4\ \Omega$$

and

$$\mathbf{Z}_T = R_T = 2\ \Omega \angle 0°$$

The total admittance is

$$Y_T = \frac{1}{Z_T} = \frac{1}{4\ \Omega} = 0.25\ \text{S}$$

and

$$\mathbf{Y}_T = 0.25\ \text{S}\ \angle 0^\circ$$

An alternate approach using Eq. 15.6 gives

$$\mathbf{Y}_1 = \frac{1}{6}\ \text{S}\ \angle 0^\circ = 0.167\ \text{S}\ \angle 0^\circ$$

$$\mathbf{Y}_2 = \frac{1}{12}\ \text{S}\ \angle 0^\circ = 0.083\ \text{S}\ \angle 0^\circ$$

$$\mathbf{Y}_T = \mathbf{Y}_1 + \mathbf{Y}_2 = 0.167\ \text{S}\ \angle 0^\circ + 0.083\ \text{S}\ \angle 0^\circ$$

$$= \mathbf{0.25\ S\ \angle 0^\circ}$$

as obtained above. The admittance diagram appears in Fig. 15.5, and

$$\mathbf{Z}_T = \frac{1}{Y_T}\angle -\theta_T = \frac{1}{0.25\ \text{S}}\angle 0^\circ = \mathbf{4\ \Omega\ \angle 0^\circ}$$

as above.

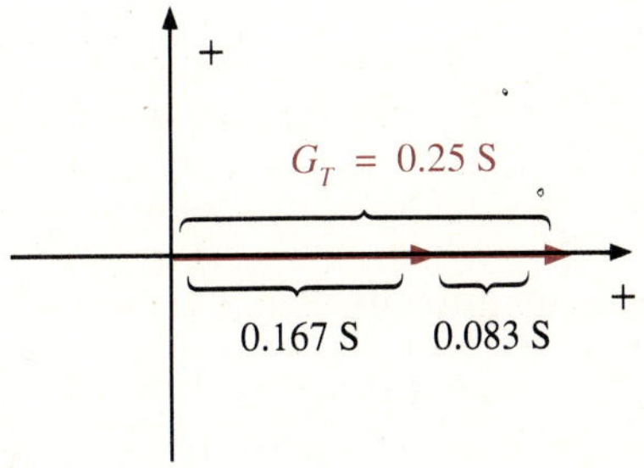

FIG. 15.5
Solution to Example 15.1.

For future reference simply keep in mind that for purely resistive networks, the angle is 0° for both $\mathbf{Z}_T$ and $\mathbf{Y}_T$ and the magnitudes are related by $Z_T = 1/Y_T$.

EXAMPLE 15.2 Determine the total admittance and impedance for the parallel network of Fig. 15.6.

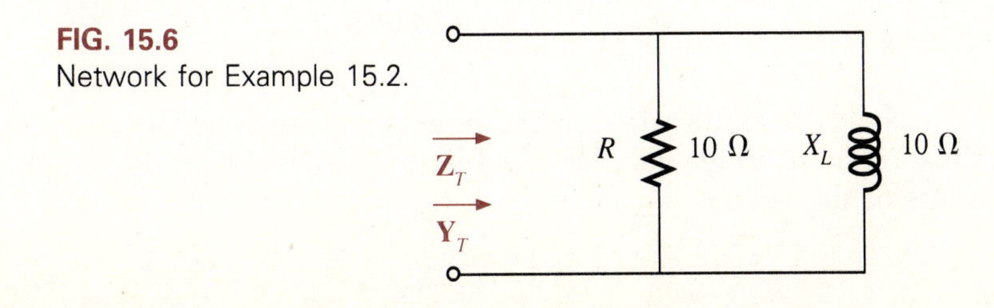

FIG. 15.6
Network for Example 15.2.

Solution

R:

$$\mathbf{G} = \frac{1}{R}\,\underline{/0^\circ} = \frac{1}{10\ \Omega}\,\underline{/0^\circ} = 0.1\ \text{S}\,\underline{/0^\circ}$$

X_L:

$$\mathbf{B}_L = \frac{1}{X_L}\,\underline{/-90^\circ} = \frac{1}{10\ \Omega}\,\underline{/-90^\circ} = 0.1\ \text{S}\,\underline{/-90^\circ}$$

Using Eq. 15.6:

$$\begin{aligned}\mathbf{Y}_T &= \mathbf{Y}_1 + \mathbf{Y}_2 = \mathbf{G} + \mathbf{B}_L\\ &= 0.1\ \text{S}\,\underline{/0^\circ} + 0.1\ \text{S}\,\underline{/-90^\circ}\end{aligned}$$

as shown in Fig. 15.7.

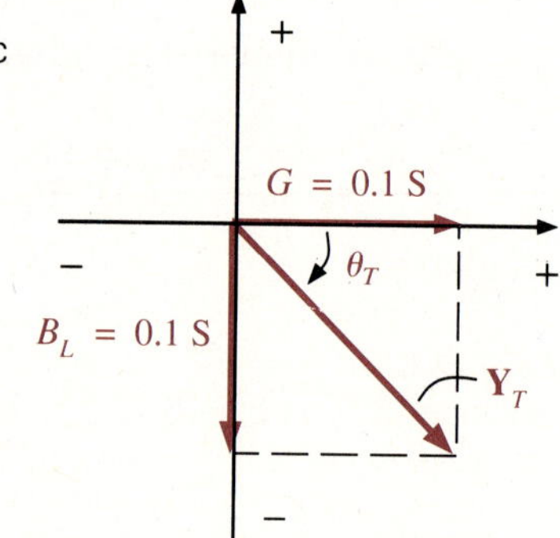

FIG. 15.7
Admittance diagram for the parallel ac network of Fig. 15.6.

The magnitude Y_T is

$$\begin{aligned}Y_T &= \sqrt{G^2 + B_L^2} = \sqrt{(0.1\ \text{S})^2 + (0.1\ \text{S})^2}\\ &= \sqrt{0.01 + 0.01} = \sqrt{0.02} = 0.1414\ \text{S}\end{aligned}$$

The magnitude of the angle θ_T is

$$\theta_T = \tan^{-1}\frac{B_L}{G} = \tan^{-1}\frac{0.1\ \text{S}}{0.1\ \text{S}} = \tan^{-1}1 = 45^\circ$$

so that

$$\mathbf{Y}_T = \mathbf{0.1414\ S}\,\underline{/\mathbf{-45^\circ}}$$

$$\mathbf{Z}_T = \frac{1}{Y_T}\,\underline{/-\theta_T} = \frac{1}{0.1414\ \text{S}}\,\underline{/-(-45^\circ)} = \mathbf{7.071\ \Omega}\,\underline{/\mathbf{45^\circ}}$$

Note the similarities to the analysis of series ac circuits, where the Pythagorean theorem and the inverse tangent function determine the components of the vector.

EXAMPLE 15.3 Determine the total admittance and impedance for the parallel network of Fig. 15.8.

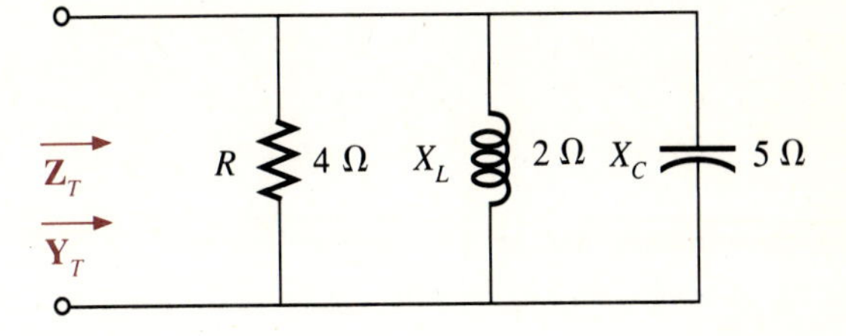

FIG. 15.8
Network for Example 15.3.

Solution

R:

$$\mathbf{G} = \frac{1}{R}\angle 0^\circ = \frac{1}{4\ \Omega}\angle 0^\circ = 0.25\ \text{S}\angle 0^\circ$$

X_L:

$$\mathbf{B}_L = \frac{1}{X_L}\angle -90^\circ = \frac{1}{2\ \Omega}\angle -90^\circ = 0.5\ \text{S}\angle -90^\circ$$

X_C:

$$\mathbf{B}_C = \frac{1}{X_C}\angle 90^\circ = \frac{1}{5\ \Omega}\angle 90^\circ = 0.2\ \text{S}\angle 90^\circ$$

Using Eq. 15.6:

$$\begin{aligned}\mathbf{Y}_T &= \mathbf{Y}_1 + \mathbf{Y}_2 + \mathbf{Y}_3 \\ &= 0.25\ \text{S}\angle 0^\circ + 0.5\ \text{S}\angle -90^\circ + 0.2\ \text{S}\angle 90^\circ\end{aligned}$$

as shown in Fig. 15.9.

$$\begin{aligned}Y_T &= \sqrt{G^2 + (B_L - B_C)^2} = \sqrt{(0.25\ \text{S})^2 + (0.5\ \text{S} - 0.2\ \text{S})^2} \\ &= \sqrt{(0.25\ \text{S})^2 + (0.3\ \text{S})^2} = \sqrt{(0.0625) + (0.09)} \\ Y_T &= \sqrt{0.1525} = 0.391\ \text{S}\end{aligned}$$

Calculator

[·] [2] [5] [x^2] [+] [(] [·] [5] [−] [·] [2] [)] [x^2] [=] [√]

with a display of 0.391.

The magnitude of angle θ_T is

$$\begin{aligned}\theta_T &= \tan^{-1}\frac{B_L - B_C}{G} = \tan^{-1}\frac{0.3\ \text{S}}{0.25\ \text{S}} \\ &= \tan^{-1}1.2 \\ &= 50.194^\circ\end{aligned}$$

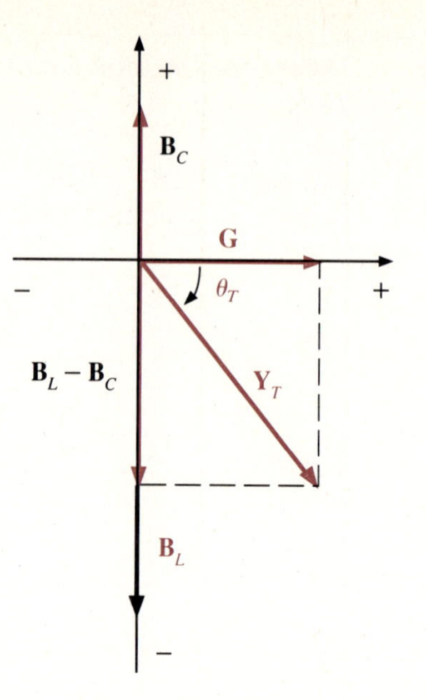

FIG. 15.9
Admittance diagram for the network of Fig. 15.8.

Calculator

(·) (3) (÷) (·) (2) (5) (=) (2nd F) (tan^{-1})

with a display of 50.194.

And

$$\mathbf{Y}_T = \mathbf{0.391\ S\ \underline{/-50.194^\circ}}$$

with

$$\mathbf{Z}_T = \frac{1}{Y_T}\underline{/-\theta_T} = \frac{1}{0.391\ \text{S}}\underline{/-(-50.194^\circ)}$$
$$= \mathbf{2.558\ \Omega\ \underline{/50.194^\circ}}$$

Before continuing, let us reflect on the results obtained thus far and what they indicate about the network. For series circuits, we found a positive angle for $\mathbf{Z}_T$ indicated an inductive network and e would lead i. The same continues here for total impedances that have a positive angle, such as in Examples 15.2 and 15.3. Note, however, that when you examine the total admittance, a negative angle reflects an inductive network. Similarly for capacitive networks, the total admittance has a positive angle and the total impedance has a negative angle.

In addition, note in Example 15.3 that the magnitude of the total impedance is more than the impedance of one of the parallel branches. For parallel resistors in a dc network, the total resistance is always less than the magnitude of the smallest resistor. In ac networks the vector relationship can result in total impedances that are more or less than the smallest or even the largest impedance (magnitude) of the network.

EXAMPLE 15.4

Determine the total admittance and impedance of the network of Fig. 15.10.

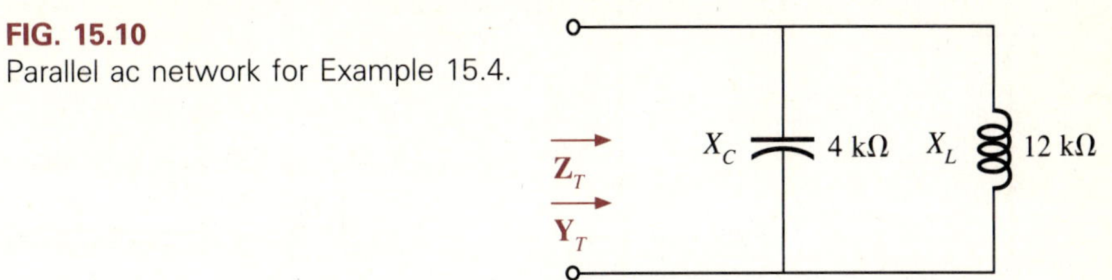

FIG. 15.10
Parallel ac network for Example 15.4.

Solution Note in this case that only reactive elements are present. The resulting impedance must therefore be totally reactive, with an angle of +90° or −90°.

$$\mathbf{Y}_1 = \mathbf{B}_C = \frac{1}{X_C}\angle 90^\circ = \frac{1}{4\text{ k}\Omega}\angle 90^\circ = 0.25 \times 10^{-3}\text{ S}\angle 90^\circ$$

$$\mathbf{Y}_2 = \mathbf{B}_L = \frac{1}{X_L}\angle -90^\circ = \frac{1}{12\text{ k}\Omega}\angle -90^\circ = 0.083 \times 10^{-3}\text{ S}\angle -90^\circ$$

From Eq. 15.6:

$$\mathbf{Y}_T = \mathbf{Y}_1 + \mathbf{Y}_2 = 0.25 \times 10^{-3}\text{ S}\angle 90^\circ + 0.083 \times 10^{-3}\text{ S}\angle -90^\circ$$

as shown in Fig. 15.11.

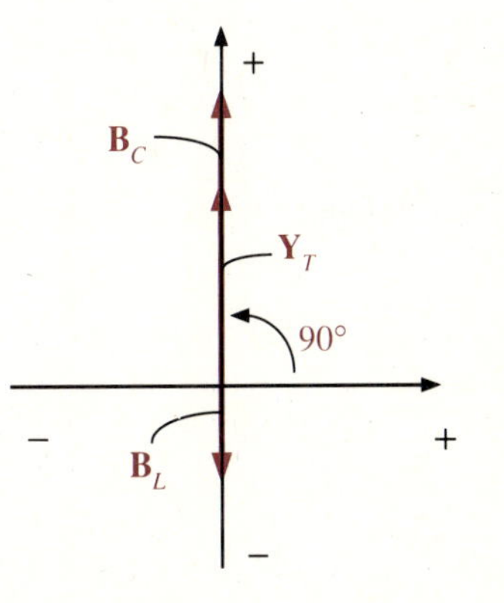

FIG. 15.11
Admittance diagram for the network of Fig. 15.10.

From Fig. 15.11 we can determine that $\mathbf{Y}_T$ is the difference of the two magnitudes, with an angle of +90°. Therefore,

$$Y_T = B_C - B_L = 0.25 \times 10^{-3}\text{ S} - 0.083 \times 10^{-3}\text{ S} = 0.167\text{ S}$$

with $\theta_T = 90°$, and

$$\mathbf{Y}_T = \mathbf{0.167\ S\angle 90^\circ}$$

$$\mathbf{Z}_T = \frac{1}{Y_T}\angle -\theta_T = \frac{1}{0.167\text{ S}}\angle -90^\circ = \mathbf{6\ \Omega\angle -90^\circ}$$

The results reveal that the smaller (magnitude) reactance of parallel reactive elements will determine whether the total impedance is capacitive or inductive—the reverse of the situation for series ac elements.

In Fig. 15.12(a) the capacitive reactance is *smaller* than the inductive reactance ($X_C < X_L$), resulting in a net capacitive impedance. For the network of Fig. 15.12(b), $X_L < X_C$ and the net inductive impedance is in parallel with the resistive element. In Fig. 15.12(c) the capacitive reactance is smaller than the parallel resistive branch by a 10:1 ratio. Any ratio of 10:1 or greater permits eliminating the larger element on an approximate basis that is acceptable for many applications. As shown in Fig. 15.12(d), if the ratio is much greater than 10:1, using the smaller impedance is assumed to be almost an exact equivalence.

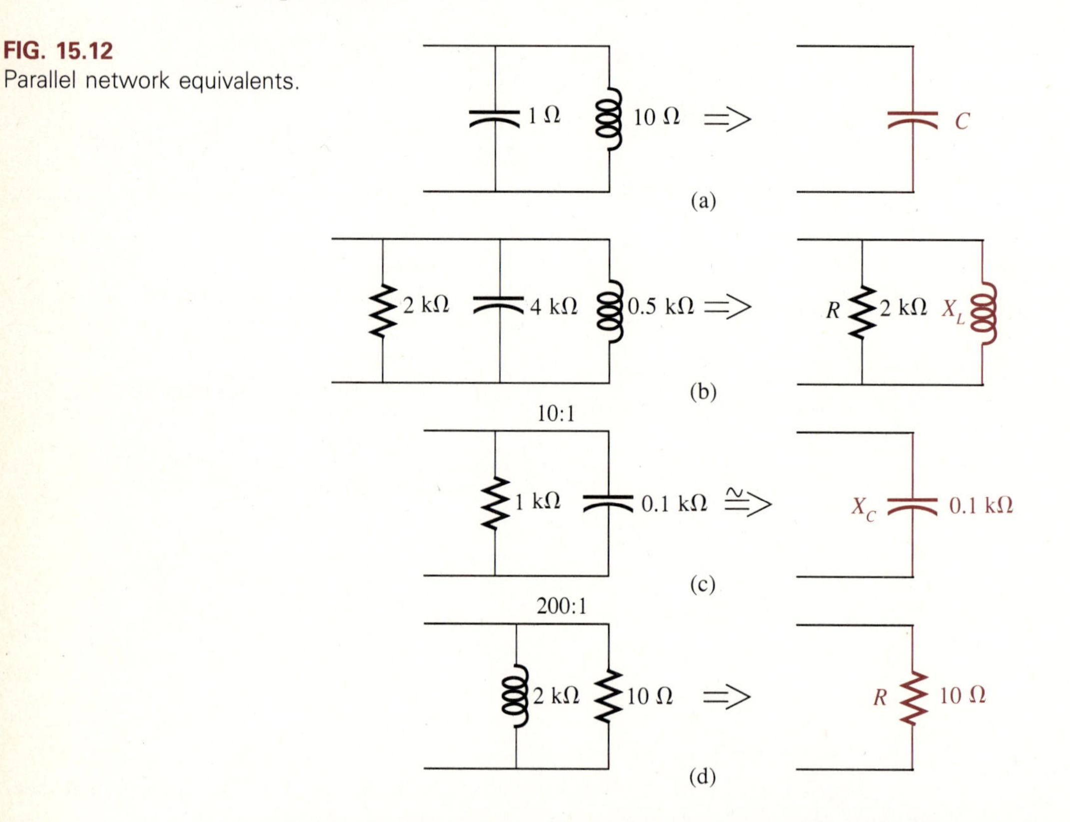

FIG. 15.12
Parallel network equivalents.

In Example 15.1, the total resistance was determined by taking the product and dividing by the sum of the resistance levels. For the ac network of Fig. 15.13, where X can be inductive or capacitive, the magnitude of the total impedance can be determined by

$$Z_T = \frac{RX}{\sqrt{R^2 + X^2}} \tag{15.9}$$

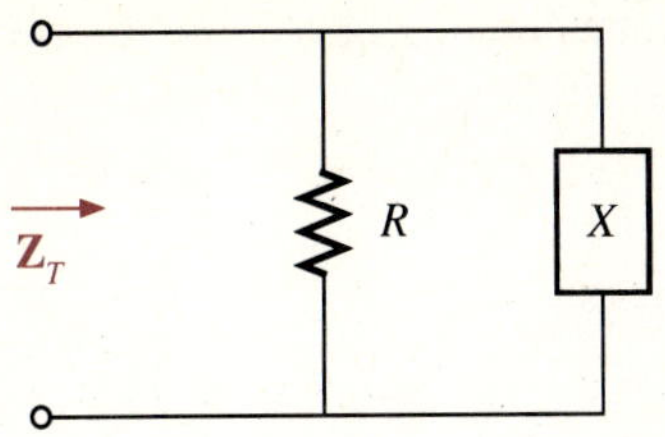

FIG. 15.13
Parallel *R*-*X* network.

The angle associated with $\mathbf{Z}_T$ can be determined from the admittance diagram when the magnitude of the angle associated with $\mathbf{Y}_T$ is determined by

$$|\theta_T| \text{ of } \mathbf{Y}_T = \left|\tan^{-1}\frac{B}{G}\right| = \left|\tan^{-1}\frac{\frac{1}{X}}{\frac{1}{R}}\right| = \left|\tan^{-1}\frac{R}{X}\right|$$

Since the magnitude of the angle of $\mathbf{Z}_T$ is equal to that of $\mathbf{Y}_T$, the magnitude of the angle associated with $\mathbf{Z}_T$ is

$$\boxed{|\theta_T| = \left|\tan^{-1}\frac{R}{X}\right|}_{\mathbf{Z}_T} \tag{15.10}$$

with a + sign applied to θ_T for inductive networks and a − sign for capacitive networks. The choice of using Eqs. 15.9 and 15.10 versus the method used in the previous examples is simply a personal preference.

For comparison, compute the total impedance of the parallel network of Fig. 15.6.

$$\mathbf{Z}_T = \frac{(R)(X)}{\sqrt{R^2 + X^2}} = \frac{(10\ \Omega)(10\ \Omega)}{\sqrt{(10\ \Omega)^2 + (10\ \Omega)^2}} = \frac{100\ \Omega}{\sqrt{200}} = \frac{100\ \Omega}{14.142}$$
$$= 7.071\ \Omega$$

and

$$|\theta_T| = \left|\tan^{-1}\frac{R}{X}\right| = \left|\tan^{-1}\frac{10\ \Omega}{10\ \Omega}\right| = 45°$$

with

$$\mathbf{Z}_T = \mathbf{7.071\ \Omega\ \angle 45°}$$

matching the result of Example 15.2.

If the 10-Ω reactance of Fig. 15.6 were capacitive, the magnitude of $\mathbf{Z}_T$ and θ_T would be the same, but a negative sign would be associated with the angle, resulting in $\mathbf{Z}_T = 7.071\ \Omega\ \angle -45°$.

15.3 *R-L* PARALLEL NETWORK

Let us now perform a detailed analysis of a parallel *R*-*L* network with a sinusoidal voltage applied at a fixed frequency.

For the network of Fig. 15.14, the resulting reactance is determined from $X_L = 2\pi fL = 20\ \Omega$.

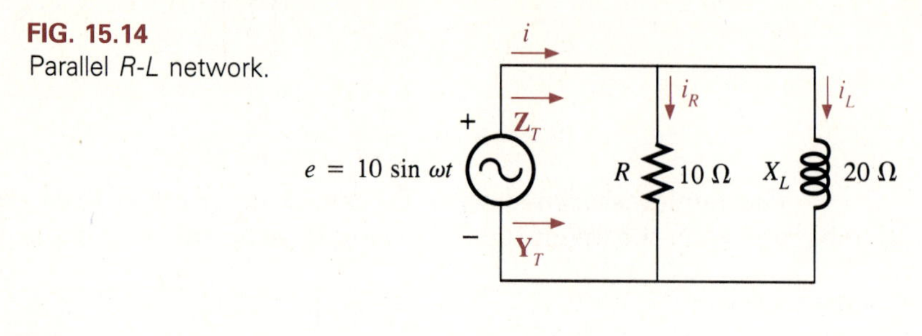

FIG. 15.14
Parallel *R-L* network.

$\mathbf{Y}_T$

$$\mathbf{Y}_T = \mathbf{Y}_1 + \mathbf{Y}_2 = \mathbf{G} + \mathbf{B}_L = \frac{1}{R}\angle 0^\circ + \frac{1}{X_L}\angle -90^\circ$$

$$= \frac{1}{10\ \Omega}\angle 0^\circ + \frac{1}{20\ \Omega}\angle -90^\circ = 0.1\ \text{S}\ \angle 0^\circ + 0.05\ \text{S}\ \angle -90^\circ$$

as shown in the admittance diagram of Fig. 15.15.

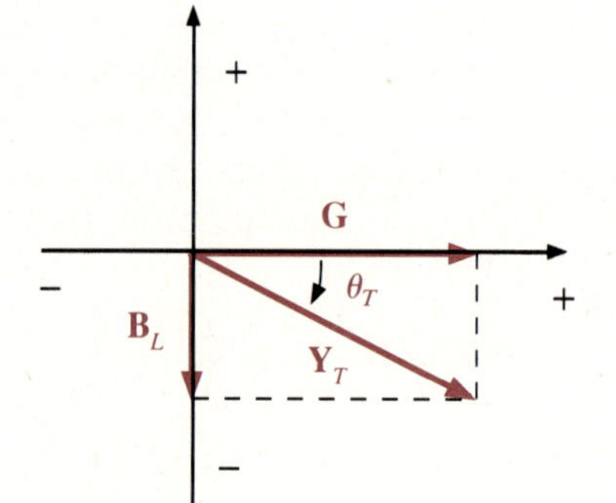

FIG. 15.15
Admittance diagram for the parallel *R-L* network of Fig. 15.14.

The magnitude of $\mathbf{Y}_T$ is determined from

$$Y_T = \sqrt{G^2 + B_L^2} = \sqrt{(0.1\ \text{S})^2 + (0.05\ \text{S})^2}$$
$$= \sqrt{0.01 + 0.0025} = \sqrt{0.0125} = 0.112\ \text{S}$$

and the magnitude of the angle θ_T is

$$|\theta_T| = \tan^{-1}\frac{B_L}{G} = \tan^{-1}\frac{0.05\ \text{S}}{0.1\ \text{S}} = \tan^{-1}0.5 = 26.57^\circ$$

so that $\mathbf{Y}_T$ = $\mathbf{0.112\ S\ \angle -26.57^\circ}$ (inductive network).

$\mathbf{Z}_T$:

$$\mathbf{Z}_T = \frac{1}{Y_T}\underline{/-\theta_T} = \frac{1}{0.112\text{ S}}\underline{/-(-26.57°)} = \mathbf{8.93\ \Omega\ \underline{/26.57°}}$$

Using Eq. 15.9:

$$Z_T = \frac{(R)(X_L)}{\sqrt{R^2 + X_L^2}} = \frac{(10\ \Omega)(20\ \Omega)}{\sqrt{(10\ \Omega)^2 + (20\ \Omega)^2}} = \frac{200\ \Omega}{22.361}$$
$$= \mathbf{8.94\ \Omega}$$

as obtained above (the slight difference is due to the level of accuracy carried through the calculations). Using Eq. 15.10:

$$|\theta_T| = \left|\tan^{-1}\frac{R}{X_L}\right| = \tan^{-1}\frac{10\ \Omega}{20\ \Omega} = \tan^{-1}0.5 = 26.57°$$

as above, with $\mathbf{Z}_T = 8.94\ \Omega\ \underline{/26.57°} = Z_T\ \underline{/\theta_T}$. Of course,

$$\mathbf{Y}_T = \frac{1}{Z_T}\underline{/-\theta_T} = \frac{1}{8.94\ \Omega}\underline{/-26.57°} = 0.112\text{ S}\ \underline{/-26.57°}$$

$\mathbf{V}_R$, $\mathbf{V}_L$: Since the voltage across parallel elements is always the same,

$$v_R = v_L = e = \mathbf{10\ sin\ \omega t}$$

The effective value is $E = (0.707)(10\text{ V}) = 7.07\text{ V}$, so

$$\mathbf{E = 7.07\ V\ \underline{/0°}}$$

with

$$\mathbf{V}_R = \mathbf{7.07\ V\ \underline{/0°}} \quad \text{and} \quad \mathbf{V}_L = \mathbf{7.07\ V\ \underline{/0°}}$$

$\mathbf{I}_R$: The peak value of the current i_R can be determined from

$$I_m = \frac{V_m}{R} = \frac{10\text{ V}}{10\ \Omega} = 1\text{ A}$$

and, since v_R and i_R are in phase,

$$i_R = \mathbf{1\ sin\ \omega t}$$

The effective value $I_{\text{eff}} = (0.707)(1\text{ A}) = 0.707\text{ A}$, and

$$\mathbf{I}_R = \mathbf{0.707\ A\ \underline{/0°}}$$

I_L: The peak value of the current i_L can be determined from

$$I_m = \frac{V_m}{X_L} = \frac{10 \text{ V}}{20 \ \Omega} = 0.5 \text{ A}$$

and, since v_L leads i_L by 90°, the angle associated with i_L must lag the voltage $v_L = e$ by 90°. Therefore,

$$i_L = \mathbf{0.5 \sin(\omega t - 90°)}$$

The effective value $I_{\text{eff}} = (0.707)(0.5 \text{ A}) = 0.354$ A, and

$$\mathbf{I}_L = \mathbf{0.354 \text{ A} \angle -90°}$$

I_T: The angle associated with the total impedance is +26.57°, revealing that the network is inductive and e leads i by 26.57°.

The magnitude of the peak value of i can be determined from

$$I_m = \frac{E_m}{Z_T} = \frac{10 \text{ V}}{8.94 \ \Omega} = 1.112 \text{ A}$$

and $\theta_T = -26.57°$ from $\mathbf{Z}_T = Z_T \angle \theta_T$, revealing that e leads i by 26.57° and $i_T = \mathbf{1.112 \sin(\omega t - 26.57°)}$.

The effective value is $I_{\text{eff}} = (0.707)(1.112 \text{ A}) = 0.786$ A, with

$$\mathbf{I}_T = \mathbf{0.786 \text{ A} \angle -26.57°}.$$

Phasor Diagram Now that all the quantities of the network have been determined in phasor form, they can all be placed on the same diagram (Fig. 15.16) to note their relative magnitudes and phase displacements.

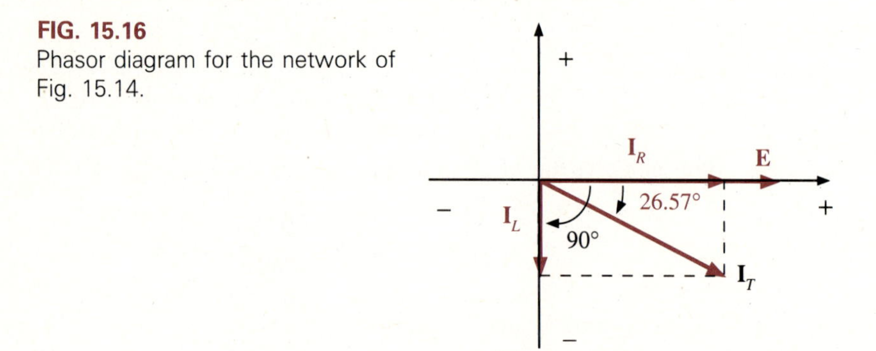

FIG. 15.16
Phasor diagram for the network of Fig. 15.14.

Figure 15.16 clearly reveals that $\mathbf{V}_R = \mathbf{E}$ is in phase with $\mathbf{I}_R$ and that $\mathbf{V}_L = \mathbf{E}$ leads $\mathbf{I}_L$ by 90°. In addition, note that the vector $\mathbf{I}_T$ equals the vector sum of $\mathbf{I}_R$ and $\mathbf{I}_L$, as required by Kirchhoff's current law. That is,

$$\mathbf{I}_T = \mathbf{I}_R + \mathbf{I}_L$$

Power The total power in watts delivered to the network is

$$P_T = EI \cos \theta_T \tag{15.11}$$

where θ_T is the magnitude (no concern for the sign) of the angle associated with the total impedance or admittance and E and I are effective values.

Substituting for the network of Fig. 15.14,

$$\begin{aligned} P_T &= (7.07\ \text{V})(0.786\ \text{A})\cos 26.57^\circ \\ &= (5.557)(0.894) \\ P_T &= \mathbf{5\ W} \end{aligned}$$

or

$$\begin{aligned} P_T &= \frac{V_R^2}{R} \\ &= \frac{(7.07\ \text{V})^2}{10\ \Omega} = \mathbf{5\ W} \end{aligned}$$

Also,

$$P_T = I_R^2 R = (0.707\ \text{A})^2\,(10\ \Omega) = \mathbf{5\ W}$$

Power Factor The power factor of the network is defined by

$$F_p = \cos \theta_T = \frac{G}{Y_T} = \frac{Z_T}{R} = \frac{P_T}{EI} \tag{15.12}$$

If θ_T of $\mathbf{Z}_T$ is a positive angle, the power factor is lagging (inductive); if negative, the power factor is leading (capacitive). If θ_T of $\mathbf{Y}_T$ is positive or negative, the resulting terminology is the reverse of that just applied.

For this example,

$$\begin{aligned} F_p &= \cos \theta_T = \cos 26.57^\circ \\ &= \mathbf{0.89\ lagging} \quad \text{(inductive network)} \end{aligned}$$

Similarly,

$$F_p = \frac{G}{Y_T} = \frac{0.1\ \text{S}}{0.112\ \text{S}} = \mathbf{0.89\ \ lagging}$$

or

$$F_P = \frac{Z_T}{R} = \frac{8.94\ \Omega}{10\ \Omega} = \mathbf{0.89\ \ lagging}$$

and

$$F_P = \frac{P_T}{EI} = \frac{5\text{ W}}{(7.07\text{ V})(0.786\text{ A})} = \mathbf{0.89 \quad lagging}$$

verifying the many forms of Eq. 15.12.

Sinusoidal Domain For this example,

$$\begin{aligned} e &= 10 \sin \omega t \\ i &= 1.118 \sin(\omega t - 26.57^\circ) \\ i_R &= 1 \sin \omega t \\ i_L &= 0.5 \sin(\omega t - 90^\circ) \end{aligned}$$

all of which are plotted in Fig. 15.17.

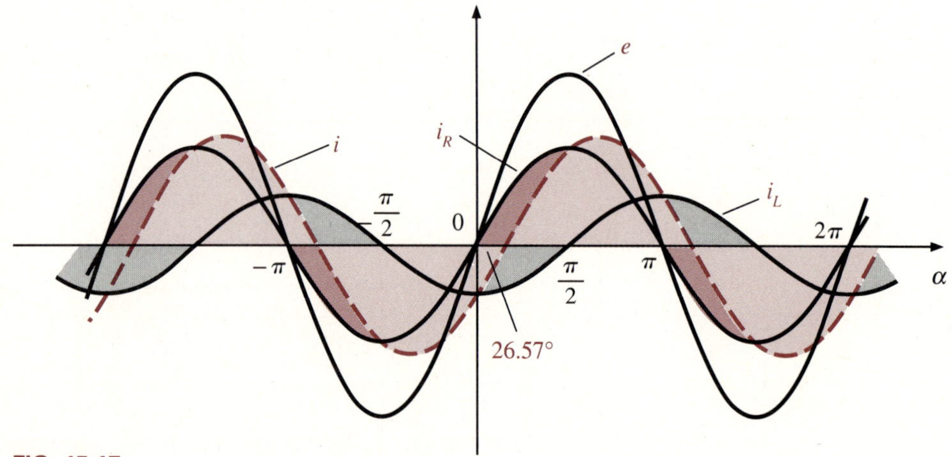

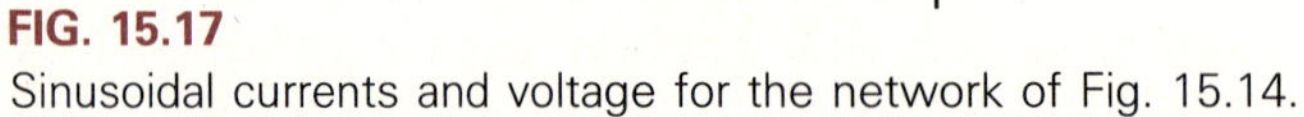
FIG. 15.17
Sinusoidal currents and voltage for the network of Fig. 15.14.

Note that v_R and i_R are in phase, v_L leads i_L by 90°, and e leads i by 26.57° due to the inductive nature of the network.

15.4 *R-C* PARALLEL NETWORK

The analysis of the parallel R-C ac network will be similar to that of the R-L parallel network, just as the analysis of the series R-L and R-C circuits were quite similar. It seems time well spent, however, at this point in the development to ensure certain fundamental concepts are understood.

At the applied frequency, the reactance of the capacitor of Fig. 15.18 is 50 Ω.

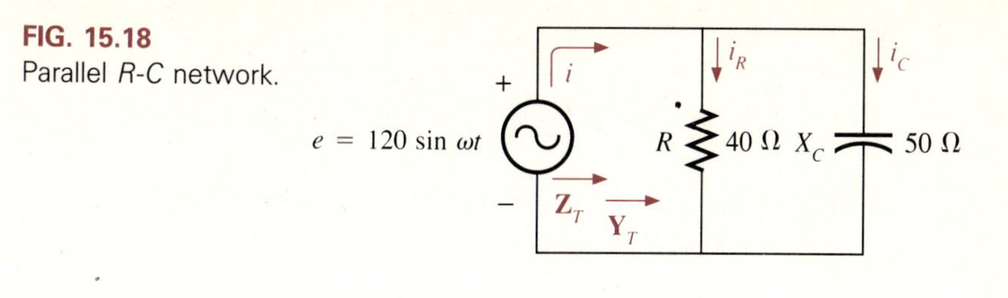

FIG. 15.18
Parallel *R-C* network.

$\mathbf{Y}_T$:

$$\mathbf{Y}_T = \mathbf{Y}_1 + \mathbf{Y}_2 = \mathbf{G} + \mathbf{B}_C = \frac{1}{R}\angle 0^\circ + \frac{1}{X_C}\angle 90^\circ$$

$$= \frac{1}{40\ \Omega}\angle 0^\circ + \frac{1}{50\ \Omega}\angle 90^\circ = 0.025\ \text{S}\angle 0^\circ + 0.02\angle 90^\circ$$

as shown in the admittance diagram of Fig. 15.19.

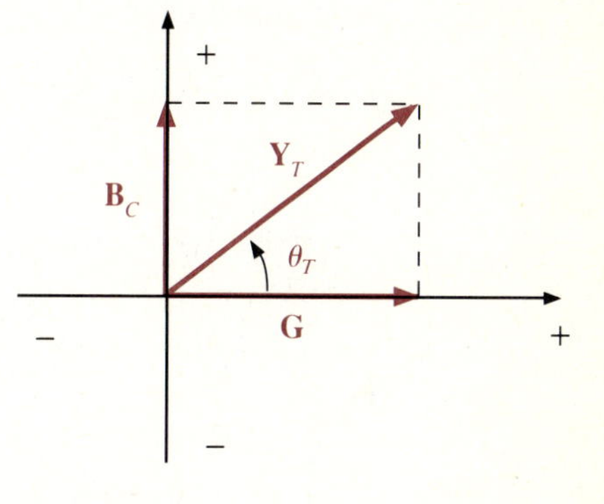

FIG. 15.19
Admittance diagram for the *R-C* network of Fig. 15.18.

$$Y_T = \sqrt{G^2 + B_C^2} = \sqrt{(0.025\ \text{S})^2 + (0.02\ \text{S})^2}$$
$$= 0.032\ \text{S}$$

Calculator

[.] [0] [2] [5] [x^2] [+] [.] [0] [2] [x^2] [=] [$\sqrt{\ }$] [=]

with a display of 0.032. The magnitude of the angle θ_T is

$$\theta_T = \tan^{-1}\frac{B_C}{G} = \tan^{-1}\frac{0.02\ \text{S}}{0.025\ \text{S}} = \tan^{-1}0.8$$
$$= 38.66^\circ$$

so that

$$\mathbf{Y}_T = \mathbf{0.032\ S\angle 38.66^\circ} \qquad \text{(capacitive network)}$$

$\mathbf{Z}_T$:

$$\mathbf{Z}_T = \frac{1}{Y_T}\angle{-\theta_T} = \frac{1}{0.032\ \text{S}}\angle{-38.66^\circ} = \mathbf{31.24\ \Omega\ \angle{-38.66^\circ}}$$

Note the negative angle to reflect a capacitive load.

Using Eq. 15.8:

$$Z_T = \frac{(R)(X_C)}{\sqrt{R^2 + X_C^2}} = \frac{(40)(50)}{\sqrt{(40)^2 + (50)^2}} = \frac{2000\ \Omega}{64.031} = 31.24\ \Omega$$

as obtained earlier.

$$|\theta_T| = \left|\tan^{-1}\frac{R}{X_C}\right| = \left|\tan^{-1}\frac{40\ \Omega}{50\ \Omega}\right| = \left|\tan^{-1}0.8\right| = 38.66^\circ$$

with a capacitive network requiring a negative sign, so that

$$\mathbf{Z}_T = \mathbf{31.24\ \Omega\ \angle{-38.66^\circ}}$$

as obtained earlier.

$\mathbf{V}_R$, $\mathbf{V}_C$: The voltage across parallel elements is the same. Therefore,

$$v_R = v_C = e = 120 \sin \omega t$$

Effective values are

$$E = (0.707)(120\ \text{V}) = 84.84\ \text{V}$$
$$V_R = V_C = 84.84\ \text{V}$$

resulting in

$$\mathbf{E} = 84.84\ \text{V}\ \angle{0^\circ}, \quad \mathbf{V}_R = 84.84\ \text{V}\ \angle{0^\circ}, \quad \mathbf{V}_C = 84.84\ \text{V}\ \angle{0^\circ}$$

$\mathbf{I}_R$: The peak value of the current i_R can be determined from

$$I_m = \frac{V_m}{R} = \frac{120\ \text{V}}{40\ \Omega} = 3\ \text{A}$$

and, since v_R and i_R are in phase,

$$i_R = \mathbf{3 \sin \omega t}$$

The effective value $I_{\text{eff}} = (0.707)(3\ \text{A}) = 2.121$ A, and $\mathbf{I}_R =$ **2.121 A** $\angle{\mathbf{0^\circ}}$.

$\mathbf{I}_C$: The peak value of the current i_C can be determined from

$$I_m = \frac{V_m}{X_C} = \frac{120\ \text{V}}{50\ \Omega} = 2.4\ \text{A}$$

and, since v_C leads i_C by 90°, the angle associated with i_C must lead the voltage $v_C = e = 120 \sin \omega t$ by 90°. Therefore,

$$i_C = \mathbf{2.4 \sin(\omega t + 90°)}$$

The effective value $I_{\text{eff}} = (0.707)(2.4 \text{ A}) = 1.697 \text{ A}$, and $\mathbf{I}_C =$ **1.697 A ∠90°.**

$\mathbf{I}_T$: The angle associated with the total impedance is −38.66°, revealing that the network is capacitive and i leads e by 38.66°.

The magnitude of the peak value of i can be determined from

$$I_m = \frac{E_m}{Z_T} = \frac{120 \text{ V}}{31.24 \ \Omega} = 3.84 \text{ A}$$

and

$$i = \mathbf{3.84 \sin(\omega t + 38.66°)}$$

since

$$e = 120 \sin(\omega t + 0°)$$

The effective value $I_{\text{eff}} = (0.707)(3.84 \text{ A}) = 2.72 \text{ A}$, with $\mathbf{I}_T =$ **2.72 A ∠38.66°**.

Phasor Diagram The phasor diagram of the currents and voltage is provided in Fig. 15.20. Note that $\mathbf{I}_C$ leads **E** by 90°, that $\mathbf{I}_R$ is in phase with **E,** and that $\mathbf{I}_T$ leads **E** due to the capacitive nature of the network. In addition, note that Kirchhoff's current law is satisfied, since

$$\mathbf{I}_T = \mathbf{I}_R + \mathbf{I}_C$$

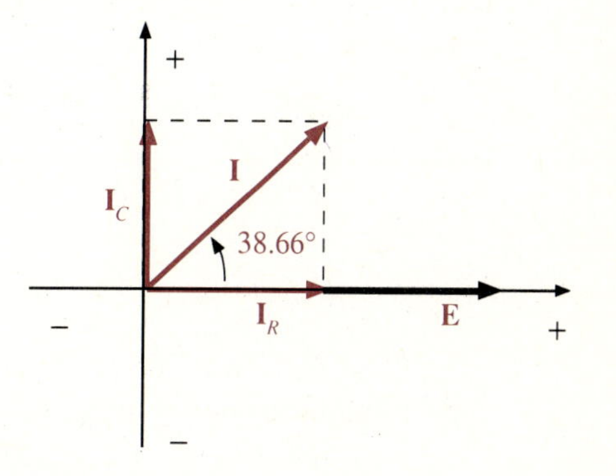

FIG. 15.20
Phasor diagram for the network of Fig. 15.18.

Power

$$\begin{aligned} P_T &= EI \cos \theta_T \\ &= (84.84 \text{ V})(2.72 \text{ A})\cos 38.66° \\ &= (229.92)(0.78) \\ P_T &\cong \mathbf{180 \text{ W}} \end{aligned}$$

or

$$\begin{aligned} P_T &= I_R^2 R \\ &= (2.121 \text{ A})^2 (40 \ \Omega) \\ P_T &\cong \mathbf{180 \ W} \end{aligned}$$

Power Factor

$$F_p = \cos \theta_T = \cos 38.66° = \mathbf{0.781 \quad leading} \qquad \text{(capacitive network)}$$

or

$$F_p = \frac{G}{Y_T} = \frac{0.025 \text{ S}}{0.032 \text{ S}} = \mathbf{0.781 \quad leading}$$

Sinusoidal Domain

$$\begin{aligned} e &= 120 \sin \omega t \\ i_T &= 3.84 \sin(\omega t + 38.66°) \\ i_R &= 3 \sin \omega t \\ i_C &= 2.4 \sin(\omega t + 90°) \end{aligned}$$

If all were plotted on the same axis, the conclusions generated by the phasor diagram would have been verified.

15.5 *R-L-C* PARALLEL NETWORK

We now conclude the detailed analysis of standard parallel configuration with the *R-L-C* configuration of Fig. 15.21.

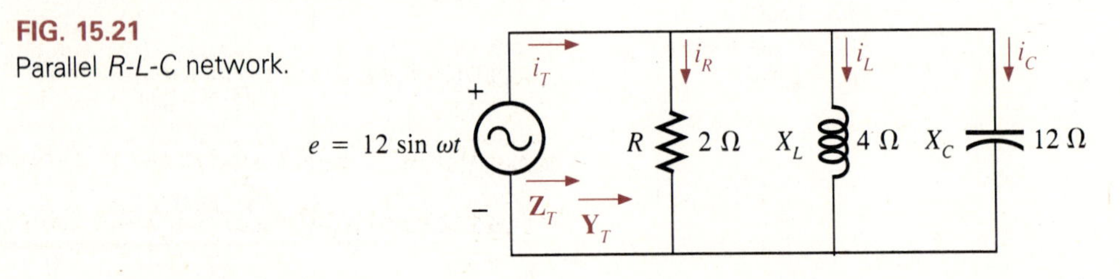

FIG. 15.21
Parallel *R-L-C* network.

The fact that $X_L < X_C$ reveals that the input impedance is inductive, and e leads the current i_T.

$\mathbf{Z}_T$:

$$\begin{aligned} \mathbf{Y}_T &= \mathbf{Y}_1 + \mathbf{Y}_2 + \mathbf{Y}_3 = \mathbf{G} + \mathbf{B}_L + \mathbf{B}_C \\ &= \frac{1}{R} \angle 0° + \frac{1}{X_L} \angle -90° + \frac{1}{X_C} \angle 90° \end{aligned}$$

$$= \frac{1}{2\ \Omega}\angle 0^\circ + \frac{1}{4\ \Omega}\angle -90^\circ + \frac{1}{12\ \Omega}\angle 90^\circ$$

$$\mathbf{Y}_T = 0.5\ \text{S}\angle 0^\circ + 0.25\ \text{S}\angle -90^\circ + 0.08\ \text{S}\angle 90^\circ$$

as shown in the admittance diagram of Fig. 15.22.

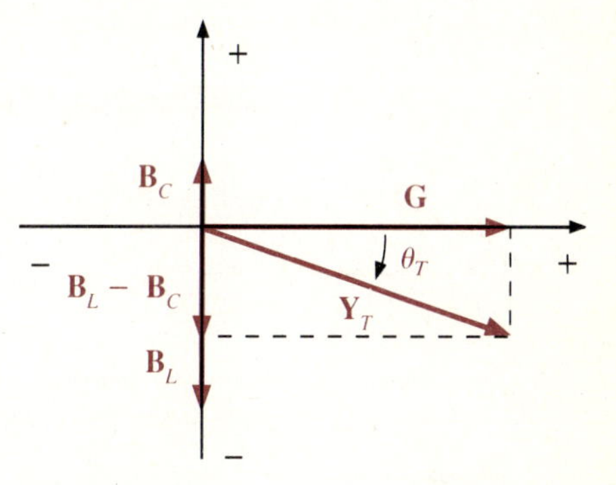

FIG. 15.22
Admittance diagram for the network of Fig. 15.21.

The magnitude of $\mathbf{Y}_T$ is determined from

$$Y_T = \sqrt{G^2 + (B_L - B_C)^2} = \sqrt{(0.5\ \text{S})^2 + (0.25\ \text{S} - 0.08\ \text{S})^2}$$
$$= \sqrt{(0.5\ \text{S})^2 + (0.167\ \text{S})^2} = 0.527\ \text{S}$$

Calculator

⊡ ⑤ x^2 ⊕ ⊡ ① ⑥ ⑦ x^2 ⊜ √

with a display of 0.527, and the magnitude of the angle θ_T is

$$|\theta_T| = \tan^{-1}\frac{B_L - B_C}{G} = \tan^{-1}\frac{0.167\ \text{S}}{0.5\ \text{S}} = \tan^{-1}0.334$$
$$= 18.47^\circ$$

Calculator

⊡ ① ⑥ ⑦ ÷ ⊡ ⑤ ⊜ 2nd F $\tan^{-1}$

with a display of 18.469, so that $\mathbf{Y}_T = 0.527\ \text{S}\angle -18.47^\circ$.

$\mathbf{Z}_T$:

$$\mathbf{Z}_T = \frac{1}{Y_T}\angle -\theta_T = \frac{1}{0.527\ \text{S}}\angle -(-18.47^\circ) = \mathbf{1.898\ \Omega\angle 18.47^\circ}$$

$\mathbf{V}_R$, $\mathbf{V}_L$, $\mathbf{V}_C$: The voltage across parallel elements is the same, and $v_R = v_L = v_C = e = \mathbf{12\ sin\ \omega t}$.

The effective value $E = (0.707)(12\ \text{V}) = 8.484\ \text{V}$, and $\mathbf{E} = 8.484\ \text{V}\angle 0^\circ$, with $\mathbf{V}_R = 8.484\ \text{V}\angle 0^\circ$, $\mathbf{V}_L = 8.484\ \text{V}\angle 0^\circ$, and $\mathbf{V}_C = 8.484\ \text{V}\angle 0^\circ$.

I_R: The peak value of the current i_R is

$$I_m = \frac{V_m}{R} = \frac{E_m}{R} = \frac{12 \text{ V}}{2 \ \Omega} = 6 \text{ A}$$

and, since v_R and i_R are in phase,

$$i_R = \mathbf{6 \sin \omega t}$$

The effective value $I_{\text{eff}} = (0.707)(6 \text{ A}) = 4.242$ A, and $\mathbf{I}_R = \mathbf{4.242 \text{ A} \angle 0°}$.

I_L: The peak value of the current i_L is

$$I_m = \frac{V_m}{X_L} = \frac{E_m}{X_L} = \frac{12 \text{ V}}{4 \ \Omega} = 3 \text{ A}$$

and, since $v_L = e$ leads i_L by 90°, the angle associated with i_L must lag the voltage v_L by 90°. Therefore,

$$i_L = \mathbf{3 \sin(\omega t - 90°)}$$

The effective value $I_{\text{eff}} = (0.707)(3 \text{ A}) = 2.121$ A, and $\mathbf{I}_L = \mathbf{2.121 \text{ A} \angle -90°}$.

I_C: The peak value of the current i_C is

$$I_m = \frac{V_m}{X_C} = \frac{E_m}{X_C} = \frac{12 \text{ V}}{12 \ \Omega} = 1 \text{ A}$$

and, since i_C leads $v_C = e$ by 90°, the angle associated with i_C must lead e by 90°. Therefore,

$$i_C = \mathbf{1 \sin(\omega t + 90°)}$$

The effective value is $I_{\text{eff}} = (0.707)(1 \text{ A}) = 0.707$ A, and $\mathbf{I}_C = \mathbf{0.707 \text{ A} \angle +90°}$.

I_T: The angle associated with the total impedance is 18.44°, supporting our earlier conclusions that the network is inductive and e leads i by 18.44°.

The magnitude of the peak value of i can be determined from

$$I_m = \frac{E_m}{Z_T} = \frac{12 \text{ V}}{1.898 \ \Omega} = 6.32 \text{ A}$$

and $i_T = \mathbf{6.32 \sin(\omega t - 18.44°)}$.

The effective value is $I_{\text{eff}} = (0.707)(6.32 \text{ A}) = 4.468$ A, and $\mathbf{I}_T = \mathbf{4.468 \text{ A} \angle -18.44°}$.

Phasor Diagram The phasor diagram of the network appears in Fig. 15.23. Note in Fig. 15.23 that **E** and $\mathbf{I}_R$ are in phase, that $\mathbf{I}_C$ leads **E** by 90°, and $\mathbf{I}_L$ lags **E** by 90°. In addition, $\mathbf{I}_C$ and $\mathbf{I}_L$ are out of phase by 180° (directly opposing) and **E** leads **I** by 18.44°, due to the inductive nature of the network.

FIG. 15.23
Phasor diagram for the network of Fig. 15.21.

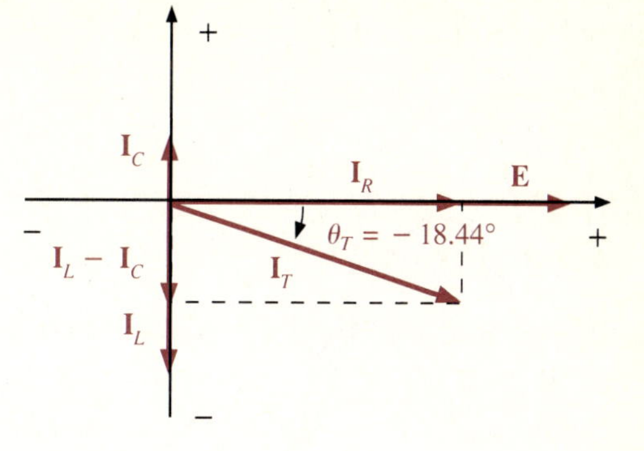

Finally, Kirchhoff's current law is satisfied, since

$$\mathbf{I}_T = \mathbf{I}_R + \mathbf{I}_L + \mathbf{I}_C$$

with the magnitude of $\mathbf{I}_T$ determined by

$$|I_T| = \sqrt{(I_R)^2 + (I_L - I_C)^2}$$

Power The total power is

$$\begin{aligned} P_T &= EI \cos \theta_T \\ &= (8.484 \text{ V})(4.468 \text{ A})\cos 18.44° \\ &= (37.91)(0.9487) \\ P_T &\cong \mathbf{36\ W} \end{aligned}$$

or

$$\begin{aligned} P_T &= I_R^2 R \\ &= (4.242 \text{ A})^2 (2\ \Omega) \\ P_T &\cong \mathbf{36\ W} \end{aligned}$$

Power Factor

$$F_p = \cos \theta_T = \cos 18.49° = \mathbf{0.949\ \ lagging} \qquad \text{(inductive network)}$$

or

$$F_p = \frac{G}{Y_T} = \frac{0.5 \text{ S}}{0.527 \text{ S}} = \mathbf{0.949\ \ lagging}$$

Sinusoidal Domain For this example,

$$\begin{aligned} e &= 12 \sin \omega t \\ i &= 6.32 \sin(\omega t - 18.44°) \\ i_R &= 6 \sin \omega t \\ i_L &= 3 \sin(\omega t - 90°) \\ i_C &= 1 \sin(\omega t + 90°) \end{aligned}$$

all of which are plotted in Fig. 15.24.

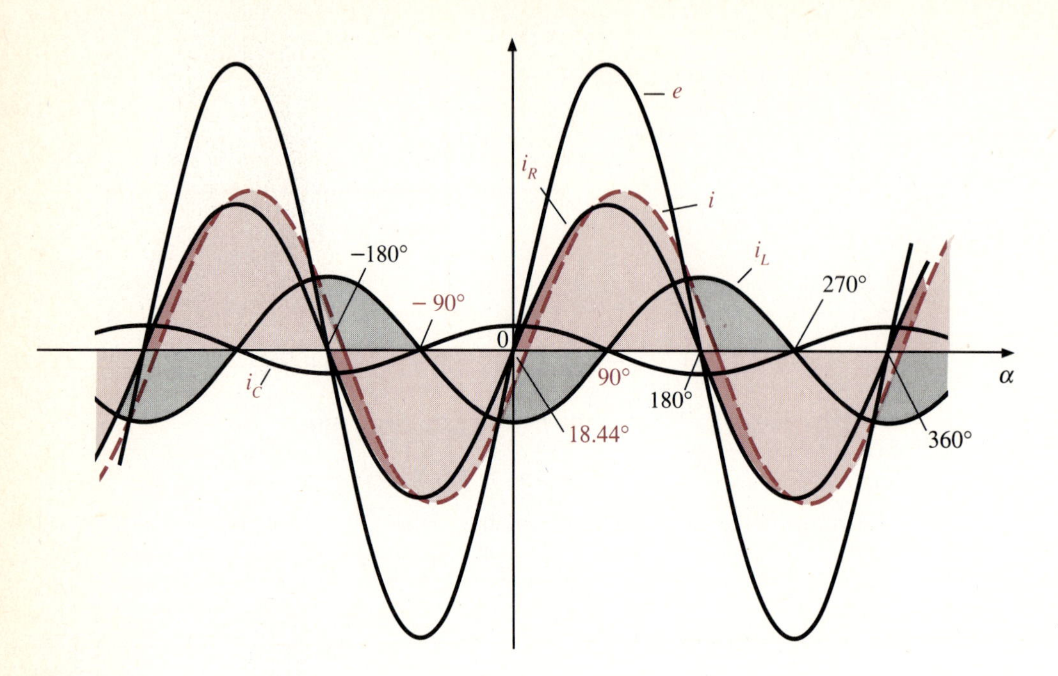

FIG. 15.24
Sinusoidal currents and voltage for the network of Fig. 15.21.

In particular, note that i_L and i_C are 180° out of phase, i_R leads i_C by 90°, and i_R lags i_L by 90°. The voltage e is in phase with i_R (resistor), leads i_L by 90° (inductor), and lags i_C by 90° (capacitor).

15.6 CURRENT SOURCES

The concept of a current source was introduced for dc systems. For sinusoidal networks, the symbol has the same appearance, and it continues to define the current in the branch in which it is located. The voltage across the current source is determined by the network to which it is applied.

EXAMPLE 15.5 For the network of Fig. 15.25, determine

a. $\mathbf{Y}_T$
b. $\mathbf{Z}_T$
c. $\mathbf{V}_S$
d. $\mathbf{I}_R$
e. $\mathbf{I}_L$

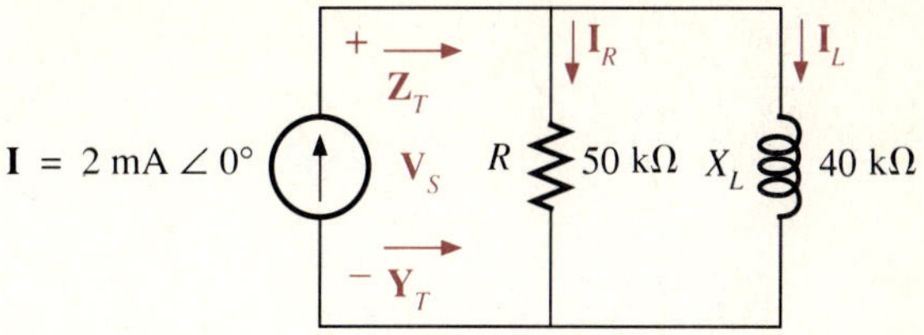

FIG. 15.25
Network for Example 15.5.

$\mathbf{I} = 2\ \text{mA} \angle 0^\circ$, $\mathbf{Z}_T$, $\mathbf{V}_S$, $\mathbf{Y}_T$, $\mathbf{I}_R$, R 50 kΩ, $\mathbf{I}_L$, X_L 40 kΩ

Solutions

a. $\mathbf{Y}_T = \mathbf{G} + \mathbf{B}_L = \dfrac{1}{50\ \text{k}\Omega}\underline{/0^\circ} + \dfrac{1}{40\ \text{k}\Omega}\underline{/-90^\circ}$

$= 0.02\ \text{mS}\ \underline{/0^\circ} + 0.025\ \text{mS}\ \underline{/-90^\circ}$

with

$$Y_T = \sqrt{(0.02\ \text{mS})^2 + (0.025\ \text{mS})^2} = 0.032\ \text{mS}$$

and

$$\theta_T = \tan^{-1}\frac{0.025\ \text{mS}}{0.02\ \text{mS}} = \tan^{-1}1.25 = 51.34^\circ$$

The admittance diagram appears in Fig. 15.26.

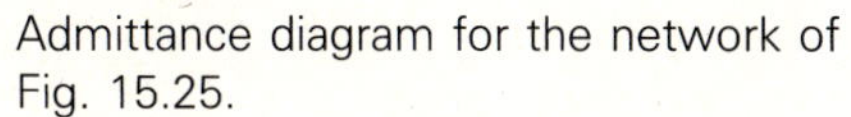

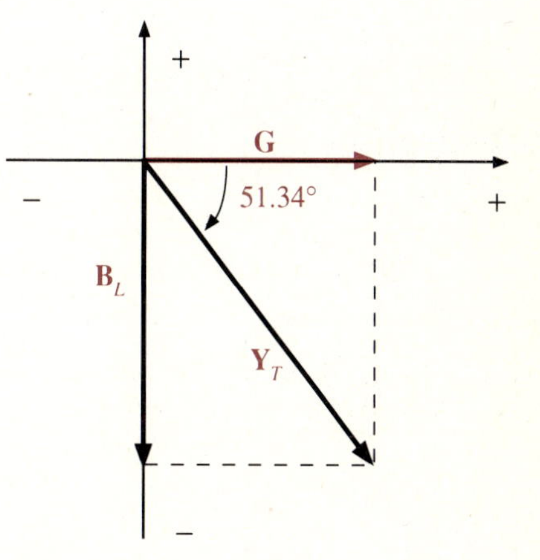

FIG. 15.26
Admittance diagram for the network of Fig. 15.25.

We conclude that

$$\mathbf{Y}_T = Y_T\,\underline{/\theta_T} = \mathbf{0.032\ mS}\ \underline{\mathbf{/-51.34^\circ}}$$

b. $Z_T = \dfrac{1}{Y_T} = \dfrac{1}{0.032\ \text{mS}} = 31.25\ \text{k}\Omega$

and $\mathbf{Z}_T = \mathbf{31.25\ k\Omega}\ \underline{\mathbf{/+51.34^\circ}}$.

c. $V_S = IZ_T = \dfrac{I}{Y_T} = \dfrac{2 \text{ mA}}{0.032 \text{ mS}} = 62.5 \text{ V}$

and, since $\mathbf{V}_S$ leads $\mathbf{I}$ by 51.34° (θ_T of $\mathbf{Z}_T = +51.34°$),

$$\mathbf{V}_S = \mathbf{62.5\ V\ \angle 51.34°}$$

d. $I_R = \dfrac{V_R}{R} = \dfrac{V_S}{R} = \dfrac{62.5 \text{ V}}{50 \text{ k}\Omega} = 1.25 \text{ mA}$

Since i_R and $v_R = v_S$ are in phase,

$$\mathbf{I}_R = \mathbf{1.25\ mA\ \angle 51.34°}$$

e. $I_L = \dfrac{V_L}{X_L} = \dfrac{V_S}{X_L} = \dfrac{62.5 \text{ V}}{40 \text{ k}\Omega} = 1.563 \text{ mA}$

Since $v_L = v_S$ leads i_L by 90°,

$$\mathbf{I}_L = 1.563 \text{ mA } \angle 51.34° - 90°$$

with $\mathbf{I}_L = \mathbf{1.563\ mA\ \angle -38.66°}$.

The phasor diagram of all the currents and voltages is provided in Fig. 15.27. Note again that $\mathbf{I} = \mathbf{I}_R + \mathbf{I}_L$ and $\mathbf{I}_R$ and $\mathbf{V}_S = \mathbf{V}_R$ are in phase. $\mathbf{V}_S = \mathbf{V}_L$ also leads $\mathbf{I}_L$ by 90° (51.34° + 38.66°).

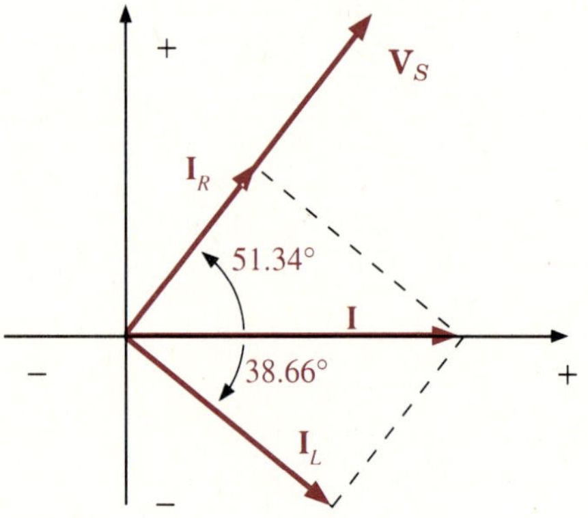

FIG. 15.27
Phasor diagram for the network of Fig. 15.25.

The currents $\mathbf{I}_R$ and $\mathbf{I}_L$ could also be determined using the current divider rule since the source current is the total current entering the parallel branches. Applying the current divider rule to ac networks is examined in the next section.

15.7 CURRENT DIVIDER RULE

The current divider rule introduced for dc circuits can also be applied to ac networks if we simply keep in mind the vector relationship between impedances and the phase relationship established by each element.

For a totally resistive configuration, such as the one appearing in Fig. 15.28, the peak value of the current i_1 can be determined by

$$I_{m_1} = \frac{R_2 I_m}{R_1 + R_2} \tag{15.13}$$

which has the same format as described for dc systems. The fact that the network is totally resistive eliminates any phase shifts, and the angle associated with i_1 is the same as that associated with i.

FIG. 15.28
Parallel resistive elements.

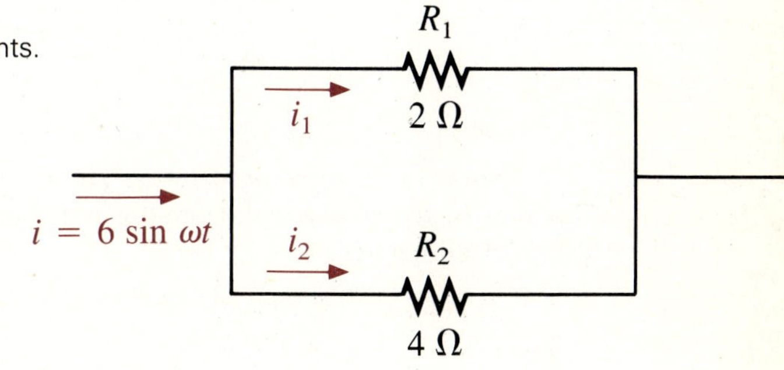

Substituting gives

$$I_{m_1} = \frac{R_2 I_m}{R_1 + R_2} = \frac{(4\ \Omega)(6\ \text{A})}{2\ \Omega + 4\ \Omega} = 4\ \text{A}$$

with i_1 = **4 sin** $\boldsymbol{\omega t}$.

For i_2:

$$I_{m_2} = \frac{R_1 I_m}{R_1 + R_2} \tag{15.14}$$

and

$$I_{m_2} = \frac{(2\ \Omega)(6\ \text{A})}{2\ \Omega + 4\ \Omega} = 2\ \text{A}$$

with i_2 = **2 sin** $\boldsymbol{\omega t}$.

Applying Kirchhoff's current law,

$$i = i_1 + i_2$$

and

$$6 \sin \omega t = 4 \sin \omega t + 2 \sin \omega t \qquad \text{(checks)}$$

If the effective values of the currents were provided in Fig. 15.28, Eqs. 15.13 and 15.14 would have the same format.

For a parallel network containing reactive elements, such as the network of Fig. 15.29, the magnitude of the effective value of $\mathbf{I}_R$ can be determined from

$$I_R = \frac{XI}{\sqrt{R^2 + X^2}} \tag{15.15}$$

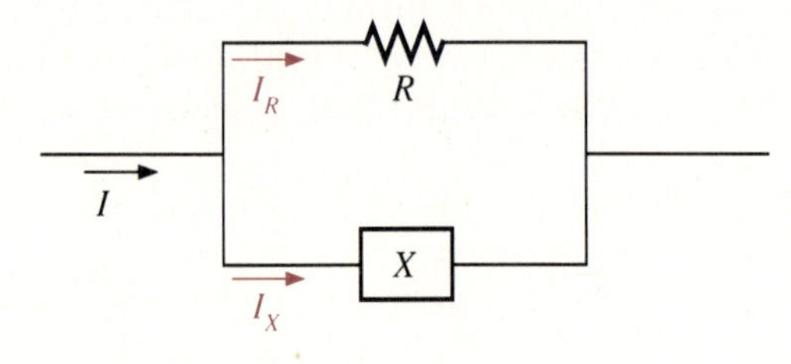

FIG. 15.29
Parallel *R-X* network.

Note the requirement to use the Pythagorean theorem in the denominator due to the vector relationship between impedances.

In addition,

$$I_X = \frac{RI}{\sqrt{R^2 + X^2}} \tag{15.16}$$

As noted for the voltage divider rule, the preceding equations provide the effective value of the unknown quantity. However, if the peak value is preferred, the peak value of i is simply substituted in place of the effective value.

For $\mathbf{I} = I\angle 0°$, the angle associated with i_R in Fig. 15.29 is the angle associated with the total impedance of the parallel branches. If X is inductive, the phase angle associated with I_X is 90° less than that of the total impedance; if capacitive, it is 90° more.

If both parallel elements are reactive, the numerator of the equation is the product of the reactances; the denominator is the difference if they are different reactances and the sum if the same. For the inductive branch, the phase angle is 90° less than the phase angle associated with the total parallel impedance, and for the capacitive branch, the phase angle is 90° more.

EXAMPLE 15.6

Determine the currents $\mathbf{I}_R$ and $\mathbf{I}_X$ for the parallel configuration of Fig. 15.30.

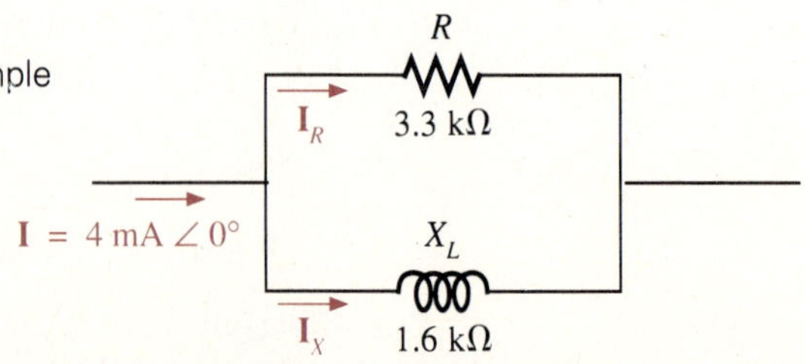

FIG. 15.30
Parallel *R-L* configuration for Example 15.6.

Solution Effective values of $\mathbf{I}_R$ and $\mathbf{I}_L$ are

$$I_R = \frac{X_L I}{\sqrt{R^2 + X_L^2}} = \frac{(1.6\ \text{k}\Omega)(4\ \text{mA})}{\sqrt{(3.3\ \text{k}\Omega)^2 + (1.6\ \text{k}\Omega)^2}}$$

$$= \frac{6.4\ \text{A}}{3.667 \times 10^3} = \mathbf{1.745\ mA}$$

$$I_L = \frac{RI}{\sqrt{R^2 + X^2}} = \frac{(3.3\ \text{k}\Omega)(4\ \text{mA})}{3.667\ \text{k}\Omega}$$

$$= \mathbf{3.6\ mA}$$

Phase angles are

$$\mathbf{Y}_T = \mathbf{G} + \mathbf{B}_L = \frac{1}{3.3\ \text{k}\Omega}\angle 0^\circ + \frac{1}{1.6\ \text{k}\Omega}\angle -90^\circ$$

$$= 0.303\ \text{mS}\angle 0^\circ + 0.625\ \text{mS}\angle -90^\circ$$

$$\mathbf{Y}_T = 0.695\ \text{mS}\angle -64.14^\circ$$

$$\mathbf{Z}_T = \frac{1}{0.695\ \text{mS}}\angle +64.14^\circ = 1.439\ \text{k}\Omega\angle +64.14^\circ$$

Therefore,

$$\mathbf{I}_R = \mathbf{1.745\ mA\angle +64.14^\circ}$$

and

$$\mathbf{I}_L = 3.6\ \text{mA}\angle +64.14^\circ - 90^\circ$$
$$= \mathbf{3.6\ mA\angle -25.86^\circ}$$

In the majority of laboratory exercises, the magnitude of a current or voltage is the quantity of primary concern. The need for the phase angle is normally reserved for more specific development activities. In the preceding analysis, therefore, the application of Eqs. 15.15 and 15.16 is normally sufficient.

EXAMPLE 15.7 Determine the reading of the rms milliammeter of Fig. 15.31.

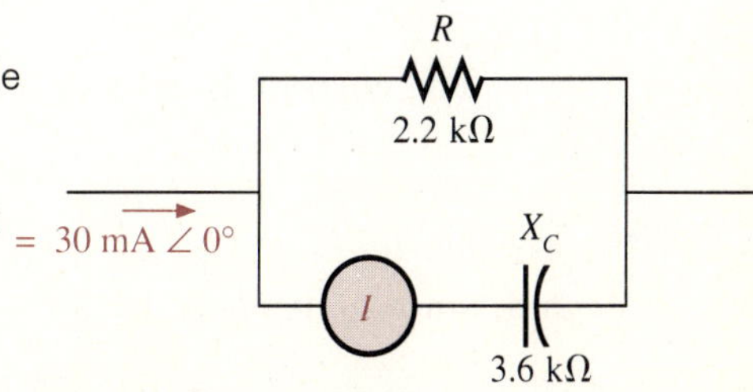

FIG. 15.31
Parallel *R-C* combination for Example 15.7.

Solution
$$I_C = \frac{X_C I}{\sqrt{R^2 + X_C^2}} = \frac{(2.2\ \text{k}\Omega)(30\ \text{mA})}{\sqrt{(2.2\ \text{k}\Omega)^2 + (3.6\ \text{k}\Omega)^2}}$$

Calculator for the Denominator

(2) (·) (2) (EXP) (3) (x^2) (+) (3) (·) (6) (EXP) (3) (x^2) (=) (√)

with a display of 4219.005.

Calculator for the Solution

(2) (·) (2) (EXP) (3) (×) (3) (0) (EXP) (+/−) (3) (=) (÷) (4) (2) (1) (9) (=)

with a display of 0.01564; I_C = **15.64 mA,** reflecting the indication of the milliammeter.

EXAMPLE 15.8 Determine the source current for the network of Fig. 15.32.

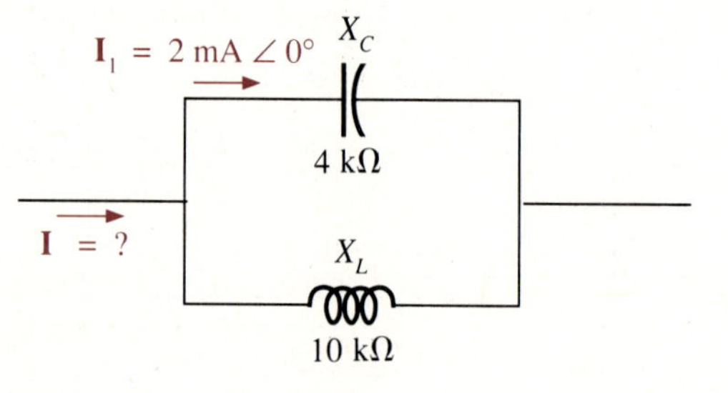

FIG. 15.32
Parallel *L-C* configuration for Example 15.8.

Solution The voltage across the 4-kΩ capacitive reactance is

$$\begin{aligned} V_C &= I_1 X_C = (2\ \text{mA})(4\ \text{k}\Omega) \\ &= 8\ \text{V} \end{aligned}$$

and, since i_C leads v_C by 90°,

$$\mathbf{V}_C = 8\ \text{V}\ \underline{/-90^\circ}$$

The voltage across X_L is the same as that across X_C because of the parallel arrangement, and

$$\mathbf{V}_L = 8\ \text{V}\ \underline{/-90^\circ}$$

The current $\mathbf{I}_L$ is

$$I_L = \frac{V_L}{X_L} = \frac{8\ \text{V}}{10\ \text{k}\Omega} = 0.8\ \text{mA}$$

and, since v_L leads i_L by 90°,

$$\mathbf{I}_L = 0.8\ \text{mA}\ \underline{/-90^\circ - 90^\circ} = 0.8\ \text{mA}\ \underline{/-180^\circ}$$

Applying Kirchhoff's current law gives

$$\begin{aligned}\mathbf{I} &= \mathbf{I}_C + \mathbf{I}_L \\ &= 2\text{ mA }\angle 0^\circ + 0.8\text{ mA }\angle -180^\circ\end{aligned}$$

Thus

$$\mathbf{I} = (2\text{ mA} - 0.8\text{ mA})\ \angle 0^\circ = \mathbf{1.2\ mA\ \angle 0^\circ}$$

15.8 FREQUENCY RESPONSE

Recall for series *R-L-C* circuits that the element with the largest impedance determines whether the network is resistive, inductive, or capacitive and can often represent an approximate measure of the total impedance.

For parallel networks, the last few sections and examples have demonstrated that the element with the smallest impedance usually determines whether the network is resistive, inductive, or capacitive; if it is small enough compared to the other elements, it may be very close in magnitude to the total impedance or admittance.

For the parallel *R-L* network of Fig. 15.33, if we begin our frequency analysis at $f = 0$ Hz, we find that $X_L = 2\pi fL = 2\pi(0)L = 0\ \Omega$, and the magnitude of the total impedance is

$$Z_T = \frac{(R)(X_L)}{\sqrt{R^2 + X_L^2}} = \frac{(R)(0\ \Omega)}{\sqrt{R^2 + (0)^2}} = 0\ \Omega$$

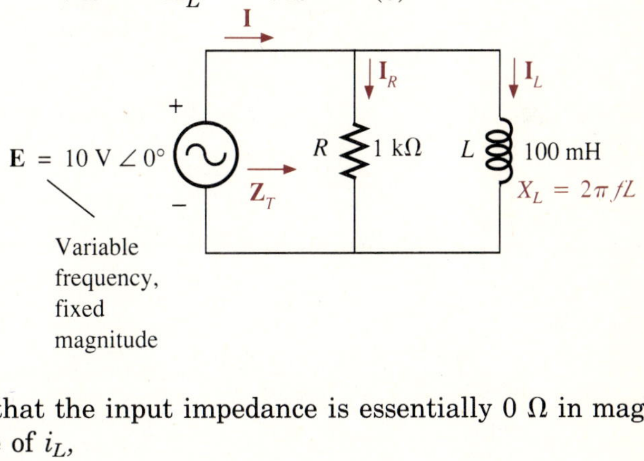

FIG. 15.33
Effect of frequency on the impedance and current levels.

The result is that the input impedance is essentially 0 Ω in magnitude and the peak value of i_L,

$$I_m = \frac{V_m}{X_L} = \frac{E_m}{X_L} = \frac{E_m}{0\ \Omega}$$

is very large.

At low frequencies the inductive reactance is relatively small in magnitude and controls the characteristics of the network and the network. That is, θ_T of $\mathbf{Z}_T = Z_T \angle \theta_T$ is essentially 90°, and e leads i by 90° (an inductive load).

At very high frequencies $X_L = 2\pi fL$ approaches levels that permit an open-circuit approximation for the coil. The result is that i_L approaches 0 A and the total resistance $\mathbf{Z}_T \cong R$. In summary, therefore, at low frequencies the network appears inductive, whereas at high frequencies it takes on resistive characteristics. In addition, at low frequencies e leads i by almost 90°, whereas at high frequencies e and i approach an in-phase relationship.

Between the low- and high-frequency regions, the magnitude of the impedance is determined by

$$Z_T = \frac{RX_L}{\sqrt{R^2 + X_L^2}}$$

and the angle θ_T is found from

$$\theta_T = \tan^{-1}\frac{R}{X_L}$$

For $f = 0.1$ kHz, substitution into the preceding equations results in $Z_T = 62.67\ \Omega$ with $\theta_T = 86.41°$.

$$\begin{aligned}
\text{At } f &= 1\text{ kHz, } Z_T = 0.532\text{ k}\Omega\text{, with } \theta_T = 57.87°.\\
\text{At } f &= 5\text{ kHz, } Z_T = 0.9528\text{ k}\Omega\text{, with } \theta_T = 17.67°.\\
\text{At } f &= 10\text{ kHz, } Z_T = 0.9876\text{ k}\Omega\text{, with } \theta_T = 9.05°.
\end{aligned}$$

Plotting the preceding results in the curves of Fig. 15.34 and Fig. 15.35.

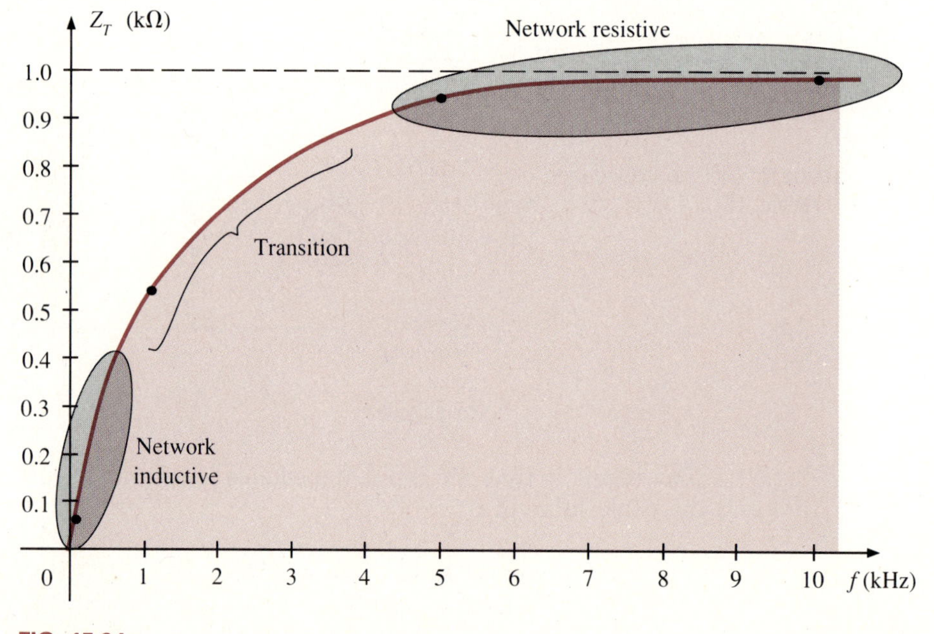

FIG. 15.34

Z_T vs. frequency for the parallel R-L network of Fig. 15.33.

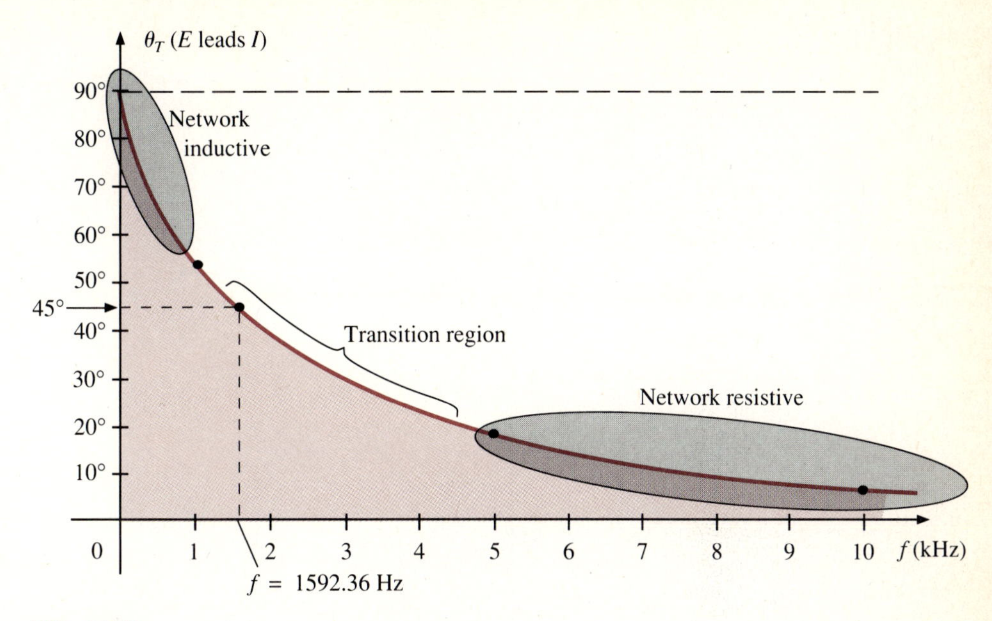

FIG. 15.35
θ_T vs. frequency for the parallel *R-L* network of Fig. 15.33.

Note in Fig. 15.34 how quickly the impedance Z_T increases toward its final resistive limit of 1 kΩ. For frequencies less than 1 kHz, the network has inductive characteristics, whereas for frequencies greater than 5 kHz, its terminal characteristics are primarily resistive. This is further evidenced by the plot of θ_T in Fig. 15.35, where the impedance angle is greater than 60° for frequencies less than 1 kHz but less than 20° for frequencies greater than 5 kHz.

If the peak value of i_R were plotted against frequency, the curve would be a horizontal straight line [Fig. 15.36(a)], since both the peak value of v_R (E_m) and the resistance R are constant. That is,

$$I_m = \frac{V_m}{R} = \frac{E_m}{R}$$

However, the peak value of the inductive current i_L is determined by

$$I_m = \frac{V_m}{X_L} = \frac{E_m}{2\pi f L} = \frac{E_m}{2\pi L(f)}$$

which is inversely related to the applied frequency. In other words, as f increases, the magnitude of I_m decreases rapidly in a parabolic manner, as shown in Fig. 15.36(b). Surely, i_L is its maximum value when X_L is its smallest (short-circuit state) and i_L is a minimum when X_L approaches its open-circuit state.

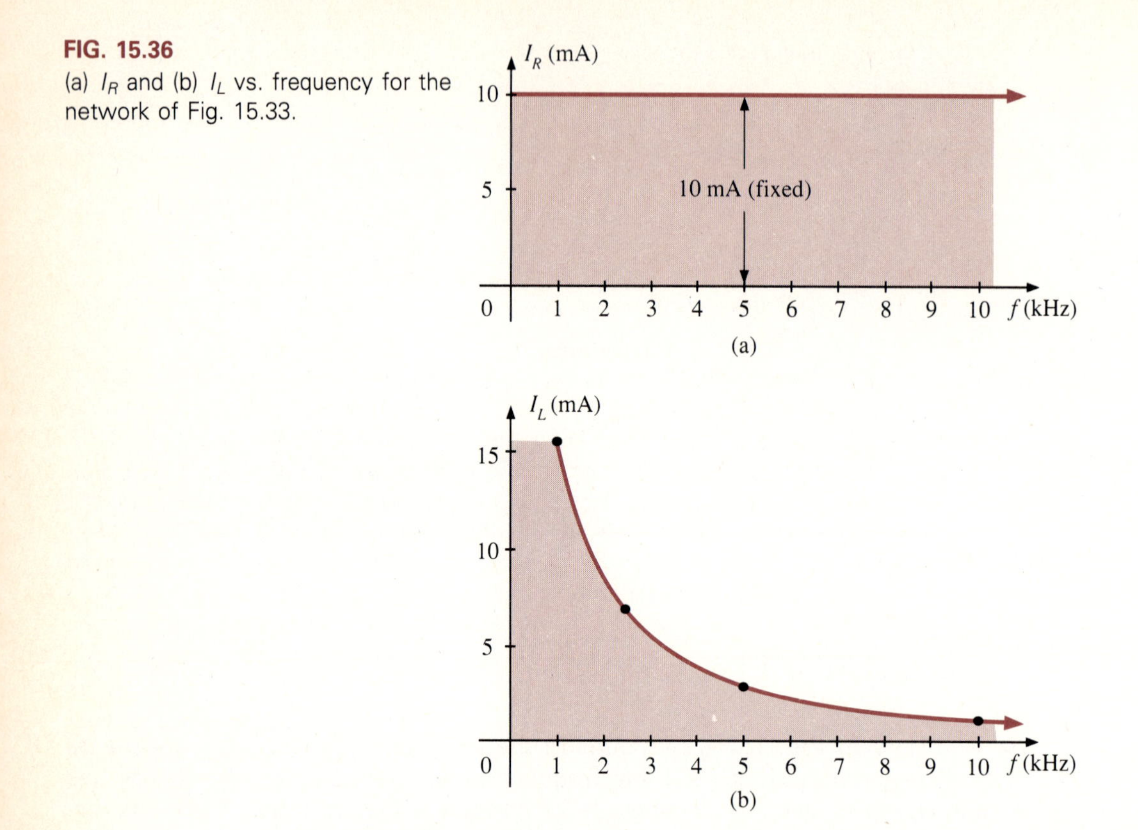

FIG. 15.36
(a) I_R and (b) I_L vs. frequency for the network of Fig. 15.33.

A plot of the source current **I** versus frequency is provided in Fig. 15.37. Note at low frequencies how the inductive reactance is the predominant factor and $\mathbf{I} \cong \mathbf{I}_L$. As the applied frequency increases, the magnitude of the current I_L

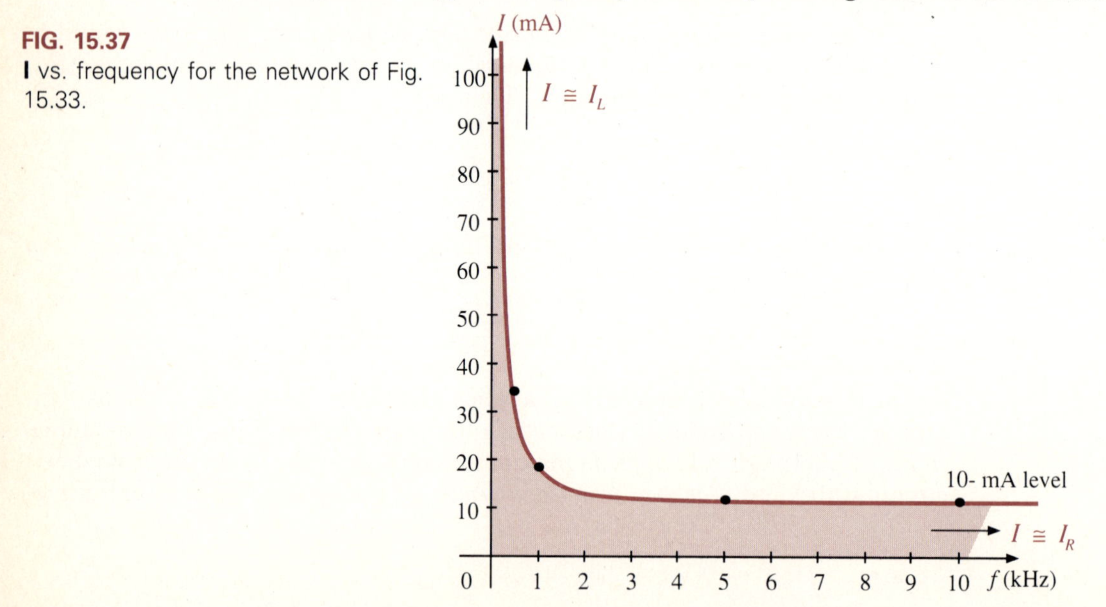

FIG. 15.37
I vs. frequency for the network of Fig. 15.33.

decreases to the point where I_R is greater than I_L and **I** approaches a constant level of $\mathbf{I} \cong \mathbf{I}_R$, which in this case is equal to $I_R = E/R = 10\text{ V}/1\text{ k}\Omega = 10\text{ mA}$. Always keep in mind, however, that the currents are related by a vector relationship $I = \sqrt{I_R^2 + I_L^2}$ and not simply an algebraic (plus or minus) equation.

For an *R*-*C* parallel network, the effect of an increasing frequency is the opposite of that obtained for the parallel *R*-*L* network. At low frequencies the capacitor is near its open-circuit equivalent ($Z_T \cong R$), and at high frequencies, it approaches its short-circuit state and $Z_T \cong 0\ \Omega$ with $\theta_T \cong 90°$ (*i* leading *e*).

The frequency response for the current of each element of the parallel *R*-*L*-*C* network of Fig. 15.38 is provided in the same figure. Note for the 0.02-μF capacitor how the current increases in an exponential manner with frequency at a rate similar to the rate at which I_L is decreasing in magnitude. Even though the magnitude of the source current is related to the branch currents by

$$I = \sqrt{I_R^2 + (I_L - I_C)^2}$$

it is possible to compare relative magnitudes to determine the general appearance of I versus frequency. At low frequencies the magnitude of I_L far outweighs the magnitude of I_R or I_C, and the shape of I is identical to that of I_L versus frequency. The same can be said for the impact of I_C at high frequencies. In the midfrequency spectrum we can expect a valley in the response curve for I versus frequency, with a minimum value occurring when $I_L = I_C$ and $I = \sqrt{I_R^2 + (0)^2} = I_R$. The frequency at which this phenomenon will occur is the same as will establish $X_L = X_C$. Solving for the frequency that establishes $X_L = X_C$ results in

$$f_p = \frac{1}{2\pi\sqrt{LC}} \tag{15.17}$$

which is similar to the value obtained for the same condition in a series *R*-*L*-*C* circuit.

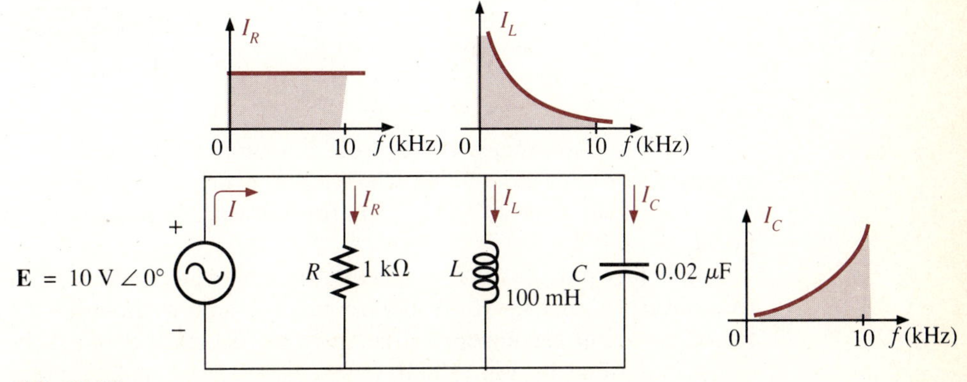

FIG. 15.38
Frequency response curves for each element of a parallel *R*-*L*-*C* network.

For this network,

$$f_p = \frac{1}{2\pi\sqrt{LC}} = \frac{1}{6.28\sqrt{(100 \times 10^{-3}\text{ H})(0.02 \times 10^{-6}\text{ F})}} = \mathbf{3560.6\ Hz}$$

Based on the preceding calculations, the curve of **I** versus frequency will have the basic shape of Fig. 15.39.

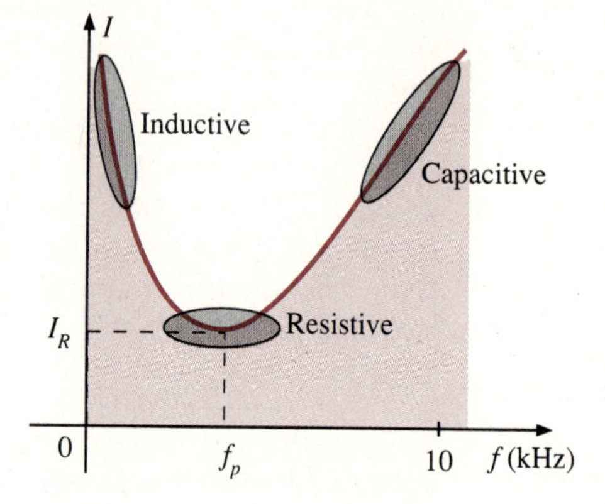

FIG. 15.39
I vs. frequency for the parallel *R-L-C* network of Fig. 15.38.

A similar approach results in the curve of the phase angle between *e* and *i* versus frequency appearing in Fig. 15.40.

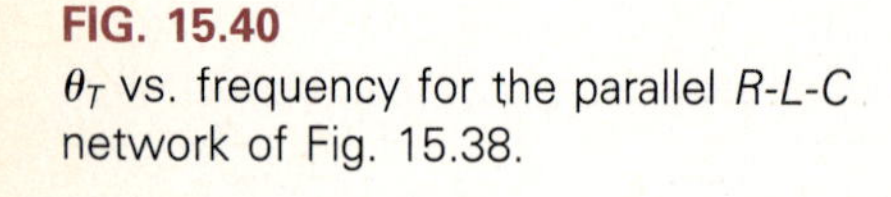

FIG. 15.40
θ_T vs. frequency for the parallel *R-L-C* network of Fig. 15.38.

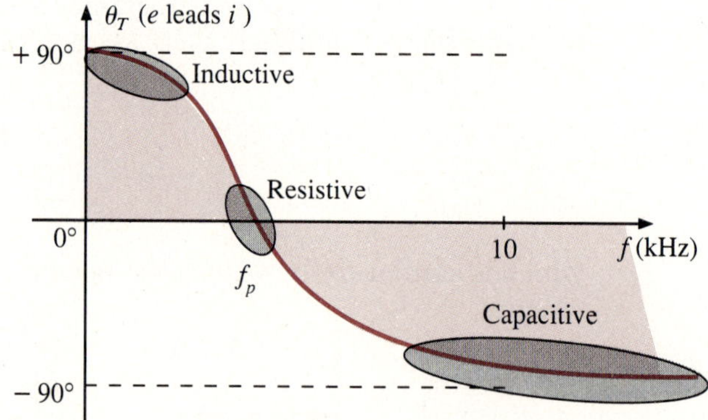

At low frequencies the system is primarily inductive (*e* leads *i* by 90°), whereas at high frequencies the system is primarily capacitive (*e* leads *i* by −90° or *i* leads *e* by 90°). At $f = f_p$ the network is resistive and *e* and *i* are in phase ($\theta_T = 0°$).

For any parallel *R-L-C* circuit the response curve for *I* will be similar to that obtained for Fig. 15.39. There will be some with wider (or narrower) valleys and higher (or lower) values of I_R at f_p, but in general the characteristic curve is defined by Fig. 15.39.

15.9 EQUIVALENT CIRCUITS

In a series ac circuit, the total impedance of two or more elements in series is often equivalent to an impedance that can be achieved with fewer elements of different values, the elements and their values being determined by the frequency applied. This is also true for parallel circuits. For example, the circuit of Fig. 15.41(a) has a total impedance at the frequency applied that is equivalent to a capacitor with a reactance of 10 Ω, as shown in Fig. 15.41(b). Always keep in mind that this equivalence is true only at the applied frequency. If the frequency changes, the reactance of each element changes, and the equivalent circuit will change—perhaps from capacitive to inductive in this example.

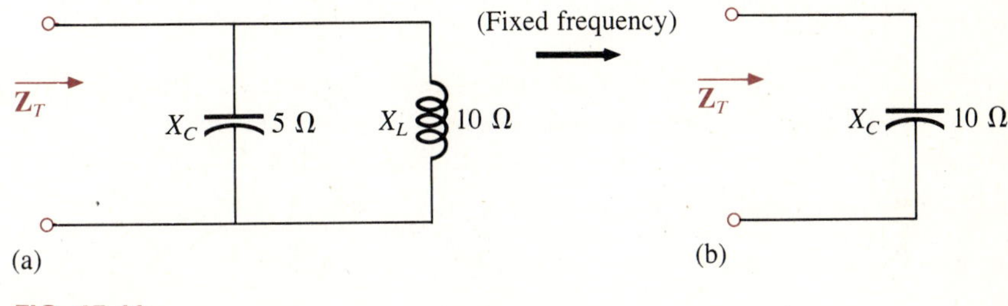

FIG. 15.41
Networks with equivalent input impedances.

In addition, keep in mind that the equivalence is true only for the total impedance of the network (at a particular frequency) that establishes the relationship between the applied voltage and the resulting current.

An equivalence also exists between a parallel R-X network and a series R-X circuit, as shown in Fig. 15.42. In other words, there exists a series combination of elements that will have the same input impedance as the parallel network of Fig. 15.42(a). Through an analysis involving both the magnitude and angle of each impedance, it can be shown that the required series elements are defined by

$$R_s = \frac{R_p X_p^2}{R_p^2 + X_p^2} \tag{15.18}$$

and

$$X_s = \frac{R_p^2 X_p}{R_p^2 + X_p^2} \tag{15.19}$$

where both R_p and X_p are defined by Fig. 15.42(a). The type of reactance X_s is also the same for both the series and parallel circuits. That is, an R-L or R-C parallel network has an equivalent R-L or R-C series circuit, respectively.

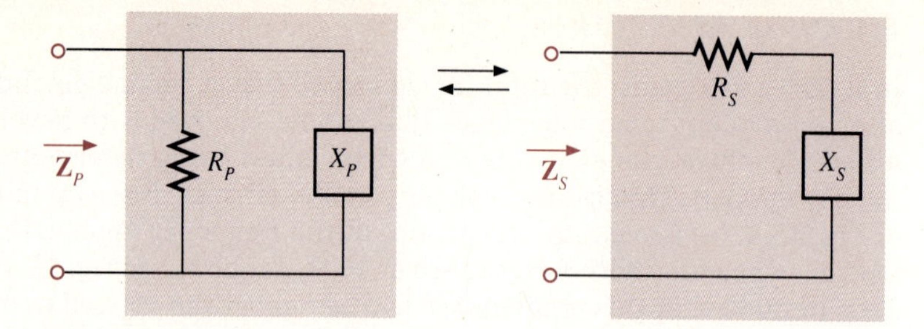

FIG. 15.42
Equivalent parallel and series networks as far as the input impedance is concerned (at a particular frequency).

The equivalence works in reverse also. For the parallel network of Fig. 15.42(a), the network parameters are defined by

$$R_p = \frac{R_s^2 + X_s^2}{R_s} \tag{15.20}$$

and

$$X_p = \frac{R_s^2 + X_s^2}{X_s} \tag{15.21}$$

EXAMPLE 15.9 Determine the equivalent series network for the parallel configuration of Fig. 15.43.

FIG. 15.43
Network for Example 15.9.

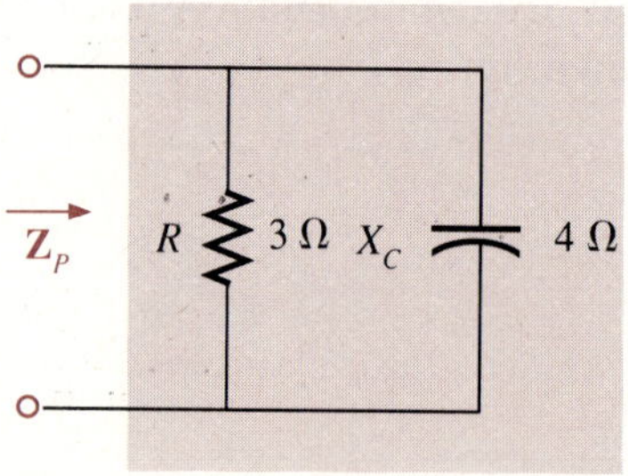

Solution From Eq. 15.18,

$$R_s = \frac{R_p X_p^2}{R_p^2 + X_p^2} = \frac{(3\ \Omega)(4\ \Omega)^2}{(3\ \Omega)^2 + (4\ \Omega)^2}$$

$$= \frac{(3)(16)}{9 + 16}\ \Omega = \frac{48}{25}\ \Omega = \mathbf{1.92\ \Omega}$$

Calculator

(3) (×) (4) (x^2) (=) (÷) (( ) (3) (x^2) (+) (4) (x^2) ( )) (=)

with a display of 1.92.

From Eq. 15.19,

$$X_s = \frac{R_p^2 X_p}{R_p^2 + X_p^2} = \frac{(3\ \Omega)^2(4\ \Omega)}{(3\ \Omega)^2 + (4\ \Omega)^2} = \frac{(9)(4)}{25}\ \Omega = \frac{36}{25}\ \Omega$$

$$= \mathbf{1.44\ \Omega} \qquad \text{(capacitive)}$$

Calculator

(3) (x^2) (×) (4) (=) (÷) (2) (5) (=)

with a display of 1.44.

The equivalent network appears in Fig. 15.44.

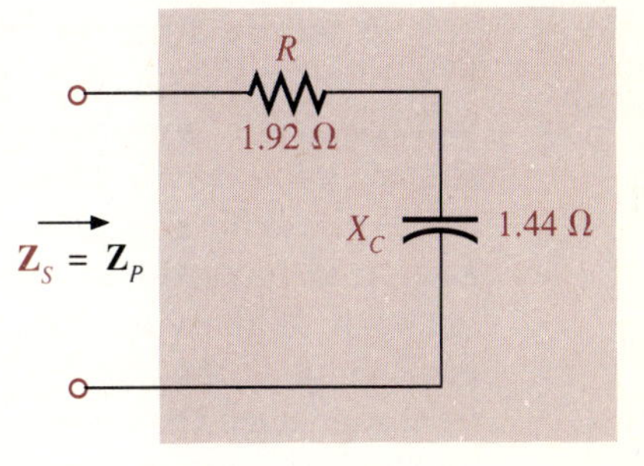

FIG. 15.44
Series circuit equivalent for the parallel network of Fig. 15.43 (as far as $\mathbf{Z}_T$ is concerned).

15.10 MEASUREMENTS—PARALLEL ac NETWORKS

A number of the general comments appearing in the discussion of measurements in a series ac network (Section 14.10) are applicable to parallel ac networks also. That is, most instruments provide the rms value of a sinusoidal voltage and current, and the applied frequency will not affect the performance of the instrument (unless very low or higher than normally encountered—check the instrument specifications). In addition, the current in a network is normally determined using a measured voltage across a resistor and Ohm's law rather than disturbing the network, as required for the insertion of ammeters.

The concerns introduced in Section 14.10 regarding the proper grounding of the vertical input of oscilloscopes are applicable to parallel ac networks also. In Fig. 15.45, since the generator, oscilloscope, and network share a common ground, v_R, v_L, and v_C can all be properly displayed on the scope. However, since $e = v_R = v_L = v_C$, all the connections indicated in Fig. 15.45 result in the same display.

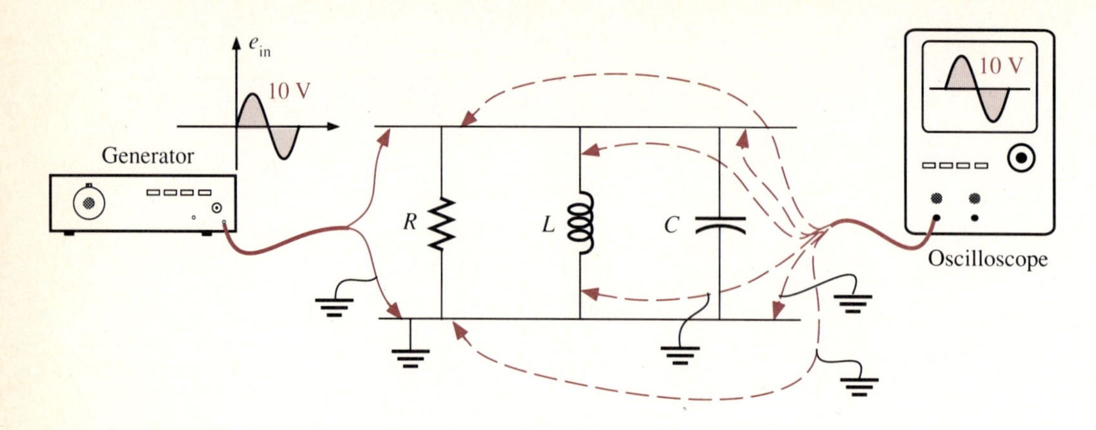

FIG. 15.45
Measuring parallel voltages with an oscilloscope.

A totally different situation arises, however, if the currents i_R and i_L of Fig. 15.46 are displayed on a scope. First, oscilloscopes display only sinusoidal voltages, so a "sensing resistor" of 10 Ω must be inserted in each branch, as shown in Fig. 15.46. The resulting waveforms will be of v_{R_s}, but since v_R and i_R are in phase for a resistor, the waveform for v_{R_s} will be the same in appearance as i_R and have the same phase relationship with the other elements of the network. When viewing v_{R_s} for each branch, one simply has to determine the peak value of the current from $I_m = V_{R_s}(\text{max})/R_s$. The sensing resistor must be chosen such that its magnitude will not approach that of the series element for the frequency range of interest. This is certainly true for the resistor where $R \gg R_s$ and for the inductive reactance where $X_L \gg R_s$. The dual-channel (two signals can be applied at once) scope will display both i_R and i_L and reveal their phase relationship, as shown on the display of Fig. 15.46. The scope is truly displaying the shape (relative magnitude) of each current and the phase relationship between i_R and i_L, since elements in series have the same current and the sensing resistor is in series with the other element of the branch. Note also the common connection between the grounds of the generator, network, and scope to avoid the problem introduced in Section 14.10.

15.11 THE *j*-OPERATOR—PARALLEL ac NETWORKS

The *j*-operator can be an effective instrument in the analysis of parallel ac networks. In general, however, it is suggested that the following guidelines be observed.

Addition and subtraction of vectors should be performed in the rectangular form.

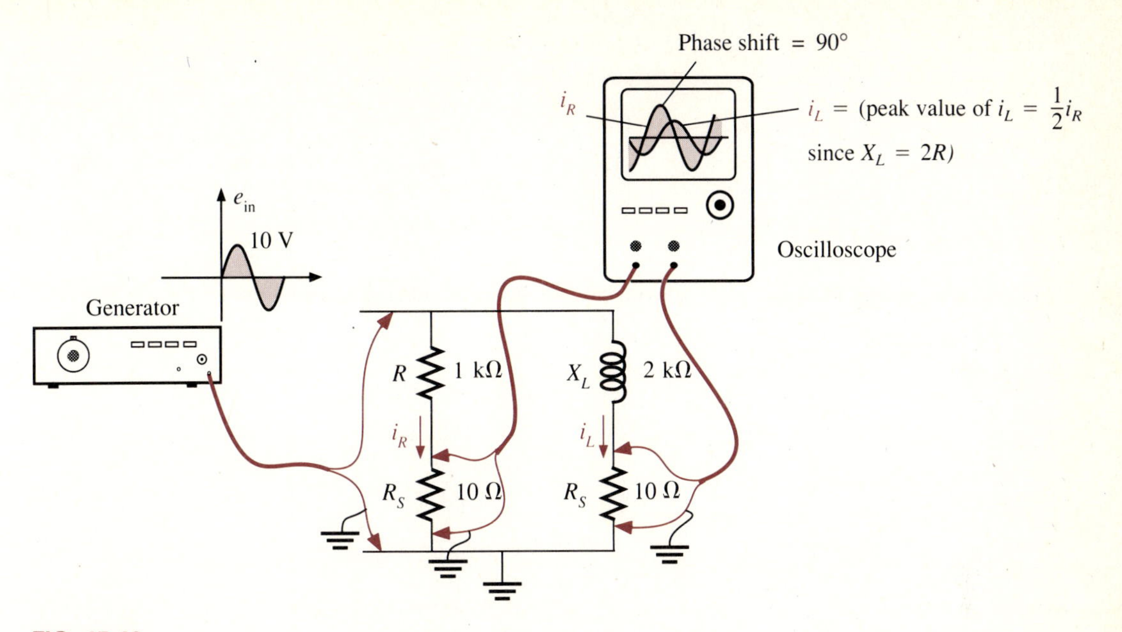

FIG. 15.46
Determining the phase relationship between i_R and i_L.

Multiplication and division should be performed in the polar form.

Since all the required operations with the *j*-operator have been introduced in past sections, the application of the *j*-operator with parallel networks is best demonstrated through a few examples.

EXAMPLE 15.10 Determine the total impedance of the network of Fig. 15.8 using the *j*-operator.

Solution From Eq. 15.7:

$$\mathbf{Y}_T = \mathbf{Y}_1 + \mathbf{Y}_2 + \mathbf{Y}_3$$

$$= \frac{1}{4\ \Omega\ \angle 0^\circ} + \frac{1}{2\ \Omega\ \angle 90^\circ} + \frac{1}{5\ \Omega\ \angle -90^\circ}$$

$$= 0.25\ \text{S}\ \angle 0^\circ + 0.5\ \text{S}\ \angle -90^\circ + 0.2\ \text{S}\ \angle +90^\circ$$

$$= 0.25\ \text{S} - j0.5\ \text{S} + j0.25\ \text{S} = 0.25\ \text{S} - j0.25\ \text{S}$$

$$\mathbf{Y}_T = 0.391\ \text{S}\ \angle -50.194^\circ$$

$$\mathbf{Z}_T = \frac{1}{\mathbf{Y}_T} = \frac{1}{0.391\ \text{S}\ \angle -50.194^\circ} = \mathbf{2.558\ \Omega\ \angle 50.194^\circ}$$

as obtained in Example 15.3.

With the j-operator at our disposal, the equation for the total impedance of two parallel impedances can be written as

$$\mathbf{Z}_T = \frac{\mathbf{Z}_1\mathbf{Z}_2}{\mathbf{Z}_1 + \mathbf{Z}_2} \qquad (15.22)$$

EXAMPLE 15.11 Repeat Example 15.2 using Eq. 15.22 and compare results.

Solution

$$\mathbf{Z}_T = \frac{\mathbf{Z}_1\mathbf{Z}_2}{\mathbf{Z}_1 + \mathbf{Z}_2} = \frac{(10\ \Omega\ \angle 0^\circ)(10\ \Omega\ \angle 90^\circ)}{10\ \Omega + j10\ \Omega}$$

$$= \frac{(10)(10)\ \Omega\ \angle 0^\circ + 90^\circ}{14.14\ \Omega\ \angle 45^\circ} = \frac{100\ \Omega\ \angle 90^\circ}{14.14\ \Omega\ \angle 45^\circ}$$

$$= 7.071\ \Omega\ \angle 90^\circ - 45^\circ$$

$$\mathbf{Z}_T = \mathbf{7.071\ \Omega\ \angle 45^\circ}$$

EXAMPLE 15.12 Determine the total impedance of the parallel R-C network of Fig. 15.18.

Solution

$$\mathbf{Z}_T = \frac{\mathbf{Z}_1\mathbf{Z}_2}{\mathbf{Z}_1 + \mathbf{Z}_2} = \frac{(40\ \Omega\ \angle 0^\circ)(50\ \Omega\ \angle -90^\circ)}{40\ \Omega - j50\ \Omega}$$

$$= \frac{(40)(50)\ \angle 0^\circ - 90^\circ}{64.03\ \angle -51.34^\circ} = \frac{2000\ \angle -90^\circ}{64.03\ \angle -51.34^\circ}$$

$$= 31.24\ \Omega\ \angle -90^\circ - (-51.34^\circ)$$

$$\mathbf{Z}_T = \mathbf{31.24\ \Omega\ \angle -38.66^\circ}$$

as obtained in Section 15.4.

EXAMPLE 15.13 Determine the currents $\mathbf{I}_R$, $\mathbf{I}_L$, and $\mathbf{I}_C$ for the parallel R-L-C network of Fig. 15.21.

Solution Since the elements are all in parallel with the applied source,

$$\mathbf{E} = \mathbf{V}_R = \mathbf{V}_L = \mathbf{V}_C$$

and

$$\mathbf{I}_R = \frac{\mathbf{V}_R}{\mathbf{R}} = \frac{8.484\ \text{V}\ \angle 0^\circ}{2\ \Omega\ \angle 0^\circ} = \mathbf{4.242\ A\ \angle 0^\circ}$$

with

$$\mathbf{I}_L = \frac{\mathbf{V}_L}{\mathbf{X}_L} = \frac{8.484\ \text{V}\ \angle 0^\circ}{4\ \Omega\ \angle 90^\circ} = 2.121\ \text{A}\ \angle 0^\circ - 90^\circ = \mathbf{2.121\ A\ \angle -90^\circ}$$

and

$$\mathbf{I}_C = \frac{\mathbf{V}_C}{\mathbf{X}_C} = \frac{8.484\text{ V}\,\angle 0^\circ}{12\,\Omega\,\angle -90^\circ} = 0.707\text{ A}\,\angle 0^\circ - (-90^\circ) = \mathbf{0.707\text{ A}\,\angle 90^\circ}$$

For two parallel impedances, the current divider rule can be written as

$$\boxed{\mathbf{I}_1 = \frac{\mathbf{Z}_2\mathbf{I}_T}{\mathbf{Z}_1 + \mathbf{Z}_2}} \tag{15.23}$$

and

$$\boxed{\mathbf{I}_2 = \frac{\mathbf{Z}_1\mathbf{I}_T}{\mathbf{Z}_1 + \mathbf{Z}_2}} \tag{15.24}$$

EXAMPLE 15.14

Repeat Example 15.6 using Eqs. 15.23 and 15.24.

Solution

$$\mathbf{I}_R = \frac{\mathbf{X}_L\mathbf{I}}{\mathbf{R} + \mathbf{X}_L} = \frac{(1.6\text{ k}\Omega\,\angle 90^\circ)(4\text{ mA}\,\angle 0^\circ)}{3.3\text{ k}\Omega + j1.6\text{ k}\Omega}$$

$$= \frac{6.4\text{ V}\,\angle 90^\circ}{3.667\text{ k}\Omega\,\angle 25.86^\circ}$$

$$\mathbf{I}_R = \mathbf{1.745\text{ mA}\,\angle 64.14^\circ}$$

$$\mathbf{I}_L = \frac{\mathbf{R}\mathbf{I}}{\mathbf{R} + \mathbf{X}_L} = \frac{(3.3\text{ k}\Omega\,\angle 0^\circ)(4\text{ mA}\,\angle 0^\circ)}{3.667\text{ k}\Omega\,\angle 25.86^\circ}$$

$$= \frac{13.2\text{ V}\,\angle 0^\circ}{3.667\text{ k}\Omega\,\angle 25.86^\circ}$$

$$\mathbf{I}_L = \mathbf{3.6\text{ mA}\,\angle -25.86^\circ}$$

FORMULA SUMMARY

$$\mathbf{G} = G\,\angle 0^\circ = \frac{1}{R}\,\angle 0^\circ$$

$$\mathbf{B}_L = B_L\,\angle -90^\circ = \frac{1}{X_L}\,\angle -90^\circ$$

$$\mathbf{B}_C = B_C\,\angle 90^\circ = \frac{1}{X_C}\,\angle 90^\circ$$

$$\mathbf{Y}_T = \mathbf{Y}_1 + \mathbf{Y}_2 + \mathbf{Y}_3 + \cdots + \mathbf{Y}_N$$

$$\mathbf{Y}_T = \frac{1}{\mathbf{Z}_T} = \frac{1}{Z_T \angle \theta_T} = \frac{1}{Z_T} \angle -\theta_T$$

R-X Parallel Branch

$$Z_T = \frac{RX}{\sqrt{R^2 + X^2}}, \qquad \theta_T = \tan^{-1}\frac{R}{X}$$

$$F_p = \cos\theta_T = \frac{G}{Y_T} = \frac{Z_T}{R} = \frac{P_T}{EI}$$

Current Divider Rule (*R-X* Parallel Branch)

$$I_R = \frac{XI}{\sqrt{R^2 + X^2}}, \qquad I_X = \frac{RI}{\sqrt{R^2 + X^2}}$$

Equivalent Circuits

$$R_s = \frac{R_p X_p^2}{R_p^2 + X_p^2}, \qquad X_s = \frac{R_p^2 X_p}{R_p^2 + X_p^2}$$

$$R_p = \frac{R_s^2 + X_s^2}{R_s}, \qquad X_p = \frac{R_s^2 + X_s^2}{X_s}$$

j-Operator

$$\mathbf{Z}_T = \frac{\mathbf{Z}_1\mathbf{Z}_2}{\mathbf{Z}_1 + \mathbf{Z}_2}$$

$$\mathbf{I}_1 = \frac{\mathbf{Z}_2\mathbf{I}_T}{\mathbf{Z}_1 + \mathbf{Z}_2}, \qquad \mathbf{I}_2 = \frac{\mathbf{Z}_1\mathbf{I}_T}{\mathbf{Z}_1 + \mathbf{Z}_2}$$

CHAPTER SUMMARY

At first glance the parallel combination of elements may appear to be more complex than series elements due to the terminology (conductance, susceptance, admittance), inverse relationships, and finding the proper angle for the usual quantities of interest. However, after carefully reviewing the examples, trying a few problems, and reviewing the material in class, you will find the analysis to be no more difficult than that encountered for series elements. In practice, it is a rare occurrence when more than three elements in parallel have to be analyzed. Consequently, a firm understanding of parallel *R-X* or *R-L-C* networks is usually a sufficient background in the subject of parallel ac networks. Also, since the analysis of parallel *R-C* networks is very similar to that of *R-L* networks and *R-L-C* networks are not that much more difficult to analyze, the basic steps to analyze parallel elements do not change significantly from configuration to configuration. Simply remember that $\mathbf{Y}_T = \mathbf{Y}_1 + \mathbf{Y}_2 + \mathbf{Y}_3$ and $\mathbf{Z}_T = 1/\mathbf{Y}_T$, with the angle of $\mathbf{Y}_T$ easily determined from the admittance diagram and the angle of $\mathbf{Z}_T$ with the opposite sign of that associated with $\mathbf{Y}_T$.

As noted for series ac networks, the analysis of parallel ac networks is very similar to that employed for parallel dc networks. Ohm's law is still applicable and the source current can be determined from an application of Kirchhoff's current law. The power equations $P = I^2R = V^2/R$ are still applicable using effective values.

In general, the major difference introduced by sinusoidal ac networks is the need to know the vector relationship between quantities. In this regard, frequent sketches of the impedance and phasor diagrams can be helpful in determining how to proceed with an analysis. The diagrams can confirm the magnitudes obtained and the sign to be associated with the angle of each quantity.

The only confusing aspect of applying the current divider rule is determining the angle for each branch current. However, since in most practical applications you are interested solely in the magnitude of the currents, the angle is reserved for special areas of concern. In fact, if the angle is required, it might be more direct to find the total impedance of the parallel elements, determine the voltage across the parallel branches using Ohm's law, and then find each current using the calculated voltage and Ohm's law. At the very least, there is more than one way to determine the angle if the need arises.

As noted in the chapter, the parallel element for R-X configurations with the least impedance has the most impact on the total parallel impedance. Therefore, a parallel network with parallel impedances of $R = 10\ \Omega$ and $X_L = 200\ \Omega$ will appear primarily resistive with a small inductive component, and the applied voltage will be very close to an in-phase relationship with the source current. However, if $R = 1\ \text{k}\Omega$ and $X_C = 50\ \Omega$, the network will appear primarily capacitive with the source current leading the applied voltage by about 90°. A change in frequency can change an R-L-C network that appears capacitive at one frequency to one that is resistive or inductive at another frequency.

At a set frequency, the total impedance obtained from a series combination of elements can be obtained from a parallel combination of elements of the same type. That is, an R-L or R-C series circuit has an R-L or R-C parallel equivalent, respectively. However, any change in frequency will change the resulting equivalent circuit. In addition, remember that the equivalence is limited solely to the terminal characteristics such as the magnitude and angle of $\mathbf{Z}_T$.

GLOSSARY

Admittance A measure of how easily a network "admits" the passage of current through that system. It is measured in siemens (S) and is represented by the capital letter Y.

Admittance diagram A vector display that clearly depicts the magnitude of the admittance of the conductance, capacitive susceptance, and inductive susceptance and the magnitude and angle of the total admittance of the system.

Current divider rule A method by which the current through either of two parallel branches can be determined in an ac network without first finding the voltage across the parallel branches.

Equivalent circuits For every series ac network there is a parallel ac network (and vice versa) that will be "equivalent" in the sense that the input current and impedance are the same.

Parallel ac circuits A connection of elements in an ac network in which all the elements have two points in common. The voltage is the same across each element.

Susceptance A measure of how "susceptible" an element is to the passage of current through it. It is measured in siemens (S) and is represented by the capital letter B.

PROBLEMS

Section 15.2

1. Find the total admittance and impedance of the circuits of Fig. 15.47.

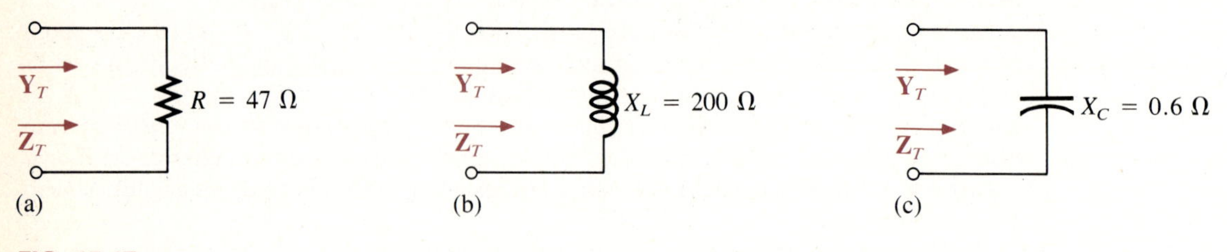

FIG. 15.47

2. Find the total admittance and impedance of the networks of Fig. 15.48.

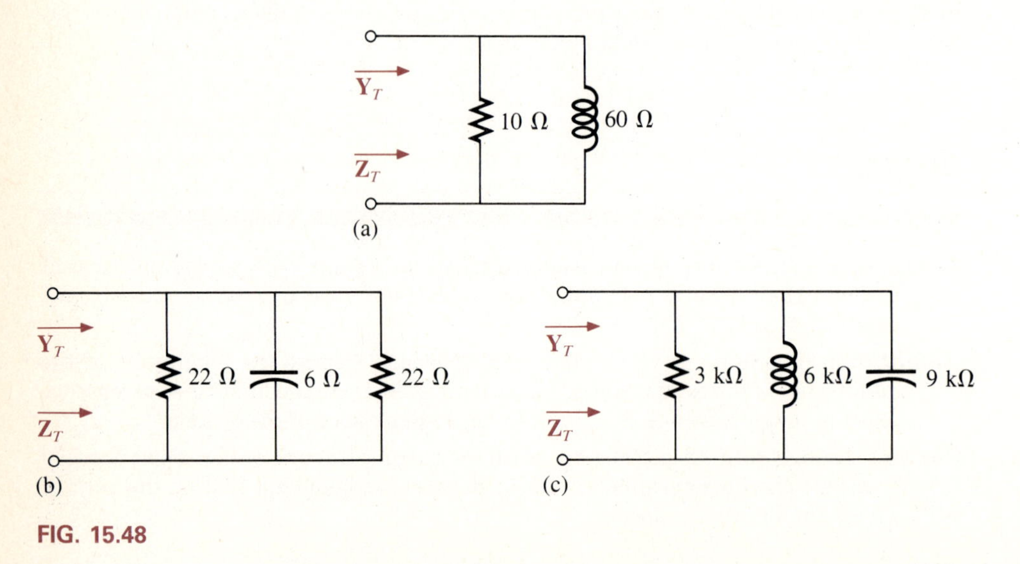

FIG. 15.48

3. Find the total impedance and admittance for the networks of Fig. 15.49.

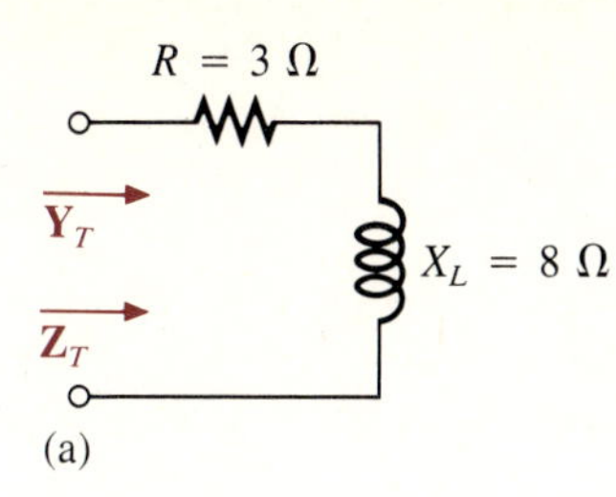

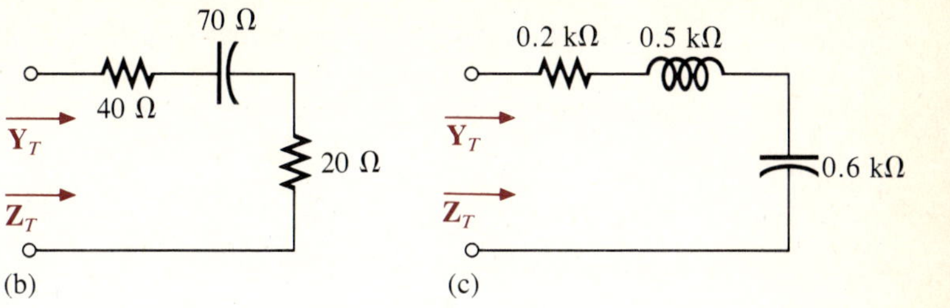

FIG. 15.49

* **4.** Find the type and impedance in ohms of the parallel elements that must be in the closed container of Fig. 15.50 in order for the indicated voltages and currents to exist at the input terminals. (Find the simplest parallel network that will satisfy the indicated conditions.)

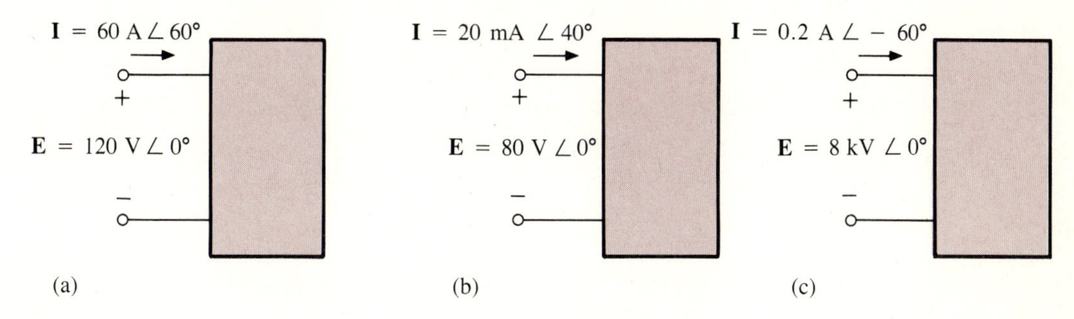

FIG. 15.50

Section 15.3

5. For the circuit of Fig. 15.51,

a. Find the total admittance $\mathbf{Y}_T$ and total impedance $\mathbf{Z}_T$.

b. Draw the admittance diagram.

c. Find the currents $\mathbf{I}_T$, $\mathbf{I}_R$, and $\mathbf{I}_L$ in phasor form.

d. Draw the phasor diagram of the currents $\mathbf{I}_T$, $\mathbf{I}_R$, and $\mathbf{I}_L$ and the voltage $\mathbf{E}$.

e. Find the average power delivered to the circuit.

f. Find the power factor of the circuit and indicate whether it is leading or lagging.

g. Find the sinusoidal expressions for the currents if the frequency is 60 Hz.

h. Plot the waveforms for the currents and voltage on the same set of axes.

FIG. 15.51

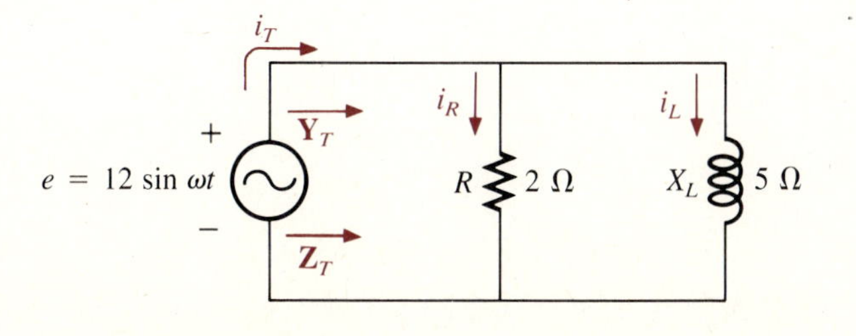

6. Repeat Problem 5 for the network of Fig. 15.52.

FIG. 15.52

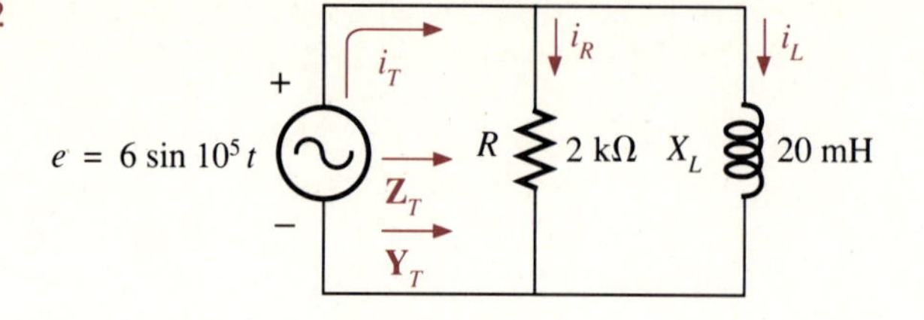

7. For the network of Fig. 15.53, determine

a. The total impedance of the parallel combination.

b. The magnitude of the peak value of the voltage v_L given the current i.

c. The magnitude of the peak value of the current i_L using the results of part b.

FIG. 15.53

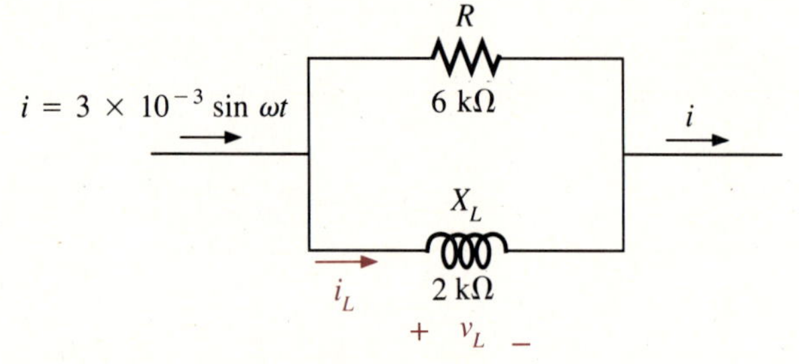

***8.** For the network of Fig. 15.54, determine

a. $\mathbf{Y}_T$

b. $\mathbf{Z}_T$

c. $\mathbf{I}_T$

d. $\mathbf{I}_2$

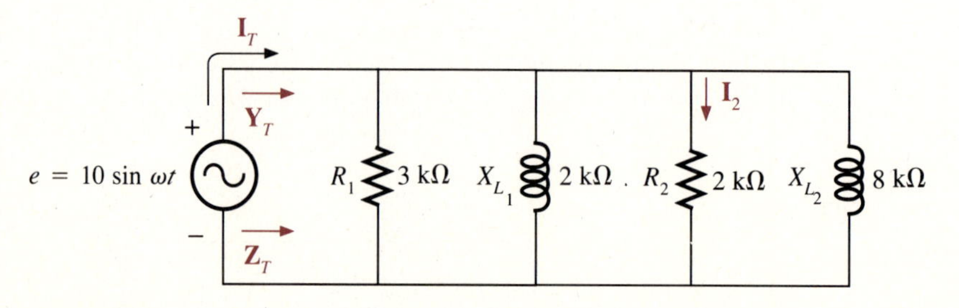

FIG. 15.54

Section 15.4

9. For the circuit of Fig. 15.55,

a. Find the total admittance $\mathbf{Y}_T$ and total impedance $\mathbf{Z}_T$.
b. Draw the admittance diagram.
c. Find the currents $\mathbf{I}_T$, $\mathbf{I}_R$, and $\mathbf{I}_C$ in phasor form.
d. Draw the phasor diagram of the currents $\mathbf{I}_T$, $\mathbf{I}_R$, and $\mathbf{I}_C$ and the voltage $\mathbf{E}$.
e. Find the average power delivered to the circuit.
f. Find the power factor of the circuit and indicate whether it is leading or lagging.
g. Find the sinusoidal expressions for the currents if $f = 60$ Hz.
h. Plot the waveforms for the currents and voltage on the same set of axes.

FIG. 15.55

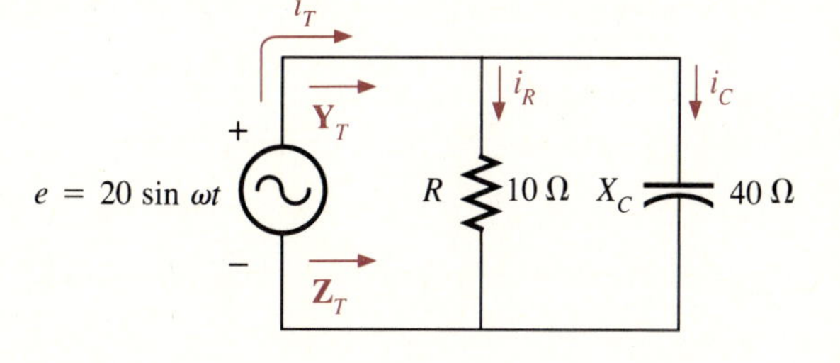

10. Repeat Problem 9 for the network of Fig. 15.56.

FIG. 15.56

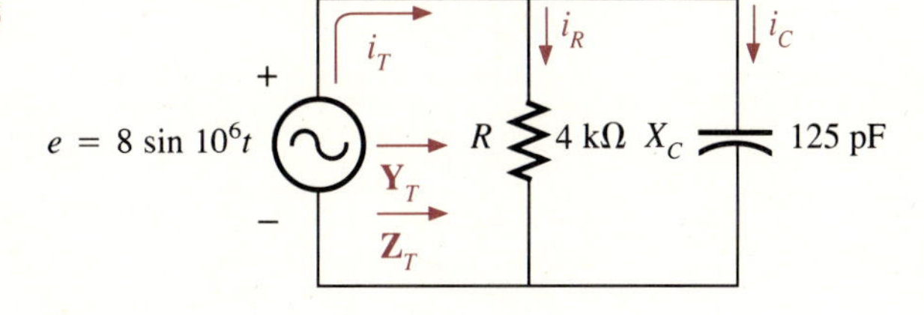

***11.** For the network of Fig. 15.57, determine

a. $\mathbf{Y}_T$
b. $\mathbf{Z}_T$
c. $\mathbf{I}_T$
d. $\mathbf{I}_R$

FIG. 15.57

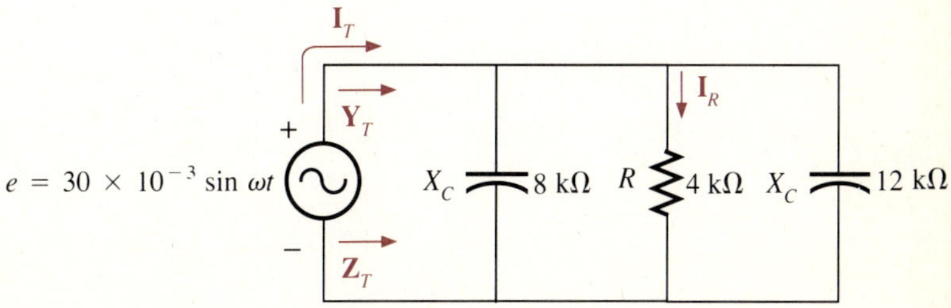

Section 15.5

12. For the circuit of Fig. 15.58,

a. Find the total admittance $\mathbf{Y}_T$ in polar form.
b. Draw the admittance diagram.
c. Find the value of C in microfarads and L in henries.
d. Find the current i and currents i_R, i_L, and i_C in phasor form.
e. Draw the phasor diagram of the currents $\mathbf{I}$, $\mathbf{I}_R$, $\mathbf{I}_L$, and $\mathbf{I}_C$ and the voltage $\mathbf{E}$.
f. Find the average power delivered to the circuit.
g. Find the power factor of the circuit and indicate whether it is leading or lagging.
h. Find the sinusoidal expressions for the currents.
i. Plot the waveforms for the currents and voltage on the same set of axes.

FIG. 15.58

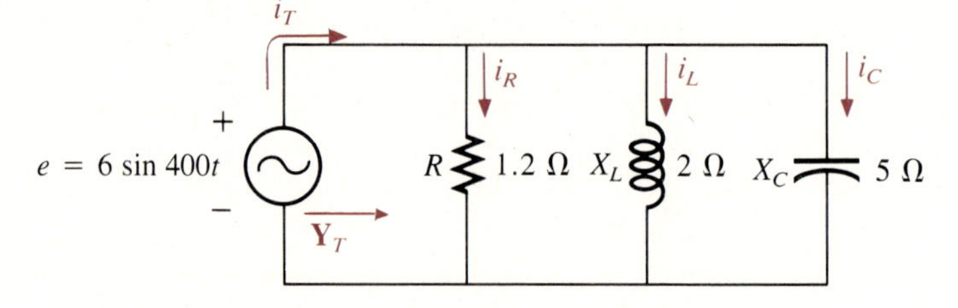

13. Repeat Problem 12 for the network of Fig. 15.59.

FIG. 15.59

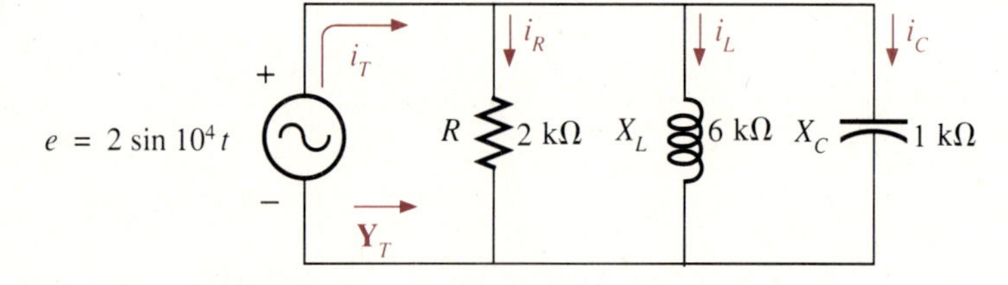

***14.** For the network of Fig. 15.60, determine

a. $\mathbf{Y}_T$
b. $\mathbf{Z}_T$
c. i (time domain)
d. i_R, i_L, and i_C (time domain)
e. Power delivered to the network
f. The power factor of the network

FIG. 15.60

i_T i_R i_L i_C + $e = 10 \sin 10^4 t$ − $\mathbf{Z}_T$ $\mathbf{Y}_T$ R 0.1 kΩ L 5 mH C 0.5 μF

***15.** For the network of Fig. 15.61, determine
a. $\mathbf{Z}_T$
b. i_T (time domain)
c. i_R (time domain)

i_T
$\mathbf{Z}_T$
i_R
+
$e = 10 \sin \omega t$
−
X_C 4 kΩ
R 10 kΩ
X_L 2 kΩ
X_C 6 kΩ

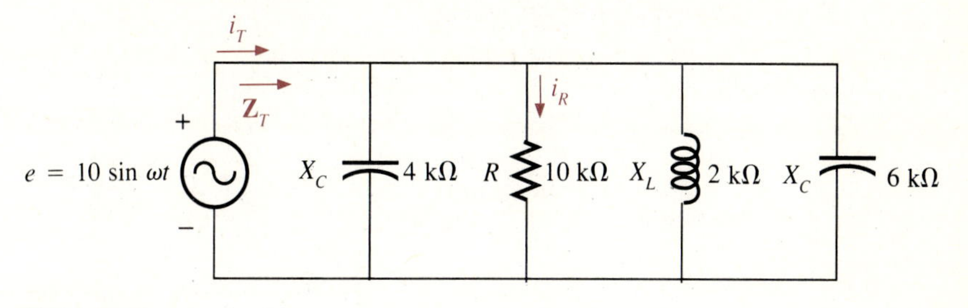

FIG. 15.61

Section 15.6

16. For the network of Fig. 15.62, determine
a. $\mathbf{Y}_T$
b. $\mathbf{Z}_T$
c. $\mathbf{V}_S$
d. $\mathbf{I}_R$
e. $\mathbf{I}_C$

FIG. 15.62

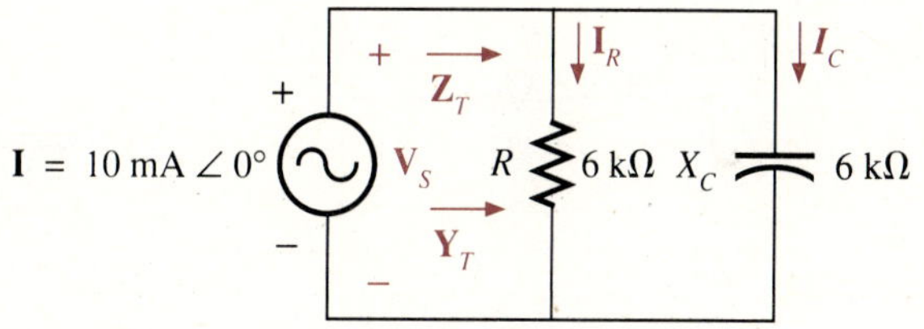

$\mathbf{I} = 10 \text{ mA} \angle 0°$
+
−
$\mathbf{V}_S$
$\mathbf{Z}_T$
$\mathbf{Y}_T$
$\mathbf{I}_R$
I_C
R 6 kΩ
X_C 6 kΩ

17. For the network of Fig. 15.63, determine
a. $\mathbf{Y}_T$
b. $\mathbf{Z}_T$
c. $\mathbf{V}_S$
d. $\mathbf{I}_L$
e. $\mathbf{I}_C$

FIG. 15.63

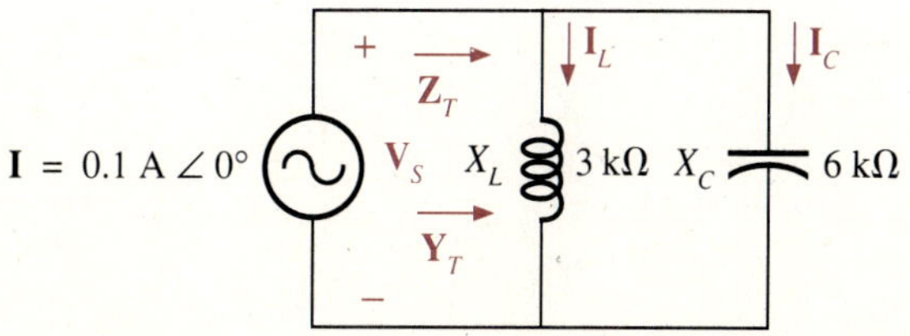

$\mathbf{I} = 0.1 \text{ A} \angle 0°$
+
−
$\mathbf{V}_S$
$\mathbf{Z}_T$
$\mathbf{Y}_T$
$\mathbf{I}_L$
$\mathbf{I}_C$
X_L 3 kΩ
X_C 6 kΩ

*18. For the network of Fig. 15.64, determine

a. $\mathbf{Z}_T$
b. i (time domain)
c. i_{R_1} (time domain)
d. i_C (time domain)
e. Power delivered by source

FIG. 15.64

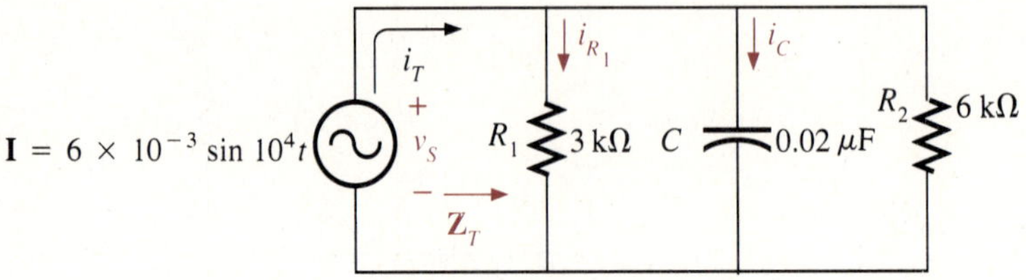

Section 15.7

19. For the networks of Fig. 15.65, determine $\mathbf{I}_1$ and $\mathbf{I}_2$ in phasor form using the current divider rule.

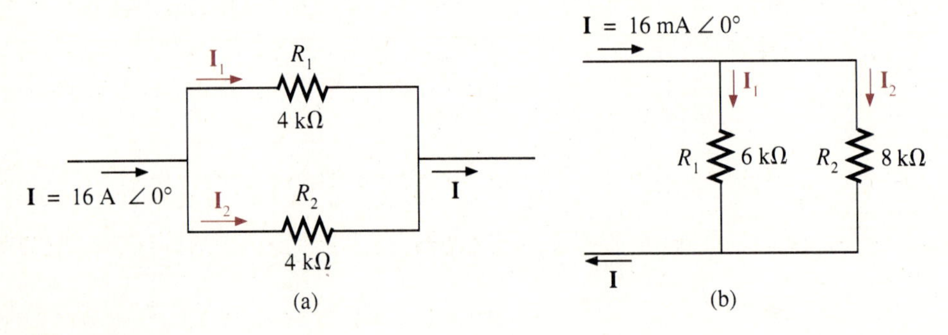

FIG. 15.65

20. Determine the currents $\mathbf{I}_1$ and $\mathbf{I}_2$ for the networks of Fig. 15.66 using the current divider rule.

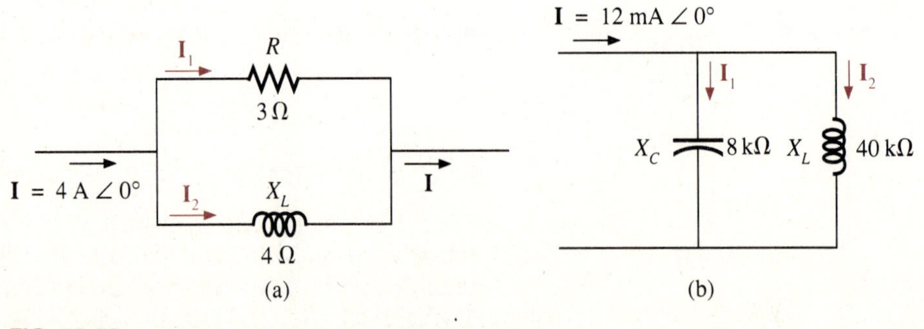

FIG. 15.66

21. Determine the rms reading of the milliammeter and voltmeter of Fig. 15.67.

FIG. 15.67

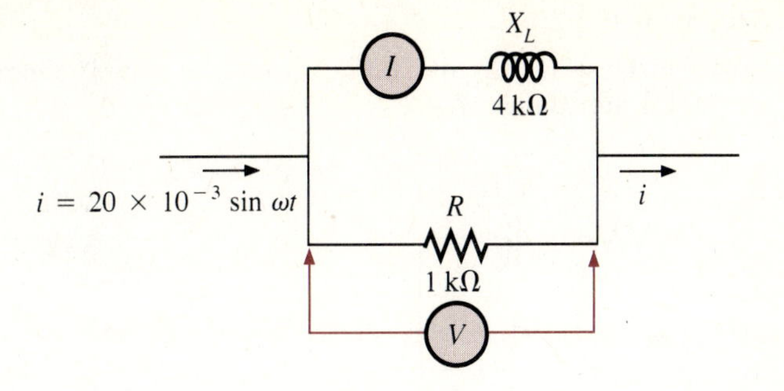

Section 15.8

22. Sketch the frequency response of Z_T, θ_T, I_T, I_R, and I_C for the network of Fig. 15.68. Use a frequency range of $f = 0$ Hz to $f = 100$ kHz and determine each at $f = 1$ kHz, 10 kHz, 50 kHz, and 100 kHz.

FIG. 15.68

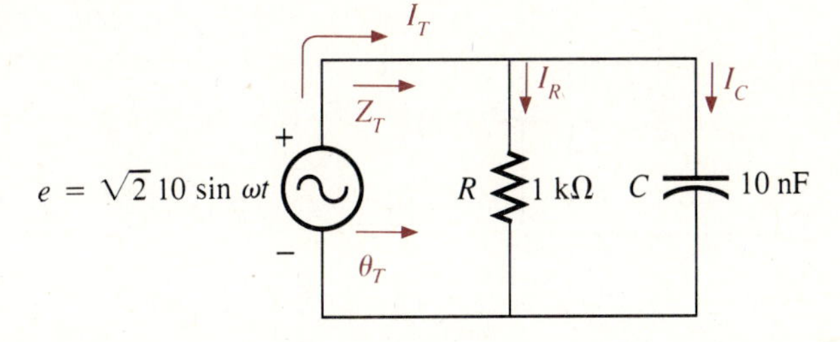

23. Sketch the frequency response of Z_T, θ_T, I_T, I_L, and I_C for the network of Fig. 15.69. Use a frequency range of $f = 0$ Hz to 100 kHz and determine each at $f = 0$ kHz, 1 kHz, 10 kHz, 50 kHz, and 100 kHz.

FIG. 15.69

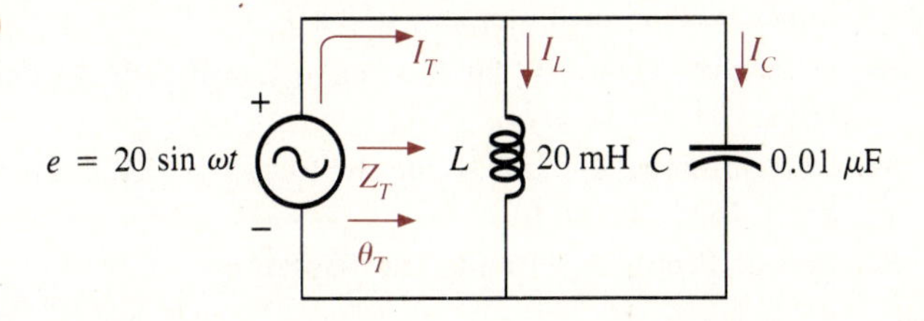

***24.** Sketch the frequency response of Z_T, θ_T, I_T, I_R, and I_C for the network of Fig. 15.70. Use a frequency range of $f = 0$ Hz to $f = 100$ kHz and determine each at $f = 1$ kHz, 10 kHz, 50 kHz, and 100 kHz.

FIG. 15.70

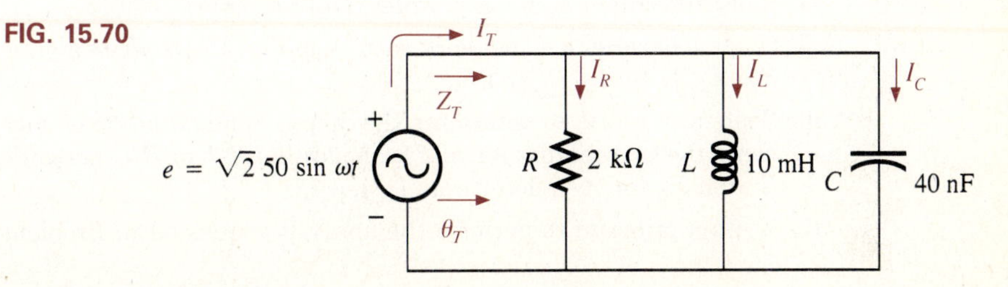

Section 15.9

25. For the series circuits of Fig. 15.71, find a parallel circuit that will have the same total impedance ($\mathbf{Z}_T$).

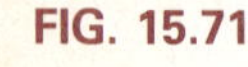

FIG. 15.71

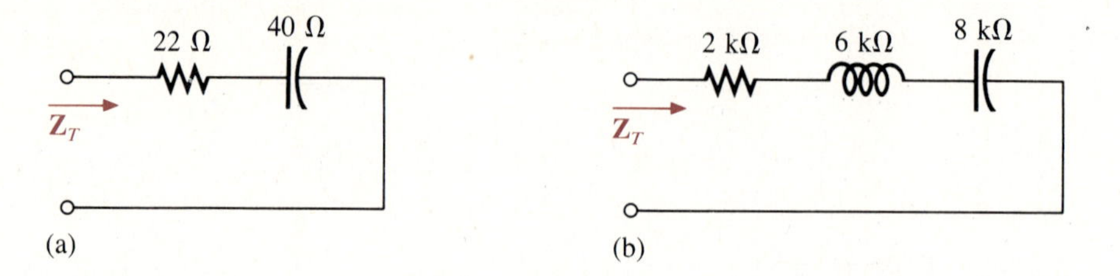

26. For the parallel circuits of Fig. 15.72, find a series circuit that will have the same total impedance.

FIG. 15.72

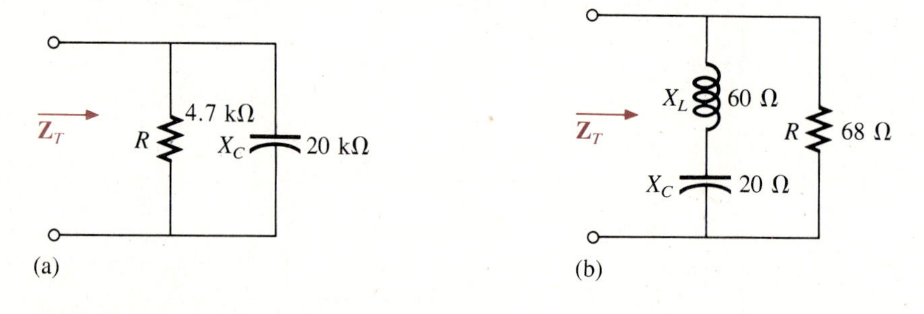

Section 15.11

27. Determine the total impedance of the network of Fig. 15.10 using the *j*-operator.

28. Using the *j*-operator approach, determine X_L in Fig. 15.12(b).

29. Determine the total impedance and admittance of a parallel *R*-*L*-*C* network if the impedance of each element is 8 kΩ.

30. Determine $\mathbf{I}_R$ and $\mathbf{I}_L$ for the network of Fig. 15.14 using *j*-operators. Then determine $\mathbf{I}_T$ from $\mathbf{I}_T = \mathbf{I}_R + \mathbf{I}_L$.

31. Determine $\mathbf{I}_R$, $\mathbf{I}_L$, and $\mathbf{I}_C$ for the network of Fig. 15.59. Then determine $\mathbf{I}_T$ from $\mathbf{I}_T = \mathbf{I}_R + \mathbf{I}_L + \mathbf{I}_C$.

32. Repeat Problem 19 using the *j*-operator.

Computer Problems

33. Given an *R*-*L* or *R*-*C* parallel network, write a program to determine the magnitude and angle of the total admittance and impedance. Assume *R* and *L* (or *C*) are given and the reactance has to be determined using the applied frequency.

34. Repeat Problem 33 for a parallel *R*-*L*-*C* network.

35. Given a parallel *R*-*C* network such as in Fig. 15.18, write a program to determine $\mathbf{Y}_T$, $\mathbf{Z}_T$, $\mathbf{I}$, $\mathbf{I}_R$, $\mathbf{I}_C$, P, and F_p.

36. Write a program to determine the magnitude and angle of each parallel current using the current divider rule for a parallel *R*-*L* or *R*-*C* network. That is, given $\mathbf{I}$, R, and X_L (or X_C), determine $\mathbf{I}_L$ and $\mathbf{I}_C$.

37. Write a program to perform the analysis requested in Problem 22.

16

Network Theorems and Methods of Analysis

OBJECTIVES

- ☐ Perform source conversions for sinusoidal ac voltage and current sources.
- ☐ Establish a procedure to analyze series-parallel ac networks.
- ☐ Be able to apply superposition to sinusoidal ac networks.
- ☐ Determine the Thevenin equivalent circuit for sinusoidal ac networks.
- ☐ Apply the maximum power transfer theorem to ac networks and determine the maximum power to the load.
- ☐ Perform Δ-Y conversions for loads with reactive elements.
- ☐ Become aware of the bridge configuration and how it can be employed to determine the magnitude of resistive and reactive elements using the balance criteria.

16.1 INTRODUCTION

A number of the network theorems and methods of analysis introduced for dc networks will now be applied to ac networks with sinusoidal inputs. The application of each is very similar to that described for dc networks, with the major differences due to the vector relationship between quantities. In some instances the method is introduced for recognition purposes only and is not intended to develop the skills beyond a level typically required in the industrial communities.

16.2 SOURCE CONVERSIONS

Some of the theorems and methods to be introduced in this and later chapters require that current sources be converted to voltage sources and vice versa. This is accomplished in much the same manner as for dc circuits, except now we shall be dealing with phasors and impedances instead of just real numbers and resistors.

In general, the format for converting from one to the other is as shown in Fig. 16.1. That is, for converting from a current source to a voltage source, the magnitude of the voltage is determined by

$$E = IZ \tag{16.1}$$

and for $\mathbf{I} = I\angle 0^\circ$, the angle associated with $\mathbf{E}$ is the same as that associated with $\mathbf{Z}$. That is, for $\mathbf{Z} = Z\angle\theta$,

$$\mathbf{E} = E\angle\theta \qquad (\mathbf{I} = I\angle 0^\circ,\ \ \mathbf{Z} = Z\angle\theta) \tag{16.2}$$

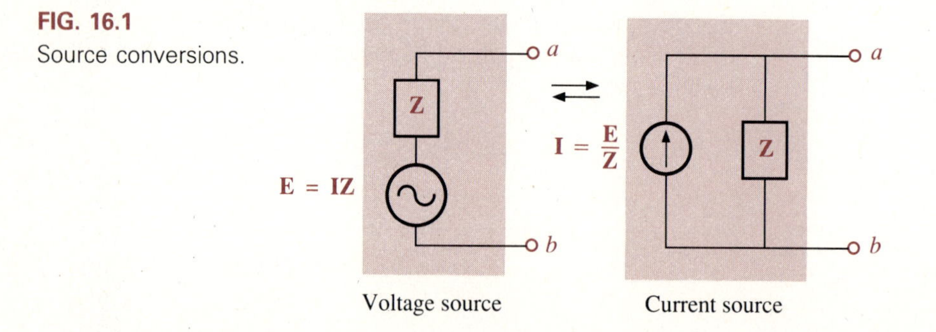

FIG. 16.1
Source conversions.

The impedance of the voltage source is in series with $\mathbf{E}$ and has the same magnitude and angle.

When converting from a voltage source to a current source, the magnitude of the current is determined by

$$I = \frac{E}{Z} \tag{16.3}$$

and for $\mathbf{E} = E \angle 0^\circ$, the angle associated with $\mathbf{I}$ is the negative of that associated with $\mathbf{Z}$. That is, for $\mathbf{Z} = Z \angle \theta$,

$$\mathbf{I} = I \angle -\theta \qquad (\mathbf{E} = E \angle 0^\circ,\ \ \mathbf{Z} = Z \angle \theta) \tag{16.4}$$

EXAMPLE 16.1 Convert the current source of Fig. 16.2 to a voltage source.

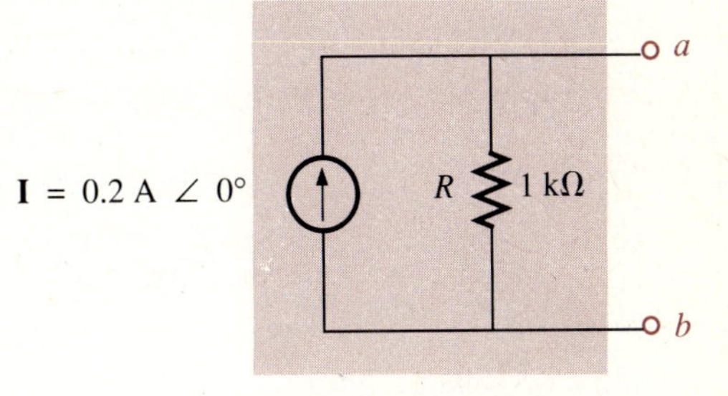

FIG. 16.2
Current source for Example 16.1.

Solution Using Eq. 16.1,

$$\begin{aligned} E &= IZ \\ &= (0.2\text{ A})(1\text{ k}\Omega) \\ &= 0.2\text{ kV} = \mathbf{200\ V} \end{aligned}$$

with

$$\mathbf{Z} = Z \angle \theta = 1\text{ k}\Omega \angle 0^\circ$$

The equivalent voltage source appears in Fig. 16.3:

R
a
1 kΩ
+
E = 200 V ∠ 0°
–
b

FIG. 16.3
Voltage source equivalent for the current source of Fig. 16.2.

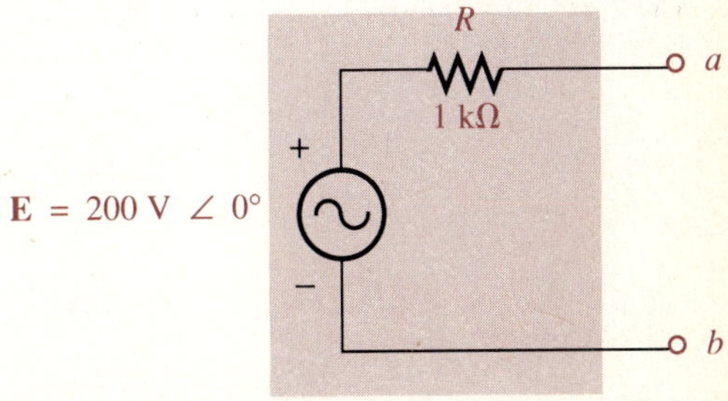

EXAMPLE 16.2

Determine the current source equivalent for the voltage source of Fig. 16.4.

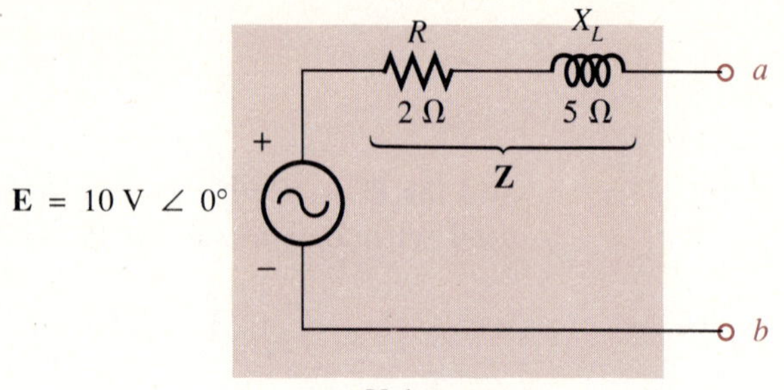

FIG. 16.4
Voltage source for Example 16.2.

Solution $\mathbf{Z} = Z\angle\theta$

with

$$Z = \sqrt{R^2 + X_L^2} = \sqrt{(2\ \Omega)^2 + (5\ \Omega)^2} = 5.385\ \Omega$$

and

$$\theta = \tan^{-1}\frac{X_L}{R} = \tan^{-1}\frac{5\ \Omega}{2\ \Omega} = \tan^{-1}2.5 = 68.2°$$

Thus,

$$\mathbf{Z} = Z\angle\theta = 5.385\ \Omega\ \angle 68.2°$$

The effective value of the current is, from Eq. 16.3,

$$I = \frac{E}{Z} = \frac{10\ \text{V}}{5.385\ \Omega} = 1.857\ \text{A}$$

$$\mathbf{I} = I\angle -\theta = \mathbf{1.857\ A\ \angle -68.2°}$$

The equivalent current source appears in Fig. 16.5.

FIG. 16.5
Current source equivalent for the voltage source of Fig. 16.4.

I = 1.875 A ∠ − 68.2°
a
Z
R 2 Ω
X_L 5 Ω
b

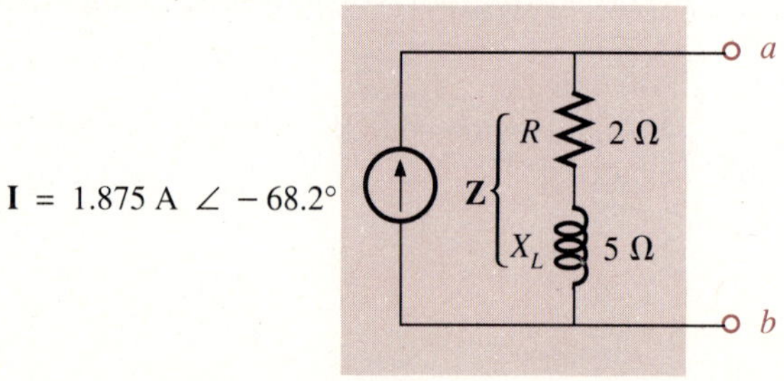

EXAMPLE 16.3 Convert the current source of Fig. 16.6 to a voltage source.

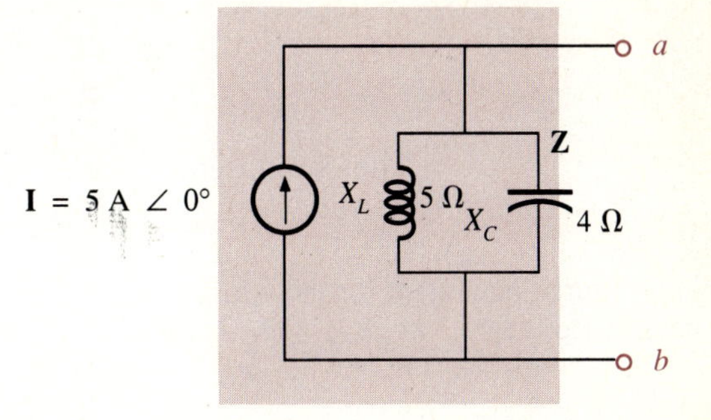

FIG. 16.6
Current source for Example 16.3.

Solution To find **Z**:

$$\mathbf{Y} = \frac{1}{\mathbf{Z}_1} + \frac{1}{\mathbf{Z}_2} = \mathbf{B}_L + \mathbf{B}_C = \frac{1}{X_L}\angle -90^\circ + \frac{1}{X_C}\angle 90^\circ$$

$$= \frac{1}{5\ \Omega}\angle -90^\circ + \frac{1}{4\ \Omega}\angle 90^\circ = 0.2\text{ S}\angle -90^\circ + 0.25\text{ S}\angle 90^\circ$$

The admittance diagram appears in Fig. 16.7, and $\mathbf{Y} = (0.25\text{ S} - 0.2\text{ S})\angle 90^\circ = 0.05\text{ S}\angle 90^\circ = Y\angle\theta$, with

$$\mathbf{Z} = \frac{1}{\mathbf{Y}} = \frac{1}{Y}\angle -\theta$$

$$= \frac{1}{0.05\text{ S}}\angle -90^\circ = \mathbf{20\ \Omega\angle -90^\circ} = X_C\angle -90^\circ$$

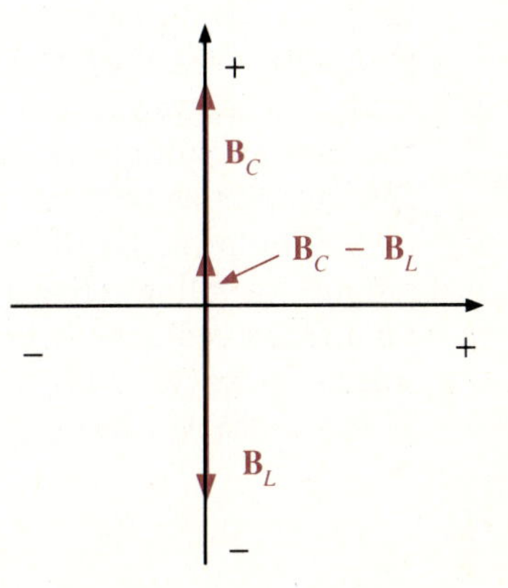

FIG. 16.7
Determining the total admittance for the parallel impedance of Fig. 16.6.

The rms value of the voltage source, from Eq. 16.1, is

$$E = IZ$$
$$= (5\text{ A})(20\ \Omega)$$
$$E = 100\text{ V}$$
$$\mathbf{E} = E\angle\theta = \mathbf{100\ V\angle{-90^\circ}}$$

The equivalent voltage source appears in Fig. 16.8.

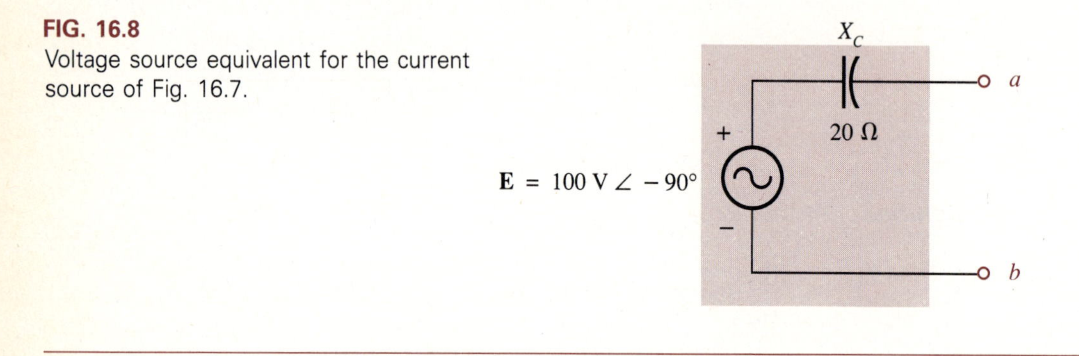

FIG. 16.8
Voltage source equivalent for the current source of Fig. 16.7.

Note in this case that the equivalent impedance of X_L in parallel with X_C is a capacitive reactance of 20 Ω. It is capacitive because the elements are in parallel and $X_C < X_L$, but also note that the total impedance turned out to be greater than either of the original reactances.

16.3 SERIES-PARALLEL ac NETWORKS

The analysis of series-parallel ac networks follows a sequence of calculations very similar to that employed for dc networks. As you will recall, the total resistance was often the first quantity determined, followed by the source current. If other quantities were required, the usual approach was to work back to the desired voltage or current using intermediate (reduced) network configurations. For ac systems the total impedance is often the first quantity determined, followed by the source current and other unknown quantities. The examples to follow clearly demonstrate that the vector relationship between quantities requires a level of mathematics beyond that required for dc networks. However, with the proper effort and patience, the magnitude and angle of the unknown quantities can be determined to a high degree of accuracy.

EXAMPLE 16.4

Determine $\mathbf{Z}_T$, $\mathbf{I}_T$, $\mathbf{V}_R$, and $\mathbf{I}_C$ for the network of Fig. 16.9.

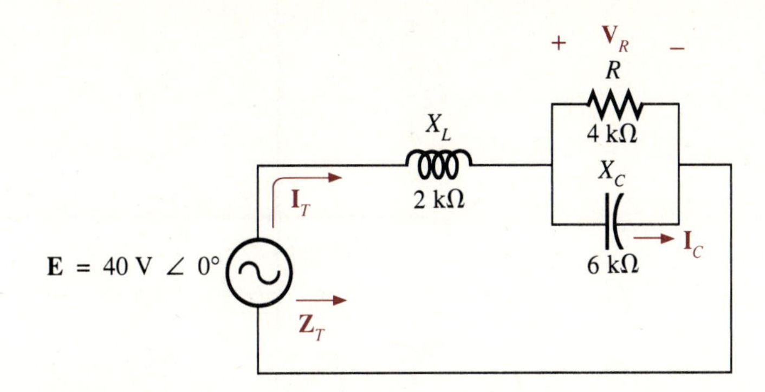

FIG. 16.9
Series-parallel ac network for Example 16.4.

Solution $\mathbf{Z}_T$: The parallel combination of the resistor and capacitive reactance is in series with the inductive reactance. For the parallel R-C combination, the total impedance $\mathbf{Z}'_T$ is determined as follows:

$$\mathbf{Y}'_T = \mathbf{G} + \mathbf{B}_C = \frac{1}{4\text{ k}\Omega}\angle 0^\circ + \frac{1}{6\text{ k}\Omega}\angle 90^\circ$$

$$= 0.25\text{ mS}\angle 0^\circ + 0.167\text{ mS}\angle 90^\circ = 0.3\text{ mS}\angle 33.74^\circ$$

with

$$\mathbf{Z}'_T = \frac{1}{\mathbf{Y}'_T} = \frac{1}{0.3\text{ mS}}\angle -33.74^\circ = 3.333\text{ k}\Omega\angle -33.74^\circ$$

The impedance $\mathbf{Z}'_T$ is broken down into its resistive and capacitive components in Fig. 16.10.

$$R' = (3.333\text{ k}\Omega)(\cos 33.74^\circ) = 2.77\text{ k}\Omega$$

$$X'_C = (3.333\text{ k}\Omega)(\sin 33.74^\circ) = 1.85\text{ k}\Omega$$

Since $\mathbf{X}_L$ and $\mathbf{Z}'_T$ are now in series,

$$\mathbf{Z}_T = \mathbf{X}_L + \mathbf{Z}'_T$$

The net reactance is

$$X_T = X_L - X_C = 4\text{ k}\Omega - 1.85\text{ k}\Omega = 2.15\text{ k}\Omega$$

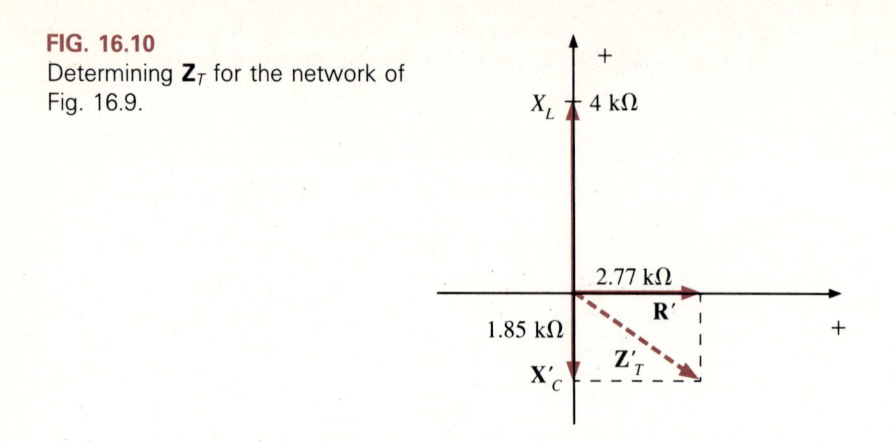

FIG. 16.10
Determining $\mathbf{Z}_T$ for the network of Fig. 16.9.

and the magnitude of the total impedance is

$$\begin{aligned} Z_T &= \sqrt{(R')^2 + (X_T)^2} = \sqrt{(2.77\ \text{k}\Omega)^2 + (2.15\ \text{k}\Omega)^2} \\ &= 3.51\ \text{k}\Omega \end{aligned}$$

Also,

$$\begin{aligned} \theta_T &= \tan^{-1}\frac{X_T}{R'} = \tan^{-1}\frac{2.15\ \text{k}\Omega}{2.77\ \text{k}\Omega} = \tan^{-1}0.776 \\ &= 37.82^\circ \end{aligned}$$

and

$$\mathbf{Z}_T = \mathbf{3.51\ k\Omega\ \angle 37.82^\circ}$$

$\mathbf{I}_T$: The magnitude is

$$I_T = \frac{E}{Z_T} = \frac{40\ \text{V}}{3.51\ \text{k}\Omega} = 11.396\ \text{mA}$$

and, since θ_T of $\mathbf{Z}_T$ is 37.82° with $\mathbf{E} = 40\ \text{V}\ \angle 0^\circ$,

$$\mathbf{I}_T = \mathbf{11.396\ mA\ \angle -37.82^\circ}$$

Recall that θ_T of $\mathbf{Z}_T$ is the angle by which **E** leads **I**.
$\mathbf{V}_R$: Using Ohm's law,

$$\begin{aligned} V_R &= I_T Z'_T = (11.396\ \text{mA})(3.333\ \text{k}\Omega) \\ &= 37.98\ \text{V} \end{aligned}$$

The angle associated with $\mathbf{V}_R$ is best determined by noting the phasor diagram of Fig. 16.11 and recalling that θ'_T of $\mathbf{Z}'_T$ is the angle by which $\mathbf{V}_R$ leads $\mathbf{I}_T$. Since θ'_T is negative, we know $\mathbf{I}_T$ in fact leads $\mathbf{V}_R$ by 33.74°, resulting in an angle of −71.56° with $\mathbf{V}_R$, so that

$$\mathbf{V}_R = \mathbf{37.98\ V\ \angle -71.56^\circ}$$

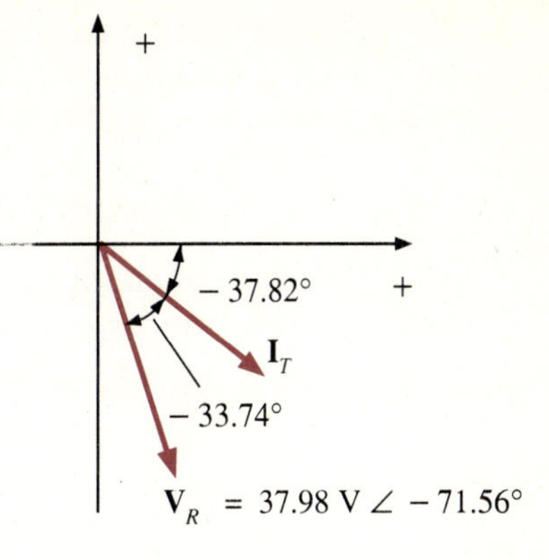

FIG. 16.11
Determining the angle associated with $\mathbf{V}_R$.

$\mathbf{I}_C$: Using Ohm's law,

$$I_C = \frac{V_C}{X_C} = \frac{V_R}{X_C} = \frac{37.98\text{ V}}{6\text{ k}\Omega} = 6.33\text{ mA}$$

and, since $\mathbf{I}_C$ leads $\mathbf{V}_C = \mathbf{V}_R$ by 90°,

$$\begin{aligned}\mathbf{I}_C &= 6.33\text{ mA}\ \underline{/-71.56^\circ + 90^\circ}\\ &= \mathbf{6.33\ mA}\ \underline{/\mathbf{18.44^\circ}}\end{aligned}$$

The network of Fig. 16.9 is "series" in nature, since a series configuration resulted once the R-C combination was reduced to an equivalent impedance. The network of the next example is "parallel" in nature, since the reduction of the R-L branch to a single equivalent impedance results in three parallel elements.

EXAMPLE 16.5 Determine $\mathbf{Z}_T$, $\mathbf{V}_S$, $\mathbf{I}_C$, and $\mathbf{V}_L$ for the network of Fig. 16.12.

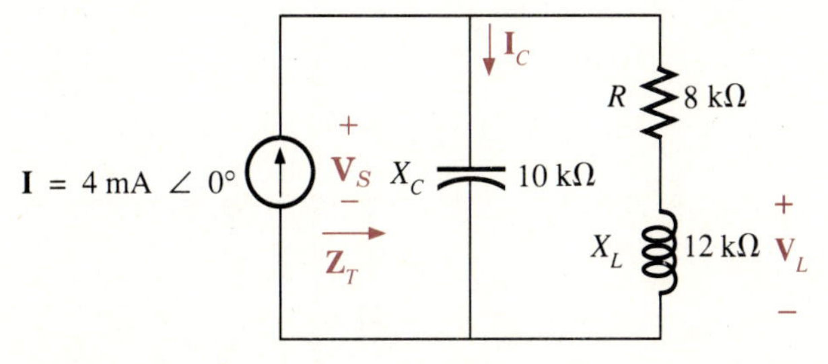

FIG. 16.12
Series-parallel ac network for Example 16.5.

Solution $\mathbf{Z}_T$: For the series R-L branch, the total impedance is

$$\begin{aligned}\mathbf{Z}'_T &= \mathbf{R} + \mathbf{X}_L = 8\text{ k}\Omega\ \underline{/0^\circ} + 12\text{ k}\Omega\ \underline{/90^\circ}\\ &= 14.42\text{ k}\Omega\ \underline{/56.31^\circ}\end{aligned}$$

The result is a parallel network of $\mathbf{X}_C$ and $\mathbf{Z}'_T$, with

$$\mathbf{Y}_T = \mathbf{B}_C + \mathbf{Y}'_T = \frac{1}{10\ \text{k}\Omega}\angle 90^\circ + \frac{1}{14.42\ \text{k}\Omega}\angle -56.31^\circ$$
$$= 0.1\ \text{mS}\ \angle 90^\circ + 0.069\ \text{mS}\ \angle -56.31^\circ$$

The various admittances are plotted in Fig. 16.13. Note that $\mathbf{Y}'_T$ was broken down into its components, and

$$Y_T = \sqrt{(G')^2 + (B_C - B'_L)^2}$$
$$= \sqrt{(0.038\ \text{mS})^2 + (0.1\ \text{mS} - 0.059\ \text{mS})^2}$$
$$= \sqrt{(0.038\ \text{mS})^2 + (0.041\ \text{mS})^2}$$
$$Y_T = 0.056\ \text{mS}$$

Also,

$$\theta_T = \tan^{-1}\frac{B_C - B'_L}{G} = \tan^{-1}\frac{0.041\ \text{mS}}{0.038\ \text{mS}} = \tan^{-1}1.079$$
$$= 47.17^\circ$$

and

$$\mathbf{Y}_T = 0.056\ \text{mS}\ \angle 47.17^\circ$$

with

$$\mathbf{Z}_T = \frac{1}{0.056\ \text{mS}}\angle -47.17^\circ$$
$$= \mathbf{17.86\ k\Omega\ \angle -47.17^\circ}$$

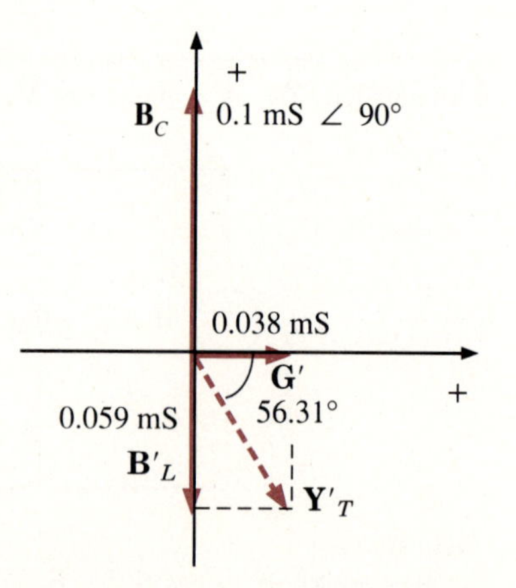

FIG. 16.13
Determining $\mathbf{Y}_T$ for the network of Fig. 16.12.

$\mathbf{V}_S$: Using Ohm's law,

$$V_S = IZ_T$$
$$= (4\ \text{mA})(17.86\ \text{k}\Omega)$$
$$V_S = 71.44\ \text{V}$$

and, since the angle of $\mathbf{Z}_T$ is the angle by which $\mathbf{V}_S$ leads **I**,

$$\mathbf{V}_S = \mathbf{71.44\ V\ \angle{-47.17^\circ}}$$

$\mathbf{I}_C$: Using Ohm's law,

$$I_C = \frac{V_S}{X_C} = \frac{71.44\ \text{V}}{10\ \text{k}\Omega} = 71.44\ \text{mA}$$

and, since $\mathbf{I}_C$ leads $\mathbf{V}_C = \mathbf{V}_S$ by 90°,

$$\begin{aligned}\mathbf{I}_C &= 7.144\ \text{mA}\ \angle{-47.17^\circ + 90^\circ} \\ &= \mathbf{7.144\ mA\ \angle{42.83^\circ}}\end{aligned}$$

$\mathbf{V}_L$: Using Ohm's law, the current through R-L branch is

$$I' = \frac{V_S}{Z'_T} = \frac{71.44\ \text{V}}{14.42\ \text{k}\Omega} = 4.95\ \text{mA}$$

The angle associated with $\mathbf{I}'$ is best determined by noting the phasor diagram of Fig. 16.14 and recalling that $\theta'_T = 56.31^\circ$ of Z'_T is the angle by which $\mathbf{V}_S$ leads $\mathbf{I}'$. Since θ'_T is positive, V_S does in fact lead $\mathbf{I}'$ by 56.31°, as shown in Fig. 16.14, with the result that

$$\mathbf{I}' = \mathbf{4.95\ mA\ \angle{-103.48^\circ}}$$

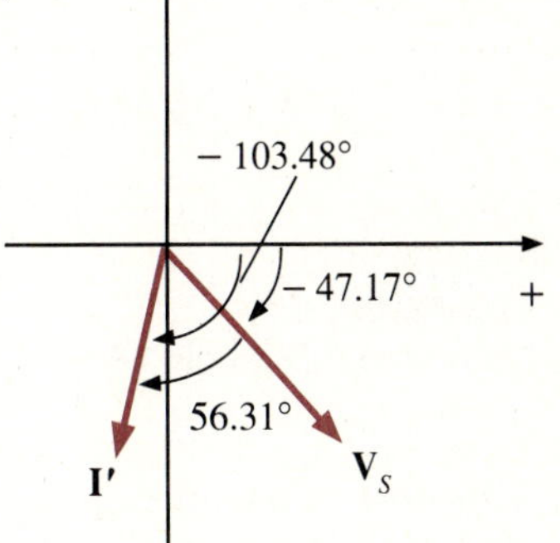

FIG. 16.14
Determining the phase angle associated with **I**′.

For $\mathbf{V}_L$:

$$\begin{aligned}V_L &= I'X_L \\ &= (4.95\ \text{mA})(12\ \text{k}\Omega) \\ V_L &= 59.4\ \text{V}\end{aligned}$$

The angle associated with $\mathbf{V}_L$ can be determined by recalling that $\mathbf{V}_L$ leads $\mathbf{I}_L = \mathbf{I}'$ by 90°. Therefore,

$$\begin{aligned}\mathbf{V}_L &= 59.4\ \text{V}\ \angle{-103.48^\circ + 90^\circ} \\ &= \mathbf{59.4\ V\ \angle{-13.48^\circ}}\end{aligned}$$

16.4 SUPERPOSITION

You will recall from Chapter 9 that the superposition theorem eliminates the need for solving simultaneous linear equations by considering the effects of each source independently. To consider the effects of each source, we had to remove the remaining sources. This was accomplished by setting voltage sources to zero (short-circuit representation) and current sources to zero (open-circuit representation). The current through or voltage across a portion of the network produced by each source was then added algebraically to find the total solution for the current or voltage.

The only variation in applying this method to ac networks is that we are now working with impedances and phasors instead of just resistors and real numbers.

The superposition theorem is not applicable to power effects in ac networks, since we are still dealing with a nonlinear relationship. It can be applied to networks with sources of different frequencies only if the total response for *each* frequency is found independently and the results are expanded in a nonsinusoidal expression.

EXAMPLE 16.6 Determine the current $\mathbf{I}_1$ of Fig. 16.15 using superposition.

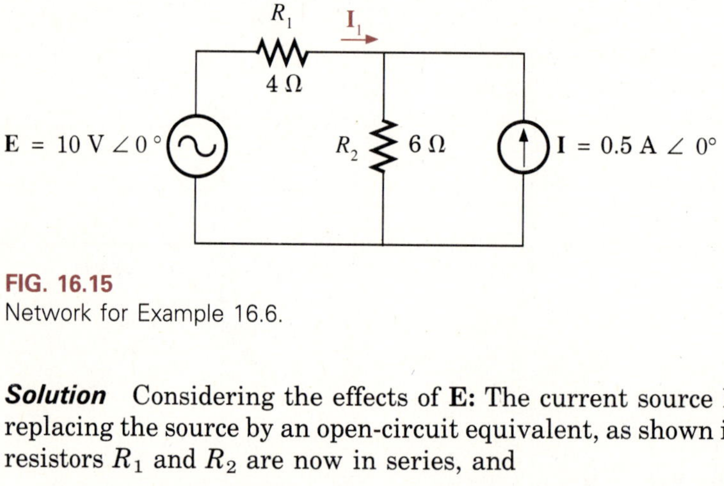

R_1 4 Ω, $\mathbf{I}_1$, E = 10 V ∠ 0°, R_2 6 Ω, I = 0.5 A ∠ 0°

FIG. 16.15
Network for Example 16.6.

Solution Considering the effects of **E:** The current source **I** is set to 0 A by replacing the source by an open-circuit equivalent, as shown in Fig. 16.16. The resistors R_1 and R_2 are now in series, and

$$\mathbf{Z}_T = \mathbf{R}_1 + \mathbf{R}_2 = 4\ \Omega\angle 0^\circ + 6\ \Omega\angle 0^\circ = 10\ \Omega\angle 0^\circ$$

The magnitude of the current $\mathbf{I}'_1$ is

$$\mathbf{I}'_1 = \frac{E}{Z_T} = \frac{10\text{ V}}{10\ \Omega} = 1\text{ A}$$

with an angle of 0° due to the resistive load, and

$$\mathbf{I}'_1 = 1\text{ A}\angle 0^\circ$$

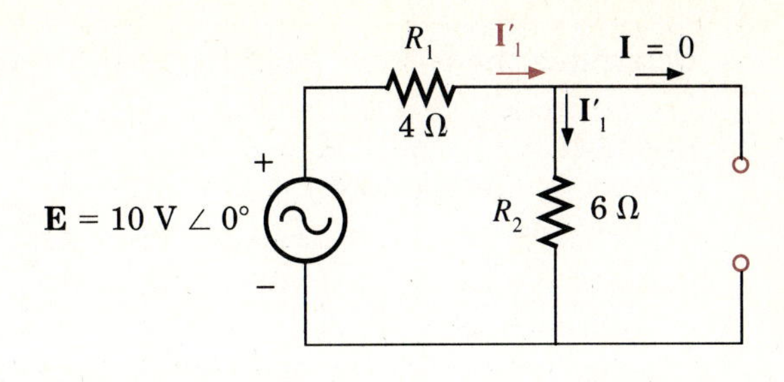

FIG. 16.16
Considering the effects of the voltage source of Fig. 16.15.

Considering the effects of **I**: The voltage source **E** is set to 0 V by replacing the source by a short-circuit equivalent, as shown in Fig. 16.17.

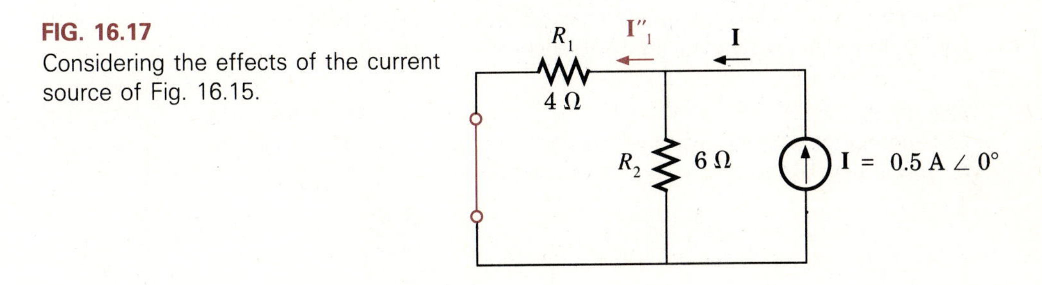

FIG. 16.17
Considering the effects of the current source of Fig. 16.15.

The magnitude of the current I''_1 (with the direction shown) can be determined by the current divider rule:

$$I''_1 = \frac{R_2(I)}{R_1 + R_2} = \frac{(6\ \Omega)(0.5\ \text{A})}{4\ \Omega + 6\ \Omega} = \frac{6}{10}(0.5\ \text{A})$$
$$= 0.3\ \text{A}$$

with an angle of 0° due to the resistive elements. Thus,

$$I''_1 = 0.3\ \text{A}\ \underline{/0^\circ}$$

Figures 16.16 and 16.17 clearly reveal that $\mathbf{I}'_1$ and $\mathbf{I}''_1$ have opposite directions. Using the direction of $\mathbf{I}_1$ appearing in Fig. 16.15,

$$\mathbf{I}_1 = \mathbf{I}'_1 - \mathbf{I}''_1$$
$$= 1\ \text{A}\ \underline{/0^\circ} - 0.3\ \text{A}\ \underline{/0^\circ}$$
$$\mathbf{I}_1 = \mathbf{0.7\ A}\ \underline{/\mathbf{0^\circ}}$$

In this example the analysis would have been quite similar if we simply assumed the voltage source was a dc supply of 10 V and the current source was a dc source of 0.5 A. However, the next example has a reactive element, which modifies the analysis due to vector relationships.

EXAMPLE 16.7 Determine the current **I** of Fig. 16.18 using superposition.

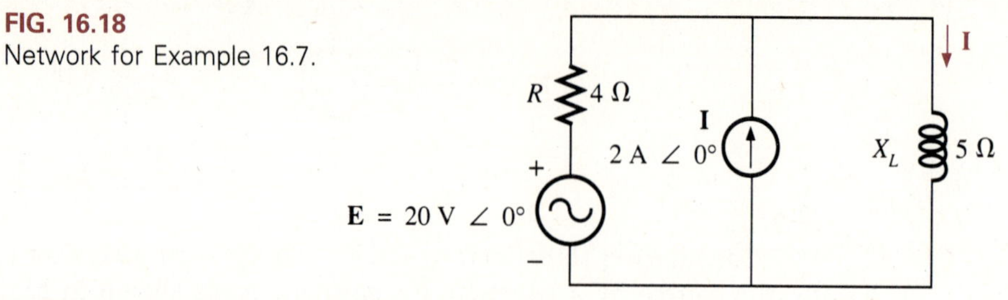

FIG. 16.18
Network for Example 16.7.

Solution Considering the effects of **E** requires that **I** be removed (open-circuit equivalent), resulting in the network of Fig. 16.19.

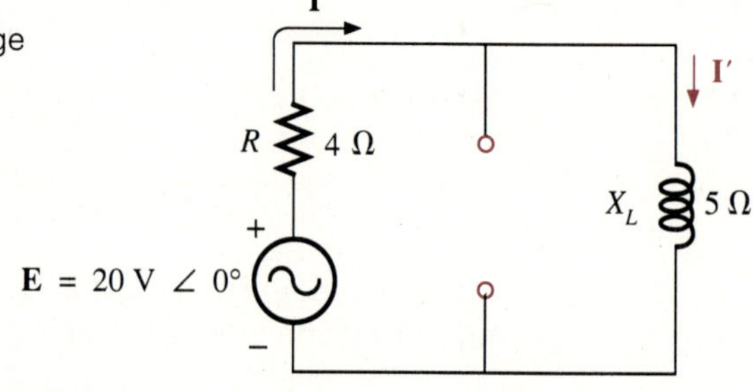

FIG. 16.19
Considering the effects of the voltage source of Fig. 16.18.

The contribution to **I** due to **E** is labeled **I**′. Note that it is now the current of the resulting series circuit. The total impedance of the network is

$$\mathbf{Z}_T = \mathbf{R} + \mathbf{X}_L = 4\ \Omega\ \underline{/0^\circ} + 5\ \Omega\ \underline{/90^\circ}$$

with

$$\begin{aligned} Z_T &= \sqrt{R^2 + X_L^2} = \sqrt{(4\ \Omega)^2 + (5\ \Omega)^2} = \sqrt{16 + 25} \\ &= \sqrt{41} = 6.403\ \Omega \end{aligned}$$

and

$$\begin{aligned} \theta_T &= \tan^{-1}\frac{X_L}{R} = \tan^{-1}\frac{5\ \Omega}{4\ \Omega} = \tan^{-1}1.25 \\ &= 51.34^\circ \\ \mathbf{Z}_T &= 6.403\ \Omega\ \underline{/51.34^\circ} \\ I' &= \frac{E}{Z_T} = \frac{20\ \text{V}}{6.403\ \Omega} = 3.124\ \text{A} \end{aligned}$$

For $\mathbf{E} = E\ \underline{/0^\circ}$ and $\mathbf{Z}_T = Z_T\ \underline{/\theta_T}$, the angle associated with **I** is $-\theta_T$. Therefore,

$$\mathbf{I}' = \mathbf{3.124\ A}\ \underline{/\mathbf{-51.34^\circ}}$$

Considering the effects of **I** requires that **E** be removed (short-circuit equivalent), resulting in the network of Fig. 16.20. The contribution due to **I** is labeled **I**″, with the rms value determined using the current divider rule:

$$I'' = \frac{R(I_1)}{\sqrt{R^2 + X_L^2}} = \frac{(4\ \Omega)(2\ \text{A})}{\sqrt{(4\ \Omega)^2 + (5\ \Omega)^2}} = \frac{8\ \text{A}}{6.403}$$
$$= 1.249\ \text{A}$$

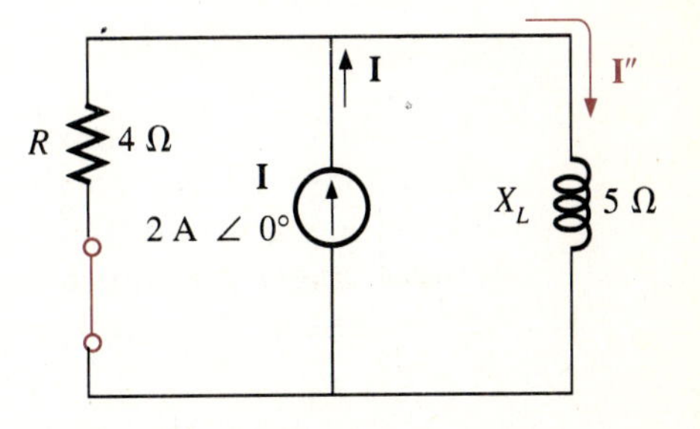

FIG. 16.20
Considering the effects of the current source of Fig. 16.18.

Recall from Section 15.6 that the angle associated with the inductive branch is ($\theta - 90°$), with θ equal the angle associated with the total impedance of the parallel branches.

$$\mathbf{Y}_T = \mathbf{G} + \mathbf{B}_L = \frac{1}{4\ \Omega}\angle 0° + \frac{1}{5\ \Omega}\angle -90°$$
$$= 0.25\ \text{S}\ \angle 0° + 0.2\ \text{S}\ \angle -90° = 0.32\ \text{S}\ \angle -38.66°$$

and

$$\mathbf{Z}_T = \frac{1}{0.32\ \text{S}}\angle 38.66° = 3.125\ \Omega\ \angle 38.66°$$

so that

$$\mathbf{I}'' = 1.249\ \text{A}\ \angle 38.66° - 90° = \mathbf{1.249\ A\ \angle -51.34°}$$

Both **I**′ and **I**″ have the same direction in their respective circuits, resulting in

$$\mathbf{I} = \mathbf{I}' + \mathbf{I}''$$
$$= 3.124\ \text{A}\ \angle -51.34° + 1.249\ \text{A}\ \angle -51.34°$$
$$\mathbf{I} = \mathbf{4.373\ A\ \angle -51.34°}$$

16.5 THEVENIN'S THEOREM

Thevenin's theorem, as stated for sinusoidal ac circuits, is changed only to include the term *impedance* instead of *resistance;* that is, *any two-terminal ac network can be replaced by an equivalent circuit consisting of a voltage source and an impedance in series,* as shown in Fig. 16.21.

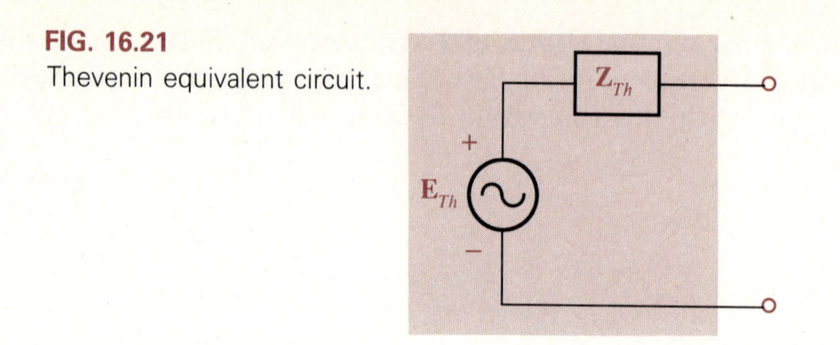

FIG. 16.21
Thevenin equivalent circuit.

Since the reactances of a circuit are frequency dependent, the Thevenin circuit found for a particular network is applicable only at *one* frequency.

The steps required to apply this method to dc circuits are repeated here with changes for sinusoidal ac circuits. As before, the only change is the replacement of the term *resistance* by *impedance*.

1. Remove that portion of the network across which the Thevenin equivalent circuit is to be found.
2. Mark (∘, •, and so on) the terminals of the remaining two-terminal network.
3. Calculate $\mathbf{Z}_{Th}$ by first setting all voltage and current sources to zero (short circuit and open circuit, respectively) and then finding the resulting impedance between the two marked terminals.
4. Calculate $\mathbf{E}_{Th}$ by first replacing the voltage and current sources and then finding the open-circuit voltage between the marked terminals.
5. Draw the Thevenin equivalent circuit with the portion of the circuit previously removed replaced between the terminals of the Thevenin equivalent circuit.

EXAMPLE 16.8

Find the Thevenin equivalent circuit for the network external to the resistor R_L in Fig. 16.22.

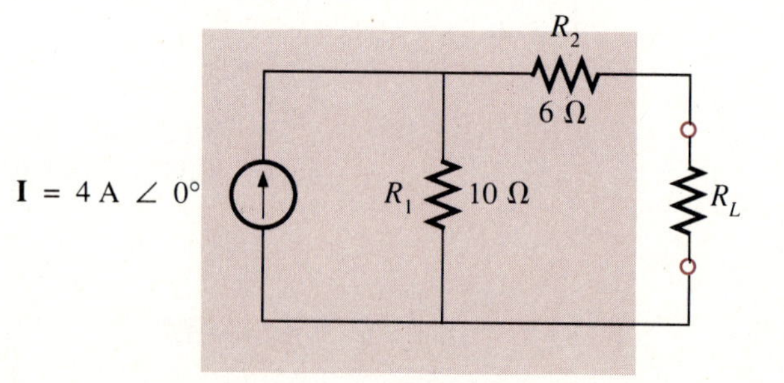

FIG. 16.22
Network for Example 16.8.

Solution Since the network is totally resistive, all the voltages and currents of the network are in phase. The analysis, therefore, can be similar to that applied to dc networks.

Steps 1–3: $\mathbf{Z}_{Th}$ is shown in Fig. 16.23, and $\mathbf{Z}_T = \mathbf{R}_1 + \mathbf{R}_2 = 10\ \Omega\ \angle 0^\circ + 6\ \Omega\ \angle 0^\circ = 16\ \Omega\ \angle 0^\circ$.

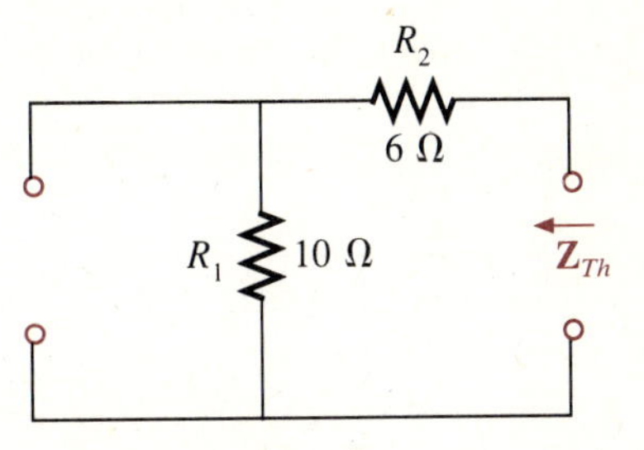

FIG. 16.23
Determining $\mathbf{Z}_{Th}$ for the network of Fig. 16.22.

Step 4: E_{Th} is shown in Fig. 16.24. Since $\mathbf{I}_2 = 0$ A due to the open circuit, the voltage across R_2 is 0 V, and $\mathbf{E}_{Th} = \mathbf{V}_1$.

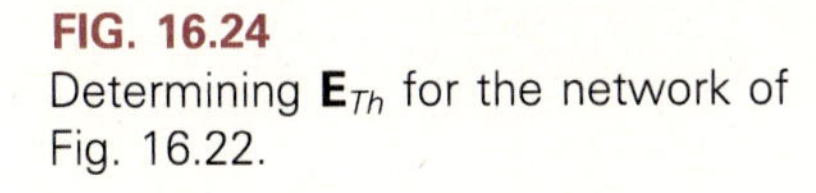
FIG. 16.24
Determining $\mathbf{E}_{Th}$ for the network of Fig. 16.22.

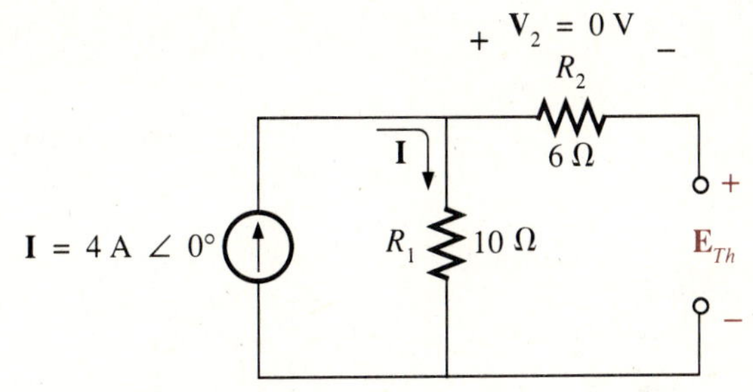

The current through R_1 is **I**, and

$$\mathbf{V}_1 = \mathbf{I}\mathbf{R}_1$$

with

$$\begin{aligned} V_1 &= IR_1 \\ &= (4\text{ A})(10\ \Omega) \\ V_1 &= 40\text{ V} \end{aligned}$$

Since v_R and i_R are in phase for a resistive element, the angle associated with $\mathbf{V}_1$ is also 0°, and

$$\mathbf{V}_1 = 40\text{ V}\ \angle 0^\circ$$
$$\mathbf{E}_{Th} = \mathbf{V}_1 = 40\text{ V}\ \angle 0^\circ$$

Step 5: The Thevenin equivalent circuit with R_L appears in Fig. 16.25.

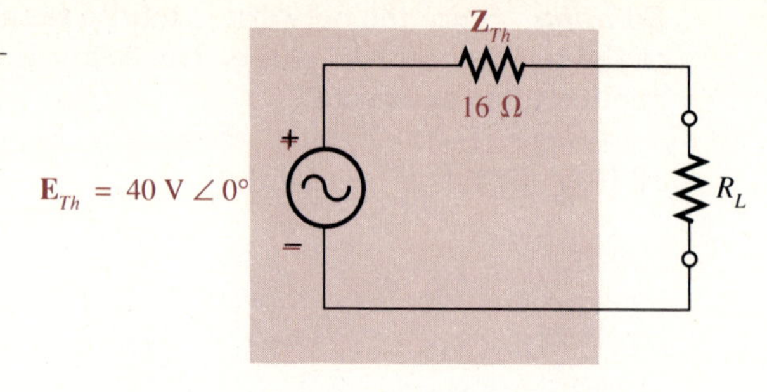

FIG. 16.25
Thevenin equivalent circuit for the network external to R_L in Fig. 16.22.

EXAMPLE 16.9 **Find the Thevenin equivalent circuit for the network external to resistor R in Fig. 16.26.**

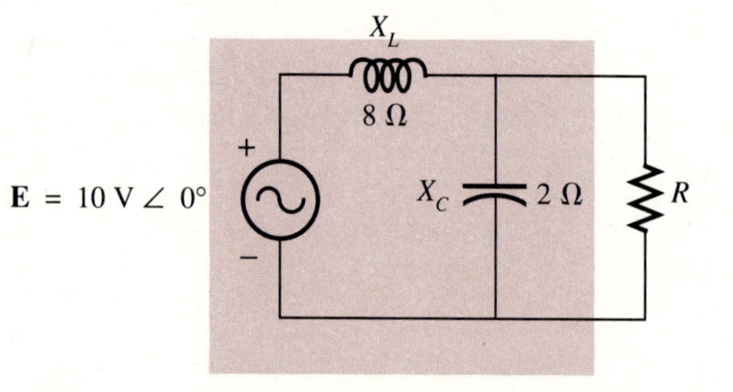

FIG. 16.26
Network for Example 16.9.

Solution

Steps 1 and 2: See Fig. 16.27.

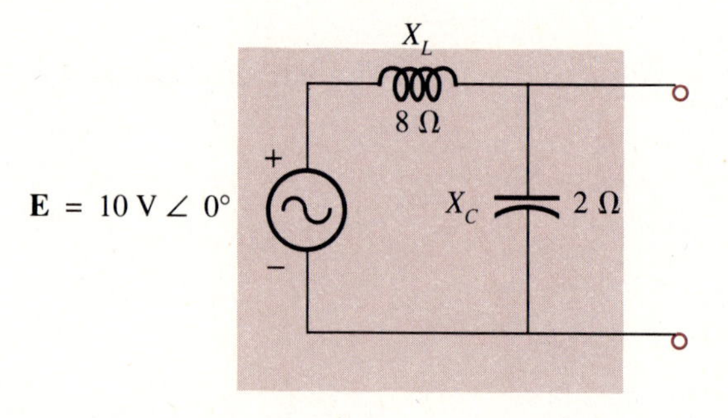

FIG. 16.27
Network of Fig. 16.26 following the application of steps 1 and 2.

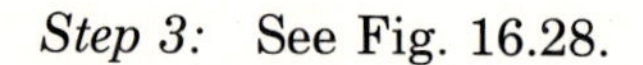

Step 3: See Fig. 16.28.

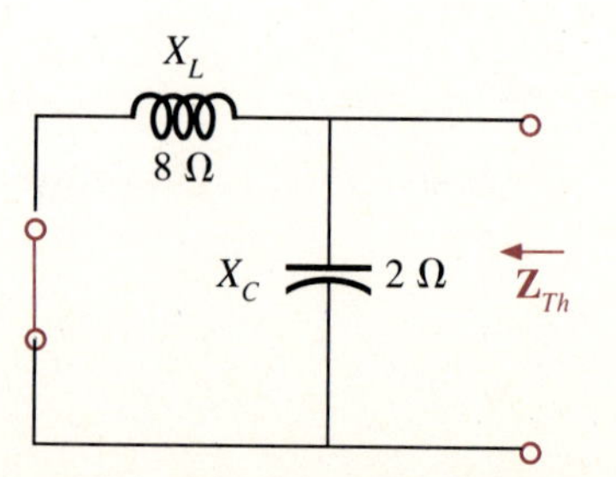

FIG. 16.28
Determining $\mathbf{Z}_{Th}$ for the network of Fig. 16.26.

Step 3 results in a parallel network for which the total admittance ca… determined. That is,

$$\begin{aligned}\mathbf{Y}_T &= \mathbf{B}_L + \mathbf{B}_C = \frac{1}{X_L}\angle -90^\circ + \frac{1}{X_C}\angle 90^\circ \\ &= \frac{1}{8\ \Omega}\angle -90^\circ + \frac{1}{2\ \Omega}\angle 90^\circ \\ &= 0.125\ \text{S}\ \angle -90^\circ + 0.5\ \text{S}\ \angle 90^\circ \\ \mathbf{Y}_T &= 0.375\ \text{S}\ \angle 90^\circ\end{aligned}$$

with

$$\mathbf{Z}_T = \mathbf{Z}_{Th} = \frac{1}{Y_T}\angle -\theta_T = \frac{1}{0.375\ \text{S}}\angle -90^\circ = \mathbf{2.667\ \Omega\ \angle -90^\circ}$$

Step 4: See Fig. 16.29. The result of this step is a series L-C combination, with $\mathbf{E}_{Th}$ being the voltage across the capacitive reactance.

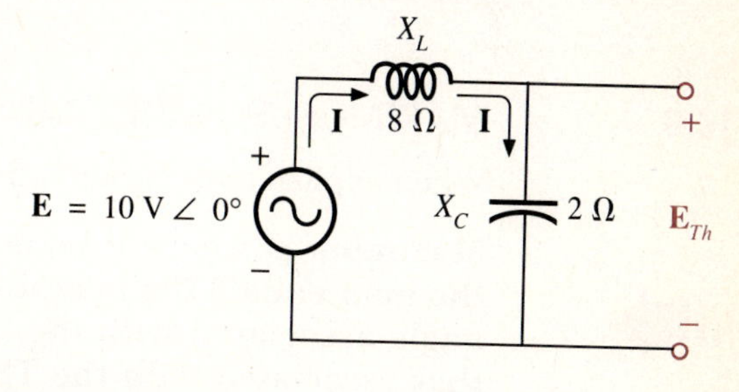

FIG. 16.29
Determining $\mathbf{E}_{Th}$ for the network of Fig. 16.26.

The total impedance of the series circuit is

$$\begin{aligned}\mathbf{Z}_T &= \mathbf{X}_L + \mathbf{X}_C = X_L\angle 90^\circ + X_C\angle -90^\circ \\ &= 8\ \Omega\ \angle 90^\circ + 2\ \Omega\ \angle -90^\circ \\ \mathbf{Z}_T &= 6\ \Omega\ \angle 90^\circ\end{aligned}$$

The magnitude of the current **I** is

$$I = \frac{E}{Z_T} = \frac{10\ \text{V}}{6\ \Omega} = 1.667\ \text{A}$$

Since the angle associated with $\mathbf{Z}_T$ is 90°, the angle associated with **I** for $\mathbf{E} = E\angle 0^\circ$ is −90°, and

$$\mathbf{I} = 1.667\ \text{A}\ \angle -90^\circ$$

The voltage across X_C is

$$\mathbf{E}_{Th} = \mathbf{IX}_C$$

with

$$\begin{aligned}E_{Th} &= IX_C \\ &= (1.667\ \text{A})(2\ \Omega) \\ E_{Th} &= 3.334\ \text{V}\end{aligned}$$

d, since v_C lags i_C by 90°, the voltage $\mathbf{E}_{Th}$ must lag $\mathbf{I}$ by 90°, so

$$\mathbf{E}_{Th} = 3.334 \text{ V} \angle -90° - 90°$$
$$= \mathbf{3.334 \text{ V} \angle -180°}$$

Step 5: The Thevenin equivalent circuit appears in Fig. 16.30.

FIG. 16.30
The Thevenin equivalent circuit for the network external to the resistor R in Fig. 16.26.

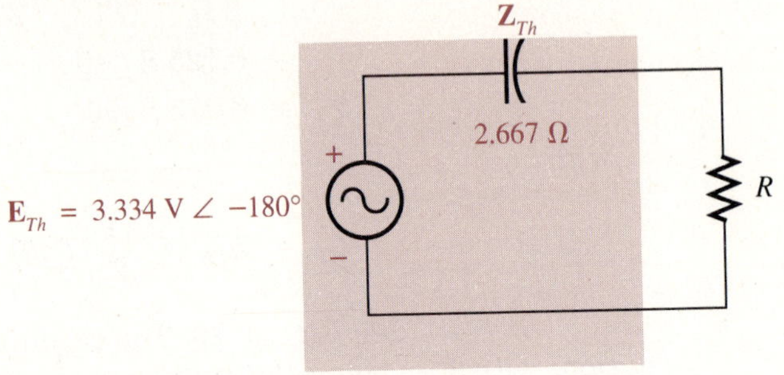

16.6 MAXIMUM POWER TRANSFER THEOREM

When applied to ac circuits, the maximum power transfer theorem states that

Maximum power will be delivered to a load when the magnitude of the load equals the magnitude of the Thevenin impedance and the angle associated with the impedance of the load is the negative of that associated with the Thevenin impedance.

Symbolically, for the network of Fig. 16.31,

$$Z_L = Z_{Th} \qquad (16.5)$$

and

$$\theta_L = -\theta_{Th} \qquad (16.6)$$

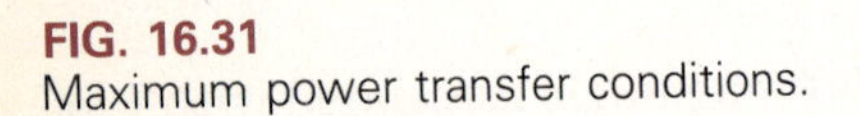

FIG. 16.31
Maximum power transfer conditions.

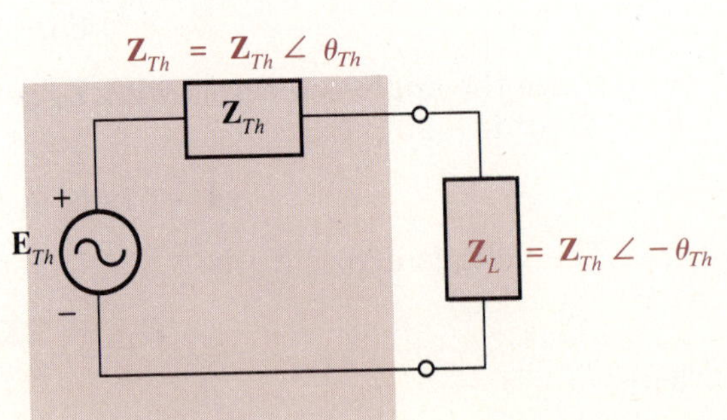

The conditions established by Eqs. 16.5 and 16.6 are satisfied by the elements of Fig. 16.32.

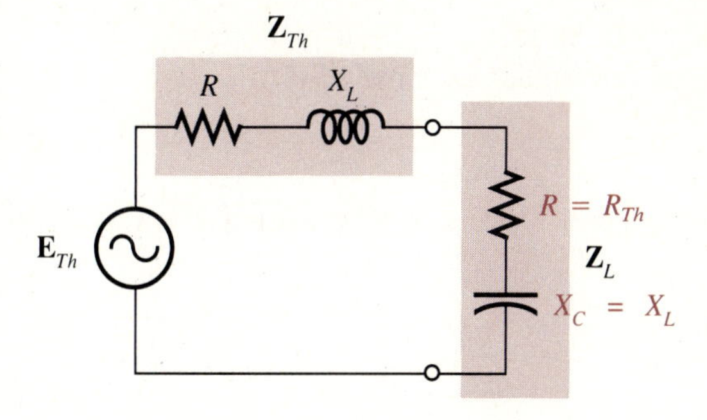

FIG. 16.32
Maximum power transfer conditions for $\mathbf{Z}_{Th} = \mathbf{R}_{Th} + \mathbf{X}_{Th}$.

Note that $X_C = X_L$, and the total impedance is

$$\begin{aligned}\mathbf{Z}_T &= \mathbf{R}_{Th} + \mathbf{X}_L + \mathbf{R}_{Th} + \mathbf{X}_C \\ &= 2R_{Th}\,\underline{/0^\circ} + X_L\,\underline{/90^\circ} + X_C\,\underline{/-90^\circ}\end{aligned}$$

and since

$$X_C = X_L$$

we find the last two terms drop out, leaving

$$\boxed{\mathbf{Z}_T = 2R_{Th}\,\underline{/0^\circ}} \qquad \textbf{(16.7)}$$

revealing that under maximum power transfer conditions, the total impedance is resistive with a phase angle of 0°. Obviously, if $\mathbf{Z}_{Th}$ includes a capacitive reactance, the load must have an inductive reactance of the same magnitude.

The magnitude of the current under these conditions is

$$I = \frac{E_{Th}}{Z_T} = \frac{E_{Th}}{2R_{Th}}$$

and the maximum power to the load is

$$\begin{aligned}P_{\max} &= I^2 R \\ &= \left(\frac{E_{Th}}{2R_{Th}}\right)^2 R_{Th} = \frac{E_{Th}^2}{4R_{Th}^2} R_{Th}\end{aligned}$$

$$\boxed{P_{\max} = \frac{E_{Th}^2}{4R_{Th}}} \qquad \textbf{(16.8)}$$

EXAMPLE 16.10 Find the load impedance in Fig. 16.33 for maximum power to the load, and find the maximum power.

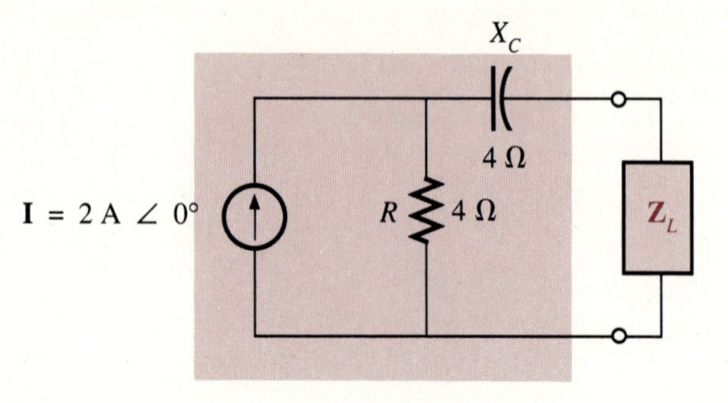

FIG. 16.33
Network for Example 16.10.

Solution $\mathbf{Z}_{Th}$ (Fig. 16.34) is

$$\mathbf{Z}_{Th} = \mathbf{R} + \mathbf{X}_C = 4\ \Omega\ \underline{/0^\circ} + 4\ \Omega\ \underline{/-90^\circ}$$

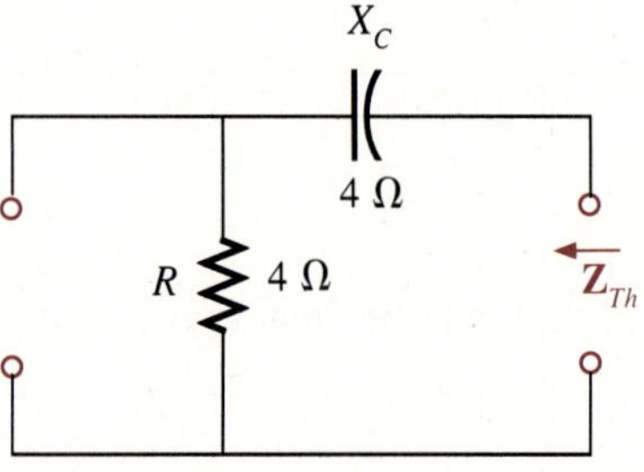

FIG. 16.34
Determining $\mathbf{Z}_{Th}$ for the network of Fig. 16.33.

For $Z_L = Z_{Th}$ and $\theta_L = -\theta_{Th}$,

$$\mathbf{Z}_L = \mathbf{R} + \mathbf{X}_L = 4\ \Omega\ \underline{/0^\circ} + 4\ \Omega\ \underline{/90^\circ}$$

For $\mathbf{E}_{Th}$, see Fig. 16.35. Since $\mathbf{V}_C = 0$ V, $\mathbf{E}_{Th} = \mathbf{V}_R$ and

$$\begin{aligned} V_R &= IR \\ &= (2\ \text{A})(4\ \Omega) \\ V_R &= 8\ \text{V} \end{aligned}$$

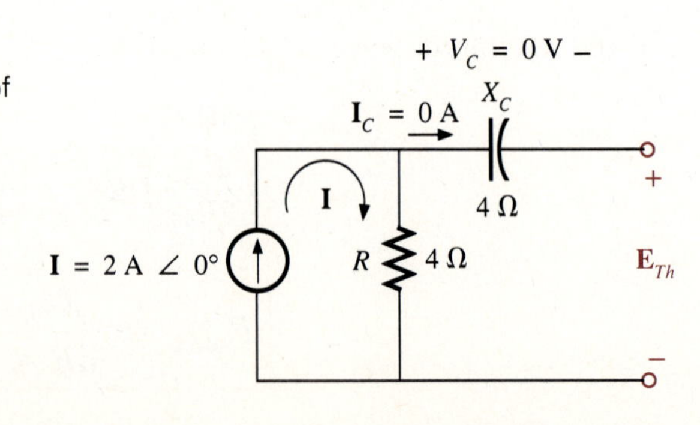

FIG. 16.35
Determining $\mathbf{E}_{Th}$ for the network of Fig. 16.33.

In addition, v_R and $i_R = i$ are in phase, so the angle associated with $\mathbf{V}_R$ is also 0°. Therefore,

$$\mathbf{E}_{Th} = \mathbf{V}_R = 8\text{ V}\,\angle 0^\circ$$

and

$$P_{\max} = \frac{E_{Th}^2}{4R} = \frac{(8\text{ V})^2}{4(4\,\Omega)} = \frac{64}{16}\text{ W}$$
$$= \mathbf{4\ W}$$

EXAMPLE 16.11 Find the load impedance in Fig. 16.36 for maximum power to the load, and find the maximum power.

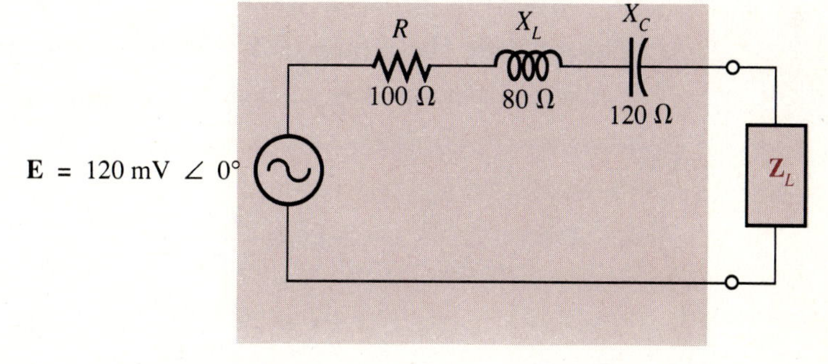

FIG. 16.36
Network for Example 16.11.

Solution For $\mathbf{Z}_{Th}$ (Fig. 16.37):

$$\begin{aligned}\mathbf{Z}_{Th} &= \mathbf{R} + \mathbf{X}_L + \mathbf{X}_C \\ &= 100\,\Omega\,\angle 0^\circ + 80\,\Omega\,\angle 90^\circ + 120\,\Omega\,\angle -90^\circ \\ &= 100\,\Omega\,\angle 0^\circ + 40\,\Omega\,\angle -90^\circ\end{aligned}$$

For maximum power transfer,

$$\mathbf{Z}_L = 100\,\Omega\,\angle 0^\circ + 40\,\Omega\,\angle +90^\circ$$

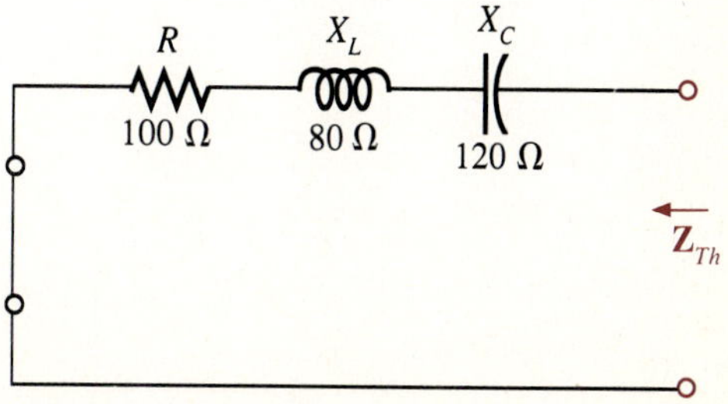

FIG. 16.37
Determining $\mathbf{Z}_{Th}$ for the network of Fig. 16.36.

For $\mathbf{E}_{Th}$, see Fig. 16.38. Since $\mathbf{E}_{Th}$ is an open-circuit voltage, the current in the series circuit must be 0 A, and $V_R = V_L = V_C = 0$ V. The result is that

$$\mathbf{E}_{Th} = \mathbf{E}$$

as shown in Fig. 16.38.

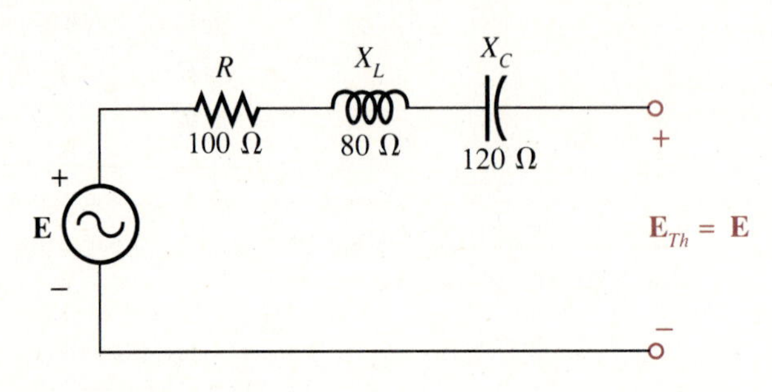

FIG. 16.38
Determining $\mathbf{E}_{Th}$ for the network of Fig. 16.36.

Substituting the required $\mathbf{Z}_L$ results in the circuit of Fig. 16.39, where

$$\mathbf{Z}_T = 2R_{Th} = \mathbf{200\ \Omega\ \angle 0^\circ}$$

and

$$\mathbf{I} = \frac{E_{Th}}{2R_{Th}}\angle 0^\circ = \frac{E}{2R_{Th}}\angle 0^\circ = \frac{120\text{ mV}}{2(100\ \Omega)}\angle 0^\circ$$

$$= \frac{120\text{ mV}}{200\ \Omega}\angle 0^\circ = \mathbf{0.6\ mA\ \angle 0^\circ}$$

with

$$P_{\max} = \frac{E_{Th}^2}{4R_{Th}} = \frac{(120\text{ mV})^2}{4(100\ \Omega)} = \frac{(120 \times 10^{-3}\text{ V})^2}{400\ \Omega}$$

$$= \frac{14{,}400 \times 10^{-6}\text{ W}}{400}$$

$$P_{\max} = \mathbf{36\ \mu W}$$

FIG. 16.39
Maximum power condition for the network of Fig. 16.36.

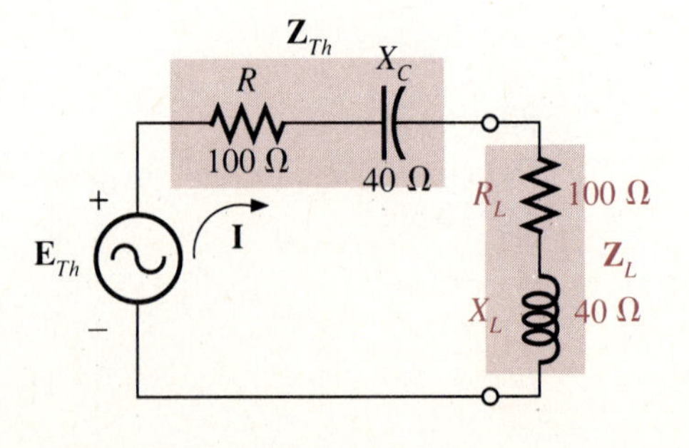

16.7 Δ-Y, Y-Δ CONVERSIONS

The Δ-Y or Y-Δ conversions for ac networks are performed in essentially the same manner as those for dc networks. The major difference again is that now we are dealing with impedances that have a vector relationship rather than resistors with a fixed value and no angle consideration.

For the delta (Δ) and wye (Y) configurations of Fig. 16.40, in which the impedances are the same for each configuration, the following equations relate the two:

$$\mathbf{Z}_\Delta = 3\mathbf{Z}_\mathrm{Y} \tag{16.9}$$

and

$$\mathbf{Z}_\mathrm{Y} = \frac{\mathbf{Z}_\Delta}{3} \tag{16.10}$$

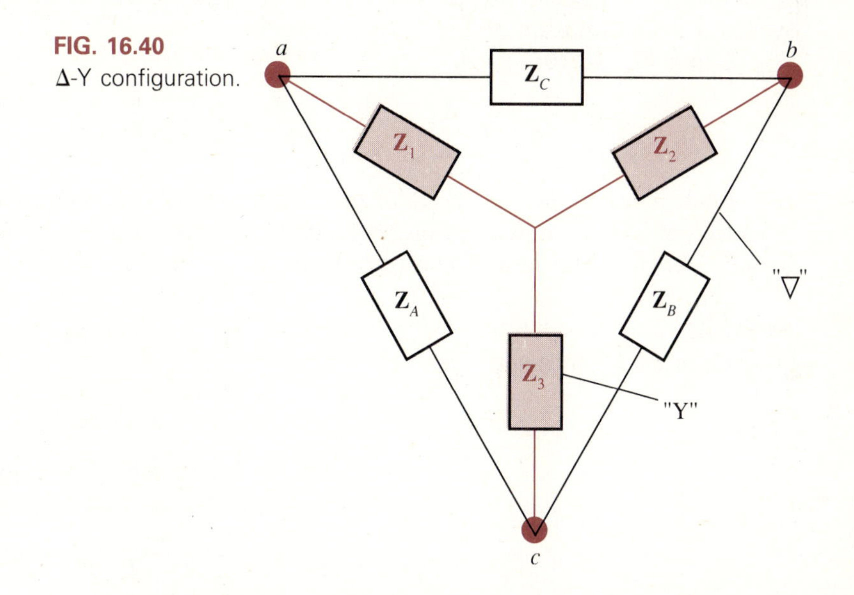

FIG. 16.40
Δ-Y configuration.

For the situation of unequal impedances in the same configuration, the resulting equations have the same format as those provided in the dc analysis, with resistive levels simply exchanged for impedances.

Keep in mind that the equations are designed to permit a choice of the Δ configuration *or* Y configuration between the same three terminals. It is not intended that both appear at the same time. They were drawn together to show how they can be connected between the same three terminals. Of course, a particular configuration may have both present at the same time, but then one is usually converted to the other to permit a network reduction to find the desired unknowns.

EXAMPLE 16.12 Find the total impedance of the bridge network of Fig. 16.41.

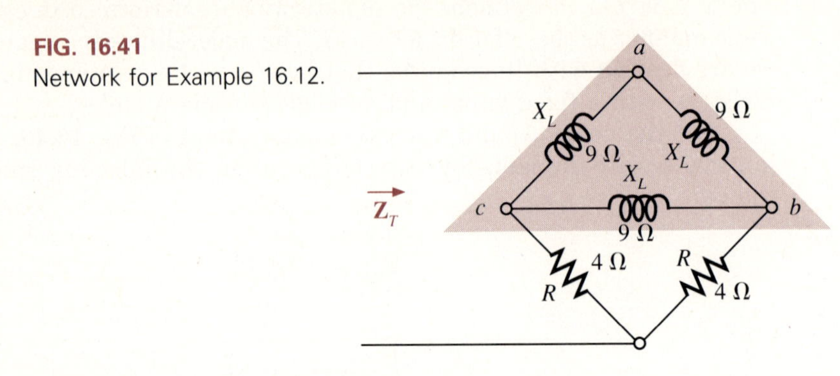

FIG. 16.41
Network for Example 16.12.

Solution To determine $\mathbf{Z}_T$, the delta (Δ) at the top of the network is first converted to a wye (Y) configuration. Note in this network that no two elements are in series or parallel, to permit an initial reduction and an eventual determination of the total impedance.

Using Eq. 16.10:

$$\mathbf{Z}_Y = \frac{\mathbf{Z}_\Delta}{3} = \frac{9\ \Omega}{3}\ \underline{/90^\circ} = 3\ \Omega\ \underline{/90^\circ}$$

Each element of the wye (Y) configuration is, therefore, an inductive reactance of 3 Ω, as shown in Fig. 16.42. Note that the same three terminals, a, b, and c, enclose the new Y configuration.

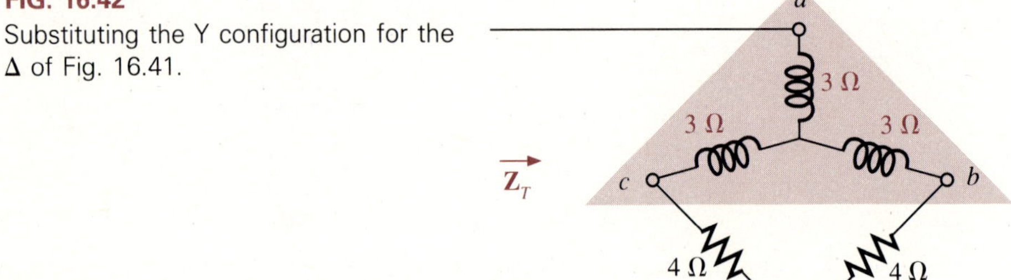

FIG. 16.42
Substituting the Y configuration for the Δ of Fig. 16.41.

The 4-Ω resistors and 3-Ω inductors are now in series, with a resulting impedance

$$\mathbf{Z} = 4\ \Omega\ \underline{/0^\circ} + 3\ \Omega\ \underline{/90^\circ}$$

with

$$Z = \sqrt{(4\ \Omega)^2 + (3\ \Omega)^2} = 5\ \Omega$$

and

$$\theta = \tan^{-1}\frac{3\ \Omega}{4\ \Omega} = 36.87^\circ$$

$$\mathbf{Z} = 5\ \Omega\ \angle 36.87^\circ$$

The parallel impedance of the two equal parallel impedances **Z** can now be determined by first finding the total parallel admittance and then the total impedance, but since the parallel impedances are the same, the equation

$$\mathbf{Z}_T = \frac{\mathbf{Z}}{N} \qquad (16.11)$$

can be applied. In Eq. 16.11, **Z** is the impedance of *one* of equal parallel branches and N is the number of equal parallel branches.

For this case

$$\mathbf{Z'}_T = \frac{\mathbf{Z}}{N} = \frac{5\ \Omega}{2}\angle 36.87^\circ$$
$$= 2.5\ \Omega\ \angle 36.87^\circ$$

as shown in Fig. 16.43.

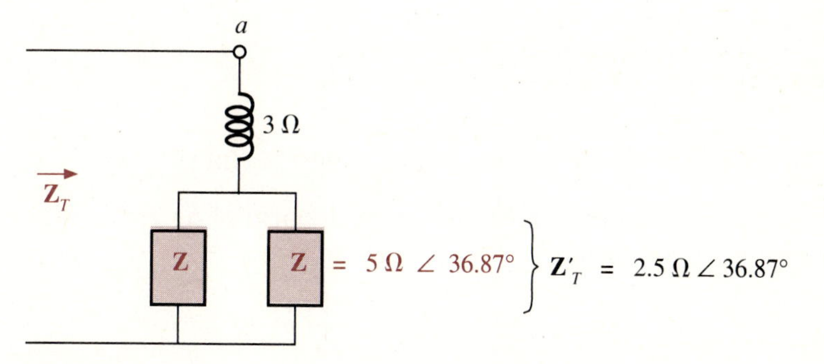

FIG. 16.43
Determining $\mathbf{Z}_T$ for the network of Fig. 16.41.

If we expand **Z′** into its components, we obtain

$$\mathbf{Z'}_T = \mathbf{R'} + \mathbf{X'}_L = 2\ \Omega\ \angle 0^\circ + 1.5\ \Omega\ \angle 90^\circ$$

Since $\mathbf{Z'}_T$ is in series with the 3-Ω inductive reactance connected to terminal a,

$$\mathbf{Z}_T = \mathbf{Z'}_T + \mathbf{X}_L$$
$$= 2\ \Omega\ \angle 0^\circ + 1.5\ \Omega\ \angle 90^\circ + 3\ \Omega\ \angle 90^\circ$$
$$\mathbf{Z}_T = 2\ \Omega\ \angle 0^\circ + 4.5\ \Omega\ \angle 90^\circ$$

with

$$Z_T = \sqrt{(2\ \Omega)^2 + (4.5\ \Omega)^2} = 4.924\ \Omega$$

$$\theta_T = \tan^{-1}\frac{4.5\ \Omega}{2\ \Omega} = \tan^{-1}2.25 = 66.04^\circ$$

and

$$\mathbf{Z}_T = \mathbf{4.924\ \Omega\ \angle 66.04^\circ}$$

EXAMPLE 16.13 Find the total impedance of the Δ-Y configuration of Fig. 16.44.

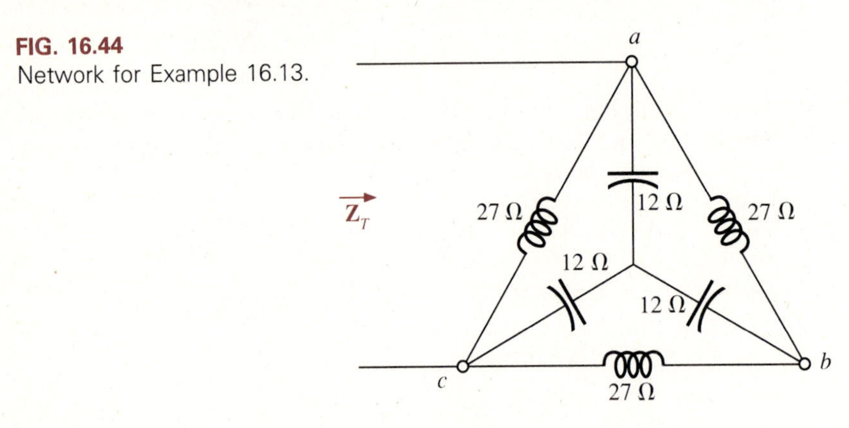

FIG. 16.44
Network for Example 16.13.

Solution Convert the Y to a Δ; use Eq. 16.9:

$$\begin{aligned}\mathbf{Z}_\Delta &= 3\mathbf{Z}_Y \\ &= 3(12\ \Omega\ \angle -90^\circ) = 36\ \Omega\ \angle -90^\circ\end{aligned}$$

Redrawing the configuration with parallel Δs yields Fig. 16.45. For each parallel combination:

$$\mathbf{Y} = \mathbf{B}_L + \mathbf{B}_C = \frac{1}{X_L}\angle -90^\circ + \frac{1}{X_C}\angle +90^\circ$$

$$= \frac{1}{27\ \Omega}\angle -90^\circ + \frac{1}{36\ \Omega}\angle +90^\circ = 0.0375\ \text{S}\ \angle -90^\circ + 0.0278\ \text{S}\ \angle +90^\circ$$

$$\mathbf{Y} = 0.0092\ \text{S}\ \angle -90^\circ$$

and

$$\mathbf{Z} = \frac{1}{\mathbf{Y}}\angle -\theta = \frac{1}{0.0092\ \text{S}}\angle -(-90^\circ) = 108.7\ \Omega\ \angle 90^\circ$$

The total impedance of the series branches a-b-c is

$$\mathbf{Z}' = \mathbf{Z} + \mathbf{Z} = 2\mathbf{Z} = 2(108.7\ \Omega\ \angle 90^\circ) = 217.4\ \Omega\ \angle 90^\circ$$

FIG. 16.45
Substituting the equivalent Δ for the Y configuration of Fig. 16.44.

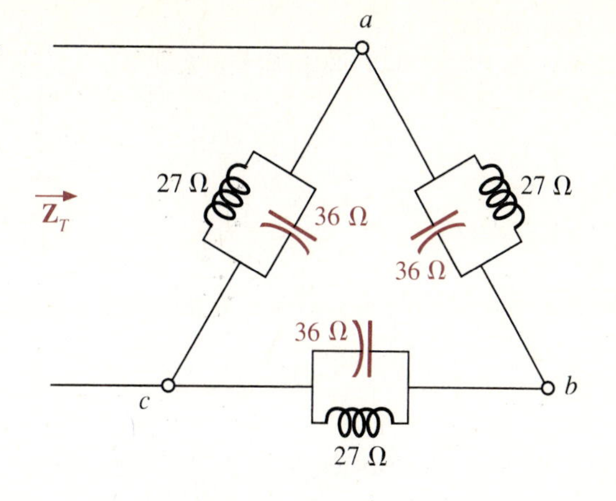

The total impedance $\mathbf{Z}_T$ is then

$$\mathbf{Z}_T = \mathbf{Z} \| 2\mathbf{Z}$$

or

$$\mathbf{Y}_T = \mathbf{Y}_1 + \mathbf{Y}_2 = \frac{1}{\mathbf{Z}} + \frac{1}{2\mathbf{Z}} = \frac{1}{108.7\ \Omega}\angle -90^\circ + \frac{1}{217.4\ \Omega}\angle -90^\circ$$

$$= 0.0092\ \text{S}\angle -90^\circ + 0.0046\ \text{S}\angle -90^\circ$$

$$\mathbf{Y}_T = 0.0138\ \text{S}\angle -90^\circ$$

and

$$\mathbf{Z}_T = \frac{1}{\mathbf{Y}_T} = \frac{1}{0.0138\ \text{S}}\angle -(-90^\circ) = \mathbf{72.46\ \Omega\angle 90^\circ}$$

16.8 BRIDGE CONFIGURATION

For the bridge configuration of Fig. 16.46, if the impedances are chosen such that

$$\boxed{\frac{\mathbf{Z}_1}{\mathbf{Z}_3} = \frac{\mathbf{Z}_2}{\mathbf{Z}_4}} \qquad \text{(16.12)}$$

the network is said to be *balanced* and the current through the bridge arm will be zero, as indicated in Fig. 16.46. Since impedances include a magnitude and an angle, it is not sufficient for the ratio of the magnitudes of $\mathbf{Z}_1/\mathbf{Z}_3$ to have the same ratio as the ratio $\mathbf{Z}_2/\mathbf{Z}_4$; the angle of one side must equal the angle of the other also.

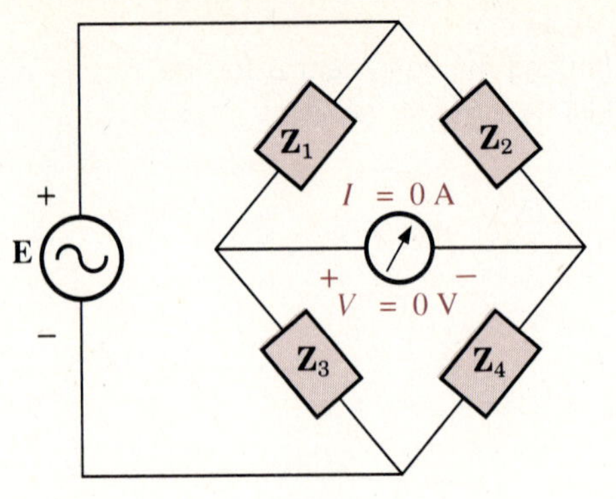

FIG. 16.46
General balanced bridge configuration.

For the network of Fig. 16.47, referred to as a *Hay bridge*, Eq. 16.12 results in

$$\frac{\mathbf{R}_1 + \mathbf{X}_{C_1}}{\mathbf{R}_3} = \frac{\mathbf{R}_2}{\mathbf{R}_4 + \mathbf{X}_{L_4}}$$

or

$$\frac{\sqrt{R_1^2 + X_{C_1}^2} \angle \theta_1}{R_3 \angle 0^\circ} = \frac{R_2 \angle 0^\circ}{\sqrt{R_4^2 + X_{L_4}^2} \angle \theta_4}$$

To satisfy the balance conditions,

$$\boxed{\frac{\sqrt{R_1^2 + X_{C_1}^2}}{R_3} = \frac{R_2}{\sqrt{R_4^2 + X_{L_4}^2}}} \tag{16.13}$$

and

$$\boxed{\theta_1 = -\theta_4} \tag{16.14}$$

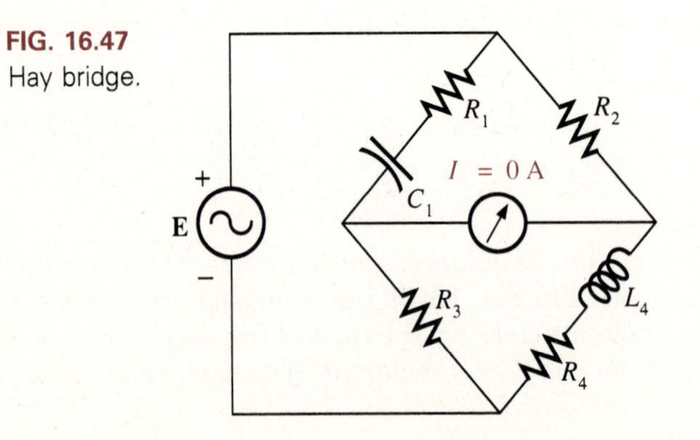

FIG. 16.47
Hay bridge.

Using a reasonable amount of algebra, equations for L_4 and R_4 can be obtained from Eqs. 16.13 and 16.14:

$$L_4 = \frac{CR_2R_3}{1 + \omega^2C^2R_1^2} \tag{16.15}$$

$$R_4 = \frac{\omega^2C^2R_1R_2R_3}{1 + \omega^2C^2R_1^2} \tag{16.16}$$

The result is that once the parameters R_1, R_2, R_3, and C_1 are adjusted until balance is obtained ($I = 0$), the unknown quantities L_4 and R_4 can be determined using Eqs. 16.15 and 16.16.

The bridge network of Fig. 16.48 is called a *Maxwell bridge;* it has the following balance equations:

$$L_4 = CR_2R_3 \tag{16.17}$$

$$R_4 = \frac{R_2R_3}{R_1} \tag{16.18}$$

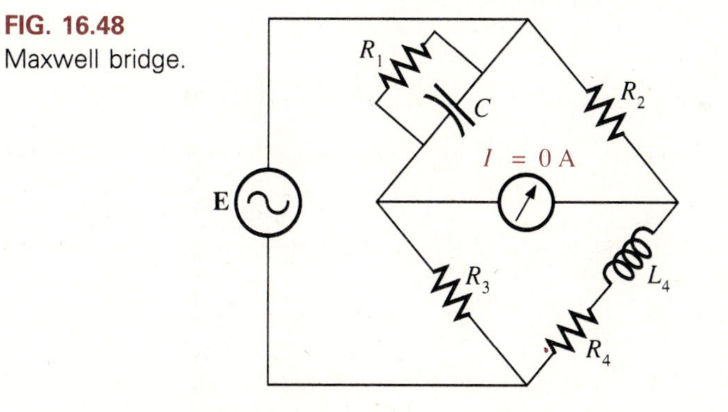

FIG. 16.48
Maxwell bridge.

Note in this case that L_4 and R_4 are not a function of frequency as was the case in Eqs. 16.15 and 16.16, where $\omega = 2\pi f$.

For the capacitance comparison bridge of Fig. 16.49, the balance equations are

$$C_4 = \frac{R_1}{R_2}C_3 \tag{16.19}$$

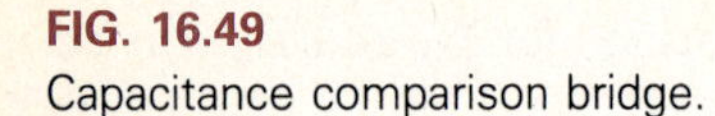

FIG. 16.49
Capacitance comparison bridge.

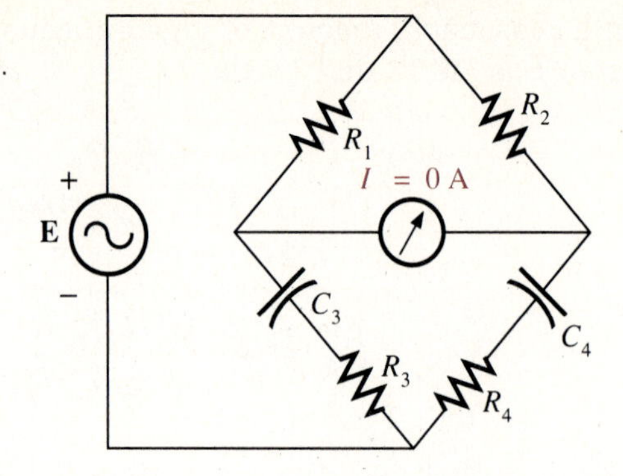

and

$$R_4 = \frac{R_2 R_3}{R_1} \tag{16.20}$$

permitting a determination of C_4 and R_4 using the other elements of the configuration that are known to a high degree of accuracy.

EXAMPLE 16.14

a. Determine whether the bridge of Fig. 16.50 is balanced ($I = 0$ A).
b. If it is balanced, determine $\mathbf{I}_1$.

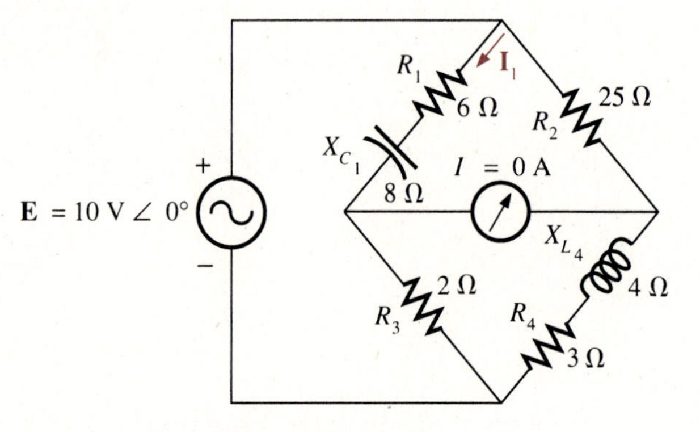

FIG. 16.50
Network for Example 16.14.

Solutions

a. $\mathbf{Z}_1 = \mathbf{R}_1 + \mathbf{X}_{C_1} = 6\ \Omega\ \angle 0^\circ + 8\ \Omega\ \angle -90^\circ = 10\ \Omega\ \angle -53.13^\circ$
$\mathbf{Z}_2 = \mathbf{R}_2 = 25\ \Omega\ \angle 0^\circ$
$\mathbf{Z}_3 = \mathbf{R}_3 = 2\ \Omega\ \angle 0^\circ$
$\mathbf{Z}_4 = \mathbf{R}_4 + \mathbf{X}_{L_4} = 3\ \Omega\ \angle 0^\circ + 4\ \Omega\ \angle 90^\circ = 5\ \Omega\ \angle 53.13^\circ$

From Eq. 16.12:

$$\frac{\mathbf{Z}_1}{\mathbf{Z}_3} = \frac{\mathbf{Z}_2}{\mathbf{Z}_4}$$

$$\frac{10\ \Omega\ \angle -53.13^\circ}{2\ \Omega\ \angle 0^\circ} = \frac{25\ \Omega\ \angle 0^\circ}{5\ \Omega\ +\ \angle 53.13^\circ}$$

The magnitude ratio of both sides is certainly satisfied, since

$$\frac{10\ \Omega}{2\ \Omega} = 5 = \frac{25\ \Omega}{5\ \Omega}$$

Using the fact that $\mathbf{Z}_T = 1/\mathbf{Y}_T$ and the angle of $\mathbf{Z}_T$ is the negative of that associated with $\mathbf{Y}_T$, we can bring the $+53.13^\circ$ to the numerator as -53.13°. The result is that

$$5\ \angle -53.13^\circ = 5\ \angle -53.13^\circ$$

satisfying all the conditions specified by Eq. 16.12.

b. If $\mathbf{I} = 0$, then $\mathbf{I}_1$ also passes through R_3 and R_3 is in series with R_1 and X_{C_1}. The total impedance of this series circuit is then

$$\begin{aligned} \mathbf{Z} &= \mathbf{R}_1 + \mathbf{X}_{C_1} + \mathbf{R}_3 \\ &= 6\ \Omega\ \angle 0^\circ + 8\ \Omega\ \angle -90^\circ + 2\ \Omega\ \angle 0^\circ \\ \mathbf{Z} &= 8\ \Omega\ \angle 0^\circ + 8\ \Omega\ \angle -90^\circ \end{aligned}$$

Thus,

$$\begin{aligned} Z &= \sqrt{(8\ \Omega)^2 + (8\ \Omega)^2} = 11.31\ \Omega \\ \theta &= -\tan^{-1}\frac{8\ \Omega}{8\ \Omega} = -\tan^{-1}1 = -45^\circ \\ \mathbf{Z} &= 11.31\ \Omega\ \angle -45^\circ \\ I_1 &= \frac{E}{Z} \\ &= \frac{10\ \text{V}}{11.31\ \Omega} = 0.884\ \text{A} \\ \mathbf{I}_1 &= 0.884\ \text{A}\ \angle -(-45^\circ) \\ &= \mathbf{0.884\ A\ \angle 45^\circ} \end{aligned}$$

A commercial bridge unit appears in Fig. 16.51. Once the dials are adjusted such that balance is established, the nameplate data of an unknown element can be determined to a high degree of accuracy.

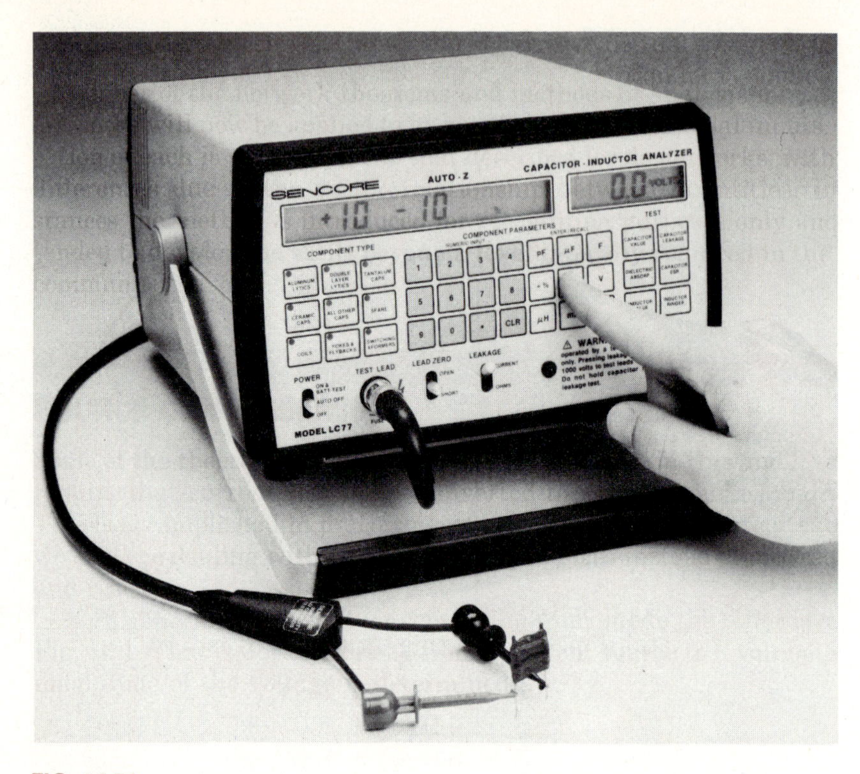

FIG. 16.51
Impedance bridge. (Courtesy of Sencore)

16.9 THE *j*-OPERATOR

The real value of the *j*-operator becomes more apparent when complex ac configurations are encountered. The operator permits the application of equations and methods that would otherwise be too lengthy and complex. In addition, the operator permits the development of a number of useful computer software packages. Before examining methods of analysis that require a *j*-operator approach, let us first repeat some of the basic calculations performed earlier to discover the impact of this mathematical technique.

EXAMPLE 16.15 Repeat Example 16.4 using the *j*-operator.

Solution For the parallel *R*-*C* combination:

$$\mathbf{Z}'_T = \mathbf{R}||\mathbf{X}_C = \frac{\mathbf{R}\mathbf{X}_C}{\mathbf{R} + \mathbf{X}_C} = \frac{(4\text{ k}\Omega \angle 0^\circ)(6\text{ k}\Omega \angle -90^\circ)}{4\text{ k}\Omega - j6\text{ k}\Omega}$$

$$= \frac{24\text{ k}\Omega \angle -90^\circ}{7.21 \angle -56.31^\circ} = 3.33\text{ k}\Omega \angle -33.69^\circ$$

and

$$\mathbf{Z}_T = \mathbf{X}_L + \mathbf{Z}'_T = j4\text{ k}\Omega + (3.33\text{ k}\Omega \angle -33.69^\circ)$$
$$= j4\text{ k}\Omega + 2.77\text{ k}\Omega - j1.85\text{ k}\Omega$$
$$= 2.77\text{ k}\Omega + j2.15\text{ k}\Omega$$
$$\mathbf{Z}_T = \mathbf{3.51\text{ k}\Omega \angle 37.82^\circ}$$

as obtained earlier.

$$\mathbf{I}_T = \frac{\mathbf{E}}{\mathbf{Z}_T} = \frac{40\text{ V} \angle 0^\circ}{3.51\text{ k}\Omega \angle 37.82^\circ} = \mathbf{11.396\text{ mA} \angle -37.82^\circ}$$
$$\mathbf{V}_R = \mathbf{I}_T\mathbf{Z}'_T = (11.396\text{ mA} \angle -37.82^\circ)(3.33\text{ k}\Omega \angle -33.69^\circ)$$
$$= \mathbf{37.95\text{ V} \angle -71.51^\circ}$$

with the differences due solely to the level of accuracy carried through the calculations.

$$\mathbf{I}_C = \frac{\mathbf{V}_C}{\mathbf{X}_C} = \frac{\mathbf{V}_R}{\mathbf{X}_C} = \frac{37.95\text{ V} \angle -71.51^\circ}{6\text{ k}\Omega \angle -90^\circ} = \mathbf{6.33\text{ mA} \angle 18.49^\circ}$$

It is fairly obvious from the preceding that the *j*-operator provides a "cleaner" route to some of the desired unknowns—particularly the angle associated with particular voltages and currents. However, there is also a tendency to lose the "personal" touch with a network due to a more mathematical approach.

EXAMPLE 16.16 Repeat Example 16.7 using the *j*-operator approach.

Solution Due to **E** (Fig. 16.19):

$$\mathbf{Z}_T = 4\ \Omega + j5\ \Omega = 6.403\ \Omega \angle 51.34^\circ$$

and

$$\mathbf{I}' = \frac{\mathbf{E}}{\mathbf{Z}_T} = \frac{20\text{ V} \angle 0^\circ}{6.403\ \Omega \angle 51.34^\circ} = 3.124\text{ A} \angle -51.34^\circ$$

Due to **I** (Fig. 16.20), using the current divider rule:

$$\mathbf{I}'' = \frac{\mathbf{RI}}{\mathbf{R} + \mathbf{X}_L} = \frac{(4\ \Omega \angle 0^\circ)(2\text{ A} \angle 0^\circ)}{4\ \Omega + j5\ \Omega}$$
$$= \frac{8\text{ A} \angle 0^\circ}{6.403 \angle 51.34^\circ} = \mathbf{1.249\text{ A} \angle -51.34^\circ}$$

with $\mathbf{I}_L = \mathbf{I}' + \mathbf{I}'' = \mathbf{4.373\text{ A} \angle -51.34^\circ}$.

The *j*-operator permits the application of mesh and nodal analysis to ac networks using the same procedure defined for dc networks.

For purely resistive networks, sources of $E \angle 0°$ and $I \angle 0°$ can be treated as dc sources of the same magnitude, and the currents and voltages can be determined as before. The phase angle of 0° can then be applied to each of the results, and the time domain expression can be determined if necessary.

For networks with reactive elements, such as the one shown in Fig. 16.52, one approach suggests labeling each impedance as $\mathbf{Z}_1$, $\mathbf{Z}_2$, $\mathbf{Z}_3$, and so on, as shown in Fig. 16.53.

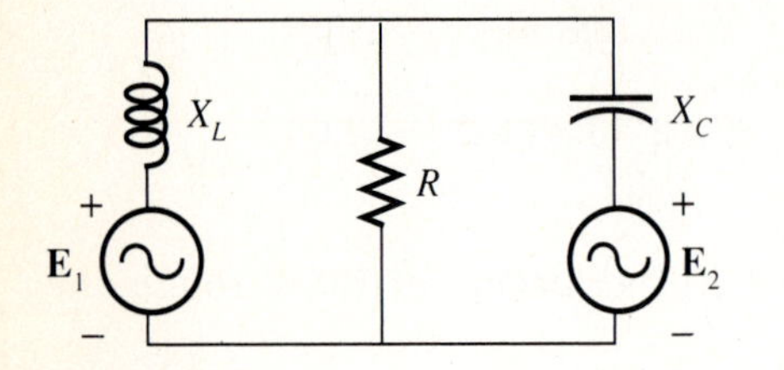

FIG. 16.52
Network to be analyzed using mesh analysis.

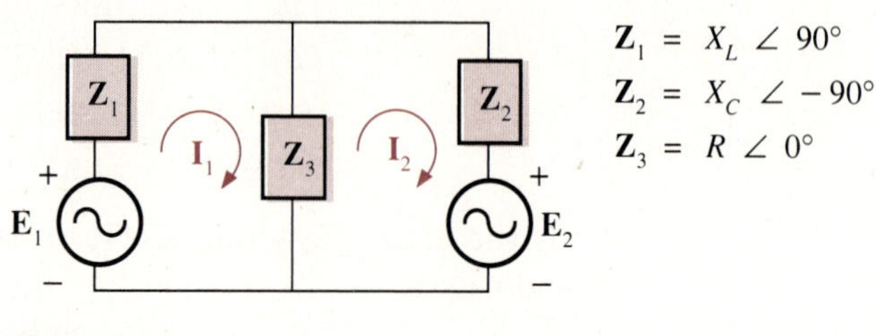

FIG. 16.53
Block diagram equivalent of Fig. 16.52.

Treating each impedance in the same manner as resistors in the dc section results in the following equations when the mesh-analysis approach is applied:

$$\begin{aligned} \mathbf{I}_1(\mathbf{Z}_1 + \mathbf{Z}_3) - \mathbf{I}_2\mathbf{Z}_3 &= \mathbf{E}_1 \\ \mathbf{I}_2(\mathbf{Z}_2 + \mathbf{Z}_3) - \mathbf{I}_1\mathbf{Z}_3 &= -\mathbf{E}_2 \end{aligned}$$

or

$$\begin{aligned} \mathbf{I}_1(\mathbf{Z}_1 + \mathbf{Z}_3) - \mathbf{I}_2\mathbf{Z}_3 &= \mathbf{E}_1 \\ -\mathbf{I}_1(\mathbf{Z}_3) + \mathbf{I}_2(\mathbf{Z}_2 + \mathbf{Z}_3) &= -\mathbf{E}_2 \end{aligned}$$

Using the substitution method described in the dc coverage or a more advanced technique such as determinants or matrices results in the following expressions for $\mathbf{I}_1$ and $\mathbf{I}_2$:

$$\mathbf{I}_1 = \frac{(\mathbf{E}_1 - \mathbf{E}_2)\mathbf{Z}_3 + \mathbf{E}_1\mathbf{Z}_2}{\mathbf{Z}_1\mathbf{Z}_2 + \mathbf{Z}_1\mathbf{Z}_3 + \mathbf{Z}_2\mathbf{Z}_3}$$

and

$$\mathbf{I}_2 = \frac{(\mathbf{E}_1 - \mathbf{E}_2)\mathbf{Z}_3 - \mathbf{E}_2\mathbf{Z}_1}{\mathbf{Z}_1\mathbf{Z}_2 + \mathbf{Z}_1\mathbf{Z}_3 + \mathbf{Z}_2\mathbf{Z}_3}$$

The equations for $\mathbf{I}_1$ and $\mathbf{I}_2$ are now in their simplest forms. The magnitude and angle of each phasor and impedance must now be substituted and the magnitude and angle of $\mathbf{I}_1$ and $\mathbf{I}_2$ determined. Using the j-operator, the mathematics required is obviously quite lengthy, but it is an approach that provides both the magnitude and angle of $\mathbf{I}_1$ and $\mathbf{I}_2$.

For nodal analysis the procedure introduced for dc networks can be applied with the use of admittance parameters, as defined in Fig. 16.55 for the network of Fig. 16.54.

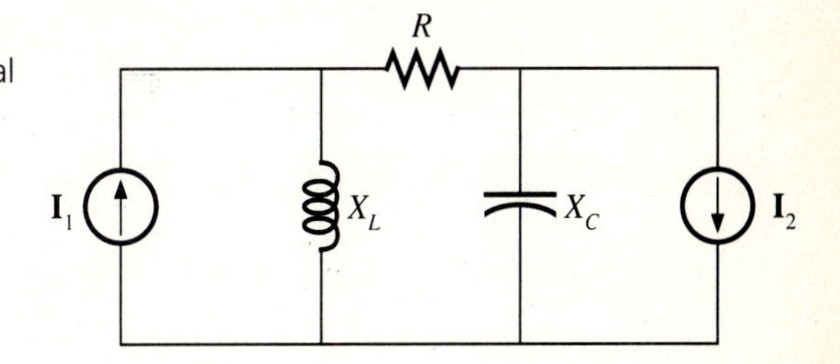

FIG. 16.54
Network to be analyzed using nodal analysis.

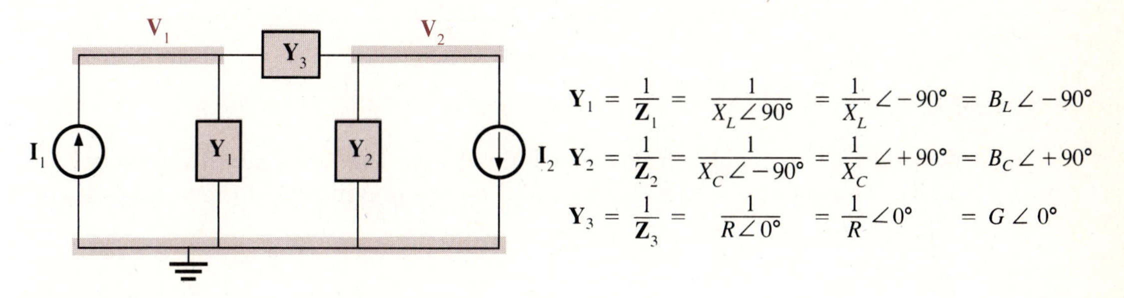

FIG. 16.55
Block diagram equivalent of Fig. 16.56.

The resulting nodal equations are

$$\begin{aligned} \mathbf{V}_1[\mathbf{Y}_1 + \mathbf{Y}_3] - \mathbf{V}_2[\mathbf{Y}_3] &= \mathbf{I}_1 \\ \mathbf{V}_2[\mathbf{Y}_2 + \mathbf{Y}_3] - \mathbf{V}_1[\mathbf{Y}_3] &= -\mathbf{I}_2 \end{aligned}$$

or

$$\begin{aligned} \mathbf{V}_1[\mathbf{Y}_1 + \mathbf{Y}_3] - \mathbf{V}_2[\mathbf{Y}_3] \qquad\qquad &= \mathbf{I}_1 \\ -\mathbf{V}_1[\mathbf{Y}_3] \qquad\qquad + \mathbf{V}_2[\mathbf{Y}_2 + \mathbf{Y}_3] &= -\mathbf{I}_2 \end{aligned}$$

Using one of the methods described earlier for mesh analysis results in the following equations for $\mathbf{V}_1$ and $\mathbf{V}_2$:

$$\mathbf{V}_1 = \frac{(\mathbf{I}_1 - \mathbf{I}_2)\mathbf{Y}_3 + \mathbf{I}_1\mathbf{Y}_2}{\mathbf{Y}_1\mathbf{Y}_2 + \mathbf{Y}_1\mathbf{Y}_3 + \mathbf{Y}_2\mathbf{Y}_3}$$

$$\mathbf{V}_2 = \frac{(\mathbf{I}_1 - \mathbf{I}_2)\mathbf{Y}_3 - \mathbf{I}_2\mathbf{Y}_1}{\mathbf{Y}_1\mathbf{Y}_2 + \mathbf{Y}_1\mathbf{Y}_3 + \mathbf{Y}_2\mathbf{Y}_3}$$

Note the similarities between the equations obtained for $\mathbf{I}_1$ and $\mathbf{I}_2$ and those obtained for $\mathbf{V}_1$ and $\mathbf{V}_2$. There is simply an interchange between **E** and **I** and **Z** and **Y**.

EXAMPLE 16.17 Determine the reading of the voltmeter of Fig. 16.56 using nodal analysis.

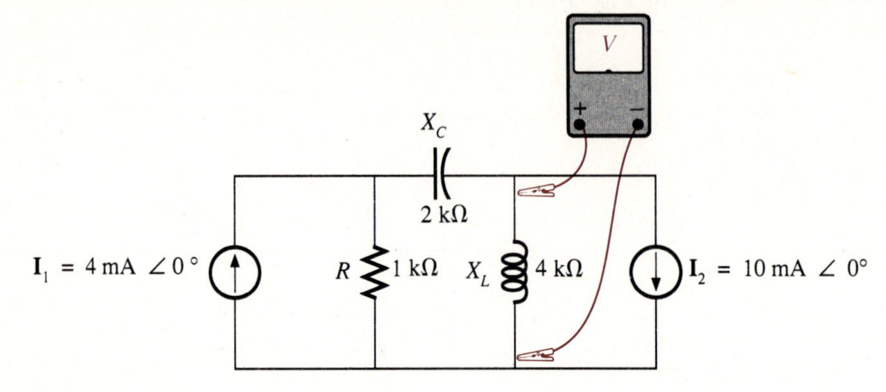

FIG. 16.56
Network for Example 16.17.

Solution The nodal voltages are defined with the block admittances in Fig. 16.57. The voltage V of Fig. 16.56 will be the magnitude of the voltage $\mathbf{V}_2$ of Fig. 16.57.

FIG. 16.57
Redrawn network of Fig. 16.56.

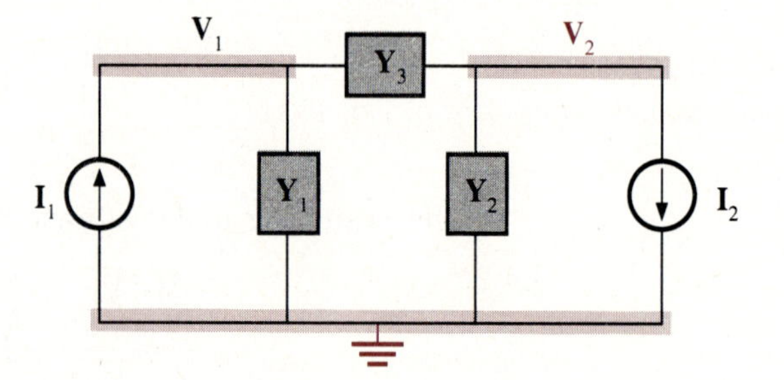

Applying nodal analysis:

$$\begin{aligned} \mathbf{V}_1(\mathbf{Y}_1 + \mathbf{Y}_3) - \mathbf{V}_2(\mathbf{Y}_3) &= \mathbf{I}_1 \\ \mathbf{V}_2(\mathbf{Y}_2 + \mathbf{Y}_3) - \mathbf{V}_1(\mathbf{Y}_3) &= -\mathbf{I}_2 \end{aligned}$$

and

$$\begin{aligned} \mathbf{V}_1(\mathbf{Y}_1 + \mathbf{Y}_3) - \mathbf{V}_2\mathbf{Y}_3 \qquad\qquad &= \mathbf{I}_1 \\ -\mathbf{V}_1(\mathbf{Y}_3) \qquad + \mathbf{V}_2(\mathbf{Y}_2 + \mathbf{Y}_3) &= -\mathbf{I}_2 \end{aligned}$$

with

$$\mathbf{V}_2 = \frac{\mathbf{I}_1\mathbf{Y}_3 - \mathbf{I}_2(\mathbf{Y}_1 + \mathbf{Y}_3)}{\mathbf{Y}_1\mathbf{Y}_2 + \mathbf{Y}_1\mathbf{Y}_3 + \mathbf{Y}_2\mathbf{Y}_3}$$

Substituting gives

$$\mathbf{Y}_1 = \frac{1}{\mathbf{Z}_1} = \frac{1}{1\ \text{k}\Omega\ \underline{/0^\circ}} = 1\ \text{mS}\ \underline{/0^\circ}$$

$$\mathbf{Y}_2 = \frac{1}{\mathbf{Z}_2} = \frac{1}{4\ \text{k}\Omega\ \underline{/90^\circ}} = 0.25\ \text{mS}\ \underline{/-90^\circ}$$

$$\mathbf{Y}_3 = \frac{1}{\mathbf{Z}_3} = \frac{1}{2\ \text{k}\Omega\ \underline{/-90^\circ}} = 0.5\ \text{mS}\ \underline{/90^\circ}$$

$$\mathbf{V}_2 = \frac{(4\ \text{mA}\ \underline{/0^\circ})(0.5\ \text{mS}\ \underline{/90^\circ}) - (10\ \text{mA}\ \underline{/0^\circ})(1\ \text{mS} + j0.5\ \text{mS})}{(1\ \text{mS}\ \underline{/0^\circ})(0.25\ \text{mS}\ \underline{/-90^\circ}) + (1\ \text{mS}\ \underline{/0^\circ})(0.5\ \text{mS}\ \underline{/90^\circ}) + (0.25\ \text{mS}\ \underline{/-90^\circ})(0.5\ \text{mS}\ \underline{/90^\circ})}$$

$$= \frac{(2 \times 10^{-6}\ \underline{/90^\circ}) - (10 \times 10^{-6} + j5 \times 10^{-6})}{(0.25 \times 10^{-6}\ \underline{/-90^\circ}) + (0.5 \times 10^{-6}\ \underline{/90^\circ}) + (0.125 \times 10^{-6}\ \underline{/0^\circ})}$$

$$= \frac{-10 \times 10^{-6} - j3 \times 10^{-6}}{0.125 \times 10^{-6} + j0.25 \times 10^{-6}} = \frac{10.44 \times 10^{-6}\ \underline{/-163.3^\circ}}{0.28 \times 10^{-6}\ \underline{/63.43^\circ}}$$

$$\mathbf{V}_2 = \mathbf{37.29\ V\ \underline{/133.27^\circ}}$$

and the reading of the voltmeter is **37.29 V.**

The mathematics required is obviously extensive, but the magnitude of voltages and currents in a sinusoidal ac network is sensitive to the vector relationship between quantities and not simply their magnitudes. In other words, substituting just the magnitudes into the preceding equation for $\mathbf{V}_2$ would result in a solution of no meaning whatsoever.

FORMULA SUMMARY

Source Conversion

$$E = IZ_p, \qquad I = \frac{E}{Z_s}, \qquad Z_s = Z_p$$

Maximum Power Transfer

$$Z_L = Z_{Th}, \qquad \theta_L = -\theta_{Th}$$

$$P_{\max} = \frac{E_{Th}^2}{4R_{Th}}$$

Δ-Y Conversion

$$\mathbf{Z}_\Delta = 3\mathbf{Z}_Y, \qquad \mathbf{Z}_Y = \frac{\mathbf{Z}_\Delta}{3}$$

Bridge Configuration

$$\frac{\mathbf{Z}_1}{\mathbf{Z}_3} = \frac{\mathbf{Z}_2}{\mathbf{Z}_4}$$

CHAPTER SUMMARY

The best preparation for the content of this chapter is a brief review of the content of Chapters 8 and 9, "Series-Parallel Networks" and "Methods of Analysis and Network Theorems." The approach for each subject area is the same for ac circuits, with the only difference being the added concern about the angle associated with each quantity of interest. Reviewing the earlier chapters will permit concentrating on the approach without being bogged down in the mathematical complexity of the vector algebra.

With regard to the angle, it is important to remember throughout this chapter that

1. The angle associated with $\mathbf{Z}_T$ is the angle by which the applied voltage leads the source current.
2. For the resistor, v_R and i_R are in phase.
   For the inductor, v_L leads i_L by 90°.
   For the capacitor, i_C leads v_C by 90°.
3. For $\mathbf{Y}_T = Y_T \angle \theta_T$, the angle associated with $\mathbf{Z}_T$ is $-\theta_T$.
4. For series elements, the angle (and magnitude) of the current is the same throughout.
5. For parallel elements, the angle (and magnitude) of the voltage across each element is the same.

For source conversions, the current or voltage of the equivalent source is determined by Ohm's law (as for dc circuits) using effective values. The magnitude and angle of the series or parallel internal impedance is the same in each equivalent source—only the location changes.

The best approach to any series-parallel configuration is first to establish mentally a plan of attack including each step required to determine the unknown quantities. It is certainly possible that a short, quick solution may result if all the avenues of attack are properly evaluated. When the actual numerical analysis commences, be sure to redraw the network as frequently as possible, carrying along as many of the unknowns as possible from the original network. The frequent redrawing of the network is the surest method of not losing an important term or changing the problem under investigation.

The superposition theorem actually reduces the complexity of the network to be analyzed by removing all the independent sources but one. However, the resulting short-circuit and open-circuit equivalents for the other sources must be treated carefully by redrawing the network, or a foolish error may result. In addition, be sure the currents or voltages determined are treated properly when determining the total current or voltage for the network—that is, deciding if the resulting currents should be added or subtracted or if the determined voltages have the same or opposite polarities.

In most cases, the Thevenin impedance is easier to determine than the Thevenin voltage, especially when the reduced networks are drawn or defined by the sequence of steps. For E_{Th}, be sure to consider carefully the impact of the fact that E_{Th} is an open-circuit voltage causing the current in series with the open circuit to be 0 A. In addition, the voltage across an element in series with the open circuit must be zero, since by Ohm's law $V = IR = (0)R = 0$ V. The maximum power transfer theorem is closely linked to Thevenin's theorem and requires only that the following relationships be put to memory: $Z_L = Z_{Th}$, $\theta_L = -\theta_{Th}$, $P_{max} = E^2_{Th}/4R_{Th}$.

The Δ-Y and Y-Δ conversions were limited solely to configurations where all the impedances of each system are the same. For any other combination of elements, simply refer to Chapter 9 for the required equations and substitute impedances with the same subscripts for the resistor elements in Chapter 9. Keep in mind that when a conversion is performed, the original configuration is replaced by the new structure between the original three terminals. The result is usually a network that can be analyzed using series-parallel techniques.

The bridge network is a special configuration permitting the establishment of a balance condition that can be used effectively in a number of practical applications. Simply ensure that the balance ratio is satisfied from a magnitude and angle aspect and the current through the bridge element will be 0 A.

When the impedances of a series-parallel ac network are replaced by the block impedance format, the mesh- and nodal-analysis techniques can be applied in much the same manner as introduced for dc networks. Usually, the writing of the required equations is fairly straightforward due to the format approach. The actual numerical calculations are an extended exercise in the use of the j-operator.

GLOSSARY

Bridge network A network configuration having the appearance of a diamond in which no two branches are in series or parallel.

Capacitance comparison bridge A bridge configuration having a galvanometer in the bridge arm that is used to determine an unknown capacitance and associated resistance.

Delta (Δ) configuration A network configuration having the appearance of the capital Greek letter delta.

Hay bridge A bridge configuration used for measuring the resistance and inductance of coils in those cases where the resistance is a small fraction of the reactance of the coil.

Maximum power transfer theorem A theorem used to determine the load impedance necessary to ensure maximum power to the load.

Maxwell bridge A bridge configuration used for inductance measurements when the resistance of the coil is large enough not to require a Hay bridge.

Mesh analysis A method through which the loop (or mesh) currents of a network can be determined. The branch currents of the network can then be determined directly from the loop currents.

Nodal analysis A method through which the node voltages of a network can be determined. The voltage across each element can then be determined through application of Kirchhoff's voltage law.

Source conversion The changing of a voltage source to a current source or vice versa that will result in the same terminal behavior of the source. In other words, the external network is unaware of the change in sources.

Superposition theorem A method of network analysis that permits considering the effects of each source independently. The resulting current and/or voltage is the phasor sum of the currents and/or voltages developed by each source independently.

Thevenin's theorem A theorem that permits the reduction of any two-terminal linear ac network to one having a single voltage source and series impedance. The resulting configuration can then be employed to determine a particular current or voltage in the original network or to examine the effects of a specific portion of the network on a particular variable.

Wye (Y) configuration A network configuration having the appearance of the capital letter Y.

PROBLEMS

Section 16.2

1. Convert the voltage sources of Fig. 16.58 to current sources.

FIG. 16.58

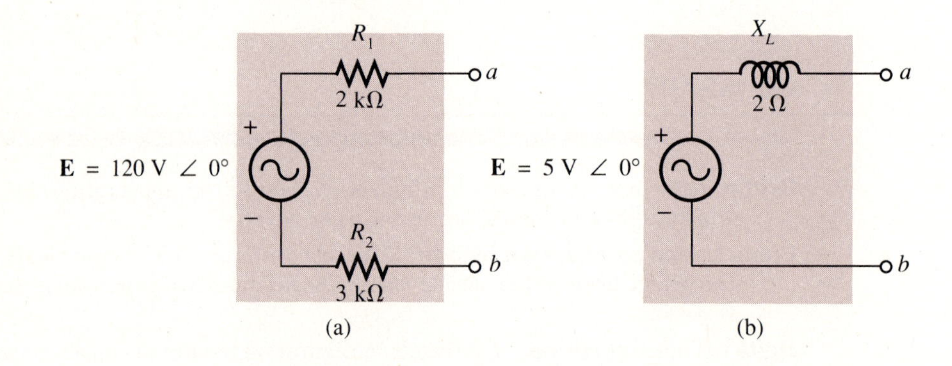

2. Convert the current sources of Fig. 16.59 to voltage sources.

FIG. 16.59

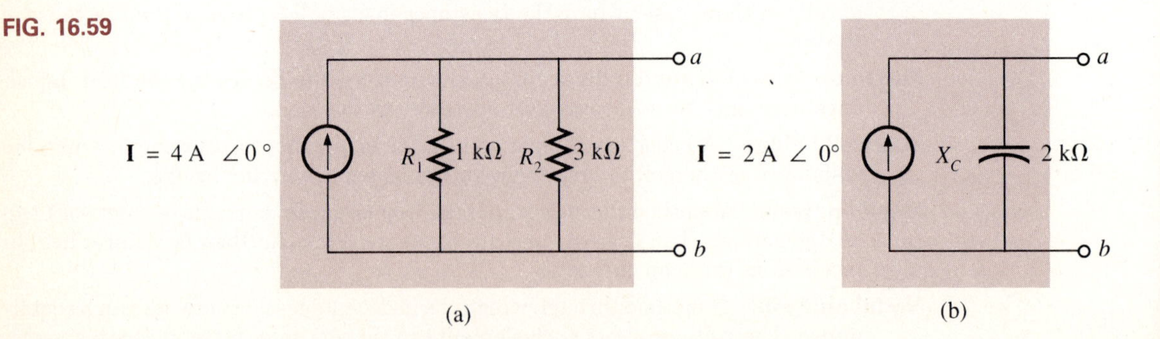

3. Convert the voltage source of Fig. 16.60 to a current source. Use effective values for the current source.

FIG. 16.60

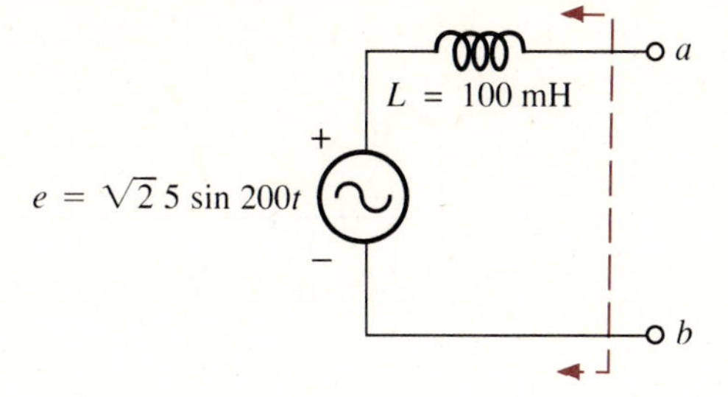

***4.** Convert the current source of Fig. 16.61 to a voltage source. Use effective values for the voltage source.

FIG. 16.61

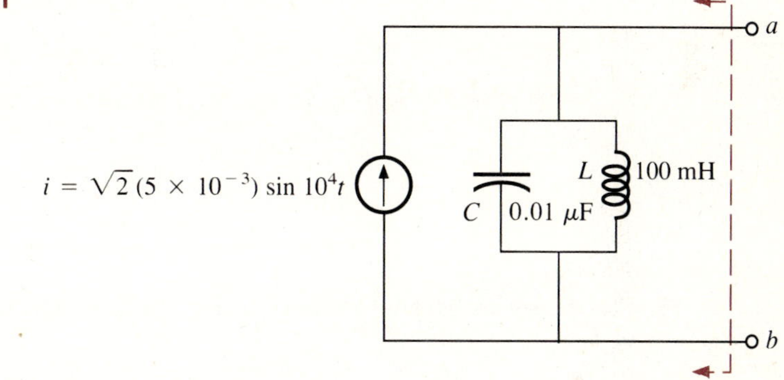

***5.** Convert the voltage source of Fig. 16.62 to a current source. Use effective values for the current source.

FIG. 16.62

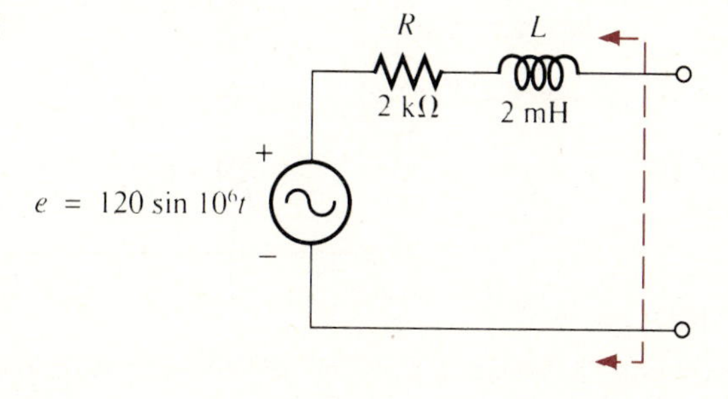

Section 16.3

6. For the series-parallel network of Fig. 16.63, determine
  a. $\mathbf{Z}_T$
  b. $\mathbf{I}$ and $\mathbf{I}_1$
  c. $\mathbf{I}_2$ and $\mathbf{I}_3$

FIG. 16.63

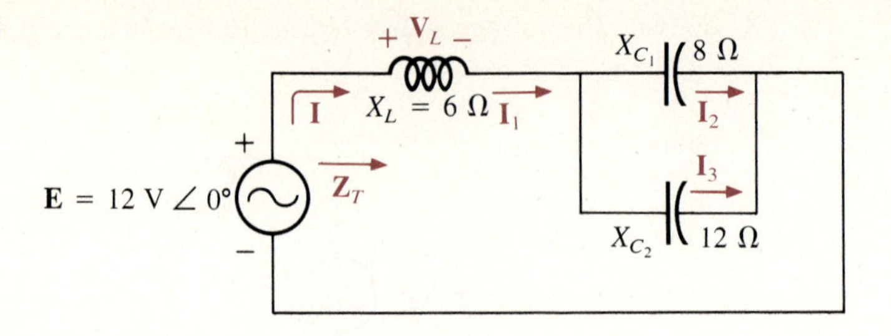

7. For the series-parallel network of Fig. 16.64, determine
a. $\mathbf{Z}_T$ and $\mathbf{Y}_T$
b. $\mathbf{I}_T$
c. $\mathbf{I}_2$
d. $\mathbf{V}_C$

FIG. 16.64

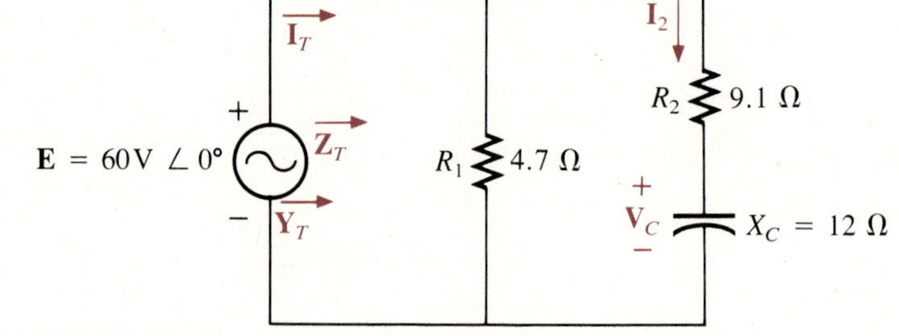

***8.** For the series-parallel network of Fig. 16.65, determine
a. $\mathbf{I}_1$
b. $\mathbf{V}_C$
c. $\mathbf{V}_{ab}$

FIG. 16.65

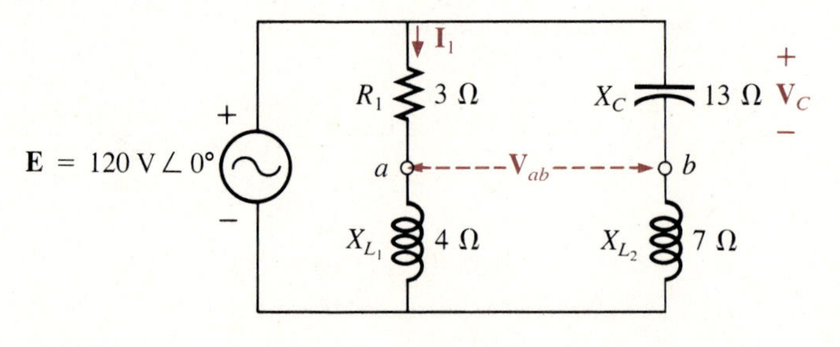

Section 16.4

9. Determine the current **I** of Fig. 16.66 using superposition.

FIG. 16.66

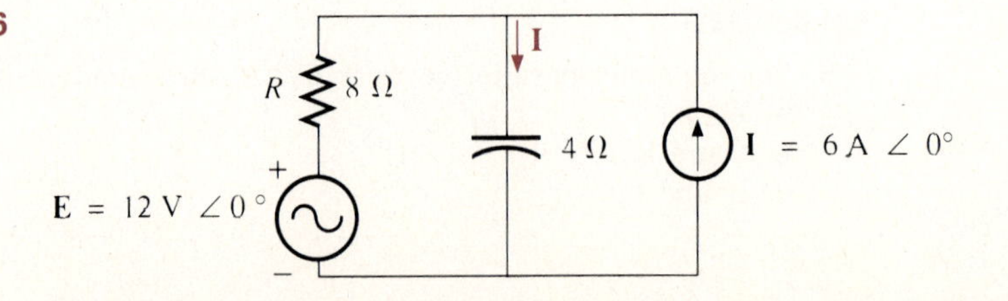

10. Determine the voltage **V** for the network of Fig. 16.67 using superposition.

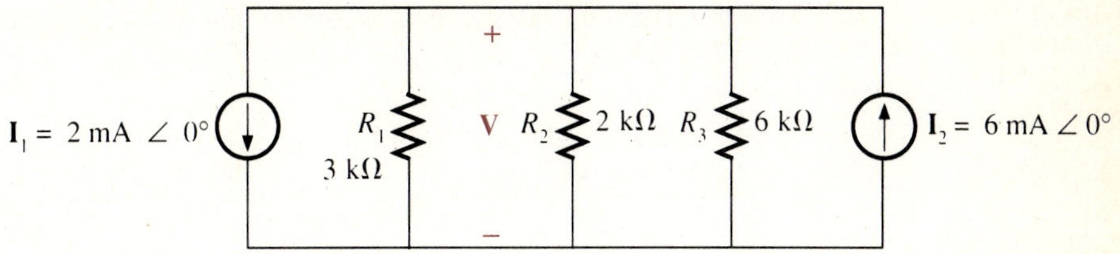

FIG. 16.67

***11.** Determine the voltage $\mathbf{V}_{R_1}$ of Fig. 16.68 using superposition.

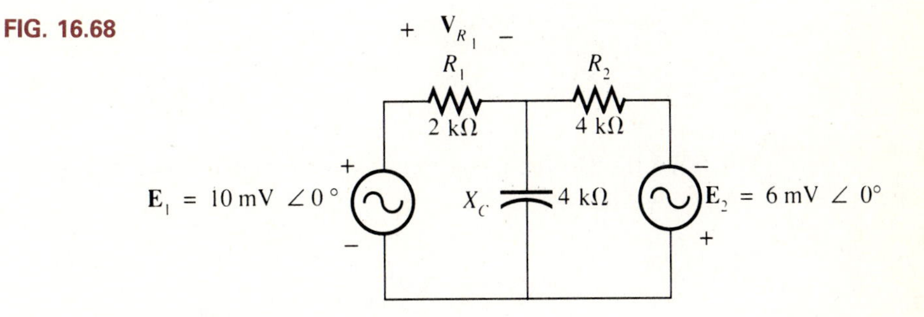

FIG. 16.68

***12.** Determine the voltage **V** for the network of Fig. 16.69 using superposition.

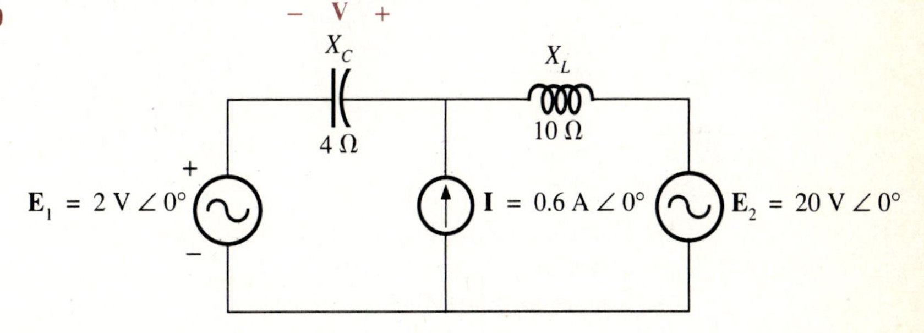

FIG. 16.69

Section 16.5

13. Find the Thevenin equivalent circuit for the network external to the impedance $\mathbf{Z}_L$ in Fig. 16.70.

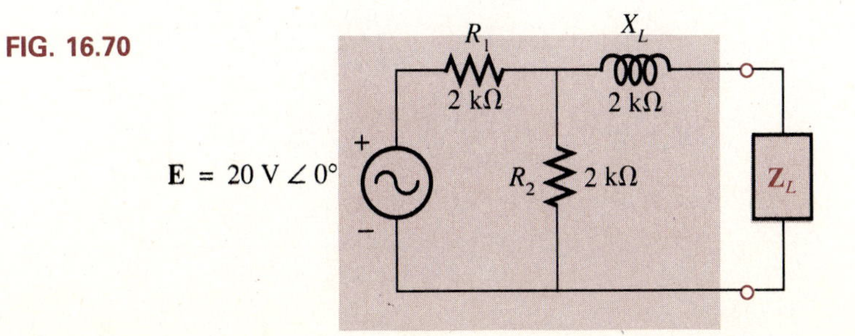

FIG. 16.70

14. Find the Thevenin equivalent circuit for the network external to the impedance $\mathbf{Z}_L$ in Fig. 16.71.

15. Find the Thevenin equivalent circuit for the network external to the impedance $\mathbf{Z}_L$ in Fig. 16.72.

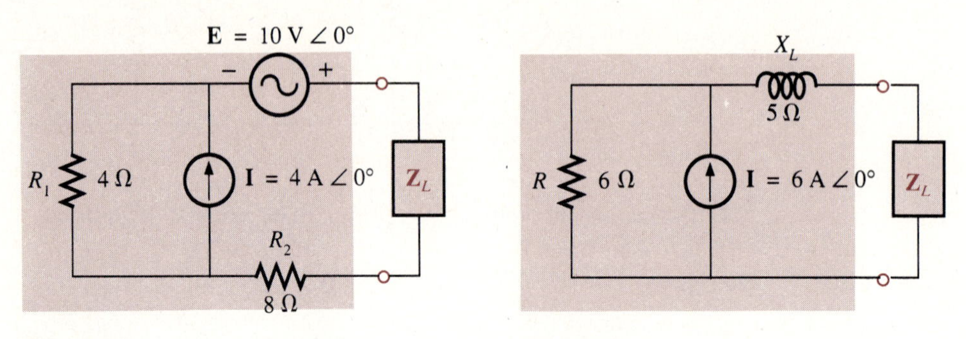

FIG. 16.71

FIG. 16.72

***16.** Find the Thevenin equivalent circuit for the network external to the impedance $\mathbf{Z}_L$ in Fig. 16.73.

FIG. 16.73

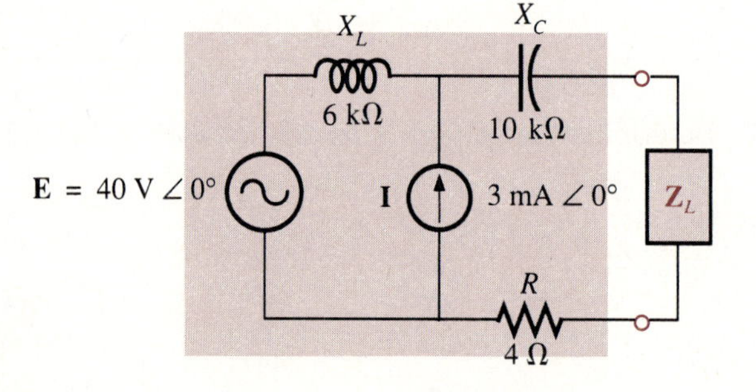

Section 16.6

17. Determine $\mathbf{Z}_L$ of Fig. 16.70 for maximum power to $\mathbf{Z}_L$ and calculate the maximum power.

18. Repeat Problem 17 for the network of Fig. 16.72.

19. **a.** Determine $\mathbf{Z}_L$ in Fig. 16.74 for maximum power to $\mathbf{Z}_L$.
b. Calculate the maximum power to $\mathbf{Z}_L$.

FIG. 16.74

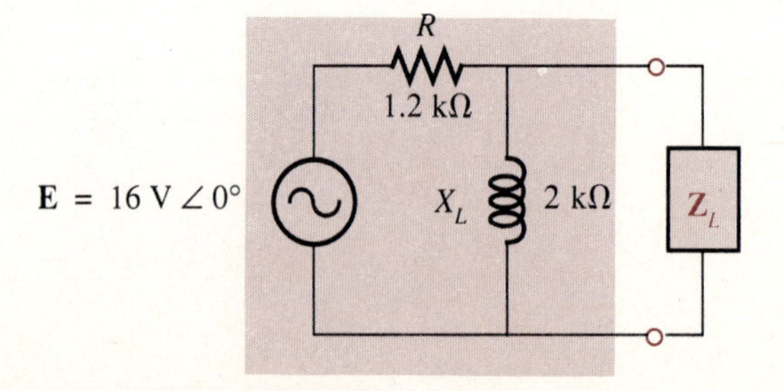

*20. Repeat Problem 19 for the network of Fig. 16.75.

FIG. 16.75

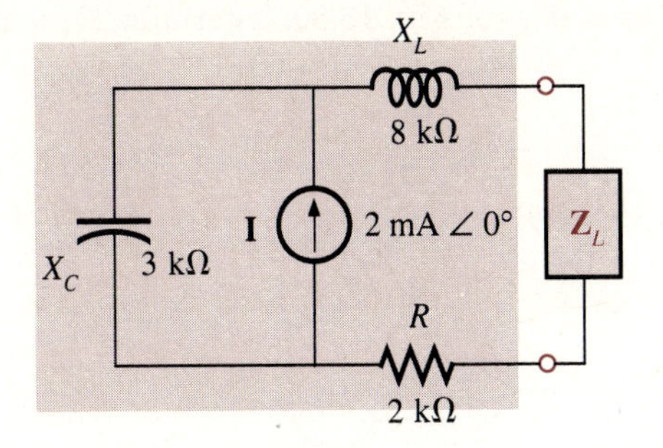

Section 16.7

21. Find the total impedance for the bridge network of Fig. 16.76.

*22. Find the total impedance for the bridge network of Fig. 16.77.

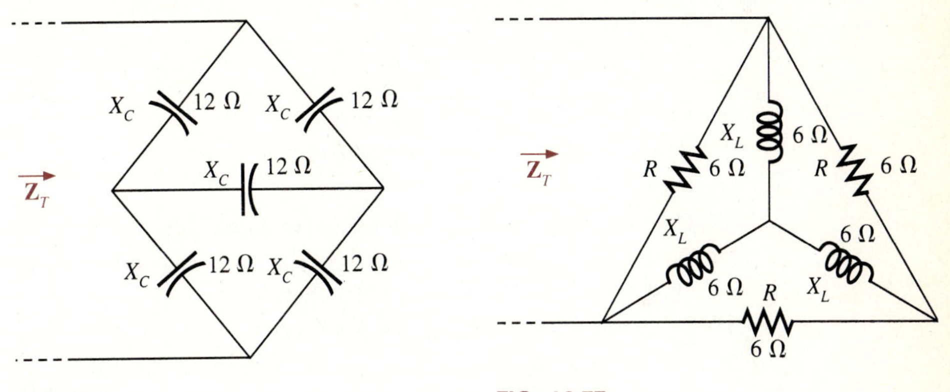

FIG. 16.76

FIG. 16.77

Section 16.8

23. Determine whether the bridge of Fig. 16.78 is balanced.

FIG. 16.78

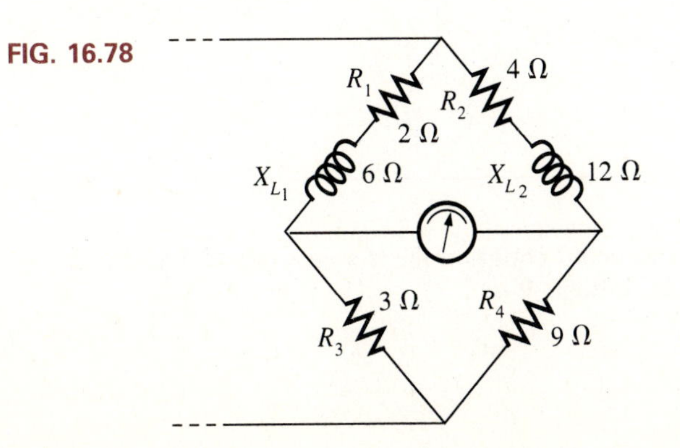

*24. **a.** Determine whether the Maxwell bridge of Fig. 16.79 is balanced.
b. If balanced, find $\mathbf{I}_2$ if $\mathbf{E} = 10 \angle 0°$ at $f = 10$ kHz.

25. For the Hay bridge of Fig. 16.80, determine R_4 and L_4 required to balance the bridge at $f = 5$ kHz.

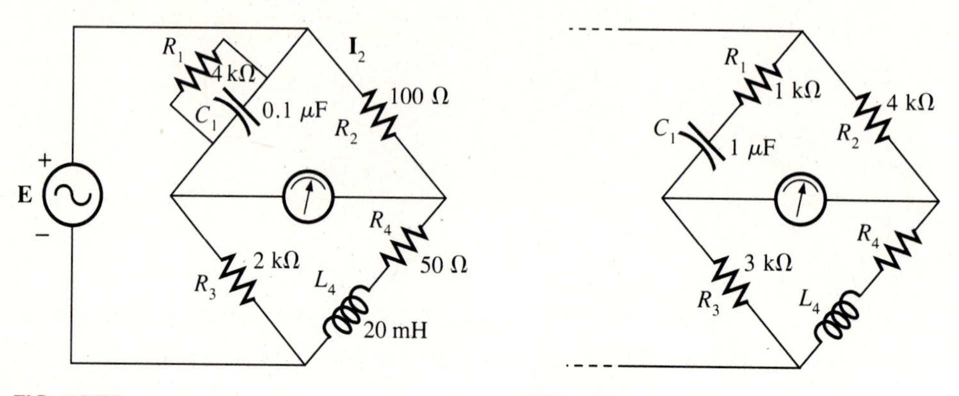

FIG. 16.79

FIG. 16.80

Section 16.9

26. Repeat Problem 4 using the *j*-operator.

27. Repeat Problem 6 using the *j*-operator.

28. Repeat Problem 8 using the *j*-operator.

29. Determine $\mathbf{V}_{R_1}$ of Fig. 16.68 using the *j*-operator.

30. Determine the Thevenin equivalent circuit for the network external to the impedance Z_L in Fig. 16.73 using the *j*-operator.

31. Repeat Problem 22 using the *j*-operator.

32. a. Write the mesh equations for the network of Fig. 16.81.
b. Determine $\mathbf{I}_L$.

FIG. 16.81

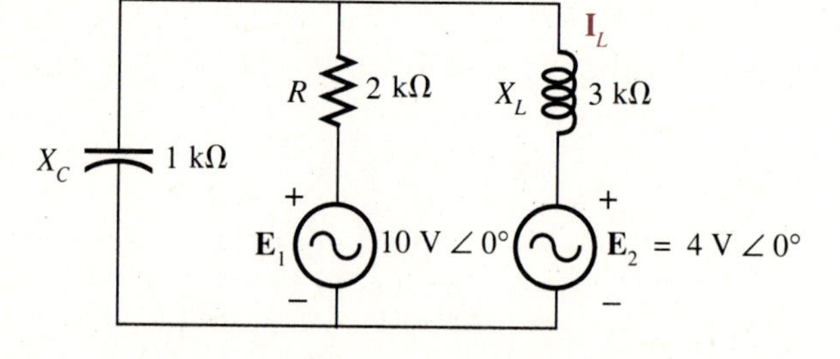

33. a. Write the nodal equation for the network of Fig. 16.82.
b. Find the voltage $\mathbf{V}_L$.

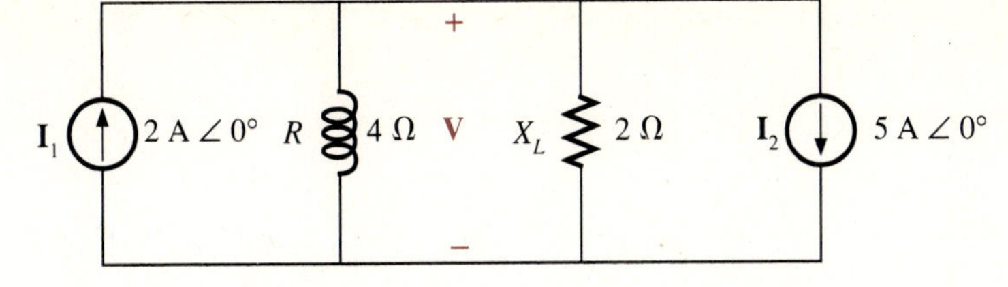

FIG. 16.82

***34.** Write the mesh equations for the network of Fig. 16.83.

***35.** Write the nodal equations for the network of Fig. 16.84.

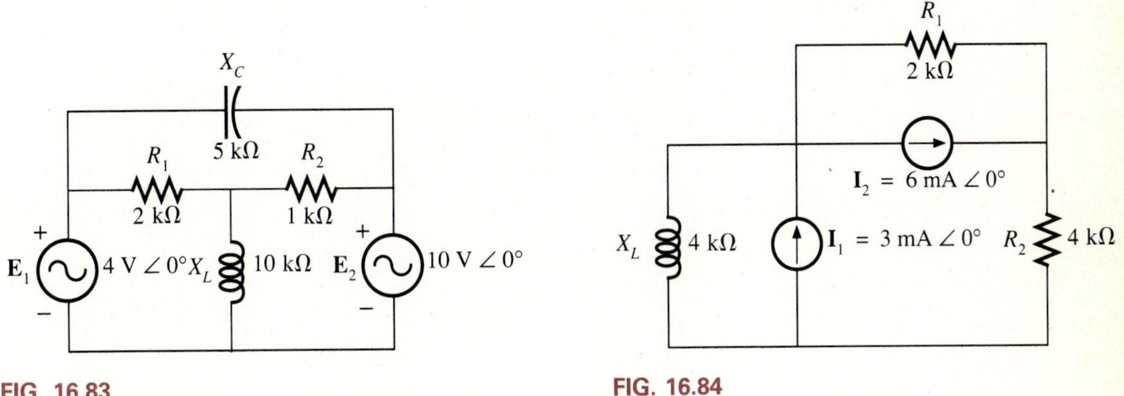

FIG. 16.83

FIG. 16.84

Computer Problems

36. Write a program that will provide the current source equivalent of a voltage source at 0° and either a series R-C or R-L branch. Of course, either R or C (or L) may not be present.

37. Write a program to provide a general solution for the current $\mathbf{I}_T$ of the network of Fig. 16.9 for any parameter values.

38. Given the network of Fig. 16.33, write a program to determine $\mathbf{Z}_L$ for maximum power to $\mathbf{Z}_L$ and determine the maximum power. The current source must have an angle of 0° associated with it, but the elements can have any magnitude.

39. Write a program that will determine the total impedance of a configuration such as the one in Fig. 16.41. The Δ or Y can be resistive, capacitive, or inductive, but each element of the Δ or Y must be the same and of equal value.

40. For the Hay bridge of Fig. 16.47, write a program to determine the value of L_4 and R_4 that will establish a balance condition given the applied frequency and all the other parameter values.

17

Power (ac)

OBJECTIVES

- ☐ Understand the concepts of reactive and apparent power and how each affects the demands on the supply.
- ☐ Be able to determine the total real, reactive, and apparent power levels of a system.
- ☐ Be able to find the total power factor of a system from the given power levels and understand how it can be used to find the series resistance and reactance that will result in the same total impedance.
- ☐ Determine the type and level of reactive element necessary to improve the power factor of a load to reduce the current demands on the supply.
- ☐ Become aware of how a wattmeter is used to measure the power to various parts of an electrical system.
- ☐ Understand the concept of effective resistance and what factors affect its level in an ac system.

17.1 INTRODUCTION

The discussion of power in Chapters 14 and 15 included only the average power delivered to an ac network—the power dissipated by the resistive elements. We now examine additional factors that affect the power demands on a voltage or current source. In particular, two additional types of power called the *reactive* and *apparent* power are introduced, along with procedures for determining one from the other two.

17.2 RESISTIVE CIRCUIT

For a purely resistive circuit, such as the one shown in Fig. 17.1, we are now aware that the voltage v and current i are in phase, as shown in Fig. 17.2. The power delivered to the resistive load *at any instant* is determined by

$$\boxed{p = vi} \tag{17.1}$$

where v and i are the instantaneous values. In particular, note that when v and i are their maximum values, the power to p_R is a maximum also. Even though

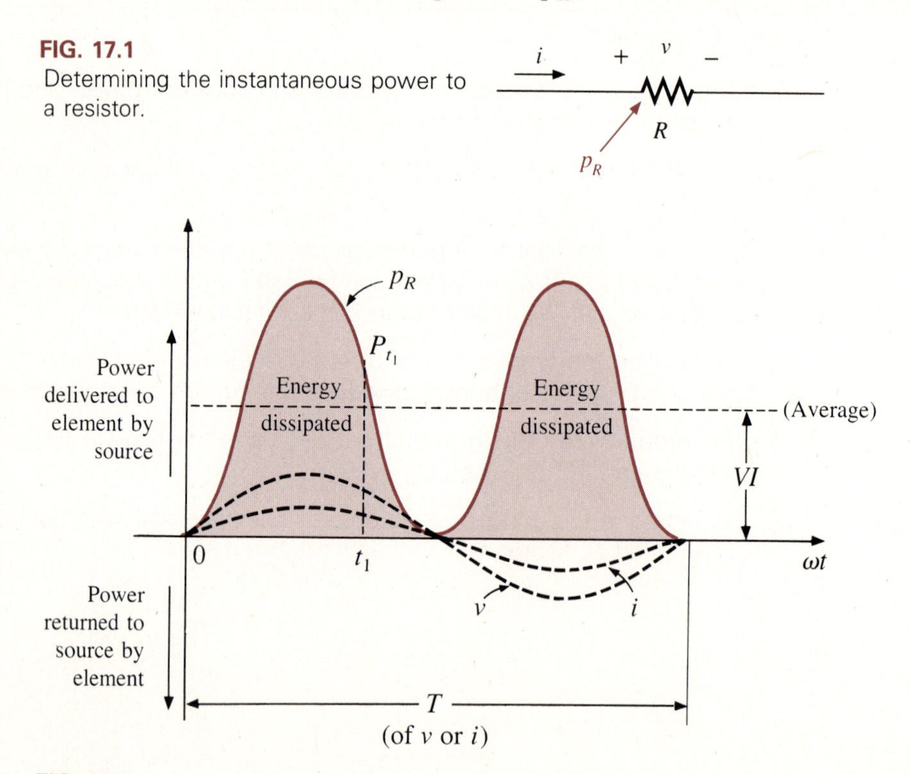

FIG. 17.1
Determining the instantaneous power to a resistor.

FIG. 17.2
Plotting the power delivered to a resistor by a sinusoidal voltage.

v and i are negative in the second half-period of v and i, the product results in a positive value for p_R. Hence, the power curve p_R is positive throughout the complete cycle of v or i. Note also that when v and i are zero, the resulting power is also 0 W. The resulting curve for p_R is actually a sine wave with an average value of VI (effective values) and a frequency twice that of v or i.

The fact that the curve for p_R is always above the axis reveals that

The total power delivered to a resistive load is dissipated.

The average power delivered to a load is the average value of the curve, which from Fig. 17.2 is

$$\boxed{P = VI} \quad \text{(watts, W)} \tag{17.2}$$

or, since $V = V_m/\sqrt{2}$ and $I = I_m/\sqrt{2}$,

$$\boxed{P = \frac{V_m I_m}{2}} \quad \text{(watts, W)} \tag{17.3}$$

Also, from before,

$$\boxed{P = I_R^2 R = \frac{V_R^2}{R}} \quad \text{(watts, W)} \tag{17.4}$$

EXAMPLE 17.1

Given $v_R = 8 \sin \omega t$ for a 2-Ω resistor, sketch the curves of v_R, i_R, and p_R and determine the average power delivered.

Solution

$$I_m = \frac{V_m}{R} = \frac{8\text{ V}}{2\ \Omega} = 4\text{ A}$$

and

$$i = 4 \sin \omega t$$

The peak value of the p_R curve is $V_m I_m = (8\text{ V})(4\text{ A}) = 32\text{ W}$, as shown in Fig. 17.3 with both v_R and i_R.

$$P = \frac{V_m I_m}{2} = \frac{(8\text{ V})(4\text{ A})}{2} = \mathbf{16\ W}$$

or

$$\begin{aligned} P = V_{\text{eff}} I_{\text{eff}} &= (0.707V_m)(0.707I_m) \\ &= (5.656\text{ V})(2.828\text{ A}) \\ P &= \mathbf{16\ W} \end{aligned}$$

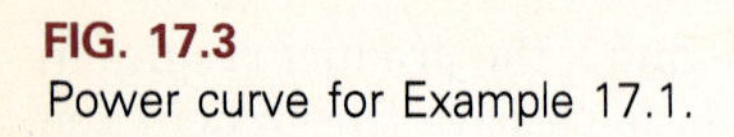

FIG. 17.3
Power curve for Example 17.1.

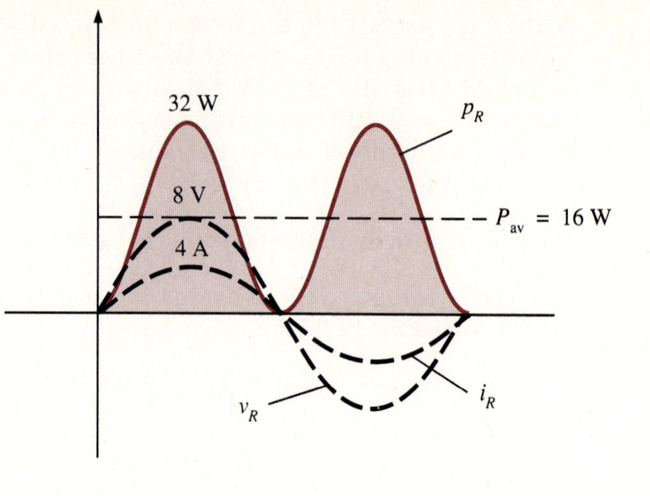

17.3 APPARENT POWER

From our analysis of dc networks (and, preceding, of resistive elements), it would seem *apparent* that the power delivered to the load of Fig. 17.4 is determined simply by the product of the applied voltage and current with no concern for the components of the load. That is, $P = VI$. However, we found in Chapter 14 that the power factor ($\cos \theta$) of the load has a pronounced effect on the power dissipated. Although the product of the voltage and current is not always the power delivered, it is a power rating of significant usefulness in the description and analysis of sinusoidal ac networks and in the maximum rating of a number of electrical components and systems. It is called the *apparent power* and is represented symbolically by S. Since it is simply the product of voltage and current, its units are *volt-amperes,* for which the abbreviation is VA. Its magnitude is determined by

$$\boxed{S = VI} \quad \text{(VA)} \tag{17.5}$$

or, since the magnitudes of V, I, and Z are related by

$$V = IZ \quad \text{and} \quad I = \frac{V}{Z}$$

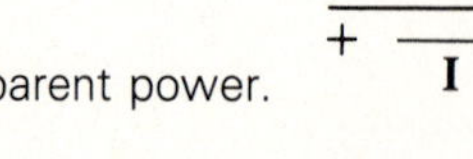

FIG. 17.4
Defining apparent power.

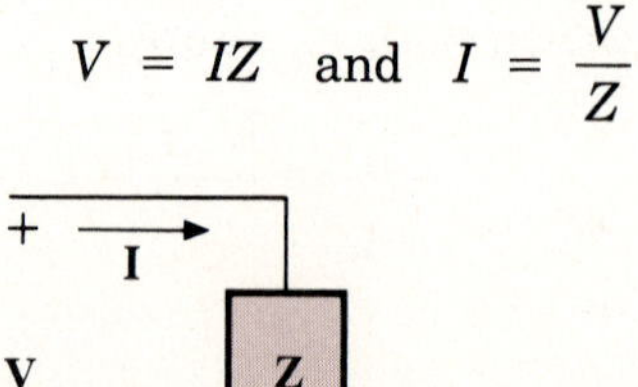

we have

$$\boxed{S = I^2 Z} \quad \text{(VA)} \tag{17.6}$$

or

$$\boxed{S = \frac{V^2}{Z}} \quad \text{(VA)} \tag{17.7}$$

The average power to a system is defined by the equation

$$P = VI \cos \theta$$

However, since

$$S = VI$$

we have

$$\boxed{P = S \cos \theta} \quad \text{(W)} \tag{17.8}$$

So, the power factor of a system F_p is

$$\boxed{F_p = \cos \theta = \frac{P}{S}} \tag{17.9}$$

The power factor of a circuit, therefore, is the ratio of the average power to the apparent power. For a purely resistive circuit, we have

$$P = VI = S$$

and

$$F_p = \cos \theta = \frac{P}{S} = 1$$

In general, electric equipment is rated in volt-amperes or in kilovolt-amperes (kVA), not in watts. By knowing the volt-ampere rating and the rated voltage of a device, we can readily determine the *maximum* current rating. For example, a device rated at 10 kVA at 200 V has a maximum current rating of $I = 10{,}000 \text{ VA}/200 \text{ V} = 50 \text{ A}$ when operated under rate conditions. The volt-ampere rating of a piece of equipment is equal to the wattage rating only when the F_p is 1. It is therefore a maximum power dissipation rating. This condition exists only when the total impedance of a system $Z \angle \theta$ is such that $\theta = 0°$.

The exact current demand of a device, when used under normal operating conditions, could be determined if the wattage rating and power factor were given instead of the volt-ampere rating. However, the power factor is sometimes not available, or it may vary with the load.

The reason for rating electrical equipment in kilovolt-amperes rather than in kilowatts is obvious from the configuration of Fig. 17.5. The load has an apparent power rating of 10 kVA and a current rating of 50 A at the applied voltage, 200 V. As indicated, the current demand is above the rated value and could damage the load element, yet the reading on the wattmeter is 0 W, since the load is purely reactive. In other words, the wattmeter reading is an indication not of the current drawn but simply of the watts dissipated.

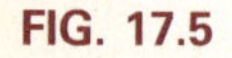

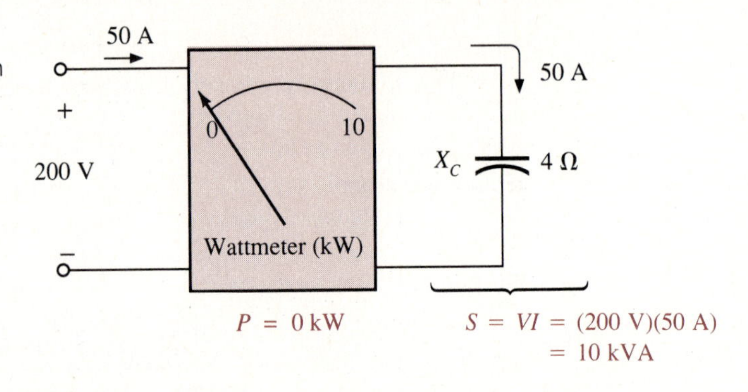

FIG. 17.5
Demonstrating the need to rate a load in kilovolt-amperes (kVA).

EXAMPLE 17.2

a. Determine the apparent power level for the load of Fig. 17.6 if $v = 16 \sin \omega t$ and $i = 0.4 \sin(\omega t + 60°)$.
b. Find the average power dissipated by the load.
c. Find the power factor of the load.

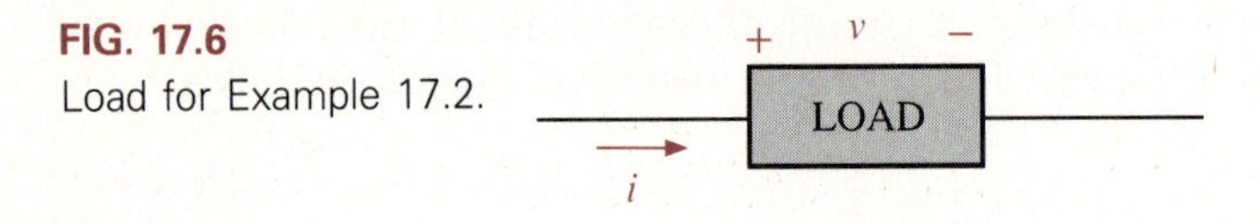

FIG. 17.6
Load for Example 17.2.

Solutions

a. $S = VI = \dfrac{V_m I_m}{2} = \dfrac{(16 \text{ V})(0.4 \text{ A})}{2} = \mathbf{3.2 \text{ VA}}$

b. $P = VI \cos\theta = S \cos\theta = (3.2 \text{ VA})(\cos 60°)$
$= (3.2 \text{ VA})(0.5)$
$P = \mathbf{1.6 \text{ W}}$

c. $F_p = \dfrac{P}{S} = \dfrac{1.6 \text{ W}}{3.2 \text{ VA}} = \mathbf{0.5} = \mathbf{\cos 60°}$

17.4 INDUCTIVE CIRCUIT AND REACTIVE POWER

For a purely inductive circuit (such as that in Fig. 17.7), v leads i by 90°, as shown in Fig. 17.8. Note for the first positive pulse of p_L that v_L and i_L

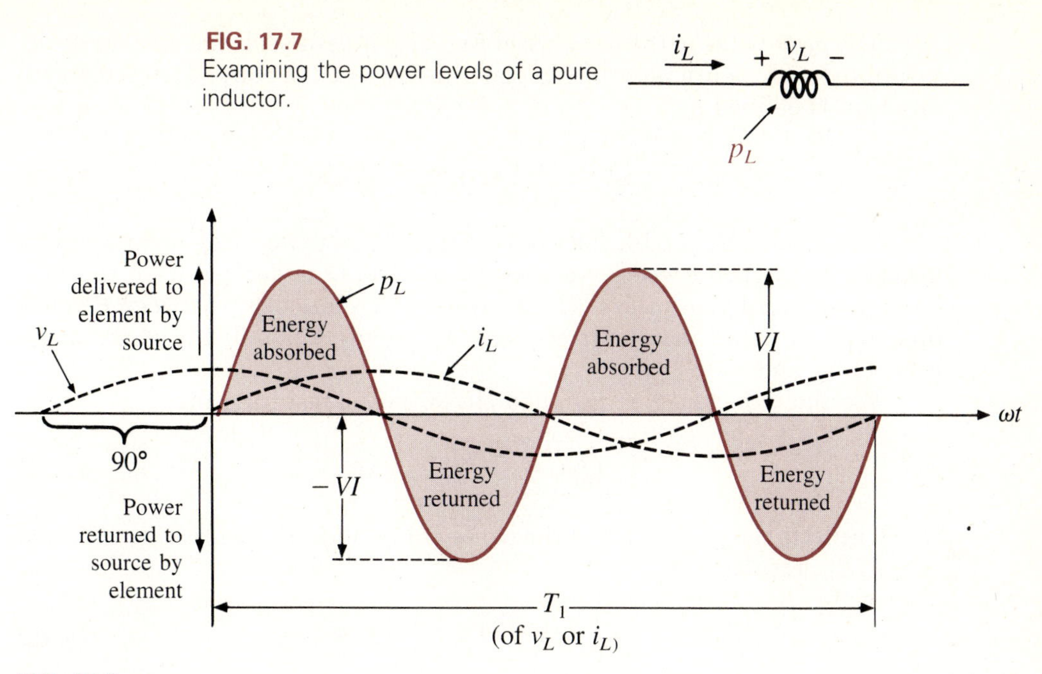

FIG. 17.7
Examining the power levels of a pure inductor.

FIG. 17.8
Plotting p_L for an applied sinusoidal voltage.

are positive, although the positive peak of p_L does not occur when v_L and i_L are at their peak values. The second pulse of p_L is negative because v_L is negative when i_L is positive and $p_L = (v_L)(i_L) = (-)(+) = -$. For the third pulse, both v_L and i_L are negative, resulting in a positive value for $p_L = (v_L)(i_L) = (-)(-) = +$. For the last pulse v_L is positive and i_L is negative, resulting in $p_L = (v_L)(i_L) = (+)(-) = -$.

The resulting curve for p_L is sinusoidal with twice the frequency of v_L and i_L. It is important to note that the area of p_L above the axis equals that below the axis for one period of v_L and i_L. The result is that the power delivered during one pulse of p_L is returned during the next pulse. Over one cycle of v_L or i_L, therefore, the net power flow in any one direction is zero. For the ideal coil, the energy absorbed during one power pulse is exactly equal to that returned during the next pulse. In total, the coil simply absorbs electrical energy during the positive pulse and converts it to a form stored as a magnetic field. During the next power pulse, the magnetic field is collapsing, returning the energy to the electrical form. Over one cycle of v_L or i_L, therefore, the net energy lost is zero (ideally), and the coil simply acts as a storage (or converting) device.

In summary,

The net flow of energy (or power) to the pure inductor is zero over one full cycle of v_L or i_L, and no energy (ideally) is lost in the transaction.

The peak value of the p_L curve of Fig. 17.8 is defined as the *reactive power* associated with a pure inductor. In general, the reactive power associated with any load is defined by

$$\boxed{Q = VI \sin \theta} \qquad \text{(VAR)} \qquad \textbf{(17.10)}$$

As noted in Eq. 17.10, the symbol for reactive power is Q, and its unit of measure is the *volt-ampere reactive* (VAR). The Q is derived from the first letter of the word *quadrature,* which reveals the relationship between P and Q (discussed in detail in a later section). The angle θ is still the phase angle between v_L and i_L.

For the inductor, $\theta = 90°$, and $\sin 90° = 1$, resulting in

$$\boxed{Q_L = VI} \qquad \text{(VAR)} \qquad \textbf{(17.11)}$$

which reflects the peak value of the curve of Fig. 17.8. However, $V = IX_L$ and $I = V/X_L$, resulting in

$$\boxed{Q_L = I^2X_L} \qquad \text{(VAR)} \qquad \textbf{(17.12)}$$

or

$$\boxed{Q_L = \frac{V_L^2}{X_L}} \qquad \text{(VAR)} \qquad \textbf{(17.13)}$$

The apparent power associated with an inductor is $S = VI$, and the average power is $P = 0$, as noted in Fig. 17.8. The power factor is, therefore,

$$F_p = \cos \theta = \frac{P}{S} = \frac{0}{VI} = 0$$

If the average power is zero and the energy supplied is returned within one cycle, why is reactive power of any significance? The reason is not obvious but can be explained using the curve of Fig. 17.8. At every instant of time along the power curve that the curve is above the axis (positive), energy must be supplied to the inductor even though it will be returned during the negative portion of the cycle. This power requirement during the positive portion of the cycle requires that the generating plant provide this energy during that interval. Therefore, the effect of reactive elements such as inductors can be to raise the power requirement of the generating plant, even though the reactive power is not dissipated but simply is "borrowed." The increased power demand during these intervals is a cost factor that must be passed on to the industrial consumer. In fact, most larger users of electrical energy pay for the apparent power demand rather than for the watts dissipated, since the volt-amperes used are sensitive to the reactive power requirement (see Section 19.7). In other words, the closer the power factor of an industrial outfit is to 1, the more efficient is the plant's operation, since it is limiting its use of borrowed power.

EXAMPLE 17.3

For an inductive load (containing both resistance and inductance), the voltage across the load is $v = 36 \sin \omega t$ and the current through the load is $i = 6 \sin(\omega t - 30°)$.

a. Determine the power dissipated.
b. Find the apparent power.
c. Calculate the reactive power to the inductive load.
d. Find the power factor of the load.

Solutions

a. Effective values:
$$V = 0.707(36 \text{ V}) = 25.452 \text{ V}$$
$$I = 0.707(6 \text{ A}) = 4.242 \text{ A}$$

$$\begin{aligned} P = VI \cos \theta &= (25.452 \text{ V})(4.242 \text{ A})\cos 30° \\ &= (107.97)(0.866) \\ P &= \mathbf{93.5\ W} \end{aligned}$$

b. $S = VI = \mathbf{107.97\ VA}$

c.
$$\begin{aligned} Q = VI \sin \theta &= (107.97)(\sin 30°) \\ &= (107.97)(0.5) \\ Q &= \mathbf{53.985\ VAR} \end{aligned}$$

d. $F_p = \dfrac{P}{S} = \dfrac{93.5 \text{ W}}{107.97 \text{ VA}} = \mathbf{0.866}$ **lagging**

17.5 CAPACITIVE CIRCUIT

For a purely capacitive circuit (such as that in Fig. 17.9), i leads v by 90°, as shown in Fig. 17.10.

Note the similarities between the power curve obtained for a capacitive element and that obtained for a pure inductor (Fig. 17.8). The power delivered by the source to the capacitor is exactly equal to that returned to the source by the capacitor over one full cycle. In other words,

The net flow of power to a pure capacitor is zero over a full cycle of v_C or i_C, and no energy (ideally) is lost in the transaction.

In this case, however, the energy is stored in the form of an electric field rather than of a magnetic field encountered for an inductor.

Substituting $\theta = 90°$ into Eq. 17.10, we find the reactive power associated with a capacitor is equal to the peak value of the p_C curve.

$$\begin{aligned} Q_C &= VI \sin \theta \\ &= VI \sin 90° \end{aligned}$$

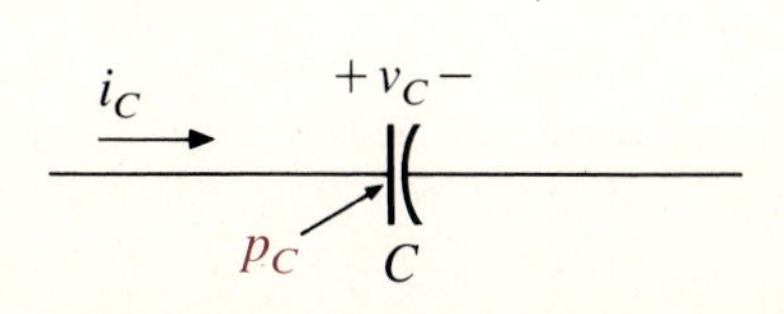

FIG. 17.9
Determining the power levels of a pure capacitor.

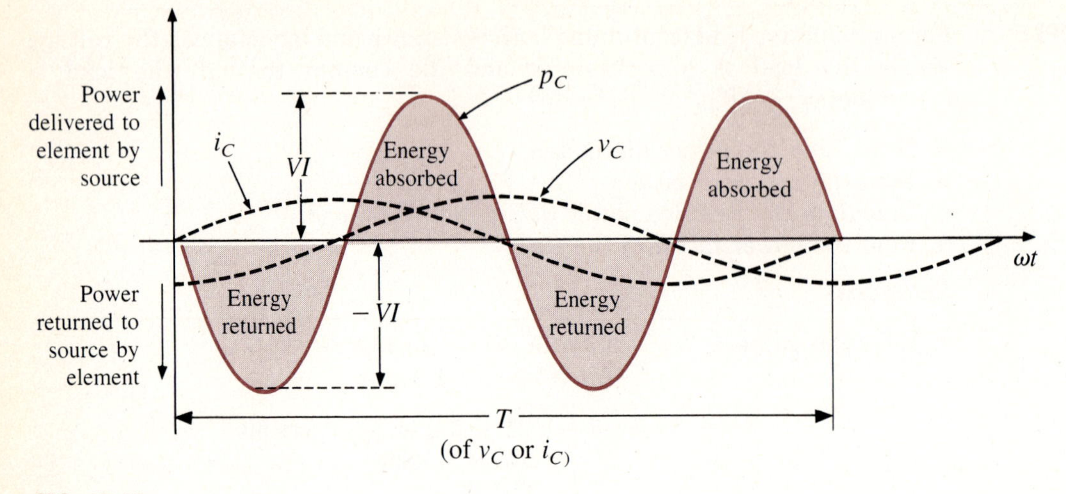

FIG. 17.10
Plotting p_C for an applied sinusoidal voltage.

and

$$\boxed{Q_C = VI} \quad \text{(VAR)} \tag{17.14}$$

Since $V = IX_C$ and $I = V/X_C$,

$$\boxed{Q_C = I^2X_C} \quad \text{(VAR)} \tag{17.15}$$

and

$$\boxed{Q_C = \frac{V^2}{X_C}} \quad \text{(VAR)} \tag{17.16}$$

The apparent power associated with the capacitor is

$$\boxed{S = VI} \quad \text{(VA)} \tag{17.17}$$

and the average power is $P = 0$, as noted from Fig. 17.10. The power factor is, therefore,

$$F_p = \cos\theta = \frac{P}{S} = \frac{0}{VI} = 0$$

17.6 THE POWER TRIANGLE

The three quantities—average power, apparent power, and reactive power—are related in the vector domain by

$$\boxed{\mathbf{S} = \mathbf{P} + \mathbf{Q}} \tag{17.18}$$

with

$$\mathbf{P} = P\,\angle 0^\circ, \qquad \mathbf{Q}_L = Q_L\,\angle 90^\circ, \qquad \mathbf{Q}_C = Q_C\,\angle -90^\circ$$

as shown in Fig. 17.11 for an inductive load and Fig. 17.12 for a capacitive load.

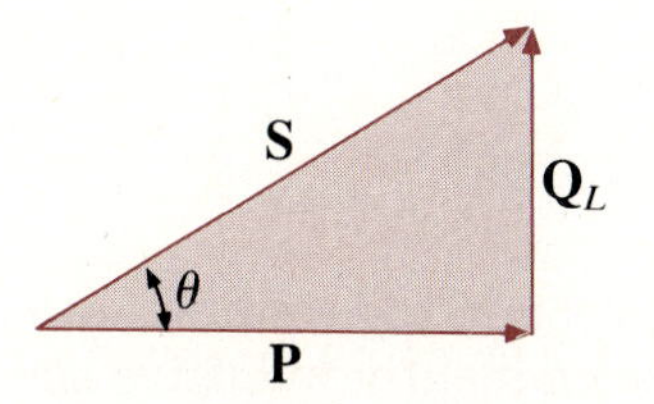

FIG. 17.11
Power diagram for inductive loads.

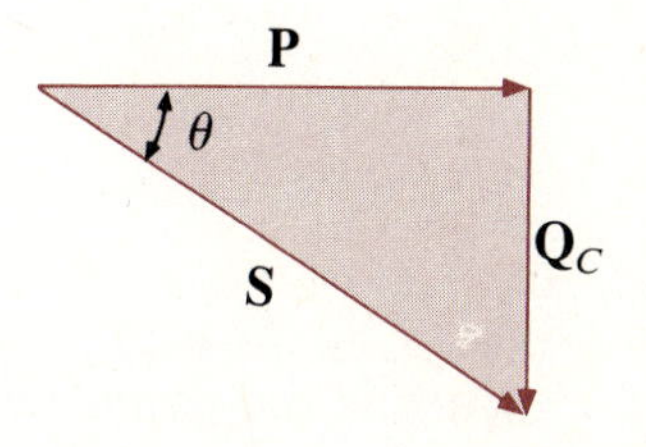

FIG. 17.12
Power diagram for capacitive loads.

If a network has both capacitive and inductive elements, the reactive component of the power triangle is determined by the *difference* between the reactive power delivered to each. If $Q_L > Q_C$, the resultant power triangle will be similar to Fig. 17.11. If $Q_C > Q_L$, the resultant power triangle will be similar to Fig. 17.12.

The preceding statement can be verified by first considering the impedance diagram of a series *R-L-C* circuit (Fig. 17.13). If we multiply each radius

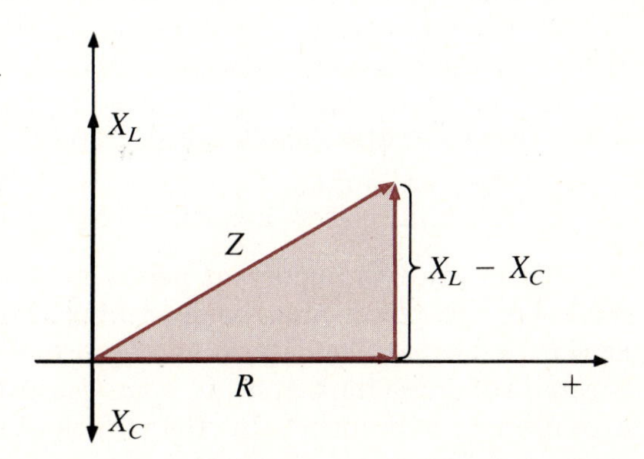

FIG. 17.13
Impedance diagram for an *R-L-C* network with $X_L > X_C$.

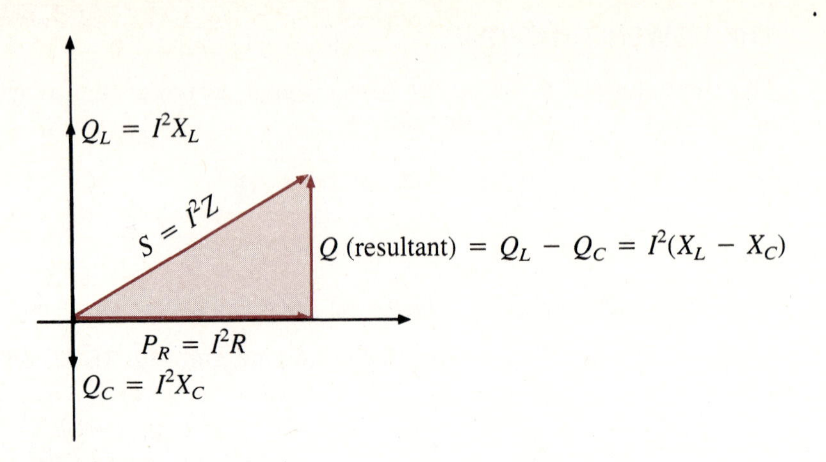

FIG. 17.14
Deriving the power triangle from the impedance diagram.

vector by the current squared (I^2), we obtain the results shown in Fig. 17.14, which is the power triangle for a predominantly inductive circuit.

Since the reactive power and average power are always at an angle of 90° to each other, the three powers are related by the Pythagorean theorem; that is,

$$S^2 = P^2 + Q^2 \tag{17.19}$$

17.7 THE TOTAL *P, Q,* AND *S*

The total number of watts, volt-amperes reactive, and volt-amperes and the power factor of any system can be found using the following procedure:

1. Find the real power and reactive power for each branch of the circuit.
2. The total real power of the system (P_T) is then the sum of the average power delivered to each branch.
3. The total reactive power (Q_T) is the difference between the reactive power of the inductive loads and that of the capacitive loads.
4. The total apparent power is $S_T = \sqrt{P_T^2 + Q_T^2}$.
5. The total power factor is P_T/S_T.

There are two important points in the preceding tabulation. First, the total apparent power must be determined from the total average and reactive powers and *cannot* be determined from the apparent powers to each branch. Second, and more important, it is *not necessary* to consider the series-parallel arrangement of branches. In other words, the total real, reactive, or apparent power is independent of whether the loads are in series, parallel, or series-parallel. The following examples demonstrate the relative ease with which all the quantities of interest can be found.

EXAMPLE 17.4 Find the total number of watts, volt-amperes reactive, and volt-amperes and the power factor F_p of the network in Fig. 17.15. Draw the power triangle and find the current in phasor form.

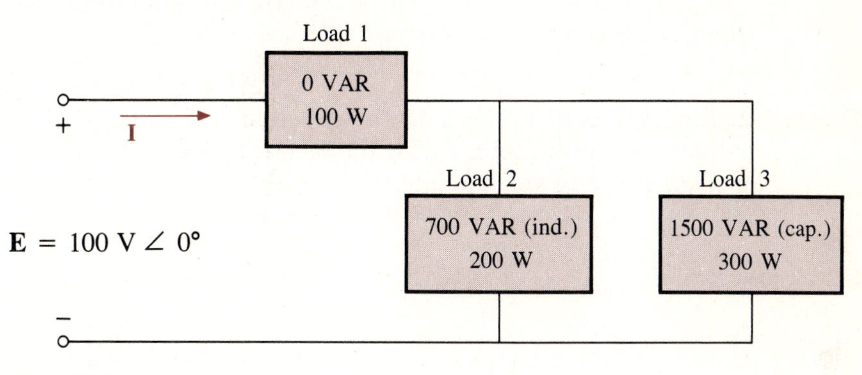

FIG. 17.15
System for Example 17.4.

Solution Using a table:

Load	W	VAR	VA
1	100	0	100
2	200	700 (ind.)	$\sqrt{(200)^2 + (700)^2} = 728.0$
3	300	1500 (cap.)	$\sqrt{(300)^2 + (1500)^2} = 1529.71$
	P_T = **600** Total power dissipated	Q_T = **800 (cap.)** Resultant reactive power of network	$S_T = \sqrt{(600)^2 + (800)^2} =$ **1000** (Note that $S_T \neq$ sum of each branch: $1000 \neq 100 + 728 + 1529.71$)

$$F_p = \frac{P_T}{S_T} = \frac{600}{1000} = \mathbf{0.6 \ \ leading\ (capacitive)}$$

The power triangle is shown in Fig. 17.16.

FIG. 17.16
Power triangle for the system of Fig. 17.14.

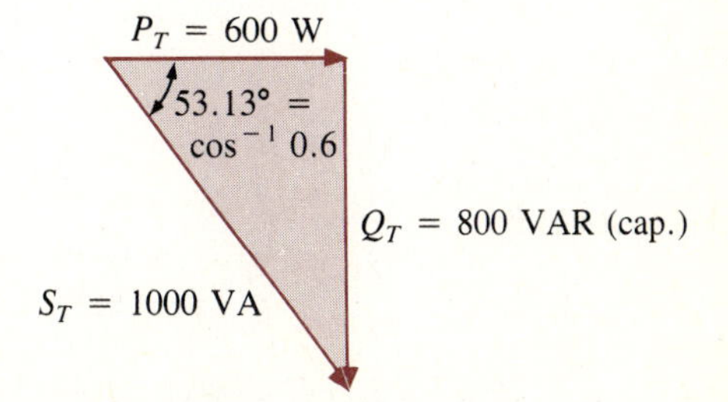

Since $S_T = VI = 1000$ VA, $I = 1000$ VA/100 V $= 10$ A; and since θ of $\cos\theta = F_p$ is the angle between the input voltage and current,

$$\mathbf{I} = \mathbf{10\ A\ \angle +53.13^\circ}$$

The plus sign is associated with the phase angle, since the circuit is predominantly capacitive.

EXAMPLE 17.5

a. Find the total number of watts, volt-amperes reactive, and volt-amperes and the power factor F_p for the network of Fig. 17.17.

b. Sketch the power triangle.

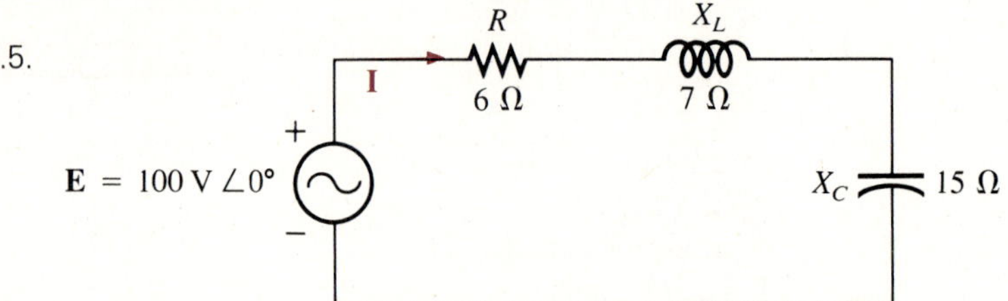

FIG. 17.17
Series *R-L-C* circuit for Example 17.5.

Solutions

a. $\mathbf{Z}_T = \mathbf{Z}_1 + \mathbf{Z}_2 + \mathbf{Z}_3 = 6\ \Omega\ \angle 0^\circ + 7\ \Omega\ \angle 90^\circ + 15\ \Omega\ \angle -90^\circ$
$= 6\ \Omega\ \angle 0^\circ + 8\ \Omega\ \angle -90^\circ$

The Pythagorean theorem gives

$$Z_T = \sqrt{(6\ \Omega)^2 + (8\ \Omega)^2} = 10\ \Omega$$

with

$$\theta_T = \tan^{-1}\frac{(X_C - X_L)}{R} = \tan^{-1}\frac{(8\ \Omega)}{(6\ \Omega)} = \tan^{-1}\frac{4}{3}$$

$$= 53.13^\circ \quad \text{(capacitive)}$$

$$I = \frac{E}{Z_T} = \frac{100\ \text{V}}{10\ \Omega} = 10\ \text{A}$$

and

$$P_R = I^2R = (10\ \text{A})^2 6\ \Omega = \mathbf{600\ W}$$

with

$$Q_L = I^2X_L = (10\ \text{A})^2 7\ \Omega = \mathbf{700\ VAR}$$
$$Q_C = I^2X_C = (10\ \text{A})^2 15\ \Omega = \mathbf{1500\ VAR}$$
$$Q_T = Q_C - Q_L = 1500\ \text{VAR} - 700\ \text{VAR} = \mathbf{800\ VAR}$$
$$S_T = \sqrt{P_T^2 + Q_T^2} = \sqrt{(600\ \text{W})^2 + (800\ \text{VAR})^2} = \mathbf{1000\ VA}$$
$$F_p = \frac{P_T}{S_T} = \frac{600\ \text{W}}{1000\ \text{VA}} = \mathbf{0.6\ \ leading\ (capacitive)}$$

b. See Fig. 17.18.

FIG. 17.18

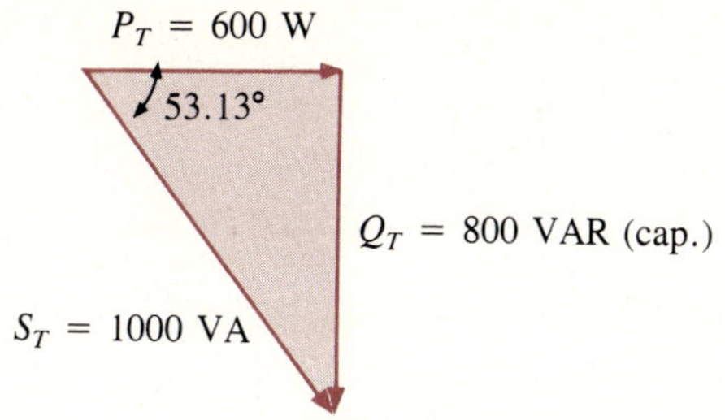

EXAMPLE 17.6

An electrical device is rated 5 kVA, 100 V at a 0.6 power-factor lag. Find the series circuit having the impedance characteristics of the device.

Solution

$$S = EI = 5000 \text{ VA}$$

Therefore,

$$I = \frac{S}{E} = \frac{5000 \text{ VA}}{100 \text{ V}} = 50 \text{ A}$$

For $F_p = 0.6$, we have

$$\theta = \cos^{-1}0.6 = 53.13°$$

Since the power factor is lagging, the circuit is predominantly inductive, and **I** lags **E**. Or, for $\mathbf{E} = 100 \text{ V} \angle 0°$,

$$\mathbf{I} = 50 \text{ A} \angle -53.13°$$

The magnitude of the total impedance is determined from

$$Z_T = \frac{E}{I} = \frac{100 \text{ V}}{50 \text{ A}} = 2 \,\Omega$$

Since the angle associated with **E** is 0° and that associated with **I** is −53.13°, the angle associated with $\mathbf{Z}_T$ is +53.13°, and

$$\mathbf{Z}_T = 2 \,\Omega \angle +53.13°$$

as drawn in Fig. 17.19.

$$R = Z_T\cos 53.13° = (2 \,\Omega)(0.6) = \mathbf{1.2 \,\Omega}$$

and

$$X_L = Z_T\sin 53.13° = (2 \,\Omega)(0.8) = \mathbf{1.6 \,\Omega}$$

The series network appears in Fig. 17.20.

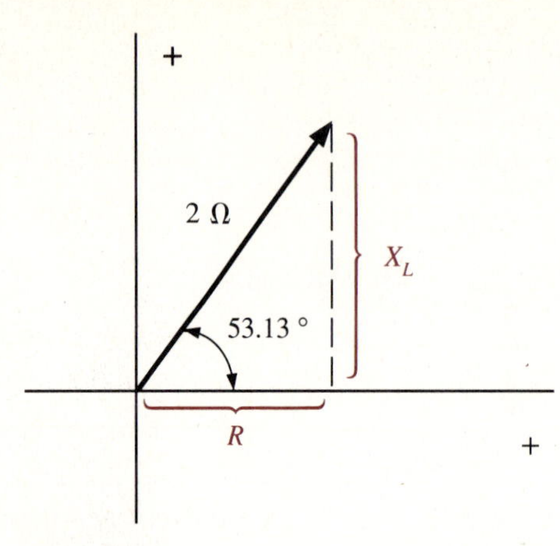

FIG. 17.19
Determining the series circuit to match a given impedance vector.

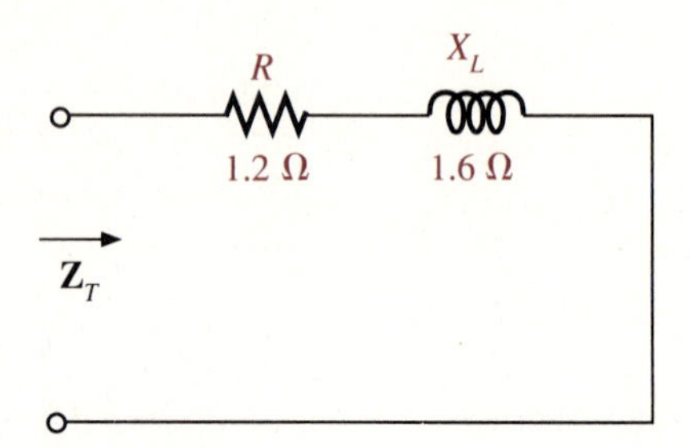

FIG. 17.20
Series circuit to match the impedance vector of Fig. 17.19.

17.8 POWER-FACTOR CORRECTION

The design of any power-transmission system is very sensitive to the magnitude of the current in the lines as determined by the applied loads. Increased currents result in increased power losses (by a squared factor, since $P = I^2R$) in the transmission lines due to the resistance of the lines. Heavier currents also require larger conductors, increasing the amount of copper needed for the system.

Every effort must therefore be made to keep current levels at a minimum. Since the line voltage of a transmission system is fixed, the apparent power is directly related to the current level. In turn, the smaller the net apparent power, the smaller the current drawn from the supply. Minimum current is therefore drawn from a supply when $S = P$ and $Q_T = 0$. Note the effect of decreasing levels of Q_T on the length (and magnitude) of S in Fig. 17.21 for the same real power. Note also that the power-factor angle approaches 0° and F_p approaches 1, revealing that the network is appearing more and more resistive at the input terminals.

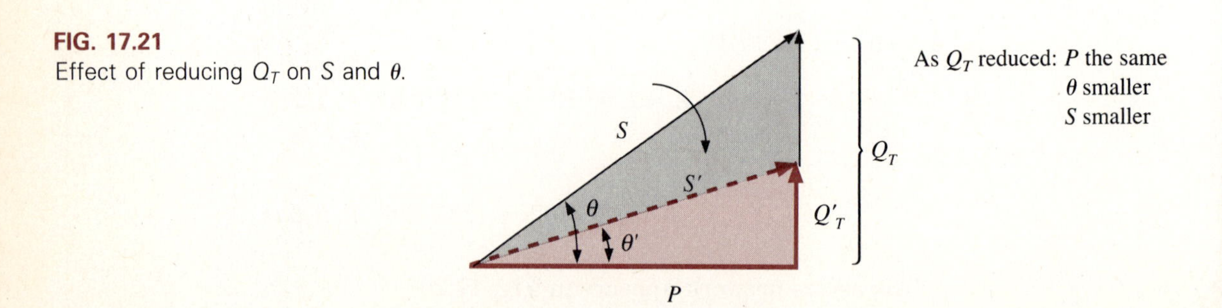

FIG. 17.21
Effect of reducing Q_T on S and θ.

The process of introducing reactive elements to bring the power factor closer to unity is called *power-factor correction.* Since most loads are inductive, the process normally involves introducing elements with capacitive terminal characteristics having the sole purpose of improving the power factor.

EXAMPLE 17.7 A 5-hp motor with a 0.6 lagging power factor and an efficiency of 92% is connected to a 208-V, 60-Hz supply. What level of capacitance in parallel with the motor will raise the power factor of the combined system to unity?

Solution First, the power triangle for the 5-hp motor is established. Since 1 hp = 746 W,

$$P_o = 5 \text{ hp} = 5(746 \text{ W}) = 3730 \text{ W}$$

and

$$P_i \text{ (drawn from the line)} = \frac{P_o}{\eta} = \frac{3730 \text{ W}}{0.92} = 4054.35 \text{ W}$$

Also,

$$F_p = \cos\theta = 0.6$$

and

$$\theta = \cos^{-1}0.6 = 53.13°$$

Then we use

$$\tan\theta = \frac{\text{Opposite}}{\text{Adjacent}} = \frac{Q_L}{P_i}$$

to obtain

$$\begin{aligned} Q_L &= P_i \tan\theta = (4054.35 \text{ W})(\tan 53.13°) \\ &= (4054.35 \text{ W})(1.333) = 5404.45 \text{ VAR} \end{aligned}$$

and

$$S = \sqrt{P_i^2 + Q_L^2} = \sqrt{(4054.35 \text{ W})^2 + (5404.45 \text{ VAR})^2} = 6756.17 \text{ VA}$$

The power triangle appears in Fig. 17.22.

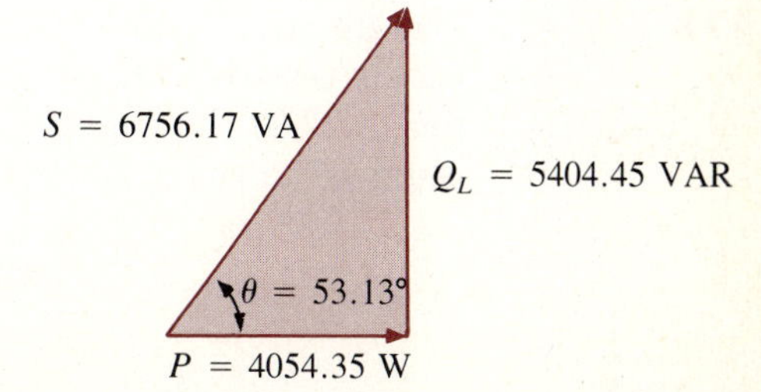

FIG. 17.22
Power triangle for the load of Example 17.7.

A net unity power-factor level is established by introducing a capacitive reactive power level of 5404.45 VAR to balance Q_L. Since

$$Q_C = \frac{V^2}{X_C}$$

then

$$X_C = \frac{V^2}{Q_C}$$

and

$$X_C = \frac{(208\ \text{V})^2}{5404.45\ \text{VAR}} = 8\ \Omega$$

and

$$C = \frac{1}{2\pi f X_C} \qquad \left(\text{from } X_C = \frac{1}{2\pi f C}\right)$$

$$= \frac{1}{(6.28)(60\ \text{Hz})(8\ \Omega)} = \mathbf{332\ \mu F}$$

At $0.6F_p$ (the original load):

$$S = VI = 6756.17\ \text{VA}$$

and

$$I = \frac{S}{V} = \frac{6756.17\ \text{VA}}{208\ \text{V}} = \mathbf{32.48\ A}$$

At unity F_p:

$$S = VI = 4054.35\ \text{VA}$$

and

$$I = \frac{S}{V} = \frac{4054.35\ \text{VA}}{208\ \text{V}} = \mathbf{19.49\ A}$$

producing a 40% reduction in supply current.

EXAMPLE 17.8 A small industrial plant has a 10-kW heating load and a 20-kVA inductive load due to a bank of induction motors. The heating elements are considered purely resistive ($F_p = 1$), and the induction motors have a lagging power factor of 0.7. If the supply is 1000 V at 60 Hz, determine the capacitive element required to raise the power factor to 0.95.

Solution For the induction motors,

$$S = VI = 20\ \text{kVA}$$
$$P = VI\cos\theta = (20 \times 10^3\ \text{VA})(0.7) = 14 \times 10^3\ \text{W}$$
$$\theta = \cos^{-1}0.7 \cong 45.6°$$

and

$$Q_L = VI\sin\theta = (20 \times 10^3\ \text{VA})(0.714) = 14.28 \times 10^3\ \text{VAR}$$

The power triangle for the total system appears in Fig. 17.23.

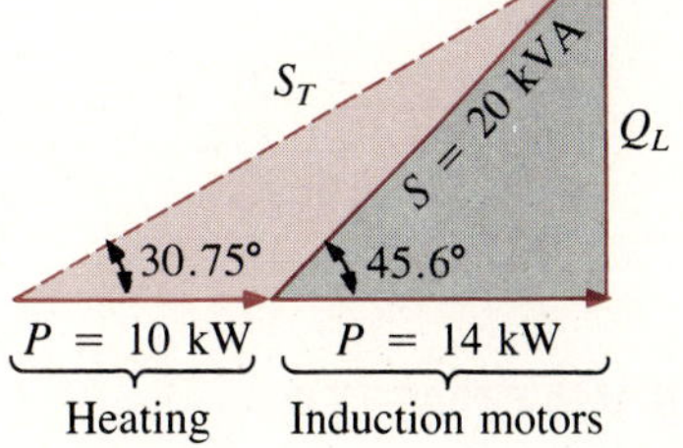

FIG. 17.23
Power triangle for Example 17.8.

Note the addition of real powers and the increased value of S:

$$S_T = \sqrt{(24\ \text{kW})^2 + (14.28\ \text{kVAR})^2} = 27.93\ \text{kVA}$$

with

$$I = \frac{S_T}{V} = \frac{27.93\ \text{kVA}}{1000\ \text{V}} = 27.93\ \text{A}$$

A power factor of 0.95 results in an angle between S and P of

$$\theta = \cos^{-1}0.95 = 18.19°$$

changing the power triangle to the following (Fig. 17.24):

$$\tan\theta = \frac{Q'_L}{P_T} \Rightarrow Q'_L = P_T\tan\theta = (24 \times 10^3\ \text{W})(\tan 18.19°)$$
$$= (24 \times 10^3\ \text{W})(0.329) = 7.9\ \text{kVAR}$$

FIG. 17.24
Resulting power triangle for Example 17.8.

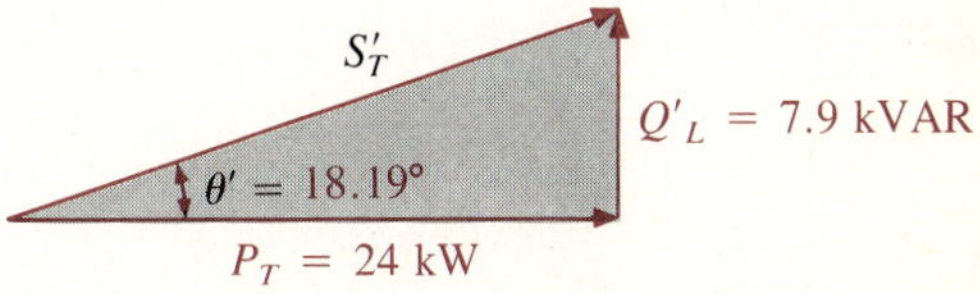

The inductive reactive power must therefore be reduced by

$$Q_L - Q'_L = 14.28\text{ kVAR} - 7.9\text{ kVAR} = 6.38\text{ kVAR}$$

Therefore, $Q_C = 6.38$ kVAR, and using

$$Q_C = \frac{V^2}{X_C}$$

we obtain

$$X_C = \frac{V^2}{Q_C} = \frac{(10^3\text{ V})^2}{6.38 \times 10^3\text{ VAR}} = 156.74\ \Omega$$

and

$$C = \frac{1}{2\pi f X_C} = \frac{1}{(6.28)(60\text{ Hz})(156.74\ \Omega)} = \mathbf{16.93\ \mu F}$$

17.9 THE WATTMETER

The wattmeter, as the name suggests, is an instrument designed to read the power to an element or network. It employs an electrodynamometer-type movement or solid-state electronic system to measure the power in a dc or an ac network. It can, in fact, be used to measure the wattage of any circuit with a periodic or nonperiodic input.

In an electrodynamometer movement, a moving coil rotates in a magnetic field produced by the current of a stationary coil. The fluxes of the stationary and movable coils interact to develop a torque on the pointer connected to the movable coil. In the wattmeter configuration (Fig. 17.25), the current in the stationary coils is the line current, whereas the current in the moving coil is derived from the line voltage. The instrument then indicates power in watts on a linear scale. A typical wattmeter using an electrodynamometer movement appears in Fig. 17.26.

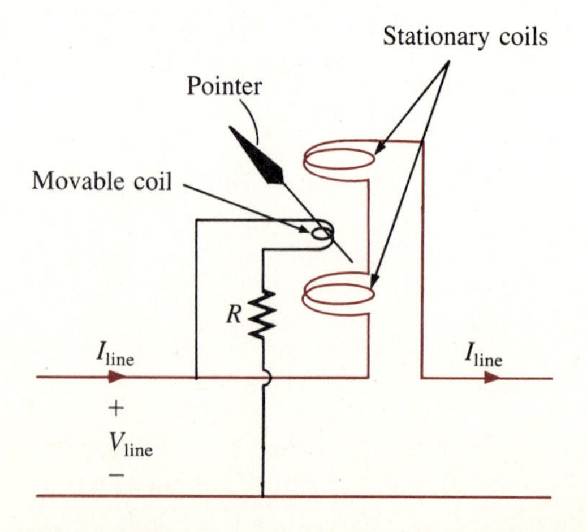

FIG. 17.25
Stationary and movable coils of an electrodynamometer movement.

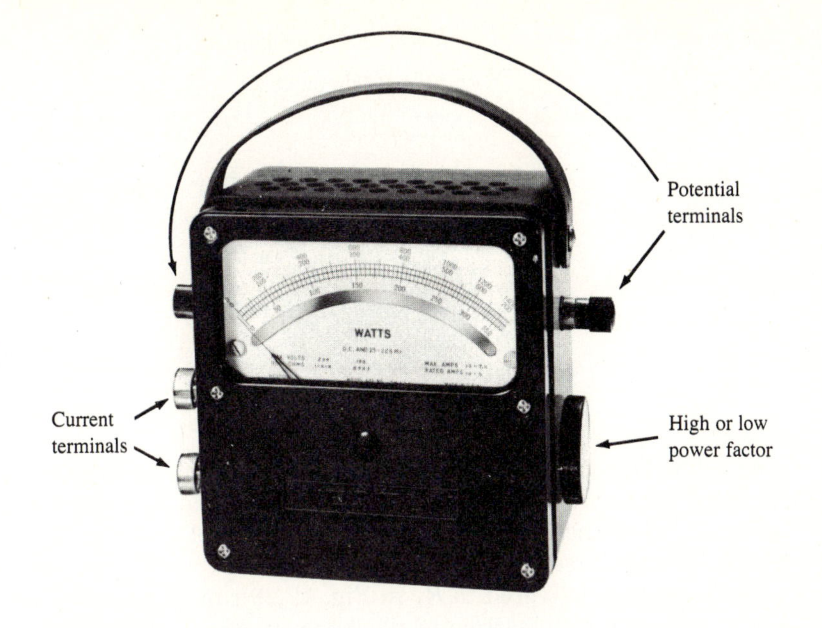

FIG. 17.26
Wattmeter. (Courtesy of Electrical Instrument Service, Inc.)

The digital display wattmeter of Fig. 17.27 employs a sophisticated electronic package to sense the voltage and current levels and, through the use of an analog-to-digital conversion unit, display the proper digits on the display.

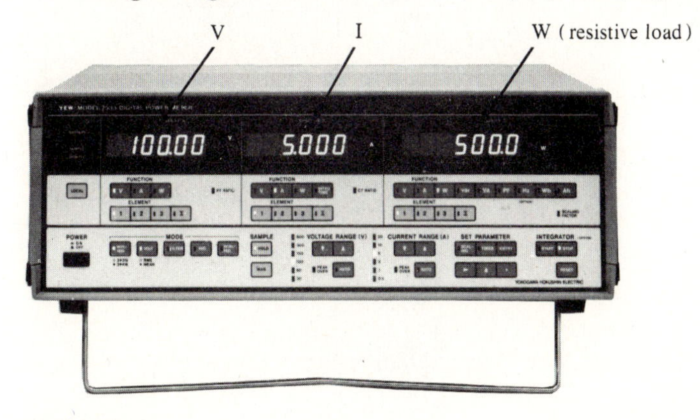

FIG. 17.27
Digital wattmeter. (Courtesy of Yokogawa Corporation of America)

For an up-scale deflection, a wattmeter is connected as shown in Fig. 17.28. Some electrodynamometer wattmeters always give a wattage reading that is higher than that actually delivered to the load. They are high by the amount of power consumed by the potential coil (V_{pc}^2/R_{pc}). This correction is important and should be considered with every set of data. Many wattmeters

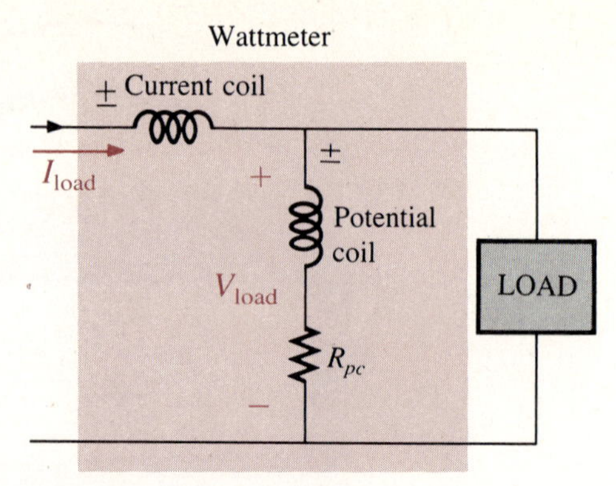

FIG. 17.28
Basic wattmeter connections.

are designed to compensate for this correction, and therefore they eliminate the need for any other adjustment in the reading. The wattmeter is always connected with the potential terminals in parallel with the load and the current terminals in series with the load to which the power is being measured.

The power delivered to R_1 in Fig. 17.29 can be found by connecting the electrodynamometer wattmeter as shown in Fig. 17.29(a). To find the power delivered to the total network, it should be connected as shown in Fig. 17.29(b). The connections for the digital meter are fundamentally the same as for the electrodynamometer meter.

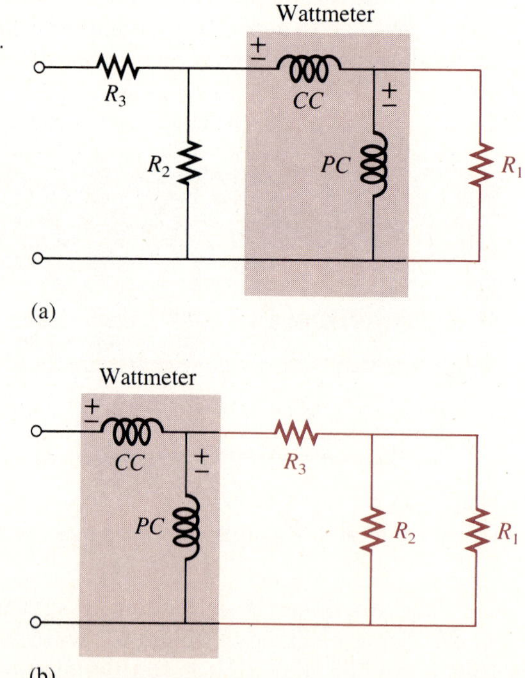

FIG. 17.29
Determining the power to specific loads.

When using a wattmeter, the operator must take care not to exceed the current, voltage, or wattage rating. The product of the voltage and current ratings may or may not equal the wattage rating. In the high-power-factor wattmeter, the product of the voltage and current ratings is usually at least 80% of the wattage rating. For a low-power-factor wattmeter, the product of the current and voltage ratings is much greater than the wattage rating. For obvious reasons, the low-power-factor meter is used only in circuits with low power factors (total impedance highly reactive). Typical ratings for high-power-factor (HPF) and low-power-factor (LPF) meters are shown in Table 17.1. Meters of both high and low power factors have an accuracy of 0.5% to 1% of full scale.

TABLE 17.1

Meter	Current Ratings	Voltage Ratings	Wattage Ratings
HPF	2.5 A	150 V	1500/750/375
	5.0 A	300 V	
LPF	2.5 A	150 V	300/150/75
	5.0 A	300 V	

17.10 EFFECTIVE RESISTANCE

The resistance of a conductor as determined by the equation $R = \rho(l/A)$ is often called the *dc, ohmic,* or *geometric* resistance. It is a constant quantity determined only by the material used and its physical dimensions. In ac circuits, the actual resistance of a conductor (called the *effective* resistance) differs from the dc resistance because of the varying currents and voltages, which introduce effects not present in dc circuits.

These effects include *radiation losses, skin effect, eddy currents,* and *hysteresis losses*. The first two effects apply to any network, whereas the latter two are concerned with the additional losses introduced by the presence of ferromagnetic materials in a changing magnetic field.

The effective resistance of an ac circuit cannot be measured by the ratio V/I, since this is now the impedance of a circuit that may have both resistance and reactance. The effective resistance can be found, however, by using the power equation $P = I^2R$, where

$$\boxed{R_{\text{eff}} = \frac{P}{I^2}} \tag{17.20}$$

A wattmeter and an ammeter are therefore necessary for measuring the effective resistance of an ac circuit.

Let us now examine the various losses in greater detail. The radiation loss is the loss of energy in the form of electromagnetic waves during the transfer of energy from one element to another. This loss in energy requires that the input power be larger to establish the same current I, causing R to increase, as determined by Eq. 17.20. At a frequency of 60 Hz, the effects of radiation losses can be completely ignored. However, at radio frequencies, this is an important effect and may in fact become the main effect in an electromagnetic device such as an antenna.

The explanation of skin effect requires the use of some basic concepts previously described. Recall from Chapter 11 that a magnetic field exists around every current-carrying conductor (Fig. 17.30). Since the amount of charge flowing in ac circuits changes with time, the magnetic field surrounding the moving charge (current) also changes. Recall also that a wire placed in a changing magnetic field has an induced voltage across its terminals as determined by Faraday's law, $e = N(d\Phi/dt)$. The higher the frequency of the changing flux as determined by an alternating current, the greater the induced voltage will be.

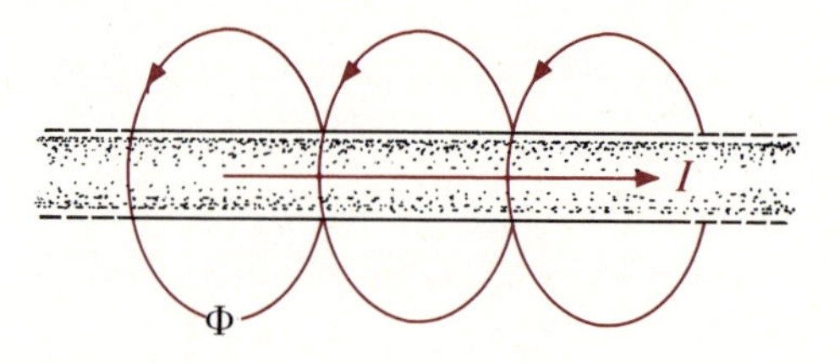

FIG. 17.30
Magnetic field induced by a current-carrying conductor.

For a conductor carrying alternating current, the changing magnetic field surrounding the wire links the wire itself, thus developing within the wire an induced voltage that opposes the original flow of charge or current. These effects are more pronounced at the center of the conductor than at the surface, since the center is linked by the changing flux inside the wire as well as that outside the wire. As the frequency of the applied signal increases, the flux linking the wire changes at a greater rate. An increase in frequency therefore increases the counterinduced voltage at the center of the wire to the point where the current will, for all practical purposes, flow on the surface of the conductor. At 60 Hz, the results of skin effect are almost noticeable. However, at radio frequencies, the skin effect is so pronounced that large conductors are frequently made hollow, since the center part is relatively ineffective. The skin effect, therefore, reduces the effective area through which the current can flow and causes the resistance of the conductor, given by the equation $R\uparrow = \rho(l/A\downarrow)$, to increase.

As mentioned earlier, hysteresis and eddy current losses appear when a ferromagnetic material is placed in the region of a changing magnetic field. To describe eddy current losses in greater detail, we consider the effects of an alternating current passing through a coil wrapped around a ferromagnetic core. As the alternating current passes through the coil, it develops a changing magnetic flux Φ linking both the coil and the core, which will develop an induced voltage within the core as determined by Faraday's law. This induced voltage and the geometric resistance of the core, $R_C = \rho(l/A)$, cause currents to be developed within the core, $i_{core} = (e_{ind}/R_C)$, called *eddy currents*. The currents flow in circular paths, as shown in Fig. 17.31, changing direction with the applied ac potential.

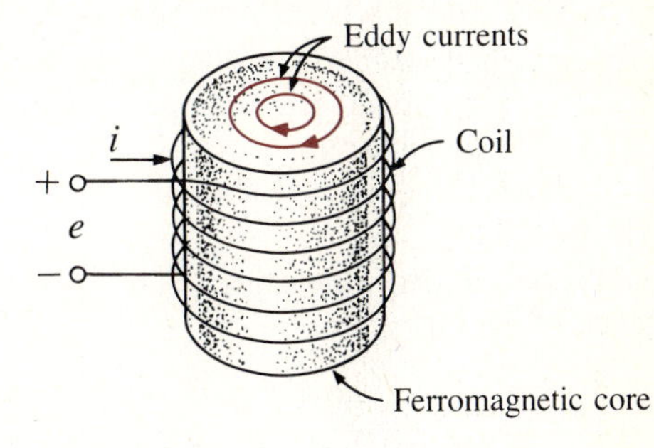

FIG. 17.31
Eddy currents induced by applied alternating currents.

The eddy current losses are determined by

$$P_{eddy} = i^2_{eddy}R_{core}$$

The magnitude of these losses is determined primarily by the type of core used. If the core is nonferromagnetic—and has a high resistivity, such as do wood or air—the eddy current losses can be neglected. In terms of the frequency of the applied signal and the magnetic field strength produced, the eddy current loss is proportional to the square of the frequency times the square of the magnetic field strength:

$$P_{eddy} \propto f^2B^2$$

Eddy current losses can be reduced if the core is constructed of thin, laminated sheets of ferromagnetic material insulated from one another and aligned parallel to the magnetic flux. Such construction reduces the magnitude of the eddy currents by placing more resistance in their path.

Hysteresis effects were described in Section 11.7. In terms of the frequency of the applied signal and the magnetic field strength produced, the hysteresis loss is proportional to the frequency to the first power times the magnetic field strength to the nth power:

$$P_{hys} \propto f^1B^n$$

when n can vary from 1.4 to 2.6, depending on the material under consideration.

Hysteresis losses can be effectively reduced by the injection of small amounts of silicon into the magnetic core, constituting some 2% or 3% of the total composition of the core. This must be done carefully, however, since too much silicon makes the core brittle and difficult to machine into the desired shape.

EXAMPLE 17.9

a. An air-core coil is connected to a 120-V, 60-Hz source, as shown in Fig. 17.32. The current is found to be 5 A, and a wattmeter reading of 75 W is observed. Find the effective resistance and the inductance of the coil.

b. A brass core is then inserted in the coil, and the ammeter reads 4 A and the wattmeter reads 80 W. Calculate the effective resistance of the core. To what do you attribute the increase in value over that of part a?

c. If a solid iron core is inserted in the coil, the current is found to be 2 A, and the wattmeter reads 52 W. Calculate the resistance and the inductance of the coil. Compare these values to those of part a, and account for the changes.

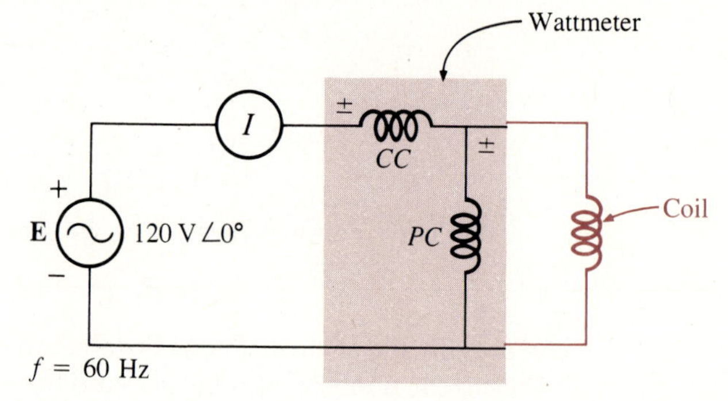

FIG. 17.32
Determining the effective resistance and inductance of the coil.

Solutions

a. $R = \dfrac{P}{I^2} = \dfrac{75\ \text{W}}{(5\ \text{A})^2} = \mathbf{3\ \Omega}$

$$Z_T = \frac{E}{I} = \frac{120\ \text{V}}{5\ \text{A}} = 24\ \Omega$$

$$X_L = \sqrt{Z_T^2 - R^2} = \sqrt{(24\ \Omega)^2 - (3\ \Omega)^2} = 23.81\ \Omega$$

Also,

$$X_L = 2\pi f L$$

or

$$L = \frac{X_L}{2\pi f} = \frac{23.81\ \Omega}{377\ \text{rad/s}} = \mathbf{63.16\ mH}$$

b. $R = \dfrac{P}{I^2} = \dfrac{80\ \text{W}}{(4\ \text{A})^2} = \dfrac{80\ \Omega}{16} = \mathbf{5\ \Omega}$

The brass core has less reluctance than the air core. Therefore, a greater magnetic flux density B will be created in it. Since $P_{\text{eddy}} \propto f^2B^2$ and $P_{\text{hys}} \propto f^1B^n$, as the flux density increases, the core losses and the effective resistance increase.

c. $R = \dfrac{P}{I^2} = \dfrac{52\ \text{W}}{(2\ \text{A})^2} = \dfrac{52\ \Omega}{4} = \mathbf{13\ \Omega}$

$$Z_T = \frac{E}{I} = \frac{120\ \text{V}}{2\ \text{A}} = 60\ \Omega$$

$$X_L = \sqrt{Z_T^2 - R^2} = \sqrt{(60\ \Omega)^2 - (13\ \Omega)^2} = 58.57\ \Omega$$

$$L = \frac{X_L}{2\pi f} = \frac{58.57\ \Omega}{377\ \text{rad/s}} = \mathbf{155.36\ mH}$$

The iron core has less reluctance than the air or brass cores. Therefore, a greater magnetic flux density B is developed in the core. Again, since $P_{\text{eddy}} \propto f^2B^2$ and $P_{\text{hys}} \propto f^1B^n$, the increased flux density causes the core losses and the effective resistance to increase.

Since the inductance L is related to the change in flux by the equation $L = N(d\Phi/di)$, the inductance will be greater for the iron core because the changing flux linking the core will increase.

17.11 COMPUTER ANALYSIS

Program 17.1 (Fig. 17.33) demonstrates how the computer can calculate the total real, reactive, and apparent power and the power factor and supply current of the system.

The program is limited to five individual loads, as defined by lines 110 through 130. Lines 260 and 270 determine the total real and reactive power using a loop routine that begins on line 210 and ends on line 280. The apparent power for each load is determined on line 340, and the total apparent power is found on line 370. The results are printed out by lines 390 through 430, and the power factor is determined by lines 440 through 530. The input current is then determined by lines 540 through 560.

A run of the program for four loads is provided with parameter values that permit a relatively easy check of the program through review of the results.

```
 10 REM *****  PROGRAM 17-1  *****
 20 REM ******************************************
 30 REM This program calculates the total real,
 40 REM reactive and apparent power of a network
 50 REM with five individual loads.
 60 REM ******************************************
 70 REM
100 DIM P(5),Q(5),S(5)
110 PRINT "This program calculates the total real,"
120 PRINT "reactive and apparent power of a network"
130 PRINT "with five individual loads."
140 PRINT
150 PRINT "Input the following data:"
160 PRINT "(use negative sign for capacitive vars)"
170 PRINT
```

```
Input  [ 180 INPUT "E=";E
       [ 190 INPUT "at an angle=";EA
         200 PRINT
Input  [ 210 FOR I=1 TO 5
       [ 220 PRINT "For";I;"   ";
       [ 230 INPUT "P(watts)=";P(I)
       [ 240 PRINT TAB(8);
       [ 250 INPUT "Q(vars)=";Q(I)
P,Pq   [ 260 PT=PT+P(I)
       [ 270 QT=QT+Q(I)
         280 NEXT I
         290 PRINT
Pa     [ 300 PRINT "The apparent power associated with each load"
       [ 310 PRINT "is the following:"
       [ 320 PRINT
       [ 330 FOR I=1 TO 5
       [ 340 S(I)=SQR(P(I)^2+Q(I)^2)
       [ 350 PRINT "S";I;"=";S(I)
       [ 360 NEXT I
       [ 370 ST=SQR(PT^2+QT^2)
         380 PRINT:PRINT
Power  [ 390 PRINT "Total real power, PT=";PT;"watts"
Output [ 400 PRINT
       [ 410 PRINT "Total reactive power, QT=";QT;"vars"
       [ 420 PRINT
       [ 430 PRINT "Total apparent power, ST=";ST;"VA"
Fp     [ 440 FP=PT/ST
       [ 450 TH=-57.296*ATN(QT/PT)
       [ 460 IF QT>0 THEN IA=EA-TH
       [ 470 IF QT<0 THEN IA=EA+TH
       [ 480 PRINT
       [ 490 PRINT "Power factor angle=";IA;"degrees"
       [ 500 PRINT
       [ 510 PRINT "Power factor=";FP;
       [ 520 IF QT>0 THEN PRINT "(lagging)"
       [ 530 IF QT<0 THEN PRINT "(leading)"
IT     [ 540 I=ST/E
       [ 550 PRINT
       [ 560 PRINT "Input current:";I;"at an angle of";IA;"degrees"
         570 END
```

```
READY


RUN


This program calculates the total real,
reactive and apparent power of a network
with five individual loads.

Input the following data:
(use negative sign for capacitive vars)

E=? 50

at an angle=? 60
```

```
For 1    P(watts)=? 200

         Q(vars)=? 100

For 2    P(watts)=? 200

         Q(vars)=? 100

For 3    P(watts)=? 100

         Q(vars)=? -200

For 4    P(watts)=? 100

         Q(vars)=? -200

For 5    P(watts)=? 0

         Q(vars)=? 0

The apparent power associated with each load
is the following:

S 1 = 224
S 2 = 224
S 3 = 224
S 4 = 224
S 5 = 0

Total real power, PT= 600 watts

Total reactive power, QT=-200 vars

Total apparent power, ST= 632 VA

Power factor angle= 78 degrees

Power factor= 1 (leading)

Input current: 13 at an angle of 78 degrees
```

FIG. 17.33
Program 17.1.

FORMULA SUMMARY

R

$$P = V_R I_R = I_R^2 R = \frac{V_R^2}{R}$$

$$S = VI = I^2 Z = \frac{V^2}{Z}$$

$$P = VI \cos \theta = S \cos \theta$$

$$F_p = \cos \theta = \frac{P}{S}$$

L

$$Q_L = V_L I_L = I_L^2 X_L = \frac{V_L^2}{X_L}$$

C

$$Q_C = V_C I_C = I_C^2 X_C = \frac{V_C^2}{X_C}$$

$$Q = VI \sin\theta = S \sin\theta$$

$$\mathbf{S} = \mathbf{P} + \mathbf{Q}, \quad S^2 = P^2 + Q^2$$

Effective Resistance

$$R_{\text{eff}} = \frac{P}{I^2}$$

CHAPTER SUMMARY

On an ideal basis:

1. Resistors dissipate all the energy delivered to them in the form of heat.
2. Inductors and capacitors store the energy delivered to them in the form of a magnetic and an electric field, respectively, and have the ability to return the energy to the electrical system where called for by the electrical design.

As noted in the formula summary, the power equations for an inductor and capacitor have the same format as those associated with a resistor. Simply interchange R, X_L, and X_C as required and apply the proper unit of measurement.

For any electrical system, the total average power is the sum of the average power levels for each load. There is *no* concern for how the various loads are connected. The total reactive power is determined by finding the difference between the total inductive and capacitive reactive power levels. Once the totals are found, the Pythagorean theorem can be applied to determine the total apparent power level. The total apparent power level is a minimum where the load is purely resistive and increases with increase in the reactive component. For a purely resistive load, $F_p = 1$ and $P = S$, and for a purely reactive load, $F_p = 0$ and $Q = S$. Any combinations of resistive and reactive components would result in F_p in the range 0 to 1 with $S = \sqrt{P^2 + Q^2}$. In general, where approaching problems of this type,

1. Determine the average and reactive power level for each load.
2. Find the total average and reactive power levels for the system.
3. Calculate the total apparent power level from the results of step 2.
4. If the source current or total impedance is required, apply the equations. $I = S_T/E$ and $Z_T = E/I$.

The first step in the application of power factor improvement techniques is usually a sketch of the system's power triangle with all its components. The desired power factor should then be employed to sketch the power triangle that will result after the improvement is made. The two sketches can then be compared to determine the type and magnitude of the capacitive componant required to improve the power factor. Diligence with each step will remove any complexity associated with the entire procedure, which must be performed with care and attention to detail.

Since power levels are sensitive to both the voltage across a load and the current through a load, a wattmeter must have terminals to cause both the voltage and current levels. The voltage terminals are always connected in parallel with the load, whereas the current terminals are in series with the load.

The effects of radiation losses, skin effect, eddy currents, and hysteresis can have a pronounced effect on the effective resistance and reactance of an element. The higher the frequency, the greater the concern about each factor, as noted in particular for eddy current and hysteresis losses. Eddy current losses increase with frequency and flux density to the second power, whereas hysteresis losses increase in a linear fashion with frequency, since it is sensitive to frequency to the first power. These phenomena help explain why a coil can seem to have an inductance level of 60 mH at one frequency and 80 mH at another. In other words, when a wide frequency range is spanned, do not expect a coil with a nameplate value of 10 mH to be exactly 10 mH at 10 Hz, 1 kHz, and 10 MHz. The preceding factors will affect its actual level.

GLOSSARY

Apparent power The power delivered to a load without consideration of the effects of a power-factor angle of the load. It is determined solely by the product of the terminal voltage and current of the load.

Average (real) power The delivered power that is dissipated in the form of heat by a network or system.

Eddy currents Small, circular currents in a paramagnetic core, causing an increase in the power losses and the effective resistance of the material.

Effective resistance The resistance value that includes the effects of radiation losses, skin effect, eddy currents, and hysteresis losses.

Hysteresis losses Losses in a magnetic material introduced by changes in the direction of the magnetic flux within the material.

Power-factor correction The addition of reactive components (typically capacitive) to establish a system power factor closer to unity.

Radiation losses The loss of energy in the form of electromagnetic waves during the transfer of energy from one element to another.

Reactive power The power associated with reactive elements that provides a measure of the energy associated with setting up the magnetic and electric fields of inductive and capacitive elements, respectively.

Skin effect At high frequencies, a counterinduced voltage builds up at the center of a conductor, resulting in an increased flow near the surface (skin) of the conductor and a great reduction near the center. As a result, resistance increases, as determined by the basic equation for resistance in terms of the geometric shape of the conductor.

PROBLEMS

Section 17.2

1. Determine the power delivered to a 2.2-kΩ resistor if the current through the resistor is $i = 30 \times 10^{-3}\sin \omega t$.
2. Find the resistance of a resistor having a voltage across it of $v_R = 8 \sin 377t$ with a power dissipation level of 28 mW.
3. Sketch the power curve (such as Fig. 17.3) for a 6-Ω resistor with a current $i_R = 3 \sin \omega t$.

Section 17.3

4. Find the apparent power of a 16-Ω resistive system due to an applied voltage of $20 \sin \omega t$. What is the power factor?
5. The power factor of a system is 0.6 lagging. What is the apparent power level if the dissipated power is 0.4 W?
6. **a.** Determine the apparent power level for a load with a voltage $v = 0.8 \sin(\omega t + 40°)$ and a current $i = 20 \times 10^{-3}(\sin \omega t + 10°)$.
   b. Find the average power dissipated by the load.
   c. Find the power factor of the load.

Sections 17.4–17.5

7. Find the reactive power of a coil with $v_L = 12 \sin \omega t$ and $X_L = 5\ \Omega$. Determine the average and apparent power levels.
8. Find the current of a capacitive load with a reactive power $Q_C = 24$ VAR and a reactance of 4 Ω.
9. Find the reactive power of a capacitor of 2 μF if the current is $5 \times 10^{-6}\sin 400t$.
10. Find the apparent power level of a 10-mH coil if the voltage across the coil is $v_L = 20 \times 10^{-3}\sin 1000t$.

Sections 17.6–17.7

11. For the network of Fig. 17.34:
    a. Find the average power delivered to each element.
    b. Find the reactive power for each element.
    c. Find the apparent power for each element.
    d. Find the total number of watts, volt-amperes reactive, and volt-amperes and the power factor F_p of the circuit.
    e. Sketch the power triangle.

FIG. 17.34

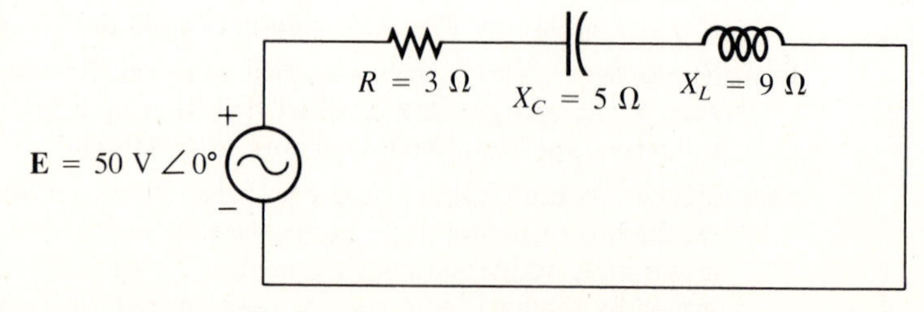

12. For the system of Fig. 17.35:

a. Find the total number of watts, volt-amperes reactive, and volt-amperes and the power factor F_p.

b. Draw the power triangle.

c. Find the current $\mathbf{I}_T$.

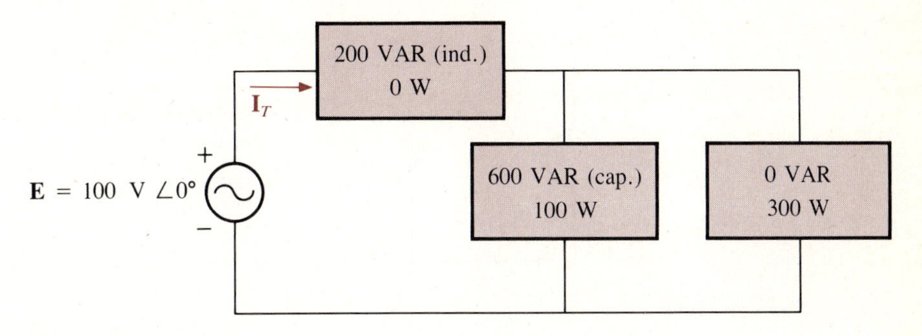

FIG. 17.35

13. Repeat Problem 12 for the system of Fig. 17.36.

FIG. 17.36

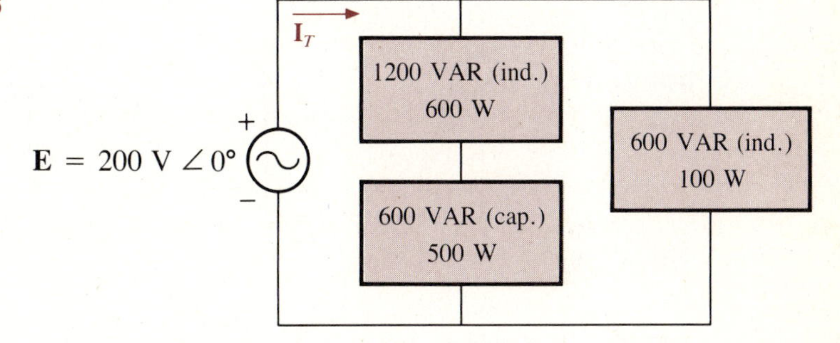

14. Repeat Problem 12 for the system of Fig. 17.37.

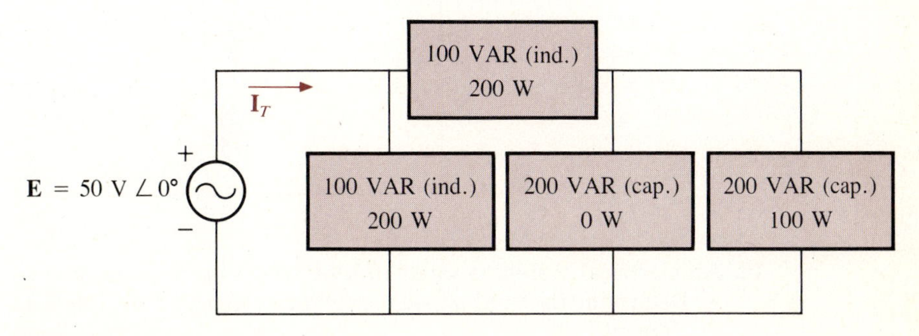

FIG. 17.37

15. For the circuit of Fig. 17.38:
  a. Find the average, reactive, and apparent power for the 20-Ω resistor.
  b. Repeat part a for the 10-Ω inductive reactance.
  c. Find the total number of watts, volt-amperes reactive, volt-amperes and the power factor F_p.
  d. Find the current $\mathbf{I}_T$.

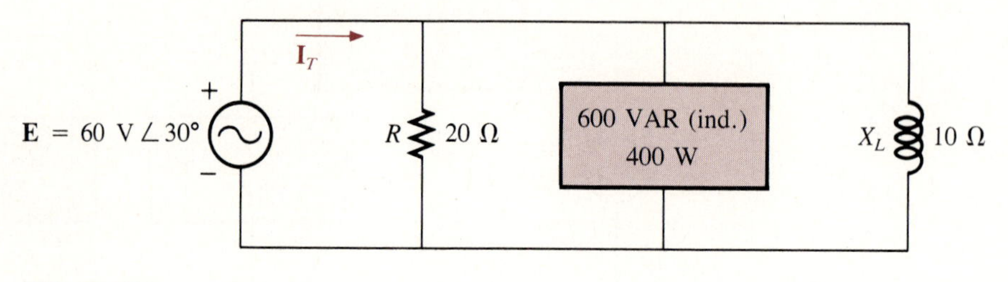

FIG. 17.38

***16.** Repeat Problem 11 for the circuit of Fig. 17.39.

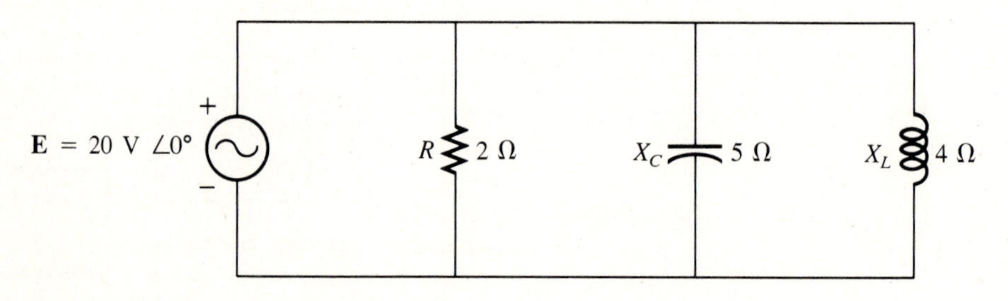

FIG. 17.39

***17.** Repeat Problem 11 for the circuit of Fig. 17.40.

FIG. 17.40

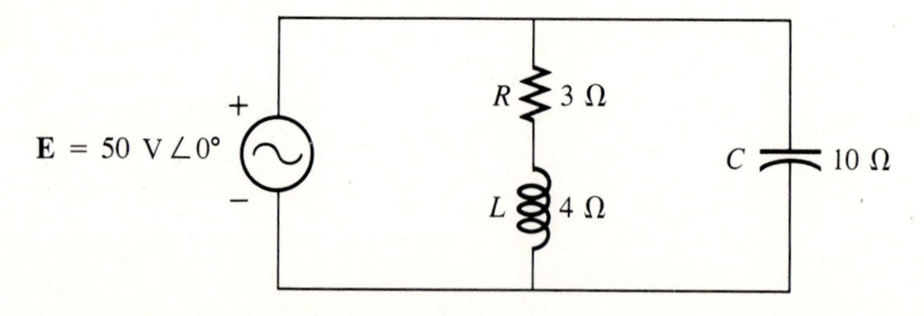

18. An electrical system is rated 10 kVA, 200 V at a 0.5 leading power factor.
  a. Determine the total impedance of the system and the impedance of each series element.
  b. Find the average power delivered to the system.

19. Repeat Problem 18 for an electrical system rated 5 kVA, 120 V at a 0.8 lagging power factor.

20. For the system of Fig. 17.41:

a. Find the total number of watts, volt-amperes reactive, and volt-amperes and F_p.

b. Find the current $\mathbf{I}_T$.

c. Draw the power triangle.

d. Find the type elements and their impedance in ohms within each electrical box. (Assume that all elements within the boxes are in series).

FIG. 17.41

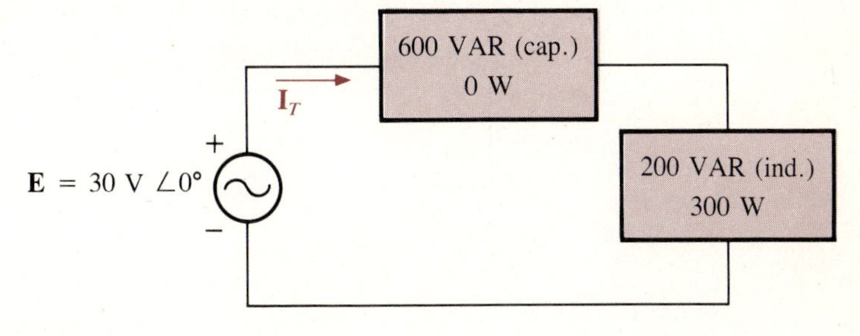

***21.** For the system of Fig. 17.42:

a. Find the total number of watts, volt-amperes reactive, and volt-amperes and F_p.

b. Find the current **I**.

c. Find the type elements and their impedance in each box. (Assume that the elements within each box are in series.)

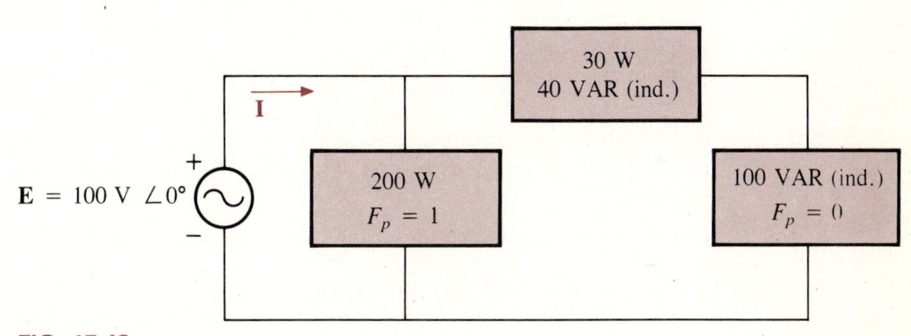

FIG. 17.42

22. For the circuit of Fig. 17.43:

a. Find the total number of watts, volt-amperes reactive, and volt-amperes and F_p.

b. Find the voltage **E**.

c. Find the type elements and their impedance in each box. (Assume that the elements within each box are in series.)

FIG. 17.43

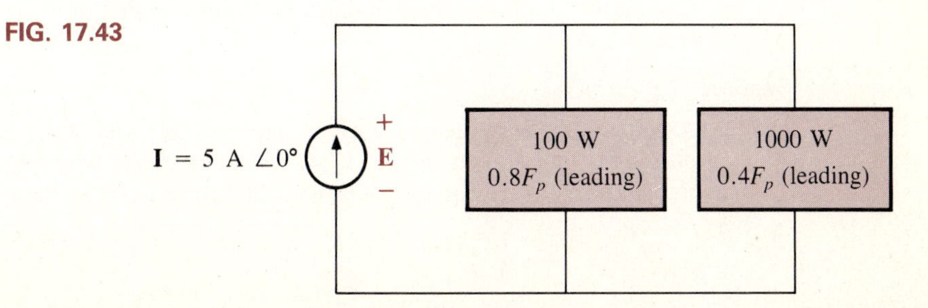

***23.** Repeat Problem 20 for the system of Fig. 17.44.

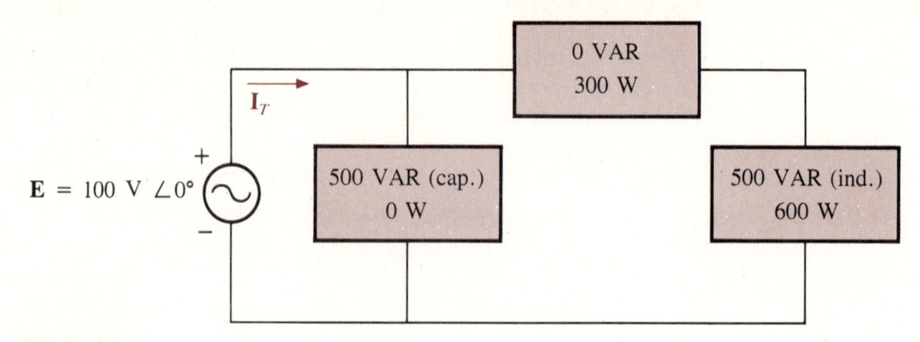

FIG. 17.44

Section 17.8

24. The lighting and motor loads of a small factory establish a 10-kVA power demand at a 0.7 lagging power factor on a 208-V, 60-Hz supply. Determine the level of capacitance in parallel with the load that will raise the power-factor level to
- **a.** Unity
- **b.** 0.9

25. The parallel load on a 1200-V, 60-Hz supply is 5 kW (resistive), 8 kVAR (inductive), and 2 kVAR (capacitive).
- **a.** Determine the F_p of the combined loads.
- **b.** Find the total kilovolt-amperes.
- **c.** Find the current drawn from the supply.
- **d.** Calculate the capacitance necessary to establish a unity power factor.
- **e.** Find the current drawn from the supply at unity power factor and compare to part c.

***26.** The loading of a factory on a 1000-V, 60-Hz system includes:

20-kW heating (unity power factor)
10-kW (P_i) induction motors (0.7 lagging power factor)
5-kW lighting (0.85 lagging power factor)

- **a.** Determine the total kilovolt-amperes.
- **b.** Find the total F_p.
- **c.** Derive the net reactive power.
- **d.** Find the current drawn from the supply.
- **e.** Calculate the capacitive contribution necessary to establish unity power factor for the total load.
- **f.** Calculate the current drawn from the supply under unity power-factor conditions.

Section 17.9

27. a. A wattmeter is connected with its current coil and the potential coil across points *f-g*, not shown in Fig. 17.45. What does the wattmeter read?
- **b.** Repeat part a with the potential coil (*PC*) across *a-b*, *b-c*, *a-c*, *a-d*, *c-d*, *d-e*, and *f-e*.

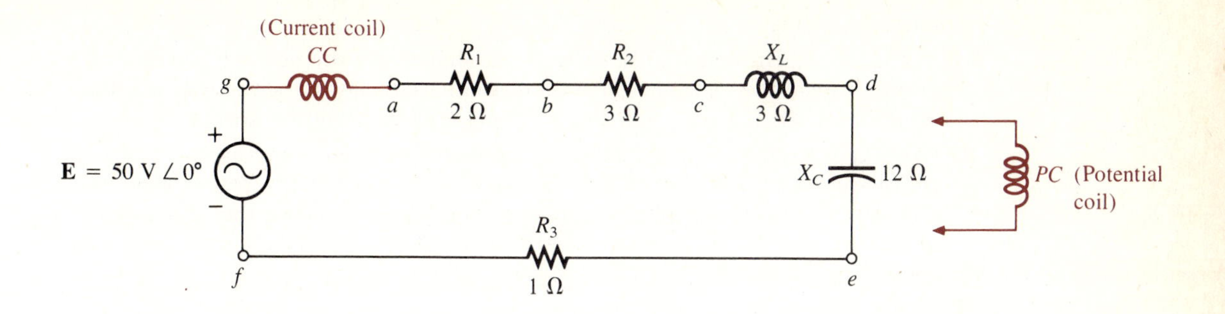

FIG. 17.45

*28. The voltage source of Fig. 17.46 delivers 640 VA at 120 V with a supply current that lags the voltage by a power factor of 0.6.

a. Determine the voltmeter, ammeter, and wattmeter readings.

b. Find the load impedance and determine the series impedances that will establish the total impedance.

FIG. 17.46

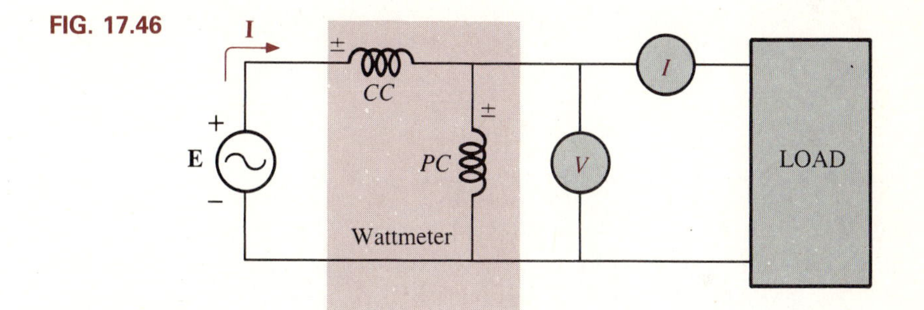

Section 17.10

29. a. An air-core coil is connected to a 200-V, 60-Hz source. The current is found to be 4 A, and a wattmeter reading of 80 W is observed. Find the effective resistance and the inductance of the coil.

b. A brass core is inserted in the coil. The ammeter reads 3 A and the wattmeter reads 90 W. Calculate the effective resistance of the core. Explain the increase over the value of part a.

c. If a solid iron core is inserted in the coil, the current is found to be 2 A, and the wattmeter reads 60 W. Calculate the resistance and inductance of the coil. Compare these values to the values of part a and account for the changes.

30. a. The inductance of an air-core coil is 0.08 H, and the effective resistance is 4 Ω when a 60-V, 50-Hz source is connected across the coil. Find the current passing through the coil and the reading of a wattmeter across the coil.

b. If a brass core is inserted in the coil, the effective resistance increases to 7 Ω, and the wattmeter reads 30 W. Find the current passing through the coil and the inductance of the coil.

c. If a solid iron core is inserted in the coil, the effective resistance of the coil increases to 10 Ω, and the current decreases to 1.7 A. Find the wattmeter reading and the inductance of the coil.

Computer Problems

31. Write a program that provides a general solution for the network of Fig. 17.17. That is, given the resistance or reactance of each element and the source voltage at 0°, calculate the real, reactive, and apparent power of the system.

32. Write a program that will demonstrate the effect of increasing reactive power on the power factor of a system. Tabulate the real power, reactive power, and power factor of the system for a fixed real power and a reactive power that starts at 10% of the real power and continues through to five times the real power in increments of 10% of the real power.

33. Write a program that will provide a general solution for systems such as that appearing in Example 17.4 (same input data).

18

Resonance

OBJECTIVES

- Become aware of the components, characteristics, and response of both series and parallel resonant circuits.
- Be able to determine the frequency response curve from the network elements for both series and parallel resonant circuits.
- Understand the effect of the quality factor on the shape of both the series and parallel resonant circuits.
- Determine the bandwidth and cutoff frequencies of the response curve for both series and parallel resonant circuits.
- Appreciate the impact of a high quality factor on the equations applied to parallel resonant circuits.
- Become familiar with log scales and the advantages associated with their use.
- Develop an understanding of the construction, characteristics, and response curves of both high- and low-pass filters.
- Be able to sketch a Bode plot of the basic *R-C* high- and low-pass filters.
- Understand how series and parallel resonant circuits are employed in the design of tuned filters.

18.1 INTRODUCTION

This chapter introduces the very important resonant (or tuned) circuit, which is fundamental to the operation of a wide variety of electrical and electronics systems in use today. The resonant circuit is a combination of R, L, and C elements having a frequency response characteristic as shown in Fig. 18.1. Note in the figure that the response is a maximum for the frequency f_r, decreasing to the right and left of this frequency. In other words, the resonant circuit selects a range of frequencies for which the response (voltage or current) is near or equal to the maximum. The frequencies to the far left or right are, for all practical purposes, nullified with respect to their effects on the system's response. A radio or television receiver has a response curve for each broadcast station of the type indicated in Fig. 18.1. When the receiver is set (or tuned) to a particular station, it is set on or near the frequency f_r of Fig. 18.1. Stations transmitting at frequencies to the far right or left of this resonant frequency are not carried through with significant power to affect the program of interest. The tuning process (setting the dial to f_r) as just described is the reason for the terminology *tuned circuit*. When the response is a maximum, the circuit is said to be in a state of *resonance* with f_r as the *resonant frequency*.

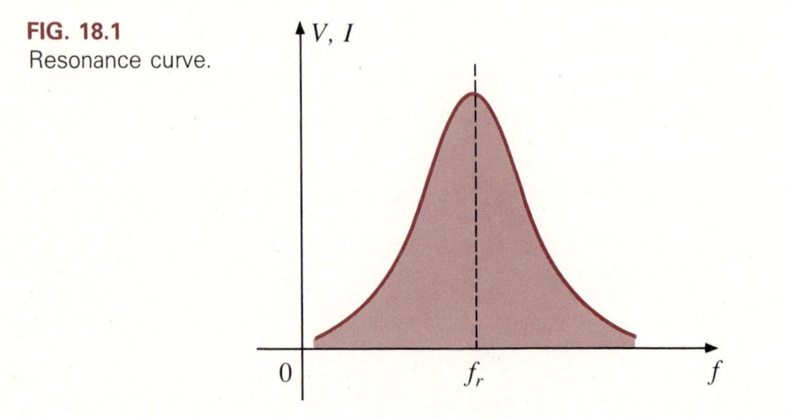

FIG. 18.1
Resonance curve.

The concept of resonance is not limited to electrical or electronics systems. If mechanical impulses are applied to a mechanical system at the power frequency, the system enters a state of resonance in which sustained vibrations of very large amplitude develop. The frequency at which this occurs is called the *natural frequency* of the system. The classic example of this effect was the Tacoma Narrows Bridge built in 1940 over Puget Sound in Washington State. It had a suspended span of 2800 ft. Four months after the bridge was completed, a 42-mi/h pulsating gale set the bridge into oscillations at its natural frequency. The amplitude of the oscillations increased to the point where the main span broke up and fell into the water below. It has since been replaced by the new Tacoma Narrows Bridge, completed in 1950.

The resonant electrical circuit *must* have both inductance and capacitance. In addition, resistance will always be present due either to the lack of ideal elements or to the control offered on the shape of the resonance curve. When resonance occurs due to the application of the proper frequency (f_r), the energy absorbed at any instant by one reactive element is exactly equal to that released by another reactive element within the system. In other words, energy pulsates from one reactive element to the other. Therefore, once the system has reached a state of resonance, it requires no further reactive power, since it is self-sustaining. The total apparent power is then simply equal to the average power dissipated by the resistive elements. *The average power absorbed by the system is also a maximum at resonance,* just as the transfer of energy to the mechanical system described earlier was a maximum at the natural frequency.

There are two types of resonant circuits: *series* and *parallel.* Each is considered in some detail in this chapter.

SERIES RESONANCE

18.2 SERIES RESONANT CIRCUIT

The basic circuit configuration for the series resonant circuit appears in Fig. 18.2. The resistance R_l is the internal resistance of the coil. The resistance R_s is the source resistance and any other resistance added in series to affect the shape of the resonance curve.

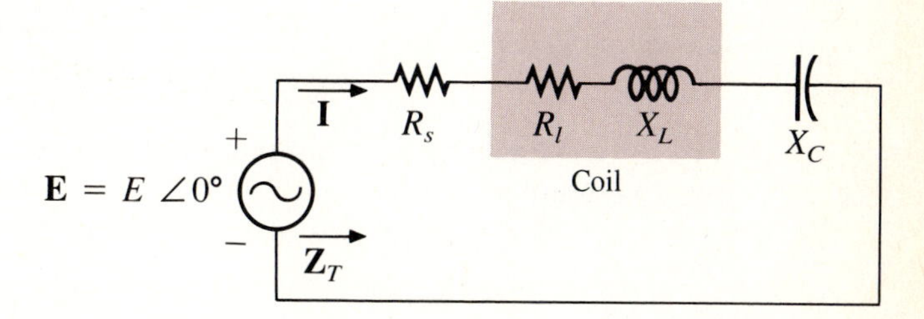

FIG. 18.2
Series resonant circuit.

Defining $R = R_s + R_l$, the magnitude of the total impedance at any frequency is determined by

$$\boxed{Z_T = \sqrt{R^2 + (X_L - X_C)^2}} \qquad (18.1)$$

Series resonance will occur when

$$\boxed{X_L = X_C} \qquad (18.2)$$

which, when substituted into Eq. 18.1, results in

$$Z_T = \sqrt{R^2 + (0)^2} = \sqrt{R^2}$$

and

$$Z_T = R \quad \text{(at resonance)} \qquad (18.3)$$

which turns out to be the minimum value of Z_T for any applied frequency.

Equation 18.2 can be used to determine the resonant frequency as follows:

$$X_L = X_C$$

$$\omega L = \frac{1}{\omega C}$$

$$\omega^2 = \frac{1}{LC}$$

$$\omega = \frac{1}{\sqrt{LC}}$$

but $\omega = 2\pi f$, so

$$f_s = \frac{1}{2\pi\sqrt{LC}} \qquad \begin{aligned} f &= \text{hertz (Hz)} \\ L &= \text{henries (H)} \\ C &= \text{farads (F)} \end{aligned} \qquad (18.4)$$

The magnitude of the current is a maximum at resonance, since Z_T is a minimum. That is,

$$I = \frac{E}{Z_T} = \frac{E}{R} \quad \text{(at resonance)} \qquad (18.5)$$

Since $\mathbf{Z}_T$ is resistive at resonance, both **E** and **I** are also *in phase* at resonance.

The magnitudes of the voltages $\mathbf{V}_L$ and $\mathbf{V}_C$ are determined by

$$V_L = IX_L \quad \text{and} \quad V_C = IX_C$$

but since $X_L = X_C$ at resonance,

$$V_L = V_C \quad \text{(at resonance)} \qquad (18.6)$$

Plotting the power curves of each element on the same set of axes (Fig. 18.3), we note that even though the total reactive power at any instant is equal to zero ($t = t'$), energy is still being absorbed and released by the inductor and capacitor at resonance.

A closer examination reveals that the energy absorbed by the inductor from time 0 to t_1 is the same as the energy being released by the capacitor from 0 to t_1. The reverse occurs from t_1 to t_2, and so on. Therefore, the total apparent power continues to be equal to the average power, even though the inductor

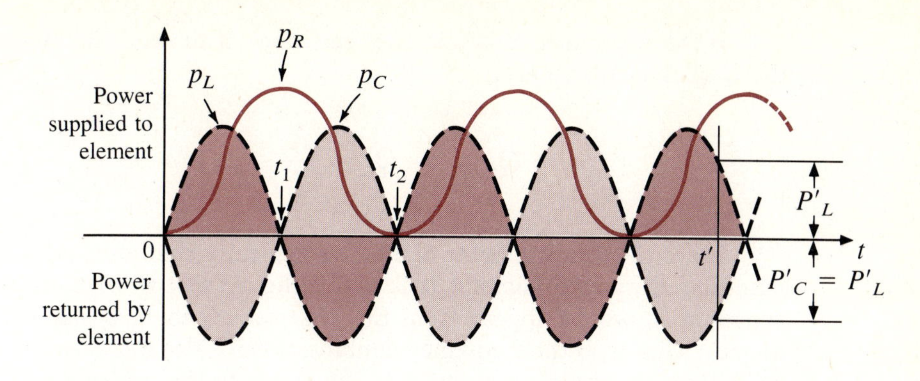

FIG. 18.3
Power curves at resonance for the series resonant circuit.

and capacitor are absorbing and releasing energy. This condition occurs only at resonance. The slightest change in frequency introduces a reactive component into the power triangle, which increases the apparent power of the system above the average power dissipation, and resonance no longer exists.

18.3 THE QUALITY FACTOR (Q)

The *quality factor* Q of a series resonant circuit is defined as the ratio of the reactive power of either the inductor or the capacitor to the average power of the resistor *at resonance;* that is,

$$\boxed{Q_s = \frac{\text{Reactive power}}{\text{Average power}}} \tag{18.7}$$

The quality factor is also an indication of how much energy is placed in storage (continual transfer from one reactive element to the other) as compared to that dissipated. The lower the level of dissipation for the same reactive power, the larger the factor Q_s and the more concentrated and intense the region of resonance.

Substituting for an inductive reactance in Eq. 18.7 at resonance gives us

$$Q_s = \frac{I^2 X_L}{I^2 R}$$

and

$$\boxed{Q_s = \frac{X_L}{R} = \frac{\omega_s L}{R}} \tag{18.8}$$

If the resistance R is just the resistance of the coil (R_l), we can speak of the Q of the coil, where

$$Q_{\text{coil}} = Q = \frac{X_L}{R_l} \qquad R = R_l \tag{18.9}$$

Since the quality factor of a coil is typically the information provided by manufacturers of inductors, it is given the symbol Q without an associated subscript. It would appear from Eq. 18.9 that Q increases linearly with frequency. That is, if the frequency doubles, then Q also increases by a factor of 2. This is approximately true for the low range to the midrange of frequencies, such as shown for the coil of Fig. 18.4. Unfortunately, however, as the fre-

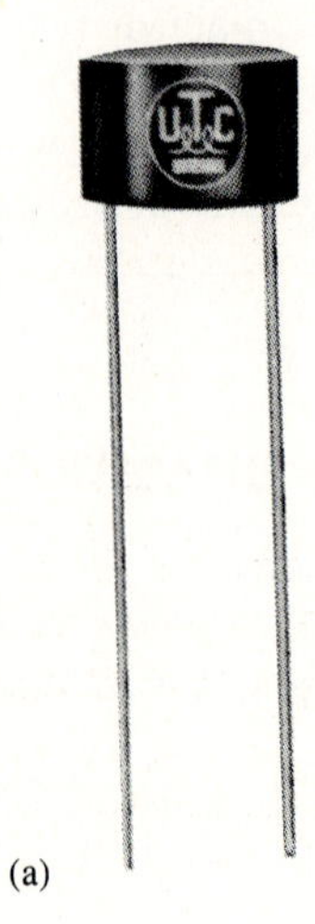

(a)

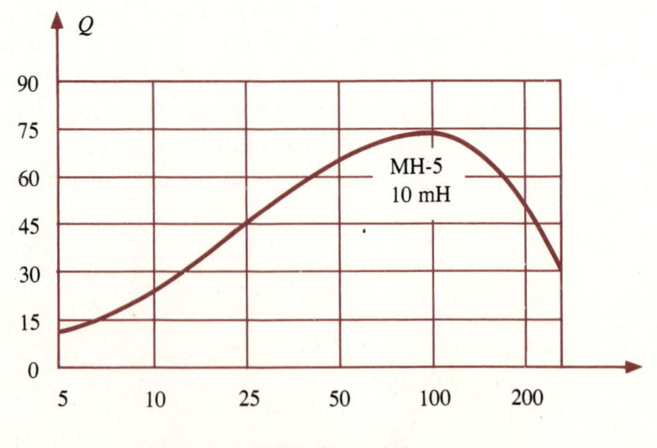

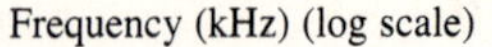

(b)

FIG. 18.4
Q vs. frequency for a TRW/UTC 10-mH coil. (Courtesy of United Transformer Corp.)

quency increases, the effective resistance of the coil also increases along with the capacitive effects between the windings, and the resulting Q decreases. For this reason, the Q of a coil must be specified at a particular frequency (usually at the maximum). For wide-frequency applications, a plot of Q versus frequency is often provided. The maximum Q for most commercially available coils approaches 100.

Applying the voltage divider rule to the circuit of Fig. 18.2 results in

$$V_L = \frac{X_L E}{Z_T} = \frac{X_L E}{R} \quad \text{(at resonance)}$$

and

$$V_L = Q_s E \quad \text{(at resonance)} \tag{18.10}$$

or

$$V_C = \frac{X_C E}{Z_T} = \frac{X_C E}{R}$$

and

$$V_C = Q_s E \quad \text{(at resonance)} \tag{18.11}$$

Since Q_s is usually greater than 1 for communication systems, the voltage across the capacitor or inductor of a series resonant circuit is usually greater than the input voltage. In fact, in many cases the Q_s is so high that careful design and handling (including adequate insulation) are mandatory with respect to the voltage across the capacitor and inductor.

In the circuit of Fig. 18.5, for example, which is in the state of resonance,

$$Q_s = \frac{X_L}{R} = \frac{480\ \Omega}{6\ \Omega} = 80$$

and

$$V_L = V_C = Q_s E = (80)(100\ \text{V}) = 8000\ \text{V}$$

which is certainly a potential to be handled with great care.

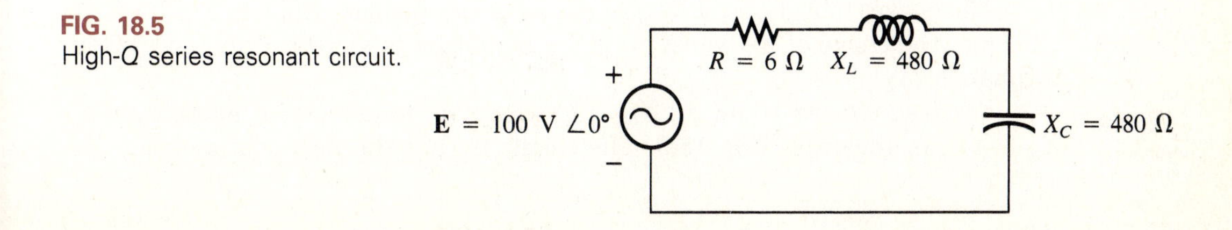

FIG. 18.5
High-Q series resonant circuit.

18.4 Z_T VERSUS FREQUENCY

The magnitude of the total impedance of the series *R-L-C* circuit of Fig. 18.2 at any frequency is given by

$$Z_T(f) = \sqrt{[R(f)]^2 + [X_L(f) - X_C(f)]^2} \tag{18.12}$$

where $R(f)$ means R as a *function of frequency;* $X_L(f)$ and $X_C(f)$ are interpreted similarly. However, we know that for the low to midrange of the frequency spectrum, R is not a function of frequency, as demonstrated by the plot of Fig. 18.6.

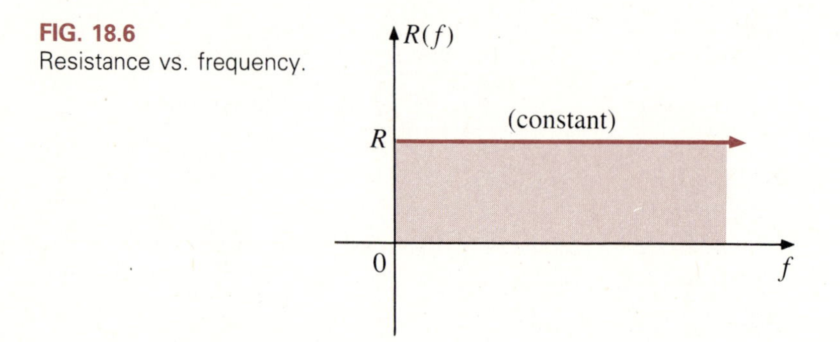

FIG. 18.6
Resistance vs. frequency.

For inductive elements, X_L increases as a straight line with a slope determined by inductance, as shown in Fig. 18.7. The larger the inductance, the steeper the curve and the more rapidly X_L increases with frequency.

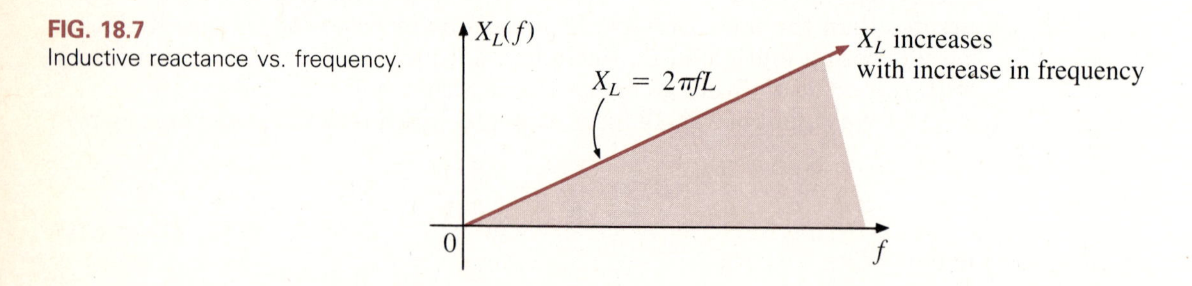

FIG. 18.7
Inductive reactance vs. frequency.

Since $X_C(f)$ in Eq. 18.12 is preceded by a negative sign, the curve of Fig. 18.8 is provided for $-X_C$. Note that it approaches zero at a fairly rapid rate as the frequency increases.

The curves of $X_L(f)$ and $X_C(f)$ are plotted on the same axis in Fig. 18.9. In addition, the curve of $X_T = (X_L - X_C)$ is plotted to show that $X_T = 0\ \Omega$ at resonance ($X_L = X_C$).

For frequencies below f_s, $X_C > X_L$, and for frequencies greater than f_s, $X_L > X_C$, as shown in Fig. 18.9. The result is that the network is capacitive

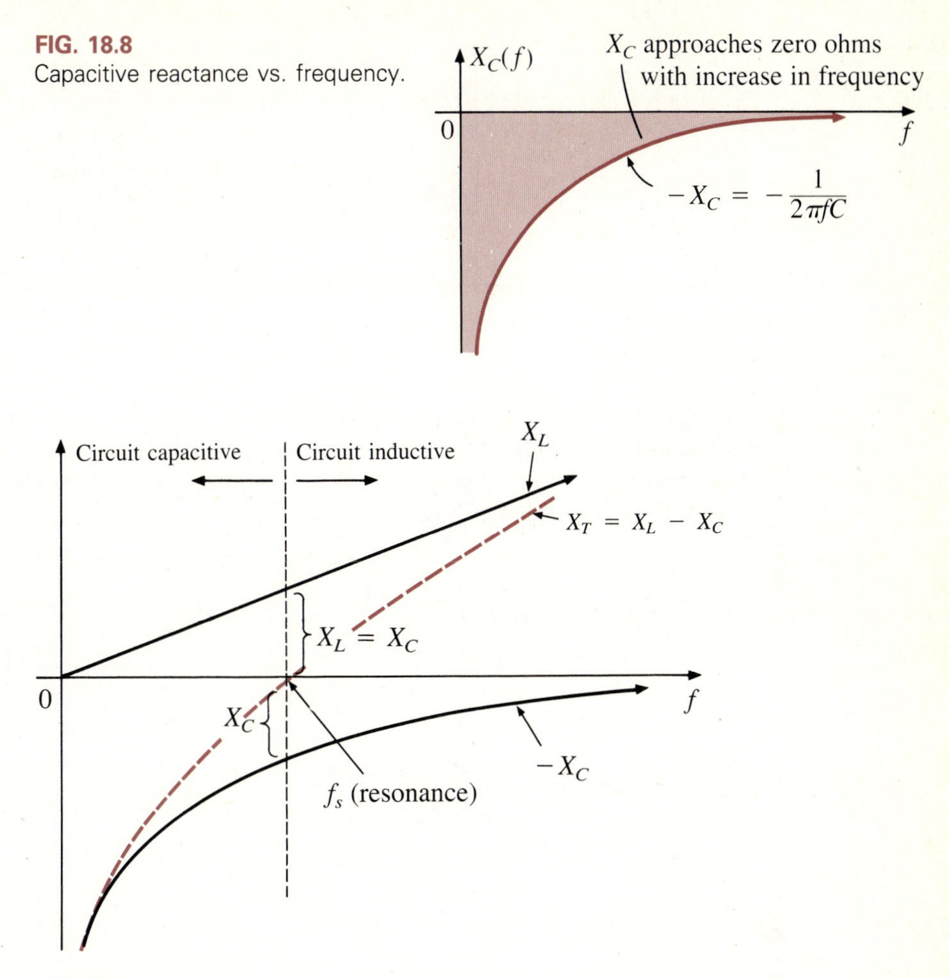

FIG. 18.8
Capacitive reactance vs. frequency.

FIG. 18.9
The total reactance $X_T = X_L - X_C$ for a series resonant circuit (vs. frequency).

(leading F_p) below f_s and inductive (lagging F_p) above f_s. Now that we have $X_T(f)$, we can modify Eq. 18.12 as follows:

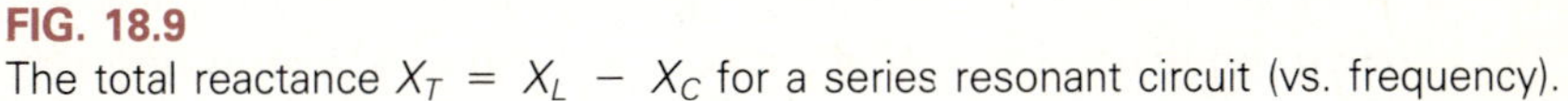

$$Z_T = \sqrt{[R(f)]^2 + [X_T(f)]^2} \qquad \textbf{(18.13)}$$

At $f = f_s$, $X_T(f) = 0\ \Omega$ and $Z_T = R(f) = R$. At very low frequencies, $X_T(f) \cong X_C(f)$ and $X_C(f) > R$, with $Z_T \cong X_C(f)$ (the minus sign is lost when $X_C(f)$ is squared, $[X_C(f)]^2$), resulting in a high level of impedance for Z_T. For very high frequencies, $X_T(f) \cong X_L(f)$ and $X_L(f) > R$, with $Z_T(f) \cong X_L(f)$, resulting in another high level of impedance. The result is the impedance curve of Fig. 18.10 for the wide range of frequencies.

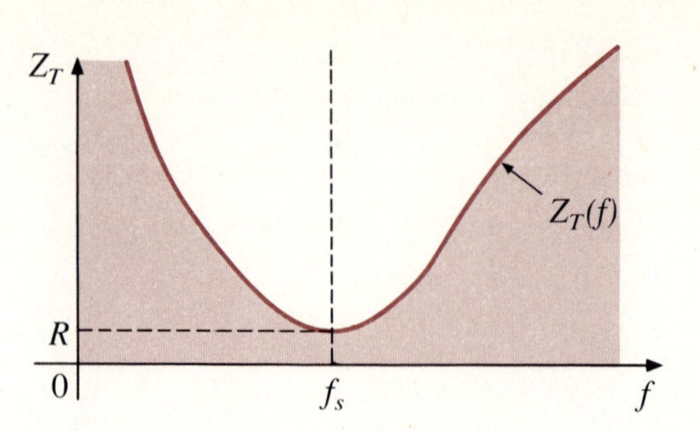

FIG. 18.10
Z_T vs. frequency for the series resonant circuit.

As just noted, the circuit is almost purely capacitive at low frequencies, and the current leads the applied voltage by 90°. At very high frequencies, the circuit is almost purely inductive, and the current lags the voltage by 90°. The applied voltage and resulting current are in phase only at resonance, as indicated by the phase plot of Fig. 18.11.

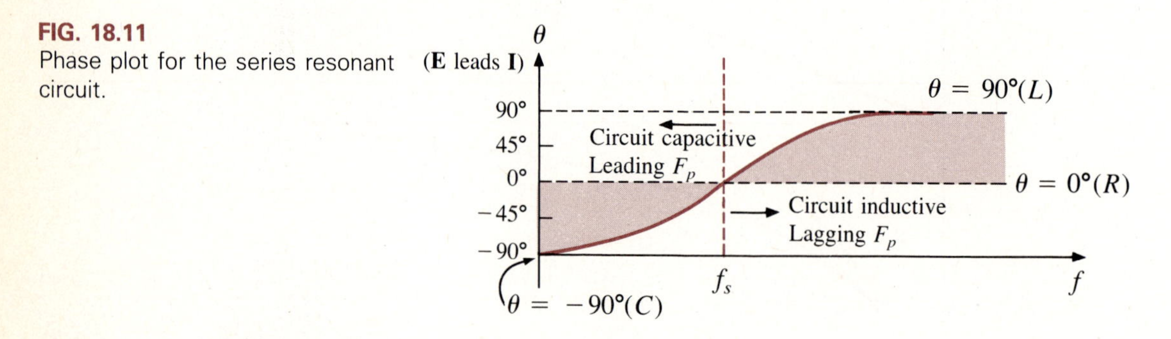

FIG. 18.11
Phase plot for the series resonant circuit.

18.5 SELECTIVITY

If we now plot the magnitude of the current $I = E/Z_T$ versus frequency for a *fixed* applied voltage E, we obtain the curve shown in Fig. 18.12, which rises

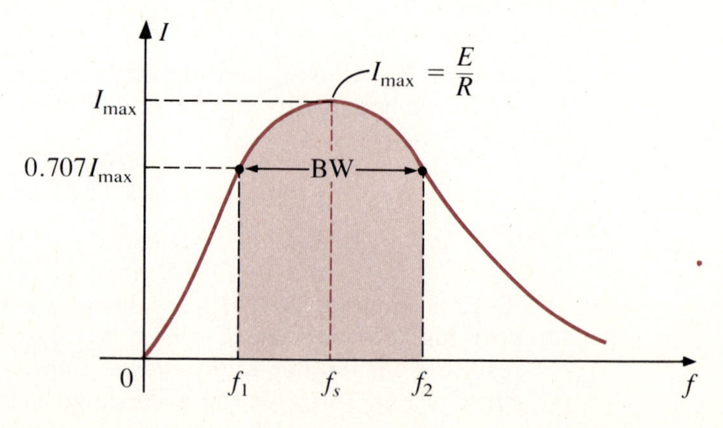

FIG. 18.12
I vs. frequency for the series resonant circuit.

from zero to a maximum value of E/R (where Z_T is a minimum) and then drops toward zero (as Z_T increases) at a slower rate than it rose to its peak value. The curve is actually the inverse of the impedance-versus-frequency curve. Since the Z_T curve is not absolutely symmetrical about the resonant frequency, the curve of the current versus frequency has the same property.

There is a definite range of frequencies at which the current is near its maximum value and the impedance is at a minimum. Those frequencies corresponding to 0.707 of the maximum current are called the *band frequencies, cutoff frequencies,* or *half-power frequencies.* They are indicated by f_1 and f_2 in Fig. 18.12. The range of frequencies between the two is referred to as the *bandwidth* (abbreviated BW) of the resonant circuit.

Half-power frequencies (denoted by HPF) are those frequencies at which the power delivered is one-half that delivered at the resonant frequency; that is,

$$P_{\text{HPF}} = \frac{1}{2}P_{\text{max}} \tag{18.14}$$

The preceding condition is derived using the fact that

$$P_{\text{max}} = I_{\text{max}}^2 R$$

and

$$P_{\text{HPF}} = I^2R = (0.707I_{\text{max}})^2R = 0.5I_{\text{max}}^2R = \frac{1}{2}P_{\text{max}}$$

Since the resonant circuit is adjusted to select a band of frequencies, the curve of Fig. 18.12 is called the *selectivity curve.* The term is derived from the fact that one must be selective in choosing the frequency to ensure that it is in the bandwidth. The smaller the bandwidth, the higher the selectivity. The shape of the curve, as shown in Fig. 18.13, depends on each element of the series R-L-C circuit. If the resistance is made smaller with a fixed inductance and capacitance, the bandwidth decreases and the selectivity increases. Similarly, if the ratio L/C increases with fixed resistance, the bandwidth again decreases with an increase in selectivity.

In terms of Q_s, if R is larger for the same X_L, then Q_s is less, as determined by the equation $Q_s = \omega_s L/R$. *A small Q_s, therefore, is associated with a resonant curve with a large bandwidth and a small selectivity, whereas a large Q_s indicates the opposite.*

For circuits where $Q_s \geq 10$, a widely accepted approximation is that the resonant frequency bisects the bandwidth and that the resonant curve is symmetrical about the resonant frequency. These conditions are shown in Fig. 18.14, indicating that the cutoff frequencies are then equidistant from the resonant frequency.

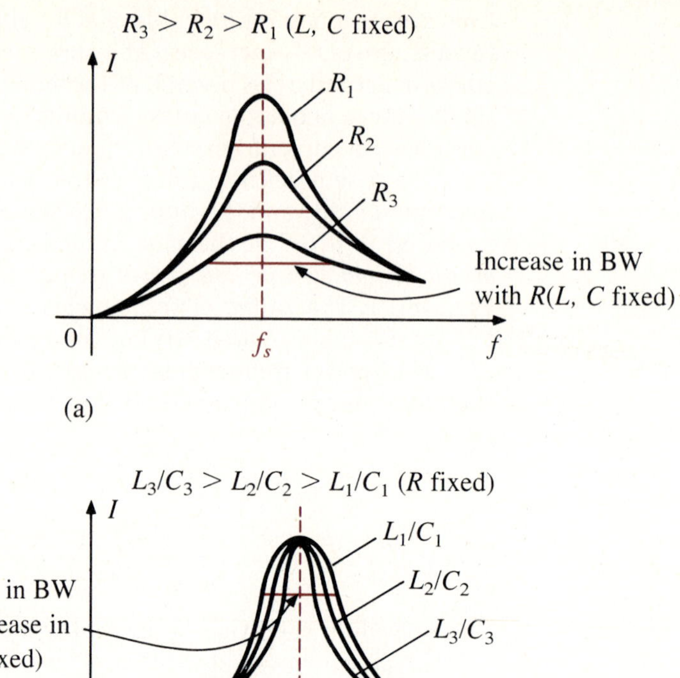

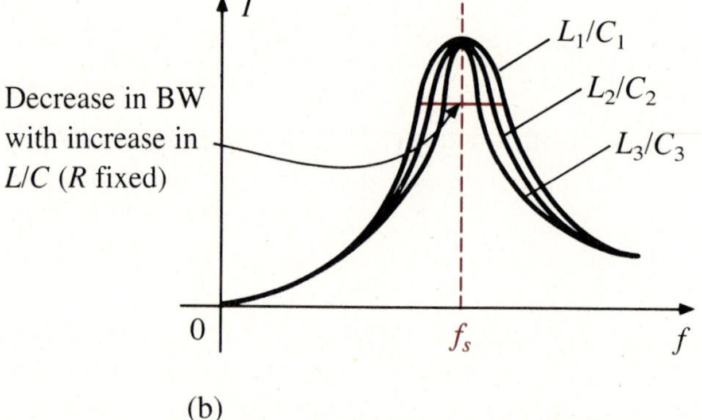

FIG. 18.13
Effect of R, L, and C on the selectivity curve for the series resonant circuit.

For $Q_s < 10$, the approximation of Fig. 18.14 can be a poor one, requiring some way to calculate f_1 and f_2. Using the general equation for Z_T (Eq. 18.1) and the fact that f_1 and f_2 are defined by 0.707 of the peak value, the following (rather complex) equations can be obtained:

$$f_1 = \frac{1}{2\pi}\left[-\frac{R}{2L} + \frac{1}{2}\sqrt{\left(\frac{R}{L}\right)^2 + \frac{4}{LC}}\right] \quad \text{(Hz)} \tag{18.15}$$

$$f_2 = \frac{1}{2\pi}\left[\frac{R}{2L} + \frac{1}{2}\sqrt{\left(\frac{R}{L}\right)^2 + \frac{4}{LC}}\right] \quad \text{(Hz)} \tag{18.16}$$

Note, however, that the equations are quite similar in format with whole terms equal in magnitude.

The *bandwidth* (BW) of the curve of Fig. 18.12 is defined by

$$\text{BW} = f_2 - f_1 \quad \text{(Hz)} \tag{18.17}$$

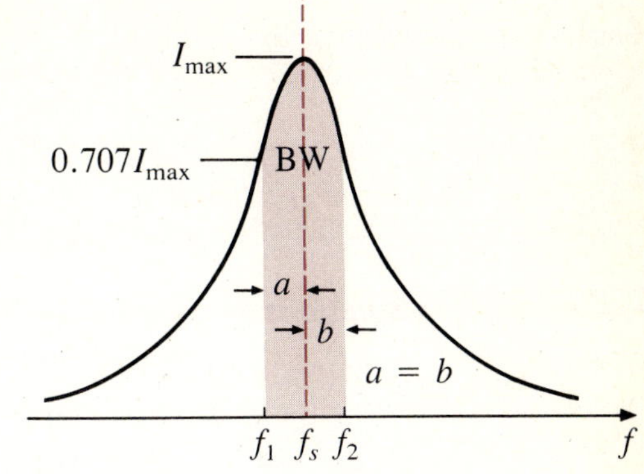

FIG. 18.14
Approximate series resonance curve for $Q_s \geq 10$.

In terms of the network parameters,

$$\text{BW} = \frac{R}{2\pi L} \quad \text{(Hz)} \tag{18.18}$$

or

$$\text{BW} = \frac{f_s}{Q_s} \quad \text{(Hz)} \tag{18.19}$$

which is a very convenient form, since it relates the bandwidth to the quality factor of the circuit. It verifies the fact that as the quality factor Q_s increases, the bandwidth decreases in width, requiring a high selectivity of frequencies to be within the bandwidth.

In a slightly different but useful form,

$$\frac{f_2 - f_1}{f_s} = \frac{1}{Q_s} \tag{18.20}$$

with the ratio $(f_2 - f_1)/f_s$ sometimes called the *fractional bandwidth.*

18.6 EXAMPLES (SERIES RESONANCE)

EXAMPLE 18.1 For the series resonant circuit of Fig. 18.15, determine

a. Z_T at resonance
b. I at resonance
c. V_R, V_L, and V_C at resonance
d. Q_s
e. BW and plot the resonance curve if $f_s = 5000$ Hz
f. P_{HPF}

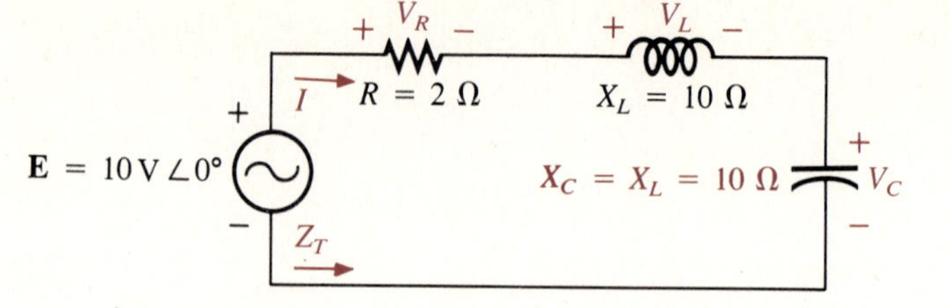

FIG. 18.15
Series resonant circuit for Example 18.1.

Solutions

a. $Z_T = \sqrt{R^2 + (X_L - X_C)^2} = \sqrt{(2\,\Omega)^2 + (10\,\Omega - 10\,\Omega)^2}$
$= \sqrt{4 + 0} = \mathbf{2\,\Omega}$

b. $I = \dfrac{E}{Z_T} = \dfrac{E}{R} = \dfrac{10\text{ V}}{2\,\Omega} = \mathbf{5\text{ A}}$

c. $V_R = IR = (5\text{ A})(2\,\Omega) = \mathbf{10\text{ V}} = E$
$V_L = IX_L = (5\text{ A})(10\,\Omega) = \mathbf{50\text{ V}}$
$V_C = IX_C = (5\text{ A})(10\,\Omega) = \mathbf{50\text{ V}} = V_L$

d. $Q_s = \dfrac{X_L}{R} = \dfrac{10\,\Omega}{2\,\Omega} = \mathbf{5}$

e. $\text{BW} = f_2 - f_1 = \dfrac{f_s}{Q_s} = \dfrac{5000\text{ Hz}}{5} = \mathbf{1\text{ kHz}}$

See Fig. 18.16.

f. $P_{\text{HPF}} = \dfrac{1}{2}P_{\max} = \dfrac{1}{2}(I^2R) = \dfrac{1}{2}[(5\text{ A})^2(2\,\Omega)]$
$= \dfrac{1}{2}(50\text{ W}) = \mathbf{25\text{ W}}$

FIG. 18.16
Resonance curve for Example 18.1.

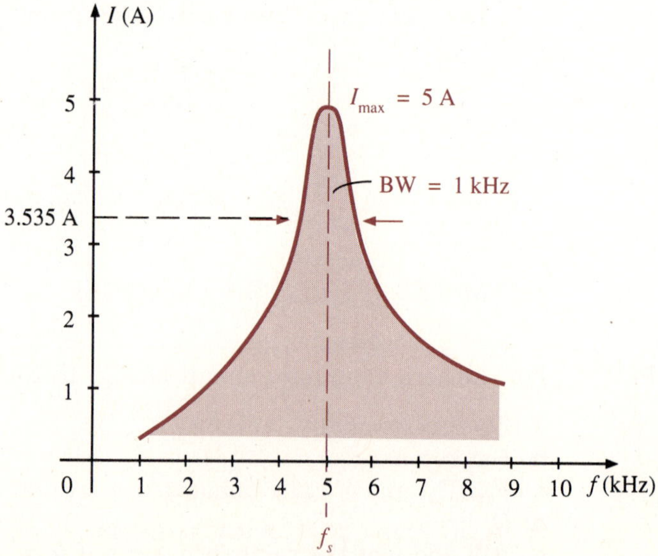

EXAMPLE 18.2

The bandwidth of a series resonant circuit is 400 Hz. If the resonant frequency is 4000 Hz, what is the value of Q_s? If $R = 10\ \Omega$, what is the value of X_L at resonance? Find the inductance L and capacitance C of the circuit.

Solution

$$\text{BW} = \frac{f_s}{Q_s} \quad \text{or} \quad Q_s = \frac{f_s}{\text{BW}} = \frac{4000\text{ Hz}}{400\text{ Hz}} = \mathbf{10}$$

$$Q_s = \frac{X_L}{R} \quad \text{or} \quad X_L = Q_s R = (10)(10\ \Omega) = \mathbf{100\ \Omega}$$

$$X_L = 2\pi f_s L \quad \text{or} \quad L = \frac{X_L}{2\pi f_s} = \frac{100\ \Omega}{(6.28)(4000\text{ Hz})} = \mathbf{3.98\ mH}$$

$$X_C = \frac{1}{2\pi f_s C} \quad \text{or} \quad C = \frac{1}{2\pi f_s X_C} = \frac{1}{(6.28)(4000\text{ Hz})(100\ \Omega)} = \mathbf{0.398\ \mu F}$$

EXAMPLE 18.3

A series R-L-C circuit has a series resonant frequency of 12 kHz. If $R = 5\ \Omega$ and X_L is 300 Ω at resonance, determine

a. BW
b. f_1 and f_2

Solutions

a. $$Q_s = \frac{X_L}{R} = \frac{300\ \Omega}{5\ \Omega} = 60$$

$$\text{BW} = \frac{f_s}{Q_s} = \frac{12{,}000\text{ Hz}}{60} = \mathbf{200\ Hz}$$

Since $Q_s \geq 10$, the bandwidth is bisected by f_s. Therefore,

$$f_1 = f_s - \frac{\text{BW}}{2} = 12{,}000\text{ Hz} - \frac{200\text{ Hz}}{2}$$
$$= 12{,}000\text{ Hz} - 100\text{ Hz} = \mathbf{11{,}900\ Hz}$$

with

$$f_2 = f_s + \frac{\text{BW}}{2} = 12{,}000\text{ Hz} + 100\text{ Hz}$$
$$= \mathbf{12{,}100\ Hz}$$

EXAMPLE 18.4

a. Determine Q_s and the bandwidth for the response curve of Fig. 18.17.
b. For $C = 0.1\ \mu\text{F}$, determine L and R for the series resonant circuit.
c. Determine I_{max} and the magnitude of the applied voltage.

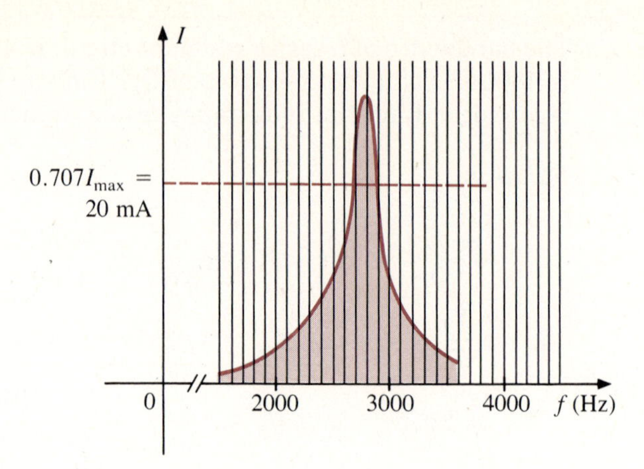

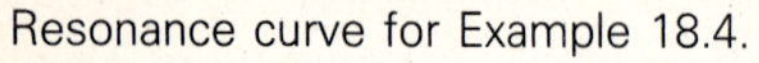
FIG. 18.17
Resonance curve for Example 18.4.

Solutions

a. The resonant frequency is 2800 Hz. At 0.707 times the peak value, BW = **200 Hz,** and

$$Q_s = \frac{f_s}{\text{BW}} = \frac{2800\text{ Hz}}{200\text{ Hz}} = \mathbf{14}$$

b. $f_s = \dfrac{1}{2\pi\sqrt{LC}}$, so

$$L = \frac{1}{4\pi^2 f_s^2 C}$$

$$= \frac{1}{4\pi^2(2.8 \times 10^3\text{ Hz})^2(0.1 \times 10^{-6}\text{ F})}$$

$$= \frac{1}{30.951} = \mathbf{32.31\text{ mH}}$$

Calculator

(4) (×) (2nd F) (π) (x²) (×) (2) (·) (8) (EXP) (3) (x²) (×) (·) (1) (EXP) (+/−) (6) (=) (1/x)

with a display of 0.032.

$$Q_s = \frac{X_L}{R}$$

or

$$R = \frac{X_L}{Q_s} = \frac{2\pi(2.8 \times 10^3\text{ Hz})(32.31 \times 10^{-3}\text{ H})}{14}$$

$$= \mathbf{40.60\ \Omega}$$

Calculator

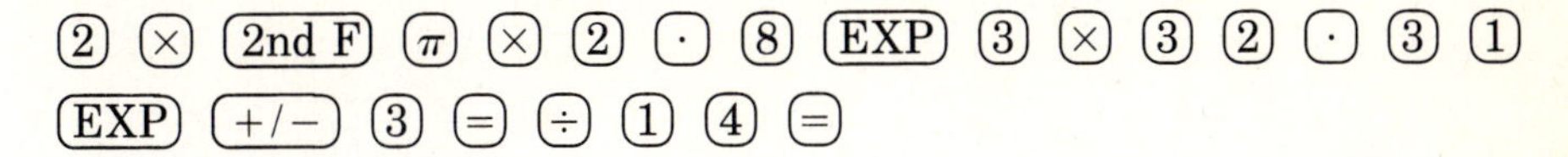

with a display of 40.602.

c. $I_{\text{max}} = \dfrac{20 \text{ mA}}{0.707} = \mathbf{28.289 \text{ mA}}$

$$E = I_{\text{max}} Z_T = I_{\text{max}} R$$
$$= (28.289 \text{ mA})(40.602 \ \Omega)$$
$$E = \mathbf{1.149 \text{ V}}$$

PARALLEL RESONANCE

18.7 PARALLEL RESONANT CIRCUIT

A parallel resonant circuit has the basic configuration of Fig. 18.18. This circuit is often called a *tank circuit,* due to the storage of energy by the inductor and capacitor. A transfer of energy similar to that discussed for a series circuit also occurs in a parallel resonant circuit. In the ideal case (no radiation losses, and so on), the capacitor absorbs energy during one half-cycle of the power curves at the same rate at which it is released by the inductor. During the next half-cycle of the power curves, the inductor absorbs energy at the same rate at which the capacitor releases it. The total reactive power at resonance is therefore 0, and the total power factor is 1.

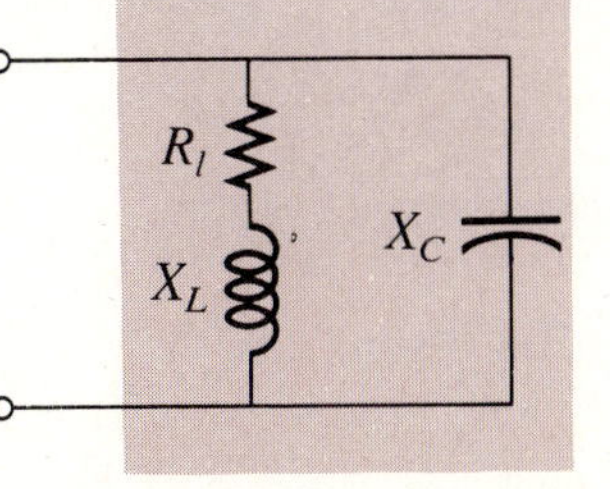

FIG. 18.18
Parallel resonant circuit.

Since tank circuits are frequently used with devices such as a transistor, which is essentially a constant-current source device, a current source is used to supply the input to the parallel resonant circuits in the following analysis, as shown in Fig. 18.19. In addition, our analysis of parallel resonant circuits will be well served if we first replace the series R-L branch by an equivalent parallel combination using a technique introduced in Chapter 15, as shown in Fig. 18.20.

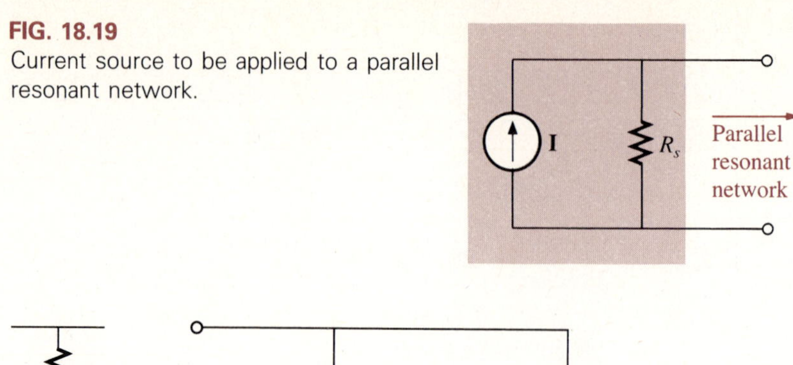

FIG. 18.19
Current source to be applied to a parallel resonant network.

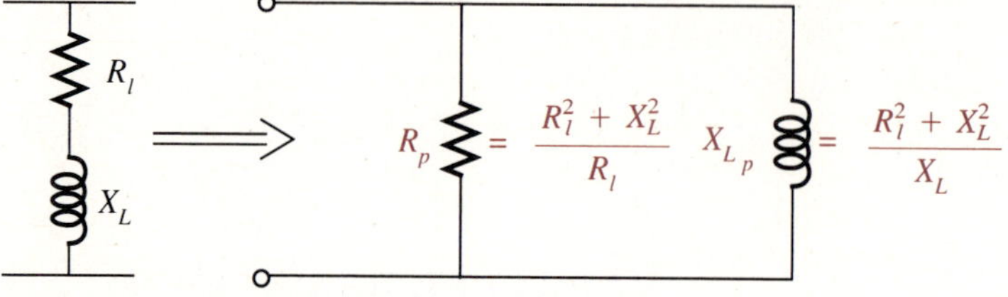

FIG. 18.20
Establishing the parallel *R-L* equivalent for the series *R-L* branch of the inductor.

The complete parallel resonant circuit with an applied current source then appears as shown in Fig. 18.21.

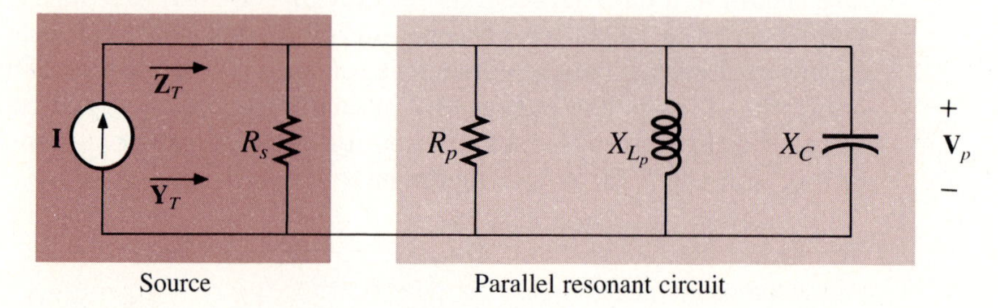

FIG. 18.21
Complete parallel resonant system.

For parallel resonance,

$$\boxed{X_{L_p} = X_C} \tag{18.21}$$

which is quite similar to the defining equation of series resonance. Substituting:

$$\boxed{\frac{R_l^2 + X_L^2}{X_L} = X_C} \tag{18.22}$$

Solving for the frequency at which Eq. 18.22 is satisfied results in the parallel resonant frequency:

$$f_p = \frac{1}{2\pi\sqrt{LC}}\sqrt{1 - \frac{R_l^2 C}{L}} \tag{18.23}$$

or, since $f_s = 1/2\pi\sqrt{LC}$ (for series resonance),

$$f_p = f_s\sqrt{1 - \frac{R_l^2 C}{L}} \tag{18.24}$$

revealing that the equation for f_p and f_s differ only by the square root factor of Eq. 18.24.

18.8 SELECTIVITY CURVE FOR PARALLEL RESONANT CIRCUITS

The total admittance of the parallel network of Fig. 18.21 is

$$\mathbf{Y}_T = \mathbf{Y}_1 + \mathbf{Y}_2 + \mathbf{Y}_3 + \mathbf{Y}_4 = \frac{1}{R_s \angle 0^\circ} + \frac{1}{R_p \angle 0^\circ} + \frac{1}{X_{L_p} \angle 90^\circ} + \frac{1}{X_C \angle -90^\circ}$$

At resonance, $X_{L_p} = X_C$ and the vector addition of the last two terms is zero.

The result is that

$$Y_T = \frac{1}{R_s} + \frac{1}{R_p} = \frac{R_p + R_s}{R_s R_p}$$

and

$$Z_T = \frac{1}{Y_T} = \frac{1}{\dfrac{R_P + R_S}{R_S R_P}}$$

with

$$Z_{T_p} = \frac{R_s R_p}{R_s + R_p} \quad \text{(at resonance)} \tag{18.25}$$

providing an equation for the total impedance of a parallel resonant circuit at resonance. Note that it is again resistive and the applied current and voltage are *in phase* at resonance.

At low frequencies, X_{L_p} is very small, resulting in a shorting effect across the parallel branches and reducing Z_T to a very small level. At high frequencies, X_C has the same effect, resulting in the Z_T-versus-frequency curve of Fig.

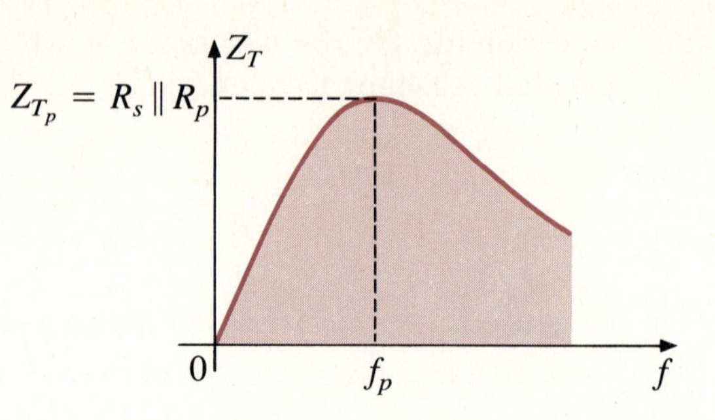

FIG. 18.22
Z_T vs. frequency for the parallel resonant circuit.

18.22. In particular, note that the impedance is now a maximum at f_p and a minimum at the high and low ends—the reverse of the series resonant circuit.

Since the current I is constant (current source) for any value of Z_T, the voltage across the parallel circuit has the same shape as the total impedance Z_T, as shown in Fig. 18.23.

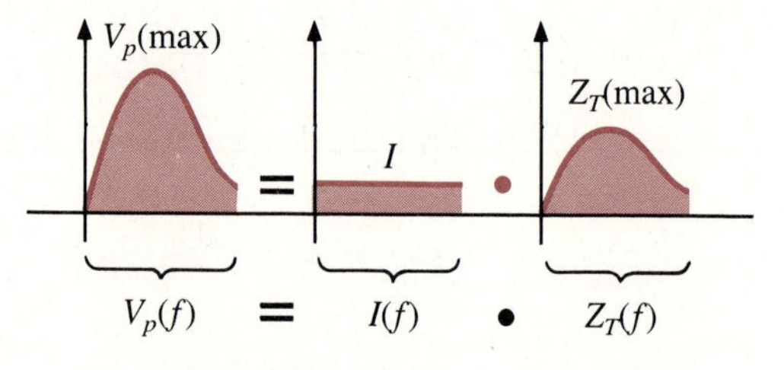

FIG. 18.23
Demonstrating the source of V_p vs. frequency resonant curve.

For the parallel circuit, the resonance curve of interest is that of the voltage V_C across the capacitor. The reason for this interest in V_C derives from electronic considerations that often place the capacitor at the input to another stage of a network.

Since the voltage across parallel elements is the same,

$$V_C = V_p = IZ_T \tag{18.26}$$

The resonant value of V_C is therefore determined by the value of Z_{T_p} and the magnitude of the current source I.

The quality factor of the parallel resonant circuit is also determined by the ratio of the reactive power to the real power and equals

$$Q_p = \frac{R_s \| R_p}{X_{L_p}} \quad \text{(at resonance)} \tag{18.27}$$

For situations where $R_s >> R_p$,

$$Q_p = \frac{X_L}{R_l} = Q \qquad R_s >> R_p \tag{18.28}$$

and the Q of the coil defines the quality factor of the parallel resonant circuit.

The bandwidth of the parallel resonant circuit is related to the resonant frequency and Q_p in the same manner as for series resonant circuits. That is,

$$\text{BW} = f_2 - f_1 = \frac{f_p}{Q_p} \tag{18.29}$$

The effect of R_l, L, and C on the shape of the parallel resonance curve, as shown in Fig. 18.24 for the input impedance, is quite similar to their effect on the series resonance curve. Whether or not R_l is zero, the parallel resonant circuit frequently appears in a network schematic, as shown in Fig. 18.24.

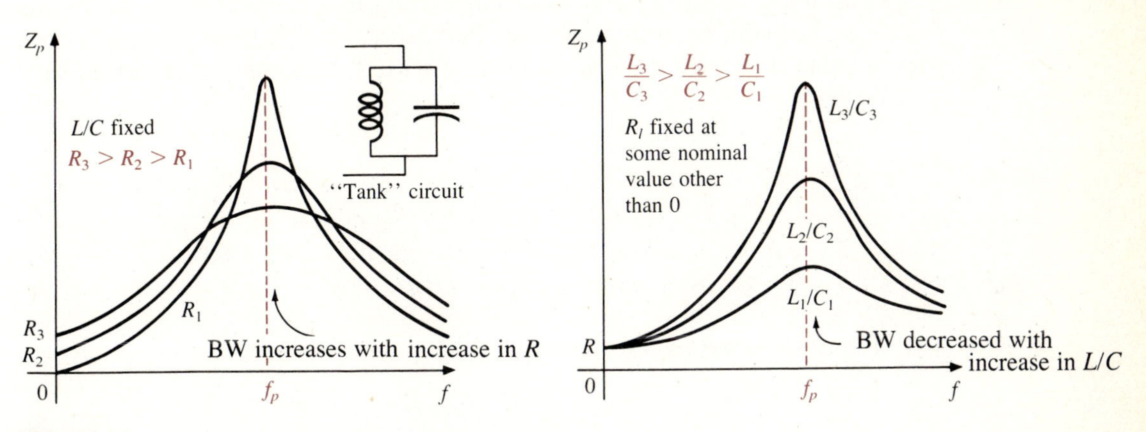

FIG. 18.24
Effect of R, L, and C on the parallel resonance curve.

At low frequencies, the capacitive reactance is quite high, and the inductive reactance is low. Since the elements are in parallel, the total impedance at low frequencies is, therefore, inductive. At high frequencies, the reverse is true, and the network is capacitive. At resonance, the network appears resistive. The facts lead to the phase plot of Fig. 18.25. Note that the plot is the inverse of that appearing for the series resonant circuit, in that at low frequencies the series resonant circuit was capacitive, and at high frequencies it was inductive.

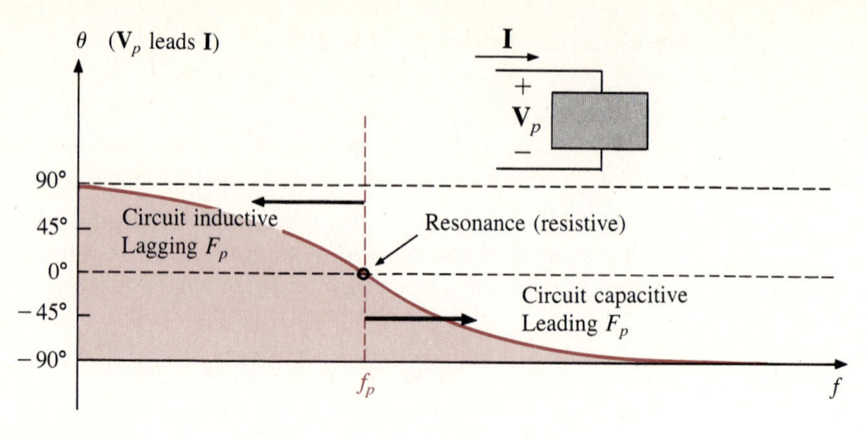

FIG. 18.25
Phase plot for the parallel resonant circuit.

18.9 EFFECT OF $Q \geq 10$

The condition $Q = X_L/R_l \geq 10$ is a common one, one that has some interesting effects on the equations just derived. The resulting equations provide a good approximation for the desired quantities with a minimum of mathematical complexity.

$\mathbf{Z}_{T_p}$

$$R_p = \frac{R_l^2 + X_L^2}{R_l} = R_l + \frac{X_L^2}{R_l}\left(\frac{R_l}{R_l}\right) = R_l + \frac{X_L^2}{R_l^2}R_l$$
$$= R_l + Q^2R_l = (1 + Q^2)R_l$$

For $Q \geq 10$, $1 + Q^2 \cong Q^2$, and

$$\boxed{R_p \cong Q^2R_l} \qquad (18.30)$$

with

$$\boxed{Z_{T_p} = R_s \| R_p \cong R_s \| Q^2R_l} \qquad \text{(at resonance)} \qquad (18.31)$$

Of course, if $R_s >> R_p$ or R_s is not included,

$$\boxed{Z_{T_p} \cong Q^2R_l} \qquad (\text{at resonance, } R_s >> R_p) \qquad (18.32)$$

an equation that frequently can be applied with excellent results.

$\mathbf{X}_{L_p}$

$$X_{L_p} = \frac{R_l^2 + X_L^2}{X_L} = \frac{R_l^2(X_L)}{X_L(X_L)} + X_L = \frac{X_L}{Q^2} + X_L$$

For $Q \geq 10$,

$$\boxed{X_{L_p} \cong X_L} \tag{18.33}$$

Resonance Condition, $X_{L_p} = X_C$ Since $X_{L_p} \cong X_L$,

$$\boxed{X_L = X_C} \tag{18.34}$$

and

$$\boxed{f_p = \frac{1}{2\pi\sqrt{LC}}} \tag{18.35}$$

as derived for series resonance.

Applying the approximations just derived to the network of Fig. 18.21 will result in the approximate equivalent circuit of Fig. 18.26.

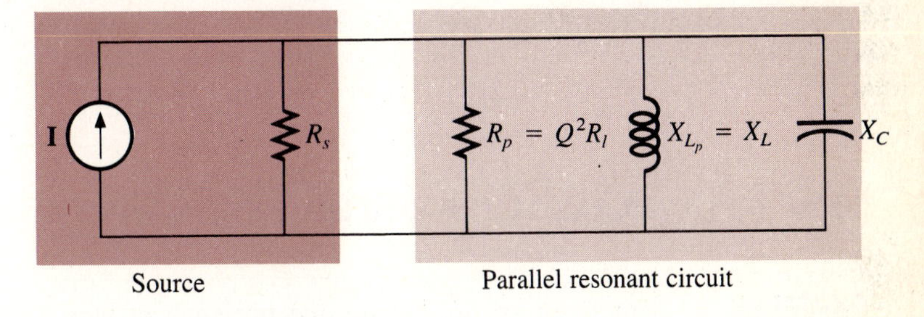

FIG. 18.26
Approximate equivalent circuit for $Q \geq 10$.

Q_p For $R_s \| R_p \cong R_p$ and $Q \geq 10$,

$$Q_p = \frac{R}{X_{L_p}} = \frac{R_p}{X_L} = \frac{Q^2R_l}{X_L} = \frac{Q^2}{X_L/R_l} = \frac{Q^2}{Q}$$

and

$$\boxed{Q_p = Q} \tag{18.36}$$

I_L, I_C You will recall that for the series resonant circuit, $V_L = V_C = QE$ at the resonant condition. For parallel resonance, if we assume $R_s >> R_p$ and $Q \geq 10$, the following relationships can be derived:

$$\boxed{I_L \cong QI} \tag{18.37}$$

and

$$\boxed{I_C \cong QI} \tag{18.38}$$

where I is the magnitude of the source current.

Table 18.1 is included as a review of the effect of $Q \geq 10$.

TABLE 18.1
Parallel resonant circuit.

	Any Q	$Q \geq 10$	$R_s = \infty\ \Omega$ $(Q \geq 10)$
Resonance	$\dfrac{R_l^2 + X_L^2}{X_L} = X_C$	$X_L = X_C$	$X_L = X_C$
f_p	$\dfrac{1}{2\pi\sqrt{LC}}\sqrt{1 - \dfrac{R_l^2 C}{L}}$	$\dfrac{1}{2\pi\sqrt{LC}}$	$\dfrac{1}{2\pi\sqrt{LC}}$
Z_{T_p}	$R_s \| \dfrac{R_l^2 + X_L^2}{R_l}, R_s \| \dfrac{L}{R_l C}$	$R_s \| Q^2 R_l$	$Q^2 R_l$
Q_p	$\dfrac{R}{X_{L_p}}, \dfrac{R}{X_C}$ $(R = R_s \| R_p)$	$\dfrac{R}{\omega_p L}$	Q
BW	$\dfrac{f_p}{Q_p}$	$I_L = I_C = Q_p I_T$ $\dfrac{f_p}{Q_p}$	$I_L = I_C = Q I_T$ $\dfrac{f_p}{Q}$

18.10 EXAMPLES (PARALLEL RESONANCE)

EXAMPLE 18.5 For the parallel resonance network of Fig. 18.27, determine

a. f_p
b. X_L (at resonance)
c. Q
d. Q_p
e. BW
f. Z_T (at resonance)
g. Magnitude of the applied voltage

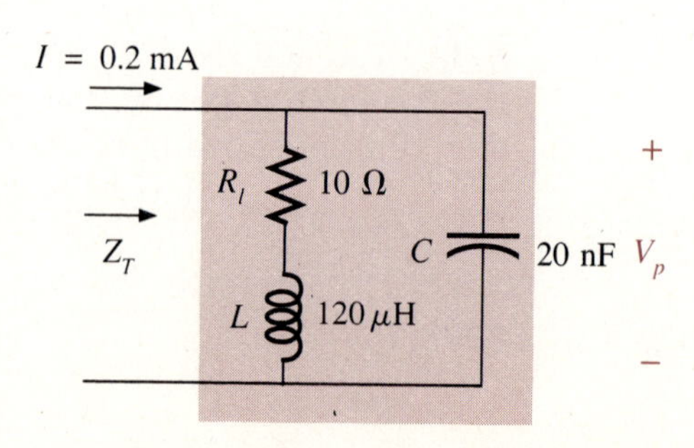

FIG. 18.27
Parallel resonant circuit for Example 18.5.

Solutions

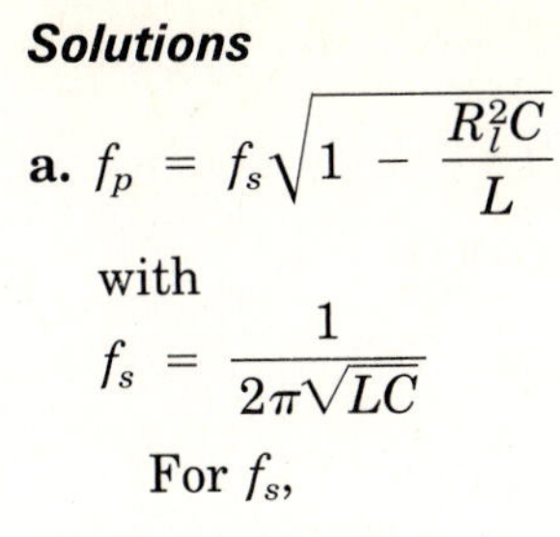

a. $f_p = f_s\sqrt{1 - \frac{R_l^2 C}{L}}$

with

$$f_s = \frac{1}{2\pi\sqrt{LC}}$$

For f_s,

Calculator (for f_s)

[1] [2] [0] [EXP] [+/−] [6] [×] [2] [0] [EXP] [+/−] [9] [=] [√] [×]
[2nd F] [π] [×] [2] [=] [1/x]

with a display of 102,734.074, so $f_s \cong$ 102.73 kHz.

For the factor $\sqrt{1 - R_l^2 C/L}$,

Calculator

[1] [0] [x^2] [×] [2] [0] [EXP] [+/−] [9] [=] [÷] [1] [2] [0] [EXP]
[+/−] [6] [=] [+/−] [+] [1] [=] [√]

with a display of 0.992, revealing that the resonant frequency of a parallel network using the same elements of a series resonant circuit will have a resonant frequency 99.2% of the series value.

That is,

$$f_p = (0.992)f_s = (0.992)(102.73 \text{ kHz})$$
$$= \mathbf{101.91\ kHz}$$

b. $X_L = 2\pi f_p L = 2\pi(101.91 \times 10^3 \text{ Hz})(120 \times 10^{-6} \text{ H})$

Calculator

[2] [×] [2nd F] [π] [×] [1] [0] [1] [.] [9] [1] [EXP] [3] [×] [1] [2] [0]
[EXP] [+/−] [6] [=]

with a display of 76.838, so $X_L \cong$ **76.84 Ω.**

c. $Q = \frac{X_L}{R_l} = \frac{76.84\ \Omega}{10\ \Omega} \cong \mathbf{7.68}$

d. Q_p: $R_s = \infty\ \Omega$, so $Q_p = Q =$ **7.68.**

e. $\text{BW} = \frac{f_p}{Q_p} = \frac{101.91 \text{ kHz}}{7.68} = \mathbf{13.27\ kHz}$

f. Since $Q < 10$,

$$Z_T = R_p = \frac{R_l^2 + X_L^2}{R_l} = \frac{(10\ \Omega)^2 + (76.84\ \Omega)^2}{10\ \Omega}$$

Calculator

[1] [0] [x^2] [+] [7] [6] [.] [8] [4] [x^2] [=] [÷] [1] [0] [=]

with a display of 600.439, and $Z_T \cong$ **0.6 kΩ.**

g. $E = IZ_T = IR_p$
$= (0.2\text{ mA})(0.6\text{ k}\Omega)$
$E = \mathbf{120\text{ mV}}$

EXAMPLE 18.6

For the network of Fig. 18.28,

a. Determine Q.
b. Determine R_p.
c. Calculate Z_{T_p}.
d. Find C at resonance.
e. Find Q_p.
f. Calculate BW.

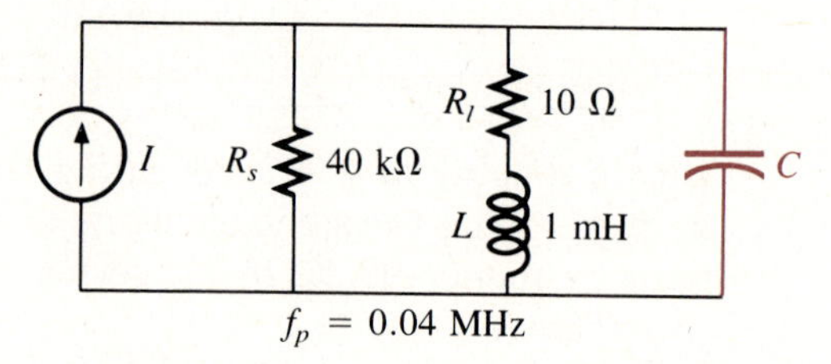

FIG. 18.28
Network for Example 18.6.

Solutions

a. $$Q = \frac{X_L}{R_l} = \frac{2\pi f_p L}{R_l} = \frac{(2\pi)(0.04 \times 10^6\text{ Hz})(1 \times 10^{-3}\text{ H})}{10\ \Omega} = \mathbf{25.13}$$

Calculator

[2] [×] [2nd F] [π] [×] [.] [0] [4] [EXP] [6] [×] [1] [EXP] [+/−] [3] [=] [÷] [1] [0] [=]

with a display of 25.132.

b. $Q \geq 10$. Therefore,

$$R_p = Q^2 R_l = (25.13)^2(10\ \Omega) = \mathbf{6.31\text{ k}\Omega}$$

c. $Z_T = R_s \| R_p = 40\text{ k}\Omega \| 6.31\text{ k}\Omega = \mathbf{5.45\text{ k}\Omega}$
d. $Q \geq 10$. Therefore,

$$f_p = \frac{1}{2\pi\sqrt{LC}}$$

and

$$C = \frac{1}{L(2\pi f)^2} = \frac{1}{(1 \times 10^{-3}\ \text{H})[2\pi(0.04 \times 10^6\ \text{Hz})]^2} = \mathbf{0.0158\ \mu F}$$

Calculator

(1) (EXP) (+/−) (3) (×) (( ) (2) (×) (2nd F) (π) (×) (·) (0) (4) (EXP) (6) ( )) (x^2) (=) (1/x) (F <−> E)

with a display of 1.583−08.

e. $Q \geq 10$. Therefore,

$$Q_p = \frac{R_s \| R_p}{X_L} = \frac{5.45\ \text{k}\Omega}{(2\pi)(0.04 \times 10^6\ \text{Hz})(1 \times 10^{-3}\ \text{Hz})}$$

Calculator

(2) (×) (2nd F) (π) (×) (·) (0) (4) (EXP) (6) (×) (1) (EXP) (+/−) (3) (=) (1/x) (×) (5) (·) (4) (5) (EXP) (3) (=)

with a display of 21.685, and $Q_p \cong$ **21.69.**

f. $\text{BW} = \frac{f_p}{Q_p} = \frac{0.04 \times 10^6\ \text{Hz}}{21.69} = \mathbf{1.84\ kHz}$

EXAMPLE 18.7 A parallel resonant network is connected directly to the output of a transistor amplifier in Fig. 18.29. The output impedance of the transistor is 50 kΩ and the magnitude of the source current is 1 mA, as shown in the equivalent circuit of Fig. 18.30.

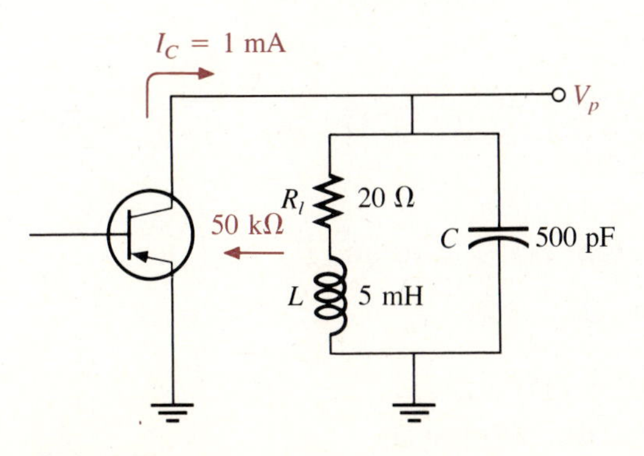

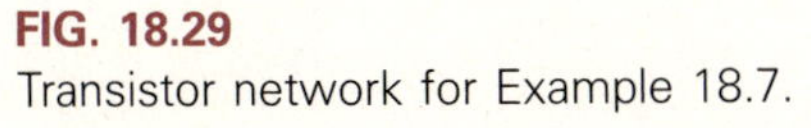

FIG. 18.29
Transistor network for Example 18.7.

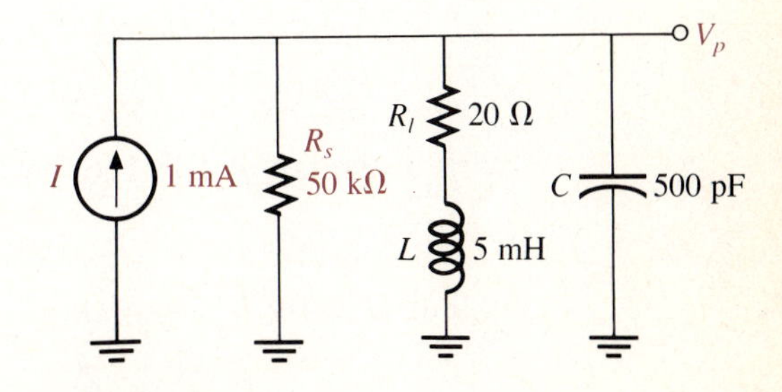

FIG. 18.30
Equivalent network for the configuration of Fig. 18.29.

Let us determine the impact of the output impedance of the current source by first finding the tuned network parameters ignoring the effects of R_s and then including its effects. Assume a high-Q coil throughout the calculations. For $R_s = \infty\ \Omega$:

a. Determine f_p.
b. Find Q and Q_p.
c. Determine Z_{T_p}.
d. Find BW.

For $R_s = 50\ \text{k}\Omega$:

e. Determine f_p.
f. Find Q.
g. Determine Z_{T_p}.
h. Calculate Q_p.
i. Find BW.

Solutions

a. Assuming $Q \geq 10$,

$$f_p = \frac{1}{2\pi\sqrt{LC}} = \frac{1}{2\pi\sqrt{(5 \times 10^{-3}\ \text{H})(500 \times 10^{-12}\ \text{F})}}$$

Calculator

(5) (EXP) (+/−) (3) (×) (5) (0) (0) (EXP) (+/−) (1) (2) (=) (√‾)
(×) (2nd F) (π) (×) (2) (=) (1/x)

with a display of 100,658.424, so $f_p \cong$ **100.66 kHz.**

b. $$Q = \frac{X_L}{R_l} = \frac{2\pi f_p L}{R_l} = \frac{2\pi(100.66 \times 10^3\ \text{Hz})(5 \times 10^{-3}\ \text{H})}{20\ \Omega}$$

Calculator

(2) (×) (2nd F) (π) (×) (1) (0) (0) (·) (6) (6) (EXP) (3) (×) (5) (EXP)
(+/−) (3) (=) (÷) (2) (0) (=)

with a display of 158.116, so $Q \cong$ **158.** For $Q \geq 10$ and $R_s = \infty\ \Omega$,

$$Q_p = Q = \mathbf{158}$$

c. For $Q \geq 10$, $R_s = \infty\ \Omega$:

$$Z_{T_p} = Q^2 R_l = (158)^2(20\ \Omega)$$

Calculator

(1) (5) (8) (x²) (×) (2) (0) (=)

with a display of 499,280, and $Z_{T_p} \cong$ **0.5 MΩ.**

d. $\text{BW} = \dfrac{f_p}{Q_p} = \dfrac{100.66 \times 10^3\ \text{Hz}}{158}$

Calculator

(1) (0) (0) (.) (6) (6) (EXP) (3) (÷) (1) (5) (8) (=)

with a display of 637.089, so BW ≅ **637.1 Hz.**

The results of ignoring the effects of R_s is a high-Q, narrow-bandwidth tuning network.

e. Assuming that f_p does not include a factor R_s results in the same solution as part a; that is, $f_p \cong$ **100.66 kHz.**

f. Q is also the same, since f_p is the same: $Q \cong$ **158.**

g. $Z_{T_p} = R_s \| R_p$
with $R_p = Q^2 R_l \cong 0.5\ \text{M}\Omega$ from part c.

Since Z_{T_p} (500 kΩ) is 10 times that of R_s, the magnitude of Z_{T_p} is now very close to R_s (rule of parallel resistive elements).

$$Z_{T_p} = \frac{(R_s)(Q^2R_l)}{R_s + Q^2R_l} = \frac{(50 \times 10^3\ \Omega)(0.5 \times 10^6\ \Omega)}{50 \times 10^3\ \Omega + 0.5 \times 10^6\ \Omega}$$

Calculator

(5) (0) (EXP) (3) (+) (.) (5) (EXP) (6) (=) (1/x) (×) (5) (0) (EXP) (3) (×) (.) (5) (EXP) (6) (=)

with a display of 45,454.545, so $Z_{T_p} \cong$ **45.5 kΩ,** which is substantially less than obtained in part c.

h. $Q_p = \dfrac{Z_{T_p}}{X_{L_p}} = \dfrac{Z_{T_p}}{X_L} \quad (Q \geq 10)$

$$= \frac{45.5 \times 10^3\ \Omega}{2\pi(100.66 \times 10^3\ \text{Hz})(5 \times 10^{-3}\ \text{H})}$$

Calculator

(2) (×) (2nd F) (π) (×) (1) (0) (0) (.) (6) (6) (EXP) (3) (×) (5) (EXP) (+/−) (3) (=) (1/x) (4) (5) (.) (5) (EXP) (3) (=)

with a display of 14.397, or $Q_p \cong$ **14.4,** which is significantly lower than the 158 obtained in part f.

i. $\text{BW} = \dfrac{f_p}{Q_p} = \dfrac{100.66 \times 10^3\ \text{Hz}}{14.4}$

Calculator

(1) (0) (0) (.) (6) (6) (EXP) (3) (÷) (1) (4) (.) (4) (=)

with a display of 6990.278, or BW ≅ **7 kHz,** compared to the 637.1 Hz of part d.

Obviously, therefore, R_s has an important impact on the shape of the resonant curve, as noted in Fig. 18.31.

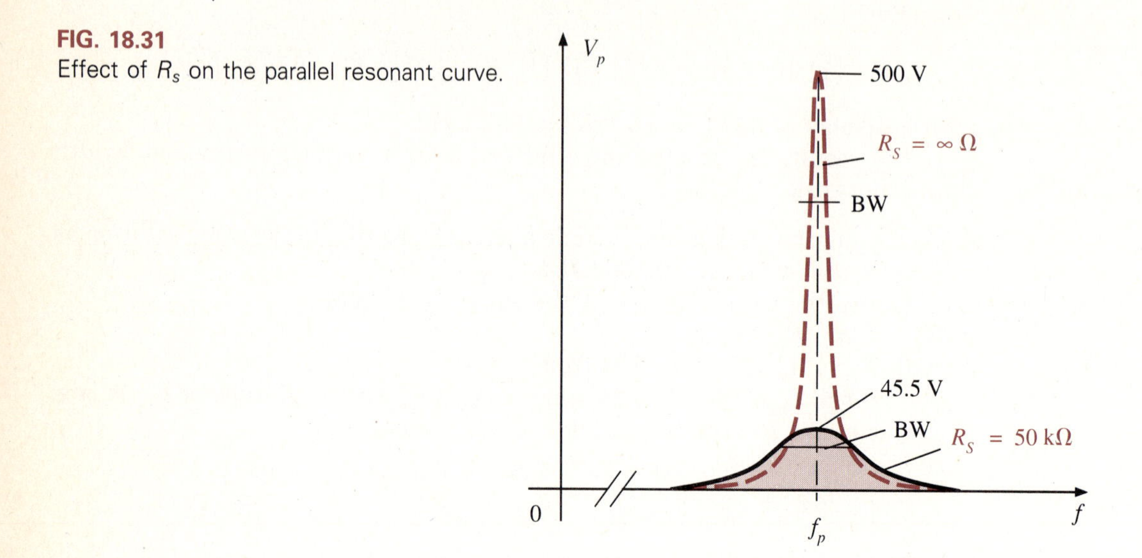

FIG. 18.31
Effect of R_s on the parallel resonant curve.

The peak value of V_p is determined by $R_s = \infty\ \Omega$:

$$V_p = IZ_{T_p} = (1\text{ mA})(0.5\text{ M}\Omega) = 500\text{ V}$$

For $R_s = 50\text{ k}\Omega$:

$$V_p = IZ_{T_p} = (1\text{ mA})(45.5\text{ k}\Omega) = 45.5\text{ V}$$

which is less than 1/10 the value obtained by ignoring the effect of R_s.

Figure 18.31 clearly reveals that the effect of R_s is to reduce severely the height of the resonant curve and increase the bandwidth by a factor of 10. It is therefore important that the magnitude of R_s be compared to the resonant impedance in the design of tank circuits.

EXAMPLE 18.8

If the network of Fig. 18.29 were approximated by the tank circuit of Fig. 18.32, what would be the impact on the network parameters? That is, determine

a. f_p
b. Q
c. Z_{T_p}
d. Q_p
e. BW

and compare the answers to the results of Example 18.7, including the effects of R_s.

FIG. 18.32
Reduced equivalent of Fig. 18.29 ($R = 0\ \Omega$).

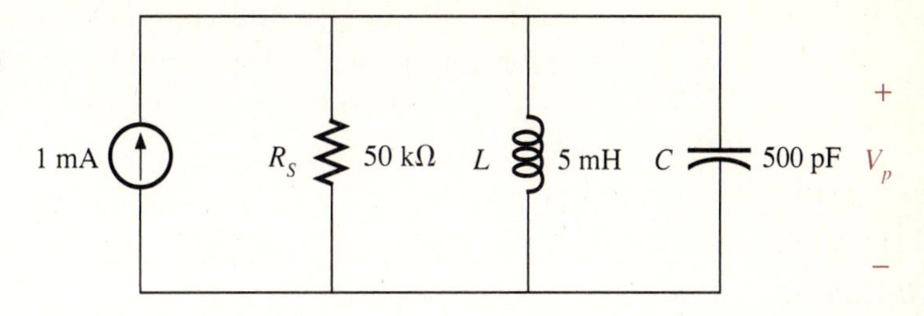

Solutions

a. Comparing Figs. 18.32 and 18.30 reveals that the only element missing is R_l. The result is that X_L and X_C are in parallel, and the condition for resonance is $X_L = X_C$. The analysis to follow demonstrates the impact of ignoring the resistance R_l in the analysis of Fig. 18.29.

The result of $X_L = X_C$ is that

$$f_p = \frac{1}{2\pi\sqrt{LC}}$$

and f_p is the same as determined in Example 18.7.

$$f_p \cong \mathbf{100.66\ kHz}$$

b. Since $Q = \dfrac{X_L}{R_l}$, if $R_l = 0\ \Omega$, the magnitude of Q is quite large.

c. $Z_{T_p} = R_s \| Q^2 R_l$, but both Q and R_l are undefined, and the total impedance should be determined by first finding the total admittance:

$$\mathbf{Y}_T = \mathbf{B}_L + \mathbf{B}_C = B_L \angle -90^\circ + B_C \angle 90^\circ$$

$$= \frac{1}{X_L} \angle -90^\circ + \frac{1}{X_C} \angle 90^\circ$$

and since $X_L = X_C$, the net admittance is 0 S, with $Z_T = 1/Y_T \Rightarrow \infty\ \Omega$, or open circuit.

At resonance, therefore, the net impedance is infinite (open circuit) for the tank circuit of Fig. 18.32. Therefore, $Z_{T_p} = R_s \| R_p = R_s \| \infty\ \Omega$, and

$$Z_{T_p} = R_s = \mathbf{50\ k\Omega}$$

which is a close match to Example 18.7(g).

d. $$Q_p = \frac{Z_{T_p}}{X_{L_p}} = \frac{Z_{T_p}}{X_L} = \frac{50\ \text{k}\Omega}{2\pi(100.66 \times 10^3\ \text{Hz})(5 \times 10^{-3}\ \text{H})}$$

Calculator

(2) (×) (2nd F) (π) (×) (1) (0) (0) (·) (6) (6) (EXP) (3) (×) (5) (EXP) (+/−) (3) (=) (1/x) (5) (0) (EXP) (3) (=)

with a display of 15.811, so $Q_p \cong \mathbf{15.81}$.

e. $$\text{BW} = \frac{f_p}{Q_p} = \frac{100.66\ \text{kHz}}{15.81} \cong \mathbf{6.4\ kHz}$$

The results are remarkably similar for the reduced effort required, revealing that for high-Q parallel resonant circuits, the important parameters of f_p, Z_{T_p}, and Q_p can be determined on an approximate basis if the effects of R_l are ignored.

EXAMPLE 18.9 Design a parallel resonant circuit to have the response curve of Fig. 18.33 using a 1-mH, 10-Ω inductor and a current source having an internal resistance of 40 kΩ.

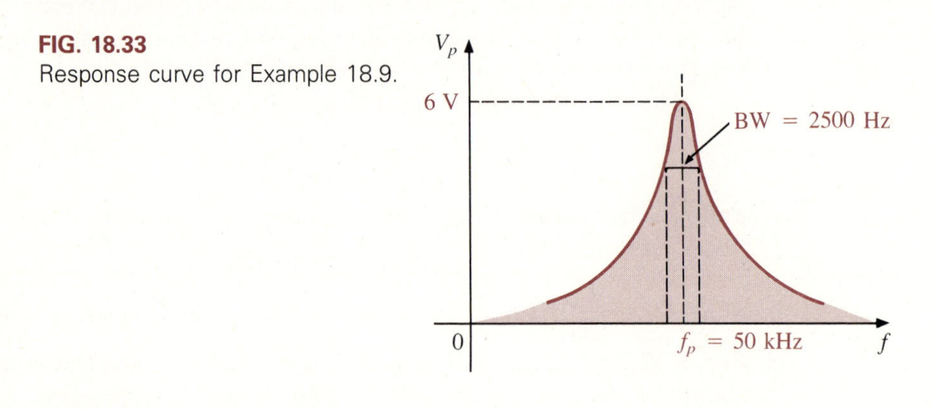

FIG. 18.33
Response curve for Example 18.9.

Solution

$$BW = \frac{f_p}{Q_p}$$

$$Q_p = \frac{f_p}{BW} = \frac{50{,}000 \text{ Hz}}{2500 \text{ Hz}} = 20$$

$$X_L = 2\pi f_p L = 2\pi(50 \times 10^3 \text{ Hz})(1 \times 10^{-3} \text{ H})$$
$$= 314.16 \ \Omega$$

$$Q = \frac{X_L}{R_L} = \frac{314.16 \ \Omega}{10 \ \Omega} = 31.4$$

For $Q \geq 10$:

$$X_C = X_L = \frac{1}{2\pi f C}$$

and

$$C = \frac{1}{2\pi f_p X_C}$$

$$= \frac{1}{2\pi(50 \times 10^3 \text{ Hz})(314.16 \ \Omega)}$$

$$C \cong \mathbf{0.01 \ \mu F}$$

From previous calculations,

$$Q_p = \frac{Z_{T_p}}{X_{L_p}} = \frac{Z_{T_p}}{X_L} = 20$$

so

$$Z_{T_p} = 20(X_L) = 20(314.16\ \Omega) \cong 6.28\ \text{k}\Omega$$

However, $Z_{T_p} = R_s||R_p = R_s||Q^2R_l = R_s(31.4)^2(10\ \Omega) = R_s||9.86\ \text{k}\Omega$ will not be 6.28 kΩ if we substitute $R_s = 40\ \text{k}\Omega$ as given.

We will, therefore, introduce a resistor R in parallel with R_s that will ensure $Z_{T_p} = 6.28\ \text{k}\Omega$.

$$Z_{T_p} = R||R_s||Q^2R_l = 6.28\ \text{k}\Omega$$
$$R_s||Q^2R_l = 40\ \text{k}\Omega||9.86\ \text{k}\Omega = 7.91\ \text{k}\Omega$$

and

$$R||7.91\ \text{k}\Omega = 6.28\ \text{k}\Omega$$

or

$$\frac{R(7.91\ \text{k}\Omega)}{R + 7.91\ \text{k}\Omega} = 6.28\ \text{k}\Omega$$
$$7.91R = 6.28R + 49.67\ \text{k}\Omega$$
$$1.63R = 49.67\ \text{k}\Omega$$
$$R = \mathbf{30.47\ k\Omega}$$

V_p is determined from $V_p = IZ_{T_p}$ and must peak at 6 V. Therefore,

$$I = \frac{V_p}{Z_{T_p}} = \frac{6\ \text{V}}{6.28\ \text{k}\Omega} = 0.955\ \text{mA} \cong \mathbf{1\ mA}$$

as shown in Fig. 18.34 with the rest of the design.

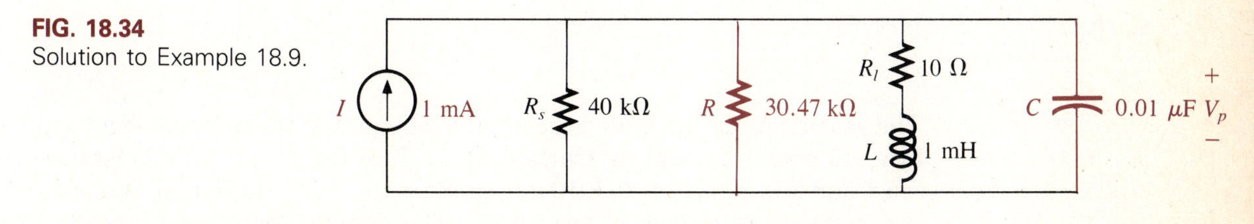

FIG. 18.34
Solution to Example 18.9.

18.11 LOG SCALES

The use of log scales permits a review of the response of a system for an extended range of frequencies. The low-, mid-, and high-frequency responses can all appear on the same graph without a significant loss in response for any frequency region.

Graph paper is available in the *semilog* or *log-log* variety. Semilog paper has only one log scale, with the other a linear scale. Both scales of log-log paper are log scales. A section of semilog paper is shown in Fig. 18.35. Note the linear (even-spaced-interval) vertical scaling and the repeating intervals of the log scale.

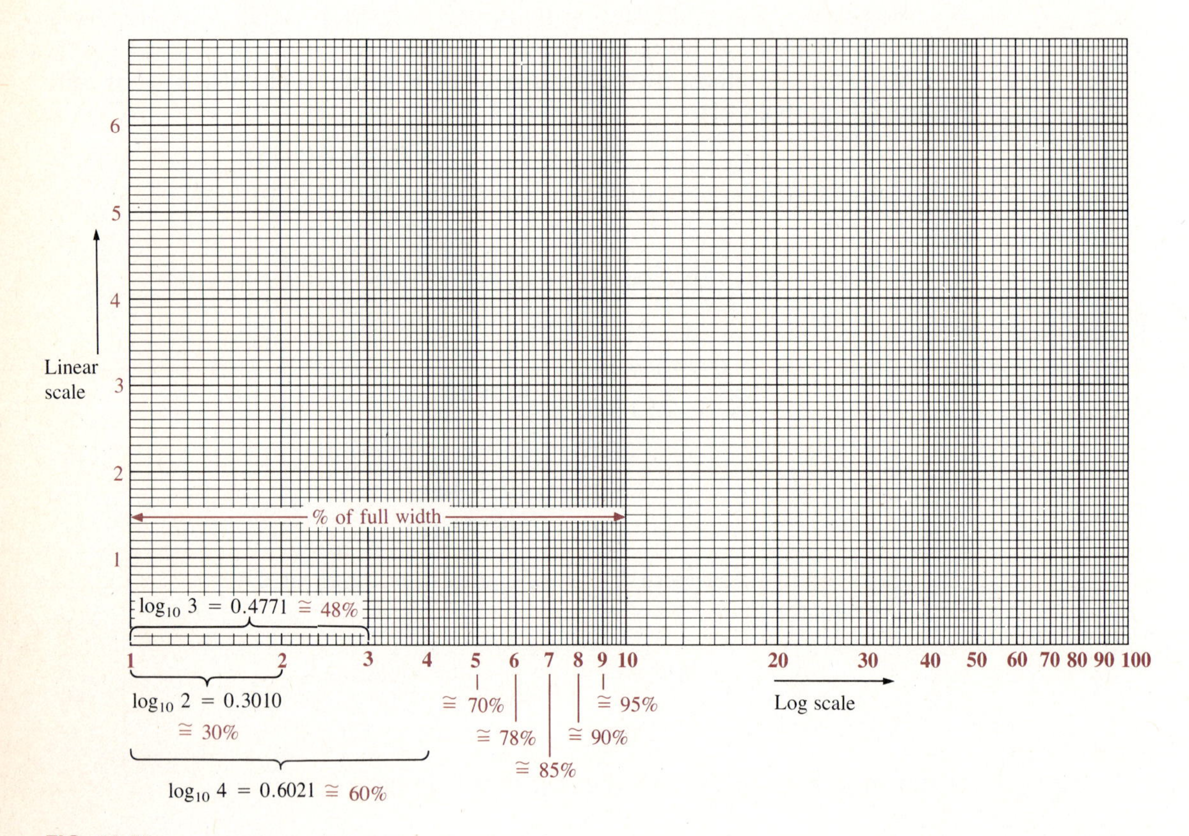

FIG. 18.35
Semilog graph paper.

The spacing of the log scale is determined by taking the common log (base 10) of the number. The scaling starts with 1, since $\log_{10}1 = 0$. The distance between 1 and 2 is $\log_{10}2 = 0.3010$, or approximately 30% of the full distance of a log interval, as shown on the graph. The distance between 1 and 3 is $\log_{10}3 = 0.4771$, or about 48% of the full width. For future reference, keep in mind that almost 50% of the width of one log interval is represented by 3 rather than by the 5 of a linear scale. This tip is particularly useful when the various lines of the graph are left unnumbered.

Note how the log scale becomes compressed at the high end of each interval. With increasing frequency levels assigned to each interval, a single graph

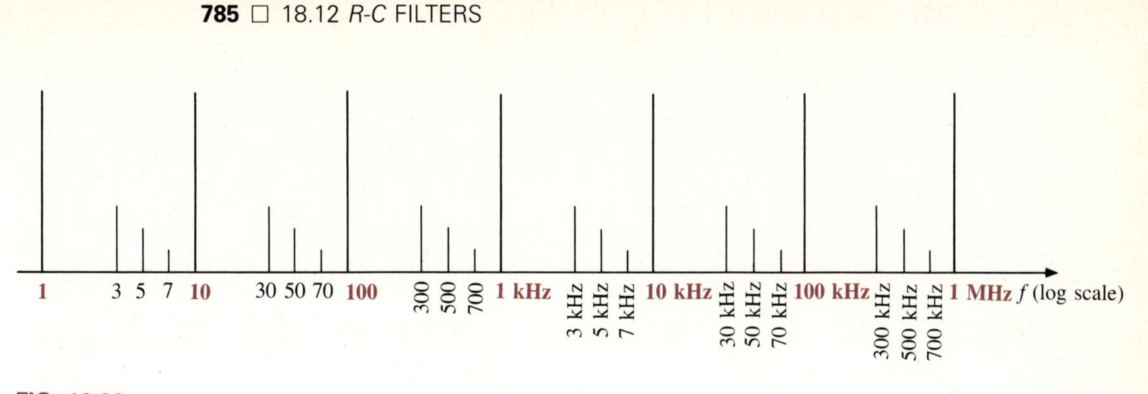

FIG. 18.36
Expanded log scale.

can provide a frequency plot extending from 1 Hz to 1 MHz, as shown in Fig. 18.36, which gives references to the 30%, 50%, and 70% levels of each interval.

The sections to follow, which cover filters, make full use of the log scale to provide a clear picture of the full response of the network to be described.

18.12 *R-C* FILTERS

The *R-C* filter, incredibly simple in design, can be used as a *low-pass* or *high-pass* filter. If the output is taken off the resistor, as shown in Fig. 18.37, the circuit is referred to as a high-pass filter—a circuit that will have the general characteristics of Fig. 18.38. The *pass band* is that range of frequencies that will result in a level of V_o equal to or greater than $0.707V_i$. In other words, the output will be at least 70.7% of the input. The *reject band* is that range of frequencies for which the output is less than 70.7% of the input. The cutoff frequency, that is, the frequency that defines the two regions, is denoted by f_1 in Fig. 18.38.

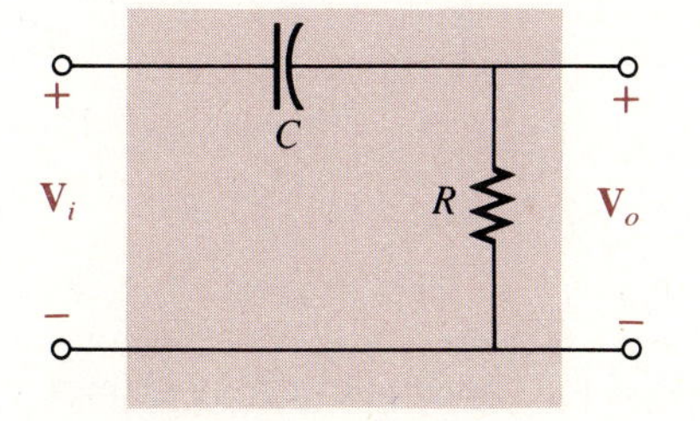

FIG. 18.37
High-pass filter.

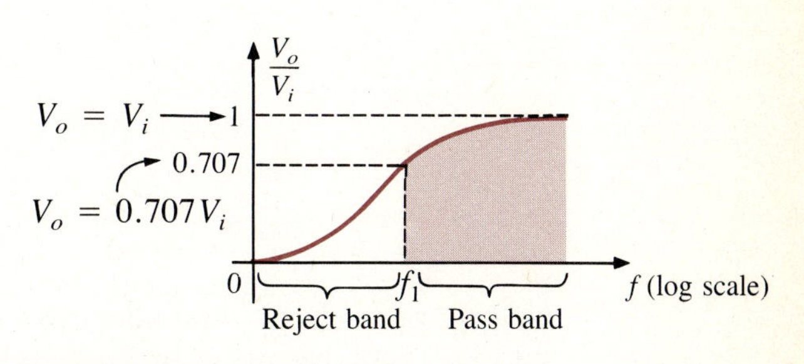

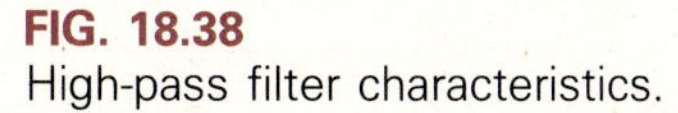
FIG. 18.38
High-pass filter characteristics.

If the output is taken off the capacitor, as shown in Fig. 18.39, the *R-C* combination is a low-pass filter, and only frequencies less than the cutoff frequency f_2 will establish V_o greater than 70.7% of V_i, as shown in Fig. 18.40.

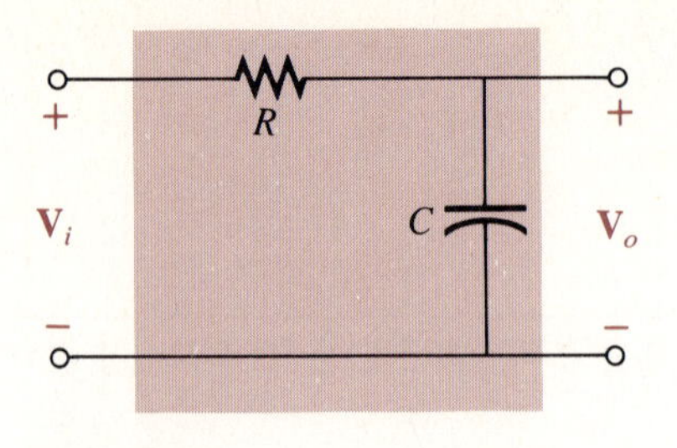

FIG. 18.39
Low-pass filter.

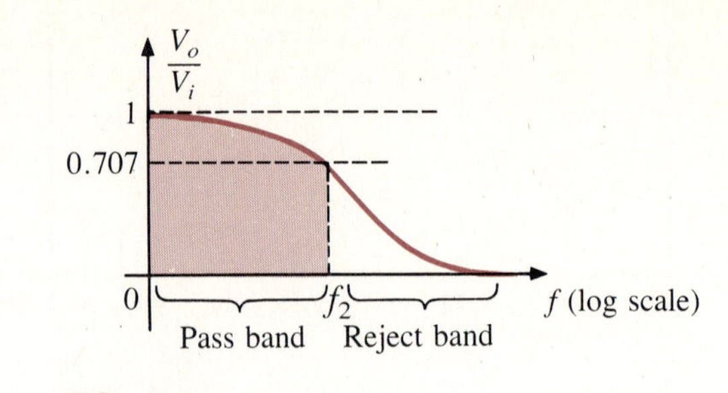

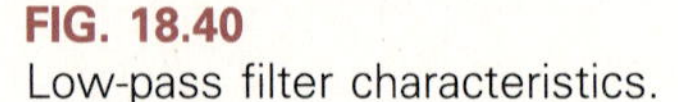
FIG. 18.40
Low-pass filter characteristics.

The R-C network of Fig. 14.39 is a low-pass filter, as noted by the high level of V_o at low frequencies and the decreasing value of V_o with increasing frequencies. The plot provided in Chapter 14 employed a linear scale. The next section demonstrates how the filter characteristics appear on a log plot.

High-pass *R-C* Filter Let us now examine the high-pass filter of Fig. 18.37 in some detail and develop a procedure for plotting the frequency characteristics of Fig. 18.38.

At very high frequencies, the reactance is

$$X_C = \frac{1}{2\pi fC} \cong 0\ \Omega$$

and the short-circuit equivalent can be substituted for the capacitor, as shown in Fig. 18.41, resulting in $V_o = V_i$, or $V_o/V_i = 1$, as shown in Fig. 18.38.

FIG. 18.41
R-C high-pass filter at very high frequencies.

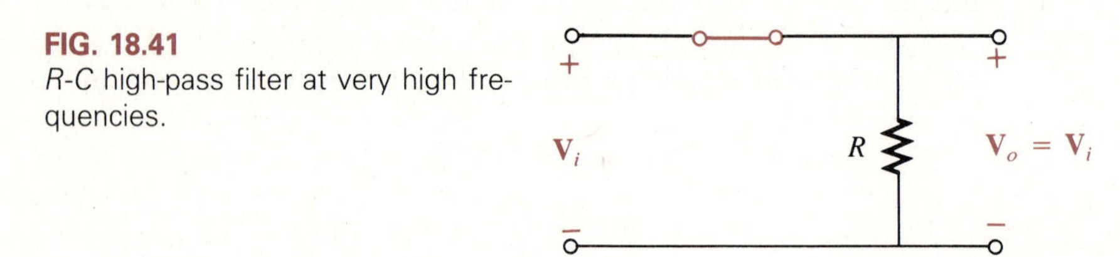

At $f = 0$ Hz,

$$X_C = \frac{1}{2\pi fC} \Rightarrow \infty\ \Omega$$

and the open-circuit equivalent can be substituted for the capacitor, as shown in Fig. 18.42, resulting in $V_o = 0$ V, as noted in Fig. 18.38.

FIG. 18.42
R-C high-pass filter at $f = 0$ Hz.

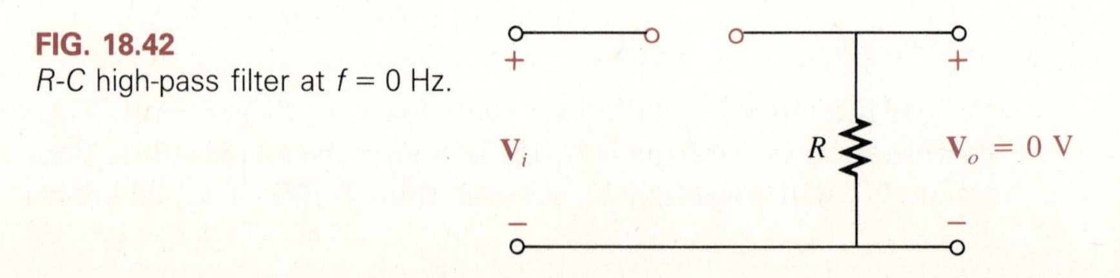

For the series circuit of Fig. 18.37, an application of the voltage divider rule as described in Chapter 15 results in the following for the magnitude of the voltage across the resistor:

$$V_o = V_R = \frac{RV_i}{\sqrt{R^2 + X_C^2}}$$

For the frequency that satisfies $X_C = 1/2\pi fC = R$,

$$V_o = \frac{RV_i}{\sqrt{R^2 + X_C^2}} = \frac{RV_i}{\sqrt{R^2 + R^2}} = \frac{RV_i}{\sqrt{2R^2}}$$

$$= \frac{RV_i}{\sqrt{2}\,R} = \frac{V_i}{\sqrt{2}}$$

$$V_o = 0.707V_i$$

or

$$\frac{V_o}{V_i} = 0.707$$

which defines the boundary between the reject and pass bands.

Solving for the frequency at which the preceding occurs,

$$X_C = \frac{1}{2\pi f_1 C} = R$$

and

$$f_1 = \frac{1}{2\pi RC} \tag{18.39}$$

The impact of Eq. 18.39 extends beyond its relative simplicity. For any high-pass *R-C* filter, the application of any frequency greater than f_1 will result in an output voltage V_o that is at least 70.7% of the magnitude of the input signal. For any frequency below f_1, the output is less than 70.7% of the applied signal.

In Chapter 15 it was also pointed out that the angle associated with $\mathbf{V}_o$ is determined by

$$\theta = \tan^{-1}\frac{X_C}{R} \tag{18.40}$$

High frequencies will result in small values of X_C, and the ratio X_C/R will approach zero with $\tan^{-1}(X_C/R)$ approaching 0°, as shown in Fig. 18.43. At low frequencies, the ratio X_C/R becomes quite large, and $\tan^{-1}(X_C/R)$ approaches 90°. For the case $X_C = R$, $\tan^{-1}(X_C/R) = \tan^{-1}1 = 45°$. Assigning a phase angle of 0° to $\mathbf{V}_i$ such that $\mathbf{V}_i = V_i \angle 0°$, the phase angle associated with $\mathbf{V}_o$ is

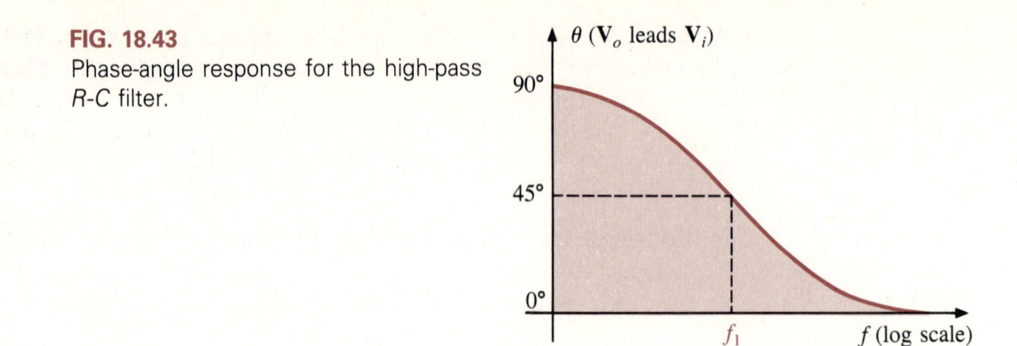

FIG. 18.43
Phase-angle response for the high-pass *R-C* filter.

θ and $\mathbf{V}_o = V_o \angle \theta$, revealing that θ is the angle by which $\mathbf{V}_o$ leads $\mathbf{V}_i$. Since the angle θ is the angle by which $\mathbf{V}_o$ leads $\mathbf{V}_i$ throughout the frequency range of Fig. 18.43, the high-pass *R-C* filter is referred to as a *leading network.*

Low-pass *R-C* Filter The analysis of the low-pass filter of Fig. 18.39 is similar in many respects to that of the high-pass filter of the same elements.

At high frequencies, $X_C \cong 0\ \Omega$ and the output approaches 0 V, as shown in Fig. 18.44. At low frequencies, X_C approaches $\infty\ \Omega$ and $\mathbf{V}_o = \mathbf{V}_i$, as shown in Fig. 18.45.

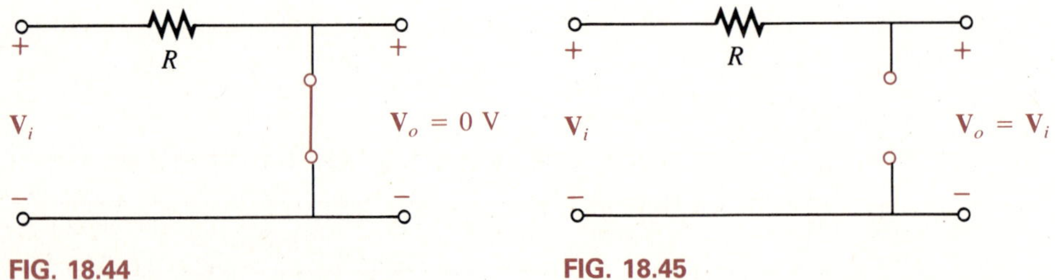

FIG. 18.44
R-C low-pass filter at high frequencies.

FIG. 18.45
R-C low-pass filter at low frequencies.

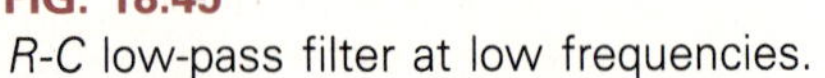

The equation for the magnitude of the voltage across the capacitive reactance can be determined using the voltage divider rule as follows:

$$V_o = V_C = \frac{X_C V_i}{\sqrt{R^2 + X_C^2}}$$

For the condition $X_C = R$, the statement

$$\frac{V_o}{V_i} = 0.707$$

is again established, and the frequency at which this occurs is determined by

$$\boxed{f_2 = \frac{1}{2\pi RC}} \qquad (18.41)$$

as shown in Fig. 18.40.

In Chapter 15 it was also determined that the angle associated with $\mathbf{V}_C = \mathbf{V}_o$ is

$$\theta = -\tan^{-1}\frac{R}{X_C} \qquad (18.42)$$

At high frequencies, X_C becomes smaller and smaller, causing R/X_C to get larger and larger, with the result that $-\tan^{-1}R/X_C$ approaches $-90°$.

At low frequencies, the ratio R/X_C gets smaller and smaller, and $-\tan^{-1}R/X_C$ approaches $0°$. At $X_C = R$, $\theta = -\tan^{-1}R/X_C = -\tan^{-1}1 = -45°$. A plot of θ versus frequency results in the phase plot of Fig. 18.46.

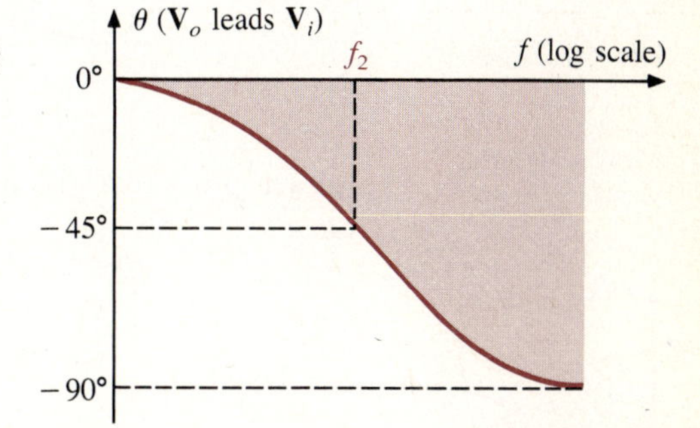

FIG. 18.46
Phase response for the low-pass *R-C* filter.

The plot is of $\mathbf{V}_o$ leading $\mathbf{V}_i$, but since the phase angle is always negative, the phase plot of Fig. 18.47 is more appropriate. Note that a change in sign requires that the vertical axis be changed to the angle by which $\mathbf{V}_o$ lags $\mathbf{V}_i$. The low-pass *R-C* filter is, therefore, a *lagging network*.

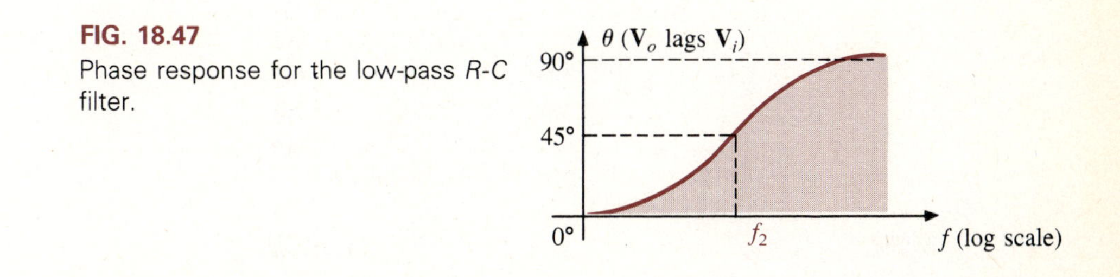

FIG. 18.47
Phase response for the low-pass *R-C* filter.

EXAMPLE 18.10

Given $R = 20\ \text{k}\Omega$ and $C = 1200\ \text{pF}$.

a. Sketch the magnitude plot if the filter is used as both a high- and a low-pass filter.
b. Sketch the phase plot for both filters of part a.
c. Determine the magnitude and phase of $\mathbf{V}_o/\mathbf{V}_i$ at $f = \frac{1}{2}f_1$ for the high-pass filter.

Solutions

a. $$f_1 = f_2 = \frac{1}{2\pi RC} = \frac{1}{(6.28)(20 \times 10^3\ \Omega)(1200 \times 10^{-12}\ \text{F})} = \mathbf{6634.82\ Hz}$$

The magnitude plots appear in Fig. 18.48.

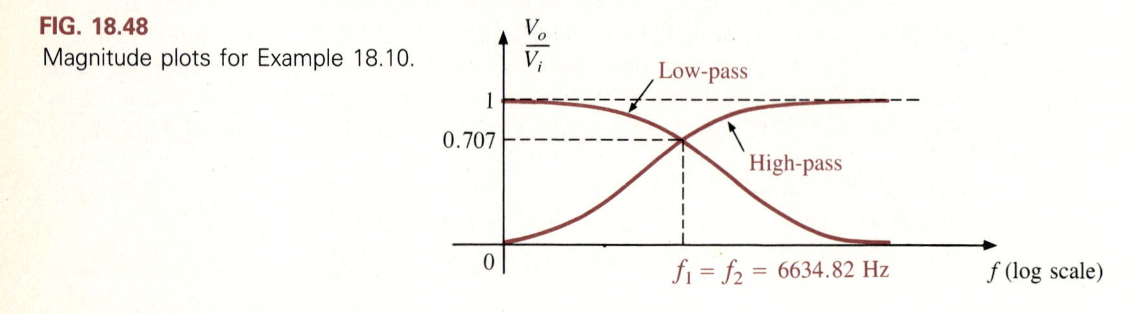

FIG. 18.48
Magnitude plots for Example 18.10.

b. The phase plots are given in Fig. 18.49.

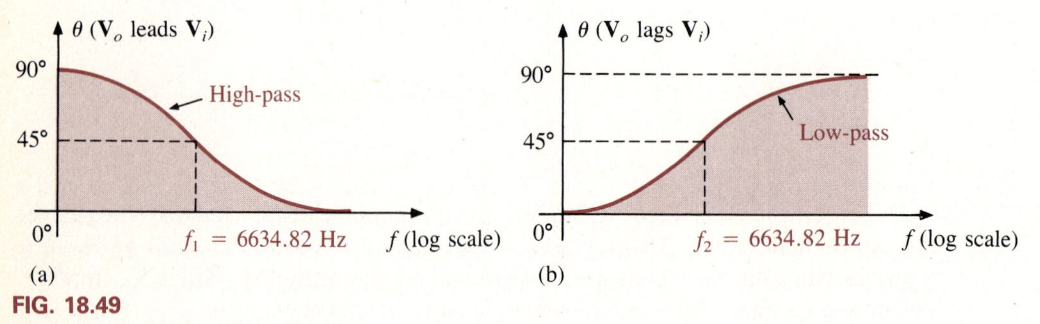

FIG. 18.49
Phase plots for Example 18.10.

c. $$f = \frac{1}{2}f_1 = \frac{1}{2}(6634.82\ \text{Hz}) = 3317.41\ \text{Hz}$$

$$X_C = \frac{1}{2\pi fC} = \frac{1}{(6.28)(3317.41\ \text{Hz})(1200 \times 10^{-12}\ \text{F})} \cong 40\ \text{k}\Omega$$

and

$$\frac{V_o}{V_i} = \frac{X_C}{\sqrt{R^2 + X_C^2}} = \frac{40\ \text{k}\Omega}{\sqrt{(20\ \text{k}\Omega)^2 + (40\ \text{k}\Omega)^2}} = \mathbf{0.4472}$$

with

$$\theta = \tan^{-1}\frac{X_C}{R} = \tan^{-1}\frac{40\ \text{k}\Omega}{20\ \text{k}\Omega} = \tan^{-1}2 = \mathbf{63.43^\circ}$$

18.13 BODE PLOTS

It is standard practice in industry to plot the frequency response of filters, amplifiers, and systems in total against a *decibel* scale defined by the following equation, where G_{dB} is the gain and P_1 and P_2 are the power levels being compared:

$$G_{dB} = 10 \log_{10} \frac{P_2}{P_1} \quad \text{(decibels, dB)} \tag{18.43}$$

Equation 18.43 permits the plotting of an extended range of interest on the same graph in much the same manner as demonstrated for frequency in Section 18.12. In addition, it permits dealing with gains of reasonable levels rather than very large magnitudes, as demonstrated in Table 18.2.

For the special case of $P_2 = 2P_i$, the gain in decibels is

$$G_{dB} = 10 \log_{10} \frac{P_2}{P_1} = 10 \log_{10} 2 = 3 \text{ dB}$$

Calculator

(2) (log) (×) (1) (0) (=)

with a display of 3.010.

Therefore, for a speaker system, a 3-dB increase in output requires that the power level be doubled. In the audio industry, it is a generally accepted rule that an increase in sound level is accomplished with 3-dB increments in the output level. In other words, a 1-dB increase is barely detectable, and a 2-dB increase just discernible. A 3-dB increase normally results in a readily detectable increase in sound level. A further increase in the sound level is normally accomplished by simply increasing the output level another 3 dB. If an 8-W system were in use, a 3-dB increase would require a 16-W output, whereas a further increase of 3 dB (a total of 6 dB) would require a 32-W system, as demonstrated by the following calculations:

$$G_{dB} = 10 \log_{10} \frac{P_2}{P_1} = 10 \log_{10} \frac{16 \text{ W}}{8 \text{ W}} = 10 \log_{10} 2 = 3 \text{ dB}$$

$$G_{dB} = 10 \log_{10} \frac{P_2}{P_1} = 10 \log_{10} \frac{32 \text{ W}}{8 \text{ W}} = 10 \log_{10} 4 = 6 \text{ dB}$$

For *similar load levels,* the power levels are defined by

$$P_2 = \frac{V_2^2}{R} \quad \text{and} \quad P_1 = \frac{V_1^2}{R}$$

Substituting into Eq. 18.43:

$$G_{dB} = 10 \log_{10} \frac{P_2}{P_1} = 10 \log_{10} \frac{V_2^2/R}{V_1^2/R}$$

$$= 10 \log_{10} \frac{V_2^2}{V_1^2} = 10 \log_{10} \left(\frac{V_2}{V_1}\right)^2$$

and

$$G_{dB} = 20 \log_{10} \frac{V_2}{V_1} \quad \text{(dB)} \tag{18.44}$$

Our current interest lies in comparing an output voltage V_o to an input level V_i, resulting in

$$G_{dB} = 20 \log_{10} \frac{V_o}{V_i} \tag{18.45}$$

Table 18.2 compares the magnitude of specific gains to the resulting decibel level. In particular, note that when voltage levels are compared, a doubling of the level results in a change of 6 dB rather than 3 dB, as obtained for power levels.

TABLE 18.2

V_o/V_i	$G_{dB} = 20 \log_{10}(V_o/V_i)$
1	0 dB
2	6 dB
10	20 dB
20	26 dB
100	40 dB
1,000	60 dB
100,000	100 dB

In addition, note that an increase in gain from 1 to 100,000 results in a change in decibels that can easily be plotted on a single graph. Also note that doubling the gain (from 1 to 2 and 10 to 20) results in a 6-dB increase in the decibel level, whereas a change of a factor of 10 (from 1 to 10, 10 to 100, and so on) always results in a 20-dB increase in the decibel level.

For any gain less than 1, the decibel level is negative, as demonstrated for $V_o/V_i = 0.707$:

$$G_{dB} = 20 \log_{10} \frac{V_o}{V_i} = 20 \log_{10} 0.707 = -3 \text{ dB}$$

Calculator

[.] [7] [0] [7] [log] [×] [2] [0] [=]

with a display of -3.012. For $V_o/V_i = (1/2)(0.707) = 0.3535$,

$$G_{dB} = 20 \log_{10} 0.3535 = -9 \text{ dB}$$

revealing, as before, that any change in level by 2:1 (whether increasing or decreasing) results in a 6-dB change in the decibel level.

High-pass *R-C* Filter If we write the gain equation for the high-pass filter in the following manner:

$$A_v = \frac{V_o}{V_i} = \frac{R}{\sqrt{R^2 + X_C^2}} = \frac{1}{\sqrt{1 + \left(\frac{X_C}{R}\right)^2}}$$

Substitute $X_C = 1/2\pi fC$:

$$A_v = \frac{1}{\sqrt{1 + \left(\frac{1}{2\pi fCR}\right)^2}}$$

This can be written as

$$A_v = \frac{1}{\sqrt{1 + \left[\left(\frac{1}{2\pi RC}\right)\frac{1}{f}\right]^2}}$$

or, since $f_1 = 1/2\pi RC$,

$$\boxed{A_v = \frac{V_o}{V_i} = \frac{1}{\sqrt{1 + \left(\frac{f_1}{f}\right)^2}}} \qquad \textbf{(18.46)}$$

which provides the gain of a high-pass *R-C* filter as a function of frequency, since f_1 is a constant determined by the network parameters.

The gain in decibels can then be determined by applying Eq. 18.45:

$$\begin{aligned} G_{\text{dB}} &= 20 \log_{10} \frac{V_o}{V_i} \\ &= 20 \log_{10} \left[\frac{1}{\sqrt{1 + \left(\frac{f_1}{f}\right)^2}}\right] \end{aligned}$$

For frequencies much lower than f_1 ($f << f_1$), the ratio $(f_1/f)^2 >> 1$, and the preceding equation can be written as

$$\boxed{G_{\text{dB}} = A_{v_{\text{dB}}} = -20 \log_{10} \frac{f_1}{f}} \qquad f << f_1 \qquad \textbf{(18.47)}$$

following a number of basic algebraic manipulations.

First note the similarities between Eq. 18.47 and the basic equation for gain in decibels: $G_{dB} = 20 \log_{10} V_o/V_i$. The comments regarding changes in decibel levels due to changes in V_o/V_i can therefore be applied here also, except now a change in frequency by a 2:1 ratio results in a −6-dB change in gain due to the negative sign in Eq. 18.47. A change in frequency by a 10:1 ratio results in a −20-dB change in gain. Any two frequencies separated by a 2:1 ratio are said to be an *octave* apart. Frequencies separated by a 10:1 ratio are said to be *decade* apart. With this terminology, Eq. 18.47 results in a 6-dB change in gain per octave and a 10-dB change per decade.

One may wonder about all the mathematical development to obtain an equation that initially appears confusing and of limited value. As specified, Eq. 18.47 is accurate only for frequency levels much less than f_1.

First, realize that the mathematical development of Eq. 18.47 does not have to be repeated for each configuration encountered. Second, the equation itself is seldom applied but is simply used to define a straight line on a log plot that permits a sketch of the frequency response of a system with a minimum of effort and a high degree of accuracy. The resulting straight-line asymptotes are called a *Bode plot* of the frequency response.

To plot Eq. 18.47, consider the following levels:

For $f = f_1$, $f_1/f = 1$ and $-20 \log_{10} 1 = 0$ dB.
For $f = f_1/2$, $f_1/f = 2$ and $-20 \log_{10} 2 = -6$ dB.

Calculator

(2) (log) (×) (2) (0) (+/−) (=)

with a display of −6.021.

For $f = f_1/4$, $f_1/f = 4$ and $-20 \log_{10} 4 = -12$ dB.
For $f = f_1/10$, $f_1/f = 10$ and $-20 \log_{10} 10 = -20$ dB.

A plot of these points on a log scale results in the decibel plot of Fig. 18.50.

For the future, note that the resulting plot is a straight line intersecting the 0-dB line at f_1. It drops off to the left at a rate of −6 dB per octave, or −10 dB per decade. In other words, once f_1 is determined, find $f_1/2$, and a plot point exists at −6 dB (or find $f_1/10$ and a plot point exists at −20 dB). The *actual* response curve then passes through the −3-dB point at f_1 and approaches the asymptote established by the straight line defined by Eq. 18.47. The curve also approaches an asymptote defined by $A_{v_{dB}} = 0$ dB derived from the midband gain of $V_o = V_i$ $(V_o/V_i = 1)$.

Using a similar set of substitutions, the equation for the phase angle (by which $\mathbf{V}_o$ leads $\mathbf{V}_i$) can be written as

$$\theta = \tan^{-1}\frac{X_C}{R} = \tan^{-1}\frac{f_1}{f} \tag{18.48}$$

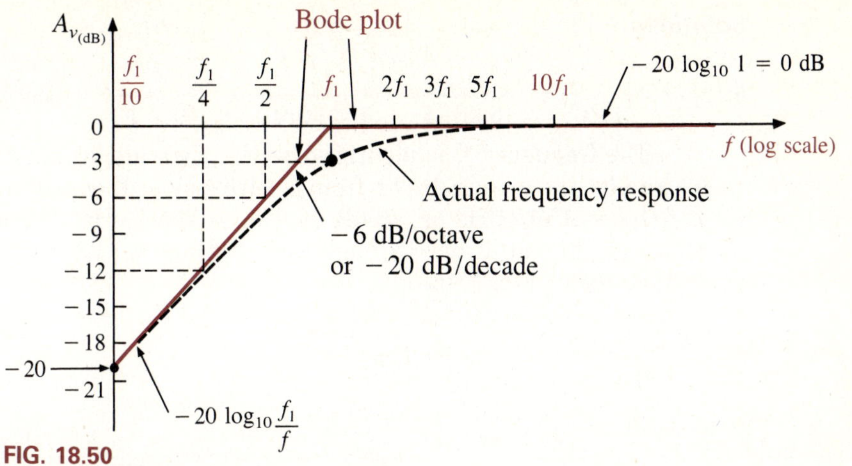

FIG. 18.50
Bode plot for the low-frequency region.

For frequencies where $f << f_1$, $\theta = \tan^{-1}(f_1/f)$ approaches 90°, and for frequencies where $f >> f_1$, $\theta = \tan^{-1}(f_1/f)$ approaches 0°. At $f = f_1$, $\theta = \tan^{-1}(f_1/f) = \tan^{-1}1 = 45°$. A plot of θ versus frequency appears in Fig. 18.51.

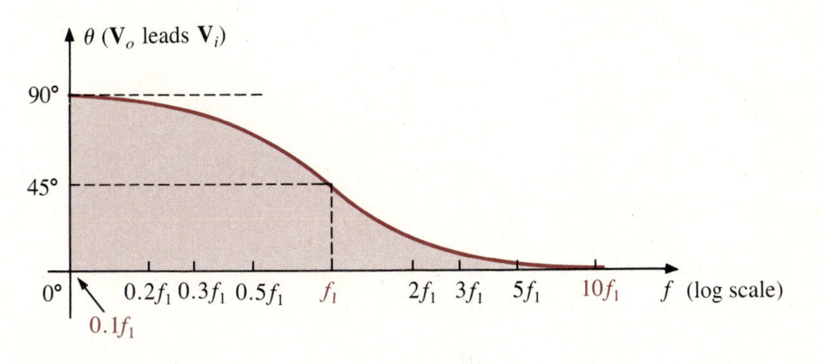

FIG. 18.51
Phase response for a high-pass *R-C* filter.

EXAMPLE 18.11

a. Sketch the frequency response for the high-pass *R-C* filter of Fig. 18.52.
b. Determine the decibel level at $f = 1$ kHz.

FIG. 18.52
High-pass filter for Example 18.11.

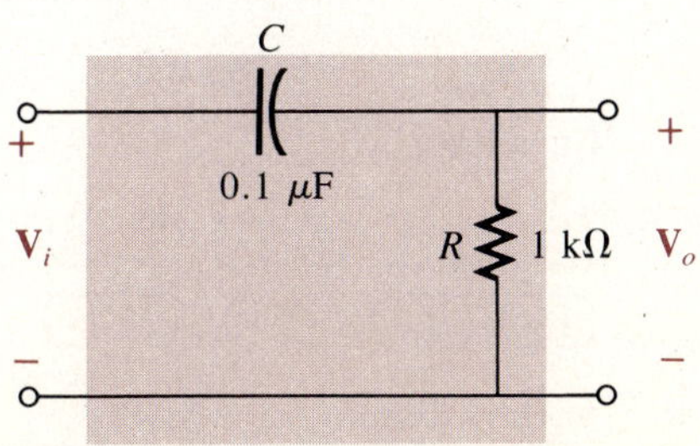

Solutions

a. $f_1 = \dfrac{1}{2\pi RC} = \dfrac{1}{(6.28)(1 \times 10^3\ \Omega)(0.1 \times 10^{-6}\ \text{F})} = 1592.36\ \text{Hz}$

The frequency f_1 is identified on the log scale, as shown in Fig. 18.51. A straight line is then drawn from f_1 with a slope that will intersect -20 dB at $f_1/10 = 159.24$ Hz or -6 dB at $f_1/2 = 796.18$ Hz. The actual response curve can then be drawn through the -3-dB level at f_1 approaching the two asymptotes of Fig. 18.53.

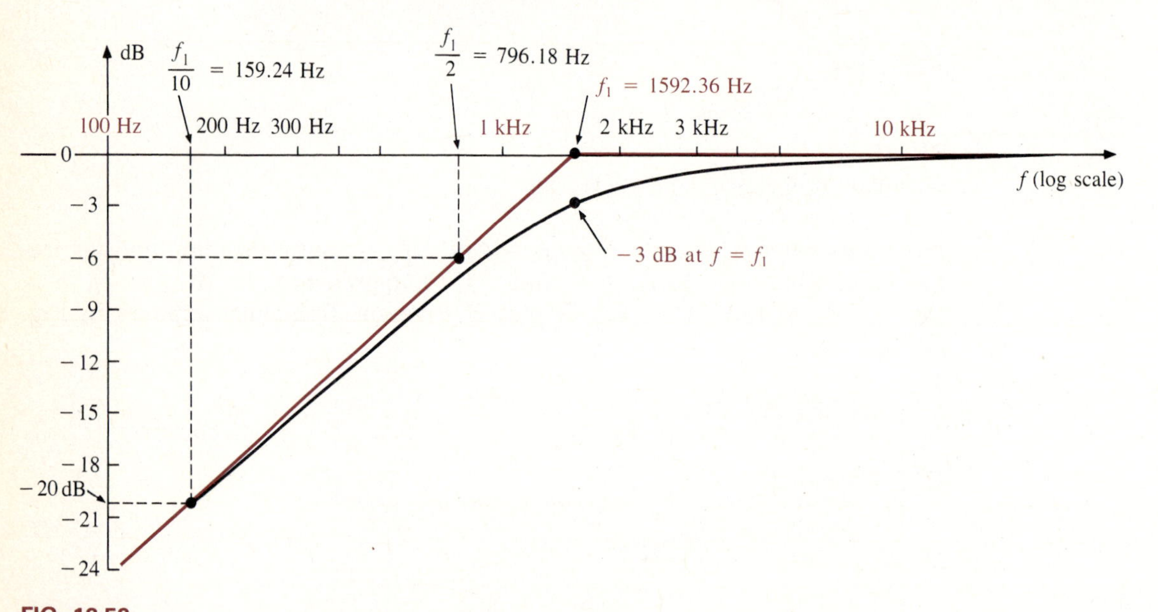

FIG. 18.53
Bode and actual frequency plot for Example 18.11.

Note in the preceding solution that there was no need to employ Eq. 18.46 or perform any extensive mathematical manipulations.

b. $|A_{v_{dB}}| = 20 \log_{10} \dfrac{1}{\sqrt{1 + \left(\dfrac{f_1}{f}\right)^2}} = 20 \log_{10} \dfrac{1}{\sqrt{1 + \left(\dfrac{1592.36\ \text{Hz}}{1000\ \text{Hz}}\right)^2}}$

$= 20 \log_{10} \dfrac{1}{\sqrt{1 + (1.592)^2}} = 20 \log_{10} 0.5318$

$|A_{v_{dB}}| = \mathbf{-5.485\ dB}$ (as verified by Fig. 18.53)

Calculator

(1) (5) (9) (2) (·) (3) (6) (÷) (1) (0) (0) (0) (=) (x^2) (+) (1) (=) (√‾)
(1/x) (log) (×) (2) (0) (=)

with a display of -5.485.

Low-pass *R-C* Filter For the low-pass filter of Fig. 18.54,

$$A_v = \frac{V_o}{V_i} = \frac{X_C}{\sqrt{R^2 + X_C^2}} = \frac{1}{\sqrt{1 + \left(\frac{R}{X_C}\right)^2}}$$

and with $f_2 = 1/2\pi RC$,

$$A_v = \frac{V_o}{V_i} = \frac{1}{\sqrt{1 + \left(\frac{f}{f_2}\right)^2}} \tag{18.49}$$

resulting in

$$A_{v_{\mathrm{dB}}} = -20 \log_{10} \frac{f}{f_2} \qquad f >> f_2 \tag{18.50}$$

for frequencies much greater than f_2 ($f >> f_2$).

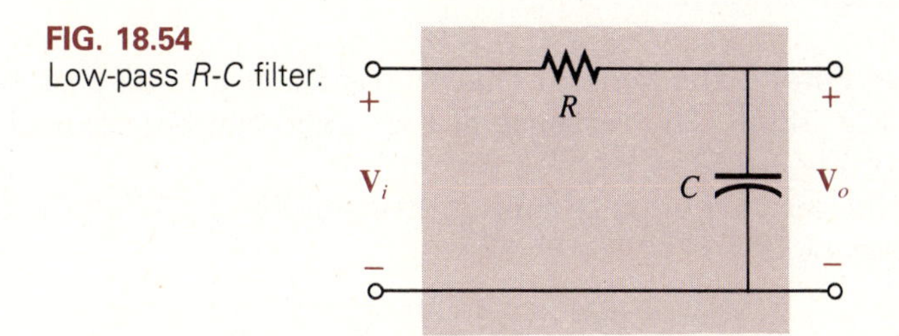

FIG. 18.54
Low-pass *R-C* filter.

The result is a decibel plot that decreases with frequency beyond f_2 at a −6-dB/octave or −20-dB/decade rate, as shown in Fig. 18.55, with f_2 the key frequency (equal to 1 kHz for Fig. 18.55). In particular, note the −6-dB drop at $f = 2f_2$ and the 20-dB drop at $f = 10f_2$.

The phase angle that $\mathbf{V}_o$ leads $\mathbf{V}_i$ by is determined by

$$\theta = -\tan^{-1}\left(\frac{f}{f_2}\right) \tag{18.51}$$

for all frequencies f.

At $f >> f_2$, the phase angle $\theta = -\tan^{-1}(f/f_2)$ approaches −90°, whereas for frequencies $f << f_2$, $\theta = -\tan^{-1}(f/f_2)$ approaches 0°. At $f = f_2$, $\theta = -\tan^{-1}1 = -45°$, confirming the plot of Fig. 18.46.

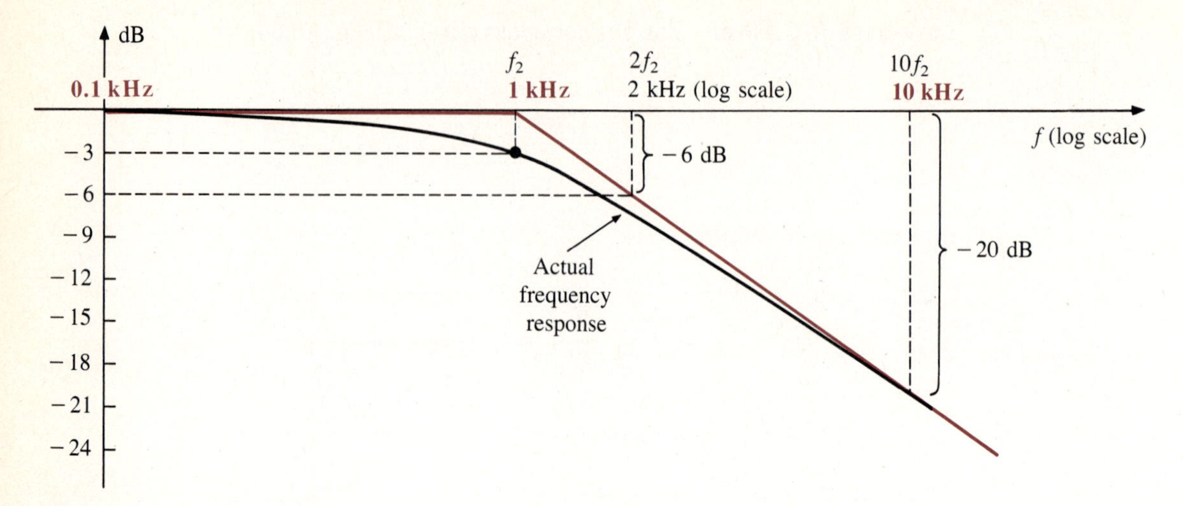

FIG. 18.55
Bode plot for the high-frequency region of a low-pass filter.

18.14 TUNED FILTERS

Series and parallel resonance can be utilized in the design of band-pass, band-stop, and double-tuned filters. Each of these filter types will be briefly described in this section.

Band-pass Filter The general characteristics of a band-pass filter are provided in Fig. 18.56. The response of Fig. 18.56 can be obtained using either network of Fig. 18.57.

For the series resonance configuration of Fig. 18.57(a), the quality factor of the network is

$$Q_s = \frac{X_L}{R} \tag{18.52}$$

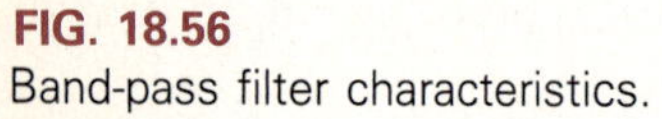

FIG. 18.56
Band-pass filter characteristics.

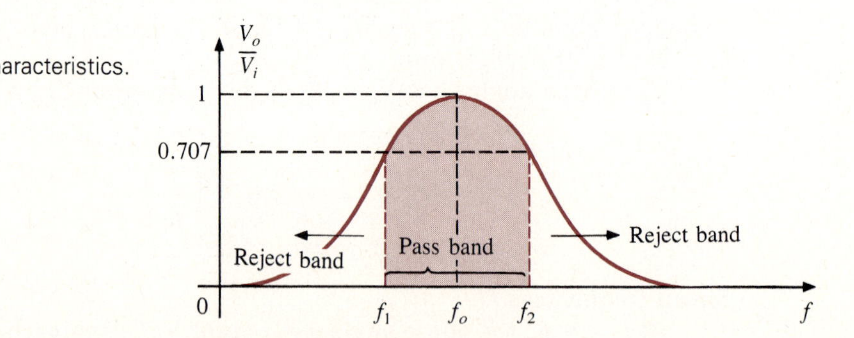

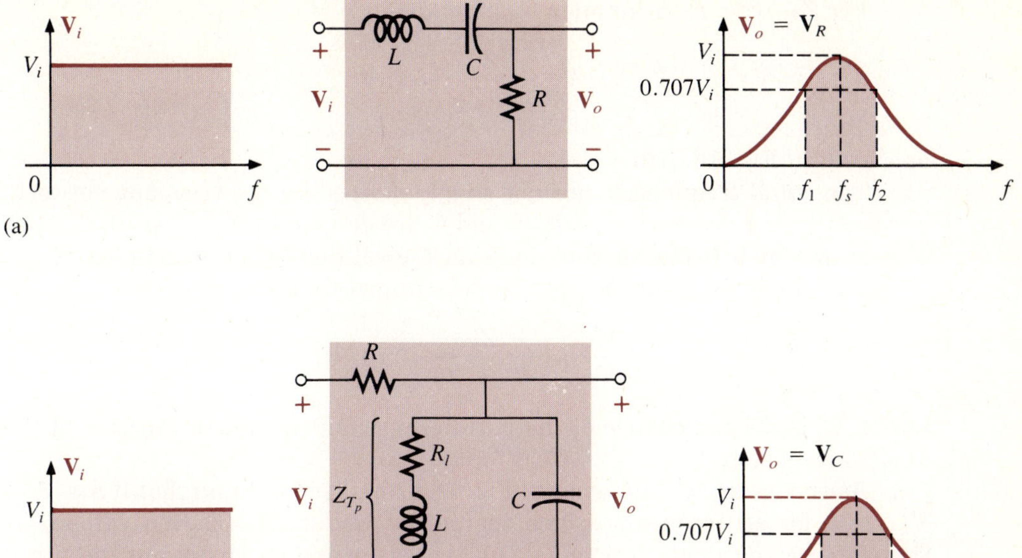

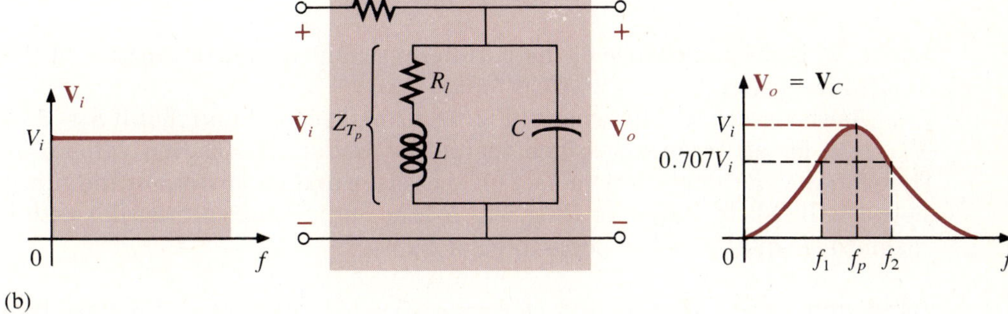

FIG. 18.57
Band-pass filters.

Applying the voltage divider rule yields

$$V_o = V_R = \frac{RV_i}{\sqrt{R^2 + (X_L - X_C)^2}}$$

At resonance, $X_L = X_C$, and

$$V_o = \frac{RV_i}{R} = V_i$$

For frequencies to the right and left of resonance, the impedance of the *L-C* combination increases over that of R, resulting in a decreasing voltage across R. The resulting response curve is shown in Fig. 18.57(a).

The parallel resonant circuit is employed in the network of Fig. 18.57(b). For this configuration, the impedance of the tank circuit is resistive at resonance, and

$$V_o = \frac{Z_{T_p} V_i}{Z_{T_p} + R} \tag{18.53}$$

For $Z_{T_p} >> R$ (a common occurrence),

$$V_o \cong \frac{Z_{T_p}V_i}{Z_{T_p}} \cong V_i$$

as shown in Fig. 18.57(b).

The cutoff frequencies are not simply defined by the resonant network. For the parallel resonant circuit, f_1 and f_2 are defined by $Z_T = 0.707Z_{T_p}$, but V_o may not be $0.707V_i$ for this impedance level due to the resistance R.

The cutoff frequencies must be determined from the condition

$$\frac{V_o}{V_i} = \frac{Z_T}{R + Z_T} = 0.707$$

where Z_T is the impedance of the parallel resonant circuit (a function of the applied frequency).

As you review the network of Fig. 18.57(b), keep in mind that if $R = 0\ \Omega$, $\mathbf{V}_o = \mathbf{V}_i$ for all frequencies, and the output is flat. The chosen value of R therefore has an important impact on the shape and bandwidth of the tuned network. If R is too large, or $R >> Q^2R_l$, the output stays almost flat with a magnitude equal to a low percentage of V_i.

Band-stop Filter The general characteristics of a band-stop filter are provided in Fig. 18.58. Since the characteristics of Fig. 18.58 are the inverse of the pattern obtained for the band-pass filters, we can employ the fact that at any frequency, the sum of the magnitude of the two waveforms to the right of the equals sign in Fig. 18.59 equals the applied voltage V_i.

For the band-pass filters of Fig. 18.57, therefore, if we take the output off the other series elements, as shown in Fig. 18.60, a band-pass characteristic will be obtained, as required by Kirchhoff's voltage law.

The band-reject and band-pass characteristics of miniature filters manufactured by TRW/UTC inductive products appear in Fig. 18.61, along with a photograph of a typical unit. Note that the band-pass characteristics are a universal set, since the resonant frequency is undefined and the horizontal axis is the ratio f/f_r. When examining the curves, remember that the vertical scale is a measure of the attenuation of the input signal. That is, at 0 dB the output is equal to the input, whereas at −30 dB the output is significantly less than the input. Since each of these units can be used for the pass-band or stop-band function, data about each application are provided. Note that the pass band is inserted by the center frequency for each unit, and the type number of the unit reflects the center frequency. The stop band is implying essentially zero response with the −35-dB criterion. Note that the center frequency does not bisect the stop band, since the resonance curve is not symmetrical about f_o.

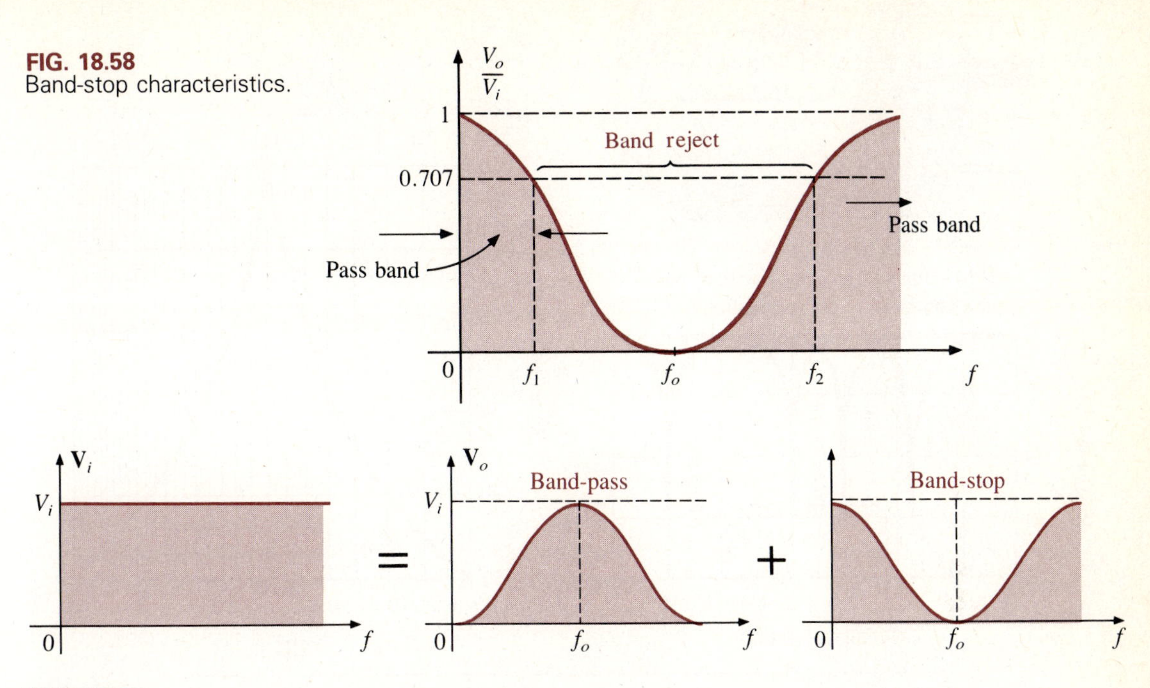

FIG. 18.58
Band-stop characteristics.

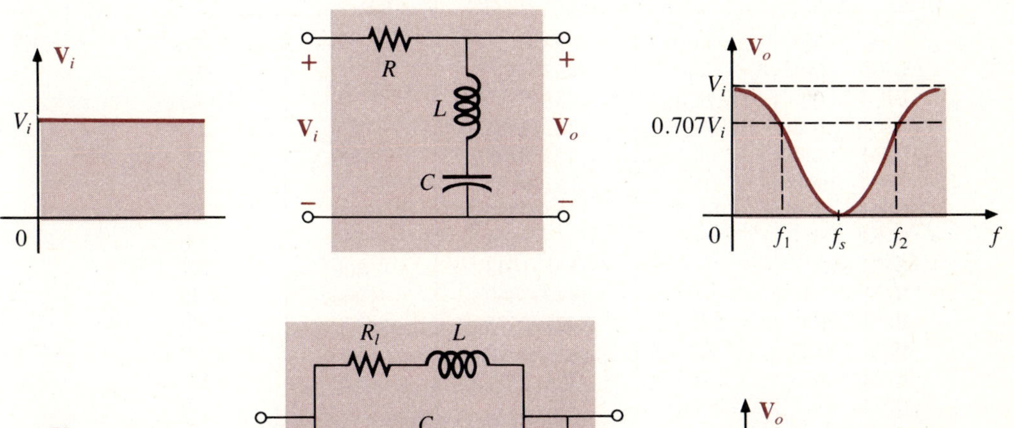

FIG. 18.59
Demonstrating the inverse relationship between band-pass and band-stop filters.

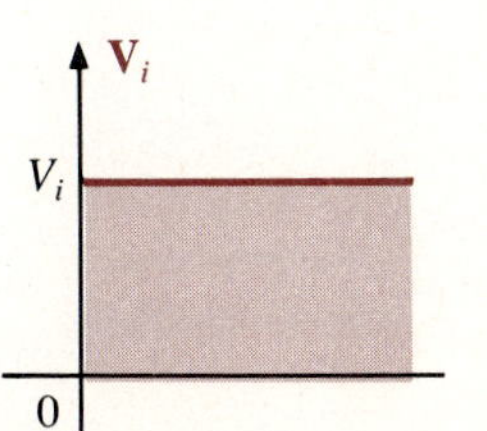
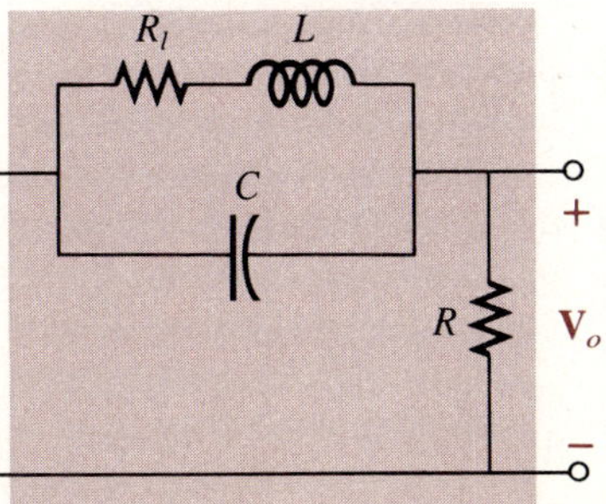
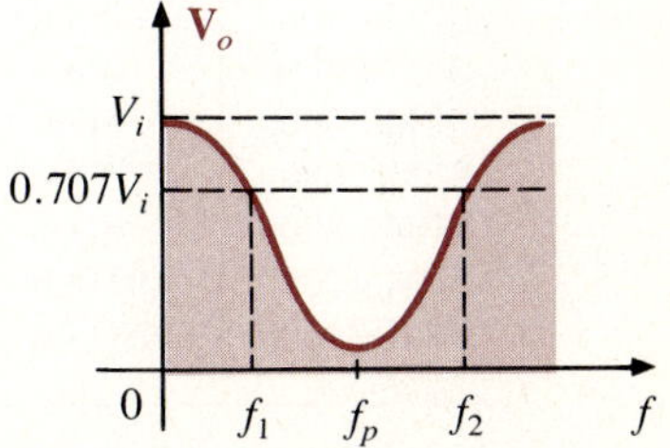

FIG. 18.60
Band-stop filters.

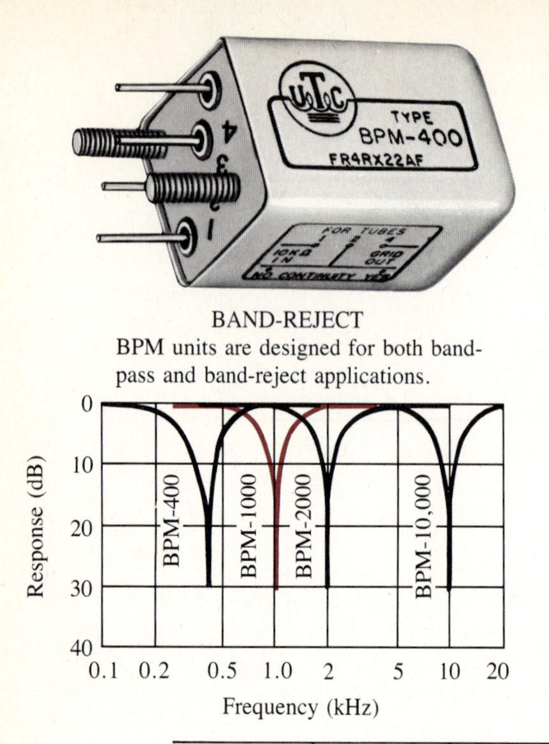

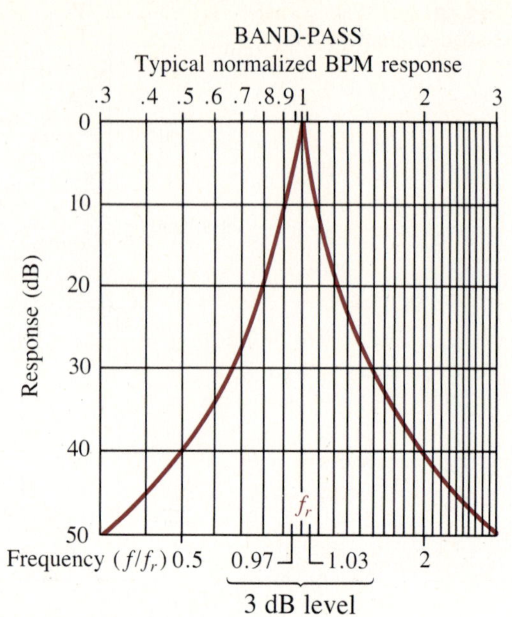

Type No.	Center Frequency (Hz)	Pass Band (Less than 2 dB) (Hz)	Stop Band (More than 35 dB) Below (Hz)	Stop Band (More than 35 dB) Above (Hz)
BPM 400	400	388–412	200	800
BPM 440	440	427–453	220	880
BPM 500	500	485–515	250	1000
BPM 600	600	582–618	300	1200
BPM 800	800	776–824	400	1600
BPM 1000	1000	970–1030	500	2000
BPM 1200	1200	1164–1236	600	2400
BPM 1500	1500	1455–1545	750	3000
BPM 1600	1600	1552–1648	800	3200
BPM 2000	2000	1940–2060	1000	4000
BPM 2500	2500	2425–2575	1250	5000
BPM 3000	3000	2910–3090	1500	6000
BPM 3200	3200	3104–3296	1600	6400
BPM 4000	4000	3880–4120	2000	8000
BPM 4800	4800	4656–4944	2400	9600
BPM 5000	5000	4850–5150	2500	10,000
BPM 6000	6000	5820–6180	3000	12,000
BPM 8000	8000	7760–8240	4000	16,000
BPM 10000	10,000	9700–10,300	5000	20,000
BPM 20000	20,000	19,400–20,600	10,000	40,000

FIG. 18.61
Band-pass and band-reject characteristics for BPM TRW/UTC filters. (Courtesy of United Transformer Corp.)

Double-tuned Filter There are some network configurations that display both a pass-band and a band-stop characteristic, such as shown in Fig. 18.62. For the network of Fig. 18.62(a), the parallel resonant circuit establishes the band stop by resonating at the frequency not permitted to establish a $\mathbf{V}_L$. The greater part of the applied voltage **E** appears across this parallel resonant circuit at this frequency due to its very high impedance compared with R_L.

For the pass band, the parallel resonant circuit is designed to be capacitive (inductive if L_s is replaced by C_s). The inductance L_s is chosen to cancel the effects of the resulting capacitive reactance at the resonant pass-band frequency of the tank circuit, thereby acting as a series resonant circuit. The applied voltage **E** then appears across R_L at this frequency.

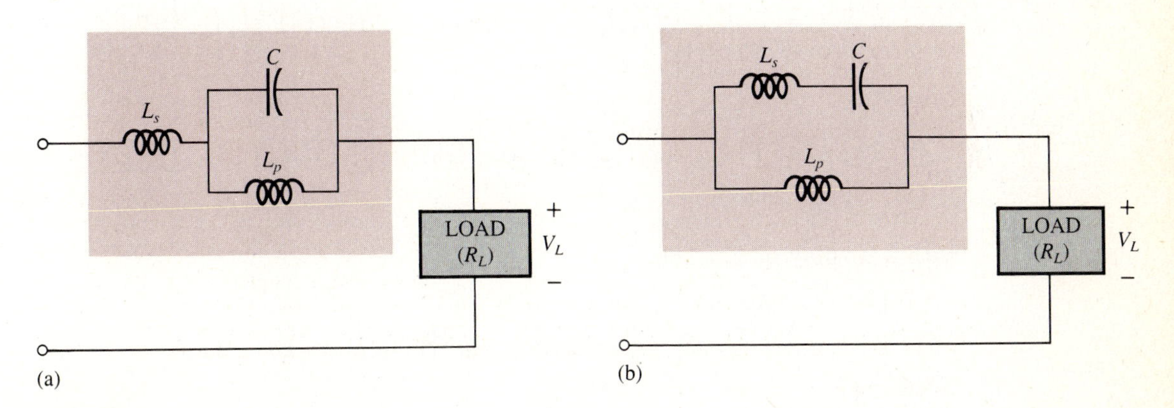

FIG. 18.62
Double-tuned networks.

For the network of Fig. 18.62(b), the series resonant circuit still determines the pass band, acting as a very low impedance across the parallel inductor at resonance, establishing $\mathbf{V}_L = \mathbf{E}$. At the desired band-stop resonant frequency, the series resonant circuit is capacitive. The inductance L_p is chosen to establish parallel resonance at the resonant band-stop frequency. The high impedance of the parallel resonant circuit results in a very low load voltage V_L.

For rejected frequencies below the pass band, the networks should appear as shown in Fig. 18.62. For the reverse situation, L_s in Fig. 18.62(a) and L_p in Fig. 18.62(b) are replaced by capacitors.

EXAMPLE 18.12 For the network of Fig. 18.62(b), determine L_s and L_p for a capacitance C of 500 pF if a frequency of 200 kHz is to be rejected and a frequency of 600 kHz accepted.

Solution For series resonance, we have

$$f_s = \frac{1}{2\pi\sqrt{LC}}$$

and

$$L_s = \frac{1}{4\pi^2 f_s^2 C} = \frac{1}{4\pi^2(600 \times 10^3 \text{ Hz})^2(500 \times 10^{-12} \text{ F})}$$
$$\cong \mathbf{141\ \mu H}$$

Calculator

(4) (×) (2nd F) (π) (x²) (×) (6) (0) (0) (EXP) (3) (x²) (×) (5) (0) (0) (EXP) (+/−) (1) (2) (=) (1/x) (F <−> E)

with a display of 1.407−04.

At 200 kHz,

$$X_L = \omega_s L = 2\pi f_s L = 2\pi(200 \times 10^3 \text{ Hz})(141 \times 10^{-6} \text{ H})$$
$$= 177.2\ \Omega$$

Calculator

(2) (×) (2nd F) (π) (×) (2) (0) (0) (EXP) (3) (×) (1) (4) (1) (EXP) (+/−) (6) (=)

with a display of 177.186.

$$X_C = \frac{1}{\omega C} = \frac{1}{2\pi(200 \times 10^3 \text{ Hz})(500 \times 10^{-12} \text{ F})}$$

Calculator

(2) (×) (2nd F) (π) (×) (2) (0) (0) (EXP) (3) (×) (5) (0) (0) (EXP) (+/−) (1) (2) (=) (1/x) (F <−> E)

with a display of 1.592 03.

For series elements,

$$X_T = X_C - X_L = 1592.52\ \Omega - 177.2\ \Omega = 1414.8\ \Omega \quad \text{(capacitive)}$$

At parallel resonance ($Q \geq 10$ assumed),

$$X_L = X_C$$

and

$$L_p = \frac{X_C}{\omega} = \frac{1414.8\ \Omega}{2\pi(200 \times 10^3 \text{ Hz})} = \mathbf{1.124\ mH}$$

Calculator

(2) (×) (2nd F) (π) (×) (2) (0) (0) (EXP) (3) (=) (1/x) (×) (1) (4) (1) (2) (·) (8) (=) (F <−> E)

with a display of 1.124−03.

FORMULA SUMMARY

Series Resonance

$$X_L = X_C, \quad f_s = \frac{1}{2\pi\sqrt{LC}}$$

$$Q_s = \frac{X_L}{R}, \quad Q_{coil} = \frac{X_L}{R_l}$$

$$V_L = Q_s E, \quad V_C = Q_s E$$

$$\text{BW} = f_2 - f_1 = \frac{f_s}{Q_s} = \frac{R}{2\pi L}$$

Parallel Resonance

Any Q:

$$\frac{R_l^2 + X_L^2}{X_L} = X_C, \quad f_p = \frac{1}{2\pi\sqrt{LC}}\sqrt{1 - \frac{R_l^2 C}{L}}$$

$$R_p = \frac{R_l^2 + X_L^2}{R_l}, \quad Z_{T_p} = R_s || R_p, \quad Q_p = \frac{R}{X_{L_p}}$$

$$\text{BW} = \frac{f_p}{Q_p}$$

$Q \geq 10$:

$$X_L = X_C, \quad f_p = \frac{1}{2\pi\sqrt{LC}}$$

$$R_p = Q^2 R_l, \quad Z_{T_p} = R_s || R_p$$

$$Q_p = \frac{R}{\omega_p L}, \quad I_L = I_C = QI_T$$

$$\text{BW} = \frac{f_p}{Q_p}$$

$Q \geq 10, R_s \Rightarrow \infty\,\Omega$:

$$X_L = X_C, \quad f_p = \frac{1}{2\pi\sqrt{LC}}$$

$$Z_{T_p} = Q^2 R_l, \quad Q_p = Q$$

$$I_L = I_C = QI_T$$

$$\text{BW} = \frac{f_p}{Q}$$

High-pass Filter

$$f_1 = \frac{1}{2\pi RC}$$

Low-pass Filter

$$f_2 = \frac{1}{2\pi RC}$$

Decibels

$$G_{dB} = 10 \log_{10} \frac{P_2}{P_1}, \qquad G_{dB} = 20 \log_{10} \frac{V_o}{V_i}$$

CHAPTER SUMMARY

This is without question one of the most extensive chapters in the text. The topics of series and parallel resonance are two of the most important within the framework of sinusoidal ac systems. Filters are an extension of the resonance phenomenon and Bode plots are a tool of the trade understood by all practicing electrical engineers. Fundamentally, the networks of this chapter "select" a range of frequencies for which a current or voltage will have a maximum value. In this way bands of frequencies can be tuned in or out as desired by the application at hand.

There is only one set of equations for series resonance that provides all the specific information necessary to plot the resonance curve. For parallel resonance, however, the conversion of a series *R-L* branch into an equivalent parallel *R-L* configuration before the analysis begins hints of a more complex system in total. However, if we simply check the magnitude of Q and note whether R_s is included in the design, the applicable set of equations in the formula review is defined and the analysis can commence using the most appropriate set of equations. However, it should be emphasized that the first set of equations for any Q are applicable for *any* network components. That is, they work for any Q or R_s level. There is no need to first determine Q or check the R_s level—simply substitute into the equations of this set. On the other side of the coin, however, most resonant networks have $Q \geq 10$, and R_s is frequently ignored in the last set of equations, which can be used for most situations. The equations are measurably simpler and easier to recall and have a close match with the equations for series resonant circuits. The choice of which to use will come with experience, but be aware that the last set will always provide a good first approximation to the desired quantities.

Log-log and semilog graph papers are quite valuable in the analysis of electronic systems because they permit the charting of the response of a system for a wide range of values not possible on a standard linear axis. The horizontal space set aside for a frequency range of 1 to 10 Hz in the same as from 1 to 10 MHz to show clearly the response at the high and low ends of the frequency spectrum on the same $8\frac{1}{2}$-in. by 11-in. sheet of graph paper. Grasp what you can from this treatment of logarithms; with exposure and time the important decibel scale and the log scales will become useful "tools of the trade."

The ability of the high- and low-pass R-C filters to perform the important filtering action they do using only two elements is indeed impressive. Through the proper choice of resistance and capacitance level, a wide band of frequencies can be permitted to pass on to a succeeding stage or blocked from further impact on the system. The Bode plot is an effective way of graphing a filter's response against a decibel scale. Once the procedure is understood, the lengthy derivations to establish the conclusions can be set on the back burner for recall only when unusual situations are encountered. The Bode plot is a technique that can provide invaluable insight into the response of a system with a minimum of time and effort.

GLOSSARY

Band (cutoff, half-power, corner, −3 dB) frequencies Frequencies that define the points on the resonance curve that are 0.707 of the peak current or voltage value. In addition, they define the frequencies at which the power transfer to the resonant circuit will be half the maximum power level (−3 dB).

Band-stop filter A network designed to reject (not pass) signals within a particular range of frequencies.

Bandwidth The range of frequencies between the band, cutoff, or half-power frequencies.

Bode plot A plot (envelope) of the frequency response of a system using straight-line segments called asymptotes.

Double-tuned filter A network having both a pass-band and a band-stop region.

Filter Networks designed either to pass or reject the transfer of signals at certain frequencies to a load.

High-pass filter A filter designed to pass high frequencies and reject low frequencies.

Low-pass filter A filter designed to pass low frequencies and reject high frequencies.

Pass-band (band-pass) filter A network designed to pass signals within a particular range of frequencies.

Quality factor (Q) A ratio that provides an immediate indication of the sharpness of the peak of a resonance curve. The higher the Q, the sharper the peak and the more quickly it drops off to the right and left of the resonant frequency.

Resonance A condition established by the application of a particular frequency (the resonant frequency) to a series or parallel R-L-C network. The transfer of power to the system is a maximum and, for frequencies above and below, the power transfer drops off to significantly lower levels.

Selectivity A characteristic of resonant networks directly related to the bandwidth of the resonant system. High selectivity is associated with small bandwidth (high Q's), and low selectivity with larger bandwidths (low Q's).

Semilog paper Graph paper with one log scale and one linear scale.

PROBLEMS

Sections 18.1–18.6

1. Find the resonant ω_s and f_s for the series circuit with the following parameters:
   a. $R = 10\ \Omega$, $L = 1\ \text{H}$, $C = 16\ \mu\text{F}$
   b. $R = 300\ \Omega$, $L = 0.5\ \text{H}$, $C = 0.16\ \mu\text{F}$
   c. $R = 20\ \Omega$, $L = 0.28\ \text{mH}$, $C = 7.46\ \mu\text{F}$
2. For the series circuit of Fig. 18.63:
   a. Find the value of X_C for resonance.
   b. Find the magnitude of the current I and the voltages V_R, V_L, and V_C at resonance.
   c. Find the Q_s of the circuit.
3. Repeat Problem 2 for the circuit of Fig. 18.64.

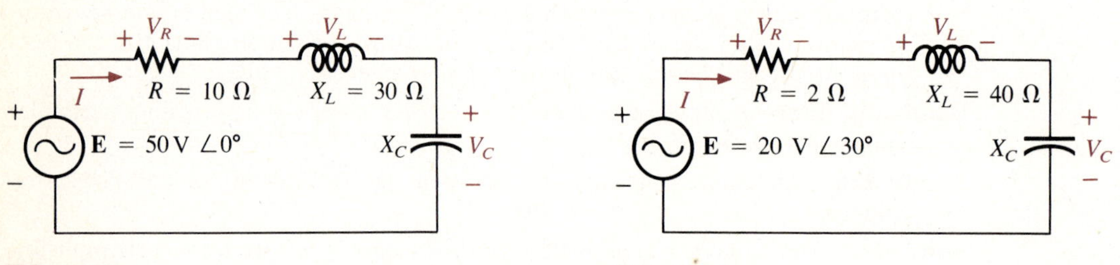

FIG. 18.63

FIG. 18.64

4. For the circuit of Fig. 18.65:
   a. Find the value of L in millihenries if the resonant frequency is 1800 Hz.
   b. Find the magnitude of the current I and the voltages V_R, V_L, and V_C at resonance.
   c. Calculate the cutoff frequencies.
   d. Find the bandwidth of the series resonant circuit.

FIG. 18.65

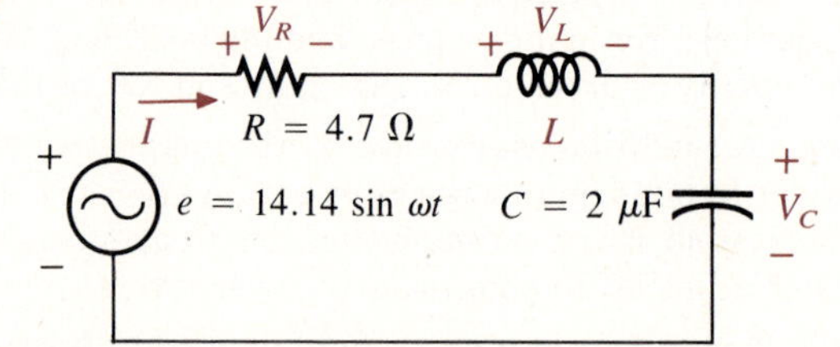

5. a. Find the bandwidth of a series resonant circuit having a resonant frequency of 6000 Hz and a Q_s of 15.
   b. Find the cutoff frequencies.
   c. If the resistance of the circuit at resonance is 3 Ω, what are the values of X_L and X_C in ohms?
   d. What is the power dissipated at the half-power frequencies if the maximum current flowing through the circuit is 0.5 A?

6. A series circuit has a resonant frequency of 10 kHz. The resistance of the circuit is 5 Ω, and X_C at resonance is 200 Ω.
   a. Find the bandwidth.
   b. Find the cutoff frequencies.
   c. Find Q_s.
   d. If the input voltage is 30 $\angle 0°$, find the magnitude of the voltage across the coil and capacitor.
   e. Find the power dissipated at resonance.
7. a. The bandwidth of a series resonant circuit is 200 Hz. If the resonant frequency is 2000 Hz, what is the value of Q_s for the circuit?
   b. If R = 2 Ω, what is the value of X_L at resonance?
   c. Find the value of L and C at resonance.
   d. Find the cutoff frequencies.
8. The cutoff frequencies of a series resonant circuit are 5400 Hz and 6000 Hz.
   a. Find the bandwidth of the circuit.
   b. If Q_s is 9.5, find the resonant frequency of the circuit.
   c. If the resistance of the circuit is 2 Ω, find the value of X_L and X_C at resonance.
   d. Find the value of L and C at resonance.

*9. Design a series resonant circuit with an input voltage of 5 V $\angle 0°$ to have the following specifications:
   a. A peak current of 500 mA at resonance
   b. A bandwidth of 120 Hz
   c. A resonant frequency of 8400 Hz
   Find the value of L and C and the cutoff frequencies.

*10. Design a series resonant circuit to have a bandwidth of 400 Hz using a coil with a Q of 20 and resistance of 2 Ω. Find the value of L and C and the cutoff frequencies.

Sections 18.7–18.10

11. For the circuit of Fig. 18.66:
    a. Find the value of X_C at resonance.
    b. Find the total impedance Z_T at resonance.
    c. If the resonant frequency is 20,000 Hz, find the value of L and C at resonance.
    d. Find Q_p and BW.

FIG. 18.66

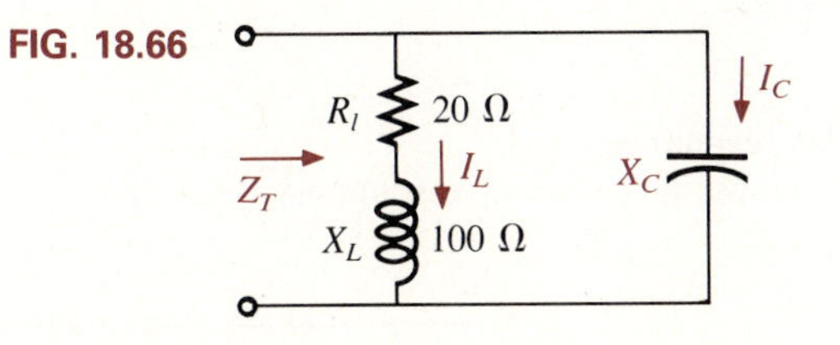

12. Repeat Problem 11 for the circuit of Fig. 18.67.

FIG. 18.67

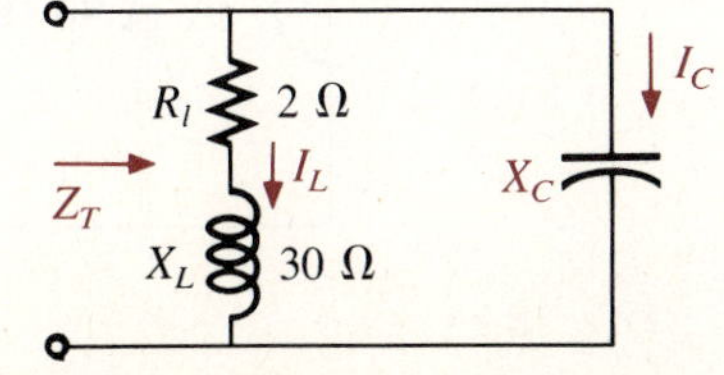

13. For the circuit of Fig. 18.68:
a. Find the resonant frequency.
b. Find the value of X_L and X_C at resonance.
c. Is the coil a high-Q or low-Q coil at resonance?
d. Find the impedance Z_{T_p} at resonance.
e. Find the currents I_L and I_C at resonance.
f. Calculate Q_p and BW.

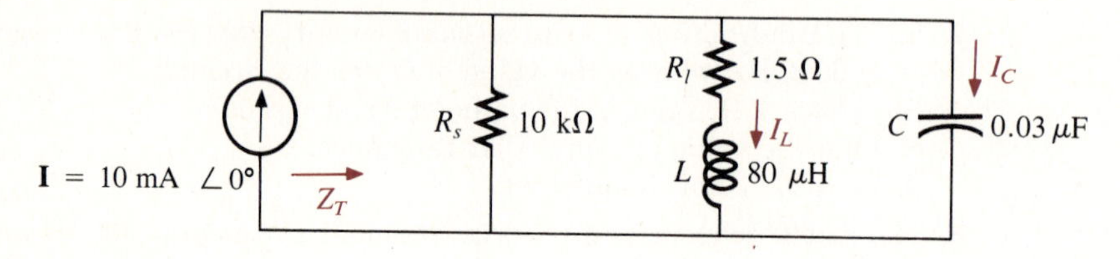

FIG. 18.68

14. Repeat Problem 13 for the circuit of Fig. 18.69.

FIG. 18.69

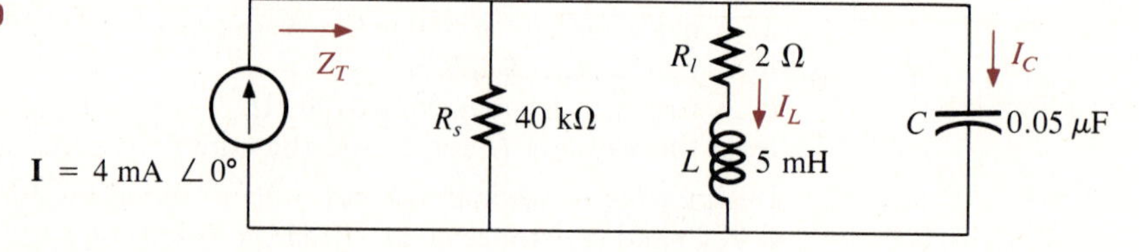

15. It is desired that the impedance Z_T of the circuit of Fig. 18.70 be a resistor of 50 kΩ at resonance.
a. Find the value of X_L.
b. Compute X_C.
c. Find the resonant frequency if $L = 16$ mH.
d. Find the value of C.

16. For the network of Fig. 18.71:
a. Find f_p.
b. Determine R_l.
c. Calculate V_C at resonance.
d. Determine the power absorbed at resonance.
e. Find BW.

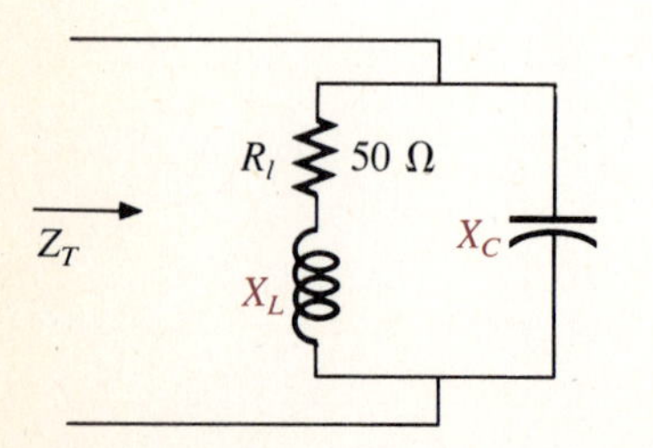

FIG. 18.70

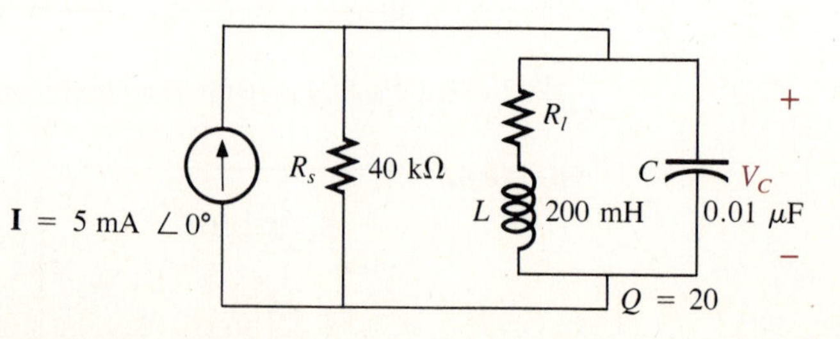

FIG. 18.71

17. Repeat Problem 16 ignoring the effects of R_l ($R_l = 0\ \Omega$) and compare solutions for parts a through e.

18. For the network of Fig. 18.72, the following are specified:

$$\begin{aligned} f_p &= 100\text{ kHz} \\ \text{BW} &= 2500\text{ Hz} \\ L &= 2\text{ mH} \\ Q &= 80 \end{aligned}$$

Find R_s and C.

FIG. 18.72

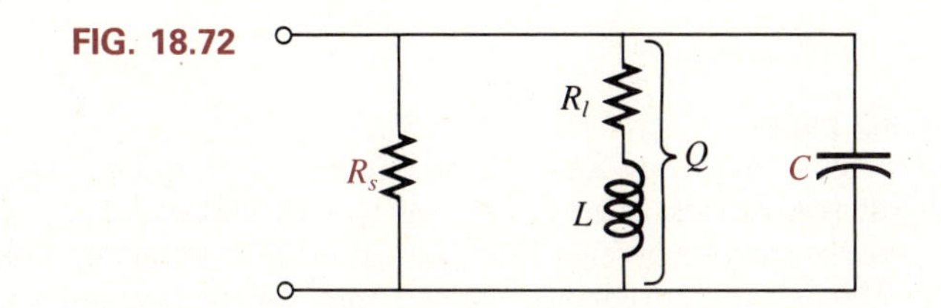

***19.** For the network of Fig. 18.73:

a. Find the value of X_L for resonance.
b. Find Q.
c. Find the resonant frequency if the bandwidth is 1000 Hz.
d. Find the maximum value of the voltage V_C.
e. Sketch the curve of V_C versus frequency. Indicate its peak value, resonant frequency, and band frequencies.

FIG. 18.73

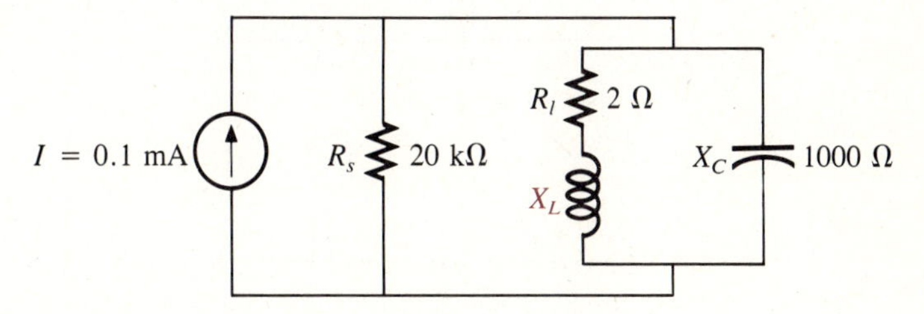

***20.** Repeat Problem 19 for the network of Fig. 18.74.

FIG. 18.74

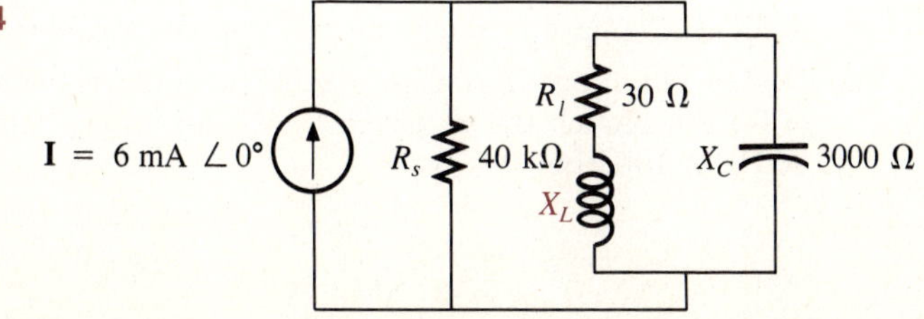

***21.** Design the network of Fig. 18.75 to have the following characteristics:
 a. A bandwidth of 500 Hz
 b. $Q_p = 30$
 c. $V_{C_{max}} = 1.8$ V

FIG. 18.75

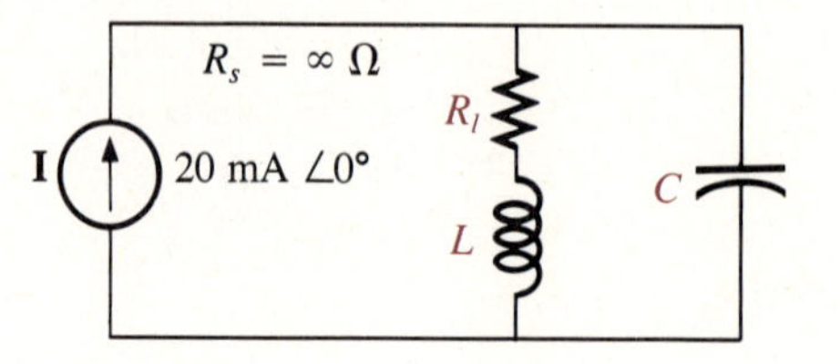

Section 18.11

22. Using semilog paper:
 a. Plot X_L versus frequency for a 10-mH coil for a frequency range of 1 Hz to 1 MHz. Set the vertical scale by the maximum value of X_L.
 b. Plot X_C versus frequency for a 1-μF capacitor for a frequency range of 10 Hz to 100 kHz. Choose an appropriate vertical scale for the frequency range of interest.

Section 18.12

23. For the high-pass R-C filter of Fig. 18.76:
 a. Determine the cutoff frequency.
 b. Plot V_o/V_i on semilog paper for a frequency range of 1 Hz to 1 MHz.
 c. Plot the phase response for the frequency range in part b.

FIG. 18.76

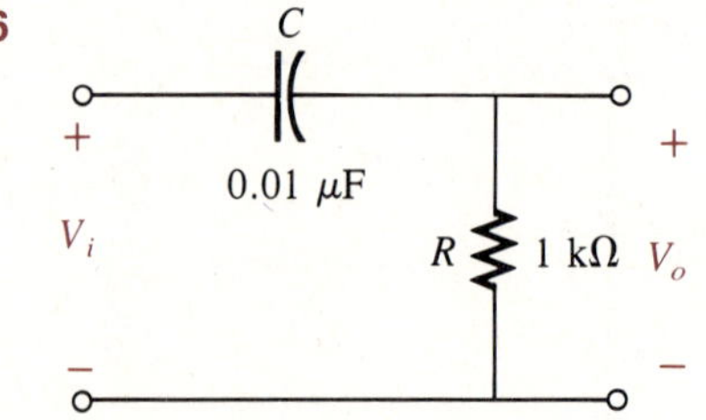

24. For a low-pass R-C filter having the same elements as in Problem 23:
 a. Sketch the network identifying V_o and V_i.
 b. Determine the cutoff frequency.
 c. Plot V_o/V_i on semilog paper for a frequency range of 1 Hz to 1 MHz.
 d. Plot the phase response for the frequency range of part c.

***25.** Design a high-pass R-C filter having a cutoff frequency of 2 kHz given a capacitor of 0.1 μF. Sketch the response V_o/V_i on semilog paper for a frequency range of 10 Hz to 100 kHz.

Section 18.13

*26. a. Sketch the Bode plot for the frequency response of a high-pass R-C filter if $R = 0.47\ \text{k}\Omega$ and $C = 0.05\ \mu\text{F}$.
b. Using the results of part a, sketch the actual frequency response for the same frequency range.
c. Determine the decibel level at frequencies equal to one-half and twice the cutoff frequency.
d. Determine the gain V_o/V_i at both frequencies of part c.
e. Sketch the phase response for the same frequency range.

*27. Repeat Problem 26 for an R-C low-pass filter if $R = 12\ \text{k}\Omega$ and $C = 0.001\ \mu\text{F}$.

Section 18.14

28. For the pass-band filter of Fig. 18.77:
a. Determine Q_s.
b. Find the cutoff frequencies.
c. Sketch the frequency characteristics.
d. Find $Q_{s(\text{loaded})}$ if a load of 200 Ω is applied.
e. Indicate on the curve of part c the change in the frequency characteristics with the load applied.

FIG. 18.77

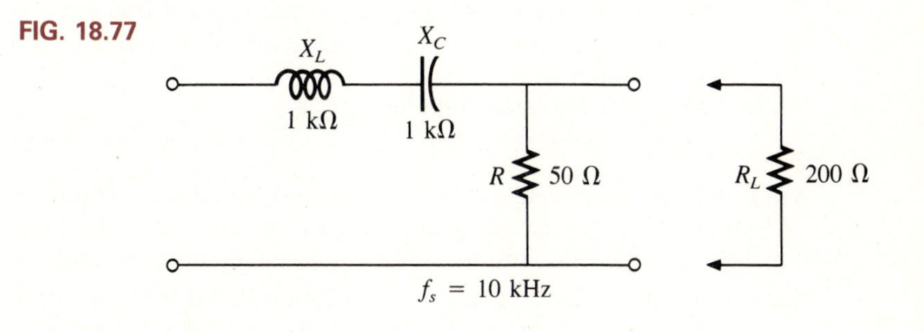

*29. For the pass-band filter of Fig. 18.78:
a. Determine Q_p ($R_L = \infty\ \Omega$, open circuit).
b. Sketch the frequency characteristics.

FIG. 18.78

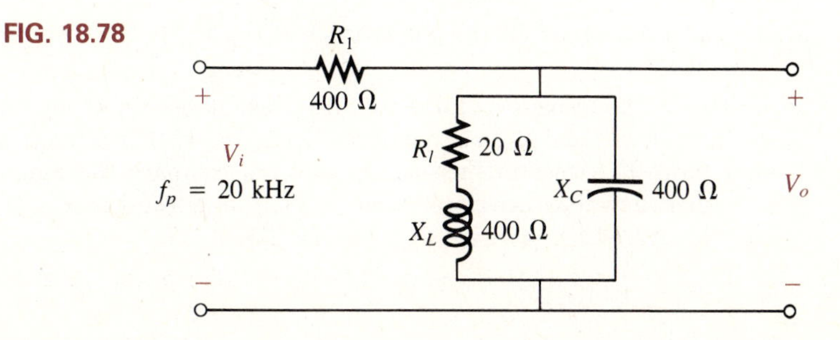

30. For the band-stop filter of Fig. 18.79:
 a. Determine Q_s.
 b. Find the bandwidth and half-power frequencies.
 c. Sketch the frequency characteristics.
 d. What is the effect on the curve of part c if a load of 2 kΩ is applied?

FIG. 18.79

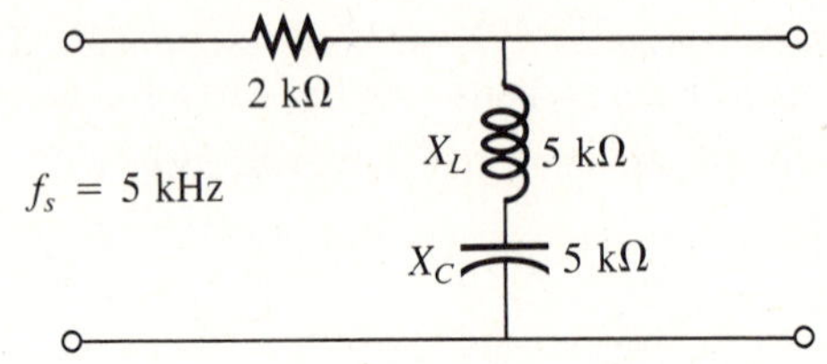

***31.** The network of Problem 29 is used as a band-stop filter.
 a. Sketch the network when used as band-stop filter.
 b. Sketch the band-stop characteristics.

32. **a.** For the network of Fig. 18.62(a), if $L_p = 400\ \mu\text{H}$ ($Q \geq 10$), $L_s = 60\ \mu\text{H}$, and $C = 120$ pF, determine the rejected and accepted frequencies.
 b. Sketch the response curve for part a.

33. **a.** For the network of Fig. 18.62(b), if the rejected frequency is 30 kHz and the accepted frequency is 100 kHz, determine the values of L_s and L_p ($Q \geq 10$) for a capacitance of 200 pF.
 b. Sketch the response curve for part a.

Computer Problems

34. Write a program to tabulate the impedance and current of the network of Fig. 18.2 versus frequency for a frequency range extending from $0.1f_s$ to $2f_s$ in increments of $0.1f_s$. For the first run, use the parameters defined by Example 18.1.

35. Write a program to provide a general solution for the network of Fig. 18.27. That is, given the parameters appearing in Fig. 18.27, determine the quantities appearing in parts a through f of Example 18.5. For the first run, use the parameters appearing in Example 18.5, and compare results.

36. Write a program that will tabulate $V_o/V_i = R/\sqrt{R^2 + X_C^2}$ versus frequency for a frequency range extending from $0.1f_1$ to $2f_1$ in increments of $0.1f_1$. Note whether $f = f_1$ when $V_o/V_i = 0.707$.

37. Repeat Problem 36 for the phase angle of Eq. 18.40. Note whether $\theta = 45°$ when $f = f_1$.

38. Write a program to tabulate $A_{v_{dB}}$ as determined by $A_{v_{dB}} = 20 \log_{10}[1/\sqrt{1 + (f_1/f)^2}]$ and $A_{v_{dB}}$ as calculated by Eq. 18.47. For a frequency range extending from $0.1f_1$ to f in increments of $0.1f_1$, compare the magnitudes, and note whether the values are closer when $f << f_1$ and whether $A_{v_{dB}} = -3$ dB at $f = f_1$ for the preceding equation and zero for Eq. 18.47.

19

Transformers

OBJECTIVES

- ☐ Become aware of the mutual inductance between coils and how it is related to the inductance of each coil and the coefficient of coupling between the coils.
- ☐ Understand how the total inductance of mutually coupled inductors is determined.
- ☐ Be able to calculate the primary and secondary voltage or current using the transformation ratio.
- ☐ Become aware of how the turns ratio affects the input impedance of a transformer and the relationship between the input and output power of the ideal transformer.
- ☐ Understand the derivation and impact of each element appearing in the equivalent circuit of the iron-core transformer on the various impedance and voltage levels.
- ☐ Appreciate the impact of applied frequency levels on the output or secondary voltage of a transformer.
- ☐ Understand how a transformer can be used as an isolation device and the true meaning of the nameplate data.
- ☐ Recognize the different types of transformers and how the area of application affects their design.
- ☐ Be able to calculate the various voltage and current levels for an autotransformer and a tapped or multiple-load transformer.

19.1 INTRODUCTION

Chapter 11 introduced the concept of the *self-inductance* of a coil. This chapter examines the *mutual inductance* that exists between coils of the same or different dimensions. Mutual inductance is a phenomenon basic to the operation of the *transformer,* an electrical device used today in almost every field of electrical engineering. This device plays an integral part in power distribution systems and can be found in many electronic circuits and measuring instruments. In this chapter, we discuss three of the basic applications of a transformer: its ability to build up or step down the voltage or current, to act as an impedance matching device, and to isolate (no physical connection) one portion of a circuit from another. In addition, the transformer equivalent circuit is introduced along with its use as an autotransformer and multiple-load device.

19.2 MUTUAL INDUCTANCE

The transformer is constructed of two coils placed so that the changing flux developed by one will link the other, as shown in Fig. 19.1. Keep in mind for the analysis to follow that the voltages induced across the coils of a transformer are directly proportional to the *rate of change* of the voltage and not simply its magnitude. The absence of a changing flux negates any transformer action. To distinguish between the coils, we will apply the transformer convention that the coil to which the source is applied is called the *primary,* and the coil to which the load is applied is called the *secondary*.

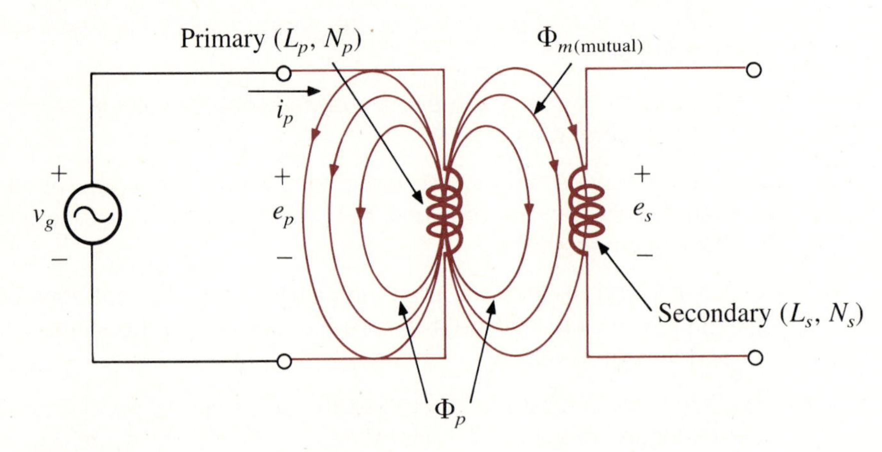

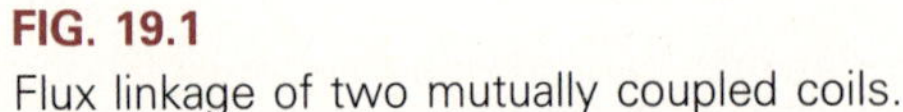
FIG. 19.1
Flux linkage of two mutually coupled coils.

The changing flux generated by the primary is denoted by Φ_p in Fig. 19.1. The mutual flux $\Phi_{m(\text{mutual})}$ is that portion of Φ_p that also links the secondary coil, as shown in same figure.

The *coefficient of coupling* between two coils is determined by

$$k \text{ (coefficient of coupling)} = \frac{\Phi_m}{\Phi_p} \tag{19.1}$$

Since the maximum changing flux that can link the secondary is Φ_p, the coefficient of coupling between two coils can never be greater than 1. The coefficient of coupling between various coils is indicated in Fig. 19.2. Note that for the iron core, $k \cong 1$, whereas for the air core, k is considerably less. Those coils with low coefficients of coupling are said to be *loosely coupled.*

FIG. 19.2
Typical levels of coefficient of coupling.

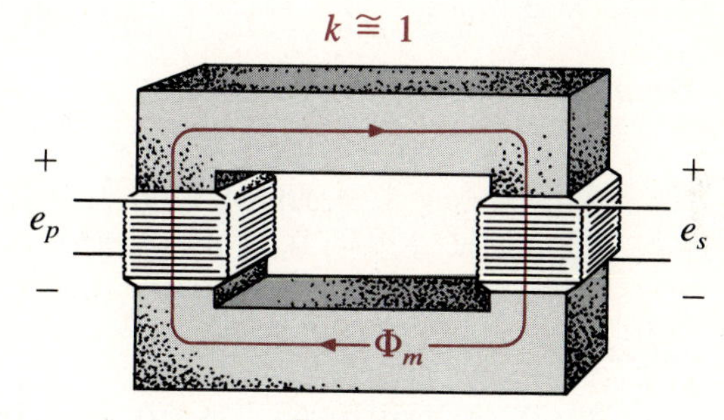

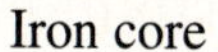

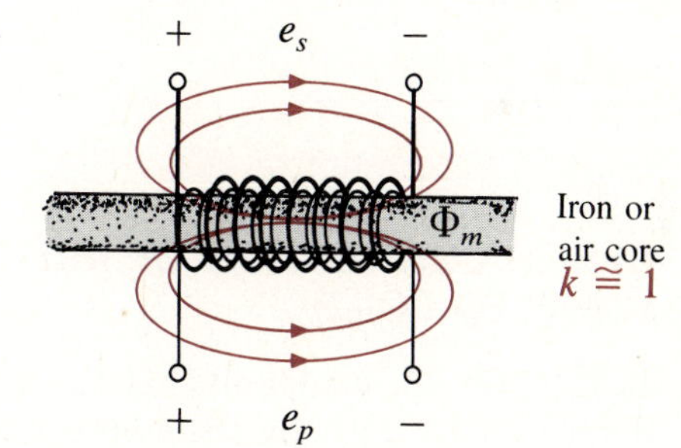

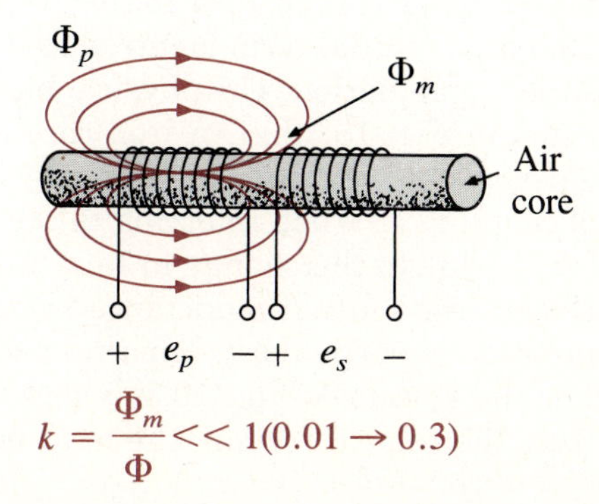

In terms of the inductance of each coil and the coefficient of coupling, the mutual inductance (measured in henries) is determined by

$$M = k\sqrt{L_p L_s} \quad \text{(henries, H)} \tag{19.2}$$

The greater the coefficient of coupling (greater flux linkages) or the greater the inductance of either coil, the higher the mutual inductance between the coils. Relate this fact to the configurations of Fig. 19.2.

EXAMPLE 19.1 For the transformer of Fig. 19.1, determine the mutual inductance if

$$k = 0.6, \quad L_p = 200 \text{ mH}, \quad \text{and} \quad L_s = 400 \text{ mH}$$

Solution Using Eq. 19.2:

$$\begin{aligned} M &= k\sqrt{L_p L_s} \\ &= (0.6)\sqrt{(200 \times 10^{-3} \text{ H})(400 \times 10^{-3})} \\ &= 0.6\sqrt{8 \times 10^{-2}} = 0.6(2.828 \times 10^{-1}) \\ M &= \mathbf{169.7 \text{ mH}} \end{aligned}$$

Calculator

(2) (0) (0) (EXP) (+/−) (3) (×) (4) (0) (0) (EXP) (+/−) (3) (=)
(√‾) (×) (·) (6) (=)

with a display of 0.1697.

19.3 SERIES CONNECTION OF MUTUALLY COUPLED COILS

In Chapter 11, we found that the total inductance of series isolated coils was determined simply by the sum of the inductances. For two coils that are connected in series but also share the same flux linkages, such as those in Fig. 19.3(a), a mutual term is introduced that will alter the total inductance of the series combination. The physical picture of how the coils are connected is indicated in Fig. 19.3(b). An iron core is included, although the equations to be developed are for any two mutually coupled coils with any value of coefficient of coupling k. When referring to the voltage induced across the inductance L_1 (or L_2) due to the change in flux linkages of the inductance L_2 (or L_1, respectively), the mutual inductance is represented by M_{12}. This type of subscript notation is particularly important when there are two or more mutual terms. For the system of Fig. 19.3, where the coils are wound such that Φ_1 and Φ_2 have the same direction, the total inductance of the mutually coupled coils is given by

$$L_{T(+)} = L_1 + L_2 + 2M_{12} \quad \text{(H)} \tag{19.3}$$

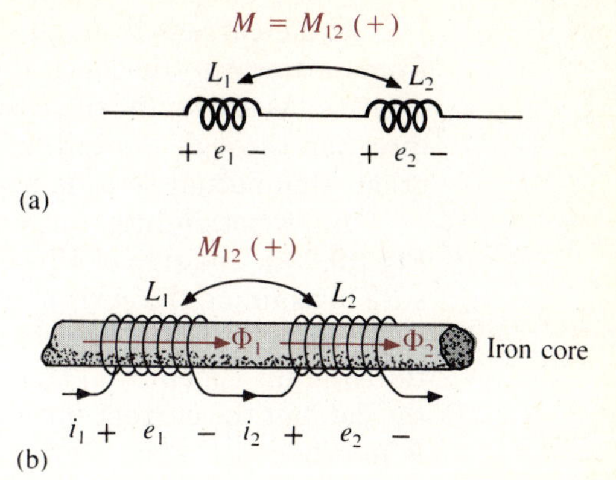

FIG. 19.3
Mutually coupled coils connected in series.

The subscript (+) is included to indicate that the mutual terms have a positive sign. If the coils were wound as shown in Fig. 19.4, where Φ_1 and Φ_2 are in opposition, the induced voltage due to the mutual term would oppose that due to the self-inductance, and the total inductance would be determined by

$$\boxed{L_{T(-)} = L_1 + L_2 - 2M_{12}} \quad \text{(H)} \qquad \textbf{(19.4)}$$

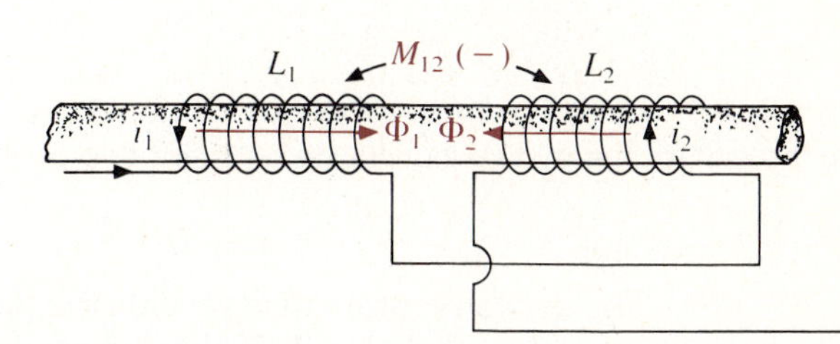

FIG. 19.4
Mutually coupled coils connected in series with negative mutual inductance.

From the preceding, it should be clear that the mutual inductance directly affects the magnitude of the voltage induced across a coil, since it determines the net inductance of the coil. Further examination reveals that the sign of the mutual term for each coil of a coupled pair is the same. For $L_{T(+)}$ they were both positive, and for $L_{T(-)}$ they were both negative. On a network schematic where it is inconvenient to indicate the windings and the flux path, a system of dots is employed that will determine whether the mutual terms are to be positive or negative. The dot convention is shown in Fig. 19.5 for the series coils of Figs. 19.2 and 19.4.

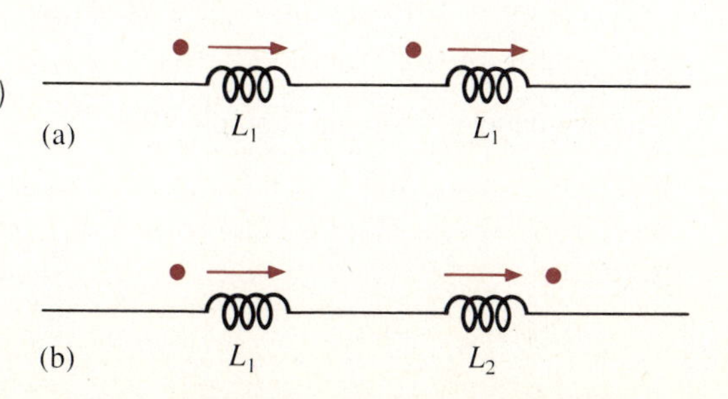

FIG. 19.5
Dot convention for the series coils of (a) Fig. 19.3 and (b) Fig. 19.4.

If the current through *each* of the mutually coupled coils is going away from (or toward) the dot as it *passes through the coil,* the mutual term will be positive, as shown for the case in Fig. 19.3(a). If the arrow indicating current direction through the coil is leaving the dot for one coil and entering for the other, the mutual term is negative.

A few possibilities for mutually coupled transformer coils are indicated in Fig. 19.6(a). The sign of M is indicated for each. When determining the sign, be sure to examine the current direction within the coil itself. In Fig. 19.6(b), one direction was indicated outside for one coil and through for the other. It initially might appear that the sign should be positive, since both currents enter the dot, but the current *through* coil 1 is leaving the dot; hence a negative sign is in order.

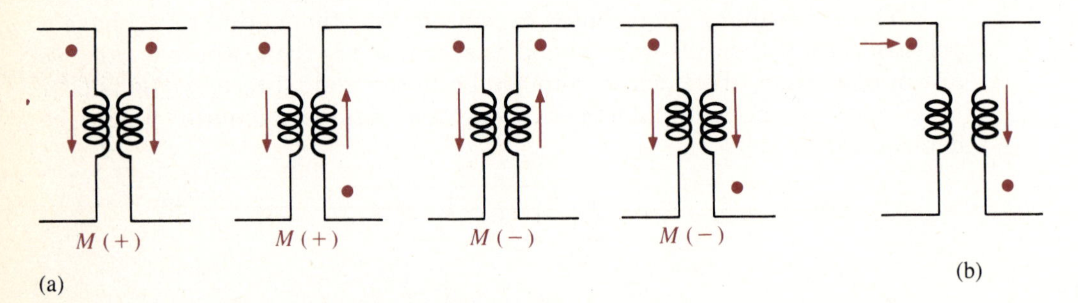

FIG. 19.6
Determining the sign of the mutual inductance between two coils using the dot convention.

The dot convention also reveals the polarity of the *induced* voltage across the mutually coupled coil. If the reference direction for the current *in* a coil leaves the dot, the polarity at the dot for the induced voltage of the mutually coupled coil is positive. In the first two figures of Fig. 19.6(a), the polarity at the dots of the induced voltages is positive. In the third figure of Fig. 19.6(a), the polarity at the dot of the right-hand coil is positive, whereas the polarity at the dot of the left-hand coil is negative, since the current enters the dot (within the coil) of the right-hand coil. The comments for the third figure of Fig. 19.6(a) can also be applied to the last figure of Fig. 19.6(a).

EXAMPLE 19.2 Find the total inductance of the series coils of Fig. 19.7.

FIG. 19.7
Mutually coupled coils for Example 19.2.

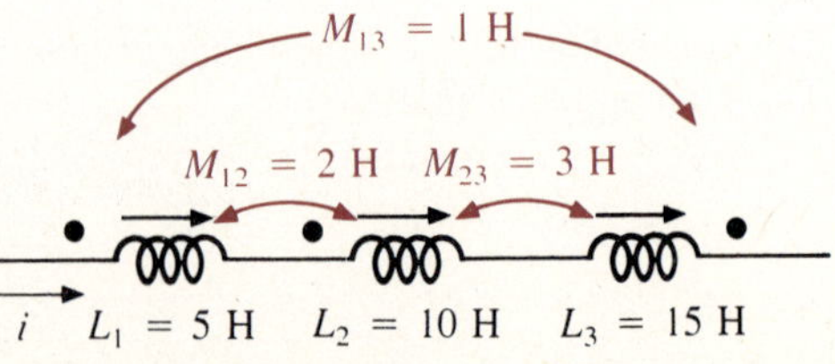

Solution

current vectors leave dot

Coil 1: $L_1 + M_{12} - M_{13}$

one current vector enters dot while one leaves

Coil 2: $L_2 + M_{12} - M_{23}$

Coil 3: $L_3 - M_{23} - M_{13}$

note that each has the same sign

and

$$\begin{aligned} L_T &= (L_1 + M_{12} - M_{13}) + (L_2 + M_{12} - M_{23}) + (L_3 - M_{23} - M_{13}) \\ &= L_1 + L_2 + L_3 + 2M_{12} - 2M_{23} - 2M_{13} \end{aligned}$$

Substituting values, we find

$$\begin{aligned} L_T &= 5\text{ H} + 10\text{ H} + 15\text{ H} + 2(2\text{ H}) - 2(3\text{ H}) - 2(1\text{ H}) \\ &= 34\text{ H} - 8\text{ H} = \mathbf{26\text{ H}} \end{aligned}$$

19.4 THE IRON-CORE TRANSFORMER

An iron-core transformer under loaded conditions is shown in Fig. 19.8. The iron core will serve to increase the coefficient of coupling between the coils by increasing the mutual flux Φ_m. Recall from Chapter 11 that magnetic flux lines always take the path of least reluctance, which in this case is the iron core.

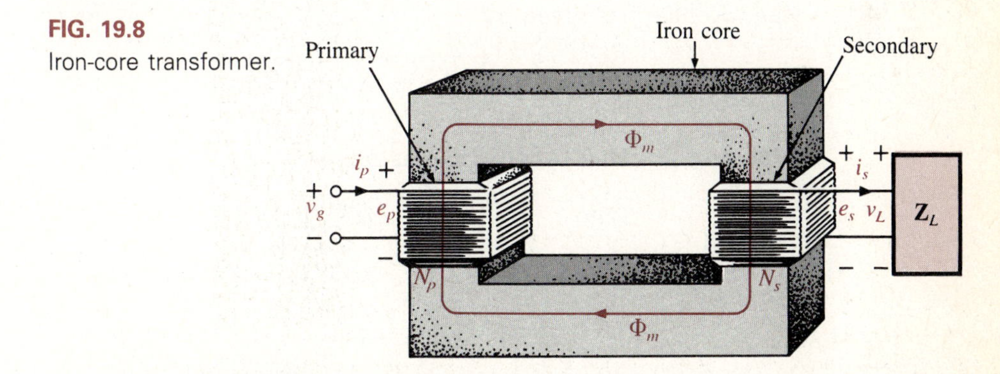

FIG. 19.8
Iron-core transformer.

We assume in the analyses to follow in this chapter that all the flux linking coil 1 links coil 2. In other words, the coefficient of coupling is its maximum value, 1, and $\Phi_m = \Phi_1$. In addition, we first analyze the transformer from an ideal viewpoint. That is, we neglect such losses as the geometric or dc resistance of the coils, the leakage reactance due to the flux linking either coil that forms no part of Φ_m, and the hysteresis and eddy current losses.

This is not to convey the impression, however, that we will be far from the actual operation of a transformer. Most transformers manufactured today can be considered almost ideal. The equations we develop under ideal conditions are, in general, a first approximation to the actual response, which will never be off by more than a few percent.

When the current i_1 through the primary circuit of the iron-core transformer is a maximum, the flux Φ_m linking both coils is also a maximum. In fact, the magnitude of the flux is directly proportional to the current through the primary windings. Therefore, the two are in phase, and for sinusoidal inputs, the magnitude of the flux varies as a sinusoid also. That is, if

$$i_p = \sqrt{2}I_p \sin \omega t$$

then

$$\Phi_m = \Phi_m \sin \omega t$$

In terms of the applied frequency, primary turns, and peak value of the flux in the core, the effective value of the primary voltage is determined by

$$\boxed{E_p = 4.44 f N_p \Phi_m} \tag{19.5}$$

In particular, note the presence of the frequency term in the equation. Recall from the discussion of inductance in Chapter 11 that the magnitude of the induced voltage is directly related to the rate of change of flux linking the coil. This fact is included in the frequency term of the equation, since the higher the frequency, the greater the rate of change of flux linking the coils.

For this situation, where it is assumed that the flux linking the secondary equals that of the primary, the effective value of the induced voltage across the secondary is given by

$$\boxed{E_s = 4.44 f N_s \Phi_m} \tag{19.6}$$

Dividing Eq. 19.5 by Eq. 19.6, as follows,

$$\frac{E_p}{E_s} = \frac{4.44 f N_p \Phi_m}{4.44 f N_s \Phi_m}$$

we obtain

$$\boxed{\frac{E_p}{E_s} = \frac{N_p}{N_s}} \tag{19.7}$$

Note that the ratio of the magnitudes of the induced voltages is the same as the ratio of the corresponding turns. The induced voltages are also in phase, and Eq. 19.7 can be changed to include phasor notation; that is,

$$\boxed{\frac{\mathbf{E}_p}{\mathbf{E}_s} = \frac{N_p}{N_s}} \tag{19.8}$$

or, since $\mathbf{V}_g = \mathbf{E}_1$ and $\mathbf{V}_L = \mathbf{E}_2$ for the ideal situation,

$$\boxed{\frac{\mathbf{V}_g}{\mathbf{V}_L} = \frac{N_p}{N_s}} \tag{19.9}$$

The ratio N_p/N_s, usually represented by the lowercase letter a, is referred to as the *transformation ratio:*

$$\boxed{a = \frac{N_p}{N_s}} \tag{19.10}$$

Equation 19.7 can then be written as

$$\frac{E_p}{E_s} = a$$

or

$$E_p = aE_s$$

If $a < 1$, the transformer is called a *step-up transformer,* since the voltage $E_s > E_p$. If $a > 1$, the transformer is called a *step-down transformer,* since $E_s < E_p$.

EXAMPLE 19.3

For the iron-core transformer of Fig. 19.9,

a. Find the maximum flux Φ_m.
b. Find the number of turns N_s.

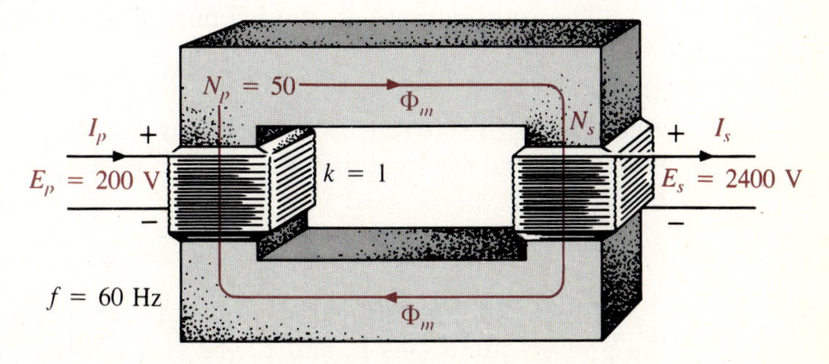

FIG. 19.9
Transformer for Example 19.3.

Solutions

a. $E_p = 4.44N_pf\Phi_m$
Therefore,

$$\Phi_m = \frac{E_p}{4.44N_pf} = \frac{200\text{ V}}{(4.44)(50\text{ turns})(60\text{ Hz})}$$

and

$$\Phi_m = \mathbf{15.02\ mWb}$$

b. $\dfrac{E_p}{E_s} = \dfrac{N_p}{N_s}$
Therefore,

$$N_s = \frac{N_pE_s}{E_p} = \frac{(50\text{ turns})(2400\text{ V})}{200\text{ V}}$$
$$= \mathbf{600\ turns}$$

The induced voltage across the secondary of the transformer of Fig. 19.8 establishes a current i_s through the load Z_L and the secondary windings. This current and the turns N_s develop an mmf N_si_s that would not be present under no-load conditions, since $i_s = 0$ and $N_si_s = 0$. Under loaded or unloaded conditions, however, the net ampere-turns on the core produced by both the primary and the secondary must remain unchanged for the same flux Φ_m to be established in the core. The flux Φ_m must remain the same to have the same induced voltage across the primary to balance the voltage impressed across the primary. In order to counteract the mmf of the secondary, which is tending to change Φ_m, an additional current must flow in the primary. This current is called the *load component of the primary current* and is represented by the notation i'_p.

For the balanced or equilibrium condition,

$$N_pi'_p = N_si_s$$

The total current in the primary under loaded conditions is

$$i_p = i'_p + i_{\phi_m}$$

where i_{ϕ_m} is the current in the primary necessary to establish the flux Φ_m. For most practical applications, $i'_p > i_{\phi_m}$ and the assumption made that $i_p \cong i'_p$, resulting in

$$N_pi_p = N_si_s$$

Since the instantaneous values of i_p and i_s are related by the turns ratio, the phasor quantities $\mathbf{I}_p$ and $\mathbf{I}_s$ are also related by the same ratio:

$$N_p\mathbf{I}_p = N_s\mathbf{I}_s$$

or

$$\boxed{\frac{\mathbf{I}_p}{\mathbf{I}_s} = \frac{N_s}{N_p}} \tag{19.11}$$

Keep in mind that Eq. 19.11 holds true only if we neglect the effects of i_{ϕ_m}.

For the step-up transformer, $a < 1$, and the current in the secondary, $I_s = aI_p$, is less in magnitude than that in the primary. For a step-down transformer, the reverse is true.

19.5 REFLECTED IMPEDANCE AND POWER

In the previous sections, we found that

$$\frac{\mathbf{V}_g}{\mathbf{V}_L} = \frac{N_p}{N_s} \quad \text{and} \quad \frac{\mathbf{I}_p}{\mathbf{I}_s} = \frac{N_s}{N_p}$$

Dividing one by the other, we have

$$\frac{\mathbf{V}_g/\mathbf{V}_L}{\mathbf{I}_p/\mathbf{I}_s} = \frac{N_p/N_s}{N_s/N_p} = \frac{a}{1/a}$$

or

$$\frac{\mathbf{V}_g/\mathbf{I}_p}{\mathbf{V}_L/\mathbf{I}_s} = a^2$$

and

$$\frac{\mathbf{V}_g}{\mathbf{I}_p} = a^2\frac{\mathbf{V}_L}{\mathbf{I}_s}$$

However, since

$$\mathbf{Z}_p = \frac{\mathbf{V}_g}{\mathbf{I}_p} \quad \text{and} \quad \mathbf{Z}_L = \frac{\mathbf{V}_L}{\mathbf{I}_s}$$

then

$$\boxed{\mathbf{Z}_p = a^2\mathbf{Z}_L} \tag{19.12}$$

which in words states that the impedance of the primary circuit of an ideal transformer is the transformation ratio squared times the impedance of the load. If a transformer is used, therefore, an impedance can be made to appear larger or smaller at the primary by placing it in the secondary of a step-down ($a > 1$) or step-up ($a < 1$) transformer, respectively. Note that if the load is capacitive or inductive, the reflected impedance will also be capacitive or inductive.

For the ideal iron-core transformer,

$$\frac{E_p}{E_s} = a = \frac{I_s}{I_p}$$

or

$$\boxed{E_pI_p = E_sI_s} \qquad (19.13)$$

and

$$\boxed{P_{\text{in}} = P_{\text{out}}} \quad \text{(ideal case)} \qquad (19.14)$$

EXAMPLE 19.4 For the iron-core transformer of Fig. 19.10,

a. Find the magnitude of the current in the primary and the impressed voltage across the primary.

b. Find the input resistance of the transformer.

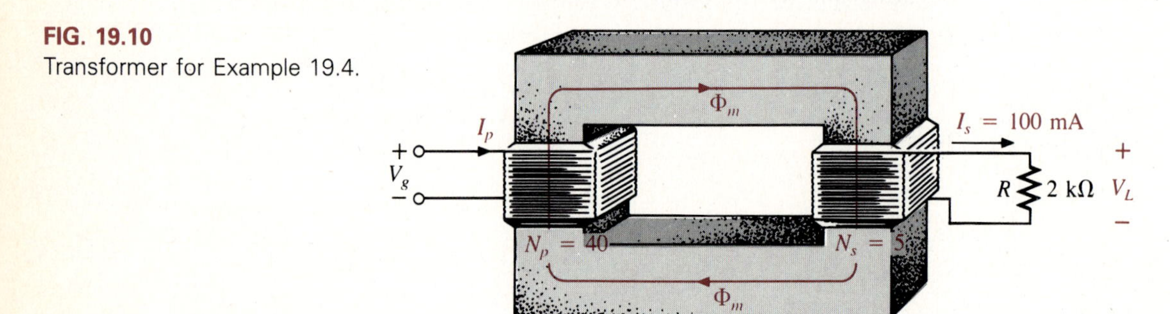

FIG. 19.10
Transformer for Example 19.4.

Solutions

a. $\dfrac{I_p}{I_s} = \dfrac{N_s}{N_p}$

or

$$I_p = \frac{N_s}{N_p}I_s = \left(\frac{5}{40}\right)(100 \text{ mA})$$

and

$$I_p = \mathbf{12.5\ mA}$$
$$V_L = I_sZ_L = (100 \text{ mA})(2 \text{ k}\Omega)$$
$$= 200 \text{ V}$$

and

$$\frac{V_g}{V_L} = \frac{N_p}{N_s}$$

or

$$V_g = \frac{N_p}{N_s}V_L = \left(\frac{40}{5}\right)(200\text{ V})$$

and

$$V_g = \mathbf{1600\ V}$$

b. $Z_p = a^2 Z_L$

$$a = \frac{N_p}{N_s} = 8$$

$$Z_p = (8)^2(2\text{ k}\Omega)$$
$$= R_p = \mathbf{128\ k\Omega}$$

EXAMPLE 19.5 For the speaker in Fig. 19.11 to receive maximum power from the circuit, the internal resistance of the speaker should be 540 Ω. If a transformer is used, the speaker resistance of 15 Ω is made to appear 540 Ω at the primary. Find the transformation ratio required and the number of turns in the primary if the secondary winding has 40 turns.

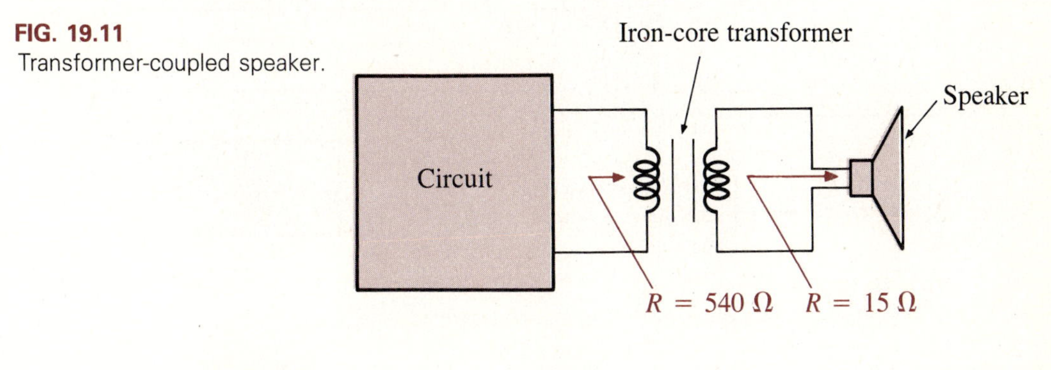

FIG. 19.11
Transformer-coupled speaker.

Solution

$$Z_p = a^2 Z_L$$

or

$$a = \sqrt{\frac{Z_p}{Z_L}} = \sqrt{\frac{540\ \Omega}{15\ \Omega}} = \sqrt{36} = 6$$

Therefore,

$$a = 6 = \frac{N_p}{N_s}$$

or

$$N_p = 6N_s = 6(40\text{ turns})$$
$$= \mathbf{240\ turns}$$

EXAMPLE 19.6 For the residential supply appearing in Fig. 19.12, determine (assuming a totally resistive load) the following:

a. The primary current I_p
b. The magnitude of I_1 and I_2
c. The total power delivered
d. The turns ratio N_1/N_2

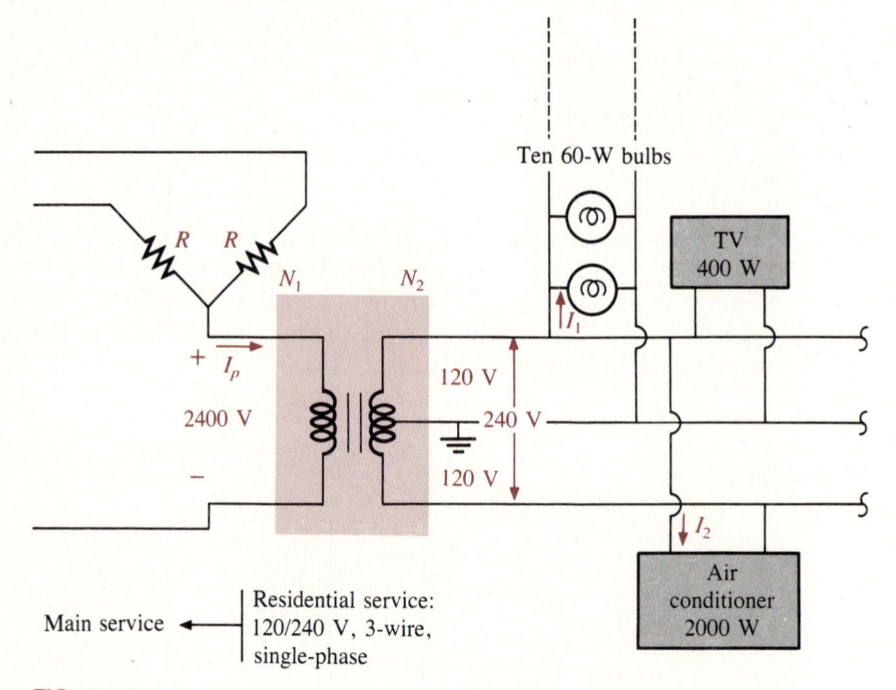

FIG. 19.12
Three-phase main service with a single-phase residential load.

Solutions

a.
$$
\begin{aligned}
P_T &= 600 \text{ W} + 400 \text{ W} + 2000 \text{ W} = 3000 \text{ W} \\
P_{\text{in}} &= P_{\text{out}} \\
V_p I_p &= V_s I_s = 3000 \text{ W (purely resistive load)} \\
2400 I_p &= 3000 \\
I_p &= \mathbf{1.25\ A}
\end{aligned}
$$

b. $P = (10)(60 \text{ W}) = 600 \text{ W} = VI_1 = 120I_1$

and

$$I_1 = \mathbf{5\ A}$$
$$P = 2000 \text{ W} = VI_2 = 240I_2$$

and

$$I_2 = \mathbf{8.333\ A}$$

c. $P_T = 3P_\phi = 3(3000\ \text{W}) = \mathbf{9\ kW}$

d. $\dfrac{N_1}{N_2} = \dfrac{V_p}{V_s} = \dfrac{2400\ \text{V}}{240\ \text{V}} = \mathbf{10}$

19.6 EQUIVALENT CIRCUIT (IRON-CORE TRANSFORMER)

For the nonideal or practical iron-core transformer, the equivalent circuit appears as in Fig. 19.13. As indicated, part of this equivalent circuit is an ideal transformer. The remaining elements of Fig. 19.13 are those elements that contribute to the nonideal characteristics of the device. The resistances R_p and R_s are simply the dc or geometric resistance of the primary and secondary coils, respectively. For the primary and secondary coils of the transformer, there is a definite amount of flux that links each coil that does not pass through the core. This situation is shown in Fig. 19.14. This *leakage* flux, serving as a definite loss to the system, since it employs an amount of input energy to be established but serves no useful purpose, is represented by an inductance L_p in the primary circuit and an inductance L_s in the secondary.

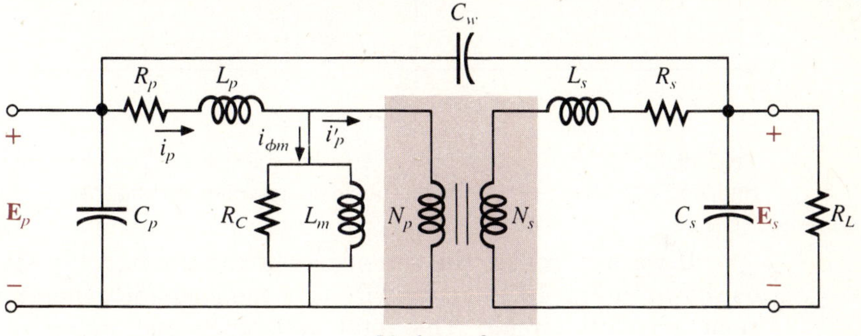

FIG. 19.13
Equivalent circuit for the practical iron-core transformer.

FIG. 19.14
Leakage flux.

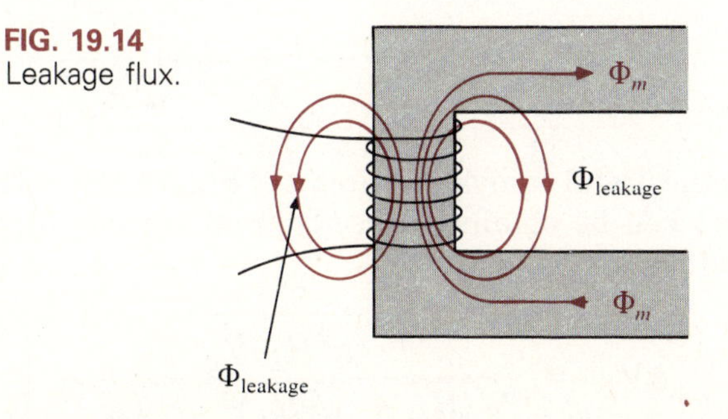

The resistance R_C represents the hysteresis and eddy current losses (core losses) within the core due to an ac flux through the core. The inductance L_m (magnetizing inductance) is the inductance associated with the magnetization of the core, that is, the establishing of the flux Φ_m in the core. The capacitances C_p and C_s are the lumped capacitances of the primary and secondary circuits, respectively, and C_w represents the equivalent lumped capacitances between the windings of the transformer.

Since i'_p is normally considerably larger than i_{Φ_m}, we ignore i_{Φ_m} for the moment (set it equal to zero), resulting in the absence of R_C and L_m in the reduced equivalent circuit of Fig. 19.15. In addition, the capacitances C_p, C_w, and C_s do not appear in the equivalent circuit of Fig. 19.15, since their reactance in the present frequency range of interest does not appreciably affect the transfer characteristics of the transformer.

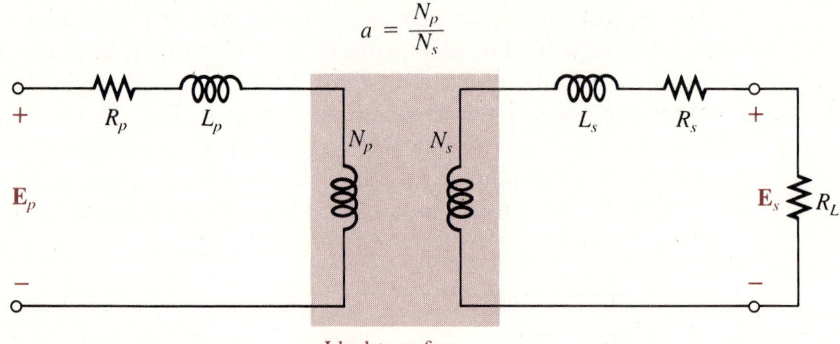

FIG. 19.15
Reduced equivalent circuit for the nonideal iron-core transformer.

If we now reflect the secondary circuit through the ideal transformer, as shown in Fig. 19.16(a), we will have the load and generator voltage in the same physical circuit. The total resistance and inductive reactance are determined by

$$R_{\text{equivalent}} = R_e = R_p + a^2 R_s \tag{19.15}$$

and

$$X_{\text{equivalent}} = X_e = X_p + a^2 X_s \tag{19.16}$$

which result in the useful equivalent circuit of Fig. 19.16(b). The magnitude of the load voltage can be obtained directly from the circuit of Fig. 19.16(b) through the voltage divider rule:

$$aV_L = \frac{(a^2 R_L)V_g}{\sqrt{(R_e + a^2 R_L)^2 + (X_e)^2}} \tag{19.17}$$

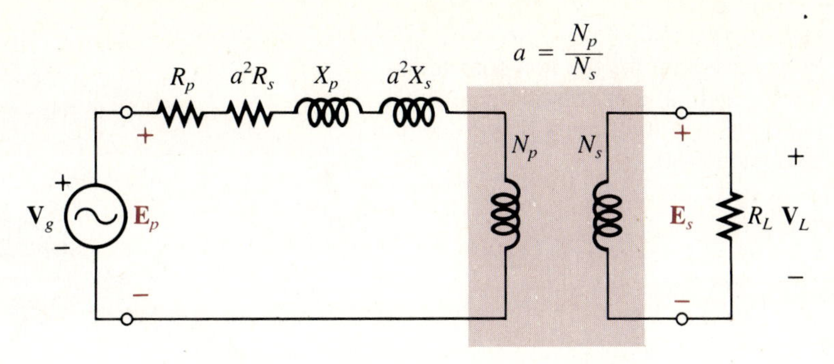

(a)

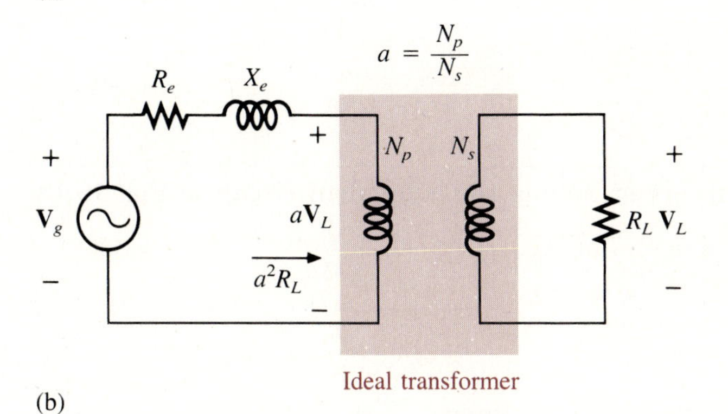

(b)

FIG. 19.16
Developing a convenient equivalent circuit for the transformer for a range of practical applications.

Equation 19.17 can also be used to determine the generator voltage necessary to establish a particular load voltage. The voltages across the elements of Fig. 19.16(b) have the phasor relationship indicated in Fig. 19.16(a). Note that the current is the reference phasor for drawing the phasor diagram. That is, the voltages across the resistive elements are *in phase* with the current phasor, whereas the voltage across the equivalent inductance leads the current by 90°. The primary voltage, by Kirchhoff's voltage law, is then the phasor sum of these voltages, as indicated in Fig. 19.17(a). For an inductive load, the phasor diagram appears in Fig. 19.17(b). Note that $a\mathbf{V}_L$ leads $\mathbf{I}$ by the power-factor angle of the load. The remainder of the diagram is then similar to that for a resistive load. (The phasor diagram for a capacitive load is left to the reader as an exercise.)

The effect of R_e and X_e on the magnitude of $\mathbf{V}_g$ for a particular $\mathbf{V}_L$ is obvious from Eq. 19.17 or Fig. 19.17. For increased values of R_e or X_e, an increase in $\mathbf{V}_g$ is required for the same load voltage. For R_e and $X_e = 0$, $\mathbf{V}_L$ and $\mathbf{V}_g$ are related by the turns ratio.

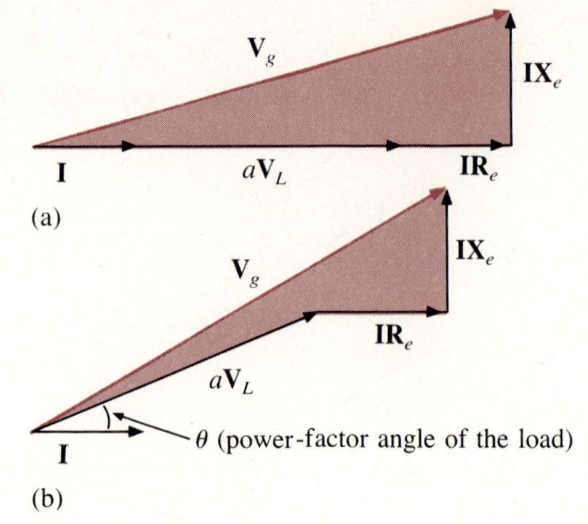

FIG. 19.17
Phasor diagram for the iron-core transformer with (a) unity power-factor load (resistive) and (b) lagging power-factor load (inductive).

EXAMPLE 19.7 For a transformer having the equivalent circuit of Fig. 19.18,

a. Determine R_e and X_e.
b. Determine V_g.

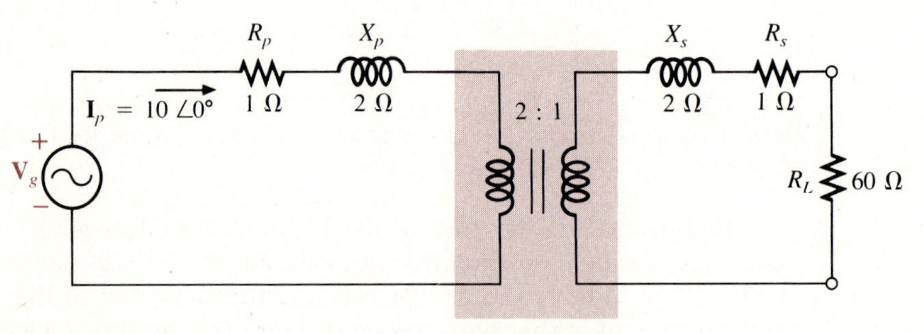

FIG. 19.18
Transformer for Example 19.7.

Solutions

a. $R_e = R_p + a^2R_s = 1\ \Omega + (2)^2(1\ \Omega) = \mathbf{5\ \Omega}$
$X_e = X_p + a^2X_s = 2\ \Omega + (2)^2(2\ \Omega) = \mathbf{10\ \Omega}$

b. The transformed equivalent circuit appears in Fig. 19.19.

$$aV_L = (I_p)(a^2R_L) = 2400\ \text{V}$$

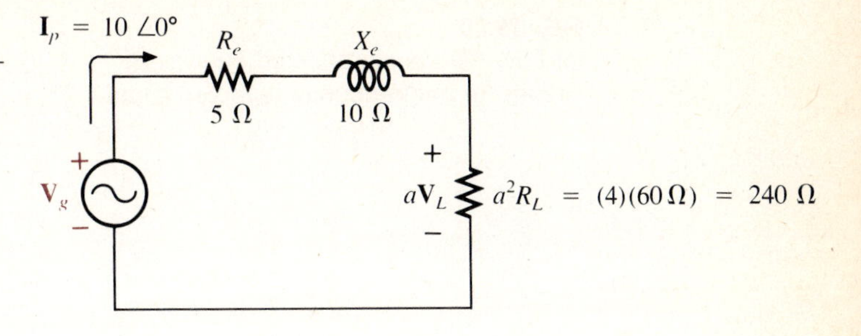

FIG. 19.19
Reduced equivalent circuit for the transformer of Fig. 19.18.

Thus,

$$V_L = \frac{2400\text{ V}}{a} = \frac{2400\text{ V}}{2} = 1200\text{ V}$$

and

$$\begin{aligned} V_g &= I_p Z_T \\ &= I_p\sqrt{(R_e + a^2R_L)^2 + (X_e)^2} \\ &= 10\text{ A}\sqrt{[5\ \Omega + (2^2)\ (60\ \Omega)]^2 + (10\ \Omega)^2} \\ &= 10\text{ A}\sqrt{(245\ \Omega)^2 + (10\ \Omega)^2} \\ &= (10\text{ A})(245.2\ \Omega) \\ V_g &= \mathbf{2452\ V} \end{aligned}$$

For R_e and $X_e = 0\ \Omega$, $V_g = aV_L = 2400$ V. Therefore, it is necessary to increase the generator voltage by 52 V (due to R_e and X_e) to obtain the same load voltage.

19.7 FREQUENCY CONSIDERATIONS

For certain frequency ranges, the effect of some parameters in the equivalent circuit of the iron-core transformer of Fig. 19.13 can be neglected. Since it is convenient to consider a low-, mid-, and high-frequency region, the equivalent circuits for each are now introduced and briefly examined.

For the low-frequency region, the reactance ($2\pi fL$) of the primary and secondary leakage reactances can be ignored, and the reflected equivalent circuit will appear as shown in Fig. 19.20(a). The magnetizing inductance must be included, since it appears in parallel with the secondary reflected circuit. As the frequency approaches zero, the reactance of the magnetizing inductance will reduce in magnitude, causing a reduction in the voltage across the secondary circuit. For $f = 0$ Hz, L_m is ideally a short circuit, and $V_L = 0$ V. As the frequency increases, the reactance of L_m will eventually be sufficiently large

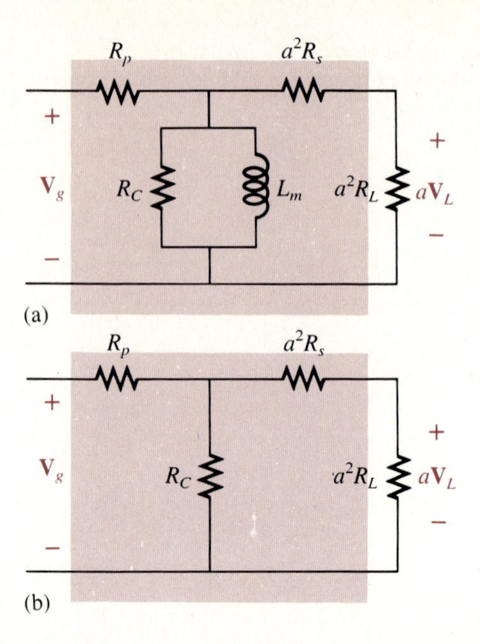

FIG. 19.20
(a) Low-frequency reflected equivalent circuit; (b) midfrequency reflected circuit.

compared with the reflected secondary impedance to be neglected. The midfrequency reflected equivalent circuit will then appear as shown in Fig. 19.20(b). Note the absence of reactive elements, resulting in an *in-phase* relationship between load and generator voltages.

For higher frequencies, the capacitive elements and primary and secondary leakage reactances must be considered, as shown in Fig. 19.21. For discussion purposes, the effects of C_w and C_s appear as a lumped capacitor C in the reflected network of Fig. 19.21; C_p does not appear, since the effect of C will predominate. As the frequency of interest increases, the capacitive reactance ($X_C = 1/2\pi fC$) will decrease to the point that it will have a shorting effect across the secondary circuit of the transformer, causing V_L to decrease in magnitude.

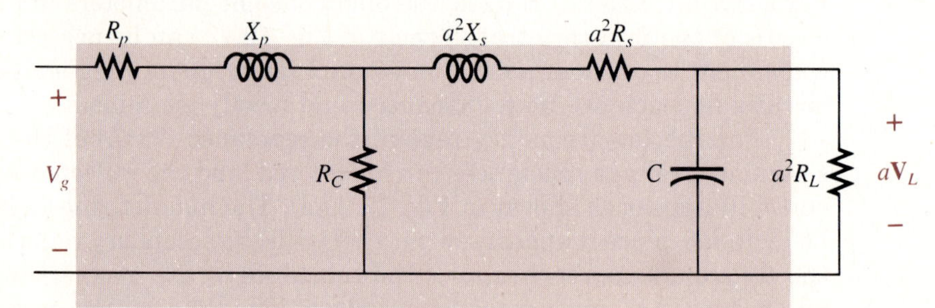

FIG. 19.21
High-frequency reflected equivalent circuit.

A typical transformer-frequency response curve appears in Fig. 19.22. For the low- and high-frequency regions, the primary element responsible for the drop-off is indicated. The peaking that occurs in the high-frequency region

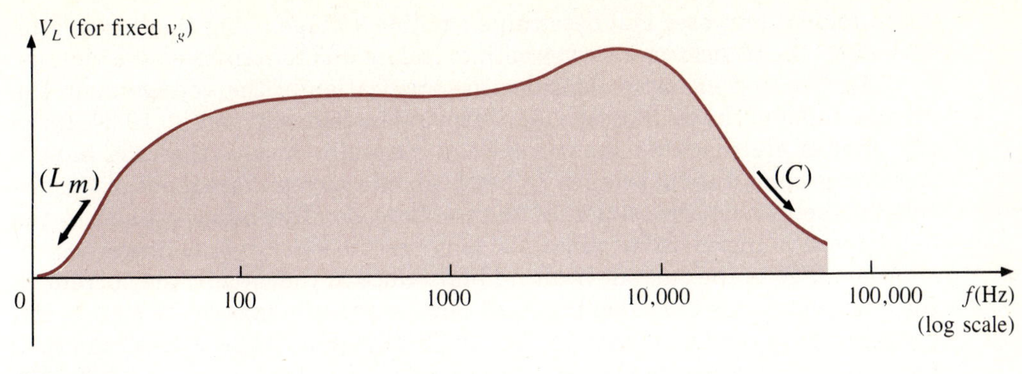

FIG. 19.22
Transformer-frequency response curve.

is due to the series resonant circuit of Fig. 19.21, composed of R_e, X_e, L, and C. In the peaking region, the series resonant circuit is in or near its resonant or tuned state.

The network discussed in some detail earlier in this chapter is for the high midfrequency region.

19.8 THE TRANSFORMER AS AN ISOLATION DEVICE

The transformer is frequently used to isolate one portion of an electrical system from another. By *isolation,* we mean the absence of any direct physical connection. As a first example of its use as an isolation device, consider the measurement of line voltages on the order of 40,000 V (Fig. 19.23).

FIG. 19.23
Transformer isolations.

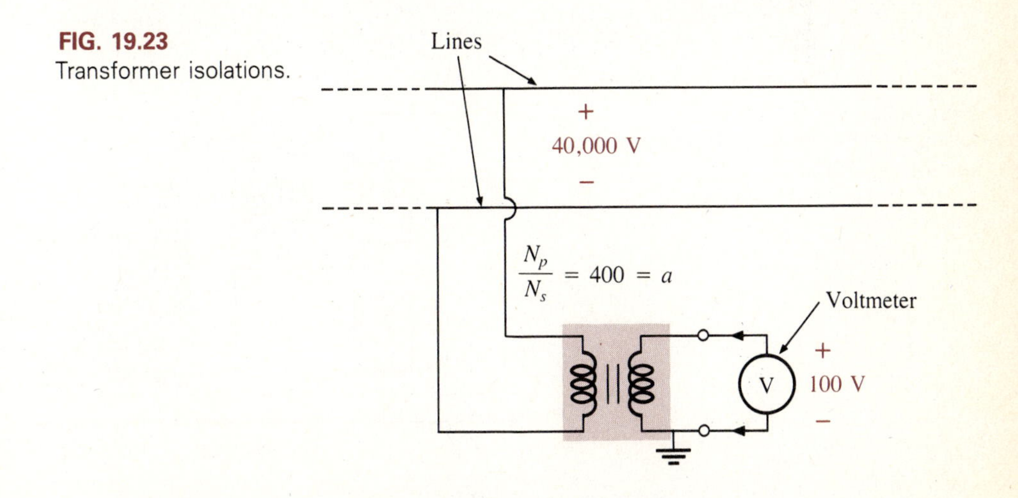

To apply a voltmeter across 40,000 V would obviously be a dangerous task due to the possibility of physical contact with the lines when making the necessary connections. By including a transformer in the transmission system as original equipment, one can bring the potential down to a safe level for measurement purposes and determine the line voltage using the turns ratio. Therefore, the transformer serves both to isolate and to step down the voltage.

As a second example, consider the application of the voltage v_x to the vertical input of the oscilloscope (a measuring instrument) in Fig. 19.24. If the connections are made as shown and the generator and oscilloscope have a common ground, the impedance $\mathbf{Z}_2$ has been effectively shorted out of the circuit by the ground connection of the oscilloscope. The input voltage to the oscilloscope is therefore meaningless as far as the voltage v_x is concerned. In addition, if $\mathbf{Z}_2$ is the current-limiting impedance in the circuit, the current in the circuit may rise to a level that will cause severe damage to the circuit. If a transformer is used as shown in Fig. 19.25, this problem will be eliminated, and the input voltage to the oscilloscope will be v_x.

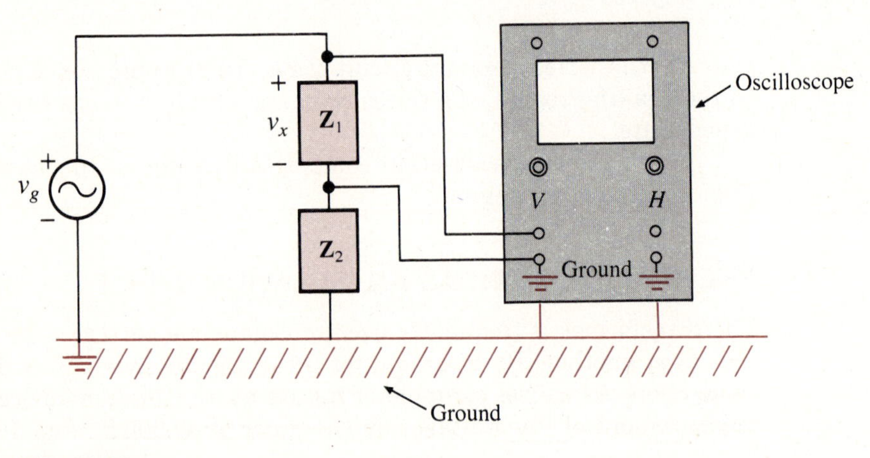

FIG. 19.24
Incorrect use of an ocilloscope to measure v_x.

FIG. 19.25
Using a transformer to correct the situation of Fig. 19.24.

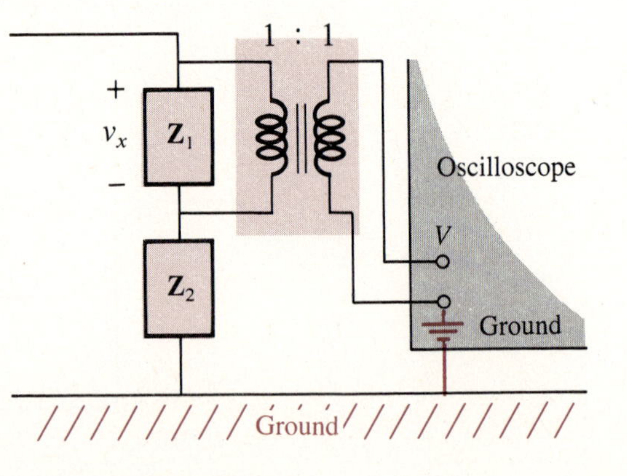

19.9 NAMEPLATE DATA

A typical power transformer rating might be the following:

5 kVA, 2000/100 V, 60 Hz

The 2000 V or the 100 V can be either the primary or the secondary voltage; that is, if 2000 V is the primary voltage, then 100 V is the secondary voltage and vice versa. The 5 kVA is the apparent power ($S = VI$) rating of the transformer. If the secondary voltage is 100 V, then the maximum load current is

$$I_L = \frac{S}{V_L} = \frac{5000}{100} = 50 \text{ A}$$

and if the secondary voltage is 2000 V, then the maximum load current is

$$I_L = \frac{S}{V_L} = \frac{5000}{2000} = 2.5 \text{ A}$$

The transformer is rated in terms of the apparent power rather than the average for the reason demonstrated by the circuit of Fig. 19.26.

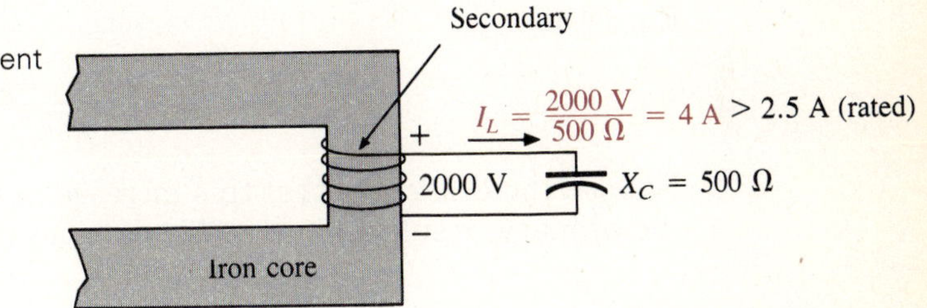

FIG. 19.26
Demonstrating the need for an apparent power rating of transformers.

Since the current through the load is greater than that determined by the apparent power rating, the transformer may be permanently damaged. Note, however, that since the load is purely capacitive, the average power to the load is zero. The wattage rating would therefore be meaningless regarding the ability of this load to damage the transformer.

The transformation ratio of the transformer under discussion can be either of two values. If the secondary voltage is 2000 V, the transformation ratio is $a = N_p/N_s = V_g/V_L = 100 \text{ V}/2000 \text{ V} = 1/20$, and the transformer is a step-up transformer. If the secondary voltage is 100 V, the transformation ratio is $a = N_p/N_s = V_g/V_L = 2000 \text{ V}/100 \text{ V} = 20$, and the transformer is a step-down transformer.

The rated primary current can be determined simply by applying Eq. 19.11:

$$I_1 = \frac{I_2}{a}$$

which is equal to [2.5 A/(1/20)] = 50 A if the secondary voltage is 2000 V and (50/20) = 2.5 A if the secondary voltage is 100 V.

To explain the necessity for including the frequency in the nameplate data, consider Eq. 19.5:

$$E_p = 4.44 f_p N_p \Phi_m$$

and the *B-H* curve for the iron core of the transformer (Fig. 19.27).

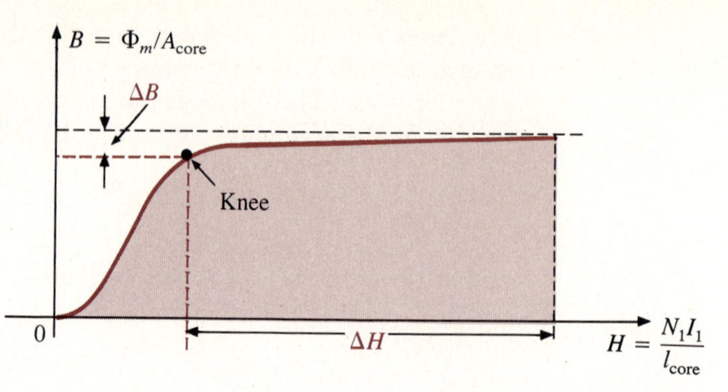

FIG. 19.27
Demonstrating the effect of an increase in flux density on the required magnetizing force when operating near saturation levels.

The point of operation on the B-H curve for most transformers is at the knee of the curve. If the frequency of the applied signal should drop and N_p and E_p remain the same, then Φ_m must increase in magnitude, as determined by Eq. 19.5:

$$\Phi_m \uparrow \;=\; \frac{E_p}{4.44 f_p \downarrow N_p}$$

Note on the B-H curve that this increase in Φ_m causes a very high current in the primary, resulting in possible damage to the transformer.

19.10 TYPES OF TRANSFORMERS

Transformers are available in many different shapes and sizes. Some of the more common types include the power transformer, audio transformer, I-F (intermediate-frequency) transformer, and R-F (radio-frequency) transformer. Each is designed to fulfill a particular requirement in a specific area of application. The symbols for some of the basic types of transformers are shown in Fig. 19.28.

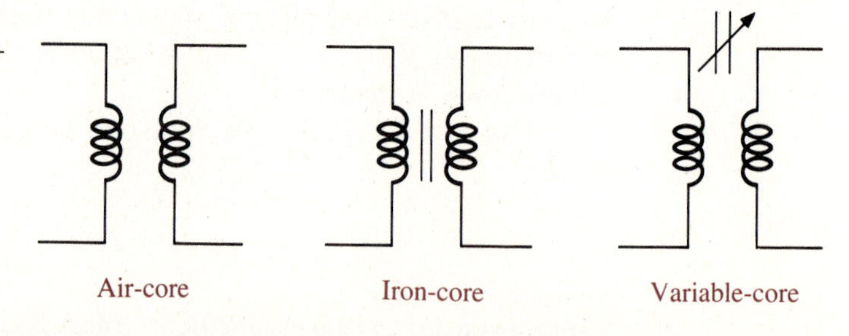

FIG. 19.28
Symbols for the various types of transformers.

The method of construction varies from one transformer to another. Two of the many different ways in which the primary and secondary coils can be wound around an iron core are shown in Fig. 19.29. In either case, the core is made of laminated sheets of ferromagnetic material separated by an insulator to reduce the eddy current losses. The sheets themselves also contain a small percentage of silicon to reduce the hysteresis losses.

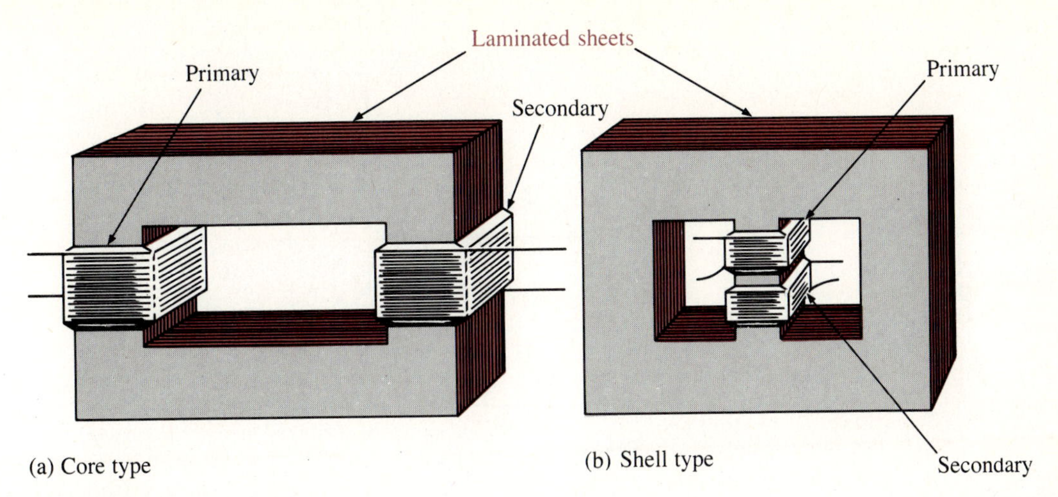

FIG. 19.29
Transformer core construction.

A shell-type power transformer with its schematic representation is shown in Fig. 19.30.

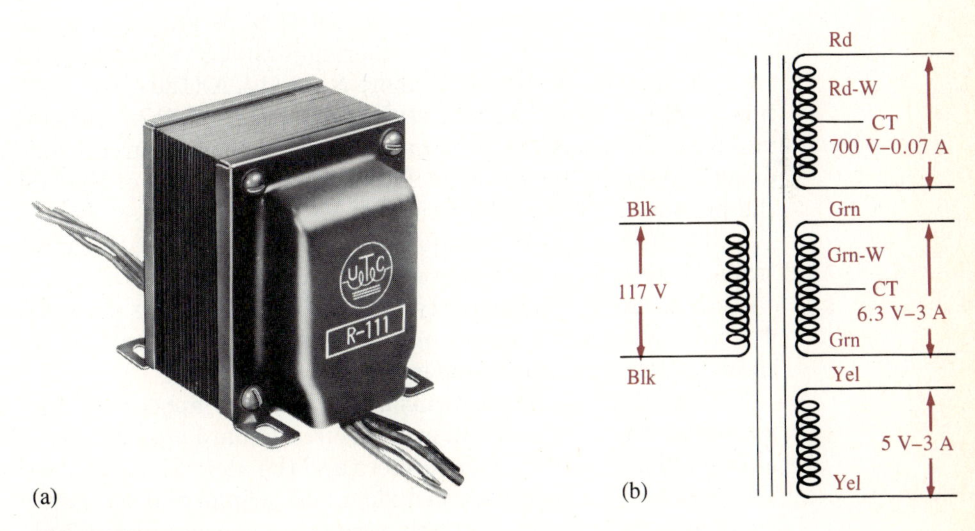

FIG. 19.30
Shell-type transformer. (a) Photograph; (b) schematic representation. (Courtesy of United Transformer Co.)

The *autotransformer* [Fig. 19.31(b)] is a type of power transformer that, instead of employing the two-circuit principle (complete isolation between coils), has one winding common to both the input and output circuits. The

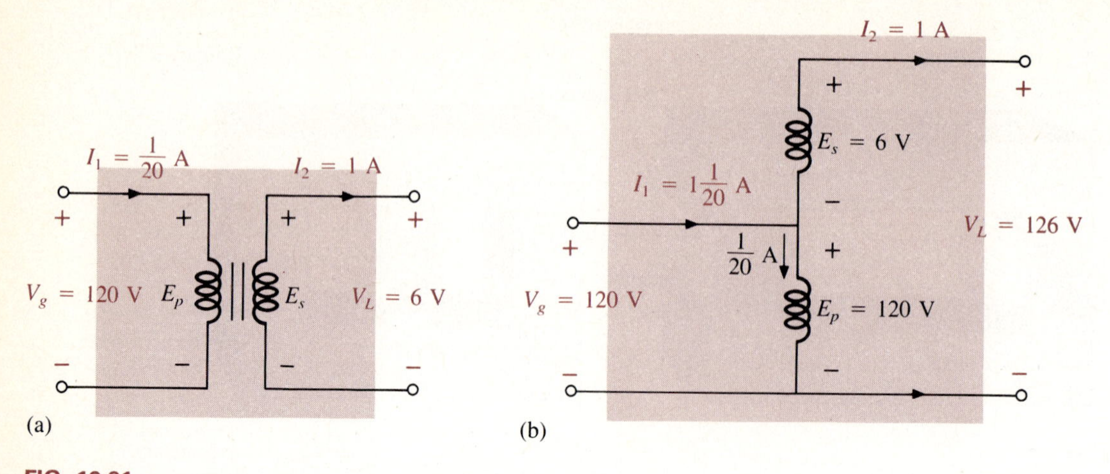

FIG. 19.31
(a) Two-circuit transformer; (b) autotransformer.

induced voltages are related to the turns ratio in the same manner as that described for the two-circuit transformer. If the proper connection is used, a two-circuit power transformer can be employed as an autotransformer. The advantage of using it as an autotransformer is that a larger apparent power can be transformed. This can be demonstrated by the two-circuit transformer of Fig. 19.31(a). It is shown in Fig. 19.31(b) as an autotransformer.

For the two-circuit transformer, note that $S = (1/20\text{ A})(120\text{ V}) = 6\text{ VA}$, whereas for the autotransformer, $S = (1\frac{1}{20}\text{ A})(120\text{ V}) = 126\text{ VA}$, which is many times that of the two-circuit transformer. Note also that the current and voltage of each coil are the same as those for the two-circuit configuration. The disadvantage of the autotransformer is obvious: loss of the isolation between the primary and secondary circuits.

The R-F and I-F transformers used extensively in radio and television transmitters and receivers are shown in Fig. 19.32. Both are available with or without a shield. Permeability-tuned I-F transformers permit the changing of Φ_m, and thereby k, by moving a ferromagnetic core within the primary and secondary coils of the transformer.

A dual-in-line pulse transformer package appears in Fig. 19.33 for use with integrated circuits and printed circuit board applications. Note the availability of four isolated transformers and the appearance of the dot convention. The data for one such unit include a 2 : 1 primary-to-secondary turns ratio; a leakage inductance of 0.50 μH; a coupling capacitance of 7 pF; a primary dc resistance of 0.18 Ω; and a secondary resistance of 0.13 Ω.

A miniature radio transformer designed for direct mounting on a printed circuit board appears in Fig. 19.34.

FIG. 19.32
I-F and R-F transformers.

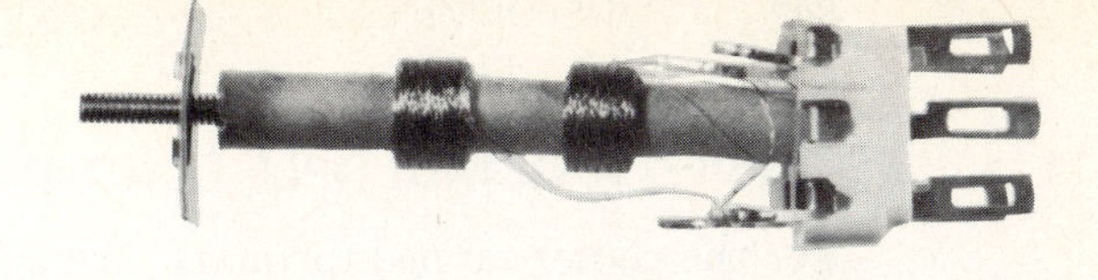

(a) I-F transformer

(b) R-F transformer

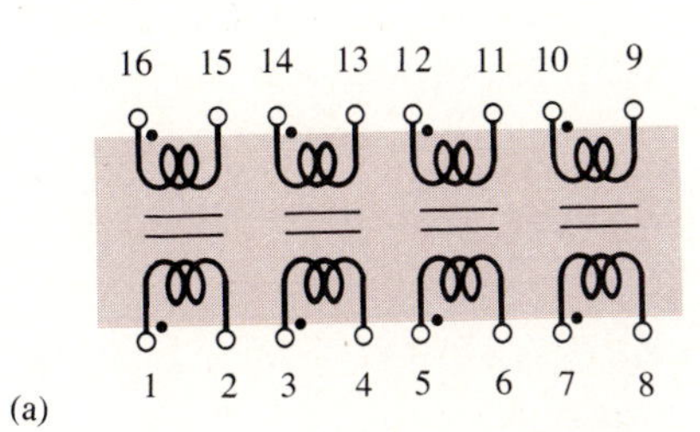

(a)

(b)

FIG. 19.33
Dual-in-line pulse transformer package. (Courtesy of Bourns®, Inc.)

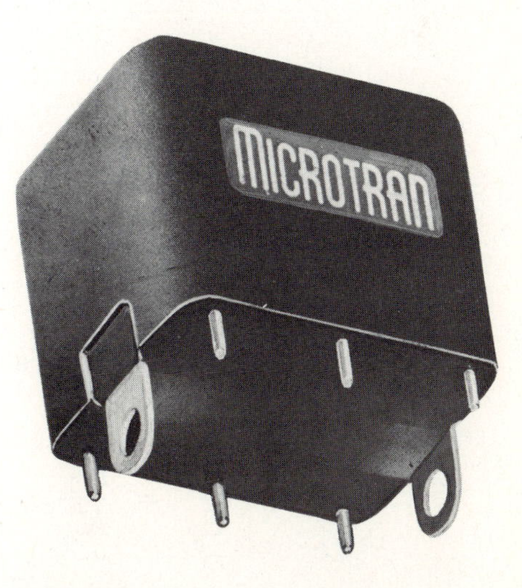

FIG. 19.34
Miniature radio transformer (for printed circuit boards). (Courtesy of Microtran Company, Inc.)

19.11 TAPPED AND MULTIPLE-LOAD TRANSFORMERS

For the center-tapped (primary) transformer of Fig. 19.35, where the voltage from the center top to either outside lead is defined as $E_p/2$, the relationship between E_p and E_s is

$$\boxed{\frac{\mathbf{E}_p}{\mathbf{E}_s} = \frac{N_p}{N_s}} \qquad (19.18)$$

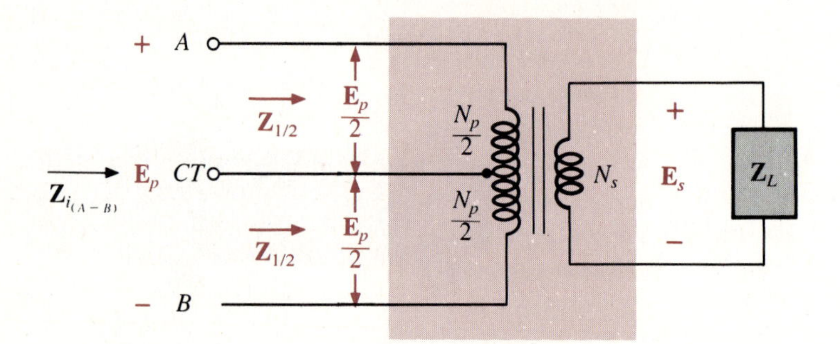

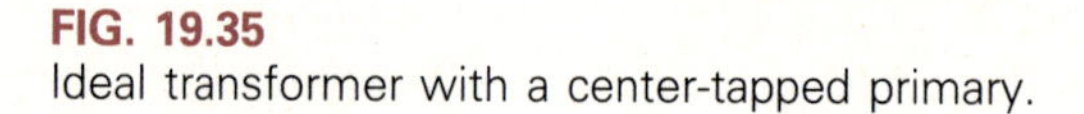

FIG. 19.35
Ideal transformer with a center-tapped primary.

For each half-section of the primary,

$$\mathbf{Z}_{1/2} = \left(\frac{N_p/2}{N_s}\right)^2 \mathbf{Z}_L$$

$$= \frac{1}{4}\left(\frac{N_p}{N_s}\right)^2 \mathbf{Z}_L$$

and

$$\boxed{\mathbf{Z}_{1/2} = \frac{1}{4}\mathbf{Z}_i} \qquad (19.19)$$

as indicated in Fig. 19.35.

For the multiple-load transformer of Fig. 19.36, the following equations apply:

$$\boxed{\frac{\mathbf{E}_1}{\mathbf{E}_2} = \frac{N_1}{N_2} \qquad \frac{\mathbf{E}_1}{\mathbf{E}_3} = \frac{N_1}{N_3} \qquad \frac{\mathbf{E}_2}{\mathbf{E}_3} = \frac{N_2}{N_3}} \qquad (19.20)$$

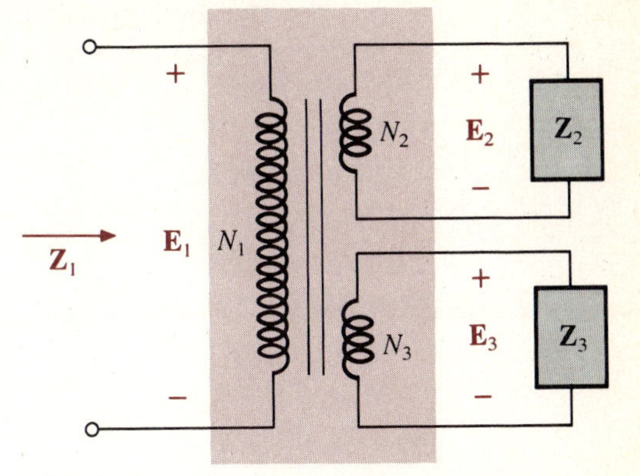

FIG. 19.36
Ideal transformer with multiple loads.

The total input impedance can be determined by first noting that for the ideal transformer, the power delivered to the primary is equal to the power dissipated by the load; that is,

$$P_1 = P_{L_2} + P_{L_3}$$

and, for resistive loads ($\mathbf{Z}_1 = R_1$, $\mathbf{Z}_2 = R_2$, and $\mathbf{Z}_3 = R_3$),

$$\frac{E_1^2}{R_1} = \frac{E_2^2}{R_2} + \frac{E_3^2}{R_3}$$

or, since

$$E_2 = \frac{N_2}{N_1} E_1 \quad \text{and} \quad E_3 = \frac{N_3}{N_1} E_1$$

then

$$\frac{E_1^2}{R_1} = \frac{[(N_2/N_1)E_1]^2}{R_2} + \frac{[(N_3/N_1)E_1]^2}{R_3}$$

and

$$\frac{E_1^2}{R_1} = \frac{E_1^2}{(N_1/N_2)^2 R_2} + \frac{E_1^2}{(N_1/N_3)^2 R_3}$$

Thus,

$$\boxed{\frac{1}{R_1} = \frac{1}{(N_1/N_2)^2 R_2} + \frac{1}{(N_1/N_3)^2 R_3}} \qquad (19.21)$$

indicating that the load resistances are reflected in parallel.

For the configuration of Fig. 19.37 with E_2 and E_3 defined as shown, Eqs. 19.20 and 19.21 are applicable.

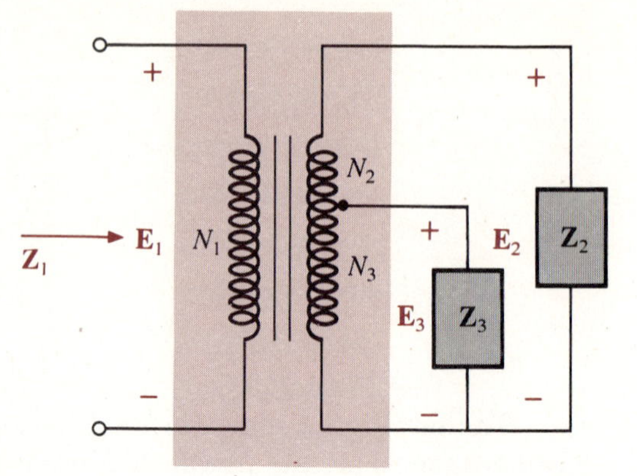

FIG. 19.37
Ideal transformer with a tapped secondary and multiple loads.

FORMULA SUMMARY

$$k = \frac{\Phi_m}{\Phi_p}, \qquad M = k\sqrt{L_p L_s}$$

$$E_p = 4.44fN_p\Phi_m, \qquad E_s = 4.44fN_s\Phi_m$$

$$\frac{\mathbf{E}_p}{\mathbf{E}_s} = \frac{N_p}{N_s}, \qquad a = \frac{N_p}{N_s}$$

$$\frac{\mathbf{I}_p}{\mathbf{I}_s} = \frac{N_s}{N_p}$$

$$\mathbf{Z}_p = a^2\mathbf{Z}_L$$

$$P_{\text{in}} = P_{\text{out}}, \qquad E_pI_p = E_sI_s$$

CHAPTER SUMMARY

Transformers play a vital role in a number of important applications. Not only can they be used to increase or decrease the voltage or current level, they can perform an important isolation or impedance matching role. For most applications, the transformer can be considered ideal, permitting the use of all the equations appearing in the formula summary. Since the majority of applications of transformers were covered in this chapter, a careful review of the examples followed by an investigation of a number of chapter problems should provide the background necessary or establish a confidence level that can handle most transformer exercises. Simply remember that the voltage levels are directly related to their respective turns ratio, whereas the current levels have an inverse relationship. Since $P_{\text{in}} = P_{\text{out}}$, you cannot expect to increase both the voltage and current levels. An increase in one must result in a decrease in the other. For impedance matching purposes, remember that the turns ratio a is squared, which can result in a significant increase or decrease in the input impedance level. Further, since a is a constant, the impedance "seen" at the input terminals has the same characteristics (R, L, or C) as the load. Transformer problems may initially appear complex, but the important equations are fairly simple in format and limited in number, permitting a rapid growth in confidence with exposure.

The iron-core equivalent model is particularly useful to describe the frequency characteristics of a transformer and reveal why it is not purely ideal. The description reveals that capacitive and inductive effects are present in all electromagnetic devices, and resonance conditions can be established in the most unexpected situations. Due to the sensitivity to frequency, the nameplate data of every transformer includes the frequency or frequency range of application.

The isolation that exists between the windings is a unique characteristic of transformers. Imagine the transfer of power from one coil to the other without a direct electrical connection. Of course, only time-varying signals can establish induced voltages due to the changing flux. The absence of a changing signal (voltage or current) at the primary will not establish a voltage or current in the secondary, even though there will be a flux of fixed magnitude in the core. The application of any dc voltage to the primary will result in 0 V at the secondary. However, there are systems manufactured that are referred to as "dc" transformers. In these systems, the dc is "chopped" to form a time-varying signal, which can be transformed and returned to dc at the output side.

A number of equations were provided for the autotransformer and multiple-load transformers. Although the autotransformer equations can be derived directly from the configuration itself, the equations for multiple-load transformers are somewhat more abstract. However, multiple-load transformers are so seldom encountered that referring to the text when necessary seems like a logical alternative to memorizing the equation for each configuration.

GLOSSARY

Autotransformer A transformer with one winding common to both the primary and the secondary circuits. A loss in isolation is balanced by the increase in its kilovolt-ampere rating.

Coefficient of coupling (k) A measure of the magnetic coupling of two coils, which ranges from a minimum of 0 to a maximum of 1.

Dot convention A technique for labeling the effect of the mutual inductance on a net inductance of a network or system.

Leakage flux The flux linking the coil that does not pass through the ferromagnetic path of the magnetic circuit.

Loosely coupled A term applied to two coils that have a low coefficient of coupling.

Multiple-load transformers Transformers having more than a single load connected to the secondary winding or windings.

Mutual inductance The inductance that exists between magnetically coupled coils of the same or different dimensions.

Nameplate data Information such as the kilovolt-ampere rating, voltage transformation ratio, and frequency of application that is of primary importance in choosing the proper transformer for a particular application.

Primary The coil or winding to which the source of electrical energy is normally applied.

Reflected impedance The impedance appearing at the primary of a transformer due to a load connected to the secondary. Its magnitude is controlled directly by the transformation ratio.

Secondary The coil or winding to which the load is normally applied.

Step-down transformer A transformer whose secondary voltage is less than its primary voltage. The transformation ratio a is greater than 1.

Step-up transformer A transformer whose secondary voltage is greater than its primary voltage. The magnitude of the transformation ratio a is less than 1.

Tapped transformer A transformer having an additional connection between the terminals of the primary or secondary windings.

Transformation ratio (a) The ratio of primary to secondary turns of a transformer.

PROBLEMS

Section 19.2

1. For the air-core transformer of Fig. 19.38:
   a. Find the value of L_s if the mutual inductance M is 40 mH.
   b. If the flux linking the primary is 4×10^{-4} Wb, what is the mutual flux or flux linking the secondary?
   c. Repeat part a if an iron core is introduced to raise the coefficient of coupling to 1.

FIG. 19.38

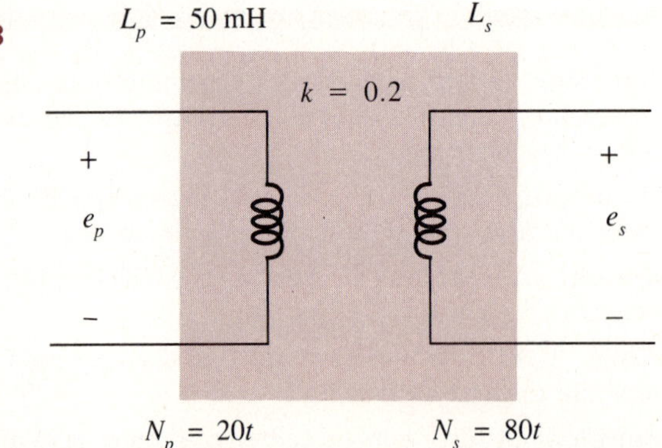

Section 19.3

2. Determine the total inductance of the series coils of Fig. 19.39.
3. Determine the total inductance of the series coils of Fig. 19.40.

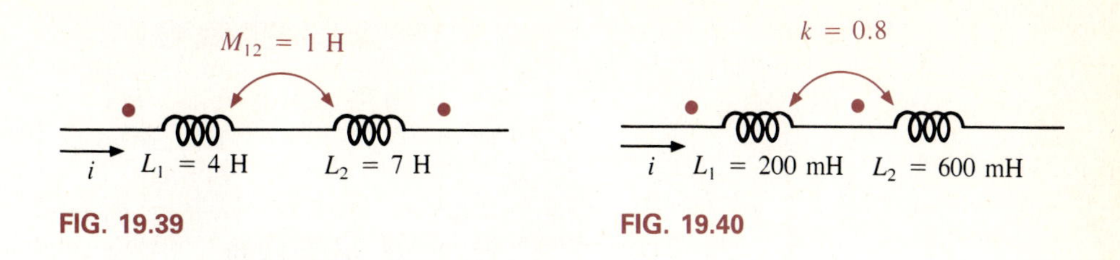

FIG. 19.39

FIG. 19.40

*4. Determine the total inductance of the series coils of Fig. 19.41.

FIG. 19.41

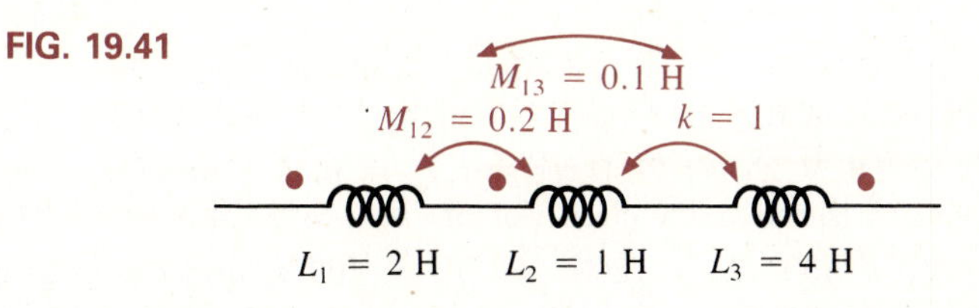

Section 19.4

5. For the iron-core transformer ($k = 1$) of Fig. 19.42:
   a. Find the magnitude of the induced voltage E_s.
   b. Find the maximum flux Φ_m.

FIG. 19.42

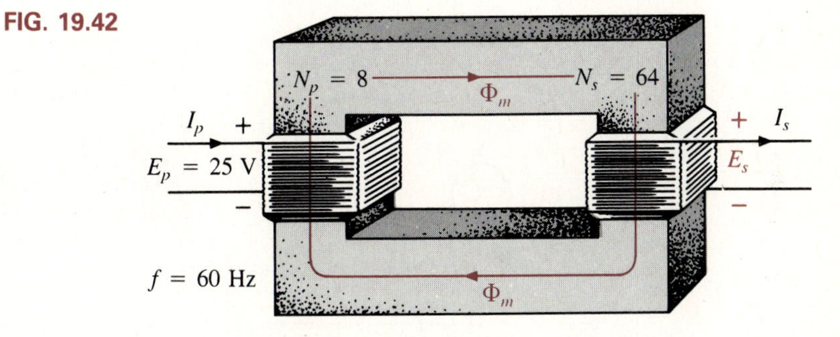

6. Repeat Problem 5 for $N_p = 240$ and $N_s = 30$.
7. Find the applied voltage of an iron-core transformer with a secondary voltage of 240 and $N_p = 60$ with $N_s = 720$.
8. If the maximum flux passing through the core of Problem 5 is 12.5 mWb, find the frequency of applied voltage.

Section 19.5

9. For the iron-core transformer of Fig. 19.43:
   a. Find the magnitude of the current I_L and the voltage V_L if $a = 1/5$, $I_p = 2$ A, and Z_L is a 2-Ω resistor.
   b. Find the input resistance for the data specified in part a.
10. Find the input impedance for the iron-core transformer of Fig. 19.43 if $a = 2$, $I_p = 4$ A, and $V_g = 1600$ V.

FIG. 19.43

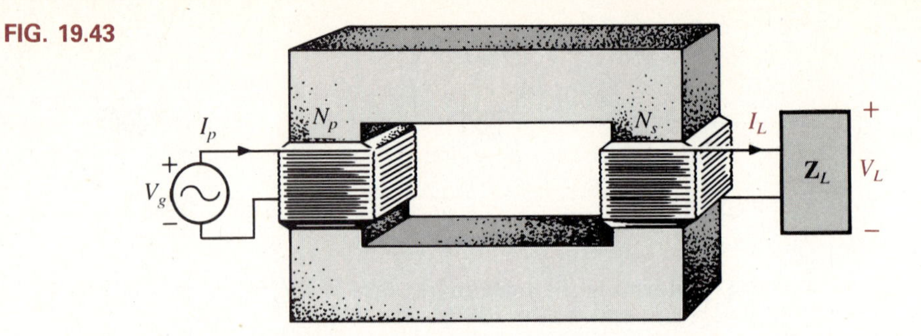

11. Find the voltage V_g and the current I_p if the input impedance of the iron-core transformer of Fig. 19.43 is 4 Ω and $V_L = 1200$ V with $a = 1/4$.

12. If $V_L = 240$ V, Z_L is a 20-Ω resistor, $I_p = 0.05$ A, and $N_s = 50$, find the number of turns in the primary circuit of the iron-core transformer of Fig. 19.43.

*13. a. If $N_p = 400$, $N_s = 1200$, and $V_g = 100$ V, find the magnitude of I_p for the iron-core transformer of Fig. 19.43 if $\mathbf{Z}_L$ is composed of a 9-Ω resistor in series with a 12-Ω inductive reactance.
  b. Find the magnitude of the voltage V_L and the current I_L for the conditions of part a.

14. a. For the circuit of Fig. 19.44, find the transformation ratio required to deliver maximum power to the speaker.
  b. Find the maximum power delivered to the speaker.

FIG. 19.44

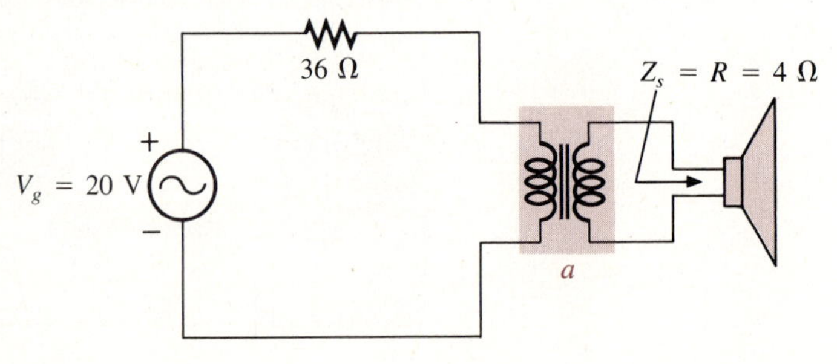

Section 19.6

15. For the transformer of Fig. 19.45, determine
  a. The equivalent resistance R_e
  b. The equivalent reactance X_e
  c. The equivalent circuit reflected to the primary
  d. The primary current for $V_g = 50 \text{ V} \angle 0°$
  e. The load voltage V_L

16. For the transformer of Fig. 19.45, if the load is replaced by an inductive reactance of 20 Ω:
  a. Determine the total reflected primary impedance.
  b. Calculate the primary current for $\mathbf{V}_g = 50 \text{ V} \angle 0°$.
  c. Determine the voltage across R_e, X_e, and the reflected load.
  d. Draw the phasor diagram.

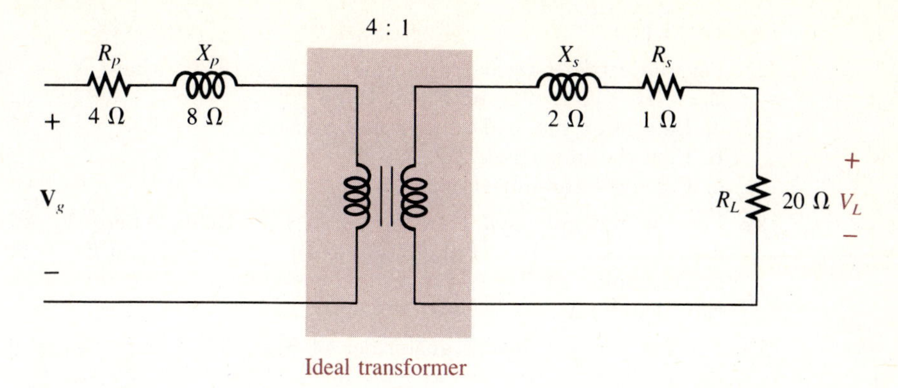

FIG. 19.45

17. Repeat Problem 16 for a capacitive load having a reactance of 20 Ω.

Section 19.7

18. Discuss in your own words the frequency characteristics of the transformer. Employ the applicable equivalent circuit and frequency characteristics appearing in this chapter.

Section 19.9

19. An ideal transformer is rated 10 kVA, 2400/120 V, 60 Hz.

a. Find the transformation ratio if the 120 V is the secondary voltage.
b. Find the current rating of the secondary if the 120 V is the secondary voltage.
c. Find the current rating of the primary if the 120 V is the secondary voltage.
d. Repeat parts a through c if the 2400 V is the secondary voltage.

Section 19.10

20. Determine the primary and secondary voltages and currents for the autotransformer of Fig. 19.46.

FIG. 19.46

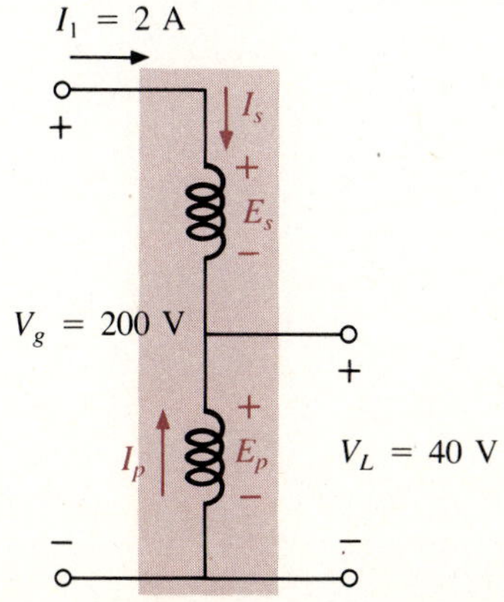

Section 19.11

21. For the center-tapped transformer of Fig. 19.35, where $N_p = 100$, $N_s = 25$, $\mathbf{Z}_L = \mathbf{R} = 5\ \Omega\ \angle 0^\circ$, and $\mathbf{E}_p = 100\ \text{V}\ \angle 0^\circ$:
a. Determine the load voltage and current.
b. Find the impedance $\mathbf{Z}_i$.
c. Calculate the impedance $\mathbf{Z}_{1/2}$.

22. For the multiple-load transformer of Fig. 19.36, where $N_1 = 90$, $N_2 = 15$, $N_3 = 45$, $\mathbf{Z}_2 = \mathbf{R} = 8\ \Omega\ \angle 0^\circ$, $\mathbf{Z}_3 = \mathbf{R}_L = 5\ \Omega\ \angle 0^\circ$, and $\mathbf{E}_1 = 60\ \text{V}\ \angle 0^\circ$:
a. Determine the load voltages and currents.
b. Calculate $\mathbf{Z}_1$.

23. For the multiple-load transformer of Fig. 19.37, where $N_1 = 120$, $N_2 = 40$, $N_3 = 30$, $\mathbf{Z}_2 = \mathbf{R}_2 = 12\ \Omega\ \angle 0^\circ$, $\mathbf{Z}_3 = \mathbf{R}_3 = 10\ \Omega\ \angle 0^\circ$, and $\mathbf{E}_1 = 120\ \text{V}\ \angle 60^\circ$:
a. Determine the load voltages and currents.
b. Calculate $\mathbf{Z}_1$.

Computer Problems

24. Write a program to provide a general solution to the problem of impedance matching as defined by Example 19.5. That is, given the speaker impedance and the internal resistance of the source, determine the required turns ratio and the power delivered to the speaker. In addition, calculate the load and source current and the primary and secondary voltages. The source voltage will have to be provided with the other parameters of the network.

25. Given the equivalent model of an iron-core transformer appearing in Fig. 19.16, write a program to calculate the magnitude of the voltage $\mathbf{V}_g$.

26. Given all the parameters of Fig. 19.37, write a program to calculate the input impedance Z_1.

20

Polyphase Systems

OBJECTIVES

- ☐ Become aware of the existence and basic operation of a multiphase generator.
- ☐ Understand the phase and magnitude relationships between the line and phase voltages of a three-phase Y- or Δ-connected generator.
- ☐ Be able to calculate the phase and line quantities for a Y-Y, Y-Δ, Δ-Y, and Δ-Δ system.
- ☐ Understand the impact of the phase sequence on the angle associated with the generated voltages and how the sequence is determined from the phasor diagram of the line or phase voltages.
- ☐ Be able to calculate the real, reactive, and apparent power of a Y- or Δ-connected load.
- ☐ Understand the procedure for measuring the total power to a three-phase load using the two- or three-wattmeter method.
- ☐ Recognize the conditions that establish an unbalanced load and how to calculate the quantities of interest for an unbalanced three-phase, four-wire, Y-connected load.

20.1 INTRODUCTION

If an ac generator is designed to develop a single sinusoidal voltage for each rotation of the shaft (rotor), it is referred to as a *single-phase generator*. If the number of coils on the rotor is increased in a specified manner, the result is a *polyphase generator,* which develops more than one ac phase voltage per rotation of the rotor. In this chapter, the three-phase system will be discussed in detail, since it is the most frequently used for power transmission.

In general, the three-phase system is more economical for transmitting power at a fixed power loss than the single-phase system. This economy is due primarily to the reduction of the I^2R losses of the transmission lines. A reduction in these losses permits the use of smaller conductors, which in turn reduces the weight of copper required.

The three-phase system is used in almost all commercial electric generators. This does not mean that single-phase and two-phase generating systems are obsolete. Most small emergency generators, such as the gasoline type, are one-phase generating systems. One of the more common applications of the two-phase system is in servomechanisms, which are self-correcting control systems capable of detecting and adjusting their own operation. Servomechanisms are used in ships and aircraft to keep them on course automatically or in simpler devices such as a thermostatic circuit, to regulate heat output. In most cases, however, where single-phase and two-phase inputs are required, they are supplied by one and two phases of a three-phase generating system rather than generated independently.

The number of phase voltages that can be produced by a polyphase generator is not limited to three. Any number of phases can be obtained by spacing the windings for each phase at the proper angular position around the rotor. Some electrical systems operate more efficiently if more than three phases are used. One such system involves the process of rectification, which is used to convert alternating current to direct current. The greater the number of phases, the smoother the dc output of the system.

20.2 THE THREE-PHASE GENERATOR

The three-phase generator of Fig. 20.1(a) has three induction coils on the rotor (armature), as shown symbolically by Fig. 20.1(b). Since the three coils have an equal number of turns and each coil rotates with the same angular velocity, the voltage induced across each coil has the same peak value, shape, and frequency. As the shaft of the generator is turned by some external means, the induced voltages e_{AN}, e_{BN}, and e_{CN} are generated simultaneously, as shown in Fig. 20.2. Note the 120° phase shift between waveforms and the similarities in appearance of the three sinusoidal functions.

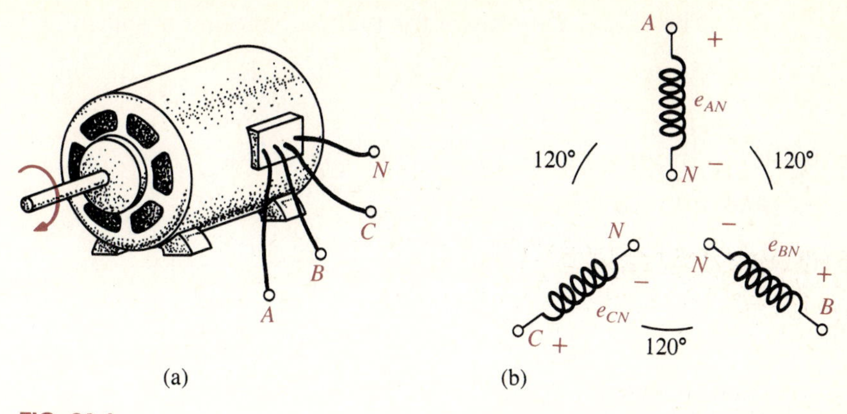

FIG. 20.1
(a) Three-phase generator; (b) induced voltages of a three-phase generator.

The sinusoidal expression for each of the induced voltages of Fig. 20.2 is

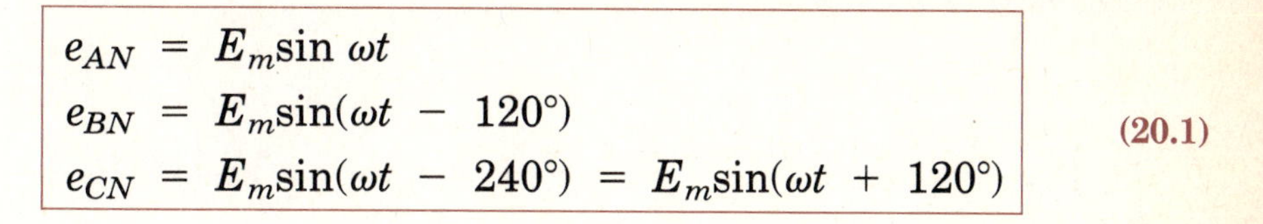

$$\begin{aligned} e_{AN} &= E_m \sin \omega t \\ e_{BN} &= E_m \sin(\omega t - 120^\circ) \\ e_{CN} &= E_m \sin(\omega t - 240^\circ) = E_m \sin(\omega t + 120^\circ) \end{aligned} \qquad \textbf{(20.1)}$$

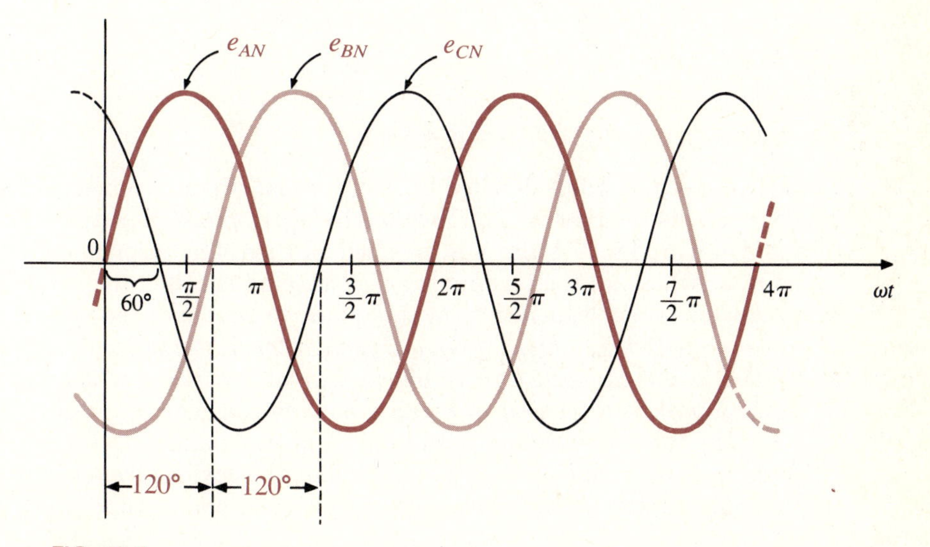

FIG. 20.2
Three-phase sinusoidal voltages.

The phasor diagram of the induced voltages is shown in Fig. 20.3, where

$$E_{AN} = 0.707E_m$$
$$E_{BN} = 0.707E_m$$
$$E_{CN} = 0.707E_m$$

and

$$\mathbf{E}_{AN} = E_{AN}\angle 0^\circ$$
$$\mathbf{E}_{BN} = E_{BN}\angle -120^\circ$$
$$\mathbf{E}_{CN} = E_{CN}\angle +120^\circ$$

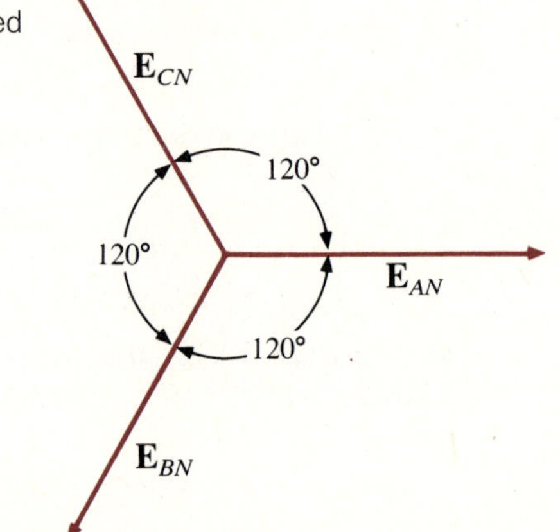

FIG. 20.3
Phasor representation of the generated voltages of a three-phase generator.

20.3 THE Y-CONNECTED GENERATOR

If the three terminals denoted by N of Fig. 20.1(b) are connected together, the generator is referred to as a *Y-connected three-phase generator* (Fig. 20.4). As indicated in Fig. 20.4, the Y is inverted for ease of notation and for clarity. The point at which all the terminals are connected is called the *neutral point.* If a conductor is not attached from this point to the load, the system is called a *Y-connected, three-phase, three-wire generator.* If the neutral is connected, the system is a *Y-connected, three-phase, four-wire generator.* The function of the neutral will be discussed in detail when we consider the load circuit.

The three conductors connected from A, B, and C to the load are called *lines.* For the Y-connected system, it should be obvious from Fig. 20.4 that the line current equals the phase current for each phase; that is,

$$\boxed{\mathbf{I}_L = \mathbf{I}_{\phi g}} \tag{20.2}$$

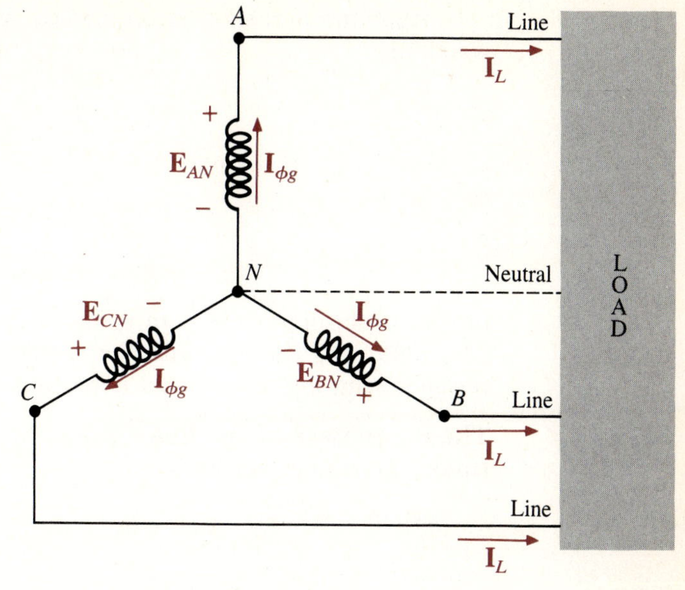

FIG. 20.4
Y-connected generator.

The voltage from one line to another is called a *line voltage*. On the phasor diagram (Fig. 20.5) it is the phasor drawn from the end of one phase to another in the counterclockwise direction.

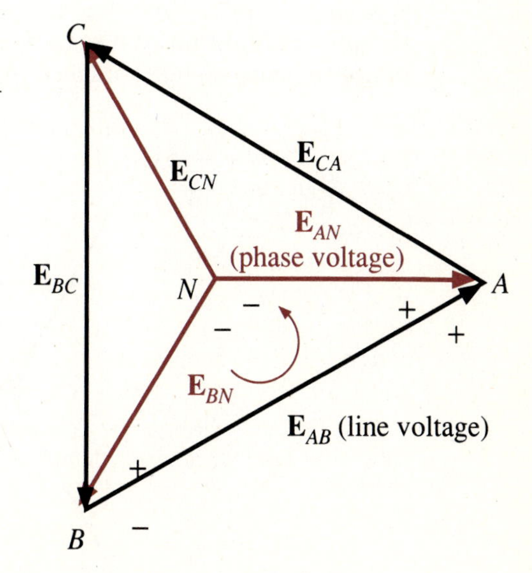

FIG. 20.5
Line and phase voltages of the Y-connected three-phase generator.

Applying Kirchhoff's voltage law around the indicated loop of Fig. 20.5, we obtain

$$\mathbf{E}_{AB} - \mathbf{E}_{AN} + \mathbf{E}_{BN} = 0$$

or

$$\mathbf{E}_{AB} = \mathbf{E}_{AN} - \mathbf{E}_{BN} = \mathbf{E}_{AN} + \mathbf{E}_{NB}$$

providing the line voltage in terms of the phase voltages.

Following through with the vector addition of the phase voltages yields the following interesting and important conclusion:

The magnitude of the line voltage of a Y-connected generator is $\sqrt{3}$ times the phase voltage.

That is,

$$\boxed{E_L = \sqrt{3}E_\phi} \qquad (20.3)$$

In other words, since $E_{AN} = E_{BN} = E_{CN} = E_\phi$ and $E_{AB} = E_{CA} = E_{BC} = E_L$, $E_L = \sqrt{3}E_\phi$ for all quantities of the Y-connected generator.

The *phase sequence* is the order in which the phasors representing the phase voltages pass through a fixed point on the phasor diagram if the phasors are rotated in a counterclockwise direction. For example, in Fig. 20.6 the phase sequence is *ABC*. However, since the fixed point can be chosen anywhere on the phasor diagram, the sequence can also be written as *BCA* or *CAB*. The phase sequence is quite important in the three-phase distribution of power. In

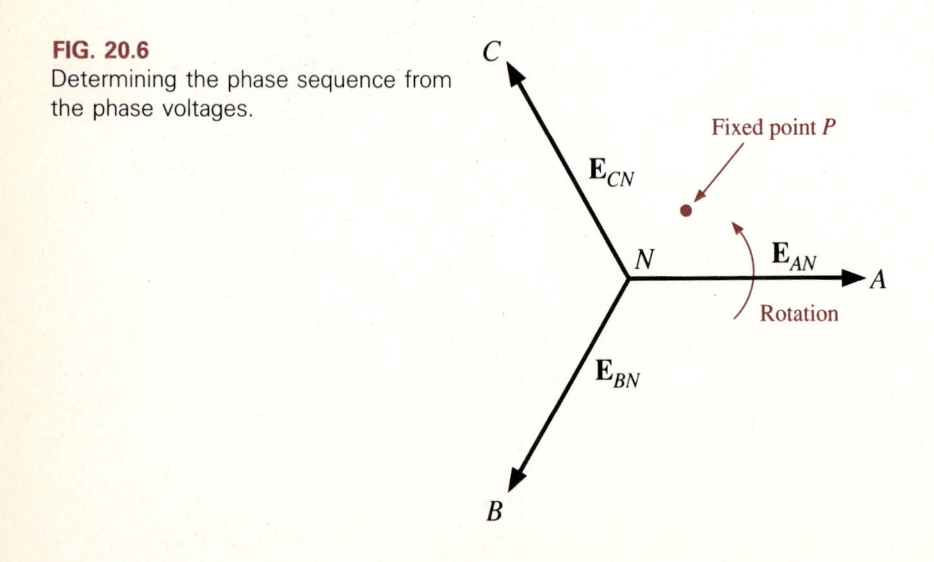

FIG. 20.6
Determining the phase sequence from the phase voltages.

a three-phase motor, for example, if two phase voltages are interchanged, the sequence will change and the direction of rotation of the motor will be reversed. Other effects will be described when we consider the loaded three-phase system.

If the sequence is given, the phasor diagram can be drawn by simply picking a reference voltage, placing it on the reference axis, and then drawing the other voltages at the proper angular position. For a sequence of *ACB*, for example, we might choose E_{AN} to be the reference voltage, as shown in Fig. 20.7 with the position of the remaining phase voltages for the indicated phase sequence. In phasor notation:

$$\mathbf{E}_{AN} = E_{AN} \angle 0^\circ \quad \text{(reference)}$$
$$\mathbf{E}_{CN} = E_{CN} \angle -120^\circ$$
$$\mathbf{E}_{BN} = E_{BN} \angle +120^\circ$$

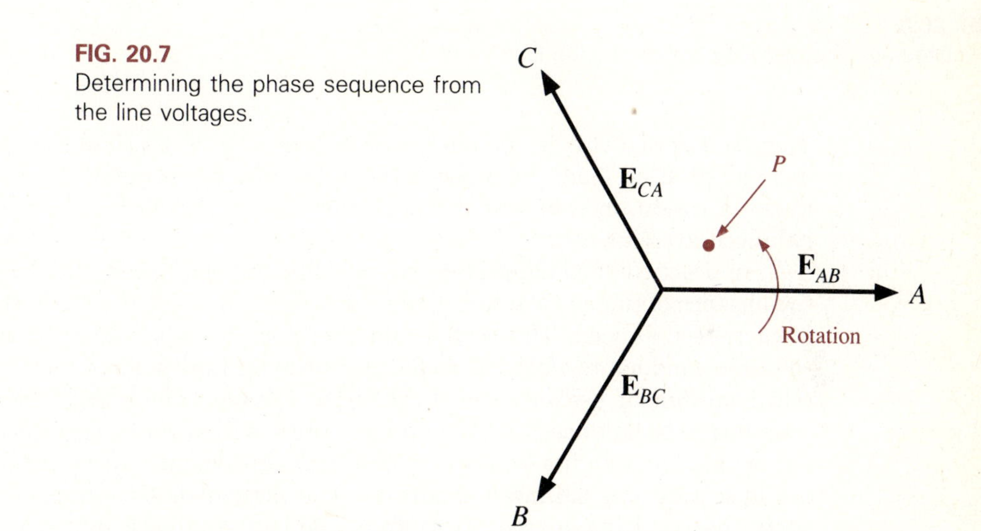

FIG. 20.7
Determining the phase sequence from the line voltages.

20.4 THE Y-CONNECTED GENERATOR WITH A Y-CONNECTED LOAD

Loads connected to three-phase supplies are of two types: the Y and the Δ. If a Y-connected load is connected to a Y-connected generator, the system is symbolically represented by Y-Y. The physical setup of such a system is shown in Fig. 20.8.

If the load is balanced, the neutral can be removed without affecting the circuit in any manner. That is, if

$$\mathbf{Z}_1 = \mathbf{Z}_2 = \mathbf{Z}_3$$

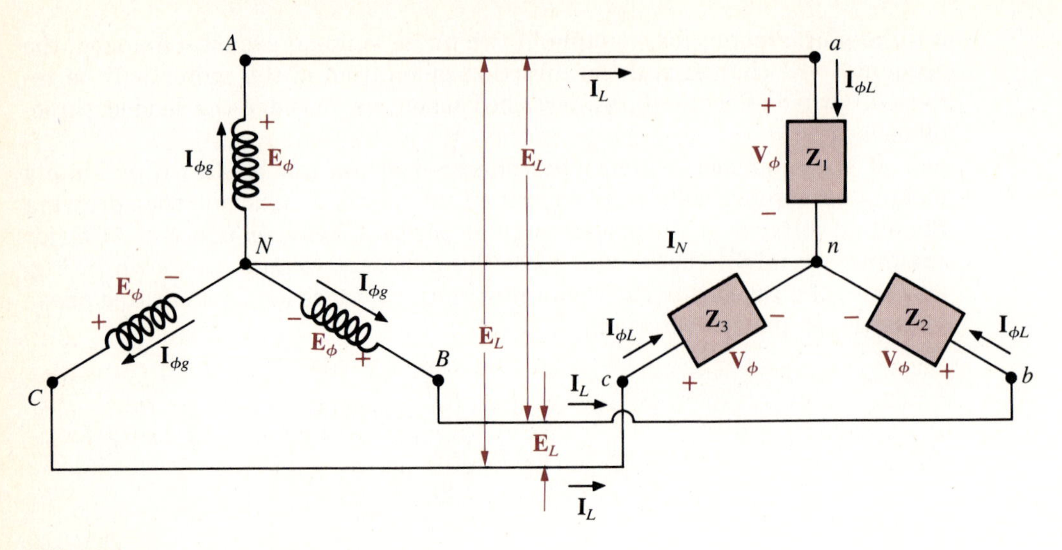

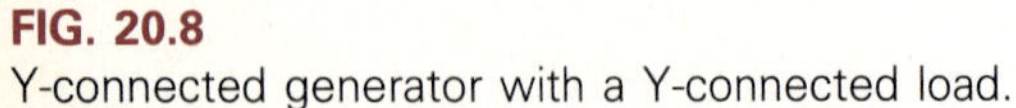

FIG. 20.8
Y-connected generator with a Y-connected load.

then I_N is zero. (This is demonstrated in Example 20.1.) Note that in order to have a balanced load, the phase angle must also be the same for each impedance—a condition that was not necessary in dc circuits when we considered balanced systems.

In practice, if a factory, for example, had only balanced three-phase loads, the absence of the neutral would have no effect, since ideally the system would always be balanced. The cost would therefore be less, since the number of required conductors would be reduced. However, lighting and most other electrical equipment use only one of the phase voltages, and even if the loading is designed to be balanced (as it should be), there will never be perfect continuous balancing, since lights and other electrical equipment will be turned on and off, upsetting the balanced condition. The neutral is therefore necessary to carry the resulting current away from the load and back to the Y-connected generator. This will be demonstrated when we consider unbalanced Y-connected systems.

We shall now examine the *four-wire Y-Y-connected system*. The current passing through each phase of the generator is the same as its corresponding line current, which, in turn, for a Y-connected load is equal to the current in the phase of the load to which it is attached:

$$\mathbf{I}_{\phi g} = \mathbf{I}_L = \mathbf{I}_{\phi L} \tag{20.4}$$

For a balanced or unbalanced load, since the generator and load have a common neutral point, then

$$\mathbf{V}_{\phi} = \mathbf{E}_{\phi} \tag{20.5}$$

In addition, since $\mathbf{I}_{\phi L} = \mathbf{V}_\phi/\mathbf{Z}_\phi$, the magnitude of the current in each phase will be equal for a balanced load and unequal for an unbalanced load. You will recall that for the Y-connected generator, the magnitude of the line voltage is equal to $\sqrt{3}$ times the phase voltage. This same relationship can be applied to a balanced or unbalanced four-wire Y-connected load:

$$\boxed{E_L = \sqrt{3}V_\phi} \qquad \textbf{(20.6)}$$

For a voltage drop across a load element, the first subscript refers to that terminal through which the current enters the load element, and the second subscript refers to the terminal from which the current leaves. In other words, the first subscript is, by definition, positive with respect to the second for a voltage drop. Note Fig. 20.9, in which the standard double subscripts for a source of voltage and a voltage drop are indicated.

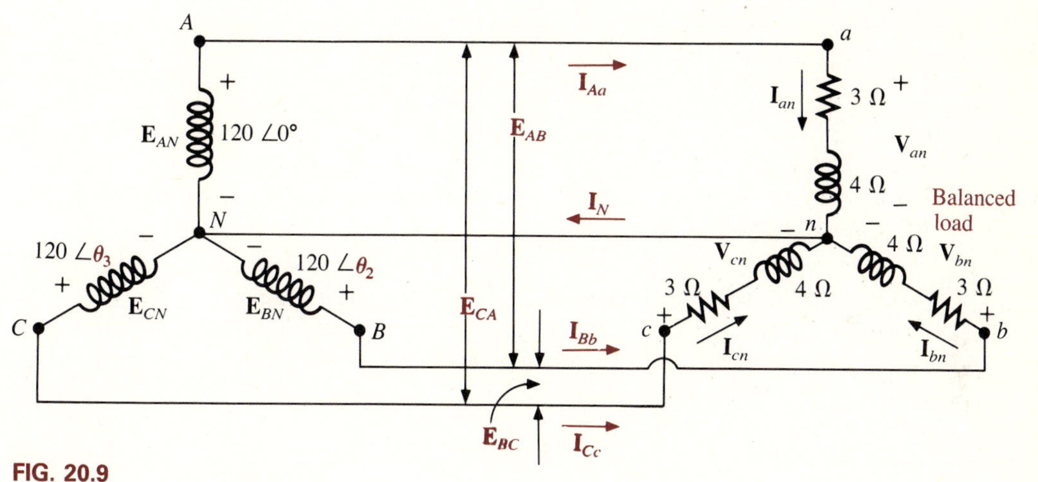

FIG. 20.9
Y-Y system for Example 20.1.

EXAMPLE 20.1 The phase sequence of the Y-connected generator in Fig. 20.9 is *ABC*.

a. Find the phase angles θ_2 and θ_3.
b. Find the magnitude of the line voltages.
c. Find the line currents.

Solutions

a. For an *ABC* phase sequence,

$$\theta_2 = \mathbf{-120°} \quad \text{and} \quad \theta_3 = \mathbf{+120°}$$

b. $E_L = \sqrt{3}E_\phi = (1.73)(120\text{ V}) = 208\text{ V}$. Therefore,

$$E_{AB} = E_{BC} = E_{CA} = \mathbf{208\ V}$$

c. $\mathbf{V}_\phi = \mathbf{E}_\phi$. Therefore,

$$\mathbf{V}_{an} = \mathbf{E}_{AN} \qquad \mathbf{V}_{bn} = \mathbf{E}_{BN} \qquad \mathbf{V}_{cn} = \mathbf{E}_{CN}$$

with

$$Z_{an} = \sqrt{R^2 + X_L^2} = \sqrt{(3\ \Omega)^2 + (4\ \Omega)^2} = 5\ \Omega$$

$$\theta = \tan^{-1}\frac{4\ \Omega}{3\ \Omega} = 53.13^\circ$$

$$I_{\phi_L} = I_{an} = \frac{V_{an}}{Z_{an}} = \frac{120\ \text{V}}{5\ \Omega} = 24\ \text{A}$$

Since the phase impedance is inductive, $\mathbf{I}_{an}$ lags $\mathbf{V}_{an}$ by the *magnitude* of the angle of the phase impedance. With $\mathbf{V}_{an} = 120\ \text{V}\ \angle 0^\circ$,

$$\mathbf{I}_{an} = 24\ \text{A}\ \angle 0^\circ - 53.13^\circ$$

and

$$\mathbf{I}_{an} = \mathbf{24\ A\ \angle -53.13^\circ}$$

$\mathbf{Z}_{bn} = 5\ \Omega\ \angle 53.13^\circ$ and $\mathbf{V}_{bn} = 120\ \text{V}\ \angle -120^\circ$, so

$$\mathbf{I}_{bn} = 24\ \text{A}\ \angle -120^\circ - 53.13^\circ$$

and

$$\mathbf{I}_{bn} = \mathbf{24\ A\ \angle -173.13^\circ}$$

$\mathbf{Z}_{cn} = 5\ \Omega\ \angle 53.13^\circ$ and $\mathbf{V}_{cn} = 120\ \text{V}\ \angle +120^\circ$, so

$$\mathbf{I}_{cn} = 24\ \text{A}\ \angle +120^\circ - 53.13^\circ$$

and

$$\mathbf{I}_{cn} = \mathbf{24\ A\ \angle +66.87^\circ}$$

Also, since $\mathbf{I}_L = \mathbf{I}_{\phi L}$,

$$\mathbf{I}_{Aa} = \mathbf{I}_{an} = \mathbf{24\ A\ \angle -53.13^\circ}$$
$$\mathbf{I}_{Bb} = \mathbf{I}_{bn} = \mathbf{24\ A\ \angle -173.13^\circ}$$
$$\mathbf{I}_{Cc} = \mathbf{I}_{cn} = \mathbf{24\ A\ \angle 66.87^\circ}$$

20.5 THE Y-Δ SYSTEM

There is no neutral connection for the Y-Δ system of Fig. 20.10. Any variation in the impedance of a phase that produces an unbalanced system will simply vary the line and phase currents of the system.

For a balanced load,

$$\mathbf{Z}_1 = \mathbf{Z}_2 = \mathbf{Z}_3 \tag{20.7}$$

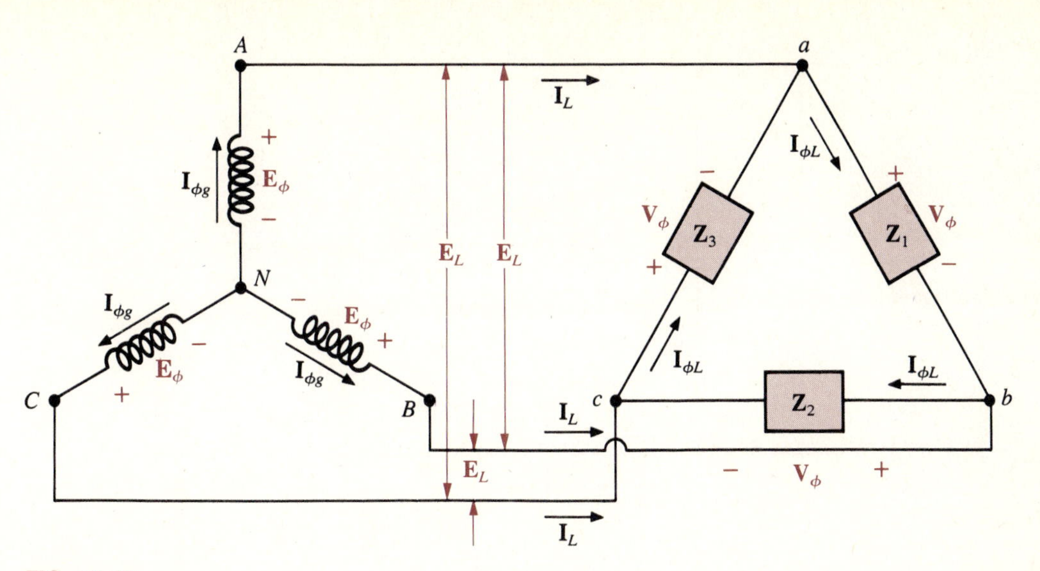

FIG. 20.10
Y-connected generator with a Δ-connected load.

The voltage across each phase of the load is equal to the line voltage of the generator for a balanced or unbalanced load:

$$\mathbf{V}_\phi = \mathbf{E}_L \quad (20.8)$$

The relationship between the magnitude of the line currents and phase currents of a balanced Δ load is the following as derived using Kirchhoff's current law in vector form.

$$I_L = \sqrt{3}I_\phi \quad (20.9)$$

For a balanced load, the line currents will be equal in magnitude, as will the phase currents.

EXAMPLE 20.2

For the three-phase system of Fig. 20.11,

a. Find the phase angles θ_2 and θ_3.
b. Find the current in each phase of the load.
c. Find the magnitude of the line currents.

Solutions

a. For an *ABC* sequence,

$$\theta_2 = \mathbf{-120°} \quad \text{and} \quad \theta_3 = \mathbf{+120°}$$

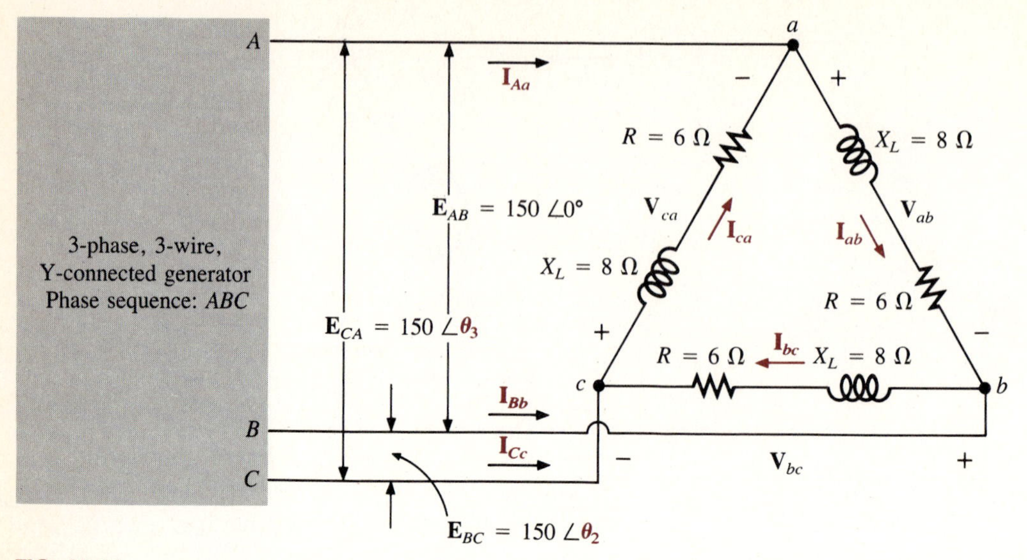

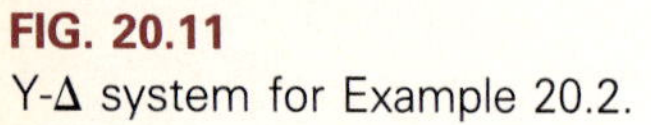
FIG. 20.11
Y-Δ system for Example 20.2.

b. $\mathbf{V}_\phi = \mathbf{E}_L$. Therefore,

$$\mathbf{V}_{ab} = \mathbf{E}_{AB} \qquad \mathbf{V}_{ca} = \mathbf{E}_{CA} \qquad \mathbf{V}_{bc} = \mathbf{E}_{BC}$$

$$Z_\phi = \sqrt{R^2 + X_L^2} = \sqrt{(6\ \Omega)^2 + (8\ \Omega)^2} = 10\ \Omega$$

$$\theta = \tan^{-1}\frac{8\ \Omega}{6\ \Omega} = 53.13°$$

and

$$I_{\phi_L} = I_{ab} = \frac{V_{ab}}{Z_{ab}} = \frac{150\text{ V}}{10\ \Omega} = 15\text{ A}$$

Since the phase impedance is inductive, $\mathbf{I}_{ab}$ lags $\mathbf{V}_{ab}$ by the *magnitude* of the angle of the phase impedance. With $\mathbf{V}_{ab} = 150\text{ V}\angle 0°$,

$$\mathbf{I}_{ab} = 15\text{ A}\angle 0° - 53.13°$$

and

$$\mathbf{I}_{ab} = \mathbf{15\ A\angle -53.13°}$$

$$\mathbf{Z}_{bc} = 10\ \Omega\angle 53.13° \text{ and } \mathbf{V}_{bc} = 150\text{ V}\angle -120°\text{, so}$$

$$\mathbf{I}_{bc} = 15\text{ A}\angle -120° - 53.13°$$

and

$$\mathbf{I}_{bc} = \mathbf{15\ A\angle -173.13°}$$

$$\mathbf{Z}_{ca} = 10\ \Omega\ \angle 53.13^\circ \text{ and } \mathbf{V}_{ca} = 150\ \text{V}\ \angle +120^\circ, \text{ so}$$

$$\mathbf{I}_{ca} = 15\ \text{A}\ \angle +120^\circ - 53.13^\circ$$

and

$$\mathbf{I}_{ca} = \mathbf{15\ A\ \angle 66.87^\circ}$$

c. $I_L = \sqrt{3}I_\phi = (1.73)(15\ \text{A}) = 25.95$ A. Therefore,

$$I_{Aa} = I_{Bb} = I_{Cc} = \mathbf{25.95\ A}$$

20.6 THE Δ-CONNECTED GENERATOR

If we rearrange the coils of the generator in Fig. 20.12(a) as shown in Fig. 20.12(b), the system is referred to as a *three-phase, three-wire, Δ-connected ac generator*. In this system, the phase and line voltages are equivalent and equal to the voltage induced across each coil of the generator. That is,

$$\left.\begin{aligned} \mathbf{E}_{AB} &= \mathbf{E}_{AN} \text{ and } e_{AN} = \sqrt{2}E_{AN}\sin \omega t \\ \mathbf{E}_{BC} &= \mathbf{E}_{BN} \text{ and } e_{BN} = \sqrt{2}E_{BN}\sin(\omega t - 120^\circ) \\ \mathbf{E}_{CA} &= \mathbf{E}_{CN} \text{ and } e_{CN} = \sqrt{2}E_{CN}\sin(\omega t + 120^\circ) \end{aligned}\right\} \begin{matrix}\text{phase}\\ \text{sequence}\\ ABC\end{matrix}$$

or

$$\boxed{\mathbf{E}_L = \mathbf{E}_{\phi g}} \qquad \textbf{(20.10)}$$

Note that only one voltage (magnitude) is available, instead of the two available in the Y-connected system.

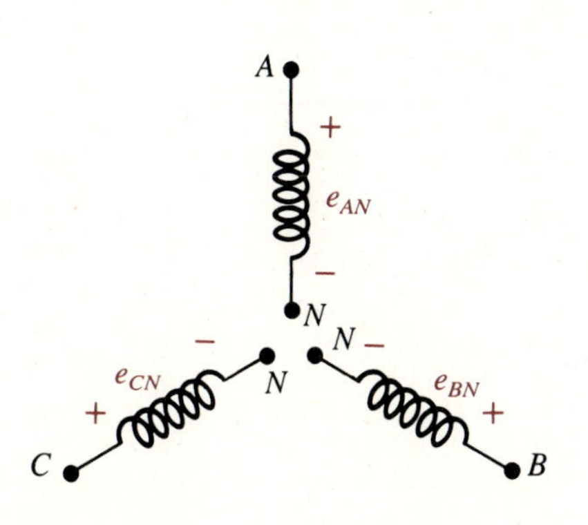

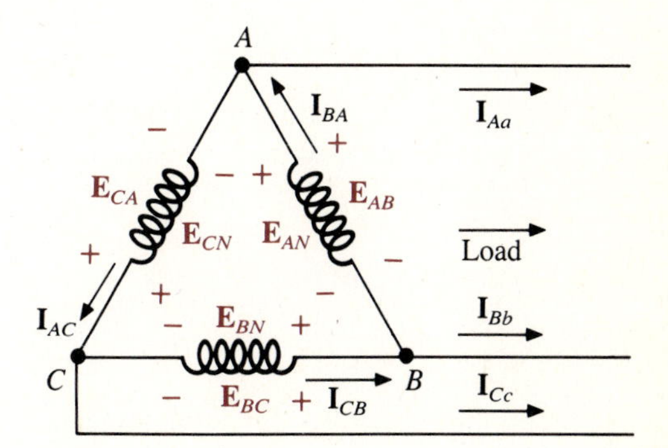

FIG. 20.12
Δ-connected generator.

Unlike the line current for the Y-connected generator, the line current for the Δ-connected system is not equal to the phase current. The relationship between the two can be found by applying Kirchhoff's current law at one of the nodes and solving for the line current in terms of the phase currents. That is, at node *A*,

$$\mathbf{I}_{BA} = \mathbf{I}_{Aa} + \mathbf{I}_{AC}$$

or

$$\mathbf{I}_{Aa} = \mathbf{I}_{BA} - \mathbf{I}_{AC} = \mathbf{I}_{BA} + \mathbf{I}_{CA}$$

The result is

$$\boxed{I_L = \sqrt{3}I_{\phi g}} \tag{20.11}$$

Even though the line and phase voltages of a Δ-connected system are the same, it is standard practice to describe the phase sequence in terms of the line voltages. The method used is the same as that described for the line voltages of the Y-connected generator. For example, the phasor diagram of the line voltages for a phase sequence *ABC* is shown in Fig. 20.13. In drawing such a diagram, one must take care to have the sequence of the first and second subscripts the same.

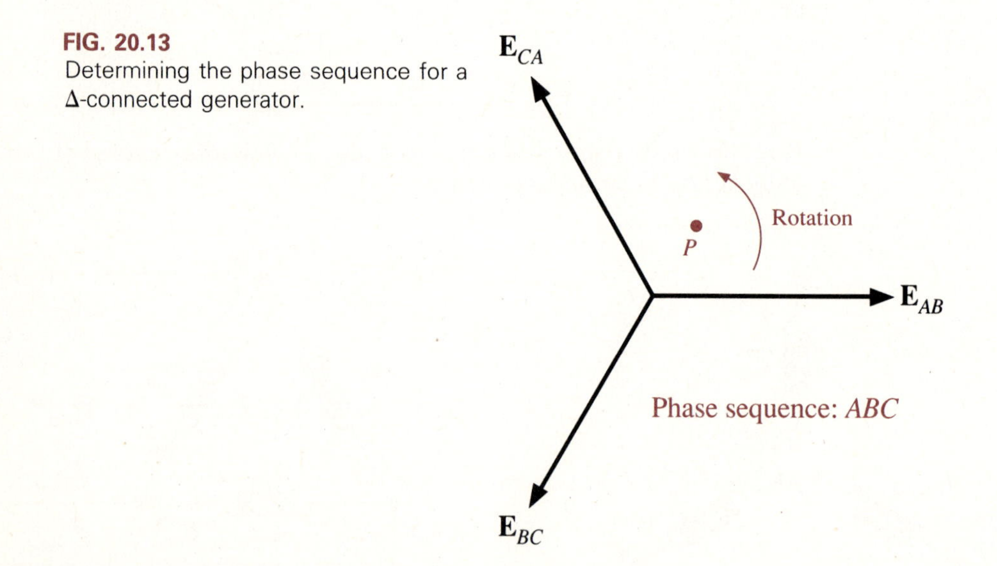

FIG. 20.13
Determining the phase sequence for a Δ-connected generator.

In phasor notation,

$$\mathbf{E}_{AB} = E_{AB}\,\underline{/0^\circ}$$
$$\mathbf{E}_{BC} = E_{BC}\,\underline{/-120^\circ}$$
$$\mathbf{E}_{CA} = E_{CA}\,\underline{/120^\circ}$$

20.7 THE Δ-Δ, Δ-Y THREE-PHASE SYSTEMS

The basic equations necessary to analyze either of the two systems (Δ-Δ, Δ-Y) have been presented at least once in this chapter. We therefore proceed directly to two descriptive examples, one with a Δ-connected load and one with a Y-connected load.

EXAMPLE 20.3 For the Δ-Δ system shown in Fig. 20.14,

a. Find the phase angles θ_2 and θ_3 for the specified phase sequence.
b. Find the current in each phase of the load.
c. Find the magnitude of the line currents.

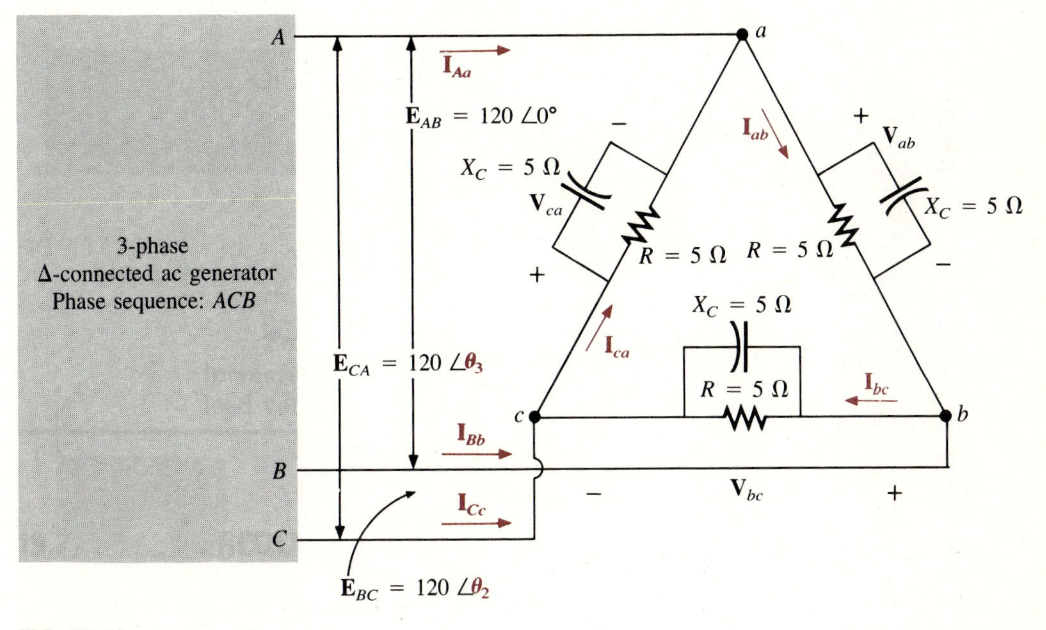

FIG. 20.14
Δ-Δ system for Example 20.3.

Solutions

a. For an *ACB* phase sequence,

$$\theta_2 = \mathbf{120°} \quad \text{and} \quad \theta_3 = \mathbf{-120°}$$

b. $\mathbf{V}_\phi = \mathbf{E}_L$. Therefore,

$$\mathbf{V}_{ab} = \mathbf{E}_{AB} \qquad \mathbf{V}_{ca} = \mathbf{E}_{CA} \qquad \mathbf{V}_{bc} = \mathbf{E}_{BC}$$

$$\mathbf{Y}_\phi = \mathbf{G} + \mathbf{B}_C = \frac{1}{R}\underline{/0°} + \frac{1}{X_C}\underline{/90°} = \frac{1}{5\ \Omega}\underline{/0°} + \frac{1}{5\ \Omega}\underline{/90°}$$

$$= 0.2\ \text{S}\ \underline{/0°} + 0.2\ \text{S}\ \underline{/90°}$$

and

$$Y_\phi = \sqrt{G^2 + B_C^2} = \sqrt{(0.2\text{ S})^2 + (0.2\text{ S})^2} = 0.283\text{ S}$$

with

$$\theta = \tan^{-1}\frac{0.2\text{ S}}{0.2\text{ S}} = \tan^{-1}1 = 45°$$

and

$$\mathbf{Y}_\phi = 0.283\text{ S}\,\underline{/45°}$$

with

$$Z_\phi = \frac{1}{Y_\phi} = \frac{1}{0.283\text{ S}} = 3.53\ \Omega$$

and

$$\mathbf{Z}_\phi = 3.53\ \Omega\,\underline{/-45°}$$

For each phase:

$$I_\phi = \frac{V_\phi}{Z_\phi} = \frac{120\text{ V}}{3.53\ \Omega} = 33.99\text{ A}$$

Since the load is capacitive, $\mathbf{I}_{ab}$ leads $\mathbf{V}_{ab}$ by the *magnitude* of the angle (45°) associated with the phase impedance. For $\mathbf{V}_{ab} = 120\text{ V}\,\underline{/0°}$,

$$\mathbf{I}_{ab} = 33.99\text{ A}\,\underline{/0° + 45°}$$

and

$$\mathbf{I}_{ab} = \mathbf{33.99\ A}\,\underline{\mathbf{/45°}}$$

$\mathbf{Z}_{bc} = 3.53\ \Omega\,\underline{/-45°}$ and $\mathbf{V}_{bc} = 120\text{ V}\,\underline{/120°}$, so

$$\mathbf{I}_{bc} = 33.99\text{ A}\,\underline{/120° + 45°}$$

and

$$\mathbf{I}_{bc} = \mathbf{33.99\ A}\,\underline{\mathbf{/165°}}$$

$\mathbf{Z}_{ca} = 3.53\ \Omega\,\underline{/-45°}$ and $\mathbf{V}_{ca} = 120\text{ V}\,\underline{/-120°}$, so

$$I_{ca} = 33.99\text{ A}\,\underline{/-120° + 45°}$$

and

$$\mathbf{I}_{ca} = \mathbf{33.99\ A}\,\underline{\mathbf{/-75°}}$$

c. $I_L = \sqrt{3}I_\phi = (1.73)(33.99\text{ A}) = \mathbf{58.8\ A}$

EXAMPLE 20.4

For the Δ-Y system shown in Fig. 20.15,

a. Find the voltage across each phase of the load.
b. Find the magnitude of the line voltages.

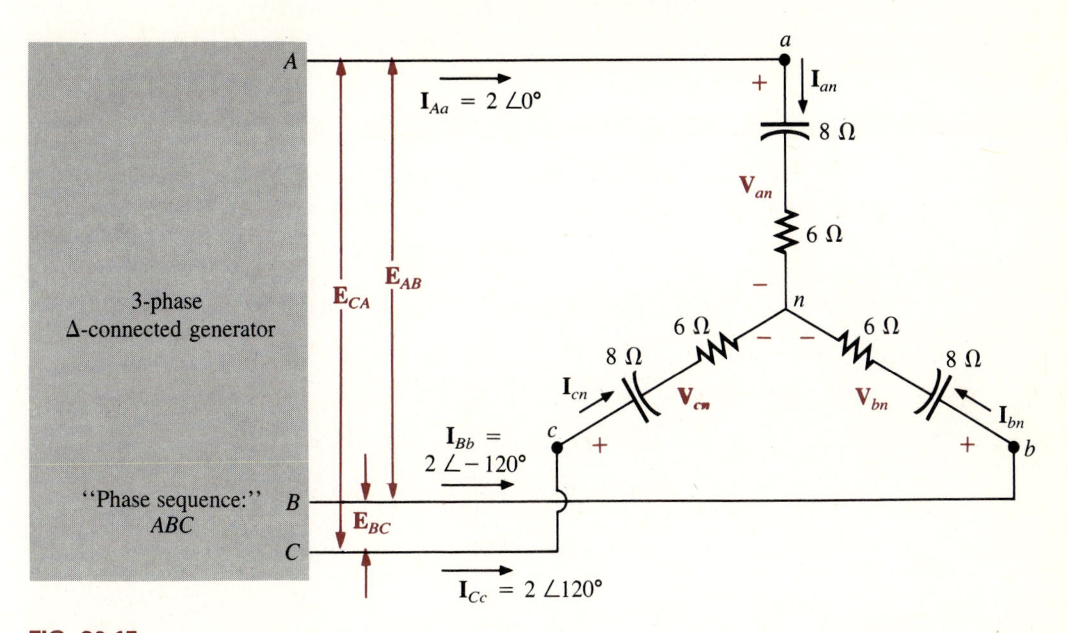

FIG. 20.15
Δ-Y system for Example 20.4.

Solutions

a. $\mathbf{I}_{\phi L} = \mathbf{I}_L$. Therefore,

$$\mathbf{I}_{an} = \mathbf{I}_{Aa} = 2\text{ A}\underline{/0^\circ}$$
$$\mathbf{I}_{bn} = \mathbf{I}_{Bb} = 2\text{ A}\underline{/-120^\circ}$$
$$\mathbf{I}_{cn} = \mathbf{I}_{Cc} = 2\text{ A}\underline{/120^\circ}$$

For each phase:

$$\mathbf{Z}_\phi = \mathbf{R} + \mathbf{X}_C = 6\ \Omega\ \underline{/0^\circ} + 8\ \Omega\ \underline{/-90^\circ}$$
$$Z_\phi = \sqrt{R^2 + X_C^2} = \sqrt{(6\ \Omega)^2 + (8\ \Omega)^2} = 10\ \Omega$$
$$\theta = \tan^{-1}\frac{X_C}{R} = \tan^{-1}\frac{8\ \Omega}{6\ \Omega} = 53.13^\circ$$

and

$$\mathbf{Z}_\phi = 10\ \Omega\ \underline{/-53.13^\circ}$$

The magnitude of each phase voltage is

$$V_\phi = I_\phi Z_\phi = (2\text{ A})(10\ \Omega) = 20\text{ V}$$

The capacitive load results in $\mathbf{I}_\phi$ leading $\mathbf{V}_\phi$ by the *magnitude* of the angle (53.13°) associated with the phase impedance.

For $\mathbf{I}_{an} = 2\text{ A}\,\angle 0°$, $\mathbf{V}_{an}$ will have an angle of $0° - 53.13° = -53.13°$ and

$$\mathbf{V}_{an} = \mathbf{20\text{ V}\,\angle{-53.13°}}$$

For $\mathbf{I}_{bn} = 2\text{ A}\,\angle{-120°}$, $\mathbf{V}_{bn}$ will have an angle of $-120° - 53.13° = -173.13°$ and

$$\mathbf{V}_{bn} = \mathbf{20\text{ V}\,\angle{-173.13°}}$$

For $\mathbf{I}_{cn} = 2\text{ A}\,\angle 120°$, $\mathbf{V}_{cn}$ will have an angle of $120° - 53.13° = 66.87°$ and

$$\mathbf{V}_{cn} = \mathbf{20\text{ V}\,\angle 66.87°}$$

b. $E_L = \sqrt{3}V_\phi = (1.73)(20\text{ V}) = 34.6\text{ V}$. Therefore,

$$E_{BA} = E_{CB} = E_{AC} = \mathbf{34.6\text{ V}}$$

20.8 POWER

Y-Connected Balanced Load See Fig. 20.16.

Average Power The average power delivered to each phase can be determined by any one of Eqs. 20.12 through 20.14.

$$P_\phi = V_\phi I_\phi \cos\theta_{I_\phi}^{V_\phi} = I_\phi^2 R_\phi = \frac{V_R^2}{R_\phi} \quad \text{(W)} \qquad (20.12)$$

where $\theta_{I_\phi}^{V_\phi}$ indicates that θ is the phase angle between V_ϕ and I_ϕ. The total power to the balanced load is

$$P_T = 3P_\phi \quad \text{(W)} \qquad (20.13)$$

In terms of the line voltage and current,

$$P_T = \sqrt{3}E_L I_L \cos\theta_{I_\phi}^{V_\phi} = 3I_L^2 R_\phi \quad \text{(W)} \qquad (20.14)$$

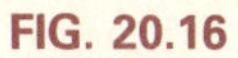

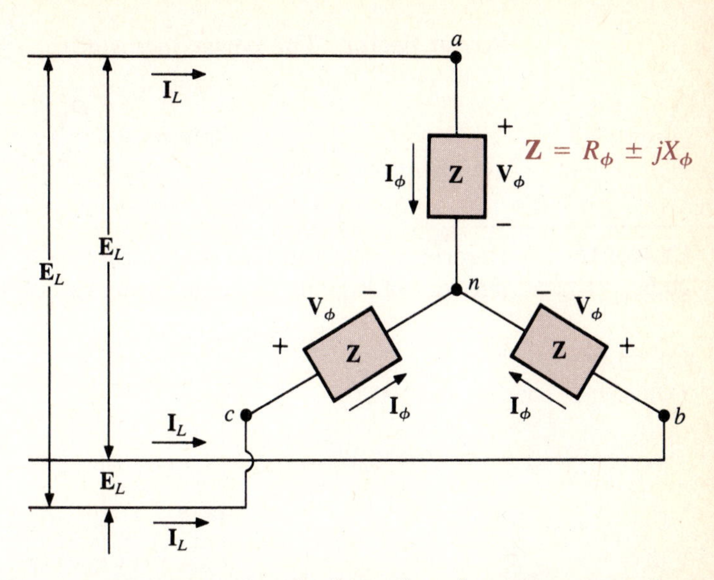

FIG. 20.16
Establishing the basic power equations for a Y-connected load.

Reactive Power The reactive power of each phase (in volt-amperes reactive) is

$$Q_\phi = V_\phi I_\phi \sin \theta_{I_\phi}^{V_\phi} = I_\phi^2 X_\phi = \frac{V_X^2}{X_\phi} \quad \text{(VAR)} \tag{20.15}$$

The total reactive power of the load is

$$Q_T = 3Q_\phi \quad \text{(VAR)} \tag{20.16}$$

or, proceeding in the same manner as before, we have

$$Q_T = \sqrt{3}E_L I_L \sin \theta_{I_\phi}^{V_\phi} = 3I_L^2 X_\phi \quad \text{(VAR)} \tag{20.17}$$

Apparent Power The apparent power of each phase is

$$S_\phi = V_\phi I_\phi \quad \text{(VA)} \tag{20.18}$$

The total apparent power of the load is

$$S_T = 3S_\phi \quad \text{(VA)} \tag{20.19}$$

or, as before,

$$S_T = \sqrt{3}E_L I_L \quad \text{(VA)} \tag{20.20}$$

Power Factor The power factor of the system is given by

$$F_p = \cos\theta = \frac{P_T}{S_T} \quad \text{(leading or lagging)} \tag{20.21}$$

EXAMPLE 20.5 Determine the total watts, volt-amperes reactive, and volt-amperes for the network of Fig. 20.17. In addition, calculate the total power factor of the load.

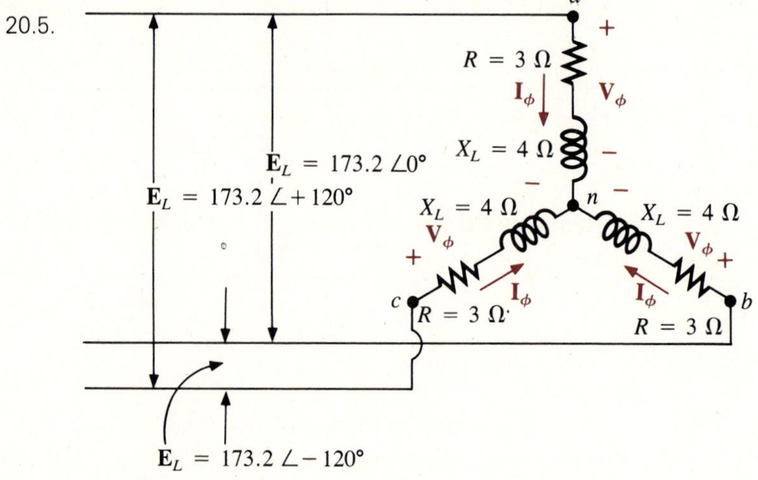

FIG. 20.17
Y-connected load for Example 20.5.

Solution See Fig. 20.17. Here,

$$V_\phi = \frac{V_L}{\sqrt{3}} = \frac{173.2}{1.732} = 100 \text{ V}$$

$$Z_\phi = \sqrt{R^2 + X_L^2} = \sqrt{(3\ \Omega)^2 + (4\ \Omega)^2} = 5\ \Omega$$

$$\theta = \tan^{-1}\frac{X_L}{R} = \tan^{-1}\frac{4\ \Omega}{3\ \Omega} = 53.13^\circ$$

$$I_\phi = \frac{V_\phi}{Z_\phi} = \frac{100 \text{ V}}{5\ \Omega} = 20 \text{ A}$$

The *average power* is

$$P_\phi = V_\phi I_\phi \cos\theta_{I_\phi}^{V_\phi} = (100 \text{ V})(20 \text{ A})\cos 53.13^\circ = (2000)(0.6) = \mathbf{1200\ W}$$

$$P_\phi = I_\phi^2 R_\phi = (20 \text{ A})^2(3\ \Omega) = (400)(3) = \mathbf{1200\ W}$$

$$P_\phi = \frac{V_R^2}{R_\phi} = \frac{(60 \text{ V})^2}{3\ \Omega} = \frac{3600}{3} = \mathbf{1200\ W}$$

$$P_T = 3P_\phi = (3)(1200 \text{ W}) = \mathbf{3600\ W}$$

or

$$P_T = \sqrt{3}E_L I_L \cos \theta_{I_\phi}^{V_\phi} = (1.732)(173.2 \text{ V})(20 \text{ A})(0.6) = \mathbf{3600 \text{ W}}$$

The *reactive power* is

$$Q_\phi = V_\phi I_\phi \sin \theta_{I_\phi}^{V_\phi} = (100 \text{ V})(20 \text{ A})\sin 53.13° = (2000)(0.8) = \mathbf{1600 \text{ VAR}}$$

or

$$Q_\phi = I_\phi^2 X_\phi = (20 \text{ A})^2(4 \ \Omega) = (400)(4) = \mathbf{1600 \text{ VAR}}$$

$$Q_T = 3Q_\phi = (3)(1600 \text{ VAR}) = \mathbf{4800 \text{ VAR}}$$

or

$$Q_T = \sqrt{3}E_L I_L \sin \theta_{I_\phi}^{V_\phi} = (1.732)(173.2 \text{ V})(20 \text{ A})(0.8) = \mathbf{4800 \text{ VAR}}$$

The *apparent power* is

$$S_\phi = V_\phi I_\phi = (100 \text{ V})(20 \text{ A}) = \mathbf{2000 \text{ VA}}$$

$$S_T = 3S_\phi = (3)(2000 \text{ VA}) = \mathbf{6000 \text{ VA}}$$

or

$$S_T = \sqrt{3}E_L I_L = (1.732)(173.2 \text{ V})(20 \text{ A}) = \mathbf{6000 \text{ VA}}$$

The *power factor* is

$$F_p = \frac{P_T}{S_T} = \frac{3600 \text{ W}}{6000 \text{ VA}} = \mathbf{0.6 \text{ lagging}}$$

Δ-Connected Balanced Load See Fig. 20.18.

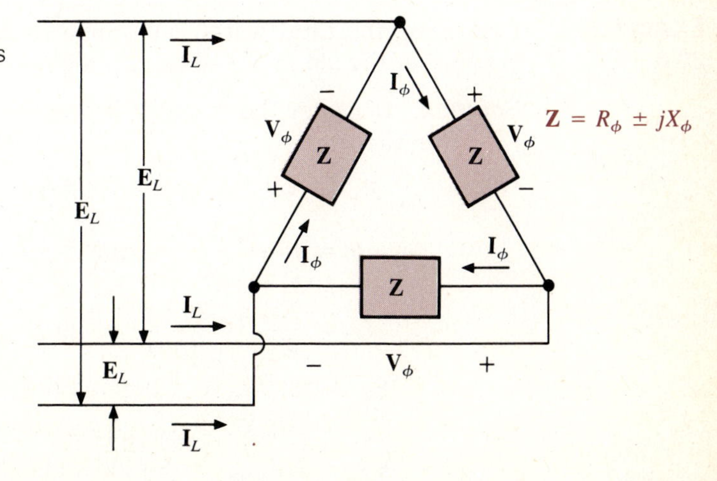

FIG. 20.18
Establishing the basic power equations for a Δ-connected load.

Average Power

$$P_\phi = V_\phi I_\phi \cos \theta_{I_\phi}^{V_\phi} = I_\phi^2 R_\phi = \frac{V_R^2}{R_\phi} \quad \text{(W)} \qquad (20.22)$$

$$P_T = 3P_\phi \quad \text{(W)} \qquad (20.23)$$

Reactive Power

$$Q_\phi = V_\phi I_\phi \sin \theta_{I_\phi}^{V_\phi} = I_\phi^2 X_\phi = \frac{V_X^2}{X_\phi} \quad \text{(VAR)} \qquad (20.24)$$

$$Q_T = 3Q_\phi \quad \text{(VAR)} \qquad (20.25)$$

Apparent Power

$$S_\phi = V_\phi I_\phi \quad \text{(VA)} \qquad (20.26)$$

$$S_T = 3S_\phi = \sqrt{3}E_L I_L \quad \text{(VA)} \qquad (20.27)$$

Power Factor

$$F_p = \cos \theta = \frac{P_T}{S_T} \qquad (20.28)$$

EXAMPLE 20.6 Determine the total watts, volt-amperes reactive, and volt-amperes for the network of Fig. 20.19. In addition, calculate the total power factor of the load.

Solution Consider the Δ and Y separately.

For the Δ:

$$Z_\Delta = \sqrt{R^2 + X_C^2} = \sqrt{(6\ \Omega)^2 + (8\ \Omega)^2} = 10\ \Omega$$

$$\theta = \tan^{-1}\frac{X_C}{R} = \tan^{-1}\frac{8\ \Omega}{6\ \Omega} = 53.13^\circ$$

$$I_\phi = \frac{E_L}{Z_\Delta} = \frac{200\ \text{V}}{10\ \Omega} = 20\ \text{A}$$

$$P_{T_\Delta} = 3I_\phi^2 R_\phi = (3)(20\ \text{A})^2(6\ \Omega) = \mathbf{7200\ W}$$

$$Q_{T_\Delta} = 3I_\phi^2 X_\phi = (3)(20\ \text{A})^2(8\ \Omega) = \mathbf{9600\ VAR}\quad \mathbf{(cap.)}$$

$$S_{T_\Delta} = 3V_\phi I_\phi = (3)(200\ \text{V})(20\ \text{A}) = \mathbf{12{,}000\ VA}$$

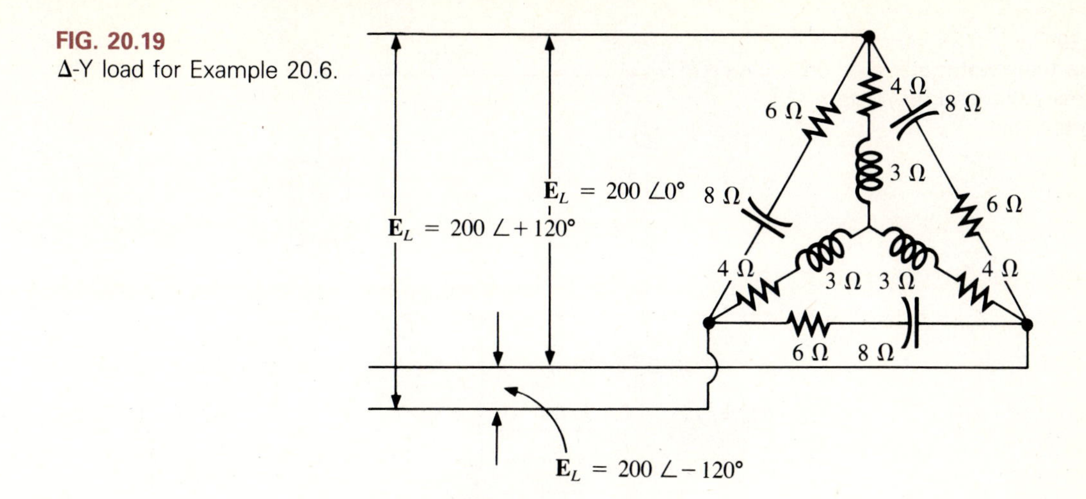

FIG. 20.19
Δ-Y load for Example 20.6.

For the Y:

$$Z_Y = \sqrt{R^2 + X_L^2} = \sqrt{(4\,\Omega)^2 + (3\,\Omega)^2} = 5\,\Omega$$

$$\theta = \tan^{-1}\frac{X_L}{R} = \tan^{-1}\frac{3\,\Omega}{4\,\Omega} = 36.87°$$

$$I_\phi = \frac{E_L/\sqrt{3}}{Z_Y} = \frac{200\text{ V}/\sqrt{3}}{5\,\Omega} = \frac{116\text{ V}}{5\,\Omega} = 23.12\text{ A}$$

$$P_{T_Y} = 3I_\phi^2 R_\phi = (3)(23.12\text{ A})^2(4\,\Omega) = \mathbf{6414.41\ W}$$

$$Q_{T_Y} = 3I_\phi^2 X_\phi = (3)(23.12\text{ A})^2(3\,\Omega) = \mathbf{4810.81\ VAR\ \ (ind.)}$$

$$S_{T_Y} = 3V_\phi I_\phi = (3)(116\text{ V})(23.12\text{ A}) = \mathbf{8045.76\ VA}$$

For the total load:

$$P_T = P_{T_\Delta} + P_{T_Y} = 7200\text{ W} + 6414.41\text{ W} = \mathbf{13{,}614.41\ W}$$

$$Q_T = Q_{T_\Delta} - Q_{T_Y} = 9600\text{ VAR (cap.)} - 4810.81\text{ VAR (ind.)}$$
$$= \mathbf{4789.19\ VAR\ \ (cap.)}$$

$$S_T = \sqrt{P_T^2 + Q_T^2} = \sqrt{(13{,}614.41\text{ W})^2 + (4789.19\text{ VAR})^2}$$
$$= \mathbf{14{,}432.2\ VA}$$

$$F_p = \frac{P_T}{S_T} = \frac{13{,}614.41\text{ W}}{14{,}432.20\text{ VA}} = \mathbf{0.943\ \ leading}$$

20.9 THE THREE-WATTMETER METHOD

The power delivered to a balanced or an unbalanced four-wire Y-connected load can be found using three wattmeters in the manner shown in Fig. 20.20. Each wattmeter measures the power delivered to each phase. The potential coil of each wattmeter is connected parallel with the load, whereas the current

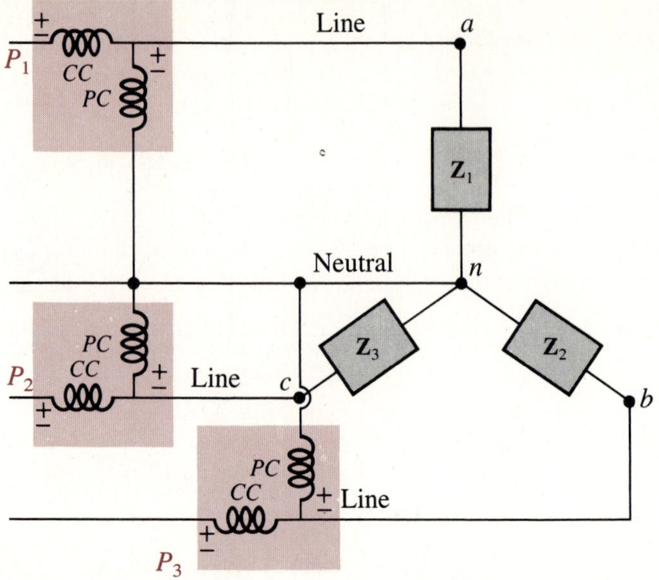

FIG. 20.20
Using the three-wattmeter method to determine the total power to a Y-connected load.

coil is in series with the load. The total average power of the system can be found by summing the three wattmeter readings; that is,

$$P_{T_Y} = P_1 + P_2 + P_3 \tag{20.29}$$

For the load (balanced or unbalanced), the wattmeters are connected as shown in Fig. 20.21. The total power is again the sum of the three wattmeter readings:

$$P_{T_\Delta} = P_1 + P_2 + P_3 \tag{20.30}$$

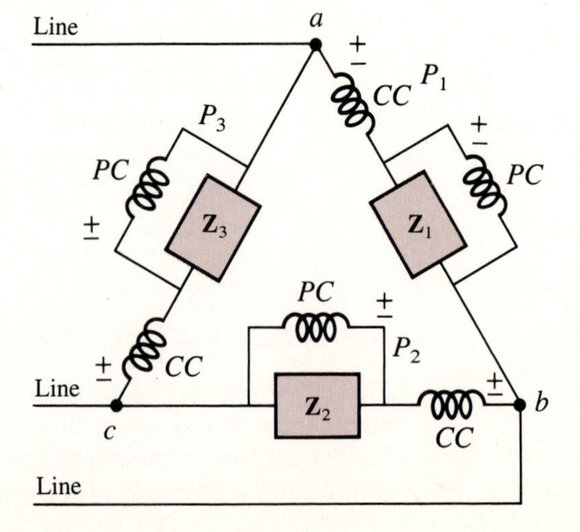

FIG. 20.21
Determining the total power to a Δ-connected load using three wattmeters.

If, in either of the cases just described, the load is balanced, the power delivered to each phase is the same. The total power is then just three times any one wattmeter reading.

20.10 THE TWO-WATTMETER METHOD

The power delivered to a three-phase, three-wire, Δ- or Y-connected balanced or unbalanced load can be found using only two wattmeters if the proper connection is employed and if the wattmeter readings are properly interpreted. The basic connections are shown in Fig. 20.22. One end of each potential coil is connected to the same line. The current coils are then placed in the remaining lines.

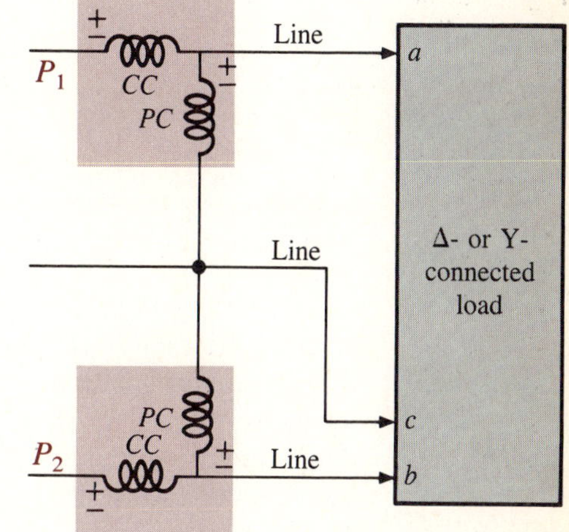

FIG. 20.22
Applying the two-wattmeter method to a Δ- or Y-connected load.

The connection shown in Fig. 20.23 also satisfies the requirements. A third hookup is also possible, but this is left to the reader as an exercise.

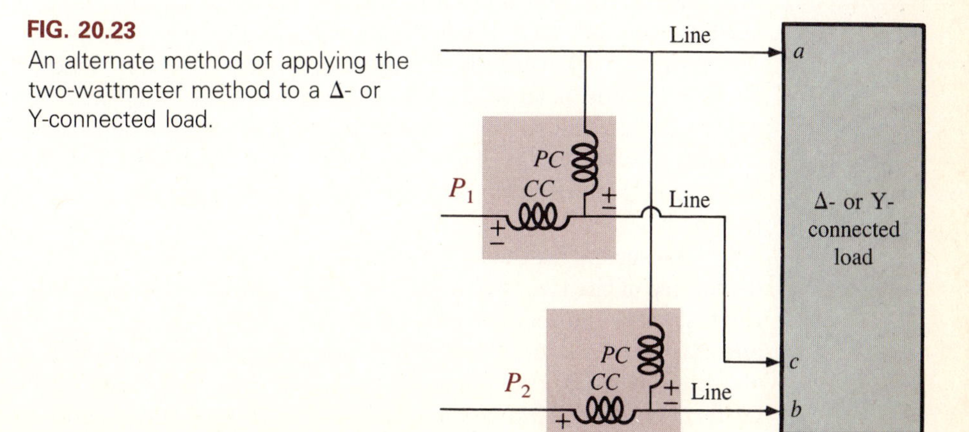

FIG. 20.23
An alternate method of applying the two-wattmeter method to a Δ- or Y-connected load.

The total power delivered to the load is the algebraic sum of the two wattmeter readings. For a *balanced* load, we now consider two methods of determining whether the total power is the sum or the difference of the two wattmeter readings. The first method to be described requires that we know or be able to find the power factor (leading or lagging) of any one phase of the load. When this information has been obtained, it can be applied directly to the curve of Fig. 20.24.

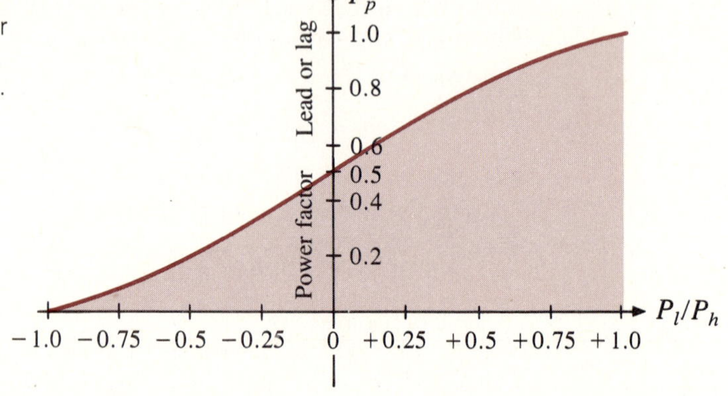

FIG. 20.24
Determining whether to find the sum or difference of the wattmeter readings when using the two-wattmeter method.

The curve in Fig. 20.24 is a plot of the power factor of the load (phase) versus the ratio P_l/P_h, where P_l and P_h are the magnitudes of the lower- and higher-reading wattmeters, respectively. Note that for a power factor (leading or lagging) greater than 0.5, the ratio has a positive value. This indicates that both wattmeters are reading positive, and the total power is the sum of the two wattmeter readings; that is, $P_T = P_l + P_h$. For a power factor less than 0.5 (leading or lagging), the ratio has a negative value. This indicates that the smaller-reading wattmeter is reading negative, and the total power is the difference of the two wattmeter readings; that is, $P_T = P_h - P_l$.

A closer examination reveals that when the power factor is 1 ($\cos 0° = 1$), corresponding to a purely resistive load, $P_l/P_h = 1$, or $P_l = P_h$, and both wattmeters have the same wattage indication. At a power factor equal to 0 ($\cos 90° = 0$), corresponding to a purely reactive load, $P_l/P_h = -1$, or $P_l = -P_h$, and both wattmeters again have the same wattage indication but with opposite signs. The transition from a negative to a positive ratio occurs when the power factor of the load is 0.5, or $\theta = \cos^{-1}0.5 = 60°$. At this power factor, $P_l/P_h = 0$, so $P_l = 0$, whereas P_h will read the total power delivered to the load.

The second method for determining whether the total power is the sum or difference of the two wattmeter readings involves a simple laboratory test. For the test to be applied, both wattmeters must first have an up-scale deflection. If one of the wattmeters has a below-zero indication, an up-scale deflection can be obtained by simply reversing the leads of the current coil of the wattmeter. To perform the test, first remove the lead of the potential coil of the *low-reading* wattmeter from the line that has no current coil in it. Take this lead and touch it to the line that has the current coil of the *high-reading* wattmeter in it. If the pointer of the low-reading wattmeter deflects upward, the two wattmeter readings should be added. If the pointer deflects downward (below 0 W), the wattage reading of the low-reading wattmeter should be subtracted from that of the high-reading wattmeter.

For a *balanced system,* since

$$P_T = P_h + P_l = \sqrt{3}E_L I_L \cos \theta_{I_\phi}^{V_\phi}$$

the power factor of the load (phase) can be found from the wattmeter readings and the magnitude of the line voltage and current:

$$F_p = \cos \theta_{I_\phi}^{V_\phi} = \frac{P_h + P_l}{\sqrt{3}E_L I_L} \tag{20.31}$$

20.11 UNBALANCED THREE-PHASE, FOUR-WIRE, Y-CONNECTED LOAD

For the three-phase, four-wire, Y-connected load of Fig. 20.25, conditions are such that *none* of the load impedances are equal. Since the neutral is a common point between the load and source, no matter what the impedance of each phase of the load, the voltage across each phase is the phase voltage of the generator:

$$\mathbf{V}_\phi = \mathbf{E}_\phi \tag{20.32}$$

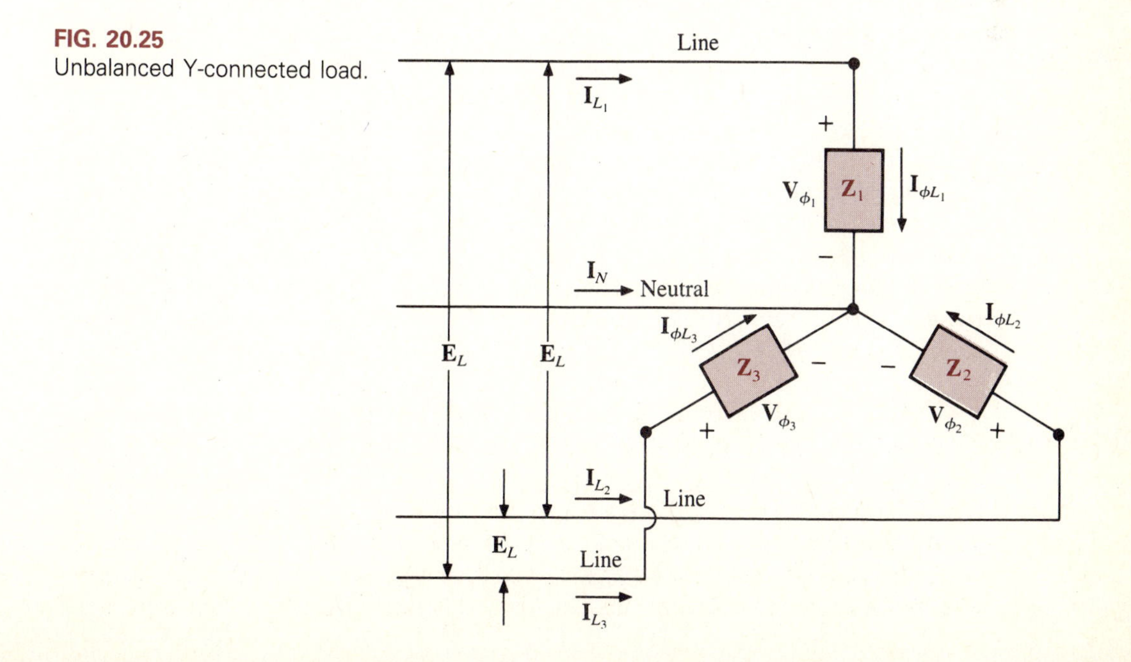

FIG. 20.25
Unbalanced Y-connected load.

Therefore, the phase currents can be determined by Ohm's law:

$$I_{\phi_1} = \frac{V_{\phi_1}}{Z_{\phi_1}} = \frac{E_{\phi_1}}{Z_{\phi_1}} \quad \text{and so on} \tag{20.33}$$

The current in the neutral for any unbalanced system can then be found by applying Kirchhoff's current law at the common point n:

$$\mathbf{I}_N = \mathbf{I}_{\phi_1} + \mathbf{I}_{\phi_2} + \mathbf{I}_{\phi_3} = \mathbf{I}_{L_1} + \mathbf{I}_{L_2} + \mathbf{I}_{L_3} \tag{20.34}$$

20.12 COMPUTER TECHNIQUES

The program of Fig. 20.27 will determine all the quantities of interest for the Δ-Y-connected load of Fig. 20.26. As indicated by the INPUT bracket, lines 130 through 220 request the line voltage and the network parameters of the series-connected phase impedances. Inductive and capacitive reactances are distinguished by the sign entered on lines 180 and 220.

The calculations for the Y-connected loads are made by lines 250 through 360 using equations introduced in Section 20.4. Lines 380 through 490 perform a detailed analysis of the Δ section. The total real and reactive power are calculated by lines 510 through 570, and the total apparent power and the power factor by lines 580 through 630.

For comparison purposes, the run employed the network parameters used in Example 20.6.

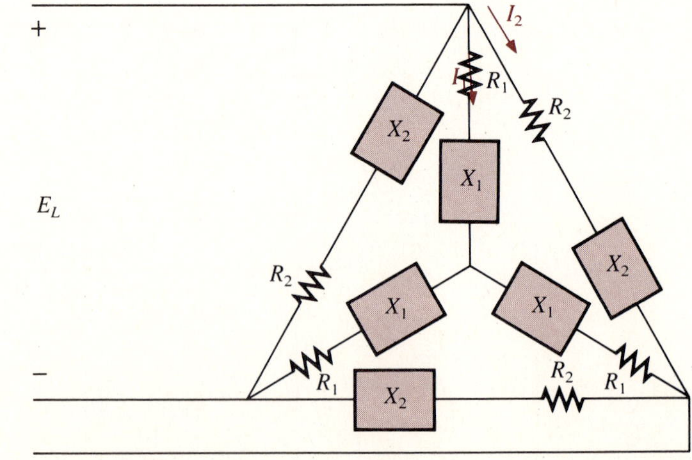

FIG. 20.26
Δ-Y configuration for computer analysis.

```
 10 REM ***** PROGRAM 20-1  *****
 20 REM ******************************
 30 REM Program to analyze a 3-phase
 40 REM delta-wye load.
 50 REM ******************************
 60 REM
100 PRINT "This program analyzes the 3-phase"
110 PRINT "delta-wye load."
120 PRINT
```

Input

```
130 PRINT "Enter the following network information:"
140 PRINT
150 INPUT "Line voltage, E=";EL
160 PRINT "For the series-connected wye load:"
170 INPUT "R=";R1
180 INPUT "X=";X1 :REM Enter negative sign if capacitive
190 PRINT
200 PRINT "And for the series-connected delta load:"
210 INPUT "R=";R2
220 INPUT "X=";X2 :REM Enter negative sign if capacitive
```

```
230 REM Now do calculations and print results
240 PRINT:PRINT
```

Y

```
250 Z1=SQR(R1^2+X1^2)
260 I1=EL/(SQR(3)*Z1)
270 P1=3*I1^2*R1
280 Q1=3*I1^2*X1
290 S1=3*EL*I1/SQR(3)
300 PRINT "For the wye connection:"
310 PRINT "The phase current I=";I1;"amps"
320 PRINT "The total power dissipated is";P1;"watts"
330 PRINT "With a net reactive power of Q=";ABS(Q1);"vars";
340 IF SGN(X1)=-1 THEN PRINT "(cap.)"
350 IF SGN(X1)=1 THEN PRINT "(ind.)"
360 PRINT "and apparent power of:";S1;"VA"
```

```
370 PRINT:PRINT
```

Δ

```
380 Z2=SQR(R2^2+X2^2)
390 I2=EL/Z2
400 P2=3*I2^2*R2
410 Q2=3*I2^2*X2
420 S2=3*EL*I2
430 PRINT "For the delta connected load:"
440 PRINT "The phase current is, I=";I2;"amps"
450 PRINT "The total power dissipated is";P2;"watts"
460 PRINT "with a net reactive power of Q=";ABS(Q2);"vars";
470 IF SGN(X2)=-1 THEN PRINT "(cap.)"
480 IF SGN(X2)=1 THEN PRINT "(ind.)"
490 PRINT "and apparent power of:";S2;"VA"
```

```
500 PRINT:PRINT
```

P_T
Q_T

```
510 PT=P1+P2
520 QT=Q1+Q2
530 PRINT "For the combined system:"
540 PRINT "The total power dissipated is";PT;"watts"
550 PRINT "and the net reactive power QT=";ABS(QT);"vars";
560 IF SGN(QT)=-1 THEN PRINT "(cap.)"
570 IF SGN(QT)=1 THEN PRINT "(ind.)"
```

S_T
F_p

```
580 ST=SQR(PT^2+QT^2)
590 FP=PT/ST
600 PRINT:PRINT "The total apparent power ST=";ST;"VA"
610 PRINT "with a network power factor Fp=";FP;
620 IF SGN(QT)=-1 THEN PRINT "(leading)"
630 IF SGN(QT)=1 THEN PRINT "(lagging)"
```

```
640 END

READY
```

```
RUN


This program analyzes the 3-phase
delta-wye load.

Enter the following network information:

Line voltage, E=? 200

For the series-connected wye load:
R=? 4

X=? 3


And for the series-connected delta load:
R=? 6

X=? -8



For the wye connection:
The phase current I= 23.094 amps
The total power dissipated is 6400 watts
With a net reactive power of Q= 4800 vars(ind.)
and apparent power of: 8000 VA


For the delta connected load:
The phase current is, I= 20 amps
The total power dissipated is 7200 watts
with a net reactive power of Q= 9600 vars(cap.)
and apparent power of: 1.2E+04 VA


For the combined system:
The total power dissipated is 1.36E+04 watts
and the net reactive power QT= 4800 vars(cap.)

The total apparent power ST= 1.4422E+04 VA
with a network power factor Fp= .943 (leading)


READY
```

FIG. 20.27
Program 20.1.

FORMULA SUMMARY

Y-Y System

$$E_L = \sqrt{3}E_\phi = \sqrt{3}V_\phi$$
$$I_L = I_\phi$$

Y-Δ System

$$E_L = \sqrt{3}E_\phi, \qquad V_\phi = E_L$$
$$I_L = \sqrt{3}I_{\text{load}}$$

Δ-Δ System

$$E_L = E_\phi = V_\phi$$
$$I_L = \sqrt{3}I_{\text{load}}$$

Δ-Y System

$$E_L = E_\phi = \sqrt{3}V_\phi$$
$$I_L = I_{\text{load}}$$

Power

Y-connected load:

$$P_\phi = V_\phi I_\phi \cos\theta = I_\phi^2 R_\phi = \frac{V_R^2}{R_\phi}$$
$$P_T = 3P_\phi = \sqrt{3}E_L I_L \cos\theta = 3I_L^2 R_\phi$$
$$Q_\phi = V_\phi I_\phi \sin\theta = I_\phi^2 X_\phi = \frac{V_X^2}{X_\phi}$$
$$Q_T = 3Q_\phi = \sqrt{3}E_L I_L \sin\theta = 3I_L^2 X_\phi$$
$$S_\phi = V_\phi I_\phi$$
$$S_T = 3S_\phi = \sqrt{3}E_L I_L$$
$$F_p = \cos\theta = \frac{P_T}{S_T}$$

Δ-connected load:

$$P_\phi = V_\phi I_\phi \cos\theta = I_\phi^2 R_\phi = \frac{V_R^2}{R_\phi}$$
$$P_T = 3P_\phi = \sqrt{3}E_L I_L \cos\theta = 3I_\phi^2 R_\phi$$
$$Q_\phi = V_\phi I_\phi \sin\theta = I_\phi^2 X_\phi = \frac{V_X^2}{X_\phi}$$
$$Q_T = 3Q_\phi = \sqrt{3}E_L I_L \sin\theta = I_\phi^2 X_\phi$$
$$S_\phi = V_\phi I_\phi$$
$$S_T = 3S_\phi = \sqrt{3}E_L I_L$$
$$F_p = \cos\theta = \frac{P_T}{S_T}$$

Two-wattmeter method:

$$P_T = P_h + P_l = \sqrt{3}E_L I_L \cos\theta$$
$$F_p = \cos\theta = \frac{P_h + P_l}{\sqrt{3}E_L I_L}$$

CHAPTER SUMMARY

There are great similarities among the analyses of the Y-Y, Y-Δ, Δ-Δ, and Δ-Y systems. In fact, the factor $\sqrt{3}$ is the same whenever two quantities are not equal. A *sketch* of the system makes it obvious whether two quantities are equal or related by the factor $\sqrt{3}$. For all configurations, the load voltage can never be greater than the line voltage; if they are not equal, the line voltage is greater by the $\sqrt{3}$ factor.

The power equations are also very similar for both the Y- and Δ-connected loads if you keep in mind that the power to a resistor is always the current through the resistor squared times the resistor, and the total power is the sum of the power levels of the system. In this case, since the power levels are the same for a balanced load, the total power is $3P_\phi$. Similarly, for reactive elements the reactive power is $I_\phi^2X_\phi$ for each element, with the current I_ϕ the current through the element and the total reactive power $3Q_\phi$. The total apparent power can be determined using an equation introduced in Chapter 17: $S_T = \sqrt{P_T^2 + Q_T^2}$ or $\sqrt{3}E_LI_L$ for each configuration. For Δ and Y configurations the power factor is $F_p = P_T/S_T$. The power delivered to a three-phase Y-connected load is fairly easy to measure with three wattmeters, since each wattmeter senses a line current and the voltage from that line to the neutral connection. For the Δ-connected load, each meter senses a phase current of the load and the voltage across a phase of the load. The two-wattmeter method requires that each wattmeter sense a line current with the "low" end of each *PC* connection hooked up to a the line without a *CC* connection. For primarily resistive loads with $F_p > 0.5$, the power levels of the two meters are added, whereas for reactive loads with $F_p < 0.5$, the difference of the readings is used to determine the total power.

For unbalanced four-wire, Y-connected loads, the phase currents can be determined simply using Ohm's law. For three-wire, Y-connected unbalanced loads, a technique such as mesh analysis must be applied to determine the various phase currents.

GLOSSARY

Delta- (Δ-) connected generator A three-phase generator having the three phases connected in the shape of the capital Greek letter delta (Δ).

Line current The current that flows from the generator to the load of a single-phase or polyphase system.

Line voltage The potential difference that exists between the lines of a single-phase or polyphase system.

Neutral connection The connection between the generator and the load that, under balanced conditions, will have zero current associated with it.

Phase current The current that flows through each phase of a single-phase or polyphase generator load.

Phase sequence The order in which the generated sinusoidal voltages of a polyphase generator will affect the load to which they are applied.

Phase voltage The voltage that appears between the line and neutral of a Y-connected generator and from line to line in a Δ-connected generator.

Polyphase ac generator An electromechanical source of ac power that generates more than one sinusoidal voltage per rotation of the rotor. The frequency generated is determined by the speed of rotation and the number of poles of the rotor.

Single-phase ac generator An electromechanical source of ac power that generates a single sinusoidal voltage having a frequency determined by the speed of rotation and the number of poles of the rotor.

Three-wattmeter method A method for determining the total power delivered to a three-phase load using three wattmeters.

Two-wattmeter method A method for determining the total power delivered to a Δ- or Y-connected three-phase load using only two wattmeters and considering the power factor of the load.

Unbalanced polyphase load A load not having the same impedance in each phase.

Wye- (Y-) connected generator A three-phase source of ac power in which the three phases are connected in the shape of the letter Y.

PROBLEMS

Section 20.4

1. A balanced Y load having a 10-Ω resistance in each leg is connected to a three-phase, four-wire, Y-connected generator having a line voltage of 208 V. Calculate the magnitude of
   a. The phase voltage of the generator
   b. The phase voltage of the load
   c. The phase current of the load
   d. The line current
2. Repeat Problem 1 if each phase impedance is changed to a 12-Ω resistor in series with a 16-Ω capacitive reactance.
3. Repeat Problem 1 if the phase impedance is changed to a 10-Ω resistor in parallel with a 10-Ω capacitive reactance.
4. The phase sequence for the Y-Y system of Fig. 20.28 is *ABC*.
   a. Find the angles θ_2 and θ_3 for the specified phase sequence.
   b. Find the voltage across each phase impedance in phasor form.
   c. Find the current through each phase impedance in phasor form.
   d. Draw the phasor diagram of the currents found in part c and show that their phasor sum is zero.
   e. Find the magnitude of the line currents.
   f. Find the magnitude of the line voltages.
5. Repeat Problem 4 if the phase impedances are changed to a 9-Ω resistor in series with a 12-Ω inductive reactance.

FIG. 20.28

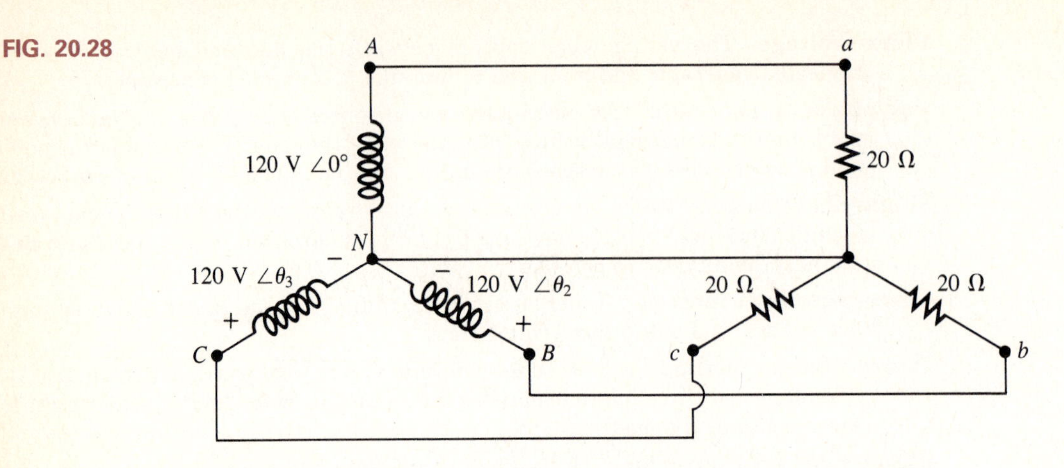

6. Repeat Problem 4 if the phase impedances are changed to a 6-Ω resistance in parallel with an 8-Ω capacitive reactance.
7. For the system of Fig. 20.29, find the magnitude of the unknown voltages and currents.

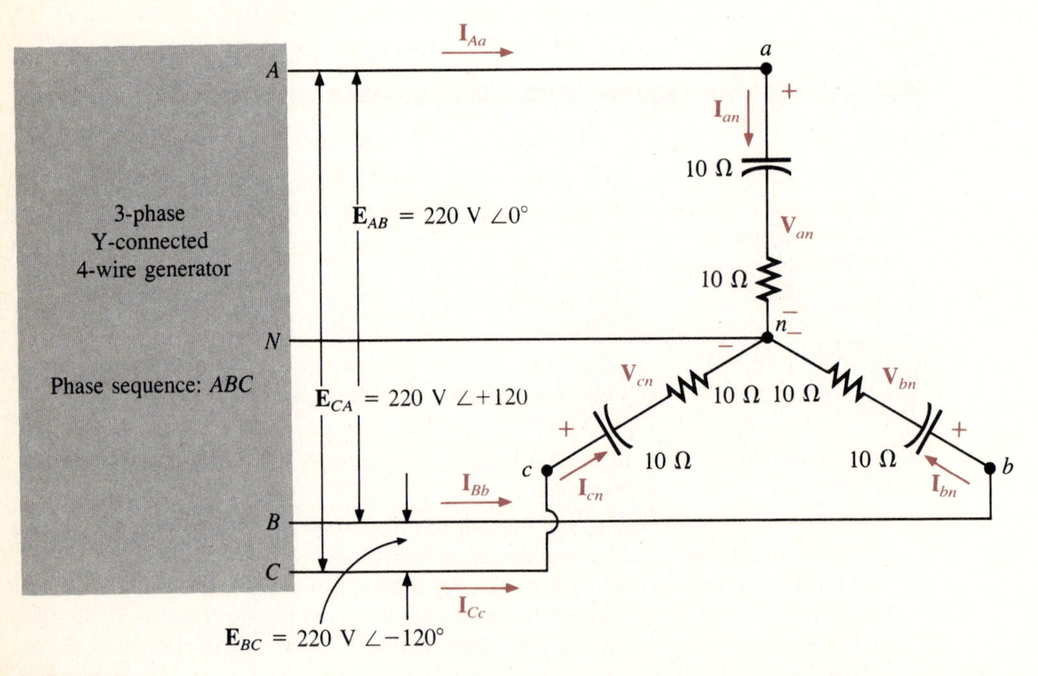

FIG. 20.29

*8. Compute the magnitude of the voltage E_{AB} for the balanced three-phase system of Fig. 20.30.

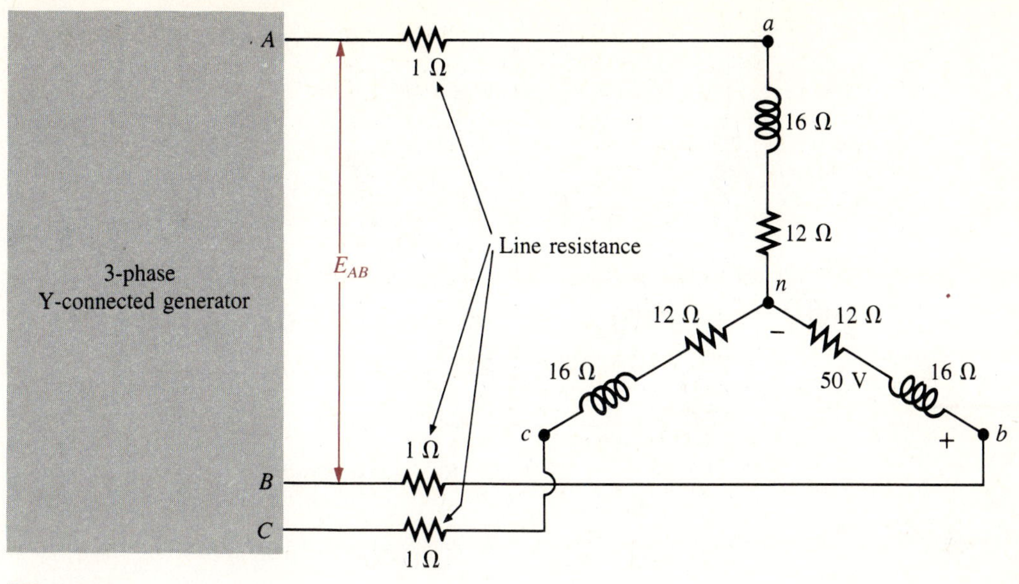

FIG. 20.30

Section 20.5

9. A balanced Δ load having a 20-Ω resistance in each leg is connected to a three-phase, three-wire, Y-connected generator having a line voltage of 208 V. Calculate the magnitude of

a. The phase voltage of the generator
b. The phase voltage of the load
c. The phase current of the load
d. The line current

10. Repeat Problem 9 if each phase impedance is changed to a 6.8-Ω resistor in series with a 14-Ω inductive reactance.

11. Repeat Problem 9 if each phase impedance is changed to an 18-Ω resistance in parallel with an 18-Ω capacitive reactance.

12. The phase sequence for the Y-Δ system of Fig. 20.31 is *ABC*.

a. Find the angles θ_2 and θ_3 for the specified phase sequence.
b. Find the voltage across each phase impedance in phasor form.
c. Draw the phasor diagram of the voltages found in part b and show that their sum is zero around the closed loop of the Δ load.
d. Find the current through each phase impedance in phasor form.
e. Find the magnitude of the line currents.
f. Find the magnitude of the generator phase voltages.

13. Repeat Problem 12 if the phase impedances are changed to a 100-Ω resistor in series with a capacitive reactance of 100 Ω.

14. Repeat Problem 13 if the phase impedances are changed to a 3-Ω resistor in parallel with an inductive reactance of 4 Ω.

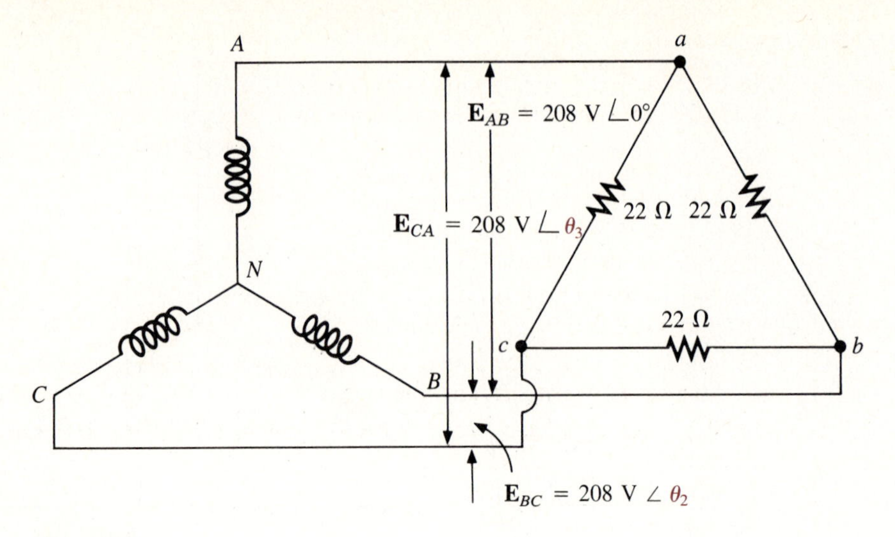

FIG. 20.31

15. For the system of Fig. 20.32, find the magnitude of the unknown voltages and currents.

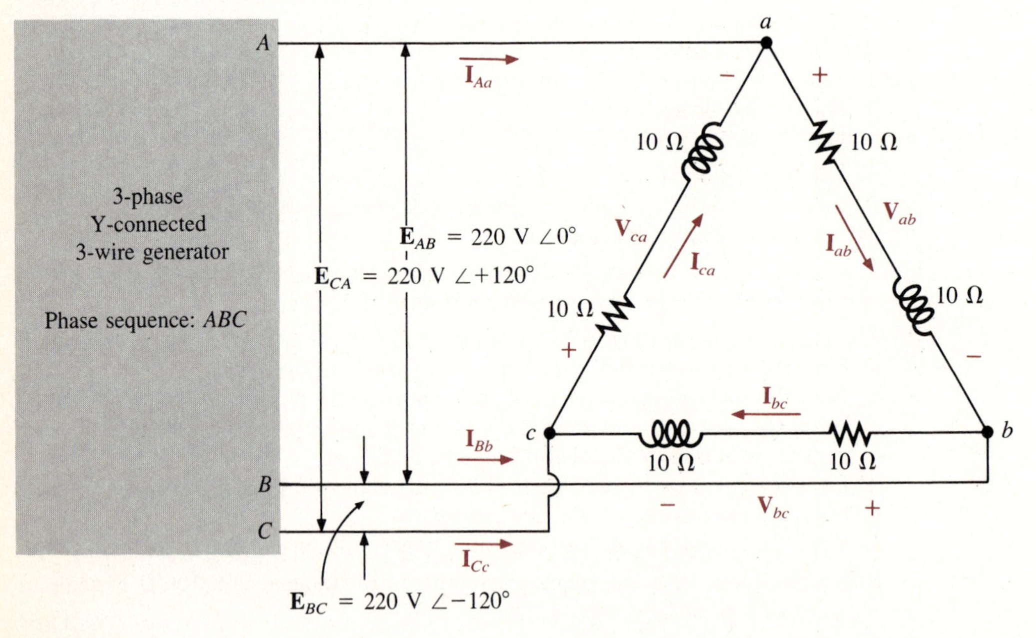

FIG. 20.32

Section 20.7

16. A balanced Y load having a 30-Ω resistance in each leg is connected to a three-phase Δ-connected generator having a line voltage of 208 V. Calculate the magnitude of
   a. The phase voltage of the generator
   b. The phase voltage of the load
   c. The phase current of the load
   d. The line current
17. Repeat Problem 16 if each phase impedance is changed to a 12-Ω resistor in series with a 12-Ω inductive reactance.
18. Repeat Problem 16 if each phase impedance is changed to a 15-Ω resistor in parallel with a 20-Ω capacitive reactance.
19. For the system of Fig. 20.33, find the magnitude of the unknown voltages and currents.

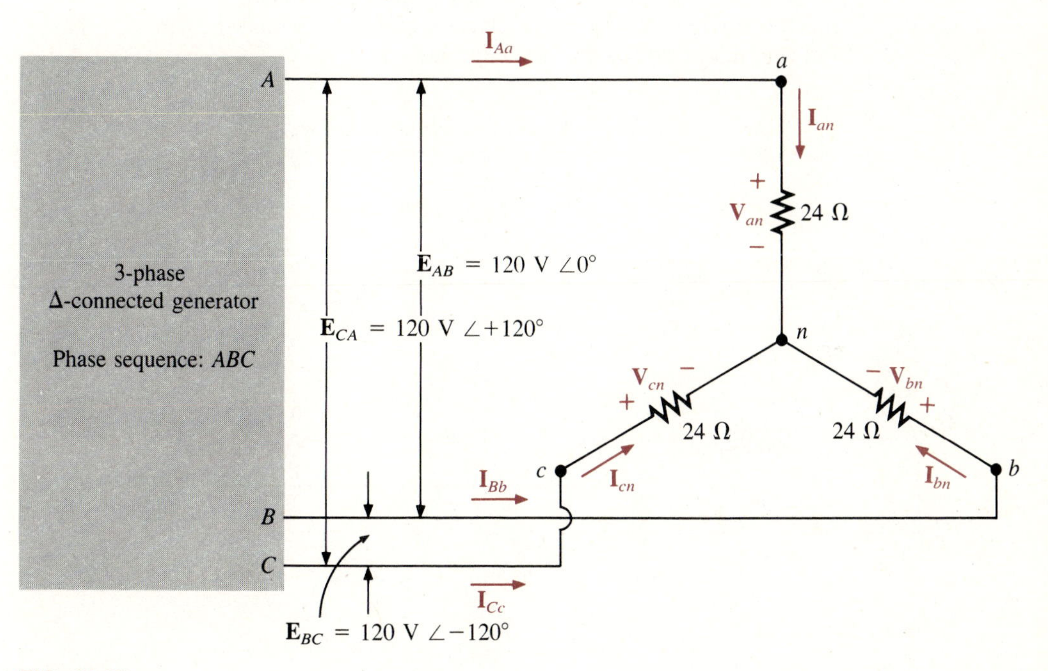

FIG. 20.33

20. Repeat Problem 19 if each phase impedance is changed to a 10-Ω resistor in series with a 20-Ω inductive reactance.
21. Repeat Problem 19 if each phase impedance is changed to a 20-Ω resistor in parallel with a 15-Ω capacitive reactance.

22. A balanced Δ load having a 220-Ω resistance in each leg is connected to a three-phase Δ-connected generator having a line voltage of 440 V. Calculate the magnitude of

a. The phase voltage of the generator
b. The phase voltage of the load
c. The phase current of the load
d. The line current

23. Repeat Problem 22 if each phase impedance is changed to a 12-Ω resistor in series with a 9-Ω capacitive reactance.

24. Repeat Problem 22 if each phase impedance is changed to a 22-Ω resistor in parallel with a 22-Ω inductive reactance.

25. The phase sequence for the Δ-Δ system of Fig. 20.34 is *ABC*.

a. Find the angles θ_2 and θ_3 for the specified phase sequence.
b. Find the voltage across each phase impedance in phasor form.
c. Draw the phasor diagram of the voltages found in part b and show that their phasor sum is zero around the closed loop of the Δ load.
d. Find the current through each phase impedance in phasor form.
e. Find the magnitude of the line currents.

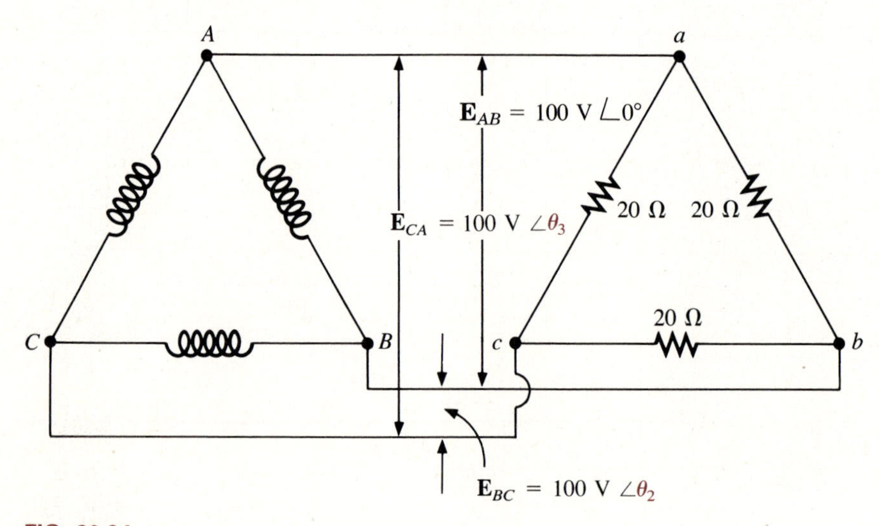

FIG. 20.34

26. Repeat Problem 25 if each phase impedance is changed to a 12-Ω resistor in series with a 16-Ω inductive reactance.

27. Repeat Problem 25 if each phase impedance is changed to a 20-Ω resistor in parallel with a 20-Ω capacitive reactance.

Section 20.8

28. Find the total watts, volt-amperes reactive, volt-amperes, and F_p of the three-phase system of Problem 2.

29. Find the total watts, volt-amperes reactive, volt-amperes, and F_p of the three-phase system of Problem 4.

30. Find the total watts, volt-amperes reactive, volt-amperes, and F_p of the three-phase system of Problem 7.

31. Find the total watts, volt-amperes reactive, volt-amperes, and F_p of the three-phase system of Problem 11.

32. Find the total watts, volt-amperes reactive, volt-amperes, and F_p of the three-phase system of Problem 13.

33. Find the total watts, volt-amperes reactive, volt-amperes, and F_p of the three-phase system of Problem 15.

34. Find the total watts, volt-amperes reactive, volt-amperes, and F_p of the three-phase system of Problem 18.

35. Find the total watts, volt-amperes reactive, volt-amperes, and F_p of the three-phase system of Problem 20.

36. Find the total watts, volt-amperes reactive, volt-amperes, and F_p of the three-phase system of Problem 24.

37. Find the total watts, volt-amperes reactive, volt-amperes, and F_p of the three-phase system of Problem 26.

***38.** A balanced three-phase, Δ-connected load has a line voltage of 200 V and a total power consumption of 4800 W at a lagging power factor of 0.8. Find the impedance of each phase in rectangular coordinates.

***39.** A balanced three-phase, Y-connected load has a line voltage of 208 V and a total power consumption of 1200 W at a leading power factor of 0.6. Find the series impedances of each phase.

***40.** Find the total watts, volt-amperes reactive, volt-amperes, and F_p of the system of Fig. 20.35.

FIG. 20.35

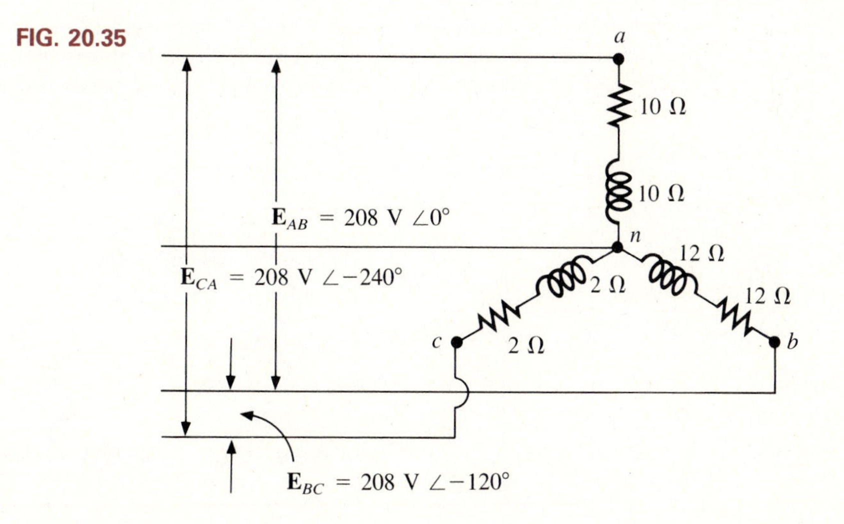

Section 20.9

***41.** **a.** Sketch the connections required to measure the total watts delivered to the load of Fig. 20.30 using three wattmeters.

b. Determine the total wattage dissipation and the reading of each wattmeter.

42. Repeat Problem 41 for the network of Fig. 20.32.

Section 20.10

43. **a.** For the three-wire system of Fig. 20.36, properly connect a second wattmeter so that the two will measure the total power delivered to the load.

b. If one wattmeter has a reading of 200 W and the other a reading of 85 W, what is the total dissipation in watts if the total power factor is 0.8 leading?

c. Repeat part b if the total power factor is 0.2 lagging.

FIG. 20.36

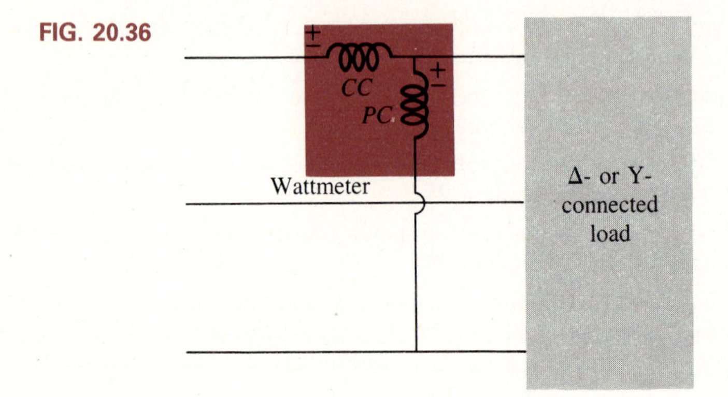

44. Sketch three different ways that two wattmeters can be connected to measure the total power delivered to the load of Problem 15.

Section 20.11

***45.** For the system of Fig. 20.37:

a. Calculate the magnitude of the voltage across each phase of the load.

b. Find the magnitude of the current through each phase of the load.

c. Find the total watts, volt-amperes reactive, volt-amperes, and F_p of the system.

FIG. 20.37

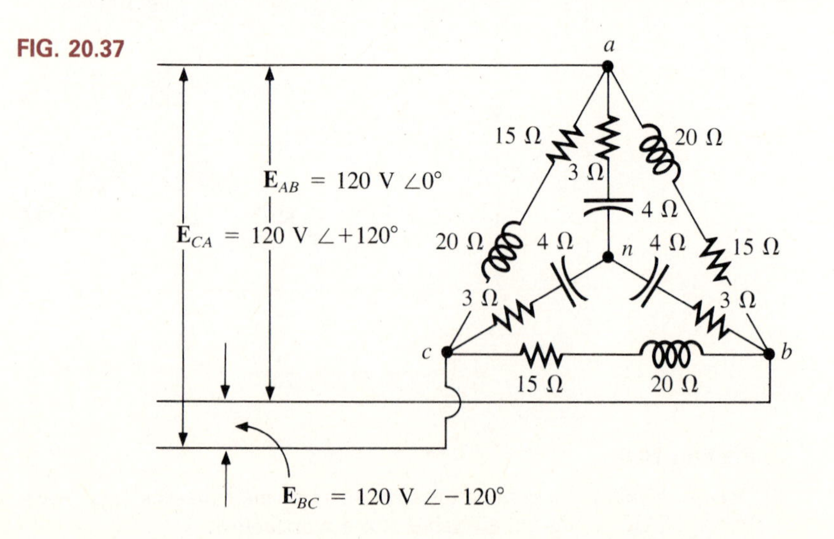

Computer Problems

46. Given the magnitude of the line voltages and the impedance of each phase (in series or parallel), write a program to determine the magnitude of all the voltages and currents of a balanced Y-connected load.

47. Repeat Problem 47 for a Δ-connected load.

48. For a balanced Y-connected load, write a program to determine
   a. The magnitude of the load currents and voltages
   b. The real, reactive, and apparent power to each phase
   c. The total real, reactive, and apparent power to the load
   d. The load power factor

49. Repeat Problem 48 for a balanced Δ-connected load.

APPENDIX A Conversion Factors

To Convert from	To	Multiply by
Btus	Calorie-grams	251.996
	Ergs	1.054×10^{10}
	Foot-pounds	777.649
	Hp-hours	0.000393
	Joules	1054.35
	Kilowatthours	0.000293
	Wattseconds	1054.35
Centimeters	Angstrom units	1×10^{8}
	Feet	0.0328
	Inches	0.3937
	Meters	0.01
	Miles (statute)	6.214×10^{-6}
	Millimeters	10
Circular mils	Square centimeters	5.067×10^{-6}
	Square inches	7.854×10^{-7}
Cubic inches	Cubic centimeters	16.387
	Gallons (U.S. liquid)	0.00433
Cubic meters	Cubic feet	35.315
Days	Hours	24
	Minutes	1440
	Seconds	86,400
Dynes	Gallons (U.S. liquid)	264.172
	Newtons	0.00001
	Pounds	2.248×10^{-6}
Electronvolts	Ergs	1.60209×10^{-12}
Ergs	Dyne-centimeters	1.0
	Electronvolts	6.242×10^{11}
	Foot-pounds	7.376×10^{-8}
	Joules	1×10^{-7}
	Kilowatthours	2.777×10^{-14}
Feet	Centimeters	30.48
	Meters	0.3048
Foot-candles	Lumens/square foot	1.0
	Lumens/square meter	10.764
Foot-pounds	Dyne-centimeters	1.3558×10^{7}
	Ergs	1.3558×10^{7}
	Horsepower-hours	5.050×10^{-7}
	Joules	1.3558
	Newton-meters	1.3558

To Convert from	To	Multiply by
Gallons (U.S. liquid)	Cubic inches	231
	Liters	3.785
	Ounces	128
	Pints	8
Gauss	Maxwells/square centimeter	1.0
	Lines/square centimeter	1.0
	Lines/square inch	6.4516
Gilberts	Ampére-turns	0.7958
Grams	Dynes	980.665
	Ounces	0.0353
	Pounds	0.0022
Horsepower	Btus/hour	2547.16
	Ergs/second	7.46×10^9
	Foot-pounds/second	550.221
	Joules/second	746
	Watts	746
Hours	Seconds	3600
Inches	Angstrom units	2.54×10^8
	Centimeters	2.54
	Feet	12
	Meters	0.0254
Joules	Btus	0.000948
	Ergs	1×10^7
	Foot-pounds	0.7376
	Horsepower-hours	3.725×10^{-7}
	Kilowatthours	2.777×10^{-7}
	Wattseconds	1.0
Kilograms	Dynes	980,665
	Ounces	35.2
	Pounds	2.2
Lines	Maxwells	1.0
Lines/square centimeter	Gauss	1.0
Lines/square inch	Gauss	0.1550
	Webers/square inch	1×10^{-8}
Liters	Cubic centimeters	1000.028
	Cubic inches	61.025
	Gallons (U.S. liquid)	0.2642
	Ounces (U.S. liquid)	33.815
	Quarts (U.S. liquid)	1.0567

To Convert from	To	Multiply by
Lumens	Candle power (spher.)	0.0796
Lumens/square centimeter	Lamberts	1.0
Lumens/square foot	Foot-candles	1.0
Maxwells	Lines	1.0
	Webers	1×10^{-8}
Meters	Angstrom units	1×10^{10}
	Centimeters	100
	Feet	3.2808
	Inches	39.370
	Miles (statute)	0.000621
Miles (statute)	Feet	5280
	Kilometers	1.609
	Meters	1609.344
Miles/hour	Kilometers/hour	1.609344
Newton-meters	Dyne-centimeters	1×10^{7}
	Kilogram-meters	0.10197
Oersteds	Ampere-turns/inch	2.0212
	Ampere-turns/meter	79.577
	Gilberts/centimeter	1.0
Quarts (U.S. liquid)	Cubic centimeters	946.353
	Cubic inches	57.75
	Gallons (U.S. liquid)	0.25
	Liters	0.9463
	Pints (U.S. liquid)	2
	Ounces (U.S. liquid)	32
Radians	Degrees	57.2958
Slugs	Kilograms	14.5939
	Pounds	32.1740
Watts	Btus/hour	3.4144
	Ergs/second	1×10^{7}
	Horsepower	0.00134
	Joules/second	1.0
Webers	Lines	1×10^{8}
	Maxwells	1×10^{8}
Years	Days	365
	Hours	8760
	Minutes	525,600
	Seconds	3.1536×10^{7}

APPENDIX B Determinants

Second-order Determinants

The use of determinants permits finding the solution for any number of simultaneous equations in a direct, time-efficient manner. All that is required is to set up the formats described below and perform a few basic algebraic manipulations in a specified manner.

For example, consider the following equations in which the variables x and y must be determined as controlled by the constants a_1, a_2, b_1, b_2, c_1, and c_2.

Col. 1		Col. 2		Col. 3	
a_1x	$+$	b_1y	$=$	c_1	(B.1a)
a_2x	$+$	b_2y	$=$	c_2	(B.1b)

Using determinants to solve for x and y requires that the following formats be established for each variable:

$$x = \frac{\begin{vmatrix} c_1 & b_1 \\ c_2 & b_2 \end{vmatrix}}{\begin{vmatrix} a_1 & b_1 \\ a_2 & b_2 \end{vmatrix}} \qquad y = \frac{\begin{vmatrix} a_1 & c_1 \\ a_2 & c_2 \end{vmatrix}}{\begin{vmatrix} a_1 & b_1 \\ a_2 & b_2 \end{vmatrix}} \tag{B.2}$$

(Columns in each numerator and denominator: Col. 1, Col. 2.)

First note that only constants appear within the vertical brackets and that the denominator of each is the same. In fact, the denominator is simply the coefficients of x and y in the same arrangement as in Eqs. (B.1a) and (B.1b). When solving for x, the coefficients of x in the numerator are replaced by the constants to the right of the equal sign in Eqs. (B.1a) and (B.1b), while the coefficients of the y variable are simply repeated. When solving for y, the y coefficients in the numerator are replaced by the constants to the right of the equal sign and the coefficients of x are repeated.

Each configuration in the numerator and denominator of Eq. (B.2) is referred to as a *determinant* (D), which can be evaluated numerically in the following manner:

$$\text{Determinant} = D = \begin{vmatrix} a_1 & b_1 \\ a_2 & b_2 \end{vmatrix} = a_1b_2 - a_2b_1 \qquad \text{(Col. 1, Col. 2)} \tag{B.3}$$

The expanded value is obtained by first multiplying the top left element by the bottom right and then subtracting the product of the lower left and upper right elements. This particular determinant is referred to as a *second-order* determinant, since it contains two rows and two columns.

It is important to remember when using determinants that the columns of the equations, as indicated in Eqs. (B.1a) and (B.1b), be placed in the same order within the determinant configuration. That is, since a_1 and a_2 are in column 1 of Eqs. (B.1a) and (B.1b), they must be in column 1 of the determinant. (The same is true for b_1 and b_2.)

Expanding the entire expression for x and y, we have the following:

$$x = \frac{\begin{vmatrix} c_1 & b_1 \\ c_2 & b_2 \end{vmatrix}}{\begin{vmatrix} a_1 & b_1 \\ a_2 & b_2 \end{vmatrix}} = \frac{c_1b_2 - c_2b_1}{a_1b_2 - a_2b_1} \tag{B.4a}$$

$$y = \frac{\begin{vmatrix} a_1 & c_1 \\ a_2 & c_2 \end{vmatrix}}{\begin{vmatrix} a_1 & b_1 \\ a_2 & b_2 \end{vmatrix}} = \frac{a_1c_2 - a_2c_1}{a_1b_2 - a_2b_1} \tag{B.4b}$$

EXAMPLE B.1 Evaluate the following determinants:

a. $\begin{vmatrix} 2 & 2 \\ 3 & 4 \end{vmatrix} = (2)(4) - (3)(2) = 8 - 6 = \mathbf{2}$

b. $\begin{vmatrix} 4 & -1 \\ 6 & 2 \end{vmatrix} = (4)(2) - (6)(-1) = 8 + 6 = \mathbf{14}$

c. $\begin{vmatrix} 0 & -2 \\ -2 & 4 \end{vmatrix} = (0)(4) - (-2)(-2) = 0 - 4 = \mathbf{-4}$

d. $\begin{vmatrix} 0 & 0 \\ 3 & 10 \end{vmatrix} = (0)(10) - (3)(0) = \mathbf{0}$

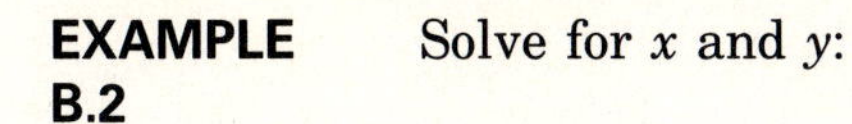

EXAMPLE B.2 Solve for x and y:

$$\begin{aligned} 2x + y &= 3 \\ 3x + 4y &= 2 \end{aligned}$$

Solution:

$$x = \frac{\begin{vmatrix} 3 & 1 \\ 2 & 4 \end{vmatrix}}{\begin{vmatrix} 2 & 1 \\ 3 & 4 \end{vmatrix}} = \frac{(3)(4) - (2)(1)}{(2)(4) - (3)(1)} = \frac{12 - 2}{8 - 3} = \frac{10}{5} = \mathbf{2}$$

$$y = \frac{\begin{vmatrix} 2 & 3 \\ 3 & 2 \end{vmatrix}}{5} = \frac{(2)(2) - (3)(3)}{5} = \frac{4 - 9}{5} = \frac{-5}{5} = \mathbf{-1}$$

Check:

$$\begin{aligned} 2x + y &= (2)(2) + (-1) \\ &= 4 - 1 = 3 \qquad \text{(checks)} \\ 3x + 4y &= (3)(2) + (4)(-1) \\ &= 6 - 4 = 2 \qquad \text{(checks)} \end{aligned}$$

EXAMPLE B.3 Solve for x and y:

$$\begin{aligned} -x + 2y &= 3 \\ 3x - 2y &= -2 \end{aligned}$$

Solution: In this example, note the effect of the minus sign and the use of parentheses to ensure the proper sign is obtained for each product:

$$x = \frac{\begin{vmatrix} 3 & 2 \\ -2 & -2 \end{vmatrix}}{\begin{vmatrix} -1 & 2 \\ 3 & -2 \end{vmatrix}} = \frac{(3)(-2) - (-2)(2)}{(-1)(-2) - (3)(2)}$$

$$= \frac{-6 + 4}{2 - 6} = \frac{-2}{-4} = \mathbf{\frac{1}{2}}$$

$$y = \frac{\begin{vmatrix} -1 & 3 \\ 3 & -2 \end{vmatrix}}{-4} = \frac{(-1)(-2) - (3)(3)}{-4}$$

$$= \frac{2 - 9}{-4} = \frac{-7}{-4} = \mathbf{\frac{7}{4}}$$

Third-order Determinants

Consider the three following simultaneous equations:

$$a_1x + b_1y + c_1z = d_1$$
$$a_2x + b_2y + c_2z = d_2$$
$$a_3x + b_3y + c_3z = d_3$$

The determinant configuration for x, y, and z can be found in a manner similar to that for two simultaneous equations. That is, to solve for x, obtain the determinant in the numerator by replacing the first column by the elements on the right of the equal sign, and the denominator is the determinant of the coefficients of the variables (the same approach applies to y and z). Again, the denominator is the same for each variable. Therefore,

$$x = \frac{\begin{vmatrix} d_1 & b_1 & c_1 \\ d_2 & b_2 & c_2 \\ d_3 & b_3 & c_3 \end{vmatrix}}{D = \begin{vmatrix} a_1 & b_1 & c_1 \\ a_2 & b_2 & c_2 \\ a_3 & b_3 & c_3 \end{vmatrix}} \qquad y = \frac{\begin{vmatrix} a_1 & d_1 & c_1 \\ a_2 & d_2 & c_2 \\ a_3 & d_3 & c_3 \end{vmatrix}}{D = \begin{vmatrix} a_1 & b_1 & c_1 \\ a_2 & b_2 & c_2 \\ a_3 & b_3 & c_3 \end{vmatrix}} \qquad z = \frac{\begin{vmatrix} a_1 & b_1 & d_1 \\ a_2 & b_2 & d_2 \\ a_3 & b_3 & d_3 \end{vmatrix}}{D = \begin{vmatrix} a_1 & b_1 & c_1 \\ a_2 & b_2 & c_2 \\ a_3 & b_3 & c_3 \end{vmatrix}}$$

There is more than one expanded format for the third-order determinant. Each, however, will give the same result. One expansion of the determinant (D) is the following:

$$D = \begin{vmatrix} a_1 & b_1 & c_1 \\ a_2 & b_2 & c_2 \\ a_3 & b_3 & c_3 \end{vmatrix} = \underset{\substack{\uparrow \\ \text{Multiplying factor}}}{a_1} \underbrace{\left(+ \underset{\text{Minor}}{\begin{vmatrix} b_2 & c_2 \\ b_3 & c_3 \end{vmatrix}} \right)}_{\text{Cofactor}} + \underset{\substack{\uparrow \\ \text{Multiplying factor}}}{b_1} \underbrace{\left(- \underset{\text{Minor}}{\begin{vmatrix} a_2 & c_2 \\ a_3 & c_3 \end{vmatrix}} \right)}_{\text{Cofactor}} + \underset{\substack{\uparrow \\ \text{Multiplying factor}}}{c_1} \underbrace{\left(+ \underset{\text{Minor}}{\begin{vmatrix} a_2 & b_2 \\ a_3 & b_3 \end{vmatrix}} \right)}_{\text{Cofactor}}$$

This expansion was obtained by multiplying the elements of the first row of D by their corresponding cofactors. It is not a requirement that the first row be used as the multiplying factors. In fact, any *row* or *column* (not diagonals) may be used to expand a third-order determinant.

The sign of each cofactor is dictated by the position of the multiplying factors (a_1, b_1, and c_1 in this case) as in the following standard format:

$$\begin{vmatrix} + \rightarrow & - & + \\ \downarrow \; - & + & - \\ + & - & + \end{vmatrix}$$

Note that the proper sign for each element can be obtained by simply assigning the upper left element a positive sign and then changing the sign as you move horizontally or vertically to the neighboring position.

For the determinant D, the elements would have the following signs:

$$\begin{vmatrix} a_1^{(+)} & b_1^{(-)} & c_1^{(+)} \\ a_2^{(-)} & b_2^{(+)} & c_2^{(-)} \\ a_3^{(+)} & b_3^{(-)} & c_3^{(+)} \end{vmatrix}$$

The minors associated with each multiplying factor are obtained by covering up the row and column in which the multiplying factor is located and writing a second-order determinant to include the remaining elements in the same relative positions that they have in the third-order determinant.

Consider the cofactors associated with a_1 and b_1 in the expansion of D. The sign is positive for a_1 and negative for b_1 as determined by the standard format. Following the procedure outlined above, we can find the minors of a_1 and b_1 as follows:

$$a_{1(\text{minor})} = \begin{vmatrix} \not{a_1} & \not{b_1} & \not{c_1} \\ \not{a_2} & b_2 & c_2 \\ \not{a_3} & b_3 & c_3 \end{vmatrix} = \begin{vmatrix} b_2 & c_2 \\ b_3 & c_3 \end{vmatrix}$$

$$b_{1(\text{minor})} = \begin{vmatrix} \not{a_1} & \not{b_1} & \not{c_1} \\ a_2 & \not{b_2} & c_2 \\ a_3 & \not{b_3} & c_3 \end{vmatrix} = \begin{vmatrix} a_2 & c_2 \\ a_3 & c_3 \end{vmatrix}$$

It was pointed out that any row or column may be used to expand the third-order determinant, and the same result will still be obtained. Using the first column of D, we obtain the expansion

$$D = \begin{vmatrix} a_1 & b_1 & c_1 \\ a_2 & b_2 & c_2 \\ a_3 & b_3 & c_3 \end{vmatrix} = a_1\left(+\begin{vmatrix} b_2 & c_2 \\ b_3 & c_3 \end{vmatrix}\right) + a_2\left(-\begin{vmatrix} b_1 & c_1 \\ b_3 & c_3 \end{vmatrix}\right) + a_3\left(+\begin{vmatrix} b_1 & c_1 \\ b_2 & c_2 \end{vmatrix}\right)$$

The proper choice of row or column can often effectively reduce the amount of work required to expand the third-order determinant. For example, in the following determinants, the first column and third row, respectively, would reduce the number of cofactors in the expansion:

$$D = \begin{vmatrix} 2 & 3 & -2 \\ 0 & 4 & 5 \\ 0 & 6 & 7 \end{vmatrix} = 2\left(+\begin{vmatrix} 4 & 5 \\ 6 & 7 \end{vmatrix}\right) + 0 + 0 = 2(28 - 30)$$

$$= \mathbf{-4}$$

$$D = \begin{vmatrix} 1 & 4 & 7 \\ 2 & 6 & 8 \\ 2 & 0 & 3 \end{vmatrix} = 2\left(+\begin{vmatrix} 4 & 7 \\ 6 & 8 \end{vmatrix}\right) + 0 + 3\left(+\begin{vmatrix} 1 & 4 \\ 2 & 6 \end{vmatrix}\right)$$

$$= 2(32 - 42) + 3(6 - 8) = 2(-10) + 3(-2)$$

$$= \mathbf{-26}$$

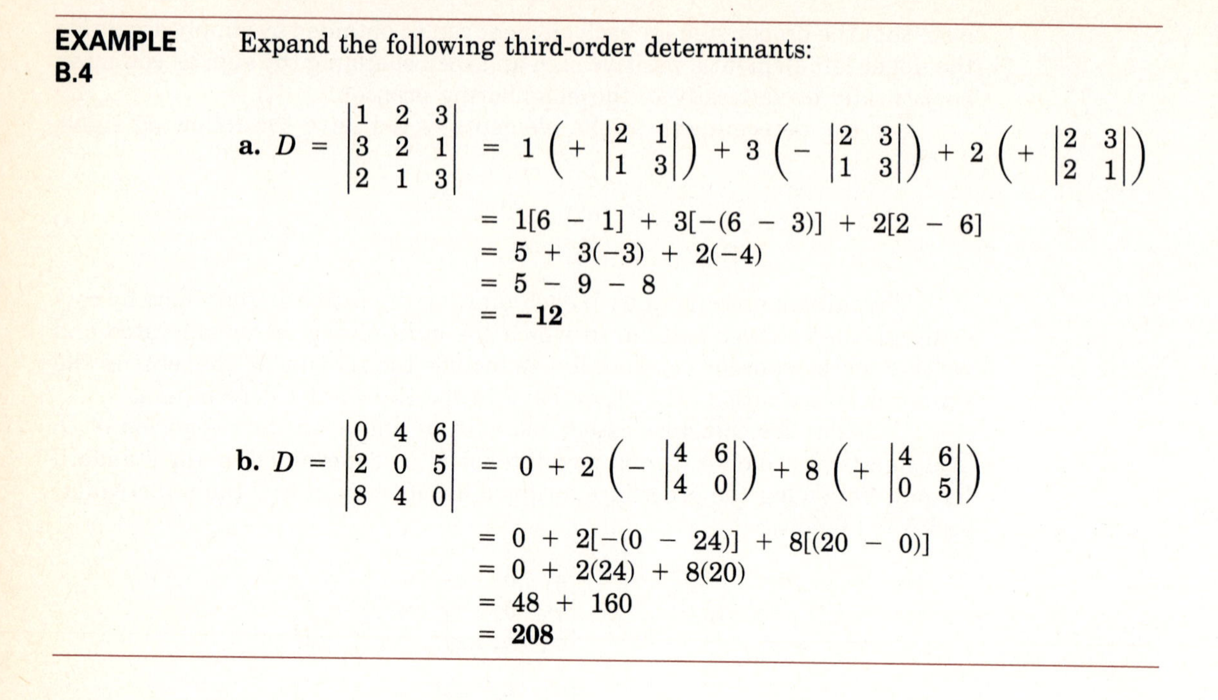

EXAMPLE B.4

Expand the following third-order determinants:

a.
$$D = \begin{vmatrix} 1 & 2 & 3 \\ 3 & 2 & 1 \\ 2 & 1 & 3 \end{vmatrix} = 1\left(+\begin{vmatrix} 2 & 1 \\ 1 & 3 \end{vmatrix}\right) + 3\left(-\begin{vmatrix} 2 & 3 \\ 1 & 3 \end{vmatrix}\right) + 2\left(+\begin{vmatrix} 2 & 3 \\ 2 & 1 \end{vmatrix}\right)$$
$$= 1[6 - 1] + 3[-(6 - 3)] + 2[2 - 6]$$
$$= 5 + 3(-3) + 2(-4)$$
$$= 5 - 9 - 8$$
$$= \mathbf{-12}$$

b.
$$D = \begin{vmatrix} 0 & 4 & 6 \\ 2 & 0 & 5 \\ 8 & 4 & 0 \end{vmatrix} = 0 + 2\left(-\begin{vmatrix} 4 & 6 \\ 4 & 0 \end{vmatrix}\right) + 8\left(+\begin{vmatrix} 4 & 6 \\ 0 & 5 \end{vmatrix}\right)$$
$$= 0 + 2[-(0 - 24)] + 8[(20 - 0)]$$
$$= 0 + 2(24) + 8(20)$$
$$= 48 + 160$$
$$= \mathbf{208}$$

APPENDIX C Color Coding of Molded Mica Capacitors (Picofarads)

RETMA *and standard* MIL *specifications*

Color	Significant Figure	Decimal Multiplier	Tolerance ±%	Class	Temp. Coeff. PPM/°C Not More than	Cap. Drift Not More than
Black	0	1	20	A	±1000	±(5% + 1 pF)
Brown	1	10	—	B	±500	±(3% + 1 pF)
Red	2	100	2	C	±200	±(0.5% + 0.5 pF)
Orange	3	1000	3	D	±100	±(0.3% + 0.1 pF)
Yellow	4	10,000	—	E	+100 − 20	±(0.1% + 0.1 pF)
Green	5	—	5	—	—	—
Blue	6	—	—	—	—	—
Violet	7	—	—	—	—	—
Gray	8	—	—	I	+150 − 50	±(0.03% + 0.2 pF)
White	9	—	—	J	+100 − 50	±(0.2% + 0.2 pF)
Gold	—	0.1	—	—	—	—
Silver	—	0.01	10	—	—	—

Courtesy of Sprague Electric Co.

NOTE: If both rows of dots are not on one face, rotate capacitor about axis of its leads to read second row on side or rear.

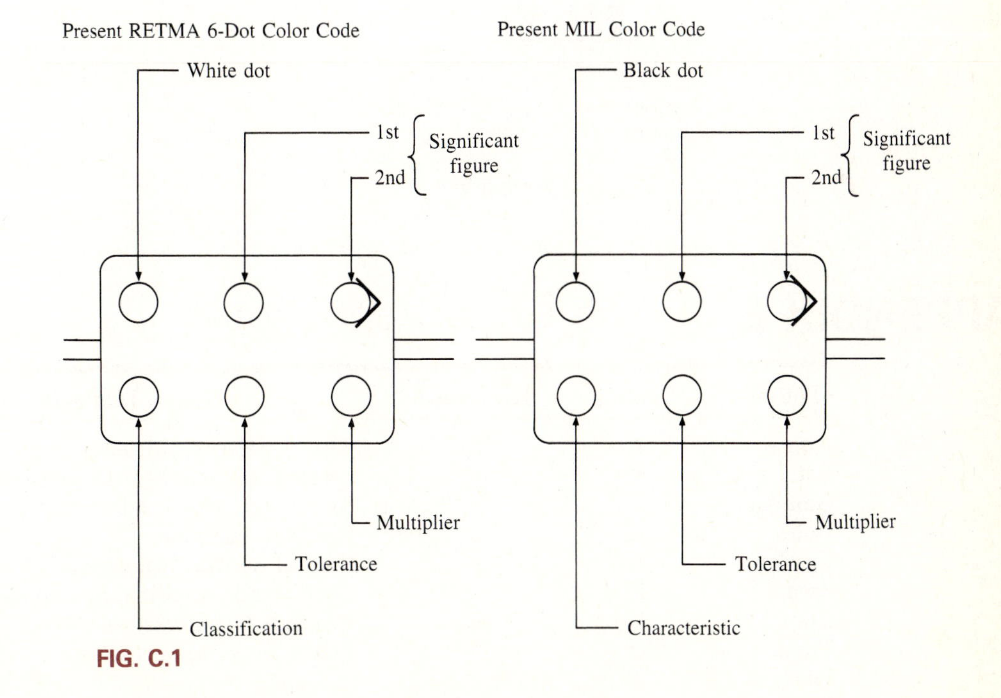

FIG. C.1

APPENDIX D Color Coding of Molded Tubular Capacitors (Picofarads)

Color	Significant Figure	Decimal Multiplier	Tolerance ±%
Black	0	1	20
Brown	1	10	—
Red	2	100	—
Orange	3	1000	30
Yellow	4	10,000	40
Green	5	10^5	5
Blue	6	10^6	—
Violet	7	—	—
Gray	8	—	—
White	9	—	10

Courtesy of Sprague Electric Co.

NOTE: Voltage rating is identified by a single-digit number for ratings up to 900 V; a two-digit number above 900 V. Two zeros follow the voltage figure.

FIG. D.1

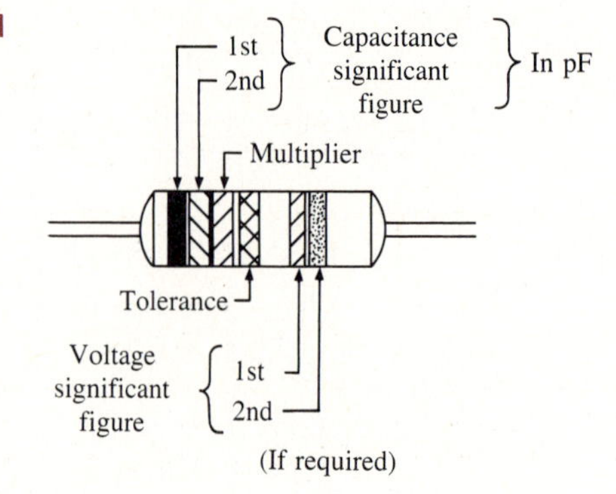

APPENDIX E The Greek Alphabet

Letter	Capital	Lowercase	Used to Designate
Alpha	A	α	Area, angles, coefficients
Beta	B	β	Angles, coefficients, flux density
Gamma	Γ	γ	Specific gravity, conductivity
Delta	Δ	δ	Density, variation
Epsilon	E	ϵ	Base of natural logarithms
Zeta	Z	ζ	Coefficients, coordinates, impedance
Eta	H	η	Efficiency, hysteresis coefficient
Theta	Θ	θ	Phase angle, temperature
Iota	I	ι	
Kappa	K	κ	Dielectric constant, susceptibility
Lambda	Λ	λ	Wavelength

APPENDIX E Greek Alphabet, *continued*

Letter	Capital	Lowercase	Used to Designate
Mu	M	μ	Amplification factor, micro, permeability
Nu	N	ν	Reluctivity
Xi	Ξ	ξ	
Omicron	O	o	
Pi	Π	π	3.1416
Rho	P	ρ	Resistivity
Sigma	Σ	σ	Summation
Tau	T	τ	Time constant
Upsilon	Υ	υ	
Phi	Φ	ϕ	Angles, magnetic flux
Chi	X	χ	
Psi	Ψ	ψ	Dielectric flux, phase difference
Omega	Ω	ω	Ohms, angular velocity

APPENDIX F Magnetic Parameter Conversions

	SI (MKS)		CGS		English
Φ	webers (Wb) 1 Wb	 =	maxwells 10^8 maxwells	 =	lines 10^8 lines
B	Wb/m^2 $1\ \text{Wb/m}^2$	 =	gauss (maxwells/cm^2) 10^4 gauss	 =	lines/in.^2 $6.452 \times 10^4\ \text{lines/in.}^2$
A	$1\ \text{m}^2$	=	$10^4\ \text{cm}^2$	=	$1550\ \text{in.}^2$
μ_o	$4\pi \times 10^{-7}$ Wb/Am	=	1 gauss/oersted	=	3.20 lines/Am
$\mathscr{F}$	NI (ampére-turns, At) 1 At	= =	$0.4\pi NI$ (gilberts) 1.257 gilberts		NI (At) 1 gilbert = 0.7958 At
H	NI/l (At/m) 1 At/m	 =	$0.4\pi NI/l$ (oersteds) 1.26×10^{-2} oersted	 =	NI/l (At/in.) 2.54×10^{-2} At/in.
H_g	$7.97 \times 10^5 B_g$ (At/m)		B_g (oersteds)		$0.313 B_g$ (At/in.)

APPENDIX G Solutions to Odd-Numbered Problems

Chapter 1

5. cm
7. 1 km = 1000 m, 1 m = 100 cm
11. time(s)
13. 1 slug = 14.6 kg = 14,600 gm
19. **(a)** 3.937 in. **(b)** 30 in. **(c)** 7.874 in.
21. 6.214 mi
23. **(a)** 15 mi/h **(b)** 7.5 mi/h
25. **(a)** 168 s **(b)** 900 s **(c)** 864 s
27. 20°C, 293.15 K
29. **(a)** 2.5 kg **(b)** 0.171 slugs

Chapter 2

1. **(a)** 4×10^3
(b) 50×10^3
(c) 104.5×10^3
(d) 1500×10^3
(e) 0.6×10^3
(f) 0.05×10^3
3. **(a)** 4,800,000
(b) 700
(c) 82,000
(d) 2
(e) 5,000,000
(f) 330
5. **(a)** 5×10^{-6} **(b)** 7000×10^{-6} **(c)** 0.081×10^{-6}
7. **(a)** 10^{-4} **(b)** 10^{+7} **(c)** 10^{-100} **(d)** $10^{+0.01}$
9. **(a)** 5×10^3
(b) 8×10^{-4}
(c) 1.5×10^6
(d) 4.2×10^4
(e) 150×10^3
(f) 2×10^{-2}
11. **(a)** 47×10^3
(b) 135×10^3
(c) 12.6×10^3
(d) 500.05×10^3
13. **(a)** 44×10^3
(b) 3.4×10^3
(c) 31.5×10^3
(d) 1075.1×10^3
15. **(a)** 1.6×10^1
(b) 4×10^{-1}
(c) 1.56×10^8
(d) 2.1×10^{-4}
(e) 5×10^{-2}
(f) 3×10^3
(g) 3.6×10^2
(h) 1.54×10^5
17. **(a)** 3.2×10^{16}
(b) 1.296×10^{-9}
(c) 2.5×10^{-3}
(d) 1.563×10^{-17}
19. **(a)** 22 km
(b) 0.5 ms
(c) 95 MHz
(d) 9 μm
(e) 4.8 nF
(f) 5080 MHz
(g) 20 ms
(h) 0.4 mm

Chapter 3

3. **(a)** 1.8×10^4 N **(b)** 1.8×10^{-2} N **(c)** 1.8×10^4 N : 1.8×10^{-2} N = 1,000,000 : 1
5. 0.3 m
7. 4.994×10^{13} electrons
9. 2.039 C, 1.019 A
11. 32 C
13. 124.84×10^{18} electrons
15. 1.111 h
17. 40 V
19. 6.242 kV
21. 60 μJ
23. 2.5 mC
29. 28 h @ 2.4 V
31. **(a)** 1.29 V **(b)** 1.2 V **(c)** 1 V
35. 7.4 : 1
37. 27 kV

Chapter 4

1. $3R$
3. $\frac{1}{2}R$
5. $16R$
7. 500 mils
9. 1200 mils
11. 256 CM
13. 155,236 CM
15. 10 mils, 0.01 in.
17. 0.797 Ω
19. 482.16 ft
21. 733.365 ft
23. 950.78 ft
25. 111.553 mils
27. 0.631 Ω
29. 0.032 in.
31. 40.82 : 1
33. **(a)** 6.385 mΩ, no **(b)** 0.065 Ω, no **(c)** 1.05 Ω, yes
35. 0.473 Ω
37. 0.1724 Ω
39. 5916.47 cm
41. 250×10^{-8} cm
43. 1.426 Ω
45. 0.022 Ω
47. **(a)** 5.213 Ω
(b) 5.426 Ω
(c) 5.640 Ω
(d) 0.213 Ω/10°C
(e) linear
49. 274.5°C
51. Cu 0.00393, Al 0.00391
53. **(a)** 2 : 1 **(b)** 4 : 1
55. 10,150 Ω
57. R_1 = 6.25 kΩ, R_2 = 18.75 kΩ
59. 198 Ω–242 Ω
61. 8 Ω–12 Ω, 12 Ω–18 Ω, continuous
63. **(a)** 25 mS **(b)** 0.1 mS **(c)** 20 S **(d)** 200 S
65. 50 S
69. −50°C: 100,000,000 Ω-cm, 50°C: 700 Ω-cm, 200°C: 9 Ω-cm
71. **(a)** 0.5 mA: 190 V, 1 mA: 205 V, 3 mA: 215 V, 5 mA: 220 V **(b)** 30 V **(c)** 5 mA : 0.5 mA = 10 : 1, 220 V : 190 V = 1.16 : 1 (fairly constant for current range)

Chapter 5

1. 15 V
3. 4 kΩ
5. 72 mV
7. 54.545 Ω
9. 1.68 kΩ
11. 12.632 Ω
13. 7.5 V
19. 20 Ω
21. 16 s
23. 5 A, 250 W
25. 4.8 W

27. 27 μW
29. 22.361 mA
31. 40 W
33. 120 V, 32 Ω
35. **(a)** 12 kW
(b) yes, 12 kW > 10.13 kW
(c) 84.471 A
37. 16.336 A
39. 14.031 A
41. 56.515 A
43. 38.4 J
45. $\eta_1 = 40\%$, $\eta_2 = 80\%$
47. **(a)** 1350 J **(b)** $W_2 = 2W_1 = 2700$ J, $P_2 = P_1$
49. 6.667 h
51. 170.38¢

Chapter 6

1. **(a)** 7 kΩ **(b)** 9 Ω **(c)** 17 kΩ **(d)** 8 kΩ
3. **(a)** 3 kΩ **(b)** 5 mA **(c)** $V_1 = 6$ V, $V_2 = 9$ V **(d)** $P_1 = 30$ mW, $P_2 = 45$ mW **(e)** $P_{del} = 75$ mW
5. **(a)** 48 Ω **(b)** 2.5 A **(c)** 30 V
7. $R = 2$ kΩ, $I = 1$ mA
9. **(a)** 16 V **(b)** 1.56 V
11. **(a)** 4.179 mA **(b)** 0.388 A
13. 20 V
15. $V = 16$ V, $V_1 = 6$ V
17. **(a)** $R = 21$ Ω, $V_1 = 2$ V, $V_2 = 1$ V, $V_3 = 21$ V, $E = 24$ V **(b)** $I = 2$ A, $R_1 = 2$ Ω, $R_2 = 13$ Ω, $E = 32$ V
19. **(a)** 4.4 V **(b)** 3.813 V
21. $V_2 = 8$ V, $V_1 = 4$ V, $E = 28$ V
23. **(a)** $V_a = 60$ V **(b)** $V_b = V_c = 20$ V **(c)** $V_{ab} = 40$ V
25. **(a)** $V_a = 50$ V, $V_b = 0$ V, $V_c = -20$ V **(b)** $V_{ac} = 70$ V, $V_{ab} = 50$ V, $V_{bc} = 20$ V **(c)** $I_1 = 35$ mA, $I_2 = 10$ mA, $I_3 = 2$ mA
27. **(a)** 27.5 Ω **(b)** 0.5 Ω
29. 1.818%

Chapter 7

3. **(a)** 6.857 Ω **(b)** 8.333 mΩ **(c)** 0.999 Ω
5. **(a)** 0.6 MΩ **(b)** 0.727 Ω
7. **(a)** 4 kΩ **(b)** $R_1 = R_2 = 8$ kΩ
9. **(a)** 2.182 kΩ **(b)** 11 mA **(c)** $V_1 = V_2 = E = 24$ V **(d)** $I_1 = 3$ mA, $I_2 = 8$ mA **(e)** $P_1 = 72$ mW, $P_2 = 192$ mW, $P_{del} = 264$ mW **(f)** yes **(g)** yes
11. **(a)** 1.714 Ω **(b)** 21 A **(c)** 36 V **(d)** $I_1 = 12$ A, $I_2 = 6$ A, $I_3 = 3$ A **(e)** yes **(f)** 108 W
13. **(a)** 66.667 mA **(b)** 0.225 Ω **(c)** 8 W **(d)** none
15. **(a)** $E = V_2 = 8$ V **(b)** $R_1 = 2$ Ω
17. **(a)** $I_1 = 3$ mA, $E = V_1 = 6$ V **(b)** $V_2 = V_3 = E = 6$ V, $I_2 = I_3 = 1$ mA, $R_2 = R_3 = 6$ kΩ
19. **(a)** $I_2 = 1$ A, $I_3 = 8$ A, $I_4 = 12$ A **(b)** $I_1 = 20$ mA, $I_2 = 60$ mA, $I_3 = 30$ mA, $I_4 = I_2 = 60$ mA
21. **(a)** $I_1 = 272.727$ mA, $I_2 = 227.273$ mA, $I_3 = 136.364$ mA, $I_4 = 90.909$ mA **(b)** $I_1 = 6$ mA, $I_2 = I_3 = 1.5$ mA
23. **(a)** $I_1 = 3$ A, $I = 4$ A, $I_2 = 4$ A **(b)** $I_2 = 2$ μA, $I_1 = 2$ μA, $I_3 = 6$ μA, $R = 9$ Ω
25. **(a)** 1.188 mA, 11.88 V **(b)** 120 mA **(c)** 12 V
27. **(a)** 4 V **(b)** 3.998 V **(c)** 3.871 V **(d)** 3 V

Chapter 8

1. **(a)** 5 Ω **(b)** 2 A **(c)** $I_1 = 2$ A, $I_2 = 1.333$ A, $I_3 = 0.667$ A **(d)** $V_1 = 6$ V, $V_2 = 4$ V, $V_3 = 4$ V **(e)** 2.667 W
3. **(a)** 12 V **(b)** $I_1 = 6$ mA, $I_3 = 1$ mA, $I_T = 7$ mA
5. **(a)** $I_1 = 1$ mA, $I_3 = 0.25$ mA **(b)** $V_1 = 1$ V, $V_3 = 0.75$ V, $V_{ab} = -0.25$ V
7. **(a)** 11 kΩ **(b)** 3 mA **(c)** $I_1 = 3$ mA, $I_2 = 1.5$ mA, $I_4 = 1$ mA **(d)** $V_a = 15$ V, $V_b = 6$ V, $V_c = 0$ V
9. **(a)** $I_2 = 3.529$ mA **(b)** $V_a = 28.236$ V
11. **(a)** $I_1 = 3$ mA, $I_3 = 2$ mA **(b)** $V = 36$ V, $V_4 = 24$ V
13. $V_2 = 6$ V, $V_5 = 4$ V
15. **(a)** 1 mA **(b)** $R_{shunt} = 5$ Ω
17. **(a)** $R_s = 299$ kΩ **(b)** 20,000 Ω/V
19. 50 nA

Chapter 9

1. $V_{ab} = 28$ V
3. **(a)** $I_T = 0.714$ A, $I_2 = 1.714$ A **(b)** $V_S = 12$ V
5. **(a)** $I = 3$ A, $R_p = 6$ Ω **(b)** $I = 4.091$ A, $R_p = 2.2$ kΩ
7. **(a)** 8 A **(b)** $E = 48$ V, $R_S = 4$ Ω, $I_L = 8$ A
9. **(a)** $I = 4$ A (up), $R_p = 2$ Ω **(b)** $I = 38$ mA (down), $R_p = 2.2$ kΩ
11. **(a)** $I_1 = 3$ A, $R_p = 3$ Ω, $I_2 = 10$ A, $R_p = 2$ Ω **(b)** $I = 1.167$ A (up)
13. **(a)** $I_1 = 2.357$ A, $I_2 = 0.762$ A, $I_3 = 1.595$ A **(b)** E_1: 16 W, E_2: 0.51 W **(c)** 22.222 W **(d)** no 16.51 W ≠ 22,222 W
15. 52.08 V
17. **(a)** $R_{Th} = 14$ Ω, $E_{Th} = -36$ V **(b)** 2 Ω: 10.125 W, 30 Ω: 20.083 W, 100 Ω: 9.972 W
19. $R_{Th} = 1.579$ kΩ, $E_{Th} = 1.147$ V
21. $R_{Th} = 4$ Ω, $E_{Th} = 36$ V
23. $R_N = 7.5$ Ω, $I_N = 1.333$ A
25. **(a)** 6 Ω **(b)** 1.5 W
27. **(a)** 2 Ω **(b)** 450 W
29. **(a)** $R_2 = \infty$ Ω (open-circuit) **(b)** no
31. $I_1 = 3.062$ A, $I_2 = 3.25$ A, $I_3 = 0.188$ A
33. $I_1 = 2.727$ mA, $I_2 = 4.818$ mA, $I_3 = 2.091$ mA
35. (same as 31)
37. (same as 33)
39. **(a)** $50I_1 - 10I_2 = 90$, $-10I_1 + 20I_2 = 0$ **(b)** 1 A
41. **(a)** $0.45V_1 - 0.2V_2 = -4$, $-0.2V_1 + 0.367V_2 = 8$ **(b)** $V_1 = 1.056$ V, $V_2 = 22.4$ V **(c)** 4.269 A
43. **(a)** $I = 5$ A, $R_p = 8$ Ω **(b)** $V_1(\frac{1}{8}\ \Omega + \frac{1}{4}\ \Omega) = 2\ \text{A} - 5\ \text{A}$ **(c)** $V_1 = -2.667$ V **(d)** 4.667 A
45. **(a)** $I_1 = 4$ A, $R_p = 2$ Ω, $I_2 = 6$ A, $R_p = 3$ Ω **(b)** $V_1[\frac{1}{4}\ \Omega + \frac{1}{2}\ \Omega + \frac{1}{3}\ \Omega] = 6\ \text{A} - 4\ \text{A}$

47. **(a)** $0.567V_1 - 0.2V_2 - 0.2V_3 = 1, -0.2V_1 + 0.5V_2 - 0.2V_3 = 0, -0.2V_1 - 0.2V_2 + 0.45V_3 = 0$ **(b)** $21I_1 - 5I_2 - 10I_3 = 6, -5I_1 + 15I_2 - 5I_3 = 0, -10I_1 - 5I_2 + 35I_3 = 0$
49. 58.335 mA
51. **(a)** no **(b)** 2.387 Ω

Chapter 10

1. 9000 N/C
3. 70 μF
5. 50 V/m
7. 8000 V/m
9. 885 pF
11. mica
13. **(a)** 10^6 V/m **(b)** 24.78 nF **(c)** 4.956 C
15. 29,035 V
17. **(a)** 0.5 s **(b)** $V_C = 20(1 - e^{-t/\tau})$ **(c)** 1τ: 12.64 V, 3τ: 19 V, 5τ: 19.86 V **(e)** $i_C = 0.2 \times 10^{-3}e^{-t/\tau}$, $v_R = 20e^{-t/\tau}$
19. **(a)** 5.5 ms **(b)** $v_C = 100(1 - e^{-t/\tau})$ **(c)** 1τ: 63.2 V, 3τ: 95 V, 5τ: 99.3 V **(e)** $i_C = 18.182 \times 10^{-3}e^{-t/\tau}$, $v_R = 60e^{-t/\tau}$
21. **(a)** $\tau_1 = 10$ ms **(c)** $v_C = 50(1 - e^{-t/\tau_1})$, $i_C = 10 \times 10^{-3}e^{-t/\tau_1}$ **(d)** $\tau_2 = 4$ ms **(f)** $v_C = 50e^{-t/\tau_2}$, $i_C = 25 \times 10^{-3}e^{-t/\tau_1}$
23. **(a)** 10 μs **(b)** 3 kA **(c)** yes
25. 1.386 μs
27. 23.026 μs
29. **(a)** $v_C = 8(1 - e^{-t/0.14})$, $i_C = 0.8 \times 10^{-3}e^{-t/0.14}$ **(b)** 0.194 s
31. 0–2 μs: +30 A, 2–4 μs: −60 A, 4–6 μs: +30 A, 6–8 μs: +20 A, 8–10 μs: −10 A, 10–15 μs: 0 A, 15–16 μs: −20 A
33. **(a)** $V_1 = 6.667$ V, $Q_1 = 40$ μC, $V_2 = 3.333$ V, $Q_2 = 40$ μC, $V_3 = 10$ V, $Q_3 = 60$ μC **(b)** $V_1 = 13.33$ V, $Q_1 = 16$ nC, $V_2 = 26.667$ V, $Q_2 = 5.333$ nC, $V_3 = 26.667$ V, $Q_3 = 10.667$ nC
35. 8.64 nJ
37. **(a)** 5 J **(b)** 0.1 C **(c)** 200 A **(d)** 10,000 W **(e)** 10 s

Chapter 11

1. CGS: 5×10^4 maxwells, 8 gauss; English: 5×10^4 lines, 51.62 lines/in.^{2}
3. **(a)** 0.4 T
5. ≅ 952,380 rels
7. 3000 At/m
9. **(a)** 3581.09 **(b)** 397.9 **(c)** 9 : 1
11. **(a)** I **(b)** I: 198,965.38 rels, II: 93,746.86 rels **(c)** I: 5.026×10^{-4} Wb, II: 0.427×10^{-4} Wb
13. 5 A
15. 5.85×10^{-3} Wb
17. 0.5 Wb/s
19. 9.867 μH
21. **(a)** 0.1696 mH **(b)** 0.339 H
23. 5 V
25. 0–3 μs: +3.333 V, 3–8 μs: 0 V, 8–9 μs: +3.2 V, 9–14 μs: −4 V, 14–17 μs: +2.933 V
27. **(a)** 2.273 μs **(b)** $i_L = 5.455 \times 10^{-3}(1 - e^{-t/\tau})$ **(c)** $v_L = 12e^{-t/\tau}$, $v_R = 12(1 - e^{-t/\tau})$ **(d)** i_L − 1τ: 3.448 mA, 3τ: 5.188 mA, 5τ: 5.417 mA, v_L − 1τ: 4.416 V, 3τ: 0.588 V, 5τ: 0.084 V
29. **(a)** $i_L = 5.455 \times 10^{-3}e^{-t/50\ \mu s}$, $v_L = 0.546e^{-t/50\ \mu s}$
31. **(a)** $i_L = 1.333 \times 10^{-3}(1 - e^{-t/55.55\ ns})$, $v_L = 48e^{-t/55.55\ ns}$ **(b)** 1.113 mA, 7.92 V
33. **(a)** 8 H **(b)** 4 H
35. $V_1 = 16$ V, $V_2 = 0$ V, $I_1 = 4$ mA
37. $W_{2H} = 16$ μJ, $W_{3H} = 24$ μJ

Chapter 12

1. **(a)** 10 ms **(b)** 20 ms **(c)** 10 V **(d)** 20 V **(e)** 100 Hz **(f)** 2.5
5. **(a)** 40 ms **(b)** 28.571 ms **(c)** 18.182 μs **(d)** 1 s
7. 0.3 ms
9. 7 Hz
11. **(a)** $\pi/4$ **(b)** $\pi/3$ **(c)** $2\pi/3$ **(d)** $3\pi/2$ **(e)** 0.989π **(f)** 1.228π
13. **(a)** 3.142 rad/s **(b)** 20.944×10^3 rad/s **(c)** 1.571×10^6 rad/s **(d)** 157.08 rad/s
15. **(a)** 120 Hz, 8.333 ms **(b)** 1.337 Hz, 0.748 s **(c)** 954.927 Hz, 1.047 ms **(d)** 0.01 Hz, 100.531 s
17. 104.72 rad/s
23. 0.476 A
25. 11.537°, 168.463°
29. **(a)** v leads i by 50° **(b)** v and i are in phase **(c)** i leads v by 80° **(d)** v and i are in phase
31. **(a)** $v = 0.01 \sin(157.08t - 120°)$ **(b)** $i = 2 \times 10^{-3} \sin(62.832 \times 10^3t + 135°)$
33. 2.5 V
35. ≅ 2.3 V
37. **(a)** 14.14 V **(b)** 5 V **(c)** 4.242 mA **(d)** 11.312 mA
39. 0.141 V
41. **(a)** 2 kΩ **(b)** 16.971 V
43. **(a)** 17.76 V **(b)** 11.312 V
45. **(a)** 8.88 V **(b)** 11.312 V

Chapter 13

1. **(a)** $30 \sin 377t$ **(b)** $6 \sin(377t + 60°)$
3. **(a)** 5 Ω **(b)** 150 Ω
5. **(a)** 1.592 H **(b)** 2.653 H **(c)** 840 mH
7. **(a)** $100 \sin(\omega t + 90°)$ **(b)** $8 \sin(\omega t + 150°)$
9. **(a)** $1 \sin(\omega t - 90°)$ **(b)** $0.6 \sin(\omega t - 70°)$ **(c)** $0.8 \sin(\omega t + 10°)$
11. **(a)** ∞ Ω **(b)** 530.516 Ω **(c)** 265.258 Ω **(d)** 17.684 Ω **(e)** 1.326 Ω
13. **(a)** 9.307 Hz **(b)** 4653.653 Hz **(c)** 18.615 Hz **(d)** 1.592 Hz
15. **(a)** $6 \sin(200t + 90°)$ **(b)** $45 \times 10^{-3} \sin(374t + 120°)$
17. **(a)** $37.135 \sin(377t - 90°)$ **(b)** $100 \sin(1600t - 170°)$
19. **(a)** 12.5 μF **(b)** 0.255 H **(c)** $R = 5$ Ω
23. 0.318 H
25. 5.066 nF
27. **(a)** 0 W **(b)** 0 W **(c)** 122.5 W
29. 192 W
31. $40 \sin(\omega t - 50°)$
33. **(a)** $2 \sin(157t - 60°)$ **(b)** 0.318 H **(c)** 0 W
35. **(a)** 100 ∠30° **(b)** 0.25 ∠−40°

(c) $70.71\angle -90°$ **(d)** $29.694\ \angle 0°$ **(e)** $4.242 \times 10^{-6}\ \angle 90°$

37. $63.2 \sin(377t + 108.4°)$

39. $85 \sin(\omega t + 93.47°)$

43. **(a)** $6.708\ \angle 26.565°$ **(b)** $6.708\ \angle 63.435°$ **(c)** $1030.776\ \angle 14.036°$ **(d)** $0.1\ \angle 53.13°$ **(e)** $11.314\ \angle -45°$ **(f)** $206.155\ \angle 75.964°$

45. see 37

47. see 39

49. $173.2 \sin \omega t$

Chapter 14

1. **(a)** $\mathbf{R} = 6.8\ \Omega\ \angle 0°$ **(b)** $\mathbf{X}_L = 754\ \Omega\ \angle 90°$ **(c)** $\mathbf{X}_L = 15.71\ \Omega\ \angle 90°$ **(d)** $\mathbf{X}_C = 265.25\ \Omega\ \angle -90°$ **(e)** $\mathbf{X}_C = 318.31\ \Omega\ \angle -90°$ **(f)** $\mathbf{R} = 200\ \Omega\ \angle 0°$

3. **(a)** $4.243\ \Omega\ \angle -45°$ **(b)** $7.02\ \text{k}\Omega\ \angle 85.91°$ **(c)** $364.13\ \Omega\ \angle 82.58°$

5. **(a)** $10\ \Omega\ \angle 36.87°$ **(c)** $\mathbf{I} = 10\ \text{A}\ \angle -36.87°$, $\mathbf{V}_R = 8\ \text{V}\ \angle -36.87°$, $\mathbf{V}_L = 60\ \text{V}\ \angle 53.13°$ **(e)** 800 W **(f)** 0.8 lagging **(g)** $i = 14.14 \sin(377t - 36.87°)$, $v_R = 113.12 \sin(377t - 36.87°)$, $v_L = 84.84 \sin(377t + 53.13°)$

7. **(a)** $9.43\ \text{k}\Omega\ \angle 35.56°$ **(c)** $2.121\ \text{mA}\ \angle -35.56°$ **(d)** $12.726\ \text{V}\ \angle -35.56°$ **(e)** 36 mW **(f)** $i = 3 \times 10^{-3} \sin(\omega t - 35.56°)$, $v_{R_2} = 18 \sin(\omega t - 35.56°)$

9. **(a)** $1659.5\ \Omega\ \angle -81.72°$ **(b)** $8.52\ \text{mA}\ \angle 81.72°$ **(c)** $\mathbf{V}_R = 4\ \text{V}\ \angle 81.72°$, $\mathbf{V}_C = 13.56\ \text{V}\ \angle -8.28°$ **(d)** 34.12 mW, 0.283 leading

11. **(a)** $4.472\ \Omega\ \angle -63.43°$ **(c)** $C = 265.25\ \mu\text{F}$, $L = 15.92$ mH **(d)** $\mathbf{I} = 11.181\ \text{A}\ \angle 63.43°$, $\mathbf{V}_R = 22.362\ \text{V}\ \angle 63.43°$, $\mathbf{V}_L = 67.09\ \text{V}\ \angle 153.43°$, $\mathbf{V}_C = 111.81\ \text{V}\ \angle -26.57°$ **(f)** 250 W **(g)** 0.447 leading **(h)** $i = 15.81 \sin(377t + 63.43°)$, $v_R = 31.62 \sin(377t + 63.43°)$, $v_L = 94.87 \sin(377t + 153.43°)$, $v_C = 158.1 \sin(377t - 26.57°)$

13. **(a)** $1.414\ \text{k}\Omega\ \angle 45°$ **(d)** $\mathbf{I} = 5\ \text{mA}\ \angle -45°$, $\mathbf{V}_R = 5\ \text{V}\ \angle -45°$, $\mathbf{V}_L = 10\ \text{V}\ \angle 45°$, $\mathbf{V}_C = 5\ \text{V}\ \angle -90°$ **(f)** 50 mW **(g)** 0.707 lagging **(h)** $i = 7.07 \times 10^{-3} \sin(10^3 t - 45°)$, $v_R = 7.07 \sin(10^3 t - 45°)$, $v_L = 14.14 \sin(10^3 t + 45°)$, $v_C = 7.07 \sin(10^3 t - 135°)$

15. **(a)** $V_1 = 8.94$ V, $V_2 = 26.83$ V **(b)** $V_1 = 112.92$ V, $V_2 = 58.66$ V

17. $R = 31.32\ \Omega$, $X_L = 22.13\ \Omega$

21. see 5

23. $\mathbf{I} = 3.795\ \text{A}\ \angle 71.57°$, $\mathbf{V}_C = 113.85\ \text{V}\ \angle -18.43°$, $\mathbf{V}_R = 37.95\ V\ \angle 71.57°$

25. $\mathbf{I} = 1.342\ \text{mA}\ \angle -18.43°$, $\mathbf{V}_R = 4.026\ \text{V}\ \angle -18.43°$, $\mathbf{V}_L = 2.684\ \text{V}\ \angle 71.57°$, $\mathbf{V}_C = 1.342\ \text{V}\ \angle -108.43°$

27. see 15

Chapter 15

1. **(a)** $\mathbf{Y}_T = 21.28\ \text{mS}\ \angle 0°$, $\mathbf{Z}_T = 47\ \Omega\ \angle 0°$ **(b)** $\mathbf{Y}_T = 5\ \text{mS}\ \angle -90°$, $\mathbf{Z}_T = 200\ \Omega\ \angle 90°$ **(c)** $\mathbf{Y}_T = 1.67\ \text{mS}\ \angle 90°$, $\mathbf{Z}_T = 0.6\ \Omega\ \angle -90°$

3. **(a)** $\mathbf{Z}_T = 8.54\ \Omega\ \angle 69.44°$, $\mathbf{Y}_T = 0.125\ \angle -69.44°$ **(b)** $\mathbf{Z}_T = 92.2\ \Omega\ \angle -49.4°$, $\mathbf{Y}_T = 10.8\ \text{mS}\ \angle 49.4°$ **(c)** $\mathbf{Z}_T = 224\ \Omega\ \angle -26.57°$, $\mathbf{Y}_T = 4.46\ \text{mS}\ \angle 26.57°$

5. **(a)** $\mathbf{Y}_T = 0.539\ \text{S}\ \angle -21.8°$, $\mathbf{Z}_T = 1.855\ \Omega\ \angle 21.8°$ **(c)** $\mathbf{I}_T = 4.574\ \text{A}\ \angle -21.8°$, $\mathbf{I}_R = 4.242\ \text{A}\ \angle 0°$, $\mathbf{I}_L = 1.697\ \text{A}\ \angle -90°$ **(e)** 36 W **(f)** 0.93 lagging **(g)** $i_T = 6.468 \sin(377t - 21.8°)$, $i_R = 6 \sin 377t$, $i_L = 2.4 \sin(377t - 90°)$

7. **(a)** $1.896\ \text{k}\Omega\ \angle 71.53°$ **(b)** 5.69 V **(c)** 2.85 mA

9. **(a)** $\mathbf{Y}_T = 0.1035\ \angle 14.04°$, $\mathbf{Z}_T = 9.71\ \Omega\ \angle -14.04°$ **(c)** $\mathbf{I}_T = 1.456\ \text{A}\ \angle 14.04°$, $\mathbf{I}_R = 1.414\ \text{A}\ \angle 0°$, $\mathbf{I}_C = 0.354\ \text{A}\ \angle +90°$ **(e)** 20 W **(f)** 0.970 leading **(g)** $i_T = 2.06 \sin(377t + 14.04°)$, $i_R = 2 \sin 377t$, $i_C = 0.5 \sin(377t + 90°)$

11. **(a)** $0.325\ \text{mS}\ \angle 39.76°$ **(b)** $3.077\ \text{k}\Omega\ \angle -39.76°$ **(c)** $6.893\ \mu\text{A}\ \angle 39.76°$ **(d)** $5.3\ \mu\text{A}\ \angle 0°$

13. **(a)** $0.972\ \text{mS}\ \angle 59.03°$ **(c)** $0.1\ \mu\text{F}$, 0.6 H **(d)** $\mathbf{I}_T = 1.374\ \text{mA}\ \angle 59.03°$, $\mathbf{I}_R = 0.707\ \text{mA}\ \angle 0°$, $\mathbf{I}_L = 0.236\ \text{mA}\ \angle -90°$, $\mathbf{I}_C = 1.414\ \text{mA}\ \angle 90°$ **(f)** 1 mW **(g)** 0.515 leading **(h)** $i_T = 1.943 \times 10^{-3} \sin(10^4 t + 59.03°)$, $i_R = 1 \times 10^{-3} \sin 10^4 t$, $i_L = 0.333 \times 10^{-3} \sin(10^4 t - 90°)$, $i_C = 2 \times 10^{-3} \sin(10^4 t + 90°)$, $i_C = 2 \times 10^{-3} \sin(10^4 t + 90°)$

15. **(a)** $7.69\ \text{k}\Omega\ \angle 39.69°$ **(b)** $I_T = 1.3 \times 10^{-3} \sin(\omega t - 39.69°)$ **(c)** $i_R = 1 \times 10^{-3} \sin \omega t$

17. **(a)** $0.166\ \text{mS}\ \angle -90°$ **(b)** $6\ \text{k}\Omega\ \angle 90°$ **(c)** $0.6\ \text{kV}\ \angle +90°$ **(d)** $0.2\ \text{A}\ \angle 0°$ **(e)** $0.1\ \text{A}\ \angle +180°$

19. **(a)** $\mathbf{I}_1 = \mathbf{I}_2 = 8\ \angle 0°$ **(b)** $\mathbf{I}_1 = 9.14\ \text{mA}\ \angle 0°$, $\mathbf{I}_2 = 6.86\ \text{mA}\ \angle 0°$

21. $I = 3.43$ mA, $V = 13.72$ V

25. **(a)** $R_p = 94.73\ \Omega$, $X_p = X_C = 52.1\ \Omega$ **(b)** $R_p = X_p = X_C = 4\ \text{k}\Omega$

27. $6\ \text{k}\Omega\ \angle -90°$

29. $\mathbf{Z}_T = 8\ \text{k}\Omega\ \angle 0°$, $\mathbf{Y}_T = 0.125\ \text{mS}\ \angle 0°$

31. $\mathbf{I}_R = 0.707\ \text{mA}\ \angle 0°$, $\mathbf{I}_L = 0.236\ \text{mA}\ \angle -90°$, $\mathbf{I}_C = 1.414\ \text{mA}\ \angle 90°$, $\mathbf{I}_T = 1.374\ \text{mA}\ \angle 59.03°$

Chapter 16

1. $R_p = 5\ \text{k}\Omega$, $\mathbf{I}_p = 24\ \text{mA}\ \angle 0°$

3. $X_{Lp} = 20\ \Omega$, $\mathbf{I}_p = 0.25\ \text{A}\ \angle 0°$

5. $\mathbf{Z}_p = 2.828\ \text{k}\Omega\ \angle 45°$, $\mathbf{I}_p = 30\ \text{A}\ \angle -45°$

7. **(a)** $\mathbf{Z}_T = 3.876\ \Omega\ \angle -11.83°$, $\mathbf{Y}_T = 0.258\ \text{S}\ \angle 11.83°$ **(b)** $\mathbf{I}_T = 15.48\ \text{A}\ \angle 11.83°$ **(c)** $\mathbf{I}_2 = 3.984\ \text{A}\ \angle 52.83°$ **(d)** $\mathbf{V}_C = 47.81\ \text{V}\ \angle -37.17°$

9. $6.7\ \text{A}\ \angle 26.57°$

11. $5.97\ \text{mV}\ \angle 13.57°$

13. $\mathbf{Z}_{Th}$ = 2.236 kΩ ∠63.44°, $\mathbf{E}_{Th}$ = 5 V ∠0°
15. $\mathbf{Z}_{Th}$ = 7.81 Ω ∠39.81°, $\mathbf{E}_{Th}$ = 36 V ∠0°
17. $\mathbf{Z}_L$ = 2.236 kΩ ∠−63.44°, P_{max} = 6.25 mW
19. **(a)** $\mathbf{Z}_L$ = 1.03 kΩ ∠−31° **(b)** P_{max} = 53.5 mW
21. 12 Ω ∠−90°
23. Not balanced
25. R_4 = 11.99 kΩ, L_4 = 12.15 mH
27. **(a)** 1.2 Ω ∠90° **(b)** $\mathbf{I} = \mathbf{I}_1$ = 10 A ∠−90° **(c)** $\mathbf{I}_2$ = 6 A ∠−90°, $\mathbf{I}_3$ = 4 A ∠−90°
29. See 11
31. 5.37 Ω ∠26.57°
33. **(a)** $\mathbf{V}[(\frac{1}{4}\ \Omega)$ ∠−90° + $(\frac{1}{2}\ \Omega$ ∠0°] = 2 A ∠0° − 5 A ∠0° **(b)** 5.37 V ∠206.57°
35. (0.559 mS ∠−26.57°)$\mathbf{V}_1$ − (0.5 mS ∠0°)$\mathbf{V}_2$ = 3 mA ∠0°, (0.75 mS ∠0°)$\mathbf{V}_2$ − (0.5 mS ∠0°)$\mathbf{V}_1$ = 6 mA ∠0°

Chapter 17

1. 0.99 W
5. 0.667 VA
7. Q_L = 14.396 VAR, P = 0 W, S = 14.396 VA
9. 1.562 × 10^{-8} VAR
11. **(a)** P_R = 300 W, $P_L = P_C$ = 0 W **(b)** Q_R = 0 VAR, Q_L = 900 VAR, Q_C = 500 VAR **(c)** S_R = 300 VA, S_L = 900 VA, S_C = 500 VA **(d)** P_T = 300 W, Q_T = 400 VAR (L), S_T = 500 VA, F_p = 0.6 lagging
13. **(a)** P_T = 1200 W, Q_T = 1200 VAR (L), S_T = 1697.056 VA, F_p = 0.707 lagging **(c)** 8.485 A
15. **(a)** P_R = 180 W, Q_R = 0 VAR, S_R = 180 VA **(b)** P_L = 0 W, Q_L = 360 VAR, S_L = 360 VA **(c)** P_T = 580 W, Q_T = 960 VAR (L), S_T = 1121.61 VA, F_p = 0.517 lagging **(d)** 18.693 A
17. **(a)** P_R = 300 W, $P_L = P_C$ = 0 W **(b)** Q_R = 0 VAR, Q_L = 400 VAR, Q_C = 250 VAR **(c)** S_R = 300 VA, S_L = 400 VA, S_C = 250 VA **(d)** P_T = 300 W, Q_T = 150 VAR (L), S_T = 335.41 VA, F_p = 0.894 lagging
19. **(a)** $\mathbf{Z}_T$ = 2.88 Ω ∠36.87°, R = 2.3 Ω, X_L = 1.73 Ω **(b)** 4000 W
21. **(a)** P_T = 230 W, Q_T = 140 VAR (L), S_T = 269.26 VA, F_p = 0.854 lagging **(b)** 2.69 A **(c)** F_p = 1: R = 50 Ω, 30 W: R = 14.67 Ω, X_L = 19.56 Ω, F_P = 0: X_L = 48.9 Ω
23. **(a)** P_T = 900 W, Q_T = 0 VAR, S_T = 900 VA, F_P = 1 **(b)** 9 A
25. **(a)** 0.640 lagging **(b)** 7.8 kVA **(c)** 6.51 A **(d)** 11.05 μF **(e)** 4.17 A
27. **(a)** 128.1 W **(b)** *a–b*: 42.69 W, *b–c*: 64.03 W, *a–c*: 106.72 W, *a–d*: 106.72 W, *c–d*: 0 W, *d–e*: 0 W, *f–e*: 21.34 W
29. **(a)** R = 5 Ω, L = 131.97 mH **(b)** 10 Ω **(c)** R = 15 Ω, L = 262.26 mH

Chapter 18

1. **(a)** ω_s = 250 rad/s, f = 39.79 Hz **(b)** ω_s = 3535.53 rad/s, f = 562.7 Hz **(c)** ω_s = 21,880.21 rad/s, f = 3482.34 Hz
3. **(a)** 40 Ω **(b)** I = 10 A, V_R = 20 V, V_L = 400 V, V_C = 400 V **(c)** 20
5. **(a)** 400 Hz **(b)** f_1 = 5800 Hz, f_2 = 6200 Hz **(c)** 45 Ω **(d)** 375 mW
7. **(a)** 10 **(b)** 20 Ω **(c)** L = 1.59 mH, C = 3.98 μF **(d)** f_1 = 1900 Hz, f_2 = 2100 Hz
9. R = 10 Ω, L = 13.26 mH, C = 27.07 nF
11. **(a)** 104 Ω **(b)** 520 Ω **(c)** L = 0.796 mH, C = 76.52 nF **(d)** Q_p = 5.2, BW = 3846.15 Hz
13. **(a)** 102.73 kHz **(b)** X_L = X_C = 51.64 Ω **(c)** Q_C = 34.43 (high) **(d)** 1510.52 Ω **(e)** I_C = 292.51 mA, I_L = 292.39 mA **(f)** Q_p = 29.25, BW = 3512.02 Hz
15. **(a)** 1580.35 Ω **(b)** 1581.93 Ω **(c)** 15.72 kHz **(d)** 6.4 nF
17. **(a)** 3558.81 Hz **(c)** 200 V **(d)** 1 W **(e)** 397.9 Hz
19. **(a)** 1000 Ω **(b)** 500 **(c)** 500 kHz **(d)** 1.923 V
21. R_l = 0.1 Ω, L = 31.83 μH, C = 3.54 μF
23. **(a)** 15.92 kHz
25. R = 795.78 Ω
27. f_2 = 13.26 kHz
29. **(a)** 20
33. **(a)** L_s = 12.67 mH, L_p = 19.21 mH

Chapter 19

1. **(a)** 0.8 H **(b)** 0.8 × 10^{-4} Wb **(c)** 32 mH
3. 1.35 H
5. **(a)** 200 V **(b)** 11.73 × 10^{-3} Wb
7. 20 V
9. **(a)** I_L = 0.4 A, V_L = 0.8 V **(b)** 0.08 Ω
11. V_g = 300 V, I_p = 75 A
13. **(a)** 60 A **(b)** V_L = 300 V, I_L = 20 A
15. **(a)** 20 Ω **(b)** 40 Ω **(c)** 340 Ω ∠0° + 40 Ω ∠90° **(d)** 146 mA **(e)** 11.68 V
17. **(a)** 280.71 Ω ∠−85.91° **(b)** 178 mA **(c)** V_{R_e} = 3.56 V, V_{X_e} = 7.12 V, V_{Z_L} = 56.96 V
19. **(a)** 20 **(b)** 83.33 A **(c)** 4.167 A **(d)** $a = \frac{1}{20}$, I_s = 4.167 A, I_p = 83.33 A
21. **(a)** V_L = 25 V, I_L = 5 A **(b)** 80 Ω **(c)** 20 Ω
23. **(a)** E_2 = 40 V, E_3 = 30 V, I_2 = 3.333 A, I_3 = 3 A **(b)** 64.47 Ω

Chapter 20

1. **(a)** 120.1 V **(b)** 120.1 V **(c)** 12.01 A **(d)** 12.01 A
3. **(a)** 120.1 V **(b)** 120.1 V **(c)** 16.98 A **(d)** 16.98 A
5. **(a)** θ_2 = −120°, θ_3 = +120° **(b)** $\mathbf{V}_{an}$ = 120 V ∠0°, $\mathbf{V}_{bn}$ = 120 V ∠−120°, $\mathbf{V}_{cn}$ = 120 V ∠+120° **(c)** $\mathbf{I}_{an}$ =

8 A $\angle 0°$, $\mathbf{I}_{bn}$ = 8 A $\angle -120°$, $\mathbf{I}_{cn}$ = 8 A $\angle +120°$ **(e)** 8 A **(f)** 207.85 V

7. $V_{an} = V_{bn} = V_{cn}$ = 127.02 V, $I_{an} = I_{bn} = I_{cn}$ = 8.983 A

9. **(a)** 120.1 V **(b)** 208 V **(c)** 10.4 A **(d)** 18.01 A

11. **(a)** 120.1 V **(b)** 208 V **(c)** 16.34 A **(d)** 28.30 A

13. **(a)** $\theta_2 = -120°$, $\theta_3 = +120°$ **(b)** $\mathbf{V}_{ab}$ = 208 V $\angle 0°$, $\mathbf{V}_{bc}$ = 208 V $\angle -120°$, $\mathbf{V}_{ca}$ = 208 V $\angle +120°$ **(d)** $\mathbf{I}_{ab}$ = 1.47 A $\angle 45°$, $\mathbf{I}_{bc}$ = 1.47 A $\angle -75°$, $\mathbf{I}_{ca}$ = 1.47 A $\angle 165°$ **(e)** 2.546 A **(f)** 120.1 V

15. I_{ϕ_L} = 15.56 A, V_{ϕ_L} = 220 V, I_L = 26.95 A

17. **(a)** 208 V **(b)** 120.1 V **(c)** 4 A **(d)** 4 A

19. V_{ϕ_L} = 69.28 V, I_{ϕ_L} = 2.89 A, I_L = 2.89 A

21. V_{ϕ_L} = 69.28 V, I_{ϕ_L} = 5.77 A, I_L = 5.77 A

23. **(a)** 440 V **(b)** 440 V **(c)** 29.33 A **(d)** 50.8 A

25. **(a)** $\theta_2 = -120°$, $\theta_3 = +120°$ **(b)** $\mathbf{V}_{ab}$ = 100 V $\angle 0°$, $\mathbf{V}_{bc}$ = 100 V $\angle -120°$, $\mathbf{V}_{ca}$ = 100 V $\angle 120°$ **(d)** $\mathbf{I}_{ab}$ = 5 A $\angle 0°$, $\mathbf{I}_{bc}$ = 5 A $\angle -120°$, $\mathbf{I}_{ca}$ = 5 A $\angle 120°$ **(e)** 8.66 A

27. **(a)** $\theta_2 = -120°$, $\theta_3 = 120°$ **(b)** $\mathbf{V}_{ab}$ = 100 V $\angle 0°$, $\mathbf{V}_{bc}$ = 100 V $\angle -120°$, $\mathbf{V}_{ca}$ = 100 V $\angle 120°$ **(d)** $\mathbf{I}_{ab}$ = 7.071 A $\angle 45°$, $\mathbf{I}_{bc}$ = 7.071 A $\angle -75°$, $\mathbf{I}_{ca}$ = 7.071 A $\angle 165°$ **(e)** 12.25 A

29. P_T = 2160 W, Q_T = 0 VAR, S_T = 2160 VA, F_p = 1

31. P_T = 7210.67 W, Q_T = 7210.67 VAR (C), S_T = 10,197.43 VA, F_p = 0.707 leading

33. P_T = 7263.41 W, Q_T = 7263.41 VAR (L), S_T = 10,272.01 VA, F_p = 0.707 lagging

35. P_T = 288.3 W, Q_T = 576.6 VAR, S_T = 644.66 VA, F_p = 0.447 lagging

37. P_T = 900 W, Q_T = 1200 VAR (L), S_T = 1500 VA, F_p = 0.6 lagging

39. R_ϕ = 38.96 Ω, X_ϕ = 51.94 Ω (C)

41. **(b)** P_T = 225 W, P_ϕ = 75 W

43. **(b)** P_T = 285 W **(c)** 115 W

45. **(a)** 120 V **(b)** I_{an} = 8.49 A, I_{bn} = 7.07 A, I_{cn} = 42.43 A **(c)** P_T = 4921.23 W, Q_T = 4921.23 VAR (L), S_T = 6958.62 VA, F_p = 0.707 lagging

Index

NOTE: Boldface page numbers refer to glossary entries.